Study Guide and Solutions Manual for McMurry's

Organic Chemistry

Fifth Edition

Susan McMurry
Cornell University

Pacific Grove • Albany • Belmont • Boston • Cincinnati • Johannesburg • London • Madrid
Melbourne • Mexico City • New York • Scottsdale • Singapore • Tokyo • Toronto

Senior Assistant Editor: *Melissa D. Henderson*
Editorial Assistant: *Dena Dowsett-Jones*
Marketing Managers: *Steve Catalano, Heather Woods*
Marketing Assistant: *Christina De Veto*
Production Coordinator: *Dorothy Bell*
Cover Design: *Vernon T. Boes*
Cover Photo: *© J. Jamsen, Natural Selection*
Manufacturing Buyer: *Nancy Panziera*
Printing and Binding: *Custom Printing, Frederick, Maryland*

For more information, contact:
BROOKS/COLE
511 Forest Lodge Road
Pacific Grove, CA 93950 USA
www.brookscole.com

Printed in the United States of America

10 9 8 7 6 5 4 3 2 1

ISBN 0-534-37192-2

Contents

Solutions to Problems

Appendices

Preface

What enters your mind when you hear the words "organic chemistry?" Some of you may think, "the chemistry of life," or "the chemistry of carbon." Other responses might include "pre-med, "pressure," "difficult," or "memorization." Although formally the study of the compounds of carbon, organic chemistry encompasses many skills that are common to other areas of study. Organic chemistry is as much a liberal art as a science, and mastery of the concepts and techniques of organic chemistry can lead to an enhanced competence in other fields.

As you proceed to solve the problems that accompany the text, you will bring to the task many problem-solving techniques. For example, planning an organic synthesis requires the skills of a chess player; you must plan your moves while looking several steps ahead, and you must keep your plan flexible. Structure-determination problems are like detective problems, in which many clues must be assembled to yield the most likely solution. Naming organic compounds is similar to the systematic naming of biological specimens; in both cases, a set of rules must be learned and then applied to the specimen or compound under study.

The problems in the text fall into two categories: drill and complex. Drill problems, which appear throughout the text and at the end of each chapter, test your knowledge of one fact or technique at a time. You may need to rely on memorization to solve these problems, which you should work on first. More complicated problems require you to recall facts from several parts of the text and then use one or more of the problem-solving techniques mentioned above. As each major type of problem—synthesis, nomenclature, or structure determination—is introduced in the text, a solution is extensively worked out in this *Solutions Manual.*

Here are several suggestions that may help you with problem solving:

1. The text is organized into chapters that describe individual functional groups. As you study each functional group, *make sure that you understand the structure and reactivity of that group*. In case your memory of a specific reaction fails you, you can rely on your general knowledge of functional groups for help.

2. *Use molecular models.* It is difficult to visualize the three-dimensional structure of an organic molecule when looking at a two-dimensional drawing. Models will help you to appreciate the structural aspects of organic chemistry and are indispensable tools for understanding stereochemistry.

3. Every effort has been made to make this *Solutions Manual* as clear, attractive, and error-free as possible. Nevertheless, you should *use the Solutions Manual in moderation*. The principal use of this book should be to check answers to problems you have already worked out. The *Solutions Manual* should not be used as a substitute for effort; at times, struggling with a problem is the only way to teach yourself.

4. *Look through the appendices at the end of the Solutions Manual.* Some of these appendices contain tables that may help you in working problems; others present information related to the history of organic chemistry.

This edition of the *Solutions Manual* is almost completely new. All but a few of the structures and drawings have been revised, and much new material has been added. Each chapter of the *Solutions Manual* begins with an outline of the text that can be used for a concise review of the text material and can also serve as a reference. After every few chapters a Review Unit has been inserted. In most cases, the chapters covered in the Review Units are related to each other, and the units are planned to appear at approximately the place in the textbook where a test might be given. Each unit lists the vocabulary for the chapters covered, the skills needed to solve problems, and several important points that might need reinforcing or that restate material in the text from a slightly different point of view. Finally, the small self-test that has been included allows you to test yourself on the material from more than one chapter.

I have tried to include many types of study aids in this *Solutions Manual.* Nevertheless, this book can only serve as an adjunct to the larger and more complete textbook. If *Organic*

Chemistry is the guidebook to your study of organic chemistry, then the *Solutions Manual* is the roadmap that shows you how to find what you need.

Acknowledgments I would like to thank my husband, John McMurry, for offering me the opportunity to write this book many years ago and for supporting my efforts while this edition was being prepared. I am indebted to C. Peter Lillya, University of Massachusetts, Amherst, and Eric Simanek, Texas A & M University, for combing this manuscript for errors and for improving the clarity of my writing. Many people at Brooks/Cole Publishing company have given me encouragement during this project; special thanks are due to Jennifer Huber, Melissa Henderson, and Beth Wilbur. Finally, I would like to thank our eleven-year-old son Paul McMurry, who helped in many small ways and who patiently watched me work on this book, hoping that his turn to use the computer would come soon.

Chapter 1 – Structure and Bonding

Chapter Outline

I. Atomic Structure (Sections 1.1 – 1.3).
   A. Introduction to atomic structure (Section 1.1).
      1. Atoms consist of a dense, positively charged nucleus surrounded by negatively charged electrons.
         a. The nucleus is made up of positively charged protons and uncharged neutrons.
         b. The nucleus contains most of the mass of the atom.
         c. Electrons move about the nucleus at a distance of about 10^{-10} m.
      2. The atomic number (Z) gives the number of protons in the nucleus.
      3. The mass number (A) gives the total number of protons and neutrons.
      4. All atoms of a given element have the same value of Z.
         a. Atoms of a given element can have different values of A.
         b. Atoms of the same element with different values of A are called isotopes.
   B. Orbitals (Section 1.2).
      1. The distribution of electrons in an atom can be described by a wave equation.
         a. The solution to a wave equation is an orbital, represented by Ψ.
         b. Ψ^2 predicts the volume of space in which an electron is likely to be found.
      2. There are four different kinds of orbitals (*s*, *p*, *d*, *f*).
         a. The *s* orbitals are spherical.
         b. The *p* orbitals are dumbbell-shaped.
         c. Four of the five *d* orbitals are cloverleaf-shaped.
      3. An atom's electrons are organized into shells.
         a. The shells differ in the numbers and kinds of orbitals they have.
         b. Electrons in different orbitals have different energies.
         c. Each orbital can hold two electrons.
      4. The two lowest-energy electrons are in the 1*s* orbital.
         a. The 2*s* orbital is the next in energy.
            Each *p* orbital has a region of zero density, called a node.
         b. The next three orbitals are $2p_x$, $2p_y$ and $2p_z$.
   C. Electron Configuration (Section 1.3).
      1. The ground-state electron configuration of an atom is a listing of the orbitals occupied by the atom.
      2. Rules for predicting the ground-state electron configuration of an atom:
         a. Orbitals with the lowest energy levels are filled first.
            The order of filling is 1*s*, 2*s*, 2*p*, 3*s*, 3*p*, 4*s*, 3*d*.
         b. Only two electrons can occupy each orbital, and they must be of opposite spin.
         c. If two or more orbitals have the same energy, one electron occupies each until all are half-full (Hund's rule). Only then does a second electron occupy one of the orbitals.
               All of the electrons in a half-filled shell have the same spin.

II. Chemical Bonding Theory (Sections 1.4 – 1.6).
   A. Development of chemical binding theory (Section 1.4).
      1. Kekulé and Couper proposed that carbon has four "affinity units" – carbon is tetravalent.
      2. Other scientists suggested that carbon can form double bonds, triple bonds and rings.

3. Van't Hoff proposed that the 4 atoms to which carbon forms bonds sit at the corners of a regular tetrahedron.
4. In a drawing of a tetrahedral carbon, a wedged line represents a bond pointing toward the viewer, and a dashed line points behind the plane of the page.

B. Covalent bonds (Section 1.5).
1. Atoms bond together because the resulting compound is more stable than the individual atoms.
   a. Atoms tend to achieve the electron configuration of the nearest noble gas.
   b. Atoms in groups 1A, 2A and 7A either lose electrons or gain electrons to form ionic compounds.
   c. Atoms in the middle of the periodic table share electrons by forming covalent bonds.
2. The number of covalent bonds formed by an atom depends on the number of electrons it has and on the number it needs to achieve an octet.
3. Covalent bonds can be represented two ways.
   a. In Lewis structures, bonds are represented as pairs of dots.
   b. In line-bond structures, bonds are represented as lines drawn between two atoms.
4. Valence electrons not used for bonding are called lone-pair electrons.
   Lone-pair electrons are represented as dots.

C. Theories of covalent bond formation (Section 1.6).
1. Valence bond theory.
   a. Covalent bonds are formed by the overlap of two atomic orbitals, each of which contains one electron. The two electrons have opposite spins.
   b. Each of the bonded atoms retains its atomic orbitals, but the electron pair of the overlapping orbitals is shared by both atoms.
   c. The greater the orbital overlap, the stronger the bond.
   d. Bonds formed by the head-on overlap of two atomic orbitals are cylindrically symmetrical and are called σ bonds.
   e. Bond strength is the measure of the amount of energy needed to break a bond.
   f. Bond length is the optimum distance between nuclei.
   g. Every bond has a characteristic bond length and bond strength.
2. Molecular orbital theory.
   a. Molecular orbitals arise from a mathematical combination of atomic orbitals and belong to the entire molecule.
   b. Two 1*s* orbitals can combine in two different ways.
      i. The additive combination is a bonding MO and is lower in energy than the two hydrogen 1*s* atomic orbitals.
      ii. The subtractive combination is an antibonding MO and is higher in energy than the two hydrogen 1*s* atomic orbitals.
   c. A node is a region between nuclei where electrons aren't found.
      If a node occurs between two nuclei, the nuclei repel each other.
   d. The number of MOs in a molecule is the same as the number of atomic orbitals combined.

III. Hybridization (Sections 1.7 – 1.11).

A. sp^3 Orbitals (Sections 1.7, 1.8).
1. Structure of methane (Section 1.7).
   a. When carbon forms 4 bonds with hydrogen, one 2*s* orbital and three 2*p* orbitals combine to form four equivalent atomic orbitals (sp^3 hybrid orbitals).
   b. These orbitals are tetrahedrally oriented.
   c. Because these orbitals are unsymmetrical, they can form stronger bonds than unhybridized orbitals can.
   d. These bonds have a specific geometry and a bond angle of 109.5°.

2. Structure of ethane (Section 1.8).
   a. Ethane has the same type of hybridization as occurs in methane.
   b. The C–C bond is formed by overlap of two sp^3 orbitals.
   c. Bond lengths, strengths and angles are very close to those of methane.

B. sp^2 Orbitals (Section 1.9).
1. If one carbon 2*s* orbital combines with two carbon 2*p* orbitals, three hybrid sp^2 orbitals are formed, and one *p* orbital remains unchanged.
2. The three sp^2 orbitals lie in a plane at angles of 120°, and the *p* orbital is perpendicular to them.
3. Two different types of bond form between two carbons.
   a. A σ bond forms from the overlap of two sp^2 orbitals.
   b. A π bond forms by sideways overlap of two *p* orbitals.
   c. This combination is known as a carbon-carbon double bond.
4. Ethylene is composed of a carbon-carbon double bond and four σ bonds formed between the remaining four sp^2 orbitals of carbon and the 1*s* orbitals of hydrogen.
   The double bond of ethylene is both shorter and stronger than the C–C bond of ethane.
5. In the molecular orbital description of ethane, both bonding and antibonding MOs can form from the combination of two *p* orbitals.

C. *sp* Orbitals (Section 1.10).
1. If one carbon 2*s* orbital combines with one carbon 2*p* orbital, two hybrid *sp* orbitals are formed, and two *p* orbitals are unchanged.
2. The two *sp* orbitals are 180° apart, and the two *p* orbitals are perpendicular to them and to each other.
3. Two different types of bonds form.
   a. A σ bond forms from the overlap of two *sp* orbitals.
   b. Two π bonds form by sideways overlap of four *p* orbitals.
   c. This combination is known as a carbon-carbon triple bond.
4. Acetylene is composed of a carbon-carbon triple bond and two σ bonds formed between the remaining two *sp* orbitals of carbon and the 1*s* orbitals of hydrogen.
   The triple bond of acetylene is the strongest carbon-carbon bond.

D. Hybridization of nitrogen and oxygen (Section 1.11).
1. Covalent bonds between other elements can be described by using hybrid orbitals.
2. Both the nitrogen atom in ammonia and the oxygen atom in water form sp^3 hybrid orbitals.
   The lone-pair electrons in these compounds occupy sp^3 orbitals.
3. The bond angles between hydrogen and the central atom is often less than 109° because the lone-pair electrons take up more room than the σ bond.

Solutions to Problems

1.1 (a) To find the ground-state electron configuration of an element, first locate its atomic number. For boron, the atomic number is 5; boron thus has 5 protons and 5 electrons. Next, assign the electrons to the proper energy levels, starting with the lowest level. Fill each level *completely* before assigning electrons to a higher energy level.

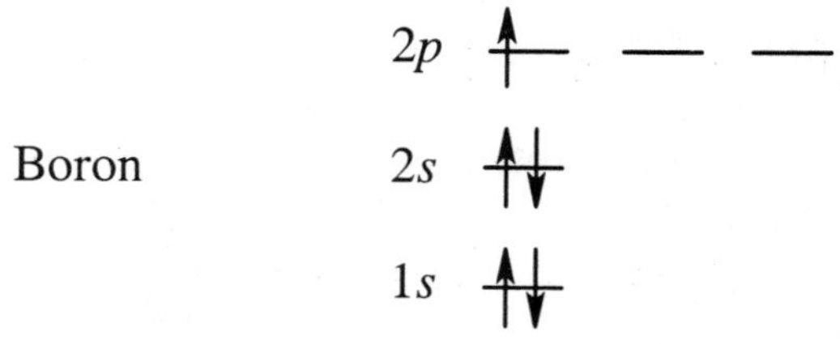

Boron

$2p$ ↑ — —
$2s$ ↑↓
$1s$ ↑↓

Remember that only two electrons can occupy the same orbital, and that they must be of opposite spin.

A different way to represent the ground-state electron configuration is to simply write down the occupied orbitals and to indicate the number of electrons in each orbital. For example, the electron configuration for boron is $1s^2\ 2s^2\ 2p^1$.

Often, we are interested only in the electrons in the outermost shell. We can then represent all filled levels by the symbol for the noble gas having the same levels filled. In the case of boron, the filled $1s$ energy level is represented by [He], and the valence shell configuration is symbolized by [He] $2s^2\ 2p^1$.

(b) Let's consider an element with many electrons. Phosphorus, with an atomic number of 15, has 15 electrons. Assigning these to energy levels:

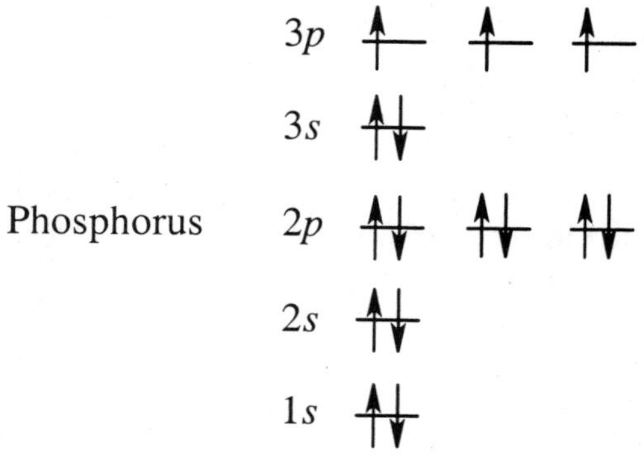

Phosphorus

$3p$ ↑ ↑ ↑
$3s$ ↑↓
$2p$ ↑↓ ↑↓ ↑↓
$2s$ ↑↓
$1s$ ↑↓

Notice that the $3p$ electrons are all in different orbitals. According to *Hund's rule*, we must place one electron into each orbital of the same energy level until all orbitals are half-filled.

The more concise way to represent ground-state electron configuration for phosphorus: $1s^2\ 2s^2\ 2p^6\ 3s^2\ 3p^3$ or [Ne] $3s^2\ 3p^3$

(c) Oxygen (atomic number 8) (d) Chlorine (atomic number 17)

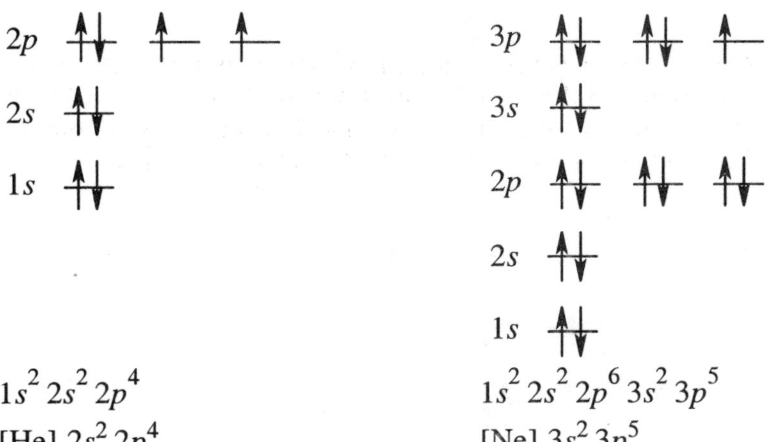

$1s^2\,2s^2\,2p^4$

[He] $2s^2\,2p^4$

$1s^2\,2s^2\,2p^6\,3s^2\,3p^5$

[Ne] $3s^2\,3p^5$

1.2 The elements of the periodic table are organized into groups that are based on the number of outer-shell electrons each element has. For example, an element in group 1A has one outer-shell electron, and an element in group 5A has five outer-shell electrons. To find the number of outer-shell electrons for a given element, use the periodic table to locate its group.

(a) Potassium (group 1A) has one electron in its outermost shell.
(b) Aluminum (group 3A) has three outer-shell electrons.
(c) Krypton is a noble gas and has eight electrons in its outermost shell.

1.3 A solid line represents a bond lying in the plane of the page, a wedged bond represents a bond pointing out of the plane of the page toward the viewer, and a dashed bond represents a bond pointing behind the plane of the page.

Chloroform

1.4

1.5 (a) Germanium (Group 4A) has four electrons in its valence shell and forms four bonds to achieve the noble-gas configuration of neon. A likely formula is $GeCl_4$.

Element	*Group*	*Likely Formula*
(b) Al	3A	AlH_3
(c) C	4A	CH_2Cl_2
(d) Si	4A	SiF_4
(e) N	5A	CH_3NH_2

1.6 Follow these three steps for drawing the Lewis structure of a molecule.

(1) Determine the number of valence, or outer-shell electrons for each atom in the molecule. For chloroform, we know that carbon has four valence electrons, hydrogen has one valence electron, and each chlorine has seven valence electrons.

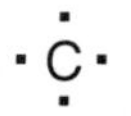 ·C· (4 dots)	4 x 1 = 4	
H·	1 x 1 = 1	
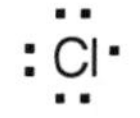 :Cl·	7 x 3 = 21	
	26	total valence electrons

(2) Next, use two electrons for each single bond.

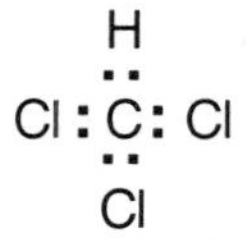

```
      H
      ..
  Cl : C : Cl
      ..
      Cl
```

(3) Finally, use the remaining electrons to achieve an noble gas configuration for all atoms.

	Molecule	*Lewis structure*	*Line-bond structure*
(a)	$CHCl_3$	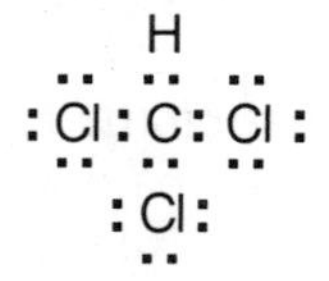 H above; :C̈l:C:C̈l: ; :C̈l: below	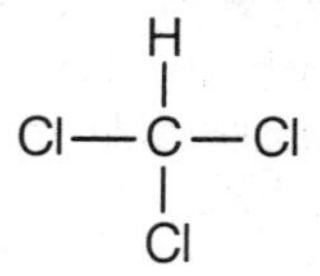 H above; Cl—C—Cl ; Cl below
(b)	H_2S 8 valence electrons	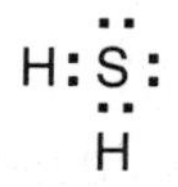 H:S̈: ; H below	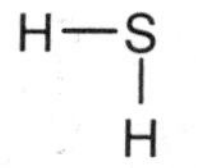 H—S ; H below
(c)	CH_3NH_2 14 valence electrons	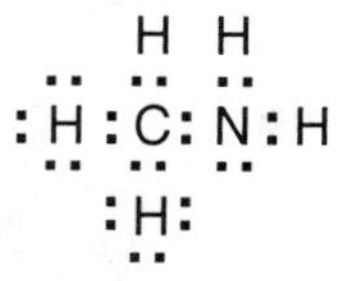 H H above; :H:C:N:H ; :H: below	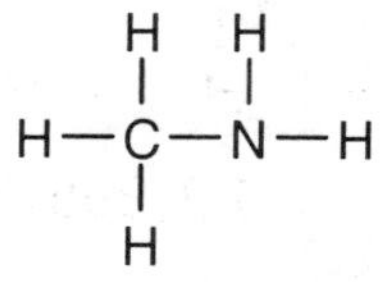 H H above; H—C—N—H ; H below
(d)	NaH 2 valence electrons	Na:H	Na—H
(e)	CH_3Li 8 valence electrons	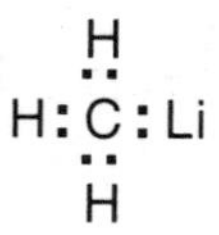 H above; H:C:Li ; H below	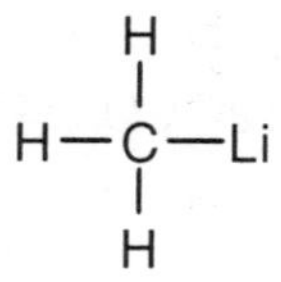 H above; H—C—Li ; H below

1.7 Each of the two carbons has 4 valence electrons. Two electrons are used to form the carbon-carbon bond, and the 6 electrons that remain can form bonds with a maximum of 6 hydrogens. Thus, the formula C_2H_7 is not possible.

1.8

Propane

The geometry around all carbon atoms is tetrahedral, and all bond angles are approximately 109°.

1.9

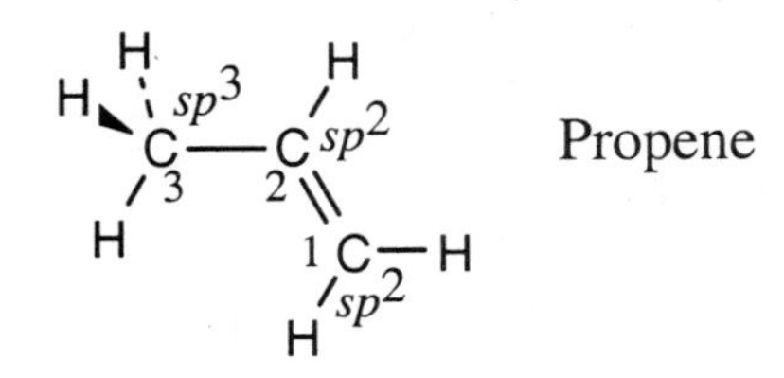

Propene

The C3–H bonds are σ bonds formed by overlap of an sp^3 orbital of carbon 3 with an *s* orbital of hydrogen.

The C2–H and C1–H bonds are σ bonds formed by overlap of an sp^2 orbital of carbon with an *s* orbital of hydrogen.

The C2–C3 bond is a σ bond formed by overlap of an sp^3 orbital of carbon 3 with an sp^2 orbital of carbon 2.

There are two C1–C2 bonds. One is a σ bond formed by overlap of an sp^2 orbital of carbon 1 with an sp^2 orbital of carbon 2. The other is a π bond formed by overlap of a *p* orbital of carbon 1 with a *p* orbital of carbon 2. All four atoms connected to the carbon-carbon double bond lie in the same plane, and all bond angles between these atoms are 120°.

1.10

All atoms lie in the same plane, and all bond angles are approximately 120°.

1,3-Butadiene

1.11 The atoms of acetaldehyde contribute 18 valence electrons. Ten electrons take part in the 5 single bonds, 4 electrons are involved in the carbon-oxygen double bond, and 4 electrons form the 2 lone pairs of electrons on oxygen.

Acetaldehyde

1.12

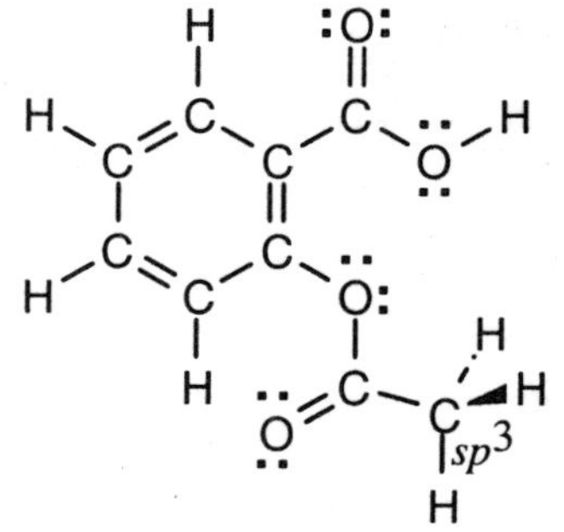

Aspirin.

All carbons are sp^2 hybridized, with the exception of the indicated carbon. All oxygen atoms have two lone pairs of electrons.

1.13

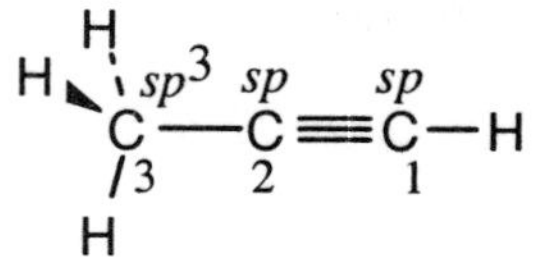

Propyne

The C3-H bonds are σ bonds formed by overlap of an sp^3 orbital of carbon 3 with an *s* orbital of hydrogen.

The C1-H bond is a σ bond formed by overlap of an *sp* orbital of carbon 1 with an *s* orbital of hydrogen.

The C2-C3 bond is a σ bond formed by overlap of an *sp* orbital of carbon 2 with an sp^3 orbital of carbon 3.

There are three C1-C2 bonds. One is a σ bond formed by overlap of an *sp* orbital of carbon 1 with an *sp* orbital of carbon 2. The other two bonds are π bonds formed by overlap of two *p* orbitals of carbon 1 with two *p* orbitals of carbon 2.

The three carbon atoms of propyne lie on a straight line; the bond angle is 180°. The $H–C_1{\equiv}C_2$ bond angle is also 180°.

1.14

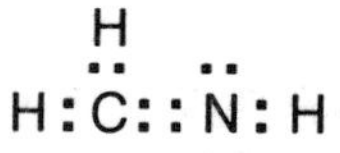

H:C::N:H Formaldimine H—C=N—H

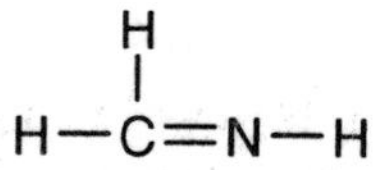

Four electrons are shared in the carbon-nitrogen double bond. The nitrogen atom is sp^2 hybridized.

1.15

(a) The sp^3-hybridized oxygen atom has tetrahedral geometry.

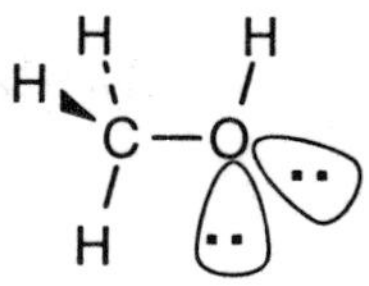

(b) Tetrahedral geometry at nitrogen and carbon.

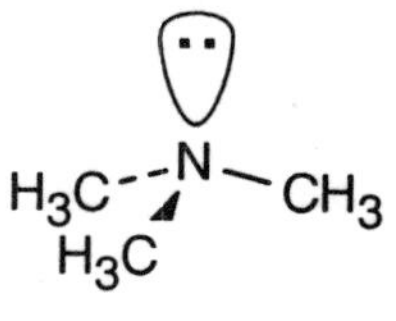

(c)

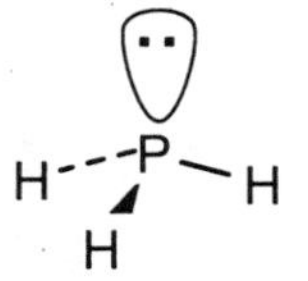

Like nitrogen, phosphorus has five outer-shell electrons. PH_3 has tetrahedral geometry.

Visualizing Chemistry

1.16

(a)

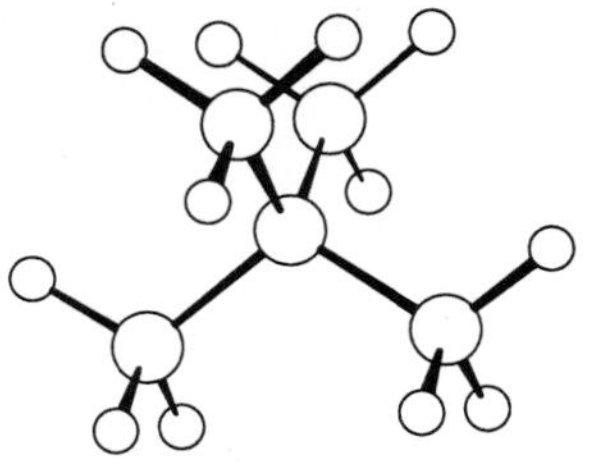

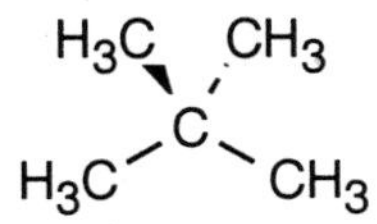

C_5H_{12}

(b)

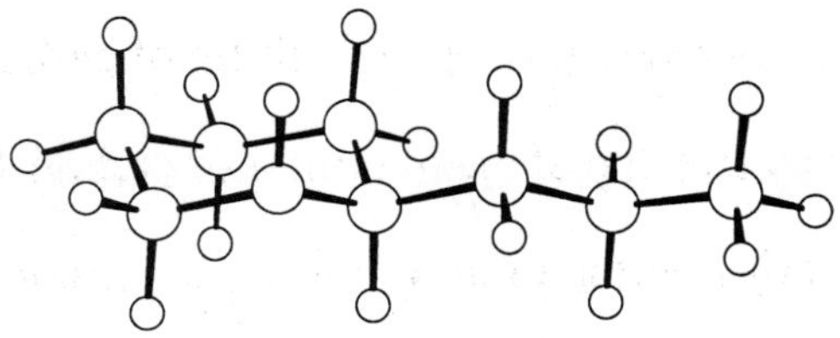

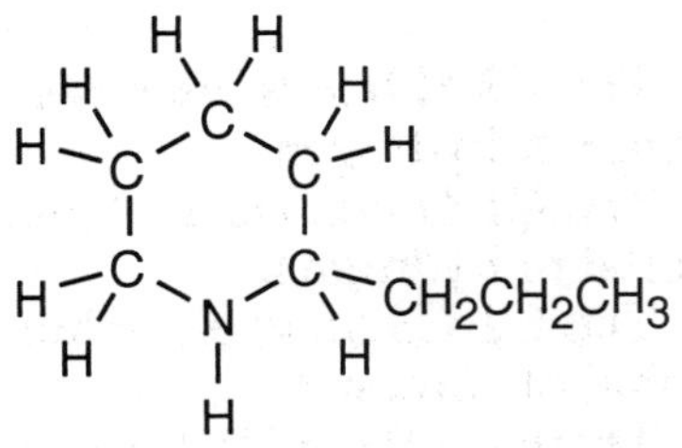

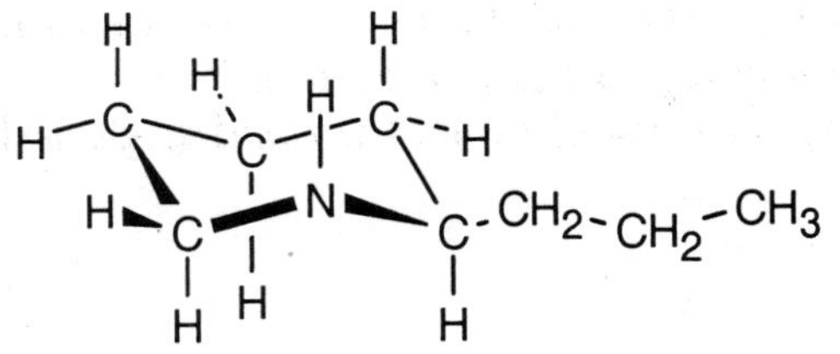

$C_8H_{17}N$

This drawing is more difficult to interpret because of the six-membered ring. Two structures are illustrated; the first one shows a flat ring, and the other shows the ring in three dimensions.

(c)

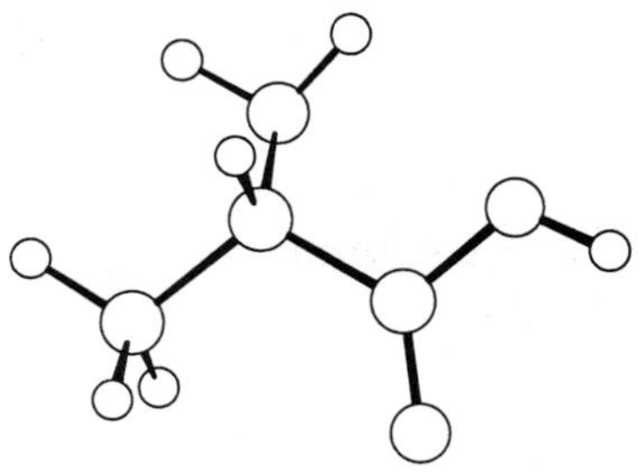

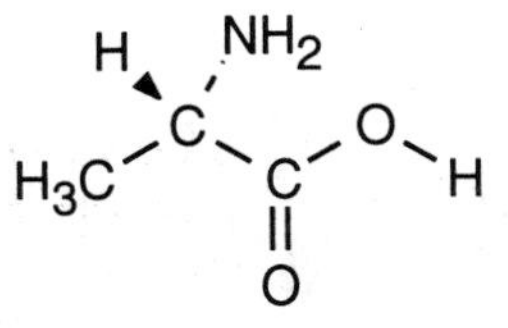

$C_3H_7NO_2$

1.17

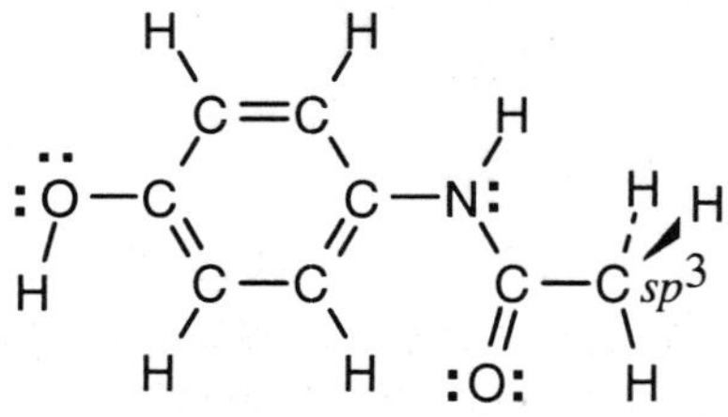

All carbons are sp^2 hybridized, except for the carbon indicated as sp^3. The two oxygen atoms and the nitrogen atom have lone pair electrons, as shown.

1.18

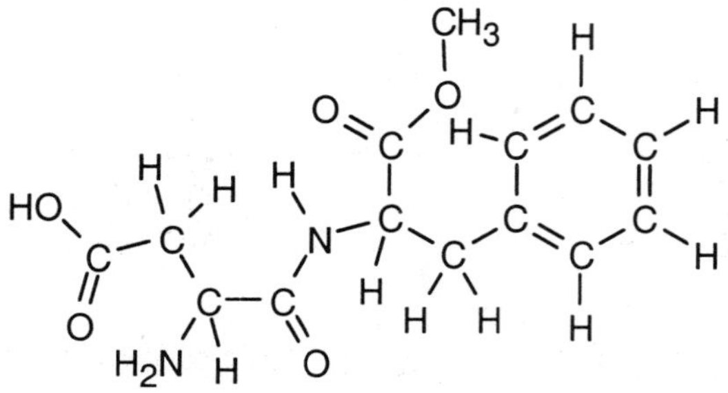

Additional Problems

1.19

Element	Atomic Number	Number of valence electrons
(a) Magnesium	12	2
(b) Sulfur	16	6
(c) Bromine	35	7

1.20

Element	Atomic Number	Ground-state Electron configuration
(a) Sodium	11	$1s^2\,2s^2\,2p^6\,3s^1$
(b) Aluminum	13	$1s^2\,2s^2\,2p^6\,3s^2\,3p^1$
(c) Silicon	14	$1s^2\,2s^2\,2p^6\,3s^2\,3p^2$
(d) Calcium	20	$1s^2\,2s^2\,2p^6\,3s^2\,3p^6\,4s^2$

1.21 (a) $AlCl_3$ (b) CF_2Cl_2 (c) NI_3

1.22

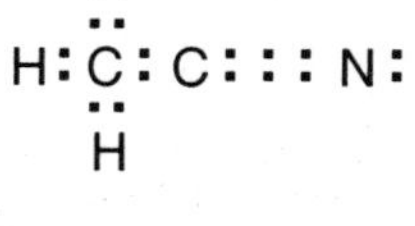

```
   H
   ..
H: C : C ::: N:
   ..
   H
```

Acetonitrile

Nitrogen has five electrons in its outer electron shell. Three are used in the carbon-nitrogen triple bond, and two are a nonbonding electron pair.

1.23 The H_3C– carbon is sp^3 hybridized, and the –CN carbon is sp hybridized.

1.24

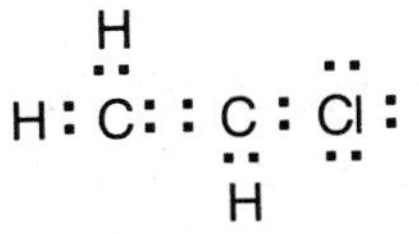

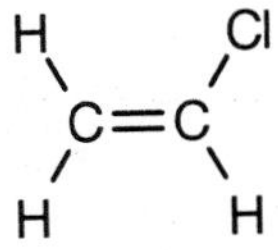

```
    H                    H       Cl
    ..   ..               \     /
H : C : : C : Cl :         C = C
         ..   ..          /     \
         H               H       H
```

Vinyl chloride has 18 valence electrons. Eight electrons are used for 4 single bonds, 4 electrons are used in the carbon-carbon double bond, and 6 electrons are in the 3 lone pairs that surround chlorine.

Note that electron pairs in Lewis structures are generally shown as either vertical or horizontal.

1.25

```
(a)                      (b)    :O:               (c)    :O:
                                 ||  ..                   ||  ..
      ..                   CH3 — C — NH2            CH3 — C — Cl:
CH3 — S — CH3                                                 ..
      ..
```

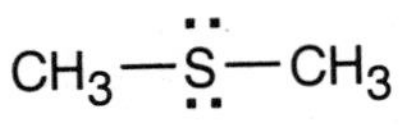

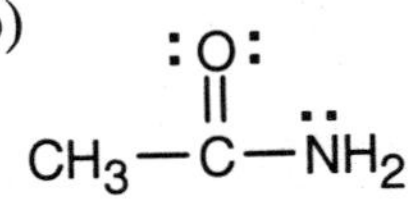

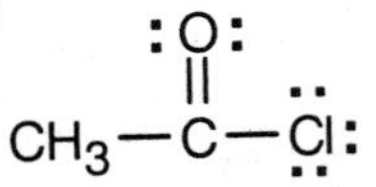

1.26 In molecular formulas of organic molecules, carbon is listed first, followed by hydrogen. All other elements are listed in alphabetical order.

Compound	*Molecular Formula*
(a) Aspirin	$C_9H_8O_4$
(b) Vitamin C	$C_6H_8O_6$
(c) Nicotine	$C_{10}H_{14}N_2$
(d) Glucose	$C_6H_{12}O_6$

1.27 To work a problem of this sort, you must examine all possible structures consistent with the rules of valence. You must systematically consider all possible attachments, including those that have branches, rings and multiple bonds.

```
(a)  H   H   H          (b)  H   H          (c)  H   H                    H       H
     |   |   |               |   |               |   |                    |       |
 H — C — C — C — H       H — C — N — H       H — C — C — O — H        H — C — O — C — H
     |   |   |               |                   |   |                    |       |
     H   H   H               H                   H   H                    H       H

(d)  H   H   H                H   Br  H
     |   |   |                |   |   |
 H — C — C — C — Br       H — C — C — C — H
     |   |   |                |   |   |
     H   H   H                H   H   H

(e)  H   O                H                        O
     |   ||               |                 H     / \     H
 H — C — C — H        H — C = C — O — H       \  C — C  /
     |                    |   |               /         \
     H                    H   H              H           H
```

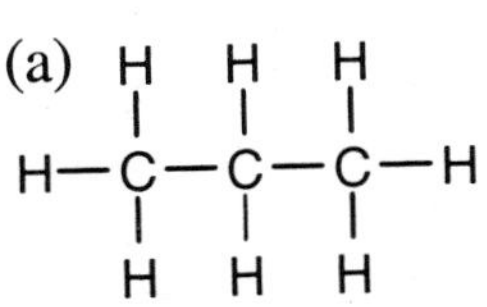

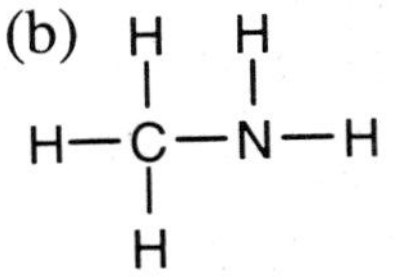

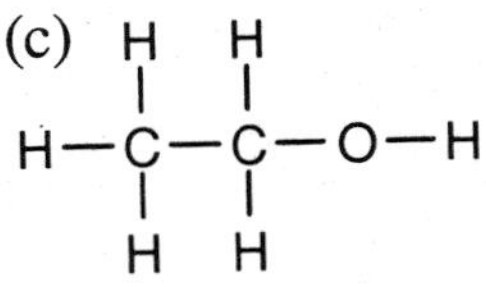

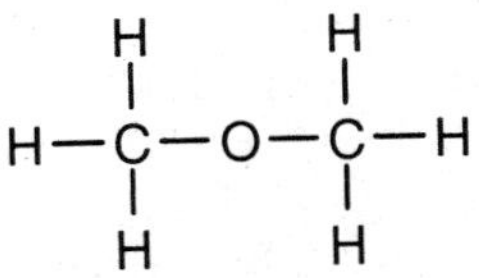

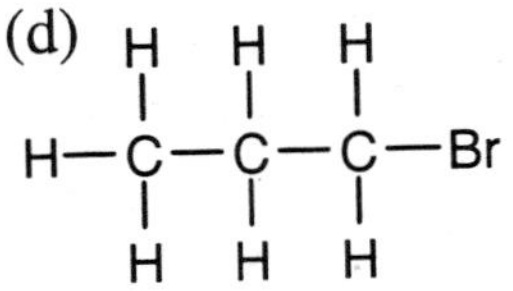

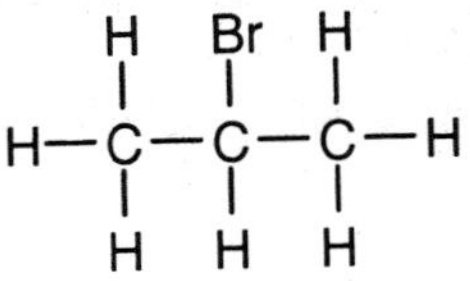

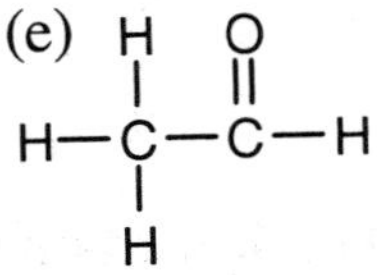

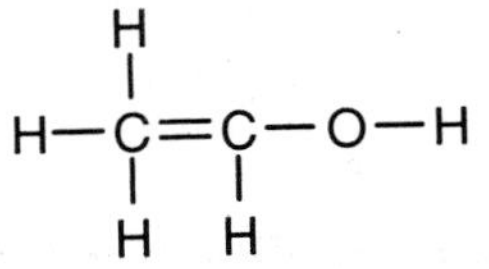

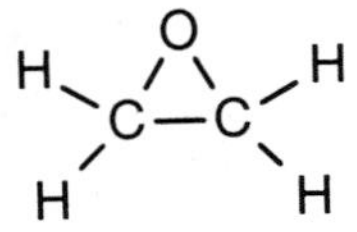

(f) $CH_3CH_2CH_2NH_2$ $CH_3CH(NH_2)CH_3$ $CH_3CH_2NHCH_3$ $(CH_3)_3N$

1.28

(a) $\overset{sp^3}{CH_3}\overset{sp^3}{CH_2}\overset{sp^3}{CH_3}$

(b) $(\overset{sp^3}{H_3C})_2\overset{sp^2}{C}=\overset{sp^2}{CH_2}$

(c) $\overset{sp^2}{H_2C}=\overset{sp^2}{CH}-\overset{sp}{C}\equiv\overset{sp}{CH}$

(d) $\overset{sp^3}{CH_3}-\overset{sp^2}{C}(=O)OH$

1.29

C_6H_6 Benzene

All carbon atoms of benzene are sp^2 hybridized, and all bond angles of benzene are 120°. Benzene is a planar molecule.

1.30 (a) The C–O–C bond angle is approximately 109°, and oxygen is sp^3–hybridized.
(b) The C–N–C bond angle is approximately 109°, and nitrogen is sp^3–hybridized.
(c) The C–N–H bond angle is approximately 109°, and nitrogen is sp^3–hybridized.
(d) The O–C–O bond angle is approximately 120°, and carbon is sp^2–hybridized.

1.31

(a) $CH_3CH_2CH=CH_2$ (b) $H_2C=CH-CH=CH_2$ (c) $H_2C=CH-C\equiv CH$

1.32 (a) The 4 valence electrons of carbon can form bonds with a maximum of 4 hydrogens. Thus, it is not possible for the compound CH_5 to exist.
(b) If you try to draw a molecule with the formula C_2H_6N, you will see that it is impossible for both carbons and nitrogen to have a complete octet of electrons. Therefore, C_2H_6N is unlikely to exist.
(c) A compound with the formula $C_3H_5Br_2$ doesn't have filled outer shells for all atoms and is thus unlikely to exist.

1.33

CH_3CH_2OH Ethanol

1.34

(a) (b) 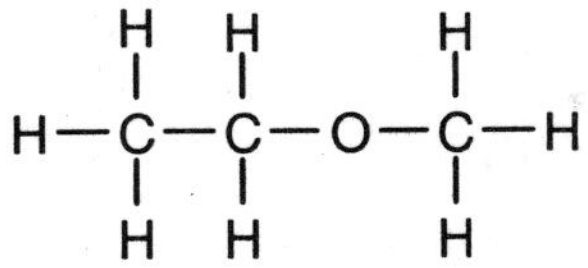 (c)

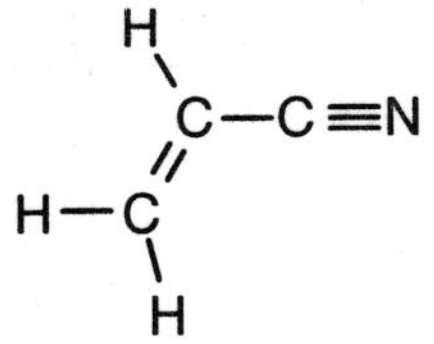

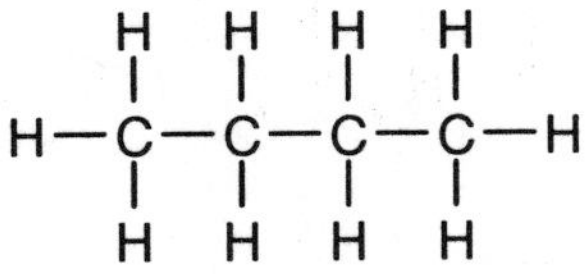

(d)

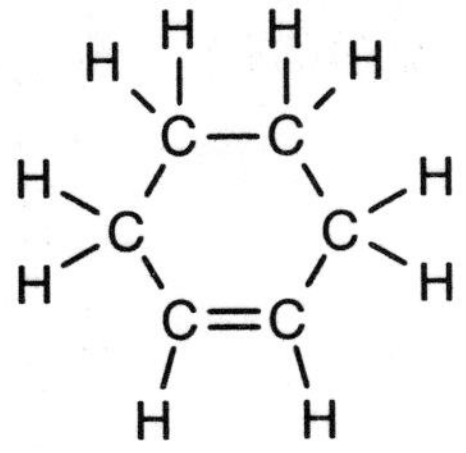

1.35

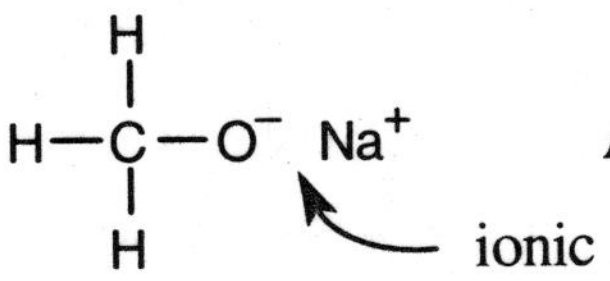

ionic

All other bonds are covalent.

1.36

(a) (b)

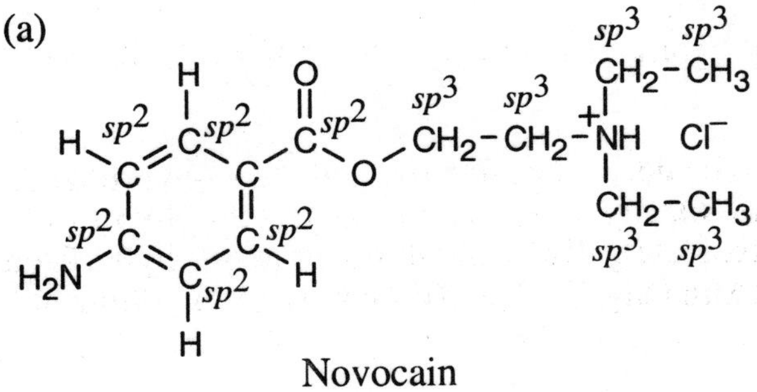

Novocain

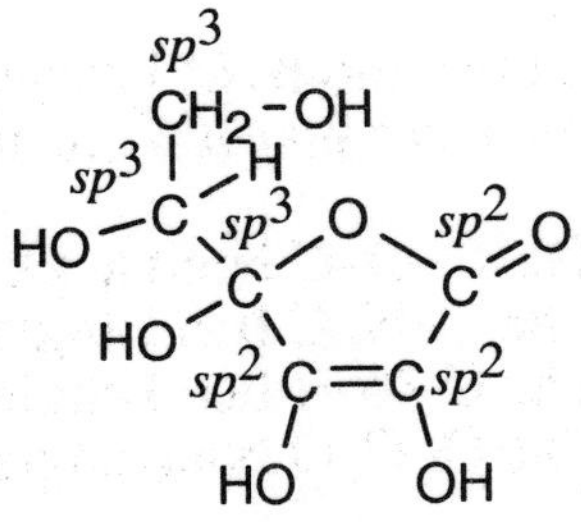

Vitamin C

1.37 All angles are approximate.
(a) 109° (b) 120° (c) 109°

1.38 Ionic: NaCl
Covalent: CH_3Cl, Cl_2, HOCl

1.39 In a compound containing a carbon-carbon triple bond, atoms bonded to the *sp*-hybridized carbons must lie in a straight line. It is not possible to form a five-membered ring if four carbons must have a linear relationship.

1.40 The π bonding molecular orbital in ethylene results from sideways overlap of two *p* atomic orbitals *with the same algebraic sign.* (A second molecular orbital can form by sideways overlap of two *p* orbitals with opposite algebraic signs, but it is an antibonding orbital.)

1.41

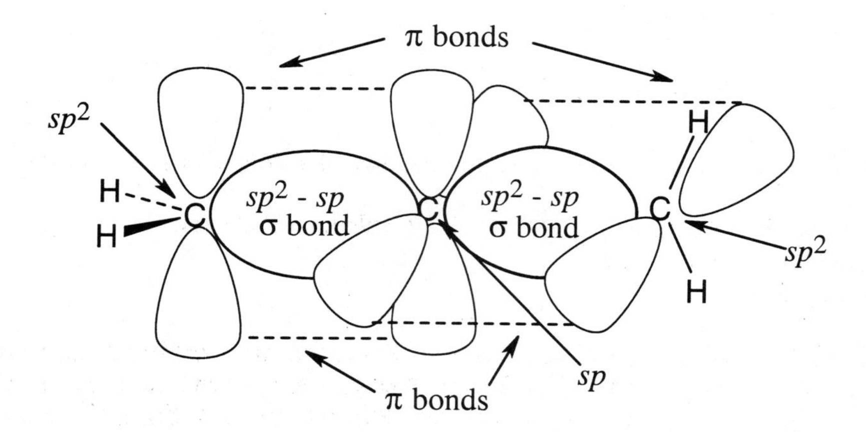

The central carbon of allene forms two σ bonds and two π bonds. The central carbon is *sp*–hybridized, and the two terminal carbons are sp^2–hybridized. The bond angle formed by the three carbons is 180°, indicating linear geometry for the carbons of allene.

1.42

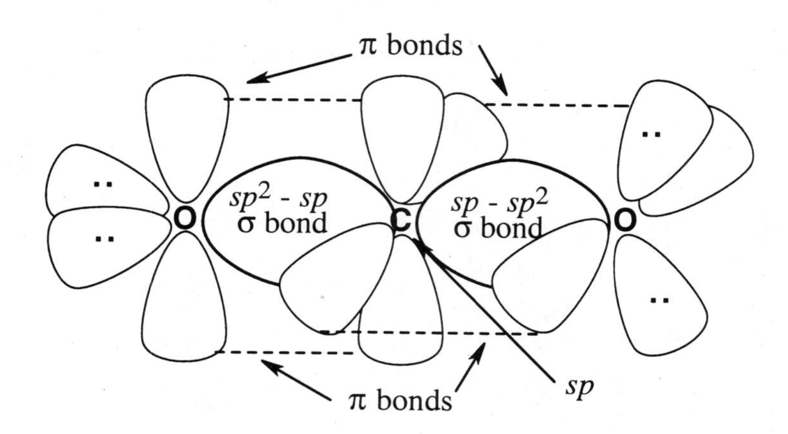

Carbon dioxide is a linear molecule.

1.43

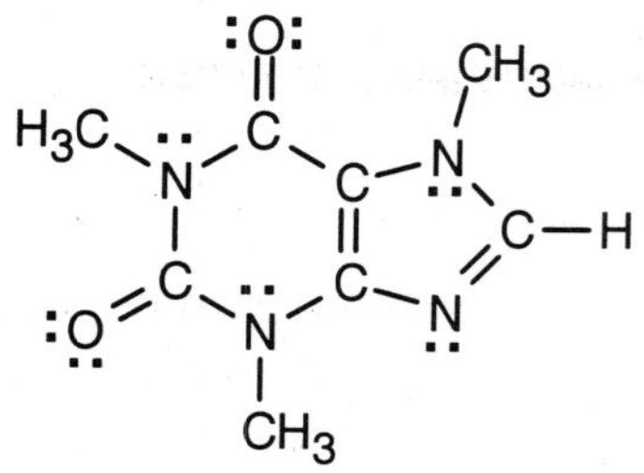

All the indicated atoms are sp^2 hybridized.

1.44 (a) The positively charged carbon atom has six valence shell electrons.
(b) A carbocation is sp^2–hybridized.
(c) A carbocation is planar.

1.45

(a) A carbanion is isoelectronic with (has the same number of electrons as) a trivalent nitrogen compound.
(b) The negatively charged carbanion carbon has eight valence electrons.
(c) The carbon atom is sp^3–hybridized.
(d) A carbanion is tetrahedral.

1.46 According to the Pauli Exclusion Principle, two electrons in the same orbital must have opposite spins. Thus, the two electrons of triplet (spin-unpaired) methylene must occupy different orbitals. In triplet methylene, *sp*–hybridized carbon forms one bond to each of two hydrogens. Each of the two unpaired electrons occupies a *p* orbital. In singlet (spin-paired) methylene the two electrons can occupy the same orbital because they have opposite spins. Including the two C–H bonds, there are a total of three occupied orbitals. We predict sp^2 hybridization and planar geometry for singlet methylene.

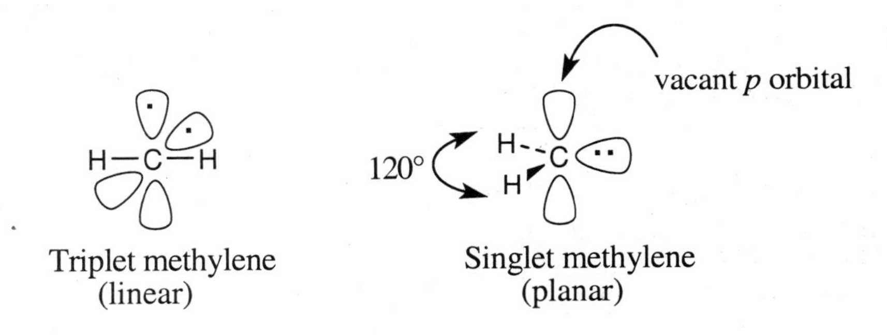

Triplet methylene (linear) Singlet methylene (planar)

A Look Ahead

1.47

$CH_3CH_2CH_2CH_3$ $CH_3CH(CH_3)CH_3$

The two compounds differ in the way that the carbon atoms are connected.

1.48

$H_2C{=}CH{-}CH_3$ (cyclopropane, C_3H_6)

One compound has a double bond, and one has a ring.

1.49

CH_3CH_2OH CH_3OCH_3

The two compounds differ in the location of the oxygen atom.

1.50

$CH_3CH_2CH{=}CH_2$ $CH_3CH{=}CHCH_3$ $H_2C{=}C(CH_3)_2$

The compounds differ in the way that the carbon atoms are connected and in the location of the double bond.

Molecular Modeling

1.51 The electron density is highest near the atomic nuclei. The "low" electron density surface most closely resembles the space-filling model.

1.52 Higher electron density is always found between O and H, rather than between O and Na. Higher electron density between O and H indicates that O–H bonds are covalent, whereas lower electron density between O and Na indicates that bonds between these two atoms are ionic.

1.53 The two highest energy molecular orbitals of acetylene have the same energy and are oriented at 90° to each other.

Chapter 2 – Polar Bonds and Their Consequences

Chapter Outline

I. Polar covalent bonds (Sections 2.1 – 2.3).
   A. Electronegativity (Section 2.1).
      1. Although some bonds are totally ionic and some are totally covalent, most chemical bonds are polar covalent bonds.
         In these bonds, electrons are attracted to one atom more than to the other atom.
      2. Bond polarity is due to differences in electronegativity (EN).
         a. Elements on the right side of the periodic table are more electronegative than elements on the left side.
         b. Carbon has an EN of 2.5.
         c. Elements with EN > 2.5 are more electronegative than carbon.
         d. Elements with EN < 2.5 are less electronegative than carbon.
      3. The difference in EN between two elements can be used to predict the polarity of a bond.
         a. If ΔEN < 0.4, a bond is nonpolar covalent.
         b. If ΔEN is between 0.4 and 2.0, a bond is polar covalent.
         c. If ΔEN > 2.0, a bond is ionic.
         d. The symbols $\delta+$ and $\delta-$ are used to indicate partial charges.
         e. A crossed arrow is used to indicate bond polarity.
      4. An inductive effect is an atom's ability to polarize a bond.
   B. Dipole moment (Section 2.2).
      1. Dipole moment is the measure of a molecule's overall polarity.
      2. Dipole moment (μ) = Q x r, where Q = charge and r = distance between charges.
         Dipole moment is measured in debyes (D).
      3. Dipole moment can be used to measure charge separation.
      4. An electrostatic potential map can be used to show the charge distribution in a molecule.
      5. Water and ammonia have large values of D; methane and ethane have D = 0.
   C. Formal charge (Section 2.3).
      1. Formal charge (FC) indicates electron "ownership" in a molecule.
      2.

$$\text{(FC)} = \left[\begin{array}{c}\text{\# of valence}\\ \text{electrons}\end{array}\right] - \left[\frac{\text{\# of bonding electrons}}{2}\right] - \left[\begin{array}{c}\text{\# nonbonding}\\ \text{electrons}\end{array}\right]$$

      3. Molecules that are neutral overall but that have + and – charges on individual atoms are dipolar.

II. Resonance (Sections 2.4 – 2.6).
   A. Chemical structures and resonance (Section 2.4).
      1. Some molecules (CH_3NO_2) can be drawn as two (or more) different Lewis structures.
         a. These structures are called resonance structures.
         b. The true structure of the molecule is intermediate between the structures.
         c. The true structure is called a resonance hybrid.
      2. Resonance structures differ only in the placement of π and nonbonding electrons.
         All atoms occupy the same positions.
      3. Resonance is an important concept in organic chemistry.

B. Rules for resonance forms (Section 2.5).
  1. Individual resonance forms are imaginary, not real.
  2. Resonance forms differ only in the placement of their π or nonbonding electrons.
     A curved arrow is used to indicate the movement of electrons, not atoms.
  3. Different resonance forms of a molecule don't have to be equivalent.
     If resonance forms are nonequivalent, the structure of the actual molecule resembles the more stable resonance form(s).
  4. Resonance forms must be valid Lewis structures and obey normal rules of valency.
  5. The resonance hybrid is more stable than any individual resonance form.

C. A useful technique for drawing resonance forms (Section 2.6).
  1. Any three-atom grouping with a multiple bond adjacent to a nonbonding *p* orbital has two resonance forms.
  2. By recognizing these three-atom pieces, resonance forms can be generated.

III. Acids and bases (Sections 2.7 – 2.11).

A. Brønsted-Lowry definition (Section 2.7).
  1. A Brønsted-Lowry acid donates an H^+ ion; a Brønsted-Lowry base accepts H^+.
  2. The product that results when a base gains H^+ is the conjugate acid of the base; the product that results when an acid loses H^+ is the conjugate base of the acid.

B. Acid and base strength (Section 2.8 – 2.10).
  1. A strong acid reacts almost completely with water (Section 2.8).
  2. The strength of an acid in water is indicated by K_a, the acidity constant.
  3. Strong acids have large acidity constants, and weaker acids have smaller acidity constants.
  4. The pK_a is normally used to express acid strength.
     a. $pK_a = -\log K_a$
     b. A strong acid has a small pK_a, and a weak acid has a large pK_a.
     c. The conjugate base of a strong acid is a weak base, and the conjugate base of a weak acid is a strong base.
  5. Predicting acid–base reactions from pK_a (Section 2.9).
     a. An acid with a low pK_a reacts with the conjugate base of an acid with a high pK_a
     b. In other words, the products of an acid–base reaction are more stable than the reactants.
  6. Organic acids and organic bases (Section 2.10).
     a. There are two main types of organic acids:
        i. Acids that contain hydrogen bonded to oxygen.
        ii. Acids that have hydrogen bonded to the carbon next to a C=O group.
     b. The main type of organic base contains a nitrogen atom with a lone electron pair.

C. Lewis acids and bases (Section 2.11).
  1. A Lewis acid accepts an electron pair.
     a. A Lewis acid may have either a vacant low-energy orbital or a polar bond to hydrogen.
     b. Examples include metal cations, halogen acids, Group 3 compounds and transition-metal compounds.
  2. A Lewis base has a pair of nonbonding electrons.
     a. Most oxygen- and nitrogen-containing organic compounds are Lewis bases.
     b. Many organic Lewis bases have more than one basic site.
  3. A curved arrow shows the movement of electrons from a Lewis base to a Lewis acid.

IV. Chemical structures (Sections 2.12 – 2.13).
- A. Drawing chemical structures (Section 2.12).
  1. Condensed structures don't show C–H bonds and don't show the bonds between CH_3, CH_2 and CH units.
  2. Skeletal structures are simpler still.
     - a. Carbon atoms aren't usually shown.
     - b. Hydrogen atoms bonded to carbon aren't usually shown.
     - c. Other atoms are shown.
- B. Molecular models (Section 2.13).
  1. Space-filling models show the crowding within a molecule.
  2. Ball-and-stick models show bonds.

Answers to Problems

2.1 After solving this problem, use Figure 2.2 to check your answers. The larger the number, the more electronegative the element.

More electronegative	*Less electronegative*
(a) H (2.1)	Li (1.0)
(b) Br (2.8)	B (2.0)
(c) Cl (3.0)	I (2.5)
(d) C (2.5)	H (2.1)

Carbon is slightly more electronegative than hydrogen.

2.2 As in Problem 2.1, use Figure 2.2.

(a) $\overset{\delta+}{H_3C}—\overset{\delta-}{Br}$ (b) $\overset{\delta+}{H_3C}—\overset{\delta-}{NH_2}$ (c) $\overset{\delta-}{H_3C}—\overset{\delta+}{Li}$ (d) $\overset{\delta-}{H_2N}—\overset{\delta+}{H}$

(e) $\overset{\delta+}{H_3C}—\overset{\delta-}{OH}$ (f) $\overset{\delta-}{H_3C}—\overset{\delta+}{MgBr}$ (g) $\overset{\delta+}{H_3C}—\overset{\delta-}{F}$

2.3

Carbon:	EN = 2.5	Carbon:	EN = 2.5	Fluorine:	EN = 4.0
Lithium:	EN = 1.0	Potassium:	EN = 0.8	Carbon:	EN = 2.5
	ΔEN = 1.5		ΔEN = 1.7		ΔEN = 1.5

Carbon:	EN = 2.5	Oxygen	EN = 3.5
Magnesium:	EN = 1.2	Carbon:	EN = 2.5
	ΔEN = 1.3		ΔEN = 1.0

The most polar bond has the largest ΔEN. Thus, in order of increasing bond polarity:

$H_3C—OH$ < $H_3C—MgBr$ < $H_3C—Li$, $H_3C—F$ < $H_3C—K$

2.4 We must look at the polarization of the individual bonds of a molecule to account for an observed dipole moment. In the case of methanol, CH_3OH, the individual bond polarities can be estimated from Figure 2.2. A polarity arrow is drawn so that the arrow points toward the more electronegative element (the larger numbers in Figure 2.2).

In addition, we must take into account the contribution of the two lone pairs of oxygen. Indicating the individual bond polarities by arrows, we can predict the direction of the dipole moment.

net dipole moment

It is difficult to calculate the dipole moment from the individual bond moments. It is often possible, however, to estimate qualitatively the direction and relative magnitude of the dipole moment by estimating the net direction of the bond polarities.

2.5

δ– δ+ δ–

:Ö=C=Ö:

The dipole moment of CO_2 is zero because the bond polarities of the two carbon-oxygen bonds cancel.

2.6

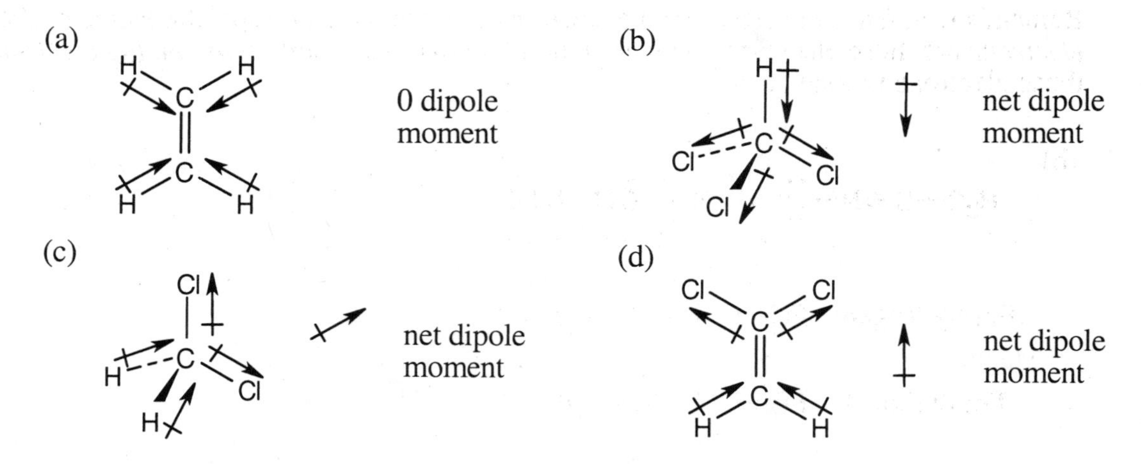

2.7 To find the formal charge of an atom in a molecule, follow these two steps:
(1) Draw a Lewis structure of the molecule.

H_3C–S^+(O^-)–CH_3 =

(2) Use the formula in Section 2.3 to determine formal charge for each atom. The periodic table shows the number of valence electrons of the element, and the Lewis structure shows the number of bonding and nonbonding electrons.

$$\text{Formal charge (FC)} = \left[\begin{matrix}\text{\# of valence}\\ \text{electrons}\end{matrix}\right] - \left[\frac{\text{\# of bonding electrons}}{2}\right] - \left[\begin{matrix}\text{\# nonbonding}\\ \text{electrons}\end{matrix}\right]$$

$$\text{For sulfur:}\quad FC = 6 - \frac{6}{2} - 2 = +1$$

$$\text{For oxygen:}\quad FC = 6 - \frac{2}{2} - 6 = -1$$

2.8

$$\text{Formal charge (FC)} = \left[\begin{matrix}\text{\# of valence}\\ \text{electrons}\end{matrix}\right] - \left[\frac{\text{\# of bonding electrons}}{2}\right] - \left[\begin{matrix}\text{\# nonbonding}\\ \text{electrons}\end{matrix}\right]$$

(a)

$H_2C{=}N{=}\ddot{N}{:}$ = H:C::N::N: (H above C; nitrogen 1 and nitrogen 2 labeled)

$$\text{For hydrogen:}\quad FC = 1 - \frac{2}{2} - 0 = 0$$

$$\text{For carbon:}\quad FC = 4 - \frac{8}{2} - 0 = 0$$

$$\text{For nitrogen 1:}\quad FC = 5 - \frac{8}{2} - 0 = +1$$

$$\text{For nitrogen 2:}\quad FC = 5 - \frac{4}{2} - 4 = -1$$

Remember: *Valence electrons* are the electrons characteristic of a specific element. *Bonding electrons* are those electrons involved in bonding to other atoms. *Nonbonding electrons* are those electrons in lone pairs.

(b)

$H_3C{-}C{\equiv}N{-}\ddot{O}{:}$ = H:C:C:::N:O: (H above and below carbon 1; carbon 1 and carbon 2 labeled)

$$\text{For hydrogen:}\quad FC = 1 - \frac{2}{2} - 0 = 0$$

$$\text{For carbon 1:}\quad FC = 4 - \frac{8}{2} - 0 = 0$$

$$\text{For carbon 2:}\quad FC = 4 - \frac{8}{2} - 0 = 0$$

$$\text{For nitrogen :}\quad FC = 5 - \frac{8}{2} - 0 = +1$$

$$\text{For oxygen:}\quad FC = 6 - \frac{2}{2} - 6 = -1$$

(c)

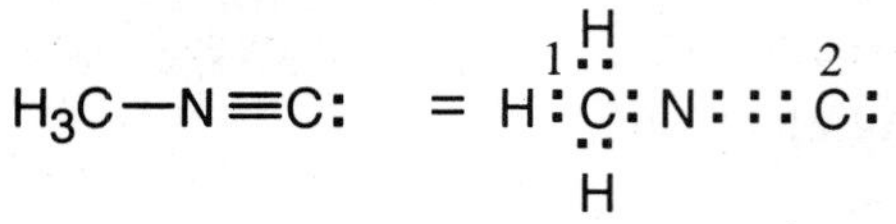

$$H_3C{-}N{\equiv}C: \;=\; H:\overset{H}{\underset{H}{\ddot{C}}}:N:::\ddot{C}:$$

For hydrogen: $FC = 1 - \frac{2}{2} - 0 = 0$

For carbon 1: $FC = 4 - \frac{8}{2} - 0 = 0$

For carbon 2: $FC = 4 - \frac{6}{2} - 2 = -1$

For nitrogen : $FC = 5 - \frac{8}{2} - 0 = +1$

2.9

(a) Nitrate anion has 3 three-atom groupings and thus has 3 resonance forms:

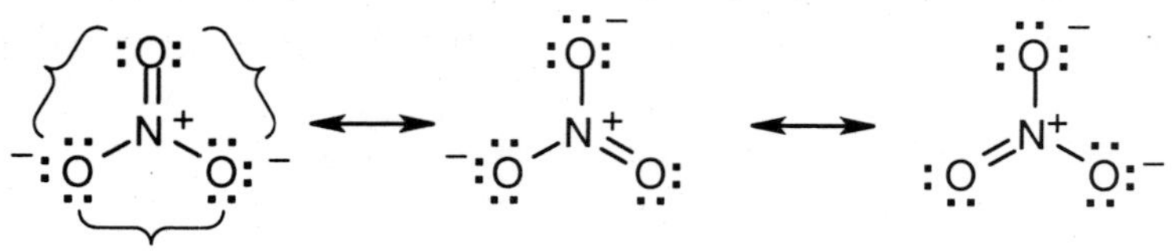

$$[O{=}N^+(O^-)(O^-)] \longleftrightarrow [^-O{-}N^+({=}O)(O^-)] \longleftrightarrow [O{=}N^+(O^-)(O^-)]$$

(b)

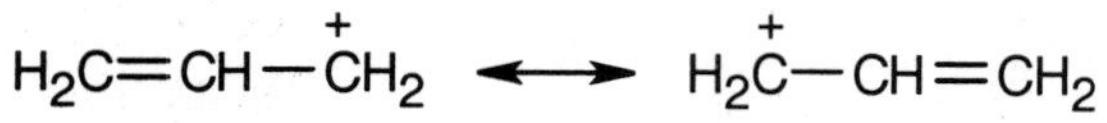

$$H_2C{=}CH{-}\overset{+}{C}H_2 \longleftrightarrow H_2\overset{+}{C}{-}CH{=}CH_2$$

(c)

$$:N{\equiv}\overset{+}{N}{-}\overset{-}{\ddot{N}}{-}H \longleftrightarrow {}^-:\ddot{N}{=}\overset{+}{N}{=}\ddot{N}{-}H$$

(d)

H H C C C Br C C C H H Br ⟷ H H C C C Br C C C H H Br

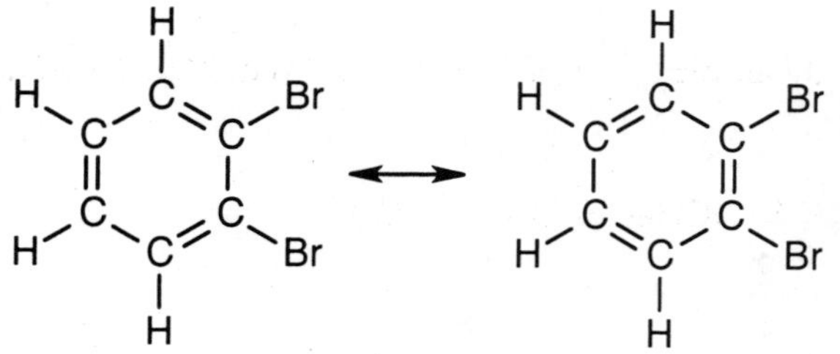

2.10 When an acid loses a proton, the product is the conjugate base of the acid. When a base gains a proton, the product is the conjugate acid of the base.

$$H{-}NO_3 + :NH_3 \rightleftharpoons NO_3^- + NH_4^+$$

Acid	Base	Conjugate base	Conjugate acid

2.11 Recall from Section 2.8 that a strong acid has a small pK_a and a weak acid has a large pK_a. Accordingly, picric acid ($pK_a = 0.38$) is a stronger acid than formic acid ($pK_a = 3.75$).

2.12 HO–H is a stronger acid than H_2N–H. Since H_2N^- is a stronger base than HO^-, the conjugate acid of H_2N^- (H_2N–H) is a weaker acid than the conjugate acid of HO^- (HO–H).

2.13

(a)

$$\text{H—CN} + CH_3COO^-\,Na^+ \xrightarrow{?} Na^+\,^-CN + CH_3COOH$$

$pK_a = 9.3$ Weaker acid (HCN); $pK_a = 4.7$ Stronger acid (CH_3COOH)

Remember that the lower the pK_a, the stronger the acid. Thus CH_3COOH, not HCN, is the stronger acid, and the above reaction will not take place in the direction written.

(b)

$$CH_3CH_2O\text{—H} + Na^+\,^-CN \xrightarrow{?} CH_3CH_2O^-\,Na^+ + HCN$$

$pK_a = 16$ Weaker acid (CH_3CH_2OH); $pK_a = 9.3$ Stronger acid (HCN)

Using the same reasoning as in part (a), we can see that the above reaction will not occur.

2.14

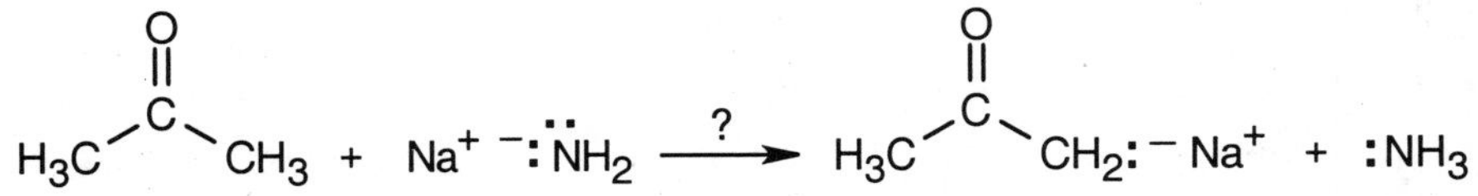

$$H_3C\text{—C(=O)—}CH_3 + Na^+\,^-{:}\ddot{N}H_2 \xrightarrow{?} H_3C\text{—C(=O)—}CH_2{:}^-\,Na^+ + {:}NH_3$$

$pK_a = 19$ Stronger acid (acetone); $pK_a = 36$ Weaker acid (NH_3)

The above reaction will take place as written.

2.15 Enter –9.31 into the calculator and use the INV LOG function to arrive at the answer $K_a = 4.9 \times 10^{-10}$.

2.16 The arrows show the movement of an electron pair from a Lewis base to a Lewis acid.

(a)

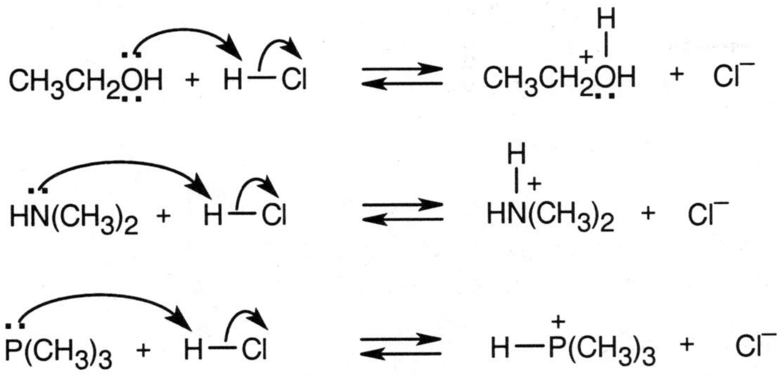

$$CH_3CH_2\ddot{O}H + \text{H—Cl} \rightleftharpoons CH_3CH_2\overset{+}{O}H_2 + Cl^-$$

$$\ddot{H}N(CH_3)_2 + \text{H—Cl} \rightleftharpoons H_2\overset{+}{N}(CH_3)_2 + Cl^-$$

$$\ddot{P}(CH_3)_3 + \text{H—Cl} \rightleftharpoons \text{H—}\overset{+}{P}(CH_3)_3 + Cl^-$$

(b)

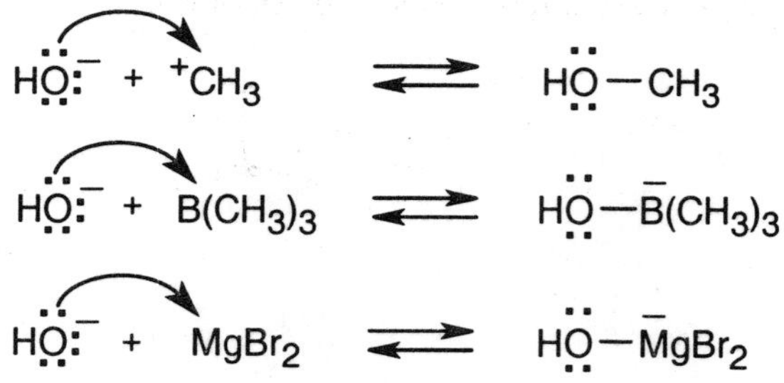

$$HO:^- + {}^+CH_3 \rightleftharpoons HO-CH_3$$

$$HO:^- + B(CH_3)_3 \rightleftharpoons HO-\bar{B}(CH_3)_3$$

$$HO:^- + MgBr_2 \rightleftharpoons HO-\bar{M}gBr_2$$

2.17

(a)

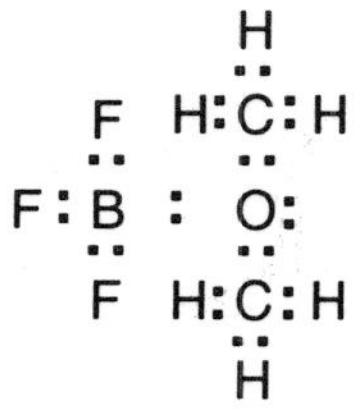

Boron: $FC = 3 - \frac{8}{2} - 0 = -1$

Oxygen: $FC = 6 - \frac{6}{2} - 2 = +1$

The formal charge of –1 for boron indicates that boron has a net negative charge; oxygen has a net positive charge.

(b)

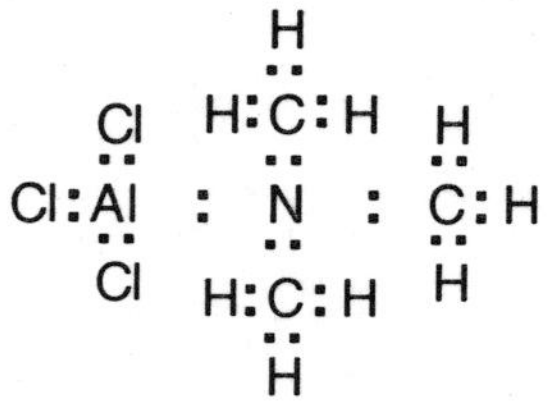

Aluminum: $FC = 3 - \frac{8}{2} - 0 = -1$

Nitrogen: $FC = 5 - \frac{8}{2} - 0 = +1$

The formal charge of –1 for aluminum indicates that it has a net negative charge; the formal charge +1 for nitrogen indicates that it has a net positive charge.
For clarity, electron dots have been left off F and Cl in the above structures.

2.18 Remember that the end of a line represents a carbon atom with 3 hydrogens, a two-way intersection represents a carbon atom with 2 hydrogens, a three-way intersection represents a carbon with 1 hydrogen and a four-way intersection represents a carbon with no hydrogens.

(a)

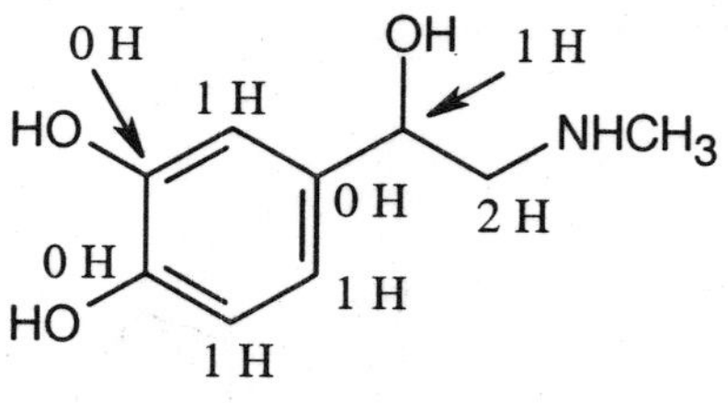

Adrenaline – $C_9H_{13}NO_3$

(b)

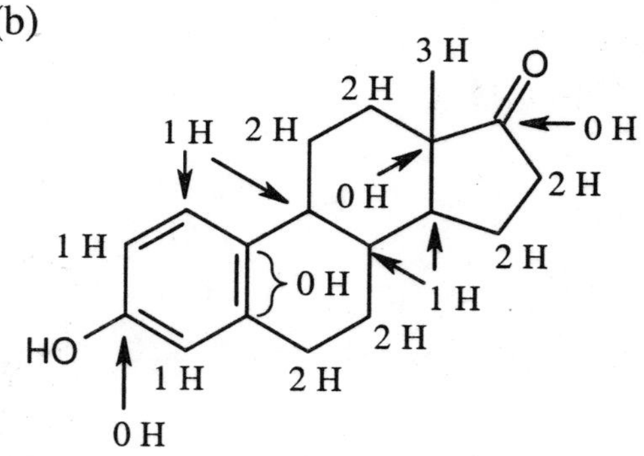

Estrone – $C_{18}H_{22}O_2$

2.19 Several possible skeletal structures can satisfy each molecular formula.

(a) C_5H_{12}

(b) C_2H_7N NH_2 NH

(c) C_3H_6O O OH OH O

OH O O O OH

(d) C_4H_9Cl Cl Cl Cl Cl

2.20

O C OH

H_2N

PABA

2.21 Two different representations of ethane are shown.

H H H C C H H H

H H H C H H H

Visualizing Chemistry

2.22

(a) (b) (c)

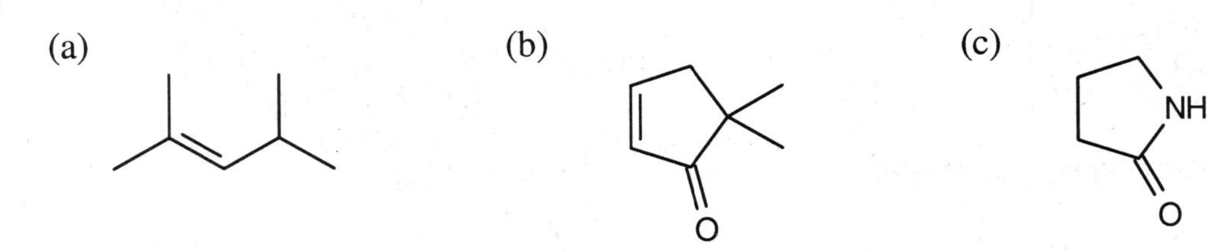

2.23 Naphthalene has 3 resonance forms.

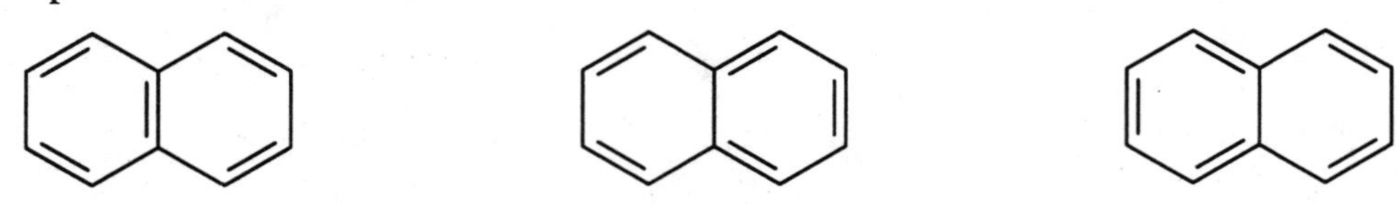

2.24

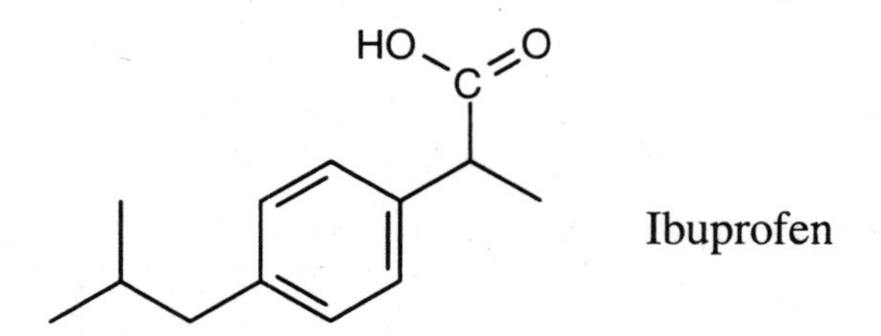

Additional Problems

2.25

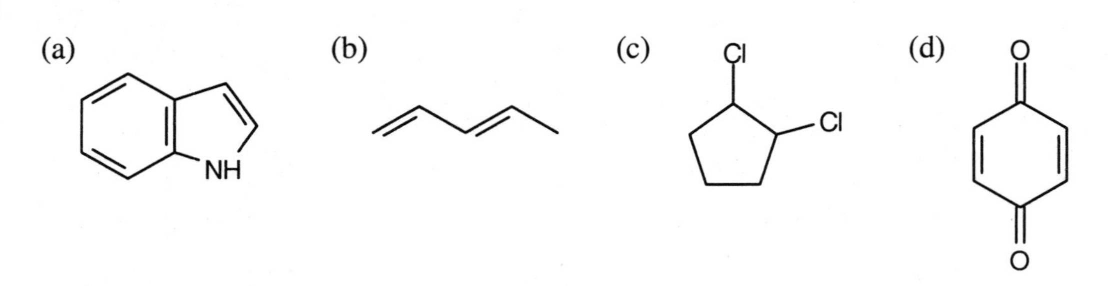

2.26

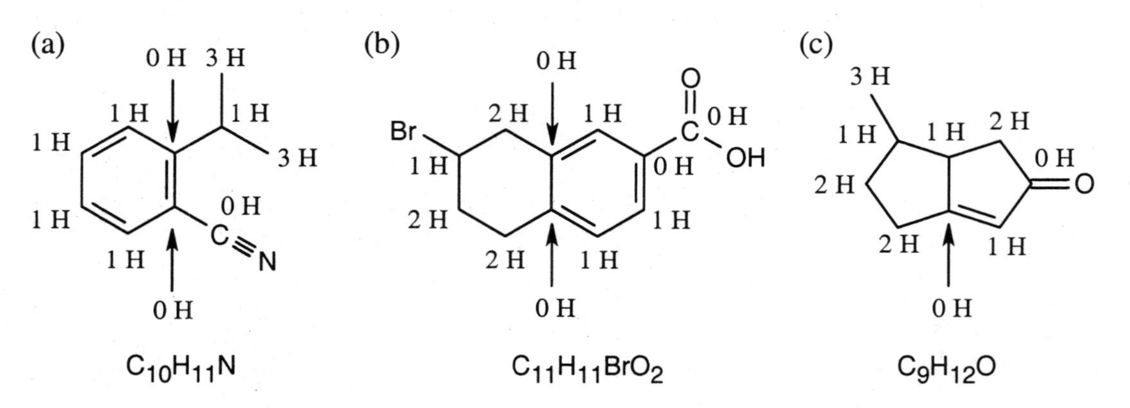

2.27 Use Figure 2.2 if you need help. The most electronegative element is starred.

(a) $CH_2\overset{*}{F}Cl$ (b) $\overset{*}{F}CH_2CH_2CH_2Br$ (c) $H\overset{*}{O}CH_2CH_2NH_2$ (d) $CH_3\overset{*}{O}CH_2Li$

2.28–2.29

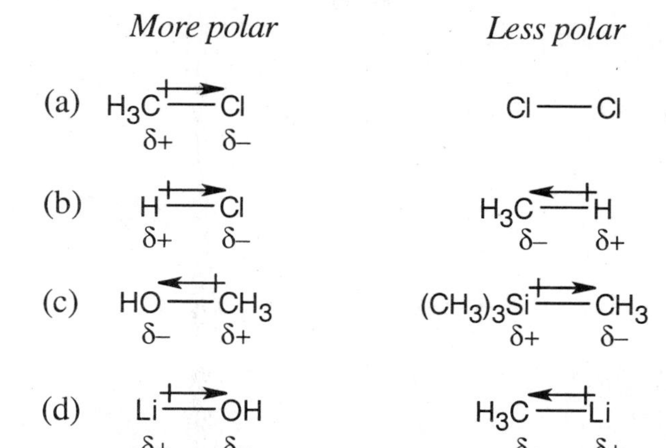

2.30

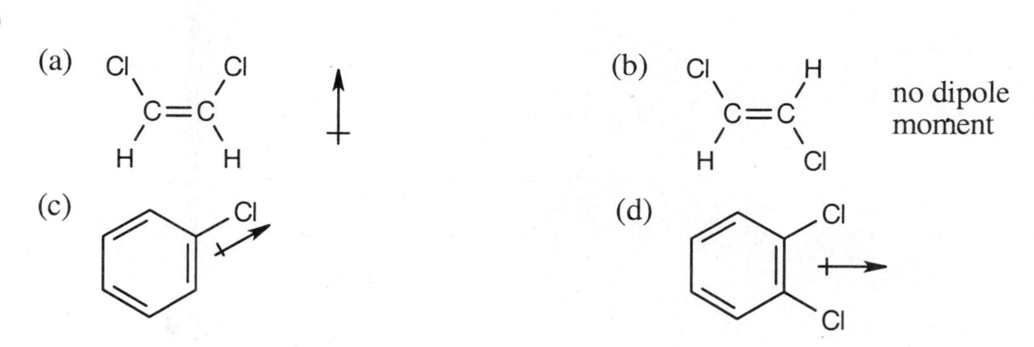

2.31 In phosgene, the individual bond polarities tend to cancel, but in formaldehyde, the bond polarities add to each other. Thus, phosgene has a smaller dipole moment.

Phosgene Formaldehyde

2.32 $\mu = Q \times r$. For a proton and an electron separated by 100 pm, μ = 4.8 D. If the two charges are separated by 136 pm, μ = 6.53 D. Since the observed dipole moment is 1.08 D, the H—Cl bond has (1.08 D / 6.53 D) x 100 % = 16.5 % ionic character.

2.33 The magnitude of a dipole moment depends on both charge and distance between atoms. Fluorine is more electronegative than chlorine, but a C–F bond is shorter than a C–Cl bond. Thus, the dipole moment of CH_3F is smaller than that of CH_3Cl.

2.34 To save space, molecules are shown as line-bond structures with lone pairs, rather than as Lewis structures.

(a) $(CH_3)_2\ddot{O}BF_3$

Oxygen: $FC = 6 - \frac{6}{2} - 2 = +1$

Boron: $FC = 3 - \frac{8}{2} - 0 = -1$

(b) $H_2\ddot{C}-\overset{1}{N}\equiv\overset{2}{N}:$

Carbon: $FC = 4 - \frac{6}{2} - 2 = -1$

Nitrogen 1: $FC = 5 - \frac{8}{2} - 0 = +1$

Nitrogen 2: $FC = 5 - \frac{6}{2} - 2 = 0$

(c) $H_2C=\overset{1}{N}=\overset{2}{\ddot{N}}:$

Carbon: $FC = 4 - \frac{8}{2} - 0 = 0$

Nitrogen 1: $FC = 5 - \frac{8}{2} - 0 = +1$

Nitrogen 2: $FC = 5 - \frac{4}{2} - 4 = -1$

(d) $:\overset{1}{\ddot{O}}=\overset{2}{\ddot{O}}-\overset{3}{\ddot{O}}:$

Oxygen 1: $FC = 6 - \frac{4}{2} - 4 = 0$

Oxygen 2: $FC = 6 - \frac{6}{2} - 2 = +1$

Oxygen 3: $FC = 6 - \frac{2}{2} - 6 = -1$

(e) $H_2\ddot{C}-P(CH_3)_3$

Carbon: $FC = 4 - \frac{6}{2} - 2 = -1$

Phosphorus: $FC = 5 - \frac{8}{2} - 0 = +1$

(f) $C_5H_5N-\ddot{O}:$ (pyridine N-oxide)

Nitrogen: $FC = 5 - \frac{8}{2} - 0 = +1$

Oxygen: $FC = 6 - \frac{2}{2} - 6 = -1$

2.35 Resonance forms do not differ in the position of nuclei. The two structures in (a) are not resonance forms because the carbon and hydrogen atoms outside the ring occupy different positions in the two forms.

not resonance structures

The pairs of structures in parts (b), (c), and (d) represent resonance forms.

2.36

(a) $H_3C-C(=\ddot{O}:)-\ddot{C}H_2^{-} \longleftrightarrow H_3C-C(-\ddot{O}:^{-})=CH_2$

(b)

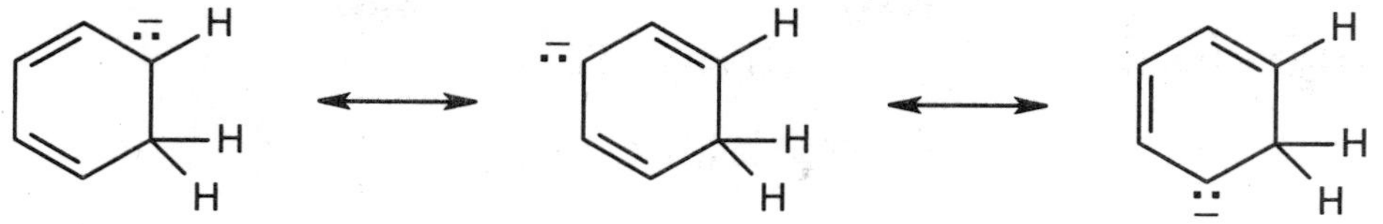

(c)

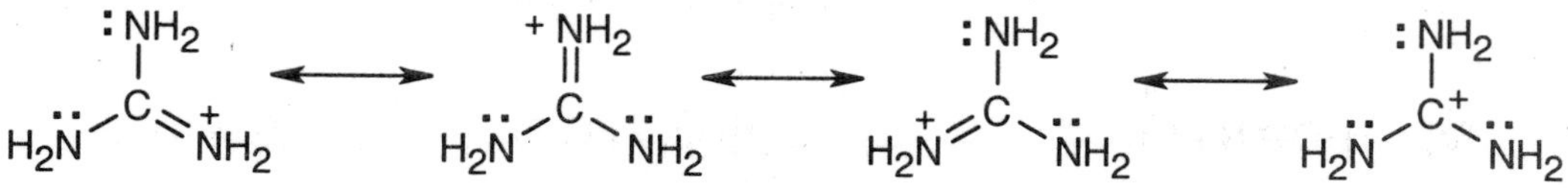

(d) $H_3C{-}\ddot{\underset{..}{S}}{-}\overset{+}{C}H_2 \longleftrightarrow H_3C{-}\overset{+}{\ddot{S}}{=}CH_2$

(e) $H_2C{=}CH{-}\overset{+}{C}H_2 \longleftrightarrow H_2\overset{+}{C}{-}CH{=}CH_2$

(f) $H_2C{=}CH{-}CH{=}CH{-}\overset{+}{C}H{-}CH_3 \longleftrightarrow H_2C{=}CH{-}\overset{+}{C}H{-}CH{=}CH{-}CH_3$

$\longleftrightarrow H_2\overset{+}{C}{-}CH{=}CH{-}CH{=}CH{-}CH_3$

2.37 The two structures are not resonance forms because the position of the carbon atoms is different in the two forms.

2.38

$$CH_3\ddot{\underset{..}{O}}H + HCl \rightleftharpoons CH_3\ddot{O}H_2^{+} + Cl^{-}$$

$$CH_3\ddot{\underset{..}{O}}H + Na^{+}\ {}^{-}\ddot{\underset{..}{N}}H_2 \rightleftharpoons CH_3\ddot{\underset{..}{O}}{:}^{-}\ Na^{+} + {:}NH_3$$

2.39

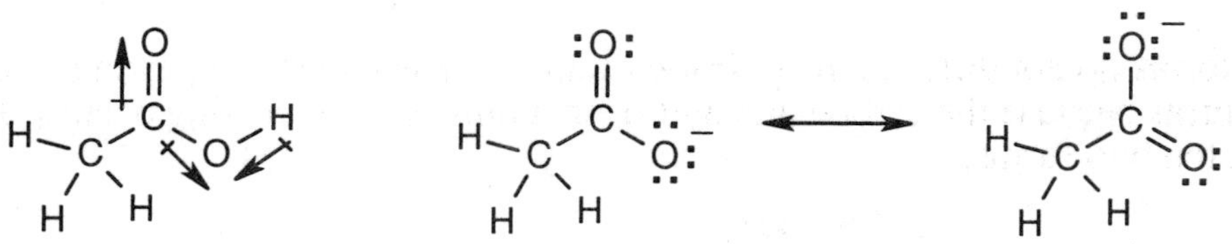

The O–H hydrogen of acetic acid is more acidic than the C–H hydrogens. The –OH oxygen is quite electronegative, and, consequently, the –O–H bond is more strongly polarized than the –C–H bonds.

2.40

Lewis acids: $AlBr_3$, BH_3, HF, $TiCl_4$

Lewis bases: $CH_3CH_2\ddot{N}H_2$, $H_3C{-}\ddot{\underset{..}{S}}{-}CH_3$

2.41

(a)

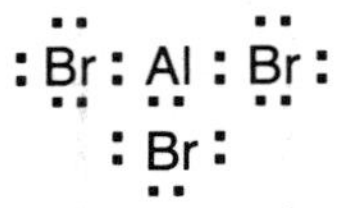

```
 ..      ..
:Br: Al :Br:
 ..  ..  ..
    :Br:
     ..
```

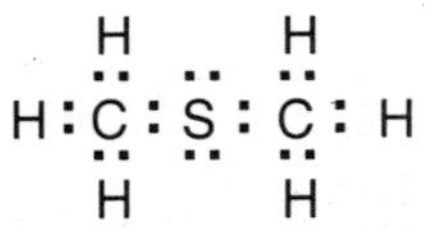

(b)

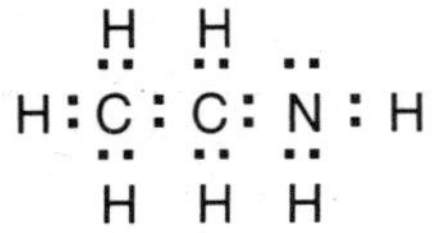

```
   H  H
   .. ..  ..
H :C :C : N : H
   .. ..  ..
   H  H   H
```

(c)

```
H : B : H
    ..
    H
```

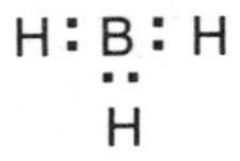

(d)

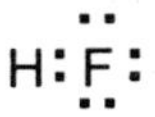

```
   ..
H :F:
   ..
```

(e)

```
   H     H
   .. .. ..
H :C :S :C : H
   .. .. ..
   H     H
```

(f)

```
       ..
      :Cl:
  ..   ..  ..
:Cl : Ti : Cl:
  ..   ..  ..
      :Cl:
       ..
```

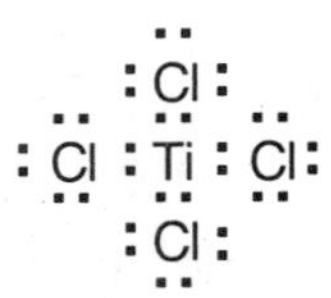

2.42

(a) CH_3OH + H_2SO_4 $\rightleftharpoons$ $CH_3OH_2^+$ + HSO_4^-

CH_3OH	H_2SO_4	$CH_3OH_2^+$	HSO_4^-
stronger base	stronger acid	weaker acid	weaker base

(b) CH_3OH + $NaNH_2$ $\rightleftharpoons$ $CH_3O^-\ Na^+$ + NH_3

CH_3OH	$NaNH_2$	$CH_3O^-\ Na^+$	NH_3
stronger acid	stronger base	weaker base	weaker acid

(c) $CH_3NH_3^+\ Cl^-$ + $NaOH$ $\rightleftharpoons$ CH_3NH_2 + H_2O + $NaCl$

$CH_3NH_3^+\ Cl^-$	$NaOH$	CH_3NH_2	H_2O
stronger acid	stronger base	weaker base	weaker acid

2.43 As in Problem 2.34, molecules are shown as line-bond structures with lone-pair electrons indicated. Only calculations for atoms with non-zero formal charge are shown.

(a)

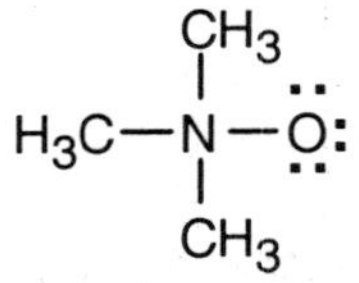

```
      CH3
      |    ..
H3C — N — O:
      |    ..
      CH3
```

Oxygen: $FC = 6 - \frac{2}{2} - 6 = -1$

Nitrogen: $FC = 5 - \frac{8}{2} - 0 = +1$

(b)

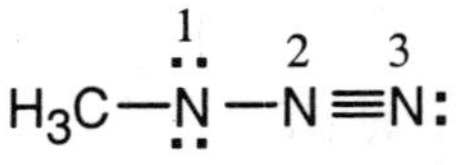

```
      1
      ..   2   3
H3C — N — N ≡ N:
      ..
```

Nitrogen 1: $FC = 5 - \frac{4}{2} - 4 = -1$

Nitrogen 2: $FC = 5 - \frac{8}{2} - 0 = +1$

Nitrogen 3: $FC = 5 - \frac{6}{2} - 2 = 0$

(c)

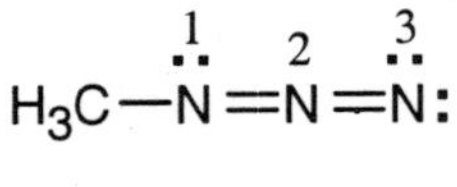

```
      1   2   3
      ..      ..
H3C — N = N = N:
```

Nitrogen 1: $FC = 5 - \frac{6}{2} - 2 = 0$

Nitrogen 2: $FC = 5 - \frac{8}{2} - 0 = +1$

Nitrogen 3: $FC = 5 - \frac{4}{2} - 4 = -1$

2.44 The substances with the largest values of pK_a are the least acidic.

Least acidic ⟶ *Most acidic*

CH_3COCH_3 < C_6H_5OH < $CH_3COCH_2COCH_3$ < CH_3COOH

pK_a = 19 pK_a = 9.9 pK_a = 9 pK_a = 4.7

2.45 To react completely (> 99.9%) with NaOH, an acid must have a pK_a at least 3 units smaller than the pK_a of H_2O. Thus, all substances in the previous problem except acetone react completely with NaOH.

2.46 The stronger the acid (smaller pK_a), the weaker its conjugate base. Since NH_4^+ is a stronger acid than $CH_3NH_3^+$, NH_3 is a weaker base than CH_3NH_2.

2.47

$$(CH_3)_3C{-}O^- K^+ + H_2O \longrightarrow (CH_3)_3C{-}OH + K^+ \ {}^-OH$$

pK_a = 15.7 (stronger acid) pK_a ≈ 18 (weaker acid)

The reaction will take place as written because water is a stronger acid than *tert*-butyl alcohol. Thus, a solution of potassium *tert*-butoxide in water can't be prepared.

2.48

$$C_5H_5N: + H{-}O{-}C(=O){-}CH_3 \rightleftharpoons C_5H_5N^+{-}H + {}^-:O{-}C(=O){-}CH_3$$

2.49 (a) Acetone: $K_a = 5 \times 10^{-20}$ (b) Formic acid: $K_a = 1.8 \times 10^{-4}$

2.50 (a) Nitromethane: pK_a = 10.30 (b) Acrylic acid: pK_a = 4.25

2.51

$$\underset{[0.050\ M]}{\text{Formic acid}} + H_2O \overset{K_a}{\rightleftharpoons} \underset{[x]}{\text{Formate}^-} + \underset{[x]}{H_3O^+}$$

$$K_a = 1.8 \times 10^{-4} = \frac{x^2}{0.050 - x}$$

If you let 0.050 – x = 0.050, then x = 3.0×10^{-3} and pH = 2.52. If you calculate x exactly, then x = 2.9×10^{-3} and pH = 2.54.

2.52 Only acetic acid will react with sodium bicarbonate. Acetic acid is the only substance in Problem 2.44 that is a stronger acid than carbonic acid.

2.53 Sodium bicarbonate reacts with acetic acid to produce carbonic acid, which breaks down to form CO_2. Bubbles of CO_2 indicate the presence of an acid stronger than carbonic acid, in this case acetic acid. Phenol does not react with sodium bicarbonate.

2.54 Reactions (a), (b) and (d) are reactions between Brønsted-Lowry acids and bases; the stronger acid and stronger base are identified. Reactions (c) and (e) occur between Lewis acids and bases.

(a)

$$CH_3OH + H^+ \longrightarrow CH_3\overset{+}{O}H_2$$

base acid

(b)

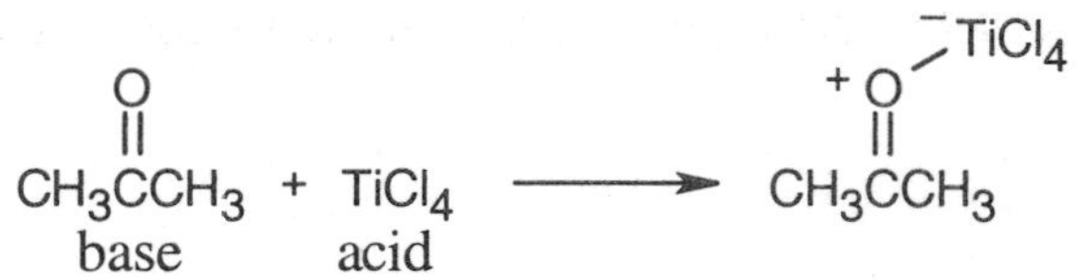

$$CH_3\overset{O}{\overset{\|}{C}}CH_3 + TiCl_4 \longrightarrow CH_3\overset{\overset{+}{O}-\overset{-}{T}iCl_4}{\overset{\|}{C}}CH_3$$

base acid

(c)

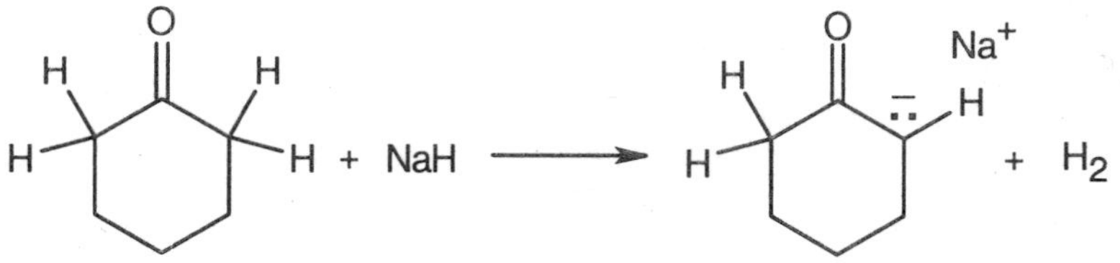

2,2,6,6-tetrahydrocyclohexanone + NaH ⟶ enolate-like anion, Na^+ + H_2

acid base

(d)

morpholine (H–N) + BH_3 ⟶ morpholine–N^+(H)–$^-BH_3$

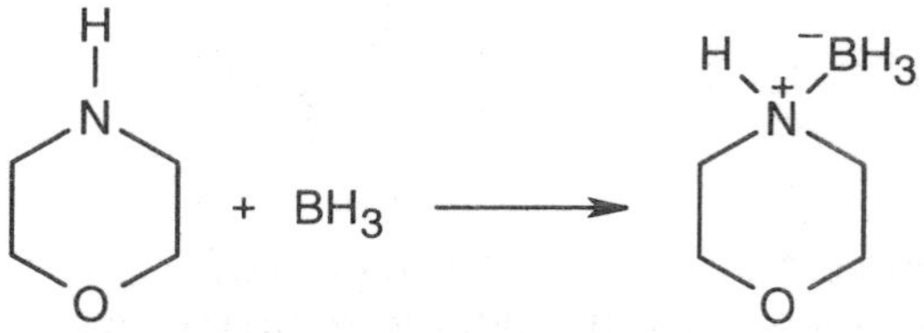

base acid

2.55 Pairs (a) and (d) represent resonance structures; pairs (b) and (c) do not. In order for two structures to be resonance forms, all atoms must be in the same position in all resonance forms.

2.56

(a)

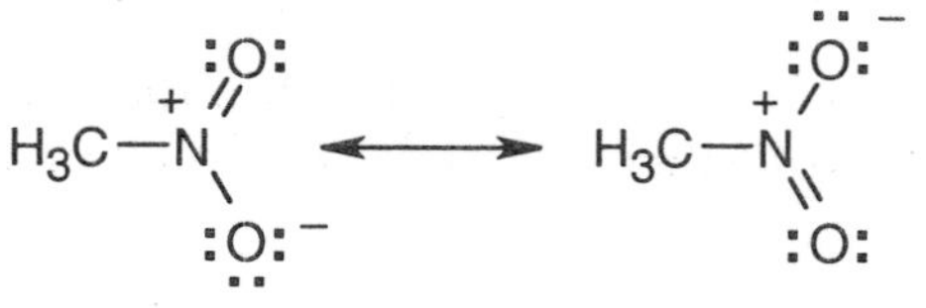

$$H_3C-\overset{+}{N}(=\ddot{O}:)(-\ddot{O}:^-) \longleftrightarrow H_3C-\overset{+}{N}(-\ddot{O}:^-)(=\ddot{O}:)$$

(b)

$$\ddot{O}=\overset{+}{\ddot{O}}-\ddot{O}:^- \longleftrightarrow {}^-:\ddot{O}-\overset{+}{\ddot{O}}=\ddot{O}$$

(c)

$$H_2C=\overset{+}{N}=\ddot{N}:^- \longleftrightarrow H_2\overset{-}{\ddot{C}}-\overset{+}{N}\equiv N: \longleftrightarrow H_2\overset{-}{\ddot{C}}-\ddot{N}=\overset{+}{N}:$$

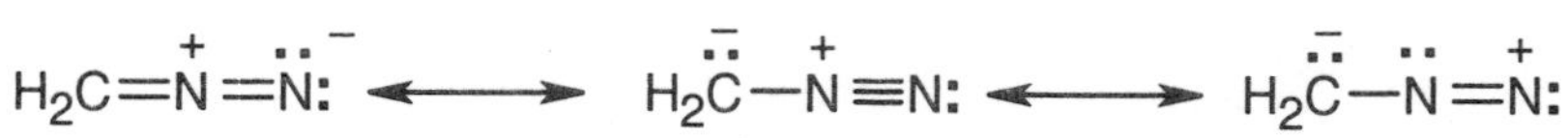

2.57

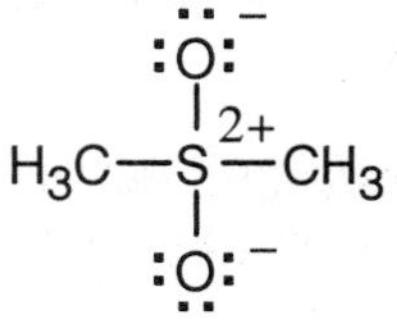

:O:⁻ | H3C—S(2+)—CH3 | :O:⁻

Each oxygen: $FC = 6 - \frac{2}{2} - 6 = -1$

Sulfur: $FC = 6 - \frac{8}{2} - 0 = +2$

The presence of formal charges indicates that electrons are not shared equally between S and O - they are strongly attracted to oxygen. The S–O bonds, therefore, are strongly polar. If the geometry of the molecule were planar (as drawn above), the dipole moments of the individual S–O bonds would cancel, resulting in a net dipole moment of zero. If dimethyl sulfone had tetrahedral geometry, however, a large dipole moment would be expected.

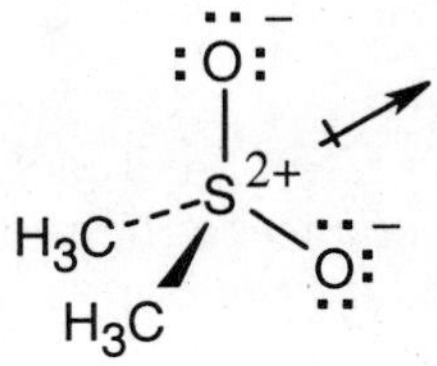

:O:⁻ | H_3C, H_3C, S^{2+}, :O:⁻

2.58

:O: || H_3C—C—O—H $\xrightleftharpoons{H_2SO_4}$:O: || H_3C—C—$\overset{+}{O}$(H)—H or $\overset{+}{O}$—H || H_3C—C—O—H

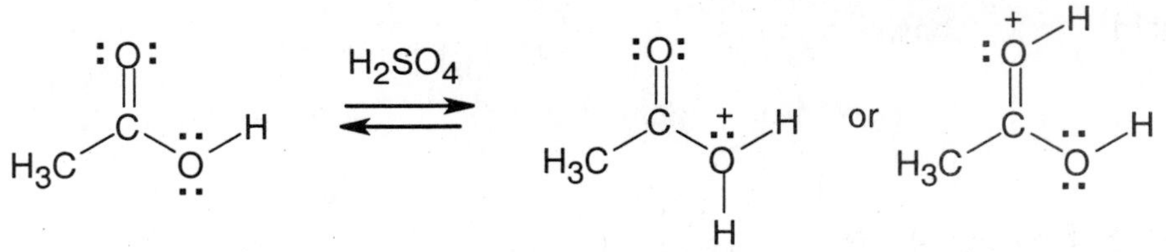

Acetic acid — protonation of single-bond oxygen — protonation of double-bond oxygen

Protonation of the single bond oxygen produces the structure illustrated above. The product formed by protonation of the double bond oxygen can be represented by the structure pictured above and by two other resonance forms.

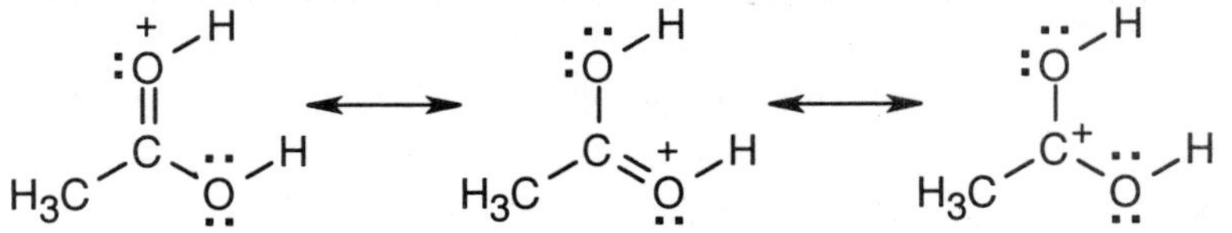

H_3C—C(=$\overset{+}{O}$—H)—O—H ⟷ H_3C—C(—O—H)=$\overset{+}{O}$—H ⟷ H_3C—C^+(—O—H)—O—H

The product formed by protonation of the double-bond oxygen is more stable because it is stabilized by resonance: the more resonance forms that can be drawn for a molecule, the more stable it is.

A Look Ahead

2.59

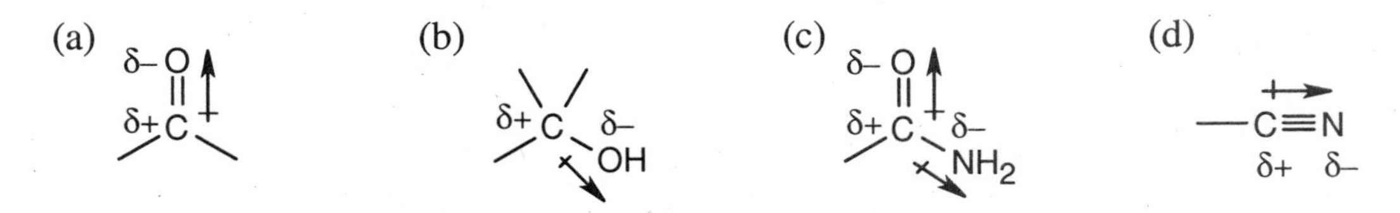

2.60

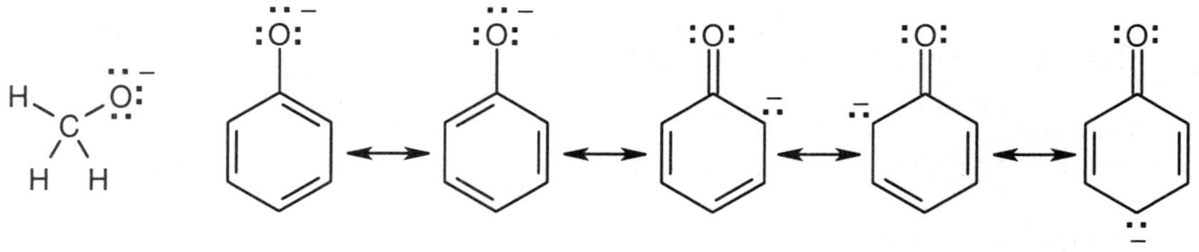

When phenol loses a proton, the resulting anion is stabilized by resonance.

Molecular Modeling

2.61 The anti conformation has zero dipole moment because the bond dipoles exactly cancel in the anti conformation.

2.62 The nitrogen atom in $CH_3NH_3^+$ and the oxygen atom in $CH_3OH_2^+$ each carry a formal charge of +1. The –NH and –OH hydrogen atoms are seen to be the most positive atoms in the electrostatic potential maps of the two ions. Formal charges don't always give the correct charges.

2.63

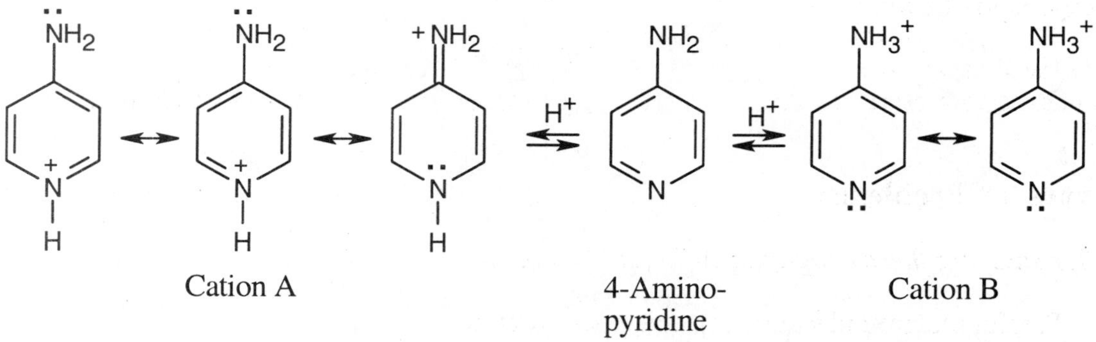

Cation A

4-Amino-
pyridine

Cation B

Cation A has greater resonance stabilization because its charge is delocalized over both the $-NH_2$ and –NH nitrogens, whereas the charge of cation B is localized on the $-NH_3$ nitrogen.

2.64 The –OH hydrogen is more positively charged than the other hydrogens.

Review Unit 1: Bonds and Bond Polarity

Major Topics Covered (with vocabulary:)

Atomic Structure:
atomic number   mass number   wave equation   orbital   shell   node   electron configuration

Chemical Bonding Theory:
covalent bond   Lewis structure   lone-pair electrons   line-bond structure   valence-bond theory   sigma (σ) bond   bond strength   bond length   molecular orbital theory   bonding MO   antibonding MO

Hybridization:
sp^3 hybrid orbital   bond angle sp^2 hybrid orbital   pi (π) bond   *sp* hybrid orbital

Polar covalent bonds:
polar covalent bond   electronegativity (EN)   inductive effect   dipole moment   formal charge

Resonance:
resonance form   resonance hybrid

Acids and Bases:
Brønsted-Lowry acid   Brønsted-Lowry base   conjugate acid   conjugate base   acidity constant   Lewis acid   Lewis base

Chemical Structures:
condensed structure   skeletal structure   space-filling models   ball-and-stick models

Types of Problems:

After studying these chapters you should be able to:

- Predict the ground state electronic configuration of atoms.
- Draw Lewis electron-dot structures of simple compounds.
- Predict and describe the hybridization of bonds in simple compounds.
- Predict bond angles and shapes of molecules.

- Predict the direction of polarity of a chemical bond, and predict the dipole moment of a simple compound.
- Calculate formal charge for atoms in a molecule.
- Draw resonance forms of molecules.
- Predict the relative acid/base strengths of Brønsted acids and bases.
- Predict the direction of Brønsted acid/base reactions.
- Calculate: pK_a from K_a, and vice versa.
  pH of a solution of a weak acid.
- Identify Lewis acids and bases.
- Draw chemical structures from molecular formulas, and vice versa.

Points to Remember:

* In order for carbon, with valence shell electron configuration of $2s^22p^2$, to form four sp^3 hybrid orbitals, it is necessary that one electron be promoted from the $2s$ subshell to the $2p$ subshell. Although this promotion requires energy, the resulting hybrid orbitals are able to form stronger bonds, and compounds containing these bonds are more stable.

* Assigning formal charge to atoms in a molecule is helpful in showing where the electrons in a bond are located. Even if a bond is polar covalent, in some molecules the electrons "belong" more to one of the atoms than the other. This "ownership" is useful for predicting the outcomes of chemical reactions, as we will see in later chapters.

* Resonance structures are representations of the distribution of π and nonbonding electrons in a molecule. Electrons don't move around in the molecule, and the molecule doesn't change back and forth, from structure to structure. Rather, resonance structures are an attempt to show, by conventional line-bond drawings, the electron distribution of a molecule that can't be represented by any one structure.

* As in general chemistry, acid-base reactions are of fundamental importance in organic chemistry. Organic acids and bases, as well as inorganic acids and bases, occur frequently in reactions, and large numbers of reactions are catalyzed by Brønsted acids and bases and Lewis acids and bases.

Self-Test:

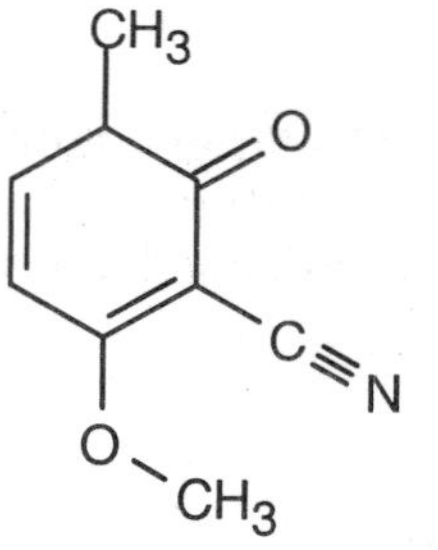

A
Ricinine
(a toxic component of castor beans)

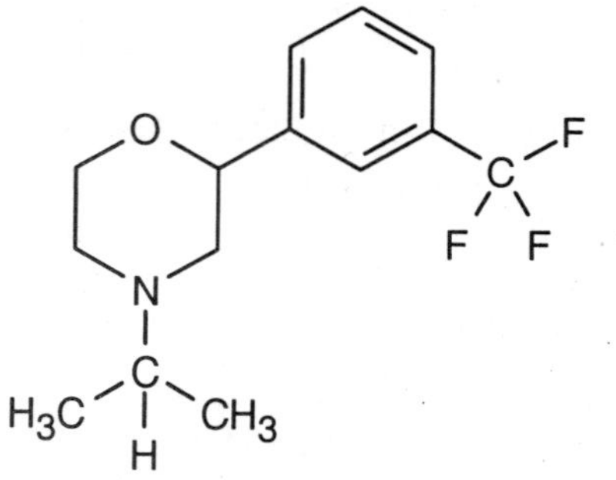

B
Oxaflozane
(an antidepressant)

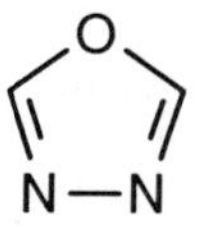

C
1,3,4-Oxadiazole

For **A** (ricinine) and **B** (oxaflozane): Add all missing electron lone pairs. Identify the hybridization of all carbons. Indicate the direction of bond polarity for all bonds with Δ EN ≥ 0.5. Which bond is the most polar? Convert **A** and **B** to molecular formulas.

Draw a resonance structure for **B**. Which atom (or atoms) of **B** can act as a Lewis base?

Add missing electron lone pairs to **C**. Is it possible to draw resonance forms for **C**? If so, draw at least one resonance form, and describe it.

Multiple Choice:

1. Which element has $4s^2 4p^2$ as its valence shell electronic configuration?
(a) Ca (b) C (c) Al (d) Ge

2. Which compound (or group of atoms) has an oxygen with a +1 formal charge?
(a) NO_3^- (b) O_3 (c) acetone anion (d) acetate anion

The following questions involve these acids: (i) HW ($pK_a = 2$); (ii) HX ($pK_a = 6$); (iii) HY ($pK_a = 10$); (iv) HZ ($pK_a = 20$).

3. Which of the above acids react almost completely with water to form hydroxide ion?
(a) none of them (b) all of them (c) HY and HZ (d) HZ

4. The conjugate bases of which of the above acids react almost completely with water to form hydroxide ion?
(a) none of them (b) all of them (c) HZ (d) HY and HZ

5. If you want to convert HX to X^-, which bases can you use?
(a) W^- (b) Y^- (c) Z^- (d) Y^- or Z^-

6. If you add equimolar amounts of HW, X^- and HY to a solution, what are the principal species in the resulting solution?
(a) HW, HX, HY (b) W^-, HX, HY (c) HW, X^-, HY (d) HW, HX, Y^-

7. What is the approximate pH difference between a solution of 1 M HX and a solution of 1 M HY?
(a) 2 (b) 3 (c) 4 (d) 6

8. If you wanted to write the structure of a molecule that shows carbon and hydrogen atoms as groups, without indicating many of the carbon-hydrogen bonds, you would draw a:
(a) molecular formula (b) Kekulé structure (c) skeletal structure (d) condensed structure

9. Which of the following molecules has zero net dipole moment?

```
(a)                (b)                (c)                (d)
 H       Cl         H       Cl         H       Cl         Cl      Cl
  \     /            \     /            \     /            \     /
   C=C                C=C                C=C                C=C
  /     \            /     \            /     \            /     \
 Cl      H          H       Cl         H       H          H       H
```

10. In which of the following bonds is carbon the more electronegative element?
(a) C—Br (b) C—I (c) C—P (d) C—S

Chapter 3 – Organic Compounds: Alkanes and Cycloalkanes

Chapter Outline

I. Functional Groups (Section 3.1).
   A. Functional groups are groups of atoms within a molecule that have a characteristic chemical behavior.
   B. The chemistry of every organic molecule is determined by its functional groups.
   C. Functional groups described in this text can be grouped into three categories:
      1. Functional groups with carbon-carbon multiple bonds.
      2. Groups in which carbon forms a single bond to an electronegative atom.
      3. Groups with a carbon-oxygen double bond.

II. Alkanes (Sections 3.2 – 3.5).
   A. Alkanes and alkane isomers (Section 3.2).
      1. Alkanes are formed by overlap of carbon sp^3 orbitals.
      2. Alkanes are described as saturated hydrocarbons.
         a. They are hydrocarbons because they contain only carbon and hydrogen.
         b. They are saturated because all bonds are single bonds.
         c. The general formula for alkanes is C_nH_{2n+2}.
      3. For alkanes with four or more carbons, the carbons can be connected in more than one way.
         a. If the carbons are in a row, the alkane is a straight-chain alkane.
         b. If the carbon chain has a branch, the alkane is a branched-chain alkane.
      4. Alkanes with the same molecular formula are isomers.
         a. Isomers whose atoms are connected differently are constitutional isomers.
            Constitutional isomers are always different compounds with different properties but with the same molecular formula.
         b. A given alkane can be drawn in many ways.
      5. Straight-chain alkanes are named according to the number of carbons in their chain.
   B. Alkyl groups (Section 3.3).
      1. An alkyl group is the partial structure that results from the removal of a hydrogen atom from an alkane.
         a. Alkyl groups are named by replacing the -ane of an alkane by -yl.
         b. *n*-Alkyl groups are formed by removal of an end carbon of a straight-chain alkane.
         c. Branched-chain alkyl groups are formed by removal of a hydrogen atom from an internal carbon.
            The prefixes *sec*- and *tert*- refer to the degree of substitution at the branching carbon atom.
      2. There are four possible degrees of alkyl substitution for carbon.
         a. A primary carbon is bonded to one other carbon.
         b. A secondary carbon is bonded to two other carbons.
         c. A tertiary carbon is bonded to three other carbons.
         d. A quaternary carbon is bonded to four other carbons.
         e. The symbol **R** refers to the rest of the molecule.
      3. Hydrogens are also described as primary, secondary and tertiary.
         a. Primary hydrogens are bonded to primary carbons (RCH_3).
         b. Secondary hydrogens are bonded to secondary carbons (R_2CH_2).
         c. Tertiary hydrogens are bonded to tertiary carbons (R_3CH).

C. Naming alkanes (Section 3.4).
1. The system of nomenclature used in this book is the IUPAC system.
In this system, a chemical name has a prefix, a parent and a suffix.
i. The parent shows the number of carbons in the principal chain.
ii. The suffix identifies the functional group family.
iii. The prefix shows the location of functional groups.
2. Naming an alkane:
a. Find the parent hydrocarbon.
i. Find the longest continuous chain of carbons, and use its name as the parent name.
ii. If two chains have the same number of carbons, choose the one with more branch points.
b. Number the atoms in the parent chain.
i. Start numbering at the end nearer the first branch point.
ii. If branching occurs an equal distance from both ends, begin numbering at the end nearer the second branch point.
c. Identify and number the substituents.
i. Give each substituent a number that corresponds to its position on the parent chain.
ii. Two substituents on the same carbon receive the same number.
d. Write the name as a single word.
i. Use hyphens to separate prefixes and commas to separate numbers.
ii. Use the prefixes, *di-*, *tri-*, *tetra-* if necessary, but don't use them for alphabetizing.
e. Name a complex substituent as if it were a compound, and set it off in parentheses.
i. Some simple branched-chain alkyl groups have common names.
ii. The prefix *iso* is used for alphabetizing, but *sec-* and *tert-* are not.
C. Properties of alkanes (Section 3.5).
1. Alkanes are chemically inert to most laboratory reagents.
2. Alkanes react with O_2 and Cl_2.
3. The boiling points and melting points of alkanes increase with increasing molecular weight.
a. This effect is due to van der Waals forces.
b. The strength of van der Waals forces increases with increasing molecular weight.
4. Increased branching lowers an alkane's boiling point.
III. Cycloalkanes (Sections 3.6 – 3.8).
A. Properties of cycloalkanes (Section 3.6).
1. Cycloalkanes have the general formula C_nH_{2n}, if they have one ring.
2. Compounds with cycloalkane rings are common in nature.
3. The effect of ring size on the melting points of cycloalkanes is irregular, but boiling points increase with increasing molecular weight.
B. Naming cycloalkanes (Section 3.7).
1. Find the parent.
a. If the number of carbon atoms in the ring is larger than the number in the largest substituent, the compound is named as an alkyl-substituted cycloalkane.
b. If the number of carbon atoms in the ring is smaller than the number in the largest substituent, the compound is named as an cycloalkyl-substituted alkane.
2. Number the substituents.
a. Start at a point of attachment and number the substituents so that the second substituent has the lowest possible number.
b. If necessary, proceed to the next substituent until a point of difference is found.

c. If two or more substituents might potentially receive the same number, number them by alphabetical priority.
d. Halogens are treated in the same way as alkyl groups.

C. Cis–trans isomerism in cycloalkanes (Section 3.8).
1. Unlike open-chain alkanes, cycloalkanes have much less rotational freedom.
   a. Very small rings are rigid.
   b. Large rings have more rotational freedom.
2. Cycloalkanes have a "top" side and a "bottom" side.
   a. If two substituents are on the same side of a ring, the ring is cis-disubstituted.
   b. If two substituents are on opposite sides of a ring,the ring is trans-disubstituted.
3. Substituents in the two types of disubstituted cycloalkanes are connected in the same order but differ in spatial orientation.
   a. These cycloalkenes are stereoisomers that are known as cis–trans isomers.
   b. Cis–trans isomers are stable compounds that can't be interconverted.

Solutions to Problems

3.1 Notice that certain functional groups have different designations if other functional groups are present in a molecule. For example, a molecule containing a carbon–carbon double bond, and no other functional group, is an alkene; if other groups are present, the group is referred to as a carbon–carbon double bond. Similarly, a compound containing a benzene ring, and only carbon- and hydrogen-containing substituents, is an arene; if other groups are present, the ring is labeled an "aromatic ring".

(a) amide; C–C double bond
(b) amine; carboxylic acid
(c) ketone; ester; C–C double bond
(d) $CH{=}CH_2$; C–C double bond; CH_2OH; aromatic ring; alcohol

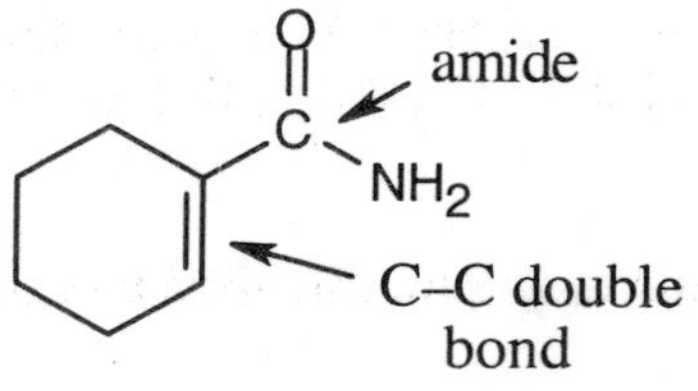

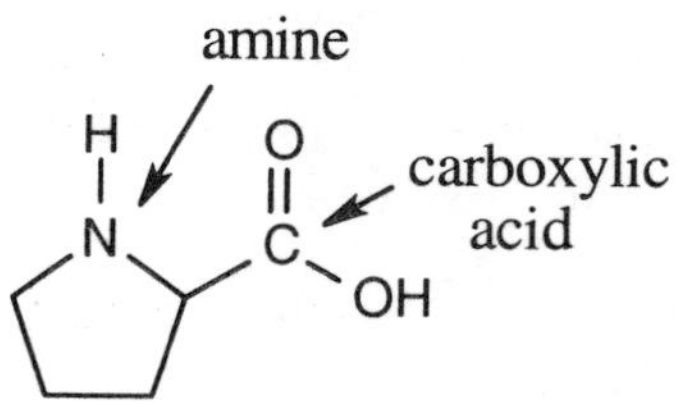

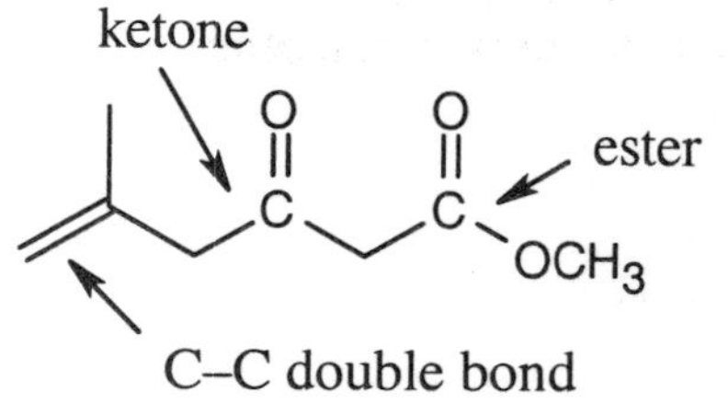

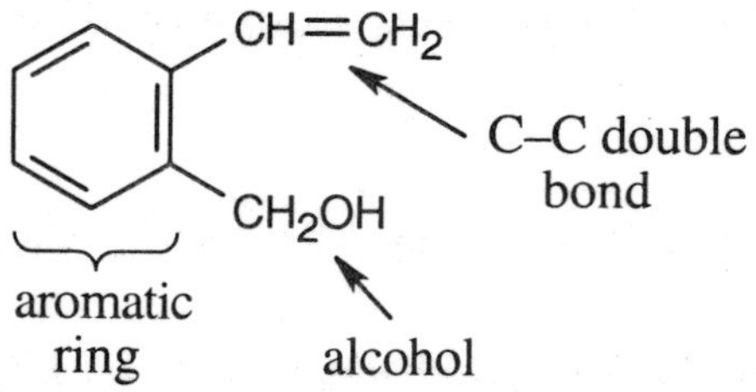

3.2

(a) CH_3OH Methanol
(b) Toluene
(c) CH_3COH Acetic acid
(d) CH_3NH_2 Methylamine
(e) $CH_3CCH_2NH_2$ Aminoacetone
(f) 1,3-Butadiene

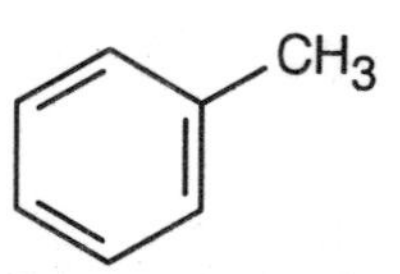

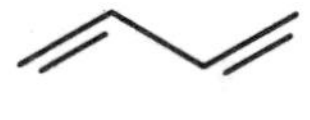

3.3

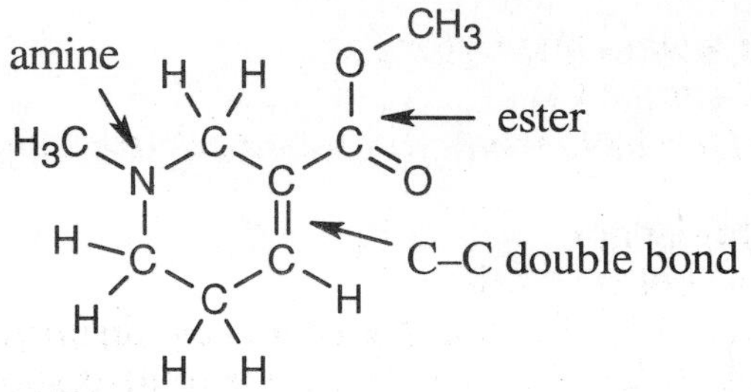
amine
ester
C–C double bond

Arecoline

$C_8H_{13}NO_2$

3.4 We know that carbon forms four bonds and hydrogen forms one bond. Thus, if you draw all possible six carbon skeletons and add hydrogens so that all carbons have four bonds, you will arrive at the following structures:

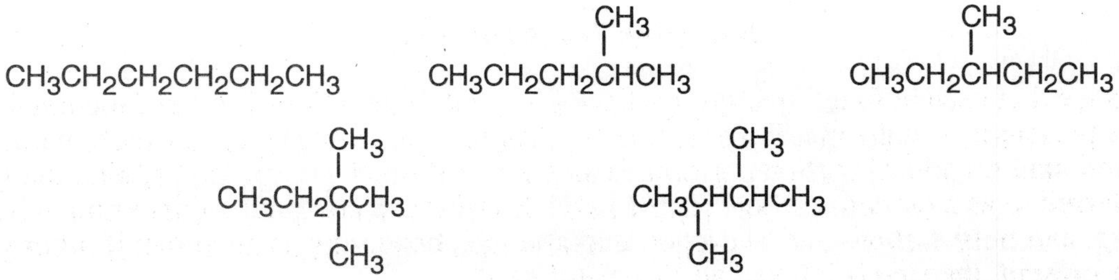

3.4 This problem becomes easier when you realize that the isomers can be alcohols and ethers. A systematic approach to this type of problem is helpful. Let's start with the alcohol isomers.

1) Draw the simplest long-chain parent alkane. In this problem, the alkane is butane, $CH_3CH_2CH_2CH_3$.
2) Find the number of different sites to which a functional group may be attached. For butane, two different sites are possible ($-CH_3$ and $-CH_2-$).
3) At each different site, replace an –H by an –OH and draw the isomer.

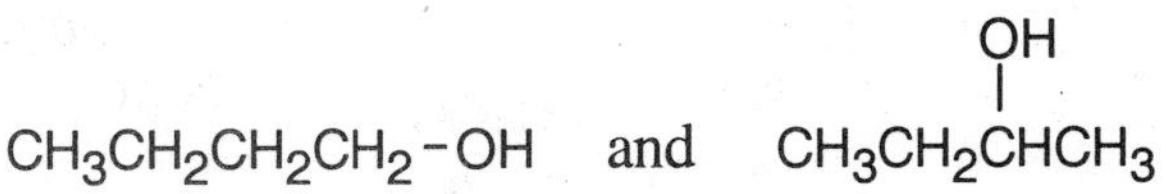

4) Draw the simplest branched C_4H_{10} alkane.

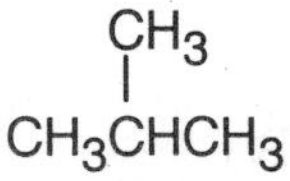

5) Find the number of different sites. (There are two for the above alkane.)
6) For each site, replace an –H with an –OH and draw the isomer.

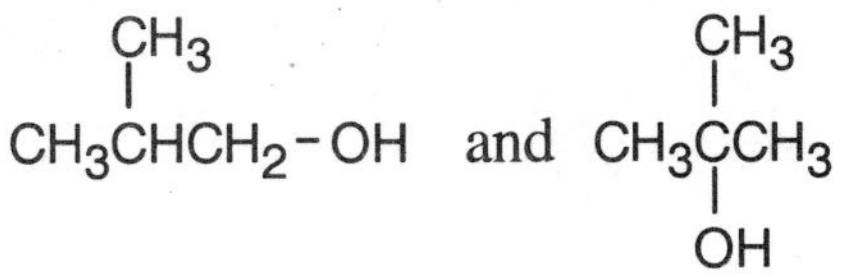

7) Proceed with the next simplest branched C_4H_{10} alkane. In this problem, we have already drawn all alcohol isomers.

For the ethers, start with the $-OCH_3$ isomers. There are two possible sites for attachment of an $-OCH_3$ group to propane ($CH_3CH_2CH_3$) and thus there are two $-OCH_3$ isomers.

$CH_3CH_2CH_2{-}OCH_3$ and $CH_3CH(OCH_3)CH_3$

Finally, there is one $-OCH_2CH_3$ ether, diethyl ether, $CH_3CH_2OCH_2CH_3$.

3.6 (a) Nine isomeric esters of formula $C_5H_{10}O_2$ can be drawn.

$CH_3CH_2CH_2C(=O)OCH_3$ $CH_3CH(CH_3)C(=O)OCH_3$ $CH_3CH_2C(=O)OCH_2CH_3$

$CH_3C(=O)OCH_2CH_2CH_3$ $CH_3C(=O)OCH(CH_3)CH_3$ $HC(=O)OCH_2CH_2CH_2CH_3$

$HC(=O)OCH(CH_3)CH_2CH_3$ $HC(=O)OCH_2CH(CH_3)CH_3$ $HC(=O)OC(CH_3)_2CH_3$

(b)

$CH_3CH_2CH_2C{\equiv}N$ and $CH_3CH(CH_3)C{\equiv}N$

3.7 (a) Two alcohols have the formula C_3H_8O.

$CH_3CH_2CH_2OH$ and $CH_3CH(OH)CH_3$

(b) Four bromoalkanes have the formula C_4H_9Br. Refer to the solution to Problem 3.5 if you need help.

$CH_3CH_2CH_2CH_2Br$ $CH_3CH_2CH(Br)CH_3$ $CH_3CH(CH_3)CH_2Br$ $CH_3C(Br)(CH_3)CH_3$

3.8

$CH_3CH_2CH_2CH_2CH_2-$ $\quad$ $CH_3CH_2CH_2\underset{|}{C}H(CH_3)-$ $\quad$ $CH_3CH_2CH(CH_2CH_3)-$ $\quad$ $CH_3CH_2CH(CH_3)CH_2-$

$CH_3CH(CH_3)CH_2CH_2-$ $\quad$ $CH_3CH_2C(CH_3)_2-$ $\quad$ $CH_3CH(CH_3)CH(CH_3)-$ $\quad$ $CH_3C(CH_3)_2CH_2-$

3.9

(a)

$CH_3CH(CH_3)CH_2CH_2CH_3$

Main chain: CH_3 (*p*), CH (*t*), CH_2 (*s*), CH_2 (*s*), CH_3 (*p*); branch CH_3 (*p*)

(b)

$CH_3CH_2CH(CH(CH_3)CH_3)CH_2CH_3$

Main chain: CH_3 (*p*), CH_2 (*s*), CH (*t*), CH_2 (*s*), CH_3 (*p*); branch CH_3CHCH_3: *p t p*

(c)

$CH_3CH(CH_3)CH_2-C(CH_3)_2-CH_3$

Main chain: CH_3 (*p*), CH (*t*), CH_2 (*s*), C (*q*), CH_3 (*p*); branches CH_3 (*p*), CH_3 (*p*), CH_3 (*p*)

p = primary; *s* = secondary; *t* = tertiary; *q* = quaternary

3.10

(a)

$CH_3CH(CH_3)CH_2CH_2CH_3$

Main chain: CH_3 (*p*), CH (*t*), CH_2 (*s*), CH_2 (*s*), CH_3 (*p*); branch CH_3 (*p*)

(b)

$CH_3CH_2CH(CH(CH_3)CH_3)CH_2CH_3$

Main chain: CH_3 (*p*), CH_2 (*s*), CH (*t*), CH_2 (*s*), CH_3 (*p*); branch CH_3CHCH_3: *p t p*

(c)

$CH_3CH(CH_3)CH_2-C(CH_3)_2-CH_3$

Main chain: CH_3 (*p*), CH (*t*), CH_2 (*s*), C, CH_3 (*p*); branches CH_3 (*p*), CH_3 (*p*), CH_3 (*p*)

3.11

(a)

$CH_3CH(CH_3)CH(CH_3)CH_3$ (the two CH carbons labelled *t*)

(b)

$CH_3CH_2CH(CH(CH_3)CH_3)CH_2CH_3$ (the CH_3CHCH_3 group is boxed)

(c)

$CH_3CH_2C(CH_3)_2CH_3$ (CH_2 labelled *s*; quaternary carbon labelled *q* with an arrow)

3.12

(a)

$CH_3CH_2CH_2CH_2CH_3$	$CH_3CH_2CH(CH_3)CH_3$	$CH_3C(CH_3)_2CH_3$
Pentane	2-Methylbutane	2,2-Dimethylpropane

(b)

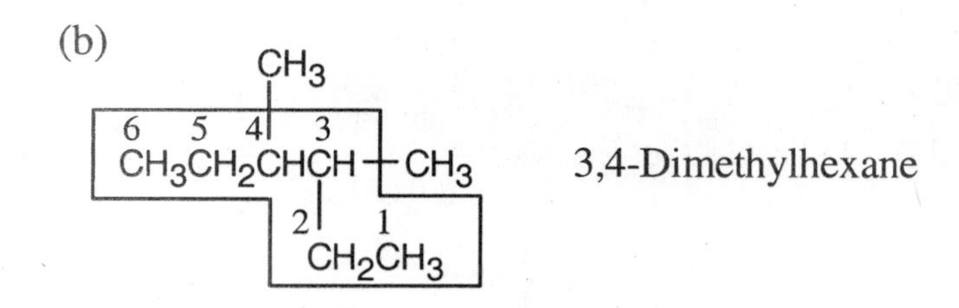

3,4-Dimethylhexane

Step 1: Find the longest continuous carbon chain and use it as the parent name. In (b), the longest chain is a hexane (boxed).
Step 2: Identify the substituents. In (b), both substituents are methyl groups.
Step 3: Number the substituents. In (b), numbering can start from either end of the carbon chain, and the methyl groups are in the 3- and 4- positions.
Step 4: Name the compound. Remember that the prefix *di-* must be used when two substituents are the same. The IUPAC name is 3,4-dimethylhexane

(c)

CH_3 (on C-4)
$(CH_3)_2CHCH_2CHCH_3$

2,4-Dimethylpentane

(d)

CH_3
$(CH_3)_3CCH_2CH_2CH$
CH_2CH_3

2,2,5–Trimethylheptane

3.13 When you are asked to draw the structure corresponding to a given name, draw the parent carbon chain, attach the specified groups to the proper carbons, and fill in the necessary hydrogens.

(a)

CH_3
$CH_3CH_2CH_2CH_2CH_2CHCHCH_2CH_3$
CH_3

3,4-Dimethylnonane

(b)

CH_3
$CH_3CH_2CH_2C—CHCH_2CH_3$
CH_3 CH_2CH_3

3-Ethyl-4,4-dimethylheptane

(c)

$CH_2CH_2CH_3$
$CH_3CH_2CH_2CH_2CHCH_2C(CH_3)_3$

2,2-Dimethyl-4-propyloctane

(d)

CH_3 CH_3
$CH_3CHCH_2CCH_3$
CH_3

2,2,4-Trimethylpentane

3.14

$CH_3CH_2CH_2CH_2CH_2-$ Pentyl

$CH_3CH_2CH_2CH(CH_3)-$ 1-Methylbutyl

$CH_3CH_2CH(CH_2CH_3)-$ 1-Ethylpropyl

$CH_3CH_2CH(CH_3)CH_2-$ 2-Methylbutyl

$CH_3CH(CH_3)CH_2CH_2-$ 3-Methylbutyl

$CH_3CH_2C(CH_3)_2-$ 1,1-Dimethylpropyl

$CH_3CH(CH_3)CH(CH_3)-$ 1,2-Dimethylpropyl

$CH_3C(CH_3)_2CH_2-$ 2,2-Dimethylpropyl

3.15

3,3,4,5-Tetramethylheptane

3.16 The steps for naming a cycloalkane are very similar to the steps used for naming an open-chain hydrocarbon.
Step 1: Name the parent cycloalkane. In (a), the parent is cyclohexane. If the compound has an alkyl substituent with more carbons than the ring size, the compound is named as a cycloalkane-substituted alkane.
Step 2: Identify the substituents. In (a), both substituents are methyl groups.
Step 3: Number the substituents so that the second substituent receives the lowest possible number. In (a), the substituents are in the 1- and 4- positions.
Step 4: Name the compound. If two different alkyl groups are present, cite them alphabetically. Halogen substituents follow the same rules as alkyl substituents.

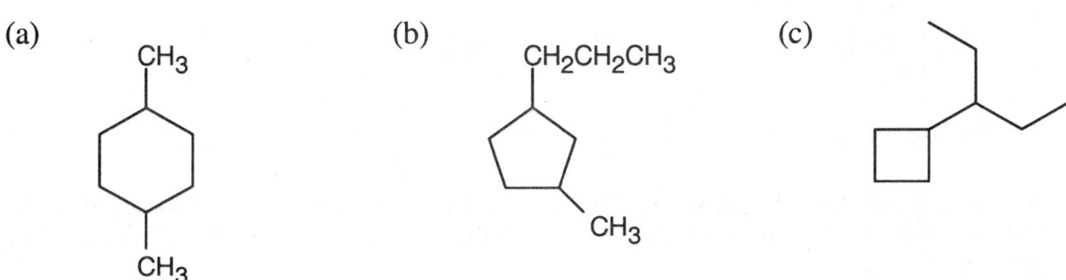

1,4-Dimethylcyclohexane 1-Methyl-3-propylcyclopentane 3-Cyclobutylpentane

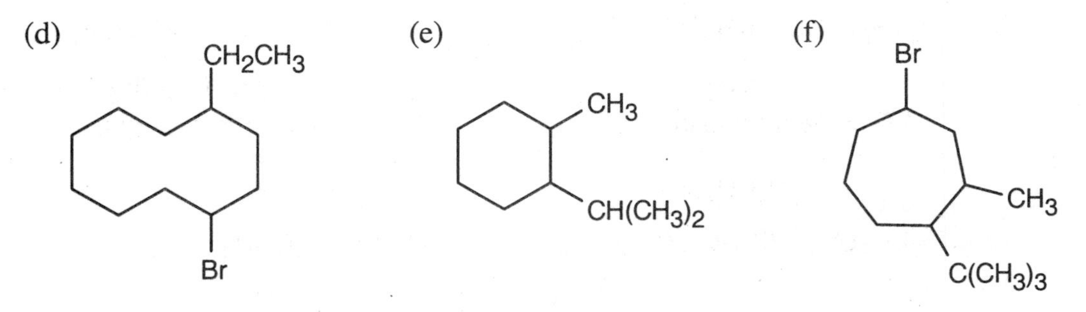

1-Bromo-4-ethylcyclodecane 1-Isopropyl-2-methyl-cyclohexane 4-Bromo-1-*tert*-butyl-2-methylcycloheptane

3.17 To draw a substituted cycloalkane, simply draw the ring and attach substituents in the specified positions.

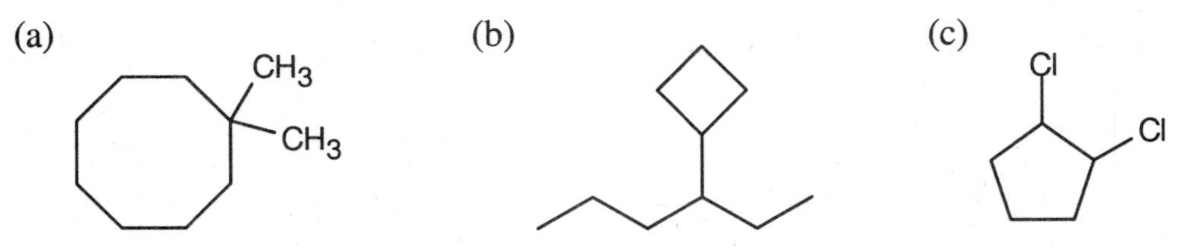

1,1-Dimethylcyclooctane 3-Cyclobutylhexane 1,2-Dimethylcyclopentane

(d)

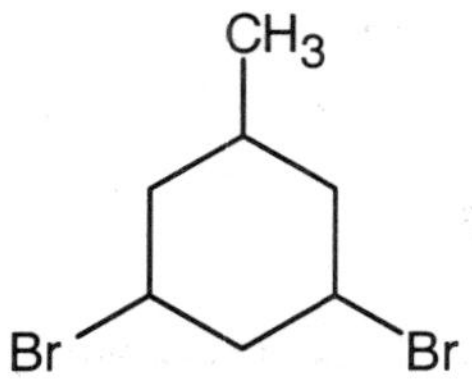

1,3-Dibromo-5-methylcyclohexane

3.18 Two substituents are cis if they both have either dashed or wedged bonds. The substituents are trans if one has a wedged bond and the other has a dashed bond.

(a) *trans*-1-Chloro-4-methylcyclohexane

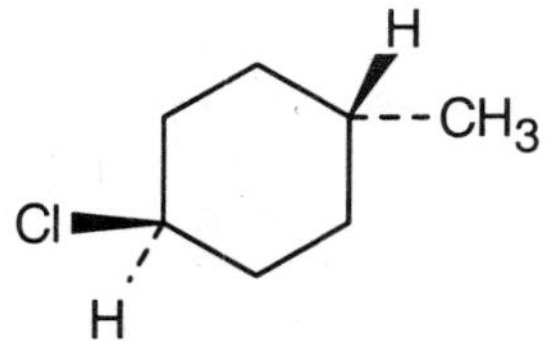

(b) *cis*-1-Ethyl-3-methylcycloheptane

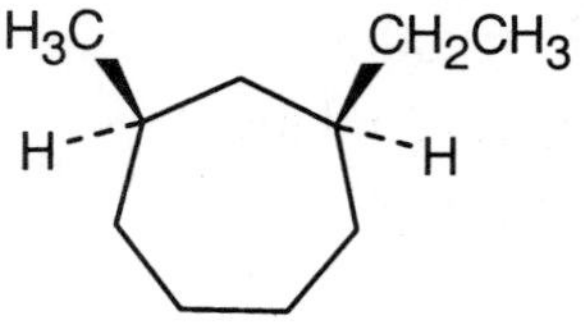

3.19

(a) *trans*-1-Bromo-3-methylcyclohexane

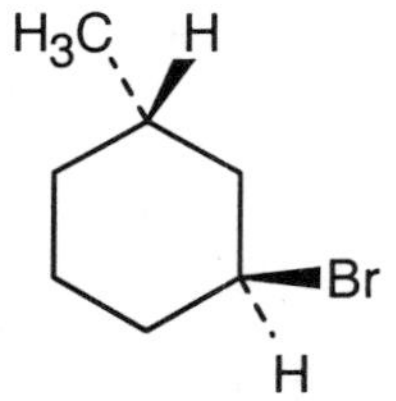

(b) *cis*-1,2-Dimethylcyclopentane

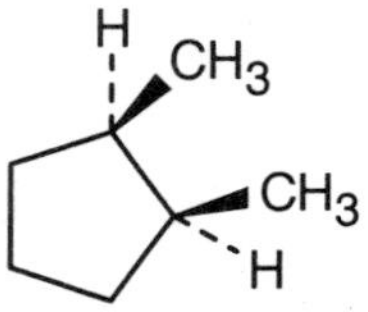

(c) *trans*-1-*tert*-Butyl-2-ethylcyclohexane

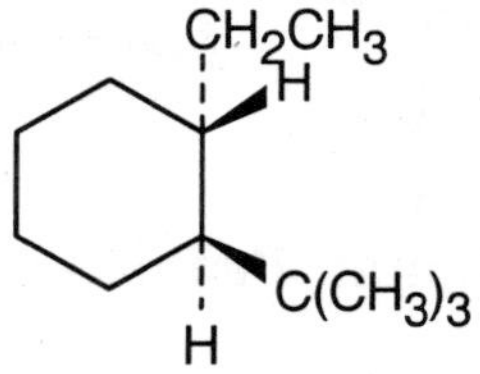

Visualizing Chemistry

3.20

(a) aromatic ring, amine, carboxylic acid

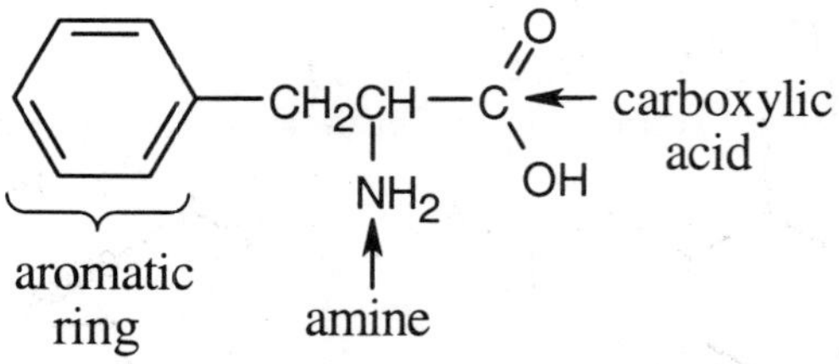

Phenylalanine $C_9H_{11}NO_2$

(b) aromatic ring, amide, amine

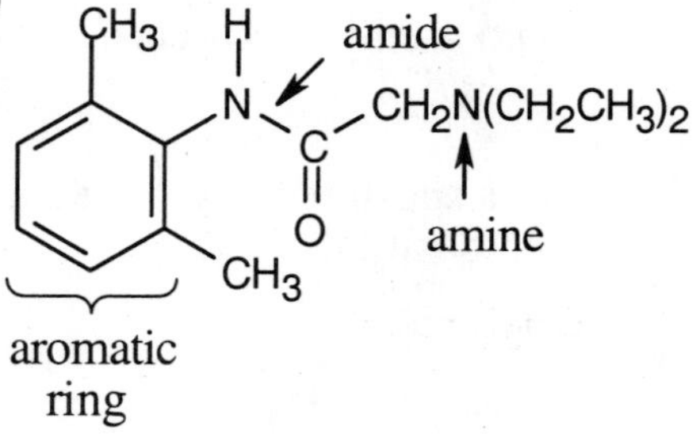

Lidocaine $C_{14}H_{22}N_2O$

3.21

(a)

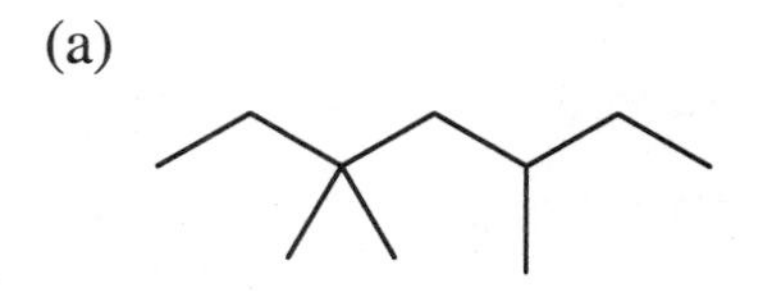

3,3,5-Trimethylheptane

(b)

H_3C H H CH_2CH_3

trans-1-Ethyl-3-methyl-cyclopentane

(c)

2,2,4-Trimethylpentane

(d)

H CH_3 Cl H

cis-1-Chloro-3-methyl-cyclohexane

3.22

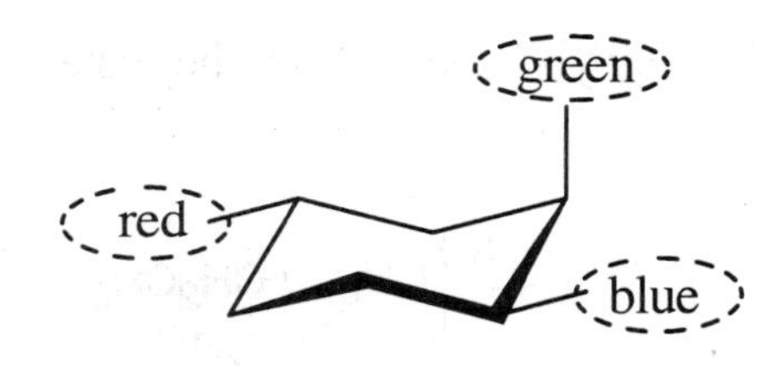

cis relationship: blue–green
trans relationship: green–red, blue–red

Additional Problems

3.23

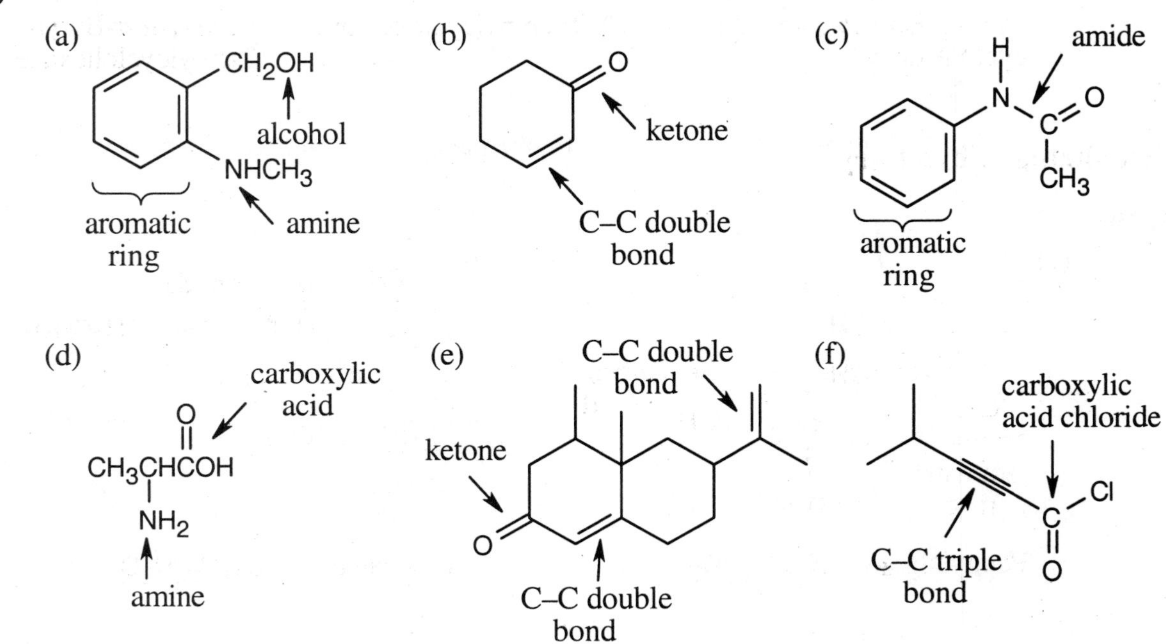

3.24 (a) Eighteen isomers have the formula C_8H_{18}. Three are pictured.

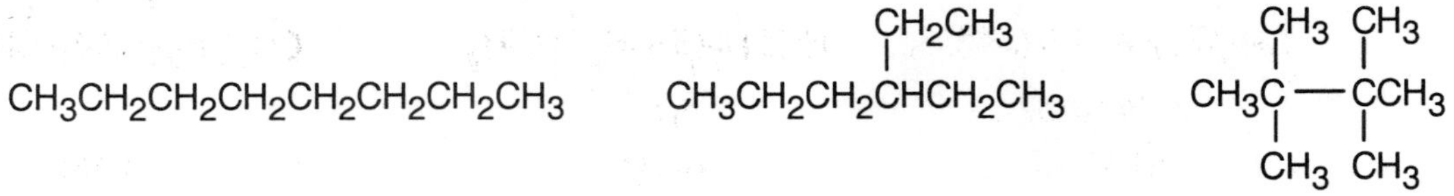

$CH_3CH_2CH_2CH_2CH_2CH_2CH_2CH_3$

CH_2CH_3 (on C-4)
$CH_3CH_2CH_2CHCH_2CH_3$

CH_3 CH_3
CH_3C—CCH_3
CH_3 CH_3

(b) Structures with the formula $C_4H_8O_2$ may represent esters, carboxylic acids or many other complicated molecules. Three possibilities:

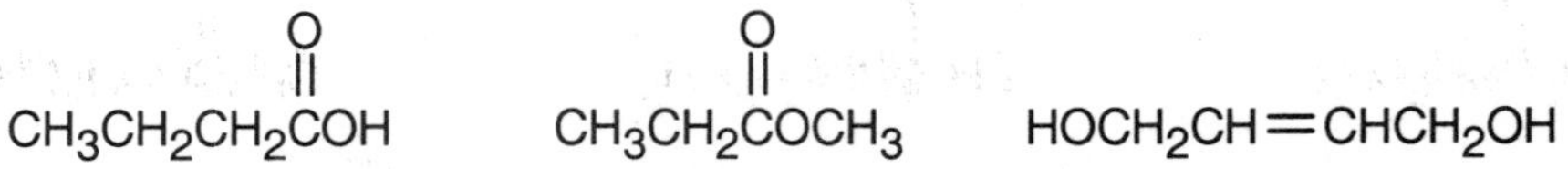

$CH_3CH_2CH_2COH$ (C=O) $CH_3CH_2COCH_3$ (C=O) $HOCH_2CH{=}CHCH_2OH$

3.25

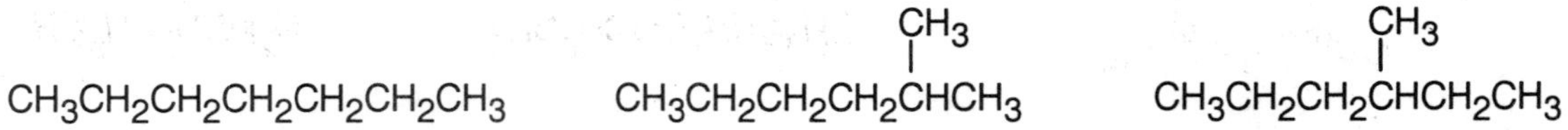

$CH_3CH_2CH_2CH_2CH_2CH_2CH_3$ Heptane	CH_3 on C-2 $CH_3CH_2CH_2CH_2CHCH_3$ 2-Methylhexane	CH_3 on C-3 $CH_3CH_2CH_2CHCH_2CH_3$ 3-Methylhexane
CH_3 $CH_3CH_2CH_2CCH_3$ CH_3 2,2-Dimethylpentane	CH_3 $CH_3CH_2CHCHCH_3$ CH_3 2,3-Dimethylpentane	CH_3 $CH_3CHCH_2CHCH_3$ CH_3 2,4-Dimethylpentane
CH_3 $CH_3CH_2CCH_2CH_3$ CH_3 3,3-Dimethylpentane	$CH_3CH_2CHCH_2CH_3$ CH_2CH_3 3-Ethylpentane	CH_3 CH_3 CH_3CH—CCH_3 CH_3 2,2,3-Trimethylbutane

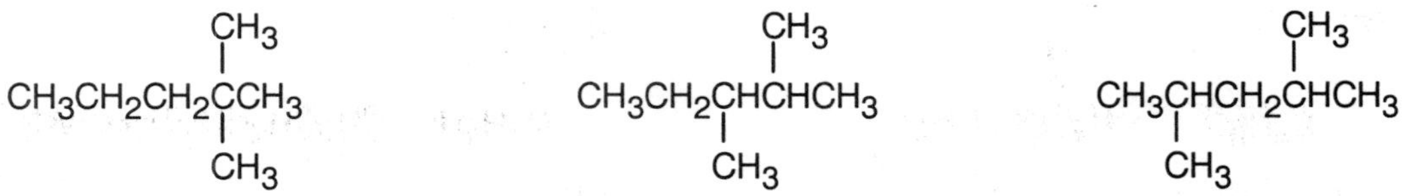

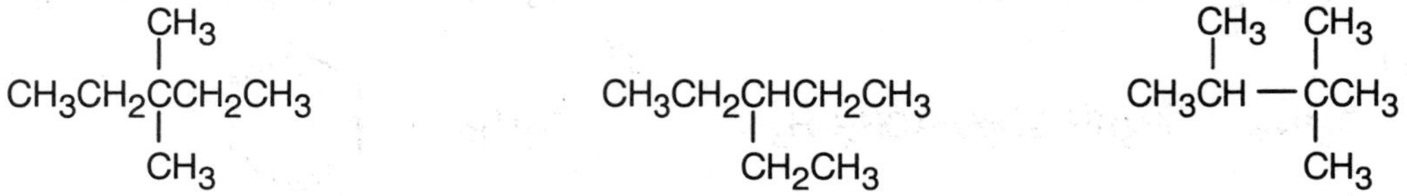

3.26

(a)

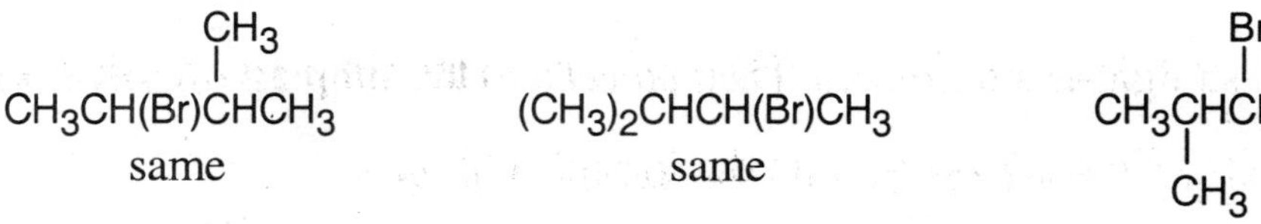

CH_3 $CH_3CH(Br)CHCH_3$ same	$(CH_3)_2CHCH(Br)CH_3$ same	Br $CH_3CHCHCH_3$ CH_3 same

(b)

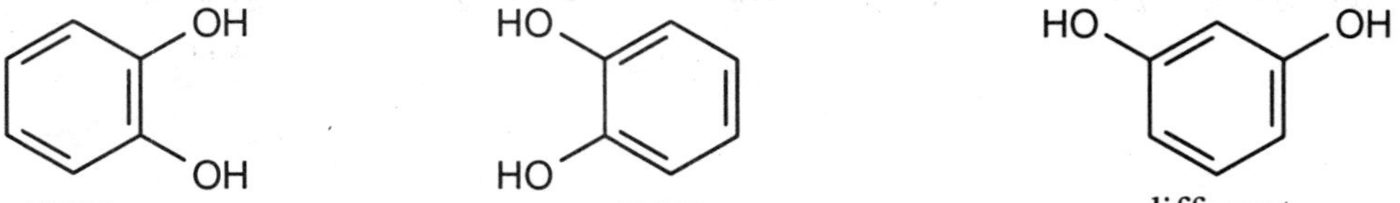

OH, OH (benzene ring) same	HO, HO (benzene ring) same	HO, OH (benzene ring) different

(c)

$$\begin{array}{c} CH_3 \\ | \\ CH_3CH_2CHCH_2CHCH_3 \\ | \\ CH_2OH \end{array}$$
different

$$\begin{array}{c} CH_2CH_3 \\ | \\ HOCH_2CHCH_2CHCH_3 \\ | \\ CH_3 \end{array}$$
same

$$\begin{array}{c} CH_3 \quad CH_3 \\ | \qquad | \\ CH_3CH_2CHCH_2CHCH_2OH \end{array}$$
same

3.27 Different answers to this problem and to Problem 3.28 are acceptable.

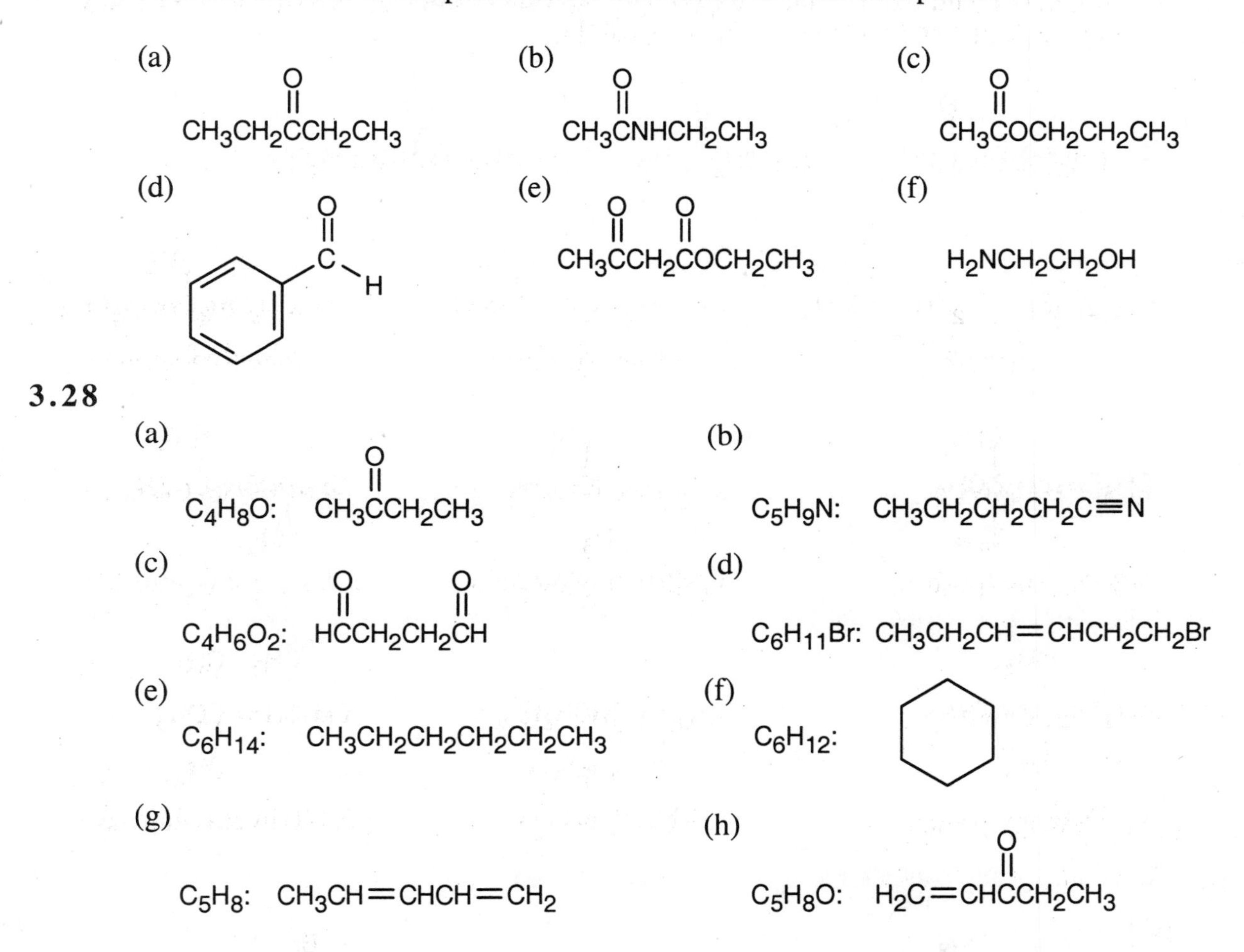

3.29 First, draw all straight-chain isomers. Then proceed to the simplest branched structure.

(a) There are four alcohol isomers with the formula $C_4H_{10}O$.

$$CH_3CH_2CH_2CH_2OH \qquad \begin{array}{c} OH \\ | \\ CH_3CH_2CHCH_3 \end{array} \qquad \begin{array}{c} CH_3 \\ | \\ CH_3CHCH_2OH \end{array} \qquad \begin{array}{c} OH \\ | \\ CH_3CCH_3 \\ | \\ CH_3 \end{array}$$

(b) There are 17 isomers of $C_5H_{13}N$. Nitrogen can be bonded to one, two or three alkyl groups.

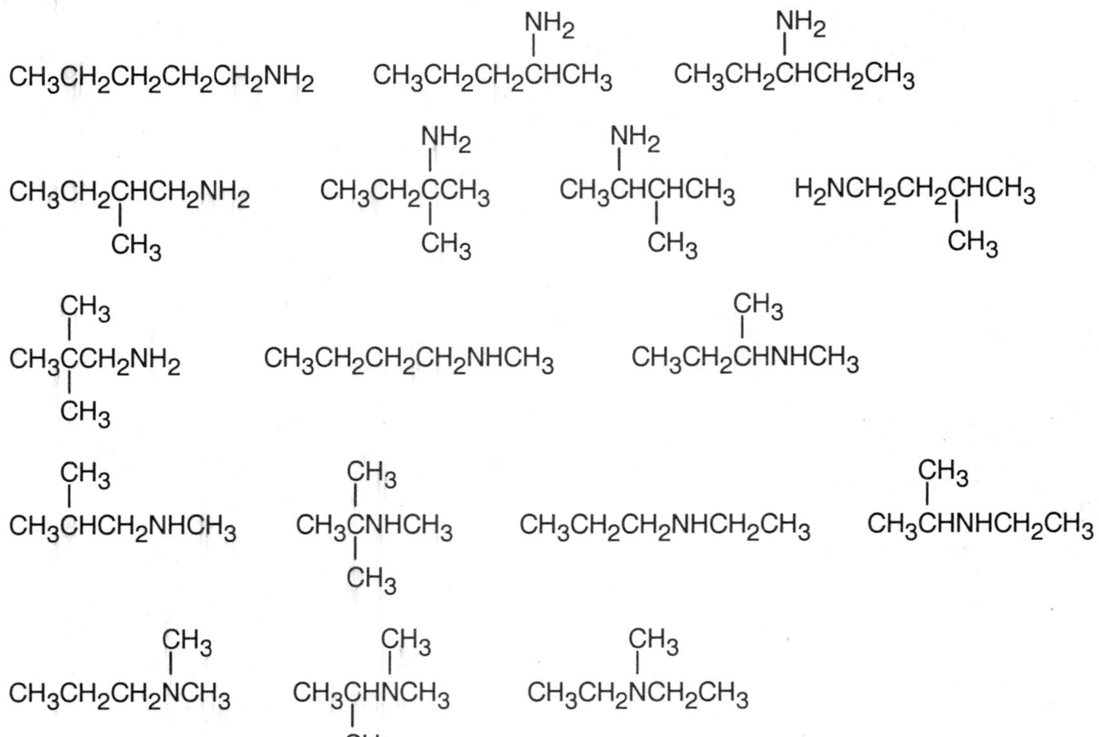

(c) There are 3 ketone isomers with the formula $C_5H_{10}O$.

$CH_3CH_2CH_2C(=O)CH_3$ $CH_3CH_2C(=O)CH_2CH_3$ $CH_3CH(CH_3)C(=O)CH_2CH_3$

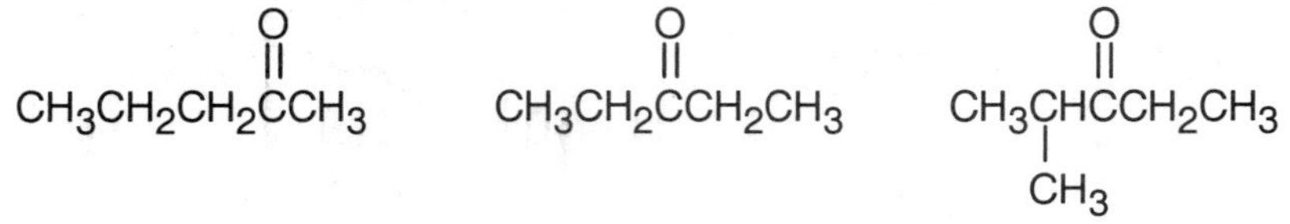

(d) There are 4 isomeric aldehydes with the formula $C_5H_{10}O$. Remember that the aldehyde functional group can occur only at the end of a chain.

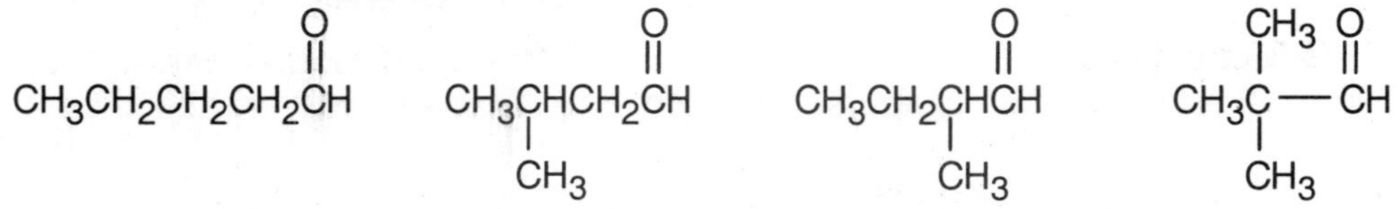

$CH_3CH_2CH_2CH_2CH(=O)$ $CH_3CH(CH_3)CH_2CH(=O)$ $CH_3CH_2CH(CH_3)CH(=O)$ $CH_3C(CH_3)_2—CH(=O)$

(e) There are 4 esters with the formula $C_4H_8O_2$.

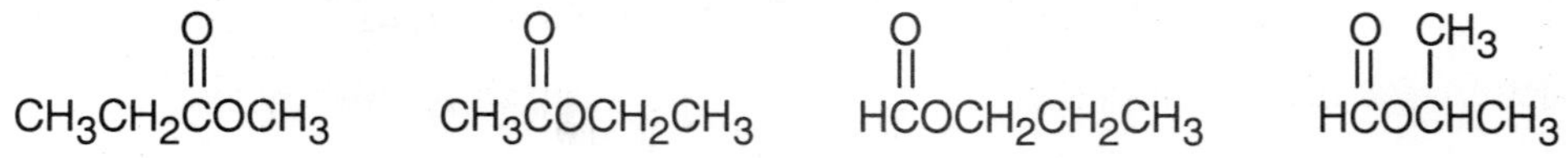

$CH_3CH_2C(=O)OCH_3$ $CH_3C(=O)OCH_2CH_3$ $HC(=O)OCH_2CH_2CH_3$ $HC(=O)OCH(CH_3)CH_3$

(f) There are 3 ethers with the formula $C_4H_{10}O$.

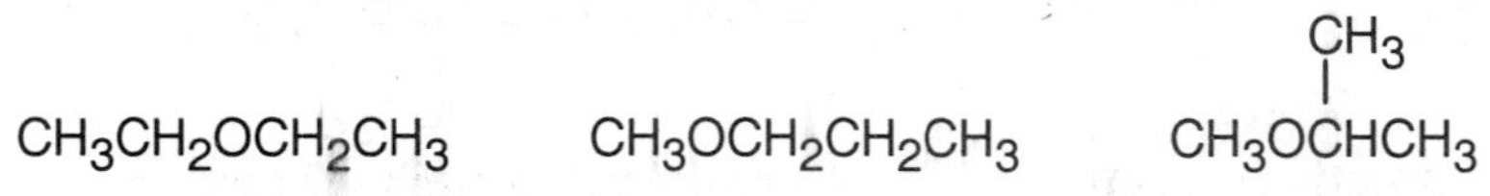

$CH_3CH_2OCH_2CH_3$ $CH_3OCH_2CH_2CH_3$ $CH_3OCH(CH_3)CH_3$

3.30

(a) CH_3CH_2OH

(b)

$$\begin{array}{c} CH_3 \\ | \\ CH_3CC{\equiv}N \\ | \\ CH_3 \end{array}$$

(c)

$$\begin{array}{c} Br \\ | \\ CH_3CHCH_3 \end{array}$$

(d)

$$\begin{array}{c} OH \\ | \\ CH_3CHCH_2OH \end{array}$$

(e)

$$\begin{array}{c} CH_3 \\ | \\ CH_3CHOCH_3 \end{array}$$

(f)

$$\begin{array}{c} CH_3 \\ | \\ CH_3CCH_3 \\ | \\ CH_3 \end{array}$$

3.31

$CH_3CH_2CH_2CH_2CH_2Br$
1-Bromopentane

$$\begin{array}{c} Br \\ | \\ CH_3CH_2CH_2CHCH_3 \end{array}$$
2-Bromopentane

$$\begin{array}{c} Br \\ | \\ CH_3CH_2CHCH_2CH_3 \end{array}$$
3-Bromopentane

3.32

$$\begin{array}{c} CH_3 \qquad\quad CH_3 \\ | \qquad\qquad | \\ CH_3CHCH_2CH_2CHCH_2Cl \end{array}$$
1-Chloro-2,5-dimethylhexane

$$\begin{array}{c} CH_3 \qquad CH_3 \\ | \qquad\quad | \\ CH_3CHCH_2CH_2CCH_3 \\ | \\ Cl \end{array}$$
2-Chloro-2,5-dimethylhexane

$$\begin{array}{c} CH_3 \qquad CH_3 \\ | \qquad\quad | \\ CH_3CHCH_2CHCHCH_3 \\ | \\ Cl \end{array}$$
3-Chloro-2,5-dimethylhexane

3.33

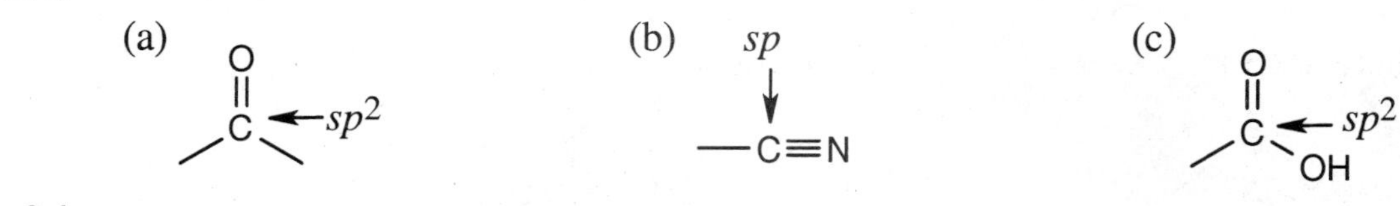

3.34

(a)

$$\begin{array}{c} CH_3 \\ | \\ CH_3CH_2CH_2CH_2CH_2CHCH_3 \end{array}$$
2-Methylheptane

(b)

$$\begin{array}{c} CH_3 \\ | \\ CH_3CH_2CH{-}CH_2CCH_3 \\ | \qquad\quad | \\ CH_2CH_3 \quad CH_3 \end{array}$$
4-Ethyl-2,2-dimethylhexane

(c)

$$\begin{array}{c} CH_3 \quad CH_3 \\ | \qquad | \\ CH_3CH_2CH_2CH_2C{-}CHCH_2CH_3 \\ | \\ CH_2CH_3 \end{array}$$
4-Ethyl-3,4-dimethyloctane

(d)

$$\begin{array}{c} CH_3 \quad CH_3 \\ | \qquad | \\ CH_3CH_2CH_2CCH_2CHCH_3 \\ | \\ CH_3 \end{array}$$
2,4,4-Trimethylheptane

(e)

$$\begin{array}{c} CH_3 \quad CH_2CH_3 \; CH_3 \\ | \qquad\quad | \qquad\quad | \\ CH_3CH_2CH_2CH_2CHCH_2C{-}CHCH_3 \\ | \\ CH_2CH_3 \end{array}$$
3,3-Diethyl-2,5-dimethylnonane

(f)

$$\begin{array}{c} CH_3CHCH_3 \\ | \\ CH_3CH_2CH_2CHCHCH_2CH_3 \\ | \\ CH_3 \end{array}$$
4-Isopropyl-3-methylheptane

3.35

(a) $CH_3CH(CH_3)CH_3$

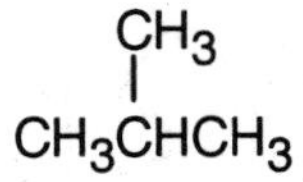

2-Methylpropane

(b) Cyclohexane

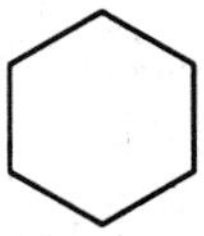

(c) $CH_3CH_2CH_2CH_2CH_2CH_3$

Hexane

3.36

(a) Cyclobutane

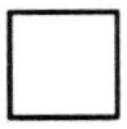

(b) $CH_3CH(CH_3)-CH(CH_3)CH_3$

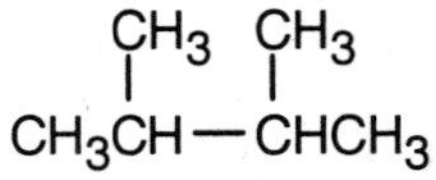

2,3-Dimethylbutane

3.37

(a) $CH_3CH_2CH_2CH_2CH_2Br$

(b) CH_2OCH_3 on a cyclobutane ring

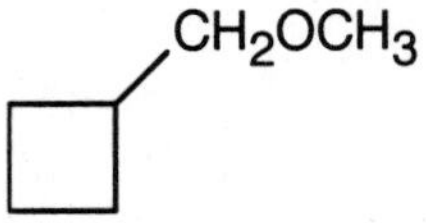

(c) $CH_3CH(CH_3)C{\equiv}N$

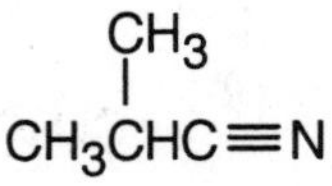

(d) CH_2OH on a cyclopentane ring

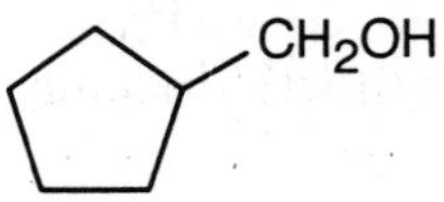

(e) There are no aldehyde isomers. However, the structure below is a ketone isomer.

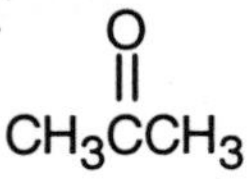

$CH_3C(=O)CH_3$

(f) Benzene ring with COOH and CH_3 on adjacent carbons

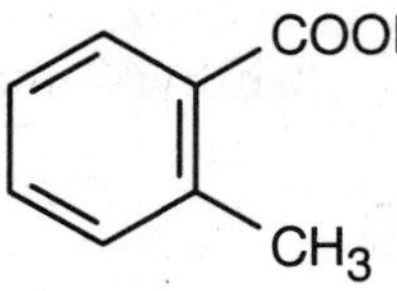

3.38

(a) *trans*-1,3-Dibromocyclopentane

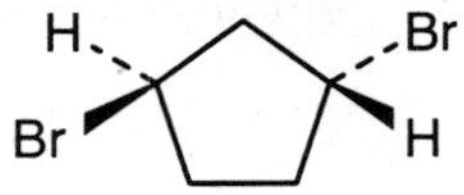

(b) *cis*-1,4-Diethylcyclohexane

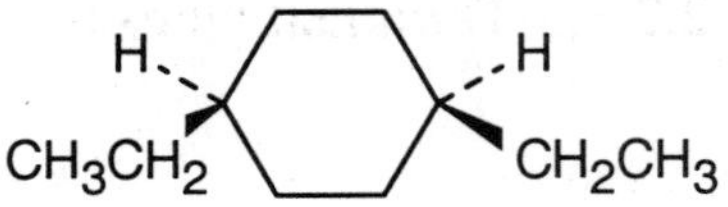

(c) *trans*-1-Isopropyl-3-methylcycloheptane

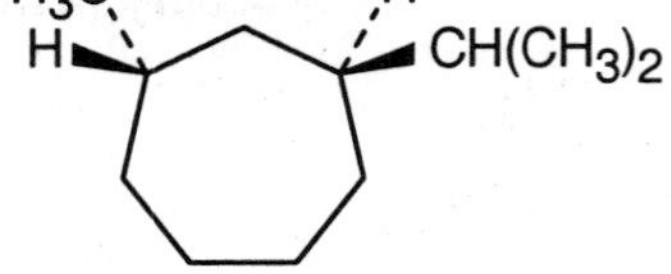

(d) Cyclohexyl$-CH_2-$cyclohexyl

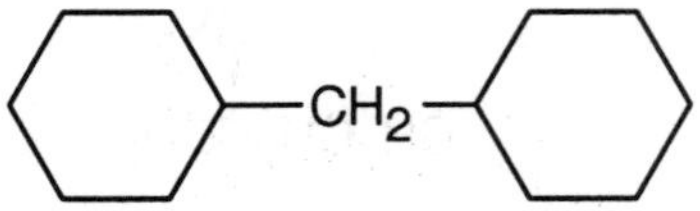

Dicyclohexylmethane

3.39

(a) $CH_3CH(CH_3)CH_2CH_3$ — labels: 1° (CH_3), 3°, 2°, 1°; branch CH_3 1°

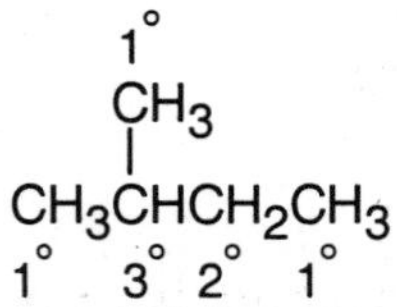

(b) $(CH_3)_2CHCH(CH_2CH_3)_2$ — labels: 1°, 3°, 3°, 2°, 1°

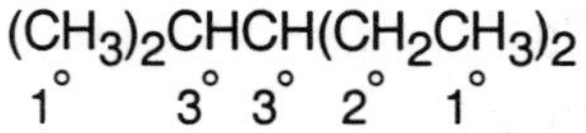

(c) $(CH_3)_3CCH_2CH_2CH(CH_3)_2$ — labels: 1°, 4°, 2°, 2°, 3° (CH), 1° (each CH_3)

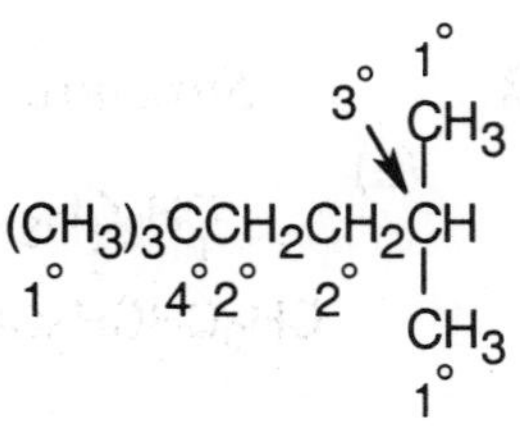

(d) Methylcyclohexane: ring CH_2 carbons 2°, ring CH 3°, CH_3 1°

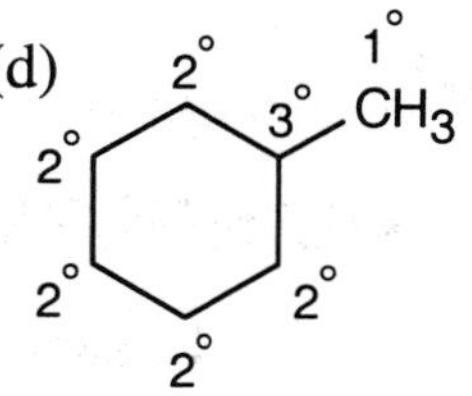

(e) Bicyclic compound: ring CH_2 carbons 2°, ring-fusion carbons 3°

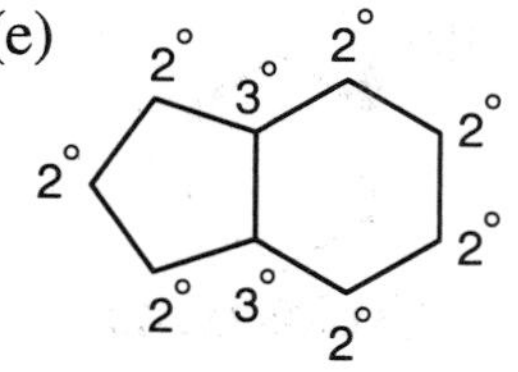

(f) Bicyclic compound: bridgehead CH 3°, CH_2 carbons 2°, bridgehead C bearing CH_3 4°, CH_3 1°

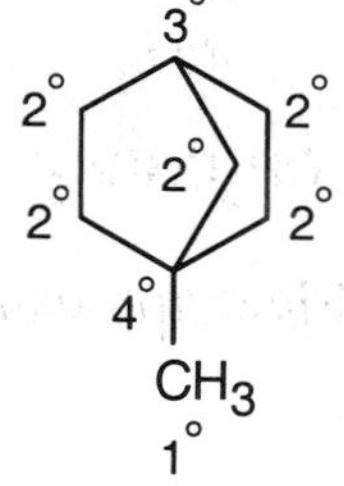

3.40

(a)

$$CH_3CH(CH_3)CH_2CH_2CH_3$$

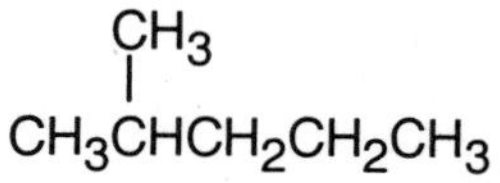

2-Methylpentane

(b)

$$CH_3CH_2C(CH_3)_2CH_3$$

2,2-Dimethylbutane

(c)

$$(CH_3)_2CHC(CH_3)_2CH_2CH_2CH_3$$

2,3,3-Trimethylhexane

(d)

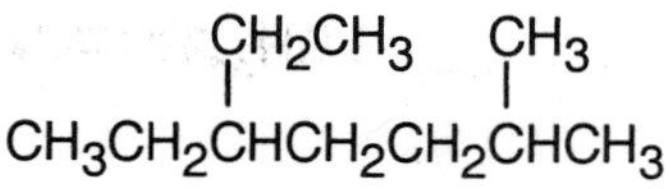

$$CH_3CH_2CH(CH_2CH_3)CH_2CH_2CH(CH_3)CH_3$$

5-Ethyl-2-methylheptane

(e)

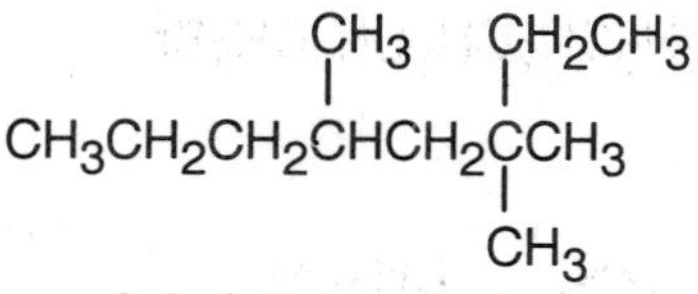

$$CH_3CH_2CH_2CH(CH_3)CH_2C(CH_3)(CH_2CH_3)CH_3$$

3,3,5-Trimethyloctane

(f)

$$(CH_3)_3CC(CH_3)_2CH_2CH_2CH_3$$

2,2,3,3-Tetramethylhexane

(g)

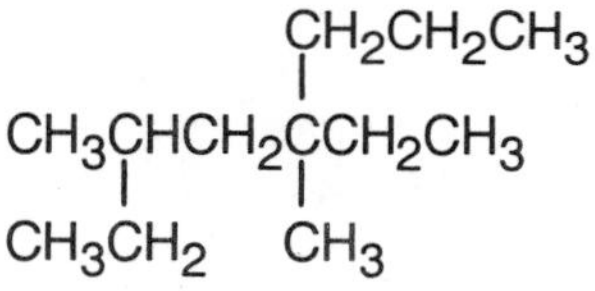

$$CH_3CH(CH_2CH_3)CH_2C(CH_2CH_2CH_3)(CH_3)CH_2CH_3$$

5-Ethyl-3,5-dimethyloctane

3.41

$$CH_3CH_2CH_2CH_2CH_2CH_3$$

Hexane

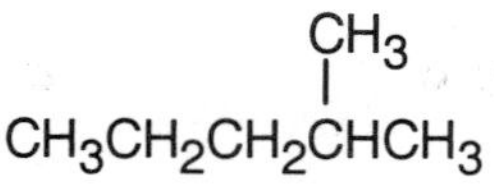

$$CH_3CH_2CH_2CH(CH_3)CH_3$$

2-Methylpentane

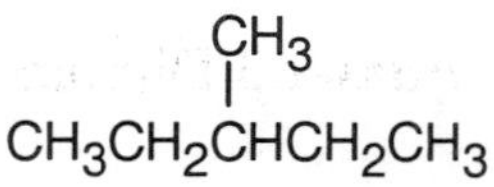

$$CH_3CH_2CH(CH_3)CH_2CH_3$$

3-Methylpentane

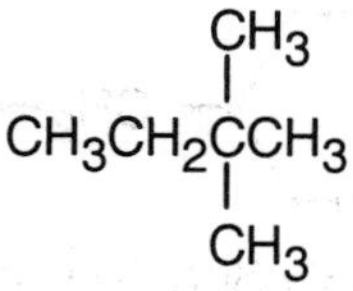

$$CH_3CH_2C(CH_3)_2CH_3$$

2,2-Dimethylbutane

$$CH_3CH(CH_3)CH(CH_3)CH_3$$

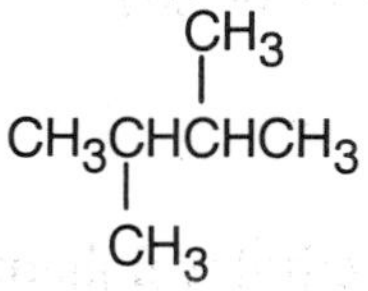

2,3-Dimethylbutane

3.42

Structure and Correct Name	*Error*
(a) $CH_3CH(CH_2CH_3)CH_2CH_2CH_2C(CH_3)_2CH_3$ (C-8 at the CH_2CH_3 end, C-1 at the $C(CH_3)_2CH_3$ end) Correct name: 2,2,6-Trimethyloctane	The longest chain is an octane.

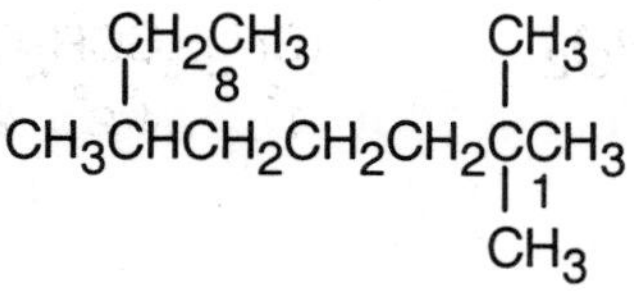

(b)

$$\begin{array}{l} \quad\;\; CH_3 \\ \quad\;\; | \\ CH_3CHCHCH_2CH_2CH_3 \\ 1 \qquad\; | \qquad\quad\;\; 6 \\ \qquad\; CH_2CH_3 \end{array}$$

Correct name: 3-Ethyl-2-methylhexane

The longest chain is a hexane. Numbering should start from the opposite end of the carbon chain, nearer the first branch.

(c)

$$\begin{array}{l} \qquad\quad\; CH_3 \\ \qquad\quad\; | \\ CH_3CH_2C{-}CHCH_2CH_3 \\ 1 \qquad\quad | \quad\;\; | \qquad\; 6 \\ \qquad\quad CH_3\;\; CH_2CH_3 \end{array}$$

Correct name: 4-Ethyl-3,3-dimethylhexane

Numbering should start from the opposite end of the carbon chain. See step 2(b) in Section 3.4.

(d)

$$\begin{array}{l} \qquad\quad\; CH_3\;\; CH_3 \\ \qquad\quad\;\; | \qquad\; | \\ CH_3CH_2CH{-}CCH_2CH_2CH_2CH_3 \\ 1 \qquad\qquad\quad | \qquad\qquad\quad 8 \\ \qquad\qquad\quad CH_3 \end{array}$$

Correct name: 3,4,4-Trimethyloctane

Numbering should start from the opposite end of the carbon chain.

(e)

$$\begin{array}{l} \qquad\qquad\quad CH_3 \\ \qquad\qquad\quad | \\ CH_3CH_2CH_2CHCH_2CHCH_3 \\ 8 \qquad\qquad\qquad\quad | \\ \qquad\qquad\quad CH_3CHCH_3 \\ \qquad\qquad\qquad\quad 1 \end{array}$$

Correct name: 2,3,5-Trimethyloctane

The longest chain is an octane.

(f)

H H
H_3C 3 1 CH_3

Correct name: *cis*-1,3-Dimethylcyclohexane

The substituents should have the lowest possible numbers.

3.43

(a)

CH_3 CH_3

1,1-Dimethylcyclooctane

(b)

$$\begin{array}{l} \qquad CH_3\;\; CH_2CH_3 \\ \qquad\; | \qquad\; | \\ CH_3CCH_2CCH_2CH_3 \\ \qquad\; | \qquad\; | \\ \qquad CH_3\;\; CH_2CH_3 \end{array}$$

4,4-Diethyl-2,2-dimethylhexane

(c)

CH_3 CH_3 CH_3

1,1,2-Trimethylcyclohexane

(d)

$$\begin{array}{l} \qquad\qquad\qquad\qquad\qquad\quad CH_3 \\ \qquad\qquad\qquad\qquad\qquad\quad | \\ \qquad\qquad\qquad\quad CH_2CH_2CHCH_3 \\ \qquad\qquad\qquad\quad | \\ CH_3CH_2CH_2CH_2CH_2CHCH_2CH_2CH_2CH_2CH_3 \end{array}$$

6-(3-Methylbutyl)-undecane

Remember that you must choose an alkane whose principal chain is long enough so that the substituent does not become part of the principal chain.

3.44

(a)

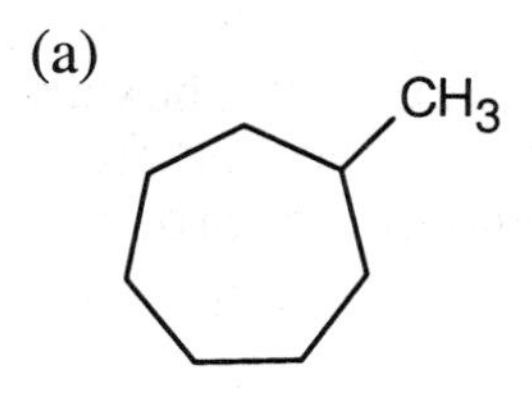

Methylcycloheptane

(b)

cis-1,3-Dimethylcyclopentane

(c)

trans-1,2-Dimethylcyclohexane

(d)

trans-1-Isopropyl-2-methylcyclobutane

(e)

1,1,4-Trimethylcyclohexane

3.45

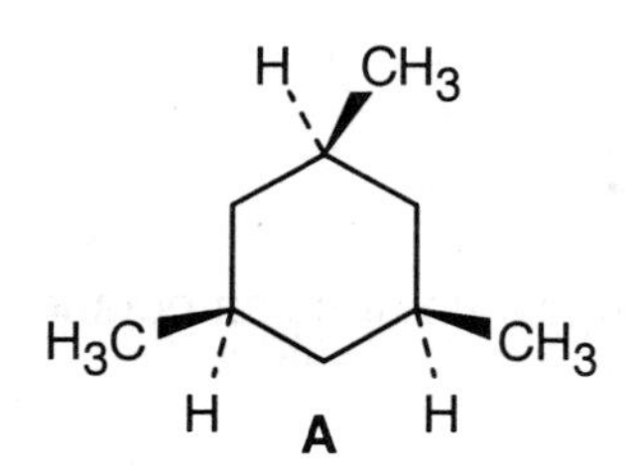

B

Two cis–trans isomers of 1,3,5–trimethylcyclohexane are possible. In one isomer (**A**), all methyl groups are cis; in **B**, one methyl group is trans to the other two.

3.46

(a)

cis-1,3-Dibromocyclohexane

trans-1,4-Dibromocyclohexane

constitutional isomers

(b)

$CH_3CH_2CH_2CH(CH_3)CH(CH_3)CH_3$

2,3-Dimethylhexane

$CH_3CH(CH_3)CH_2CH_2CH(CH_3)CH_3$

2,5,5-Trimethylpentane
(correct name: 2,5-Dimethylhexane)

constitutional isomers

(c)

identical

3.47

cis-1,3-Dibromocyclopentane *trans*-1,3-Dibromocyclopentane

3.48

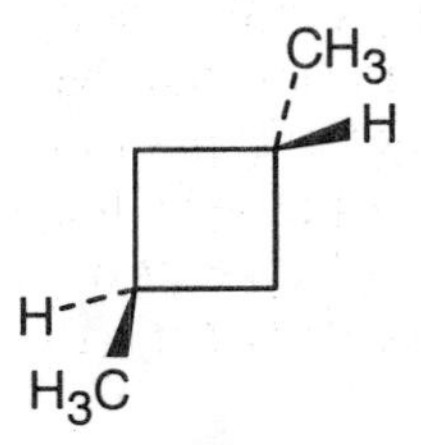

trans-1,3-Dimethylcyclobutane *cis*-1,3-Dimethylcyclobutane

3.49 (a) Because malic acid has two –COOH groups, the formula for the rest of the molecule is C_2H_4O. Possible structures for malic acid are:

$HOC(O)CH(CH_2OH)COH$ — primary alcohol

$HOC(O)CH(OH)CH_2COH(O)$ — secondary alcohol

$HOC(O)C(OH)(CH_3)COH(O)$ — tertiary alcohol

$HOC(O)CH_2OCH_2COH(O)$

$HOC(O)OCH_2CH_2COH(O)$

$HOC(O)OCH(CH_3)COH(O)$

(b) Because only one of these compounds (the second one) is also a secondary alcohol, it must be malic acid.

3.50

If $H_2C(CH_2Br)_2$ + 2 Na → cyclopropane + 2 NaBr

Then $C(CH_2Br)_4$ + 4 Na → spiropentane + 4 NaBr

The two rings are perpendicular in order to keep the geometry of the central carbon as close to tetrahedral as possible.

3.51 To solve this type of problem, read the problem carefully, word for word. Then try to interpret parts of the problem. For example:

1) Formaldehyde is an aldehyde, H–C(=O)–H
2) It trimerizes — that is, 3 formaldehydes come together to form a compound $C_3H_6O_3$. Because no atoms are eliminated, all of the original atoms are still present.
3) There are no carbonyl groups. This means that trioxane cannot contain any –C=O functional groups. If you look back to Table 3.1, you can see that the only oxygen-containing functional groups that can be present are either ethers or alcohols.
4) A monobromo derivative is a compound in which one of the –H's has been replaced by a –Br. Because only one monobromo derivative is possible, we know that there can only be one type of hydrogen in trioxane. The only possibility for trioxane is:

Trioxane

3.52

A B C

Menthol Isomers of menthol

The substituents on the ring have the following relationships:

	Menthol	Isomer A	Isomer B	Isomer C
—CH(CH$_3$)$_2$,—CH$_3$	trans	trans	cis	cis
—CH(CH$_3$)$_2$,—OH	trans	cis	trans	cis
—CH$_3$,—OH	cis	trans	trans	cis

3.53

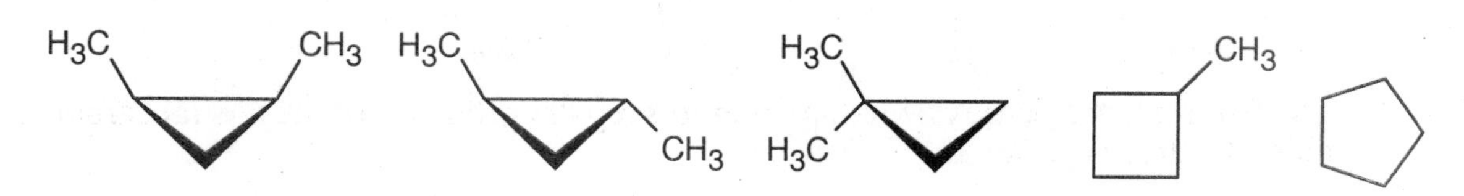

The first two structures are cis-trans isomers.

A Look Ahead

3.54

H_3C H H CH_3

H CH_3 H_3C H

The two *trans*–1,2–dimethylcyclopentanes are mirror images.

3.55 A puckered ring allows all the bonds in the ring to have a nearly tetrahedral bond angle.

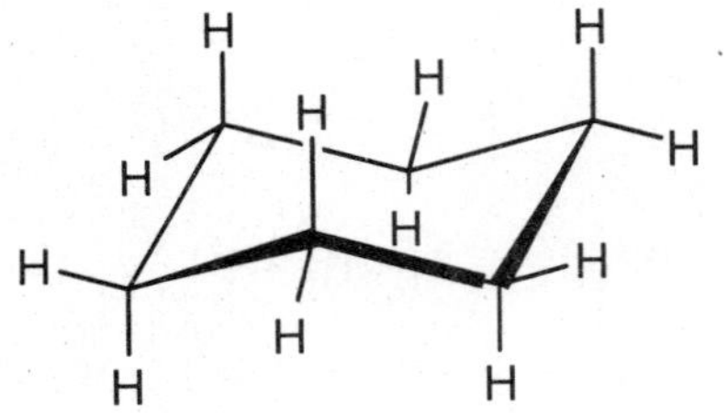

Molecular Modeling

3.56 Only cyclooctadecane contains a cavity.

3.57 (no answer)

3.58

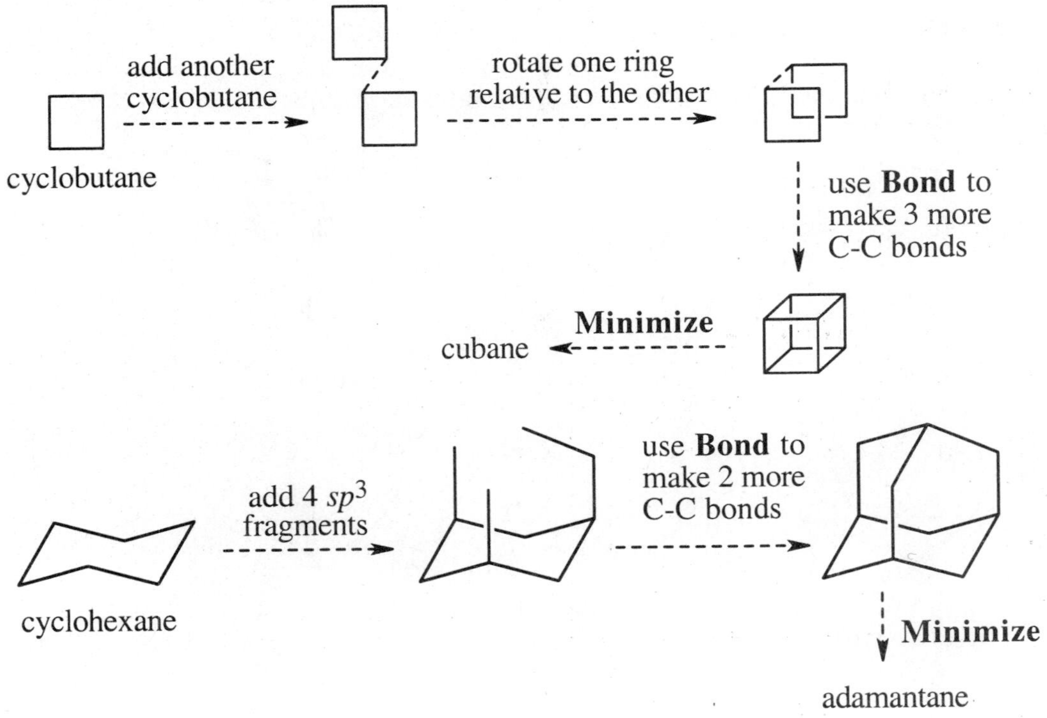

cyclobutane → add another cyclobutane → rotate one ring relative to the other → use **Bond** to make 3 more C-C bonds → **Minimize** → cubane

cyclohexane → add 4 sp^3 fragments → use **Bond** to make 2 more C-C bonds → **Minimize** → adamantane

Chapter 4 – Stereochemistry of Alkanes and Cycloalkanes

Chapter Outline

I. Conformations of straight-chain alkanes (Sections 4.1 – 4.3).
  A. Conformations of ethane (Section 4.1).
    1. Rotation about a single bond produces isomers that differ in conformation.
      These isomers (conformers) have the same connections of atoms and can't be isolated.
    2. These isomers can be represented in two ways:
      a. Sawhorse representations view the C–C bond from an oblique angle.
      b. Newman projections represent the two carbons as a circle.
    3. There is a barrier to rotation that makes some conformers of lower energy than others.
      a. The lowest energy conformer (staggered conformation) occurs when all C–H bonds are as far from each other as possible.
      b. The highest energy conformer (eclipsed conformation) occurs when all C–H bonds are as close to each other as possible.
      c. Between these two conformations lie an infinite number of other conformations.
    4. The staggered conformation is 12 kJ/mol lower in energy than the eclipsed conformation.
      a. This energy difference is due to torsional strain from the repulsion of electron clouds in the C–H bond.
      b. The torsional strain resulting from a single C–H interaction is 4.0 kJ/mol.
      c. The barrier to rotation can be represented on a graph of potential energy vs. angle of rotation.
  B. Conformations of propane (Section 4.2).
    1. Propane also shows a barrier to rotation that is 14 kJ/mol.
    2. The eclipsing interaction between a C–C bond and a C–H bond is 6.0 kJ/mol.
  C. Conformations of butane (Section 4.3).
    1. Not all staggered conformations of butane have the same energy; not all eclipsed conformations have the same energy.
      a. In the lowest energy conformation (anti) the two large methyl groups are as far from each other as possible).
      b. The eclipsed conformation that has two methyl–hydrogen interactions and a H–H interaction is 16 kJ/mol higher in energy than the anti conformation.
      c. The conformation with two methyl groups 60° apart (gauche conformation) is 3.8 kJ/mol higher in energy than the anti conformation.
        This energy difference is due to steric strain – the repulsive interaction that results from forcing atoms to be closer together than their atomic radii allow.
      d. The highest energy conformations occurs when the two methyl groups are eclipsed.
        i. This conformation is 19 kJ/mol less stable than the anti conformation.
        ii. The value of a methyl-methyl eclipsing interaction is 11 kJ/mol.
    2. The most favored conformation for any straight-chain alkane has carbon-carbon bonds in staggered arrangements and large substituents anti to each other.
    3. At room temperature, bond rotation occurs rapidly, but a majority of molecules adopt the most stable conformation.

II. Conformations of cycloalkanes (Sections 4.4 – 4.15).
  A. General principles (Sections 4.4 – 4.6).
    1. Baeyer strain theory (Section 4.4).
      a. A. von Baeyer suggested that rings other than those of 5 or 6 carbons were too strained to exist.
      b. This concept of angle strain is true for smaller rings, but larger rings can be easily prepared.
    2. Heats of combustion of cycloalkanes (Section 4.5).
      a. To measure strain, it is necessary to measure the total energy of a compound and compare it to a strain-free reference compound.
      b. Heat of combustion measures the amount of heat released when a compound is burned in oxygen.
        i. The more strained the compound, the higher the heat of combustion.
        ii. Strain per CH_2 unit can be calculated and plotted as a function of ring size.
      c. Graphs show that only small rings have serious strain.
    3. The nature of ring strain (Section 4.6).
      a. Rings tend to adopt puckered conformations.
      b. Several factors account for ring strain.
        i. Angle strain occurs when bond angles are distorted from their normal values.
        ii. Torsional strain is due to eclipsing of bonds.
        iii. Steric strain results when atoms approach too closely.
  B. Conformations of small rings (Sections 4.7 – 4.8).
    1. Cyclopropane (Section 4.7).
      a. Cyclopropane has bent bonds.
      b. Because of bent bonds, cyclopropane is more reactive than other cycloalkanes.
    2. Cyclobutane (Section 4.8).
      a. Cyclobutane has less angle strain than cyclopropane but has more torsional strain.
      b. Cyclobutane has almost the same total strain as cyclopropane.
      c. Cyclobutane is slightly bent to relieve torsional strain, but this increases angle strain.
    3. Cyclopentane
      a. Cyclopentane has little angle strain but considerable torsional strain.
      b. To relieve torsional strain, cyclopentane adopts a puckered conformation.
        In this conformation, one carbon is bent out of plane; hydrogens are nearly staggered.
  C. Conformations of cyclohexane (Sections 4.9 – 4.14).
    1. Chair cyclohexane (Section 4.9).
      a. The chair conformation of cyclohexane is strain-free.
      b. In a standard drawing of cyclohexane the lower bond is in front.
    2. Axial and equatorial bonds in cyclohexane (Section 4.10).
      a. There are two kinds of positions on a cyclohexane ring.
        i. Six axial hydrogens are perpendicular to the plane of the ring.
        ii. Six equatorial hydrogens are roughly in the plane of the ring.
      b. Each carbon has one axial hydrogen and one equatorial hydrogen.
      c. Each side of the ring has alternating axial and equatorial hydrogens.
      d. All hydrogens on the same side of the ring are cis.
    3. Conformational mobility of cyclohexanes (Section 4.11).
      a. Different chair conformations of cyclohexanes interconvert by a ring-flip.
      b. After a ring-flip, an axial bond becomes an equatorial bond, and vice versa.
      c. The energy barrier to interconversion is 45 kJ/mol, making interconversion rapid at room temperature.

4. Conformations of monosubstituted cyclohexanes (Section 4.12).
    a. Both conformations aren't equally stable at room temperature.
        In methylcyclohexane, 95% of molecules have the methyl group in the equatorial position.
    b. The energy difference is due to 1,3-diaxial interactions.
        i. These interactions are due to steric strain.
        ii. They are the same interactions as occur in gauche butane.
    c. Axial methylcyclohexane has two gauche interactions that cause it to be 7.6 kJ/mol less stable than equatorial methylcyclohexane.
    d. All substituents are more stable in the equatorial position.
        The size of the strain depends on the size of the group.
5. Conformations of disubstituted cyclohexanes (Section 4.13).
    a. In *cis*-dimethylcyclohexane, one methyl group is axial and one is equatorial in both chair conformations, which are of equal energy.
    b. In *trans*-dimethylcyclohexane, both methyl groups are either both axial or both equatorial.
        i. The conformation with both methyl groups axial is 15.2 kJ/mol less stable than the conformation with both groups equatorial.
        ii. The trans isomer exists almost exclusively in the diequatorial conformation.
    c. This type of conformational analysis can be carried out for most substituted cyclohexanes.
6. Boat cyclohexane (Section 4.14).
    a. Boat cyclohexane has two types of hydrogen.
    b. Boat cyclohexane shows both steric strain and torsional strain.
    c. Boat cyclohexane is 29 kJ/mol less stable than chair cyclohexane.
        i. Twisting can reduce some of this strain.
        ii. Twist-boat cyclohexane is still much more strained than chair cyclohexane.

D. Conformations of polycyclic molecules (Section 4.15).

1. Decalin has two rings that can be either cis-fused or trans-fused.
    The two decalins are nonconvertible.
2. Steroids have four fused rings.
3. Bicyclic ring systems have rings that are connected by bridges.
    In norbornane, the six-membered ring is locked into a boat conformation.

Solutions to Problems

4.1 Make models of staggered and eclipsed ethane and measure the distance between a hydrogen atom on carbon 1 and a hydrogen atom on carbon 2. You will find that the distance between hydrogens (**A**) in the staggered conformation is 10% – 15% greater than the distance between hydrogens in the eclipsed conformation (**B**). (The calculated percent difference of the distance between hydrogens in the two conformations is 11.1%).

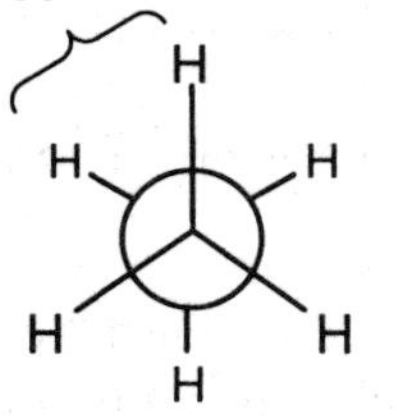

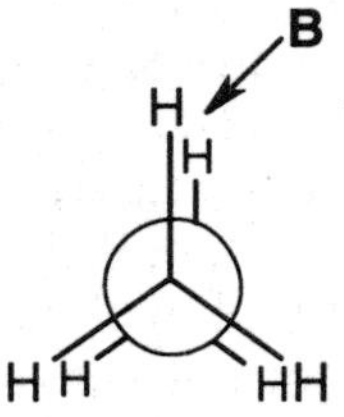

4.2

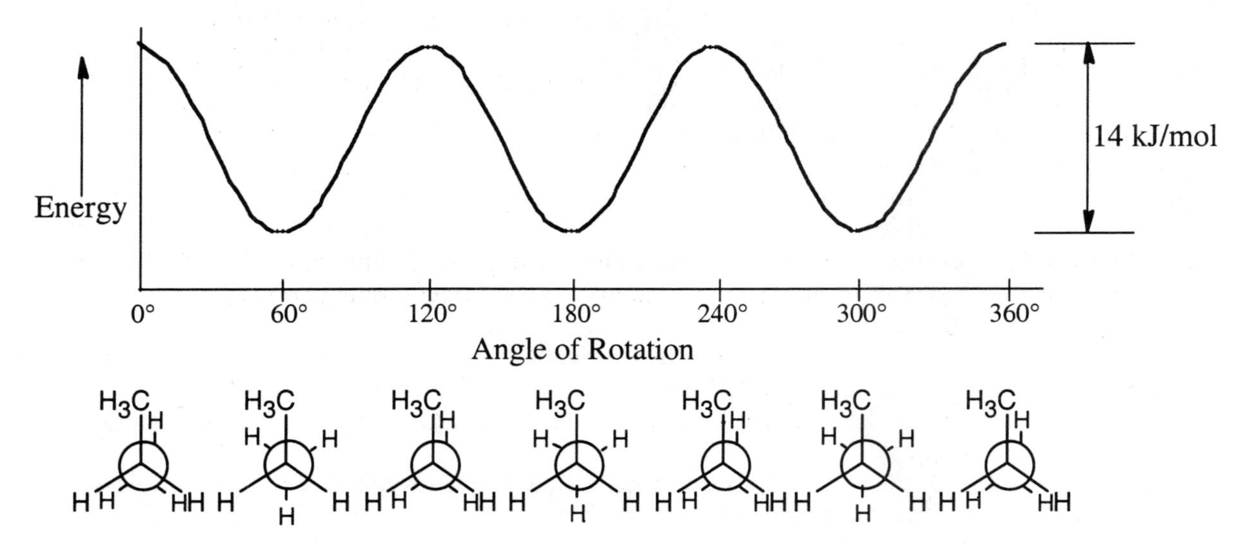

4.3

(a)

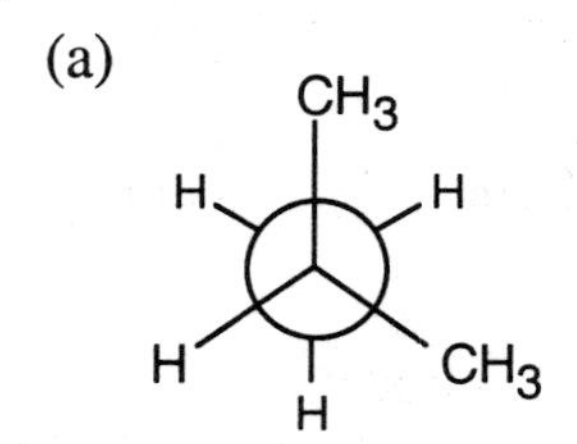

The *most stable* conformation is staggered and occurs at 60°, 180°, and 300°.

(b)

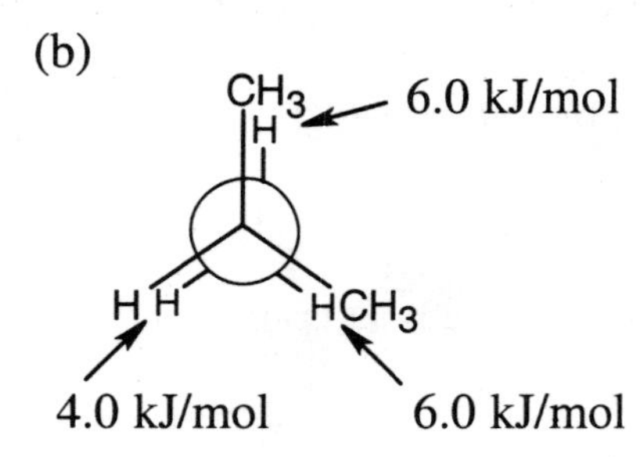

The *least stable* conformation is eclipsed and occurs at 0°, 120°, 240°, and 360°.

(c),(d)

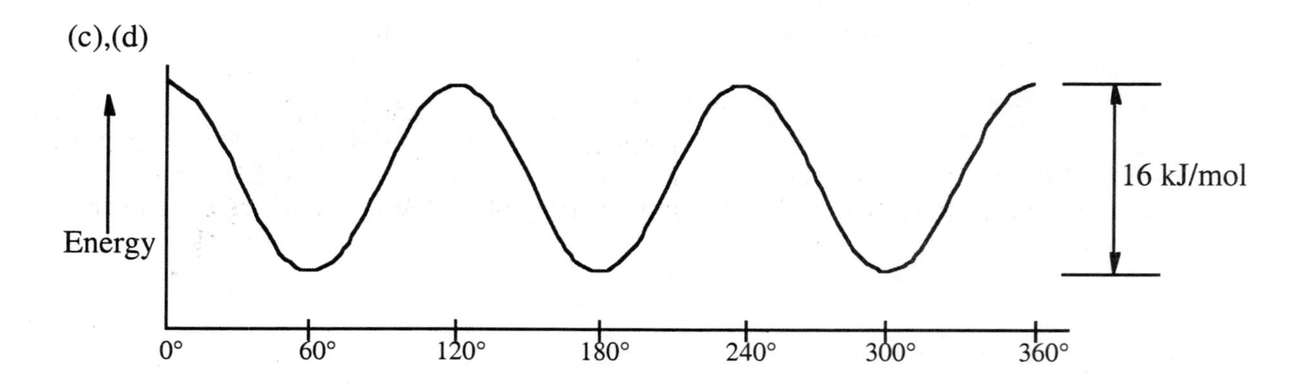

4.4

This conformation of 2,3-dimethylbutane is the most stable because it is staggered and has the fewest CH_3–CH_3 gauche interactions.

4.5 The conformation shown in the stereo view is a staggered conformation in which the hydrogens on carbons 2 and 3 are 60° apart. Draw the Newman projection.

3.8 kJ/mol

3.8 kJ/mol

3.8 kJ/mol

The Newman projection shows three gauche interactions, each of which has an energy cost of 3.8 kJ/mol. The total strain energy is 11.4 kJ/mol.

4.6 Cyclopropane has a higher heat of combustion than cyclohexane. Combustion of cyclopropane releases more heat per gram than combustion of cyclohexane because of the high strain energy of the cyclopropane ring.

4.7

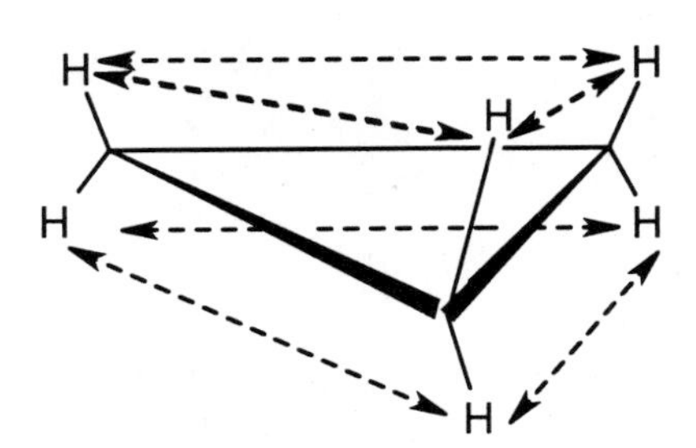

All hydrogen atoms on the same side of the cyclopropane ring are eclipsed by each other. If we draw each hydrogen-hydrogen interaction, we count six eclipsing interactions. Since each of these interactions "costs" 4.0 kJ/mol, all six cost 24.0 kJ/mol. 24 kJ/mol ÷115 kJ/mol = 0.21; thus, 21% of the total strain energy of cyclopropane is due to eclipsing strain.

4.8

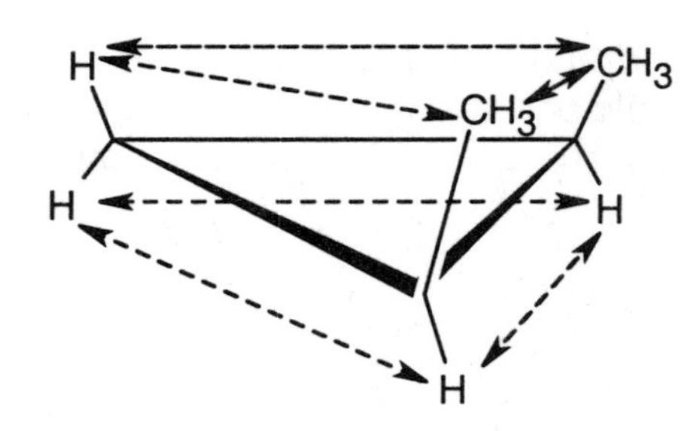

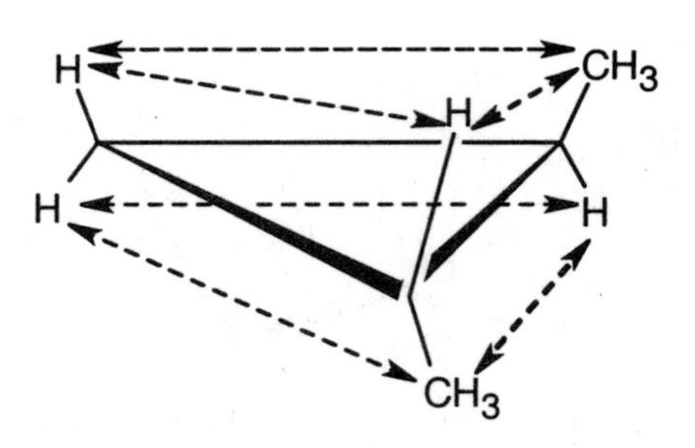

Eclipsing interaction	Energy cost (kJ/mol)	*cis isomer* # of interactions	*cis isomer* Total energy cost (kJ/mol)	*trans isomer* # of interactions	*trans isomer* Total energy cost (kJ/mol)
H–H	4.0	3	12.0	2	8.0
H–CH_3	6.0	2	12.0	4	24.0
CH_3–CH_3	11	1	11	0	0
			35		32

The added energy cost of eclipsing interactions causes *cis*-1,2-dimethylcyclopropane to be of higher energy, and to be less stable than the trans isomer. Since the cis isomer is of higher energy, its heat of combustion is also greater.

4.9

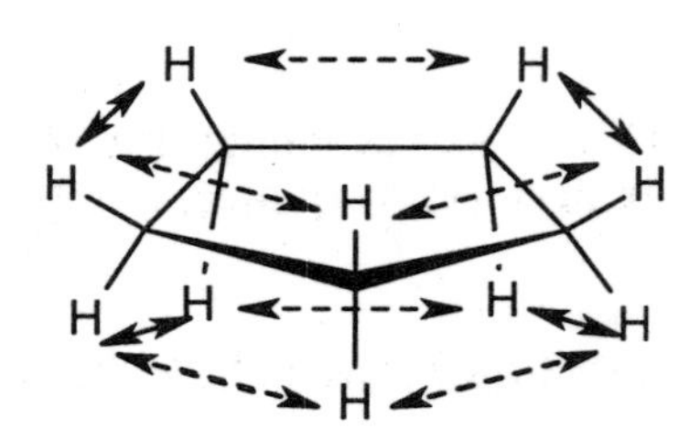

If cyclopentane were planar, it would have ten hydrogen–hydrogen interactions with a total energy cost of 40 kJ/mol. The measured total strain energy of 26 kJ/mol indicates that 14 kJ/mol of eclipsing strain in cyclopentane (35%) has been relieved by puckering.

4.10

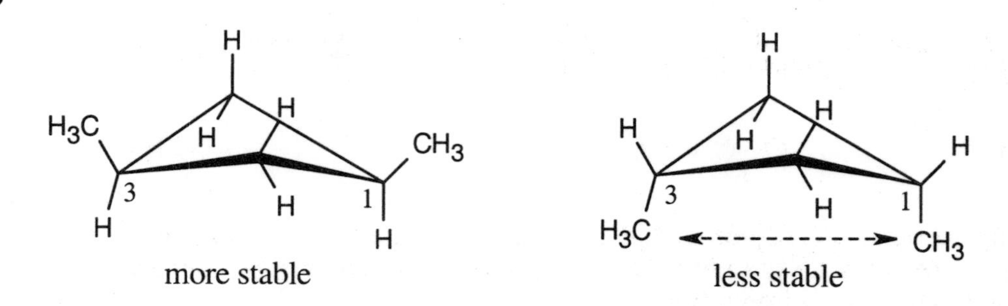

The methyl groups are farther apart in the more stable conformation of *cis*-1,3-dimethylcyclobutane.

4.11

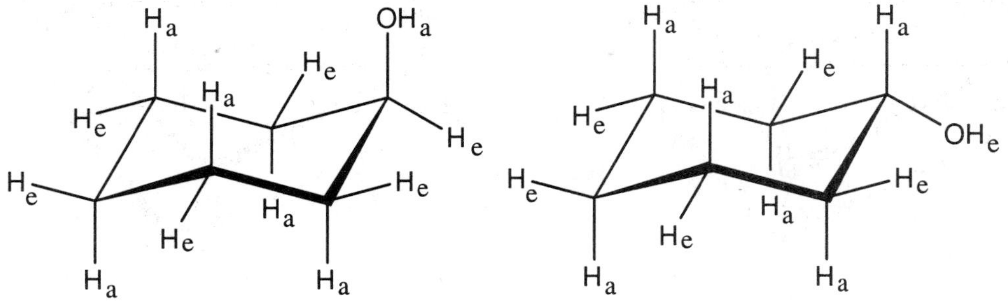

The conformation with –OH in the equatorial position is more stable.

4.12 Make a model of *cis*-1,2-dichlorocyclohexane. Notice that all cis substituents are on the same side of the ring and that two adjacent cis substituents have an axial-equatorial relationship. Now, perform a ring-flip on the cyclohexane.

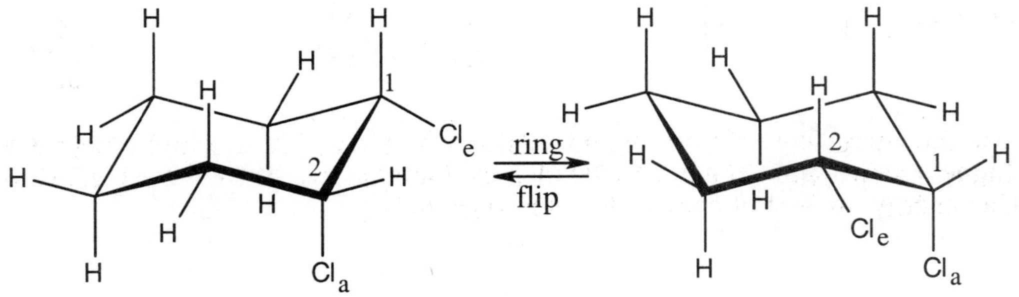

After the ring-flip, the relationship of the two substituents is still axial-equatorial. No two adjacent cis substituents can be converted to being both axial or both equatorial without breaking bonds.

4.13 For a *trans*-1,2-disubstituted cyclohexane, two adjacent substituents must be either both axial or both equatorial.

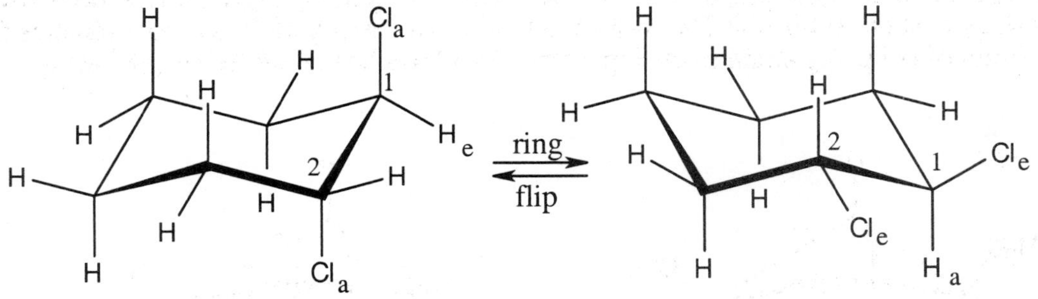

A ring flip converts two adjacent axial substituents to equatorial substituents, and vice versa. As in Problem 4.12, no two adjacent trans substituents can be converted to an axial-equatorial relationship without bond breaking.

4.14 In trans-1,4-disubstituted cyclohexanes, the methyl substituents are either both axial or both equatorial.

trans-1,4-Dimethylcyclohexane

4.15

The most stable conformation of axial *tert*-butylcyclohexane is pictured. One methyl group is positioned above the ring and competes for space with two axial ring protons. In the other axial alkylcyclohexanes, this methyl group is replaced by hydrogen, which has a much smaller space requirement. The steric strain caused by an axial *tert*-butyl group is therefore higher than the strain caused by axial methyl, ethyl or isopropyl groups.

4.16 The energy difference between an axial and an equatorial cyano group is very small because there are no 1–3 diaxial interactions for a cyano group.

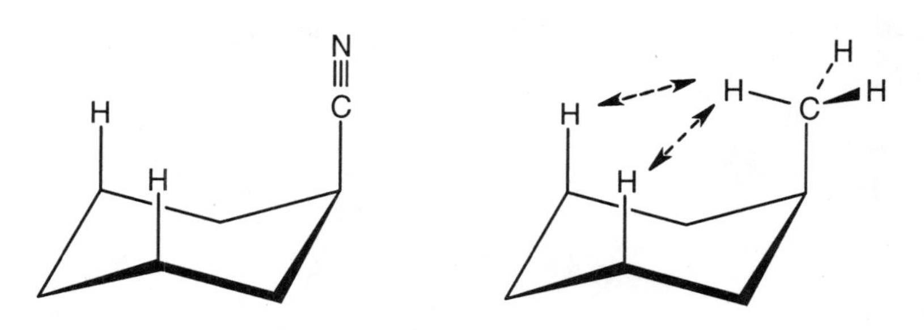

Cyclohexanecarbonitrile Methylcyclohexane

4.17 Table 4.2 shows that an axial bromine causes 2 x 1.0 kJ/mol of steric strain. Thus, the energy difference between axial and equatorial bromocyclohexane is 2.0 kJ/mol. According to Figure 4.19, this energy difference corresponds approximately to a 70:30 ratio of more stable : less stable conformer. Thus, 70% of bromocyclohexane molecules are in the equatorial conformation, and 30% are in the axial conformation at any given moment.

4.18

(a)

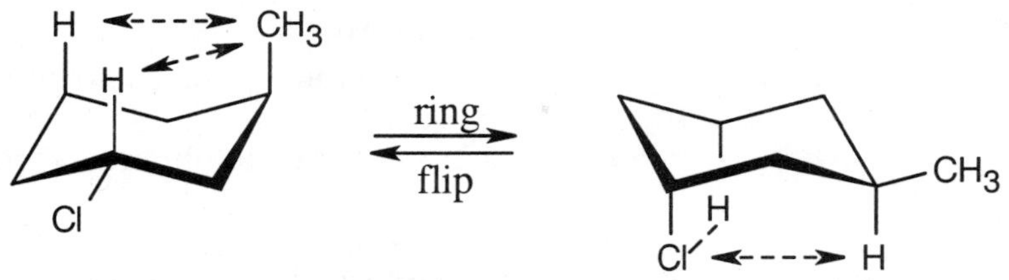

ring flip

trans-1-Chloro-3-methylcyclohexane

2 $(H–CH_3)$ = 7.6 kJ/mol   2 (H–Cl) = 2.0 kJ/mol

The second conformation is more stable than the first.

(b)

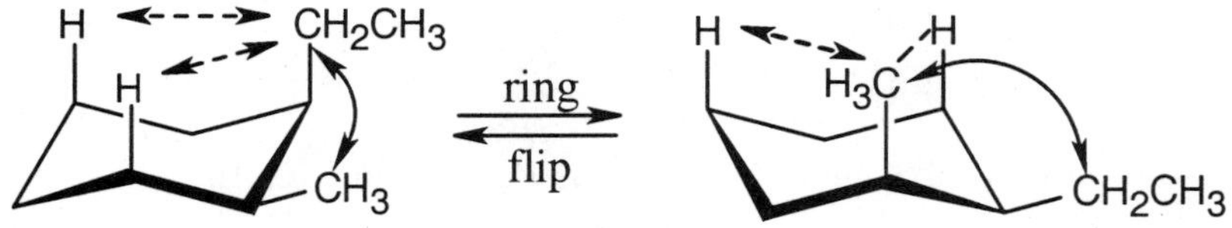

ring flip

cis-1-Ethyl-2-methylcyclohexane

First conformation	Second conformation
one $CH_3–CH_2CH_3$ gauche interaction = 3.8 kJ/mol	one $CH_3–CH_2CH_3$ gauche interaction = 3.8 kJ/mol
2 $(H–CH_2CH_3)$ = 8.0 kJ/mol	2 $(H–CH_3)$ = 8.0 kJ/mol
Total = 11.8 kJ/mol	Total = 11.4 kJ/mol

The second conformation is more stable than the first.

(c)

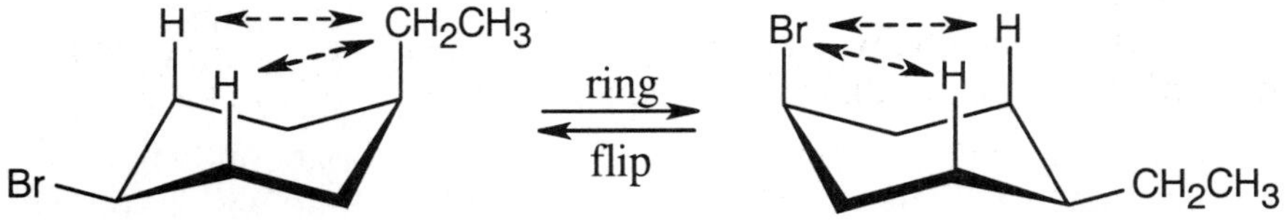

ring flip

cis-1-Bromo-4-ethylcyclohexane

2 $(H–CH_2CH_3)$ = 8.0 kJ/mol   2 (H–Br) = 2.0 kJ/mol

The second conformation is more stable than the first.

(d)

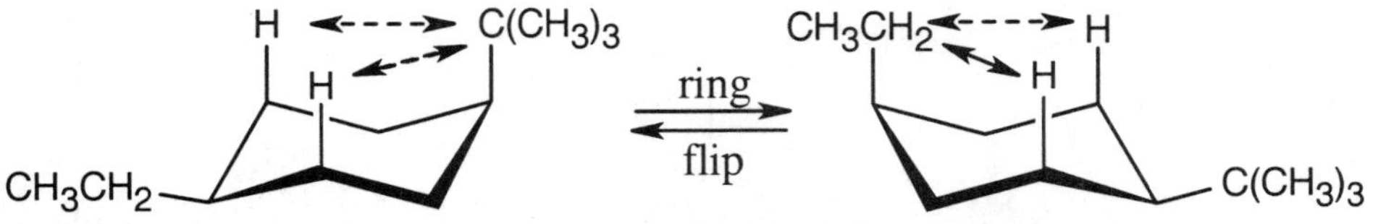

ring flip

cis-1-*tert*-Butyl-4-ethylcyclohexane

2 $[H–C(CH_3)_3]$ = 22.8 kJ/mol   2 $(H–CH_2CH_3)$ = 8.0 kJ/mol

The second conformation is more stable than the first.

4.19 The compound pictured is *cis*-1-chloro-2-methyl-*trans*-4-methylcyclohexane, and the three substituents have the orientations shown in the first structure. To decide if the conformation shown is the more stable conformation or the less stable conformation, perform a ring-flip on the illustrated conformation and do a calculation like those in the previous problem. Notice that each conformation has a Cl–CH_3 gauche interaction, but we don't need to know its energy cost because it is present in both conformations.

ring flip

2 [H–CH_3] = 7.6 kJ/mol
2 [H–Cl] = 2.0 kJ/mol
= 9.6 kJ/mol

2 [H–CH_3] = 7.6 kJ/mol

The illustrated conformation is the less stable chair form.

4.20

In the chair conformation of *trans*-1,3-di-*tert*-butylcyclohexane, one *tert*–butyl group must be axial and the other equatorial. The 1,3–diaxial interactions of the axial *tert*-butyl group make the chair form of *trans*-1,3-di-*tert*-butylcyclohexane 22.8 kJ/mol less stable than a cyclohexane with no axial substituents. Since a twist boat conformation is 23 kJ/mol less stable than a chair, it is almost as likely that the compound will assume a twist-boat conformation as a chair conformation. The twist-boat removes the 1,3–diaxial interaction present in the chair form of *trans*-1,3-di-*tert*-butylcyclohexane.

4.21

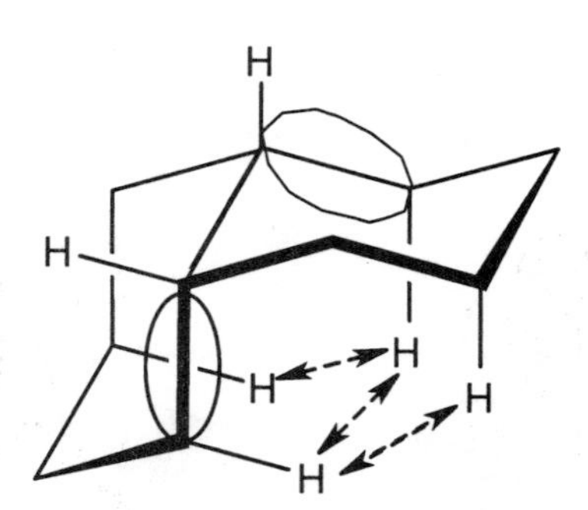

Trans-decalin is more stable than *cis*-decalin. Three 1,3–diaxial interactions cause *cis*–decalin to be of higher energy than *trans*-decalin. You may be able to visualize these interactions by thinking of the circled parts of *cis*-decalin as similar to axial methyl groups. The gauche interactions that occur with axial methyl groups also occur in *cis*-decalin.

Visualizing Chemistry

4.22

trans-1-Chloro-3-methylcyclohexane

2 (H–CH_3) = 7.6 kJ/mol 2 (H–Cl) = 2.0 kJ/mol

The conformation shown (the left structure) is the less stable conformation.

4.23

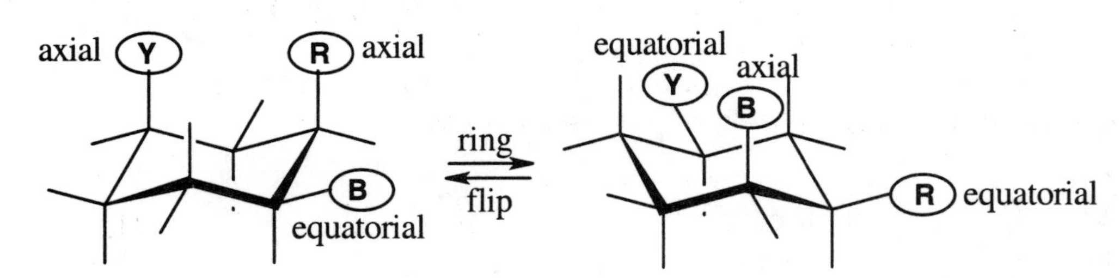

4.24

α-Glucose β-Glucose

The only difference between α-glucose and β-glucose is in the orientation of the –OH group at carbon 1; the –OH group is axial in α-glucose, and it is equatorial in β-glucose. You would expect β-glucose to be more stable because it has all substituents in the equatorial position.

Additional Problems

4.25

(a), (b)

2-Methylbutane — C2–C3 bond

Most stable conformation: 3.8 kJ/mol

Least stable comformation: 6.0 kJ/mol, 4.0 kJ/mol, 11 kJ/mol

The energy difference between the two conformations is (11 + 6.0 + 4.0) –3.8 = 17 kJ/mol.

(c) Consider the least stable conformation to be at zero degrees. Keeping the "front" of the projection unchanged, rotate the "back" by 60° to obtain each conformation.

3.8 kJ/mol 6.0 kJ/mol 6.0 kJ/mol 6.0 kJ/mol 3.8 kJ/mol

at 60°: energy = 3.8 kJ/mol at 120°: energy = 18.0 kJ/mol at 180°: energy = 3.8 kJ/mol

11 kJ/mol 3.8 kJ/mol 3.8 kJ/mol

4.0 kJ/mol 6.0 kJ/mol

at 240°: energy = 21 kJ/mol at 300°: energy = 7.6 kJ/mol

Use the lowest energy conformation as the energy minimum. The highest energy conformation is 17 kJ/mol higher in energy than the lowest energy conformation.

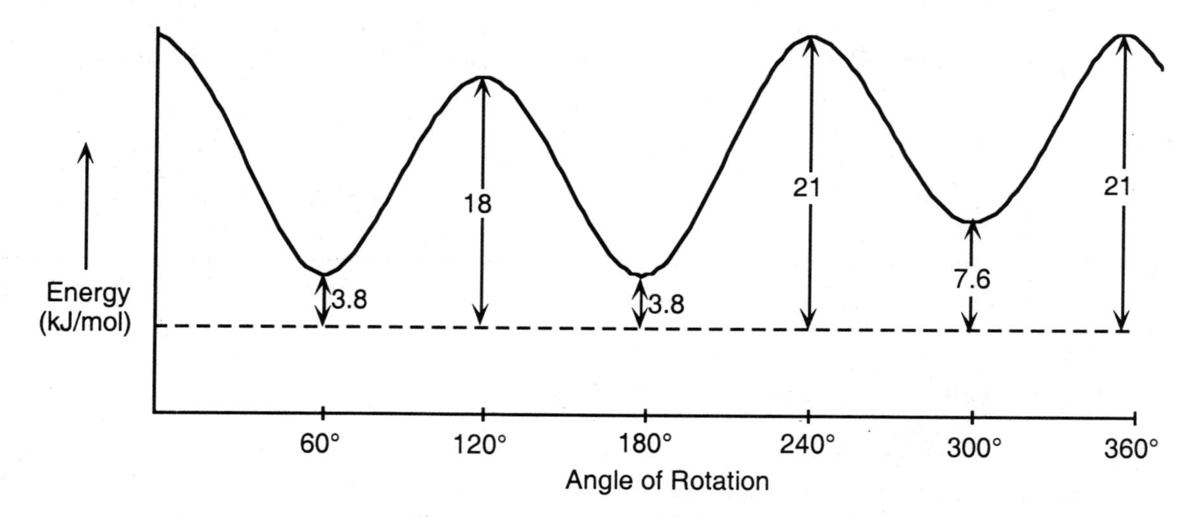

4.26

2 CH_3–CH_3 gauche	3 CH_3–CH_3 gauche	3 CH_3–CH_3 gauche
= 2(3.8 kJ/mol)	= 3(3.8 kJ/mol)	= 3(3.8 kJ/mol)
= 7.6 kJ/mol	= 11.4 kJ/mol	= 11.4 kJ/mol

4.27 Since we are not told the values of the interactions for 1,2–dibromoethane, the diagram can only be qualitative.

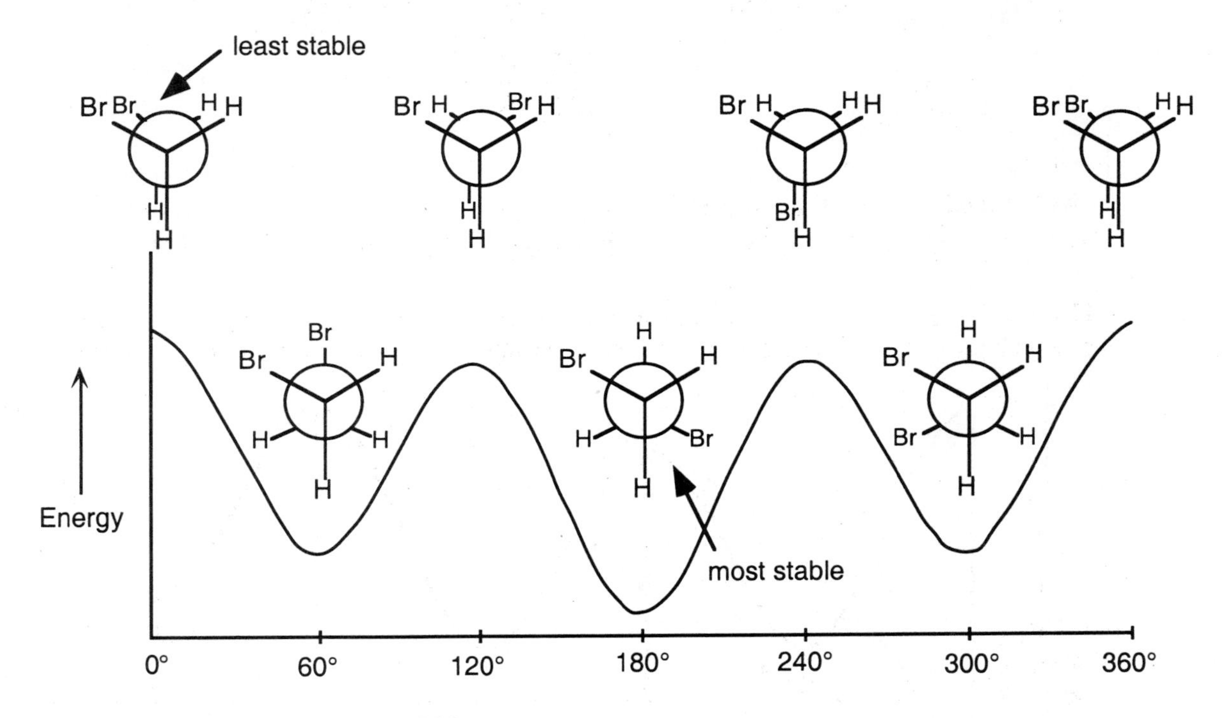

The anti conformation is at 180°.
The gauche conformations are at 60°, 300°.

4.28 The anti conformation has no net dipole moment because the polarities of the individual bonds cancel. The gauche conformation, however, has a dipole moment. Because the observed dipole moment is 1.0 D at room temperature, a mixture of conformations must be present.

4.29 The highest energy conformation of bromoethane is 15 kJ/mol. Because this includes two H–H eclipsing interactions of 4.0 kJ/mol each, the value of an H–Br eclipsing interaction is 15-2(4.0) = 7 kJ/mol.

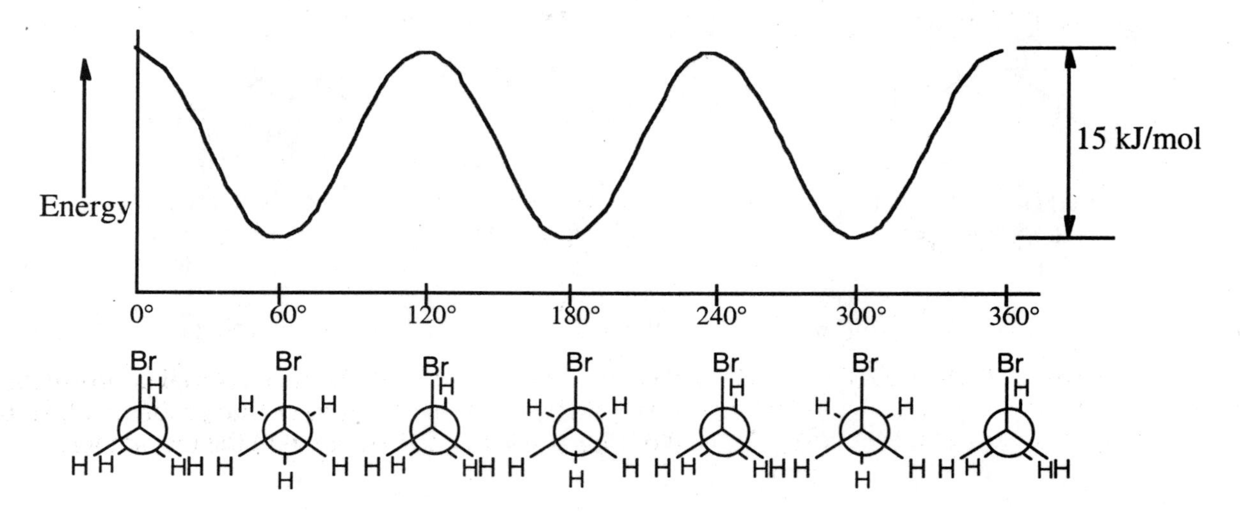

4.30 The best way to draw pentane is to make a model and to copy it onto the page. A model shows the relationship among atoms, and its drawing shows how these relationships appear in two dimensions. From your model, you should be able to see that all atoms are staggered in the drawing.

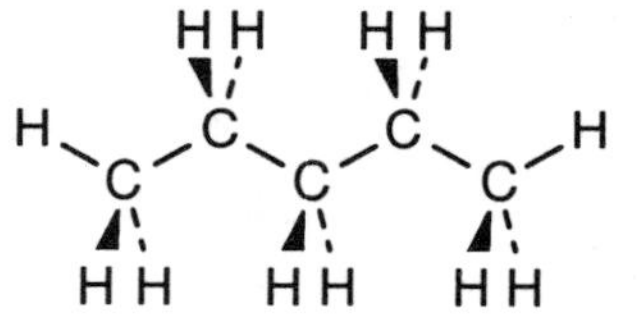

4.31

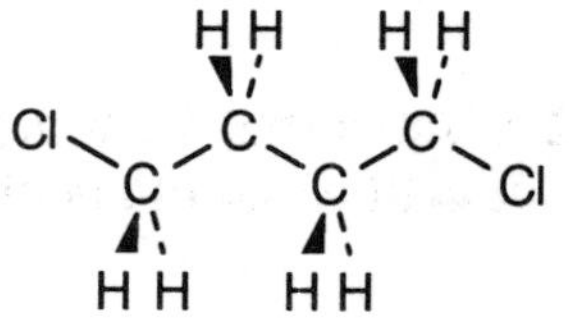

4.32

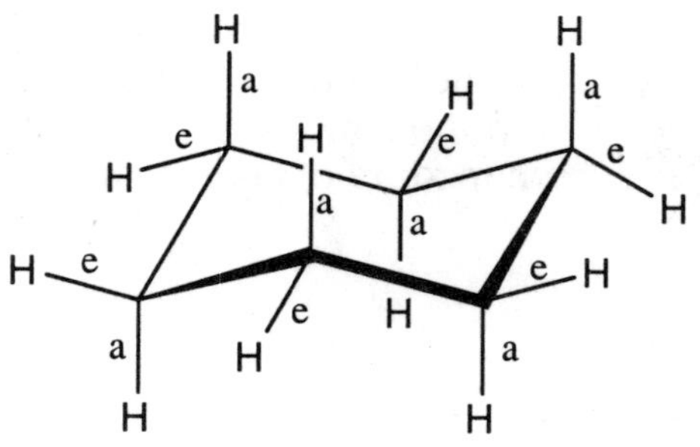

4.33

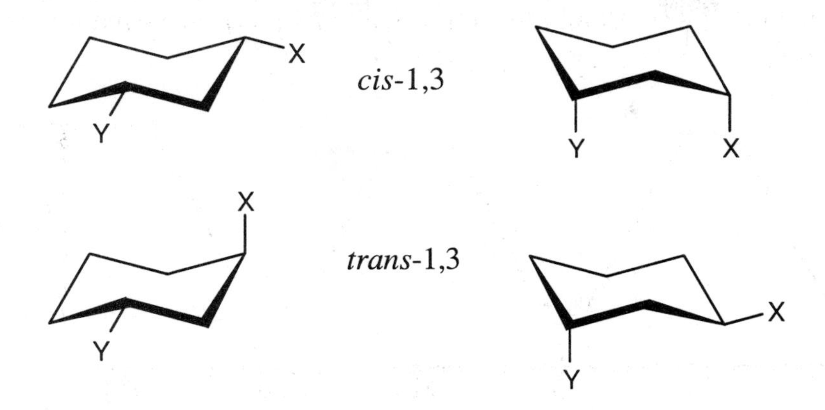

A cis–1,3–disubstituted isomer exists almost exclusively in the diequatorial conformation, which has no 1,3–diaxial interactions. The trans isomer must have one group axial, leading to 1,3–diaxial interactions. Thus, the trans isomer is less stable than the cis isomer.

4.34

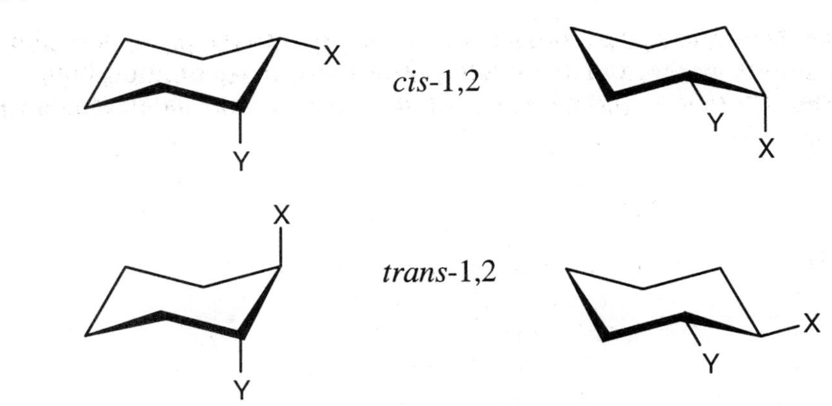

Reasoning similar to that in the previous problem can be used to show that the trans–1,2 disubstituted isomer, which exists principally in the diequatorial conformation, is more stable than the cis isomer.

4.35

The *trans*–1,4–isomer is more stable.

4.36

less stable
cis

more stable
trans

Two types of interaction are present in *cis*-1,2-dimethylcyclobutane. One interaction occurs between the two methyl groups, which are almost eclipsed. The other is an across-the-ring interaction between methyl group at position 1 of the ring and a hydrogen at position 3. Because neither of these interactions are present in trans isomer, it is more stable than the cis isomer.

more stable
cis

less stable
trans

In *trans*-1,3-dimethylcyclobutane an across-the-ring interaction occurs between the methyl group at position 3 of the ring and a hydrogen at position 1. Because no interactions are present in the cis isomer, it is more stable than the trans isomer.

4.37 Since the methyl group of *N*-methylpiperidine prefers an equatorial conformation, the steric requirements of a methyl group must be greater than those of an electron lone pair.

4.38

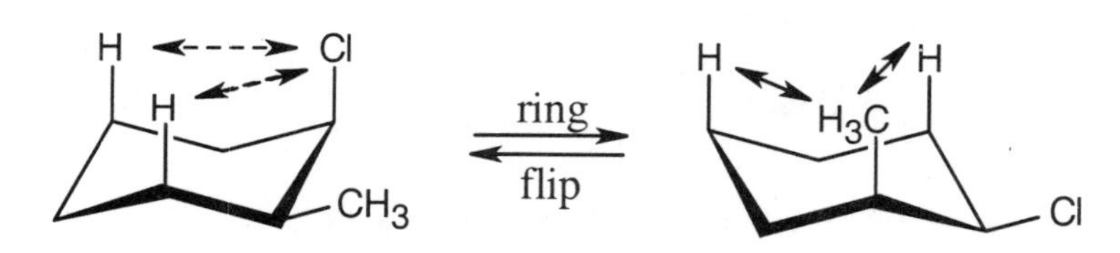

cis-1-Chloro-2-methylcyclohexane

Use Table 4.2 to find the values of 1,3–diaxial interactions. For the first conformation, the steric strain is 2 x 1.0 kJ/mol = 2.0 kJ/mol. The steric strain in the second conformation is 2 x 3.8 kJ/mol, or 7.6 kJ/mol. The first conformation is more stable than the second conformation by 5.6 kJ/mol.

4.39

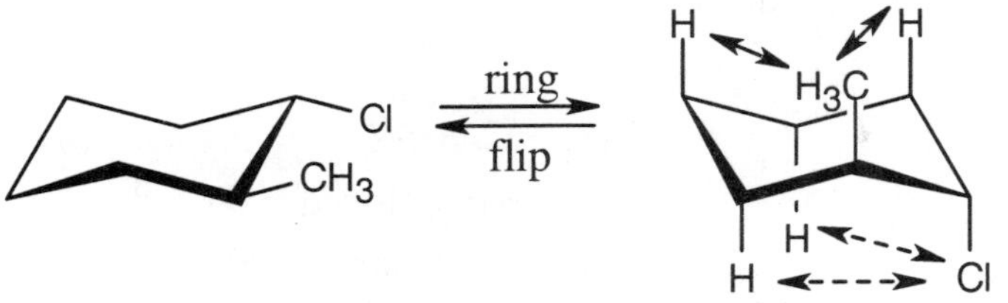

ring flip

no 1,3-diaxial interactions

2 x 3.8 kJ/mol = 7.6 kJ/mol
2 x 1.0 kJ/mol = 2.0 kJ/mol
= 9.6 kJ/mol

trans-1-Chloro-2-methylcyclohexane

The first conformation is more stable than the second conformation by a maximum of 9.6 kJ/mol. (A gauche interaction between the two substituents in the diequatorial conformation reduces the value of the energy difference.)

4.40

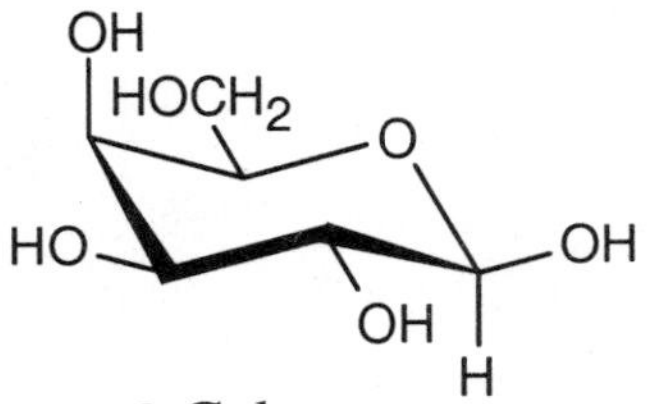

β-Galactose

In this conformation, all substituents, except for one hydroxyl group, are equatorial.

4.41 From the flat-ring drawing you can see that the methyl group and the –OH group have a cis relationship, and the isopropyl group has a trans relationship to both of these groups. Draw a chair cyclohexane ring and attach the groups with the correct relationship.

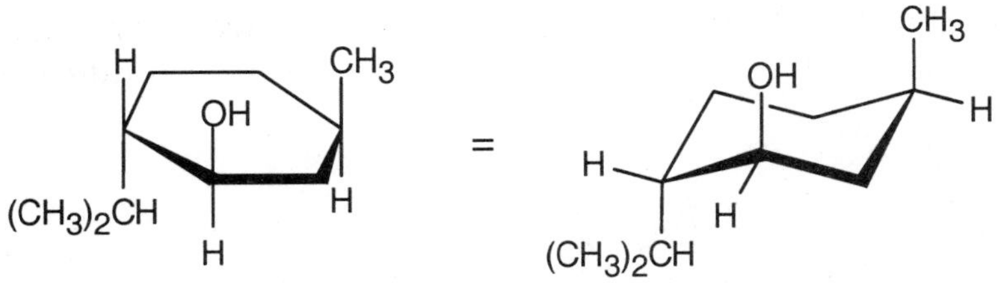

In this conformation, all substituents are axial. Now, perform a ring flip.

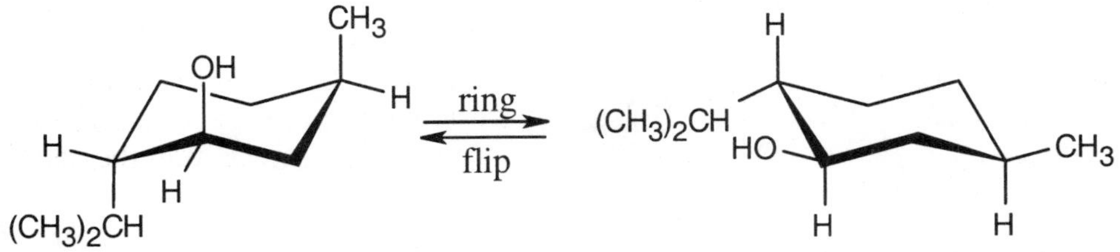

ring flip

The second conformation is more stable because all substituents are equatorial.

4.42 To solve this problem: (1) Find the energy cost of a 1–3 diaxial interaction by using Table 4.2. (2) Convert this energy difference into a percent by using Figure 4.19.

(a)

2 (H – $CH(CH_3)_2$ = 9.2 kJ/mol
% equatorial = 97.5
% axial = 2.5

(b)

2 (H – F) = 1.0 kJ/mol
% equatorial = 63
% axial = 37

(c)

2 (H – CN) = 0.8 kJ/mol
% equatorial = 60
% axial = 40

(d)

2 (H – OH) = 4.2 kJ/mol
% equatorial = 85
% axial = 15

4.43 Make sure you know the difference between axial–equatorial and cis–trans. Axial substituents are parallel to the axis of the ring; equatorial substituents lie around the "equator" of the ring. Cis substituents are on the same side of the ring; trans substituents are on opposite side of the ring.

(a) 1,3–trans

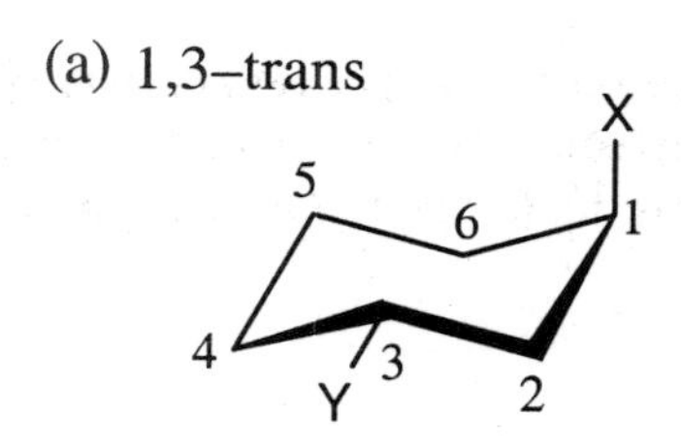

ring flip

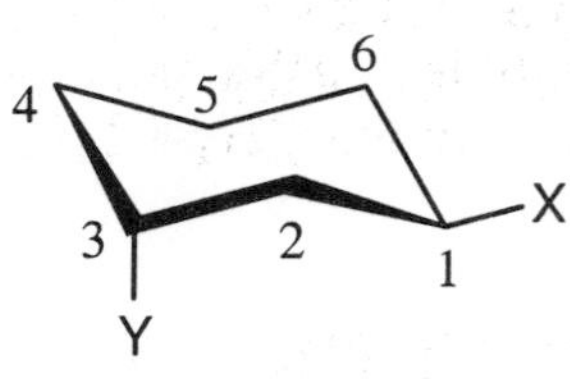

axial, equatorial

equatorial, axial

(b) 1,4–cis

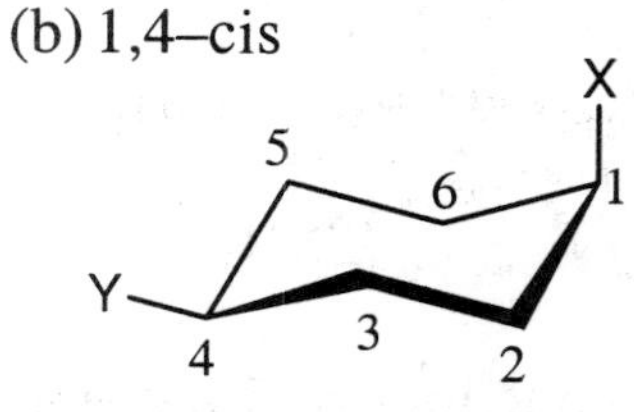

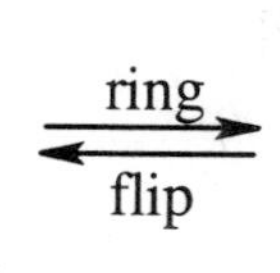

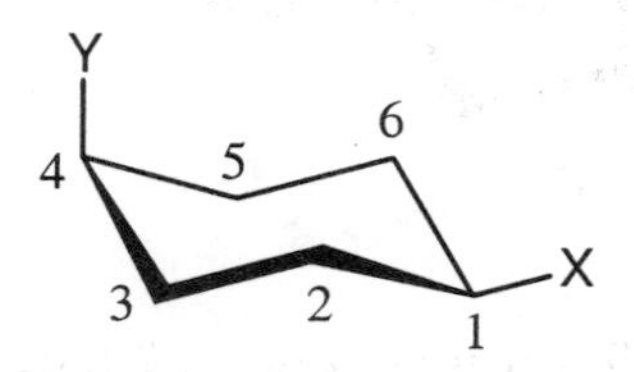

axial, equatorial

equatorial, axial

(c) 1,3–cis

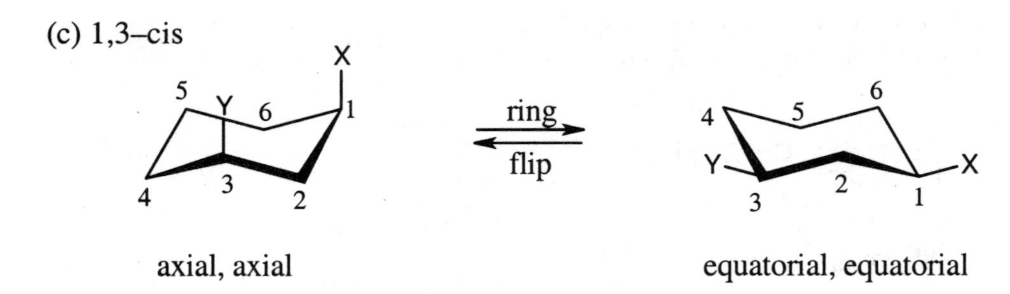

axial, axial — equatorial, equatorial

(d) 1,5–trans is the same as 1,3–trans

(e) 1,5–cis is the same as 1,3–cis

(f) 1,6–trans

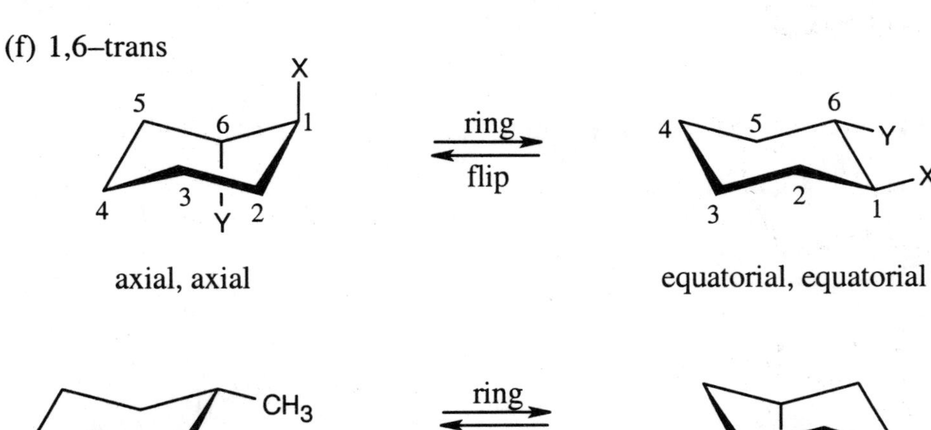

axial, axial — equatorial, equatorial

4.44

diequatorial — diaxial

The large energy difference between conformations is due to the severe 1,3–diaxial interaction between the two methyl groups.

4.45 Diaxial cis-1,3-dimethylcyclohexane contains three 1,3–diaxial interactions -- two H–CH_3 interactions of 3.8 kJ/mol each, and one CH_3–CH_3 interaction. If the diaxial conformation is 23 kJ/mol less stable than the diequatorial, 23 – 2(3.8) ≈ 15 kJ/mol of this strain energy must be due to the CH_3–CH_3 interaction.

4.46

A — **B**

2 H – CH_3 interactions = 7.6 kJ/mol

2 H – CH_3 interactions =	7.6 kJ/mol
1 CH_3 – CH_3 interaction =	15 kJ/mol
	≈ 23 kJ/mol

Conformation **A** is favored because it is 15 kJ/mol lower in energy than conformation **B**.

4.47

There are two cis-trans stereoisomers of 1,3,5-trimethylcyclohexane. In one isomer, all methyl groups are cis to each other; in the other isomer, one methyl group is trans to the other two. The isomer with all substituents cis to each other is more stable because it has no 1,3- diaxial interactions in its chair conformation.

4.48 Note: In working with decalins, it is essential to use models. Many structural features of decalins that are obvious with models are not easily visualized with drawings.

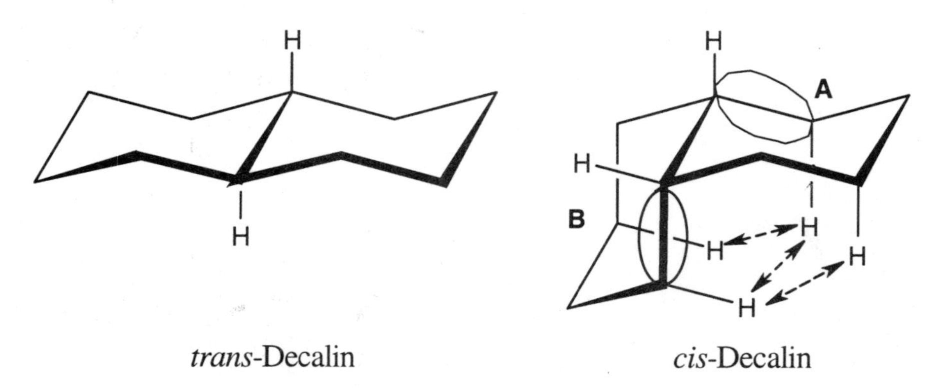

trans-Decalin *cis*-Decalin

No 1,3–diaxial interactions are present in *trans*–decalin.

At the ring junction of *cis*-decalin, one ring acts as an axial substituent of the other (see circled bonds). The circled part of ring **B** has two 1,3–diaxial interactions with ring **A** (indicated by arrows). Similarly, the circled part of ring **A** has two 1,3–diaxial interactions with ring **B**; one of these interactions is the same as an interaction of part of the **B** ring with ring **A**. These three 1,3–diaxial interactions have a total energy cost of 3 x 3.8 kJ/mol = 11.4 kJ/mol. *Cis*-decalin is therefore less stable than *trans*-decalin by 11.4 kJ/mol.

4.49 A ring-flip converts an axial substituent into an equatorial substituent and vice versa. At the ring junction of *trans*-decalin, each ring is a trans–trans diequatorial substituent of the other. If a ring-flip were to occur, the two rings would become axial substituents of each other. You can see with models that a diaxial ring junction is impossibly strained. Consequently, *trans*-decalin does not ring-flip.

The rings of *cis*-decalin are joined by an axial bond and an equatorial bond. After a ring-flip, the rings are still linked by an equatorial and an axial bond. No additional strain or interaction is introduced by a ring-flip of *cis*-decalin.

4.50

The first isomer is the most stable because all chlorine atoms can assume an equatorial conformation.

4.51

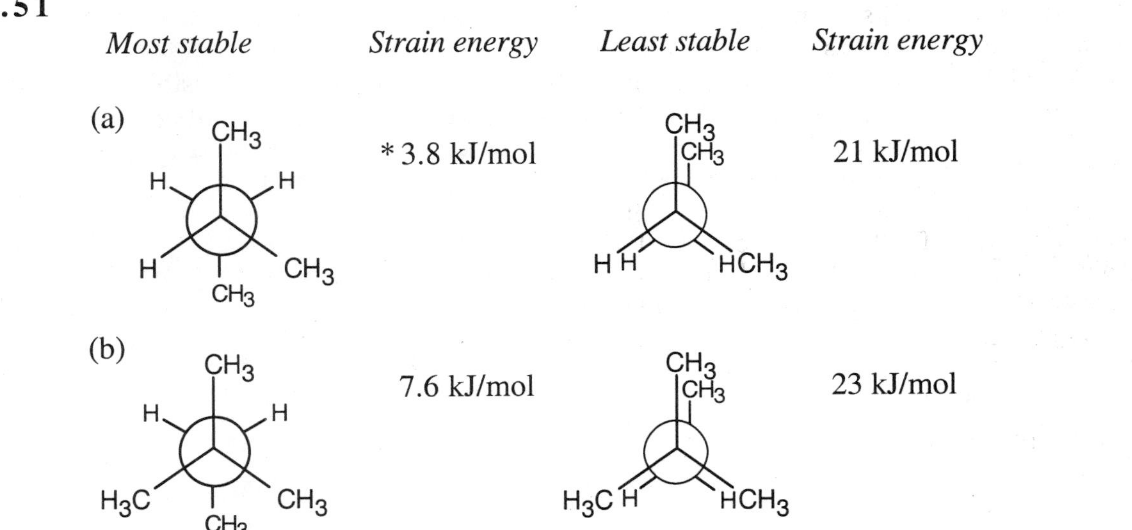

(c) 7.6 kJ/mol 26 kJ/mol

(d) 15.2 kJ/mol ** 28 kJ/mol

*most stable overall (least strain)
**least stable overall (most strain)

4.52

Conformation **A** of *cis*-1-chloro-3-methylcyclohexane has no 1,3–diaxial interactions and is the more stable conformation. Steric strain in **B** is due to one CH_3–H interaction (3.8 kJ/mol), one Cl–H interaction (1.0 kJ/mol) and one CH_3–Cl interaction. Since the total-strain energy of B is 15.5 kJ/mol, 15.5 - 3.8 - 1.0 = 10.7 kJ/mol of strain is caused by a CH_3–Cl interaction.

4.53

If you build a model of 1-norbornene, you will find that it is almost impossible to form the bridgehead double bond. sp^2-Hybridization at the double bond requires all carbons bonded to the starred carbons to lie in a common plane, yet the angle strain caused by sp^2 hybridization is too severe to allow a bridgehead double bond to exist.

4.54 A steroid ring system is fused, and ring-flips don't occur. Thus, substituents such as the methyl groups shown remain axial. Substituents on the same side of the ring system as the methyl groups are in alternating axial and equatorial positions. Thus, an "up" substituent at C3 (a) is equatorial.

Substituents on the bottom side of the ring system also alternate axial and equatorial positions. A substituent at C7 (b) is axial, and one at C11 (c) is equatorial

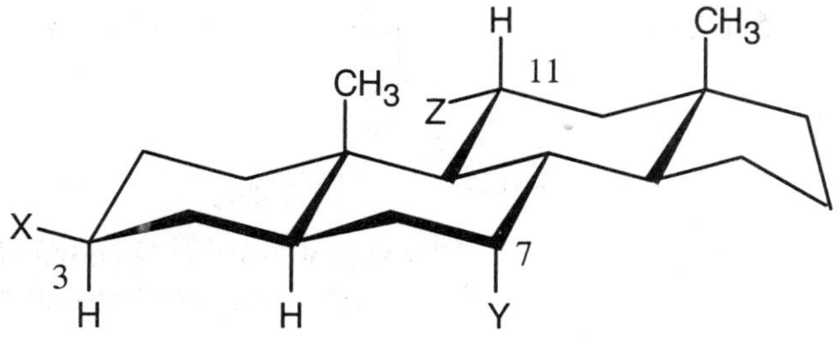

4.55

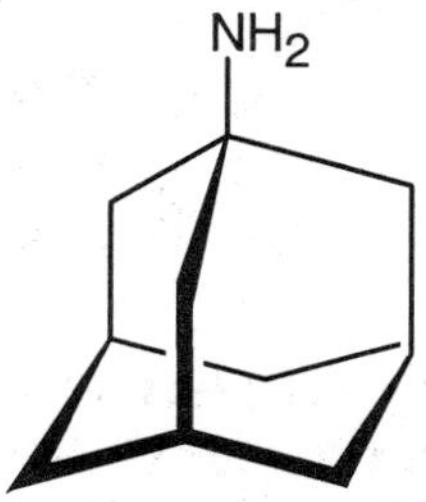

A Look Ahead

4.56

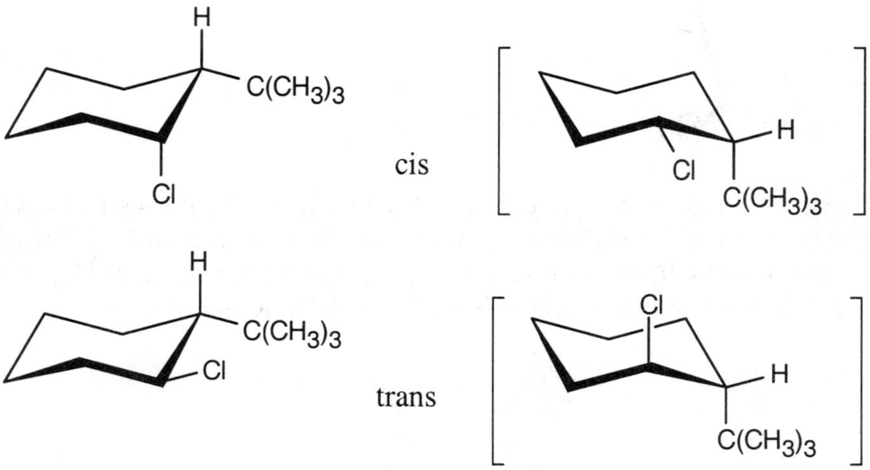

cis

trans

All four conformations of the two isomers are illustrated. The second conformation of each pair has a high degree of steric strain, and thus each isomer adopts the first conformation. Since only the cis isomer has chlorine in the necessary axial position, it is more reactive than the trans isomer.

4.57 Draw the four possible isomers of 4-*tert*-butylcyclohexane-1,3-diol. Make models of these isomers also. The bulky *tert*-butyl group determines the stable conformation because of its strong preference for the equatorial position.

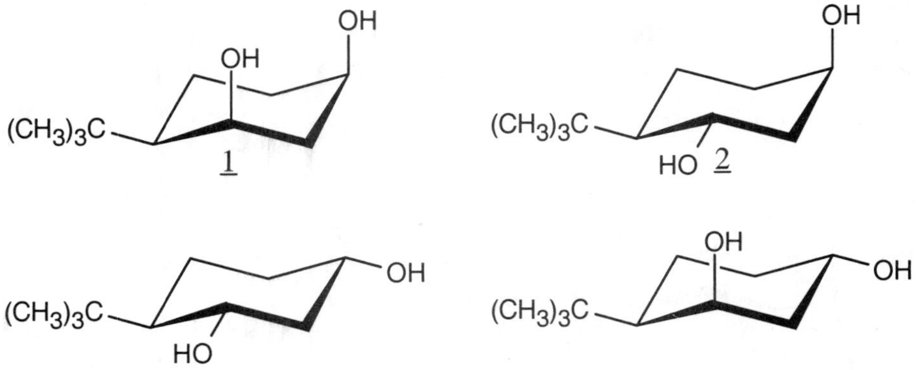

Only when the two hydroxyl groups are cis–diaxial (structure 1) can the acetal ring form. In any other conformation, the oxygen atoms are too far apart to be incorporated into a six-membered ring.

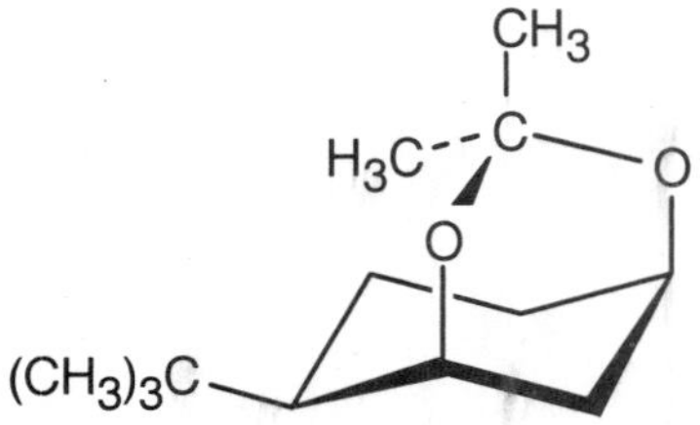

Molecular Modeling

4.58 The energy difference between staggered and eclipsed conformations is larger for 2,2-dimethylpropane (18 kJ/mol) than for ethane (12 kJ/mol). For ethane, bond rotation creates three hydrogen-hydrogen eclipsing interactions. For 2,2-dimethylpropane, bond rotation creates three methyl-hydrogen eclipsing interactions.

4.59 The gauche conformer is shown in frames 3 and 11, and the anti conformer is shown in frame 7. The central C–C bond distance ranges from 153.3 pm (eclipsed methyls) to 153.0 pm (anti conformer). The C–C–C bond angle ranges from 114.6° (eclipsed methyls) to 113.1° (anti conformer). Both changes reduce steric interactions by moving the methyl groups farther apart in the eclipsed methyl conformer.

4.60 Only bond rotation about the C3–C4 bond gives staggered conformers with different geometries. A is lower in energy than B by 12 kJ/mol because B contains two 1,3 methyl-methyl interactions and A contains only one. (Note that the right side of the molecule twists somewhat to reduce 1,3 interactions, and the structures aren't perfectly staggered.)

4.61

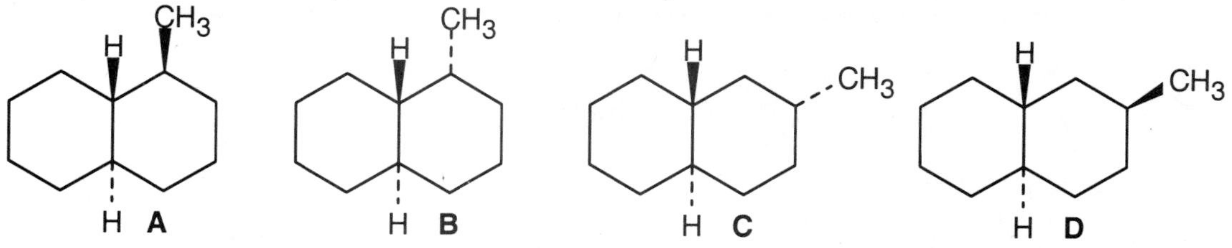

Four isomers are possible: A (49.3 kJ/mol), B (54.7 kJ/mol), C (43.9 kJ/mol), D (49.8 kJ/mol). 1,3-Methyl-hydrogen interactions are responsible for the energy differences: A (1 interaction), B (3 interactions), C (0 interactions), D (2 interactions).

4.62 When groups are axial, both compounds are distorted, with alkyl groups leaning away from axial hydrogens; this is especially noticeable for the *tert*-butyl group. When groups are equatorial, there is no strain. Because an axial *tert*-butyl group introduces more strain than an axial methyl group, *tert*-butyl has a stronger equatorial preference.

Chapter 5 – An Overview of Organic Reactions

Chapter Outline

I. Organic Reactions (Sections 5.1 – 5.6).
  A. Kinds of organic reactions (Section 5.1).
    1. Addition reactions occur when two reactants add to form one product, with no atoms left over.
    2. Elimination reactions occur when a single reactant splits into two products.
    3. Substitution reactions occur when two reactants exchange parts to yield two new products.
    4. Rearrangement reactions occur when a single product undergoes a rearrangement of bonds to yield an isomeric product.
  B. Reaction mechanisms -general information (Section 5.2).
    1. A reaction mechanism describes the bonds broken and formed in a chemical reaction and its rate, and accounts for all reactants and products.
    2. Bond breaking and formation in chemical reactions.
      a. Bond breaking is heterolytic if one electron remains with each fragment.
      b. Bond breaking is homolytic if both electrons remain with one fragment and the other fragment has a vacant orbital.
      c. Bond formation is homogenic if one electron in a covalent bond comes from each reactant.
      d. Bond formation is heterogenic if both electrons in a covalent bond come from one reactant.
    3. Types of reactions.
      a. Radical reactions involve symmetrical bond breaking and bond formation.
      b. Polar reactions involve unsymmetrical bond breaking and bond formation.
      c. Pericyclic reactions will be studied later.
  C. Radical reactions (Section 5.3).
    1. Radicals are highly reactive because they contain an atom with an unpaired electron.
    2. A substitution reaction occurs when a radical abstracts an electron from another molecule.
    3. An addition reaction occurs when a radical adds to a double bond.
    4. Steps in a radical reaction.
      a. The *initiation step* produces radicals by the homolytic cleavage of a bond.
      b. The *propagation steps* occur when a radical abstracts an atom to produce a new radical and a stable molecule.
        This sequence of steps is a chain reaction.
      c. A *termination step* occurs when two radicals combine.
    5. In radical reactions, all bonds are broken and formed by reactions of species with odd numbers of electrons.
  D. Polar reactions (Sections 5.4 – 5.6).
    1. Characteristics of polar reactions (Section 5.4).
      a. Polar reactions occur as a result of positive and negative charges within molecules.
      b. These charge differences are usually due to electronegativity differences between atoms.
        i. Differences may also be due to interactions of functional groups with solvents, as well as with Lewis acids or bases.
        ii. Some bonds in which one atom is polarizable may also behave as polar bonds.

c. In polar reactions, electron-rich sites in one molecule react with electron-poor sites in another molecule.
d. The movement of an electron pair in a polar reaction is shown by a curved arrow.
e. The reacting species:
i. A nucleophile is a compound with an electron-rich atom.
ii. An electrophile is a compound with an electron-poor atom.
iii. Some compounds can behave as both nucleophiles and as electrophiles
f. Many polar reactions can be explained in terms of acid-base reactions.
2. An example of a polar reaction: addition of HBr to ethylene (Section 5.5).
a. This reaction is known as an electrophilic addition.
b. The π electrons in ethylene behave as a nucleophile.
c. The reaction begins by the addition of the electrophile H^+ to the double bond.
d. The resulting intermediate carbocation reacts with Br^- to form bromoethane.
3. Rules for using curved arrows in polar reaction mechanisms (Section 5.6).
a. Electrons must move from a nucleophilic source to an electrophilic sink.
b. The nucleophile can be either negatively charged or neutral.
c. The electrophile can be either positively charged or neutral.
d. The octet rule must be followed.

II. Describing a reaction (Sections 5.7 – 5.10).
A. Equilibria, rates, and energy changes (Section 5.7).
1. All chemical reactions are equilibria that can be expressed by an equilibrium constant K_{eq} that shows the ratio of products to reactants.
a. If $K_{eq} > 1$, [products] > [reactants].
b. If $K_{eq} < 1$, [reactants] > [products].
2. For a reaction to proceed as written, the energy of the products must be lower than the energy of the reactants.
a. The energy change that occurs during a reaction is described by $\Delta G°$, the Gibbs free-energy change.
b. Favorable reactions have negative $\Delta G°$ and are exergonic.
c. Unfavorable reactions have positive $\Delta G°$ and are endergonic.
d. $\Delta G° = -RT \ln K_{eq}$.
3. $\Delta G°$ is composed of two terms – $\Delta H°$, and $\Delta S°$, which is temperature-dependent.
a. $\Delta H°$ is a measure of the change in total bonding energy during a reaction.
i. If $\Delta H°$ is negative, a reaction is exothermic.
ii. If $\Delta H°$ is positive, a reaction is endothermic.
b. $\Delta S°$ (entropy) is a measure of the freedom of motion of a reaction.
i. A reaction that produces two product molecules from one reactant molecule has positive entropy.
ii. A reaction that produces one product molecule from two reactant molecules has negative entropy.
c. $\Delta G° = \Delta H° - T\,\Delta S°$
4. None of these expressions predict the rate of a reaction
B. Bond dissociation energies (Section 5.8).
1. The bond dissociation energy (D) measures the heat needed to break a bond.
2. Each bond has a characteristic strength.
3. It is possible to calculate $\Delta H°$ for a reaction by using values of D.
4. Values of D are of limited usefulness.
a. Calculations of $\Delta H°$ can't provide values of $\Delta S°$ or $\Delta G°$.
b. Calculations of $\Delta H°$ can't give information about reaction rates.
c. Values of D are for gas-phase reactions and don't account for the effects of solvent.

C. Energy diagrams and transition states (Section 5.9).
   1. Reaction energy diagrams show the energy changes that occur during a reaction.
      The vertical axis represents energy changes, and the horizontal axis represents the progress of a reaction.
   2. The transition state is the highest-energy species in this reaction.
      It is possible for a reaction to have more than one transition state.
   3. The difference in energy between the reactants and the transition state is the energy of activation $\Delta G^\ddagger$.
      Values of $\Delta G^\ddagger$ range from 40 – 150 kJ/mol.
   4. After reaching the transition state, a molecule can go on to form products or can revert to starting material.
   5. Every reaction has its own energy profile.

D. Intermediates (Section 5.10).
   1. In a reaction of at least two steps, an intermediate is the species that lies at the energy minimum between two transition states.
   2. Even though an intermediate lies at an energy minimum between two transition states, it is a high-energy species and usually can't be isolated.
   3. Each step of a reaction has its own $\Delta G^\ddagger$ and $\Delta G°$, but the total reaction has an overall $\Delta G°$.

Solutions to Problems

5.1

(a) $CH_3Br + KOH \longrightarrow CH_3OH + KBr$ substitution

(b) $CH_3CH_2OH \longrightarrow H_2C{=}CH_2 + H_2O$ elimination

(c) $H_2C{=}CH_2 + H_2 \longrightarrow CH_3CH_3$ addition

5.2

2-Methylpentane ($CH_3CH_2CH_2CH(CH_3)CH_3$) $\xrightarrow[h\nu]{Cl_2}$

1-Chloro-2-methylpentane ($CH_3CH_2CH_2CH(CH_3)CH_2Cl$) + 2-Chloro-2-methylpentane ($CH_3CH_2CH_2CCl(CH_3)CH_3$) + 3-Chloro-2-methylpentane ($CH_3CH_2CHClCH(CH_3)CH_3$) + 2-Chloro-4-methylpentane ($CH_3CHClCH_2CH(CH_3)CH_3$) + 1-Chloro-4-methylpentane ($ClCH_2CH_2CH_2CH(CH_3)CH_3$)

5.3 Pentane has three types of hydrogen atoms, $\overset{a}{CH_3}\overset{b}{CH_2}\overset{c}{CH_2}\overset{b}{CH_2}\overset{a}{CH_3}$. Although monochlorination produces $CH_3CH_2CH_2CH_2CH_2Cl$, it is not possible to avoid producing $CH_3CH_2CH_2CHClCH_3$ and $CH_3CH_2CHClCH_2CH_3$ as well. Since neopentane has only one type of hydrogen, monochlorination yields a single product.

5.4 Keep in mind:
(1) An electrophile is electron-poor, either because it is positively charged or because it has a functional group that is positively polarized.
(2) A nucleophile is electron-rich, either because it has a negative charge, because it has a functional group containing a lone electron pair, or because it has a functional group that is negatively polarized.
(3) Some molecules can act as both nucleophiles and electrophiles, depending on the reaction conditions.
(a) HCl can act as both a nucleophile and as an electrophile.
(b) CH_3NH_2 is a nucleophile because of the lone-pair electrons of nitrogen.
(c) CH_3SH is both a nucleophile and an electrophile – a nucleophile because of the sulfur lone-pair electrons, and an electrophile because of the polarized S–H bond.
(d) CH_3CHO is both a nucleophile and an electrophile because of its polar C=O bond.

(a) $\overset{\delta+}{H}-\overset{\delta-}{\ddot{C}l}:$ — H: electrophilic (as H^+); Cl: nucleophilic (as Cl^-)

(b) $CH_3-\ddot{N}H_2$ — N: nucleophilic

(c) $CH_3-\overset{\delta-}{\ddot{S}}-\overset{\delta+}{H}$ — S: nucleophilic; H: electrophilic

(d) $H_3C-\overset{\delta+}{C}(=\overset{\delta-}{\ddot{O}}:)-H$ — O: nucleophilic; C: electrophilic

5.5 Reaction of cyclohexene with HCl or HBr is an electrophilic addition reaction in which a halogen acid adds to a double bond to produce a haloalkane.

Cyclohexene + H—Br ⟶ Bromocyclohexane

Cyclohexene + H—Cl ⟶ Chlorocyclohexane

5.6 For curved arrow problems, follow these steps:
(1) Locate the bonding changes. In (a), a bond from nitrogen to chlorine has formed, and a Cl–Cl bond has broken.
(2) Identify the nucleophile and electrophile (in (a), the nucleophile is ammonia and the electrophile is one Cl in the Cl_2 molecule), and draw a curved arrow whose tail is near the nucleophile and whose head is near the electrophile.
(3) Check to see that all bonding changes are accounted for. In (a), we must draw a second arrow to show the heterolytic bond-breaking of Cl_2 to form Cl^-.

(a)

$$NH_3 + :\ddot{Cl}-\ddot{Cl}: \longrightarrow H_3\overset{+}{N}-\ddot{Cl}: + :\ddot{Cl}:^-$$

(b)

$$H_3C-\ddot{O}:^- + H_3C-\ddot{Br}: \longrightarrow H_3C-\ddot{O}-CH_3 + :\ddot{Br}:^-$$

A bond has formed between oxygen and the carbon of bromomethane. The bond between carbon and bromine has broken. CH_3O^- is the nucleophile and bromomethane is the electrophile.

(c)

$$H_3C-C(O^-)(Cl)-OCH_3 \longrightarrow H_3C-C(=O)-OCH_3 + :\ddot{Cl}:^-$$

A double bond has formed between oxygen and carbon, and a carbon – chlorine bond has broken. Electrons move from oxygen to the double bond and from carbon to chlorine.

5.7 (1) The nucleophile HO^- forms a bond with a hydrogen atom of bromoethane. (2) The electrons from the former C–H bond are used to form a C–C double bond, and (3) bromide ion is eliminated

$$H\ddot{O}:^- + CH_3CH_2Br \longrightarrow H_2O + H_2C{=}CH_2 + :\ddot{Br}:^-$$

5.8 According to Table 5.3, a negative value of $\Delta G°$ indicates that a reaction is favorable. Thus, a reaction with $\Delta G° = -44$ kJ/mol (-11 kcal/mol) is more favorable than a reaction with $\Delta G° = +44$ kJ/mol ($+11$ kcal/mol).

5.9 From the expression $\Delta G° = -RT\ln K_{eq}$, we can see that a large K_{eq} is related to a large negative $\Delta G°$. Consequently, a reaction with $K_{eq} = 1000$ is more exergonic than a reaction with $K_{eq} = 0.001$.

5.10 $\Delta G° = -RT\ln K_{eq}$; $R = 0.008315$ kJ/(K·mol); $T = 298$ K

$= -[0.008315 \text{ kJ/(K·mol)}]\,(298\text{ K})\,(\ln K_{eq}) = (-2.478 \text{ kJ/mol})\,(\ln K_{eq})$

If $K_{eq} = 1000$, then $\ln K_{eq} = 6.91$ and $\Delta G° = (-2.478 \text{ kJ/mol})\,(6.91) = -17.1$ kJ/mol

If $K_{eq} = 1$, then $\ln K_{eq} = 0$ and $\Delta G° = 0$

If $K_{eq} = 0.001$, then $\Delta G° = +17.1$ kJ/mol

$\Delta G° = -RT\ln K_{eq}$; $\ln K_{eq} = -\Delta G°/RT$; $T = 298$ K

$\ln K_{eq} = (-\Delta G°) / (2.478 \text{ kJ/mol})$

If $\Delta G° = -40$ kJ/mol, then $\ln K_{eq} = -(-40 \text{ kJ/mol}) / (2.478 \text{ kJ/mol}) = 16.14$; $K_{eq} = 1.0 \times 10^{7}$

If $\Delta G° = 0$, then $K_{eq} = 1$

If $\Delta G° = +40$ kJ/mol, then $\ln K_{eq} = (-40 \text{ kJ/mol}) / (2.478 \text{ kJ/mol}) = -16.14$; $K_{eq} = 1.0 \times 10^{-7}$

5.11

$$CH_4 + Br_2 \longrightarrow CH_3Br + HBr$$

Reactant bonds broken	*D*	*Product bonds formed*	*D*
$CH_3—H$	438 kJ/mol	$CH_3—Br$	293 kJ/mol
$Br—Br$	193 kJ/mol	$H—Br$	366 kJ/mol
	631 kJ/mol		659 kJ/mol

$$\Delta H° = D_{\text{bonds broken}} - D_{\text{bonds formed}}$$
$$= 631 \text{ kJ/mol} - 659 \text{ kJ/mol} = -28 \text{ kJ/mol}$$

The reaction of bromine with methane is less exothermic than the reaction of chlorine with methane because the value of $\Delta H°$ is a smaller negative number.

5.12

(a) $CH_3CH_2OCH_3 + HI \longrightarrow CH_3CH_2OH + CH_3I$

Reactant bonds broken	*D*	*Product bonds formed*	*D*
$CH_3CH_2O—CH_3$	339 kJ/mol	$CH_3CH_2O—H$	436 kJ/mol
$H—I$	298 kJ/mol	$CH_3—I$	234 kJ/mol
	637 kJ/mol		670 kJ/mol

$$\Delta H° = D_{\text{bonds broken}} - D_{\text{bonds formed}}$$
$$= 637 \text{ kJ/mol} - 670 \text{ kJ/mol} = -33 \text{ kJ/mol}$$

(b) $CH_3Cl + NH_3 \longrightarrow CH_3NH_2 + HCl$

Reactant bonds broken	*D*	*Product bonds formed*	*D*
CH_3—Cl	351 kJ/mol	CH_3—NH_2	335 kJ/mol
NH_2—H	449 kJ/mol	H—Cl	432 kJ/mol
	800 kJ/mol		767 kJ/mol

$$\Delta H^\circ = D_{\text{bonds broken}} - D_{\text{bonds formed}}$$
$$= 800 \text{ kJ/mol} - 767 \text{ kJ/mol} = +33 \text{ kJ/mol}$$

5.13 A reaction with $\Delta G^\ddagger = 45$ kJ/mol is faster than a reaction with $\Delta G^\ddagger = 70$ kJ/mol. It is not possible to measure the size of K_{eq} from $\Delta G^\ddagger$ because $\Delta G^\ddagger$ measures the energy difference between reactant and *transition state* rather than between reactant and *product*. The energy difference between reactant and product is described by ΔG°, and thus also by K_{eq}.

5.14

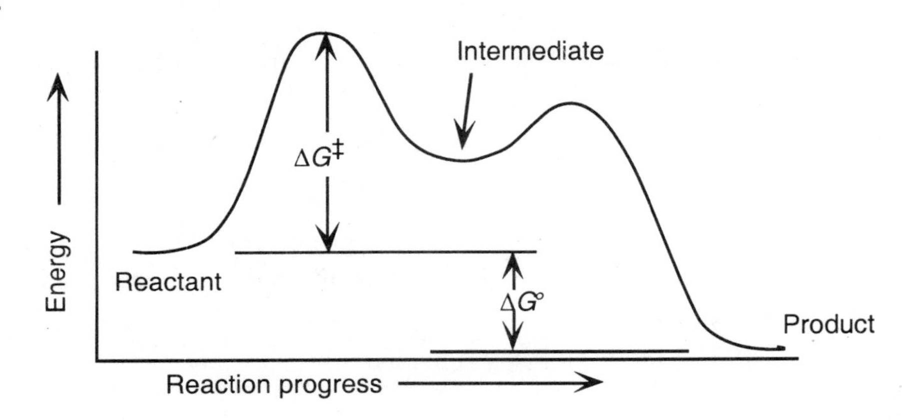

5.15

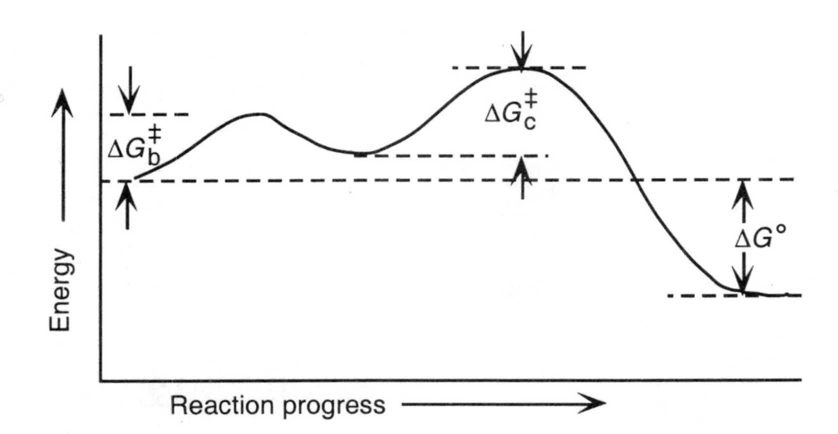

Step b: CH_3—H + Cl· ⟶ CH_3· + H—Cl

Step c: CH_3· + Cl_2 ⟶ CH_3—Cl + Cl·

Since $\Delta H^\circ_{\text{overall}}$ is negative, ΔG° is also negative.

Visualizing Chemistry

5.16

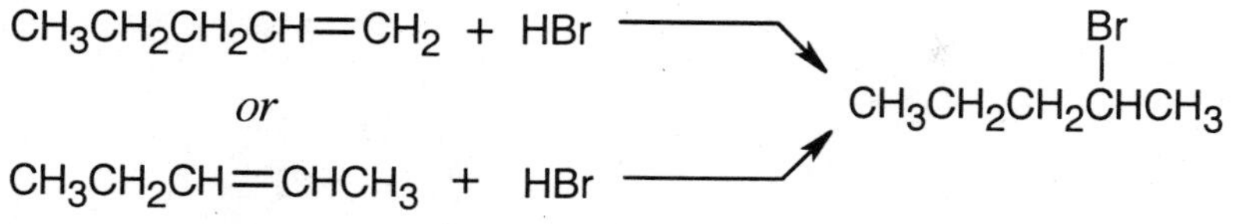

$$CH_3CH_2CH_2CH{=}CH_2 + HBr \longrightarrow CH_3CH_2CH_2CH(Br)CH_3$$

or

$$CH_3CH_2CH{=}CHCH_3 + HBr \longrightarrow CH_3CH_2CH_2CH(Br)CH_3$$

5.17

CH_2 H—Cl:

or

CH_3 H—Cl:

$\longrightarrow$ [:Cl:$^-$ $^+$ CH_3] $\longrightarrow$ Cl CH_3

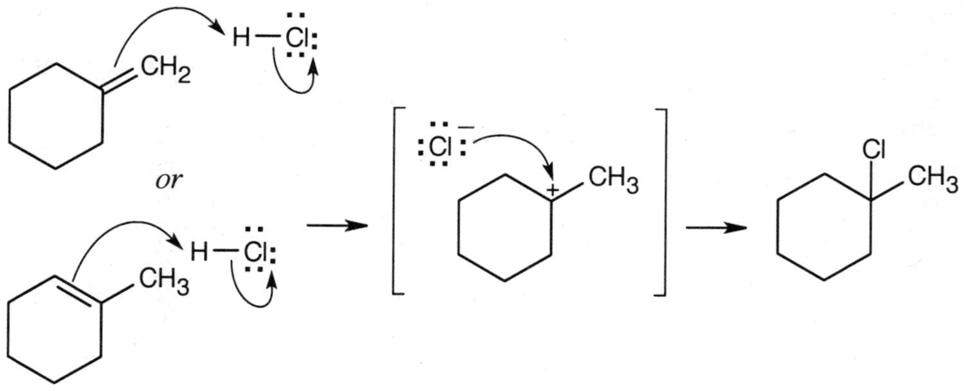

Additional Problems

5.18 – 5.19

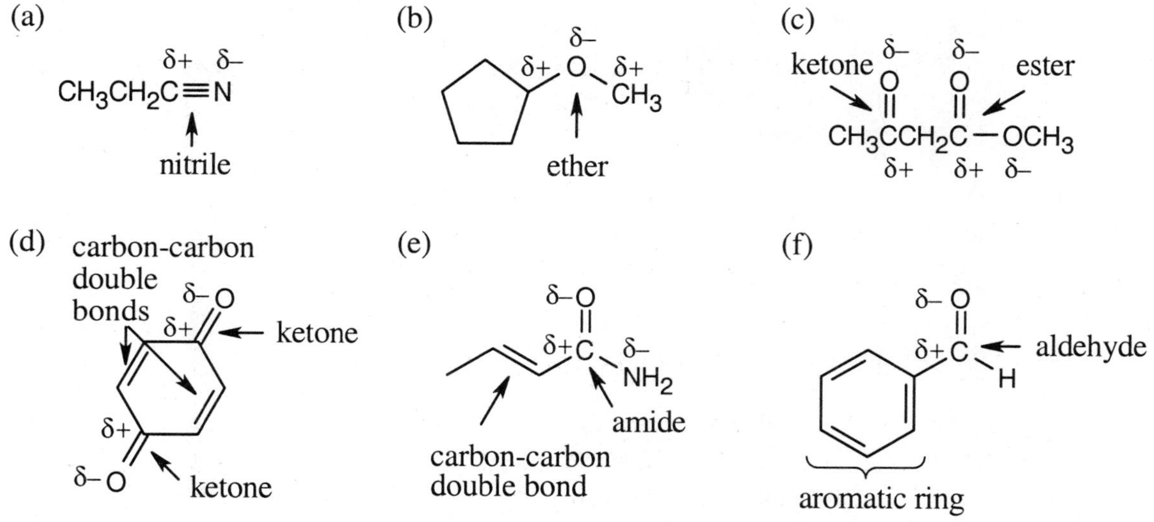

(a) δ+ δ– $CH_3CH_2C{\equiv}N$ nitrile

(b) δ– δ+ O δ+ CH_3 ether

(c) ketone δ– δ– ester $CH_3CCH_2C{-}OCH_3$ (O above each C; δ+ δ+ δ–)

(d) carbon-carbon double bonds; δ–O δ+ ketone; δ+ δ–O ketone

(e) δ–O δ+C δ– NH_2 amide; carbon-carbon double bond

(f) δ– O δ+C H aldehyde; aromatic ring

5.20 (a) The reaction between bromoethane and sodium cyanide is a substitution because two reagents exchange parts.
(b) This reaction is an elimination because two products (cyclohexene and H_2O) are produced from one reactant
(c) Two reactants form one product in this addition reaction.
(d) This is a substitution reaction.

5.21

(a) $:\ddot{\text{Br}}:^-$ is a nucleophile. (b) H^+ is an electrophile.

(c) (d)

$$CH_3{-}Br + H\ddot{O}:^- \longrightarrow CH_3OH + :\ddot{Br}:^-$$

The above reaction is both a substitution reaction (because two reactants change parts) and a polar reaction (because the carbon-bromine bond is broken heterolytically).

(e)

$$H\ddot{O}{-}H \longrightarrow H\ddot{O}:^- + H^+$$

Both electrons stay with the hydroxide anion.

(f)

$$Cl{-}Cl \longrightarrow 2\ :\ddot{Cl}\cdot$$

Each chlorine radical receives an electron from the former Cl–Cl bond.

5.22 (a) Cl^- is a nucleophile because it is electron-rich.
(b) BF_3 is an electrophile because it is an electron-poor Lewis acid.
(c) The illustrated amine is a nucleophile because of its lone electron pair.

5.23 A transition state represents a structure occurring at an energy maximum. An intermediate occurs at an energy minimum between two transition states. Even though an intermediate may be of such high energy that it cannot be isolated, it is still of lower energy than all transition states in the reaction under consideration.

5.24

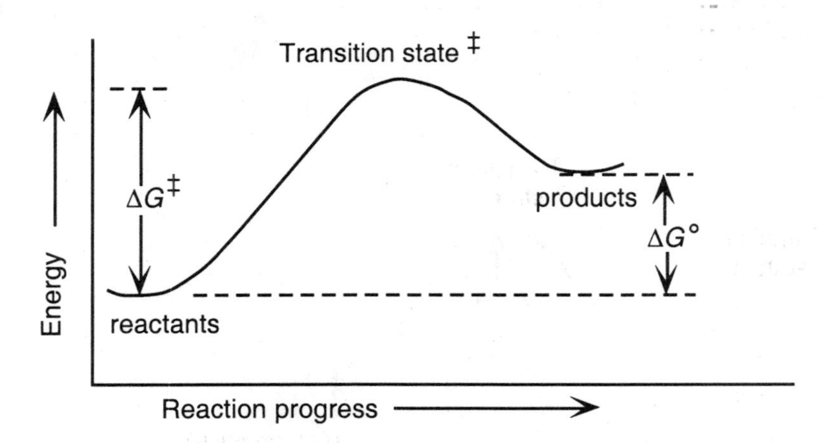

$\Delta G°$ is positive.

5.25

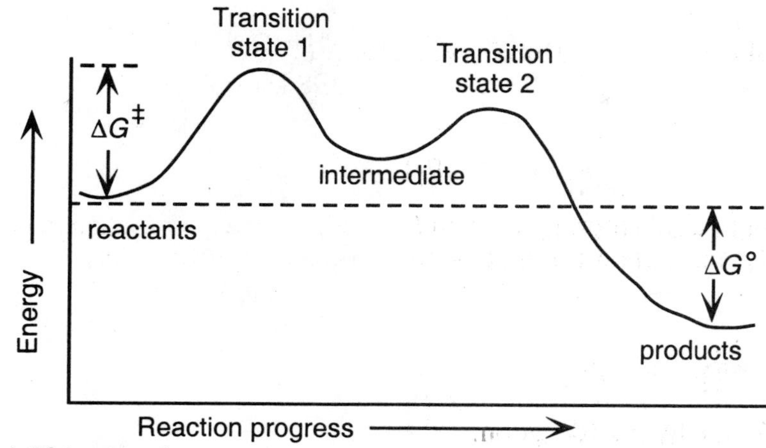

$\Delta G°$ is negative.

5.26 Problem 5.25 shows a reaction energy diagram of a two-step exothermic reaction. Step 2 is faster than step 1 because $\Delta G^{\ddagger}{}_2 < \Delta G^{\ddagger}{}_1$.

5.27

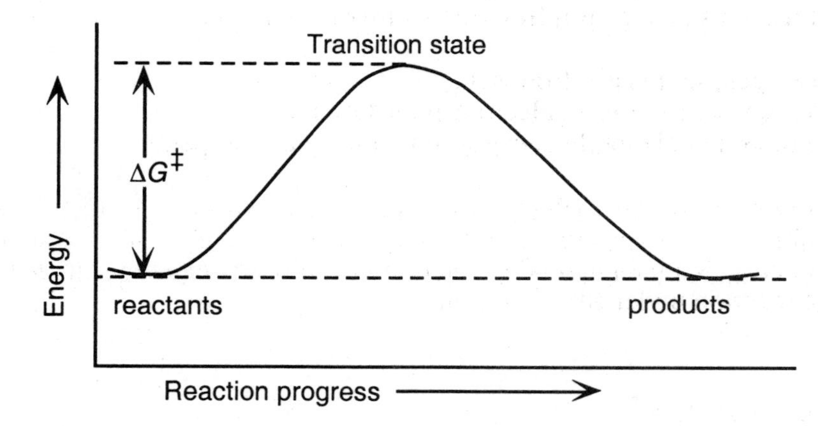

A reaction with $K_{eq} = 1$ has $\Delta G° = 0$.

5.28

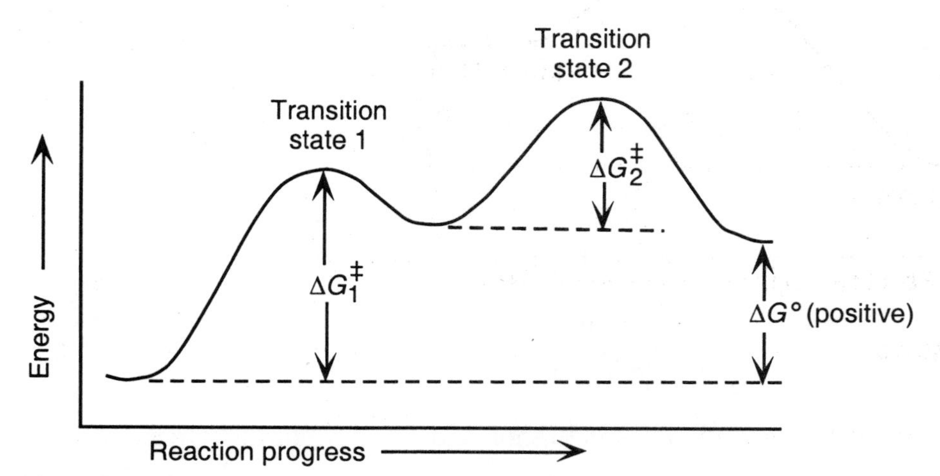

(a) $\Delta G°$ is positive.
(b) There are two steps in the reaction.
(c) Step 2 is faster because $\Delta G^{\ddagger}{}_2 < \Delta G^{\ddagger}{}_1$
(d) There are two transition states, as indicated on the diagram.

5.29

(a)

$$CH_3{-}OH + H{-}Br \longrightarrow CH_3{-}Br + H{-}OH$$

380 kJ/mol   366 kJ/mol   293 kJ/mol   498 kJ/mol

$$\Delta H^\circ = D_{\text{bonds broken}} - D_{\text{bonds formed}}$$
$$= 746 \text{ kJ/mol} - 791 \text{ kJ/mol} = -45 \text{ kJ/mol}$$

(b)

$$CH_3CH_2O{-}H + CH_3{-}Cl \longrightarrow CH_3CH_2O{-}CH_3 + H{-}Cl$$

436 kJ/mol   351 kJ/mol   339 kJ/mol   432 kJ/mol

$$\Delta H^\circ = D_{\text{bonds broken}} - D_{\text{bonds formed}}$$
$$= 787 \text{ kJ/mol} - 771 \text{ kJ/mol} = +16 \text{ kJ/mol}$$

5.30

(a)

$$CH_3CH_2{-}H + Cl{-}Cl \longrightarrow CH_3CH_2{-}Cl + H{-}Cl$$

420 kJ/mol   243 kJ/mol   338 kJ/mol   432 kJ/mol

$$\Delta H^\circ = D_{\text{bonds broken}} - D_{\text{bonds formed}}$$
$$= 663 \text{ kJ/mol} - 770 \text{ kJ/mol} = -107 \text{ kJ/mol}$$

(b)

$$CH_3CH_2{-}H + Br{-}Br \longrightarrow CH_3CH_2{-}Br + H{-}Br$$

420 kJ/mol   193 kJ/mol   285 kJ/mol   366 kJ/mol

$$\Delta H^\circ = D_{\text{bonds broken}} - D_{\text{bonds formed}}$$
$$= 613 \text{ kJ/mol} - 651 \text{ kJ/mol} = -38 \text{ kJ/mol}$$

(c)

$$CH_3CH_2{-}H + I{-}I \longrightarrow CH_3CH_2{-}I + H{-}I$$

420 kJ/mol   151 kJ/mol   222 kJ/mol   298 kJ/mol

$$\Delta H^\circ = D_{\text{bonds broken}} - D_{\text{bonds formed}}$$
$$= 571 \text{ kJ/mol} - 520 \text{ kJ/mol} = +51 \text{ kJ/mol}$$

Of the three halogenation reactions, chlorination is energetically the most favorable.

5.31

$$H_3C{-}CH_3 + Br{-}Br \longrightarrow CH_3{-}Br + CH_3{-}Br$$

376 kJ/mol   193 kJ/mol   293 kJ/mol   293 kJ/mol

$$\Delta H^\circ = D_{\text{bonds broken}} - D_{\text{bonds formed}}$$
$$= 569 \text{ kJ/mol} - 586 \text{ kJ/mol} = -17 \text{ kJ/mol}$$

ΔH° for bromoethane formation is –38 kJ/mol, and ΔH° for bromomethane formation is –17 kJ/mol. Although both of these reactions have negative ΔH°, the reaction that forms bromoethane is more favorable.

5.32 Irradiation initiates the chlorination reaction by producing chlorine radicals. Although these radicals are consumed in the propagation steps, new Cl· radicals are formed to carry on the reaction. After irradiation stops, chlorine radicals are still present to carry on the propagation steps, but, as time goes on, radicals combine with other radicals in termination reactions that remove radicals from the reaction mixture. Because the number of radicals decreases, fewer propagation cycles occur, and the reaction gradually slows down and stops.

5.33, 5.34

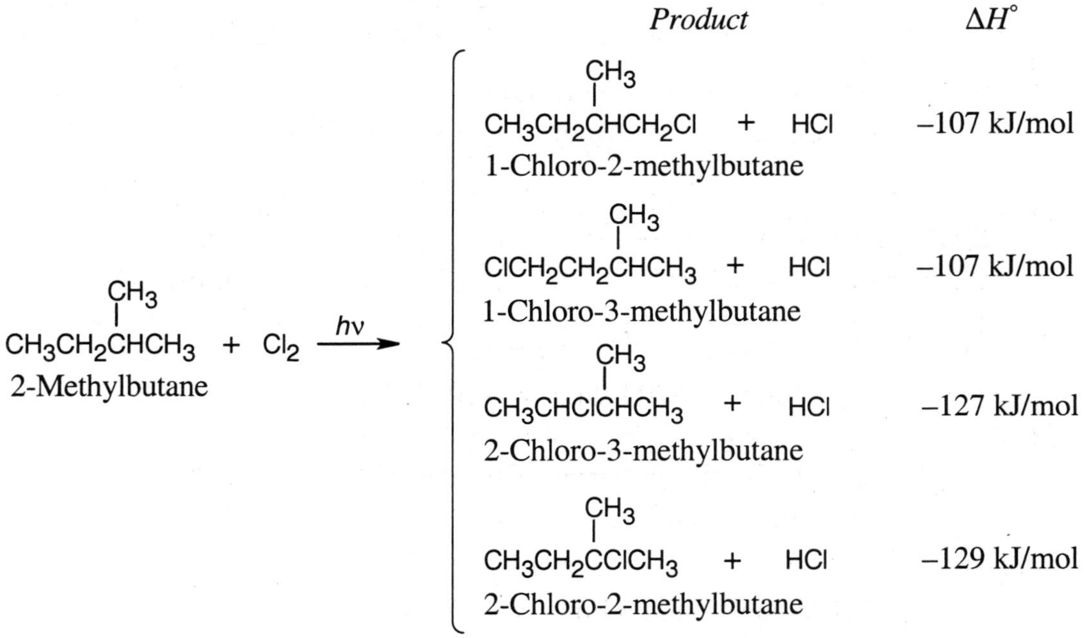

$CH_3CH_2CH(CH_3)CH_3 + Cl_2 \xrightarrow{h\nu}$
2-Methylbutane

Product	ΔH°
$CH_3CH_2CH(CH_3)CH_2Cl$ + HCl 1-Chloro-2-methylbutane	–107 kJ/mol
$ClCH_2CH_2CH(CH_3)CH_3$ + HCl 1-Chloro-3-methylbutane	–107 kJ/mol
$CH_3CHClCH(CH_3)CH_3$ + HCl 2-Chloro-3-methylbutane	–127 kJ/mol
$CH_3CH_2CCl(CH_3)CH_3$ + HCl 2-Chloro-2-methylbutane	–129 kJ/mol

Formation of the tertiary chloroalkane is favored, but a mixture of products is expected.

5.35 The following compounds yield single monohalogenation products because each has only one kind of hydrogen atom.

(a) CH_3CH_3 (c) 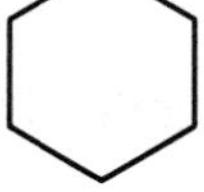 cyclohexane (f) $CH_3C{\equiv}CCH_3$

5.36 For the first series of steps:

(a) $\Delta H^\circ = +243$ kJ/mol
(b) $\Delta H^\circ = +6$ kJ/mol
(c) $\Delta H^\circ = -108$ kJ/mol

$\Delta H^\circ_{overall} = -102$ kJ/mol

For the alternate series:

(a) $\Delta H^\circ = +243$ kJ/mol
(b) $\Delta H^\circ = +87$ kJ/mol
(c) $\Delta H^\circ = -189$ kJ/mol

$\Delta H^\circ_{overall} = -102$ kJ/mol

For both series of reactions, the radical-producing initiation step has $\Delta H° = +243$ kJ/mol and the propagation steps b + c have $\Delta H° = -102$ kJ/mol. In series 2, however, one step has $\Delta H° = +87$ kJ/mol; this step is energetically much less favorable than any step in series 1 and disfavors the second series as a whole. The first route for chlorination of methane is thus more likely to occur.

5.37

(a)

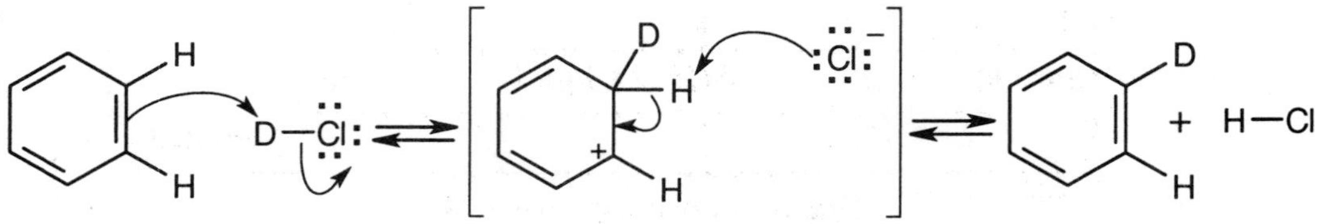

(b)

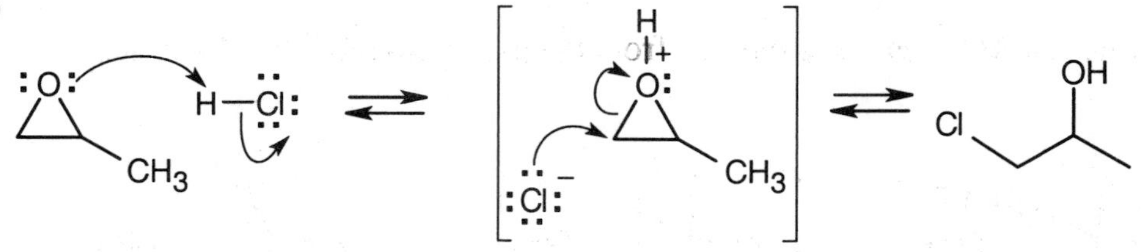

5.38

(a)

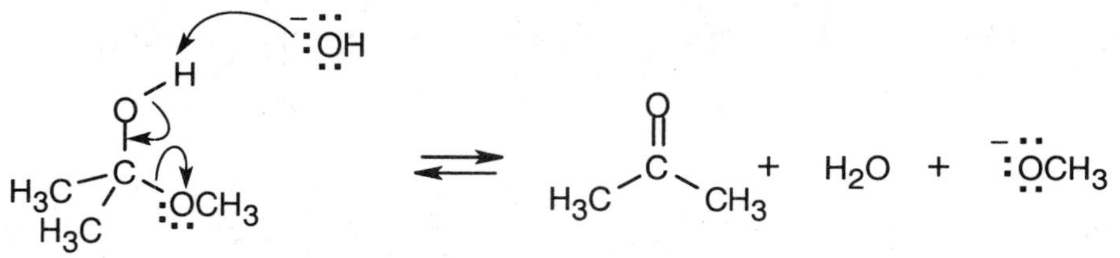

(b)

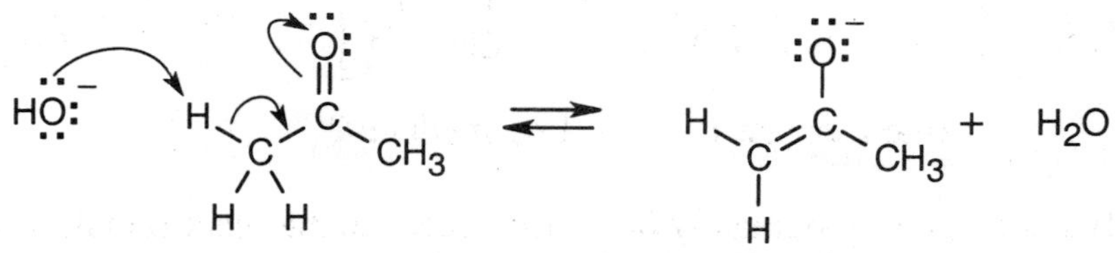

5.39

(a)

$$K_{eq} = \frac{[\text{Products}]}{[\text{Reactants}]} = \frac{0.70}{0.30} = 2.3$$

(b) Section 5.9 states that reactions that occur spontaneously have $\Delta G^{\ddagger}$ of less than 80 kJ/mol at room temperature. Since this reaction proceeds slowly at room temperature, $\Delta G^{\ddagger}$ is probably close to 80 kJ/mol.

(c)

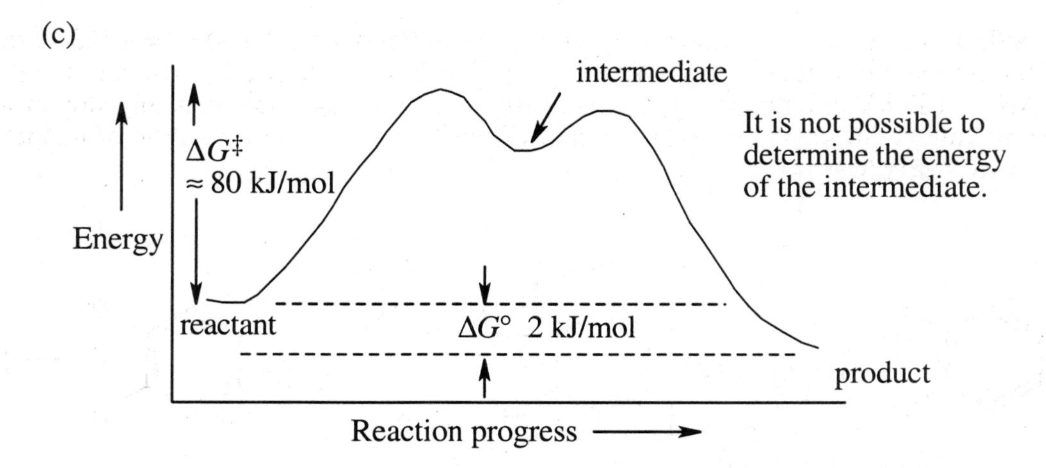

The value of $\Delta G°$ can be calculated from the expression $\Delta G° = -RT \ln K_{eq}$.

5.40

(Isopropylidenecyclohexane + H^+ ⇌ [tertiary carbocation: cyclohexyl cation bearing $C(CH_3)_2H$] ⇌ 1-isopropylcyclohexene + H^+)

5.41

$(CH_3)_3C{-}Cl$ ⇌ (T.S. 1) [$(CH_3)_3C^+$ (Intermediate 1) ⇌ (H_2O, T.S. 2) $(CH_3)_3C{-}OH_2^+$ (Intermediate 2)] ⇌ (H_2O, T.S. 3) $(CH_3)_3C{-}O{-}H$ + H_3O^+ + Cl^-

Intermediate 1 Intermediate 2

(a) $\Delta G^‡$ for the first step is approximately 80 kJ/mol because the reaction takes place slowly at room temperature. $\Delta G^‡$ values for the second and third steps are smaller - perhaps 60 kJ/mol for Step 2, and 40 kJ/mol for Step 3. $\Delta G°$ is approximately zero.
(b)

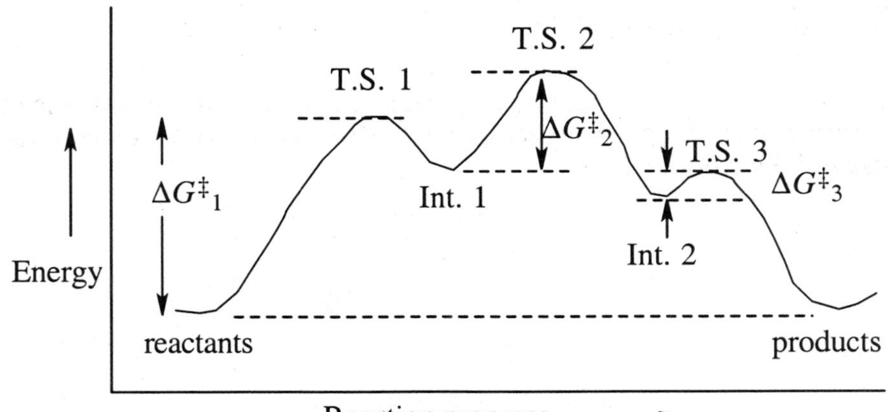

5.42

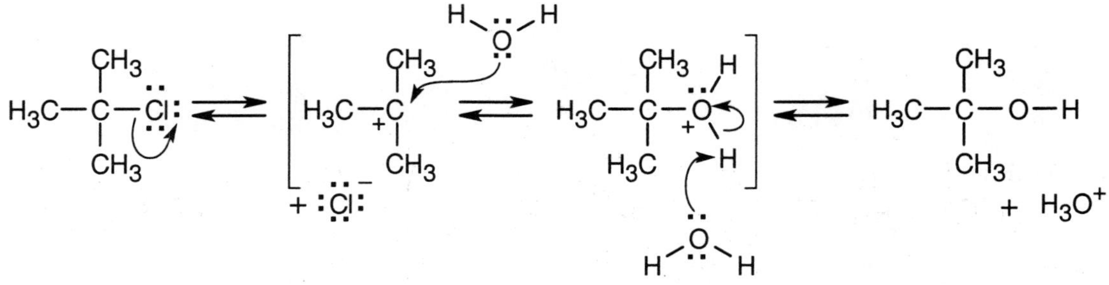

$$(CH_3)_3C{-}Cl \rightleftharpoons [(CH_3)_3C^+ + :Cl:^-] \xrightarrow{H_2O} \rightleftharpoons [(CH_3)_3C{-}OH_2^+] \rightleftharpoons (CH_3)_3C{-}O{-}H + H_3O^+$$

5.43 $\Delta G° = \Delta H° - T\Delta S°$

$= -75\ \text{kJ/mol} - (298\ \text{K})\ [0.054\ \text{kJ/(K·mol)}]$

$= -75\ \text{kJ/mol} - 16\ \text{kJ/mol}$

$= -91\ \text{kJ/mol}$

The reaction is exothermic because $\Delta H°$ is negative, and it is endergonic because $\Delta G°$ is negative..

5.44 $\Delta G° = -RT\ln K_{eq}$

$= [-0.008315\ \text{kJ/(K·mol)}]\ (298\ \text{K})\ (\ln K_{eq}) = 91\ \text{kJ/mol}$

$\ln K_{eq} = (-91\ \text{kJ/mol}) / (-2.48\ \text{kJ/mol}) = 36.7$

$K_{eq} = 8.7 \times 10^{15}$

A Look Ahead

5.45

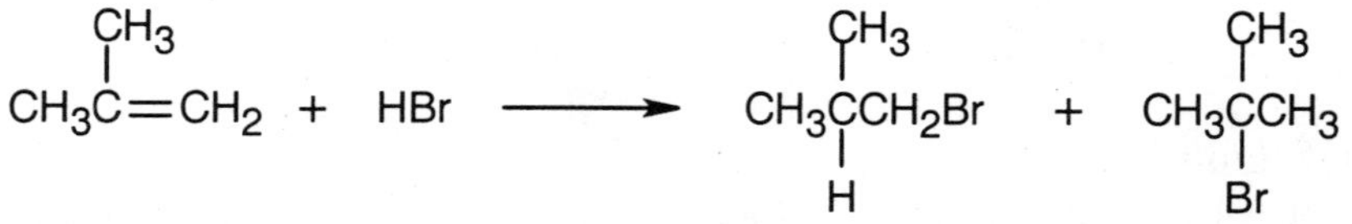

$$CH_3C(CH_3){=}CH_2 + HBr \longrightarrow CH_3CH(CH_3)CH_2Br + CH_3CBr(CH_3)CH_3$$

2-Methylpropene → 1-Bromo-2-methylpropane + 2-Bromo-2-methylpropane

5.46

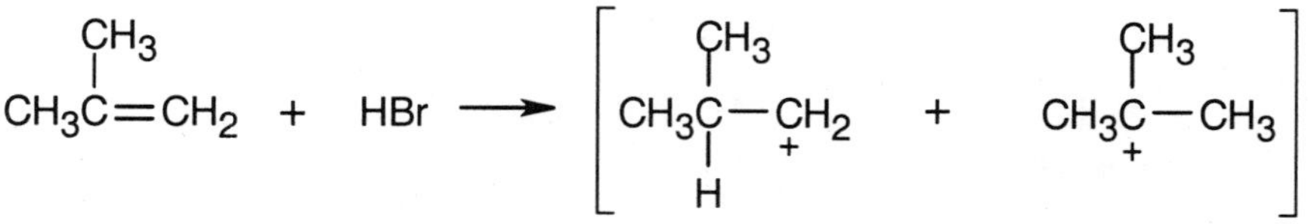

$$CH_3C(CH_3){=}CH_2 + HBr \longrightarrow [CH_3CH(CH_3){-}\overset{+}{C}H_2 + CH_3\overset{+}{C}(CH_3){-}CH_3]$$

The second carbocation is more stable because more alkyl substituents are bonded to the positively charged carbon.

5.47

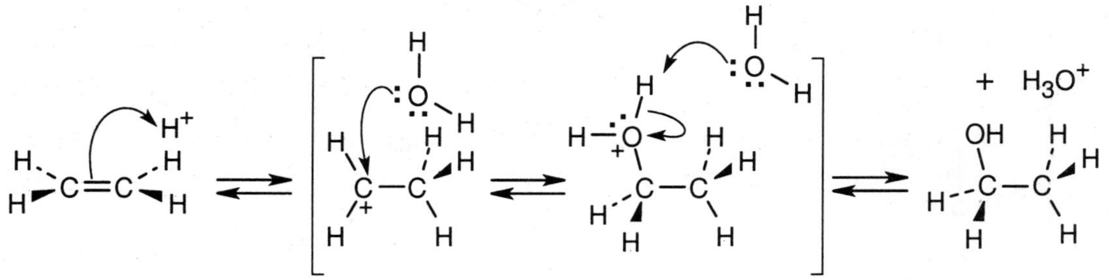

$+ \ H_3O^+$

Molecular Modeling

5.48 The most electrophilic atoms are starred.

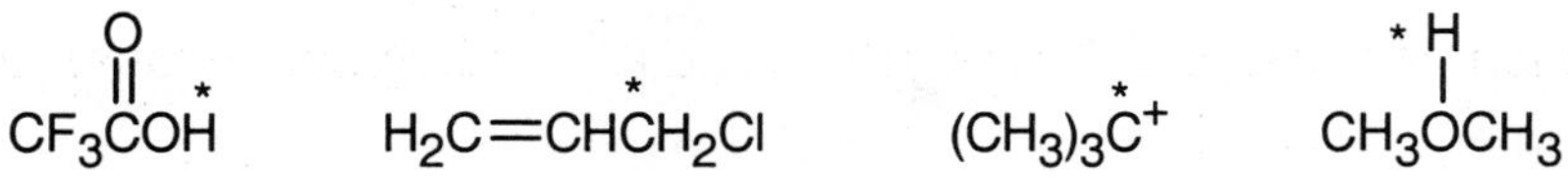

$CF_3\overset{*}{C}OH$ (C=O) $H_2C{=}CH\overset{*}{C}H_2Cl$ $(CH_3)_3\overset{*}{C}^+$ CH_3OCH_3 (starred H on O)

5.49

$H_2\overset{*}{C}{-}C{\equiv}N: \longleftrightarrow H_2C{=}C{=}\overset{*}{N}:$

H_3C–N, N*, CH_3

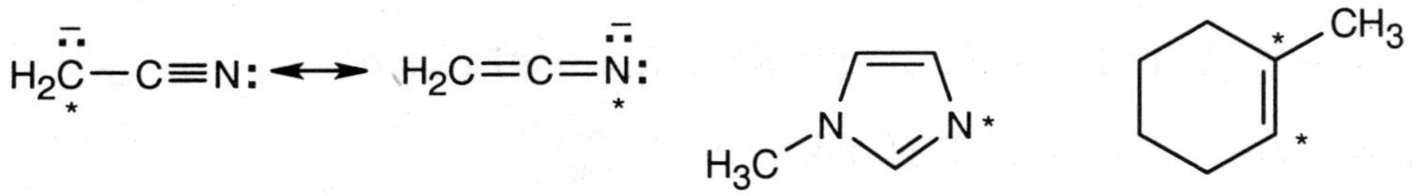

5.50 $\Delta H° = -331$ kJ/mol. The reaction is exothermic.

5.51 $\Delta H^{\ddagger} = 100$ kJ/mol for (a) and $\Delta H^{\ddagger} = 84$ kJ/mol for (b). Reaction (b) will be faster if both reactions are carried out under identical conditions. Notice that there is a statistical factor that also favors (b).

Review Unit 2: Alkenes and Their Stereochemistry: Organic Reactions

Major Topics Covered (with Vocabulary):

Functional Groups.

Alkanes:
saturated aliphatic straight-chain alkane branched-chain alkane isomer constitutional isomer alkyl group primary, secondary, tertiary, quaternary carbon paraffin cycloalkane primary, secondary, tertiary hydrogen cis-trans isomer stereoisomer

Alkane Stereochemistry:
conformer sawhorse representation Newman projection staggered conformation eclipsed conformation torsional strain anti conformation gauche conformation steric strain angle strain heat of combustion chair conformation axial group equatorial group ring-flip 1,3-diaxial interaction conformational analysis boat conformation twist-boat conformation bicycloalkane

Organic Reactions:
addition reaction elimination reaction substitution reaction rearrangement reaction reaction mechanism homolytic heterolytic homogenic heterogenic radical reaction polar reaction initiation propagation termination electronegativity polarizability curved arrow electrophile nucleophile carbocation

Describing a Reaction:
K_{eq} $\Delta G°$ exergonic endergonic enthalpy entropy heat of reaction exothermic endothermic bond dissociation energy reaction energy diagram transition state activation energy reaction intermediate

Types of Problems:

After studying these chapters, you should be able to:

- Identify functional groups, and draw molecules containing a given functional group.
- Draw all isomers of a given molecular formula.
- Name and draw alkanes, alkyl groups, and cycloalkanes, including cis-trans isomers.
- Identify carbons and hydrogens as being primary, secondary or tertiary.

- Draw energy vs. angle of rotation graphs for single bond conformations.
- Draw Newman projections of bond conformations and predict their relative stability.
- Understand the geometry of, and predict the stability of, cycloalkanes having fewer than 6 carbons.
- Draw and name substituted cyclohexanes.
- Predict the stability of substituted cyclohexanes by estimating steric interactions.

- Identify reactions as polar, radical, substitution, elimination, addition, or rearrangement reactions.
- Understand the mechanism of radical reactions.

- Identify reagents as electrophiles or nucleophiles.
- Understand the concepts of equilibrium and rate.
- Calculate K_{eq} and $\Delta G°$ of reactions, and use bond dissociation energies to calculate $\Delta H°$ of reactions.
- Draw reaction energy diagrams and label them properly.

Points to Remember:

* In identifying the functional groups in a compound, some groups have different designations that depend on the number and importance of other groups in the molecule. For example, a compound containing an –OH group and few other groups is probably named as an alcohol, but when several other groups are present , the –OH group is referred to as a hydroxyl group. There is a priority list of functional groups in the Appendix of this book, and this priority order will become more apparent as you progress through the text.

* It is surprising how many errors can be made in naming compounds as simple as alkanes. Why is this? Often the problem is a result of just not paying attention. It is very easy to undercount or overcount the $–CH_2–$ groups in a chain and to misnumber substituents. Let's work through a problem, using the rules in Section 3.4.

```
 CH3  CH2CH3
  |    |
CH3CCH2CHCH2CH3
  |
 CH3
```

Find the longest chain. In the above compound, the longest chain is a hexane (Try all possibilities; there are two different six-carbon chains in the compound.) Identify the substituents. The compound has two methyl groups and an ethyl group. It's a good idea to list these groups to keep track of them. Number the chain and the groups. Try both possible sets of numbers, and see which results in the lower combination of numbers. The compound might be named either as a 2,2,4-trisubstituted hexane or a 3,5,5-trisubstituted hexane, but the first name has a lower combination of numbers. Name the compound, remembering the prefix *di-* and remembering to list substituents in alphabetical order. The correct name for the above compound is 4-ethyl-2,2-dimethylhexane.
The acronym FINN (from the first letters of each step listed above) may be helpful.

* When performing a ring-flip on a cyclohexane ring, keep track of the positions on the ring.

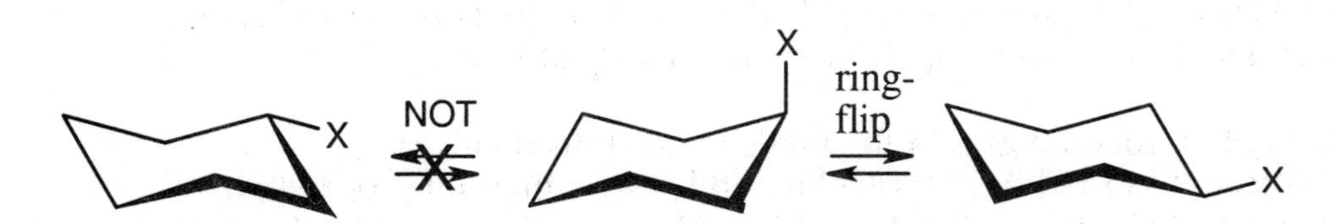

* In virtually all cases, a compound is of lower energy than the free elements of which it is composed. Thus, energy is released when a compound is formed from its component elements, and energy is required when bonds are broken. Entropy decreases when a compound is formed from its component elements (because disorder decreases). For two compounds of similar structure, less energy is required to break all bonds of the higher energy compound than is required to break all bonds of the lower energy compound.

Self-test

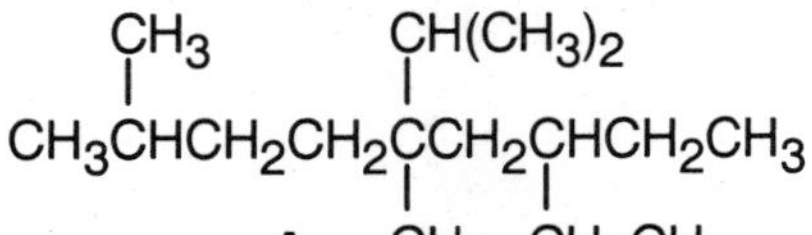

$$\begin{array}{c} \quad CH_3 \qquad\qquad\quad CH(CH_3)_2 \\ CH_3CHCH_2CH_2CCH_2CHCH_2CH_3 \\ \qquad\qquad\quad CH_3 \;\; CH_2CH_3 \end{array}$$

A

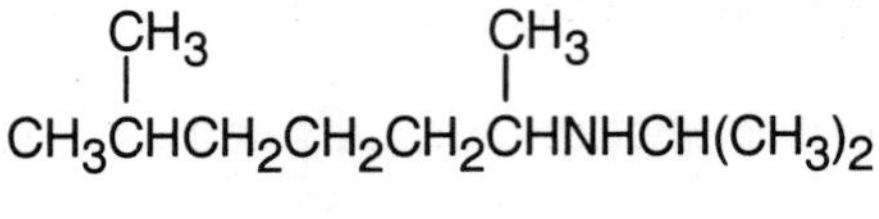

$$\begin{array}{c} CH_3 \qquad\qquad\qquad CH_3 \\ CH_3CHCH_2CH_2CH_2CHNHCH(CH_3)_2 \end{array}$$

B

Metron S (an antihistamine)

Name **A**, and identify carbons as primary, secondary, tertiary or quaternary.

B is an amine with two alkyl groups. Name these groups and identify alkyl hydrogens as primary, secondary or tertiary.

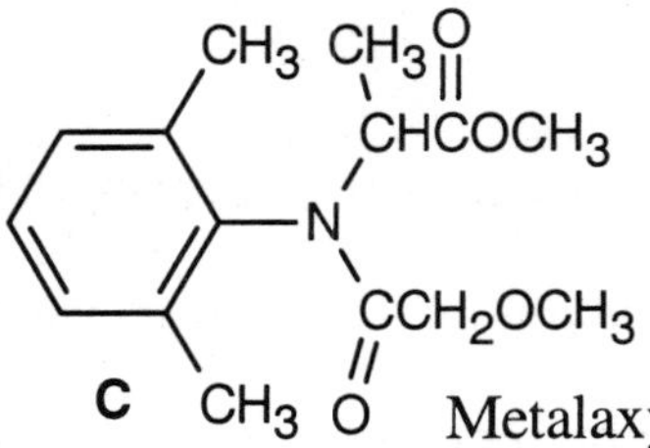

CH3 CH3 O, CHCOCH3, N, CCH2OCH3, CH3 O

C Metalaxyl (a fungicide)

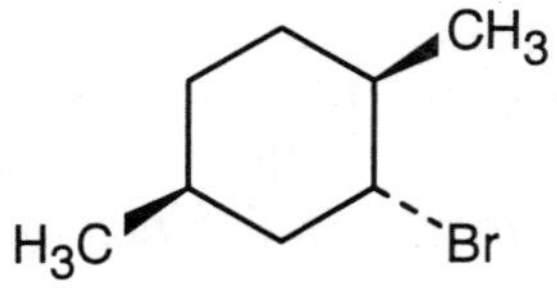

CH3, H3C, Br

D

Identify all functional groups of **C** (metalaxyl).

Name **D** and indicate the cis/trans relationship of the substituents. Draw both possible chair conformations, and calculate the energy difference between them.

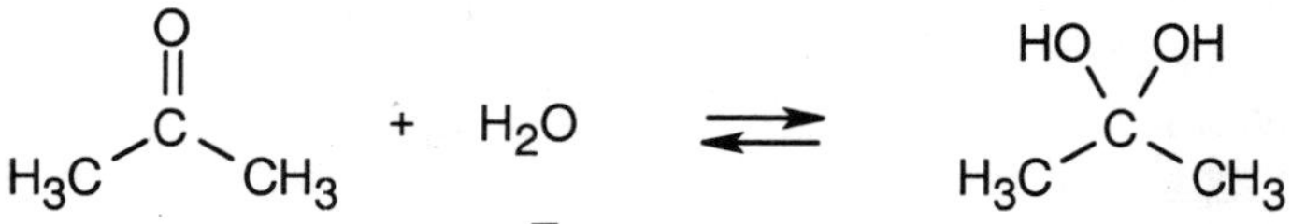

$$H_3C{-}C(=O){-}CH_3 \; + \; H_2O \; \rightleftharpoons \; H_3C{-}C(OH)_2{-}CH_3$$

E

What type of reaction is occurring in **E**? Would you expect that the reaction occurs by a polar or a radical mechanism? If K_{eq} for the reaction at 298 K is 10^{-3}, what sign do you expect for $\Delta G°$? Would you expect $\Delta S°$ to be negative or positive? What about $\Delta H°$?

Multiple Choice

1. Which of the following functional groups doesn't contain a carbonyl group?
   (a) aldehyde (b) ester (c) ether (d) ketone

2. Which of the following compounds contains primary, secondary, tertiary and quaternary carbons?
   (a) 2,2,4-Trimethylhexane (b) Ethylcyclohexane (c) 2-Methyl-4-ethylcyclohexane
   (d) 2,2-Dimethylcyclohexane

3. How many isomers of the formula $C_4H_8Br_2$ are there?
   (a) 4 (b) 6 (c) 8 (d) 9

4. The lowest energy conformation of 2-methylbutane occurs:
(a) when all methyl groups are anti (b) when all methyl groups are gauche
(c) when two methyl groups are anti (d) when two methyl groups are eclipsed

5. The strain in a cyclopentane ring is due to:
(a) angle strain (b) torsional strain (c) steric stain (d) angle strain and torsional strain

6. In which molecule do the substituents in the more stable conformation have a diequatorial relationship?
(a) cis-1,2 disubstituted (b) cis-1,3 disubstituted (c) trans-1,3-disubstituted
(d) cis-1,4 disubstituted

7. Which of the following molecules is not a nucleophile?
(a) BH_3 (b) NH_3 (c) HO^- (d) $H_2C{=}CH_2$

8. Which of the following reactions probably has the greatest entropy increase?
(a) addition reaction (b) elimination reaction (c) substitution reaction (d) rearrangement

9. At a specific temperature T, a reaction has negative $\Delta S°$ and $K_{eq} > 1$. What can you say about $\Delta G°$ and $\Delta H°$?
(a) $\Delta G°$ is negative and $\Delta H°$ is positive (b) $\Delta G°$ and $\Delta H°$ are both positive (c) $\Delta G°$ and $\Delta H°$ are both negative (d) $\Delta G°$ is negative but you can't predict the sign of $\Delta H°$.

10. In which of the following situations is $\Delta G^{\ddagger}$ likely to be smallest?
(a) a slow exergonic reaction (b) a fast exergonic reaction (c) a fast endergonic reaction
(d) a slow endergonic reaction

Chapter 6 – Alkenes: Structure and Reactivity

Chapter Outline

I. Introduction to alkene chemistry (Sections 6.1 – 6.7).
   A. Industrial preparation and use of alkenes (Section 6.1).
      1. Ethylene and propylene are the two most important organic chemicals produced industrially.
      2. Ethylene, propylene and butene are synthesized by thermal cracking.
         a. Thermal cracking involves homolytic breaking of C–H and C–C bonds.
         b. Thermal cracking reactions are dominated by entropy.
   B. Calculating a molecule's degree of unsaturation (Section 6.2).
      1. The degree of unsaturation of a molecule describes the number of multiple bonds and/or rings a molecule has.
      2. To calculate degree of unsaturation of a compound, first determine the equivalent hydrocarbon formula of the compound.
         a. Add the number of halogens to the number of hydrogens.
         b. Subtract one hydrogen for every nitrogen.
         c. Ignore the number of oxygens.
      3. Calculate the number of pairs of hydrogens that would be present in an alkane C_nH_{2n+2} that has the same number of carbons as the equivalent hydrocarbon of the compound of interest. This is the degree of unsaturation.
   C. Naming alkenes (Section 6.3).
      1. Find the longest chain containing the double bond, and name it.
      2. Number the carbon atoms in the chain.
      3. Number the substituents and write the name.
         a. Name the substituents alphabetically.
         b. Indicate the position of the double bond.
         c. Use the suffixes -diene, -triene, etc. if more than one double bond is present.
   D. Electronic structure of alkenes (Section 6.4).
      1. Carbon atoms in a double bond are sp^2-hybridized.
      2. The two carbons in a double bond form one σ bond and one π bond.
      3. Free rotation doesn't occur around double bonds.
      4. 268 kJ/mol of energy is required to break a π bond.
   E. Double bond geometry (Sections 6.5 – 6.6).
      1. Cis–trans isomerism (Section 6.5).
         a. A disubstituted alkene can have substituents either on the same side of the double bond (cis) or on opposite sides (trans).
         b. These isomers don't interconvert because free rotation about a double bond isn't possible.
         c. Cis–trans isomerism doesn't occur if one carbon in the double bond is bonded to identical substituents.
      2. *E,Z* isomerism (Section 6.6).
         a. The *E,Z* system is used to describe the arrangement of substituents around a double bond that can't be described by the cis–trans system.
         b. Sequence rules for *E,Z* isomers:
            i. For each double bond carbon, rank the substituents by atomic number. An atom with a high atomic number receives a higher priority than an atom with a lower atomic number.
            ii. If a decision can't be reached, look at the second or third atom until a difference is found.

iii. Multiple-bonded atoms are equivalent to the same number of single-bonded atoms.

F. Alkene stability (Section 6.7).
1. Cis alkenes are less stable than trans alkenes because of steric strain between double bond substituents.
2. Stabilities of alkenes can be determined experimentally by measuring:
a. Heats of combustion.
b. Cis–trans equilibrium constants.
c. Heats of hydrogenation – the most useful method.
3. The heat of hydrogenation of a cis isomer is a larger negative number than the heat of hydrogenation of a trans isomer.
This indicates that a cis isomer is of higher energy and is less stable than a trans isomer.
4. Alkene double bonds become more stable with increasing substitution for two reasons:
a. Hyperconjugation – a stabilizing interaction between the antibonding π orbital and a filled C–H σ orbital on an adjacent substituent.
b. More substituted double bonds have more of the stronger sp^2–sp^3 bonds.

II. Electrophilic addition reactions (Sections 6.8 – 6.12).
A. Addition of H–X to alkenes (Sections 6.8 – 6.9).
1. Mechanism of addition (Section 6.8).
a. The electrons of the nucleophilic π bond attack the electrophile H–X (X = Cl, Br, I).
b. Two electrons from the π bond form a new σ bond between –H and an alkene carbon.
c. The carbocation intermediate reacts with –X to form a C–X bond.
2. An energy diagram has two peaks separated by a valley (carbocation intermediate).
a. The reaction is exergonic.
b. The first step is slower than the second step.
(3. Organic reactions are often written different ways to emphasize different points.)
4. Orientation of addition: Markovnikov's Rule (Section 6.9).
a. In the addition of HX to a double bond, H attaches to the carbon with fewer substituents, and X attaches to the carbon with more substituents.
b. If the carbons have the same number of substituents, a mixture of products results.
B. Carbocation structure and stability (Section 6.10).
1. Carbocations are planar; the unoccupied *p* orbital extends above and below the plane.
2. The stability of carbocations increases with increasing substitution.
a. Carbocation stability can be measured by studying gas-phase dissociation enthalpies.
b. Carbocations can be stabilized by inductive effects of neighboring alkyl groups.
c. Carbocation can be stabilized by hyperconjugation: The more alkyl groups on the carbocation, the more opportunities there are for hyperconjugation.
C. The Hammond Postulate (Section 6.11).
1. The transition state for an endergonic reaction step resembles the product of that step.
2. The transition state for an exergonic reaction step resembles the reactant for that step.
3. In an electrophilic addition reaction, the transition state for alkene protonation resembles the carbocation intermediate.
4. More stable carbocations form faster because their transition states are also stabilized.

D. Carbocation rearrangements (Section 6.12).
1. In some electrophilic addition reactions, products from carbocation rearrangements are formed.
2. The appearance of these products supports the two-step electrophilic addition mechanism, in which an intermediate carbocation is formed.
3. Intermediate carbocations can rearrange to more stable carbocations by either a hydride shift or by an alkyl shift.
4. In both cases a group moves to an adjacent positively charged carbon, taking its bonding electron pair with it.

Solutions to Problems

6.1 The expression "degree of unsaturation" refers to the number of rings, double bonds or triple bonds that a compound contains. For example, consider the compound in (a), C_8H_{14}. A saturated alkane with eight carbons has the formula C_8H_{18}. C_8H_{14}, which has four fewer (or two pairs fewer) hydrogens, may have two double bonds, or two rings, or one of each, or a triple bond. C_8H_{14} thus has a degree of unsaturation of 2.

Compound	*Degree of unsaturation*	*Compound*	*Degree of unsaturation*
(a) C_8H_{14}	2	(b) C_5H_6	3
(c) $C_{12}H_{20}$	3	(d) $C_{20}H_{32}$	5
(e) $C_{40}H_{56}$	13		

6.2

Compound	*Degree of Unsaturation*	*Structures*
(a) C_4H_8	1	$CH_3CH_2{=}CH_2CH_3$ $CH_3CH_2CH_2{=}CH_2$ $(CH_3)_2C{=}CH_2$
		[cyclobutane] [methylcyclopropane: $-CH_3$]
(b) C_4H_6	2	$CH_2{=}CHCH{=}CH_3$ $CH_3CH{=}C{=}CH_2$ $CH_3C{\equiv}CCH_3$
		$CH_3CH_2C{\equiv}CH$ [bicyclobutane] [cyclobutene]
		[methylcyclopropene: $-CH_3$] [methylcyclopropene: $-CH_3$] [methylenecyclopropane: $=CH_2$]
(c) C_3H_4	2	$H_2C{=}C{=}CH_2$ $CH_3C{\equiv}CH$ [cyclopropene]

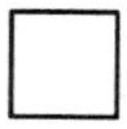

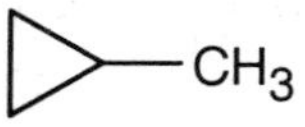

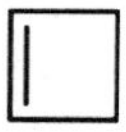

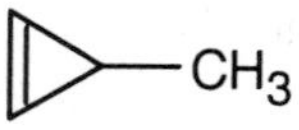

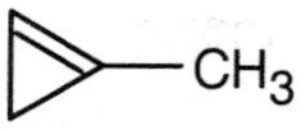

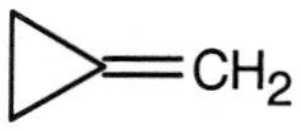

6.3 Unlike the hydrocarbons in the previous problems, the compounds in this problem contain additional elements. Review the rules for these elements.

(a) Subtract one hydrogen for each nitrogen present to find the formula of the equivalent hydrocarbon – C_6H_4. Compared to the alkane C_6H_{14}, the compound of formula C_6H_4 has 10 fewer hydrogens, or 5 fewer hydrogen pairs, and has a degree of unsaturation of 5.

(b) $C_6H_5NO_2$ also has 5 degrees of unsaturation because oxygen doesn't affect the equivalent hydrocarbon formula of a compound.

(c) A halogen atom is equivalent to a hydrogen atom in calculating the equivalent hydrocarbon formula. In this problem, the equivalent hydrocarbon formula is C_8H_{12}, and the degree of unsaturation is 3.

(d) $C_9H_{16}Br_2$ – one degree of unsaturation.
(e) $C_{10}H_{12}N_2O_3$ – 6 degrees of unsaturation.
(f) $C_{20}H_{32}ClN$ – 5 degrees of unsaturation.

6.4 (1) Find the longest chain containing the double bond and name it. In (a), the longest chain is a pentene.
(2) Identify the substituents. There are three methyl groups in (a).
(3) Number the substituents, remembering that the double bond receives the lowest possible number. The methyl groups are attached to C3 and C4 (two methyl groups).
(4) Name the compound, remembering to use the prefix "tri-" before "methyl" and remembering to use a number to signify the location of the double bond. The name of the compound in (a) is 3,4,4-trimethyl-1-hexene.

(a)

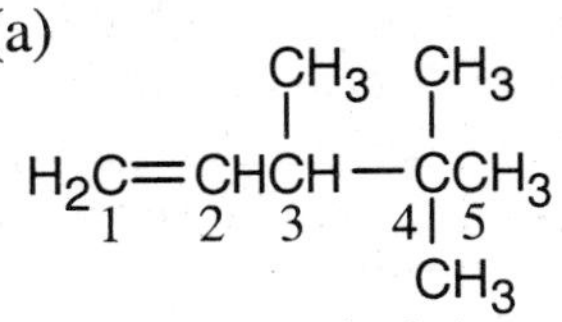

CH_3 CH_3

$H_2C{=}CHCH{-}CCH_3$ (1 2 3 4 5)

CH_3

3,4,4-Trimethyl-1-pentene

(b)

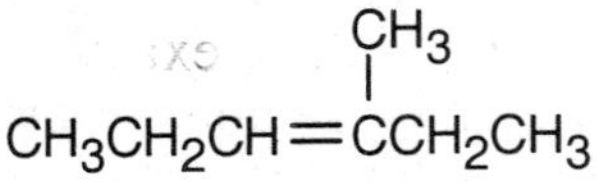

CH_3

$CH_3CH_2CH{=}CCH_2CH_3$

3-Methyl-3-hexene

(c)

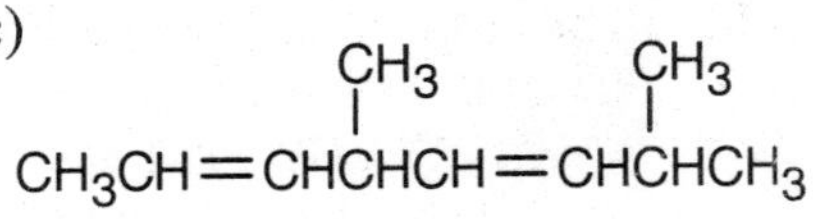

CH_3 CH_3

$CH_3CH{=}CHCHCH{=}CHCHCH_3$

4,7-Dimethyl-2,5-octadiene

6.5 It's much easier to draw a structure from a given name than it is to name a structure. First, draw the carbon chain, placing the double bond or bonds in the designated locations. Then attach the cited groups in the proper positions.

(a)

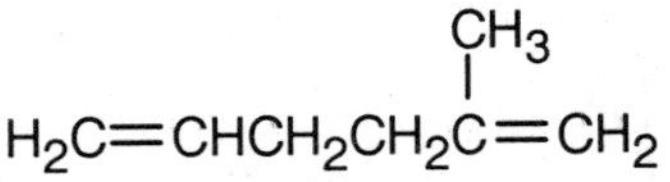

CH_3

$H_2C{=}CHCH_2CH_2C{=}CH_2$

2-Methyl-1,5-hexadiene

(b)

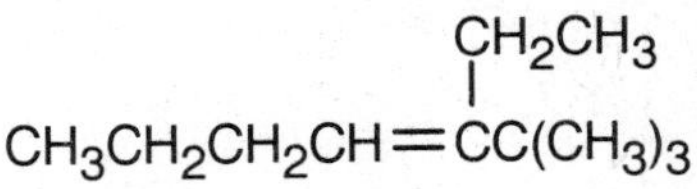

CH_2CH_3

$CH_3CH_2CH_2CH{=}CC(CH_3)_3$

3-Ethyl-2,2-dimethyl-3-heptene

(c)

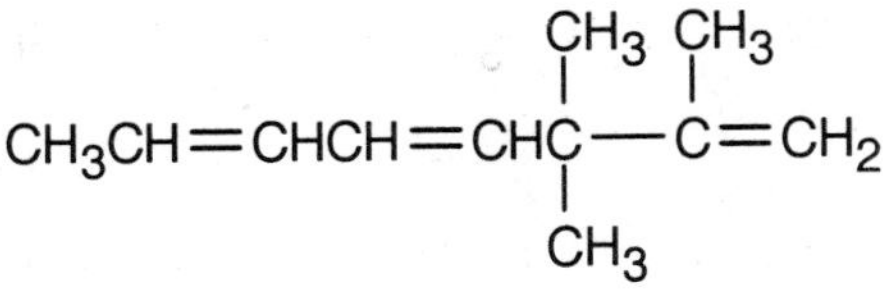

CH_3 CH_3

$CH_3CH{=}CHCH{=}CHC{-}C{=}CH_2$

CH_3

2,3,3-Trimethyl-1,4,6-octatriene

(d)

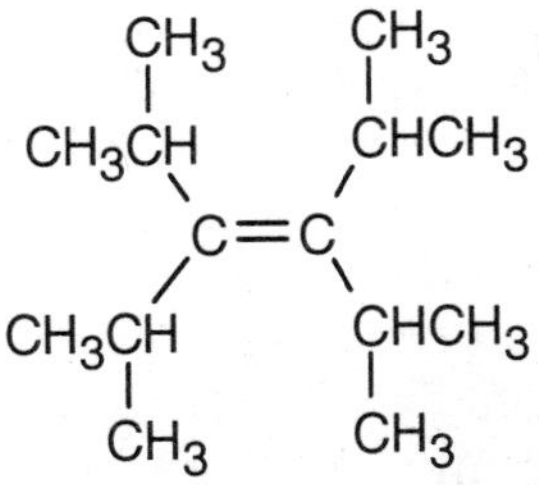

CH_3 CH_3

CH_3CH $CHCH_3$

$C{=}C$

CH_3CH $CHCH_3$

CH_3 CH_3

3,4-Diisopropyl-2,5-dimethyl-3-hexene

(e)

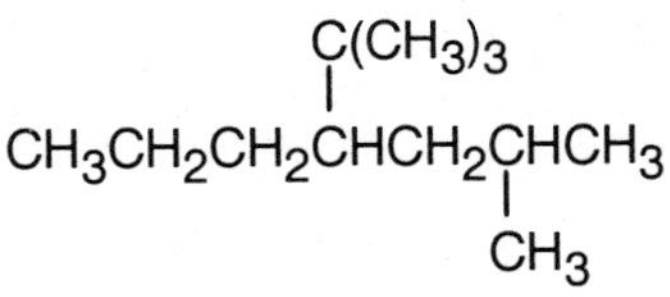

$C(CH_3)_3$

$CH_3CH_2CH_2CHCH_2CHCH_3$

CH_3

4-*tert*-Butyl-2-methylheptane

6.6

(a)

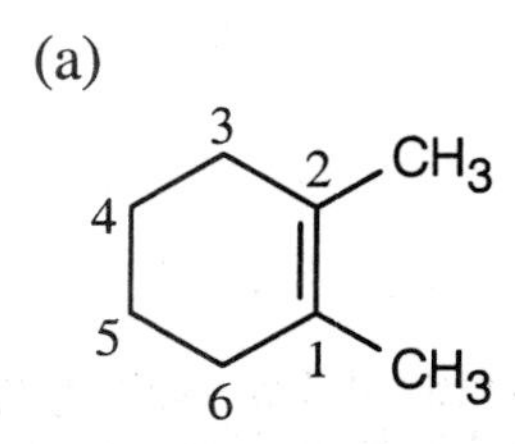

1,2-Dimethylcyclohexene

(b) 4,4-Dimethylcycloheptene

(c) 3-Isopropylcyclopentene

6.7 Compounds in (c), (e), and (f) can exist as cis-trans isomers.

	cis	trans
(c) $CH_3CH_2CH{=}CHCH_3$	H, H / C=C / CH_3CH_2, CH_3	H, CH_3 / C=C / CH_3CH_2, H
(e) $ClCH{=}ClCH$	H, H / C=C / Cl, Cl	H, Cl / C=C / Cl, H
(f) $BrCH{=}CHCl$	H, H / C=C / Br, Cl	H, Cl / C=C / Br, H

6.8 A model of cyclohexene shows that a six-membered ring is too small to contain a trans double bond without causing severe strain to the ring. A ten-membered ring is flexible enough to accommodate either a cis or a trans double bond, although the cis isomer has less strain than the trans isomer.

6.9 Review the sequence rules presented in Section 6.6. A summary:

Rule 1: An atom with a high atomic number has priority over an atom with a low atomic number.
Rule 2: If a decision can't be reached by using Rule 1, look at the second, third, or fourth atom away from the double-bond carbon until a decision can be made.
Rule 3: Multiple-bonded atoms are equivalent to the same number of single-bonded atoms.

High	*Low*	*Rule*	*High*	*Low*	*Rule*
(a) –Br	–H	1	(b) –Br	–Cl	1
(c) $–CH_2CH_3$	$–CH_3$	2	(d) –OH	$–NH_2$	1
(e) $–CH_2OH$	$–CH_3$	2	(f) –CH=O	$–CH_2OH$	3

6.10 *Highest priority* ————> *Lowest Priority*

(a) –Cl, –OH, $–CH_3$, –H

(b) $–CH_2OH$, $–CH{=}CH_2$, $–CH_2CH_3$, $–CH_3$

(c) –COOH, $–CH_2OH$, –C≡N, $–CH_2NH_2$

(d) $–CH_2OCH_3$, –C≡N, –C≡CH, $–CH_2CH_3$

6.11

(a)

Low H_3C ... CH_2OH Low

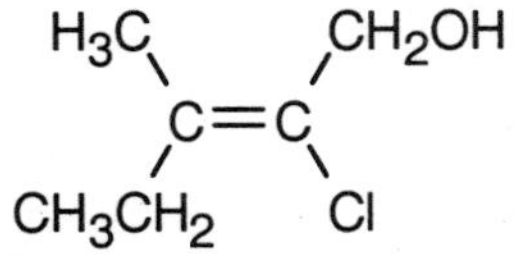

$C{=}C$ *Z*

High CH_3CH_2 ... Cl High

First, consider the substituents on the right side of the double bond. –Cl ranks higher than $-CH_2OH$ by Rule 1 of the Cahn-Ingold-Prelog Rules. On the left side of the double bond, $-CH_2CH_3$ ranks higher than $-CH_3$

(b)

High Cl ... CH_2CH_3 Low

$C{=}C$ *E*

Low CH_3O ... $CH_2CH_2CH_3$ High

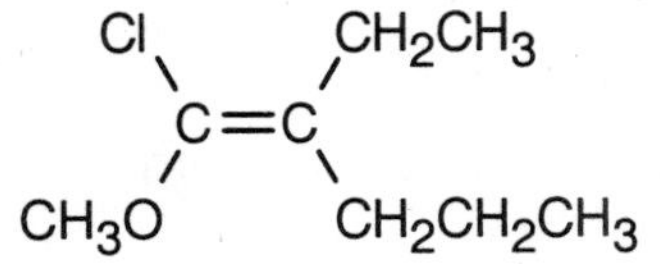

(c)

High H_3C– (ring) ... COOH High

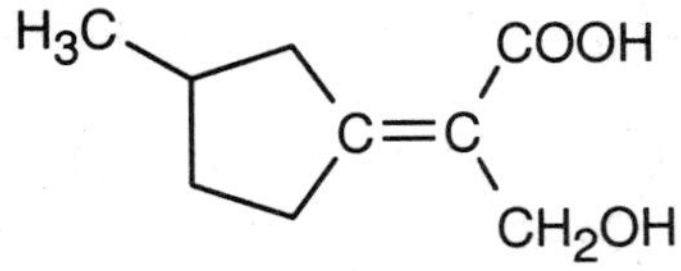

$C{=}C$ *Z*

Low ... CH_2OH Low

Notice that the upper substituent on the left side of the double bond is of higher priority because of the methyl group attached to the ring.

(d)

Low H ... CN High

$C{=}C$ *E*

High H_3C ... CH_2NH_2 Low

6.12

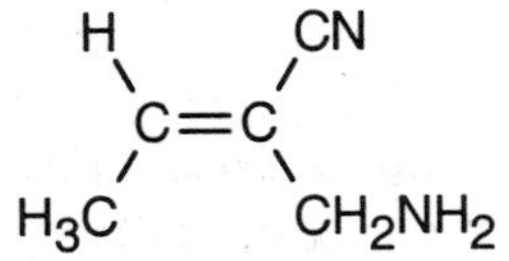

High ... $COOCH_3$ High

Z

Low ... CH_2OH Low

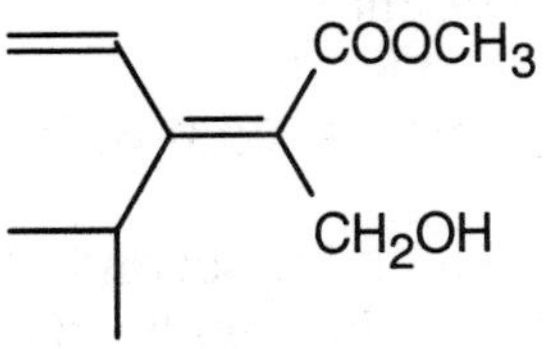

6.13

More stable *Less stable*

(a)

H_3C H CH_3CH_2 H

$C{=}C$ $C{=}C$

H_3C H H H

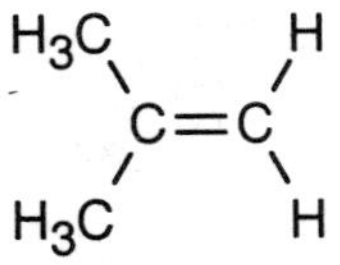

disubstituted double bond monosubstituted double bond

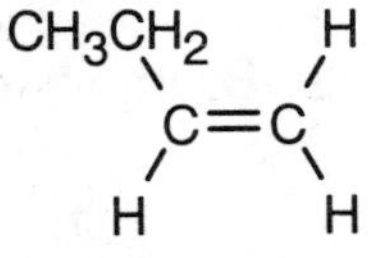

(b)

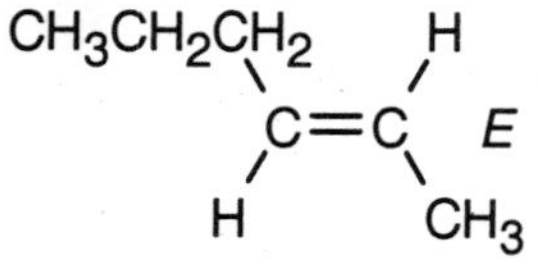

$$CH_3CH_2CH_2(H)C{=}C(H)(CH_3) \quad E$$

no steric strain

$$CH_3CH_2CH_2(H)C{=}C(CH_3)(H) \quad Z$$

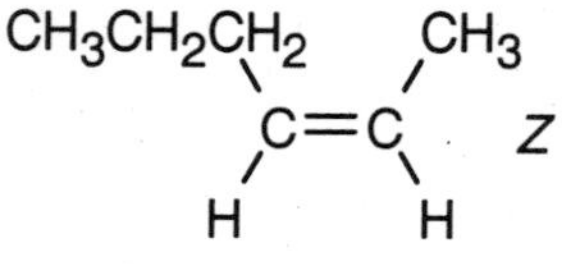

steric strain of groups on the same side of the double bond

(c)

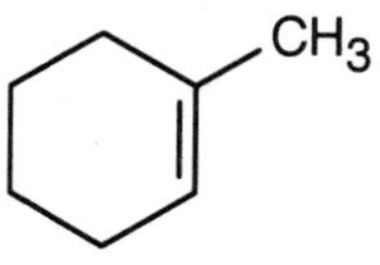

trisubstituted double bond

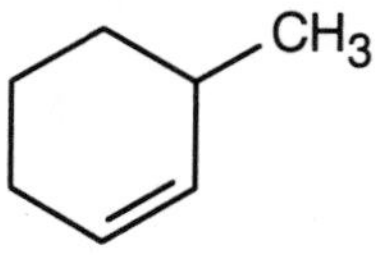

disubstituted double bond

6.14 All these reactions are electrophilic additions of HX to an alkene. Use Markovnikov's rule to predict orientation.

(a)

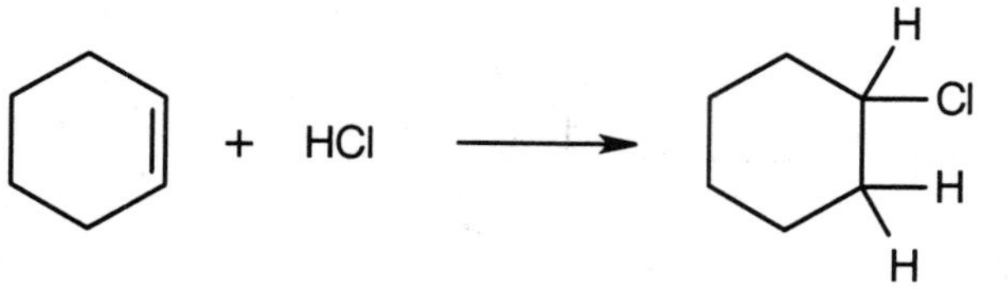

Cyclohexene + HCl ⟶ Chlorocyclohexane

(b)

$$(CH_3)_2C{=}CHCH_2CH_3 + HBr \longrightarrow (CH_3)_2CBrCH_2CH_2CH_3$$

2-Bromo-2-methylpentane

In accordance with Markovnikov's Rule, H forms a bond to the carbon with fewer substituents, and Br forms a bond to the carbon with more substituents.

(c)

$$CH_3CH_2CH_2CH{=}CH_2 \xrightarrow[H_3PO_4]{KI} CH_3CH_2CH_2CHICH_3$$

2-Iodopentane

(d)

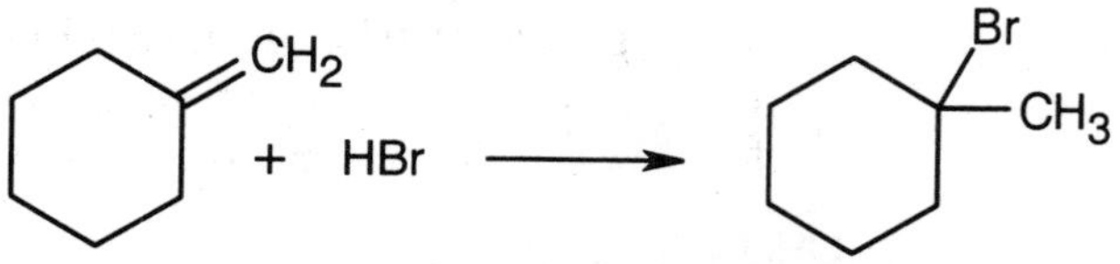

Methylenecyclohexane + HBr ⟶ 1-Bromo-1-methylcyclohexane

6.15 Think backward in choosing the alkene starting material for synthesis of the desired haloalkanes. Remember that halogen is bonded to one end of the double bond and that more than one starting material can give rise to the desired product.

(a)

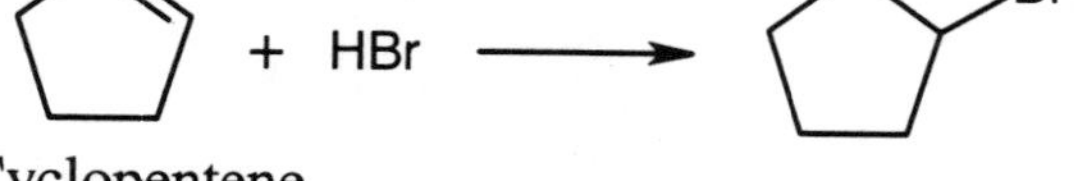

Cyclopentene + HBr ⟶ Bromocyclopentane

Cyclopentene

(b)

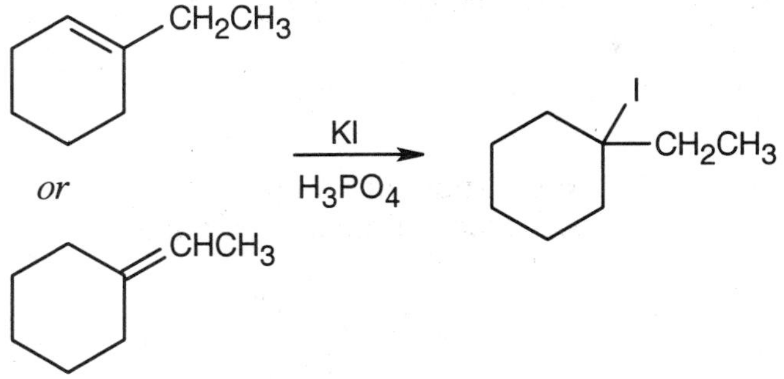

(1-ethylcyclohexene) CH_2CH_3 *or* (ethylidenecyclohexane) $CHCH_3$ $\xrightarrow[H_3PO_4]{KI}$ 1-iodo-1-ethylcyclohexane (I, CH_2CH_3)

(c)

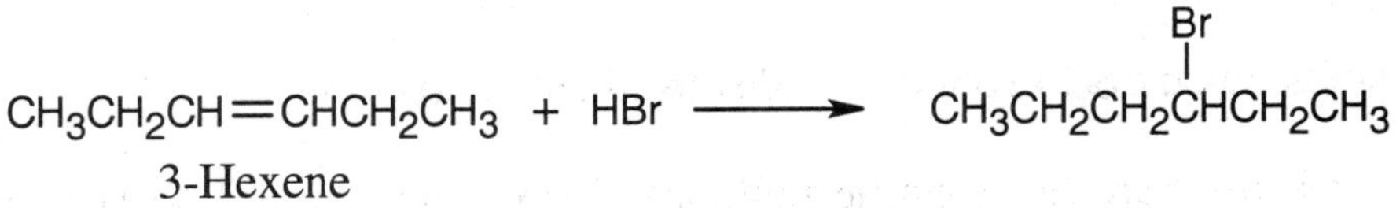

$$CH_3CH_2CH{=}CHCH_2CH_3 + HBr \longrightarrow CH_3CH_2CH_2CH(Br)CH_2CH_3$$

3-Hexene

(d)

(vinylcyclohexane) + HCl $\longrightarrow$ (1-chloroethyl)cyclohexane (Cl)

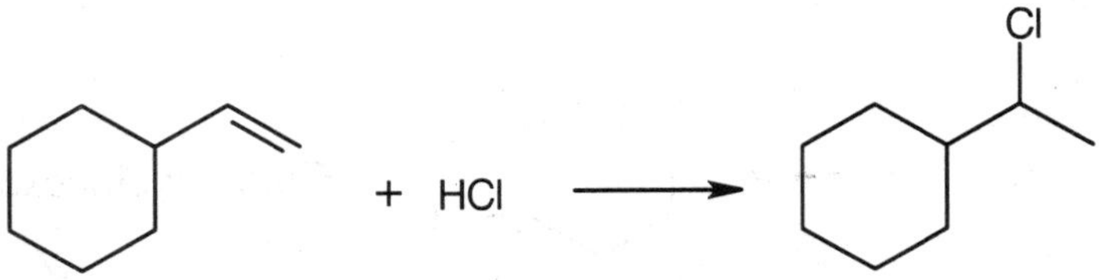

6.16 The more stable carbocation is formed

(a)

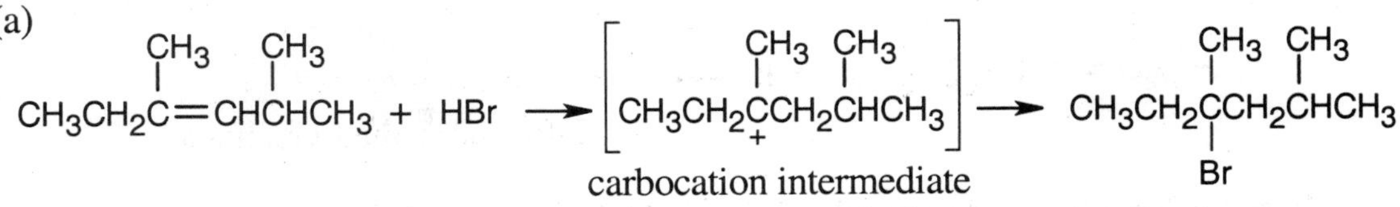

$$CH_3CH_2C(CH_3){=}CHCH(CH_3)CH_3 + HBr \longrightarrow \left[CH_3CH_2\overset{+}{C}(CH_3)CH_2CH(CH_3)CH_3\right] \longrightarrow CH_3CH_2C(CH_3)(Br)CH_2CH(CH_3)CH_3$$

carbocation intermediate

(b)

(ethylidenecyclopentane) $=CHCH_3$ $\xrightarrow[H_3PO_4]{KI}$ [(cyclopentyl cation)$^{+}$–CH_2CH_3] $\longrightarrow$ 1-iodo-1-ethylcyclopentane (I, CH_2CH_3)

carbocation intermediate

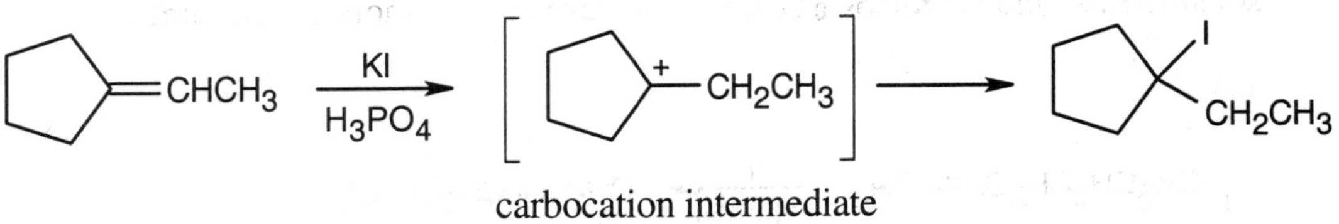

6.17 Two representations of the secondary carbocation are shown on the left below. This secondary carbocation can experience hyperconjugative overlap with two hydrogens under normal circumstances. However, in the alignment shown in the stereodrawing, only one hydrogen (circled) is in the correct position for hyperconjugative overlap with the carbocation carbon.

Because there is rotation about the carbon-carbon bonds, all of the hydrogens starred in the representation on the far right can be involved in hyperconjugation at some time.

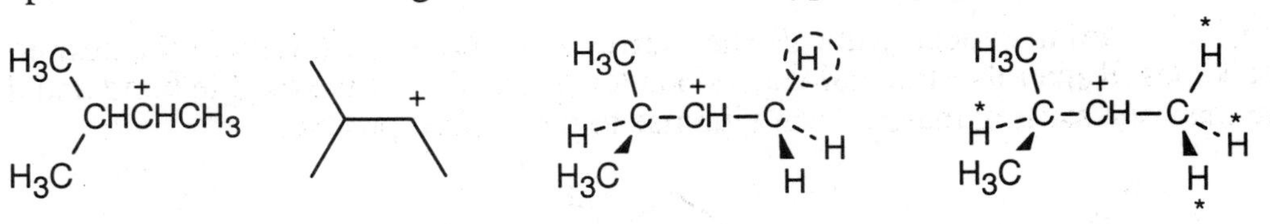

$(H_3C)_2\overset{+}{C}HCHCH_3$ — [skeletal structure] — H_3C, H–C(H$_3$C)–$\overset{+}{C}$H–C(H)(H)(H, circled) — H_3C, H*–C(H$_3$C)–$\overset{+}{C}$H–C(H*)(H*)(H*)

6.18 The second step in the electrophilic addition of HCl to alkenes is exergonic. According to the Hammond Postulate, the transition state should resemble the carbocation intermediate.

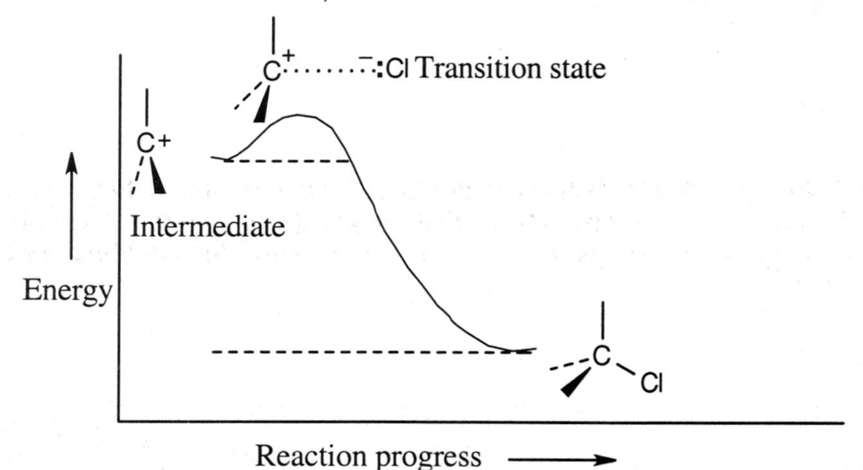

6.19

H—Br

CHCH$_3$

H

:Br:

CH$_2$CH$_3$

Br

CH$_2$CH$_3$

electrophilic attack by π electrons

hydride shift

reaction of carbocation with Br$^-$

Visualizing Chemistry

6.20

(a)

2,4,5-Trimethyl-2-hexene

(b)

1-Ethyl-3,3-dimethylcyclohexene

6.21

(a)

High Cl Low

Low C=O High

H

E

(b)

High OCH$_3$ High

C=O

Low OH Low

Z

6.22

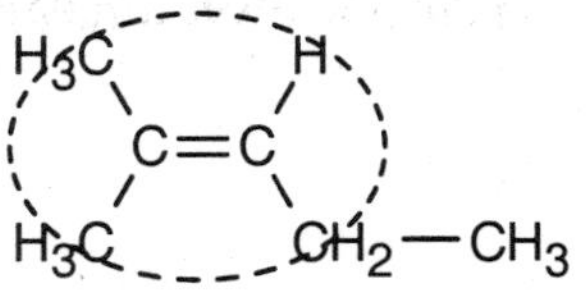

All of the atoms attached to a double bond (circled) must lie in the same plane. If you look at this drawing with a stereoviewer, you can see that the groups on one of the double bond carbons are at a 90° angle to the groups on the other carbon. Thus, the structure doesn't represent a stable molecule.

6.23

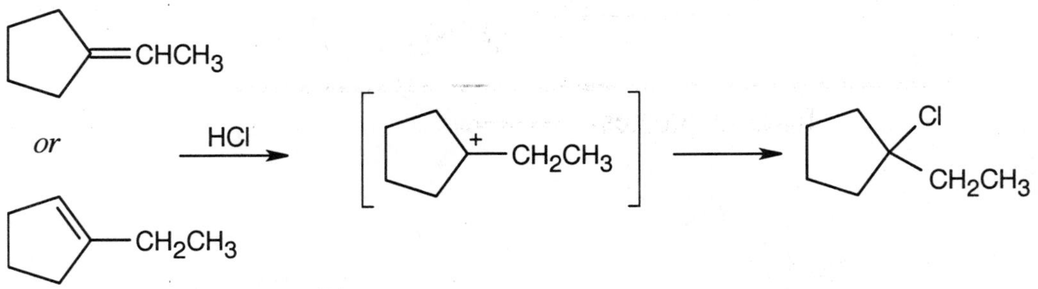

Either of the two compounds shown can form the illustrated tertiary carbocation when they react with HCl. If you make a model of this intermediate, you can see that the three circled hydrogens are aligned for maximum overlap with the vacant *p* orbital. Because of conformational mobility, the three starred hydrogens are also able to be involved in hyperconjugation.

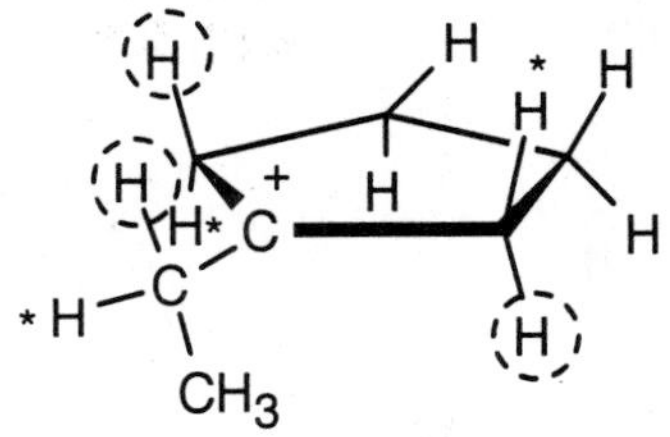

Additional Problems

6.24 The purpose of this problem is to give you experience in calculating the number of double bonds and/or rings in a formula. Additionally, you will learn to draw structures containing various functional groups. Remember that any formula that satisfies the rules of valency is acceptable. Try to identify functional groups in the structures that you draw.

(a) $C_{10}H_{16}$ – 3 degrees of unsaturation. Examples:

$CH_3CH_2CH{=}CHCH{=}CHCH{=}CHCH_2CH_3$

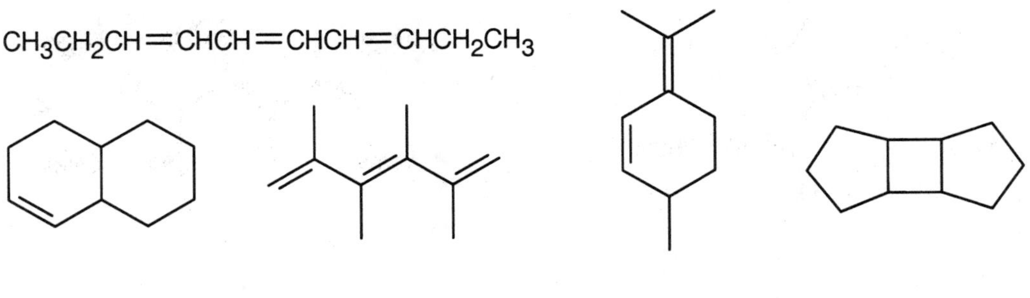

(b) C_8H_8O. The equivalent hydrocarbon is C_8H_8, which has 5 degrees of unsaturation.

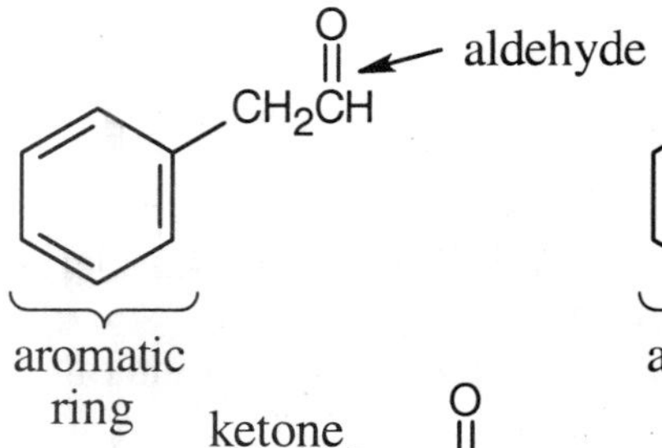

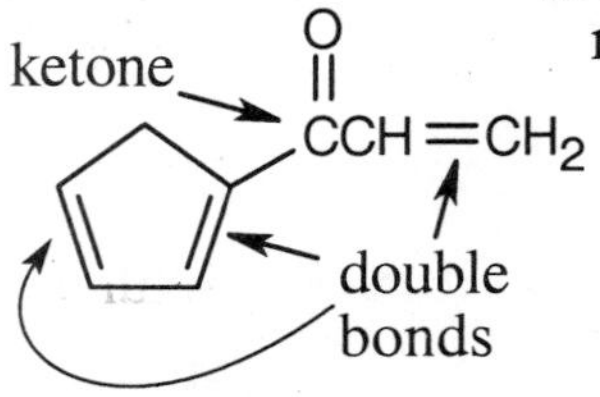

ketone
aromatic ring
aldehyde
aromatic ring
aldehyde
aromatic ring
ketone
double bonds
ketone
double bonds

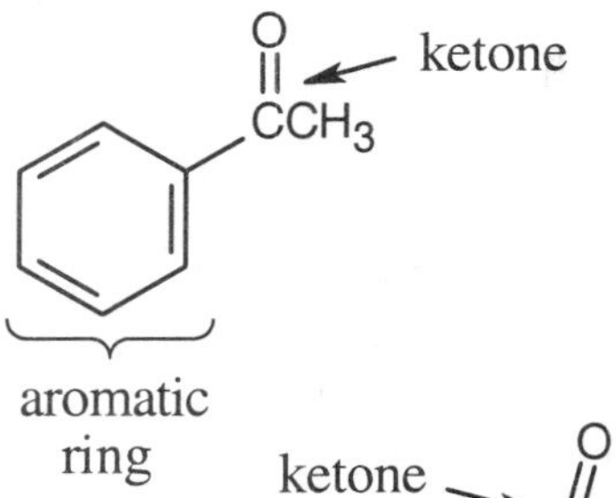

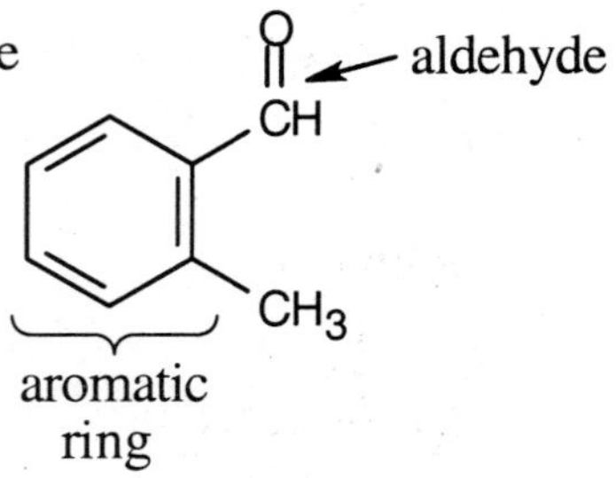

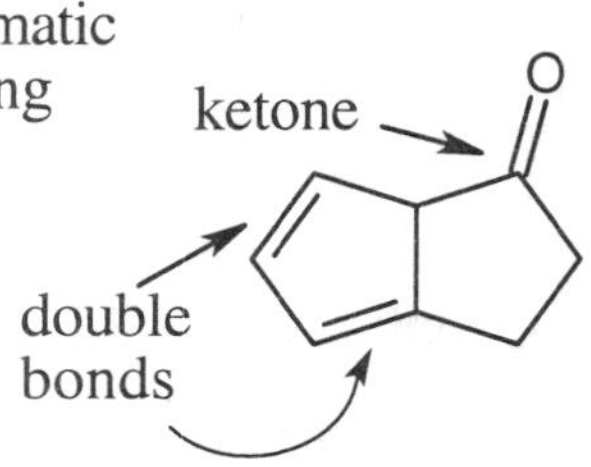

(c) $C_7H_{10}Cl_2$ has C_7H_{12} as its equivalent hydrocarbon formula. $C_7H_{10}Cl_2$ has two degrees of unsaturation.

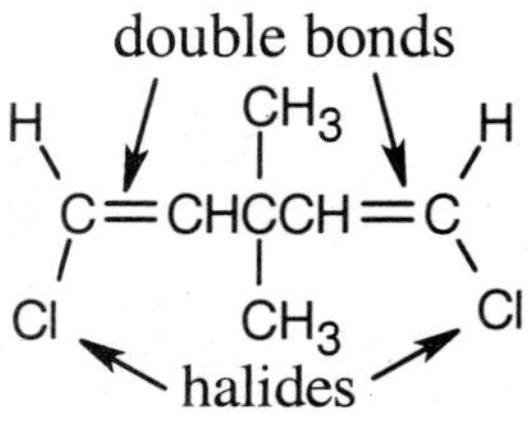

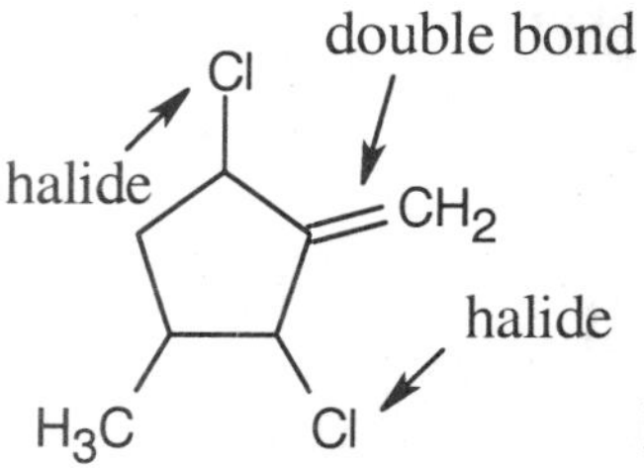

triple bond
halide
halide
double bonds
halides
double bond
halide
halide
double bond
halide
halide
halides

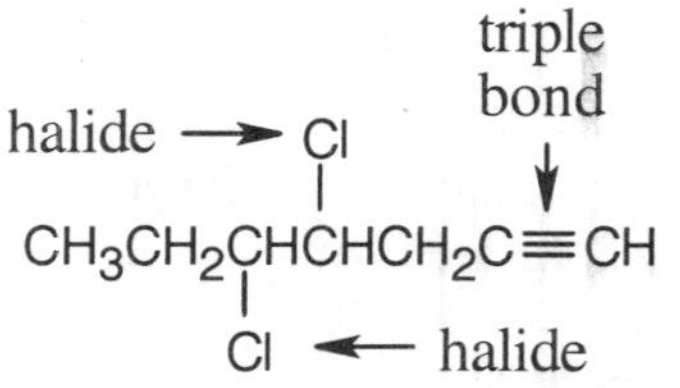

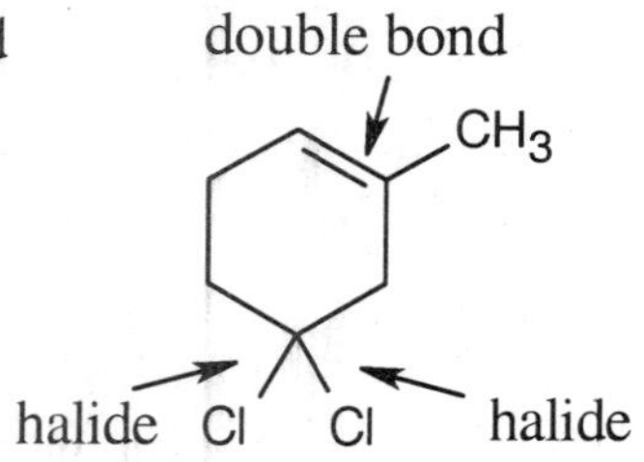

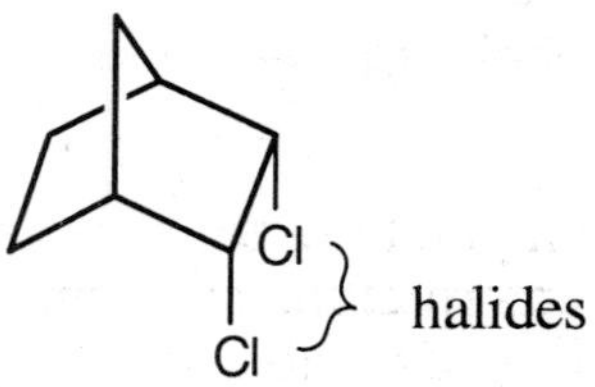

(d) $C_{10}H_{16}O_2$ – 3 degrees of unsaturation.

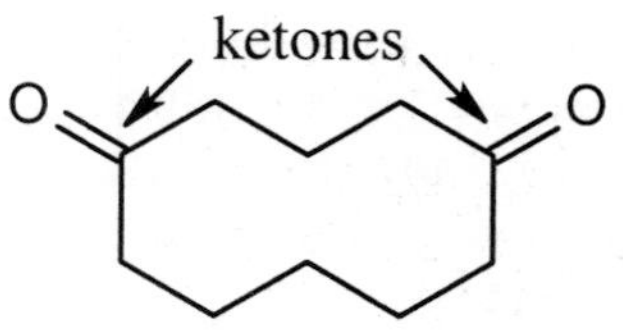

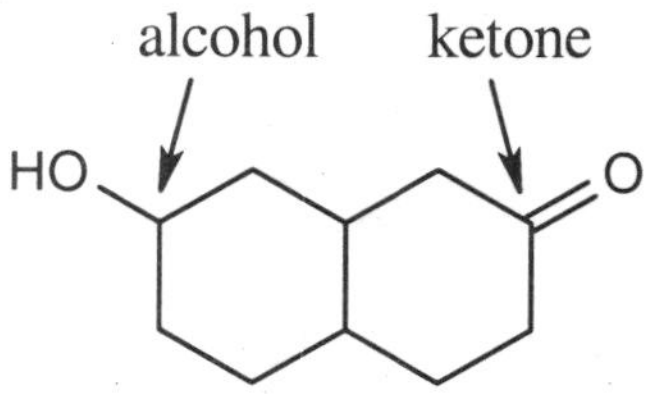

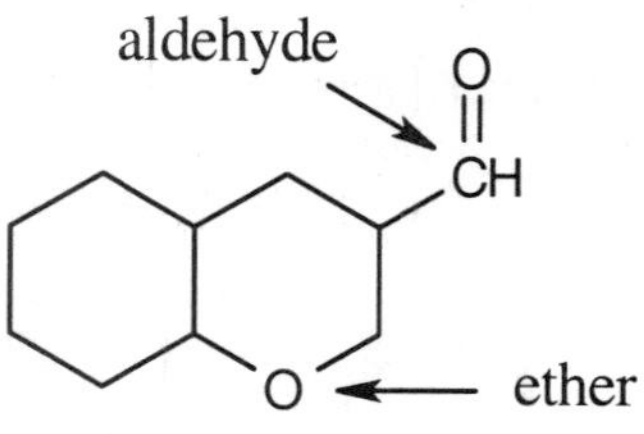

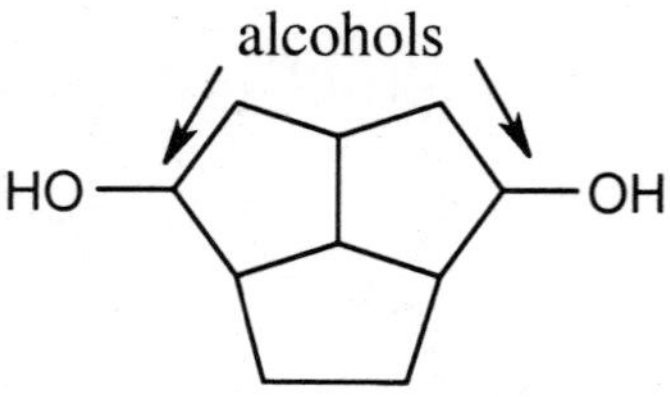

double bonds
ester
ketones
alcohol
ketone
aldehyde
ether
alcohols

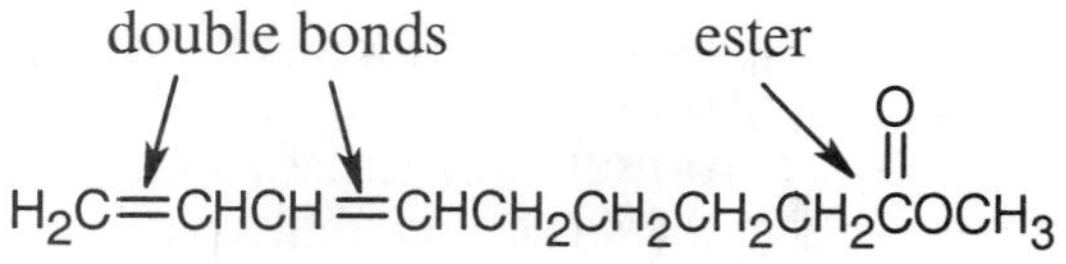

(e) $C_5H_9NO_2$ – 2 degrees of unsaturation.

ketone, amide: $CH_3CH_2C(=O)CH_2C(=O)NH_2$

double bond, nitro group: $H_2C{=}CHCH_2CH_2CH_2N^+(=O){-}O^-$

double bond, carboxylic acid: $H_2C{=}CHCH_2NHCH_2C(=O)OH$

HO–, N–H; amide, alcohol

O, NH_2; ketone, amine, ether

(f) $C_8H_{10}ClNO$ – 4 degrees of unsaturation.

double bonds, amide, halide: H(Cl)$C{=}CHCH{=}CHCH{=}CHCH_2C(=O)NH_2$

double bonds, ether, halide, amine: O, Cl, N–H

aromatic ring, alcohol, halide, amine: CH_3, HO, Cl, CH_2NH_2

halide, double bond, amine, ketone: Cl, H_2N, O

ketone, amine, halide: O, N–H, Cl

6.25 When you have to solve a word problem such as this one, train yourself to read every word and interpret the problem phrase by phrase. For example, "A compound of formula $C_{10}H_{14}$" describes a compound with four degrees of unsaturation. The phrase, " undergoes catalytic hydrogenation", means that H_2 is added to the double bonds. The phrase, " absorbs only 2 molar equivalents of hydrogen", means that only two of the degrees of unsaturation are double bonds (or a triple bond). The other two must be rings.

6.26 A compound of formula $C_{12}H_{13}N$ has as its equivalent hydrocarbon formula $C_{12}H_{12}$, which has 7 degrees of unsaturation. Since two of the degrees of unsaturation are due to rings, the other five are due to double or triple bonds. Thus, $C_{12}H_{13}N$ absorbs 5 equivalents of hydrogen.

6.27

(a) $H_3C(H)C{=}C(H)CH(CH_3)CH_2CH_3$

E-4-Methyl-2-hexene

(b) $CH_3CH(CH_3)CH_2CH_2CH(CH_2CH_3)(H)C{=}C(H)CH_3$

Z-4-Ethyl-7-methyl-2-octene

(c) $H_2C{=}C(CH_2CH_3)CH_2CH_3$

2-Ethyl-1-butene

(d)

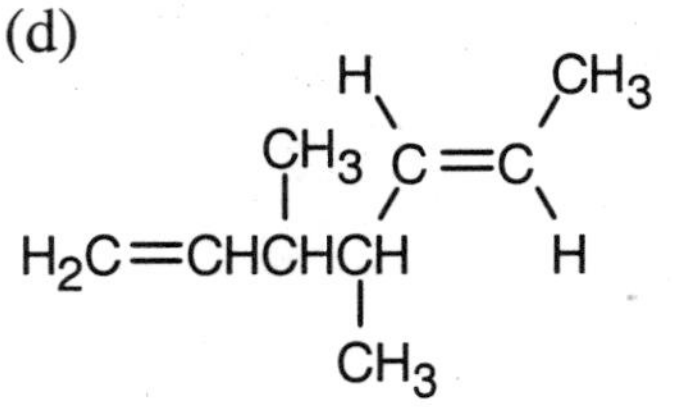

H CH$_3$ / CH$_3$ C=C / H$_2$C=CHCHCH H / CH$_3$

(5*E*)-3,4-Dimethyl-1,5-heptadiene

(e)

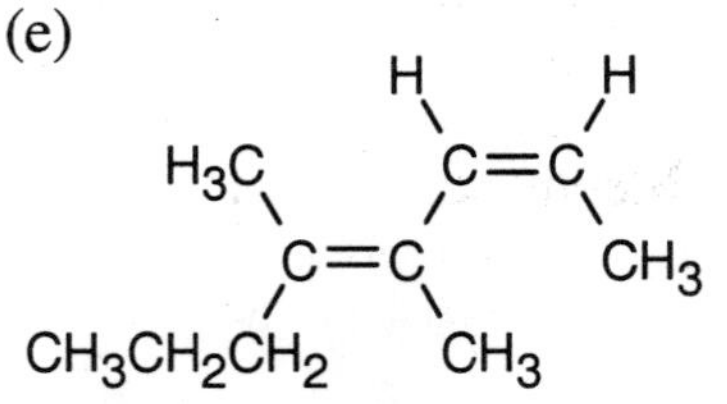

H H / H$_3$C C=C / C=C CH$_3$ / CH$_3$CH$_2$CH$_2$ CH$_3$

(2*Z*,4*E*)-4,5-Dimethyl-2,4-octadiene

(f)

H$_2$C=C=CHCH$_3$

1,2-Butadiene

6.28 Because the longest carbon chain contains 8 carbons and 3 double bonds, ocimene is an *octatriene*. Start numbering at the end that will give the lower number to the first double bond (1,3, 6 is lower than 2,5,7). Number the methyl substituents and, finally, name the compound.

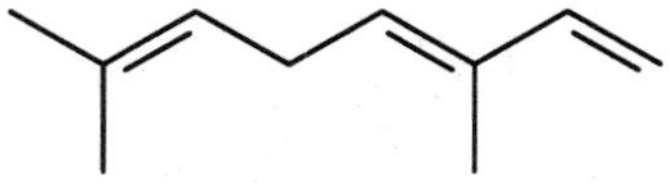

(3*E*)-3,7-Dimethyl-1,3,6-octatriene

6.29

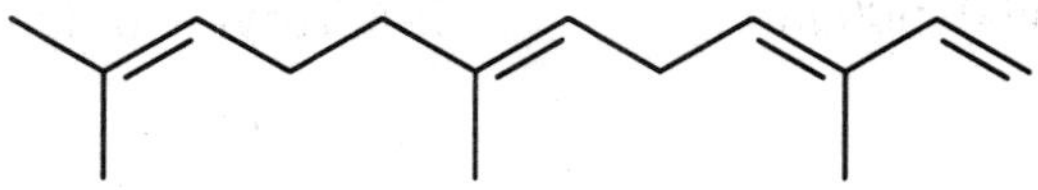

(3*E*,6*E*)-3,7,11-Trimethyl-1,3,6,10-dodecatetraene

6.30

(a)

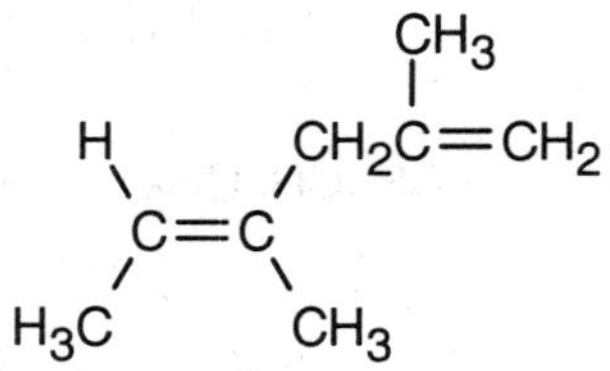

CH$_3$ / H CH$_2$C=CH$_2$ / C=C / H$_3$C CH$_3$

(4*E*)-2,4-Dimethyl-1,4-hexadiene

(b)

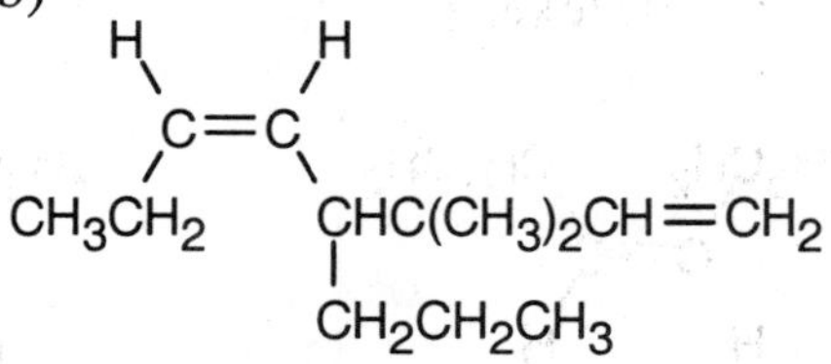

H H / C=C / CH$_3$CH$_2$ CHC(CH$_3$)$_2$CH=CH$_2$ / CH$_2$CH$_2$CH$_3$

cis-3,3-Dimethyl-4-propyl-1,5-octadiene

(c)

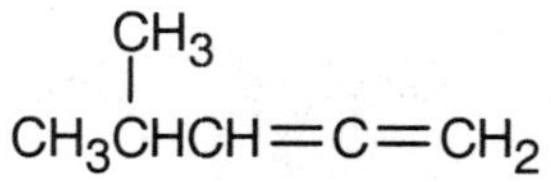

CH$_3$ / CH$_3$CHCH=C=CH$_2$

4-Methyl-1,2-pentadiene

(d)

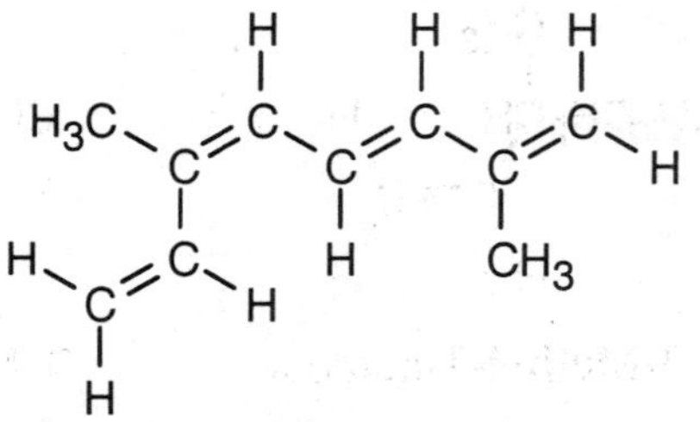

H H H / H$_3$C C=C C=C C=C H / H C=C H H CH$_3$ / H

(3*E*,5*Z*)-2,6-Dimethyl-1,3,5,7-octatetraene

(e)

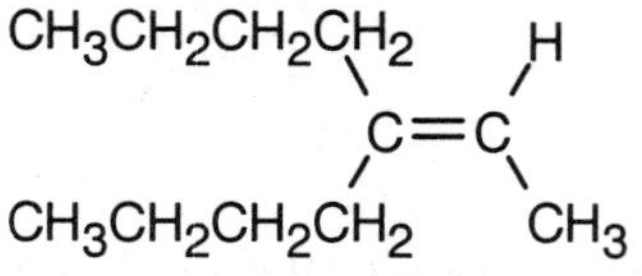

CH$_3$CH$_2$CH$_2$CH$_2$ H / C=C / CH$_3$CH$_2$CH$_2$CH$_2$ CH$_3$

3-Butyl-2-heptene

(f)

(CH$_3$)$_3$C H / C=C / H C(CH$_3$)$_3$

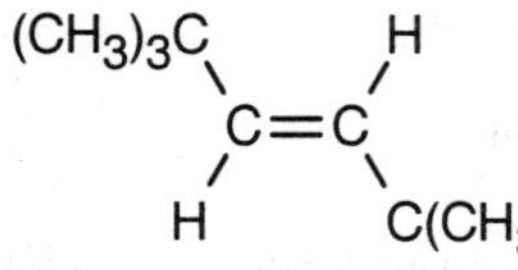

trans-2,2,5,5-Tetramethyl-3-hexene

6.31

Menthene

6.32

$CH_3CH_2CH_2CH{=}CH_2$

1-Pentene

CH_3CH_2 CH_3
C=C
H H

(*Z*)-2-Pentene

CH_3CH_2 H
C=C
H CH_3

(*E*)-2-Pentene

CH_3CH_2 H
C=C
H_3C H

2-Methyl-1-butene

CH_3
CH_3CH H
C=C
H H

3-Methyl-1-butene

H_3C CH_3
C=C
H_3C H

2-Methyl-2-butene

6.33

$CH_3CH_2CH_2CH_2CH{=}CH_2$

1-Hexene

$CH_3CH_2CH_2$ CH_3
C=C
H H

(*Z*)-2-Hexene

$CH_3CH_2CH_2$ H
C=C
H CH_3

(*E*)-2-Hexene

CH_3CH_2 CH_2CH_3
C=C
H H

(*Z*)-3-Hexene

CH_3CH_2 H
C=C
H CH_2CH_3

(*E*)-3-Hexene

$CH_3CH_2CH_2$ H
C=C
H_3C H

2-Methyl-1-pentene

CH_3
CH_3CH_2CH H
C=C
H H

3-Methyl-1-pentene

CH_3
CH_3CHCH_2 H
C=C
H H

4-Methyl-1-pentene

CH_3CH_2 CH_3
C=C
H CH_3

2-Methyl-2-pentene

CH_3CH_2 CH_3
C=C
H_3C H

(*Z*)-3-Methyl-2-pentene

CH_3CH_2 H
C=C
H_3C CH_3

(*E*)-3-Methyl-2-pentene

CH_3
CH_3CH CH_3
C=C
H H

(*Z*)-4-Methyl-2-pentene

```
     CH3                      CH3                      CH3
     |                        |                        |
   CH3CH     H              CH3CH     H              CH3CCH3  H
       \    /                   \    /                   \    /
        C=C                      C=C                      C=C
       /    \                   /    \                   /    \
      H      CH3              H3C     H                 H      H
```

(*E*)-4-Methyl-2-pentene   2,3-Dimethyl-1-butene   3,3-Dimethyl-1-butene

```
   CH3CH2    H              H3C       CH3
        \   /                  \     /
         C=C                    C=C
        /   \                  /     \
   CH3CH2    H              H3C       CH3
```

2-Ethyl-1-butene   2,3-Dimethyl-2-butene

6.34 As expected, the two trans compounds are more stable than their cis counterparts. The cis–trans difference is much more pronounced for the tetramethyl compound, however. Build a model of *cis*-2,2,5,5-tetramethyl-3-hexene and notice the extreme crowding of the methyl groups. Steric strain makes the cis isomer much less stable than the trans isomer and causes cis ΔH°_{hydrog} to have a much larger negative value than trans ΔH°_{hydrog} for the hexene isomers.

6.35 Build models of the two cyclooctenes and notice the large amount of strain in *trans*-cyclooctene relative to *cis*-cyclooctene. This strain causes the trans isomer to be of higher energy and to have a ΔH°_{hydrog} larger than the ΔH°_{hydrog} of the cis isomer.

6.36 Models show that the difference in strain between the two cyclononene isomers is smaller than the difference between the two cyclooctene isomers.This reduced strain is reflected in the fact that the values of ΔH°_{hydrog} for the two cyclononene isomers are relatively close. Nevertheless, the trans isomer is still more strained than the cis isomer.

6.37 The central carbon of allene forms two σ bonds and two π bonds. The central carbon is *sp*-hybridized, and the carbon-carbon bond angle is 180°, indicating linear geometry for the carbons of allene. The terminal $=CH_2$ units are twisted 90° with respect to each other.

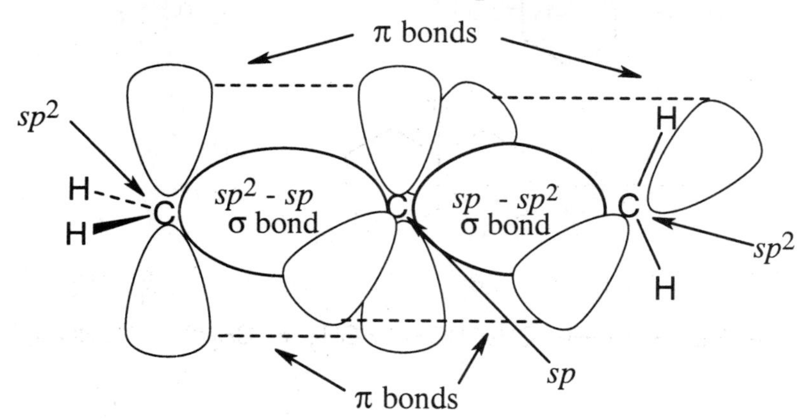

6.38 The heat of hydrogenation for a typical diene is 2 x (ΔH°_{hydrog} of an alkene) = –252 kJ/mol. Thus, allene, with ΔH°_{hydrog} = –295 kJ/mol is 43 kJ/mol higher in energy than a typical diene. Allene is less stable than a typical diene because more heat is released on hydrogenation of allene, whose initial energy level is higher than that of a typical diene.

6.39

(a)

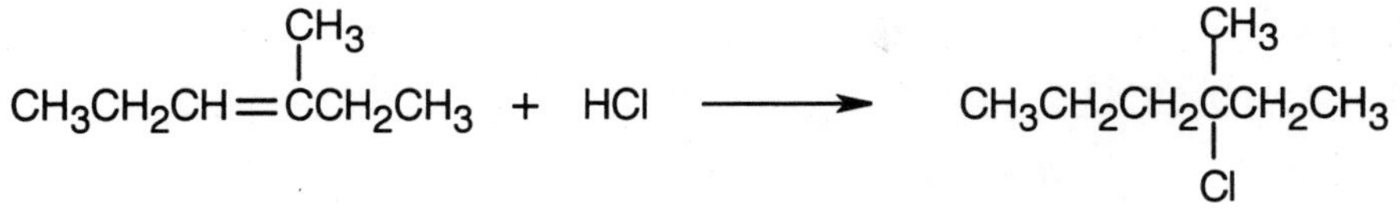

$$CH_3CH_2CH{=}C(CH_3)CH_2CH_3 + HCl \longrightarrow CH_3CH_2CH_2C(CH_3)(Cl)CH_2CH_3$$

(b)

1-Ethylcyclopentene + HBr ⟶ 1-bromo-1-ethylcyclopentane (CH_2CH_3, Br)

(c)

$$CH_3CH_2C(CH_3){=}CHC(CH_3)_3 + HI \longrightarrow CH_3CH_2C(CH_3)(I)CH_2C(CH_3)_3$$

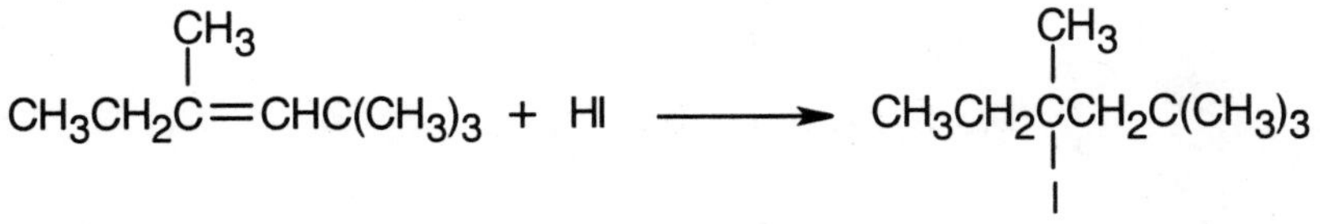

(d)

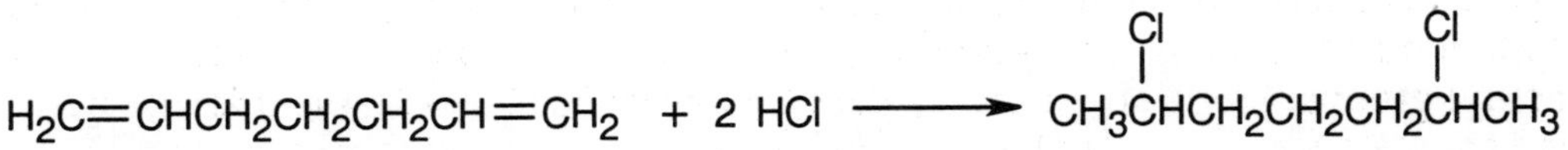

$$H_2C{=}CHCH_2CH_2CH_2CH{=}CH_2 + 2\ HCl \longrightarrow CH_3CH(Cl)CH_2CH_2CH_2CH(Cl)CH_3$$

(e)

3-Methylcyclohexene (CH_3) + HBr ⟶ 1-bromo-3-methylcyclohexane (Br, CH_3) + 1-bromo-2-methylcyclohexane (Br, CH_3)

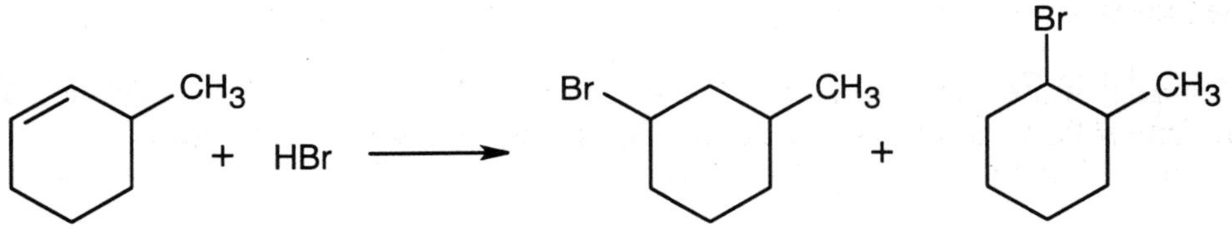

Two products are formed because the two possible carbocations formed are of similar stability.

6.40

(a)

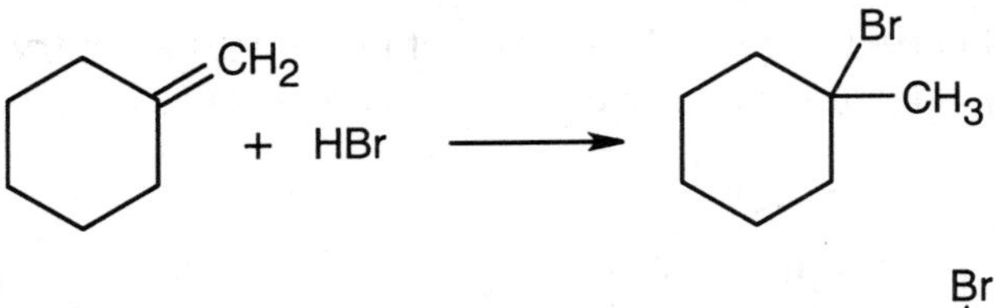

Methylenecyclohexane (CH_2) + HBr ⟶ 1-bromo-1-methylcyclohexane (Br, CH_3)

(b)

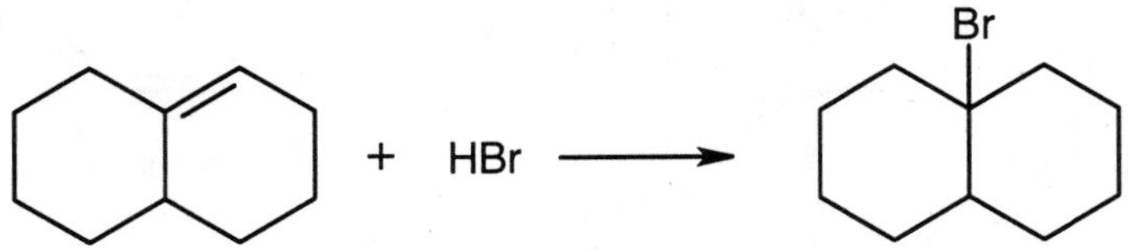

Bicyclic alkene + HBr ⟶ bicyclic bromide (Br at ring-fusion carbon)

(c)

$$CH_3CH{=}CHCH(CH_3)CH_3 + HBr \longrightarrow CH_3CH(Br)CH_2CH(CH_3)CH_3 + CH_3CH_2CH(Br)CH(CH_3)CH_3$$

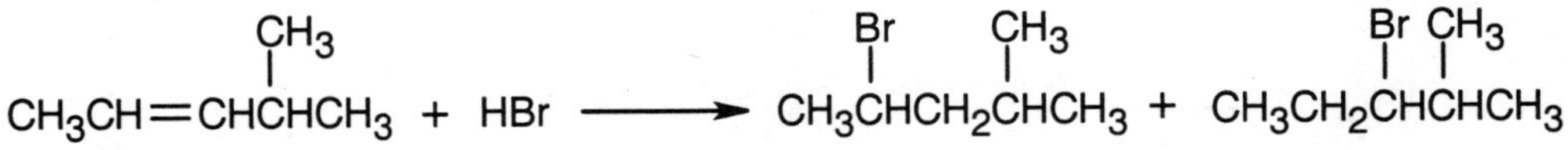

6.41 *Highest priority ————> Lowest Priority*

(a) –I, –Br, $–CH_3$, –H

(b) $–OCH_3$, –OH, –COOH, –H

(c) $–COOCH_3$, –COOH, $–CH_2OH$, $–CH_3$

(d) $–COCH_3$, $–CH_2CH_2OH$, $–CH_2CH_3$, $–CH_3$

(e) $–CH_2Br$, –C≡N, $–CH_2NH_2$, $–CH{=}CH_2$

(f) $–CH_2OCH_3$, $–CH_2OH$, $–CH{=}CH_2$, CH_2CH_3

6.42

(a) High $HOCH_2$, CH_3 High; C=C; Low H_3C, H Low — *Z*

(b) Low HOOC, H Low; C=C; High Cl, OCH_3 High — *Z*

(c) High NC, CH_3 Low; C=C; Low CH_3CH_2, CH_2OH High — *E*

(d) High H_3COOC, $CH{=}CH_2$ High; C=C; Low HOOC, CH_2CH_3 Low — *Z*

6.43

(a)

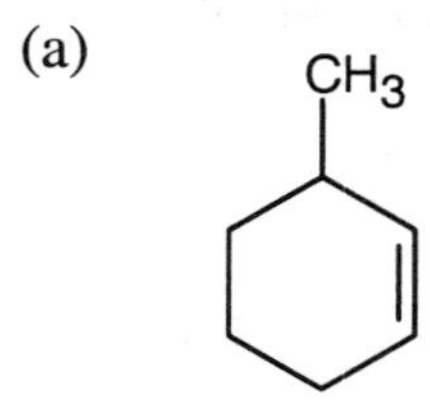

3-Methylcyclohexene

(b)

2,3-Dimethylcyclopentene

(c)

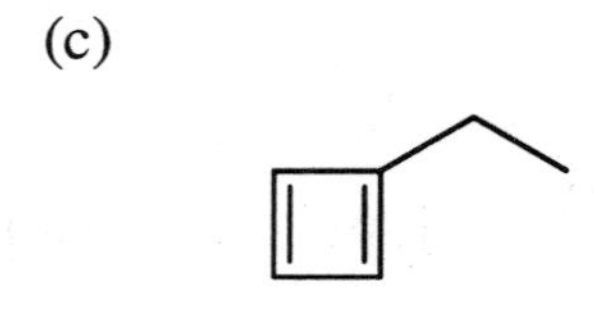

Ethylcyclobutadiene

(d)

1,2-Dimethyl-1,4-cyclohexadiene

(e)

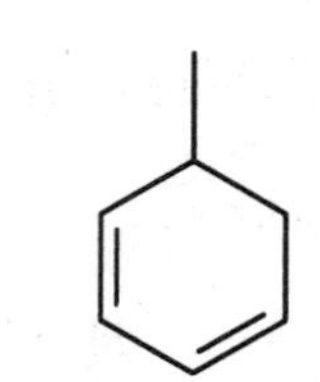

5-Methyl-1,3-cyclohexadiene

(f)

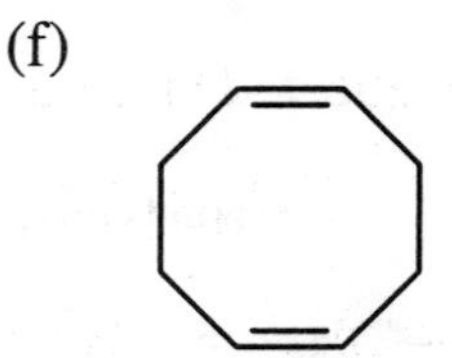

1,5-Cyclooctadiene

6.44

(a) High H_3C (on cyclohexylidene ring) / Low; $C{=}C$; $COOH$ High / H Low — Z (correct)

(b) Low H, High H_3C; $C{=}C$; $CH_2CH{=}CH_2$ High, $CH_2CH(CH_3)_2$ Low — E (correct)

(c) High Br, Low H; $C{=}C$; CH_2NH_2 Low, CH_2NHCH_3 High — Z (incorrect)

(d) High NC, Low $(CH_3)_2NCH_2$; $C{=}C$; CH_3 Low, CH_2CH_3 High — E (correct)

(e) High Br, Low H; $C{=}C$ (cyclopentylidene) — This compound doesn't show *E-Z* isomerism.

(f) Low $HOCH_2$, High H_3COCH_2; $C{=}C$; $COOH$ High, $COCH_3$ Low — E (correct)

6.45 The value of D for the π bond of ethylene is given in Section 6.4.

$$\Delta H^\circ = D_{\text{bonds broken}} - D_{\text{bonds formed}}$$

$$\underset{235\text{ kJ/mol}}{H_2C{=}CH_2} + \underset{432\text{ kJ/mol}}{H{-}Cl} \longrightarrow \underset{420\text{ kJ/mol}}{H{-}CH_2}\underset{338\text{ kJ/mol}}{CH_2{-}Cl} \quad \Delta H^\circ = -91\text{ kJ/mol}$$

$$\underset{235\text{ kJ/mol}}{H_2C{=}CH_2} + \underset{366\text{ kJ/mol}}{H{-}Br} \longrightarrow \underset{420\text{ kJ/mol}}{H{-}CH_2}\underset{285\text{ kJ/mol}}{CH_2{-}Br} \quad \Delta H^\circ = -104\text{ kJ/mol}$$

$$\underset{235\text{ kJ/mol}}{H_2C{=}CH_2} + \underset{298\text{ kJ/mol}}{H{-}I} \longrightarrow \underset{420\text{ kJ/mol}}{H{-}CH_2}\underset{222\text{ kJ/mol}}{CH_2{-}I} \quad \Delta H^\circ = -109\text{ kJ/mol}$$

The reaction between ethylene and HI is the most favorable because its ΔH° has the largest negative value.

6.46

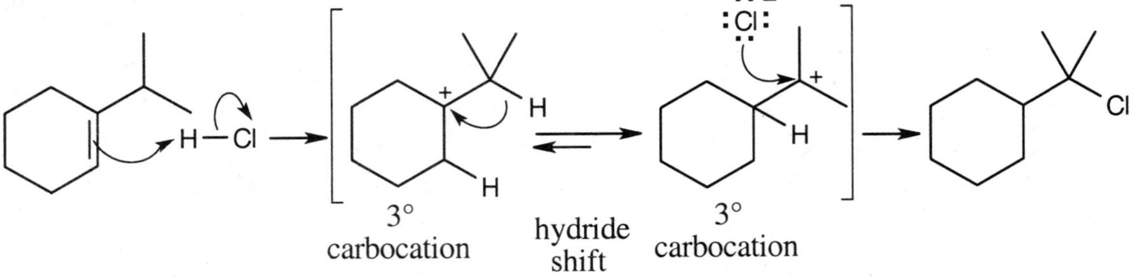

H—Cl

H

H

H

:Cl:⁻

Cl

3° carbocation

hydride shift

3° carbocation

6.47

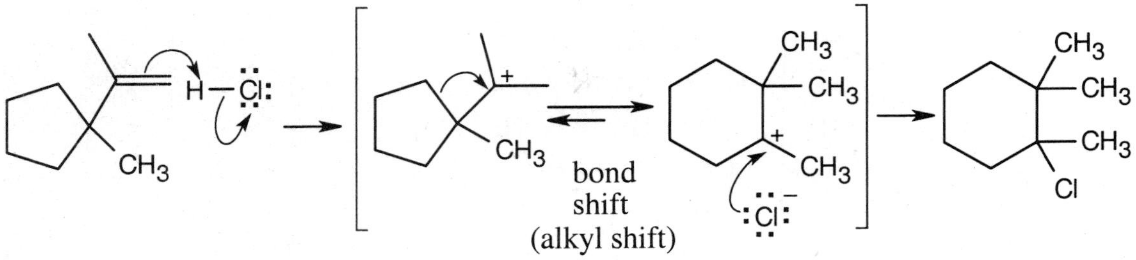

H—Cl:

CH_3

CH_3

CH_3

CH_3

CH_3

CH_3

CH_3

CH_3

Cl

:Cl:⁻

bond shift (alkyl shift)

6.48

H—Br

CH_3

Br

Br^-

CH_3

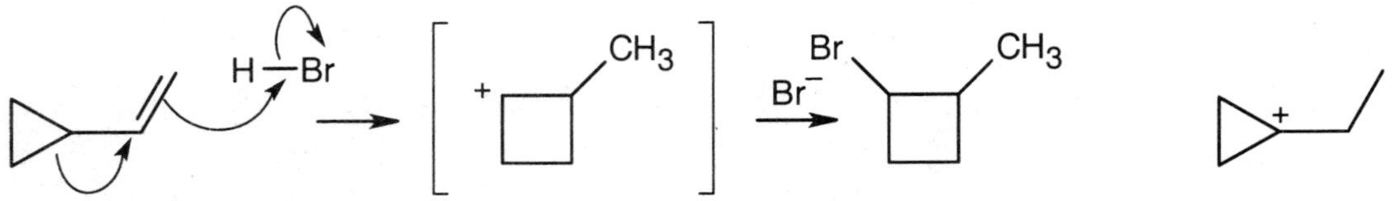

Attack of the π electrons on H^+ yields the carbocation pictured on the far right. A bond shift (alkyl shift) produces the bracketed intermediate, which reacts with Br^- to yield 1-bromo-2-methylcyclobutane.

6.49

(a) $C_{27}H_{46}O$ 5 degrees of unsaturation
(b) $C_{14}H_9Cl_5$ 8 degrees of unsaturation
(c) $C_{20}H_{34}O_5$ 4 degrees of unsaturation
(d) $C_8H_{10}N_4O_2$ 6 degrees of unsaturation
(e) $C_{21}H_{28}O_5$ 8 degrees of unsaturation
(f) $C_{17}H_{23}NO_3$ 7 degrees of unsaturation

6.50 The reaction is exergonic because it is spontaneous. According to the Hammond Postulate, the transition state should resemble the isobutyl cation.

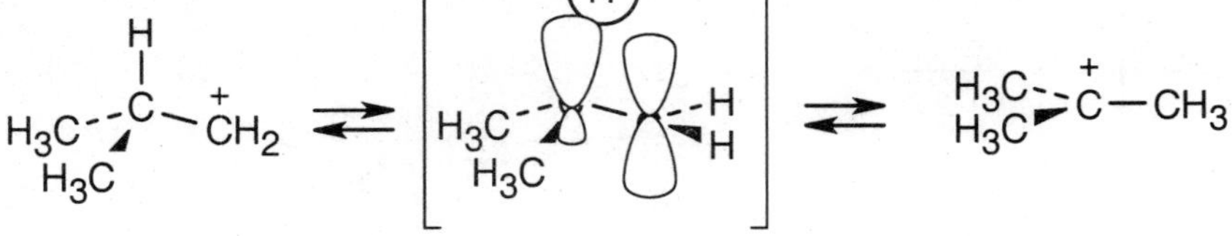

H

H_3C—C—$\overset{+}{C}H_2$

H_3C

H

H_3C

H_3C

H

H

H_3C

H_3C—$\overset{+}{C}$—CH_3

6.51

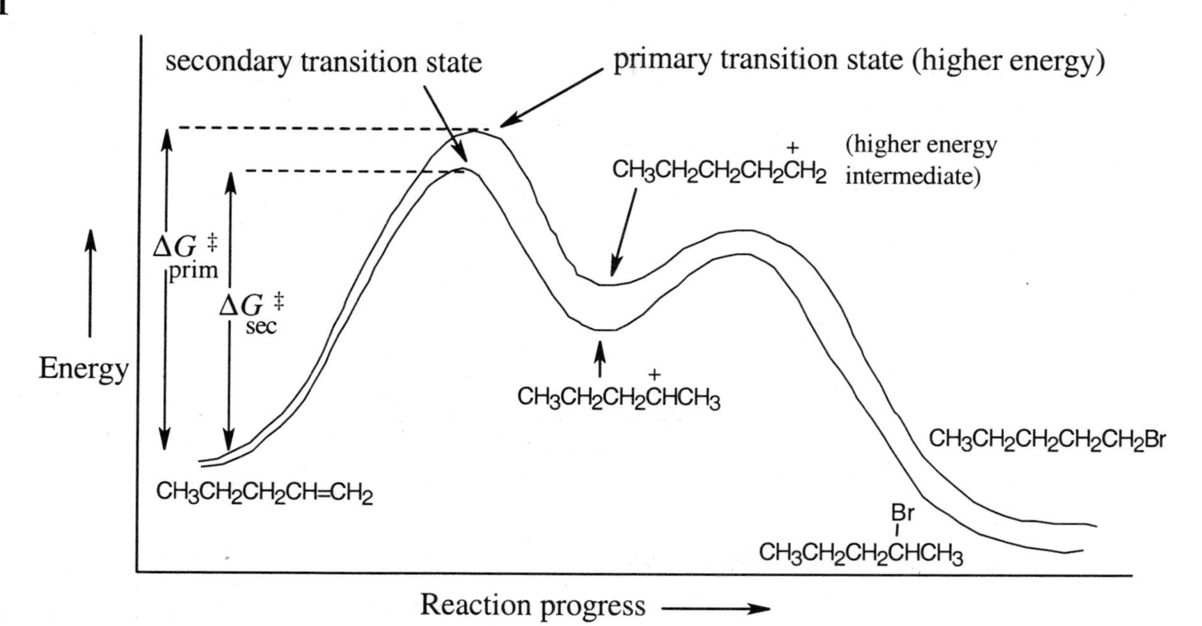

6.52

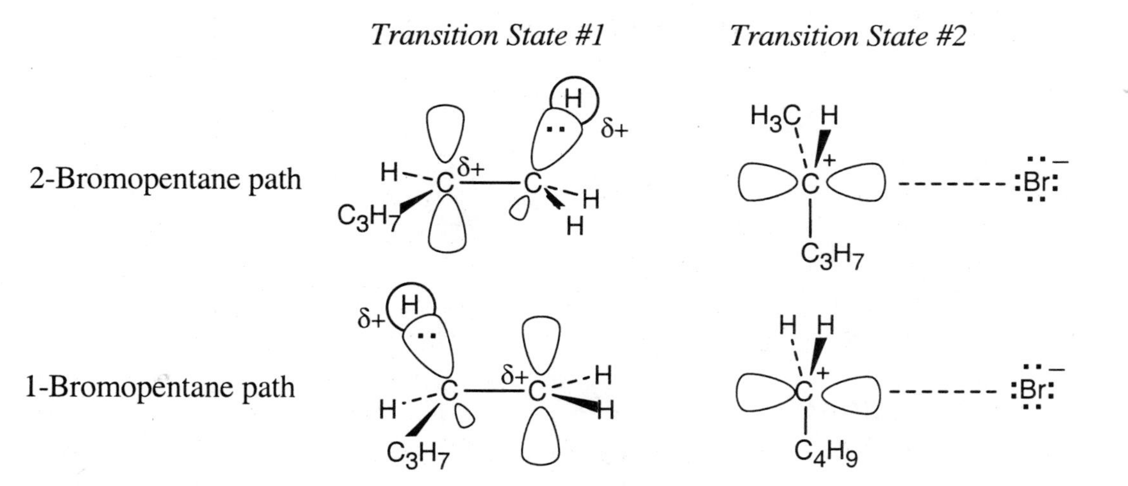

The first step (carbocation formation) is endergonic for both reaction paths, and both transition states resemble the carbocation intermediates. Transition states for the exergonic second step also resemble the carbocation intermediate. Transition state #1 for 1-bromopropane is more like the carbocation intermediate than is transition state #1 for 2-bromopentane.

A Look Ahead

6.53 Reaction of 1-chloropropane with the Lewis acid $AlCl_3$ forms a carbocation. The less stable propylcarbocation undergoes a hydride shift to produce the more stable isopropyl carbocation, which reacts with benzene to give isopropylbenzene.

$$CH_3CH_2CH_2{-}Cl + AlCl_3 \rightleftharpoons [CH_3CH_2CH_2^+ \; AlCl_4^-]$$

$$CH_3CH(H){-}\overset{+}{C}H_2 \rightleftharpoons CH_3\overset{+}{C}HCH_3$$

$$CH_3\overset{+}{C}HCH_3 + C_6H_6 \longrightarrow C_6H_5CH(CH_3)CH_3$$

6.54

(a)

$$CH_3CH_2C(CH_3){=}CHCH_3 + H_2O \xrightarrow[\text{catalyst}]{\text{acid}} CH_3CH_2C(CH_3)(OH)CH_2CH_3$$

(b)

$$\text{methylenecyclohexane } (C_6H_{10}{=}CH_2) + H_2O \xrightarrow[\text{catalyst}]{\text{acid}} \text{1-methylcyclohexanol (OH, } CH_3\text{)}$$

(c)

$$CH_3CH(CH_3)CH_2CH{=}CH_2 + H_2O \xrightarrow[\text{catalyst}]{\text{acid}} CH_3CH(CH_3)CH_2CH(OH)CH_3$$

6.55

$$CH_3CH(CH_3){-}C(CH_3){=}CH_2 \xrightarrow{HBr} CH_3CH(CH_3){-}C(CH_3)(Br){-}CH_3 \xrightarrow[CH_3OH]{KOH} CH_3C(CH_3){=}C(CH_3){-}CH_3$$

2,3-Dimethyl-1-butene — 2-Bromo-2,3-dimethylbutane — 2,3-Dimethyl-2-butene

The product, 2,3-dimethyl-2-butene, is formed by elimination of HBr from 2-bromo-2,3-dimethylbutane. The product has the more substituted double bond.

Molecular Modeling

6.56

```
H       CH3            (CH3)3C     CH3
 \     /                      \   /
  C=C                          C=C
 /     \                      /   \
(CH3)3C  H                   H     H
```

(E)-4,4-Dimethyl-2-pentene (Z)-4,4-Dimethyl-2-pentene

E,Z energy differences depend on the conformations you select for your molecules. A "reasonable" choice of conformation should show that (E)-4,4-dimethyl-2-pentene is the lower energy conformation. Steric repulsion between the methyl and *tert*-butyl groups in the Z isomer leads to large values of C–C=C angles (129° and 133°), compared to values of 124° and 128° in the E isomer.

6.57 The positive carbon becomes less positively charged as the number of alkyl groups bonded to it is increased. This implies transfer of electron density from alkyl groups to the positive carbon or, equivalently, delocalization of positive charge from carbon to the alkyl groups.

6.58 Carbon-carbon distances in the cation are shorter (147.3 pm vs. 153.1 pm in 2-methylpropane), consistent with some double-bond character in these bonds. Out-of-plane C–H distances in the cation are longer (109.0 pm vs. an average distance of 108.7 pm in 2-methylpropane), consistent with bond weakening due to hyperconjugation. In-plane C–H distances in the cation are shorter (107.9 pm), indicating that these bonds do not participate in hyperconjugation.

6.59

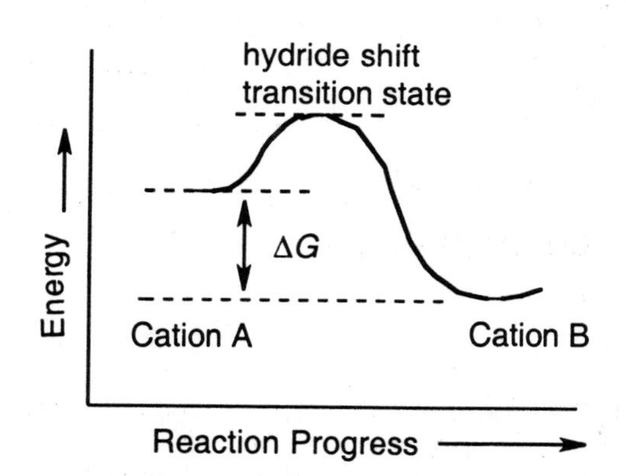

The transition state's geometry is more like that of the reactant carbocation, exactly the result that the Hammond Postulate predicts for an exothermic reaction.

Chapter 7 – Alkenes: Reactions and Synthesis

Chapter Outline

I. Preparation of alkenes (Section 7.1).
  A. Dehydrohalogenation.
    Reaction of an alkyl halide with a strong base forms an alkene, with loss of HX.
  B. Dehydration.
    Treatment of an alcohol with a strong acid forms an alkene, with loss of H_2O.
II. Addition reactions of alkenes (Sections 7.2 – 7.6).
  A. Addition of halogens (Section 7.2).
    1. Br_2 and Cl_2 react with alkenes to yield 1,2-dihaloalkanes.
    2. Reaction occurs with anti stereochemistry – both bromines come from opposite sides of the molecule.
    3. The reaction intermediate is a cyclic bromonium intermediate that is formed in a single step by interaction of an alkene with Br^+.
  B. Halohydrin formation (Section 7.3).
    1. Alkenes add HO–X (X = Br or Cl) when they react with halogens in the presence of H_2O.
    2. The added nucleophile (H_2O) intercepts the bromonium ion to yield a bromohydrin.
    3. Bromohydrin formation is usually achieved by NBS in aqueous DMSO.
    4. Aromatic rings are inert to halohydrin reagents.
  C. Addition of water to alkenes (Sections 7.4).
    1. Hydration.
      a. Water adds to alkenes to yield alcohols in the presence of a strong acid catalyst.
      b. Although this reaction is important industrially, reaction conditions are too severe for most molecules.
    2. Oxymercuration.
      a. Addition of $Hg(OAc)_2$, followed by $NaBH_4$, converts an alkene to an alcohol.
      b. The mechanism of addition proceeds through a mercurinium ion.
      c. The reaction follows Markovnikov regiochemistry.
  D. Hydroboration/oxidation (Section 7.5).
    1. BH_3 adds to an alkene to produce an organoborane.
      Three molecules of alkene add to BH_3 to produce a trialkylborane.
    2. Treatment of the trialkylborane with H_2O_2 forms 3 molecules of an alcohol.
    3. Addition occurs with syn stereochemistry.
    4. Addition occurs with non-Markovnikov regiochemistry.
      Hydroboration is complementary to oxymercuration/reduction.
    5. The mechanism of hydroboration involves a four-center, cyclic transition state.
      a. This transition state explains syn addition.
      b. Stabilization of the transition state by a substituted double-bond carbon also explains non-Markovnikov regiochemistry.
  E. Addition of carbenes (Section 7.6).
    1. A carbene (R_2C:) adds to an alkene to give a cyclopropane.
    2. The reaction occurs in a single step, without intermediates.
    3. Treatment of $HCCl_3$ with KOH forms dichlorocarbene.
      Addition of dichlorocarbene to a double bond is stereospecific, and only cis-dichlorocyclopropanes are formed.
    4. The Simmons-Smith reaction (CH_2I_2, Zn-Cu) produces an unsubstituted cyclopropane via a carbenoid reagent.

III. Reduction and oxidation of alkenes (Sections 7.7 – 7.8).
  A. Reduction of alkenes (Section 7.7).
    1. Catalytic hydrogenation reduces alkenes to saturated hydrocarbons.
    2. The catalysts used are Pt and Pd.
       Catalytic hydrogenation is a heterogeneous process that takes place on the surface of the catalyst.
    3. Hydrogenation occurs with syn stereochemistry.
       The reaction is sensitive to the steric environment around the double bond.
    4. Alkenes are much more reactive than other functional groups.
  B. Oxidation of alkenes (Section 7.8).
    1. Hydroxylation.
      a. OsO_4 causes the addition of two –OH groups to an alkene to form a diol.
      b. Hydroxylation occurs through a cyclic osmate.
      c. The reaction occurs with syn stereochemistry.
    2. Cleavage.
      a. O_3 causes cleavage of an alkene to produce aldehyde and/or ketone fragments.
         The reaction proceeds through a cyclic molozonide, which rearranges to an ozonide that is reduced by Zn.
      b. $KMnO_4$ cleaves alkenes to yield ketones, carboxylic acids or CO_2.
      c. Diols can be cleaved with HIO_4 to produce carbonyl compounds.

IV. Biological alkene addition reactions (Section 7.9).
  A. Many biochemical reactions are additions to alkenes.
  B. These reactions are enzyme-catalyzed.

V. Alkene polymers (Section 7.10).
  A. Many types of polymers can be formed by radical polymerization of alkene monomers.
    1. There are 3 steps in a polymerization reaction.
      a. Initiation involves homolytic cleavage of a weak bond to form a radical
         The radical adds to an alkene to generate an alkyl radical.
      b. The alkyl radical adds to another alkene molecule to yield a second radical.
         This step is repeated many,many times.
      c. Termination occurs when two radical fragments combine.
    2. Mechanisms of radical reactions are shown by using fishhook arrows.
    3. As in electrophilic addition reactions, the more stable radical (more substituted) is formed in preference to the less stable radical.
    4. Chain branches can form when the radical end of a growing chain abstracts a hydrogen atom from the middle of a chain.
      a. Short-chain branching occurs when a hydrogen is abstracted from a position 4 carbons from the end of the chain.
      b. Long-chain branching occurs when a radical at the end of a chain abstracts a hydrogen from the middle of a second chain.
      c. Short-chain branching is much more common than long-chain branching.
  B. Some alkene monomers can form polymers by cationic polymerization.
    1. A strong protic or Lewis catalyst is needed.
    2. Cationic polymerization is more effective when a tertiary carbocation is involved.

Solutions to Problems

7.1

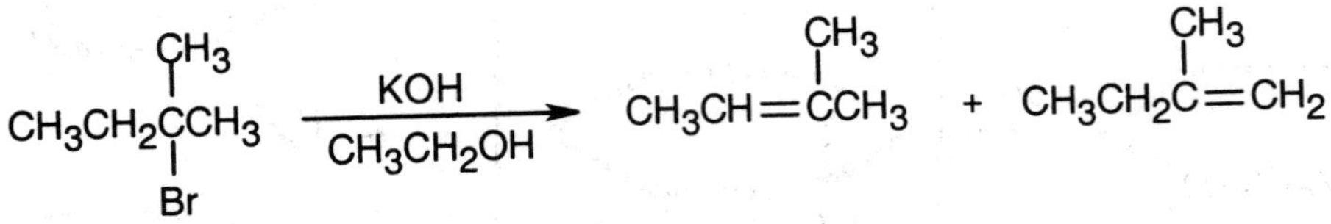

$$CH_3CH_2C(CH_3)(Br)CH_3 \xrightarrow[CH_3CH_2OH]{KOH} CH_3CH{=}C(CH_3)CH_3 + CH_3CH_2C(CH_3){=}CH_2$$

Dehydrobromination can occur in two directions to yield a mixture of products.

7.2

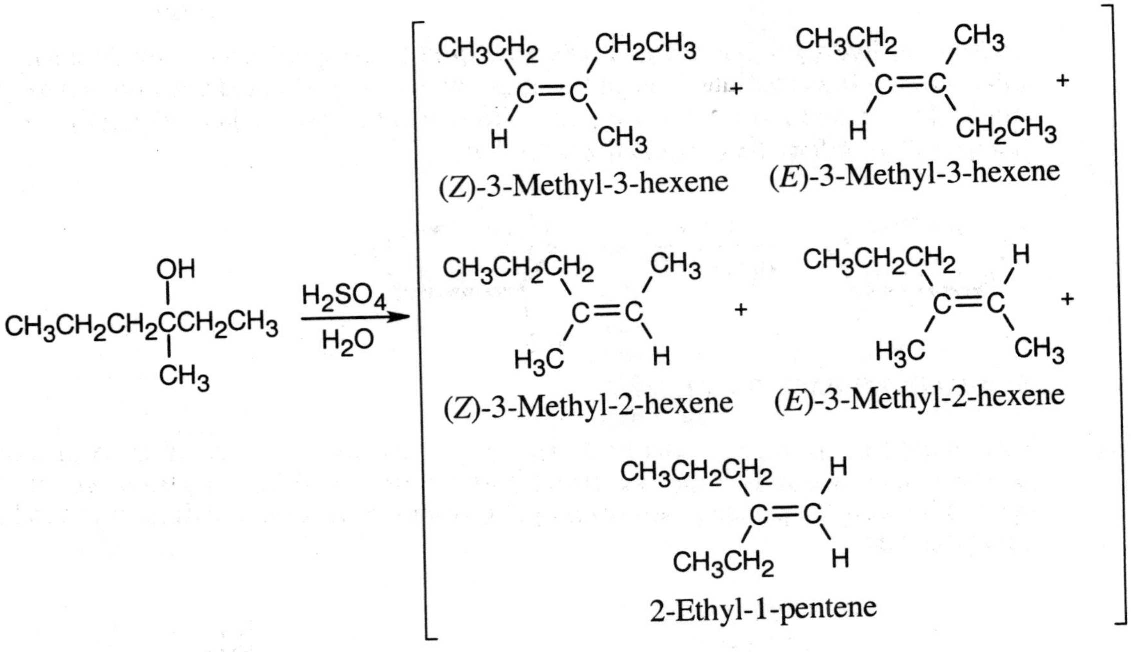

$$CH_3CH_2CH_2C(OH)(CH_3)CH_2CH_3 \xrightarrow[H_2O]{H_2SO_4}$$

(*Z*)-3-Methyl-3-hexene + (*E*)-3-Methyl-3-hexene + (*Z*)-3-Methyl-2-hexene + (*E*)-3-Methyl-2-hexene + 2-Ethyl-1-pentene

Five alkene products, including *E*, *Z* isomers, might be obtained by dehydration of 3–methyl–3–hexanol.

7.3

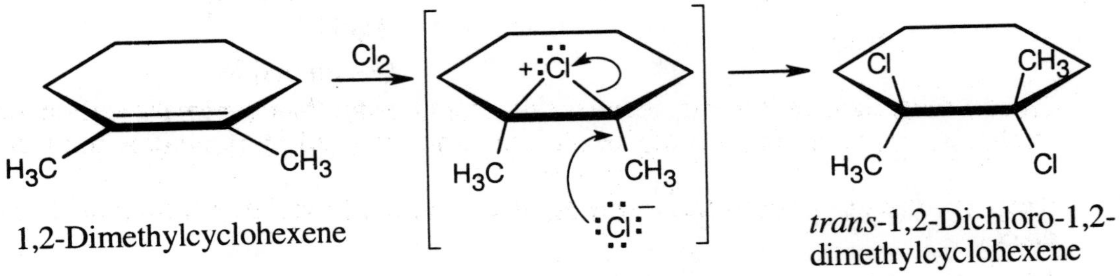

1,2-Dimethylcyclohexene $\xrightarrow{Cl_2}$ [chloronium ion intermediate, attacked by $:\ddot{Cl}:^-$] $\rightarrow$ *trans*-1,2-Dichloro-1,2-dimethylcyclohexene

The chlorines are trans to one another in the product, as are the methyl groups.

7.4

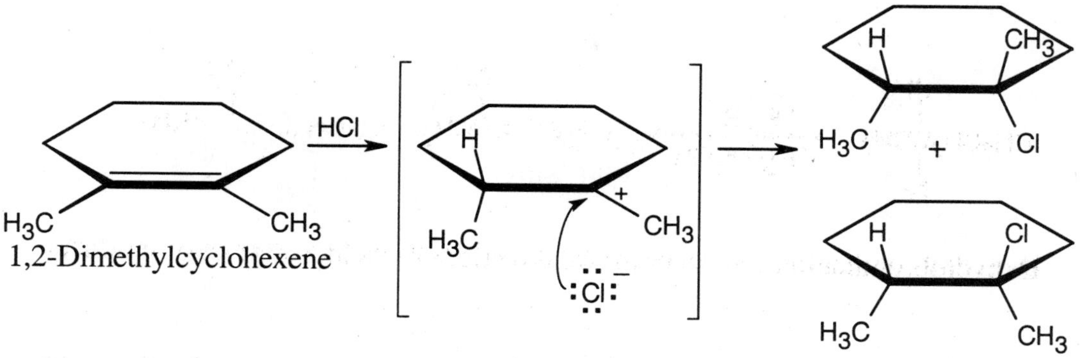

1,2-Dimethylcyclohexene

HCl

Addition of hydrogen halides involves formation of an open carbocation, not a cyclic halonium ion intermediate. The carbocation, which is sp^2-hybridized and planar, can be attacked by chloride from either top or bottom, yielding products in which the two methyl groups can be either cis or trans to each other.

7.5

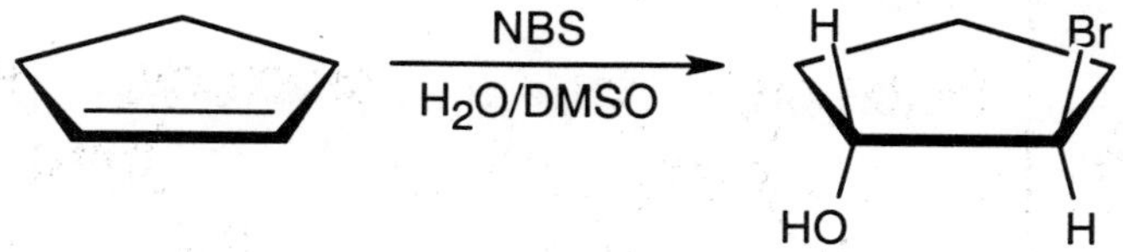

NBS, H_2O/DMSO

Br and OH are trans in the product.

7.6 NBS is the source of the electrophilic Br^+ ion in bromohydrin formation. Attack of the alkene π electrons on Br^+ forms a cyclic bromonium ion. When this bromonium ion is opened by water, a partial positive charge develops at the carbon whose bond to bromine is being cleaved.

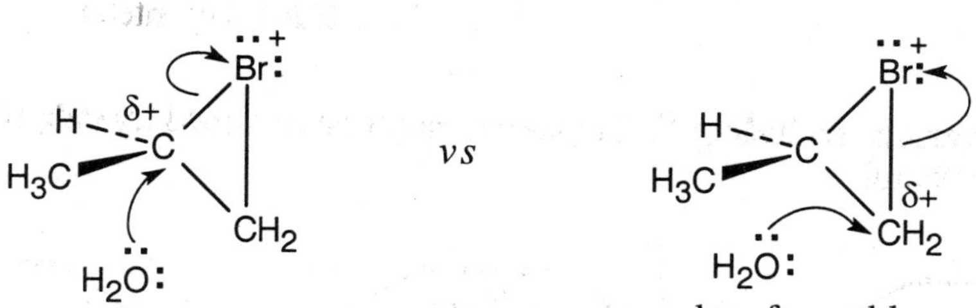

vs

less favorable

Since a secondary carbon can stabilize this charge better than a primary carbon, opening of the bromonium ion occurs at the secondary carbon to yield Markovnikov product.

7.7 Keep in mind that oxymercuration is equivalent to Markovnikov addition of H_2O to an alkene.

(a)

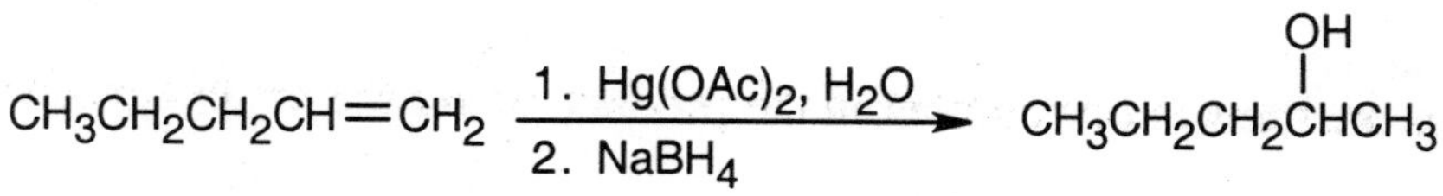

$$CH_3CH_2CH_2CH{=}CH_2 \xrightarrow[\text{2. NaBH}_4]{\text{1. Hg(OAc)}_2,\ H_2O} CH_3CH_2CH_2CH(OH)CH_3$$

(b)

$$CH_3CH_2CH{=}C(CH_3)CH_3 \xrightarrow[\text{2. NaBH}_4]{\text{1. Hg(OAc)}_2,\ H_2O} CH_3CH_2CH_2C(CH_3)(OH)CH_3$$

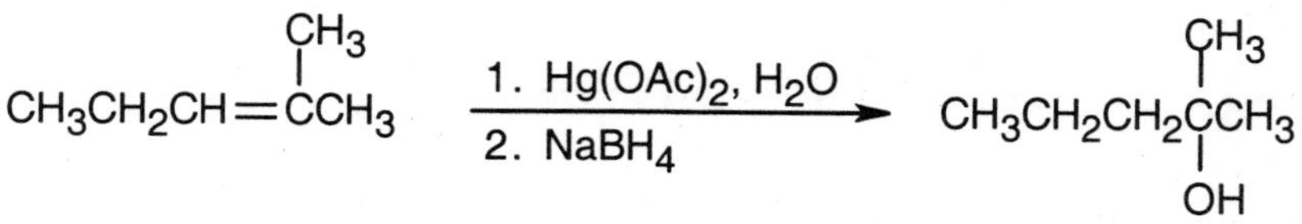

7.8 Remember to think backwards to determine the possible alkene starting materials for the alcohols pictured.

(a)

$CH_3C(CH_3){=}CHCH_2CH_2CH_3$ *or* $H_2C{=}C(CH_3)CH_2CH_2CH_2CH_3$ $\xrightarrow[\text{2. NaBH}_4]{\text{1. Hg(OAc)}_2\text{, H}_2\text{O}}$ $CH_3C(OH)(CH_3)CH_2CH_2CH_2CH_3$

(b)

vinylcyclohexane $\xrightarrow[\text{2. NaBH}_4]{\text{1. Hg(OAc)}_2\text{, H}_2\text{O}}$ 1-cyclohexylethanol (OH)

Oxymercuration occurs with Markovnikov orientation.

7.9 Recall the mechanism of hydroboration and note that the hydrogen added to the double bond comes from borane. The product of hydroboration with BD_3 has deuterium bonded to the more substituted carbon; –D and –OH are cis to one another.

1-methylcyclopentene (CH_3) $\xrightarrow[\text{2. H}_2\text{O}_2\text{, }^{-}\text{OH}]{\text{1. BD}_3\text{, THF}}$ product with HO, D (cis), H, CH_3

7.10 As described in Practice Problem 7.2, the strategy in this sort of problem begins with a look backward. In more complicated syntheses this approach is essential, but even in problems in which the functional group(s) in the starting material and the reagents are known, this approach is productive.

All the products in this problem result from hydroboration of a double bond. The –OH group is bonded to the less substituted carbon of the double bond in the starting material.

(a)

$(CH_3)_2CHCH{=}CH_2$ $\xrightarrow[\text{2. H}_2\text{O}_2\text{, }^{-}\text{OH}]{\text{1. BH}_3\text{, THF}}$ $(CH_3)_2CHCH_2CH_2OH$

(b)

$(CH_3)_2C{=}CHCH_3$ $\xrightarrow[\text{2. H}_2\text{O}_2\text{, }^{-}\text{OH}]{\text{1. BH}_3\text{, THF}}$ $(CH_3)_2CHCH(OH)CH_3$

This product can also result from oxymercuration of the starting material in (a).

(c)

methylenecyclohexane ($=CH_2$) $\xrightarrow[\text{2. H}_2\text{O}_2\text{, }^{-}\text{OH}]{\text{1. BH}_3\text{, THF}}$ cyclohexyl–CH_2OH

7.11 The drawings below show the transition states resulting from addition of BH_3 to the double bond of the cycloalkene. Addition can occur on either side of the double bond.

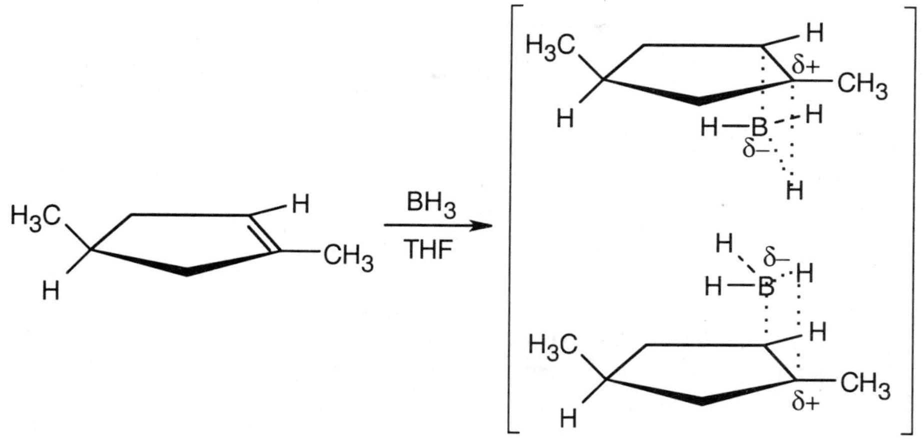

H_3C H H CH_3 $\xrightarrow[THF]{BH_3}$ [transition states: H_3C, H, δ+, CH_3, H—B, δ−, H]

Reaction of the two neutral alkylborane adducts with hydrogen peroxide gives two alcohol isomers. In one isomer, the two methyl groups have a cis relationship, and in the other isomer they have a trans relationship.

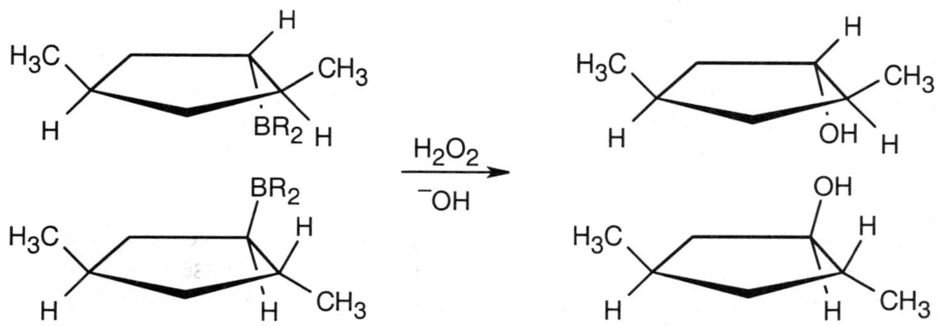

$\xrightarrow[^{-}OH]{H_2O_2}$

7.12 Reaction of a double bond with chloroform under basic conditions gives a product with a cyclopropane ring in which one of the carbons has two chlorine atoms bonded to it. Reaction of a double bond with CH_2I_2 yields a product with a cyclopropane ring that has a $-CH_2-$ group.

(a)

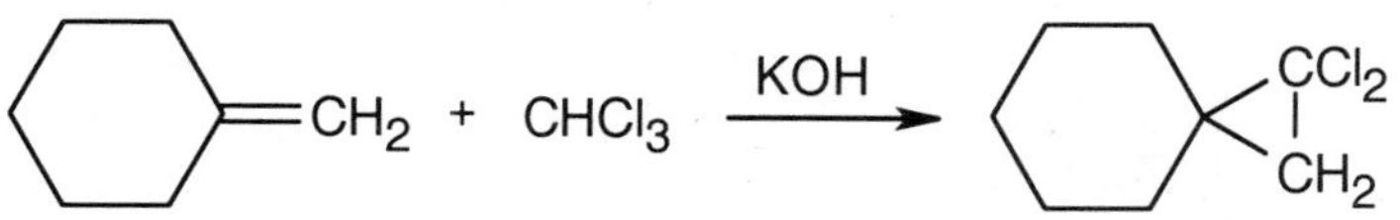

$=CH_2$ + $CHCl_3$ $\xrightarrow{KOH}$ CCl_2, CH_2

(b)

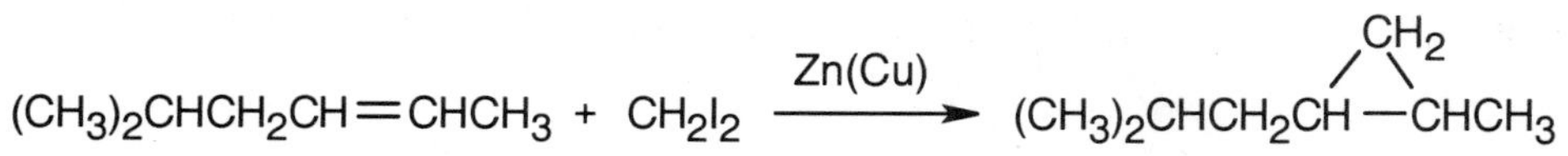

$(CH_3)_2CHCH_2CH{=}CHCH_3 + CH_2I_2 \xrightarrow{Zn(Cu)} (CH_3)_2CHCH_2CH{-}CHCH_3$ (bridged by CH_2)

Depending on the stereochemistry of the double bond of the alkene in (b), two different isomers can be formed.

7.13 Catalytic hydrogenation produces alkanes from alkenes.

(a)

$$(CH_3)_2C{=}CHCH_2CH_3 \xrightarrow[\text{Pd/C in ethanol}]{H_2} (CH_3)_2CHCH_2CH_2CH_3$$

2-Methyl-2-pentene → 2-Methylpentane

(b)

3,3-Dimethylcyclopentene $\xrightarrow[\text{Pd/C in ethanol}]{H_2}$ 1,1-Dimethylcyclopentane

7.14 Reaction of an alkene with OsO_4, followed by treatment with $NaHSO_3$, yields a diol product. To choose a starting material for these products, pick an alkene that has a double bond between the diol carbons.

(a)

1-Methylcyclohexene $\xrightarrow[\text{pyridine}]{OsO_4}$ [cyclic osmate ester] $\xrightarrow[H_2O]{NaHSO_3}$ cis-1-methyl-1,2-cyclohexanediol

(b)

$$CH_3CH_2CH{=}C(CH_3)_2 \xrightarrow[\text{2. } NaHSO_3,\ H_2O]{\text{1. } OsO_4,\text{ pyridine}} CH_3CH_2CH(OH){-}C(OH)(CH_3)CH_3$$

2-Methyl-2-pentene

(c)

$$CH_2{=}CHCH{=}CH_2 \xrightarrow[\text{2. } NaHSO_3,\ H_2O]{\text{1. } OsO_4,\text{ pyridine}} HOCH_2CH(OH)CH(OH)CH_2OH$$

1,3-Butadiene

7.15 Both sets of reactants cleave double bonds. Aqueous $KMnO_4$ produces a carboxylic acid from a double bond carbon that is monosubstituted and a ketone from a double bond carbon that is disubstituted. Ozone produces an aldehyde from a double bond carbon that is monosubstituted and a ketone from a double bond carbon that is disubstituted. If the double bond is part of a ring, both carbonyl groups occur in the same product molecule.

(a)

1-Methylcyclohexene $\xrightarrow[H_3O^+]{KMnO_4}$ $CH_3C(=O)CH_2CH_2CH_2CH_2C(=O)OH$

(b)

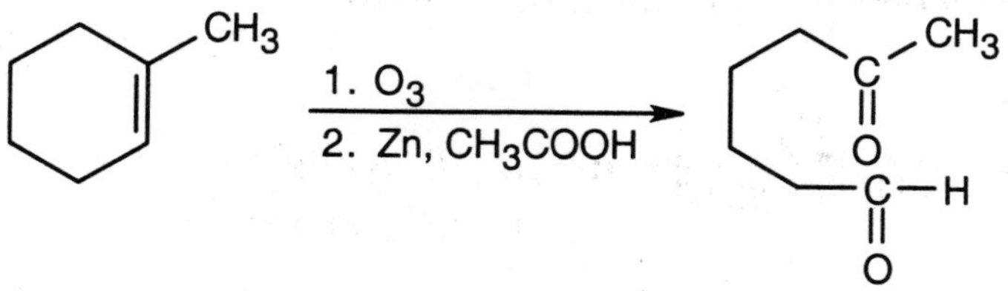

$$\text{1-methylcyclohexene} \xrightarrow[\text{2. Zn, } CH_3COOH]{\text{1. } O_3} \text{CH}_3\text{C(=O)CH}_2\text{CH}_2\text{CH}_2\text{CH}_2\text{C(=O)H}$$

7.16 Orient the fragments so that the oxygens point toward each other. Remove the oxygens, and draw a double bond between the remaining carbons.

(a)

$$(CH_3)_2C{=}CH_2 \xrightarrow[\text{2. Zn, } H_3O^+]{\text{1. } O_3} (CH_3)_2C{=}O \; + \; O{=}CH_2$$

(b)

$$CH_3CH_2CH{=}CHCH_2CH_3 \xrightarrow[\text{2. Zn, } H_3O^+]{\text{1. } O_3} CH_3CH_2CH{=}O \; + \; O{=}CHCH_2CH_3$$

7.17 Find the smallest repeating unit in each polymer. This is the monomer unit.

	Monomer	*Polymer*
(a)	$H_2C{=}CHOCH_3$	$\sim CH_2{-}CH(OCH_3){-}CH_2{-}CH(OCH_3){-}CH_2{-}CH(OCH_3) \sim$
(b)	$ClHC{=}CHCl$	$\sim CH(Cl){-}CH(Cl){-}CH(Cl){-}CH(Cl){-}CH(Cl){-}CH(Cl) \sim$

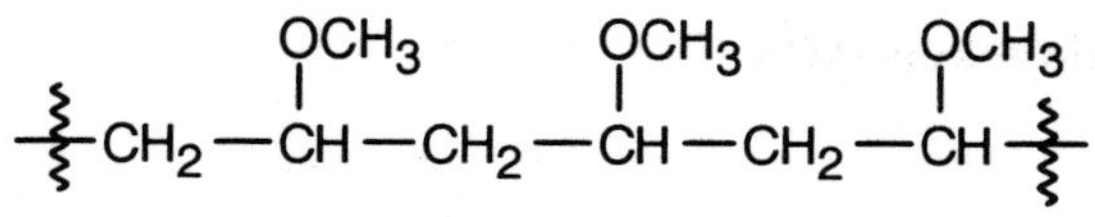

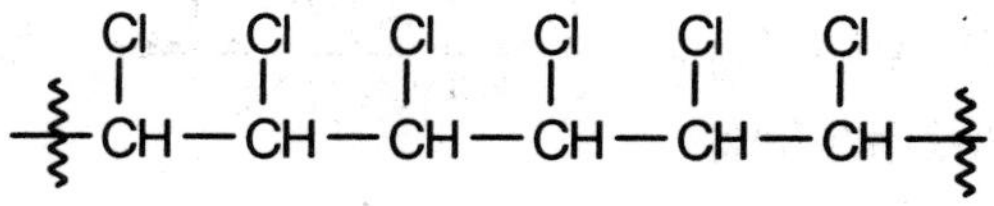

7.18 One radical abstracts a hydrogen atom from another radical, and the remaining electrons create a double bond.

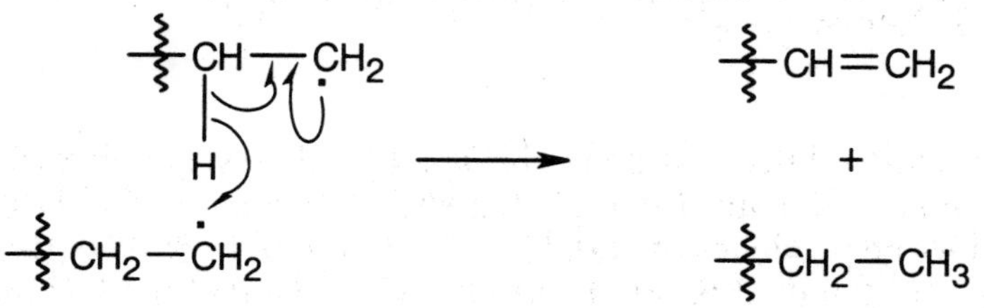

$$\sim CH(H){-}\dot{C}H_2 \; + \; \sim CH_2{-}\dot{C}H_2 \longrightarrow \sim CH{=}CH_2 \; + \; \sim CH_2{-}CH_3$$

7.19 An acid catalyst forms the initial carbocation, which is attacked by the π electrons of an alkene monomer. The resulting carbocation can be attacked by the π electrons of a second alkene monomer. The process is repeated a vast number of times to generate the polymer.

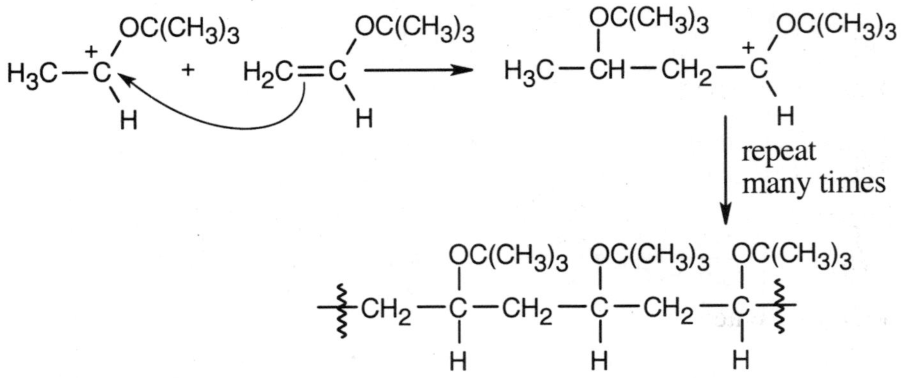

$$H_3C-\overset{+}{C}H(OC(CH_3)_3) + H_2C{=}CH(OC(CH_3)_3) \longrightarrow H_3C-CH(OC(CH_3)_3)-CH_2-\overset{+}{C}H(OC(CH_3)_3)$$

repeat many times

$$-CH_2-CH(OC(CH_3)_3)-CH_2-CH(OC(CH_3)_3)-CH_2-CH(OC(CH_3)_3)-$$

Visualizing Chemistry

7.20

(a)

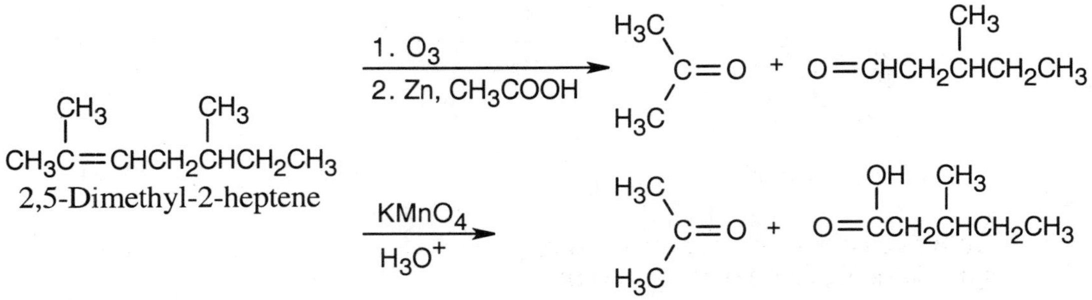

2,5-Dimethyl-2-heptene: $CH_3C(CH_3){=}CHCH_2CH(CH_3)CH_2CH_3$

$\xrightarrow{\text{1. } O_3;\ \text{2. Zn, } CH_3COOH}$ $(H_3C)_2C{=}O + O{=}CHCH_2CH(CH_3)CH_2CH_3$

$\xrightarrow{KMnO_4,\ H_3O^+}$ $(H_3C)_2C{=}O + O{=}C(OH)CH_2CH(CH_3)CH_2CH_3$

(b)

3,3-Dimethylcyclopentene

$\xrightarrow{\text{1. } O_3;\ \text{2. Zn, } CH_3COOH}$ $HC(=O)C(CH_3)_2CH_2CH_2C(=O)H$

$\xrightarrow{KMnO_4,\ H_3O^+}$ $HOC(=O)C(CH_3)_2CH_2CH_2C(=O)OH$

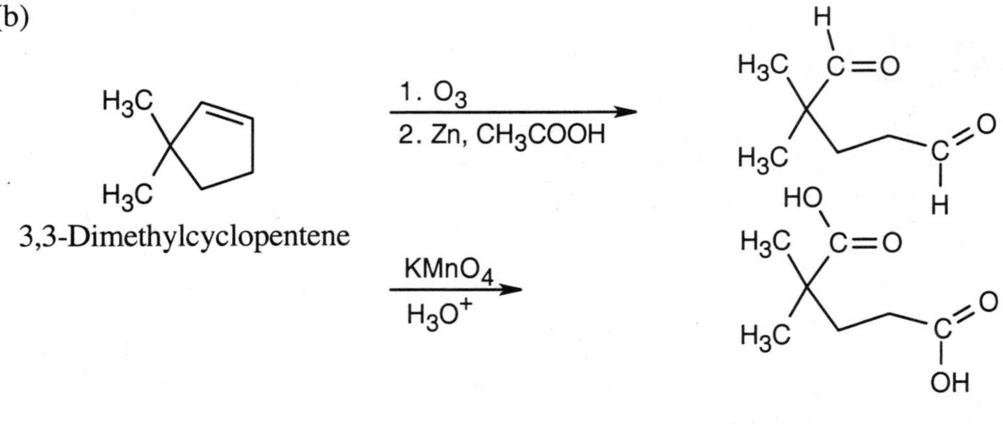

7.21

(a)

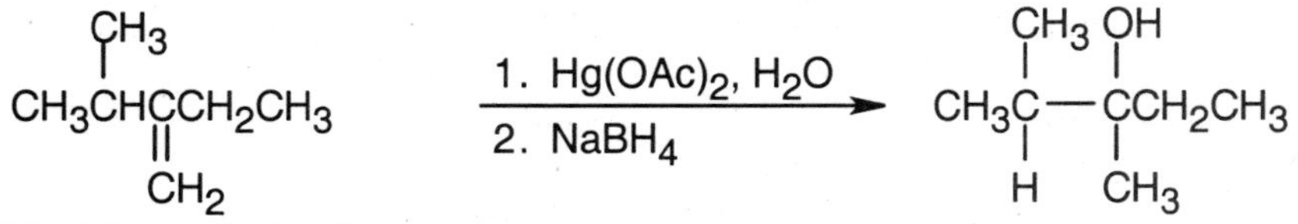

$CH_3CHC(=CH_2)CH_2CH_3$ (2-Ethyl-3-methyl-1-butene) $\xrightarrow[\text{2. } NaBH_4]{\text{1. } Hg(OAc)_2,\ H_2O}$ $CH_3CH(CH_3)C(OH)(CH_3)CH_2CH_3$

or

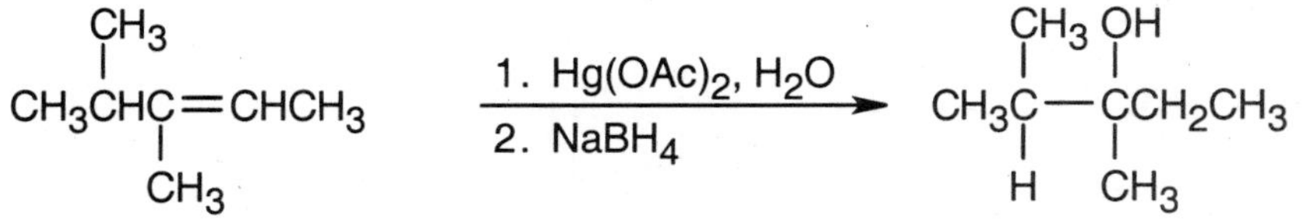

$CH_3CH(CH_3)C(CH_3){=}CHCH_3$ (3,4-Dimethyl-2-pentene) $\xrightarrow[\text{2. } NaBH_4]{\text{1. } Hg(OAc)_2,\ H_2O}$ $CH_3CH(CH_3)C(OH)(CH_3)CH_2CH_3$

Only oxymercuration/reduction can be used to produce an alcohol that has –OH bonded to the more substituted carbon. A third alkene, 2,3-dimethyl-2-pentene, gives a mixture of tertiary alcohols when treated with either BH_3 or $Hg(OAc)_2$.

(b)

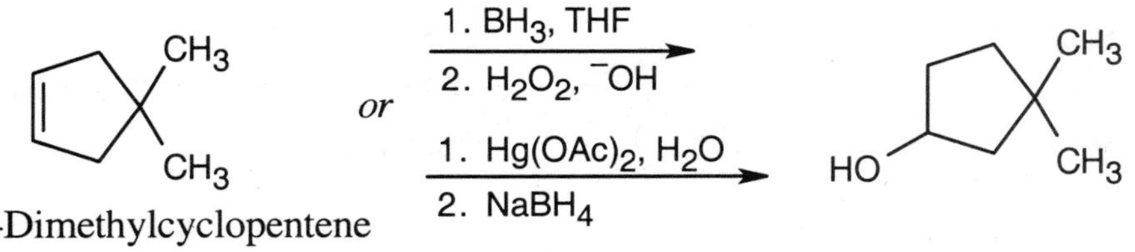

4,4-Dimethylcyclopentene $\xrightarrow[\text{2. } H_2O_2,\ ^-OH]{\text{1. } BH_3,\ THF}$ *or* $\xrightarrow[\text{2. } NaBH_4]{\text{1. } Hg(OAc)_2,\ H_2O}$ 3,3-dimethylcyclopentanol (HO, CH_3, CH_3)

Both hydroboration/oxidation and oxymercuration yield the same alcohol product from the symmetrical alkene starting material.

7.22

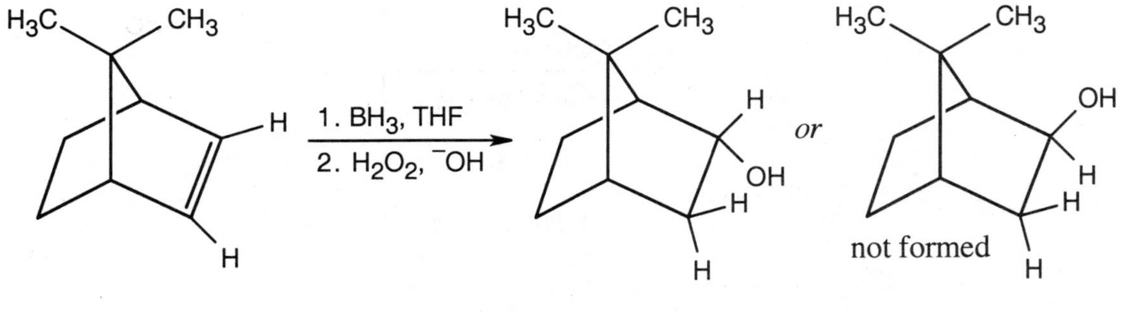

(bicyclic alkene with H_3C, CH_3 on bridge; H, H on double bond) $\xrightarrow[\text{2. } H_2O_2,\ ^-OH]{\text{1. } BH_3,\ THF}$ (alcohol: H, OH, H, H) *or* (alcohol: OH, H, H, H) not formed

On paper, it is possible to draw two possible alcohols that might be formed by hydroboration/oxidation of the alkene shown. One product results from addition to the top face of the double bond (not formed), and the other product results from addition to the bottom face of the double bond (formed). The second product shown above is not produced because a methyl group on the bridge of the bicyclic ring system blocks formation of the bulky alkylborane intermediate.

Additional Problems

7.23

(a) $C_6H_5CH{=}CH_2 \xrightarrow{H_2/Pd} C_6H_5CH_2CH_3$

(b) $C_6H_5CH{=}CH_2 \xrightarrow{Br_2} C_6H_5CHBrCH_2Br$

(c) $C_6H_5CH{=}CH_2 \xrightarrow{HBr} C_6H_5CHBrCH_3$

(d) $C_6H_5CH{=}CH_2 \xrightarrow[\text{2. NaHSO}_3\text{, H}_2\text{O}]{\text{1. OsO}_4\text{, pyridine}} C_6H_5CH(OH)CH_2OH$

(e) $C_6H_5CH{=}CH_2 \xrightarrow{D_2/Pd} C_6H_5CHDCH_2D$

7.24

(a)

$CH_3CH_2CH_2CH_2C(CH_3){=}CH_2$	2-Methyl-1-hexene
$CH_3CH_2CH_2CH{=}C(CH_3)_2$	2-Methyl-2-hexene
$CH_3CH_2CH{=}CHCH(CH_3)_2$	2-Methyl-3-hexene
$CH_3CH{=}CHCH_2CH(CH_3)_2$	5-Methyl-2-hexene
$H_2C{=}CHCH_2CH_2CH(CH_3)_2$	5-Methyl-1-hexene

$\xrightarrow{H_2/Pd}$ $CH_3CH_2CH_2CH_2CH(CH_3)_2$
2-Methylhexane

(b)

3,3-Dimethylcyclohexene

4,4-Dimethylcyclohexene

$\xrightarrow{H_2/Pd}$

1,1-Dimethylcyclohexane

(c)

$$\underset{\text{5-Methyl-2-hexene}}{CH_3CH{=}CHCH_2CH(CH_3)_2} \xrightarrow{Br_2/CH_2Cl_2} \underset{\text{2,3-Dibromo-5-methylhexane}}{CH_3CH(Br){-}CH(Br)CH_2CH(CH_3)_2}$$

(d)

$$\underset{\text{1-Pentene}}{CH_3CH_2CH_2CH{=}CH_2} \xrightarrow[\text{2. } NaBH_4]{\text{1. } Hg(OAc)_2,\ H_2O} \underset{\text{2-Pentanol}}{CH_3CH_2CH_2CH(OH)CH_3}$$

(e)

$$\underset{\text{3-Methyl-1-heptene}}{CH_3CH_2CH_2CH_2CH(CH_3)CH{=}CH_2} \xrightarrow{HCl,\ ether} \underset{\text{2-Chloro-3-methylheptane}}{CH_3CH_2CH_2CH_2CH(CH_3)CH(Cl)CH_3}$$

7.25

(a)

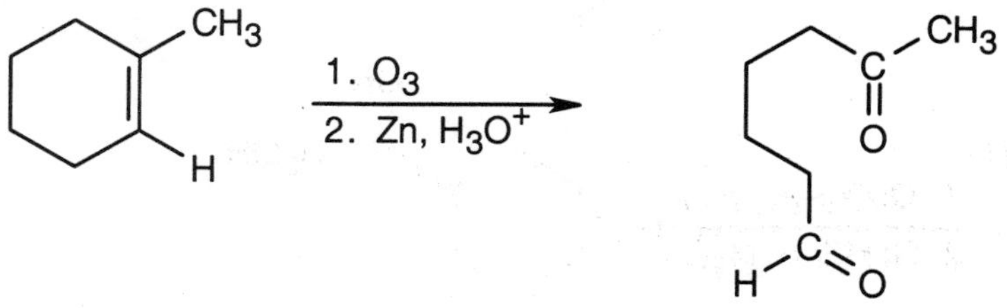

$$\text{1-methylcyclohexene} \xrightarrow[\text{2. Zn, } H_3O^+]{\text{1. } O_3} CH_3C(=O)CH_2CH_2CH_2CH_2CHO$$

(b)

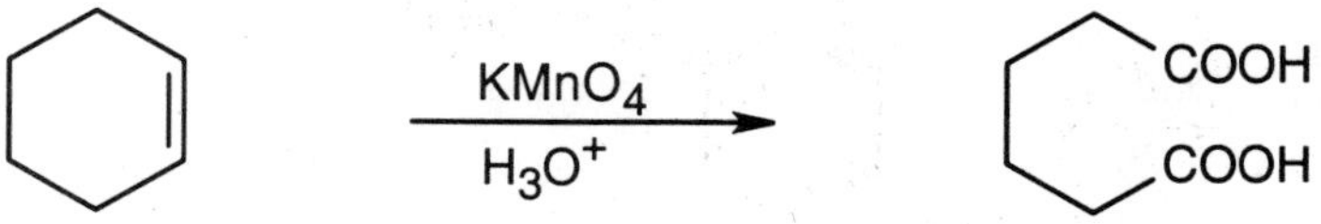

$$\text{cyclohexene} \xrightarrow[H_3O^+]{KMnO_4} HOOC(CH_2)_4COOH$$

(c)

$$\text{1-methylcyclohexene} \xrightarrow[\text{2. } H_2O_2,\ ^-OH]{\text{1. } BH_3,\ THF} \text{trans-2-methylcyclohexanol}$$

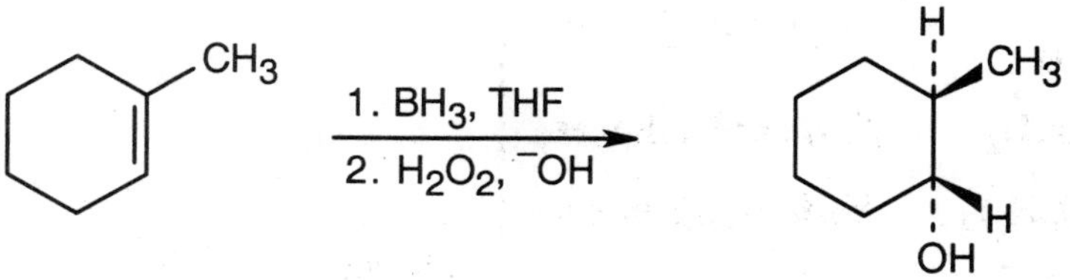

Remember that –H and –OH add *syn* across the double bond.

(d)

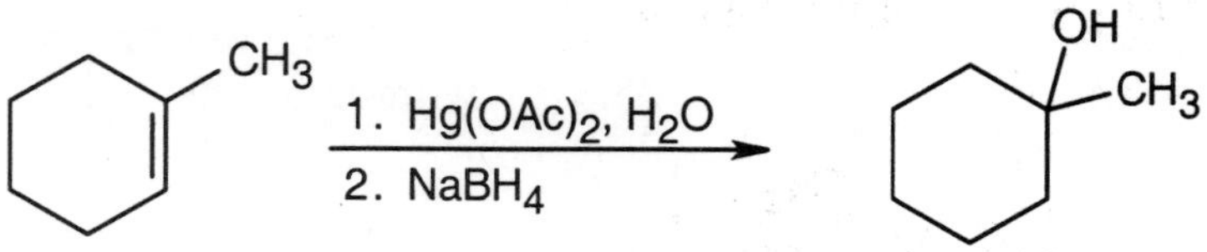

$$\text{1-methylcyclohexene} \xrightarrow[\text{2. } NaBH_4]{\text{1. } Hg(OAc)_2,\ H_2O} \text{1-methylcyclohexanol (OH, } CH_3\text{)}$$

7.26

(a)

$$\text{cyclopentene} \xrightarrow[\text{2. } NaHSO_3,\ H_2O]{\text{1. } OsO_4,\ pyridine} \text{cis-1,2-cyclopentanediol (H, OH; OH, H)}$$

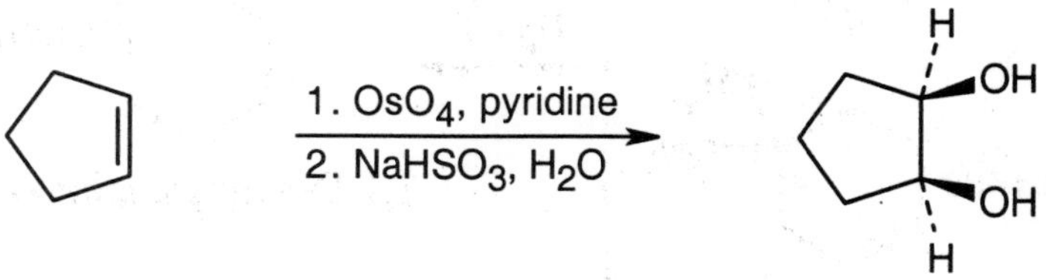

(b)

$$\text{cyclopentene} \xrightarrow[\text{2. NaBH}_4]{\text{1. Hg(OAc)}_2\text{, H}_2\text{O}} \text{cyclopentanol (OH)}$$

Hydroboration/oxidation is another route to this product.

(c)

$$\text{cyclopentene} \xrightarrow{\text{CHCl}_3\text{, KOH}} \text{6,6-dichlorobicyclo[3.1.0]hexane (H, H, Cl, Cl)}$$

(d)

$$\text{1-methylcyclohexanol (CH}_3\text{, OH)} \xrightarrow[\text{heat}]{\text{H}_2\text{SO}_4\text{, H}_2\text{O}} \text{1-methylcyclohexene (CH}_3\text{)}$$

(e)

$$CH_3CH{=}CHCH(CH_3)CH_3 \xrightarrow[\text{2. Zn, H}_3\text{O}^+]{\text{1. O}_3} CH_3CHO + CH_3CH(CH_3)CHO$$

(f)

$$CH_3C(CH_3){=}CH_2 \xrightarrow[\text{2. H}_2\text{O}_2\text{, }^-\text{OH}]{\text{1. BH}_3\text{, THF}} CH_3CH(CH_3)CH_2OH$$

7.27 Because ozonolysis gives only one product, we can assume that the alkene is symmetrical.

$$(H_3C)_2C{=}C(CH_3)_2 \xrightarrow[\text{2. Zn, H}_3\text{O}^+]{\text{1. O}_3} (H_3C)_2C{=}O + O{=}C(CH_3)_2$$

2,3-Dimethyl-2-butene

7.28 If the hydrocarbon reacts with only one equivalent of hydrogen, it has only one double bond.
If only one type of aldehyde is produced on ozonolysis, the alkene must be symmetrical. These two facts allow us to conclude that the unknown hydrocarbon is $CH_3CH_2CH_2CH_2CH{=}CHCH_2CH_2CH_2CH_3$.

$$CH_3CH_2CH_2CH_2CH{=}CHCH_2CH_2CH_2CH_3 \xrightarrow[\text{2. Zn, H}_3\text{O}^+]{\text{1. O}_3} 2\ CH_3CH_2CH_2CH_2CHO$$

5-Decene → Pentanal

$$\downarrow \text{H}_2\text{/Pd}$$

$$CH_3CH_2CH_2CH_2CH_2CH_2CH_2CH_2CH_2CH_3$$

Decane

7.29 Remember that alkenes can give ketones, carboxylic acids, and CO_2 on oxidative cleavage with $KMnO_4$ in acidic solution.

(a)

$$CH_3CH_2CH{=}CH_2 \xrightarrow[H_3O^+]{KMnO_4} CH_3CH_2COOH + CO_2$$

(b)

$$CH_3CH_2CH_2CH{=}C(CH_3)_2 \xrightarrow[H_3O^+]{KMnO_4} CH_3CH_2CH_2COOH + (CH_3)_2C{=}O$$

(c)

$$\text{(cyclohexylidene)}{=}C(CH_3)_2 \xrightarrow[H_3O^+]{KMnO_4} \text{cyclohexanone }({=}O) + (CH_3)_2C{=}O$$

7.30 Compound **A** has three degrees of unsaturation. Because compound **A** contains only one double bond, the other two degrees of unsaturation must be rings.

B $\xleftarrow[2.\ Zn,\ H_3O^+]{1.\ O_3}$ **A** $\xrightarrow{H_2/Pd}$ (decalin, H, H)

Other structures having two fused rings are possible.

7.31 Don't get discouraged by the amount of information in this problem. Read slowly and interpret the information phrase by phrase. We know the following:

1) Hydrocarbon **A** (C_6H_{12}) has one double bond or ring.
2) Because **A** reacts with one equivalent of H_2, it has one double bond and no ring.
3) Compound **A** forms a diol (**B**) when reacted with OsO_4.
4) When alkenes are oxidized with $KMnO_4$ they give either carboxylic acids or ketones, depending on the substitution pattern of the double bond.
   a) A ketone is produced from what was originally a disubstituted carbon in the double bond.
   b) A carboxylic acid is produced from what was originally a monosubstituted carbon in the double bond.
5) One fragment from $KMnO_4$ oxidation is a carboxylic acid, CH_3CH_2COOH.
   a) This fragment was $CH_3CH_2CH{=}$ (a monosubstituted double bond) in compound **A**.
   b) It contains three of the six carbons of compound **A**.
6) a) The other fragment contains three carbons.
   b) It forms ketone **C** on oxidation.
   c) The only three carbon ketone is acetone, $O{=}C(CH_3)_2$.
   d) This fragment was $=C(CH_3)_2$ in compound **A**.
7) If we join the fragment in 5a with the one in 6d, we get:

$CH_3CH_2CH{=}C(CH_3)_2$ **A** $\qquad C_6H_{12}$

The complete scheme:

$$CH_3CH_2CH{=}C(CH_3)CH_3 \ (\mathbf{A}) \xrightarrow[2.\ NaHSO_3,\ H_2O]{1.\ OsO_4,\ pyridine} CH_3CH_2CH(OH){-}C(OH)(CH_3)CH_3 \ (\mathbf{B})$$

$$\mathbf{A} \xrightarrow{H_2/Pd} CH_3CH_2CH_2CH(CH_3)CH_3$$

$$\mathbf{A} \xrightarrow[H_3O^+]{KMnO_4} CH_3CH_2COOH + O{=}C(CH_3)_2 \ (\mathbf{C})$$

7.32 The oxidative cleavage reaction of alkenes with O_3, followed by Zn in acid, produces aldehyde and ketone functional groups at sites where double bonds used to be. On ozonolysis, these two dienes yield only aldehydes because all double bonds are monosubstituted.

$$\text{1,4-cyclohexadiene} \xrightarrow[2.\ Zn,\ H_3O^+]{1.\ O_3} OHC{-}CH_2{-}CH_2{-}CHO + OHC{-}CHO$$

Because the other diene is symmetrical, only one dialdehyde, $OCHCH_2CHO$, is produced.

7.33 Try to solve this problem phrase by phrase.

1) $C_{10}H_{18}O$ has two double bonds and/or rings.
2) $C_{10}H_{18}O$ must be an alcohol because it undergoes reaction with H_2SO_4 to yield alkenes.
3) When $C_{10}H_{18}O$ is treated with dilute H_2SO_4, a mixture of alkenes of the formula $C_{10}H_{16}$ is produced.
4) Since the major alkene product **B** yields only cyclopentanone, C_5H_8O, on ozonolysis, **B** and **A** contain two rings. **A** therefore has no double bonds.

$$\text{1-cyclopentylcyclopentanol} \ (\mathbf{A}) \xrightarrow[H_2SO_4]{dilute} \text{1-cyclopentylcyclopentene} + \text{cyclopentylidenecyclopentane} \ (\mathbf{B})$$

$$\mathbf{B} \xrightarrow[2.\ Zn,\ H_3O^+]{1.\ O_3} 2 \ \text{cyclopentanone}$$

7.34

$$\text{Cyclohexene} \xrightarrow{HBr} \text{Bromocyclohexane}$$

$$\text{1-Methylcyclohexene} \xrightarrow{HBr} \text{1-Bromo-1-methylcyclohexane}$$

Cyclohexene 1-Methylcyclohexene

Addition to 1–methylcyclohexene occurs at a faster rate. Positive charge generated during the reaction is better stabilized by a tertiary carbocation intermediate than by a secondary carbocation intermediate.

7.35

(a)

$$\underset{\text{2-Butene}}{CH_3CH{=}CHCH_3} \xrightarrow{HBr} CH_3CH_2\overset{Br}{\overset{|}{C}}HCH_3$$

(b)

$$3\ \underset{\text{2-Butene}}{CH_3CH{=}CHCH_3} \xrightarrow[THF]{BH_3} (CH_3CH_2\overset{CH_3}{\overset{|}{C}}H)_3B \xrightarrow[^-OH]{H_2O_2} 3\ CH_3CH_2\overset{OH}{\overset{|}{C}}HCH_3$$

(c)

$$\text{H(H}_3\text{C)C=C(H)CH}_3 \xrightarrow[\text{Zn–Cu}]{CH_2I_2} \text{cyclopropane}$$

cis-2-Butene *cis*-1,2-Dimethylcyclopropane

(d)

$$\text{H}_3\text{C(H)C=C(H)CH}_3 \xrightarrow[\text{Zn–Cu}]{CH_2I_2} \text{cyclopropane}$$

trans-2-Butene *trans*-1,2-Dimethylcyclopropane

The Simmons-Smith reaction, in (c) and (d), occurs with syn stereochemistry. Only *cis*-1,2 -dimethylcyclopropane is produced from *cis*-2-butene, and only *trans*-1,2-dimethylcyclopropane is produced from *trans*-2-butene.

7.36

(a)

$$:\ddot{\underset{..}{I}}{-}\ddot{N}{=}N{=}\ddot{N}: \longleftrightarrow\ :\ddot{\underset{..}{I}}{-}\ddot{\underset{..}{N}}{-}N{\equiv}N:$$

(b)

$$(FC) = \left[\begin{matrix}\text{\# of valence}\\ \text{electrons}\end{matrix}\right] - \left[\frac{\text{\# of bonding electrons}}{2}\right] - \left[\begin{matrix}\text{\# nonbonding}\\ \text{electrons}\end{matrix}\right]$$

A: $:\ddot{I}-\ddot{N}=N=\ddot{N}:$ (N 1 2 3) ⟷ B: $:\ddot{I}-\ddot{N}-N\equiv N:$ (N 1 2 3)

A: $I-N=\overset{+1}{N}=\overset{-1}{N}$ B: $I-\overset{-1}{N}-\overset{+1}{N}\equiv N$

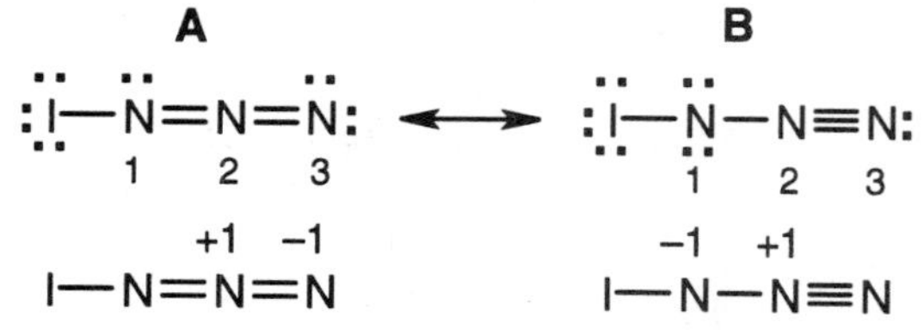

Formal Charge

	I	N1	N2	N3
A	0	0	+1	–1
B	0	–1	+1	0

Formal charge calculations show a partial negative charge on N1.

(c) Addition of IN_3 to the alkene yields a product in which I is bonded to the primary carbon and N_3 is bonded to the secondary carbon. If addition occurs with Markovnikov orientation, I^+ must be the electrophile, and the reaction must proceed through an iodonium ion intermediate. Opening of the iodonium ion gives Markovnikov product for the reasons discussed in Problem 7.6. The bond polarity of iodine azide is:

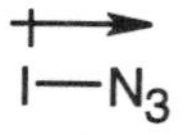

$$\overset{+\longrightarrow}{I-N_3}$$

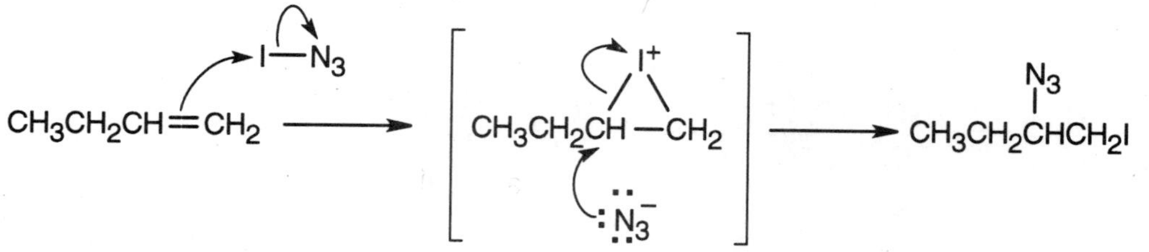

$CH_3CH_2CH=CH_2$ + $I-N_3$ ⟶ $[CH_3CH_2CH-CH_2$ (bridged by I^+) + $:\ddot{N}_3^-]$ ⟶ $CH_3CH_2CH(N_3)CH_2I$

7.37

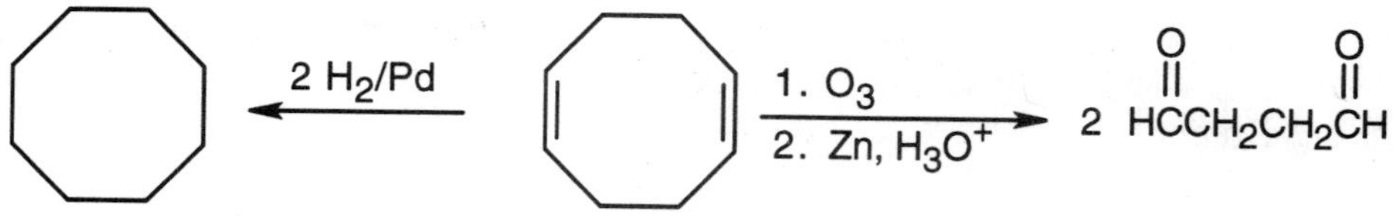

Cyclooctane ← (2 H_2/Pd) — 1,5-Cyclooctadiene —(1. O_3; 2. Zn, H_3O^+)→ 2 $HCCH_2CH_2CH$ (O=, =O)

7.38

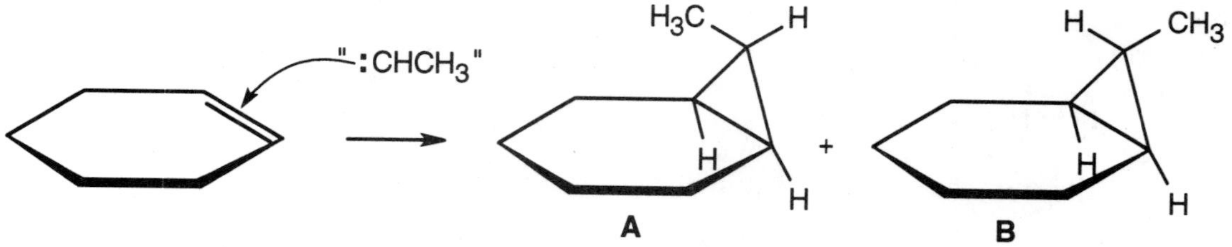

Cyclohexene + "$:CHCH_3$" ⟶ **A** (H_3C, H on cyclopropane carbon; H, H at ring fusion) + **B** (H, CH_3 on cyclopropane carbon; H, H at ring fusion)

Focus on the stereochemistry of the three-membered ring. Simmons-Smith reaction of 1,1–diiodoethane with the double bond occurs with syn stereochemistry and can produce two isomers. In one of these isomers (**A**), the methyl group is on the same side of the three-membered ring as the cyclohexane ring carbons. In **B**, the methyl group is on the side of the three-membered ring opposite to the cyclohexane ring carbons.

7.39 (a) Addition of HI occurs with Markovnikov regiochemistry – iodine adds to the more substituted carbon.
(b) Hydroxylation of double bonds produces cis, not trans, diols.
(c) Ozone reacts with both double bonds of 1,4–cyclohexadiene.
(d) Because hydroboration is a syn addition, the –H and the –OH added to the double bond must be cis to each other.

7.40 (a) This alcohol can't be synthesized selectively by hydroboration/oxidation. Consider the two possible starting materials.

1.

$$CH_3CH_2CH_2CH{=}CH_2 \xrightarrow[2.\ H_2O_2,\ ^{-}OH]{1.\ BH_3,\ THF} CH_3CH_2CH_2CH_2CH_2OH$$

1-Pentene yields only the primary alcohol.

2.

$$CH_3CH_2CH{=}CHCH_3 \xrightarrow[2.\ H_2O_2,\ ^{-}OH]{1.\ BH_3,\ THF} CH_3CH_2\underset{OH}{C}HCH_2CH_3 + CH_3CH_2CH_2\underset{OH}{C}HCH_3$$

2-Pentene yields a mixture of alcohols.

(b)

$$(CH_3)_2C{=}C(CH_3)_2 \xrightarrow[2.\ H_2O_2,\ ^{-}OH]{1.\ BH_3,\ THF} (CH_3)_2CH\underset{OH}{C}(CH_3)_2$$

2,3-Dimethyl-2-butene yields the desired alcohol exclusively.

(c) This alcohol can't be formed cleanly by a hydroboration reaction. The –H and –OH added to a double bond must be cis to each other.

(d) The product shown is not a hydroboration product; hydroboration yields an alcohol in which –OH is bonded to the less substituted carbon.

7.41

(a)

$$H_2C{=}CHCH(CH_3)_2 \xrightarrow[Zn(Cu)]{CH_2I_2} H_2C\overset{CH_2}{—}CHCH(CH_3)_2$$

3-Methyl-1-butene

(b)

$$\text{Cycloheptene} + CHCl_3 \xrightarrow{KOH} \text{7,7-dichlorobicyclo[4.1.0]heptane (ring with } CCl_2 \text{)}$$

Cycloheptene

7.42

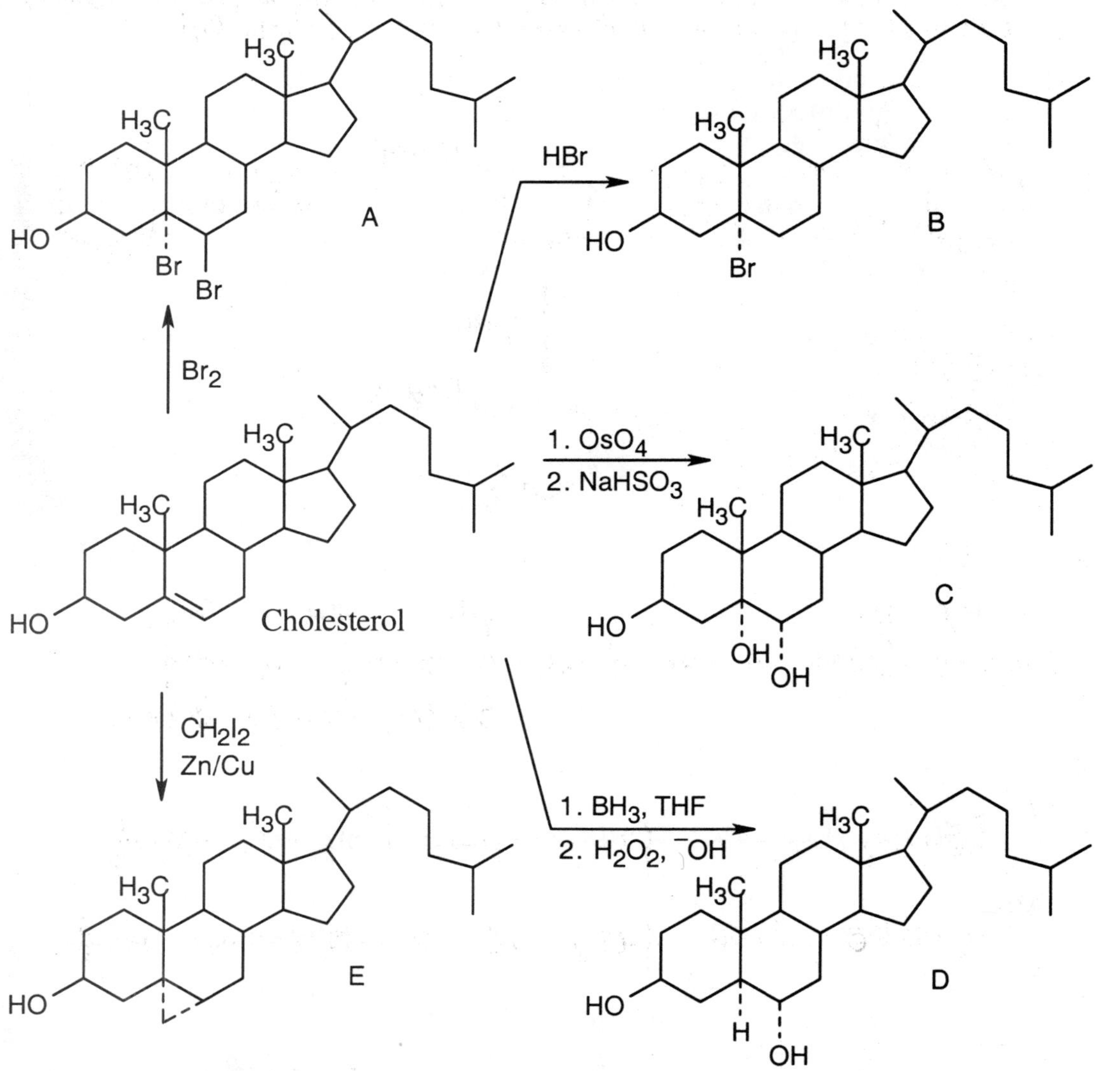

HBr

Br₂ → A

1. OsO_4
2. $NaHSO_3$ → C

CH_2I_2
Zn/Cu → E

1. BH_3, THF
2. H_2O_2, ^{-}OH → D

Cholesterol

7.43

$$CH_3(CH_2)_{12}CH{=}CH(CH_2)_7CH_3 \xrightarrow[H_3O^+]{KMnO_4} CH_3(CH_2)_{12}COOH + CH_3(CH_2)_7COOH$$

7.44 C_8H_8 has five double bonds and/or rings. One of these double bonds reacts with H_2/Pd. Stronger conditions cause the uptake of four equivalents of H_2. C_8H_8 thus contains four double bonds, three of which are in an aromatic ring, and one C=C double bond. A good guess for C_8H_8 at this point is:

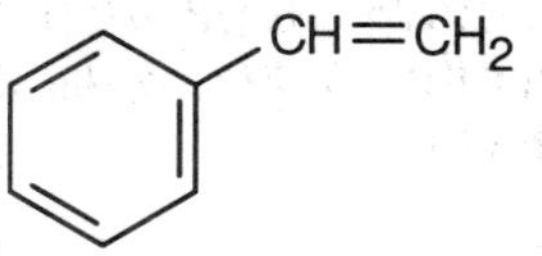

$C_6H_5CH{=}CH_2$

Reaction of a double bond with $KMnO_4$ yields cleavage products of the highest possible degree of oxidation. In this case, the products are $CO_2 + C_6H_5CO_2H$.

$$\text{B (}C_6H_5COOH\text{)} + CO_2 \xleftarrow[H_3O^+]{KMnO_4} \text{A (}C_6H_5CH{=}CH_2\text{)} \xrightarrow{H_2/Pd} C_6H_5CH_2CH_3$$

$$\text{A} \xrightarrow{H_2/Rh} \text{C (ethylcyclohexane, } C_6H_{11}CH_2CH_3\text{)}$$

7.45

$$H_2C{=}C(CH_3){-}COCH_3 \text{ (}C{=}O\text{)} \longrightarrow {-}CH_2{-}C(CH_3)(COOCH_3){-}CH_2{-}C(CH_3)(COOCH_3){-}CH_2{-}C(CH_3)(COOCH_3){-}$$

7.46

(a)

$$F_2C{=}CF_2 \longrightarrow {-}CF_2{-}CF_2{-}CF_2{-}CF_2{-}CF_2{-}CF_2{-}$$

(b)

$$H_2C{=}CHCl \longrightarrow {-}CH_2{-}CHCl{-}CH_2{-}CHCl{-}CH_2{-}CHCl{-}$$

7.47

$$(H_3C)_2C{=}CH_2 + H{-}A \rightleftharpoons \left[(H_3C)_2\overset{+}{C}{-}CH_3 + CH_3\ddot{O}{-}H \rightleftharpoons CH_3\overset{+}{O}(H){-}C(CH_3)_2CH_3 \;\; :A^- \right] \rightleftharpoons HA + CH_3OC(CH_3)_3$$

protonation of double bond — nucleophilic attack of methanol on carbocation — loss of proton

The above mechanism is the same as the mechanism shown in Section 7.4 with one exception: In this problem, methanol, rather than water, is the nucleophile, and an ether, rather than an alcohol, is the observed product.

7.48

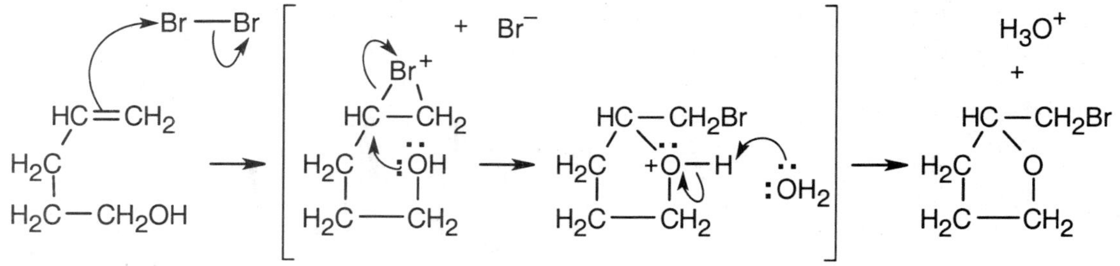

formation of cyclic bromonium ion

nucleophilic attack of –OH on bromonium ion

loss of H^+

The above mechanism is the same as that for halohydrin formation, shown in Section 7.3. In this case, the nucleophile is the hydroxyl group of 4–penten–1–ol.

7.49 (a) Bromine dissolved in CH_2Cl_2 has a reddish-brown color. When an alkene is dissolved in Br_2/CH_2Cl_2, the double bond reacts with bromine, and the color disappears. This test distinguishes cyclopentene from cyclopentane, which does not react with Br_2. Alternatively, each compound can be treated with H_2/Pd. The alkene takes up H_2, and the alkane is unreactive.
(b) An aromatic compound such as benzene is unreactive to the Br_2/CH_2Cl_2 reagent and can be distinguished from 2–hexene, which decolorizes Br_2/CH_2Cl_2. Also, an aromatic compound doesn't take up H_2 under reaction conditions used for hydrogenation of alkenes.

7.50

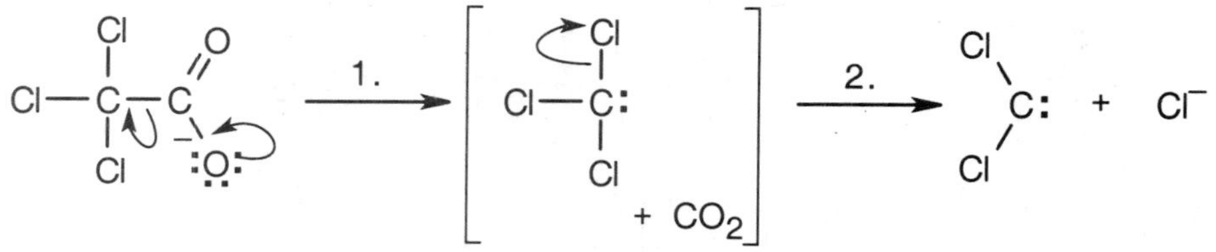

In step 1, carbon dioxide is lost from the trichloroacetate anion. In step 2, elimination of chloride anion produces dichlorocarbene. Step 2 is the same for both the above reaction and the base-induced elimination of HCl from chloroform.

7.51 (a) α-Terpinene, $C_{10}H_{16}$, has three degrees of unsaturation.
(b) Hydrogenation removes only two degrees of saturation, producing a hydrocarbon $C_{10}H_{20}$, that has one ring. α-Terpinene thus has two double bonds and one ring.
(c)

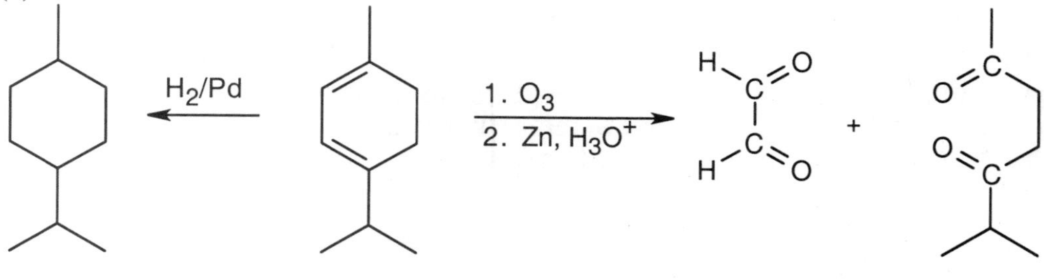

α-Terpinene

Glyoxal

6-Methylheptane-2,5-dione

7.52 Make models of the cis and trans diols. Notice that it is much easier to form a five-membered cyclic periodate from the cis diol A than from the trans diol B. We therefore predict that the cis periodate intermediate is of lower energy than the trans periodate intermediate because of the lack of strain in the cis periodate ring.

Because any factor that lowers the energy of a transition state or intermediate also lowers $\Delta G^{\ddagger}$ and increases the rate of reaction, we predict that diol cleavage should proceed more rapidly for cis diols than for trans diols.

7.53

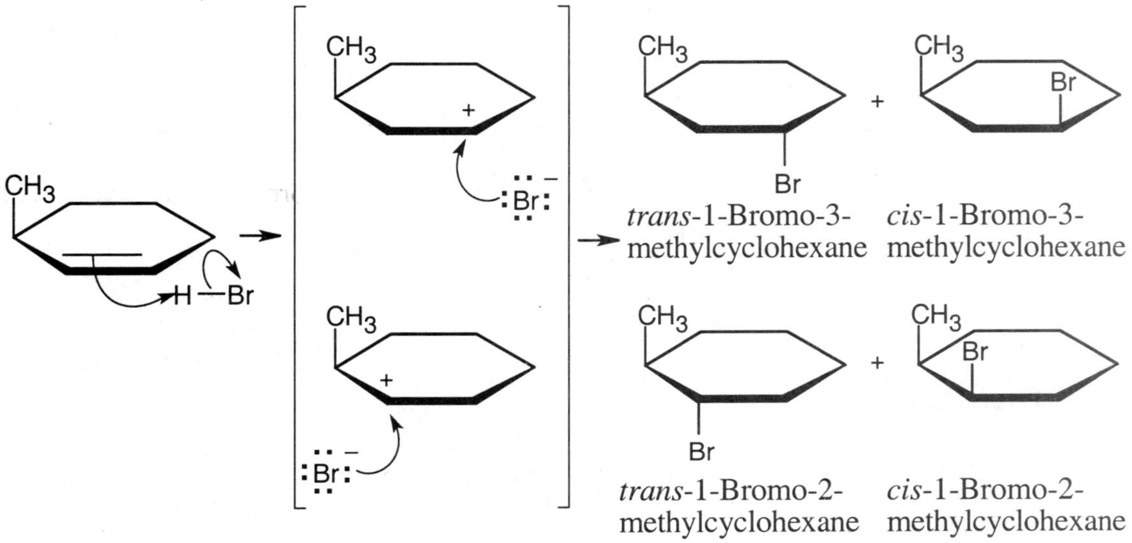

trans-1-Bromo-3-methylcyclohexane + *cis*-1-Bromo-3-methylcyclohexane

trans-1-Bromo-2-methylcyclohexane + *cis*-1-Bromo-2-methylcyclohexane

In the reaction of 3-methylcyclohexene with HBr, there are two intermediate carbocations of approximately equal stability that are attacked by bromide ion on either side of each carbocation to give four different products.

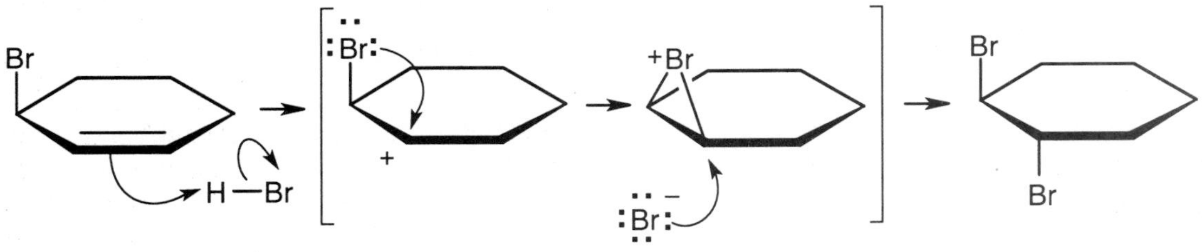

The most stable cation intermediate from protonation of 3-bromocyclohexene is a cyclic bromonium ion, which is attacked by Br^- from the opposite side to yield anti product.

7.54

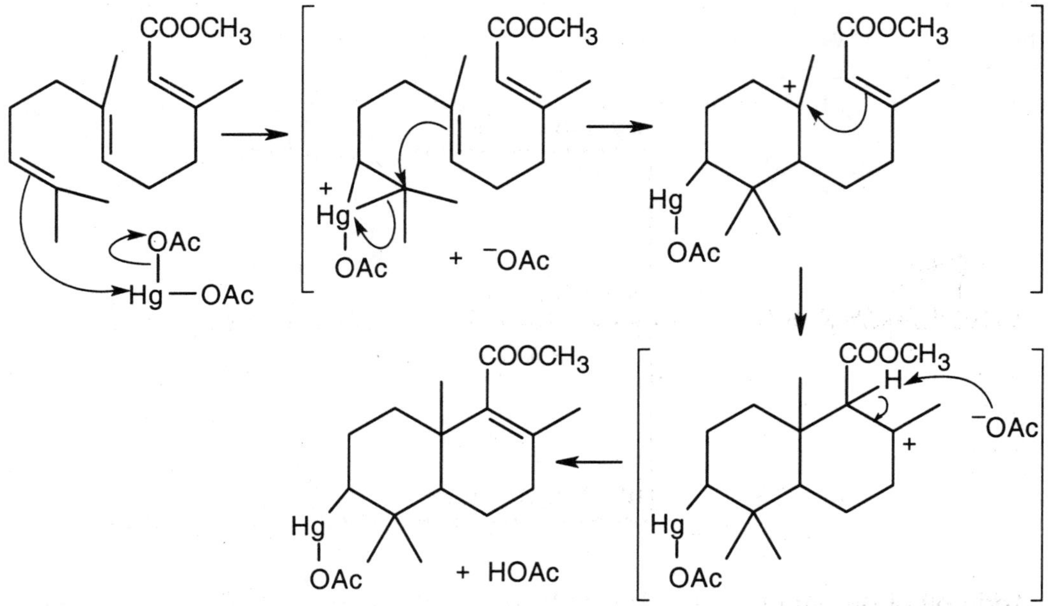

COOCH3 OAc Hg—OAc + Hg OAc + ⁻OAc Hg OAc + ⁻OAc HOAc

7.55 Hydroboration of 2-methyl-2-pentene at 160°C is reversible. The initial organoborane intermediate can eliminate BH_3 in either of two ways, yielding either 2-methyl-2-pentene or 4-methyl-2-pentene, which in turn can undergo reversible hydroboration to yield either 4-methyl-2-pentene or 4-methyl-1-pentene. The effect of these reversible reactions is to migrate the double bond along the carbon chain. A final hydroboration then yields the most stable (primary) organoborane, which is oxidized to form 4-methyl-1-pentanol.

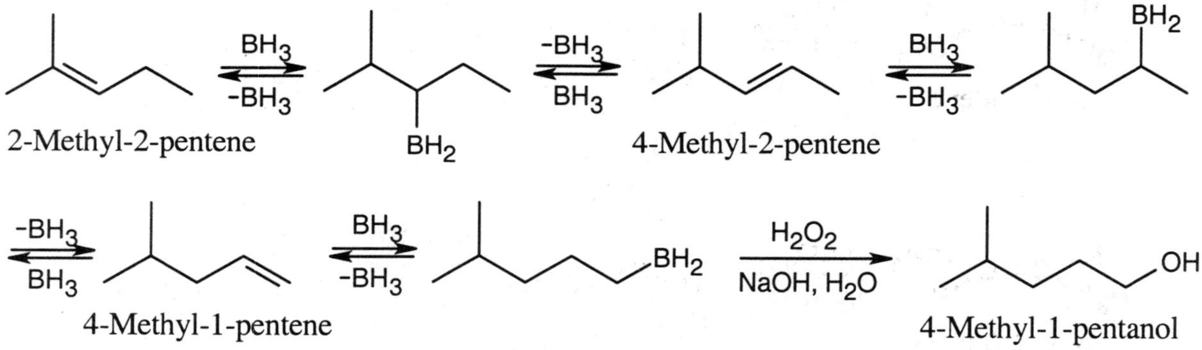

2-Methyl-2-pentene ⇌ (BH_3 / $-BH_3$) BH_2 ⇌ ($-BH_3$ / BH_3) 4-Methyl-2-pentene ⇌ (BH_3 / $-BH_3$) BH_2

⇌ ($-BH_3$ / BH_3) 4-Methyl-1-pentene ⇌ (BH_3 / $-BH_3$) BH_2 → (H_2O_2 / NaOH, H_2O) OH 4-Methyl-1-pentanol

A Look Ahead

7.56

$CH_3CHCH_2CH_2C{\equiv}CH$ (with CH_3 on C2)

(a) 1 equiv Br_2 → $CH_3CH(CH_3)CH_2CH_2(Br)C{=}C(Br)H$

(b) 2 equiv H_2, Pd/C → $CH_3CH(CH_3)CH_2CH_2CH_2CH_3$

(c) 1 equiv HBr → $CH_3CH(CH_3)CH_2CH_2C(Br){=}CH_2$

Addition of one equivalent of HX or X_2 to a triple bond occurs with Markovnikov regiochemistry to yield a product in which the two added atoms usually have a trans-relationship across the double bond.

7.57

cis-2-Butene —OsO_4, pyridine→ cyclic osmate —$NaHSO_3$, H_2O→ diol

trans-2-Butene —OsO_4, pyridine→ cyclic osmate —$NaHSO_3$, H_2O→ diol

Formation of the cyclic osmate, which occurs with syn stereochemistry, retains the cis-trans stereochemistry of the double bond because osmate formation is a single-step reaction. Treatment of the osmate ester with $NaHSO_3$ does not affect the stereochemistry of the carbon-oxygen bond. The diol produced from *cis*–2–butene is a stereoisomer of the diol produced from *trans*–2–butene.

7.58

The above mechanism is the same as the mechanism shown in Section 7.4 with one exception: In this problem, methanol, rather than water, is the nucleophile, and an ether, rather than an alcohol, is the observed product.

Molecular Modeling

7.59 The C–Br bond distances in the bromonium ions show that one C–Br is shorter and stronger than the other (203.1 pm vs. 217.0 pm in propene, and 196.4 vs. 269.4 pm in styrene). The shorter bond becomes the C–Br bond in the major product, and the longer bond is broken when a water molecule attacks the bromonium ion.

7.60 There are two electrophilic sites above and below the carbon plane. These coincide with the lobes of the empty carbon *p* orbital. The carbene initially acts as an electrophile. Its initial approach brings one of its electrophilic sites together with a nucleophilic site (the π electrons) of the alkene.

7.61 Transition state B is 14 kJ/mol lower in energy than transition state A, and transition state B leads to the observed product. The B–H bond length of the bond that breaks is 126.5 pm and is only slightly longer than the other B–H bonds ($\approx$ 119 pm). The C–H bond length of the bond that forms is 168.6 pm and is much longer than the other C–H bonds ($\approx$ 110 pm). Thus, the transition state is reactant-like, consistent with the Hammond Postulate.

Chapter 8 – Alkynes: An Introduction to Organic Synthesis

Chapter Outline

I. Introduction to alkynes (Section 8.1 – 8.3).
  A. Electronic structure of alkynes (Section 8.1).
    1. Alkyne triple bonds result from the overlap of two *sp* hybridized carbon atoms. One σ bond and two π bonds are formed.
    2. The length (120 pm) and strength (835 kJ/mol) of a –C≡C– bond make it the strongest carbon–carbon bond.
    3. Alkynes are somewhat less reactive than alkenes in electrophilic addition reactions
  B. Naming alkynes (Section 8.2).
    The rules for naming alkynes are like the rules for alkenes (Sec. 6.3), with a few exceptions.
      a. The suffix *-yne* is used for an alkyne.
      b. Compounds with both double bonds and triple bonds are *enynes.*
      c. When there is a choice in numbering, double bonds receive lower numbers than triple bonds.
      d. Compounds can also contain alkynyl groups.
  C. Preparation of alkynes (Section 8.3).
    1. Alkynes can be prepared by elimination reactions of 1,2-dihalides, using a strong base.
    2. The dihalides are formed by addition of X_2 to alkenes.
    3. Vinylic halides give alkynes when treated with a strong base.

II. Reactions of alkynes (Sections 8.4 – 8.7).
  A. Addition of X_2 and HX (Section 8.4).
    1. HX adds to alkynes by an electrophilic addition mechanism.
      a. Addition of two equivalents of HX occurs if the acid is in excess.
      b. Addition occurs with Markovnikov regiochemistry and with trans stereochemistry.
    2. X_2 also adds in the same manner, and trans stereochemistry is observed.
    3. The intermediate in addition reactions is a vinylic carbocation, which forms less readily than an alkyl carbocation.
    4. Mechanisms of some alkyne addition reactions are complex.
  B. Hydration reactions of alkynes (Section 8.5).
    1. Hg(II)-catalyzed additions.
      a. The –OH group adds to the more substituted carbon to give Markovnikov product.
      b. The intermediate enol product tautomerizes to a ketone.
      c. The mechanism is similar to that of addition to alkenes, but no $NaBH_4$ is necessary for removal of Hg.
      d. A mixture of products is formed from an internal alkyne, but a terminal alkyne yields a methyl ketone.
    2. Hydroboration/oxidation of alkynes.
      a. Hydroboration/ oxidation of alkynes gives an intermediate enol product that tautomerizes to a carbonyl product.
        i. Hydroboration of a terminal alkyne gives an aldehyde.
        ii. Hydroboration of an internal alkyne gives a ketone.
      b. Hydroboration/ oxidation is complementary to Hg(II)-catalyzed hydration.
  C. Reduction of alkynes (Section 8.6).
    1. Complete reduction to an alkane occurs when H_2/Pd is used.
    2. Partial reduction to a cis alkene occurs with H_2 and a Lindlar catalyst.

3. Partial reduction with Li in NH_3 produces a trans alkene.
   a. The reaction proceeds through an anion radical —> vinylic radical —> vinylic anion.
   b. The more stable trans vinylic anion is formed.

D. Oxidative cleavage of alkynes (Section 8.7).
1. O_3 or $KMnO_4$ cleave alkyne bonds to produce carboxylic acids or CO_2 (terminal alkyne).
2. Oxidative cleavage reactions were formerly used for structure determinations.

III. Alkyne acidity (Sections 8.8 – 8.9).

A. Formation of acetylide anions (Section 8.8).
1. Terminal alkynes are weakly acidic ($pK_a = 25$).
2. Very strong bases ($^{-}NH_2$) can deprotonate terminal alkynes.
3. Acetylide anions are stabilized by the large amount of "*s* character" of the orbital that holds the electron.

B. Alkylation of acetylide anions (Section 8.9).
1. Acetylide anions can react with haloalkanes to form substitution products.
   a. The nucleophilic acetylide anion attacks the electrophilic carbon of a haloalkane to produce a new alkyne.
   b. This reaction is called an alkylation reaction.
   c. Any terminal alkyne can form an alkylation product.
2. Acetylide alkylations are limited to primary alkyl bromides and iodides.
   Acetylide ions cause dehydrohalogenation reactions with secondary and tertiary halides.

IV. Organic synthesis (Section 8.10).

A. Reasons for the study of organic synthesis.
1. In the pharmaceutical and chemical industries, synthesis produces new molecules, or better routes to important molecules.
2. In academic laboratories, synthesis is done for creative reasons.
3. In the classroom, synthesis is a tool for teaching the logic of organic chemistry.

B. Strategies for organic synthesis.
1. Work backward, but -
2. Keep the starting material in mind.

Solutions to Problems

8.1 The rules for naming alkynes are almost the same as the rules for naming alkenes. The suffix *-yne* is used, and compounds containing both double bonds and triple bonds are -*enynes*.

(a)

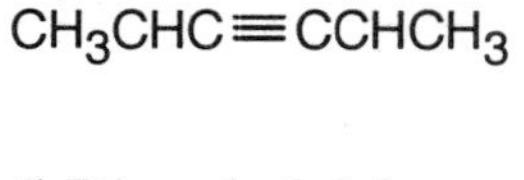

$$\begin{array}{c} \quad CH_3 \quad\;\; CH_3 \\ \quad | \qquad\quad | \\ CH_3CHC{\equiv}CCHCH_3 \end{array}$$

2,5-Dimethyl-3-hexyne

(b)

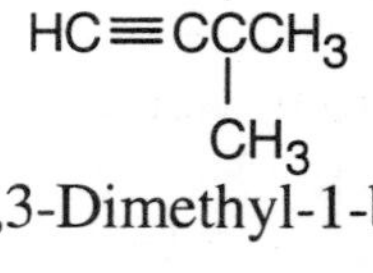

$$\begin{array}{c} \qquad\quad CH_3 \\ \qquad\quad | \\ HC{\equiv}CCCH_3 \\ \qquad\quad | \\ \qquad\quad CH_3 \end{array}$$

3,3-Dimethyl-1-butyne

(c)

$$CH_3CH{=}CHCH{=}CHC{\equiv}CCH_3$$

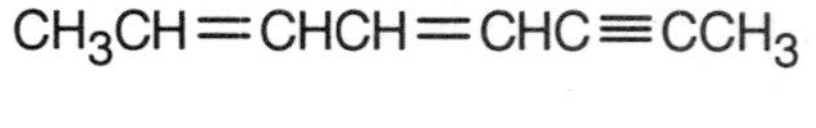

2,4-Octadien-6-yne
(not 4,6-Octadien-2-yne)

(d)

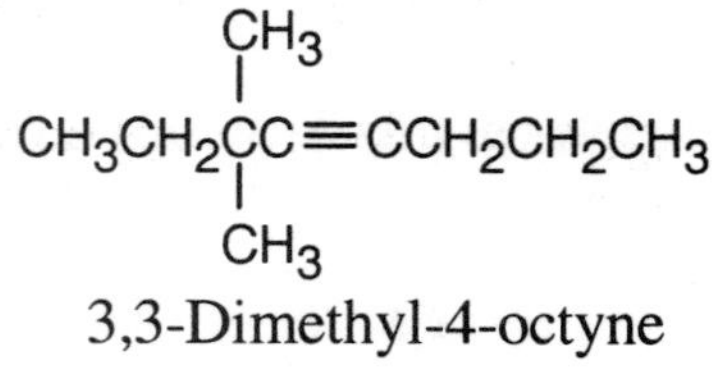

$$\begin{array}{c} CH_3 \\ | \\ CH_3CH_2CC{\equiv}CCH_2CH_2CH_3 \\ | \\ CH_3 \end{array}$$

3,3-Dimethyl-4-octyne

(e)

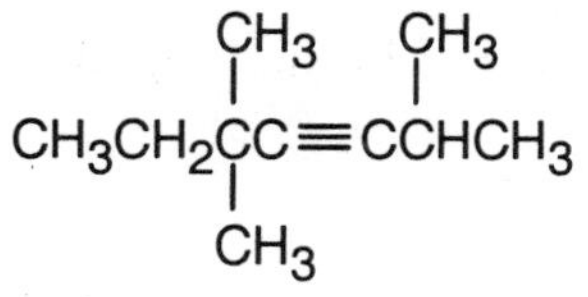

$$CH_3CH_2C(CH_3)_2C{\equiv}CCH(CH_3)CH_3$$

2,5,5-Trimethyl-3-heptyne

(f)

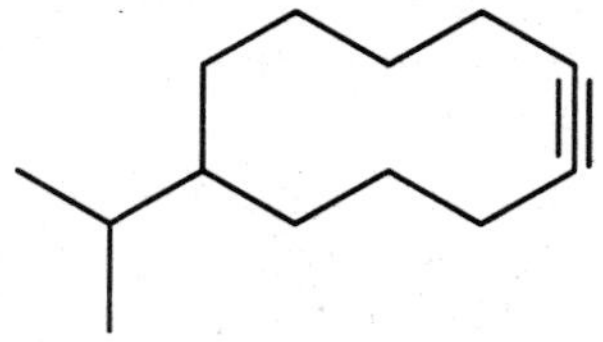

6-Isopropylcyclodecyne

8.2

$CH_3CH_2CH_2CH_2C{\equiv}CH$
1-Hexyne

$CH_3CH_2CH_2C{\equiv}CCH_3$
2-Hexyne

$CH_3CH_2C{\equiv}CCH_2CH_3$
3-Hexyne

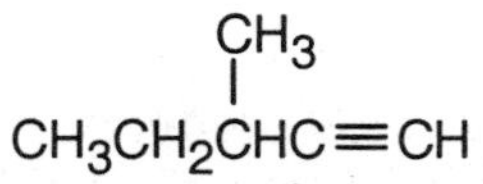

$CH_3CH_2CH(CH_3)C{\equiv}CH$
3-Methyl-1-pentyne

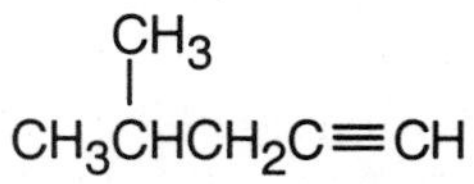

$CH_3CH(CH_3)CH_2C{\equiv}CH$
4-Methyl-1-pentyne

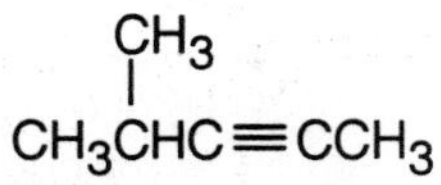

$CH_3CH(CH_3)C{\equiv}CCH_3$
4-Methyl-2-pentyne

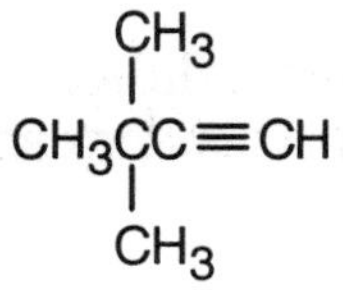

$CH_3C(CH_3)_2C{\equiv}CH$
3,3-Dimethyl-1-butyne

8.3 Markovnikov addition is observed with alkynes as well as with alkenes.

(a)

$$CH_3CH_2CH_2C{\equiv}CH \;+\; 2\;Cl_2 \longrightarrow CH_3CH_2CH_2CCl_2CHCl_2$$

(b)

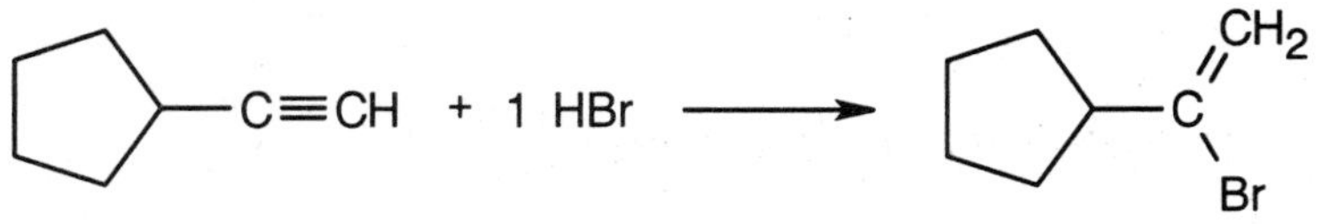

$$C_5H_9{-}C{\equiv}CH \;+\; 1\;HBr \longrightarrow C_5H_9{-}C(Br){=}CH_2$$

(c)

$$CH_3CH_2CH_2CH_2C{\equiv}CCH_3 \;+\; 1\;HBr \longrightarrow (CH_3CH_2CH_2CH_2)(H)C{=}C(Br)(CH_3) \;+\; (CH_3CH_2CH_2CH_2)(Br)C{=}C(H)(CH_3)$$

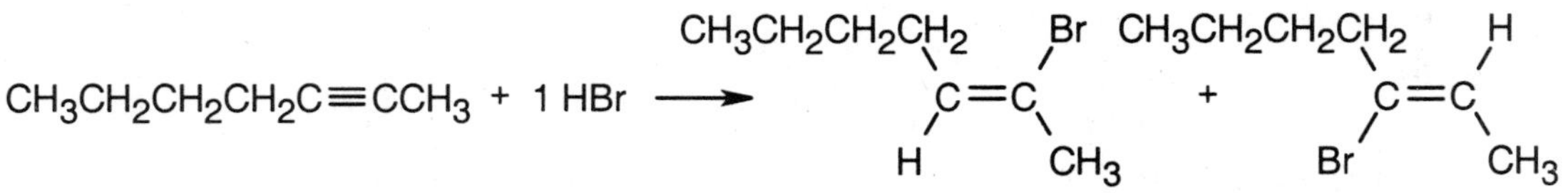

Two products result from addition to an internal alkyne.

8.4

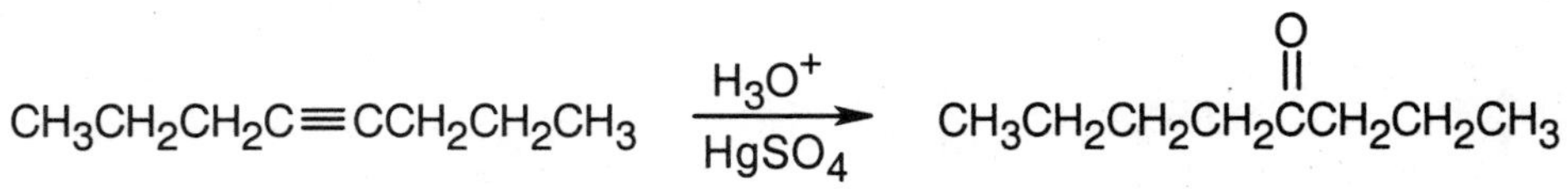

$$CH_3CH_2CH_2C{\equiv}CCH_2CH_2CH_3 \xrightarrow[HgSO_4]{H_3O^+} CH_3CH_2CH_2CH_2\overset{O}{\overset{\|}{C}}CH_2CH_2CH_3$$

This symmetrical alkyne yields only one product.

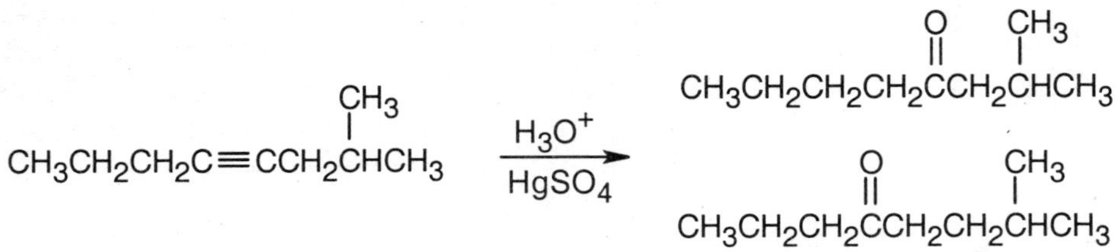

$$CH_3CH_2CH_2C{\equiv}CCH_2CH(CH_3)CH_3 \xrightarrow[HgSO_4]{H_3O^+} CH_3CH_2CH_2CH_2C(=O)CH_2CH(CH_3)CH_3 + CH_3CH_2CH_2C(=O)CH_2CH_2CH(CH_3)CH_3$$

Two ketone products result from hydration of 2-methyl-4-octyne.

8.5

(a)

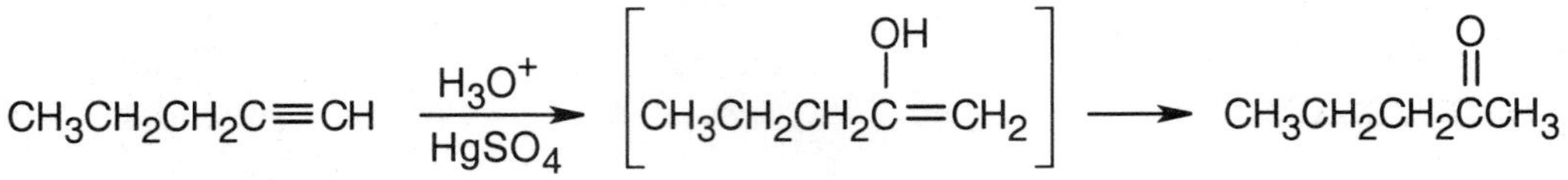

$$CH_3CH_2CH_2C{\equiv}CH \xrightarrow[HgSO_4]{H_3O^+} \left[CH_3CH_2CH_2C(OH){=}CH_2\right] \longrightarrow CH_3CH_2CH_2C(=O)CH_3$$

(b)

$$CH_3CH_2C{\equiv}CCH_3 \xrightarrow[HgSO_4]{H_3O^+} CH_3CH_2C(=O)CH_2CH_3 + CH_3CH_2CH_2C(=O)CH_3$$

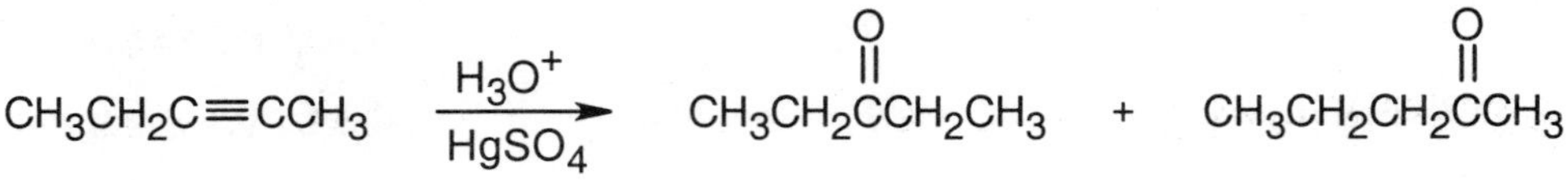

The desired ketone can be prepared only as part of a product mixture.

8.6 Remember that hydroboration yields aldehydes from terminal alkynes and ketones from internal alkynes.

(a)

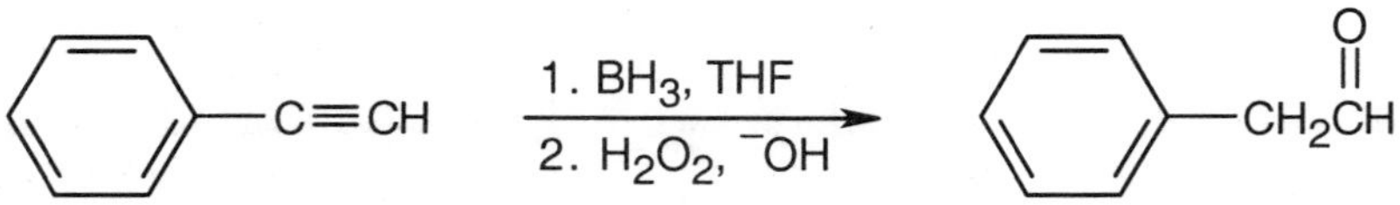

$$C_6H_5{-}C{\equiv}CH \xrightarrow[2.\ H_2O_2,\ ^-OH]{1.\ BH_3,\ THF} C_6H_5{-}CH_2CH{=}O$$

(b)

$$(CH_3)_2CHC{\equiv}CCH(CH_3)_2 \xrightarrow[2.\ H_2O_2,\ ^-OH]{1.\ BH_3,\ THF} (CH_3)_2CHCH_2C(=O)CH(CH_3)_2$$

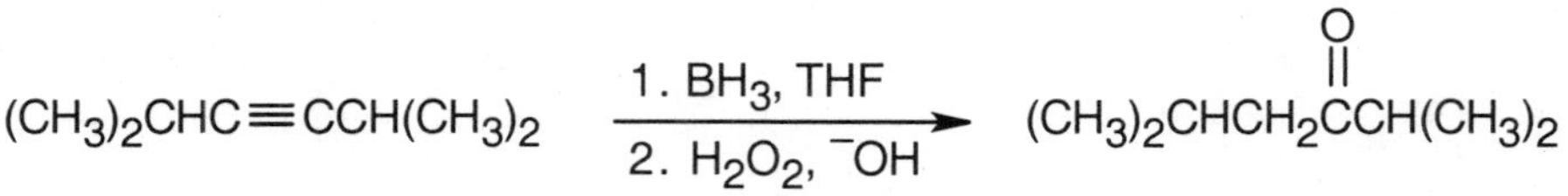

8.7 Choice of the correct reducing reagent gives a double bond with the desired geometry.

(a)

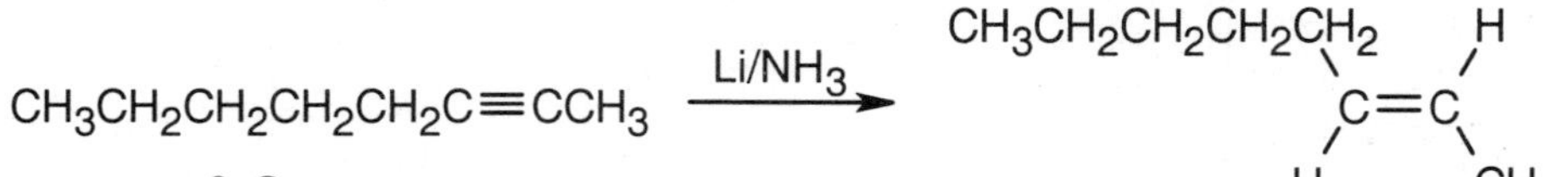

$$\underset{\text{2-Octyne}}{CH_3CH_2CH_2CH_2CH_2C{\equiv}CCH_3} \xrightarrow{Li/NH_3} \underset{\textit{trans}\text{-2-Octene}}{(CH_3CH_2CH_2CH_2CH_2)HC{=}CH(CH_3)}$$

(b)

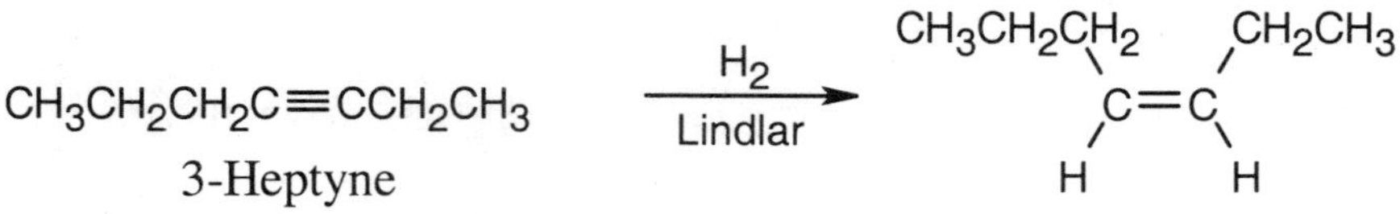

$$\underset{\text{3-Heptyne}}{CH_3CH_2CH_2C{\equiv}CCH_2CH_3} \xrightarrow[Lindlar]{H_2} \underset{\textit{cis}\text{-3-Heptyne}}{(CH_3CH_2CH_2)HC{=}CH(CH_2CH_3)}$$

(c)

$$CH_3CH_2CH(CH_3)C{\equiv}CH \xrightarrow[\text{or } H_2,\ \text{Lindlar}]{Li/NH_3} CH_3CH_2CH(CH_3)CH{=}CH_2$$

3-Methyl-1-pentyne → 3-Methyl-1-pentene

8.8 Oxidative cleavage reactions of alkynes yield carboxylic acids or CO_2 (with terminal alkynes).

(a)

$$C_6H_5C{\equiv}CH \xrightarrow[H_3O^+]{KMnO_4} C_6H_5COOH + CO_2$$

(b)

$$CH_3(CH_2)_7C{\equiv}C(CH_2)_7C{\equiv}C(CH_2)_7CH_3 \xrightarrow[H_3O^+]{KMnO_4} 2\ CH_3(CH_2)_7COOH + HOOC(CH_2)_7COOH$$

8.9 A base that is strong enough to deprotonate acetone must be the conjugate base of an acid weaker than acetone. In this problem, only $Na^+\ ^-C{\equiv}CH$ is a base strong enough to deprotonate acetone.

8.10 Remember that the desired alkyne must be a terminal alkyne and the desired halide must be primary. More than one combination of terminal alkyne and halide may be possible.

	Alkyne	*R'X (X=Br or I)*	*Product*
(a)	$CH_3CH_2CH_2C{\equiv}CH$	CH_3X	$CH_3CH_2CH_2C{\equiv}CCH_3$ 2-Hexyne
	or $HC{\equiv}CCH_3$	$CH_3CH_2CH_2X$	
(b)	$(CH_3)_2CHC{\equiv}CH$	CH_3CH_2X	$(CH_3)_2CHC{\equiv}CCH_2CH_3$ 2-Methyl-3-hexyne
(c)	$C_6H_{11}C{\equiv}CH$ (cyclohexyl)	CH_3X	$C_6H_{11}C{\equiv}CCH_3$ (cyclohexyl)
(d)	$(CH_3)_2CHCH_2C{\equiv}CH$	CH_3X	$(CH_3)_2CHCH_2C{\equiv}CCH_3$ 5-Methyl-2-hexyne
	or $HC{\equiv}CCH_3$	$(CH_3)_2CHCH_2X$	

(e)

$HC{\equiv}CC(CH_3)_3$ $\quad$ CH_3CH_2X $\quad$ $CH_3CH_2C{\equiv}CC(CH_3)_3$
2,2-Dimethyl-3-hexyne

Products (b), (c), and (e) can be synthesized by only one route because only primary halides can be used for acetylide alkylations.

8.11 The cis double bond can be formed by hydrogenation of an alkyne, which can be synthesized by an alkylation reaction of a terminal alkyne.

$$CH_3C{\equiv}CH \xrightarrow[\text{2. } CH_3Br,\ THF]{\text{1. } NaNH_2,\ NH_3} CH_3C{\equiv}CCH_3 \xrightarrow[\text{Lindlar catalyst}]{H_2} \text{(H)(H}_3\text{C)C=C(H)(CH}_3\text{)}$$

cis-2-Butene

8.12 The starting material is $CH_3CH_2CH_2C{\equiv}CCH_2CH_2CH_3$.Look at the functional groups in the target molecule and work backward to 4-octyne.

(a) $KMnO_4$ cleaves 4-octyne into two four-carbon fragments.

$$CH_3CH_2CH_2C{\equiv}CCH_2CH_2CH_3 \xrightarrow[H_3O^+]{KMnO_4} 2\ CH_3CH_2CH_2COOH$$

Butanoic acid

(b) To reduce a triple bond to a double bond with *cis* stereochemistry use H_2 with Lindlar catalyst.

$$CH_3CH_2CH_2C{\equiv}CCH_2CH_2CH_3 \xrightarrow[\text{Lindlar catalyst}]{H_2} \text{(CH}_3\text{CH}_2\text{CH}_2\text{)(H)C=C(H)(CH}_2\text{CH}_2\text{CH}_3\text{)}$$

cis-4-Octene

(c) Addition of HBr to *cis*-4-octene (part b) yields 4-bromooctane.

$$\text{(CH}_3\text{CH}_2\text{CH}_2\text{)(H)C=C(H)(CH}_2\text{CH}_2\text{CH}_3\text{)} \xrightarrow{HBr} CH_3CH_2CH_2CHBrCH_2CH_2CH_2CH_3$$

from (b) $\quad$ 4-Bromooctane

Alternatively, lithium/ammonia reduction of 4-octyne, followed by addition of HBr, gives 4-bromooctane.

(d) Hydration or hydroboration/oxidation of *cis*-4-octene (part b) yields 4-hydroxyoctane (4-octanol).

$$\underset{\text{from (b)}}{(CH_3CH_2CH_2)HC{=}CH(CH_2CH_2CH_3)} \xrightarrow[\substack{or \\ 1.\ BH_3,\ THF \\ 2.\ H_2O_2,\ ^-OH}]{\substack{1.\ Hg(OAc)_2,\ H_2O \\ 2.\ NaBH_4}} \underset{\text{4-Hydroxyoctane}}{CH_3CH_2CH_2CH(OH)CH_2CH_2CH_2CH_3}$$

(e) Addition of Cl_2 to 4-octene (part b) yields 4,5-dichlorooctane.

$$\underset{\text{from (b)}}{(CH_3CH_2CH_2)HC{=}CH(CH_2CH_2CH_3)} \xrightarrow{Cl_2} \underset{\text{4,5-Dichlorooctane}}{CH_3CH_2CH_2CH(Cl)CH(Cl)CH_2CH_2CH_3}$$

8.13 The following syntheses are explained in detail in order to illustrate retrosynthetic logic -- the system of planning syntheses by working backwards.

(a) 1. An immediate precursor to $CH_3CH_2CH_2CH_2CH_2CH_2CH_2CH_2CH_2CH_3$ might be an alkene or alkyne. Try $C_8H_{17}C{\equiv}CH$, which can be reduced to decane by H_2/Pd.

2. The alkyne $C_8H_{17}C{\equiv}CH$ can be formed by alkylation of $HC{\equiv}C{:}^-\,Na^+$ by $C_8H_{17}Br$, 1-bromooctane.

3. $HC{\equiv}C{:}^-\,Na^+$ can be formed by treatment of $HC{\equiv}CH$ with $NaNH_2$, NH_3.

The complete sequence:

$$HC{\equiv}CH \xrightarrow[NH_3]{NaNH_2} HC{\equiv}C{:}^-\,Na^+ \xrightarrow[THF]{C_8H_{17}Br} C_8H_{17}C{\equiv}CH \xrightarrow{H_2/Pd} \text{Decane}$$

$C_8H_{17}Br$ = 1-Bromooctane

(b) 1. An immediate precursor to $CH_3CH_2CH_2CH_2C(CH_3)_3$ might be $HC{\equiv}CCH_2CH_2C(CH_3)_3$, which, when hydrogenated, yields 2,2–dimethylhexane.

2. $HC{\equiv}CCH_2CH_2C(CH_3)_3$ can be formed by alkylation of $HC{\equiv}C{:}^-\,Na^+$ with $BrCH_2CH_2C(CH_3)_3$.

The complete sequence:

$$HC{\equiv}CH \xrightarrow[NH_3]{NaNH_2} HC{\equiv}C{:}^-\,Na^+$$

$$BrCH_2CH_2C(CH_3)_3 \xrightarrow[THF]{HC{\equiv}C{:}^-\,Na^+} HC{\equiv}CCH_2CH_2C(CH_3)_3 \xrightarrow[Pd]{2\ H_2} \underset{\text{2,2-Dimethylhexane}}{CH_3CH_2CH_2CH_2C(CH_3)_3}$$

(c) 1. $CH_3CH_2CH_2CH_2CH_2CHO$ can be made by treating $CH_3CH_2CH_2CH_2C{\equiv}CH$ with borane, followed by H_2O_2.

2. $CH_3CH_2CH_2CH_2C{\equiv}CH$ can be synthesized from $CH_3CH_2CH_2CH_2Br$ and $HC{\equiv}C{:}^-\,Na^+$.

The complete sequence:

$$HC{\equiv}CH \xrightarrow[NH_3]{NaNH_2} HC{\equiv}C{:}^-\,Na^+$$

$$CH_3CH_2CH_2CH_2Br + HC{\equiv}C{:}^-\,Na^+ \xrightarrow{THF} CH_3CH_2CH_2CH_2C{\equiv}CH \xrightarrow[2.\ H_2O_2,\ ^-OH]{1.\ BH_3,\ THF} CH_3CH_2CH_2CH_2CH_2CH{=}O$$

Hexanal

(d) 1. The desired ketone can be formed by mercuric ion-catalyzed hydration of 1-heptyne.

2. 1-Heptyne can be synthesized by an alkylation of sodium acetylide by 1-bromopentane.

The complete sequence:

$$HC{\equiv}CH \xrightarrow[NH_3]{NaNH_2} HC{\equiv}C{:}^-\,Na^+$$

$$HC{\equiv}C{:}^-\,Na^+ + CH_3CH_2CH_2CH_2CH_2Br \xrightarrow{THF} CH_3CH_2CH_2CH_2CH_2C{\equiv}CH$$

$$CH_3CH_2CH_2CH_2CH_2C{\equiv}CH \xrightarrow[HgSO_4]{H_2SO_4,\ H_2O} CH_3CH_2CH_2CH_2CH_2C(=O)CH_3$$

2-Heptanone

Visualizing Chemistry

8.14

(a)

$$HC{\equiv}CCH_2C(CH_3)_2CH_2CH_3$$

4,4-Dimethyl-1-hexyne

(i) H_2, Lindlar catalyst $\rightarrow$ $H_2C{=}CHCH_2C(CH_3)_2CH_2CH_3$

(ii) H_3O^+, $HgSO_4$ $\rightarrow$ $CH_3C(=O)CH_2C(CH_3)_2CH_2CH_3$

(b)

$$CH_3CHCH_2C{\equiv}CCH_2CHCH_3 \text{ (with } CH_3 \text{ on C-2 and C-7)}$$

2,7-Dimethyl-4-octyne

(i) H_2, Lindlar catalyst → $(CH_3)_2CHCH_2CH{=}CHCH_2CH(CH_3)_2$ (cis; H and H on the same side)

(ii) H_3O^+, $HgSO_4$ → $CH_3CH(CH_3)CH_2CH_2C(=O)CH_2CH(CH_3)CH_3$

8.15

(a)

$$CH_3CH(CH_3)CH_2C{\equiv}CH \xrightarrow[2.\ H_2O_2,\ ^-OH]{1.\ BH_3,\ THF} CH_3CH(CH_3)CH_2CH_2CH{=}O$$

4-Methyl-1-pentyne

An aldehyde is formed by reacting a terminal alkyne with borane, followed by oxidation.

(b)

$$\text{(CH}_3\text{-cyclohexyl)}C{\equiv}CH \xrightarrow[\text{ether}]{2\ HCl} \text{(CH}_3\text{-cyclohexyl)}CCl_2CH_3$$

HCl must be used if both chlorines are to be on the same carbon.

8.16 First, draw the structure of each target compound. Then, analyze the structures for a synthetic route.

(a) $CH_3CH(OH)CH_2CH{-}CH_2$ (cyclopropyl ring closed by CH_2)

(b) $H_2C{=}CHCH_2CH_2C(=O)CH_3$

(a) The left side and the right side might have double bonds as immediate precursors; the right side may result from a Simmons-Smith carbenoid addition to an alkene, and the left side may result from hydration of an alkene. Let's start with 3-bromo-1-propene.

$$BrCH_2CH{=}CH_2 \xrightarrow[Zn(Cu)]{CH_2I_2} BrCH_2CH{-}CH_2\ (\text{cyclopropane with } CH_2) \xrightarrow{HC{\equiv}C{:}^-\ Na^+} HC{\equiv}CCH_2CH{-}CH_2\ (\text{cyclopropane with } CH_2)$$

$$\xrightarrow[\text{catalyst}]{H_2,\ \text{Lindlar}} H_2C{=}CHCH_2CH{-}CH_2\ (\text{cyclopropane with } CH_2) \xrightarrow[2.\ NaBH_4]{1.\ Hg(OAc)_2,\ H_2O} CH_3CH(OH)CH_2CH{-}CH_2\ (\text{cyclopropane with } CH_2)$$

(b) The right side can result from Hg-catalyzed addition of H_2O to a terminal alkyne.

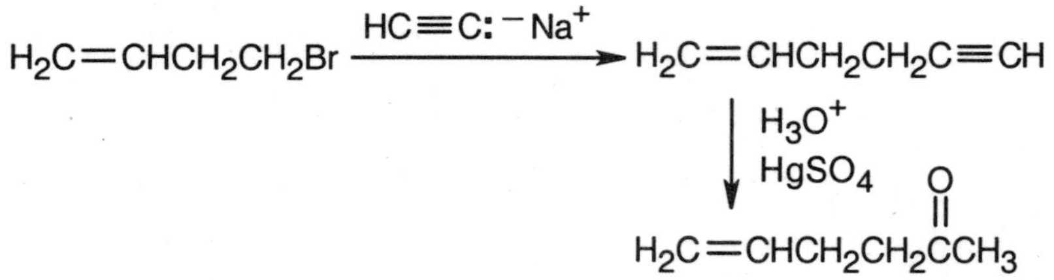

$$H_2C{=}CHCH_2CH_2Br \xrightarrow{HC{\equiv}C{:}^{-}\ Na^{+}} H_2C{=}CHCH_2CH_2C{\equiv}CH$$

$$\downarrow H_3O^+,\ HgSO_4$$

$$H_2C{=}CHCH_2CH_2C({=}O)CH_3$$

8.17 It's not possible to form a small ring containing a triple bond because the angle strain that would result from bending the bonds of an *sp*-hybridized carbon to form a small ring is too great.

Additional Problems

8.18

(a) $CH_3CH_2C{\equiv}CC(CH_3)_2CH_3$

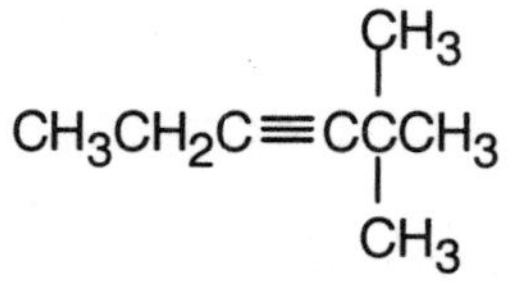

2,2-Dimethyl-3-hexyne

(b) $CH_3C{\equiv}CCH_2C{\equiv}CCH_2CH_3$

2,5-Octadiyne

(c) $CH_3CH{=}C(CH_3)C{\equiv}CCH(CH_3)CH_3$

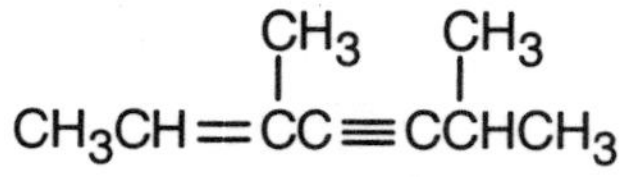

3,6-Dimethyl-2-hepten-4-yne

(d) $HC{\equiv}CC(CH_3)_2CH_2C{\equiv}CH$

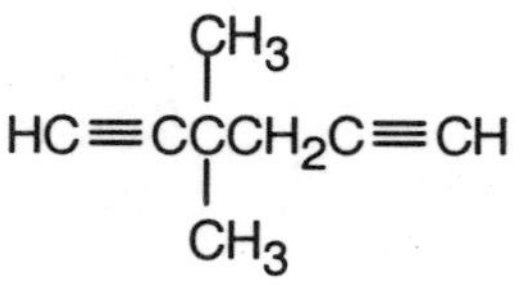

3,3-Dimethyl-1,5-hexadiyne

(e) $H_2C{=}CHCH{=}CHC{\equiv}CH$

1,3-Hexadien-5-yne

(f) $CH_3CH_2CH(CH_2CH_3)C{\equiv}CCH(CH_2CH_3)CH(CH_3)CH_3$

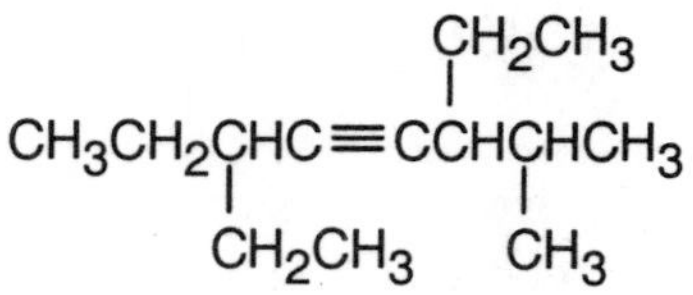

3,6-Diethyl-2-methyl-4-octyne

8.19

(a) $CH_3CH_2CH_2C{\equiv}CC(CH_3)_2CH_2CH_3$

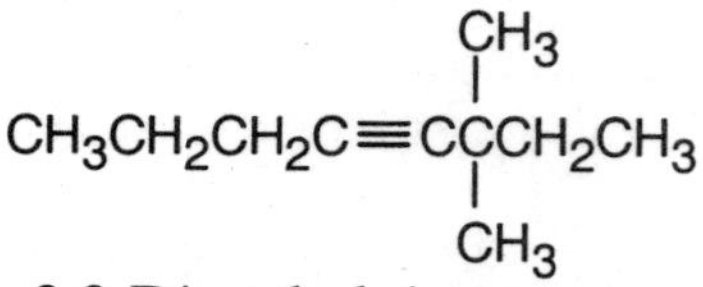

3,3-Dimethyl-4-octyne

(b) $CH_3C{\equiv}CC{\equiv}CCH(CH_3)CH_2CH(CH_2CH_3)C{\equiv}CH$

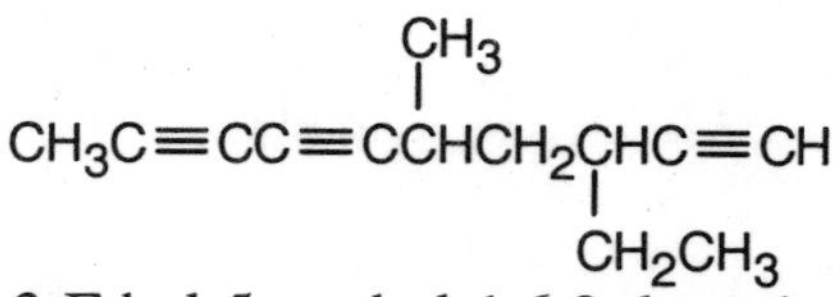

3-Ethyl-5-methyl-1,6,8-decatriyne

(c) $(CH_3)_3CC{\equiv}CC(CH_3)_3$

2,2,5,5-Tetramethyl-3-hexyne

(d) Ring: H_2C–$CH_2C{\equiv}CCH(CH_3)$–$CHCH_3$–$CH_2CH_2CH_2CH_2$–(back to H_2C)

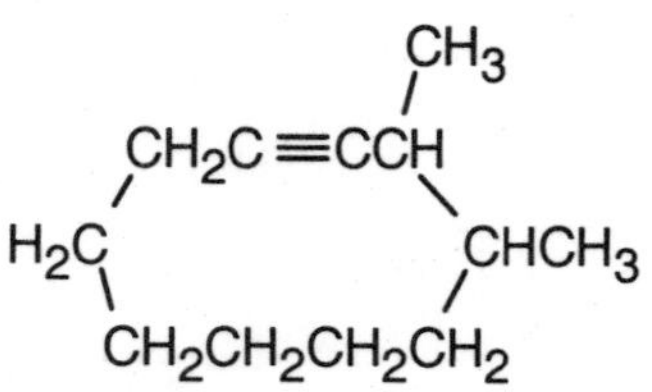

3,4-Dimethylcyclodecyne

(e)

$CH_3CH{=}CHCH{=}CHC{\equiv}CH$

3,5-Heptadien-1-yne

(f)

$CH_3CH_2C{\equiv}CCH_2C(CH_3)_2{-}CHClCH{=}CH_2$

3-Chloro-4,4-dimethyl-1-nonen-6-yne

(g)

$CH_3CH_2CH_2CH_2CH(CH(CH_3)CH_2CH_3)C{\equiv}CH$

3-*sec*-Butyl-1-heptyne

(h)

$CH_3CH_2CH_2CH(C(CH_3)_3)C{\equiv}CCH(CH_3)CH_3$

5-*tert*-Butyl-2-methyl-3-octyne

8.20 (a) $CH_3CH{=}CHC{\equiv}CC{\equiv}CCH{=}CHCH{=}CHCH{=}CH_2$.
1,3,5,11–Tridecatetraen-7,9–diyne

Using *E–Z* notation: (3*E*,5*E*,11*E*)–1,3,5,11–Tridecatetraen–7,9–diyne
The parent alkane of this hydrocarbon is tridecane.

(b) $CH_3C{\equiv}CC{\equiv}CC{\equiv}CC{\equiv}CC{\equiv}CCH{=}CH_2$. 1–Tridecen–3,5,7,9,11–pentayne
This hydrocarbon is also of the tridecane family.

8.21

Starting compound: $C_6H_5{-}CH{=}CH{-}C{\equiv}CH$

H_2, Pd/C → **A**: $C_6H_5{-}CH_2CH_2CH_2CH_3$

H_2, Lindlar → **B**: $C_6H_5{-}CH{=}CH{-}CH{=}CH_2$

8.22 (a) An acyclic alkane with eight carbons has the formula C_8H_{18}. C_8H_{10} has eight fewer hydrogens, or four fewer pairs of hydrogens, than C_8H_{18}. Thus, C_8H_{10} contains four degrees of unsaturation (rings/double bonds/ triple bonds).
(b) Because only one equivalent of H_2 is absorbed over the Lindlar catalyst, *one* triple bond is present.
(c) Three equivalents of H_2 are absorbed when reduction is done over a palladium catalyst; two of them hydrogenate the triple bond already found to be present. Therefore, one *double* bond must also be present.
(d) C_8H_{10} must contain one ring.
(e) Many structures are possible.

1-ethynylcyclohexene ($C{\equiv}CH$ on cyclohexene ring)

$H_2C{=}CH{-}$(cyclobutane-1,3-diyl)${-}C{\equiv}CH$

$H_2C{=}CHCH(C_3H_5)C{\equiv}CH$ (cyclopropyl substituent)

8.23

(a)

$$CH_3CH_2CH_2CH_2C{\equiv}CH \xrightarrow[\text{HBr}]{\text{1 equiv}} CH_3CH_2CH_2CH_2C(Br){=}CH_2$$

(b)

$$CH_3CH_2CH_2CH_2C{\equiv}CH \xrightarrow[\text{Cl}_2]{\text{1 equiv}} (CH_3CH_2CH_2CH_2)(Cl)C{=}C(Cl)(H)$$

(c)

$$CH_3CH_2CH_2CH_2C{\equiv}CH \xrightarrow[\text{Lindlar catalyst}]{H_2} CH_3CH_2CH_2CH_2CH{=}CH_2$$

(d)

$$CH_3CH_2CH_2CH_2C{\equiv}CH \xrightarrow[\text{2. }CH_3Br]{\text{1. }NaNH_2,\ NH_3} CH_3CH_2CH_2CH_2C{\equiv}CCH_3$$

(e)

$$CH_3CH_2CH_2CH_2C{\equiv}CH \xrightarrow[HgSO_4]{H_2O,\ H_2SO_4} CH_3CH_2CH_2CH_2C({=}O)CH_3$$

(f)

$$CH_3CH_2CH_2CH_2C{\equiv}CH \xrightarrow[\text{HCl}]{\text{2 equiv}} CH_3CH_2CH_2CH_2CCl_2CH_3$$

8.24

(a)

$$CH_3(CH_2)_3C{\equiv}C(CH_2)_3CH_3 \xrightarrow[\text{Lindlar catalyst}]{H_2} (CH_3CH_2CH_2CH_2)(H)C{=}C(CH_2CH_2CH_2CH_3)(H)$$

(b)

$$CH_3(CH_2)_3C{\equiv}C(CH_2)_3CH_3 \xrightarrow{\text{Li in }NH_3} (CH_3CH_2CH_2CH_2)(H)C{=}C(H)(CH_2CH_2CH_2CH_3)$$

(c)

$$CH_3(CH_2)_3C{\equiv}C(CH_2)_3CH_3 \xrightarrow[Br_2]{\text{1 equiv}} (CH_3CH_2CH_2CH_2)(Br)C{=}C(Br)(CH_2CH_2CH_2CH_3)$$

(d)

$$CH_3(CH_2)_3C{\equiv}C(CH_2)_3CH_3 \xrightarrow[\text{2. }H_2O_2,\ ^-OH]{\text{1. }BH_3,\ THF} CH_3CH_2CH_2CH_2CH_2C({=}O)CH_2CH_2CH_2CH_3$$

(e)

$$CH_3(CH_2)_3C{\equiv}C(CH_2)_3CH_3 \xrightarrow[HgSO_4]{H_2O,\ H_2SO_4} CH_3CH_2CH_2CH_2CH_2C({=}O)CH_2CH_2CH_2CH_3$$

(f)

$$CH_3(CH_2)_3C{\equiv}C(CH_2)_3CH_3 \xrightarrow[\text{Pd/C}]{\text{excess } H_2} CH_3(CH_2)_8CH_3$$

8.25

(a)

$$CH_3CH_2CH_2C{\equiv}CCH_3 \xrightarrow[Br_2]{\text{2 equiv}} CH_3CH_2CH_2CBr_2CBr_2CH_3$$

(b)

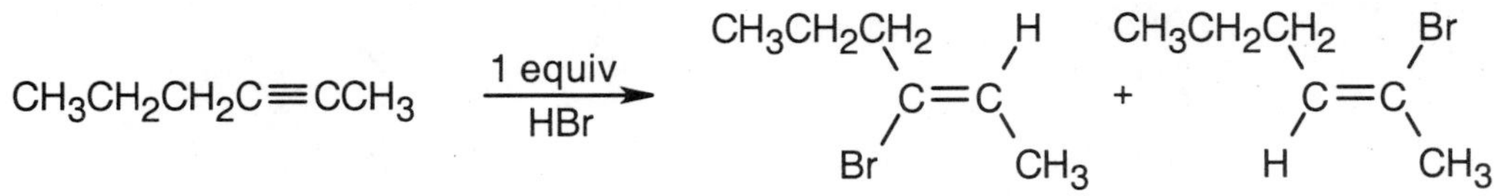

$$CH_3CH_2CH_2C{\equiv}CCH_3 \xrightarrow[HBr]{\text{1 equiv}} (CH_3CH_2CH_2)(Br)C{=}C(H)(CH_3) + (CH_3CH_2CH_2)(H)C{=}C(Br)(CH_3)$$

(c)

$$CH_3CH_2CH_2C{\equiv}CCH_3 \xrightarrow[HBr]{\text{excess}} CH_3CH_2CH_2CBr_2CH_2CH_3 + CH_3CH_2CH_2CH_2CBr_2CH_3$$

(d)

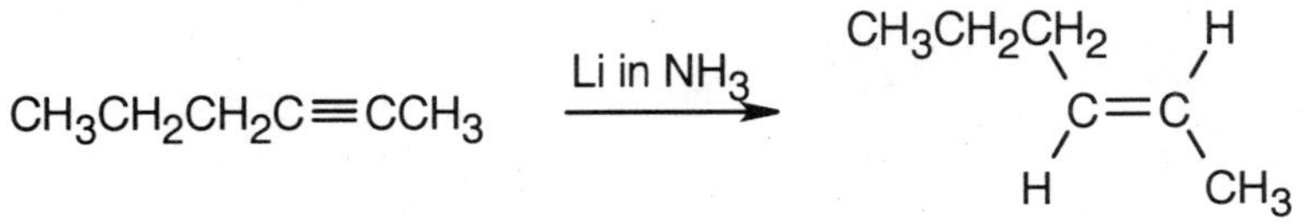

$$CH_3CH_2CH_2C{\equiv}CCH_3 \xrightarrow{\text{Li in } NH_3} (CH_3CH_2CH_2)(H)C{=}C(H)(CH_3)$$

(e)

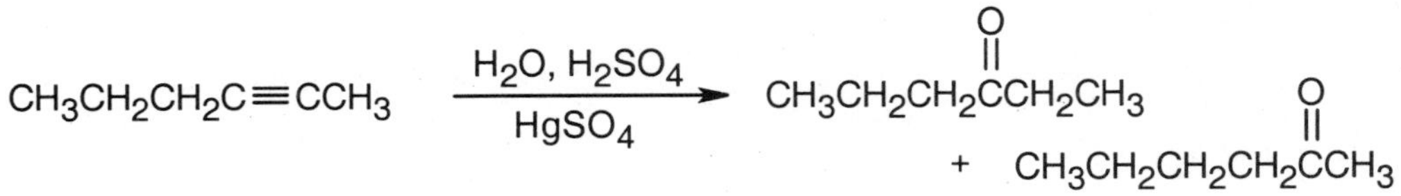

$$CH_3CH_2CH_2C{\equiv}CCH_3 \xrightarrow[HgSO_4]{H_2O,\ H_2SO_4} CH_3CH_2CH_2\overset{O}{\overset{\|}{C}}CH_2CH_3 + CH_3CH_2CH_2CH_2\overset{O}{\overset{\|}{C}}CH_3$$

8.26

(a)

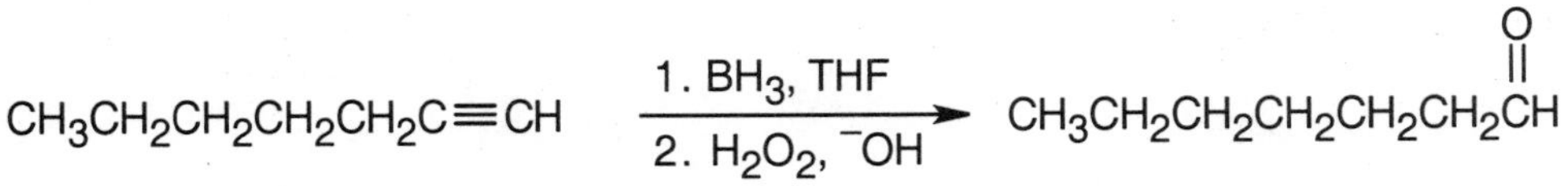

$$CH_3CH_2CH_2CH_2CH_2C{\equiv}CH \xrightarrow[\text{2. } H_2O_2,\ ^-OH]{\text{1. } BH_3,\ THF} CH_3CH_2CH_2CH_2CH_2CH_2\overset{O}{\overset{\|}{C}}H$$

(b)

$$C_6H_5C(Br){=}CH_2 \xrightarrow{NaNH_2,\ NH_3} C_6H_5C{\equiv}CH$$

(c)

$$C_6H_5CBr_2CH_3 \xrightarrow{2\ NaNH_2,\ NH_3} C_6H_5C{\equiv}CH$$

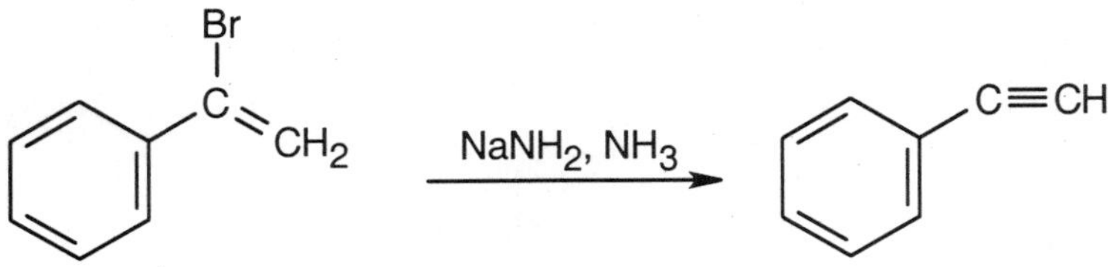

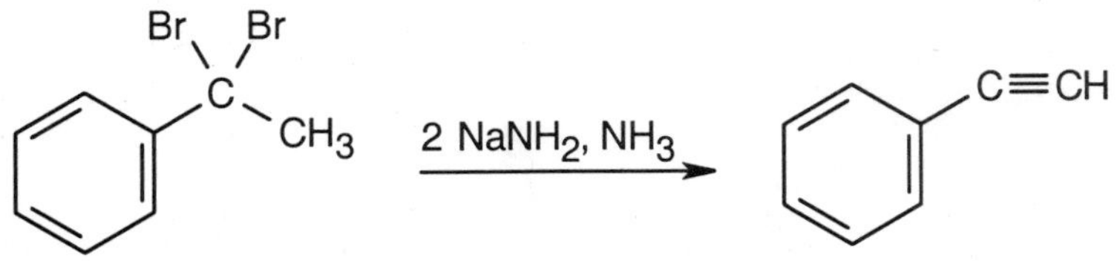

8.27

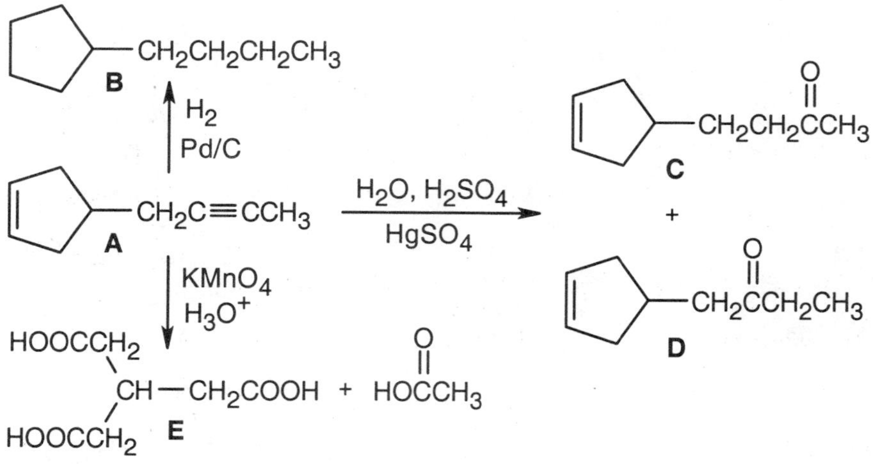

B: cyclopentyl–$CH_2CH_2CH_2CH_3$

$\uparrow$ H_2, Pd/C

A: cyclopentenyl–$CH_2C{\equiv}CCH_3$ $\xrightarrow[HgSO_4]{H_2O,\ H_2SO_4}$ C: cyclopentenyl–$CH_2CH_2\overset{O}{\overset{\|}{C}}CH_3$ + D: cyclopentenyl–$CH_2\overset{O}{\overset{\|}{C}}CH_2CH_3$

$\downarrow$ $KMnO_4$, H_3O^+

E: $(HOOCCH_2)_2CH{-}CH_2COOH$ + $HO\overset{O}{\overset{\|}{C}}CH_3$

8.28

(a)

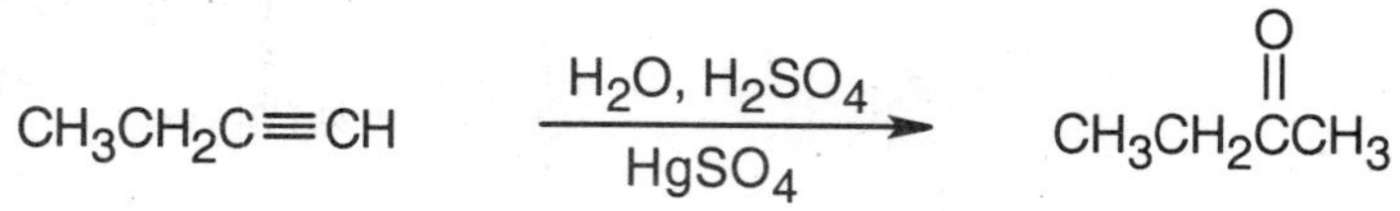

$$CH_3CH_2C{\equiv}CH \xrightarrow[HgSO_4]{H_2O,\ H_2SO_4} CH_3CH_2\overset{O}{\overset{\|}{C}}CH_3$$

(b)

$$CH_3CH_2C{\equiv}CH \xrightarrow[2.\ H_2O_2,\ ^-OH]{1.\ BH_3,\ THF} CH_3CH_2CH_2CHO$$

(c)

$$C_6H_5C{\equiv}CH \xrightarrow[2.\ CH_3Br]{1.\ NaNH_2,\ NH_3} C_6H_5C{\equiv}CCH_3$$

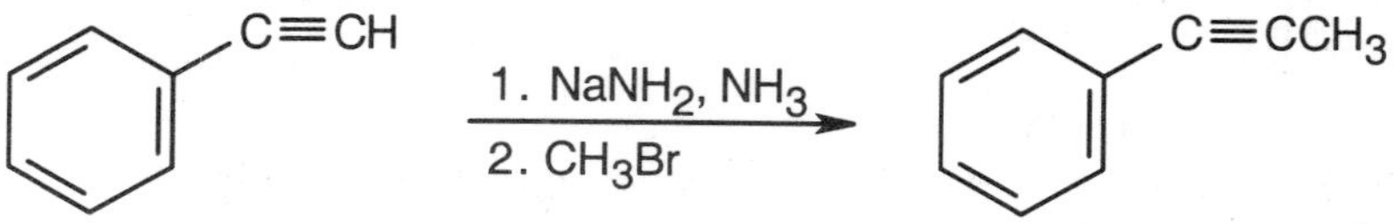

(d)

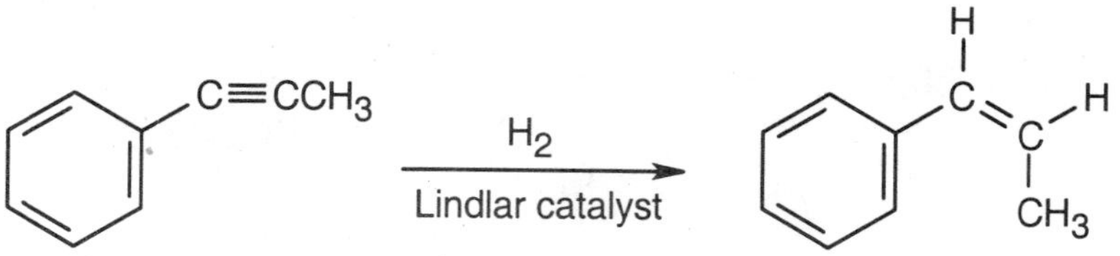

$$C_6H_5C{\equiv}CCH_3 \xrightarrow[\text{Lindlar catalyst}]{H_2} C_6H_5(H)C{=}C(H)CH_3$$

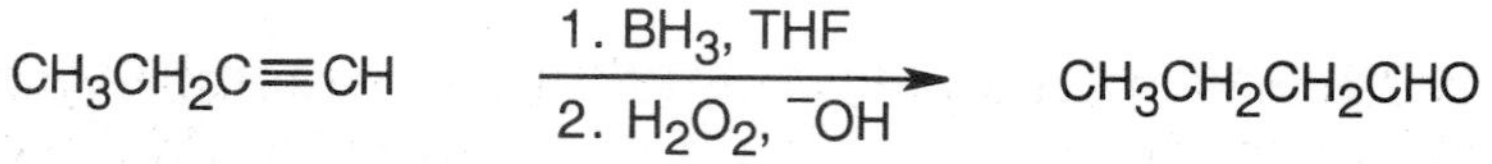

(e)

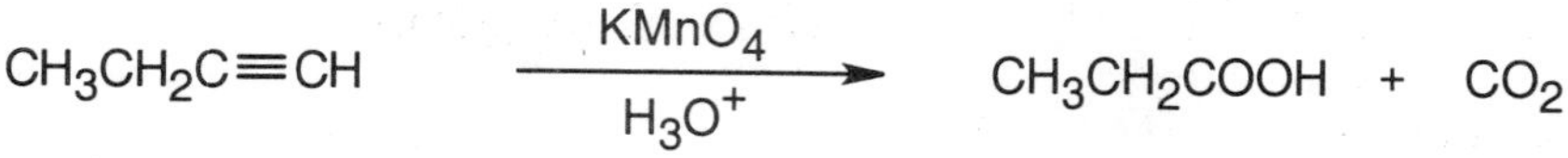

$$CH_3CH_2C{\equiv}CH \xrightarrow[H_3O^+]{KMnO_4} CH_3CH_2COOH + CO_2$$

(f)

$$CH_3CH_2CH_2CH_2CH{=}CH_2 \xrightarrow[CH_2Cl_2]{Br_2} CH_3CH_2CH_2CH_2CH(Br)CH_2Br$$

$\downarrow$ 1. 2 $NaNH_2$, NH_3; 2. H_3O^+

$$CH_3CH_2CH_2CH_2C{\equiv}CH$$

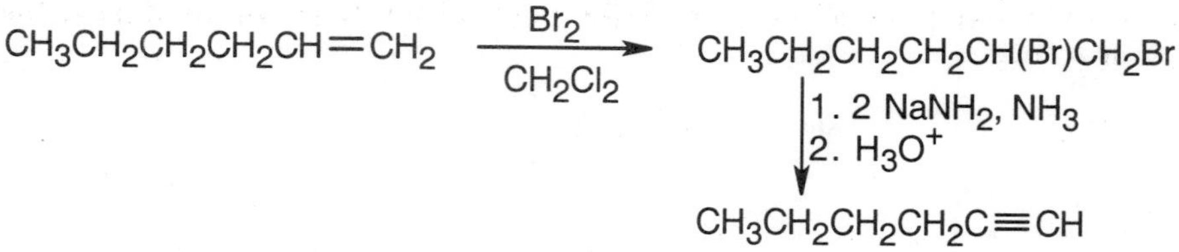

8.29

(a)

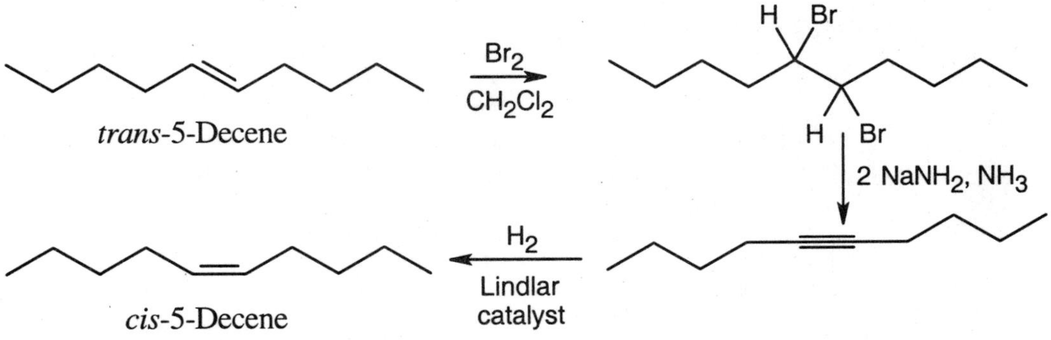

trans-5-Decene $\xrightarrow[CH_2Cl_2]{Br_2}$ 5,6-dibromodecane (H, Br; H, Br) $\xrightarrow{2\ NaNH_2,\ NH_3}$ 5-decyne $\xrightarrow[\text{Lindlar catalyst}]{H_2}$ *cis*-5-Decene

(b)

cis-5-Decene $\xrightarrow[CH_2Cl_2]{Br_2}$ 5,6-dibromodecane (H, Br; H, Br) $\xrightarrow{2\ NaNH_2,\ NH_3}$ 5-decyne $\xrightarrow{\text{Li in } NH_3}$ *trans*-5-Decene

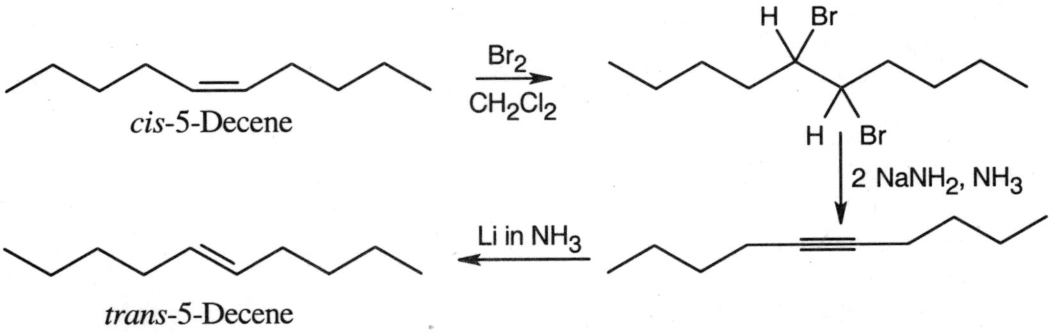

8.30 Both $KMnO_4$ and O_3 oxidation of alkynes yield carboxylic acids; terminal alkynes give CO_2 also. In (a), (b), and (c), the observed products can be formed by $KMnO_4$ oxidation of the corresponding alkenes.

(a)

$$CH_3(CH_2)_5C{\equiv}CH \xrightarrow[H_3O^+]{KMnO_4} CH_3(CH_2)_5COOH + CO_2$$

(b)

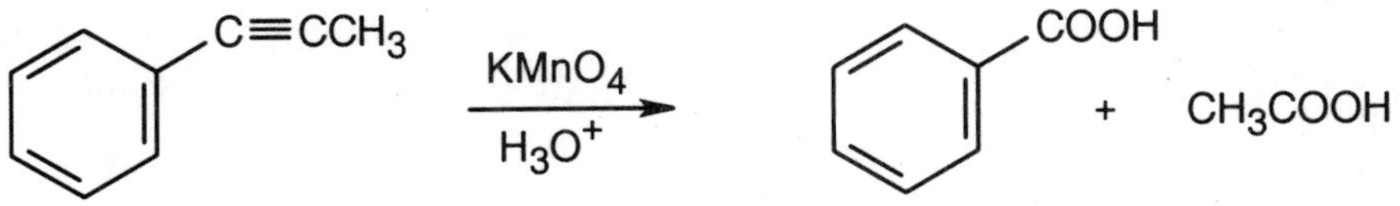

$$C_6H_5C{\equiv}CCH_3 \xrightarrow[H_3O^+]{KMnO_4} C_6H_5COOH + CH_3COOH$$

(c)

$$\text{cyclodecyne} \xrightarrow[H_3O^+]{KMnO_4} HOOC(CH_2)_8COOH$$

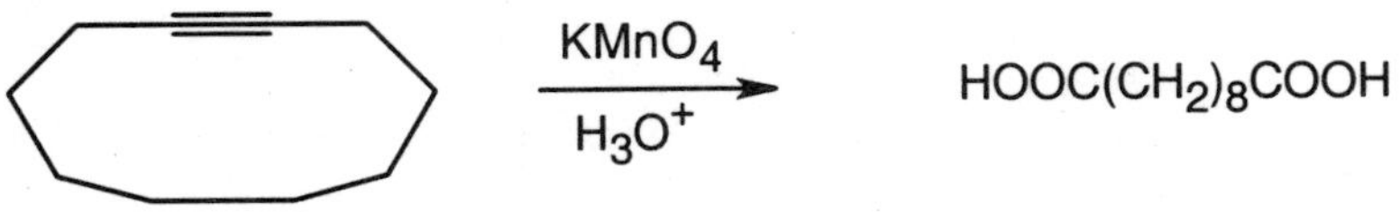

Since only one cleavage product is formed, the parent hydrocarbon must have had a triple bond as part of a ring.

(d)

$$CH_3CH{=}C(CH_3)CH_2CH_2C{\equiv}CH \xrightarrow[2.\ Zn,\ H_3O^+]{1.\ O_3} CH_3CHO + CH_3C(=O)CH_2CH_2COOH + CO_2$$

Notice that the products of this ozonolysis contain aldehyde and ketone functional groups, as well as a carboxylic acid and CO_2. The parent hydrocarbon must thus contain a double and a triple bond.

(e)

$$\text{1-ethynylcyclohexene } (C{\equiv}CH) \xrightarrow[2.\ Zn,\ H_3O^+]{1.\ O_3} OHCCH_2CH_2CH_2CH_2C(=O)COOH + CO_2$$

8.31

(a)

$$CH_3CH_2CH_2C{\equiv}CH \xrightarrow[\text{Lindlar catalyst}]{H_2} CH_3CH_2CH_2CH{=}CH_2 \xrightarrow[2.\ Zn,\ H_3O^+]{1.\ O_3} CH_3CH_2CH_2CHO + CH_2O$$

(b)

$$(CH_3)_2CHCH_2C{\equiv}CH \xrightarrow[2.\ CH_3CH_2Br]{1.\ NaNH_2,\ NH_3} (CH_3)_2CHCH_2C{\equiv}CCH_2CH_3$$

$$\xrightarrow{\text{Li in } NH_3} \text{trans-}(CH_3)_2CHCH_2CH{=}CHCH_2CH_3 \text{ (H, } CH_2CH_3 \text{ / } (CH_3)_2CHCH_2\text{, H)}$$

8.32 Synthetic analysis: The product contains a cis-disubstituted cyclopropane ring, which can be formed from a Simmons-Smith reaction of CH_2I_2 with a cis alkene. The alkene with a cis bond can be produced from an alkyne by hydrogenation using a Lindlar catalyst. The needed alkyne can be formed from the starting material shown by an alkylation using bromomethane.

$$CH_3CH_2CH_2CH_2C{\equiv}CH \xrightarrow[2.\ CH_3Br]{1.\ NaNH_2,\ NH_3} CH_3CH_2CH_2CH_2C{\equiv}CCH_3$$

$$\xrightarrow[\text{Lindlar catalyst}]{H_2} \text{cis-}CH_3CH_2CH_2CH_2CH{=}CHCH_3 \text{ (H, H / } CH_3CH_2CH_2CH_2\text{, } CH_3\text{)}$$

$$\xrightarrow[Zn\ (Cu)]{CH_2I_2} \text{cis-1-butyl-2-methylcyclopropane (ring } CH_2\text{; H, H; } CH_3CH_2CH_2CH_2\text{, } CH_3\text{)}$$

8.33

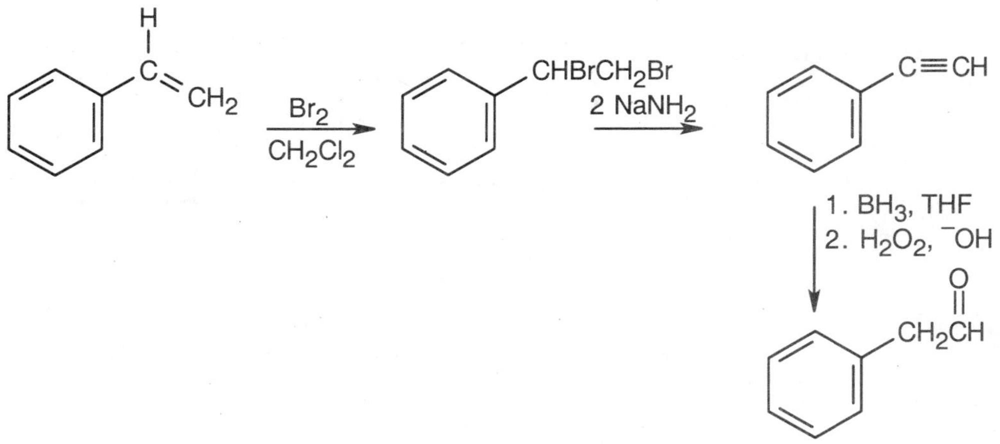

Styrene ($C_6H_5CH{=}CH_2$) $\xrightarrow[CH_2Cl_2]{Br_2}$ $C_6H_5CHBrCH_2Br$ $\xrightarrow{2\ NaNH_2}$ $C_6H_5C{\equiv}CH$ $\xrightarrow[2.\ H_2O_2,\ ^-OH]{1.\ BH_3,\ THF}$ $C_6H_5CH_2CHO$

8.34 Synthetic analysis: The trans double bond in the target molecule is a product of reduction of a triple bond with Li in NH_3. The alkyne was formed by an alkylation of a terminal alkyne with bromomethane. The terminal alkyne was synthesized from the starting alkene by bromination, followed by dehydrohalogenation.

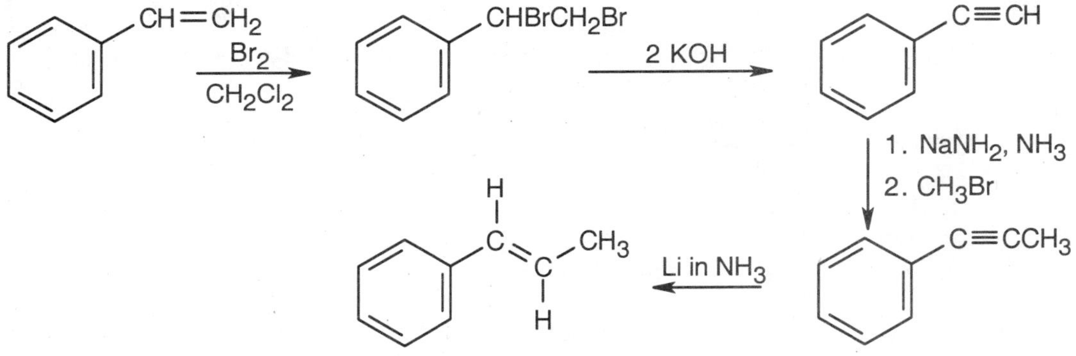

$C_6H_5CH{=}CH_2$ $\xrightarrow[CH_2Cl_2]{Br_2}$ $C_6H_5CHBrCH_2Br$ $\xrightarrow{2\ KOH}$ $C_6H_5C{\equiv}CH$ $\xrightarrow[2.\ CH_3Br]{1.\ NaNH_2,\ NH_3}$ $C_6H_5C{\equiv}CCH_3$ $\xrightarrow{Li\ in\ NH_3}$ trans-$C_6H_5CH{=}CHCH_3$

8.35

(a)

$$CH_3CH_2C{\equiv}CH \xrightarrow[CH_2Cl_2]{2\ Cl_2} CH_3CH_2CCl_2CHCl_2$$

1,1,2,2-Tetrachlorobutane

(b)

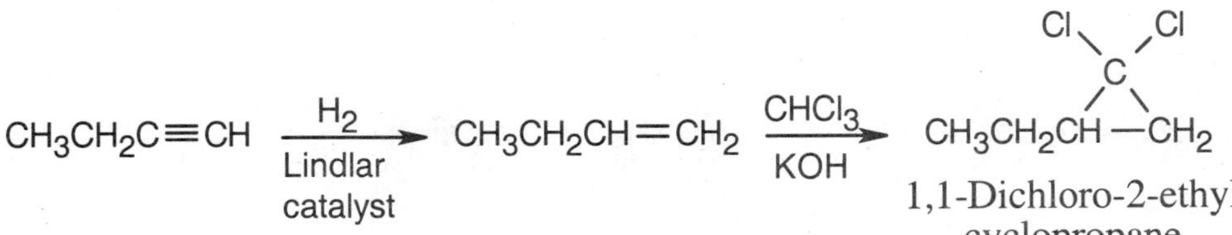

$$CH_3CH_2C{\equiv}CH \xrightarrow[\text{Lindlar catalyst}]{H_2} CH_3CH_2CH{=}CH_2 \xrightarrow[KOH]{CHCl_3} CH_3CH_2CH{-}CH_2 \text{ (bridged by } CCl_2)$$

1,1-Dichloro-2-ethyl-cyclopropane

(c)

$$CH_3CH_2C{\equiv}CH \xrightarrow[2.\ H_2O_2,\ ^-OH]{1.\ BH_3,\ THF} CH_3CH_2CH_2CHO$$

Butanal

8.36 In all of these problems, an acetylide ion (or an anion of a terminal alkyne) is alkylated by a haloalkane.

(a)

$$HC\equiv CH \xrightarrow[\text{2. } CH_3CH_2CH_2Br]{\text{1. } NaNH_2,\ NH_3} CH_3CH_2CH_2C\equiv CH$$

(b)

$$HC\equiv CH \xrightarrow[\text{2. } CH_3CH_2Br]{\text{1. } NaNH_2,\ NH_3} CH_3CH_2C\equiv CH \xrightarrow[\text{2. } CH_3CH_2Br]{\text{1. } NaNH_2,\ NH_3} CH_3CH_2C\equiv CCH_2CH_3$$

(c)

$$HC\equiv CH \xrightarrow[\text{2. } (CH_3)_2CHCH_2Br]{\text{1. } NaNH_2,\ NH_3} (CH_3)_2CHCH_2C\equiv CH$$

$$\downarrow H_2,\ \text{Lindlar catalyst } \textit{or } \text{Li in } NH_3$$

$$(CH_3)_2CHCH_2CH{=}CH_2$$

(d)

$$\underset{\text{from (a)}}{CH_3CH_2CH_2C\equiv CH} \xrightarrow[\text{2. } CH_3CH_2CH_2Br]{\text{1. } NaNH_2,\ NH_3} CH_3CH_2CH_2C\equiv CCH_2CH_2CH_3$$

$$\downarrow H_2O,\ H_2SO_4,\ HgSO_4$$

$$CH_3CH_2CH_2\overset{O}{\overset{\|}{C}}CH_2CH_2CH_2CH_3$$

(e)

$$HC\equiv CH \xrightarrow[\text{2. } CH_3CH_2CH_2CH_2Br]{\text{1. } NaNH_2,\ NH_3} CH_3CH_2CH_2CH_2C\equiv CH$$

$$\downarrow \text{1. } BH_3,\ THF;\ \text{2. } H_2O_2,\ ^-OH$$

$$CH_3CH_2CH_2CH_2CH_2CHO$$

8.37

(a)

$$CH_3CH_2C\equiv CCH_2CH_3 \xrightarrow[\text{Lindlar catalyst}]{D_2} \text{(D)(CH}_3\text{CH}_2\text{)C=C(D)(CH}_2\text{CH}_3\text{)}$$

(cis: both D on the same side; CH_3CH_2 and CH_2CH_3 on the same side)

(b)

$$CH_3CH_2C\equiv CCH_2CH_3 \xrightarrow{\text{Li in } ND_3} \text{(D)(CH}_3\text{CH}_2\text{)C=C(CH}_2\text{CH}_3\text{)(D)}$$

(trans: D and D on opposite sides)

(c)

$$CH_3CH_2CH_2C\equiv CH \xrightarrow[NH_3]{NaNH_2} [CH_3CH_2CH_2C\equiv C{:}^-\ Na^+] \xrightarrow{D_3O^+} CH_3CH_2CH_2C\equiv CD$$

(d)

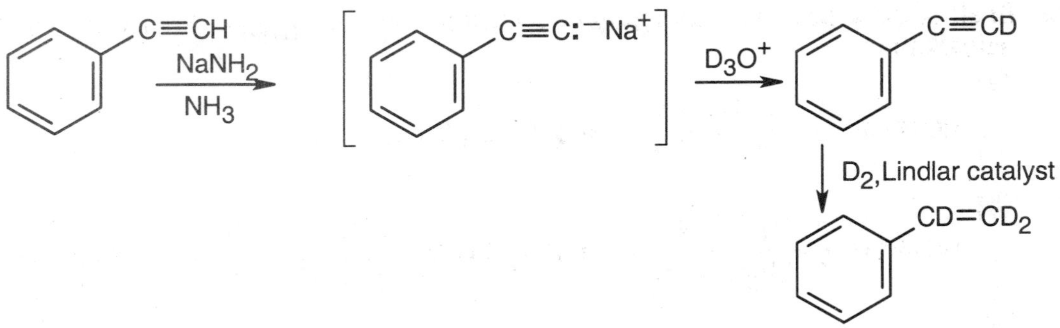

$$C_6H_5C{\equiv}CH \xrightarrow[NH_3]{NaNH_2} \left[C_6H_5C{\equiv}C{:}^- Na^+\right] \xrightarrow{D_3O^+} C_6H_5C{\equiv}CD \xrightarrow{D_2,\ \text{Lindlar catalyst}} C_6H_5CD{=}CD_2$$

8.38

$$HC{\equiv}CH \xrightarrow[2.\ BrCH_2(CH_2)_6CH_2Br]{1.\ NaNH_2,\ NH_3} \text{cyclodecyne}$$

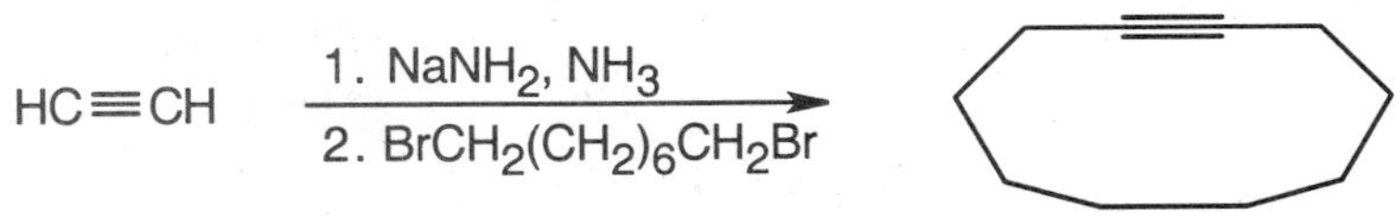

8.39 Synthetic analysis: Muscalure is a C_{23} alkene. The only functional group present is the double bond between C_9 and C_{10}. Since our synthesis begins with acetylene, we can assume that the double bond can be produced by hydrogenation of a triple bond.

$$HC{\equiv}CH \xrightarrow[2.\ CH_3(CH_2)_6CH_2Br]{1.\ NaNH_2,\ NH_3} CH_3(CH_2)_7C{\equiv}CH \xrightarrow[2.\ CH_3(CH_2)_{11}CH_2Br]{1.\ NaNH_2,\ NH_3}$$

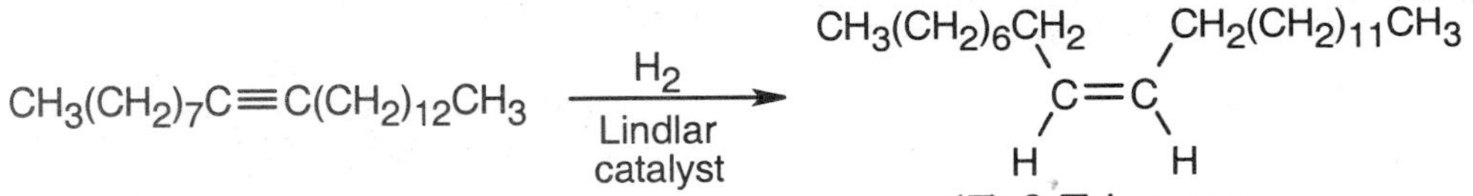

$$CH_3(CH_2)_7C{\equiv}C(CH_2)_{12}CH_3 \xrightarrow[\text{Lindlar catalyst}]{H_2} \underset{H}{\overset{CH_3(CH_2)_6CH_2}{}}C{=}C\underset{H}{\overset{CH_2(CH_2)_{11}CH_3}{}}$$

(*Z*)-9-Tricosene

8.40

A (cyclohexylidene–CH–C≡CH):

$$\mathbf{A} \xrightarrow[Pd/C]{3\ H_2} \mathbf{B}\ (\text{cyclohexyl}-CH_2CH_2CH_3)$$

$$\mathbf{A} \xrightarrow[2.\ Zn,\ H_3O^+]{1.\ O_3} \text{cyclohexanone } (=O) \text{ plus other fragments}$$

$$\mathbf{A} \xrightarrow[2.\ CH_3I]{1.\ NaNH_2,\ NH_3} \mathbf{C}\ (\text{cyclohexylidene}=CH-C{\equiv}C-CH_3)$$

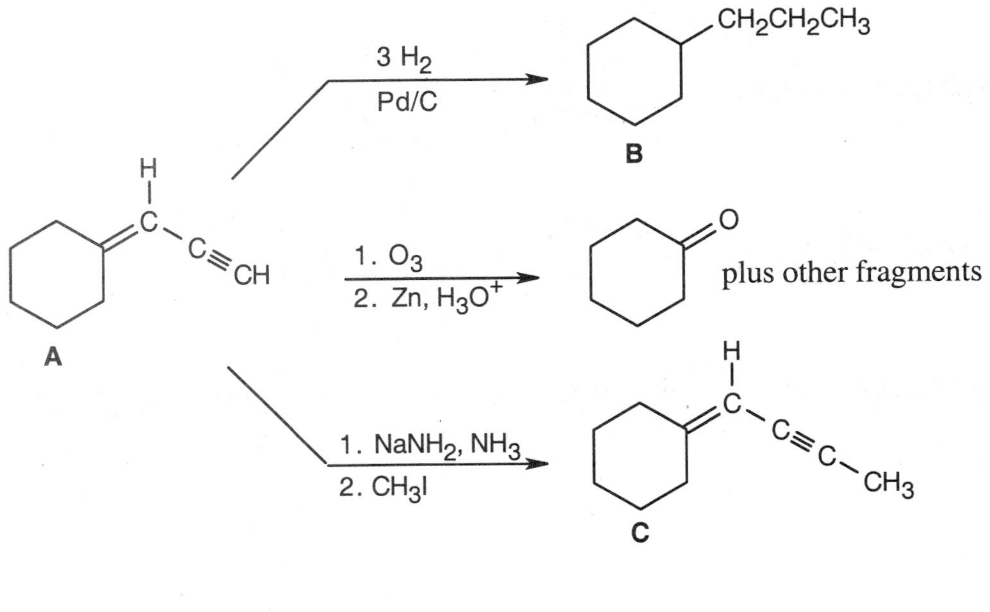

8.41

$$\begin{array}{l} H_2C-C\equiv C-C\equiv C-CH_2 \\ \quad | \qquad\qquad\qquad\qquad | \\ H_2C-C\equiv C-C\equiv C-CH_2 \end{array}$$

A

$\xrightarrow[\text{Pd/C}]{H_2}$ (cyclic product)

$\xrightarrow[\text{2. Zn, } H_3O^+]{\text{1. } O_3}$ 2 $HOOCCH_2CH_2COOH$ + 2 $HOOC-COOH$

8.42

$$CH_3\overset{O}{\overset{\|}{C}}CH_3 \xrightarrow[\text{2. } H_3O^+]{\text{1. } HC\equiv C:^- Na^+} CH_3\underset{CH_3}{\overset{OH}{C}}C\equiv CH \xrightarrow[H_2SO_4]{H_2O} H_2C=\overset{CH_3}{C}C\equiv CH$$

$$\downarrow H_2, \text{ Lindlar catalyst}$$

$$H_2C=\overset{CH_3}{C}CH=CH_2$$

2-Methyl-
1,3-butadiene

8.43 (1) Erythrogenic acid contains six degrees of unsaturation (see Sec. 6.2 for the method of calculating unsaturation equivalents for compounds containing elements other than C and H).

(2) One of these double bonds is contained in the carboxylic acid functional group –COOH; thus, five other degrees of unsaturation are present.

(3) Because five equivalents of H_2 are absorbed on catalytic hydrogenation, erythrogenic acid contains no rings.

(4) The presence of both aldehyde and carboxylic acid products of ozonolysis indicates that both double and triple bonds are present in erythrogenic acid.

(5) Only two ozonolysis products contain aldehyde functional groups; these fragments must have been double-bonded to each other in erythrogenic acid.

$H_2C=CH(CH_2)_4C\equiv$

(6) The other ozonolysis products result from cleavage of triple bonds. However, not enough information is available to tell in which order the fragments were attached. The two possible structures:

A $H_2C=CH(CH_2)_4C\equiv C-C\equiv C(CH_2)_7COOH$

B $H_2C=CH(CH_2)_4C\equiv C(CH_2)_7C\equiv CCOOH$

One method of distinguishing between the two possible structures is to treat erythrogenic acid with two equivalents of H_2, using Lindlar catalyst. The resulting trialkene can then be ozonized. The fragment that originally contained the carboxylic acid can then be identified.

8.44 This reaction mechanism is similar to the mechanism of halohydrin formation.

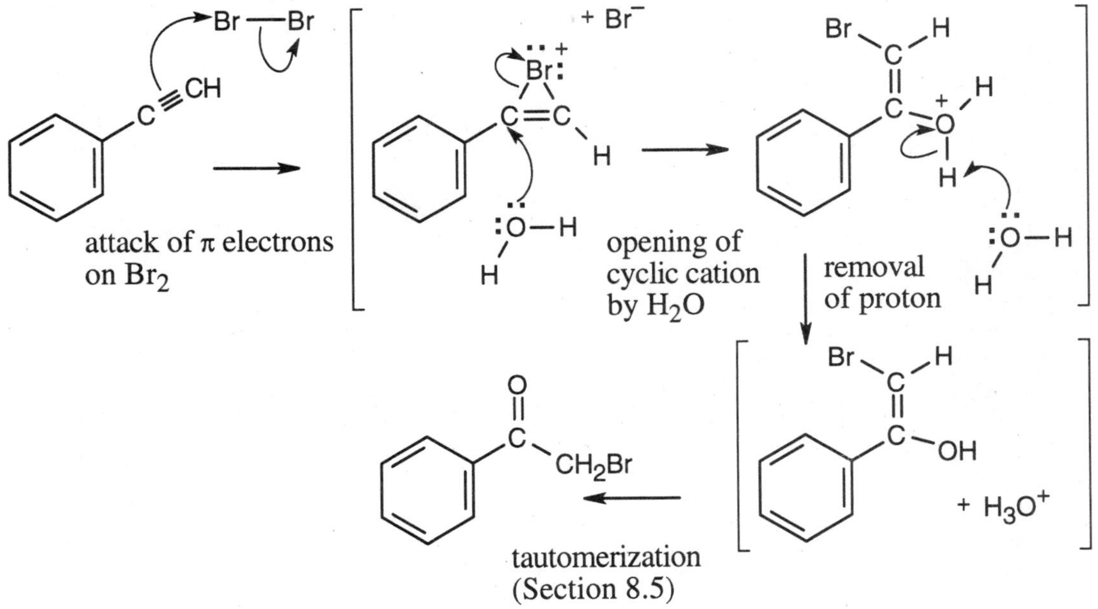

attack of π electrons on Br_2

opening of cyclic cation by H_2O

removal of proton

tautomerization (Section 8.5)

8.45

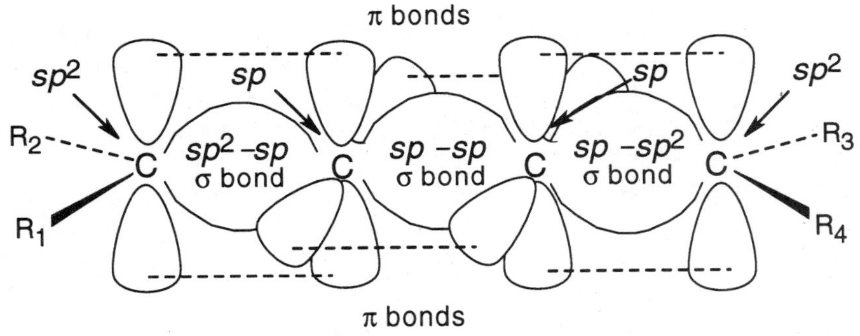

This simplest cumulene is pictured above. The carbons at the end of the cumulated double bonds are sp^2-hybridized and form one π bond to the "interior" carbons. The interior carbons are *sp*-hybridized; each carbon forms two π bonds – one to an "exterior" carbon and one to the other interior carbon. If you build a model of this cumulene, you can see that the substituents all lie in the same plane. This cumulene can thus exhibit cis–trans isomerism, just as simple alkenes can.

In general, the substituents of any compound with an odd number of adjacent double bonds lie in a plane; these compounds can exhibit cis–trans isomerism. The relationship of substituents at the ends of any compound with an even number of adjacent double bonds will be explained in the next chapter.

8.46

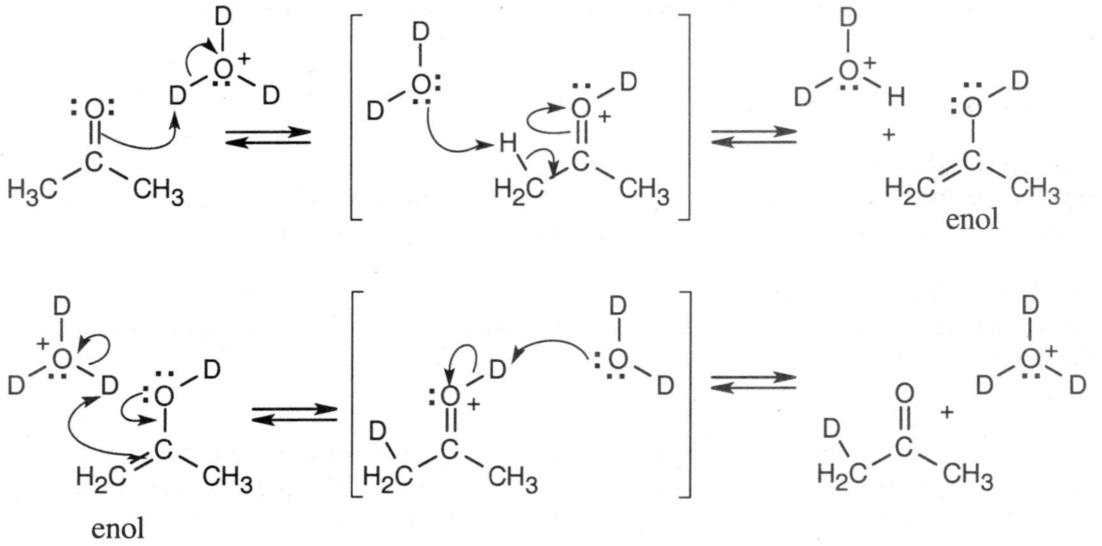
D
D
D
:O:
H3C
CH3
D
D
D
H
H2C
CH3
D
D
H
D
H2C
CH3
enol
D
D
D
D
H2C
CH3
enol
D
D
D
D
H2C
CH3
D
D
D
O
D
H2C
CH3

Repeating this process several times replaces all hydrogen atoms with deuterium atoms.

Molecular Modeling

8.47 Cycloheptyne is highly reactive because it is highly strained; the C–C≡C bond angle is 144°, instead of the usual value of 180° for a triple bond. Additions to the triple bond relieve strain by generating a relatively unstrained cycloheptene derivative, and these additions are faster than additions to noncyclic alkynes and alkenes.

8.48 The alkyne hydrogen is most positive.

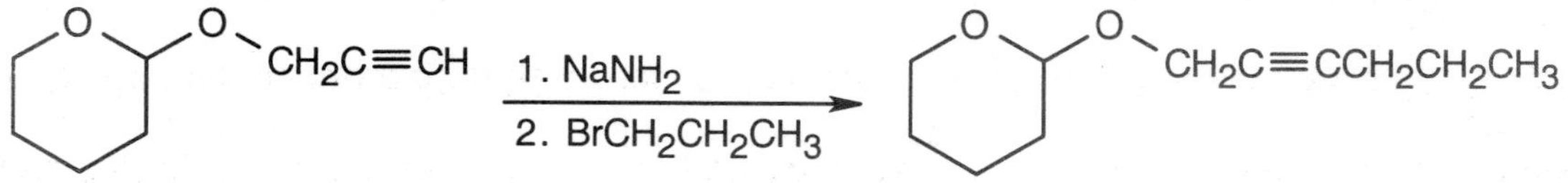
O
O
CH2C≡CH
1. NaNH2
2. BrCH2CH2CH3
O
O
CH2C≡CCH2CH2CH3

8.49 The reaction gives the lower energy product. 2-Butyn-1-ol is more stable than 2,3-butadien-1-ol by 15 kJ/mol.

8.50 The two radical conformations have nearly the same energy (ΔE = 3.3 kJ/mol). The anion shows a significant preference for the trans conformation (ΔE = 15 kJ/mol).

Review Unit 3: Alkenes and Alkynes

Major Topics Covered (with vocabulary):

Introduction to alkenes:
degree of unsaturation methylene group vinyl group allyl group cis–trans isomerism *E,Z* isomerism sequence rules heat of hydrogenation hyperconjugation

Electrophilic addition reactions:
electrophilic addition reaction regiospecific Markovnikov's rule Hammond Postulate hydride shift

Other reactions of alkenes:
dehydrohalogenation dehydration anti stereochemistry bromonium ion halohydrin hydration oxymercuration hydroboration syn stereochemistry carbene stereospecific Simmons–Smith reaction hydrogenation diol osmate molozonide ozonide

Polymerization reactions:
polymer monomer chain branching radical polymerization cationic polymerization

Alkynes:
alkyne enyne vicinal tautomer Lindlar catalyst acetylide anion alkylation

Organic Synthesis.

Types of Problems:

After studying these chapters you should be able to:

- Calculate the degree of unsaturation of any compound, including those containing N, O, and halogen.
- Name acyclic and cyclic alkenes and alkynes, and draw structures corresponding to names.
- Assign *E,Z* priorities to groups.
- Assign cis-trans and *E,Z* designations to double bonds.
- Predict the relative stability of alkene double bonds.

- Formulate mechanisms of electrophilic addition reactions.
- Predict the products of reactions involving alkenes and alkynes.
- Choose the correct alkene or alkyne starting material to yield a given product.
- Deduce the structure of an alkene or alkyne from its molecular formula and products of cleavage.
- Carry out syntheses involving alkenes and alkynes.

Points to Remember:

* Calculating the degree of unsaturation is an absolutely essential technique in the structure determination of all organic compounds. It is the starting point for deciding which functional groups are or aren't present in a given compound, and eliminates many possibilities. When a structure determination problem is given, always calculate the degree of unsaturation first.

* All cis–trans isomers can also be described by the *E,Z* designation, but not all *E,Z* isomers can be described by the cis–trans designation.

* Bond dissociation energies, described in Chapter 5, measure the energy required to homolytically break a bond. They are not the same as dissociation enthalpies, which measure the ability of a compound to dissociate heterolytically. Bond dissociation energies can be used to calculate dissociation enthalpies in the gas phase if other quantities are also known.

* Not all hydrogens bonded to carbons adjacent to a carbocation can take part in hyperconjugation at the same time. At any given instant, some of the hydrogens have C-H bonds that lie in the plane of the carbocation and are not suitably oriented for hyperconjugative overlap

* Although it is very important to work backwards when planning an organic synthesis, don't forget to pay attention to the starting material, also. Planning a synthesis is like solving a maze from the middle outward: keeping your eye on the starting material can keep you from running into a dead end.

Self-Test:

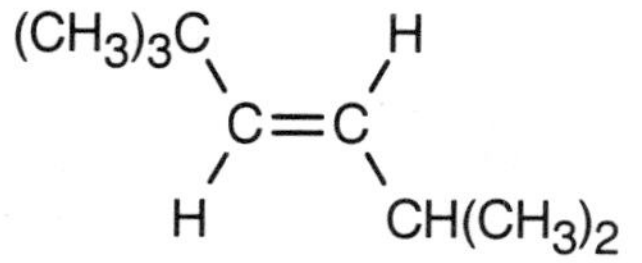

$(CH_3)_3C(H)C{=}C(H)CH(CH_3)_2$

A

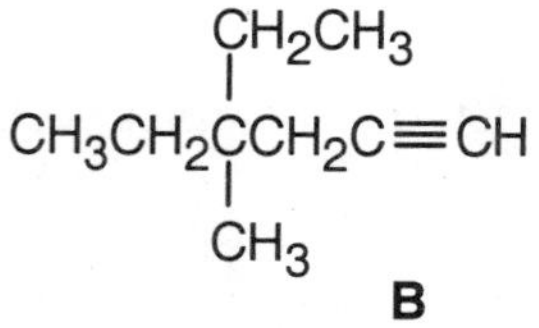

$CH_3CH_2C(CH_2CH_3)(CH_3)CH_2C{\equiv}CH$

B

Provide names for **A** and **B** (include bond stereochemistry). Predict the products of reaction of **A** and **B** with (a) 1 equiv HBr (b) H_2, Pd/C (c) BH_3, THF, then H_2O_2, HO^- (d) O_3, then Zn, H_3O^+. Give a reagent that reacts with **A** but not **B**. Give a reagent that reacts with **B**, but not **A**.

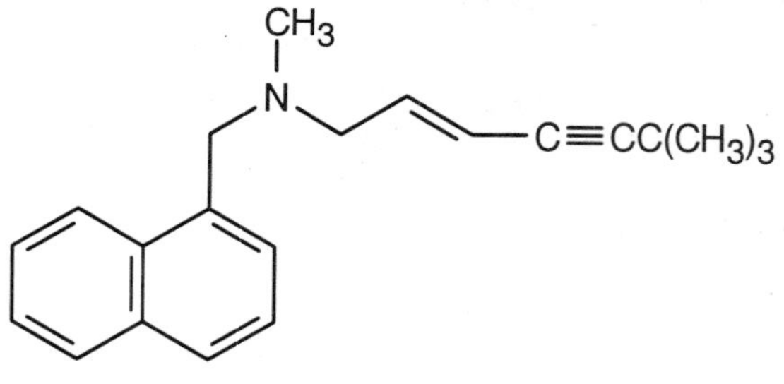

C Terbinafine (an antifungal)

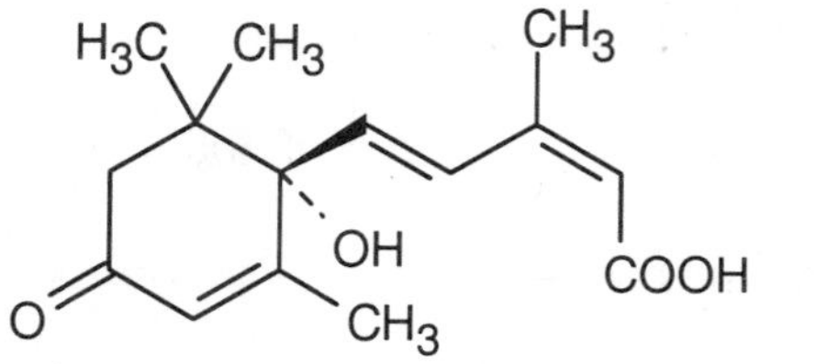

D Abscisic acid (a plant hormone)

What is the configuration of the double bond in **C**? What products result from treatment of **C** with $KMnO_4$, H_3O^+ (neither the aromatic ring nor the amine are affected)? How might the triple bond be introduced?

Give *E,Z* configurations for the double bonds in **D**.

Multiple Choice:

1. What is the degree of unsaturation of a compound whose molecular formula is $C_{11}H_{13}N$?
(a) 4 (b) 5 (c) 6 (d) 7

2. Two equivalents of H_2 are needed to hydrogenate a hydrocarbon. It is also known that the compound contains two rings and has 15 carbons. What is its molecular formula?
(a) $C_{15}H_{22}$ (b) $C_{15}H_{24}$ (c) $C_{15}H_{28}$ (d) $C_{15}H_{32}$

3. Which group is of lower priority than $-CH{=}CH_2$?
(a) $-CH(CH_3)_2$ (b) $-CH{=}C(CH_3)_2$ (c) $-C{\equiv}CH$ (d) $-C(CH_3)_3$

4. What is the usual relationship between the heats of hydrogenation of a pair of cis/trans alkene isomers?
(a) Both have positive heats of hydrogenation (b) Both have negative heats of hydrogenation, and ΔH_{hydrog} for the cis isomer has a greater negative value (c) Both have negative heats of hydrogenation, and ΔH_{hydrog} for the trans isomer has a greater negative value (d) Both have negative heats of hydrogenation, but the relationship between the two values of ΔH_{hydrog} can't be predicted.

5. In a two-step exergonic reaction, what is the relationship of the two transition states?
(a) both resemble the intermediate (b) the first resembles the starting material, and the second resembles the product (c) the first resembles the intermediate and the second resembles the product (d) there is no predictable relationship between the two transition states

6. For synthesis of an alcohol, acid-catalyzed hydration of an alkene is useful in all of the following instances except:
(a) when an alkene has no acid-sensitive groups (b) when an alkene is symmetrical
(c) when a large amount of the alcohol is needed
(d) when two possible carbocation intermediates are of similar stability.

7. A reaction that produces a diol from an alcohol is a:
(a) hydration (b) hydrogenation (c) hydroboration (d) hydroxylation

8. Which of these cyclic intermediates is not part of an oxidative cleavage reaction?
(a) ozonide (b) molozonide (c) osmate (d) periodate

9. An enol is a tautomer of an:
(a) alcohol (b) alkyne (c) alkene (d) ketone

10. Which reaction proceeds through a vinylic radical?
(a) Hg-catalyzed hydration of an alkyne (b) Li/NH_3 reduction of an alkyne
(c) catalytic hydrogenation of an alkyne (d) treatment of an alkyne with a strong base

Chapter 9 – Stereochemistry

Chapter Outline

I. Handedness (Sections 9.1 – 9.5).
  A. Enantiomers and tetrahedral carbon (Section 9.1).
    When four different groups are bonded to a carbon atom, two different arrangements are possible.
    a. These arrangements are mirror images.
    b. The two mirror-image molecules are enantiomers.
  B. The reason for handedness in molecules: Chirality (Section 9.2).
    1. Molecules that are not superimposable on their mirror-images are chiral.
      a. A molecule is not chiral if it contains a plane of symmetry.
      b. A molecule with no plane of symmetry is chiral.
    2. A carbon bonded to four different groups is a chirality center.
    3. Any $–CH_2–$ or $–CH_3$ carbon is achiral.
  C. Optical activity (Section 9.3).
    1. Solutions of certain substances rotate the plane of plane-polarized light. These substances are said to be optically active.
    2. The amount of rotation can be measured with a polarimeter.
    3. The degree of rotation can also be measured.
      a. A compound whose solution rotates plane-polarized light to the right is dextrorotatory.
      b. A compound whose solution rotates plane-polarized light to the left is levorotatory.
  D. Specific rotation (Section 9.4).
    1. The amount of rotation depends on concentration, path length and wavelength.
    2. Specific rotation is the observed rotation of a sample with concentration = 1 g/mL, sample pathlength of 1 dm, and light of wavelength = 589 nm.
    3. Specific rotation is a physical constant characteristic of a given optically active compound.
  E. Pasteur's discovery of enantiomerism (Section 9.5).
    1. Pasteur discovered two different types of crystals in a solution that he was evaporating.
    2. The crystals were mirror images.
    3. Solutions of each of the two types of crystals were optically active, and their specific rotations were equal in magnitude but opposite in sign.
    4. Pasteur postulated that some molecules are handed and thus discovered the phenomenon of enantiomerism.

II. Stereoisomers and configurations (Sections 9.6 – 9.11).
  A. Specification of configurations of stereoisomers (Section 9.6).
    1. Rules for assigning configurations at a chirality center:
      a. Assign priorities to each group bonded to the carbon by using Cahn-Ingold-Prelog rules (Section 6.6).
      b. Orient the molecule so that the group of lowest priority is pointing to the rear.
      c. Draw a curved arrow from group 1 to group 2 to group 3.
      d. If the arrow is clockwise, the chirality center is *R*, and if the arrow is counterclockwise, the chirality center is *S*.
    2. The sign of rotation is not related to *R,S* designation.
    3. X-ray experiments have proven *R,S* conventions to be correct.

B. Diastereomers (Section 9.7).
  1. A molecule with two chirality centers can have four possible stereoisomers.
    a. The stereoisomers group into two pairs of enantiomers.
    b. A stereoisomer from one pair is the diastereomer of a stereoisomer from the other pair.
  2. Diastereomers are stereoisomers that are not mirror images.

C. Meso compounds (Section 9.8).
  1. A meso compound occurs when a compound with two chirality centers possesses a plane of symmetry.
  2. A meso compound is achiral despite having two chirality centers.

D. Molecules with more than two chirality centers (Section 9.9).
  A molecule with n chirality centers can have a maximum of 2^n stereoisomers.

E. Racemic mixtures and their resolution (Section 9.10).
  1. A racemic mixture is a 50:50 mixture of two enantiomers.
    Racemic mixtures show zero optical rotation.
  2. Some racemic mixtures can be resolved into their component enantiomers.
    a. If a racemic mixture of a carboxylic acid reacts with a chiral amine, the product ammonium salts are diastereomers.
    b. The diastereomeric salts differ in chemical and physical properties and can be separated.
    c. The original enantiomers can be recovered by acidification.

F. Physical properties of stereoisomers (Section 9.11).
  1. Enantiomers have identical physical properties, except for the sign of their specific rotations.
  2. The physical properties of meso compounds, diastereomers and racemic mixtures differ from each other and from the properties of enantiomers.

III. A review of isomerism (Section 9.12).
  A. Constitutional isomers differ in connections between atoms.
    1. Skeletal isomers have different carbon skeletons.
    2. Functional isomers contain different functional groups.
    3. Positional isomers have functional groups in different positions.
  B. Stereoisomers have the same connections between atoms, but different geometry.
    1. Enantiomers have a mirror-image relationship.
    2. Diastereomers are non-mirror-image stereoisomers.
      a. Configurational diastereomers.
      b. Cis–trans isomers differ in the arrangement of substituents in a double bond or ring.

IV. Fischer projections (Sections 9.13 – 9.14).
  A. Working with Fischer projections (Section 9.13).
    1. A Fischer projection represents a tetrahedral carbon as a pair of perpendicular lines.
      a. The horizontal line represents bonds coming out of the page.
      b. The vertical line represents bonds going into the page.
    2. An atom can be represented by many Fischer projections.
    3. To manipulate, and thus to compare, Fischer projections, there are two allowable motions.
      a. A Fischer projection can be rotated on the page by 180°, but not by 90° or 270°.
      b. Holding one group steady, the other three groups can be rotated clockwise or counterclockwise.
    4. To see if two Fischer projections represent the same enantiomer, perform allowed motions on it until two groups of one projection are superimposed with two groups on the other projection.

a. If the other two groups also coincide, the two Fischer projections represent the same enantiomer.
b. If the other two groups don't coincide, the two Fischer projections represent different enantiomers.

B. Assigning *R,S* configurations to Fischer projections (Section 9.14).
1. Rules for assigning *R,S* configurations.
a. Assign priorities to the substituents in the usual way.
b. Perform one of the two allowed motions to place the lowest priority group at the top of the Fischer projection.
c. Determine the direction of rotation of the arrow that travels from group 1 to group 2 to group 3, and assign *R* or *S* configuration.
2. Fischer projections can be used to specify more than one chirality center.

V. Stereochemistry of reactions (Sections 9.15 – 9.17).
A. Addition of HBr to alkenes (Section 9.15).
1. When HBr adds to an achiral alkene, a racemic mixture of products is formed.
2. The achiral cationic intermediate can be attacked by Br^- from either side to produce a racemic mixture.
3. Alternatively, the transition states for topside attack and bottomside attack are enantiomers and have the same energy.
B. Addition of Br_2 to alkenes (Section 9.16).
1. Br_2 can add to either face of a double bond to form a bromonium ion.
2. In the case of *cis*-2-butene, addition of Br^- to the bromonium ions forms the (2*S*,3*S*) and (2*R*,3*R*) enantiomers in equal yield – a racemic mix.
3. For *trans*-2-butene, addition of Br^- to the bromonium ion yields *meso*-2,3-dibromobutane.
4. Reaction between two achiral partners always leads to an optically inactive product, either racemic or meso.
C. Addition of HBr to a chiral alkene (Section 9.17)..
1. When H^+ adds to a chiral alkene, the intermediate carbocation is chiral.
2. The original chirality center is unaffected by the reaction.
3. Attack of Br^- on the carbocation doesn't occur with equal probability from either side, and the resulting product is an optically active mixture of diastereomeric bromoalkanes.
4. Reaction of a chiral reactant with an achiral reactant leads to unequal amounts of diastereomeric products.

VI. Chirality at atoms other than carbon (Section 9.18).
A. Other elements with tetrahedral atoms can be chirality centers.
B. Trivalent nitrogen can, theoretically, be chiral, but rapid inversion of the nitrogen lone pair interconverts the enantiomers.

VII. Chirality in nature (Section 9.19).
A. Different enantiomers of a chiral molecule have different properties in nature.
1. (+)-Limonene has the odor of oranges, and (–)-limonene has the odor of lemons.
2. Racemic fluoxetine is an antidepressant, but the *S* enantiomer is effective against migraine.
B. In nature, a molecule must fit into a chiral receptor, and only one enantiomer usually fits.

Solutions to Problems

9.1 Objects having a plane of symmetry are achiral.
Chiral: screw, beanstalk, shoe.
Achiral: screwdriver, hammer.

9.2 Use the following rules to locate centers that are *not* chirality centers.

1. All $-CH_3$ and $-CX_3$ carbons are not chirality centers.
2. All $-CH_2$ and $-CX_2-$ carbons are not chirality centers.
3. All $—C{=}C—$ and $—C{\equiv}C—$ carbons are not chirality centers

By rule 3, all aromatic ring carbons are not chirality centers.

(a) CH_3 — Toluene, achiral

(b) * $CH_2CH_2CH_3$, N, H — Coniine, chiral

(c) CH_3CH_2, O, O, N, N, H, H, O — Phenobarbital, achiral

9.3 Use the rules in Problem 9.2 to identify chirality centers.

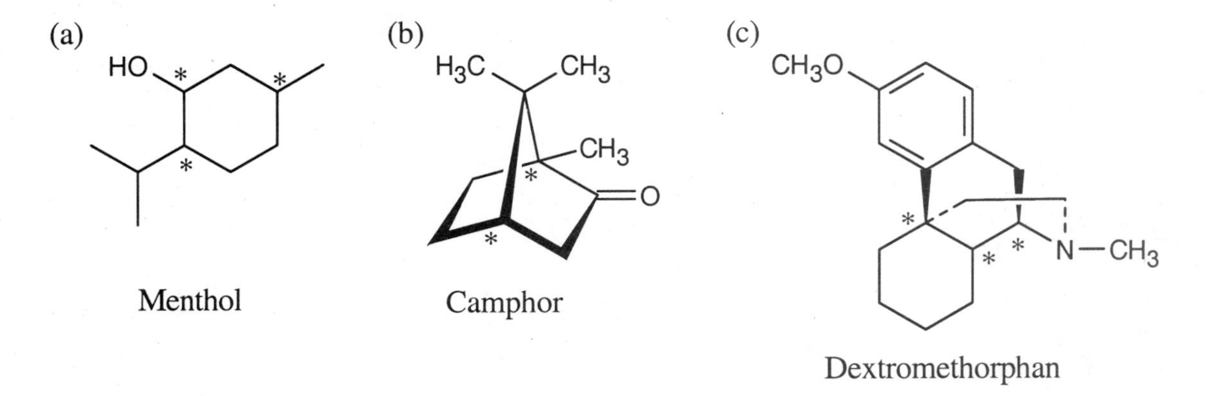

Menthol

Camphor

Dextromethorphan

9.4

COOH, H, C, NH_2, H_3C — COOH, H_3C, C, NH_2, H — Alanine

9.5

Use the formula $[\alpha]_D = \dfrac{\alpha}{l \times C}$, where

$[\alpha]_D$ = specific rotation

α = observed rotation

l = path length of cell (in dm)

C = concentration (in g/mL)

In this problem: α = 1.21°

l = 5.00 cm = 0.500 dm

C = 1.50 g/10.0 mL = 0.150 g/mL

$$[\alpha]_D = \frac{+1.21°}{0.500 \text{ dm} \times 0.150 \text{ g/mL}} = +16.1°$$

9.6 Use the sequence rules in Section 9.6.

(a) By Rule 1, –H is of lowest priority, and –Br is of highest priority. By Rule 2, $–CH_2CH_2OH$ is of higher priority than $–CH_2CH_3$.

Highest ⟶ *Lowest*

–Br, $–CH_2CH_2OH$, $–CH_2CH_3$, –H

(b) By Rule 3, –COOH is considered as $—\overset{—O\ \ O—}{\underset{}{C}}—OH$. Because 3 oxygens are attached to a –COOH carbon and only one oxygen is attached to $–CH_2OH$, –COOH is of higher priority than $–CH_2OH$. $–CO_2CH_3$ is of higher priority than –COOH by Rule 2, and –OH is of highest priority by Rule 1.

Highest ⟶ *Lowest*

—OH, $—CO_2CH_3$. —COOH, $—CH_2OH$

(c) $—NH_2$, —CN, $—CH_2NHCH_3$, $—CH_2NH_2$

(d) —Br, —Cl, $—CH_2Br$, $—CH_2Cl$

9.7 All stereochemistry problems are easier if you use models. Part (a) will be solved by two methods – with models and without models.
(a) *With models:* Build a model of (a). Orient the model so that group 4 is pointing to the rear. Note the direction of rotation of arrows that go from group 1 to group 2 to group 3. The arrows point counterclockwise, and the configuration is *S*.

Without models: Imagine yourself looking at the molecule, with the group of lowest priority pointing to the back. Your viewpoint would be at the upper right of the molecule, and you would see group 1 on the left, group 3 on the right and group 2 at the bottom. The arrow of rotation travels counterclockwise, and the configuration is *S*.

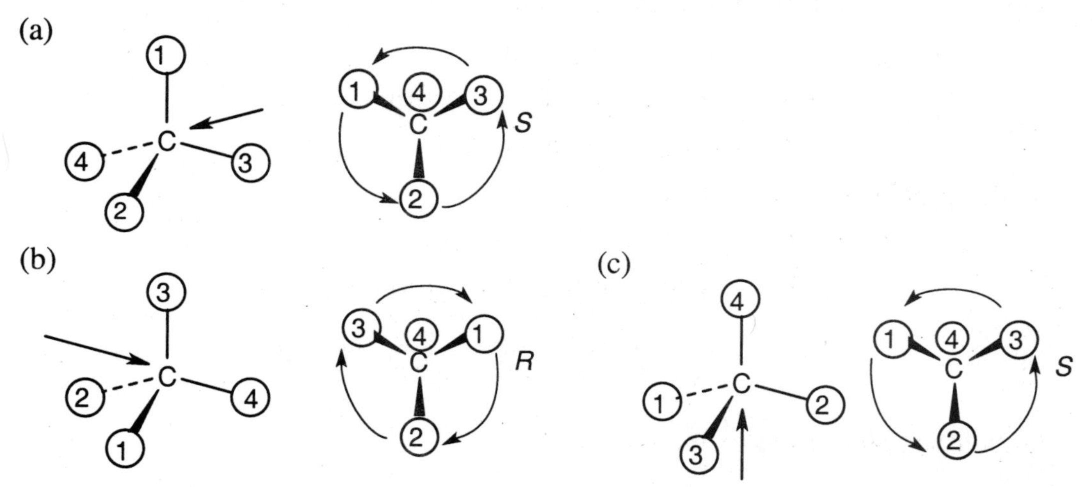

9.8 The following scheme may be used to assign *R,S* configurations to chirality centers:

Step 1. For each chirality center, rank substituents by the Cahn–Ingold–Prelog system; give the number 4 to the lowest priority substituent. For part (a):

Substituent	*Priority*
–Br	1
–COOH	2
$–CH_3$	3
–H	4

Step 2. As in the previous problem, orient yourself so that you are 180° from the lowest priority group (indicated by the arrow in the drawing). From that viewpoint, draw the molecule as it looks when you face it. Draw the arrow that travels from group 1 to group 2 to group 3, and note its direction of rotation. The molecule in (a) has *S* configuration.

(b)

(c)

In (b), the observer is behind the page, looking out and down toward the right. In (c), the observer is behind the page looking out and up to the left

9.9 As in the previous problem, assign priorities to the substituents, giving the number 4 to the lowest priority substituent. Orient the lowest priority group toward the back, and arrange the three other groups, as we have done previously, as spokes on a steering wheel, with a counterclockwise rotation (because we need to draw an *S* enantiomer). Then, tilt the drawing until it is a tetrahedral representation.

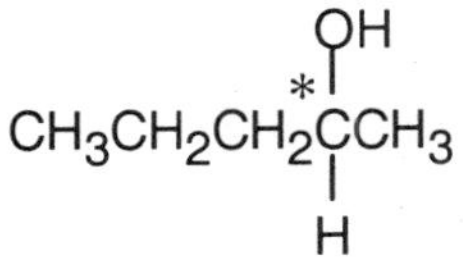

$CH_3CH_2CH_2\overset{*}{C}(OH)(H)CH_3$ (*S*)-2-Pentanol

Substituent	*Priority*
—OH	1
$—CH_2CH_2CH_3$	2
$—CH_3$	3
—H	4

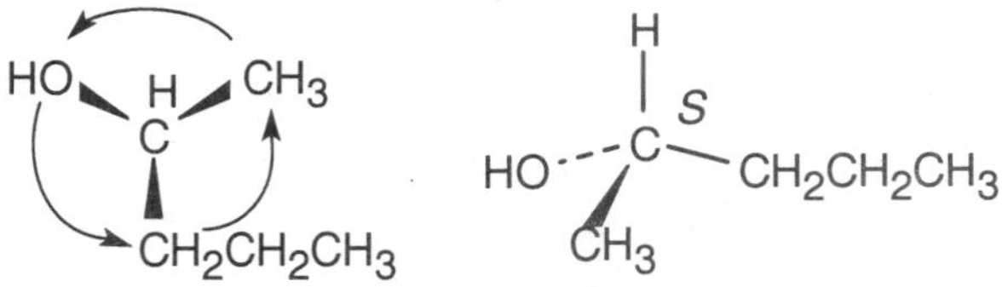

HO, H, CH_3, C, $CH_2CH_2CH_3$

H, *S*, HO, C, $CH_2CH_2CH_3$, CH_3

9.10 Fortunately, methionine is represented in the correct configuration.

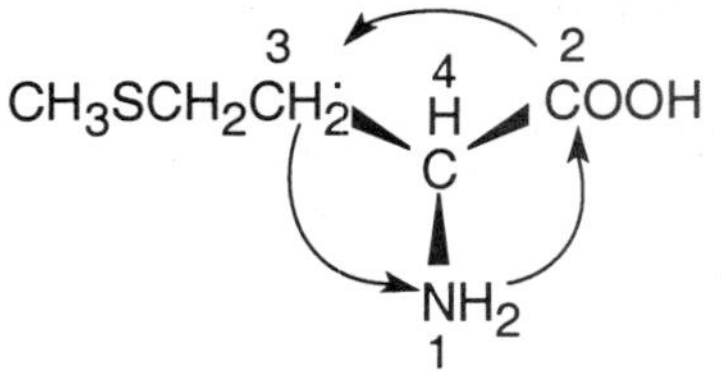

3 $CH_3SCH_2CH_2$, 4 H, C, 2 COOH (*S*)-Methionine, NH_2 1

9.11 (*R,S*) assignments for more complicated molecules can be made using the same method as in Problem 9.8, taking one carbon atom at a time. It's especially important to use molecular models when a compound has more than one chirality center. For part (a):

Step 1. Assign priorities to groups of the top chirality center.

Substituent	*Priority*
–Br	1
–CH(OH)CH_3	2
$–CH_3$	3
–H	4

Step 2. Orient the model so that the lowest priority group of the first chirality center points to the rear. If you aren't using a model, orient yourself so that you are 180° from the lowest priority group. You would be looking out of the page, upward to the left.

Step 3. Note the direction of rotation of the arrow that travels from to group 1 to group 2 to group 3. The rotation is clockwise, and the configuration is *R*.

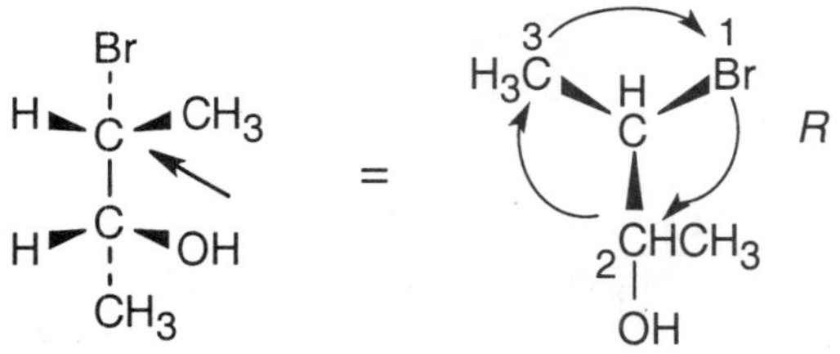

Br, H, C, CH_3, H, C, OH, CH_3 = 3 H_3C, H, C, 1 Br, *R*, 2 $CHCH_3$, OH

Step 4. Repeat steps 1-3 for the next chirality center.

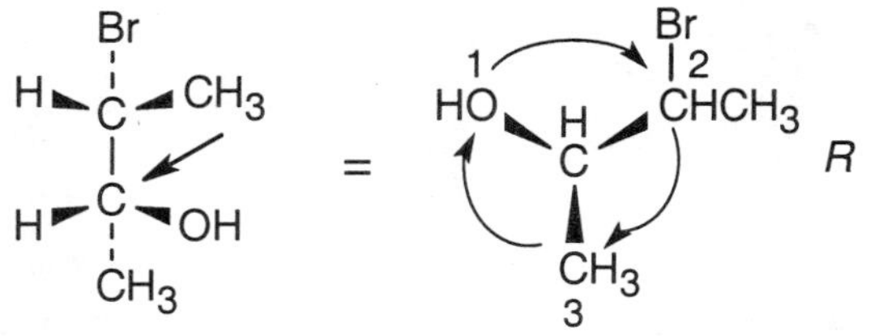

Br
H C CH_3
C OH
H CH_3
=
1 HO H C Br 2 $CHCH_3$ CH_3 3
R

(b) *S*, *R*
(c) *R*, *S*
(d) *S*, *S*

a, d are enantiomers and are diastereomeric with b, c.
b, c are enantiomers and are diastereomeric with a, d.

9.12

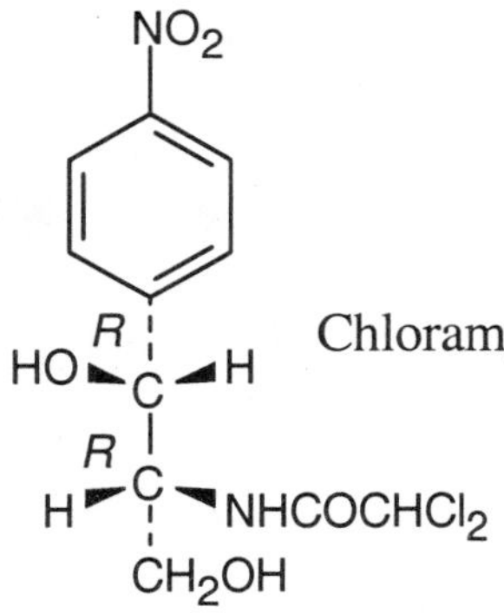

NO_2
R HO C H
R H C $NHCOCHCl_2$
CH_2OH

Chloramphenicol

9.13

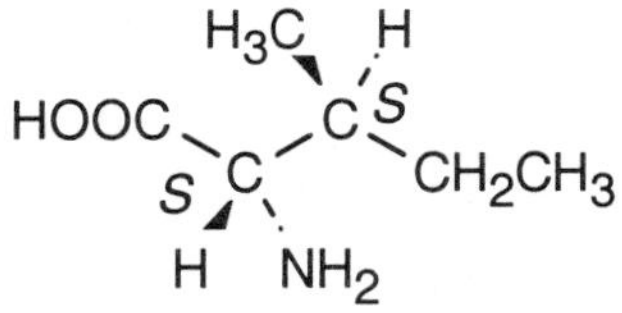

H_3C H
HOOC C *S*
S C CH_2CH_3
H NH_2

Isoleucine

9.14 To decide if a structure represents a meso compound, try to locate a plane of symmetry that divides the molecule into two halves that are mirror images. Molecular models may be helpful.

(a)

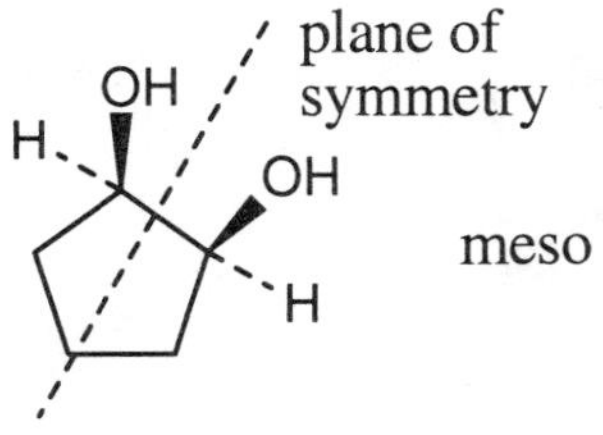

OH
H
OH
H
plane of symmetry

meso

(b) and (c) are not meso structures.

(d)

Br, H, CH_3, C, C, H_3C, H, Br ≡ Br, H, CH_3, C, C, Br, H, CH_3

plane of symmetry

meso

9.15 For a molecule to exist as a meso form, it must possess a plane of symmetry. 2,3-Dibromobutane can exist as a pair of enantiomers *or* as a meso compound, depending on the configurations at carbons 2 and 3.

(a)

Br, H, CH_3, C *S*, *S* C, H_3C, H, Br — H, H_3C, Br, C *R*, *R* C, Br, H, CH_3 — Br, H, CH_3, C *S*, *R* C, Br, H, CH_3

not meso

plane of symmetry

meso

(b) 2,3-Dibromopentane has no symmetry plane and thus can't exist in a meso form.

(c) 2,4-Dibromopentane can exist in a meso form.

Br, H, CH_3, C *S*, CH_2, *R* C, Br, H, CH_3

plane of symmetry

2,4–Dibromopentane can also exist as a pair of enantiomers (2*R*,4*R*) and (2*S*,4*S*) that are not meso compounds.

9.16 The molecule represents a meso compound. The symmetry plane passes through the carbon bearing the –OH group and between thr two ring carbons that are bonded to methyl groups.

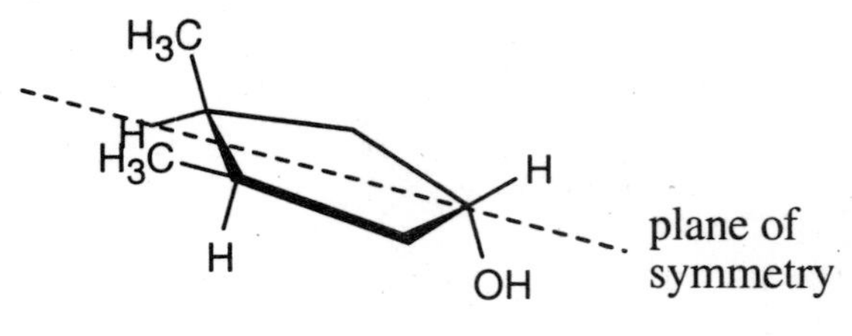

9.17

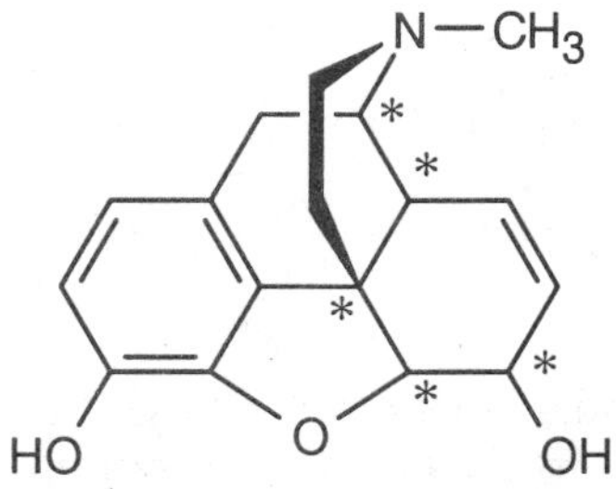

N—CH_3

Morphine

HO O OH

Morphine has five chirality centers and, in principle, can have $2^5 = 32$ stereoisomers. Many of these stereoisomers are too strained to exist.

9.18

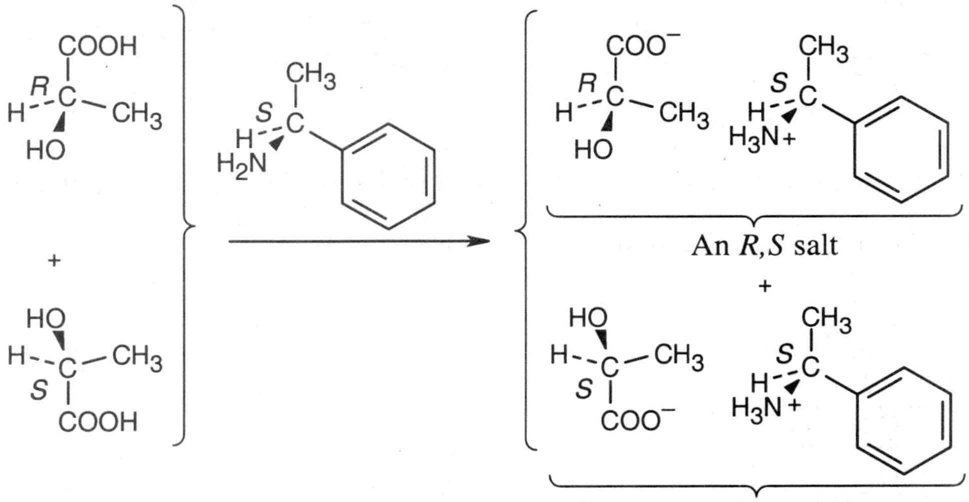

COOH, R, H, C, CH_3, HO + HO, H, C, CH_3, S, COOH; CH_3, S, C, H, H_2N → COO^-, R, H, C, CH_3, HO; CH_3, S, C, H, H_3N+

An *R*,*S* salt

+

HO, H, C, CH_3, S, COO^-; CH_3, S, C, H, H_3N+

An *S*,*S* salt

The two product salts have the configurations (*R*,*S*) and (*S*,*S*) and are diastereomers.

9.19 (a)

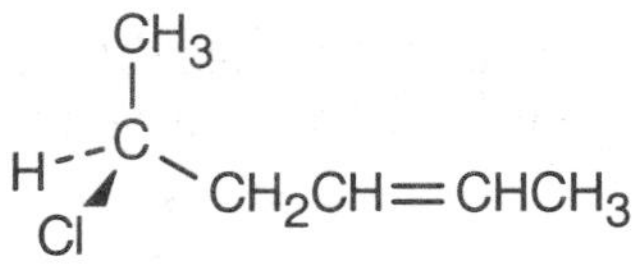

CH_3, H, C, Cl, $CH_2CH{=}CHCH_3$; Cl

(*S*)-5-Chloro-2-hexene 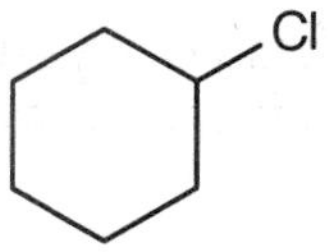 Chlorocyclohexane

These two compounds are constitutional isomers.

(b) The two dibromopentane stereoisomers are diastereomers.

9.20 Two manipulations of Fischer projections are allowable.

1. A Fischer projection may be rotated on the page by 180°, but not by 90° or 270°.
2. Holding one group of a Fischer projection steady, we may rotate the other three groups either clockwise or counterclockwise.

If we hold the –COOH group of projection A steady and rotate the other three groups by 120°, we arrive at a projection identical with projection B. Thus A and B are identical.

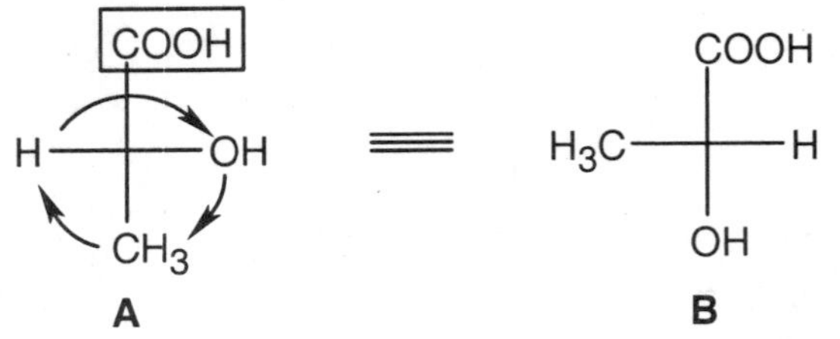

If projection C is rotated 180°, the resulting projection has two groups superimposable with projection B. However, no manipulation can make all four groups superimposable; C is thus an enantiomer of A and B.

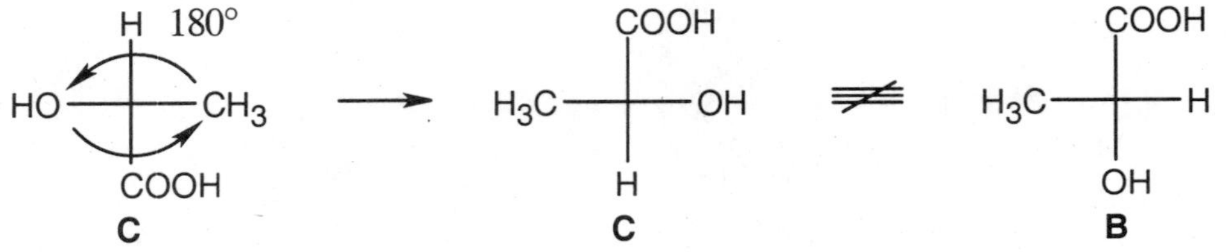

Likewise, projection D is not superimposable on projection B but is identical to projection C. Thus, A and B are identical and enantiomeric with C and D, which are also identical.

9.21 (a) By an allowable rotation, manipulate two groups in the first structure to the positions they occupy in the second structure.

For example:

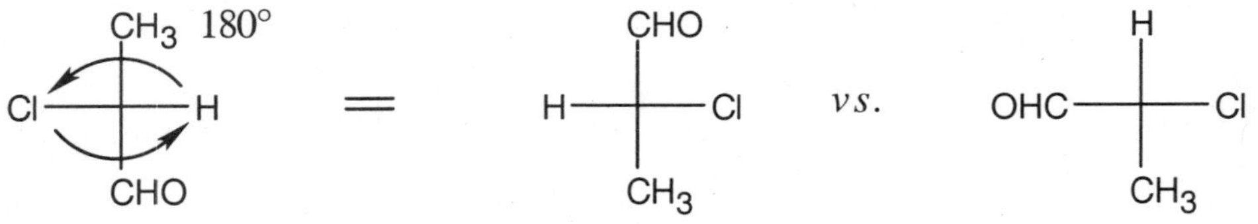

Although this projection resembles the second one in placement of $-CH_3$ and $-Cl$, $-H$ and $-CHO$ are interchanged. The two projections thus represent enantiomers.

(b) As above, rotate the first projection by 180°.

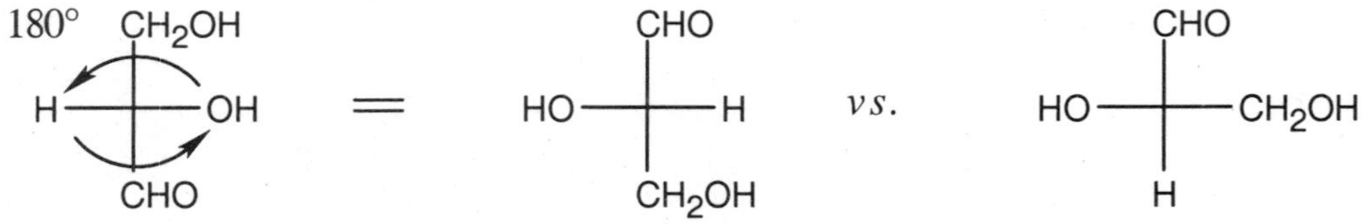

If we now compare this structure to the second structure of the pair, two of the groups occupy the same relative position. There is, however, no rotation that can make these two structures superimposable; they are enantiomers.

9.22 As in Practice Problem 9.4, rotate the molecule so that the two horizontal groups are pointing toward you, and the two vertical bonds are receding from you. Then flatten the molecule and draw the groups at the ends of the crossed bonds.

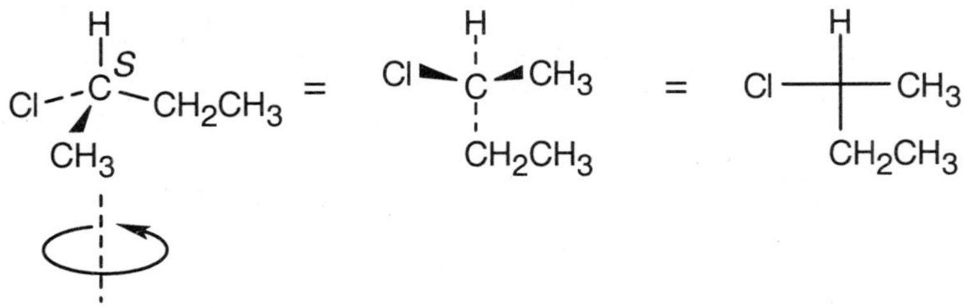

H, S, Cl, C, CH_2CH_3, CH_3 = H, Cl, C, CH_3, CH_2CH_3 = H, Cl, CH_3, CH_2CH_3

9.23 One of the hardest spatial problems in organic chemistry is to visualize two-dimensional drawings as three-dimensional chemical structures. This difficulty becomes particularly troublesome when it is necessary to assign (*R,S*) configurations to structures, especially if they are drawn as Fischer projections. Working out these assignments is easier when models are used, but it is still possible to determine a configuration from a two-dimensional drawing.

The following system may be used for assigning (*R,S*) configurations. Part (a) is used as an example.

Step 1. Rank substituents in priority order.

Substituents	*Priority*
–Br	1 high
–COOH	2
$-CH_3$	3
–H	4 low

Step 2. Use either of the allowable manipulations of Fischer projections to bring the group of lowest priority to the top of the projection. In this case, hold –Br steady and rotate the other three groups clockwise.

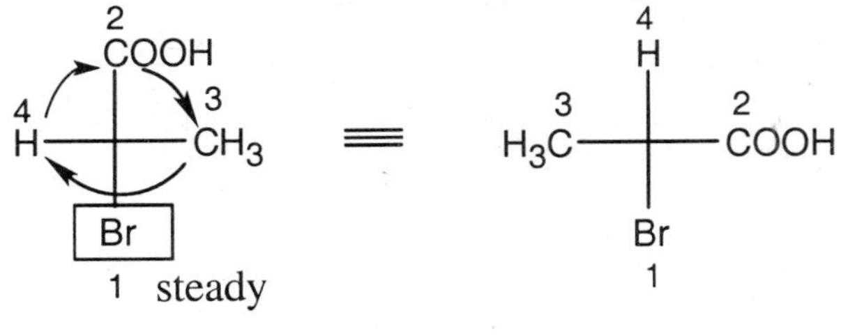

2 COOH, 4 H, 3 CH_3, Br 1 steady ≡ 4 H, 3 H_3C, 2 COOH, Br 1

Step 3. Indicate on the rotated projection the direction of the arrows that proceed from group 1 to group 2 to group 3. If the direction of the arrows is clockwise, the configuration is *R*; if the direction is counterclockwise, the configuration is *S*. Here the configuration is *S*.

4 H, 3 H_3C, 2 COOH, Br 1 *S*

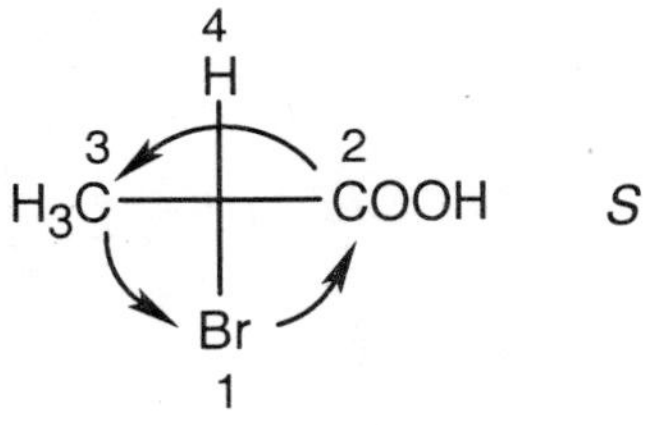

(b)

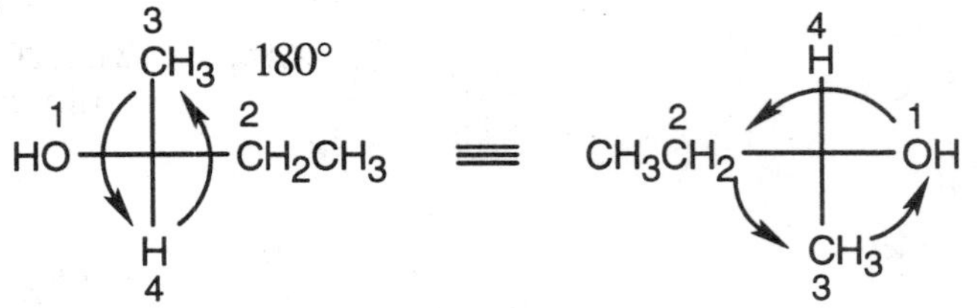

3 CH3 180°
1 HO — 2 CH2CH3
H 4
≡
4 H
2 CH3CH2 — 1 OH
CH3 3
S

(c)

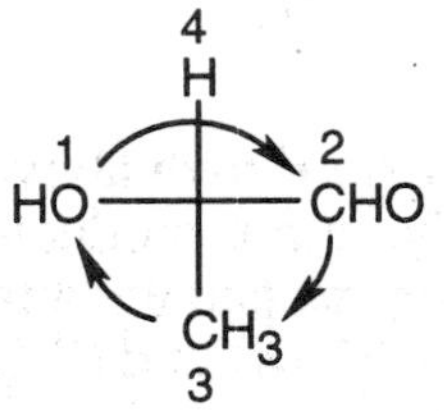

4 H
1 HO — 2 CHO
CH3 3
R

9.24 The easiest way to solve this problem is to build a model, assign *R* or *S* configuration to the chirality center, manipulate the model so that two horizontal groups are pointing out and two vertical groups are pointing back, and draw the Fischer projection of the model. Without a model: Bring the methyl group slightly forward, so that you have the following structure:

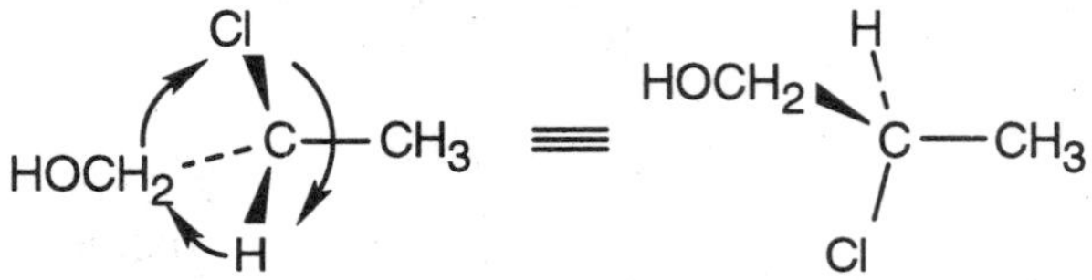

Cl
HOCH2 - - - C — CH3
H
≡
H
HOCH2 — C — CH3
Cl

Perform the indicated rotation to arrive at the second structure shown above. This structure is close in structure to the steering wheel structures that we used to assign configurations in Problems 9.6 and 9.7. Redraw the second structure above as a steering wheel structure, assign priorities to the four substituents, and determine the configuration (*R*).

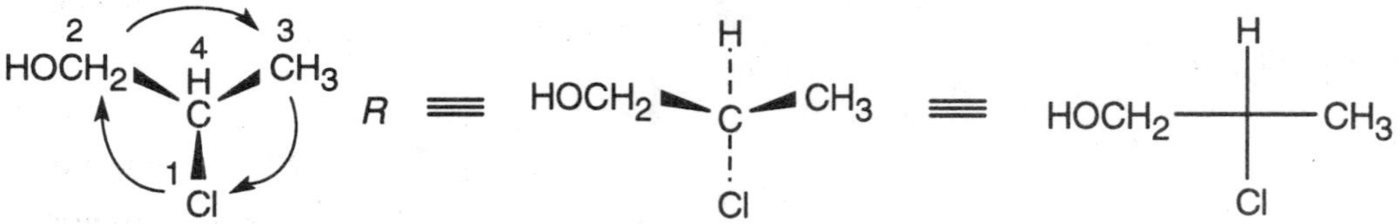

2 HOCH2 — 4 H C — 3 CH3
1 Cl
R ≡
H
HOCH2 — C — CH3
Cl
≡
H
HOCH2 — CH3
Cl

A small tilt of the first structure above produces the second structure, which can easily be drawn as a Fischer projection.

9.25 Possible bromonium ion intermediates are shown below. In this problem, assume that attack of Br^- is somewhat more likely at carbon 2.

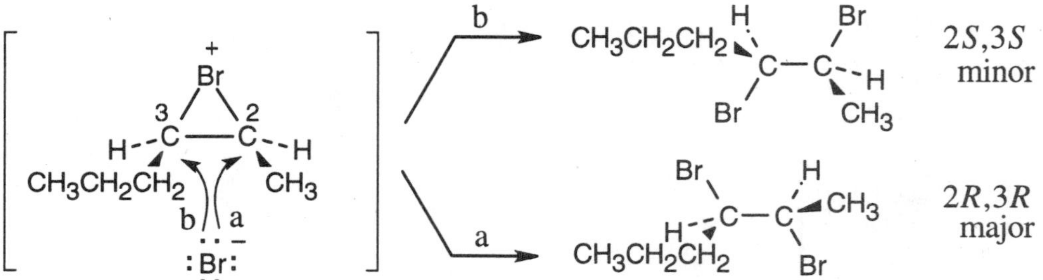

+ Br
3 H - - C — 2 C - - H
CH3CH2CH2 CH3
b a
:Br:⁻

b → CH3CH2CH2 H C — C Br H Br CH3 2*S*,3*S* minor

a → Br H H - - C — C CH3 CH3CH2CH2 Br 2*R*,3*R* major

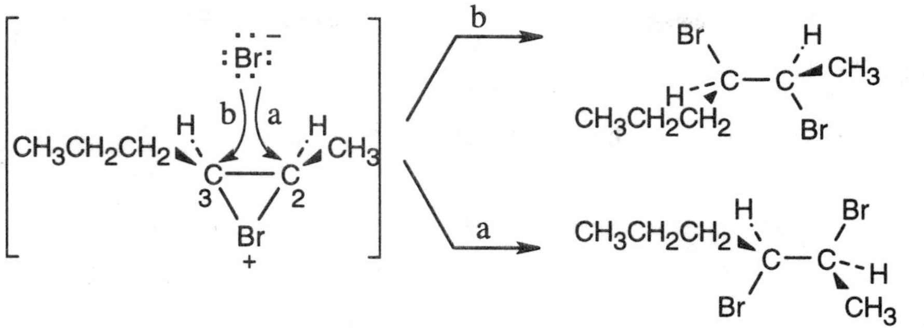

2R,3R minor

2S,3S major

The products of attack of bromide ion on each bromonium ion are shown above. Notice that the major products are enantiomers of each other, as are the minor products. Because the bromonium ions are formed in a 50:50 mixture, and because the percent attack at carbon 2 is the same for each bromonium ion, the amount of (*2R,3R*) and (*2S,3S*)-dibromohexanes is equal, and the product is a racemic mixture.

9.26 As in Problem 9.25, let's assume that attack at carbon 2 is be more likely.

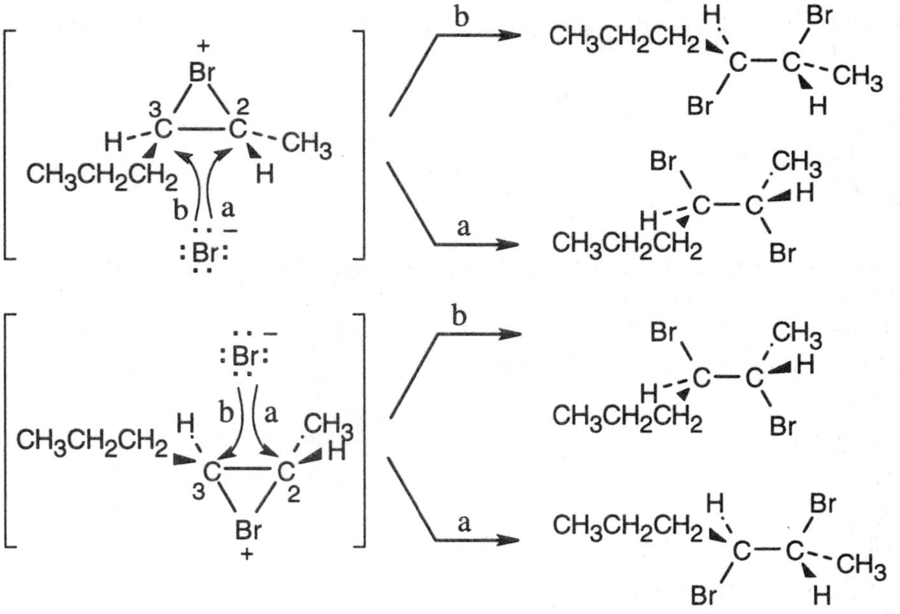

2R,3S minor

2S,3R major

2S,3R minor

2R,3S major

The product of bromination of *trans*-2-hexene is a racemic mixture of (*2S,3R*)-2,3-dibromohexane and (*2R,3S*)-2,3-dibromohexane. The reasoning is explained in Problem 9.25. The racemic mixture formed in this problem is diastereomeric to the mixture formed from *cis*-2-hexene (Problem 9.25).

9.27 Look back to Figure 9.19, which shows the reaction of (*R*)-4-methyl-1-hexene with HBr. In a similar way, we can write a reaction mechanism for the reaction of HBr with (*S*)-4-methyl-1-hexene.

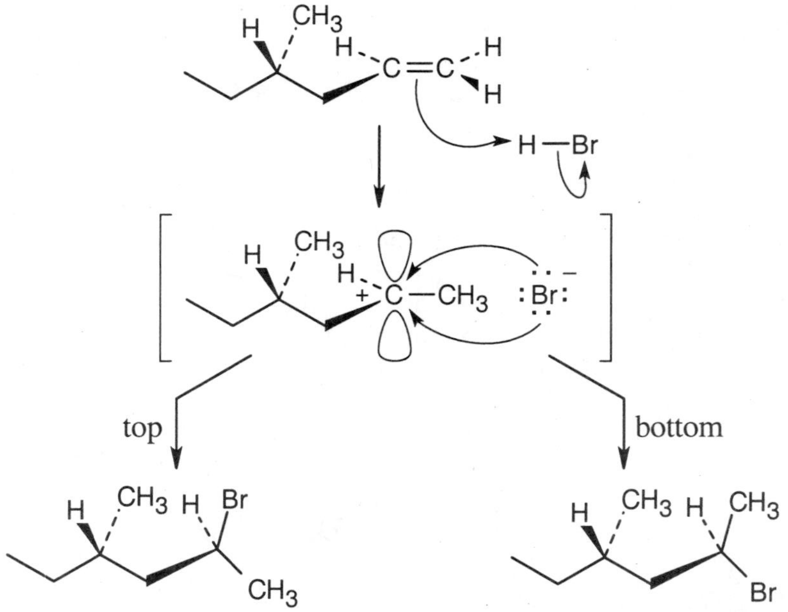

top

bottom

(2*S*,4*S*)-2-Bromo-4-methylhexane (2*R*,4*S*)-2-Bromo-4-methylhexane

The (2*S*,4*S*) stereoisomer is the enantiomer of the (2*R*,4*R*) isomer, and the transition states leading to the formation of these two isomers are enantiomeric and of equal energy. Thus, the (2*S*,4*S*) and (2*R*,4*R*) enantiomers are formed in equal amounts. A similar argument can be used to show that the (2*R*,4*S*)and 2(*S*,4*R*) isomers are formed in equal amounts. The product of the reaction of HBr with racemic starting material is thus a racemic mixture of the four possible stereoisomers and is optically inactive.

Note that the ratio of (2*R*,4*R* + 2*S*,4*S*):(2*R*,4*S* + 2*S*,4*R*) is not 50:50. Nevertheless, the product mixture is optically inactive because the enantiomers are formed in equal amounts.

9.28

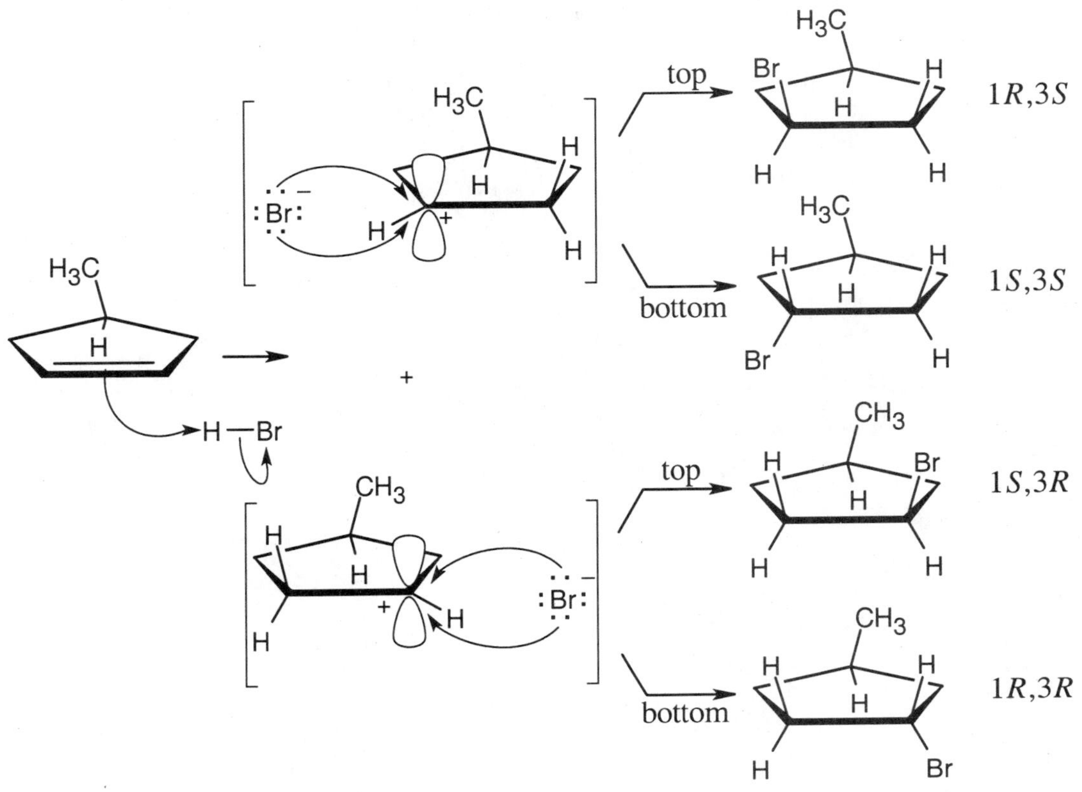

Two enantiomeric carbocations are formed. Each carbocation can be attacked by bromide ion from either the top or the bottom to yield four stereoisomers of 1-bromo-3-methylcyclopentane. The same argument used in Problem 9.27 can be used to show that the (1*S*,3*R*) and (1*R*,3*S*) enantiomers are formed in equal amounts, and the (1*S*,3*S*) and (1*R*,3*R*) isomers are formed in equal amounts. The product mixture is optically inactive (racemic). The result is a non-50:50 mixture of two racemic pairs.

Visualizing Chemistry

9.29 Structures (a), (b), and (d) are identical (*R* enantiomer), and (c) represents the *S* enantiomer.

9.30

(a)

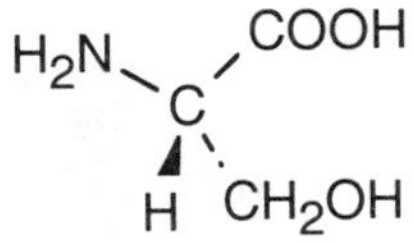

(*S*)-Serine

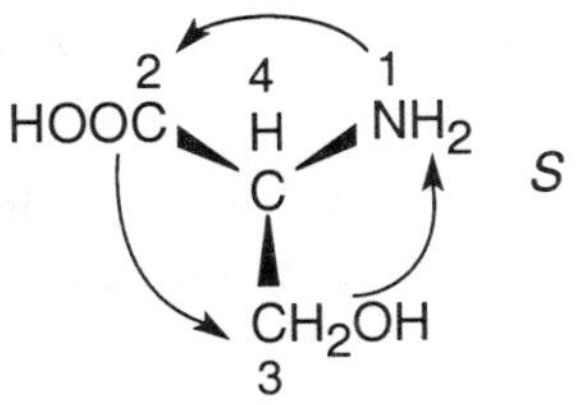

(b)

(*R*)-Adrenaline

R = ring

9.31

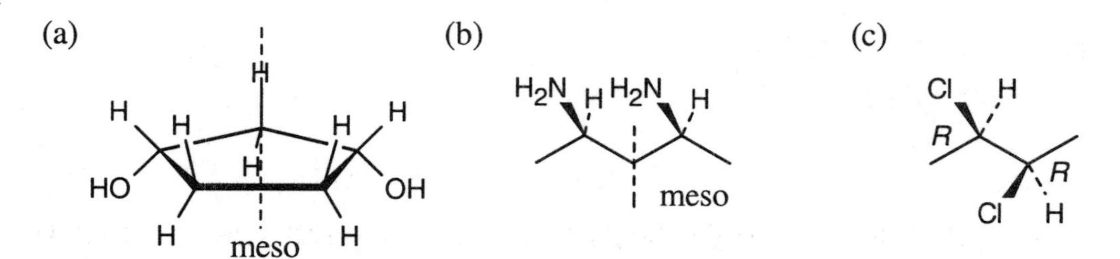

9.32

Pseudoephedrine

Additional Problems

9.33 Refer to Problem 9.5 for the formula for calculating $[\alpha]_D$. In this problem, C = 3.00 g ÷ 5.00 mL = 0.600 g/mL, and l = 1.00 cm = 0.100 dm.

For cholic acid: $$[\alpha]_D = \frac{+2.22°}{0.100 \text{ dm} \times 0.600 \text{ g/mL}} = \frac{+2.22°}{0.0600} = +37.0°$$

9.34

For ecdysone: $$[\alpha]_D = \frac{+0.087°}{0.200 \text{ dm} \times 0.00700 \text{ g/mL}} = +62°$$

9.35

(a) $CH_3CH_2CH_2\overset{*}{C}H(CH_3)CH_2CH(CH_3)_2$ 2,4-Dimethylheptane has one chirality center.

(b) $CH_3CH_2C(CH_3)_2CH_2CH(CH_2CH_3)_2$ 3-Ethyl-5,5-dimethylheptane is achiral.

(c) *cis*-1,4-Dichlorocyclohexane is achiral. Notice the plane of symmetry that passes through the –Cl groups.

(d) $CH_3C{\equiv}C\overset{*}{C}H(CH_3)\overset{*}{C}H(CH_3)C{\equiv}CCH_3$ 4,5-Dimethyl-2,6-octadiyne has two chirality centers.

The chirality of (d) depends on the configuration at each of the chirality centers. The (*R,R*) and (*S,S*) isomers are chiral enantiomers; the (*R,S*) isomer is an achiral meso compound.

9.36

(a)

Cl

$CH_3CH_2CH_2\overset{*}{C}HCH_3$

2-Chloropentane

(b)

OH

$CH_3CH_2CH_2CH_2\overset{*}{C}HCH_3$

2-Hexanol

(c)

CH_3

$CH_3CH_2\overset{*}{C}HCH{=}CH_2$

3-Methyl-1-pentene

(d)

CH_3

$CH_3CH_2CH_2CH_2\overset{*}{C}HCH_2CH_3$

3-Methylheptane

9.37

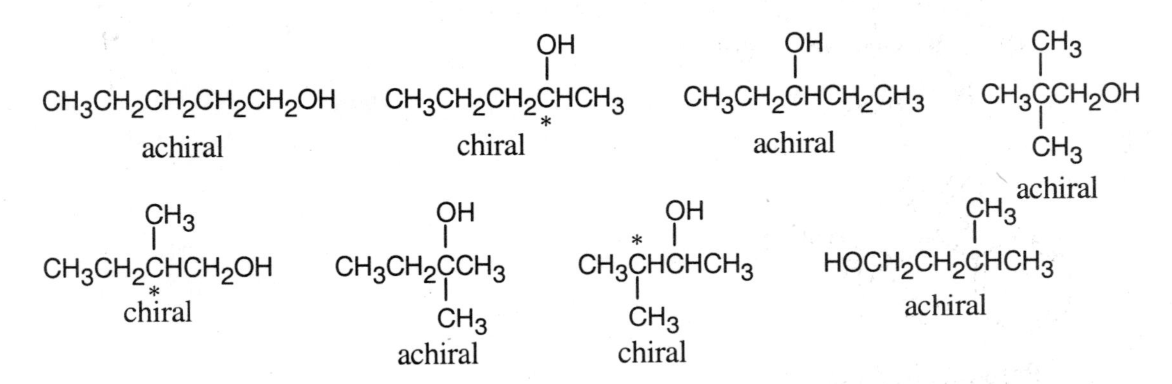

9.38 Draw the five C_6H_{14} hexanes.

$CH_3CH_2CH_2CH_2CH_2CH_3$
3 kinds of –H

$CH_3CH_2CH_2CH(CH_3)_2$
5 kinds of –H

$CH_3CH_2CH(CH_3)CH_2CH_3$
4 kinds of –H

$(CH_3)_2CHCH(CH_3)_2$
2 kinds of –H

$CH_3CH_2C(CH_3)_3$
3 kinds of –H

17 monobromohexanes can be formed from the hexane isomers. You may need to draw all the bromohexanes to find the nine that are chiral:

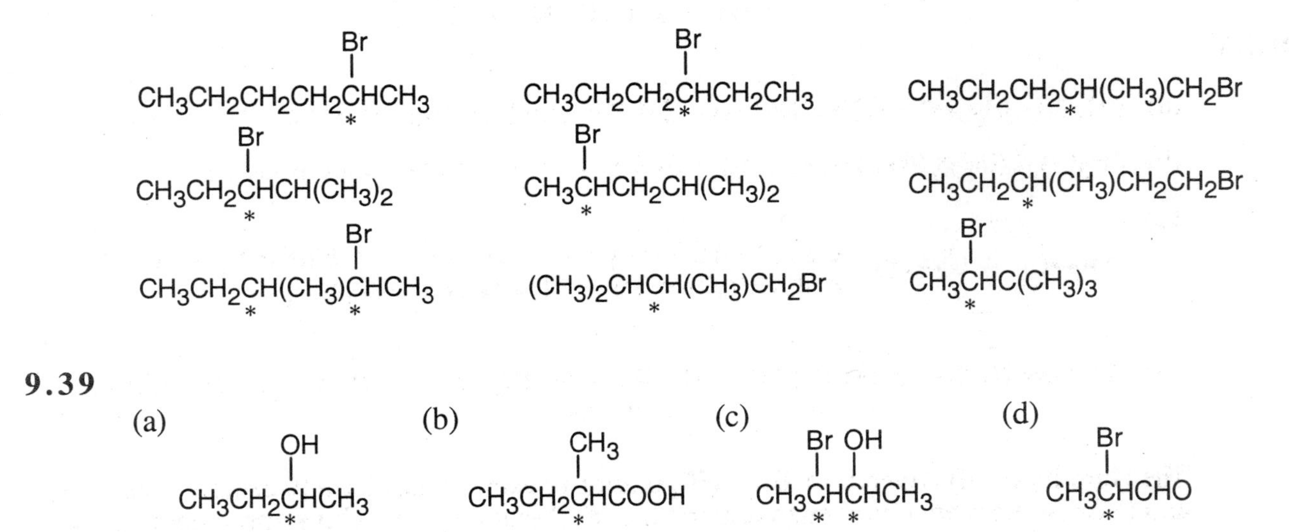

9.40 Chiral: (d) golf club, (e) monkey wrench (because of the threads).
Achiral: (a) basketball, (b) fork, (c) wine glass, (f) snowflake.

9.41

Penicillin V
3 chirality centers

9.42

(a) symmetry plane

(b) symmetry plane

(c) This compound is also a meso compound.

9.43 The specific rotation of (2*R*,3*R*)-dichloropentane is equal in magnitude and opposite in sign to the specific rotation of (2*S*,3*S*)-dichloropentane because the compounds are enantiomers. There is no predictable relationship between the specific rotations of the (2*R*,3*S*) and (2*R*,3*R*) isomers because they are diastereomers.

9.44–9.45

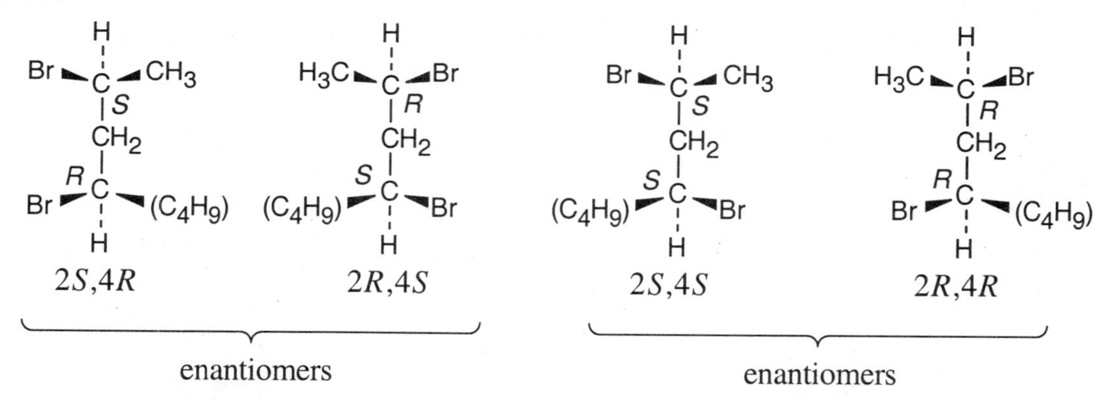

The (2*R*,4*S*) stereoisomer is the enantiomer of the (2*S*,4*R*) stereoisomer.
The (2*S*,4*S*) and (2*R*,4*R*) stereoisomers are diastereomers of the (2*S*,4*R*) stereoisomer.

9.46

(a)

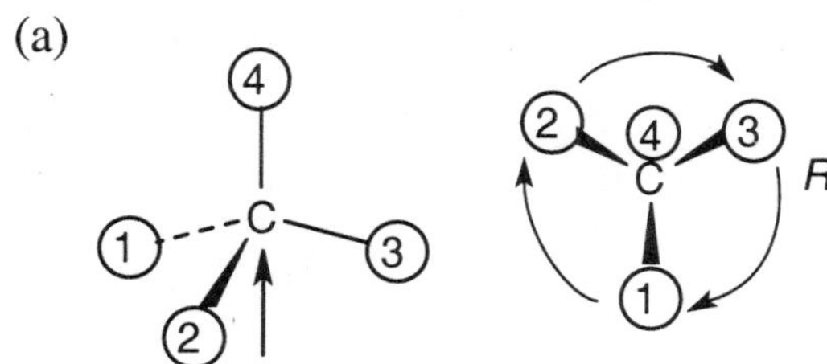

(b)

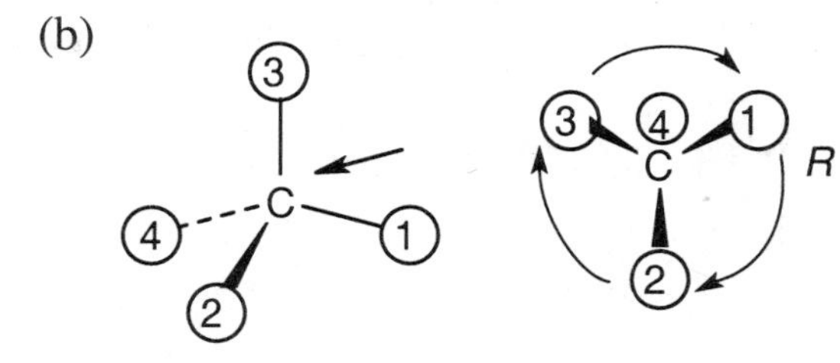

(c)

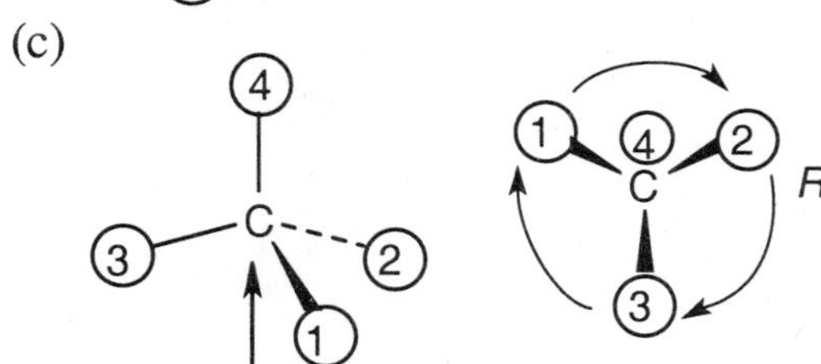

9.47

Highest ⟶ *Lowest*

(a) —$C(CH_3)_3$, —$CH{=}CH_2$, —$CH(CH_3)_2$, —CH_2CH_3

(b) , —$C{\equiv}CH$, —$C(CH_3)_3$, —$CH{=}CH_2$

(c) —$COOCH_3$, —$COCH_3$, —CH_2OCH_3, —CH_2CH_3

(d) —Br, —CH_2Br, —CN, —CH_2CH_2Br

9.48

(a) H OH *S*

(b) Cl H *S*

(c) H OCH_3 *S* $HOCH_2$ COOH

9.49

(a) OH H *S* *S* Cl H

(b) H *S* *S* CH_3 CH_3CH_2 H

(c) HO *R* *S* OH H_3C CH_3

9.50 Problem 9.9 shows a method of solution.

(a)

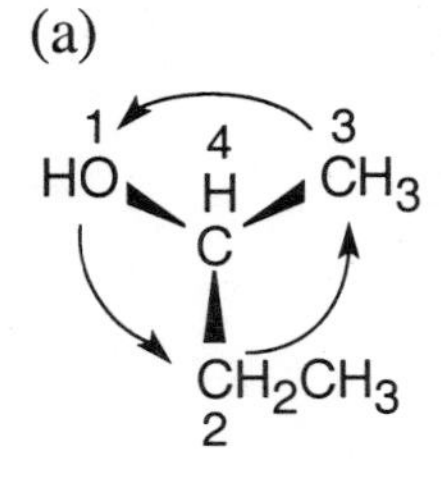

H *S* HO C CH_2CH_2 CH_3

(b)

3 CH_3CH_2 4 H C 1 Cl $CH{=}CH_2$ 2

H *R* $H_2C{=}CH$ C Cl CH_2CH_3

9.51

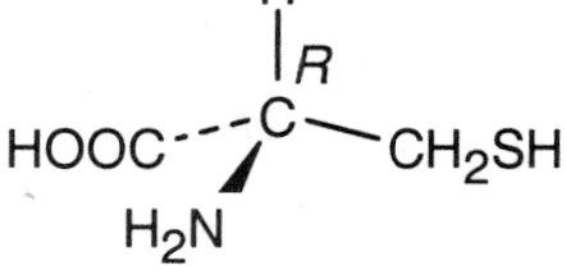

H
R
HOOC–C–CH_2SH
H_2N

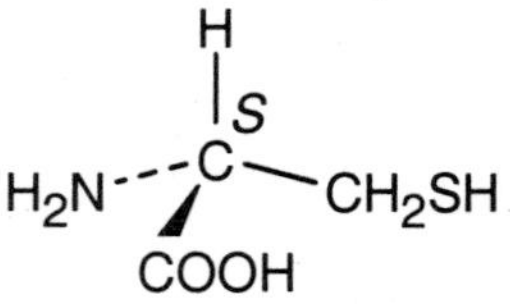

H
S
H_2N–C–CH_2SH
COOH

9.52 Identical molecules: b (*R* enantiomer), c (*R* enantiomer), d (*S* enantiomer).
Pair of enantiomers: a.

9.53

(a)
CN 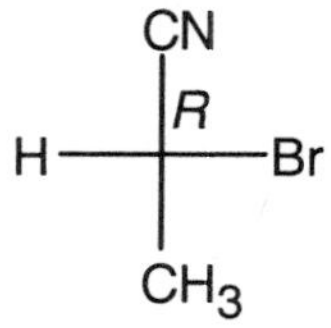
R
H–C–Br
CH_3

(b)
$CH{=}CH_2$
S
H–C–CH_2CH_3
COOH

(c)
Br
S
H–C–(phenyl)
CH_2CH_3

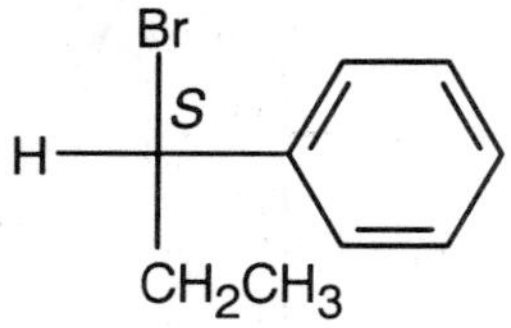

9.54

(a)
H 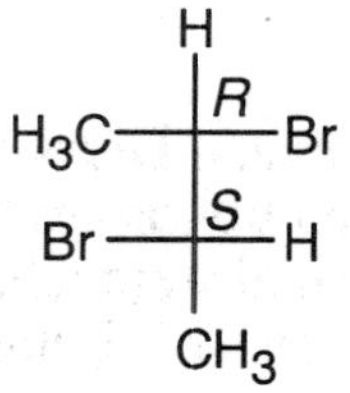
R
H_3C–C–Br
S
Br–C–H
CH_3

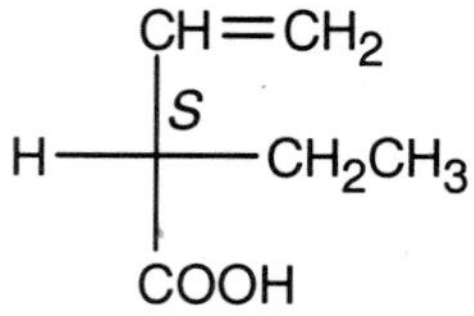

(b) 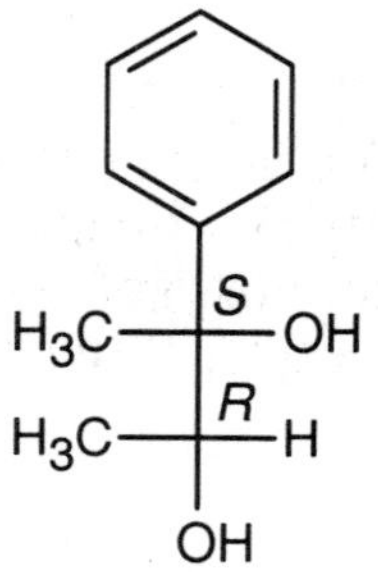
(phenyl)
S
H_3C–C–OH
R
H_3C–C–H
OH

(c)
COOH
S
HO–C–H
R
H–C–OH
R
H–C–OH
CH_2OH

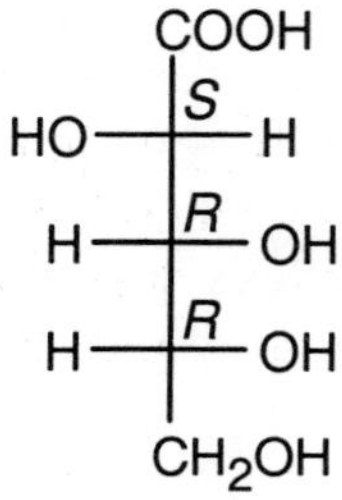

(d)
NH_2
S
H–C–COOH
R
H–C–OH
H–C–H
(phenyl)

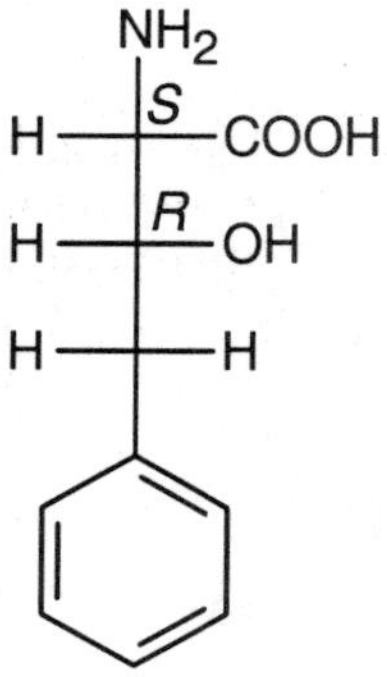

9.55

(a) 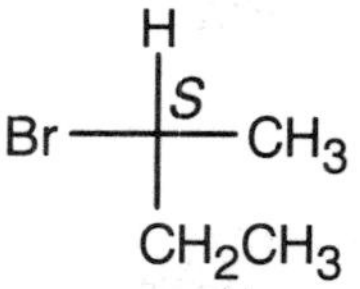
H
S
Br–C–CH_3
CH_2CH_3
(*S*)-2-Bromobutane
$CH_3CH_2CH(Br)CH_3$

(b) 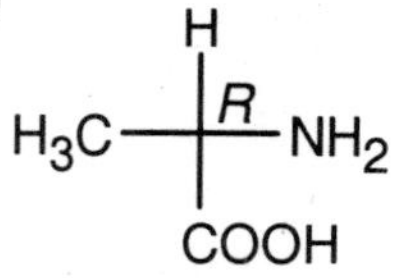
H
R
H_3C–C–NH_2
COOH
(*R*)-Alanine
$CH_3CH(NH_2)COOH$

(c) 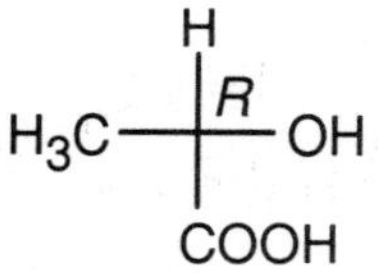
H
R
H_3C–C–OH
COOH
(*R*)-2-Hydroxypropanoic acid
$CH_3CH(OH)COOH$

(d) 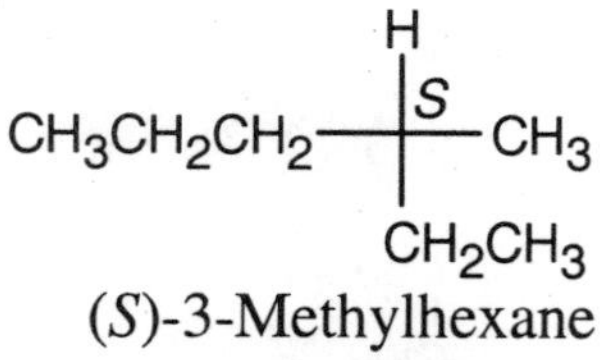
H
S
$CH_3CH_2CH_2$–C–CH_3
CH_2CH_3
(*S*)-3-Methylhexane
$CH_3CH_2CH_2CH(CH_3)CH_2CH_3$

9.56

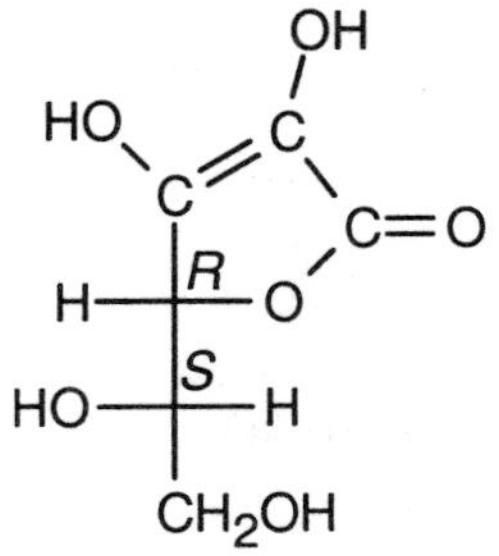

Ascorbic acid

9.57

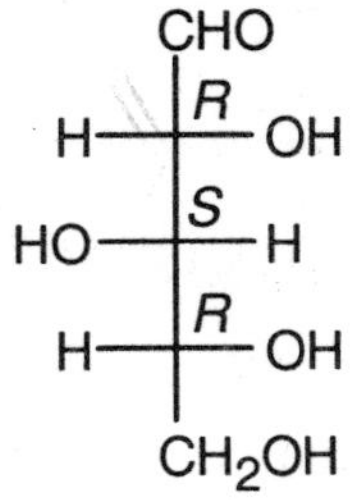

(+)-Xylose

9.58 The initial product of hydroxylation of a double bond is a cyclic *osmate*. Drawings of the osmate for *cis*-2-butene are shown below. No carbon-oxygen bonds are broken in the cleavage step; cleavage occurs at osmium-oxygen bonds. The final stereochemistry, therefore, is the same as that of the initial adduct. The product is *meso*-2,3-butanediol.

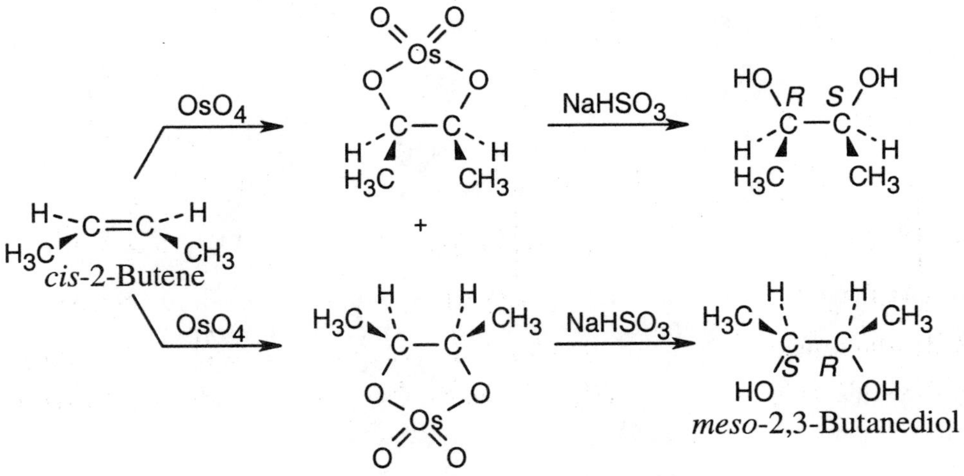

cis-2-Butene

OsO_4

$NaHSO_3$

meso-2,3-Butanediol

9.59 The osmates and the diol products of hydroxylation of *trans*-2-butene are shown below. The product is a racemic mixture of the (*R*,*R*) and (*S*,*S*) enantiomers.

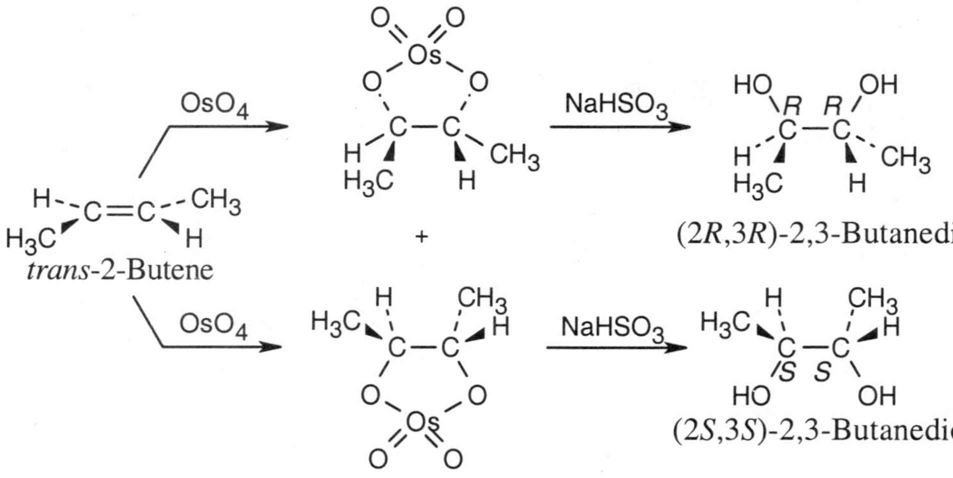

H ‑ C=C ‑ CH_3 / H_3C H

trans-2-Butene

OsO_4 → (osmate: O, O, Os, O, O; C—C; H, H_3C, CH_3, H) — $NaHSO_3$ → HO, OH, *R* *R*, C—C, H, H_3C, CH_3, H

(2*R*,3*R*)-2,3-Butanediol

+

OsO_4 → (osmate: H_3C, H, CH_3, H; C—C; O, O, Os, O, O) — $NaHSO_3$ → H_3C, H, CH_3, H, C—C, *S* *S*, HO, OH

(2*S*,3*S*)-2,3-Butanediol

9.60

O; *R* C—C *S*; H, H; $CH_3CH_2CH_2$, $CH_2CH_2CH_3$

$CH_3CH_2CH_2$ H, H $CH_2CH_2CH_3$; *S* C—C *R*; O

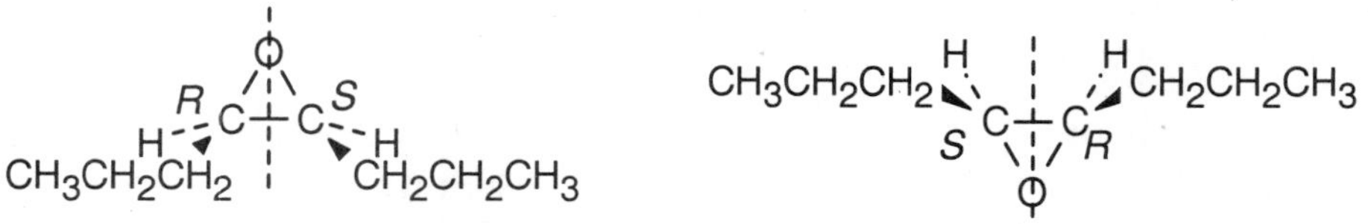

Peroxycarboxylic acids can attack either the "top" side or the "bottom" side of a double bond. The epoxide resulting from "top-side" attack on *cis*-4-octene has two chirality centers, but because it has a plane of symmetry, it is a meso compound. The two epoxides are identical.

9.61

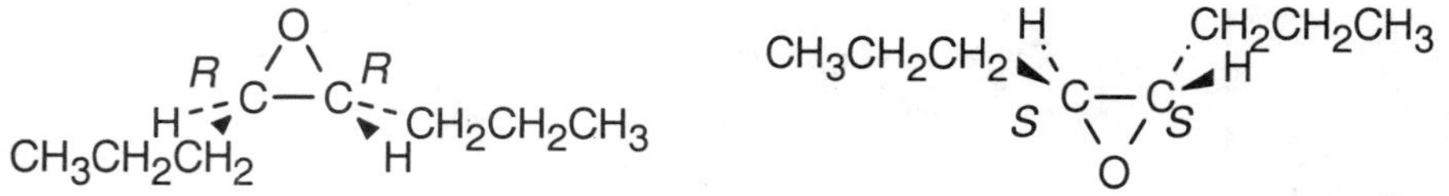

O; *R* C—C *R*; H, $CH_2CH_2CH_3$; $CH_3CH_2CH_2$, H

$CH_3CH_2CH_2$ H, $CH_2CH_2CH_3$ H; *S* C—C *S*; O

The epoxide formed by "top-side" attack of a peroxyacid on *trans*-4-octene has two chirality centers of *R* configuration. The epoxide formed by "bottom-side" attack has (*S*,*S*) configuration. The two epoxide enantiomers are formed in equal amounts and constitute a racemic mixture.

9.62

(a) Br_2, H_2O / DMSO → Br, H, *S* *S*, H, OH + H, OH, *R* *R*, Br, H

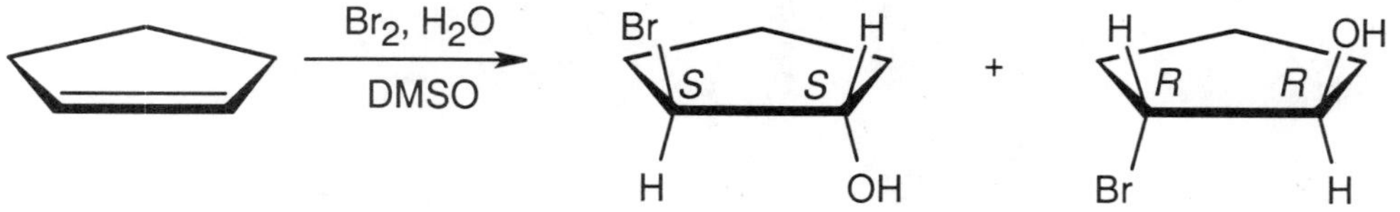

racemic mixture

(b) Br_2 / CH_2Cl_2 → Br, H, *S* *S*, H, Br + H, Br, *R* *R*, Br, H

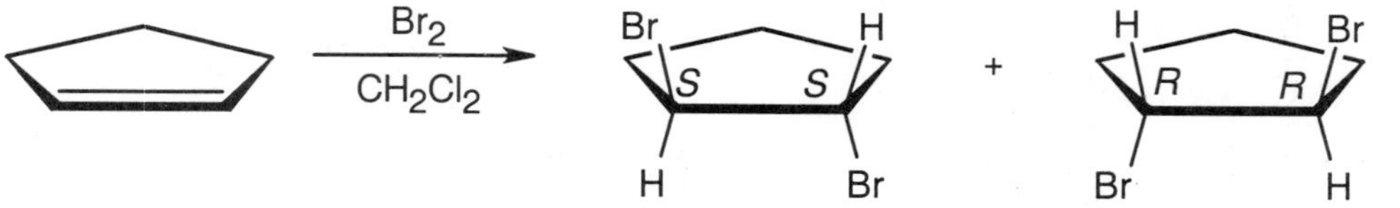

racemic mixture

(c)

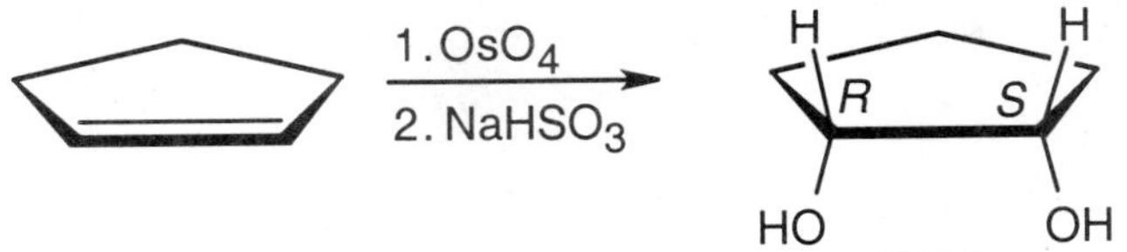

1. OsO_4
2. $NaHSO_3$

meso

9.63

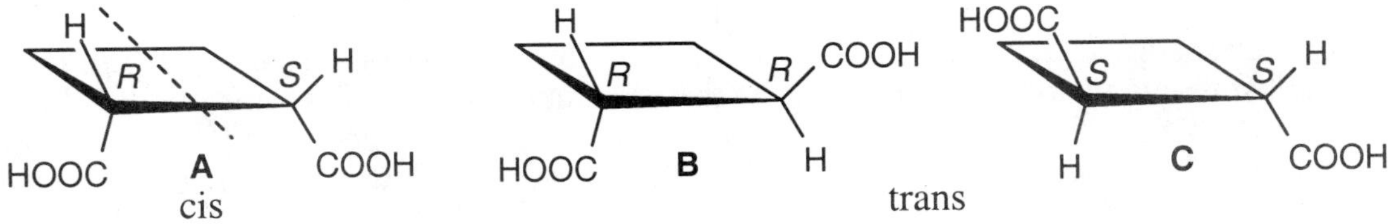

A cis meso

B trans C

B and **C** are enantiomers and are optically active. Compound **A** is their diastereomer and is a *meso* compound, which is not optically active.

The two isomeric cyclobutane-1,3-dicarboxylic acids are diastereomers and are both *meso* compounds. The cis isomer has two planes of symmetry.

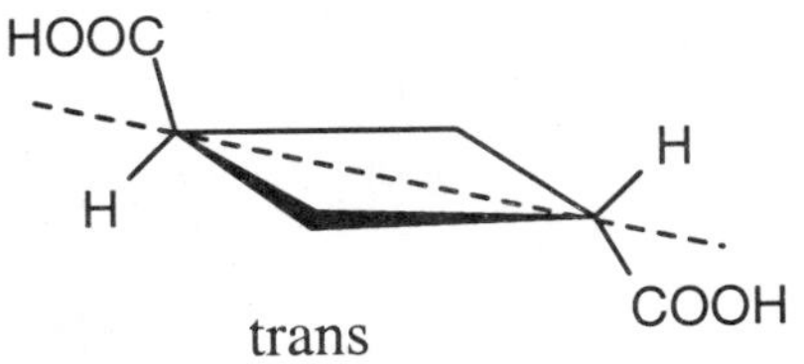

trans 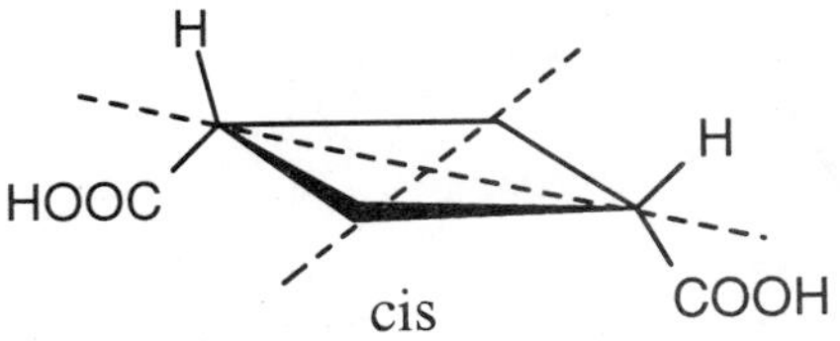 cis

9.64

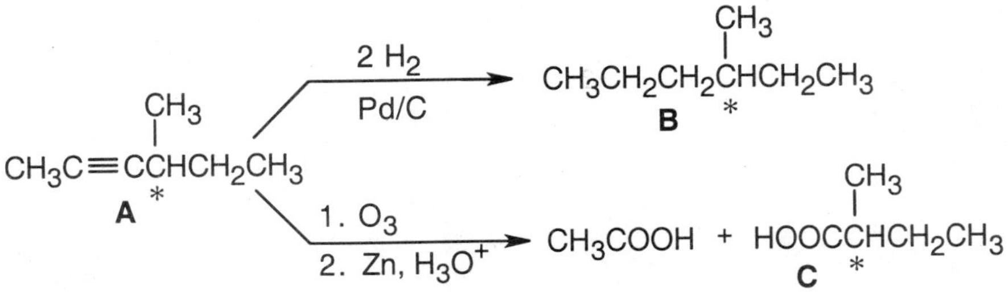

$CH_3C{\equiv}CCH(CH_3)CH_2CH_3$ (**A**, *) —2 H_2, Pd/C→ $CH_3CH_2CH_2CH(CH_3)CH_2CH_3$ (**B**, *)

A —1. O_3; 2. Zn, H_3O^+→ CH_3COOH + $HOOCCH(CH_3)CH_2CH_3$ (**C**, *)

9.65 **A** has four multiple bonds/rings.

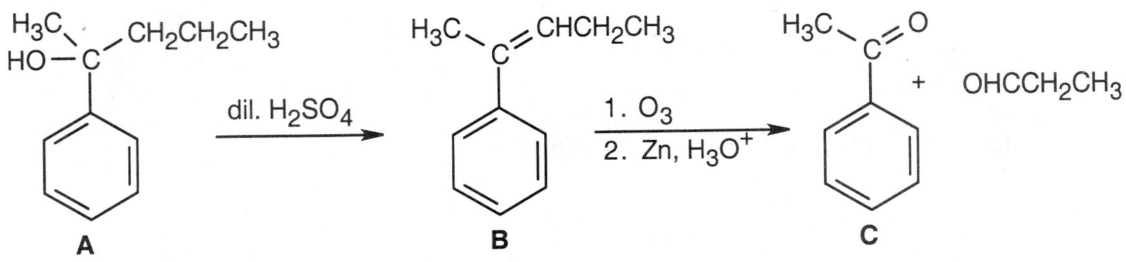

A (2-phenyl-2-pentanol: H_3C, HO, $CH_2CH_2CH_3$ on C bearing phenyl) —dil. H_2SO_4→ **B** (H_3C–C(Ph)=$CHCH_2CH_3$) —1. O_3; 2. Zn, H_3O^+→ **C** (H_3C–C(Ph)=O) + $OHCCH_2CH_3$

2-Phenyl-3-pentanol is also an acceptable answer.

9.66

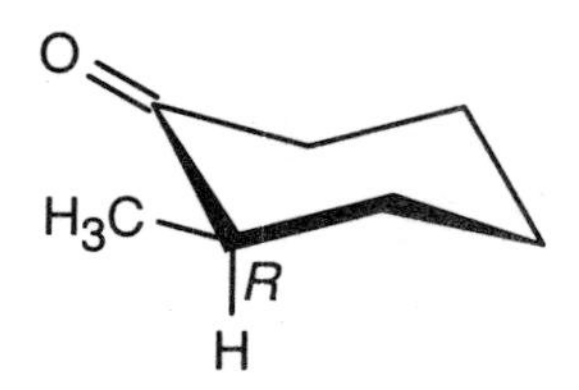

(*R*)-2-Methylcyclohexanone

9.67 A tetrahedrane can be chiral. Notice that the orientations of the four tetrahedrane substituents in space are the same as the orientations of the four substituents of a tetrasubstituted carbon atom. If you were able to make a model of a tetrasubstituted tetrahedrane (without having your models fall apart), you would also be able to make a model of its mirror image.

9.68 Mycomycin contains no chiral carbon atoms, yet is chiral. To see why, make a model of mycomycin. For simplicity, call –CH=CHCH=CHCH$_2$COOH "A" and –C≡CC≡CH "B". Remember from Chapter 6 that the carbon atoms of an allene are linear and that the π bonds formed are perpendicular to each other. Attach substituents at the *sp*^{2} carbons.

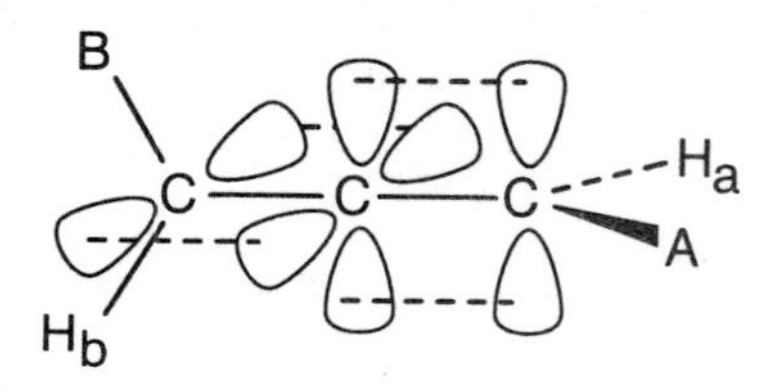

Notice that the substituents A, H$_a$, and all carbon atoms lie in a plane that is perpendicular to the plane that contains B, H$_b$, and all carbon atoms.

Now, make another model identical to the first, except for an exchange of A and H$_a$. This new allene is not superimposable on the original allene. The two allenes are enantiomers and are chiral because they possess no plane of symmetry.

9.69 4-Methylcyclohexylideneacetic acid is chiral for the same reason that mycomycin (Problem 9.68) is chiral: It possesses no plane of symmetry and is not superimposable on its mirror image. As in the case of allenes, the two functional groups at one end of the molecule lie in a plane perpendicular to the plane that contains the two functional groups at the other end.

9.70

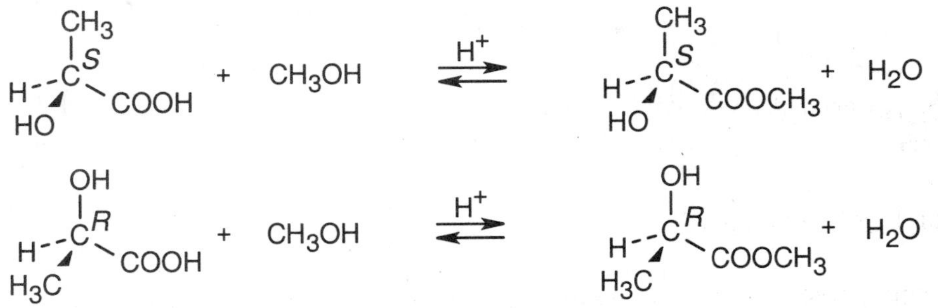

The product is a racemic mixture of (*R*)- and (*S*)- methyl lactates.

9.71

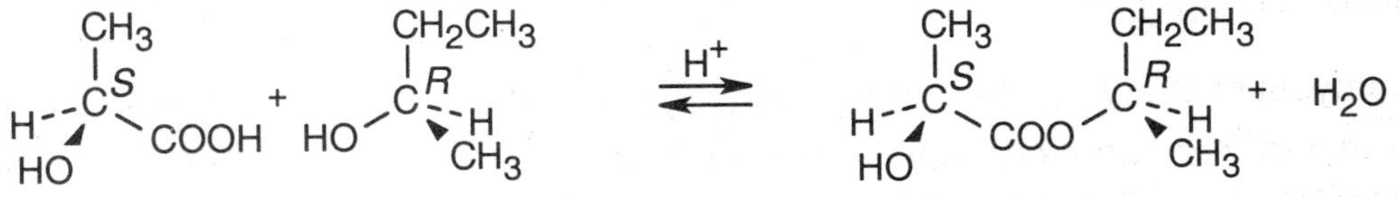

The product is (*R*)-*sec*–butyl (*S*)-lactate. As in the previous problem, the stereochemistry at the chirality centers of the product is the same as the stereochemistry of the reactants because no bonds were formed or broken at the chirality center.

9.72

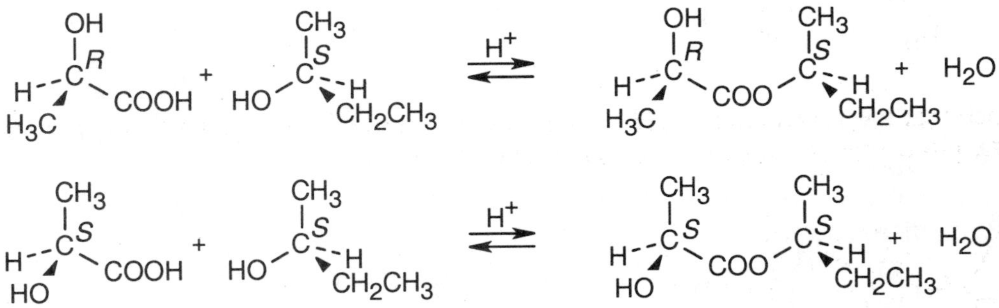

The product esters are diastereomers and differ in physical properties and chemical behavior. It should be possible to separate them by a technique such as distillation, fractional crystallization or chromatography. After separation, the esters can be converted back into (*R*)- or (*S*)-lactic acid and (*S*)-2-butanol.

9.73

(a) (*S*)-1-Chloro-2-methylbutane $\xrightarrow[h\nu]{Cl_2}$ (*S*)-1,4-Dichloro-2-methylbutane + (*R*)-1,2-Dichloro-2-methylbutane + (*S*)-1,2-Dichloro-2-methylbutane + other products

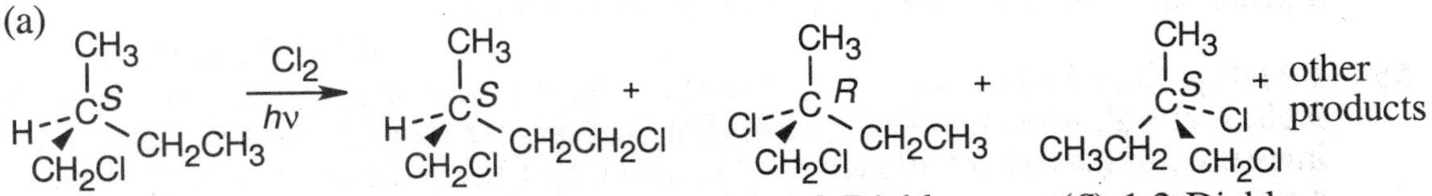

(*R*)- and (*S*)-1,2-Dichloro-2-methylbutane: 1:1 mixture

(b) Chlorination at carbon 4 yields optically active product. Chlorination at carbon 2 yields optically inactive product.

(c) Radical chlorination reactions taking place at a chirality center occur with racemization; radical chlorination reactions at a site other than the chirality center do not affect the stereochemistry of the chirality center.

9.74 Both of the diastereomers shown below are meso compounds with three chirality centers. Each is a meso compound because it has a symmetry plane, and in each structure the central carbon is bonded to four different groups (a group with *R* configuration, a group with *S* configuration, –OH, and –H).

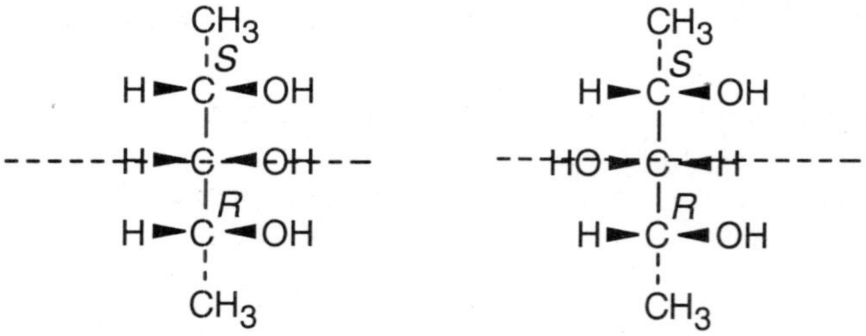

CH_3 S H–C–OH H–C–OH R H–C–OH CH_3 CH_3 S H–C–OH HO–C–H R H–C–OH CH_3

9.75

CH_3 S H–C–Br H–C–Cl R H–C–Br CH_3 **A** CH_3 S H–C–Br Cl–C–H R H–C–Br CH_3 **B** CH_3 S H–C–Br Cl–C–H S Br–C–H CH_3 **C** CH_3 R Br–C–H H–C–Cl R H–C–Br CH_3 **D**

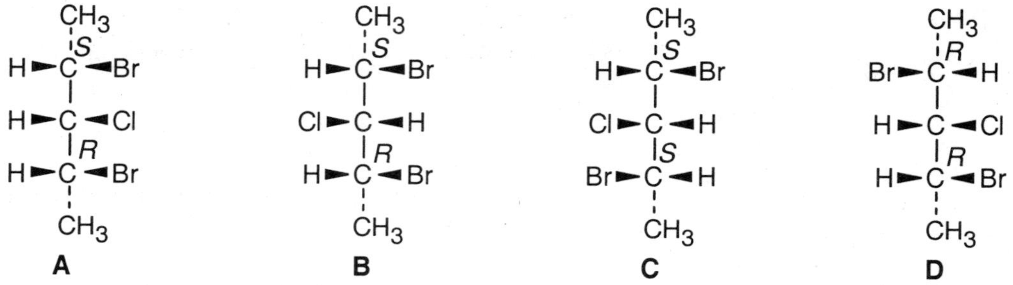

There are four stereoisomers of 2,4-dibromo-3-chloropentane. **C** and **D** are enantiomers and are optically active. **A** and **B** are optically inactive meso compounds and are diastereomers.

9.76

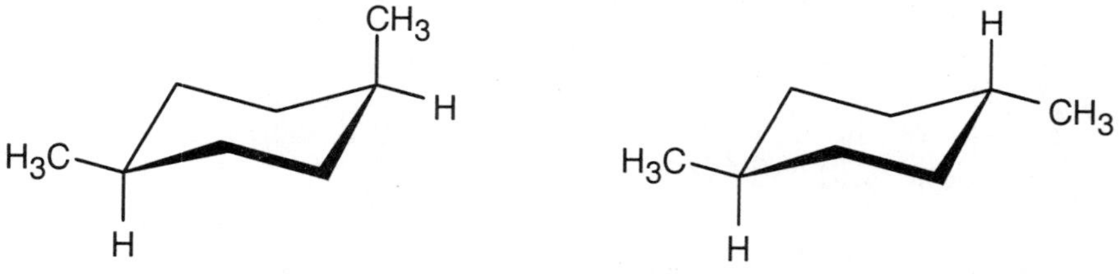

CH_3 H H_3C H *cis*-1,4-Dimethylcyclohexane H CH_3 H_3C H *trans*-1,4-Dimethylcyclohexane

(a) There is only one stereoisomer of each of the 1,4-dimethylcyclohexanes.
(b) Neither 1,4-dimethylcyclohexane is chiral; both are meso compounds.
(c) The two 1,4-dimethylcyclohexanes are diastereomers.

9.77

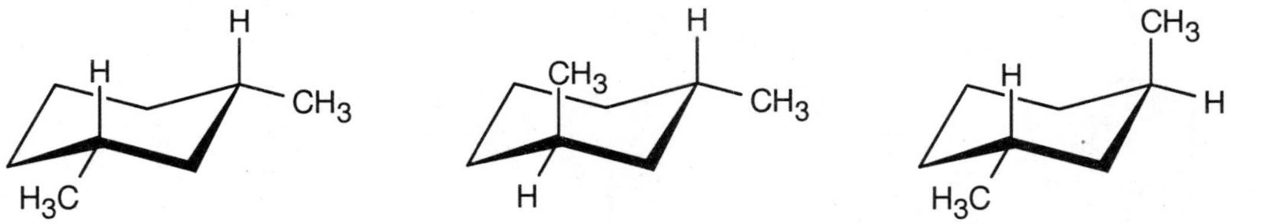

cis-1,3-Dimethylcyclohexane

trans-1,3-Dimethylcyclohexane

(a) There is one stereoisomer of *cis*-1,3-dimethylcyclohexane, and there are two stereoisomers of *trans*-1,3-dimethylcyclohexane.
(b) *trans*-1,3-Dimethylcyclohexane exists as a pair of chiral enantiomers. *cis*-1,3-Dimethylcyclohexane is a achiral meso compound.
(c) The two trans stereoisomers are enantiomers, and are diastereomers of the cis stereoisomer.

9.78

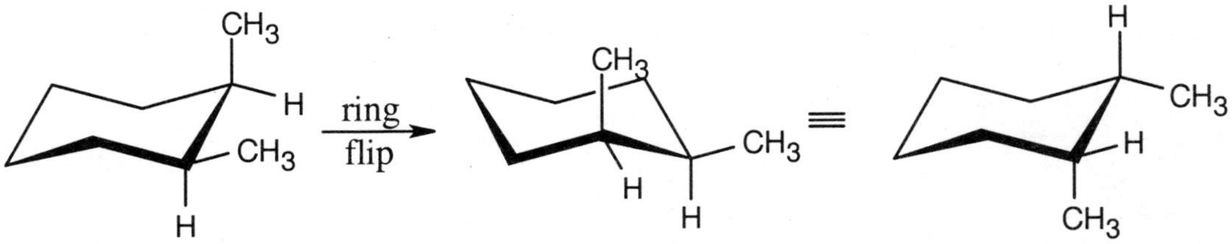

ring flip

cis-1,2-Dimethylcyclohexane

The two *cis*-1,2-dimethylcyclohexane enantiomers can interconvert by a ring flip and thus exist as a racemic mixture.

A Look Ahead

9.79

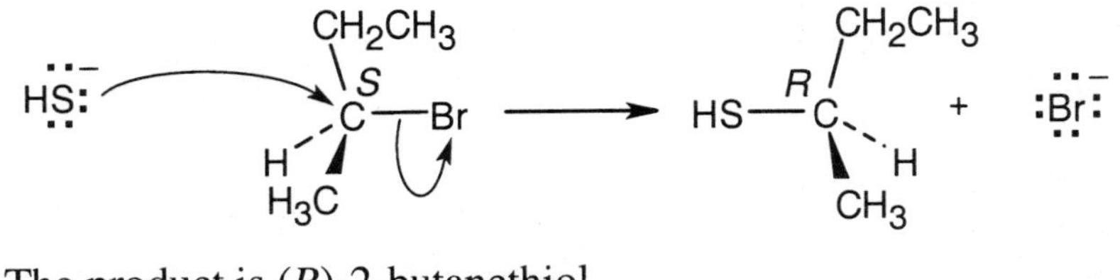

The product is (*R*)-2-butanethiol.

9.80

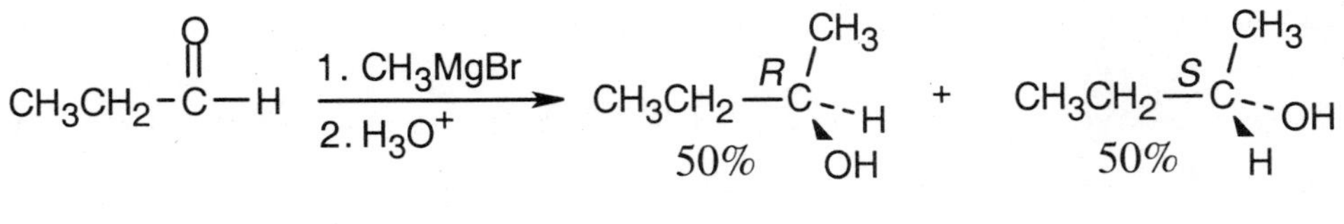

50% 50%

(a) Reaction of a Grignard reagent with an achiral starting material, such as propanal, yields racemic product.
(b) The product consists of a 50:50 mixture of (*R*)-2-butanol and its enantiomer, (*S*)-2-butanol.

9.81

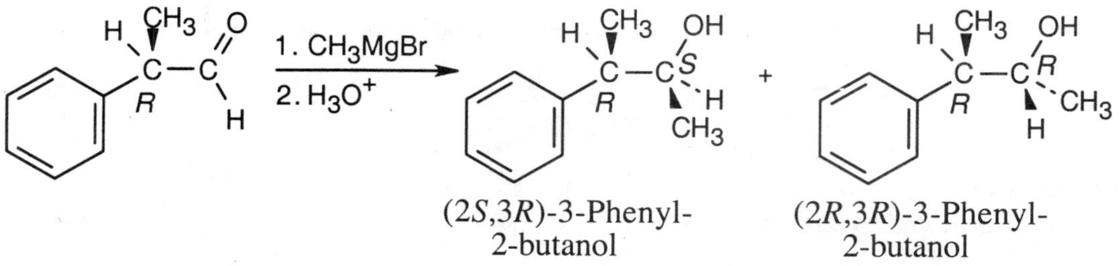

$$\text{C}_6\text{H}_5\text{–CH(CH}_3\text{)–CHO} \xrightarrow[\text{2. } H_3O^+]{\text{1. } CH_3MgBr} \text{(2}S\text{,3}R\text{)} + \text{(2}R\text{,3}R\text{)}$$

(2*S*,3*R*)-3-Phenyl-2-butanol (2*R*,3*R*)-3-Phenyl-2-butanol

(a) Reaction of a Grignard reagent with a chiral starting material yields chiral products; the product mixture is optically active.
(b) The two products are a mixture of the (2*S*,3*R*) and (2*R*,3*R*) diastereomers of 3-phenyl-2-butanol. The product ratio can't be predicted, but it is not 50:50.

Molecular Modeling

9.82

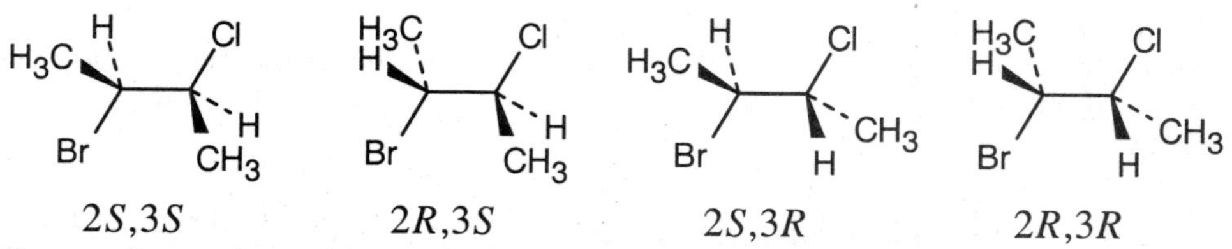

2*S*,3*S* 2*R*,3*S* 2*S*,3*R* 2*R*,3*R*

The conformation with anti halogens has the lowest strain energy for each stereoisomer. Strain energies for these conformations are 40.8 kJ/mol for the (2*R*,3*R*) and (2*S*,3*S*) enantiomers and 43.0 kJ/mol for the (2*R*,3*S*) and (2*S*,3*R*) enantiomers. Notice that enantiomers have identical energies, and diastereomers have different energies.

9.83 The pyramidal form has the lowest energy for all three molecules. ΔE = 45 kJ/mol for *N*-ethyl-*N*-methylpropylamine, ΔE = 199 kJ/mol for *P*-ethyl-*P*-methylpropylphosphine, and ΔE = 225 kJ/mol for ethyl methyl sulfoxide. The amine racemizes much more rapidly than the other two molecules; the phosphine and sulfoxide maintain their configurations indefinitely at room temperature.

Chapter 10 – Alkyl Halides

Chapter Outline

I. Introduction to alkyl halides (Sections 10.1 – 10.2).
   A. Naming alkyl halides (Section 10.1).
      1. Rules for naming alkyl halides:
         a. Find the longest chain and name it as the parent.
            If a double or triple bond is present, the parent chain must contain it.
         b. Number the carbon atoms of the parent chain, beginning at the end nearer the first substituent, whether alkyl or halo.
         c. Number each substituent.
            i. If more than one of the same kind of substituent is present, number each, and use the prefixes *di-*, *tri-*, *tetra-* and so on.
            ii. If different halogens are present, number all and list them in alphabetical order.
         d. If the parent chain can be numbered from either end, start at the end nearer the substituent that has alphabetical priority.
      2. Some alkyl halides are named by first citing the name of the alkyl group and then citing the halogen.
   B. Structure of alkyl halides (Section 10.2).
      1. Alkyl halides have approximately tetrahedral geometry.
      2. Bond lengths increase with increasing size of the halogen bonded to carbon.
      3. Bond strengths decrease with increasing size of the halogen bonded to carbon.
      4. Carbon–halogen bonds are polar, and halomethanes have dipole moments.
      5. Alkyl halides behave as electrophiles in polar reactions.

II. Preparation of alkyl halides (Sections 10.3 – 10.7).
   A. Alkyl halides can be prepared by reacting alkenes with HX or X_2 (Section 10.3).
   B. Alkyl halides can also be formed by radical reaction of alkanes with Cl_2 or Br_2.
      1. The sequence of steps: initiation, propagation, termination.
      2. Complications of radical halogenation (Section 10.4).
         a. The reaction continues on to produce di- and polysubstituted products.
         b. If more than one type of hydrogen is present, more than one type of monosubstituted product is formed.
         c. The reactivity order of different types of hydrogen towards chlorination is: primary < secondary < tertiary.
            i. This reactivity order is due to the bond dissociation energies for formation of the alkyl radicals.
            ii. The stability order of alkyl radicals: primary < secondary < tertiary.
         d. According to the Hammond Postulate, the product with the more stable transition state (here, the more stable radical intermediate) forms faster.
         e. Bromination is more selective than chlorination.
            The stability of the radical intermediate is even more important for bromination because radical formation is less endergonic.
   C. Allylic bromination of alkenes (Sections 10.5 – 10.6).
      1. Reaction of an alkene with NBS causes bromination at the position allylic to the double bond (Section 10.5).
      2. This reaction occurs by a radical chain mechanism.
         a. Br· abstracts an allylic hydrogen.
         b. The allylic radical reacts with Br_2 to form an allylic bromide, plus Br·.
      3. Reaction occurs at the allylic position because an allylic C–H bond is weaker than most other C–H bonds, and an allylic radical is more stable.

4. Reasons for stability of an allylic radical (Section 10.6).
   a. The carbon with the unpaired electron is sp^2-hybridized, and its *p* orbital can overlap with the *p* orbitals of the double-bond carbons.
   b. The radical intermediate is thus stabilized by resonance.
      This stability is due to delocalization of the unpaired electron over an extended π network.
   c. Reaction of the allylic radical with Br_2 can occur at either end of the π orbital system.
      i. A mixture of products may be formed if the alkene is unsymmetrical.
      ii. These products aren't usually formed in equal quantities: reaction to form the more substituted double bond is favored.
   d. Products of allylic bromination can be dehydrohalogenated to form dienes.

D. Alkyl halides from alcohols (Section 10.7).
   1. Tertiary alkyl chlorides, bromides or iodides can be prepared by the reaction of a tertiary alcohol with HCl, HBr or HI.
      Reaction of secondary or primary alcohols occurs under more drastic conditions, which may destroy other acid-sensitive functional groups.
   2. Primary and secondary alkyl halides can be formed by treatment of the corresponding alcohols with $SOCl_2$ or PBr_3.
      Reaction conditions are mild, and the reagents don't interfere with other functional groups.

III. Reactions of alkyl halides (Sections 10.8 –10.9).
   A. Grignard reagents (Section 10.8).
      1. Organohalides react with Mg to produce organomagnesium halides, RMgX.
         These compounds are known as Grignard reagents.
      2. Grignard reagents can be formed from alkyl, alkenyl and aryl halides.
         Steric hindrance is no barrier to formation of Grignard reagents.
      3. The carbon bonded to Mg is negatively polarized and is nucleophilic.
      4. Grignard reagents react with weak acids to form hydrocarbons.
   B. Organometallic coupling reagents (Section 10.9).
      1. Alkyl halides can react with Li to form alkyllithiums.
      2. These alkyllithiums can combine with CuI to form lithium diorganocopper compounds (R_2CuLi), which are known as Gilman reagents.
      3. R_2CuLi compounds can react with alkyl halides (except for fluorides) to form hydrocarbon products.
      4. Organometallic coupling reactions are useful for forming large molecules from small pieces.
         a. The reaction can be carried out on alkyl, vinyl and aryl halides.
         b. A radical mechanism is probably involved.

IV. Oxidation and reduction in organic chemistry (Section 10.10).
   A. In organic chemistry, an oxidation is a reaction that results in a loss in electron density by carbon.
      1. This loss may be due to two kinds of reactions:
         a. Bond formation between carbon and a more electronegative atom (usually O, N or halogen).
         b. Bond breaking between carbon and a less electronegative atom (usually H).
      2. Examples include chlorination of alkanes and reaction of alkenes with Br_2.
   B. A reduction is a reaction that results in a gain of electron density by carbon.
      1. This gain may be due to two kinds of reactions:
         a. Bond formation between carbon and a less electronegative atom.
         b. Bond breaking between carbon and a more electronegative atom.
      2. Examples include conversion of a Grignard reagent to an alkane, and reduction of an alkene with H_2.
   C. Alkanes are at the lowest oxidation level, and CO_2 is at the highest level.

D. A reaction that converts a compound from a lower oxidation level to a higher oxidation level is an oxidation.
E. A reaction that converts a compound from a higher oxidation level to a lower oxidation level is an reduction.

Solutions to Problems

10.1 The rules that were given for naming alkanes in Section 3.6 are used for alkyl halides. A halogen is treated as an alkyl substituent.

(a) $CH_3CH_2CH_2CH_2I$

1-Iodobutane

(b) $CH_3CH(CH_3)CH_2CH_2Cl$

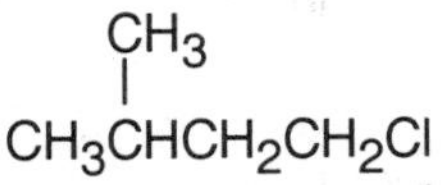

1-Chloro-3-methylbutane

(c) $BrCH_2CH_2CH_2C(CH_3)_2CH_2Br$

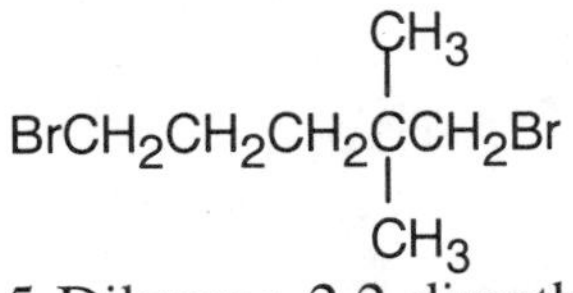

1,5-Dibromo-2,2-dimethylpentane

(d) $CH_3C(CH_3)(Cl)CH_2CH_2Cl$

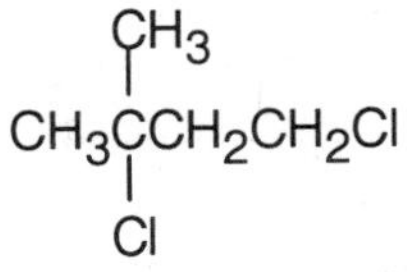

1,3-Dichloro-3-methylbutane

(e) $CH_3CH(I)CH(CH_2CH_2Cl)CH_2CH_3$

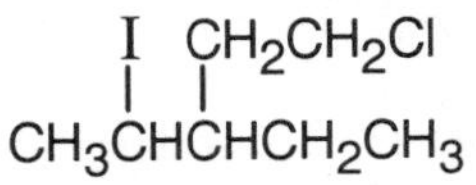

1-Chloro-3-ethyl-4-iodopentane

(f) $CH_3CH(Br)CH_2CH_2CH(Cl)CH_3$

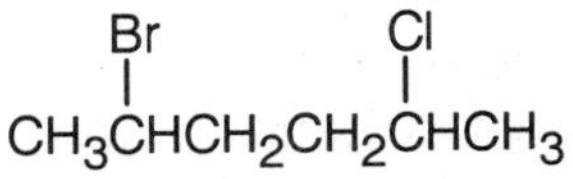

2-Bromo-5-chlorohexane

10.2

(a) $CH_3CH_2CH_2C(CH_3)_2-CH(Cl)CH_3$

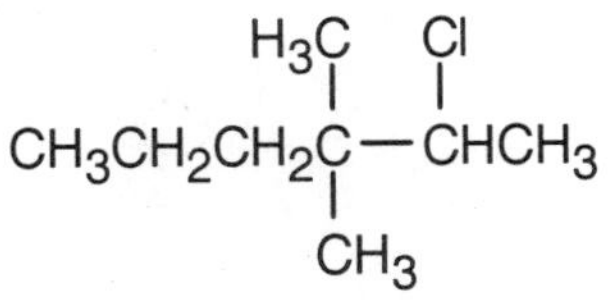

2-Chloro-3,3-dimethylhexane

(b) $CH_3CH_2CH_2C(Cl)_2-CH(CH_3)CH_3$

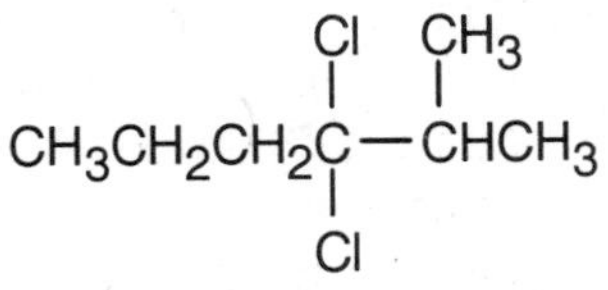

3,3-Dichloro-2-methylhexane

(c) $CH_3CH_2C(Br)(CH_2CH_3)CH_2CH_3$

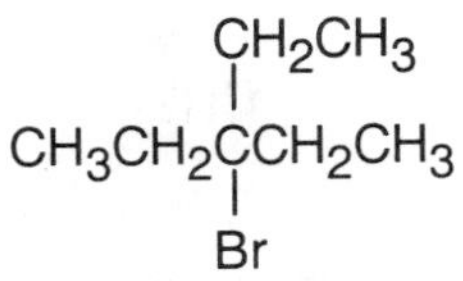

3-Bromo-3-ethylpentane

(d) $(H_3C)_2HC-C_6H_{10}Br_2$ (isopropyl group on cyclohexane ring, two Br on ring carbon)

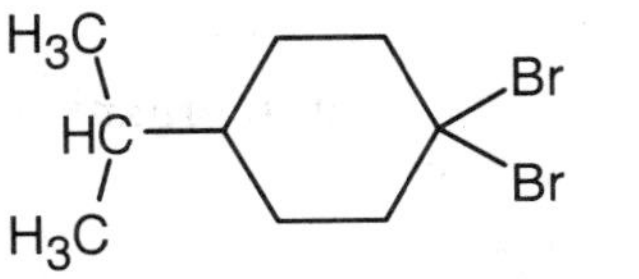

1,1-Dibromo-4-isopropylcyclohexane

(e) $CH_3CH_2CH_2CH_2CH_2CH(CH(CH_3)CH_2CH_3)CH_2CH(Cl)CH_3$

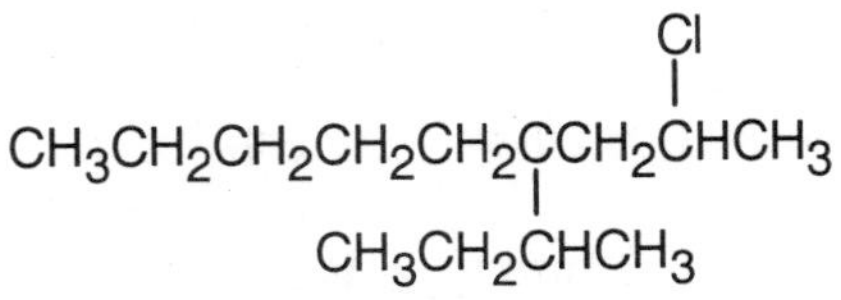

4-*sec*-Butyl-2-chlorononane

(f) $(CH_3)_3C-C_6H_{10}Br_2$ (tert-butyl group on cyclohexane ring, two Br on ring carbon)

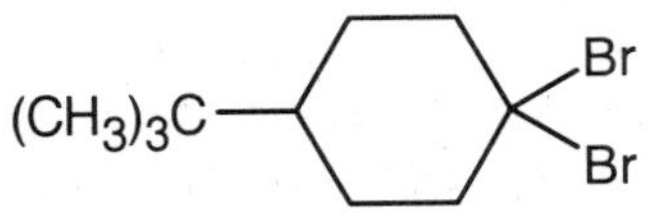

1,1-Dibromo-4-*tert*-butylcyclohexane

10.3

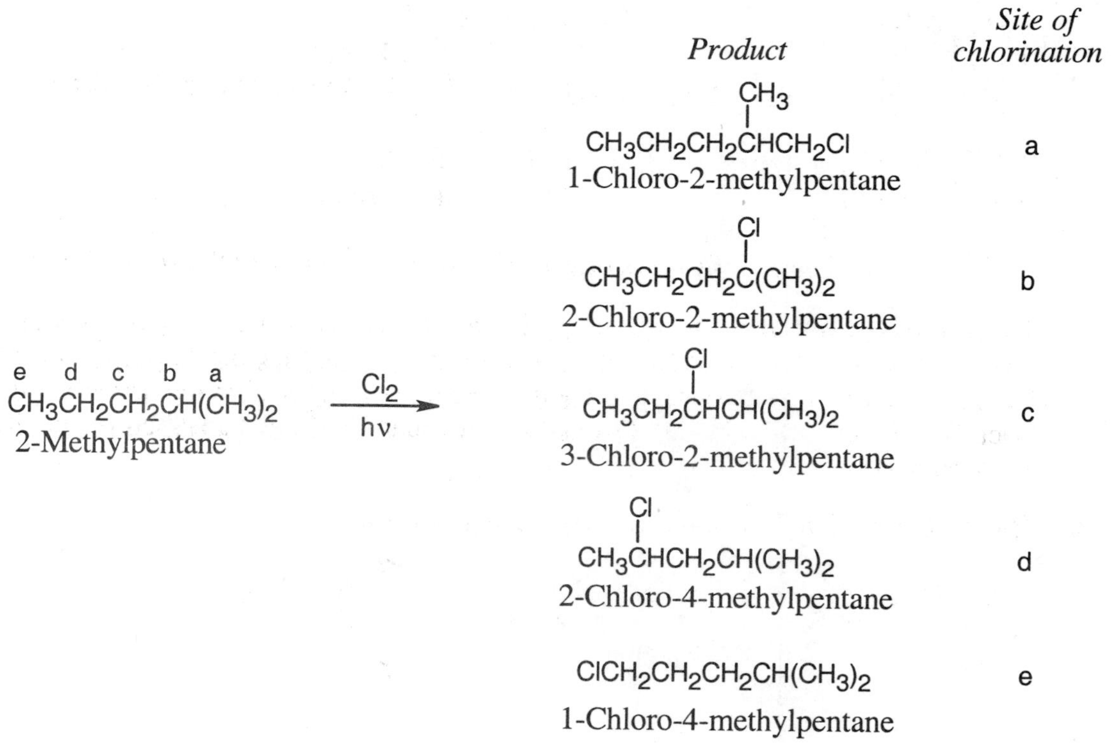

$$\underset{\text{2-Methylpentane}}{\overset{e\;\;d\;\;c\;\;b\;\;a}{CH_3CH_2CH_2CH(CH_3)_2}} \xrightarrow[h\nu]{Cl_2}$$

Product	*Site of chlorination*
$CH_3CH_2CH_2CH(CH_3)CH_2Cl$ 1-Chloro-2-methylpentane	a
$CH_3CH_2CH_2CCl(CH_3)_2$ 2-Chloro-2-methylpentane	b
$CH_3CH_2CHClCH(CH_3)_2$ 3-Chloro-2-methylpentane	c
$CH_3CHClCH_2CH(CH_3)_2$ 2-Chloro-4-methylpentane	d
$ClCH_2CH_2CH_2CH(CH_3)_2$ 1-Chloro-4-methylpentane	e

Chlorination at sites b and e yields achiral products. The products of chlorination at sites a, c and d are chiral; each product is formed as a racemic mixture of enantiomers.

10.4

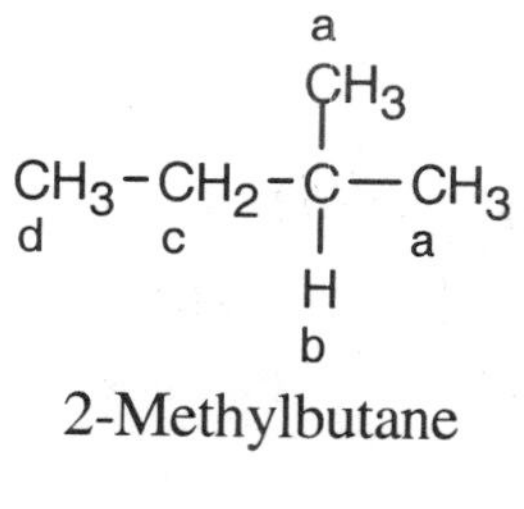

$\overset{d}{CH_3}-\overset{c}{CH_2}-\underset{\overset{|}{H\,(b)}}{C}(\overset{a}{CH_3})-\overset{a}{CH_3}$

2-Methylbutane

Type of –H	a	b	c	d
Number of –H of each type	6	1	2	3
Relative reactivity	1.0	5.0	3.5	1.0
Number times reactivity	6.0	5.0	7.0	3.0
Percent chlorination	29%	24%	33%	14%

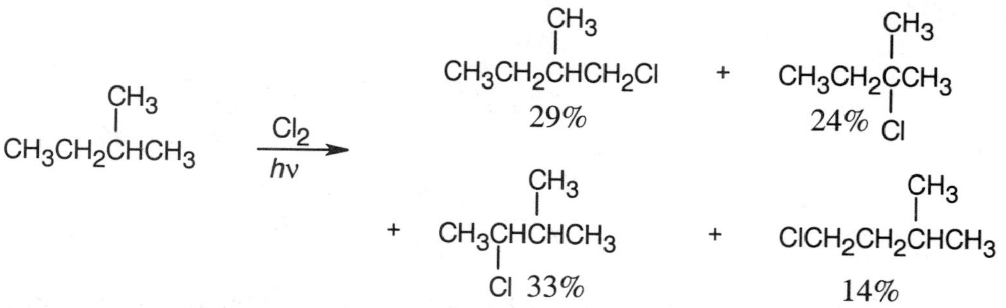

$$CH_3CH_2CH(CH_3)CH_3 \xrightarrow[h\nu]{Cl_2} \underset{29\%}{CH_3CH_2CH(CH_3)CH_2Cl} + \underset{24\%}{CH_3CH_2CCl(CH_3)CH_3} + \underset{33\%}{CH_3CHClCH(CH_3)CH_3} + \underset{14\%}{ClCH_2CH_2CH(CH_3)CH_3}$$

10.5 Review Sections 5.7 – 5.9 if you need help.

$(CH_3)_2CH\text{–}H \quad + \quad Cl\cdot \quad \rightarrow \quad (CH_3)_2CH\cdot \quad + \quad H\text{–}Cl$

$D = 401$ kJ/mol $\qquad\qquad\qquad\qquad D = 432$ kJ/mol

$\Delta H° = D_{\text{bonds broken}} - D_{\text{bonds formed}} = 401 \text{ kJ/mol} - 432 \text{ kJ/mol} = -31 \text{ kJ/mol}.$

$(CH_3)_2CH\text{–}H \quad + \quad Br\cdot \quad \rightarrow \quad (CH_3)_2CH\cdot \quad + \quad H\text{–}Br$

$D = 401$ kJ/mol $\qquad\qquad\qquad\qquad D = 366$ kJ/mol

$\Delta H° = D_{\text{bonds broken}} - D_{\text{bonds formed}} = 401 \text{ kJ/mol} - 366 \text{ kJ/mol} = +35 \text{ kJ/mol}.$

The reaction of Br• with a secondary hydrogen atom is more selective. In the endergonic reaction of $(CH_3)_2CH_2$ with Br•, the transition state resembles the isopropyl radical. Because the secondary isopropyl radical is much more stable than the primary radical, the reaction is much more likely to proceed by secondary hydrogen abstraction and thus to be more selective.

10.6 The cyclohexadienyl radical has three resonance forms.

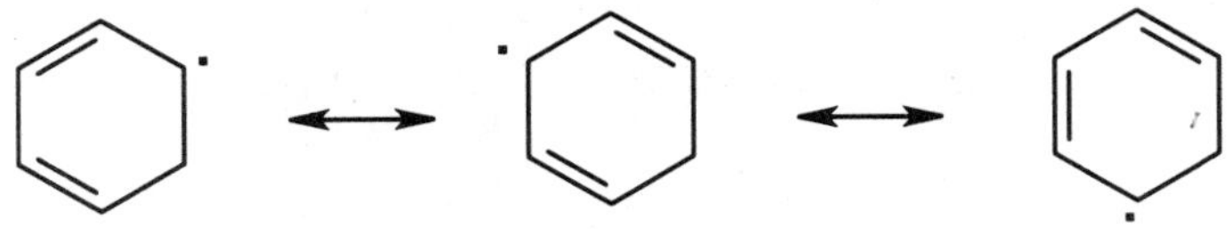

10.7

CH_2

Abstraction of hydrogen by a bromine radical yields an allylic radical.

Br·

CH_2 $\longleftrightarrow$ $\dot{C}H_2$ + HBr

The allylic radical reacts with Br_2 to produce **A** and **B**.

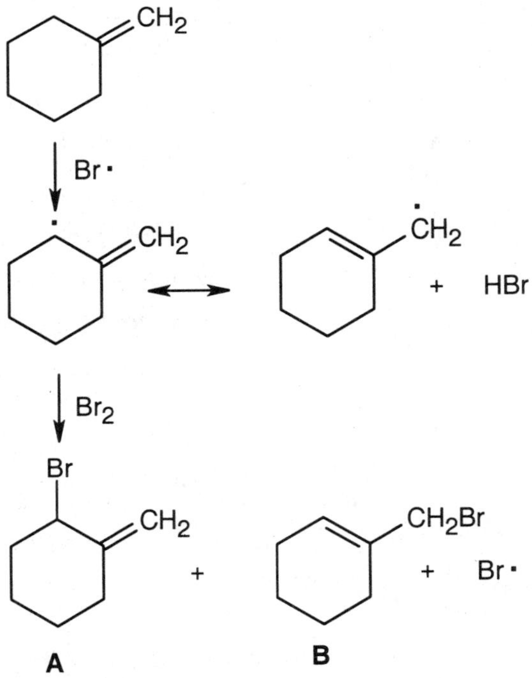

Br_2

Br

CH_2 + CH_2Br + Br·

A **B**

Product **B**, which has a trisubstituted double bond, forms in preference to product **A**, which has a disubstituted double bond. In addition, product **B** is favored because reaction at the primary end of the allylic radical is favored.

10.8

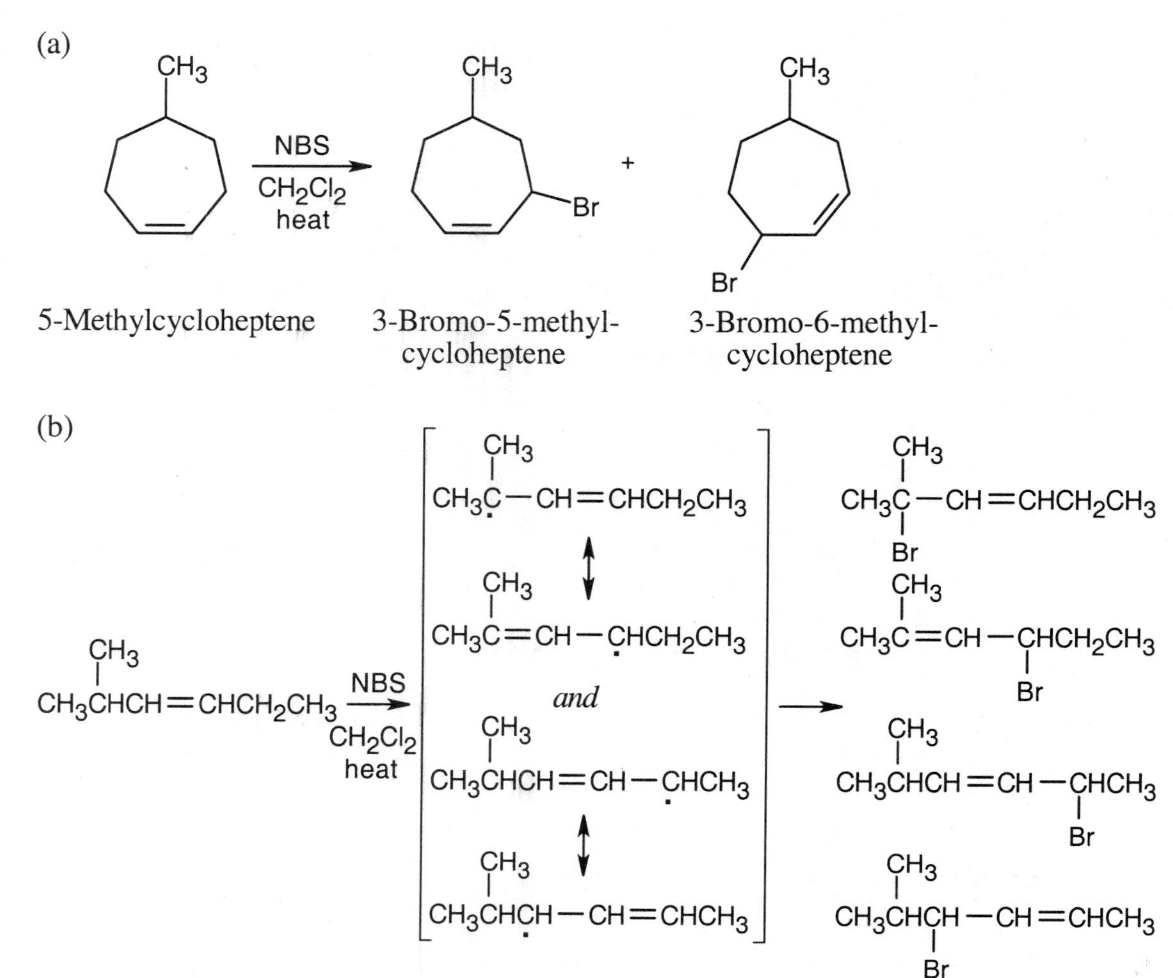

Two different allylic radicals can form, and four different bromohexenes can be produced.

10.9 Remember that halogen acids are used for converting tertiary alcohols to alkyl halides. PBr_3 and $SOCl_2$ are used for converting secondary and primary alcohols to alkyl halides.

(a)

$$\underset{\text{OH}}{\overset{\text{CH}_3}{\text{CH}_3\text{CCH}_3}} \xrightarrow[\text{ether}]{\text{HCl}} \underset{\text{Cl}}{\overset{\text{CH}_3}{\text{CH}_3\text{CCH}_3}}$$

(b)

$$\text{CH}_3\overset{\text{CH}_3}{\text{CH}}\text{CH}_2\overset{\text{OH}}{\text{CH}}\text{CH}_3 \xrightarrow[\text{ether}]{\text{PBr}_3} \text{CH}_3\overset{\text{CH}_3}{\text{CH}}\text{CH}_2\overset{\text{Br}}{\text{CH}}\text{CH}_3$$

(c)

$$\text{HOCH}_2\text{CH}_2\text{CH}_2\text{CH}_2\overset{\text{CH}_3}{\text{CH}}\text{CH}_3 \xrightarrow[\text{ether}]{\text{PBr}_3} \text{BrCH}_2\text{CH}_2\text{CH}_2\text{CH}_2\overset{\text{CH}_3}{\text{CH}}\text{CH}_3$$

(d)

$$\underset{\text{(CH}_3\text{ at C3; OH and CH}_3\text{ at C5)}}{CH_3CH_2CHCH_2CCH_3} \xrightarrow[\text{ether}]{HCl} \underset{\text{(CH}_3\text{ at C3; Cl and CH}_3\text{ at C5)}}{CH_3CH_2CHCH_2CCH_3}$$

10.10 Table 8.1 shows that the pK_a of CH_3–H is 60. Since CH_4 is a very weak acid, $^-{:}CH_3$ is a very strong base. Alkyl Grignard reagents are similar in base strength to $^-{:}CH_3$, but alkynyl Grignard reagents are somewhat weaker bases. Both reactions (a) and (b) occur as written.

(a)

$$\underset{\text{stronger base}}{CH_3MgBr} + \underset{\text{stronger acid}}{H{-}C{\equiv}C{-}H} \longrightarrow \underset{\text{weaker acid}}{CH_4} + \underset{\text{weaker base}}{H{-}C{\equiv}C{-}MgBr}$$

(b)

$$\underset{\text{stronger base}}{CH_3MgBr} + \underset{\text{stronger acid}}{NH_3} \longrightarrow \underset{\text{weaker acid}}{CH_4} + \underset{\text{weaker base}}{H_2N{-}MgBr}$$

10.11 Just as Grignard reagents react with *proton* donors to convert R–MgX into R–H, they also react with *deuterium* donors to convert R–MgX into R–D. In this case:

$$\underset{\text{(Br at C2)}}{CH_3CHCH_2CH_3} \xrightarrow[\text{ether}]{Mg} \underset{\text{(MgBr at C2)}}{CH_3CHCH_2CH_3} \xrightarrow{D_2O} \underset{\text{(D at C2)}}{CH_3CHCH_2CH_3}$$

10.12 (a) Synthetic analysis: The methyl group has an allylic relationship to the double bond. Thus, an organometallic coupling reaction between 3-bromocyclohexane and lithium dimethylcopper gives the desired product. 3-Bromohexane can be formed by allylic bromination of cyclohexene with NBS.

$$\text{cyclohexene} \xrightarrow[CH_2Cl_2]{NBS} \text{3-bromocyclohexene} \xrightarrow{(CH_3)_2CuLi} \text{3-methylcyclohexene}$$

3-Methylcyclohexene

(b) Synthetic analysis: We are asked to synthesize an eight-carbon product from a four-carbon starting material. Thus, an organometallic coupling reaction between 1-bromobutane and lithium dibutylcopper gives octane as the product. The Gilman reagent is formed from 1-bromobutane.

$$2\ CH_3CH_2CH_2CH_2Br \xrightarrow[\text{pentane}]{4\ Li} 2\ CH_3CH_2CH_2CH_2Li + 2\ LiBr \xrightarrow[\text{ether}]{CuI} (CH_3CH_2CH_2CH_2)_2Cu^-Li^+ + LiI$$

$$CH_3CH_2CH_2CH_2Br + (CH_3CH_2CH_2CH_2)_2Cu^-Li^+ \xrightarrow{\text{ether}} \underset{\text{Octane}}{CH_3(CH_2)_6CH_3} + LiBr + CH_3CH_2CH_2CH_2Cu$$

(c) Synthetic analysis: The synthesis in (b) suggests a way to proceed in this problem. Decane can be synthesized from 1-bromopentane and lithium dipentylcopper. 1-Bromopentane is formed by hydroboration of 1-pentene, followed by treatment of the resulting alcohol with PBr_3.

$$CH_3CH_2CH_2CH{=}CH_2 \xrightarrow[\text{2. } H_2O_2,\ ^-OH]{\text{1. } BH_3,\ THF} CH_3CH_2CH_2CH_2CH_2OH \xrightarrow{PBr_3,\ \text{ether}} CH_3CH_2CH_2CH_2CH_2Br$$

$$2\ CH_3CH_2CH_2CH_2CH_2Br \xrightarrow[\text{pentane}]{4\ Li} 2\ CH_3CH_2CH_2CH_2CH_2Li + 2\ LiBr$$

$$\xrightarrow{CuI,\ \text{ether}} (CH_3CH_2CH_2CH_2CH_2)_2Cu^-Li^+ + LiI$$

$$CH_3CH_2CH_2CH_2CH_2Br + (CH_3CH_2CH_2CH_2CH_2)_2Cu^-Li^+ \xrightarrow{\text{ether}} \underset{\text{Decane}}{CH_3(CH_2)_8CH_3}$$

$$+ LiBr + CH_3CH_2CH_2CH_2CH_2Cu$$

10.13 (a) As described in Practice Problem 10.2, the oxidation level of a compound can be found by adding the number of C–O, C–N, and C–X bonds and subtracting the number of C–H bonds. Cyclohexane, the first compound shown, has 12 C–H bonds, and has an oxidation level of –12. Cyclohexanone has 2 C–O bonds (from the double bond) and 10 C–H bonds, for an oxidation level of –8. 1-Chlorocyclohexene has one C–Cl bond and 9 C–H bonds, and also has an oxidation level of –8. Benzene has 6 C–H bonds, for an oxidation level of –6. Thus, in order of increasing oxidation level:

cyclohexane < cyclohexanone = 1-chlorocyclohexene < benzene

(b)

$CH_3CH_2NH_2$	<	$H_2NCH_2CH_2NH_2$	<	CH_3CN
oxidation level = –4		oxidation level = –2		oxidation level = 0

10.14 (a) The aldehyde carbon of the reactant has an oxidation level of 1 (2 C–O bonds minus 1 C–H bond). The alcohol carbon of the product has an oxidation level of –1 (1 C–O bond minus 2 C–H bonds). The reaction is a reduction because the oxidation level of the product is lower than the oxidation level of the reactant.

(b) The oxidation level of the upper carbon of the double bond in the reactant changes from 0 to +1 in the product; the oxidation level of the lower carbon of the double bond changes from 0 to –1. The total oxidation level, however, is the same for both product and reactant, and the reaction is neither an oxidation nor a reduction.

Visualizing Chemistry

10.15

(a)

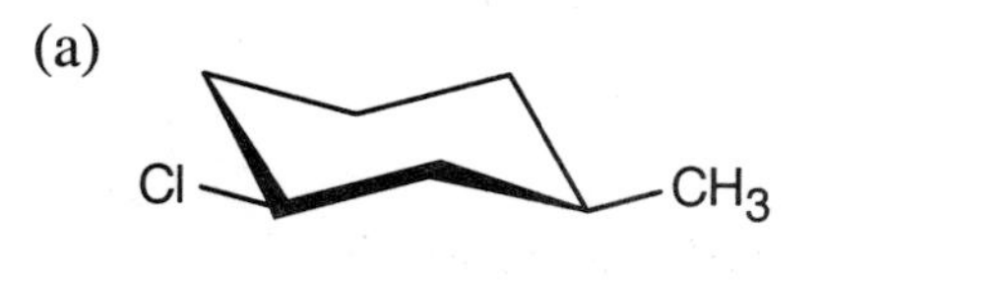

cis-1-Chloro-3-methylcyclohexane

(b)

CH_3 Cl
$CH_3C{=}CHCHCH_2CH_2CH_3$

4-Chloro-2-methyl-2-heptene

10.16

(a)

CH_3
$CH_3C{=}CHCH_2CH_2CH_3$ $\xrightarrow{NBS}$
2-Methyl-2-hexene

[CH_3 / $CH_3C{=}CH\dot{C}HCH_2CH_3$ ↔ CH_3 / $CH_3\dot{C}CH{=}CHCH_2CH_3$; $\cdot CH_2$ / $CH_3C{=}CHCH_2CH_2CH_3$ ↔ CH_2 / $CH_3C{-}\dot{C}HCH_2CH_2CH_3$] ⟶

CH_3 Br / $CH_3C{=}CHCHCH_2CH_3$
+
CH_3 / $CH_3CCH{=}CHCH_2CH_3$ / Br
+
CH_2Br / $CH_3C{=}CHCH_2CH_2CH_3$
+
H_2C Br / $CH_3C{-}CHCH_2CH_2CH_3$

(b)

NBS ⟶ [↔] ⟶

Br

+

Br

10.17 Recall the method of assigning configuration from Chapter 9. First, assign priorities to the groups around the chiral carbon. If you aren't using a model, imagine yourself looking at the chirality center, with the lowest priority group (4) pointing 180° away from you. (You would be looking out from behind the plane of the page.) Note the direction of rotation of the arrows that travel from group 1 to group 2 to group 3. The rotation is clockwise, and the configuration is *R*. The name of the compound is (*R*)-2-bromopentane.

Reaction of (*S*)-2-pentanol with PBr_3 to form (*R*)-2-bromopentane occurs with a change in stereochemistry because the configuration at the chirality center changes from *S* to *R*.

Additional Problems

10.18

(a) $CH_3CH(CH_3)CH(Br)CH(Br)CH_2CH(CH_3)CH_3$

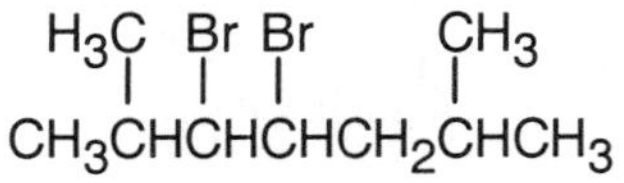

3,4-Dibromo-2,6-dimethylheptane

(b) $CH_3CH{=}CHCH_2CH(I)CH_3$

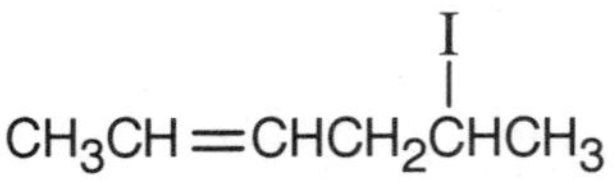

5-Iodo-2-hexene

(c) $CH_3C(Br)(CH_3)CH_2CH(Cl)CH(CH_3)CH_3$

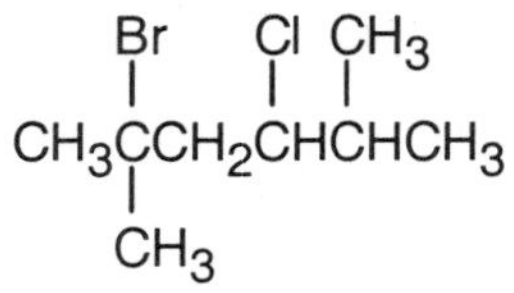

2-Bromo-4-chloro-2,5-dimethylhexane

(d) $CH_3CH_2CH(CH_2Br)CH_2CH_2CH_3$

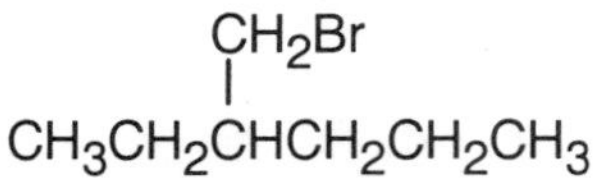

3-(Bromomethyl)hexane

(e) $ClCH_2CH_2CH_2C{\equiv}CCH_2Br$

1-Bromo-6-chloro-2-hexyne

10.19

(a) $CH_3CH_2CH(CH_3)CH(Cl)CH(Cl)CH_3$

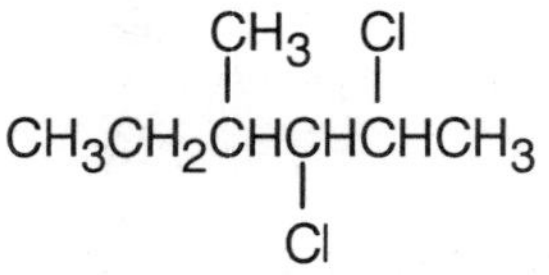

2,3-Dichloro-4-methylhexane

(b) $CH_3CH_2C(Br)(CH_2CH_3)CH_2CH(CH_3)CH_3$

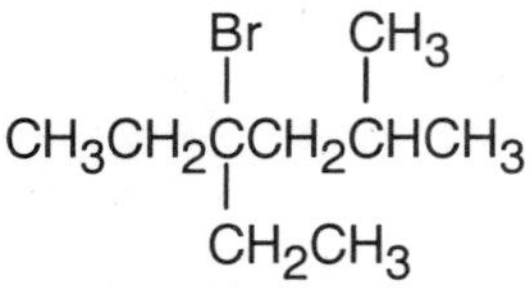

4-Bromo-4-ethyl-2-methylhexane

(c) $CH_3C(CH_3)_2CH(I)C(CH_3)_2CH_3$

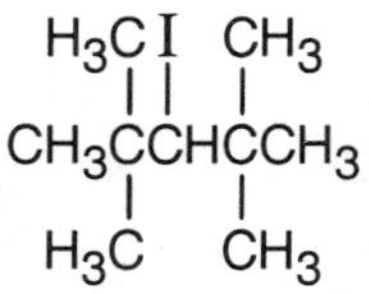

3-Iodo-2,2,4,4-tetramethylpentane

(d) Br, H and CH_2CH_3, H on cyclopentane (cis)

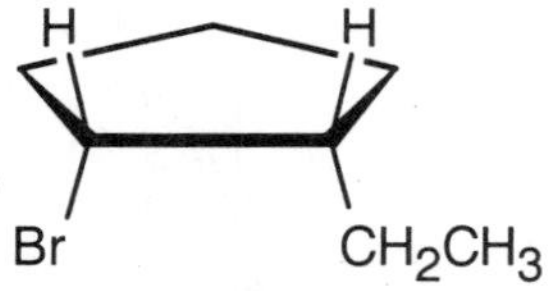

cis-1-Bromo-2-ethylcyclopentane

10.20

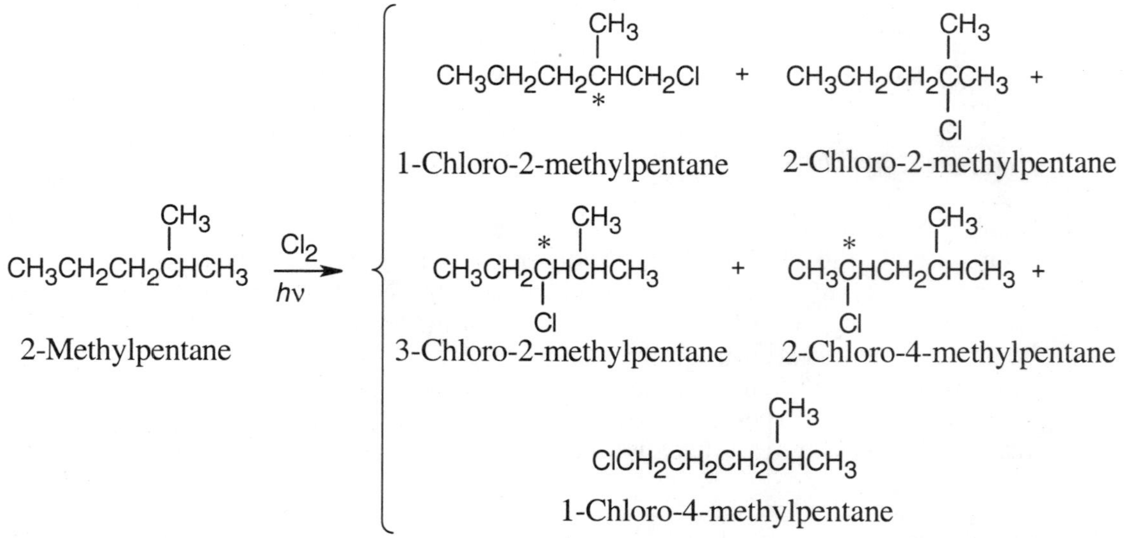

$$CH_3CH_2CH_2CH(CH_3)CH_3 \xrightarrow[h\nu]{Cl_2}$$

2-Methylpentane

$CH_3CH_2CH_2\overset{*}{C}H(CH_3)CH_2Cl$ — 1-Chloro-2-methylpentane

+ $CH_3CH_2CH_2C(Cl)(CH_3)CH_3$ — 2-Chloro-2-methylpentane

+ $CH_3CH_2\overset{*}{C}H(Cl)CH(CH_3)CH_3$ — 3-Chloro-2-methylpentane

+ $CH_3\overset{*}{C}H(Cl)CH_2CH(CH_3)CH_3$ — 2-Chloro-4-methylpentane

+ $ClCH_2CH_2CH_2CH(CH_3)CH_3$ — 1-Chloro-4-methylpentane

Three of the above products are chiral (chirality centers are starred). None of the products are optically active; each chiral product is a racemic mixture.

10.21 Abstraction of hydrogen by Br• can produce either of two allylic radicals. The first radical, resulting from abstraction of a secondary hydrogen, is more likely to be formed.

$$CH_3\dot{C}HCH{=}CHCH_3 \longleftrightarrow CH_3CH{=}CH\dot{C}HCH_3 \quad \text{(identical resonance forms)}$$

and

$$CH_3CH_2CH{=}CH\dot{C}H_2 \longleftrightarrow CH_3CH_2\dot{C}HCH{=}CH_2$$

Reaction of the radical intermediates with a bromine source leads to a mixture of products:

$CH_3CH(Br)CH{=}CHCH_3$ — *cis-* and *trans*-4-Bromo-2-pentene

and

$CH_3CH_2CH{=}CHCH_2Br$ — *cis-* and *trans*-1-Bromo-2-pentene

and

$CH_3CH_2CH(Br)CH{=}CH_2$ — 3-Bromo-1-pentene

The major product is 4–bromo–2–pentene, instead of the desired product, 1–bromo–2–pentene.

10.22 Three different allylic radical intermediates can be formed. Bromination of these intermediates can yield as many as five bromoalkenes. This is definitely not a good reaction to use in a synthesis.

CH_3 CH_3 NBS → Br CH_3

(allylic – secondary hydrogen abstracted) 3-Bromo-2-methylcyclohexene

CH_3 CH_3 NBS → CH_3 + CH_3 Br

Br

(allylic – secondary hydrogen abstracted) 3-Bromo-1-methyl-cyclohexene 3-Bromo-3-methyl-cyclohexene

CH_2 CH_2 NBS → CH_2Br + CH_2 Br

(allylic – primary hydrogen abstracted) 1-(Bromomethyl)-cyclohexene 2-Bromomethylene-cyclohexane

10.23

(a)

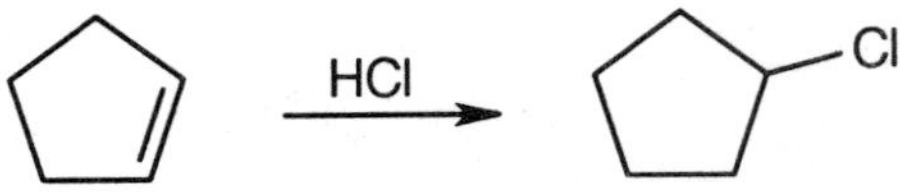
HCl
Cl

Chlorocyclopentane

(b)

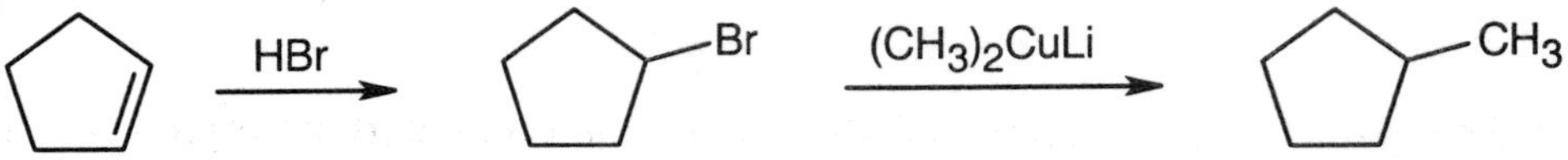
HBr
Br
$(CH_3)_2CuLi$
CH_3

Methylcyclopentane

(c)

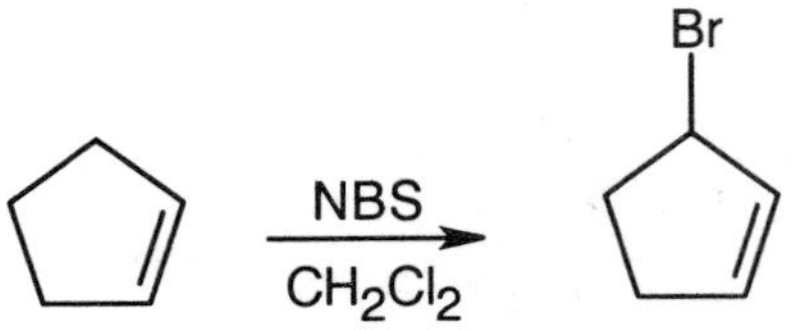
NBS
CH_2Cl_2
Br

3-Bromocyclopentene

(d)

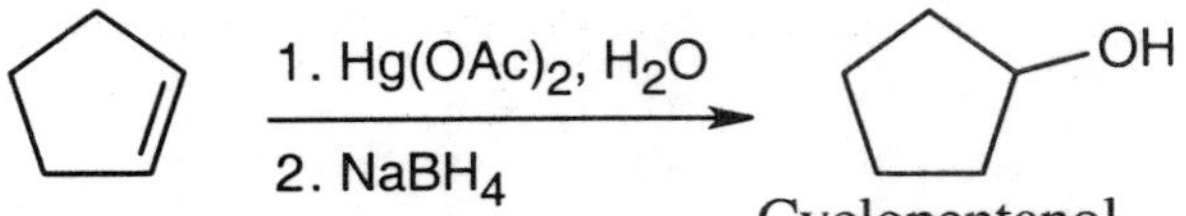
1. $Hg(OAc)_2$, H_2O
2. $NaBH_4$
OH

Cyclopentanol

(e)

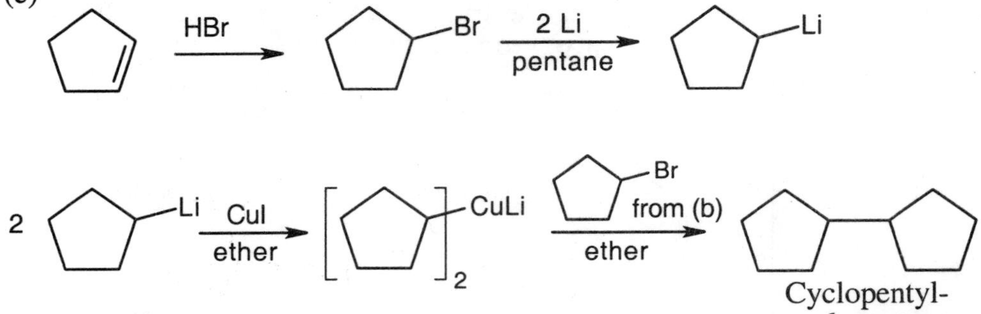
HBr
Br
2 Li
pentane
Li
2
Li
CuI
ether
CuLi
2
Br
from (b)
ether

Cyclopentyl-
cyclopentane

(f)

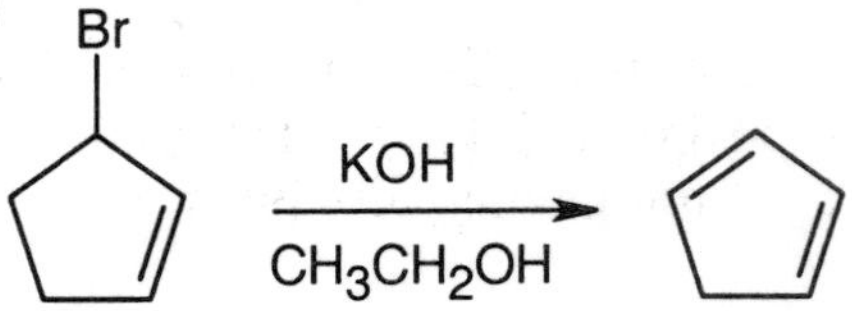
Br
KOH
CH_3CH_2OH

from (c) 1,3-Cyclopentadiene

10.24

(a)

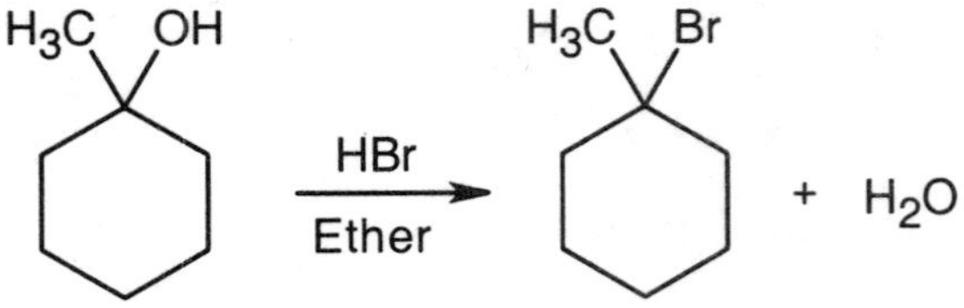
H_3C OH
HBr
Ether
H_3C Br
+ H_2O

(b)

$$CH_3CH_2CH_2CH_2OH \xrightarrow{SOCl_2} CH_3CH_2CH_2CH_2Cl + SO_2 + HCl$$

(c)

$$\text{octahydronaphthalene (tetrasubstituted C=C)} \xrightarrow[\text{CH}_2\text{Cl}_2]{\text{NBS}} \text{allylic bromide (Br on ring carbon next to C=C)} + \text{bromide with Br at ring fusion carbon (trisubstituted C=C)} + \text{succinimide (NH)}$$

The major product contains a tetrasubstituted double bond, and the minor product contains a trisubstituted double bond.

(d)

$$\text{cyclohexanol (C}_6\text{H}_{11}\text{OH)} \xrightarrow[\text{Ether}]{\text{PBr}_3} \text{bromocyclohexane (C}_6\text{H}_{11}\text{Br)} + \text{P(OH)}_3$$

(e)

$$\text{CH}_3\text{CH}_2\text{CH(Br)CH}_3 \xrightarrow[\text{Ether}]{\text{Mg}} \underset{\textbf{A}}{\text{CH}_3\text{CH}_2\text{CH(MgBr)CH}_3} \xrightarrow{\text{H}_2\text{O}} \underset{\textbf{B}}{\text{CH}_3\text{CH}_2\text{CH}_2\text{CH}_3} + \text{MgBr(OH)}$$

(f)

$$2\ \text{CH}_3\text{CH}_2\text{CH}_2\text{CH}_2\text{Br} \xrightarrow[\text{Pentane}]{4\ \text{Li}} 2\ \underset{\textbf{A}}{\text{CH}_3\text{CH}_2\text{CH}_2\text{CH}_2\text{Li}} \xrightarrow{\text{CuI}} \underset{\textbf{B}}{(\text{CH}_3\text{CH}_2\text{CH}_2\text{CH}_2)_2\text{CuLi}}$$

(g)

$$\text{CH}_3\text{CH}_2\text{CH}_2\text{CH}_2\text{Br} + (\text{CH}_3)_2\text{CuLi} \xrightarrow{\text{Ether}} \text{CH}_3\text{CH}_2\text{CH}_2\text{CH}_2\text{CH}_3 + \text{CH}_3\text{Cu} + \text{LiBr}$$

10.25

$$(S)\text{-}\text{C(H)(CH}_3\text{)(CH}_2\text{CH}_3\text{)(CH}_2\text{CH}_2\text{CH}_3\text{)} \xrightarrow[h\nu]{\text{Br}_2} (R)\text{-}\text{C(Br)(CH}_3\text{)(CH}_2\text{CH}_3\text{)(CH}_2\text{CH}_2\text{CH}_3\text{)} + (S)\text{-}\text{C(Br)(CH}_3\text{)(CH}_2\text{CH}_3\text{)(CH}_2\text{CH}_2\text{CH}_3\text{)}$$

Abstraction of a hydrogen atom from the chirality center of (*S*)–3–methyl–hexane produces an achiral radical intermediate, which reacts with bromine to form a 1:1 mixture of *R* and *S* enantiomeric, chiral bromoalkanes. The product mixture is optically inactive.

10.26

$$\text{H}_3\text{C–CH}_2\text{–CH}_2\text{–}\overset{R}{\text{C}}\text{H(Cl)CH}_3 \xrightarrow[h\nu]{\text{Cl}_2} \text{H}_3\text{C–}\overset{R}{\text{C}}\text{H(Cl)–CH}_2\text{–}\overset{R}{\text{C}}\text{H(Cl)CH}_3 + \text{H}_3\text{C–}\overset{S}{\text{C}}\text{H(Cl)–CH}_2\text{–}\overset{R}{\text{C}}\text{H(Cl)CH}_3$$

+ other products

Abstraction of a hydrogen atom from carbon 4 yields a chiral radical intermediate. Reaction of this intermediate with chlorine does not occur with equal probability from each side, and the two diastereomeric products are not formed in 1:1 ratio. The first product is optically active, and the second product is a meso compound.

10.27

for Cl·: Cl· + CH_3—H ⟶ CH_3· + H—Cl
438 kJ/mol 432 kJ/mol

CH_3· + Cl—Cl ⟶ CH_3—Cl + Cl·
243 kJ/mol 351 kJ/mol

$$\Delta H° = D_{\text{bonds broken}} - D_{\text{bonds formed}}$$
$$= 438\text{ kJ/mol} + 243\text{ kJ/mol} - 432\text{ kJ/mol} - 351\text{ kJ/mol} = -102\text{ kJ/mol}$$

for Br·: Br· + CH_3—H ⟶ CH_3· + H—Br
438 kJ/mol 366 kJ/mol

CH_3· + Br—Br ⟶ CH_3—Br + Br·
193 kJ/mol 293 kJ/mol

$$\Delta H° = D_{\text{bonds broken}} - D_{\text{bonds formed}}$$
$$= 438\text{ kJ/mol} + 193\text{ kJ/mol} - 366\text{ kJ/mol} - 293\text{ kJ/mol} = -28\text{ kJ/mol}$$

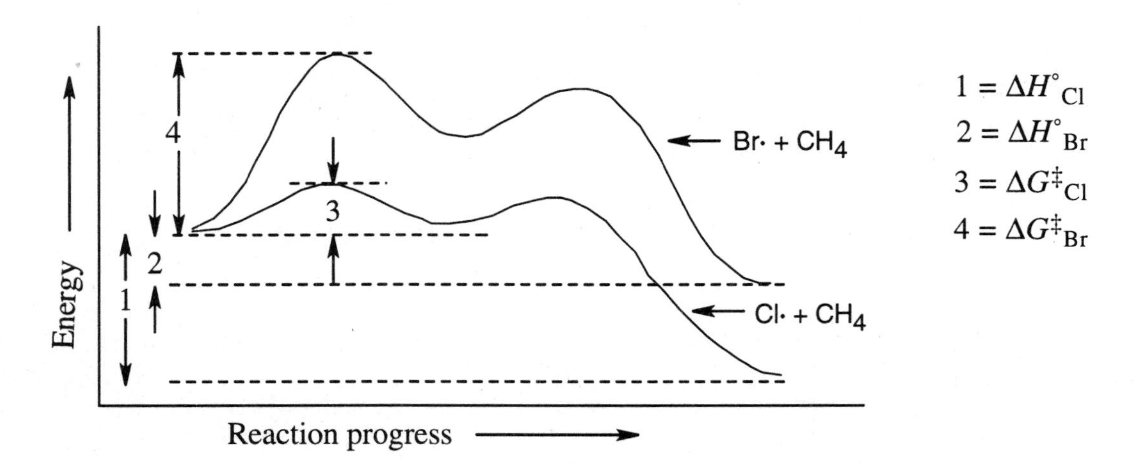

Reaction between Cl• and CH_4 is likely to be faster. ΔH° for formation of the CH_3• intermediate is lower for chlorination than for bromination, and thus $\Delta G^{\ddagger}$ is likely to be lower also, according to the Hammond Postulate.

10.28

$CH_3CH{=}CHCH_2CH{=}CH_2$ —NBS→ (left and right)

Left intermediate:

$$\left[\dot{C}H_2CH{=}CHCH_2CH{=}CH_2 \longleftrightarrow H_2C{=}CH\dot{C}HCH_2CH{=}CH_2\right]$$

↓ Br_2

$BrCH_2CH{=}CHCH_2CH{=}CH_2$
6-Bromo-1,4-hexadiene

\+ $H_2C{=}CHCH(Br)CH_2CH{=}CH_2$
3-Bromo-1,5-hexadiene

Right intermediate:

$$\left[CH_3CH{=}CH\dot{C}HCH{=}CH_2 \longleftrightarrow CH_3CH{=}CHCH{=}CH\dot{C}H_2 \longleftrightarrow CH_3\dot{C}HCH{=}CHCH{=}CH_2\right]$$

↓ Br_2

$CH_3CH{=}CHCH{=}CHCH_2Br$
1-Bromo-2,4-hexadiene

\+ $CH_3CH{=}CHCH(Br)CH{=}CH_2$
3-Bromo-1,4-hexadiene

\+ $CH_3CH(Br)CH{=}CHCH{=}CH_2$
5-Bromo-1,3-hexadiene

The intermediate on the right is the more stable, because the unpaired electron is delocalized over more atoms.

10.29 Table 5.3 shows that the bond dissociation energy for $C_6H_5CH_2$–H is 368 kJ/mol. This value is comparable in size to the bond dissociation energy for allylic hydrogens, and thus it is relatively easy to form the $C_6H_5CH_2\cdot$ radical. The high bond dissociation energy for formation of $C_6H_5\cdot$, 464 kJ/mol, indicates the bromination on the benzene ring will not occur. The only product of reaction with NBS is $C_6H_5CH_2Br$.

10.30

10.31 Two allylic radicals can form:

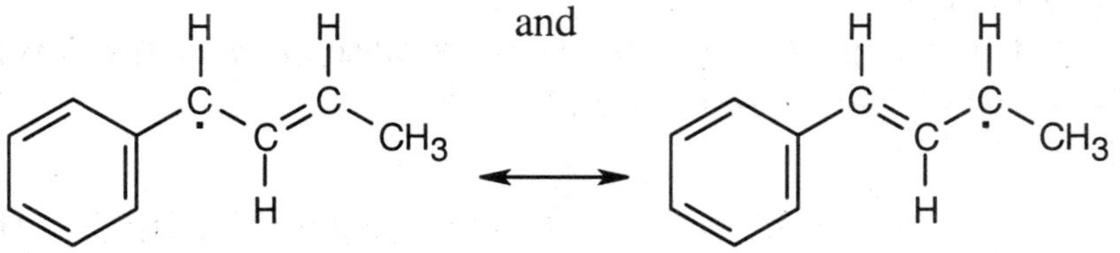

The second radical is much more likely to form because it is both allylic and benzylic, and it yields the following products:

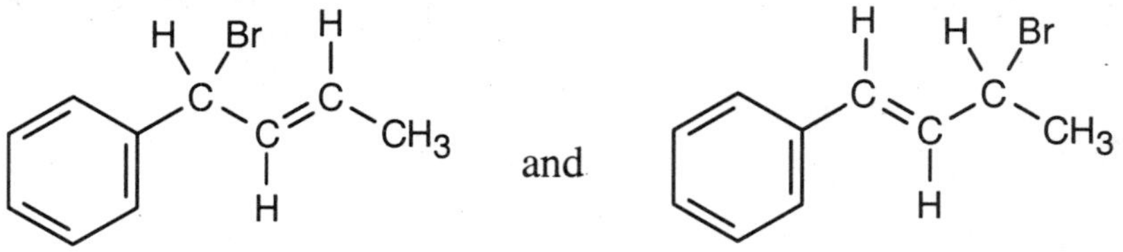

10.32

(a)

$CH_3CH{=}CHCH{=}CHCH{=}CH\overset{+}{C}H_2$

$CH_3CH{=}CHCH{=}CH\overset{+}{C}HCH{=}CH_2$

$CH_3CH{=}CH\overset{+}{C}HCH{=}CHCH{=}CH_2$

$CH_3\overset{+}{C}HCH{=}CHCH{=}CHCH{=}CH_2$

(b)

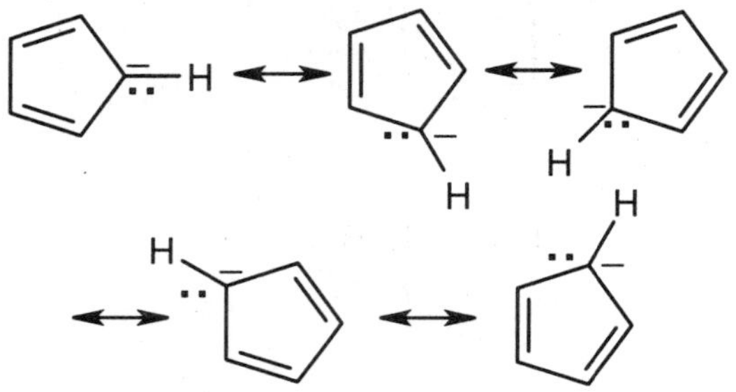

(c)

$CH_3C{\equiv}\overset{+}{N}{-}\ddot{\underset{..}{O}}{:}^{-} \longleftrightarrow CH_3\bar{C}{=}\overset{+}{N}{=}\ddot{O} \longleftrightarrow CH_3\overset{+}{C}{=}N{-}\ddot{\underset{..}{O}}{:}^{-}$

10.33 Remember that the oxidation level is found by subtracting the number of C–H bonds from the number of C–O, C–N, and C–X bonds. The oxidation levels are shown beneath the structures. In order of increasing oxidation level:

(a)

$CH_3CH{=}CHCH_3$, $CH_3CH_2CH{=}CH_2$ < $CH_3CH_2CH_2CHO$ < $CH_3CH_2CH_2COOH$

−8, −8, −6, −4

(b)

$CH_3CH_2CH_2NH_2$, $CH_3CH_2CH_2Br$ < $BrCH_2CH_2CH_2Cl$ < CH_3COCH_2Cl

−6, −6, −4, −2

10.34

OH

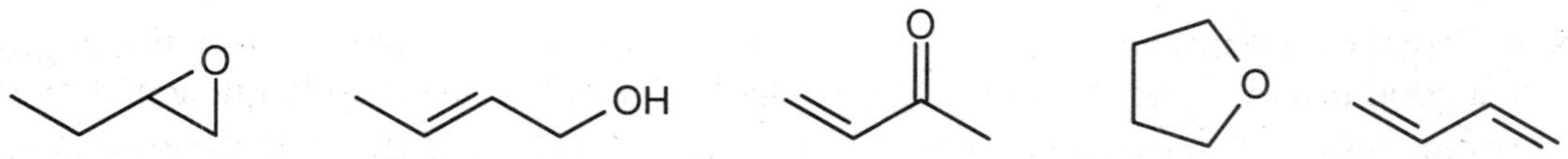

−6 −6 −4 −6 −6

All of the compounds, except for butenone, have the same oxidation level.

10.35 (a) This reaction is an oxidation.
(b) The reaction is neither an oxidation nor a reduction because the oxidation level of the reactant is the same as the oxidation level of the product.
(c) This reaction is a reduction.

10.36 All these reactions involve addition of a dialkylcopper reagent ($(CH_3CH_2CH_2CH_2)_2CuLi$) to an alkyl halide. The dialkylcopper is prepared by treating 1–bromobutane with lithium, followed by addition of CuI:

$$2\ CH_3CH_2CH_2CH_2Br \xrightarrow[\text{Pentane}]{2\ Li} 2\ CH_3CH_2CH_2CH_2Li \xrightarrow[\text{Ether}]{CuI} (CH_3CH_2CH_2CH_2)_2CuLi$$

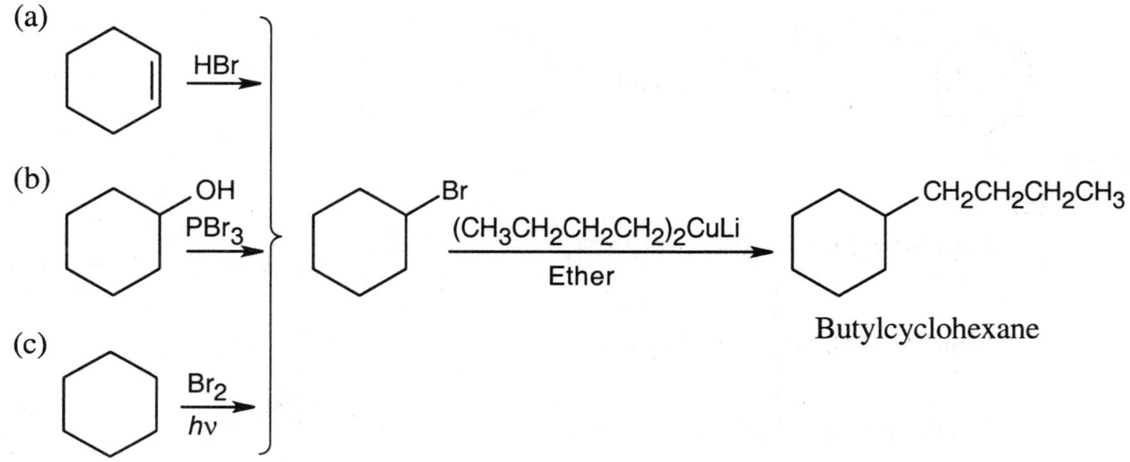

(a) cyclohexene $\xrightarrow{HBr}$

(b) cyclohexanol (OH) $\xrightarrow{PBr_3}$

(c) cyclohexane $\xrightarrow[h\nu]{Br_2}$

(a)–(c) → bromocyclohexane (Br) $\xrightarrow[\text{Ether}]{(CH_3CH_2CH_2CH_2)_2CuLi}$ cyclohexyl–$CH_2CH_2CH_2CH_3$

Butylcyclohexane

10.37 (a) Fluoroalkanes don't usually form Grignard reagents.

(b) Two allylic radicals can be produced.

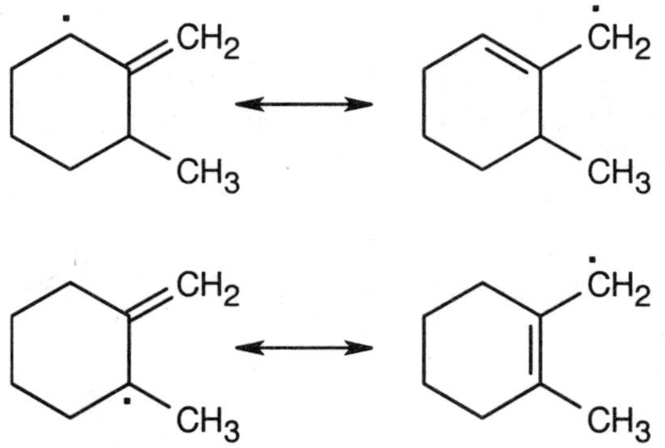

(radical on ring carbon adjacent to $=CH_2$, with CH_3) ⟷ (ring double bond, $\dot{C}H_2$, CH_3)

($=CH_2$, radical on ring carbon bearing CH_3) ⟷ (ring double bond, $\dot{C}H_2$, CH_3)

Instead of a single product, as many as four bromoalkene products may result.

(c) Dialkylcopper reagents don't react with fluoroalkanes.

10.38 A Grignard reagent can't be prepared from a compound containing a functional group that is a good proton donor because the Grignard reagent is immediately quenched by the proton source. For example, the $-COOH$, $-OH$, $-NH_2$, and $RC{\equiv}CH$ functional groups are too acidic to be used for preparation of a Grignard reagent. $BrCH_2CH_2CH_2NH_2$ is another compound that doesn't form a Grignard reagent.

10.39

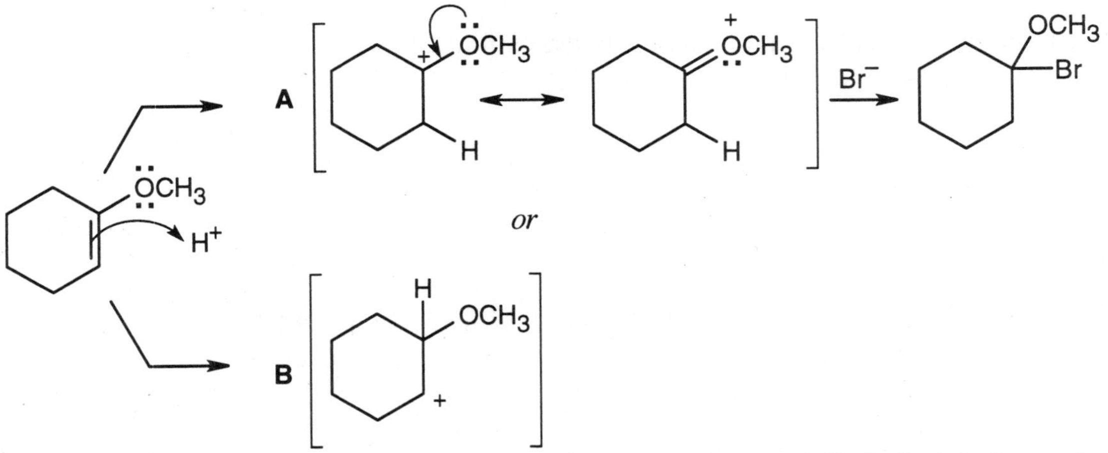

Reaction of the ether with HBr can occur by either path **A** or path **B**. Path **A** is favored because its cation intermediate can be stabilized by resonance.

10.40

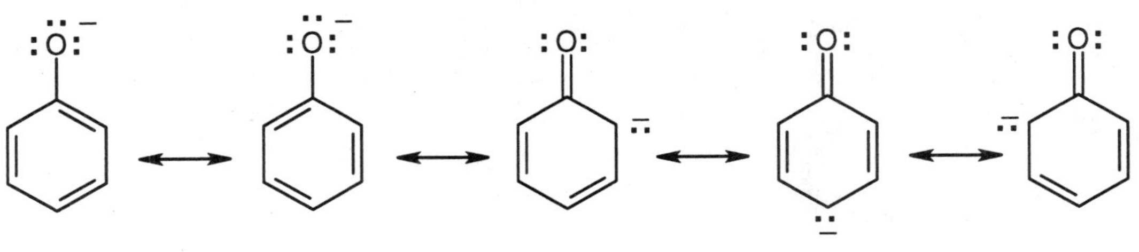

10.41

$$(C_4H_9)_3SnH \xrightarrow{h\nu} (C_4H_9)_3Sn\cdot + H\cdot$$

$$(C_4H_9)_3Sn\cdot + RX \longrightarrow (C_4H_9)_3SnX + R\cdot$$

$$(C_4H_9)_3SnH + R\cdot \longrightarrow (C_4H_9)_3Sn\cdot + RH$$

A Look Ahead

10.42 As we saw in Chapter 6, tertiary carbocations (R_3C^+) are more stable than either secondary or primary carbocations, due to the ability of the three alkyl groups to stabilize positive charge. If the substrate is also allylic, as in the case of $H_2C{=}CHC(CH_3)_2Br$, positive charge can be further delocalized. Thus, $H_2C{=}CHC(CH_3)_2Br$ should form a carbocation faster than $(CH_3)_3CBr$ because the resulting carbocation is more stable.

10.43

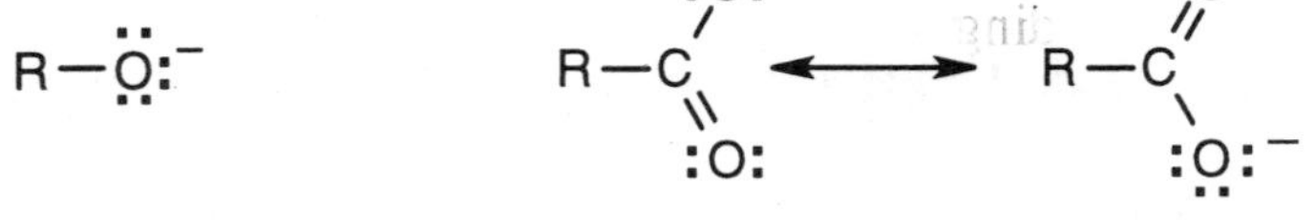

alkoxide ion carboxylate ion

Carboxylic acids are more acidic than alcohols because the negative charge of carboxylate ions is stabilized by resonance. This resonance stabilization favors formation of carboxylate anions over alkoxide anions, and increases K_a for carboxylic acids.

Molecular Modeling

10.44 The unpaired electron in the methyl radical is localized on carbon in a *p* orbital. The unpaired electron in the trichloromethyl radical is delocalized over all *p* orbitals.

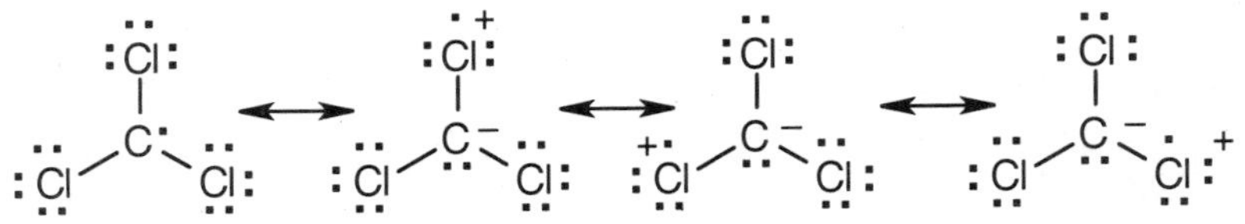

10.45 The complex shows that oxygen atoms in the solvent approach the magnesium atom in the Grignard reagent, changing bond distances and bond angles around magnesium. The oxygen-magnesium interaction is mainly electrostatic. Oxygen has a very negative potential in diethyl ether, and magnesium has a very positive potential in the Grignard reagent. Solvation enhances the ionic nature of the carbon-magnesium bond; The methyl group is more negative (red) in the complex and is presumably more reactive.

10.46 The breaking C–H bond distances decrease in the following order: 150.3 pm (CH_4), 148.3 pm [$HC(CH_3)_3$], and 134.7 pm ($H_2C{=}CHCH_3$). C–H bond strengths also decrease in this order, and the reactions progressively change from endothermic to exothermic.According to the Hammond Postulate, the transition states for a family of reactions are more reactant-like for more exothermic reactions, as can be observed in this series of reactions.

10.47

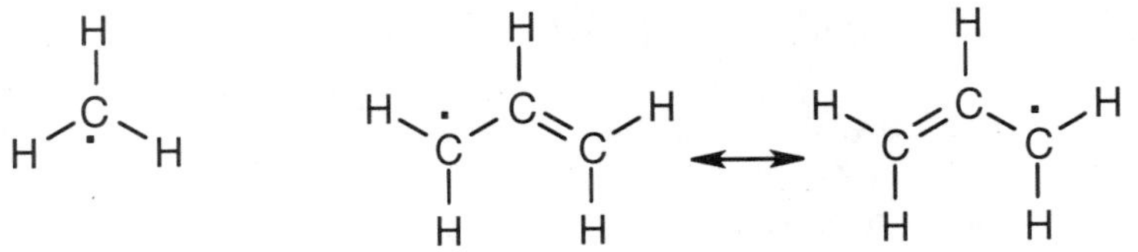

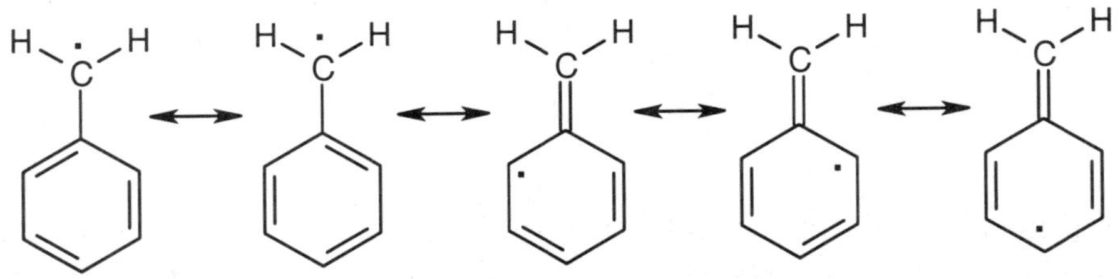

The unpaired electron in the transition state is delocalized in the same way as it is delocalized in the product radical, and the transition state has some of the characteristics of the product radical.

10.48 Radical B, which leads to the observed product, is formed more rapidly because it is lower in energy, and the transition state leading to B is also lower in energy. Radical B can react with Br_2 to yield 4-bromo-2-heptene, the observed product, or 2-bromo-3-heptene, which is not formed.

Chapter 11 – Reactions of Alkyl Halides: Nucleophilic Substitutions and Eliminations

Chapter Outline

I. Substitution Reactions (Sections 11.1 – 11.9)
  A. S_N2 reactions (Sections 11.1 – 11.5)
    1. Background of S_N2 reactions (Sections 11.1 – 11.4).
      a. Walden discovered that (+) malic acid and (–) malic acid could be interconverted (Section 11.1).
      b. This discovery meant that one or more reactions must have occurred with inversion of configuration at the chirality center.
    2. Stereochemistry of S_N2 reactions (Section 11.2).
      a. Kenyon and Phillips discovered that the nucleophilic substitution of tosylate ion by acetate ion occurred with inversion of configuration.
      b. They concluded that the nucleophilic substitution reactions of primary and secondary alkyl halides always proceed with inversion of configuration.
    3. Kinetics of S_N2 reactions (Section 11.3).
      a. Reaction rate is the rate at which reactants are converted to products.
      b. The kinetics of a reaction measure the relationship between reactant concentrations and product concentrations.
      c. In an S_N2 reaction, reaction rate depends on the concentration of both alkyl halide and nucleophile.
         This type of reaction is a second-order reaction.
      d. In a second-order reaction, rate = k x [RX] x [Nu]
         The constant k is the rate constant; its units are (L/mol·s).
    4. Mechanism of the S_N2 reaction (Section 11.4).
      a. The mechanism of an S_N2 reaction must account for both the observed stereochemistry and the observed kinetics.
      b. The reaction takes place in a single step, without intermediates.
        i. The nucleophile attacks the substrate from a direction directly opposite to the leaving group.
        ii. This type of attack accounts for inversion of configuration.
        iii. In the transition state, the new bond forms at the same time as the old bond breaks.
        iv. Negative charge is shared between the attacking nucleophile and the leaving group.
        v. The three remaining bonds to carbon are in a planar arrangement.
        vi. Both substrate and nucleophile are involved in the step whose rate is measured.
    5. Characteristics of the S_N2 reaction (Section 11.5).
      a. Changes in the energy levels of reactants or of the transition state affect the reaction rate.
      b. Changes in the substrate.
        i. Reaction rate is decreased if the substrate is bulky.
        ii. Substrates, in order of increasing reactivity: tertiary, neopentyl, secondary, primary, methyl.
        iii. S_N2 reactions can occur only at relatively unhindered sites.
        iv. Vinyl and aryl halides are unreactive to S_N2 substitutions.

c. Changes in the nucleophile.
   i. Any species can act as a nucleophile if it has an unshared electron pair.
      If the nucleophile has a negative charge, the product is neutral.
      If the nucleophile is neutral, the product is positively charged.
   ii. The reactivity of a nucleophile is dependent on reaction conditions.
   iii. In general, nucleophilicity parallels basicity.
   iv. Nucleophilicity increases going down a column of the periodic table.
   v. Negatively charged nucleophiles are usually more reactive than neutral nucleophiles.
d. Changes in the leaving group.
   i. In general, the best leaving groups are best able to stabilize negative charge.
   ii. Usually, the best leaving groups are the weakest bases.
   iii. Good leaving groups lower the energy of the transition state.
   iv. Poor leaving groups include F^-, HO^-. RO^-, and H_2N^-.
e. Changes in the solvent.
   i. Polar, protic solvents slow S_N2 reactions by lowering the reactivity of the nucleophile .
   ii. Polar, aprotic solvents raise the ground-state energy of the nucleophile and make it more reactive.
f. A summary:
   i. Steric hindrance in the substrate raises the energy of the transition state, increasing $\Delta G^\ddagger$, and deceasing the reaction rate.
   ii. More reactive nucleophiles have a higher ground-state energy, decreasing $\Delta G^\ddagger$, and increasing the reaction rate.
   iii. Good leaving groups decrease the energy of the transition state, decreasing $\Delta G^\ddagger$, and increasing the reaction rate.
   iv. Polar protic solvents solvate the nucleophile, lowering the ground-state energy, increasing $\Delta G^\ddagger$, and decreasing the reaction rate. Polar aprotic solvents don't solvate the nucleophile, raising the ground-state energy, decreasing $\Delta G^\ddagger$, and deceasing the reaction rate.

B. S_N1 Reactions (Sections 11.6 – 11.9).
1. Background of S_N1 reactions (Section 11.6).
   a. Under certain reaction conditions, tertiary halides are much more reactive than primary and methyl halides.
   b. These reactions must be occurring by a different mechanism.
2. Kinetics of the S_N1 reaction (Section 11.7).
   a. The rate of reaction of a tertiary alkyl halide with water depends only on the concentration of the alkyl halide.
   b. The reaction is a first order process, with reaction rate = k x [RX].
   c. The rate expression shows that only RX is involved in the slowest, or rate-limiting, step, and the nucleophile is involved in a different, faster step.
   d. The rate expression also shows that there must be at least two steps in the reaction.
   e. In an S_N1 reaction, slow dissociation of the substrate is followed by rapid reaction with the nucleophile.
3. Stereochemistry of S_N1 reactions (Section 11.8).
   a. An S_N1 reaction of an enantiomer produces racemic product because an S_N1 reaction proceeds through a planar, achiral intermediate.

b. Few S_N1 reactions proceed with complete racemization.
    The ion pair formed by the leaving group and the carbocation sometimes shields one side of the carbocation from attack before the leaving group can diffuse away.

4. Characteristics of the S_N1 reaction (Section 11.9).
    a. As in S_N2 reactions, factors that lower $\Delta G^{\ddagger}$ favor faster reactions.
    b. Changes in the substrate.
        i. The more stable the carbocation intermediate, the faster the S_N1 reaction.
        ii. Substrates, in order of increasing reactivity: methyl, primary, secondary and allyl and benzyl, tertiary.
        iii. Allylic and benzylic substrates are also reactive in S_N2 reactions.
    c. Changes in the leaving group.
        i. The best leaving groups are the conjugate bases of strong acids.
        ii. In S_N1 reactions, water can act as a leaving group.
    d. Changes in the nucleophile have no effect on S_N1 reactions.
    e. Changes in the solvent.
        i. Polar solvents (high dielectric constant) increase the rates of S_N1 reactions.
        ii. Polar solvents stabilize the carbocation intermediate more than the reactants and lower $\Delta G^{\ddagger}$.
        iii. Polar solvents stabilize by orienting themselves around the carbocation, with electron-rich ends facing the positive charge.
    f. A summary;
        i. The best substrates are those that form stable carbocations.
        ii. Good leaving groups lower the energy of the transition state leading to carbocation formation and increase the reaction rate.
        iii. The nucleophile doesn't affect the reaction rate, but it must be nonbasic.
        iv. Polar solvents stabilize the carbocation intermediate and increase the reaction rate.

II. Elimination reactions (Sections 11.10 – 11.14).
  A. Introduction (Section 11.10).
    1. In addition to bringing about substitution, a basis nucleophile can also cause elimination of HX from an alkyl halide to form a carbon–carbon double bond.
    2. A mixture of double-bond products is usually formed, but the product with the more substituted double bond is the major product.
        This observation is the basis of Zaitsev's rule.
    3. Double-bond formation can occur by several mechanistic routes, but at this point, we will study only two mechanisms.
  B. The E2 reaction (Sections 11.11 – 11.13)
    1. General features (Section 11.11).
        a. An E2 reaction occurs when an alkyl halide is treated with strong base.
        b. The reaction occurs in one step, without intermediates.
        c. E2 reactions follow second-order kinetics.
        d. E2 reactions always occur with periplanar geometry.
            i. Periplanar geometry is required because of the need for overlap of the sp^3 orbitals of the reactant as they become π orbitals in the product.
            ii. Anti periplanar geometry is preferred because it allows the substituents of the two reacting carbons to assume a staggered relationship.
            iii. Syn periplanar geometry occurs only when anti periplanar geometry isn't possible.
        e. The preference for anti periplanar geometry results in the formation of double bonds with specific *E,Z* conformations.

2. Elimination reactions and cyclohexane conformations (Section 11.12).
   a. The chemistry of substituted cyclohexanes is controlled by their conformations.
   b. The preference of antiperiplanar geometry for E2 reactions can be met only if the atoms to be eliminated have a trans-diaxial relationship.
   c. Neomenthyl chloride reacts 200x faster than menthyl chloride because the groups to be eliminated are trans diaxial in the most favorable conformation, and the Zaitsev product is formed.
   d. For menthyl chloride, reaction must proceed through a higher energy conformation, and non-Zaitsev product is formed.
3. The deuterium isotope effect (Section 11.13).
   a. In a reaction in which a C–H bond is cleaved in the rate-limiting step, substitution of –D for –H results in a decrease in rate.
   b. Because this effect is observed in E2 reactions, these reactions must involve C–H bond breaking in the rate-limiting step.

C. The E1 reaction (Section 11.14).
1. An E1 reaction occurs when the intermediate carbocation of an S_N1 loses H^+ to form a C=C bond.
2. E1 reactions usually occur in competition with S_N1 reactions.
3. E1 reactions show first-order kinetics.
4. There is no geometric requirement for the groups to be eliminated, and the most stable (Zaitsev) product is formed.
5. No deuterium isotope effect is observed: C–H bond-breaking occurs after the rate-limiting step.

III. Summary of reactivity (Section 11.15).

A. Primary halides.
1. S_N2 reaction is usually observed.
2. E1 reaction occurs if a strong, bulky base is used.

B. Secondary halides.
1. S_N2 and E2 reaction occur in competition.
2. Strong bases promote E2 elimination.
3. Secondary halides (especially allylic and benzylic halides) can react by S_N1 and E1 routes if weakly basic nucleophiles and protic solvents are used.

C. Tertiary halides.
1. Under basic conditions, E2 elimination is favored.
2. S_N1 and E1 products are formed under nonbasic conditions.

IV. Substitution reactions in synthesis (Section 11.16).

A. Alkylation of acetylide anions is an S_N2 reaction.
   E2 products are formed if the substrate is a secondary or tertiary halide.

B. Preparation of alkyl halides from alcohols and HX.
1. Reaction of tertiary alcohols occurs by an S_N1 mechanism.
2. Reaction of secondary and primary alcohols occurs by an S_N2 mechanism.

C. Preparation of alkyl bromides from alcohols and PBr_3.
   PBr_3 converts a poor (–OH) leaving group into a good leaving group, which can be displaced by Br^-.

Solutions to Problems

11.1 As described in Practice Problem 11.1, identify the leaving group and the chirality center. Draw the product carbon skeleton, inverting the configuration at the chirality center, and replace the leaving group with the nucleophilic reactant.

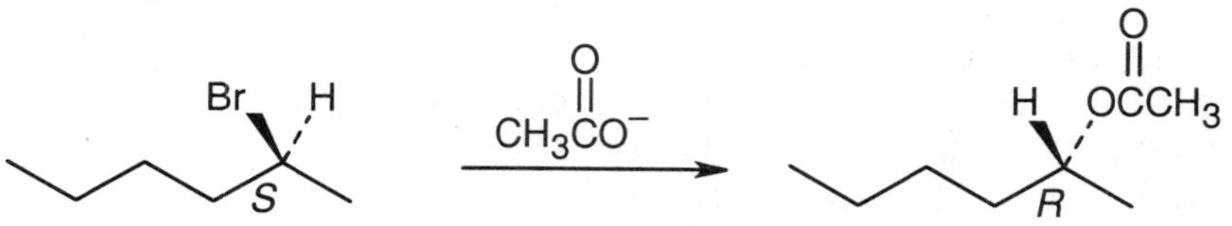

Br H (*S*) $\xrightarrow{CH_3CO_2^-}$ H $OCCH_3$ (O) (*R*)

11.2 Use the suggestions in the previous problem to draw the correct product.

HO^- + H_3C, H, CH_2CH_3 C—Br (*R*) $\longrightarrow$ HO—C (CH_3, H, CH_2CH_3) (*S*) + $:Br:^-$

11.3

HS^- + H_3C—C(H, CH_3)—CH_2—C(Br, H)(*S*)—CH_3 $\longrightarrow$ H_3C—C(H, CH_3)—CH_2—C(H, SH)(*R*)—CH_3 + Br^-

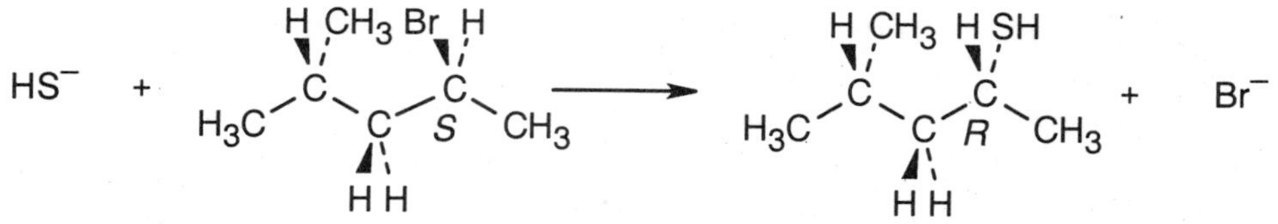

11.4 If back-side attack were necessary for S_N2 reaction, a molecule with a hindered back-side would not be able to react by an S_N2 mechanism. In this problem, approach by the hydroxide ion from the back side of the bromoalkane is blocked by the rigid ring system, and displacement can't occur.

11.5 All of the nucleophiles in this problem are relatively reactive.

(a)
$$CH_3CH_2CH_2CH_2Br + NaI \longrightarrow CH_3CH_2CH_2CH_2I + NaBr$$

(b)
$$CH_3CH_2CH_2CH_2Br + KOH \longrightarrow CH_3CH_2CH_2CH_2OH + KBr$$

(c)
$$CH_3CH_2CH_2CH_2Br + HC{\equiv}C^-\ Li^+ \longrightarrow CH_3CH_2CH_2CH_2C{\equiv}CH + LiBr$$

(d)
$$CH_3CH_2CH_2CH_2Br + NH_3 \longrightarrow CH_3CH_2CH_2CH_2NH_3^+\ Br^-$$

11.6 (a) $(CH_3)_2N^-$ is more nucleophilic because it is more basic than $(CH_3)_2NH$ and because a negatively charged nucleophile is more nucleophilic than a neutral nucleophile.

(b) $(CH_3)_3N$ is more nucleophilic than $(CH_3)_3B$. $(CH_3)_3B$ is non-nucleophilic because it has no lone electron pair.

(c) H_2S is more nucleophilic than H_2O because nucleophilicity increases in going down a column of the periodic table.

11.7 In this problem, we are comparing two effects – the effect of the substrate and the effect of the leaving group. Tertiary substrates are less reactive than secondary substrates, which are less reactive than primary substrates.

Least reactive ——————→ *Most reactive*

$(CH_3)_3CCl$ < $(CH_3)_2CHCl$ < CH_3Br < CH_3OTos

tertiary carbon	secondary carbon	good leaving group	excellent leaving group

11.8

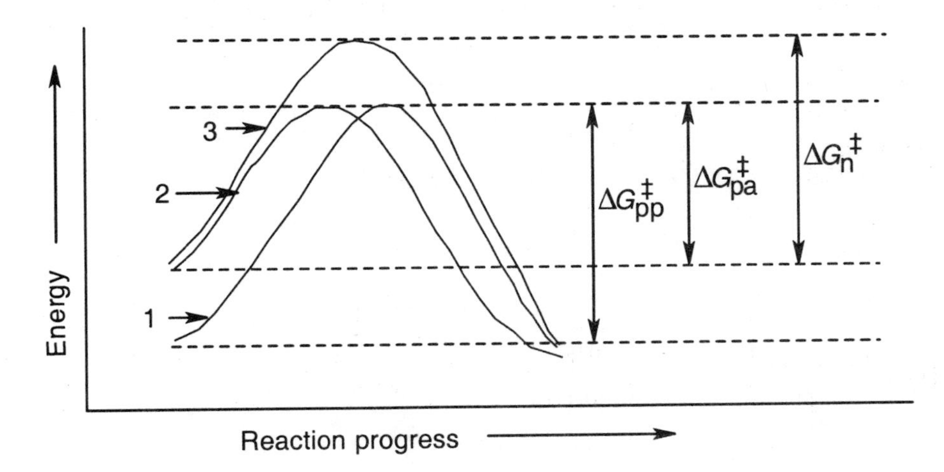

Polar protic solvents (curve 1) stabilize the charged transition state by solvation and also stabilize the nucleophile by hydrogen bonding.

Polar aprotic solvents (curve 2) stabilize the charged transition state by solvation, but do not hydrogen-bond to the nucleophile. Since the energy level of the nucleophile is higher, $\Delta G^{\ddagger}$ is smaller and the reaction is faster in polar aprotic solvents than in polar protic solvents.

Nonpolar solvents (curve 3) stabilize neither the nucleophile nor the transition state. $\Delta G^{\ddagger}$ is therefore higher in nonpolar solvents than in polar solvents, and the reaction rate is slower.

11.9

S-$CH_3(CH_2)_3CH_2C(CH_3)(CH_2CH_3)$–Cl → [$CH_3(CH_2)_3CH_2C^{+}(CH_3)(CH_2CH_3)$ + Cl^-] $\xrightarrow{CH_3CO_2^-}$ *S*-$CH_3(CH_2)_3CH_2C(CH_3)(CH_2CH_3)$–$OCCH_3$ (O) + *R*-CH_3CO(O)–$C(CH_3)(CH_2CH_3)CH_2(CH_2)_3CH_3$

Attack by acetate can occur on either side of the planar, achiral carbocation intermediate, resulting in a mixture of both the *R* and *S* enantiomeric acetates. The ratio of enantiomers is probably close to 50:50.

11.10 The *S* substrate reacts with water to form a mixture of *R* and *S* alcohols. The ratio of enantiomers is close to 50:50.

$$\text{(}S\text{)-}C_6H_5C(Br)(CH_3)CH_2CH_3 \xrightarrow{H_2O} \text{(}S\text{)-}C_6H_5C(OH)(CH_3)CH_2CH_3 + \text{(}R\text{)-}C_6H_5C(OH)(CH_3)CH_2CH_3$$

11.11 If reaction had proceeded with complete inversion, the product would have had a specific rotation of +53.6°. If complete racemization had occurred, $[\alpha]_D$ would have been zero. The observed rotation was +5.3°. Since $\frac{+5.3°}{+53.6°} = 0.099$, 9.9% of the original tosylate was inverted. The remaining 90.1% of the product must have been racemized.

11.12 S_N1 reactivity is related to carbocation stability. Thus, substrates that form the most stable carbocations are the most reactive in S_N1 reactions.

Least reactive ⟶ *Most reactive*

$$\underset{\text{vinylic}}{H_2C{=}CHBr} \ll \underset{\text{primary}}{CH_3CH_2Br} < \underset{\text{secondary}}{CH_3CH(Br)CH_3} < \underset{\text{allylic}}{H_2C{=}CHCH(Br)CH_3}$$

11.13

$$CH_3CH(Br)CH{=}CH_2 \underset{}{\overset{-Br^-}{\rightleftharpoons}} \left[CH_3\overset{+}{C}HCH{=}CH_2 \longleftrightarrow CH_3CH{=}CH\overset{+}{C}H_2\right] \overset{Br^-}{\rightleftharpoons} CH_3CH{=}CHCH_2Br$$

The two bromobutenes form the same allylic carbocation.

11.14

$$C_6H_5CH_2CH(Cl)C_6H_5$$

1-Chloro-1,2-diphenylethane

Nucleophilic substitution of 1-chloro-1,2-diphenylethane proceeds via an S_N1 mechanism because of stabilization of the carbocation intermediate by the phenyl group at C1. In an S_N1 reaction, the rate-limiting step is carbocation formation; all subsequent steps occur at a faster rate and do not affect the rate of reaction. After carbocation formation, the rate does not depend on the identity of the nucleophile that combines with the carbocation. In this example, F^- and $(CH_3CH_2)_3N$**:** react at the same rate.

11.15 Both substrates have allylic groups and can react either by a S_N1 or a S_N2 route. The reaction mechanism is determined by the leaving group, the solvent, or the nucleophile.
(a) This reaction probably occurs by a S_N1 mechanism. HCl converts the poor –OH leaving group into an excellent $-OH_2^+$ leaving group, and the polar solvent stabilizes the carbocation intermediate, which is a resonance-stabilized allylic carbocation.
(b) This reaction takes place in a polar, aprotic solvent, and the nucleophile and leaving group are both reactive. It is very likely that the reaction occurs by a S_N2 mechanism.

11.16 The major elimination product in each case has the most substituted double bond.

(a)

Br CH_3 (substituents)

$CH_3CH_2CHCHCH_3$ —Base→ $CH_3CH_2CH{=}CCH_3$ (with CH_3) + $CH_3CH{=}CHCHCH_3$ (with CH_3)

major (trisubstituted double bond) minor (disubstituted double bond)

(b)

CH_3 Cl CH_3 (substituents), CH_3 below

$CH_3CHCH_2C-CHCH_3$ —Base→ $CH_3CHCH_2C{=}CCH_3$ (CH_3, CH_3, CH_3) + $CH_3CHCH{=}CCHCH_3$ (CH_3, CH_3, CH_3)

major (tetrasubstituted double bond) minor (trisubstituted double bond)

+ $CH_3CHCH_2CCHCH_3$ (CH_3, CH_3; $=CH_2$)

minor (disubstituted double bond)

(c)

cyclohexyl–$CHCH_3$ (Br) —Base→ cyclohexylidene$={}CHCH_3$ + cyclohexyl–$CH{=}CH_2$

major (trisubstituted double bond) minor (monosubstituted double bond)

11.17

H, Ph, Br on C (*R*); C (*R*) with Br, Ph, H

(1*R*,2*R*)-1,2-Dibromo-1,2-diphenylethane

Convert this drawing into a Newman projection, and draw the conformation having anti periplanar geometry for –H and –Br.

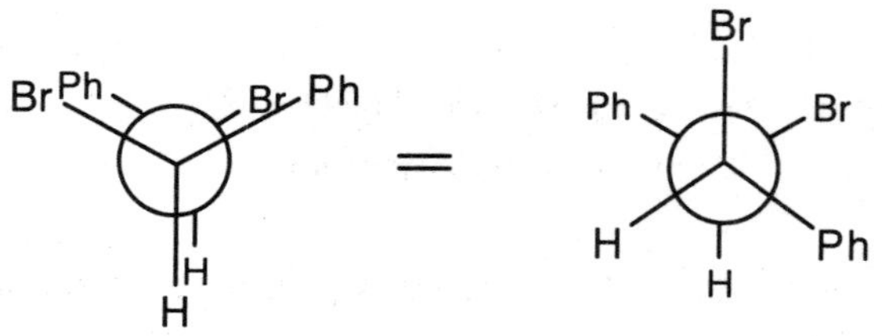

The alkene resulting from dehydrohalogenation is (*Z*)-1-bromo-1,2-diphenylethylene.

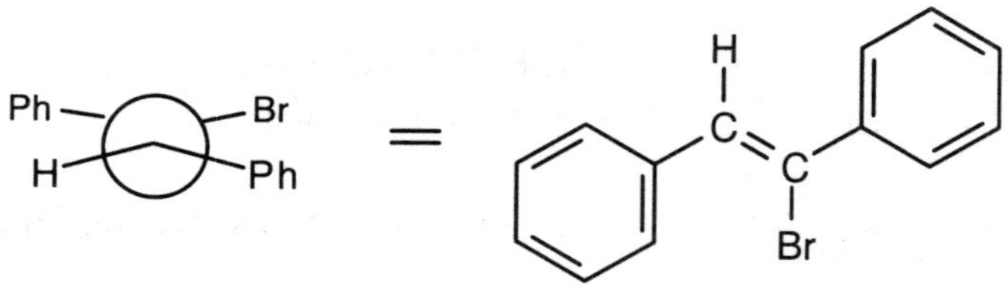

11.18 As in the previous problem, draw the structure, convert it to a Newman projection, and rotate the groups so that the –H and –Br to be eliminated have an anti periplanar relationship.

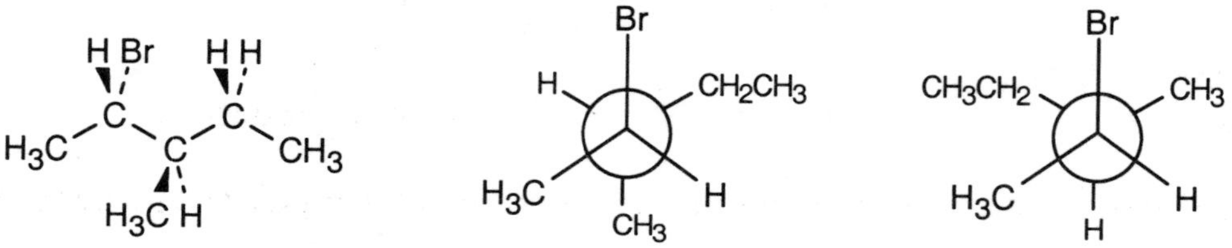

The major product is (*Z*)-3-methyl-2-pentene. A small amount of 3-methyl-1-pentene is also formed.

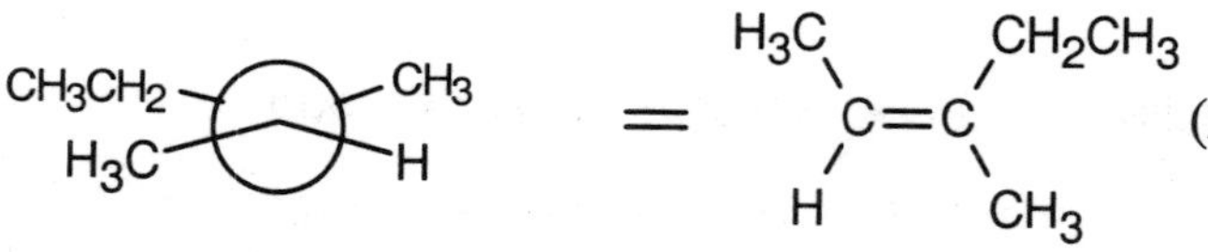

(*Z*)-3-Methyl-2-pentene

11.19

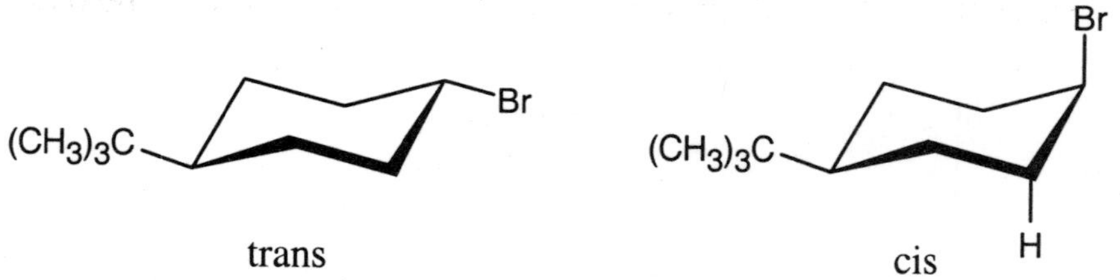

trans

cis

The more stable conformations of each of the two isomers are pictured above; the larger *tert*–butyl group is always equatorial in the more stable conformation. The cis isomer reacts faster under E2 conditions because –Br and –H are in the anti-periplanar arrangement that favors E2 elimination.

11.20

(a)

$$CH_3CH_2CH_2CH_2Br + NaN_3 \longrightarrow CH_3CH_2CH_2CH_2N_3$$

primary — substitution product

The reaction occurs by a S_N2 mechanism because the substrate is primary, the nucleophile is nonbasic, and the product is a substitution product.

(b)

$$CH_3CH_2CH(Cl)CH_2CH_3 + KOH \longrightarrow CH_3CH_2CH{=}CHCH_3$$

secondary — strong base — elimination product

This is an E2 reaction since a secondary halide reacts with a strong base to yield an elimination product.

(c)

$$\text{1-chloro-1-methylcyclohexane} + CH_3COOH \longrightarrow \text{1-methylcyclohexyl acetate } (C(CH_3)(OCCH_3{=}O))$$

tertiary — substitution product

This is an S_N1 reaction. Tertiary substrates form substitution products only by the S_N1 route.

Visualizing Chemistry

11.21 (a)

(i) $CH_3CH_2Cl + Na^+ \ {}^-SCH_3 \longrightarrow CH_3CH_2SCH_3 + NaCl$

(ii) $CH_3CH_2Cl + Na^+ \ {}^-OH \longrightarrow CH_3CH_2OH + NaCl$

Both reactions yield S_N2 substitution products because the substrate is primary and both nucleophiles are good. E2 elimination product is unlikely to form because the base ^-OH isn't bulky.

(b)

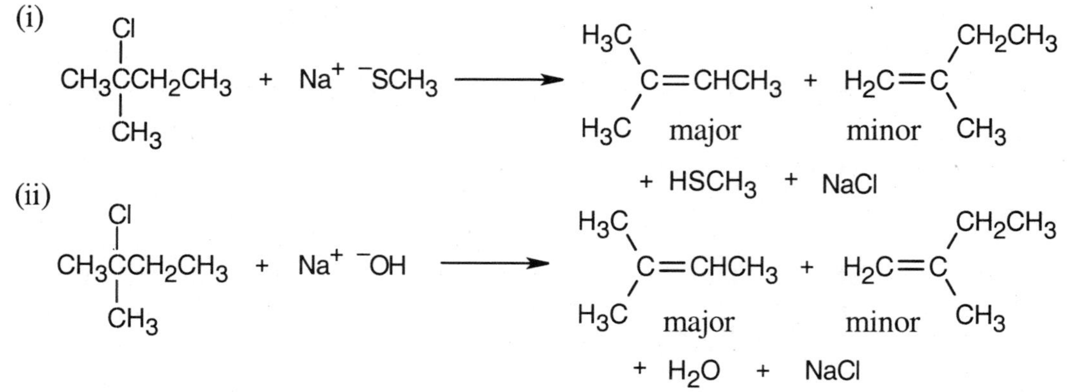

(i)

$$CH_3C(Cl)(CH_3)CH_2CH_3 + Na^+ \ {}^-SCH_3 \longrightarrow (H_3C)_2C{=}CHCH_3 \text{ (major)} + H_2C{=}C(CH_2CH_3)CH_3 \text{ (minor)} + HSCH_3 + NaCl$$

(ii)

$$CH_3C(Cl)(CH_3)CH_2CH_3 + Na^+ \ {}^-OH \longrightarrow (H_3C)_2C{=}CHCH_3 \text{ (major)} + H_2C{=}C(CH_2CH_3)CH_3 \text{ (minor)} + H_2O + NaCl$$

The substrate is tertiary, and the nucleophiles are basic. Two elimination products are expected; the major product has the more substituted double bond, in accordance with Zaitsev's rule.

(c)

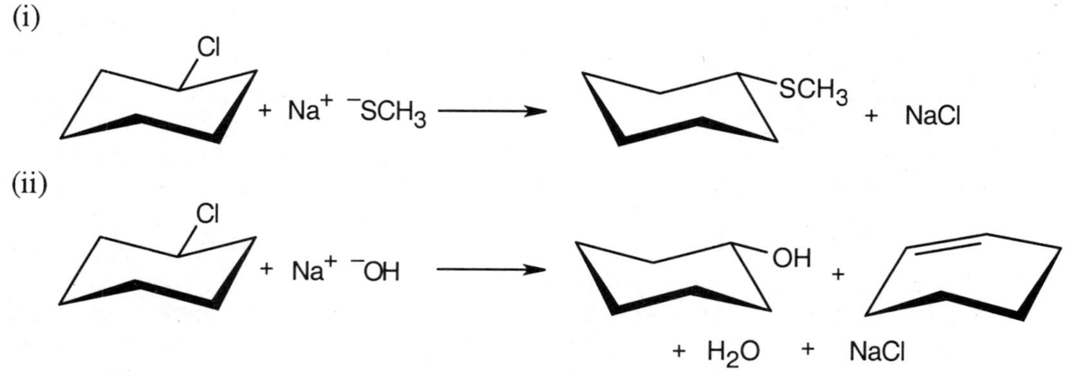

(i)

Cl (cyclohexane) + $Na^+ \ {}^-SCH_3$ ⟶ SCH_3 (cyclohexane) + NaCl

(ii)

Cl (cyclohexane) + $Na^+ \ {}^-OH$ ⟶ OH (cyclohexane) + (cyclohexene) + H_2O + NaCl

In (i), the secondary substrate reacts with the good, but weakly basic, nucleophile to yield substitution product. In (ii), NaOH is a poorer nucleophile but a stronger base, and both substitution and elimination product are formed.

11.22

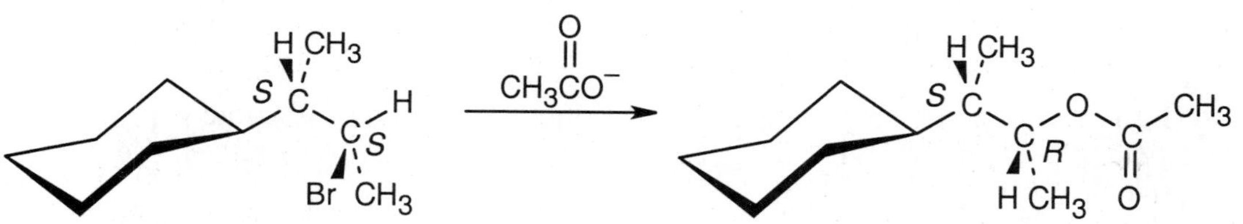

(cyclohexyl)–C(H)(CH₃) *S* –C(H)(Br)(CH₃) *S* $\xrightarrow{CH_3CO^-\ (O)}$ (cyclohexyl)–C(H)(CH₃) *S* –C(H)(CH₃)–O–C(=O)–CH₃ *R*

Reaction of the secondary bromide with the weakly basic acetate nucleophile occurs by an S_N2 route, with inversion of configuration, to produce the *R* acetate.

11.23

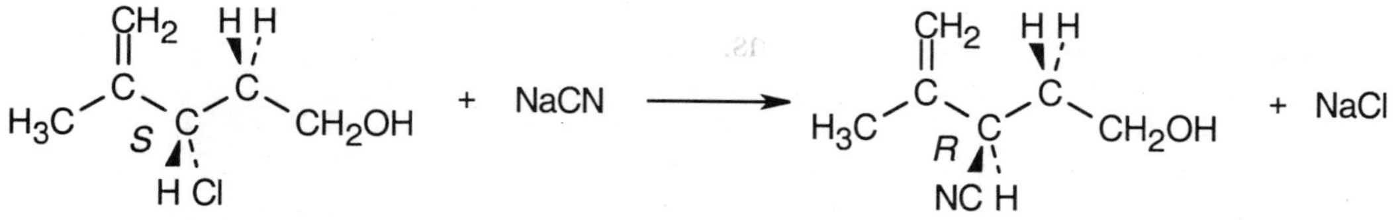

H_3C–C(=CH_2)–C(H)(Cl) *S* –C(H)(H)–CH_2OH + NaCN ⟶ H_3C–C(=CH_2)–C(NC)(H) *R* –C(H)(H)–CH_2OH + NaCl

The *S* substrate has a secondary allylic chloride group and a primary hydroxyl group. S_N2 reaction occurs at the secondary carbon to give the *R* cyano product because hydroxide is a poor leaving group.

11.24

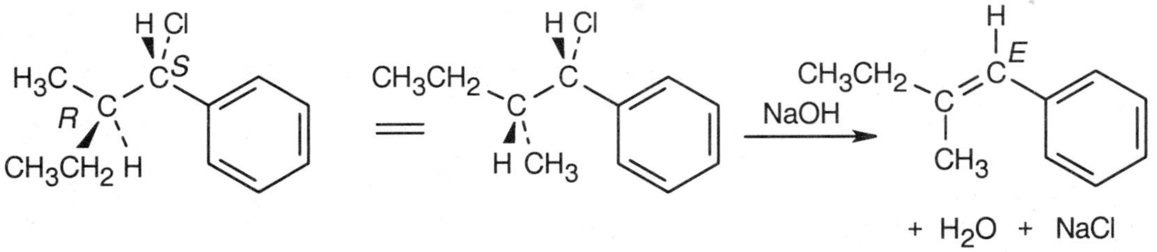

$\xrightarrow{NaOH}$ + H_2O + NaCl

Rotate the left side of the molecule so that the groups to be eliminated have an anti periplanar relationship. The double bond in the product has an *E* configuration.

Additional Problems

11.25 (a) The rates of both S_N1 and S_N2 reactions are affected by the use of polar solvents. S_N1 reactions are accelerated because polar solvents stabilize developing charges in the transition state. Most S_N2 reactions, however, are slowed down by polar *protic* solvents because these solvents hydrogen-bond to the nucleophile and decrease its reactivity. Polar aprotic solvents solvate nucleophiles without hydrogen bonding and increase nucleophile reactivity in S_N2 reactions.

(b) Good leaving groups (weak bases whose negative charge can be delocalized) increase the rates of S_N1 and S_N2 reactions.

(c) A good attacking nucleophile accelerates the rate of an S_N2 reaction. Since the nucleophile is involved in the rate-limiting step of an S_N2 reaction, a good attacking nucleophile lowers the energy of the transition state and increases the rate of reaction. Choice of nucleophile has no effect on the rate of a S_N1 reaction because attack of the nucleophile occurs after the rate-limiting step.

(d) Because the rate-limiting step in an S_N2 reaction involves attack of the nucleophile on the substrate, any factor that makes approach of the nucleophile more difficult slows down the rate of reaction. Especially important is the degree of crowding at the reacting carbon atom. Tertiary carbon atoms are too crowded to allow S_N2 substitution to occur. Even steric hindrance one carbon atom away from the reacting site causes a drastic slowdown in rate of reaction.

The rate-limiting step in an S_N1 reaction involves formation of a carbocation. Any structural factor in the substrate that stabilizes a carbocation increases the rate of reaction. Substrates that are tertiary, allylic, or benzylic react the fastest.

11.26 (a) CH_3I reacts faster than CH_3Br because I^- is a better leaving group than Br^-.

(b) CH_3CH_2I reacts faster with OH^- in dimethylsulfoxide (DMSO) than in ethanol. Ethanol, a protic solvent, hydrogen-bonds with hydroxide ion and decreases its reactivity.

(c) Under the S_N2 conditions of this reaction, CH_3Cl reacts faster than $(CH_3)_3CCl$. Approach of the nucleophile to the bulky $(CH_3)_3CCl$ molecule is hindered.

(d) $H_2C{=}CHCH_2Br$ reacts faster because vinylic halides such as $H_2C{=}CHBr$ are unreactive to substitution reactions.

11.27

$$\underset{\text{primary halide}}{CH_3CH_2\overset{\overset{CH_3}{|}}{C}HCH_2I} + {}^{-}CN \longrightarrow CH_3CH_2\overset{\overset{CH_3}{|}}{C}HCH_2CN + I^{-}$$

This is a S_N2 reaction, whose rate depends on the concentration of both alkyl halide and nucleophile. Rate = k x [RX] x [Nu:$^-$]

(a) Halving the concentration of cyanide ion and doubling the concentration of alkyl halide doesn't change the reaction rate.
(b) Tripling the concentrations of both cyanide ion and alkyl halide causes a ninefold increase in reaction rate.

11.28

$$\underset{\text{tertiary halide}}{CH_3CH_2\overset{\overset{CH_3}{|}}{\underset{\underset{I}{|}}{C}}CH_3} + CH_3CH_2OH \longrightarrow CH_3CH_2\overset{\overset{CH_3}{|}}{\underset{\underset{OCH_2CH_3}{|}}{C}}CH_3 + HI$$

This is an S_N1 reaction, whose rate depends only on the concentration of 2-iodo-2-methylbutane. Rate = k x [RX].
(a) Tripling the concentration of alkyl halide triples the rate of reaction.
(b) Halving the concentration of ethanol by dilution with diethyl ether reduces the dielectric polarization of the solvent and decreases the rate.

11.29

(a)

$$CH_3Br + Na^+ \ {}^{-}C{\equiv}CH(CH_3)_2 \longrightarrow CH_3C{\equiv}CH(CH_3)_2 + NaBr$$

Not $CH_3C{\equiv}C^- Na^+ + BrCH(CH_3)_2$. The strong base $CH_3C{\equiv}C^-$ brings about elimination, producing $CH_3C{\equiv}CH$ and $H_2C{=}CHCH_3$.

(b)

$$CH_3CH_2CH_2CH_2Br + NaCN \longrightarrow CH_3CH_2CH_2CH_2CN + NaBr$$

(c)

$$H_3C{-}Br + {}^{-}OC(CH_3)_3 \longrightarrow H_3C{-}OC(CH_3)_3 + NaBr$$

or

$$(CH_3)CCl + CH_3OH \longrightarrow \underset{\text{(major)}}{(CH_3)C{-}OCH_3} + HCl + \underset{\text{(minor)}}{(CH_3)_2C{=}CH_2}$$

(d)

$$CH_3CH_2CH_2Br + \text{excess } NH_3 \longrightarrow \underset{\text{major}}{CH_3CH_2CH_2NH_2} + NH_4{}^+Br^-$$

(e)

$$(C_6H_5)_3P: + CH_3Br \longrightarrow (C_6H_5)_3\overset{+}{P}CH_3\ Br^-$$

(f)

$$\text{1-methylcyclohexanol (OH, } CH_3\text{)} + HBr \longrightarrow \text{1-bromo-1-methylcyclohexane (Br, } CH_3\text{)} + H_2O$$

11.30 (a) The difference in this pair of reactions is in the leaving group. Since $^-$OTos is a better leaving group than $^-$Cl (see Section 11.5), S_N2 displacement by iodide on CH_3–OTos proceeds faster.

(b) The substrates in these two reactions are different. Bromoethane is a primary bromoalkane, and bromocyclohexane is a secondary bromoalkane. Since S_N2 reactions proceed faster at primary than secondary carbon atoms, S_N2 displacement on bromoethane is a faster reaction.

(c) Ethoxide ion and cyanide ion are different nucleophiles. Since CN^- is more reactive than $CH_3CH_2O^-$ in S_N2 reactions, S_N2 displacement on 2-bromopropane by CN^- proceeds at a faster rate.

(d) The solvent in each reaction is different. S_N2 reactions in hexamethylphosphoramide (HMPA) proceed faster than those in other solvents. Thus, S_N2 displacement by acetylide ion on bromomethane proceeds faster in HMPA than in benzene.

11.31 Because 1–bromopropane is a primary haloalkane, the reaction proceeds by either a S_N2 or E2 mechanism, depending on the basicity and the amount of steric hindrance in the nucleophile.

(a)

$$CH_3CH_2CH_2Br + NaNH_2 \longrightarrow CH_3CH_2CH_2NH_2 + NaBr$$

(b)

$$CH_3CH_2CH_2Br + K^+\ ^-OC(CH_3)_3 \longrightarrow CH_3CH{=}CH_2 + HOC(CH_3)_3 + KBr$$

$K^+\ ^-OC(CH_3)_3$ is a strong, bulky base that brings about elimination, not substitution.

(c)

$$CH_3CH_2CH_2Br + NaI \longrightarrow CH_3CH_2CH_2I + NaBr$$

(d)

$$CH_3CH_2CH_2Br + NaCN \longrightarrow CH_3CH_2CH_2CN + NaBr$$

(e)

$$CH_3CH_2CH_2Br + Na^+\ ^-C{\equiv}CH \longrightarrow CH_3CH_2CH_2C{\equiv}CH + NaBr$$

(f)

$$CH_3CH_2CH_2Br \xrightarrow{Mg} CH_3CH_2CH_2MgBr \xrightarrow{H_2O} CH_3CH_2CH_3$$

11.32 To predict nucleophilicity, remember these rules:

1) In comparing nucleophiles that have the same attacking atom, nucleophilicity parallels basicity. In other words, a more basic nucleophile is a more effective nucleophile.
2) Nucleophilicity increases in going down a column of the periodic table.
3) A negatively charged nucleophile is usually more reactive than a neutral nucleophile.

	More Nucleophilic	*Less Nucleophilic*	*Reason*
(a)	$^-NH_2$	NH_3	Rule 1 or 3
(b)	CH_3COO^-	H_2O	Rule 1 or 3
(c)	F^-	BF_3	BF_3 is not a nucleophile
(d)	$(CH_3)_3P$	$(CH_3)_3N$	Rule 2
(e)	I^-	Cl^-	Rule 2
(f)	$^-C{\equiv}N$	$^-OCH_3$	Reactivity chart, Section 11.5

11.33 An alcohol is converted to an ether by two different routes in this series of reactions. The two resulting ethers have identical structural formulas but differ in sign of specific rotation. Therefore, at some step or steps in these reaction sequences, inversion of configuration at the chiral carbon must have occurred. Let's study each step of the Phillips and Kenyon series to find where inversion is occurring.

$CH_3CH(OH)CH_2C_6H_5$, $[\alpha]_D = +33.0°$ —③ TosCl→ $CH_3CH(OTos)CH_2C_6H_5$, $[\alpha]_D = +31.1°$

① K ↓ (from the alcohol): $[CH_3CH(O^-K^+)CH_2C_6H_5]$

② CH_3CH_2Br ↓: $CH_3CH(OCH_2CH_3)CH_2C_6H_5$, $[\alpha]_D = +23.5°$

④ CH_3CH_2OH, heat ↓ (from the tosylate): $CH_3CH(OCH_2CH_3)CH_2C_6H_5$, $[\alpha]_D = -19.9°$

In step 1, the alcohol reacts with potassium metal to produce a potassium alkoxide. Since the bond between carbon and oxygen has not been broken, no inversion occurs in this step.

The potassium alkoxide acts as a nucleophile in the S_N2 displacement on CH_3CH_2Br in step 2. It is the C–Br bond of bromoethane, however, not the C–O bond of the alkoxide, that is broken. No inversion at the carbon chirality center occurs in step 2.

The starting alcohol reacts with tosyl chloride in step 3. Again, because the O–H bond, rather than the C–O bond, of the alcohol is broken, no inversion occurs at this step.

Inversion does occur at step 4 when the $^-$OTos group is displaced by CH_3CH_2OH. The C–O bond of the tosylate (–OTos) is broken, and a new C–O bond is formed.

Notice the specific rotations of the two enantiomeric products. The product of steps 1 and 2 should be enantiomerically pure because neither reaction has affected the C–O bond. Reaction 4 proceeds with some racemization at the chirality center to give a smaller absolute value of $[\alpha]_D$.

11.34 (a) Substitution does not take place with secondary alkyl halides when a strong, bulky base is used. Elimination occurs instead, and produces $H_2C{=}CHCH_2CH_3$ and $CH_3CH{=}CHCH_3$.
(b) Fluoroalkanes don't undergo S_N2 reactions, and no reaction occurs.
(c) $SOCl_2$ in pyridine converts primary and secondary alcohols to chlorides by an S_N2 mechanism. 1–Methyl–1–cyclohexanol is a tertiary alcohol, and does not undergo S_N2 substitution. Instead, E2 elimination occurs to give 1–methylcyclohexene.

11.35 S_N1 reactivity:

Least reactive ⟶ *Most reactive*

(a) $CH_3CH_2CH(NH_2)CH_3$ < $(CH_3)_3C{-}Cl$ < $C_6H_5C(CH_3)_2Cl$

most stable carbocation

(b) $(CH_3)_3COH$ < $(CH_3)_3CF$ < $(CH_3)_3CBr$

best leaving group

(c) $C_6H_5CH_2Br$ < $C_6H_5CHBrCH_3$ < $(C_6H_5)_3CBr$

most stable carboction

11.36 S_N2 reactivity:

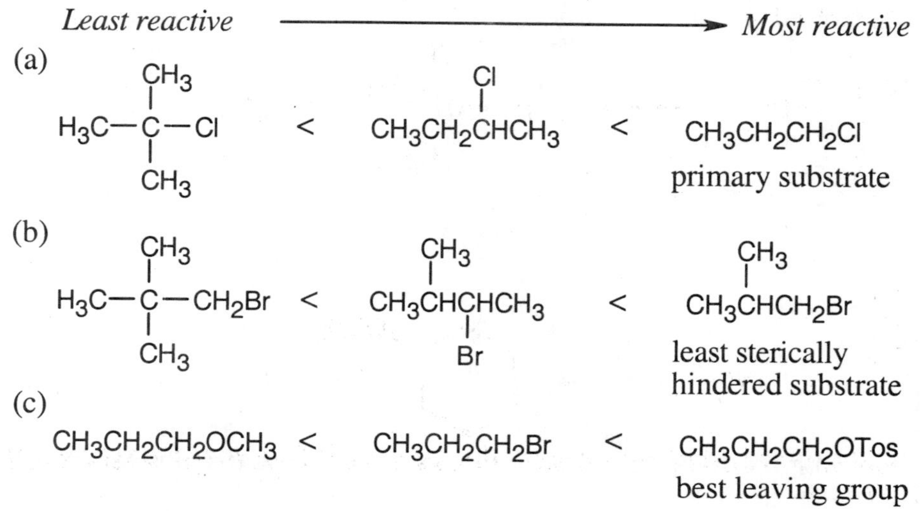

Least reactive ⟶ *Most reactive*

(a)

$(CH_3)_3C–Cl$ < $CH_3CH_2CH(Cl)CH_3$ < $CH_3CH_2CH_2Cl$
primary substrate

(b)

$(CH_3)_3C–CH_2Br$ < $CH_3CH(CH_3)CH(Br)CH_3$ < $CH_3CH(CH_3)CH_2Br$
least sterically hindered substrate

(c)

$CH_3CH_2CH_2OCH_3$ < $CH_3CH_2CH_2Br$ < $CH_3CH_2CH_2OTos$
best leaving group

11.37

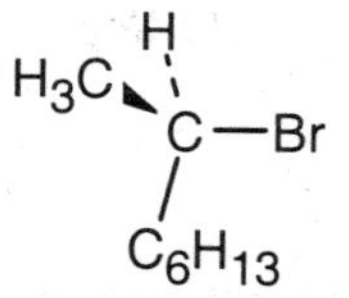

H_3C, H, $C–Br$, C_6H_{13}

(*R*)-2-Bromooctane

(*R*)-2-Bromooctane is a secondary bromoalkane, which undergoes S_N2 substitution. Since S_N2 reactions proceed with inversion of configuration, the configuration at the carbon chirality center is inverted. (This does not necessarily mean that all *R* isomers become *S* isomers after an S_N2 reaction. The *R,S* designation refers to the priorities of groups, which may change when the nucleophile is varied.)

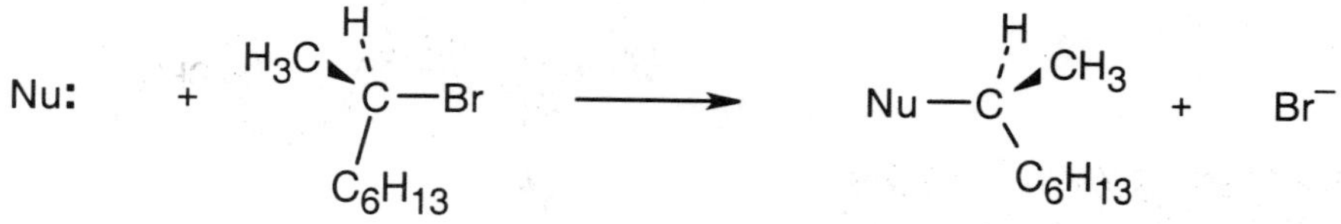

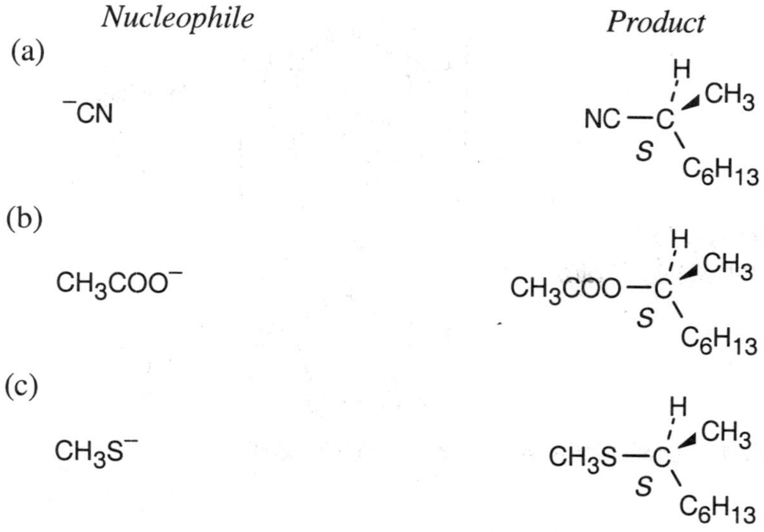

$$Nu{:} + (R)\text{-}H_3C(H)(C_6H_{13})C–Br \longrightarrow Nu–C(H)(CH_3)(C_6H_{13}) + Br^-$$

	Nucleophile	*Product*
(a)	^-CN	$NC–C(H)(CH_3)(C_6H_{13})$ *S*
(b)	CH_3COO^-	$CH_3COO–C(H)(CH_3)(C_6H_{13})$ *S*
(c)	CH_3S^-	$CH_3S–C(H)(CH_3)(C_6H_{13})$ *S*

11.38 After 50% of the starting material has reacted, the reaction mixture consists of 50% (*R*)-2-bromooctane and 50% (*S*)-2-bromooctane. At this point, the *R* starting material is completely racemized.

Br^- + H_3C(H)(*R*)C—Br(C_6H_{13}) ⟶ Br—C(H)(*S*)(CH_3)(C_6H_{13}) + Br^-

11.39

Cyclohexyl—$\ddot{O}$:$^-$ + CH_3I ⟶ Cyclohexyl—OCH_3 + I^-

This is an excellent method of ether preparation because iodomethane is very reactive in S_N2 displacements.

Cyclohexyl—I + $CH_3\ddot{O}$:$^-$ ⟶ Cyclohexyl—OCH_3 + Cyclohexene + CH_3OH + I^-

Reaction of a secondary haloalkane with a basic nucleophile yields both substitution and elimination products. This is a less satisfactory method of ether preparation.

11.40

Methoxide removes a proton from the hydroxyl group of 4-bromo-1-butanol.

$BrCH_2CH_2CH_2CH_2$—O—H + :$\ddot{O}CH_3^-$ ⇌ [$BrCH_2CH_2CH_2CH_2$—$\ddot{O}$:$^-$] + $HOCH_3$

S_N2 displacement of Br^- by O^- yields the cyclic ether tetrahydrofuran.

⟶ Tetrahydrofuran + :$\ddot{Br}$:$^-$

$CH_3OCH_2CH_2CH_2CH_2OH$ is also formed.

11.41

$$BrCH_2CH_2Br \; + \; 2\ NaOH \longrightarrow HOCH_2CH_2OH \; + \; 2\ NaBr$$

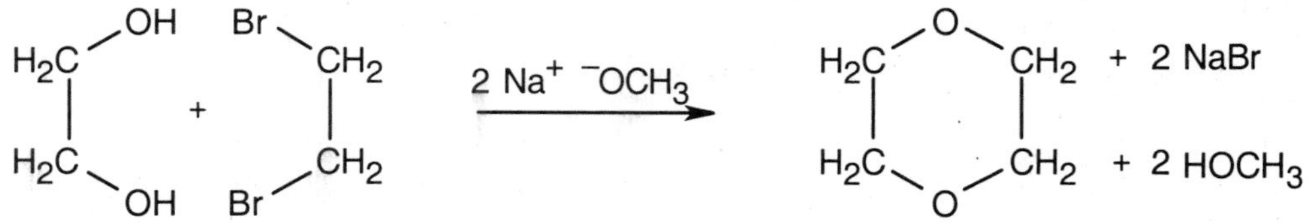

$$HOCH_2CH_2OH \; + \; BrCH_2CH_2Br \xrightarrow{2\ Na^+\ ^-OCH_3} \text{1,4-dioxane} \; + \; 2\ NaBr \; + \; 2\ HOCH_3$$

1,4-Dioxane

11.42 The first step in an S_N1 displacement is dissociation of the substrate to form a planar, sp^2–hybridized carbocation and a leaving group. The carbocation that would form from dissociation of this haloalkane can't become planar because of the rigid structure of the rest of the molecule. Because it's not possible to form the necessary carbocation, an S_N1 reaction can't occur.

11.43

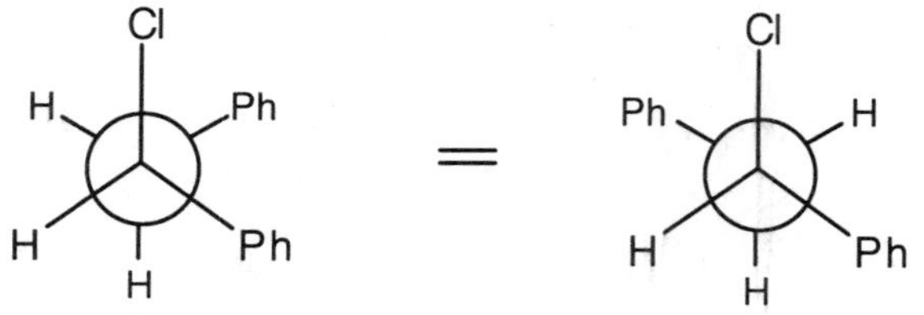

A = **B**

Both Newman projections place –H and –Cl in the correct anti periplanar geometry for E2 elimination.

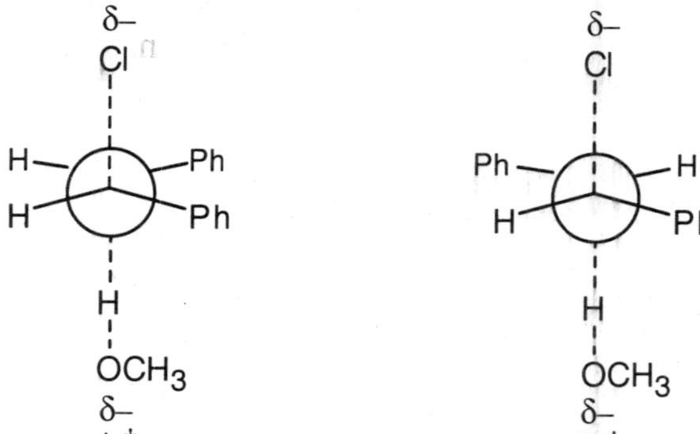

A‡ **B‡**

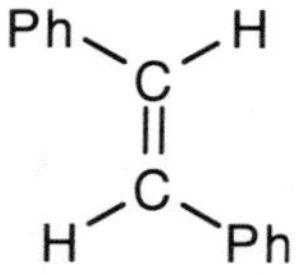

trans-1,2-Diphenylethylene

Either transition state **A**‡ or **B**‡ can form when 1-chloro-1,2-diphenylethane undergoes E2 elimination. Crowding of the two phenyl groups in T.S. **A**‡ make this transition state (and the product resulting from it) of higher energy than transition state **B**‡. Formation of the product from **B**‡ is therefore favored, and *trans*-1,2-diphenylethylene is the major product.

11.44

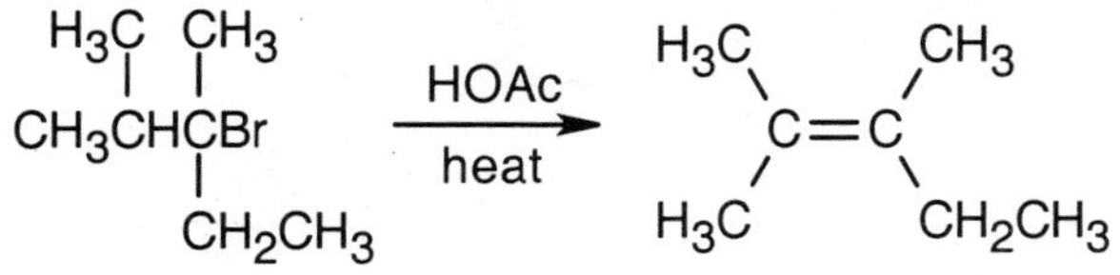

$CH_3CH(CH_3)C(CH_3)(Br)CH_2CH_3 \xrightarrow[\text{heat}]{\text{HOAc}} (H_3C)_2C{=}C(CH_3)CH_2CH_3$

The alkene shown above has the most highly substituted double bond, and, according to Zaitsev's rule, is the major product. The following minor products may also form. Smaller amounts of these products form, either because they have less substituted double bonds or because of steric strain.

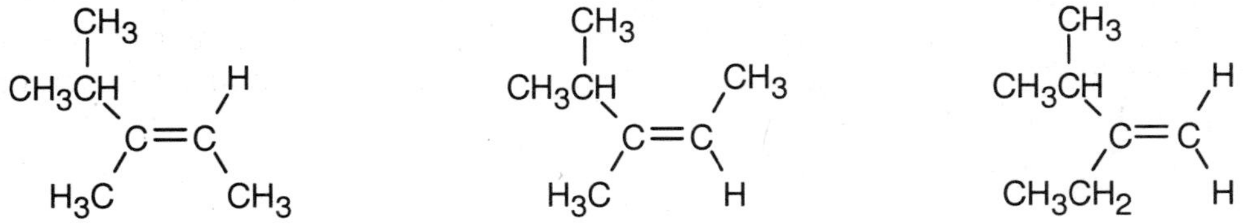

$(CH_3)_2CH(H_3C)C{=}C(H)CH_3$ $(CH_3)_2CH(H_3C)C{=}C(CH_3)H$ $(CH_3)_2CH(CH_3CH_2)C{=}CH_2$

11.45

H3C–C(H)(OTos) *R* / Ph–C(H)(CH3) *S* = Newman projection = Newman projection

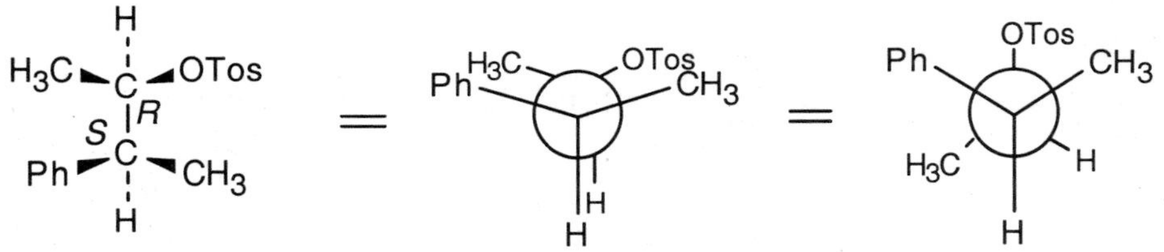

Draw a Newman projection of the tosylate of (2*R*,3*S*)-3-phenyl-2-butanol, and rotate the projection until the –OTos and the –H on the adjoining carbon atom are anti periplanar. Even though this conformation has several gauche interactions, it is the only conformation in which –OTos and –H are 180° apart.

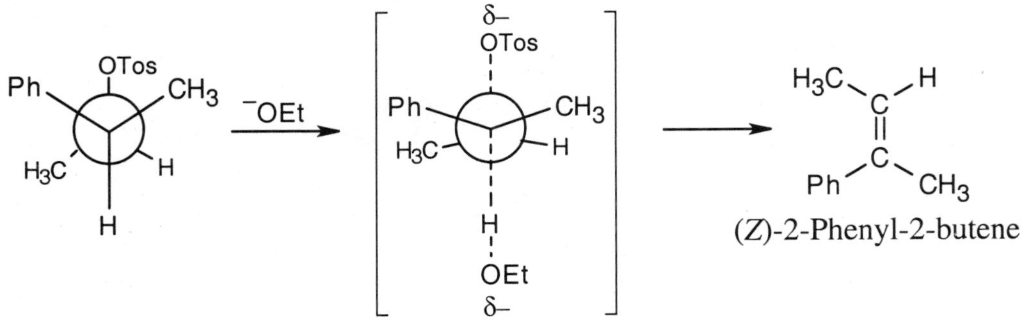

$\xrightarrow{^-OEt}$ [δ– OTos ··· H ··· OEt δ–] ⟶ (*Z*)-2-Phenyl-2-butene

Elimination yields the *Z* isomer of 2-phenyl-2-butene. Refer to Chapter 6 for the method of assigning *E, Z* designation.

11.46 By the same argument used in the previous problem, you can show that elimination from the tosylate of (2*R*,3*R*)-3-phenyl-2-butanol gives the *E*–alkene.

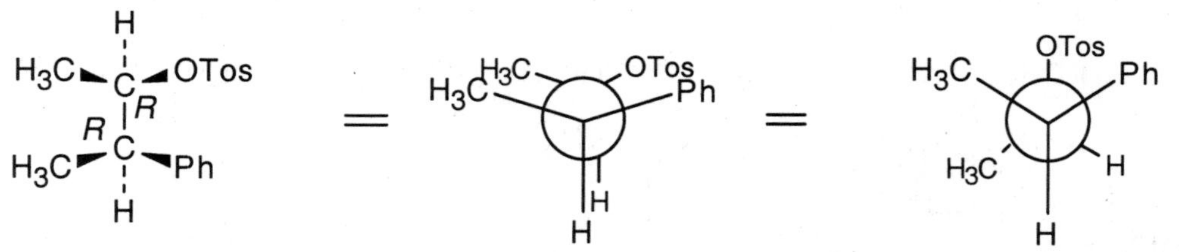

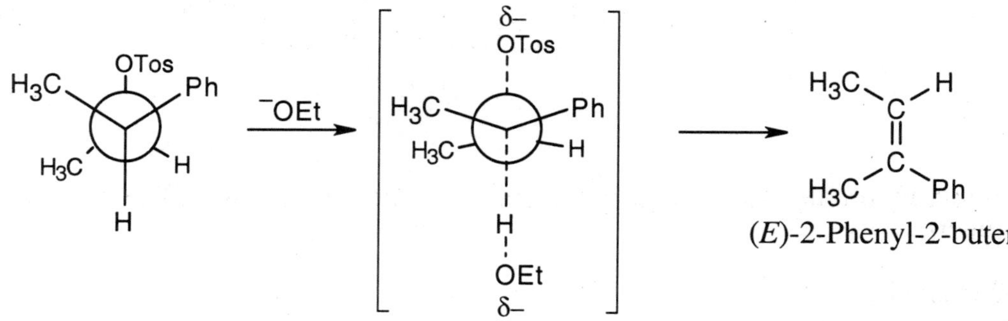

(*E*)-2-Phenyl-2-butene

The (2*S*,3*S*) isomer also forms the *E*–alkene; the (2*S*,3*R*) isomer yields *Z*–alkene.

11.47

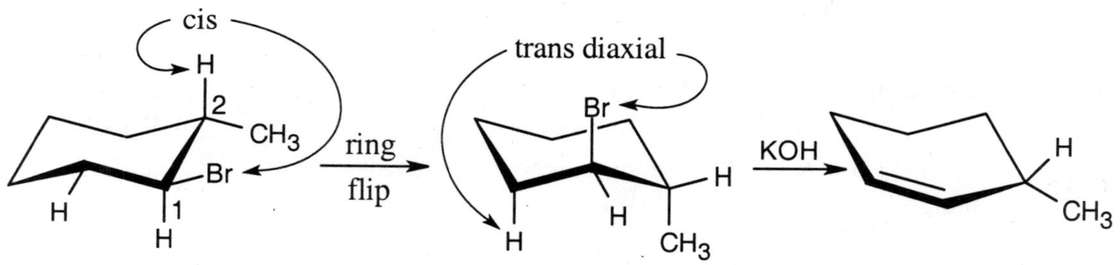

E2 reactions require that the two atoms to be eliminated have a trans diaxial relationship. Since it's impossible for bromine and the hydrogen at C2 to be trans diaxial, elimination occurs in the opposite direction to yield 3-methylcyclohexene, the non-Zaitsev product.

11.48

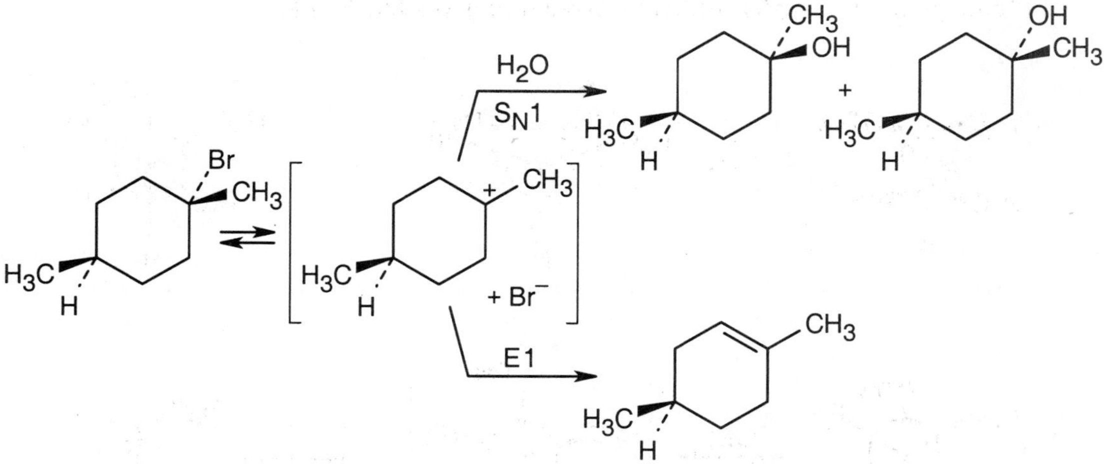

This tertiary bromoalkane reacts by S_N1 and E1 routes to yield alcohol and alkene products.

11.49 S_N2 reactivity:

Most reactive ⟶ *Least reactive*

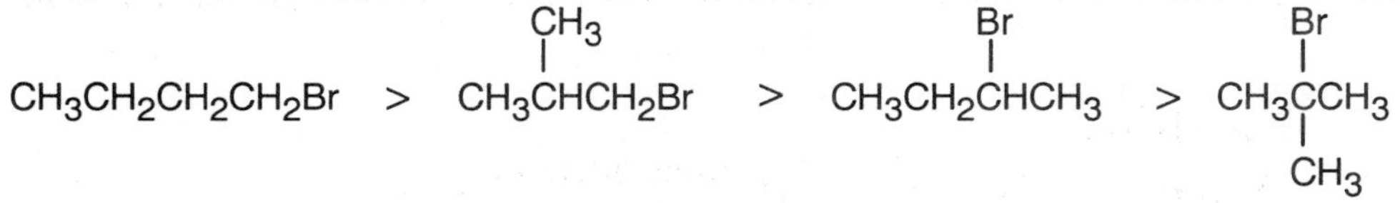

$CH_3CH_2CH_2CH_2Br$ > $CH_3CH(CH_3)CH_2Br$ > $CH_3CH_2CH(Br)CH_3$ > $CH_3C(Br)(CH_3)CH_3$

1-Bromobutane	1-Bromo-2-methylpropane	2-Bromobutane	2-Bromo-2-methylpropane

11.50

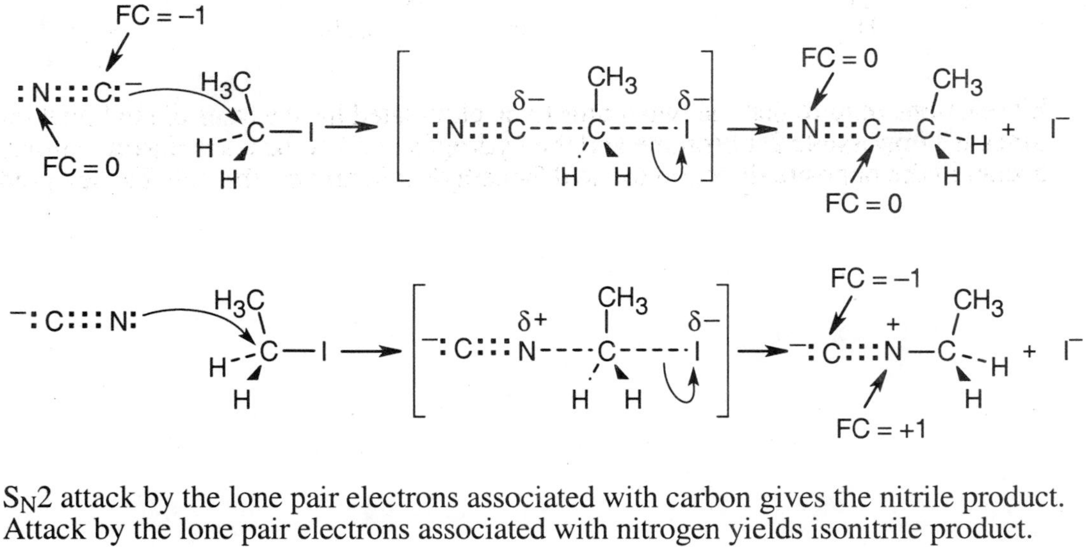

S_N2 attack by the lone pair electrons associated with carbon gives the nitrile product. Attack by the lone pair electrons associated with nitrogen yields isonitrile product.

11.51

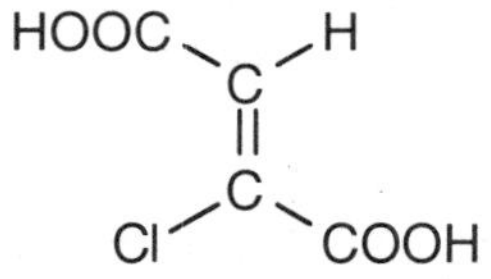

HOOC–C(H)=C(Cl)–COOH

(*Z*)-2-Chloro-2-butene-1,4-dioic acid

HOOC–C(H)=C(Cl)–COOH

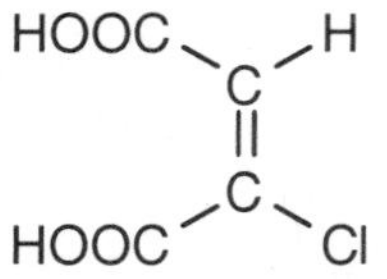

(*E*)-2-Chloro-2-butene-1,4-dioic acid

Hydrogen and chlorine are anti to each other in the *Z* isomer and are syn in the *E* isomer. Since the *Z* isomer reacts fifty times faster than the *E* isomer, elimination must proceed more favorably when the substituents to be eliminated are anti to one another. This is the same stereochemical result as occurs in E2 eliminations of alkyl halides.

11.52 Since 2-butanol is a secondary alcohol, substitution can occur by either an S_N1 or S_N2 route, depending on reaction conditions. Two factors favor an S_N1 mechanism in this case. (1) The reaction is run under solvolysis (solvent as nucleophile) conditions in a polar, protic solvent. (2) Dilute acid converts a poor leaving group (^-OH) into a good leaving group (OH_2), which dissociates easily.

Protonation of the hydroxyl oxygen..

is followed by loss of water to form a planar carbocation.

Attack of water from either side of the planar carbocation yields racemic product.

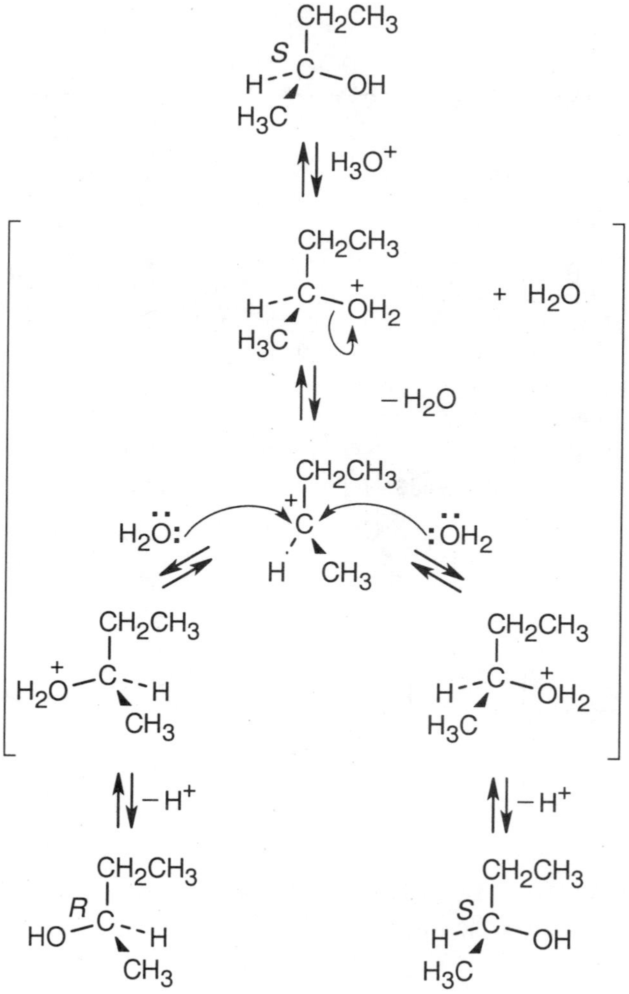

CH_2CH_3, *S*, H, C, OH, H_3C

H_3O^+

CH_2CH_3, H, C, $\overset{+}{O}H_2$, H_3C + H_2O

$-H_2O$

CH_2CH_3, $\overset{+}{C}$, H, CH_3, $H_2\ddot{O}$:, :$\ddot{O}H_2$

CH_2CH_3, $H_2\overset{+}{O}$–C, H, CH_3 | CH_2CH_3, H, C, $\overset{+}{O}H_2$, H_3C

$-H^+$ | $-H^+$

CH_2CH_3, *R*, HO–C, H, CH_3 | CH_2CH_3, *S*, H, C, OH, H_3C

11.53 The chiral tertiary alcohol (*R*)-3-methyl-3-hexanol reacts with HBr by an S_N1 pathway. HBr protonates the hydroxyl group, which dissociates to yield a planar, achiral carbocation. Attack by the nucleophilic bromide anion can occur from either side of the carbocation to produce (±)3-bromo-3-methylhexane.

11.54 Since carbon-deuterium bonds are slightly stronger than carbon-hydrogen bonds, more energy is required to break a C–D bond than to break a C–H bond. In a reaction where either a carbon-deuterium or a carbon-hydrogen bond can be broken in the rate-limiting step, a higher percentage of C–H bond-breaking occurs because the energy of activation for C–H breakage is lower.

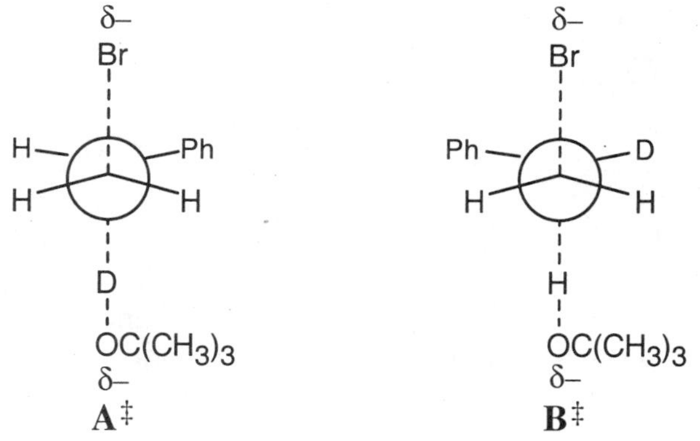

δ– Br H Ph H H D OC(CH3)3 δ– **A‡**

δ– Br Ph D H H H OC(CH3)3 δ– **B‡**

Transition state **A**‡ is of higher energy than transition state **B**‡ because more energy is required to break the C–D bond. The product that results from transition state **B**‡ is thus formed in greater abundance.

11.55 One of the steric requirements of E2 elimination is the need for periplanar geometry, which optimizes orbital overlap in the transition state leading to alkene product. Two types of periplanar arrangements of substituents are possible — syn and anti.

A model of the deuterated bromo compound shows that the deuterium, bromine, and the two carbon atoms that will constitute the double bond all lie in a plane. This arrangement of atoms leads to syn elimination. Even though anti elimination is usually preferred, it doesn't occur for this compound because the bromine, hydrogen, and two carbons can't achieve the necessary geometry.

11.56

Cl H H A Cl

H Cl H B Cl

We concluded in Problem 11.55 that E2 elimination in compounds of this bicyclic structure occurs with syn-periplanar geometry. In compound **A**, –H and –Cl can be eliminated via the syn-periplanar route. Since neither syn nor anti-periplanar elimination is possible for **B**, elimination occurs by a slower, E1 route.

11.57

(a)

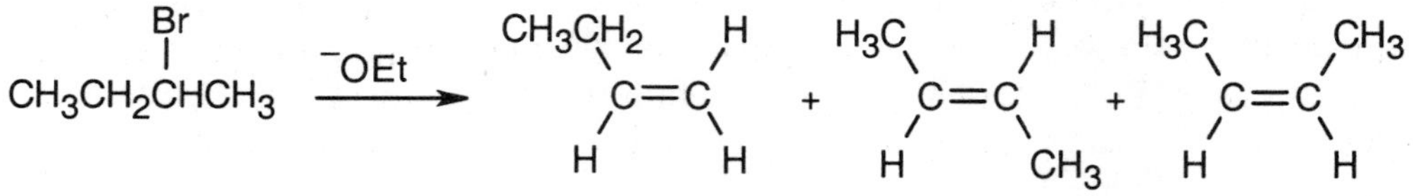

$$CH_3CH_2CHBrCH_3 \xrightarrow{^{-}OEt} CH_3CH_2CH{=}CH_2 + \textit{trans-}CH_3CH{=}CHCH_3 + \textit{cis-}CH_3CH{=}CHCH_3$$

(b)

$H_2C{=}CHBr$, like other vinylic organohalides, does not undergo nucleophilic substitutions.

(c)

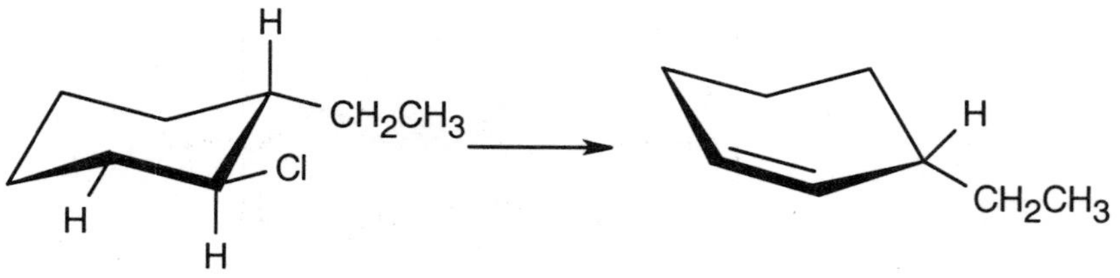

This alkyl halide gives the less substituted cycloalkene (non-Zaitsev product). Elimination to form Zaitsev product is not likely to occur because the –Cl and –H involved cannot assume the anti-periplanar geometry preferred for E2 elimination.

(d)

$$(CH_3)_3C{-}OH + HCl \xrightarrow{0°} (CH_3)_3C{-}Cl + H_2O$$

11.58

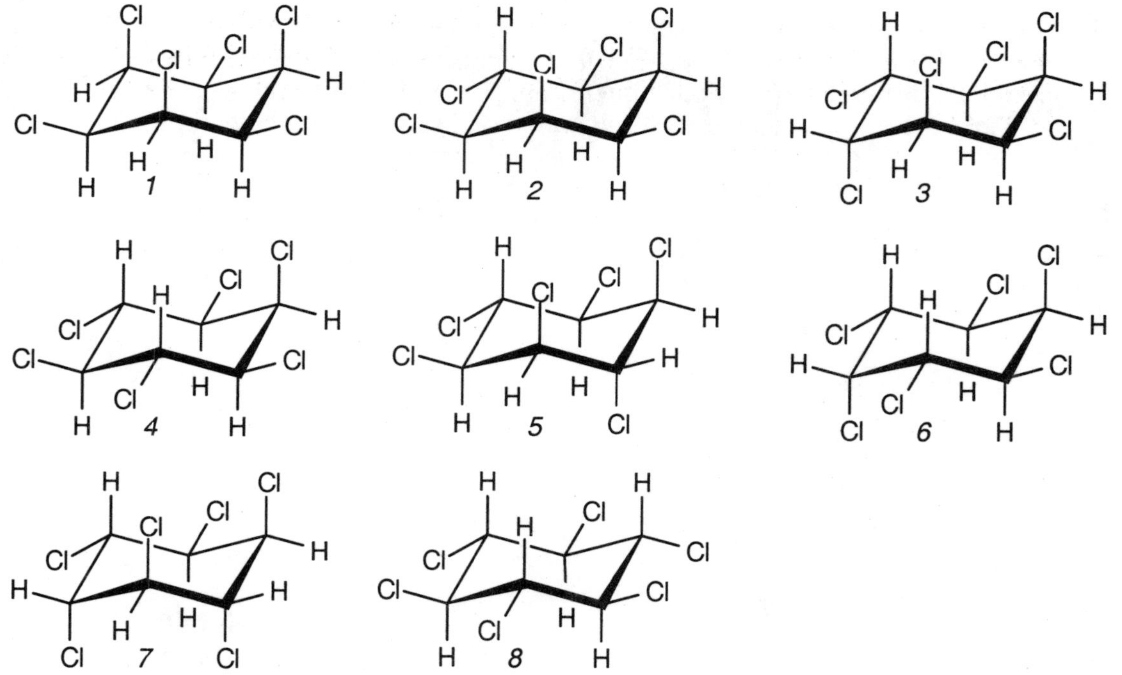

Diastereomer *8* reacts much more slowly than other isomers in an E2 reaction. No pair of hydrogen and chlorine atoms can assume the anti-periplanar orientation preferred for E2 elimination.

11.59 Build molecular models of triethylamine and quinuclidine. A model of the most stable conformation of triethylamine shows that the ethyl groups interfere with approach of the nitrogen lone pair electrons to iodomethane. In quinuclidine, however, the hydrocarbon framework is rigidly held back from the nitrogen lone pair. It is sterically easier for quinuclidine to approach methyl iodide, and reaction therefore occurs at a faster rate.

11.60 The two pieces of evidence indicate that the reaction proceeds by an S_N2 mechanism: S_N2 reactions proceed much faster in polar aprotic solvents such as DMF, and methyl esters react faster than ethyl esters. This reaction is an S_N2 displacement on a methyl ester by iodide ion.

Other experiments can provide additional evidence for an S_N2 mechanism. We can determine if the reaction is second-order by varying the concentration of LiI. We can also vary the type of nucleophile to distinguish an S_N2 mechanism from an S_N1 mechanism, which does not depend on the identity of the nucleophile.

11.61 Because Cl^- is a relatively poor leaving group and acetate is a relatively poor nucleophile, a substitution reaction involving these two groups proceeds at a very slow rate. I^-, however, is both a good nucleophile and a good leaving group. 1-Chlorooctane thus reacts preferentially with iodide to form 1-iodooctane. Only a small amount of 1-iodooctane is formed (because of the low concentration of iodide ion), but 1-iodooctane is more reactive than 1-chlorooctane toward substitution by acetate. Reaction with acetate produces 1-octyl acetate and regenerates iodide ion. The whole process can now be repeated with another molecule of 1-chlorooctane. The net result is production of 1-octyl acetate, and no iodide is consumed.

11.62 Two optically inactive compounds are possible for compound **X**. The racemates of these compounds are also correct answers.

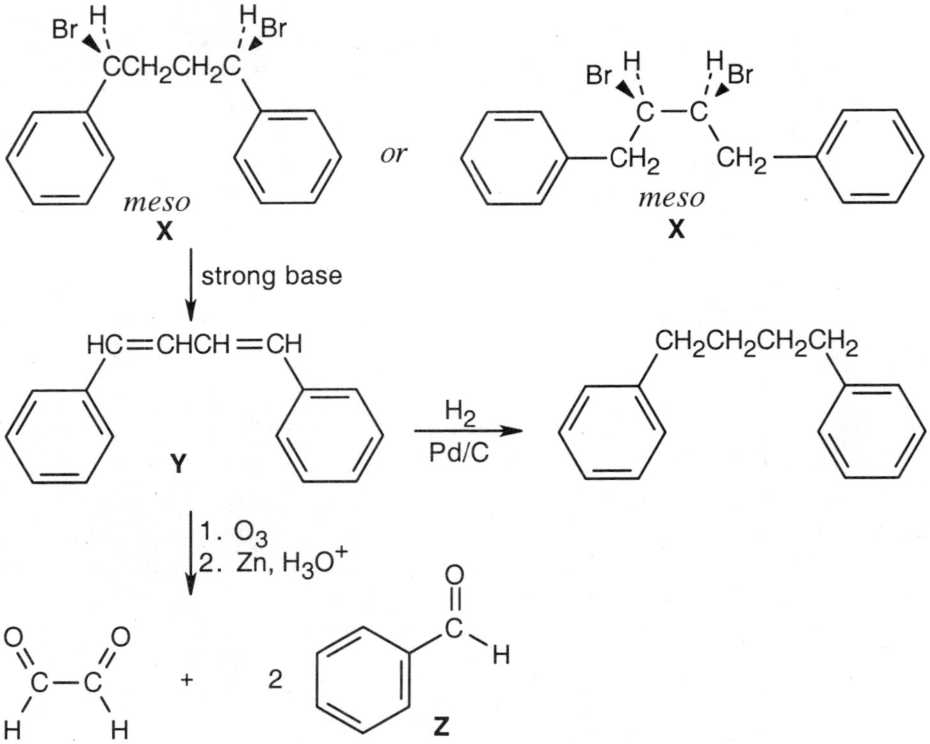

meso **X** *or* *meso* **X**

strong base

Y

H_2, Pd/C

1. O_3
2. Zn, H_3O^+

+ 2 **Z**

11.63

E2 Elimination

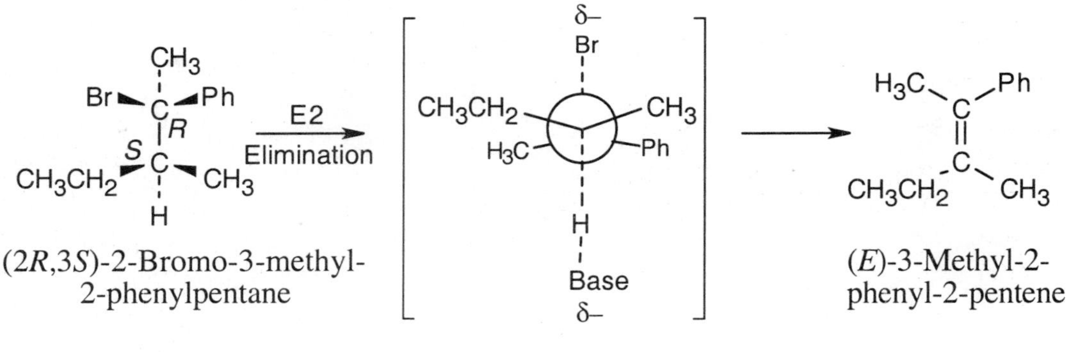

(2*R*,3*S*)-2-Bromo-3-methyl-2-phenylpentane

(*E*)-3-Methyl-2-phenyl-2-pentene

The (2*S*,3*R*) isomer also yields *E* product.

11.64

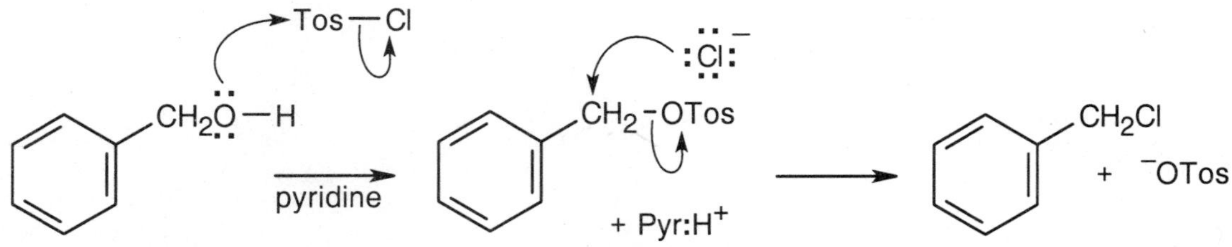

pyridine

+ Pyr:H^+

+ $^-$OTos

At lower temperatures, a tosylate is formed from the reaction of *p*-toluenesulfonyl chloride and an alcohol. The new bond is formed between the toluenesulfonyl group and the oxygen of the alcohol. At higher temperatures, the chloride anion can displace the –OTos group, which is an excellent leaving group, to form an organochloride.

11.65

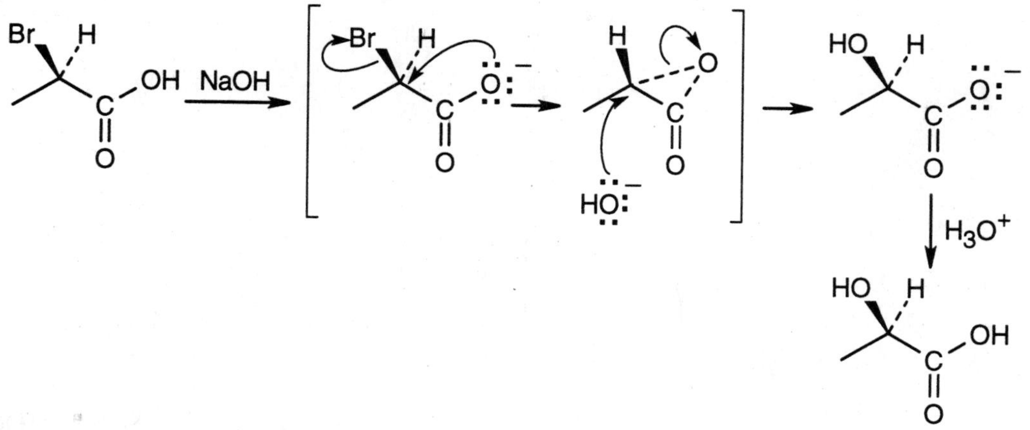

Br H OH NaOH → [Br H O⁻ → H O HO⁻] → HO H O⁻ → H_3O^+ → HO H OH

Two inversions of configuration equal a net retention of configuration.

11.66

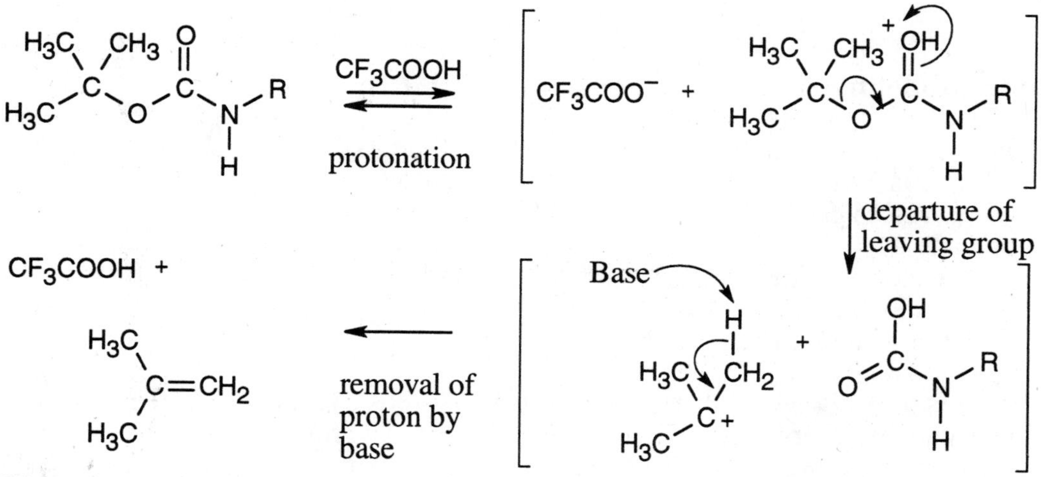

H_3C CH_3 O R H ⇌ CF_3COOH protonation [CF_3COO^- + H_3C CH_3 $\overset{+}{O}H$ R H]

departure of leaving group

CF_3COOH + $(H_3C)_2C{=}CH_2$ ← removal of proton by base [Base H H_3C CH_2 C+ H_3C + OH O=C N R H]

This reaction proceeds by an E1 mechanism.

A Look Ahead

11.67

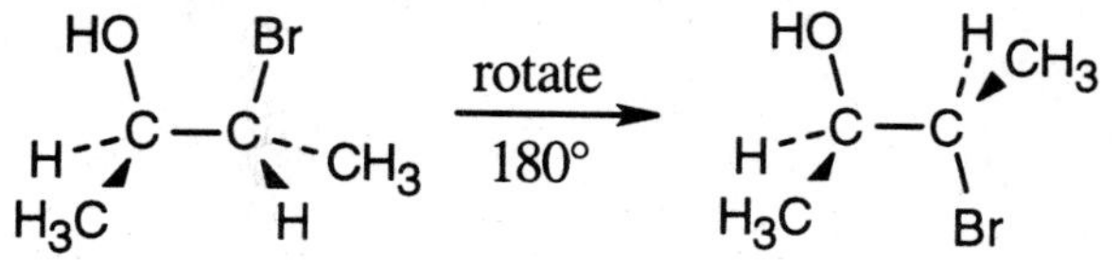

HO Br H C C CH_3 H_3C H → rotate 180° → HO H CH_3 H C C H_3C Br

Rotate around the C–C bond so that –OH and –Br are 180° apart.

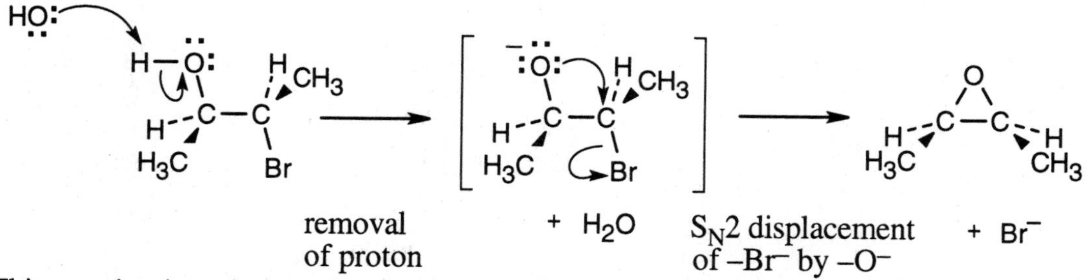

HO^- H–O: H CH_3 H C C H_3C Br → [$^-$:O: H CH_3 H C C H_3C Br] → O H C C H H_3C CH_3

removal of proton + H_2O S_N2 displacement of $-Br^-$ by $-O^-$ + Br^-

This reaction is an intramolecular S_N2 displacement.

11.68

$$\text{H}_3\text{C}(\text{H})\text{C}{=}\text{C}(\text{H})\text{CH}_3 \xrightarrow[\text{H}_2\text{O}]{\text{Br}_2} \text{H}_3\text{C}(\text{H})(\text{HO})\text{C}{-}\text{C}(\text{Br})(\text{H})\text{CH}_3 \xrightarrow[\text{Ethanol}]{\text{NaOH}} \text{epoxide} + \text{Br}^-$$

11.69

$$CH_3CH_2CH_2CH_2CH_2NH_2 \xrightarrow{\text{excess } CH_3I} CH_3CH_2CH_2CH_2CH_2\overset{+}{N}(CH_3)_3 \ I^-$$

$$CH_3CH_2CH_2CH_2CH_2\overset{+}{N}(CH_3)_3 \xrightarrow{Ag_2O,\ H_2O} CH_3CH_2CH_2CH{=}CH_2 + :N(CH_3)_3$$

The intermediate is a charged quaternary ammonium compound that results from S_N2 substitutions on three CH_3I molecules by the amine nitrogen. E2 elimination occurs because the neutral $N(CH_3)_3$ molecule is a good leaving group.

Molecular Modeling

11.70 The transition state in the S_N2 reaction of Cl^- and CH_3Br has a Cl–C–Br bond angle of 180° and is closest to ideal. In the other reactions with Cl^-, the bond angle changes from 164° (CH_3CH_2Br) to 158° ($(CH_3)_2CHBr$) to 147° ($(CH_3)_3CCH_2Br$). These distortions of bond angle are due to steric repulsions between methyl hydrogens and Cl and Br. The reacting atoms in the transition state of the reaction of Cl^- with $(CH_3)_3Br$ have a linear relationship, but the C–Cl and C–Br distances are much longer than in the transition state of the reaction of Cl^- with CH_3Br.

11.71 Water is the only one of these solvents that has a positive atom (H) that can cluster around the negatively charged nucleophile and solvate it, lowering its energy and slowing the rate of reaction. All three solvents have negative atoms that can solvate K^+. Both DMSO and acetonitrile should promote rapid reaction because they solvate only cations, leaving the nucleophilic anion unsolvated and reactive.

11.72 The reactive conformer places bromine in an axial orientation. In the cis isomer, two hydrogens are anti to the axial bromine, and elimination yields two cycloalkenes. The trans isomer has only one hydrogen anti to the axial bromine, and elimination produces only one cycloalkene. The lower energy cis conformer is also the reactive conformer, but the lower energy trans conformer (equatorial bromine) is not reactive. Thus, it is expected that the cis isomer will undergo elimination more rapidly.

cis isomer:
reactive conformer (lower energy)

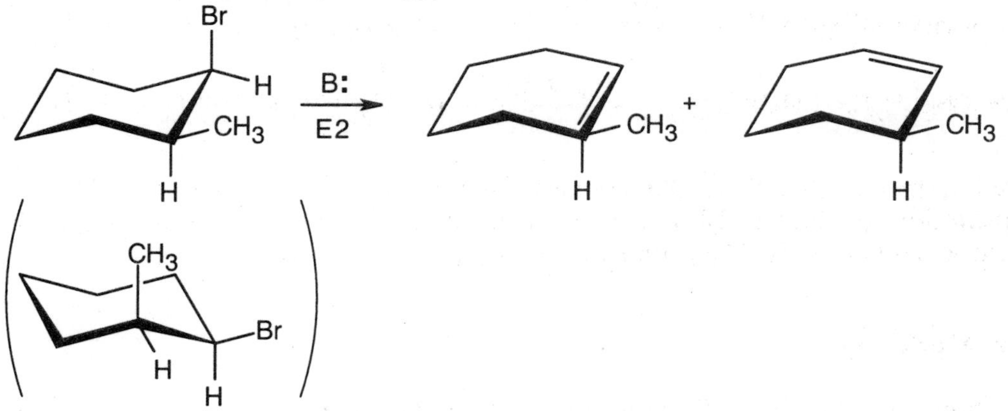

trans isomer (one enantiomer):
reactive conformer (higher energy)

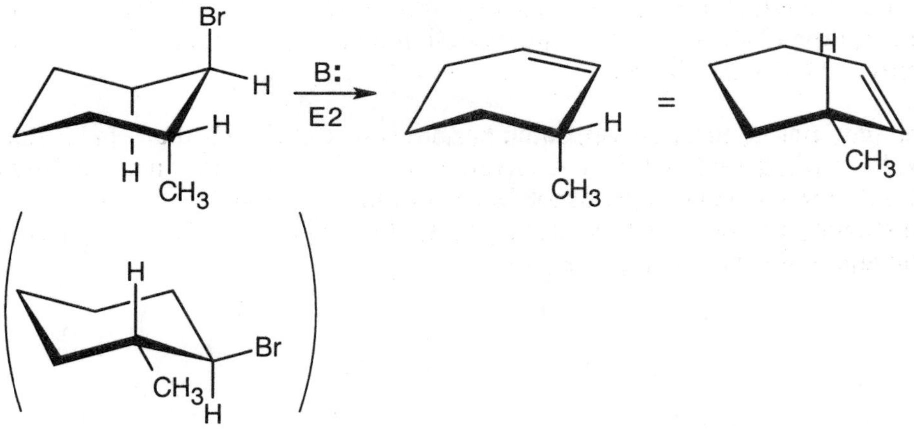

Review Unit 4: Stereochemistry; Alkyl Halides; Substitutions and Eliminations

Major Topics Covered (with vocabulary):

Handedness:
stereoisomer enantiomer chiral plane of symmetry achiral chirality center plane-polarized light optically active levorotatory specific rotation

Stereoisomers and configuration:
configuration absolute configuration diastereomer meso compound racemate resolution Fischer projection

Stereochemistry of reactions.

Alkyl halides:
allylic position delocalization Grignard reagent Gilman reagent

Oxidation and reduction in organic chemistry:
oxidation reduction

Substitution reactions:
nucleophilic substitution reaction Walden inversion reaction rate kinetics second-order reaction rate constant S_N2 reaction nucleophilicity leaving group solvation S_N1 reaction first-order reaction rate-limiting step ion pair dielectric polarization

Elimination reactions:
Zaitsev's rule E2 reaction syn periplanar geometry anti periplanar geometry deuterium isotope effect E1 reaction

Types of Problems:

After studying these chapters, you should be able to:

- Calculate the specific rotation of an optically active compound.
- Locate chirality centers, assign priorities to substituents, and assign *R,S* designations to chirality centers.
- Given a stereoisomer, draw its enantiomer and/or diastereomers.
- Locate the symmetry plane of a meso compound.
- Manipulate Fischer projections to see if they are identical.
- Assign *R,S* designations to Fischer projections.
- Predict the stereochemistry of reaction products.

- Draw, name, and synthesize alkyl halides.
- Understand the mechanism of radical halogenation and the stability order of radicals.
- Prepare Grignard reagents and dialkylcopper reagents and use them in synthesis.
- Predict the oxidation level of a compound.

- Formulate the mechanisms of S_N2, S_N1 and elimination reactions.
- Predict the effect of substrate, nucleophile, leaving group and solvent on substitution and elimination reactions.
- Predict the products of substitution and elimination reactions.
- Classify substitution and elimination reactions by type.

Points to Remember:

* A helpful strategy for assigning *R,S* designations: Using models, build two enantiomers by adding four groups to each of two tetrahedral carbons. Number the groups 1–4, to represent priorities of groups at a tetrahedral carbon, and assign a configuration to each carbon. Attach a label that indicates configuration to each enantiomer. Keep these two enantiomers, and use them to check your answer every time that you need to assign *R,S* configurations to a chiral atom.

* In naming alkyl halides by the IUPAC system, remember that a halogen is named as a substituent on an alkane. When numbering the alkyl halide, the halogens are numbered in the same way as alkyl groups and are cited alphabetically.

* The definition of oxidation and reduction given in Chapter 10 expands the concept to reactions that you might not have considered to be oxidations or reductions. As you learn new reactions, try to classify them as oxidations, reductions or neither.

* Predicting the outcome of substitutions and eliminations is only straightforward in certain cases. For primary halides, S_N2 and E2 reactions are predicted. For tertiary halides, S_N1, E2 and E1(to a certain extent) are the choices. The possibilities for secondary halides are more complicated. In addition, many reactions yield both substitution and elimination products, and both inversion and retention of configuration may occur in the same reaction.

Self-Test:

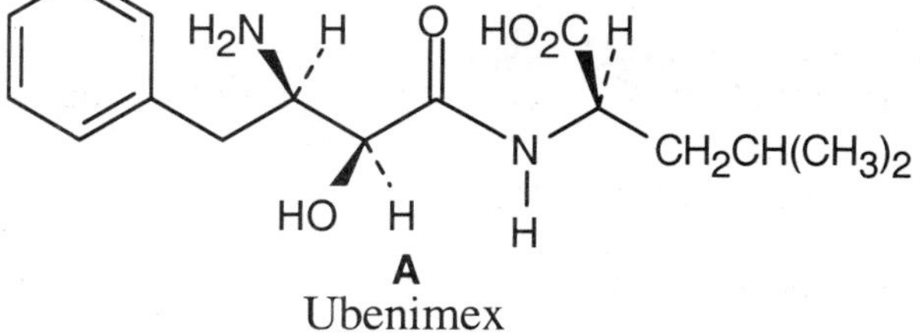

A
Ubenimex
(an antitumor drug)

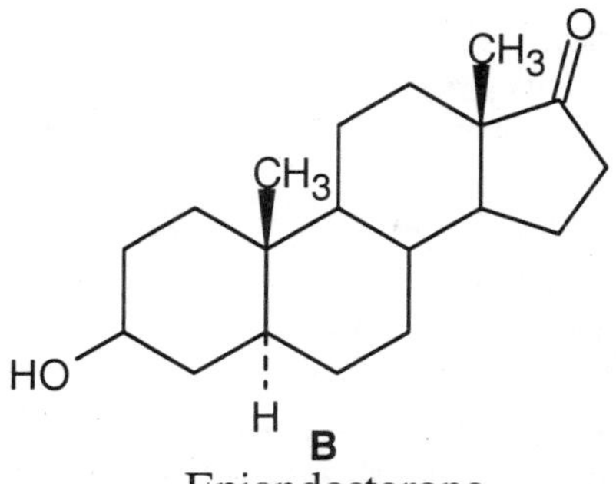

B
Epiandosterone
(an androgen)

Assign *R,S* designations to the chiral carbons in **A**. Indicate the chirality centers in **B**. How many possible stereoisomers of **B** are there?

$$\begin{array}{c} CH_3 \\ | \\ CH_3CH_2CHCHCH_3 \\ | \\ Br \end{array}$$

C

$$\begin{array}{c} CH_2CH_3 \\ | \\ HC{\equiv}CCCH{=}CHCl \\ | \\ OH \end{array}$$

D

Ethchlorvynol
(a sedative)

Name **C**. Draw all stereoisomers of **C**, label them, and describe their relationship. Predict the products of reaction of **C** with:(a) NaOH; (b) Mg, then H_2O; (c) product of (b) + Br_2, $h\nu$ (show the major product); (d) $(CH_3CH_2)_2CuLi$.

Draw the *R* enantiomer of **D**. Predict the products of reaction of **D** with: (a) HBr; (b) product of (a) + aqueous ethanol. Describe the reactivity of the –Cl atom in substitution and elimination reactions.

Multiple Choice:

1. A meso compound and a racemate are identical in all respects except:
(a) molecular formula (b) degree of rotation of plane-polarized light
(c) connectivity of atoms (d) physical properties

2. Which of the following Fischer projections represents an *R* enantiomer?

3. In the reaction of (2*R*,3*S*)-3-methyl-2-pentanol with tosyl chloride, what is the configuration of the product?
(a) a mixture of all four possible stereoisomers (b) (2*R*,3*S*) and (2*S*,3*S*) (c) (2*R*,3*S*)
(d) (2*S*,3*S*)

4. Monochlorination of 2,3-dimethylbutane yields what percent of 2-chloro-2,3-dimethylbutane?
(a) 16% (b) 35% (c) 45% (d) 55%

5. How many monobromination products can be formed by NBS bromination of 2-ethyl-1-pentene? Include double-bond isomers.
(a) 3 (b) 4 (c) 5 (d) 6

6. Which of the following reactions is an oxidation?
(a) hydroxylation (b) hydration (c) hydrogenation (d) addition of HBr

7. All of the following are true of S_N2 reactions except:
(a) The rate varies with the concentration of nucleophile (b) The rate varies with the type of nucleophile (c) The nucleophile is involved in the rate-determining step (d) The rate of the S_N2 reaction of a substrate and a nucleophile is the same as the rate of the E2 reaction of the same two compounds.

8. Which of the following is true of S_N1 reactions?
(a) The rate varies with the concentration of nucleophile (b) The rate varies with the type of nucleophile (c) The rate is increased by use of a polar solvent. (d) The nucleophile is involved in the rate-determining step.

9. Which base is best for converting 1-bromohexane to 1-hexene?
(a) $(CH_3)_3CO^-$ (b) ^-CN (c) ^-OH (d) $^-C{\equiv}CH$

10. Which of the following is both a good nucleophile and a good leaving group?
(a) ^-OH (b) ^-CN (c) ^-Cl (d) ^-I

Chapter 12 – Structure Determination. Mass Spectrometry and Infrared Spectroscopy

Chapter Outline

I. Mass Spectrometry (Sections 12.1 – 12.4).
  A. General features of mass spectrometry (Section 12.1).
    1. Purpose of mass spectrometry.
      a. Mass spectrometry is used to measure the molecular weight of a compound.
      b. Mass spectrometry can also provide information on the structure of an unknown compound.
    2. Technique of mass spectrometry.
      a. A small amount of vaporized sample is bombarded by a stream of high-energy electrons.
      b. An electron is dislodged, producing a cation radical.
      c. Most of the cation radicals fragment; the fragments may be positively charged or neutral.
      d. A strong magnetic field deflects the positively charged fragments, which are separated by *m/z* ratio.
      e. A detector records the fragments as peaks.
    3. Important terms.
      a. The mass spectrum is presented as a bar graph, with masses (*m/z*) on the *x* axis and intensity on the *y* axis.
      b. The base peak is the tallest peak and is assigned an intensity of 100%.
      c. The parent peak, or molecular ion (M^+), corresponds to the unfragmented cation radical.
        In large molecules, the base peak is often not the molecular ion.
  B. Interpreting mass spectra (Sections 12.2 – 12.4).
    1. Molecular weight (Section 12.2).
      a. Mass spectra can frequently provide the molecular weight of a sample.
        i. Double-focusing mass spectrometers can provide mass measurements accurate to 0.0001 amu.
        ii. Some samples fragment so easily that M^+ is not seen.
      b. If you know the molecular weight of the sample, you can often deduce its molecular formula.
      c. There is often a peak at M+1 that is due to contributions from ^{13}C and ^2H.
    2. Fragmentation patterns of hydrocarbons (Section 12.3).
      a. Fragmentation patterns can be used to identify a known compound, because a given compound has a unique fragmentation "fingerprint".
      b. Fragmentation patterns can also provide structural information.
        i. Most hydrocarbons fragment into carbocations and radicals.
        ii. The positive charge remains with the fragment most able to stabilize it.
        iii. It is often difficult to assign structures to fragments.
        iv. For hexane, major fragments correspond to the loss of methyl, ethyl, propyl, and butyl radicals.
    3. Fragmentation patterns of common functional groups (Section 12.4).
      a. Alcohols.
        i. Alcohols can fragment by alpha cleavage, in which a C–C bond next to the –OH group is broken.
          The products are a cation and a radical.
        ii. Alcohols can also dehydrate, leaving an alkene cation radical with a mass 18 units less than M^+.

b. Amines also undergo alpha cleavage, forming a cation and a radical.
c. Carbonyl compounds.
   i. Aldehydes and ketones with a hydrogen 3 carbons from the carbonyl group can undergo the McLafferty rearrangement.
      The products are a cation radical and a neutral alkene.
   ii. Aldehydes and ketones also undergo alpha cleavage, which breaks a bond between the carbonyl group and a neighboring carbon.
      The products are a cation and a radical.

II. Spectroscopy and the electromagnetic spectrum (Section 12.5).
  A. The nature of radiant energy.
    1. The different types of electromagnetic radiation make up the electromagnetic spectrum.
    2. Electromagnetic radiation behaves both as a particle and as a wave.
    3. Electromagnetic radiation can be characterized by three variables.
       a. The wavelength (λ) measures the distance from one maximum to the next.
       b. The frequency (ν) measures the number of wave maxima that pass a fixed point per unit time.
       c. The amplitude is the height measured from the midpoint to the maximum.
    4. Wavelength times frequency equals the speed of light.
    5. Electromagnetic energy is transmitted in discrete energy bundles called quanta.
       a. $\varepsilon = h \times \nu$
       b. Energy varies directly with frequency but inversely with wavelength.
       c. $E = 1.20 \times 10^{-2}$ kJ/mol $\div\ \lambda$ (cm) for a "mole" of photons.
  B. Electromagnetic radiation and organic molecules.
    1. When an organic compound is struck by a beam of electromagnetic radiation, it absorbs radiation of certain wavelengths, and transmits radiation of other wavelengths.
    2. If we determine which wavelengths are absorbed and which are transmitted, we can obtain an absorption spectrum of the compound.
       For an infrared spectrum:
       i. The horizontal axis records wavelength.
       ii. The vertical axis records percent transmittance.
       iii. The baseline runs across the top of the spectrum.
       iv. Energy absorption is a downward spike.
    3. The energy a molecule absorbs is distributed over the molecule.
    4. There are many types of spectroscopies that differ in the region of the electromagnetic spectrum that is being used.

III. Infrared Spectroscopy (Sections 12.6 – 12.9).
  A. Infrared radiation (Section 12.6).
    1. The infrared (IR) region of the electromagnetic spectrum extends from 7.8×10^{-7} m to 10^{-4} m.
       a. Organic chemists use the region from 2.5×10^{-6} m to 2.5×10^{-5} m.
       b. Wavelength is usually given in μm, and frequency is expressed in wavenumber, which is the reciprocal of wavelength.
       c. The useful range of IR radiation is 4000 cm^{-1} – 400 cm^{-1}; this corresponds to energies of 48.0 kJ/mol – 4.80 kJ/mol.
    2. IR radiation causes bonds to stretch and bend and causes other molecular vibrations.
    3. Energy is absorbed at a specific frequency that corresponds to the frequency of the vibrational motion of a bond.
    4. If we measure the frequencies at which IR energy is absorbed, we can find out the kinds of bonds a compound contains and identify functional groups.

B. Interpreting IR spectra (Sections 12.7 – 12.9).
1. General principles (Section 12.7).
a. Most molecules have very complex IR spectra.
i. This complexity means that each molecule has a unique fingerprint that allows it to be identified by IR spectroscopy.
ii. Complexity also means that not all absorptions can be identified.
b. Most functional groups have characteristic IR absorption bands that don't change from one compound to another.
c. The significant regions of IR absorptions :
i. 4000 cm^{-1}–2500 cm^{-1} corresponds to absorptions by C–H, O–H, and N–H bonds.
ii. 2500 cm^{-1}–2000 cm^{-1} corresponds to triple-bond stretches.
iii. 2000 cm^{-1}–1500 cm^{-1} corresponds to double bond stretches.
iv. The region below 1500 cm^{-1} is the fingerprint region, where many complex bond vibrations occur.
d. The frequency of absorption of different bonds depends on two factors:
i. The strength of the bond.
ii. The difference in mass between the two atoms in the bond.
2. Interpreting IR spectra of hydrocarbons (Section 12.8).
a. Alkanes.
i. C–C absorbs at 800–1300 cm^{-1}.
ii C–H absorbs at 2850–2960 cm^{-1}.
b. Alkenes.
i. =C–H absorbs at 3020–3100 cm^{-1}.
ii. C=C absorbs at 1650–1670 cm^{-1}.
iii. $RCH=CH_2$ absorbs at 910 and 990 cm^{-1}.
iv. $R_2C=CH_2$ absorbs at 890 cm^{-1}.
c. Alkynes.
i. –C≡C– absorbs at 2100–2260 cm^{-1}.
ii. ≡C–H absorbs at 3300 cm^{-1}.
3. Interpreting IR spectra of some other functional groups.
a. The alcohol O–H bond absorbs at 3400–3650 cm^{-1}.
b. The N–H bond of amines absorbs at 3300–3500 cm^{-1}.
c. Aromatic compounds.
i. =C–H absorbs at 3030 cm^{-1}.
ii. Ring absorptions occur at 1660–2000 cm^{-1} and at 1450–1600 cm^{-1}.
d. Carbonyl compounds.
i. Saturated aldehydes absorb at 1730 cm^{-1}; unsaturated aldehydes absorb at 1705 cm^{-1}.
ii. Saturated ketones absorb at 1715 cm^{-1}; unsaturated ketones absorb at 1690 cm^{-1}.
iii. Saturated esters absorb at 1735 cm^{-1}; unsaturated esters absorb at 1715 cm^{-1}.

Solutions to Problems

12.1 The following systematic approach may be helpful.

(a) $M^+ = 86$

1. Find the compound of molecular weight 86 that contains only C and H. In Practice Problem 12.1, a useful technique is described. First, divide the molecular weight of the compound by 12 (the molecular weight of carbon). $86 \div 12 = 7$ (remainder 2). The simplest hydrocarbon formula is C_7H_2.
2. Replace one carbon with 12 hydrogens to generate the next formula, C_6H_{14}. Remember that a hydrocarbon with n carbon atoms can have no more than $2n + 2$ hydrogen atoms. We have thus listed all hydrocarbon formulas.
3. Find the formula corresponding to $M^+ = 86$ that contains carbon, hydrogen, and one oxygen atom. If one oxygen atom (atomic weight = 16) is added to the base formula from step 1, one carbon atom (atomic weight 12) and four hydrogen atoms (atomic weight 4) must be removed.Since the formula in Step 1, however, has only 2 hydrogens, we must apply this technique to the formula in Step 2.The resulting formula is $C_5H_{10}O$.
4. Proceed to find the remaining molecular formulas. Each time one oxygen is added, one carbon and four hydrogens must be removed. The remaining formulas for $M^+ = 86$ are $C_4H_6O_2$ and $C_3H_2O_3$.

(b) $M^+ = 128$. The procedure is the same as in part (a). Two hydrocarbons having $M^+ = 128$ are C_9H_{20} and $C_{10}H_8$. The formulas containing one oxygen are $C_8H_{16}O$ and C_9H_4O. The remaining formulas are $C_7H_{12}O_2$, $C_6H_8O_3$, $C_5H_4O_4$.

(c) $M^+ = 156$. Possible formulas are $C_{11}H_{24}$, $C_{12}H_{12}$, $C_{11}H_8O$, $C_{10}H_{20}O$, $C_{10}H_4O_2$, $C_9H_{16}O_2$, $C_8H_{12}O_3$, $C_7H_8O_4$, $C_6H_4O_5$.

12.2 Use the method described in Problem 12.1(a). The hydrocarbons having $M^+ = 218$ are $C_{16}H_{26}$ and $C_{17}H_{14}$ ($C_{18}H_2$ is also possible, but unlikely). Since nootkatone also contains oxygen, we must consider only those formulas that include oxygen. Using the previous procedure, we can determine that $C_{15}H_{22}O$, $C_{14}H_{18}O_2$, $C_{13}H_{14}O_3$, $C_{12}H_{10}O_4$, $C_{11}H_6O_5$, $C_{16}H_{10}O$ and $C_{15}H_6O_2$ are possible formulas for nootkatone. The actual formula of nootkatone is $C_{15}H_{22}O$.

12.3

$CH_3CH_2CH{=}C(CH_3)_2$ — 2-Methyl-2-pentene

$CH_3CH_2CH_2CH{=}CHCH_3$ — 2-Hexene

Fragmentation occurs to a greater extent at the weakest carbon-carbon bonds, and the positive charge remains with the fragment that is more able to stabilize it. A table of bond-dissociation energies (Table 5.4) shows that allylic bonds have lower bond-dissociation energies than the other bonds in these two compounds. Thus, the principal fragmentations of these compounds yield allylic cations.

$^+CH_2CH{=}C(CH_3)_2$ — $m/z = 69$

$^+CH_2CH{=}CHCH_3$ — $m/z = 55$

Spectrum (b), which has $m/z = 55$ as its base peak, corresponds to 2–hexene. Spectrum (a), which has an abundant peak at $m/z = 69$, corresponds to 2–methyl–2–pentene.

12.4 In a mass spectrum, the molecular ion is both a cation and a radical. When it fragments, two kinds of cleavage can occur. (1) Cleavage can form a radical and a cation (the species observed in the mass spectrum). Alpha cleavage shows this type of pattern. (2) Cleavage can form a neutral molecule and a different radical cation (the species observed in the mass spectrum). Alcohol dehydration and the McLafferty rearrangement show this cleavage pattern.

(a)

$$[H_3C{-}C(=O){-}CH_2CH_2CH_3]^{+\cdot} \xrightarrow[\text{cleavage}]{\text{Alpha}} [H_3C{-}C{=}O]^{+} + \cdot CH_2CH_2CH_3$$

$M^+ = 86$ → $m/z = 43$

$$[H_3C{-}C(=O){-}CH_2CH_2CH_3]^{+\cdot} \xrightarrow[\text{cleavage}]{\text{Alpha}} H_3C\cdot + [O{=}C{-}CH_2CH_2CH_3]^{+}$$

$M^+ = 86$ → $m/z = 71$

In theory, alpha cleavage can take place on either side of the carbonyl group, to produce the cations with $m/z = 43$ and $m/z = 71$. In practice, cleavage occurs on the more substituted side of the carbonyl group, and the first cation, with $m/z = 43$ is observed.

(b)

$$[\text{cyclohexanol}]^{+\cdot} \xrightarrow{\text{Dehydration}} [\text{cyclohexene}]^{+\cdot} + H_2O$$

$M^+ = 100$ → $m/z = 82$

Dehydration of cyclohexanol produces a cation radical with $m/z = 82$.

(c)

$$[CH_3CH(CH_3)CH_2C(=O)CH_3]^{+\cdot} \xrightarrow[\text{rearrangement}]{\text{McLafferty}} [\ldots]^{+\cdot} \rightarrow [H_2C{=}C(OH){-}CH_3]^{+\cdot} + CH_2{=}C(H){-}CH_3$$

$M^+ = 100$ → $m/z = 58$

The cation radical fragment resulting from McLafferty rearrangement has $m/z = 58$.

(d)

$$[H_3C{-}CH_2{-}N(CH_2CH_3){-}CH_2CH_3]^{+\cdot} \xrightarrow[\text{cleavage}]{\text{Alpha}} H_3C\cdot + [CH_2{-}N(CH_2CH_3){-}CH_2CH_3]^{+}$$

$M^+ = 101$ → $m/z = 86$

Alpha cleavage of the above amine yields a cation with $m/z = 86$.

12.5 At first glance, we know that: (1) energy increases as wavelength decreases, and (2) the wavelength of X-radiation is smaller than the wavelength of infrared radiation. Thus, we estimate that an X ray is of higher energy than an infrared ray.
An exact solution:

$E = h\nu = hc/\lambda$; $h = 6.62 \times 10^{-34}$ J·s; $c = 3.00 \times 10^{8}$ m/s

for $\lambda = 10^{-6}$ m (infrared radiation):

$$E = \frac{(6.62 \times 10^{-34}\ \text{J·s})(3.00 \times 10^{8}\ \text{m/s})}{1.0 \times 10^{-6}\ \text{m}} = 2.0 \times 10^{-19}\ \text{J}$$

for $\lambda = 3.0 \times 10^{-9}$ m (X radiation):

$$E = \frac{(6.62 \times 10^{-34}\ \text{J·s})(3.00 \times 10^{8}\ \text{m/s})}{3.0 \times 10^{-9}\ \text{m}} = 6.6 \times 10^{-17}\ \text{J}$$

Confirming our estimate, the calculation shows that an X ray is of higher energy than infrared radiation.

12.6 First, convert radiation in m to radiation in Hz by the equation:

$$\nu = \frac{c}{\lambda} = \frac{3.00 \times 10^{8}\ \text{m/s}}{9.0 \times 10^{-6}\ \text{m}} = 3.3 \times 10^{13}\ \text{Hz}$$

The equation $E = h\nu$ shows that the greater the value of ν, the greater the energy. Thus, radiation with $\nu = 3.3 \times 10^{13}$ Hz ($\lambda = 9.0 \times 10^{-6}$ m) is higher in energy than radiation with $\nu = 4.0 \times 10^{9}$ Hz.

12.7

(a) $$E = \frac{1.20 \times 10^{-4}\ \text{kJ/mol}}{\lambda\ (\text{in m})} = \frac{1.20 \times 10^{-4}\text{kJ/mol}}{5.0 \times 10^{-11}}$$
$= 2.4 \times 10^{6}$ kJ/mol for a gamma ray.

(b) $E = 4.0 \times 10^{4}$ kJ/mol for an X ray.

(c) $$\nu = \frac{c}{\lambda};\ \lambda = \frac{c}{\nu} = \frac{3.0 \times 10^{8}\text{m/s}}{6.0 \times 10^{15}\text{Hz}} = 5.0 \times 10^{-8}\text{m}$$

$$E = \frac{1.20 \times 10^{-4}\text{kJ/mol}}{5.0 \times 10^{-8}} = 2.4 \times 10^{3}\ \text{kJ/mol for ultraviolet light}$$

(d) $E = 2.8 \times 10^{2}$ kJ/mol for visible light.

(e) $E = 6.0$ kJ/mol for infrared radiation

(f) $E = 4.0 \times 10^{-2}$ kJ/mol for microwave radiation.

12.8 The wavenumber is the reciprocal of the wavelength, which is expressed in centimeters. The conversion 1 μm = 10^{-4} cm is useful.

(a) $3.10\ \mu m = 3.10 \times 10^{-4}$ cm ; $\dfrac{1}{3.10 \times 10^{-4}\ cm} = 3225\ cm^{-1}$

(b) 5.85 μm ; 1710 cm^{-1}

(c) $\dfrac{1}{2250\ cm^{-1}} = 4.44 \times 10^{-4}\ cm = 4.44 \mu m$

(d) 970 cm^{-1} ; 10.3 μm

12.9 (a) A compound with a strong absorption at 1710 cm^{-1} contains a carbonyl group and is either a ketone or aldehyde.

(b) A compound with a nitro group has a strong absorption at 1540 cm^{-1}.

(c) A compound showing both carbonyl (1720 cm^{-1}) and –OH (2500-3000 cm^{-1} broad) absorptions is a carboxylic acid.

12.10 To use IR spectroscopy to distinguish between isomers, find a strong IR absorption that is present in one isomer but absent in the other.

(a)

CH_3CH_2OH
Strong hydroxyl band
at 3400 – 3640 cm^{-1}

CH_3OCH_3
No band in the region
3400 – 3640 cm^{-1}

(b)

$CH_3CH_2CH_2CH_2CH{=}CH_2$
Alkene bands at
3020–3100 cm^{-1} and
at 1650–1670^{-1}.

No bands in alkene region.

(c)

CH_3CH_2COOH
Strong, broad band
at 2500–3100 cm^{-1}

$HOCH_2CH_2CHO$
Strong band at
3400–3640 cm^{-1}

12.11 Based on what we know at this point, we can identify four absorptions in this spectrum.

(a) Absorptions in the region 1450 cm^{-1} – 1600 cm^{-1} are due to aromatic ring –C=C– motions.
(b) The absorption at 2100 cm^{-1} is due to a –C≡C– stretch.
(c) Absorptions in the range 3000 cm^{-1} – 3100 cm^{-1} are due to aromatic ring =C–H stretches.
(d) The absorption at 3300 cm^{-1} is due to a ≡C–H stretch.

12.12 (a) An ester next to a double bond absorbs at 1715 cm^{-1}. The alkene double bond absorbs at 1640–1680 cm^{-1} and at 3020–3100 cm^{-1}.

(b) The aldehyde carbonyl group absorbs at 1730 cm^{-1}. The alkyne C≡C bond absorbs at 2100–2260 cm^{-1}, and the alkyne H–C bond absorbs at 3300 cm^{-1}.

(c) The most important absorptions for this compound are due to the alcohol group (a broad, intense band at 3400–3650 cm^{-1}) and to the carboxylic acid group, which has a C=O absorption in the range 1710 –1760 cm^{-1} and a broad O–H absorption in the range 2500–3100 cm^{-1}. Absorptions due to the aromatic ring [3030 cm^{-1} (w) and 1450–1600 cm^{-1} (m)] may also be seen.

12.13

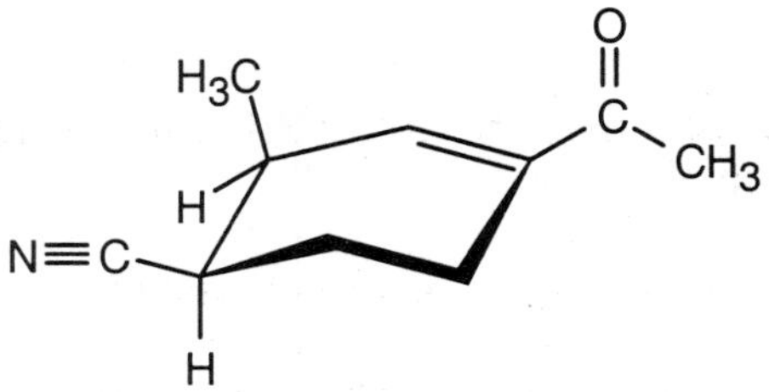

The compound contains nitrile and ketone groups, as well as a carbon-carbon double bond. The nitrile absorption occurs at 2210–2260 cm^{-1}. The ketone shows an absorption at 1690 cm^{-1}; this value is lower than the usual value because the ketone is next to the double bond. The double bond absorptions occur at 3020–3100 cm^{-1} and at 1640–1680 cm^{-1}.

Visualizing Chemistry

12.14

Compound	*Significant IR Absorption*	*Due to:*
(a)	1540 cm^{-1}	nitro group (1)
	1730 cm^{-1}	aldehyde (2)
	3030 cm^{-1},	aromatic ring C–H(3)
	1450–1600 cm^{-1}	aromatic ring C=C(3)
(b)	1735 cm^{-1}	ester (1)
	3020–3100 cm^{-1}	vinylic stretch C–H(2)
	910 cm^{-1}, 990 cm^{-1}	$C=CH_2$ bend(3)
(c)	1715 cm^{-1}	ketone (1)
	3400–3650 cm^{-1}	alcohol (2)

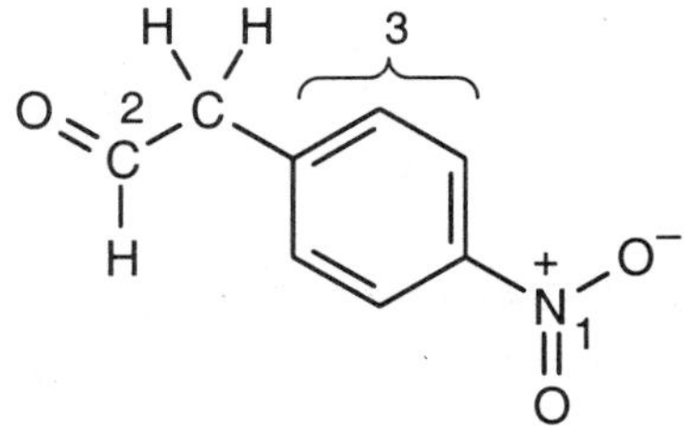

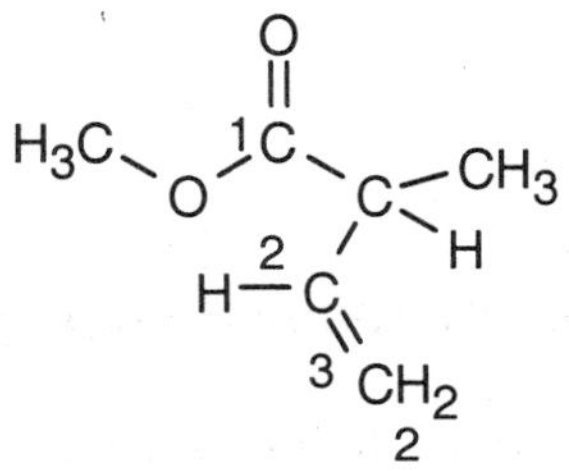

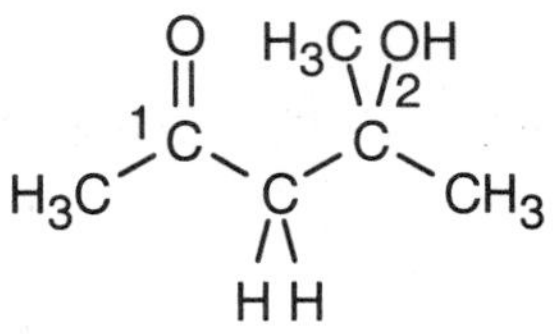

12.15 (a) The mass spectrum of this ketone shows fragments resulting from both McLafferty rearrangement and alpha cleavage.
McLafferty rearrangement:

$$\left[CH_3CH_2\underset{\displaystyle CH_3}{\underset{|}{C}H}\overset{\displaystyle O}{\overset{\|}{C}}CH_2CH_3\right]^{+\cdot} \xrightarrow[\text{rearrangement}]{\text{McLafferty}} [\,\cdots\,]^{+\cdot} \longrightarrow H_2C{=}CH_2 + \left[H_3C{-}CH{=}C(OH){-}CH_2CH_3\right]^{+\cdot}$$

$M^+ = 114$ $\qquad$ $m/z = 86$

Alpha cleavage:

$$\left[CH_3CH_2\underset{\displaystyle CH_3}{\underset{|}{C}H}{-}\overset{\displaystyle O}{\overset{\|}{C}}{+}CH_2CH_3\right]^{+\cdot} \xrightarrow[\text{cleavage}]{\text{Alpha}} \left[CH_3CH_2\underset{\displaystyle CH_3}{\underset{|}{C}H}{-}\overset{\displaystyle O}{\overset{\|}{C}}\right]^{+} + \cdot CH_2CH_3$$

$M^+ = 114$ $\qquad$ $m/z = 85$

$$\left[CH_3CH_2\underset{\displaystyle CH_3}{\underset{|}{C}H}{+}\overset{\displaystyle O}{\overset{\|}{C}}{-}CH_2CH_3\right]^{+\cdot} \xrightarrow[\text{cleavage}]{\text{Alpha}} CH_3CH_2\underset{\displaystyle CH_3}{\underset{|}{C}H^{\cdot}} + \left[\overset{\displaystyle O}{\overset{\|}{C}}{-}CH_2CH_3\right]^{+}$$

$M^+ = 114$ $\qquad$ $m/z = 57$

(b) Two different fragments can arise from alpha cleavage of this amine:

$$\left[\text{pyrrolidinyl-}N{-}CH(CH_3){-}CH_3\right]^{+\cdot} \xrightarrow[\text{cleavage}]{\text{Alpha}} CH_3^{\cdot} + \left[\text{pyrrolidinyl-}N{-}CH(CH_3)\right]^{+}$$

$M^+ = 113$ $\qquad$ $m/z = 98$

$$\left[\text{pyrrolidinyl-}N{-}CH(CH_3)_2\right]^{+\cdot} \xrightarrow[\text{cleavage}]{\text{Alpha}} \left[CH_2CH_2CH_2{-}N(=CH_2){-}CH(CH_3)_2\right]^{+\cdot}$$

$M^+ = 113$ $\qquad$ $m/z = 113$

The second product results from cleavage of a bond in the five-membered ring.

Additional Problems

12.16

M^+	Molecular Formula	Degree of Unsaturation
(a) 86	C_6H_{14}	0
(b) 110	C_8H_{14}	2
(c) 146	$C_{11}H_{14}$	5
(d) 190	$C_{14}H_{22}$	4
	$C_{15}H_{10}$	11

12.17

M^+	Molecular Formula	Degree of Unsaturation	Possible Structure
(a) 132	$C_{10}H_{12}$	5	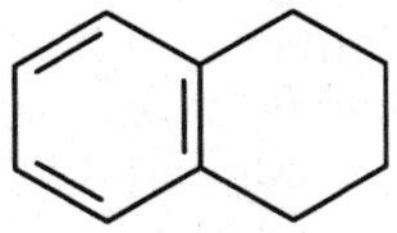
(b) 166	$C_{13}H_{10}$	9	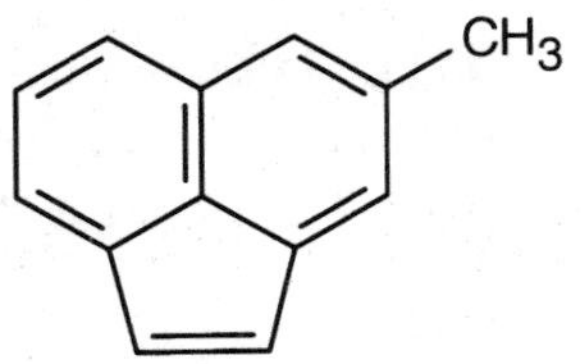 CH_3
	$C_{12}H_{22}$	2	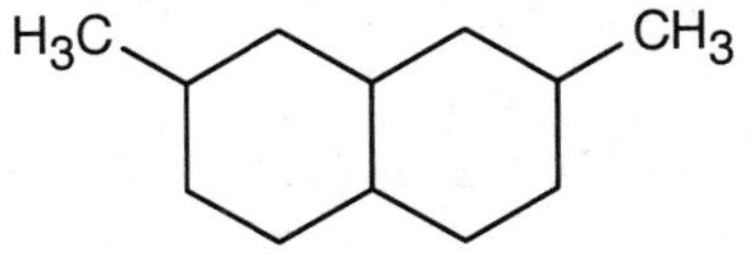 H_3C CH_3
(c) 84	C_6H_{12}	1	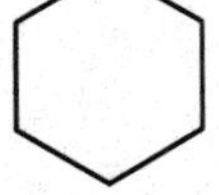

12.18 Remember that compounds in this problem may contain carbon, hydrogen, oxygen, and nitrogen. In addition, the molecular ions of many compounds may have the same value of M^+. Some of the less likely molecular formulas -- those with few carbon or hydrogen atoms -- have been omitted.

a) $M^+ = 74$. Any nitrogen-containing compound that has a molecular ion at $M^+ = 74$ must have an even number of nitrogen atoms.
Compounds containing:
C, H; C_6H_2
C, H, O; $C_4H_{10}O$, $C_3H_6O_2$, $C_2H_2O_3$
C, H, N; $C_3H_{10}N_2$, CH_6N_4
C, H, N, O; $C_2H_6N_2O$, $CH_2N_2O_2$

b) $M^+ = 131$ has an odd number of nitrogen atoms; no hydrocarbons correspond to this molecular ion.

C, H, N; C_9H_9N, $C_7H_5N_3$, $C_6H_{17}N_3$, $C_4H_{13}N_5$

C, H, N, O; $C_7H_{17}NO$, C_8H_5NO, $C_6H_{13}NO_2$, $C_5H_9NO_3$, $C_4H_5NO_4$, $C_5H_{13}N_3O$, $C_4H_9N_3O_2$, $C_3H_5N_3O_3$, $C_3H_9N_5O$

12.19 Reasonable molecular formulas for camphor are $C_{10}H_{16}O$, $C_9H_{12}O_2$, $C_8H_8O_3$. The actual formula, $C_{10}H_{16}O$, corresponds to three degrees of unsaturation. The ketone functional group accounts for one of these. Since camphor is a saturated compound, the other two degrees of unsaturation are due to two rings.

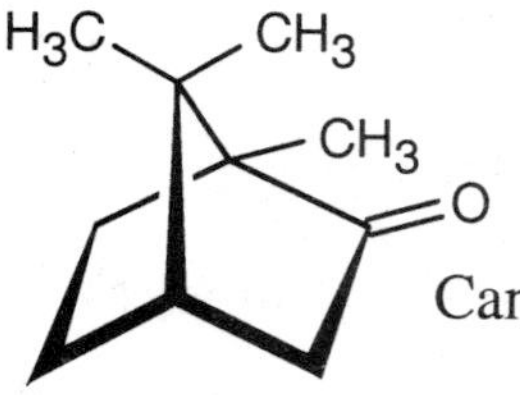

H_3C CH_3 CH_3 O

Camphor

12.20 Carbon is tetravalent, and nitrogen is trivalent. If a C–H unit (formula weight 13) is replaced by an N atom (formula weight 14), the molecular weight of the resulting compound increases by one. Since all neutral hydrocarbons have even-numbered molecular weights (C_nH_{2n+2}, C_nH_{2n}, and so forth) the resulting nitrogen-containing compounds have odd-numbered molecular weights. If two C–H units are replaced by two N atoms, the molecular weight of the resulting compound increases by two and remains an even number.

12.21 Because M^+ is an odd number, pyridine contains an odd number of nitrogen atoms. If pyridine contained one nitrogen atom (atomic weight 14) the remaining atoms would have a formula weight of 65, corresponding to $-C_5H_5$. C_5H_5N is, in fact, the molecular formula of pyridine.

12.22 The molecular formula of nicotine is $C_{10}H_{14}N_2$. To find the equivalent hydrocarbon formula, subtract the number of nitrogens from the number of hydrogens. The equivalent hydrocarbon formula of nicotine, $C_{10}H_{12}$, indicates five degrees of unsaturation — two of them due to the two rings and the other three due to three double bonds.

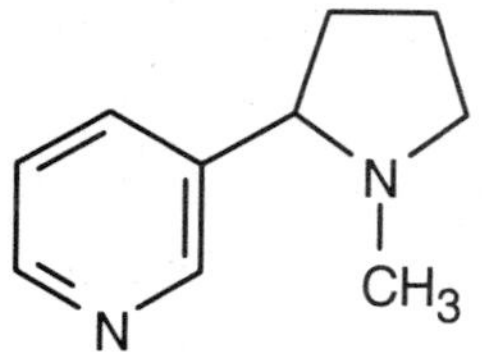

N N CH_3

Nicotine

12.23 In order to simplify this problem, neglect the ^{13}C and ^{2}H isotopes in determining the molecular ions of these compounds.

(a) The formula weight of $-CH_3$ is 15, and the atomic masses of the two bromine isotopes are 79 and 81. The two molecular ions of bromoethane occur at $M^+ = 96$ (49.3%) and $M^+ = 94$ (50.7%).

(b) The formula weight of $-C_6H_{13}$ is 85, and the atomic masses of the two chlorine isotopes are 35 and 37. The two molecular ions of 1-chlorohexane occur at $M^+ = 122$ (24.2%) and $M^+ = 120$ (75.8%).

12.24 Again, neglect ^{13}C and ^{2}H in these calculations.

(a) Finding the molecular ions of chloroform is a statistical exercise.

1. The probability that all three chlorine atoms are ^{37}Cl is $(0.242)^3 = 0.014$.
2. The probability that two chlorine atoms are ^{37}C and one is ^{35}Cl is $3(0.242)(0.242)(0.758) = 3(0.0444) = 0.133$. The factor *3* enters the calculations because three permutations of two ^{37}Cl's and one ^{35}Cl are possible.
3. The probability that one chlorine atom is ^{37}Cl and two are ^{35}Cl is $3(0.242)(0.758)(0.758) = 3(0.139) = 0.417$.
4. The probability that all chlorine atoms are ^{35}Cl is $(0.758)^3 = 0.436$.
5. The mass of: $CH^{37}Cl^{37}Cl^{37}Cl = 124$
   $CH^{37}Cl^{37}Cl^{35}Cl = 122$
   $CH^{37}Cl^{35}Cl^{35}Cl = 120$
   $CH^{35}Cl^{35}Cl^{35}Cl = 118$.
6. Thus, molecular ions for chloroform occur at:

M^+	124	122	120	118
Abundance	1.4%	13.3%	41.7%	43.6%

(b) The molecular ions for Freon 12:

M^+	124	122	120
Abundance	5.9%	36.7%	57.4%

12.25 Each carbon atom has a 1.10% probability of being ^{13}C and a 98.90% probability of being ^{12}C. The ratio of the height of the ^{13}C peak to the height of the ^{12}C peak for a one-carbon compound is (1.10/98.9) x 100% = 1.11%. For a six-carbon compound, the contribution to $(M+1)^+$ from ^{13}C is 6 x (1.10/98.9) x 100% = 6.66%. For benzene, the relative height of $(M+1)^+$ is 6.66% of the height of M^+.

A similar line of reasoning can be used to calculate the contribution to $(M+1)^+$ from ^{2}H. The natural abundance of ^{2}H is 0.015%, so the ratio of a ^{2}H peak to a ^{1}H peak for a one-hydrogen compound is 0.015%. For a six-hydrogen compound, the contribution to $(M+1)^+$ from ^{2}H is 6 x 0.015% = 0.09%.

For benzene, $(M+1)^+$ is 6.75% of M^+. Notice that ^{2}H contributes very little to the size of $(M+1)^+$.

12.26 (a) The molecular formula of the ketone is $C_5H_{10}O$, and the fragments correspond to the products of alpha cleavage (McLafferty rearrangement fragments have even-numbered values of *m/z*). Draw all possible ketone structures, show the charged products of alpha cleavage, and note which fragments correspond to those listed.

$$[H_3C-C(=O)-CH_2CH_2CH_3]^{+\bullet} \xrightarrow{\text{Alpha cleavage}} [H_3C-C{=}O]^+ + [O{=}C-CH_2CH_2CH_3]^+$$

$M^+ = 86$; $m/z = 43$; $m/z = 71$

$$[H_3C-C(=O)-CH(CH_3)CH_3]^{+\bullet} \xrightarrow{\text{Alpha cleavage}} [H_3C-C{=}O]^+ + [O{=}C-CH(CH_3)CH_3]^+$$

$M^+ = 86$; $m/z = 43$; $m/z = 71$

$$[CH_3CH_2-C(=O)-CH_2CH_3] \xrightarrow{\text{Alpha cleavage}} [CH_3CH_2-C{=}O] + [O{=}C-CH_2CH_3]$$

$M^+ = 86$; $m/z = 57$; $m/z = 57$

Either of the first two compounds shows the observed fragments in its mass spectrum.

(b) $C_5H_{12}O$ is the formula of an alcohol with $M^+ = 88$. The fragment at $m/z = 70$ is due to the product of dehydration of M^+. The other two fragments are a result of alpha cleavage. Draw the possible C_5 alcohol isomers, and draw their products of alpha cleavage. The tertiary alcohol shown fits the data.

$$\left[CH_3CH_2-\overset{\displaystyle OH}{\underset{\displaystyle CH_3}{C}}-CH_3\right]^{+\bullet} \xrightarrow{\text{Alpha cleavage}} \left[\overset{\displaystyle OH}{\underset{CH_3CH_2\quad CH_3}{C}}\right]^{+} + \left[\overset{\displaystyle OH}{\underset{H_3C\quad CH_3}{C}}\right]^{+}$$

$M^+ = 88$ $\qquad$ $m/z = 73$ $\qquad$ $m/z = 59$

$$\left[CH_3CH_2-\overset{\displaystyle OH}{\underset{\displaystyle CH_3}{C}}-CH_3\right]^{+\bullet} \xrightarrow{\text{Dehydration}} \left[(H_3C)(H)C{=}C(CH_3)_2\right]^{+\bullet}$$

$M^+ = 88$ $\qquad$ $m/z = 70$

12.27

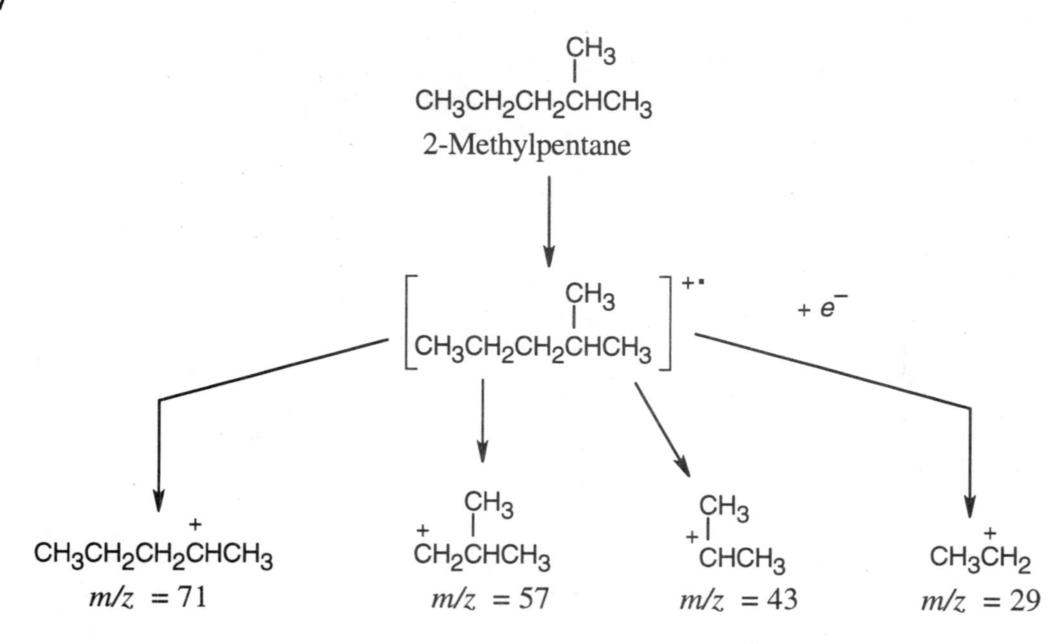

The molecular ion, at $m/z = 86$, is present in very low abundance. The base peak, at $m/z = 43$, represents a stable secondary carbocation.

12.28 Before doing the hydrogenation, familiarize yourself with the mass spectra of cyclohexene and cyclohexane. Note that M^+ is different for each compound. After the reaction is underway, inject a sample from the reaction mixture into the mass spectrometer. If the reaction is finished, the mass spectrum of the reaction mixture should be superimposable with the mass spectrum of cyclohexane.

12.29 See Problem 12.8 for the method of solution.
(a) 3360 cm^{-1} (b) 1720 cm^{-1} (c) 2030 cm^{-1}

12.30 (a) 5.70 µm (b) 3.08 µm (c) 5.80 µm (d) 5.62 µm

12.31 $CH_3CH_2C{\equiv}CH$ shows absorptions at 2100-2260 cm^{-1} ($C{\equiv}C$) and at 3300 cm^{-1} ($C{\equiv}C{-}H$) that are due to the terminal alkyne bond.

$H_2C{=}CHCH{=}CH_2$ has absorptions in the regions 1650-1670 cm^{-1} and 3020-3100 that are due to the double bonds. It also shows absorptions at 910 cm^{-1} and 990 cm^{-1} that are due to monosubstituted alkene bonds. No absorptions occur in the alkyne region.

$CH_3C{\equiv}CCH_3$. For reasons we won't discuss, symmetrically substituted alkynes such as 2–butyne do not show a $C{\equiv}C$ bond absorption in the IR. This alkyne is distinguished from the other isomers in that it shows no absorptions in either the alkyne or alkene regions.

12.32 Two enantiomers have identical physical properties (other than the sign of specific rotation). Thus, their IR spectra are also identical.

12.33 Diastereomers have different physical properties and chemical behavior and their IR spectra are also different.

12.34 (a) Absorptions at 3300 cm^{-1} and 2150 cm^{-1} are due to a terminal triple bond. Possible structures:

$CH_3CH_2CH_2C{\equiv}CH$ $(CH_3)_2CHC{\equiv}CH$

(b) An IR absorption at 3400 cm^{-1} is due to a hydroxyl group. Since no double bond absorption is present, the compound must be a cyclic alcohol.

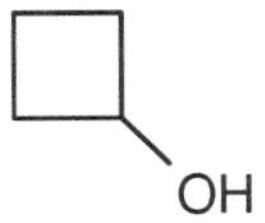

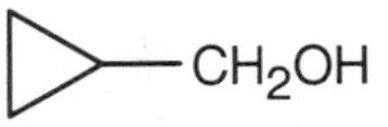

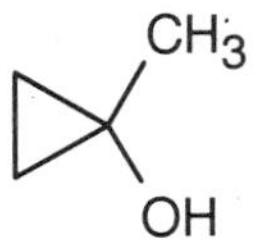

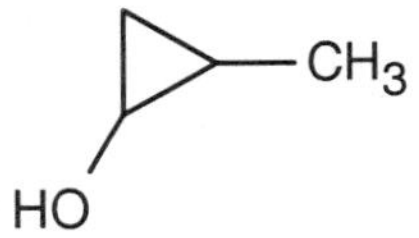

(c) An absorption at 1715 cm^{-1} is due to a ketone. The only possible structure is $CH_3CH_2COCH_3$.

(d) Absorptions at 1600 cm^{-1} and 1500 cm^{-1} are due to an aromatic ring. Possible structures:

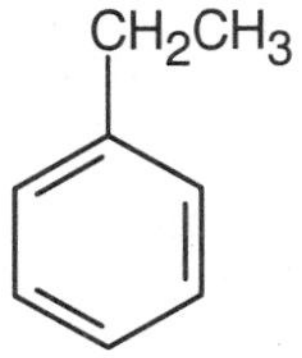

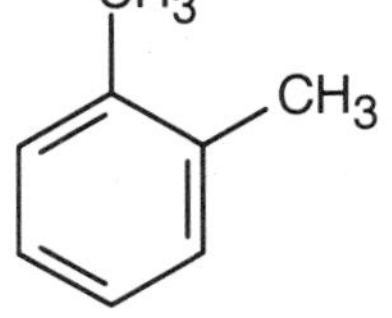

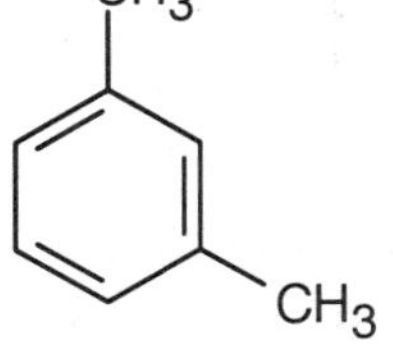

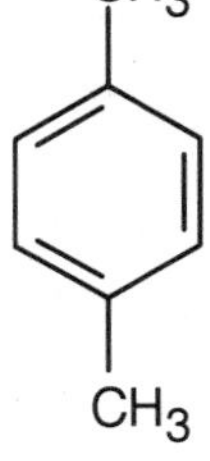

12.35 (a) $HC{\equiv}CCH_2NH_2$
Alkyne absorptions at 3300 cm^{-1}, 2100-2260 cm^{-1}
Amine absorption at 3300-3500 cm^{-1}

$CH_3CH_2C{\equiv}N$
Nitrile absorption at 2210-2260 cm^{-1}

(b) CH_3COCH_3
Strong ketone absorption at 1715 cm^{-1}

CH_3CH_2CHO
Strong aldehyde absorption at 1730 cm^{-1}

12.36 Spectrum (b) differs from spectrum (a) in several respects. Note in particular the absorptions at 715 cm^{-1} (strong), 1140 cm^{-1} (strong), 1650 cm^{-1} (medium), and 3000 cm^{-1} (medium) in spectrum (b). The absorptions at 1650 cm^{-1} (C=C stretch) and 3000 cm^{-1} (=C–H stretch) can be found in Table 12.1. They allow us to assign spectrum (b) to cyclohexene and spectrum (a) to cyclohexane.

12.37 (a) $CH_3C{\equiv}CCH_3$ exhibits no terminal ≡C–H stretching vibration at 3300 cm^{-1}, as $CH_3CH_2C{\equiv}CH$ does.

(b) $CH_3COCH{=}CHCH_3$, a conjugated unsaturated ketone, shows a strong ketone absorption at 1690 cm^{-1}; $CH_3COCH_2CH{=}CH_2$, a nonconjugated ketone, shows a ketone absorption at 1715 cm^{-1} and monosubstituted alkene absorptions at 910 cm^{-1} and 990 cm^{-1}.

(c) CH_3CH_2CHO exhibits an aldehyde band at 1730 cm^{-1}; $H_2C{=}CHOCH_3$ shows characteristic monosubstituted alkene absorptions at 910 cm^{-1} and 990 cm^{-1}.

12.38 If the isotopic masses of the atoms C, H, and O had integral values of 12 amu, 1 amu and 16 amu, many molecular formulas would correspond to a molecular weight of 360 amu. Because isotopic masses are not integral, however, only one molecular formula is associated with a molecular ion at 360.1937 amu.

To reduce the number of possible formulas, assume that the difference in molecular weight between 360 and 360.1937 is due mainly to hydrogen. Divide 0.1937 by 0.00783, the amount by which the atomic weight of one ^1H atom differs from 1. The answer, 24.8, gives a "ballpark" estimate of the number of hydrogens in cortisone. Then make a list of molecular formulas containing C, H and O whose mass is 360 and which contain 20-30 hydrogens. Tabulate these, and calculate their exact masses using the values in the text.

	Isotopic mass			
Molecular formula	Mass of carbons	Mass of hydrogens	Mass of oxygens	Mass of molecular ion
$C_{27}H_{20}O$	324.0000 amu	20.1566 amu	15.9949 amu	360.1515 amu
$C_{25}H_{28}O_2$	300.0000	28.2192	31.9898	360.2090
$C_{24}H_{24}O_3$	288.0000	24.1879	47.9847	360.1726
$C_{21}H_{28}O_5$	252.0000	28.2192	79.9745	360.1937

The molecular weight of $C_{21}H_{28}O_5$ corresponds to the observed molecular weight of cortisone. (Note that only the last formula has the correct degree of unsaturation, 8).

12.39

$$\text{1-Methylcyclohexanol} \xrightarrow{H_3O^+} \text{1-Methylcyclohexene}$$

The infrared spectrum of the starting alcohol shows a broad absorption at 3400-3640 cm^{-1}, due to an O–H stretch. The alkene product exhibits medium intensity absorbances at 1645-1670 cm^{-1} and at 3000-3100 cm^{-1}. Monitoring the disappearance of the alcohol absorptions makes it possible to decide when reaction is complete. It is also possible to monitor the *appearance* of the alkene absorbances.

12.40

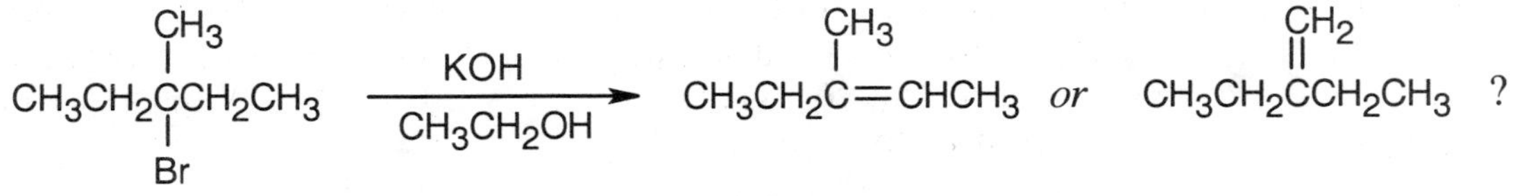

$$\underset{\text{3-Bromo-3-methylpentane}}{CH_3CH_2C(CH_3)(Br)CH_2CH_3} \xrightarrow[CH_3CH_2OH]{KOH} \underset{\text{3-Methyl-2-pentene}}{CH_3CH_2C(CH_3){=}CHCH_3} \text{ or } \underset{\text{2-Ethyl-1-butene}}{CH_3CH_2C({=}CH_2)CH_2CH_3} \text{ ?}$$

The IR spectra of both products show the characteristic absorptions of alkenes in the regions 3020-3100 cm^{-1} and 1650 cm^{-1}. However, in the region 700-1000 cm^{-1}, 2–ethyl–1–butene shows a strong absorption at 890 cm^{-1} that is typical of 2,2–disubstituted $R_2C{=}CH_2$ alkenes. The presence or absence of this peak should help to identify the product. (3–Methyl–2–pentene is the major product of the dehydrobromination reaction.)

12.41

	Compound	*Distinguishing Absorption*	*Due to:*
(a)	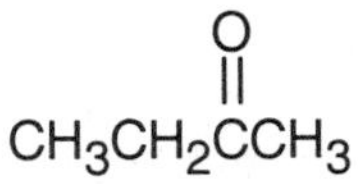 $CH_3CH_2C({=}O)CH_3$	1715 cm^{-1}	C=O (ketone)
(b)	$(CH_3)_2CHCH_2C{\equiv}CH$	2100–2260 cm^{-1} 3300 cm^{-1}	C≡C C≡C—H
(c)	$(CH_3)_2CHCH_2CH{=}CH_2$	910 cm^{-1}, 990 cm^{-1} 1650–1670 cm^{-1} 3020–3100 cm^{-1}	$RCH{=}CH_2$ C=C =C—H
(d)	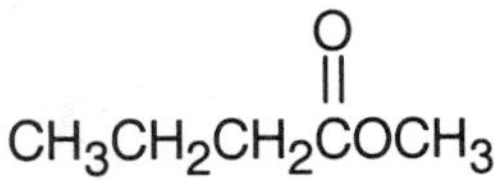 $CH_3CH_2CH_2C({=}O)OCH_3$	1735 cm^{-1}	C=O (ester)
(e)	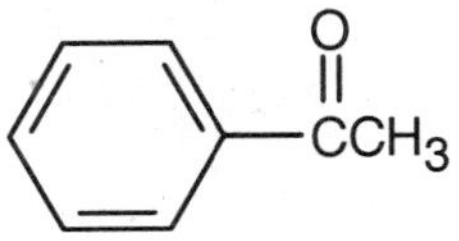 $C_6H_5C({=}O)CH_3$	1690 cm^{-1} 1450–1600 cm^{-1}	ketone next to aromatic ring aromatic ring

12.42 (a) This ketone shows mass spectrum fragments that are due to alpha cleavage and to the McLafferty rearrangement. The molecular ion occurs at $M^+ = 148$, and major fragments have m/z = 120, 105, and 71. (Note that only charged species are shown.)

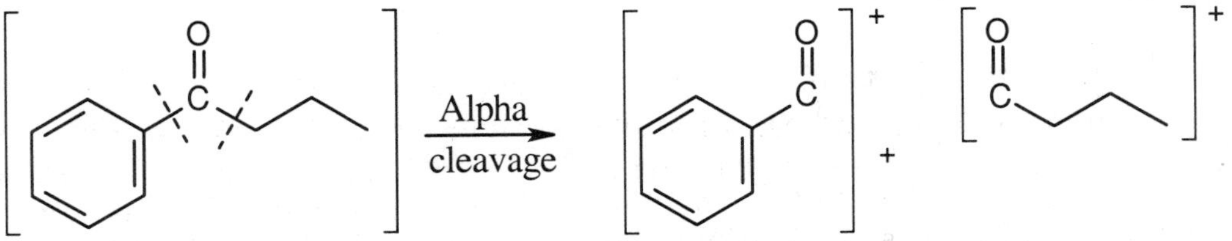

$$[C_6H_5C({=}O)CH_2CH_2CH_3] \xrightarrow[\text{cleavage}]{\text{Alpha}} [C_6H_5C{=}O]^+ + [O{=}CCH_2CH_2CH_3]^+$$

$C_{10}H_{12}O$ $M^+ = 148$ m/z = 105 m/z = 71

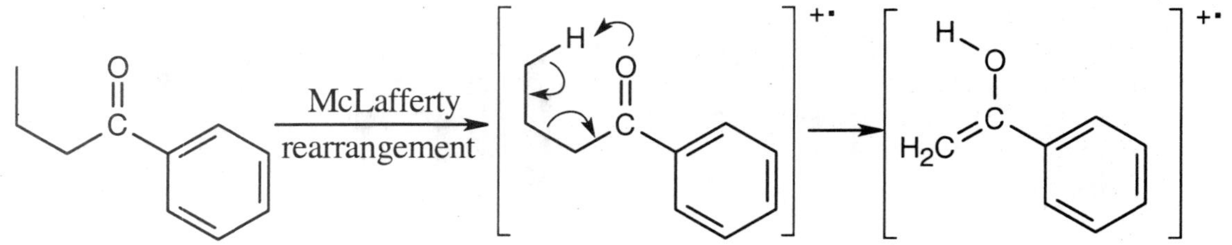

O C McLafferty rearrangement H O C H O H_2C=C

m/z = 120

(b) The fragments in the mass spectrum of this alcohol ($C_8H_{16}O$) result from dehydration and alpha cleavage. Major fragments have m/z values of 128 (the same value as the molecular ion), 110. and 99.

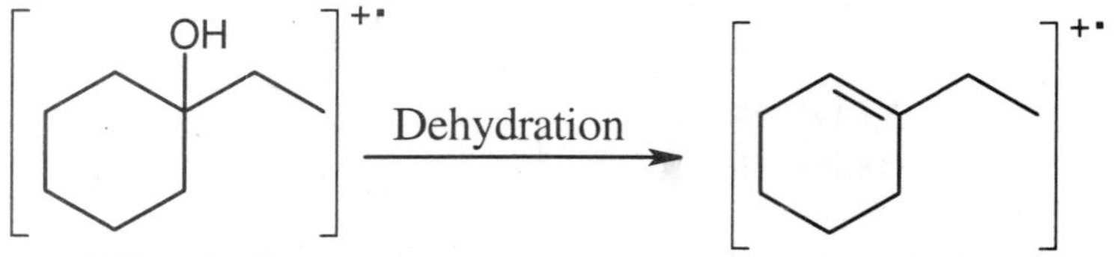

OH Dehydration

$M^+ = 128$ m/z = 110

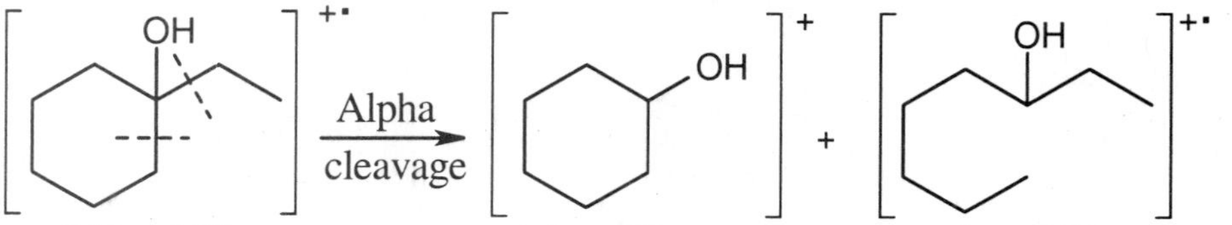

OH Alpha cleavage OH + OH

$M^+ = 128$ m/z = 99 m/z = 128

(c) Amines fragment by alpha cleavage. In this problem, cleavage occurs in the ring, producing a fragment with the same value of m/z as the molecular ion (99).

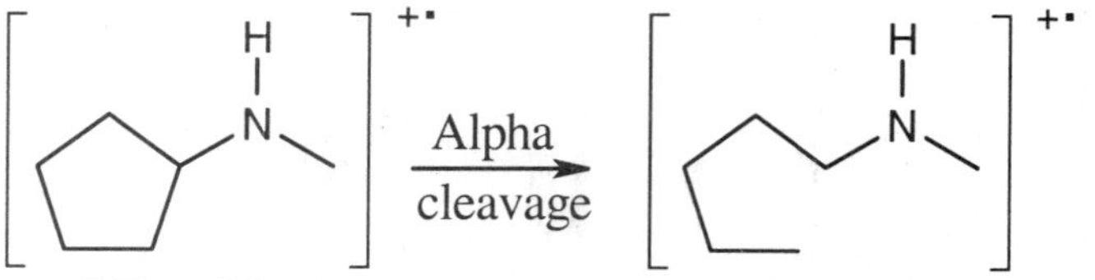

H N Alpha cleavage H N

$M^+ = 99$ m/z = 99

12.43 The following expressions are needed:

$E = h\nu = hc/\lambda = hc\tilde{\nu}$ where $\tilde{\nu}$ is the wavenumber. The last expression shows that, as $\tilde{\nu}$ increases, the energy needed to cause IR absorption increases, indicating greater bond strength. Thus, an ester C=O bond ($\tilde{\nu}$ = 1735 cm^{-1}) is stronger than a ketone C=O bond ($\tilde{\nu}$ = 1715 cm^{-1}).

12.44 Possible molecular formulas containing carbon, hydrogen, and oxygen and having M^+ = 150 are $C_{10}H_{14}O$, $C_9H_{10}O_2$, and $C_8H_6O_3$. The first formula has four degrees of unsaturation, the second has five degrees of unsaturation, and the third has six degrees of unsaturation. Since carvone has three double bonds (including the ketone) and one ring, $C_{10}H_{14}O$ is the correct molecular formula for carvone.

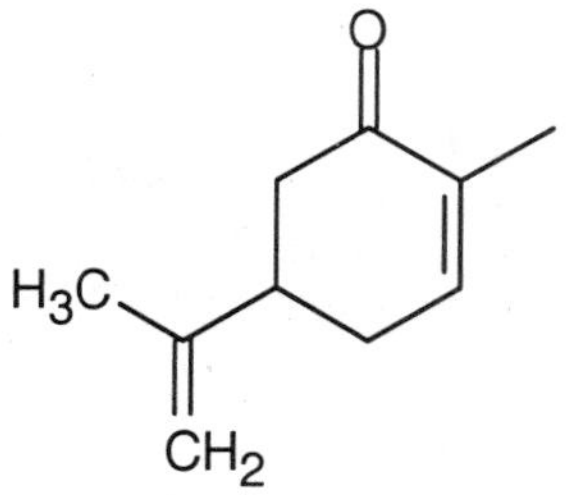

O

H_3C

CH_2

Carvone

12.45 The intense absorption at 1690 cm^{-1} is due to a ketone next to a double bond.

12.46 The peak of maximum intensity (base peak) in the mass spectrum occurs at *m/z* = 67. This peak does *not* represent the molecular ion, however, because M^+ of a hydrocarbon must be an even number. Careful inspection reveals the molecular ion peak at *m/z* = 68. M^+ = 68 corresponds to a hydrocarbon of molecular formula C_5H_8 with a degree of unsaturation of two.

Fairly intense peaks in the mass spectrum occur at *m/z* = 67, 53, 40, 39, and 27. The peak at *m/z* = 67 corresponds to loss of one hydrogen atom, and the peak at *m/z* = 53 represents loss of a methyl group. The unknown hydrocarbon thus contains a methyl group.

Significant IR absorptions occur at 2130 cm^{-1} (–C≡C– stretch) and at 3320 cm^{-1} (≡C–H stretch). These bands indicate that the unknown hydrocarbon is a terminal alkyne. Possible structures for C_5H_8 are $CH_3CH_2CH_2C{\equiv}CH$ and $(CH_3)_2CHC{\equiv}CH$. [1–Pentyne is correct.]

12.47 The molecular ion, M^+ = 70, corresponds to the molecular formula C_5H_{10}. This compound has one double bond or ring.

The base peak in the mass spectrum occurs at *m/z* = 55. This peak represents loss of a methyl group from the molecular ion and indicates the presence of a methyl group in the unknown hydrocarbon. All other peaks occur with low intensity.

In the IR spectrum, it is possible to distinguish absorptions at 1660 cm^{-1} and at 3000 cm^{-1} due to a double bond. (The 2960 cm^{-1} absorption is rather hard to detect because it occurs as a shoulder on the alkane C–H stretch at 2850-2960 cm^{-1}.)

Since no absorptions occur in the region 890 cm^{-1} – 990 cm^{-1}, we can exclude terminal alkenes as possible structures. The remaining possibilities for C_5H_{10} are $CH_3CH_2CH{=}CHCH_3$ and $(CH_3)_2C{=}CHCH_3$. [2–Methyl–2–butene is correct.]

12.48

(a)

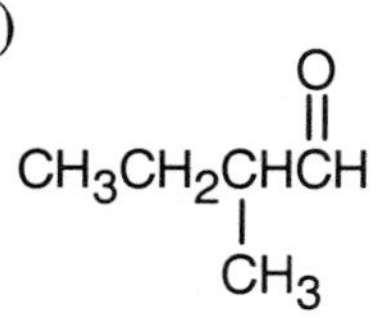

$CH_3CH_2CH(CH_3)CH{=}O$ (drawn: CH_3CH_2CHCH with O above the terminal carbon and CH_3 below the third carbon)

(b)

$CH_3CH_2CH_2CH_2C{\equiv}N$ $\quad$ $CH_3C(CH_3)_2C{\equiv}N$ $\quad$ $CH_3CH(CH_3)CH_2C{\equiv}N$

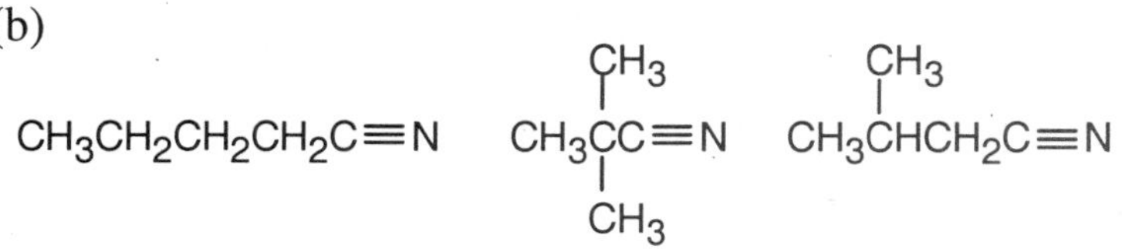

12.49 The simplest way to distinguish between the two isomers is by taking their IR spectra. The aldehyde carbonyl group absorbs at 1730 cm^{-1}, and the ketone carbonyl group absorbs at 1715 cm^{-1}.

The mass spectra of the two isomers also differ. Like ketones, aldehydes also undergo alpha cleavage and McLafferty rearrangements.

McLafferty rearrangement:

$[H_2C{=}C(OH)CH_3]^{+\cdot}$ m/z = 58

$[H_2C{=}C(OH)H]^{+\cdot}$ m/z = 44

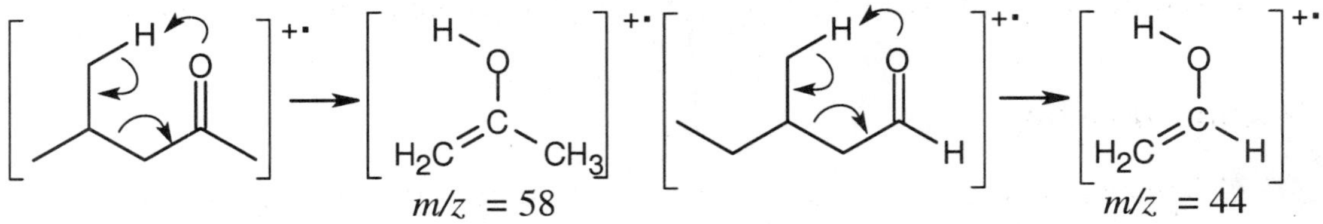

The fragments from the McLafferty rearrangements differ in values of m/z.

Alpha cleavage:

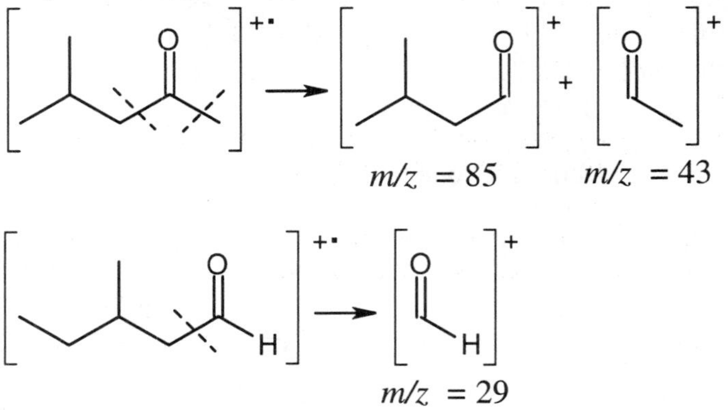

m/z = 85 m/z = 43

m/z = 29

The fragments resulting from alpha cleavage also differ in values of m/z.

A Look Ahead

12.50

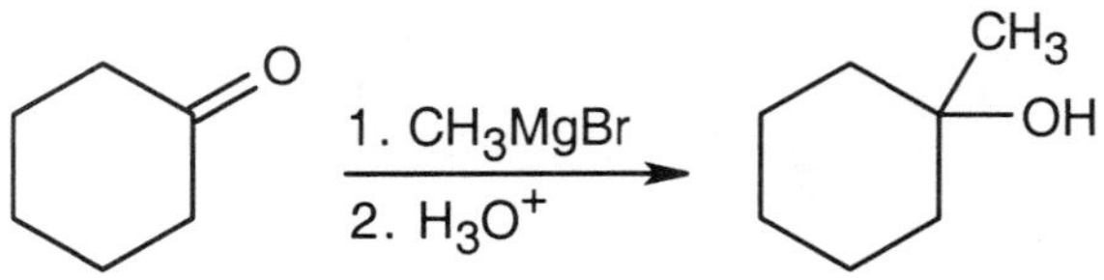

cyclohexanone → (1. CH_3MgBr; 2. H_3O^+) → 1-methylcyclohexanol (CH_3, OH)

The absorption at 3400 cm^{-1} is due to a hydroxyl group.

12.51

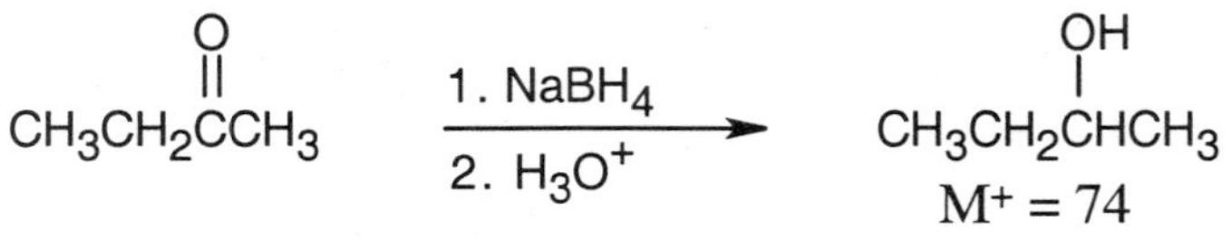

$CH_3CH_2COCH_3$ → (1. $NaBH_4$; 2. H_3O^+) → $CH_3CH_2CH(OH)CH_3$

$M^+ = 74$

12.52

$$CH_3CH_2C\equiv N \xrightarrow[\text{heat}]{H_3O^+} CH_3CH_2\overset{O}{\overset{\|}{C}}OH$$

$M^+ = 74$

The absorption at 1710 cm^{-1} is due to the carbonyl group of a carboxylic acid, and the absorption at 2500–3100 cm^{-1} is due to the –OH group of the carboxylic acid.

Molecular Modeling

12.53 Calculated frequencies increase with bond strength: 1580 cm^{-1} (ethane, C–C bond), 1856 cm^{-1} (ethylene, C=C bond), 2234 cm^{-1} (acetylene, C≡C bond).

12.54 The C=O stretching frequencies are 1940 cm^{-1} (acetone), 1920 cm^{-1} (methyl benzoate), and 1876 cm^{-1} (*N,N*-dimethylformamide). This vibration is a useful diagnostic tool for two reasons: (1) The vibration involves the carbon-oxygen bond (other bonds are not involved); (2) The frequency is unique and is well-separated from the vibrational frequencies of other functional groups.

12.55 Both bonds are equally involved in the C=O stretching vibrations of carbon dioxide: 2463 cm^{-1} (asymmetric stretch) and 1428 cm^{-1} (symmetric stretch).

12.56 The O–H stretching frequencies: 3870 cm^{-1} (acetic acid), 3366 cm^{-1} (acetic acid + water), 3347 cm^{-1} (acetic acid dimer). Hydrogen bonding substantially lowers O–H stretching frequencies.

Chapter 13 – Structure Determination Nuclear Magnetic Resonance Spectroscopy

Chapter Outline

I. Principles of Nuclear Magnetic Resonance Spectroscopy (Sections 13.1 – 13.3)
- A. Theory of NMR Spectroscopy (Section 13.1)
  1. Many nuclei behave as if they were spinning about an axis.
     - a. The positively charged nuclei produce a magnetic field that can interact with an externally applied magnetic field.
     - b. The ^{13}C nucleus and the ^{1}H nucleus behave in this manner.
     - c. In the absence of an external magnetic field the spins of magnetic nuclei are randomly oriented.
  2. When a sample containing these nuclei is placed between the poles of a strong magnet, the nuclei align themselves either with the applied field or against the applied field.
     The parallel orientation is slightly lower in energy and is slightly favored.
  3. If the sample is irradiated with radiofrequency energy of the correct frequency, the nuclei of lower energy absorb energy and "spin-flip" to the higher energy state.
     - a. The magnetic nuclei are in resonance with the applied radiation.
     - b. The frequency of the rf radiation needed for resonance depends on the magnetic field strength and on the identity of the magnetic nuclei.
       - i. In a strong magnetic field, higher frequency rf energy is needed.
       - ii. At a magnetic field strength of 1.41 T, rf energy of 60 MHz is needed to bring a ^{1}H nucleus into resonance, and energy of 15 MHz is needed for ^{13}C.
  4. Nuclei with an odd number of protons and nuclei with an odd number of neutrons show magnetic properties.
- B. The nature of NMR absorptions (Section 13.2).
  1. Not all ^{13}C nuclei and not all ^{1}H nuclei absorb at the same frequency.
     - a. Each magnetic nucleus is surrounded by electrons that set up their own magnetic fields.
     - b. These little fields oppose the applied field and shield the magnetic nuclei.
       - i. $\boldsymbol{B}_{\text{effective}} = \boldsymbol{B}_{\text{applied}} - \boldsymbol{B}_{\text{local}}$
       - ii. This expression shows that the magnetic field felt by a nucleus is less than the applied field.
     - c. These shielded nuclei absorb at slightly different values of magnetic field strength.
     - d. A sensitive NMR spectrometer can detect these small differences.
     - e. Thus, NMR spectra can be used to map the carbon–hydrogen framework of a molecule.
  2. NMR spectra.
     - a. The horizontal axis shows effective field strength, and the vertical axis shows intensity of absorption.
     - b. Each peak corresponds to a chemically distinct nucleus.
     - c. Zero absorption is at the bottom.
     - d. Absorptions due to both ^{13}C and ^{1}H can't both be observed at the same time.
  3. Operation of an NMR spectrometer
     - a. A solution of a sample is placed in a thin glass tube between the poles of a magnet.
     - b. The strong magnetic field causes the nuclei to align in either of the two possible orientations.

c. The strength of the applied magnetic field is varied, holding rf frequency constant.
d. Chemically distinct nuclei come into resonance at slightly different values of ***B***.
e. A detector monitors the absorption of rf energy
f. The signal is amplified and recorded.

4. Time scale of NMR absorptions.
   a. The time scale (10^{-3} s) of NMR spectra is much slower than that of most other spectra.
   b. If a process occurs faster than the time scale of NMR, absorptions are observed as "time-averaged processes.
      NMR records only a single spectrum.
   c. NMR can be used to measure rates and activation energies of fast processes.
      i. Because cyclohexane ring-flips are very fast at room temperature, only a single peak is observed for equatorial and axial hydrogens at room temperature.
      ii At –90°, both axial and equatorial hydrogens can be identified.

C. Chemical Shifts (Section 13.3).
1. Field strength increases from left (downfield) to right (upfield).
   a. Nuclei that absorb downfield require a lower field strength for resonance and are deshielded.
   b. Nuclei that absorb upfield require a higher field strength and are shielded.
2. TMS is used as a reference point in both ^{13}C NMR and ^{1}H NMR.
   The TMS absorption occurs upfield of other absorptions, and is set as the zero point.
3. The chemical shift is the position on the chart where a nucleus absorbs.
4. NMR charts are calibrated by using an arbitrary scale – the delta scale.
   a. One δ equals 1 ppm of the spectrometer operating frequency.
   b. By using this system, all chemical shifts occur at the same value of δ, regardless of the spectrometer operating frequency.
5. NMR absorptions occur over a narrow range.
   a. ^{1}H absorptions occur 0–10 δ downfield from TMS.
   b. ^{13}C absorptions occur 1–220 δ downfield from TMS.
   c. Accidental overlap can be avoided by using an instrument with a higher field strength.

II. ^{13}C NMR spectroscopy (Sections 13.4 – 13.7)

A. Signal averaging and FT-NMR (Section 13.4).
1. The low natural abundance of ^{13}C (1.1%) makes it difficult to observe ^{13}C peaks because of background noise.
2. If hundreds of individual runs are averaged, the background noise cancels.
   This technique takes a long time.
3. In FT-NMR, all signals are recorded simultaneously.
   a. The sample is irradiated with a pulse of rf energy that covers all useful frequencies.
   b. The resulting complex signal must be mathematically manipulated before display.
   c. FT-NMR takes only a few seconds per spectrum.
4. FT-NMR and signal averaging provide increased speed and sensitivity.
   a. Only a few mg of sample are needed for ^{13}C NMR spectra.
   b. Only a few μg of sample are needed for ^{1}H NMR spectra.

B. Characteristics of ^{13}C NMR spectroscopy (Section 13.5).
1. Each distinct carbon shows a single line.
2. The chemical shift depends on the electronic environment within a molecule.
   a. Carbons bonded to electronegative atoms absorb downfield.
   b. Carbons with sp^3 hybridization absorb in the range 0–90 δ.

c. Carbons with sp^2 hybridization absorb in the range 110–220 δ.
Carbonyl carbons absorb in the range 160–220 δ.
3. Symmetry reduces the number of absorptions.
4. Peaks aren't uniform in size.

C. DEPT ^{13}C NMR spectra (Section 13.6).
1. With DEPT experiments, the number of hydrogens bonded to each carbon can be determined.
2. DEPT experiments are run in three stages.
a. A broadband decoupled spectrum gives the chemical shifts of all carbons.
b. A DEPT–90 spectrum shows signals due only to CH carbons.
c. A DEPT–135 spectrum shows CH_3 and CH resonances as positive signals, and CH_2 resonances as negative signals.
3. Interpretation of DEPT spectra.
a. Subtract all peaks in the DEPT–135 spectrum from the broadband-decoupled spectrum to find C.
b. Use DEPT–90 spectrum to identify CH.
c. Use negative DEPT–135 peaks to identify CH_2.
d. Subtract DEPT–90 peaks from positive DEPT–135 peaks to identify CH_3.

D. Uses of ^{13}C NMR spectroscopy (Section 13.7).
^{13}C NMR spectroscopy can show the number of nonequivalent carbons in a molecule and can identify symmetry in a molecule.

III. ^{1}H NMR Spectroscopy (Sections 13.8 – 13.13).

A. Proton equivalence (Section 13.8).
1. ^{1}H NMR can be used to determine the number of nonequivalent protons in a molecule.
2. If it is not possible to quickly decide if two protons are equivalent, try the following test:
a. Replace one of the protons by the group X.
b. If the protons are equivalent, the same product will form, regardless of which proton is replaced.
c. If the protons aren't equivalent, different products will form.

B. Chemical shifts in ^{1}H NMR spectroscopy (Section 13.9).
1. Chemical shifts are determined by the local magnetic fields surrounding magnetic nuclei.
a. More strongly shielded nuclei absorb upfield.
b. Less shielded nuclei absorb downfield.
2. Most ^{1}H NMR chemical shifts are in the range 0–10 δ.
a. Protons that are sp^2-hybridized absorb at higher field strength.
b. Protons that are sp^2-hybridized absorb at lower field strength.
c. Protons on carbons that are bonded to electronegative atoms absorb at lower field strength.
3. The ^{1}H NMR spectrum can be divided into 5 regions:
a. Saturated (0–1.5 δ).
b. Allylic (1.5–2.5 δ).
c. H bonded to C next to an electronegative atom (2,5–4.5 δ).
d. Vinylic (4.5–6.5 δ).
e. Aromatic (6.5–8.0).
f. Aldehyde and carboxylic acid protons absorb even farther downfield.

C. Integration of ^{1}H NMR signals: proton counting (Section 13.10).
1. The area of a peak is proportional to the number of protons causing the peak.
2. Integrated peak areas are superimposed over a spectrum as a stair-step line.
3. To compare two peaks, measure their relative heights.

D. Spin-spin splitting (Section 13.11).
   1. The tiny magnetic field produced by one nucleus can affect the magnetic field felt by other nuclei.
   2. Protons that have n equivalent neighboring protons show a peak in their ^{1}H NMR spectrum that is split into $n + 1$ smaller peaks (a multiplet).
   3. This splitting is caused by the coupling of spins of neighboring nuclei.
   4. The distance between peaks in a multiplet is called the coupling constant (J).
      a. The value of J is usually 0–18 δ.
      b. The value of J is determined by the geometry of the molecule and is independent of the spectrometer operating frequency.
      c. The value of J is shared between both groups of hydrogens whose spins are coupled.
      d. By comparing values of J, it is possible to know the atoms whose spins are coupled.
   5. Three rules for spin-spin splitting in ^{1}H NMR:
      a. Chemically identical protons don't show spin-spin splitting.
      b. The signal of a proton with n equivalent neighboring protons is split into a multiplet of $n + 1$ peaks with coupling constant J.
      c. Two groups of coupled protons have the same value of J.
   6. Spin-spin splitting isn't seen in ^{13}C NMR.
      a. Although spin-spin splitting can occur between carbon and other magnetic nuclei, the spectrometer operating conditions suppress it.
      b. Coupling between the spins of two ^{13}C nuclei isn't seen because of the low probability that two ^{13}C nuclei would be adjacent.

E. Complex spin-spin splitting (Section 13.12).
   1. At times the signals in a ^{1}H NMR absorption overlap accidentally.
   2. Also, signals may be split by two or more nonequivalent kinds of protons.
      a. To understand the effect of multiple coupling, it helps to draw a tree diagram.
      b. In this type of multiplet, the peaks on one side of the multiplet may be larger than those on the other side.
         i. The larger peaks are on the side nearer to the coupled partner.
         ii. This helps identify the nuclei whose spins are coupled.

F. Uses of ^{1}H NMR spectroscopy (Section 13.13).
   ^{1}H NMR can be used to identify the products of reactions.

Solutions to Problems

13.1

$$E = \frac{1.20 \times 10^{-4}\ \text{kJ/mol}}{\lambda\ (\text{in m})}$$

$$\lambda = \frac{c}{\nu} = \frac{3.0 \times 10^{8}\ \text{m/s}}{\nu}; \quad \nu = 56\ \text{MHz} = 5.6 \times 10^{7}\ \text{Hz}$$

$$\lambda = \frac{3.0 \times 10^{8}\ \text{m/s}}{5.6 \times 10^{7}\ \text{Hz}} = 5.4\ \text{m}$$

$$E = \frac{1.20 \times 10^{-4}\ \text{kJ/mol}}{5.4} = 2.2 \times 10^{-5}\ \text{kJ/mol}$$

Compare this value with $E = 2.4 \times 10^{-5}$ kJ/mol for ^{1}H. It takes slightly less energy to spin-flip a ^{19}F nucleus than to spin-flip a ^{1}H nucleus.

13.2

$$\lambda = \frac{c}{\nu} = \frac{3.0 \times 10^8 \text{ m/s}}{\nu}; \quad \nu = 100 \text{ MHz} = 1.0 \times 10^8 \text{ Hz}$$

$$\lambda = \frac{3.0 \times 10^8 \text{ m/s}}{1.0 \times 10^8 \text{ Hz}} = 3.0 \text{ m}$$

$$E = \frac{1.20 \times 10^{-4} \text{ kJ/mol}}{3.0} = 4.0 \times 10^{-5} \text{ kJ/mol}$$

Increasing the spectrometer frequency increases the amount of energy needed for resonance.

13.3

```
b H         CH3 a
   \       /
    C=C
   /       \
c H         Cl
```

2–Chloropropene has three kinds of protons. Protons b and c differ because one is cis to the chlorine and the other is trans.

13.4

(a)

$$\delta = \frac{\text{Observed chemical shift (\# Hz away from TMS)}}{\text{Spectrometer frequency in MHz}}$$

Units of δ are parts per million. In this problem, $\delta = 2.1$ ppm

$$2.1 \text{ ppm} = \frac{\text{Observed chemical shift}}{60 \text{ (MHz)}}$$

$$126 \text{ Hz} = \text{Observed chemical shift}$$

(b) If the ^{1}H NMR spectrum of acetone were recorded at 100 MHz, the position of absorption would still be 2.1 δ because measurements given in ppm or δ units are independent of the operating frequency of the NMR spectrometer.

(c) $$2.1\ \delta = \frac{\text{Observed chemical shift}}{100 \text{ (MHz)}}; \quad \text{Observed chemical shift} = 210 \text{ Hz}$$

13.5

$$\delta = \frac{\text{Observed chemical shift (in Hz)}}{60 \text{ MHz}}$$

(a) $\delta = \frac{436 \text{ Hz}}{60 \text{ MHz}} = 7.27\ \delta$ for $CHCl_3$ (b) $\delta = \frac{183 \text{ Hz}}{60 \text{ MHz}} = 3.05\ \delta$ for CH_3Cl

(c) $\delta = \frac{208 \text{ Hz}}{60 \text{ MHz}} = 3.47\ \delta$ for CH_3OH (d) $\delta = \frac{318 \text{ Hz}}{60 \text{ MHz}} = 5.30\ \delta$ for CH_2Cl_2

13.6 Methyl propanoate has 4 distinct carbons, and each one absorbs in a specific region of the ^{13}C spectrum. The absorption (4) has the lowest value of δ and occurs in the $-CH_3$ region of the ^{13}C spectrum. Absorption (3) occurs in the $-CH_2-$ region. The methyl group (1) is next to an electronegative atom and absorbs downfield from the other two absorptions. The carbonyl carbon (2) absorbs far downfield.

$CH_3CH_2COCH_3$ (C=O; carbons numbered 4, 3, 2, 1)

δ (ppm)	Assignment
9.3	4
27.6	3
51.4	1
174.6	2

13.7

(a) Methylcyclopentane (plane of symmetry; carbons numbered 1–4)

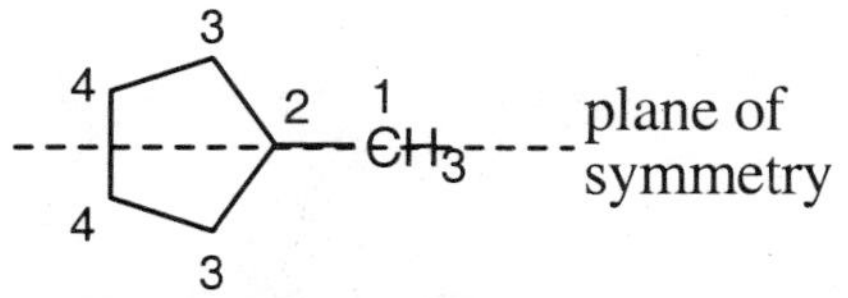

Four resonance lines are observed because of symmetry.

(b) 1-Methylcyclohexene (carbons numbered 1–7)

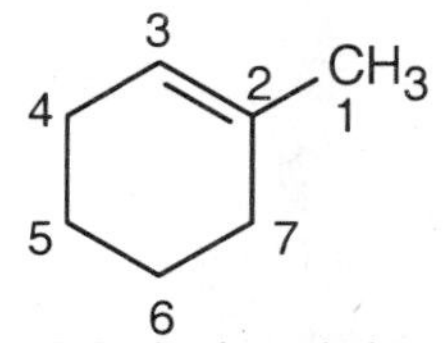

Seven lines are seen because no two carbons are equivalent

(c) 1,2-Dimethylbenzene (plane of symmetry; carbons numbered 1–4)

Four resonance lines are seen.

(d) 2-Methyl-2-butene (carbons numbered 1–5)

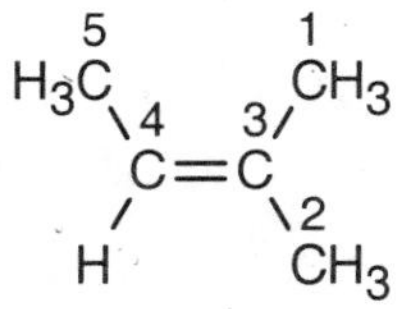

Five resonance lines are observed. Carbons 1 and 2 are nonequivalent because of the double bond stereochemistry.

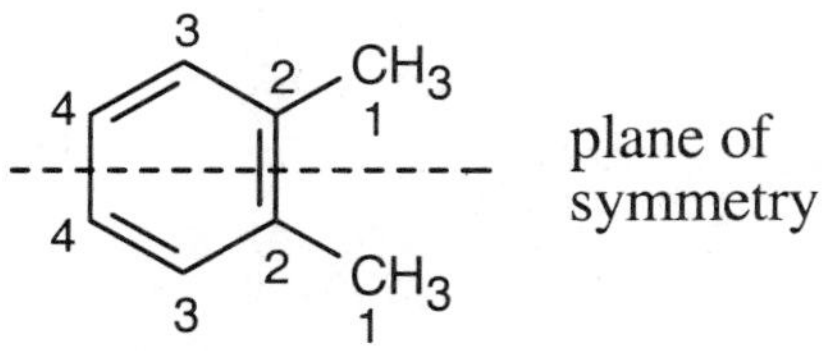

13.8

(a) 1-Methylcyclohexene (carbons 1–7); 1-methyl-4-methylcyclopentene (carbons 1–7, H_3C 6)

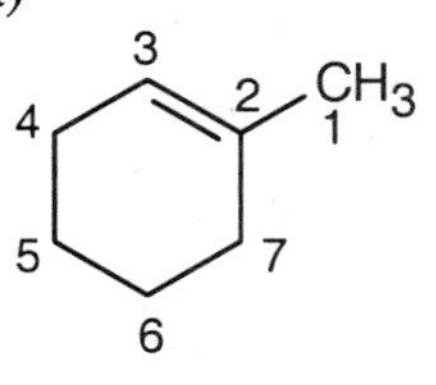

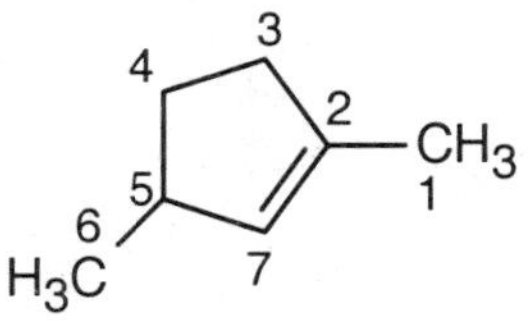

(b) $(H_3C)_2CHCH_2CH_2CH_3$ (carbons numbered 5, 5, 4, 3, 2, 1)

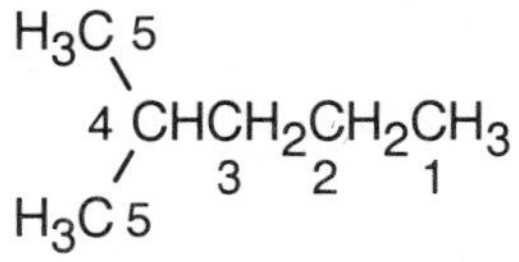

Two of the 6 carbons are equivalent.

(c) $(H_3C)_2CHCH_2Cl$ (carbons numbered 3, 3, 2, 1)

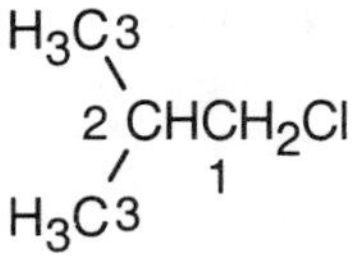

Two of the 4 carbons are equivalent.

13.9 The top spectrum shows all eight ^{13}C NMR peaks. The middle spectrum (DEPT-90) shows only peaks due to CH carbons. From the DEPT-90 spectrum, the absorption at 124 δ can be assigned to the vinyl carbon (5), and the absorption at 68 δ can be assigned to the –OH carbon (2).

The DEPT-135 spectrum shows all but the quaternary carbon (6), which appears in the top spectrum at 132 δ. The top half of the DEPT-135 spectrum shows absorptions due to CH_3 carbons and CH carbons (which we have already identified). The 3 remaining peaks on the top of the DEPT-135 spectrum are due to methyl groups. The peak at 18 δ is due to carbon (1). The other two peaks arise from carbons (7) and (8); their environments are too similar to allow distinguishing between them (23 δ and 26 δ).

The bottom half of the DEPT-135 shows the two CH_2 carbons. Carbon (3) absorbs at 24 δ (negative), and carbon (4) absorbs at 39 δ negative).

In summary:

6-Methyl-5-hepten-2-ol

Carbon	*Chemical Shift (δ)*
1	18
2	68
3	24 (negative)
4	39 (negative)
5	124
6	132
7, 8	23, 26

13.10 Identify the carbons as primary, secondary, tertiary or quaternary, and use Figure 13.7 to find approximate values for chemical shifts. (When an actual spectrum is given, it is easier to assign the carbons to the chemical shifts.) Remember: DEPT-90 spectra identify tertiary carbons, and DEPT-135 spectra identify primary carbons (positive peaks) and secondary carbons (negative peaks). Quaternary carbons are identified in the broadband-decoupled spectrum, in which all peaks appear.

Carbon	*Chemical Shift (δ)*	*DEPT-90?*	*DEPT-135?*
1	10–30	no	yes (positive)
2	30–50	no	yes (negative)
3	160–220	no	no
4	110–150	yes	yes (positive)
5	110–150	no	no
6	10–30	no	yes (positive)
7	50–90	no	yes (positive)

13.11 Always start this type of problem by calculating the degree of unsaturation of the unknown compound. $C_{11}H_{16}$ has 4 degrees of unsaturation. Since the unknown hydrocarbon is aromatic, a benzene ring accounts for all four degrees of unsaturation.

Next, look for elements of symmetry. Although the molecular formula indicates 11 carbons, only 7 peaks appear in the ^{13}C NMR spectrum, indicating a plane of symmetry. Four of the 7 peaks are due to aromatic carbons, indicating a benzene ring that is probably monosubstituted. (Prove to yourself that a monosubstituted benzene ring has 4 different kinds of carbons).

The DEPT-90 spectrum shows that 3 of the kinds of carbons in the aromatic ring are CH carbons. The positive peaks in the DEPT-135 spectrum include these three peaks, along with the peak at 29.5 δ, which is due to a CH_3 carbon. The negative peak in the DEPT-135 spectrum is due to a CH_2 carbon.

Two peaks remain unidentified and are thus quaternary carbons; one of them is aromatic.

At this point, the unknown structure is a monosubstituted benzene ring with a substituent that contains CH_2, C, and CH_3 carbons. A structure for the unknown compound that satisfies all data:

$C_6H_5-CH_2-C(CH_3)_2-CH_3$

13.12

$CH_3CH_2CH_2CH_2C{\equiv}CH$ + HBr ⟶ $CH_3CH_2CH_2CH_2C(Br){=}CH_2$ 2-Bromo-1-hexene *or* $CH_3CH_2CH_2CH_2CH{=}CHBr$? 1-Bromo-1-hexene

The two possible products are easy to distinguish by using ^{13}C NMR. 2-Bromo-1-hexene, the actual product formed, shows no peaks in its DEPT-90 ^{13}C NMR spectrum because it has no CH carbons. The other possible product, 1-bromo-1-hexene, shows 2 peaks in its DEPT-90 spectrum.

13.13

Compound	Kinds of non-equivalent protons	Compound	Kinds of non-equivalent protons
(a) $\overset{1}{CH_3}\overset{2}{CH_2}Br$	2	(b) $\overset{1}{CH_3}O\overset{2}{CH_2}\overset{3}{CH}(\overset{4}{CH_3})\overset{4}{CH_3}$	4
(c) $\overset{1}{CH_3}\overset{2}{CH_2}\overset{3}{CH_2}NO_2$	3	(d) p-xylene-type ring: $\overset{1}{CH_3}$ on ring, ring H's labeled 2, 2, 3, 3, and H labeled 4	4
(e) $(\overset{3}{H_3C})(\overset{1}{CH_3}\overset{2}{CH_2})C{=}C(\overset{4}{H})(\overset{5}{H})$ The two vinylic protons are nonequivalent.	5	(f) $(\overset{1}{CH_3}\overset{2}{CH_2})(\overset{3}{H})C{=}C(\overset{3}{H})(\overset{2}{CH_2}\overset{1}{CH_3})$ plane of symmetry	3

13.14

5 signals

13.15

	Compound	δ	*Kind of proton*
(a)	Cyclohexane	1.43	secondary alkyl
(b)	CH_3COCH_3	2.17	methyl ketone
(c)	C_6H_6	7.37	aromatic
(d)	Glyoxal	9.70	aldehyde
(e)	CH_2Cl_2	5.30	protons adjacent to two halogens
(f)	$(CH_3)_3N$	2.12	methyl protons adjacent to nitrogen

13.16

Proton	δ	*Kind of proton*
1	1.0	primary alkyl
2	1.8	allylic
3	6.1	vinylic
4	6.3	vinylic (different from proton 3)
5	7.2	aromatic
6	6.8	aromatic
7	3.8	ether

This compound has seven different kinds of protons. Notice that the two "5" protons are equivalent to each other, as are the two "6" protons, because of rapid rotation around the bond joining the aromatic ring and the alkenyl side chain.

13.17

H_3C–C_6H_4–CH_3 *p*-Xylene

There are two absorptions in the ^1H NMR spectrum of *p*–xylene. The four ring protons absorb at 7.0 δ, and the six methyl-group protons absorb at 2.3 δ. The peak ratio of methyl protons:ring protons is 3:2.

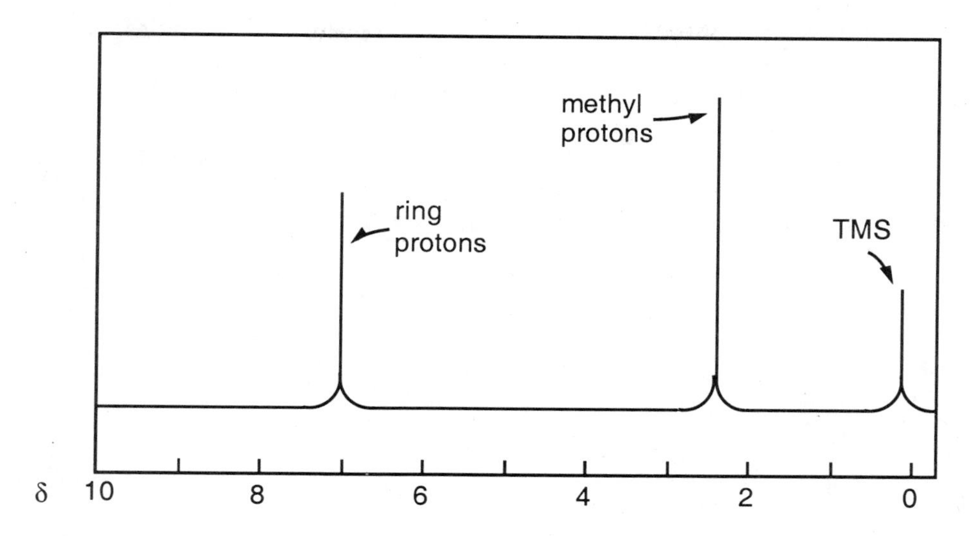

13.18

	Compound	Proton	Number of Adjacent Protons	Splitting
(a)	$\overset{1}{C}HBr_2\overset{2}{C}H_3$	1	3	quartet
		2	1	doublet
(b)	$\overset{1}{C}H_3O\overset{2}{C}H_2\overset{3}{C}H_2Br$	1	0	singlet
		2	2	triplet
		3	2	triplet
(c)	$Cl\overset{1}{C}H_2\overset{2}{C}H_2\overset{1}{C}H_2Cl$	1	2	triplet
		2	4	quartet
(d)	$\overset{1}{C}H_3\overset{2}{C}H(\overset{1}{C}H_3)C(=O)O\overset{3}{C}H_2\overset{4}{C}H_3$	1	1	doublet
		2	6	septet
		3	3	quartet
		4	2	triplet
(e)	$\overset{1}{C}H_3\overset{2}{C}H_2C(=O)O\overset{3}{C}H(\overset{4}{C}H_3)\overset{4}{C}H_3$	1	2	triplet
		2	3	quartet
		3	6	septet
		4	1	doublet
(f)		1	2	triplet
		2	1	doublet
		3	1	doublet
		4	1	doublet

13.19 (a) This compound has no degrees of saturation and only one kind of hydrogen. The only possible structure is CH_3OCH_3.
(b) Again, this compound has no degrees of unsaturation and has two kinds of hydrogens. The compound is 2-chloropropane.
(c) This compound, with no degrees of unsaturation, has two different kinds of hydrogen, each of which has two neighboring hydrogens.
(d) $C_4H_8O_2$; one degree of unsaturation and 3 different kinds of hydrogen.

(a) CH_3OCH_3

(b) $CH_3CH(Cl)CH_3$

(c) $ClCH_2CH_2OCH_2CH_2Cl$

(d) $CH_3CH_2C(=O)OCH_3$ *or* $CH_3C(=O)OCH_2CH_3$

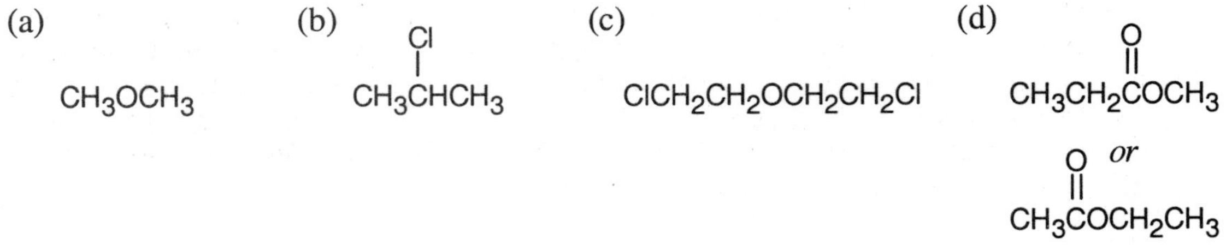

13.20 The molecular formula reveals a compound with a degree of unsaturation of zero (no multiple bonds or rings). The ^1H NMR spectrum shows two signals, corresponding to two types of hydrogens in the ratio 33:50, or 2:3. Since the unknown contains 10 hydrogens, four protons are of one type and six are of the other type.

The upfield signal at 1.2 δ is due to saturated primary protons. The downfield signal at 3.5 δ is due to protons on carbon adjacent to an electronegative atom — in this case, oxygen.

The signal at 1.2 δ is a triplet, indicating two neighboring protons. The signal at 3.5 δ is a quartet, indicating three neighboring protons. The compound is diethyl ether, $CH_3CH_2OCH_2CH_3$.

13.21

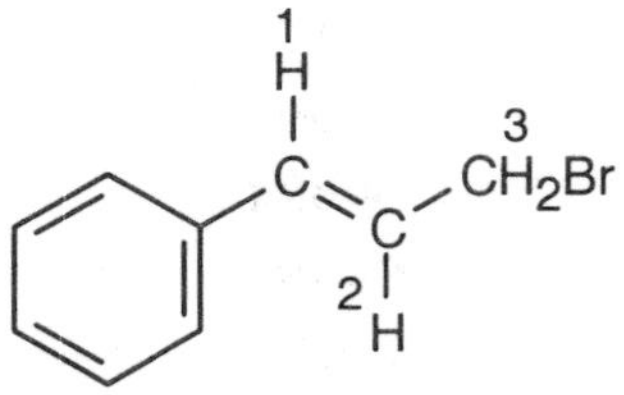

(*E*)-3-Bromo-1-phenyl-1-propene

Coupling of the C2 proton to the Cl vinylic proton occurs with $J = 16$ Hz and causes the signal of the C2 proton to be split into a doublet. The C2 proton is also coupled to the two C3 protons with $J = 8$ Hz. This splitting causes each leg of the C2 proton doublet to be split into a triplet, producing six lines in all. Because of the size of the coupling constants, two of the lines coincide, and a quintet is observed.

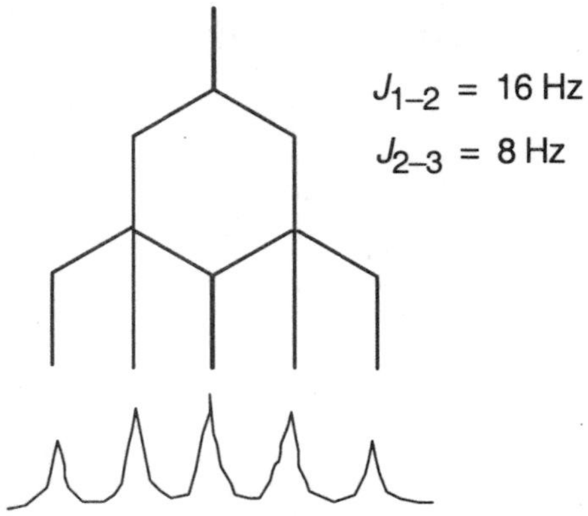

$J_{1-2} = 16$ Hz

$J_{2-3} = 8$ Hz

13.22

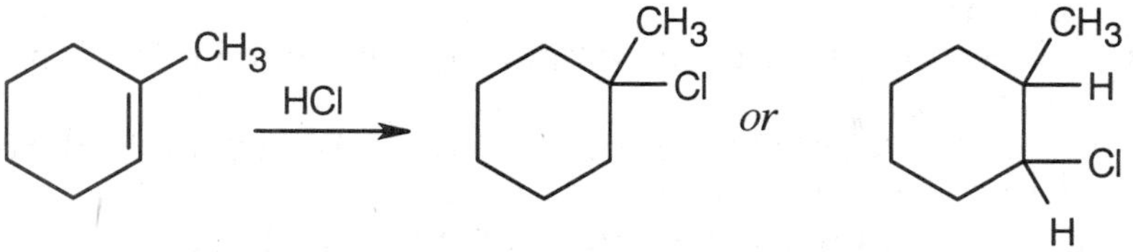

HCl

or

Focus on the ^{1}H NMR methyl group absorption. In the first product, the methyl group signal is unsplit; in the other product, it appears as a doublet. In addition, the second product shows a downfield absorption in the 2.5 δ – 4.0 δ region due to the proton bonded to a carbon that is also bonded to an electronegative atom. If you were to take the ^{1}H NMR spectrum of the reaction product, you would find an unsplit methyl group, and you could conclude that the product was 1-chloro-1-methylcyclohexane.

Visualizing Chemistry

13.23

(a)

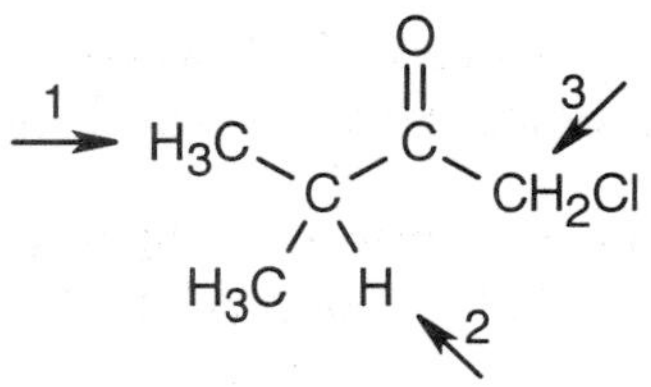

1. doublet
2. septet
3. singlet

(b)

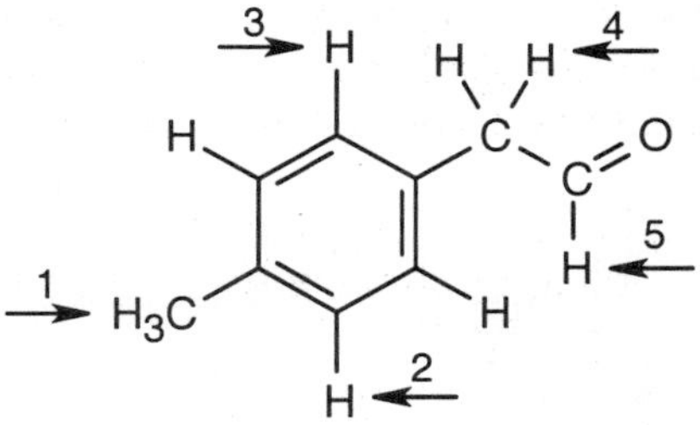

1. singlet
2. doublet
3. doublet
4. doublet
5. triplet

13.24 Surprisingly, this compound has only two different kinds of carbons. In the ^{13}C NMR spectrum, the $-CH_2-$ carbons absorb at 30–50 δ, and the =C– carbons absorb at 110–150 δ.

13.25 The compound has 5 different types of carbons and 4 different types of hydrogens.

^{13}C

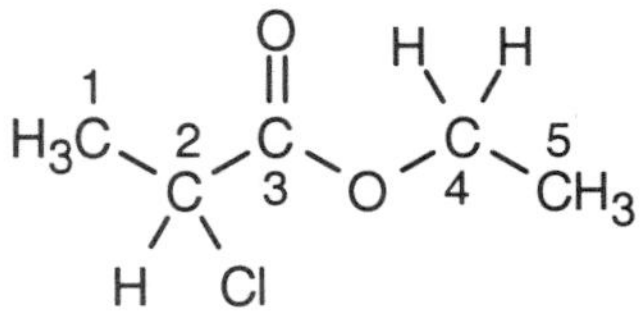

^{1}H

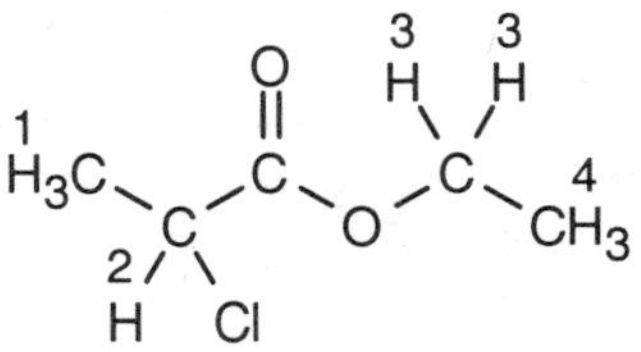

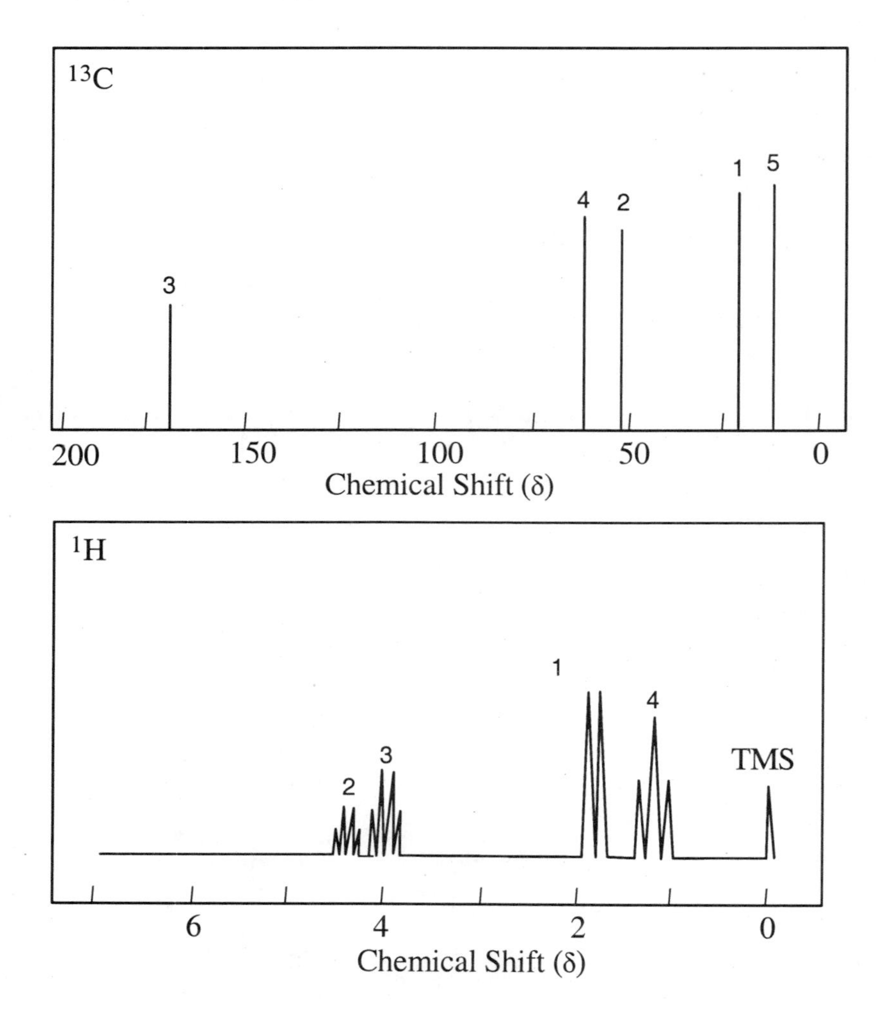

13.26 *cis*-1,2-Dimethylcyclohexane has two chirality centers, and, because of them, all ring protons and all ring carbons are different, as are the methyl groups. However, the cyclohexane ring can ring-flip, interconverting the isomer shown with its enantiomer (Problem 9.78). Since the rate of interconversion is faster than the time scale of NMR spectra, the spectra of *cis*-1,2-dimethylcyclohexane show "time-averaged" peaks for nuclei in molecules that undergo ring-flips. Spectroscopically, *cis*-1,2-dimethylcyclohexane behaves as a compound with a plane of symmetry. Both the ^{13}C NMR spectrum and the ^{1}H NMR spectrum show 4 peaks.

^{1}H ^{13}C

Additional Problems

13.27

$$\delta = \frac{\text{Observed chemical shift (in Hz)}}{100 \text{ MHz}}$$

(a) 2.18 δ (b) 4.78 δ (c) 7.52 δ

13.28 δ x 300 MHz = Observed chemical shift (in Hz)
(a) 630 Hz (b) 1035 Hz (c) 1890 Hz (d) 2310 Hz

13.29 (a) Since the symbol "δ" indicates ppm downfield from TMS, chloroform absorbs at 7.3 ppm.

(b)
$$\delta = \frac{\text{Observed chemical shift (in Hz)}}{\text{Spectrometer frequency in MHz}}$$

$$7.3 \text{ ppm} = \frac{\text{chemical shift}}{360 \text{ MHz}};\quad 7.3 \text{ ppm} \times 360 \text{ MHz} = \text{chemical shift}$$

$$2600 \text{ Hz} = \text{chemical shift}$$

(c) The value of δ is still 7.3 because the chemical shift measured in δ is independent of the operating frequency of the spectrometer.

13.30

(a)

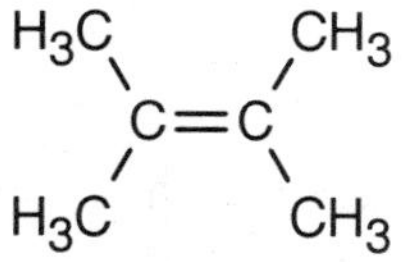

^{13}C: 2 absorptions
^{1}H: 1 absorption

(b)

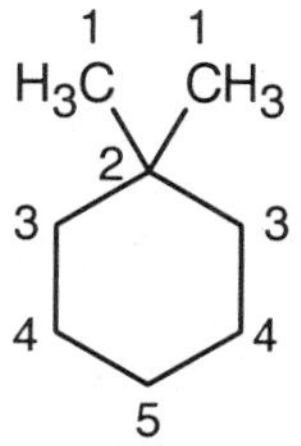

^{13}C: 5 absorptions
(at room temperature)

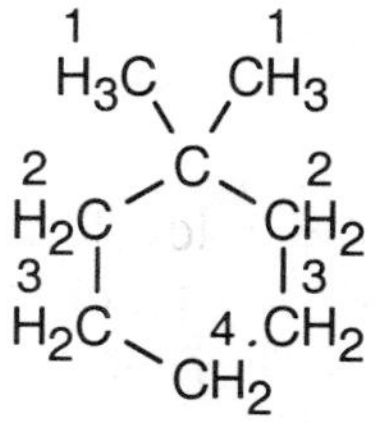

^{1}H: 4 absorptions
(at room temperature)

(c)

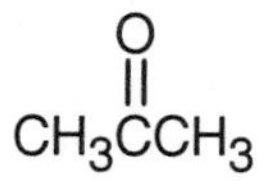

^{13}C: 2 absorptions
^{1}H: 1 absorption

(d)

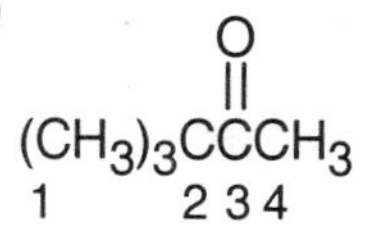

^{13}C: 4 absorptions

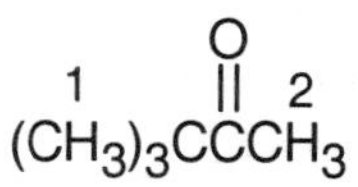

^{1}H: 2 absorptions

(e)

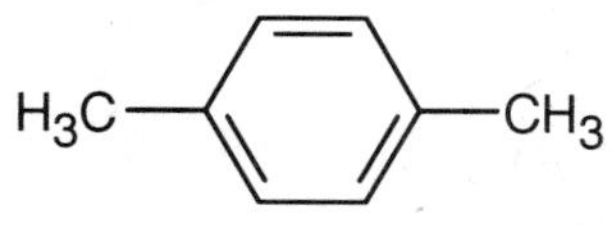

^{13}C: 3 absorptions
^{1}H: 2 absorptions

(f)

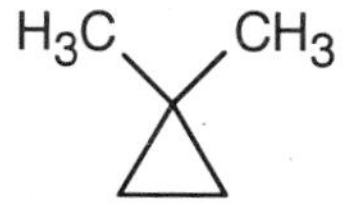

^{13}C: 3 absorptions
^{1}H: 2 absorptions

13.31–13.32

Compound	Number of ^{13}C Absorptions	*Carbons Showing Peaks in DEPT-135 ^{13}C NMR Spectrum* Positive Peaks	Negative Peaks	No Peaks
(a) 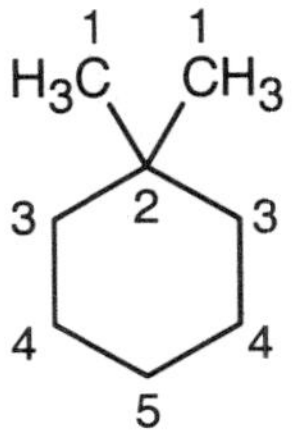	5	carbon 1	carbons 3,4,5	carbon 2
(b) 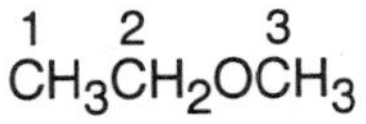	3	carbons 1,3	carbon 2	
(c) 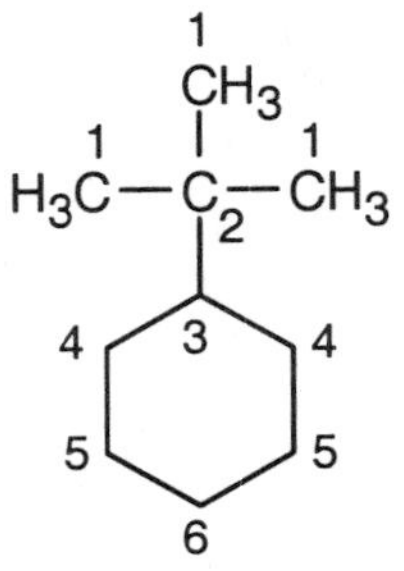	6	carbons 1,3	carbons 4,5,6	carbon 2
(d) 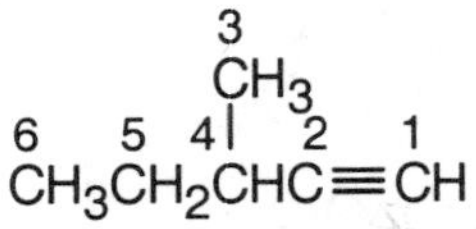	6	carbons 1,3,4,6	carbon 5	carbon 2
(e) 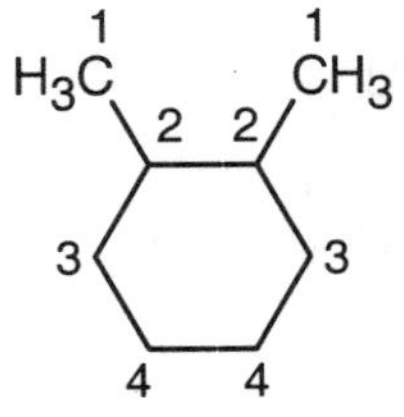	4	carbons 1,2	carbons 3,4	
(f) 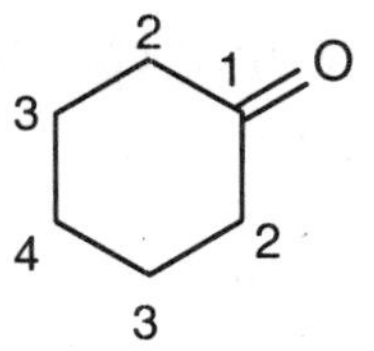	4		carbons 2,3,4	carbon 1

13.33 ^{13}C NMR absorptions occur over a range of 250 ppm, while ^{1}H NMR absorptions generally occur over a range of only 10 ppm. The spread of peaks in ^{13}C NMR is therefore much greater, so accidental overlap is less likely. In addition, normal ^{13}C NMR spectra are uncomplicated by spin-splitting, and the total number of lines is smaller.

13.34 A nucleus that absorbs at 6.50 δ is less shielded than a nucleus that absorbs at 3.20 δ and thus requires a weaker applied field to come into resonance. A shielded nucleus feels a smaller effective field, and a stronger applied field is needed to bring it into resonance.

13.35

Compound	*Kinds of non-equivalent protons*	*Compound*	*Kinds of non-equivalent protons*
(a) 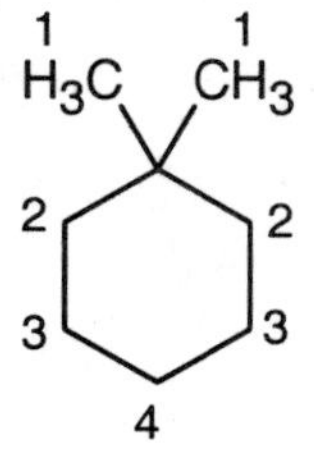 1,1-dimethylcyclohexane: 1 H_3C, 1 CH_3; ring positions 2, 2, 3, 3, 4	4	(b) 1 2 3 4 $CH_3CH_2CH_2OCH_3$	4
(c) 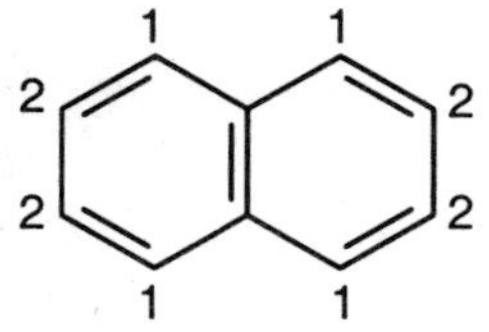 naphthalene: 1, 1, 2, 2, 2, 2, 1, 1	2	(d) 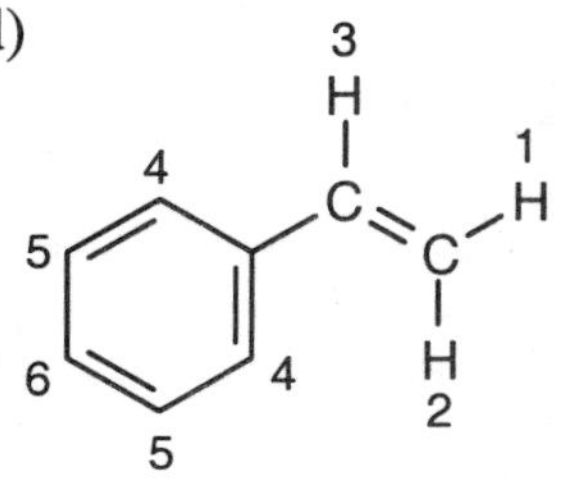 styrene: vinyl H 1, 2, 3; ring 4, 5, 6, 4, 5	6
(e) 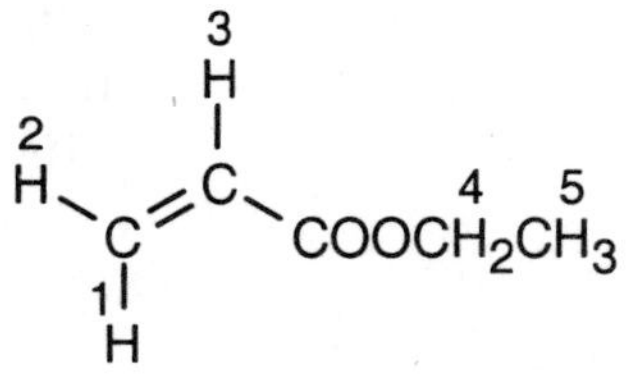 H(2), H(1) C=C H(3) $COOCH_2CH_3$ (4, 5)	5		

13.36

Lowest Chemical Shift ——————→ *Highest Chemical Shift*

CH_4 < Cyclohexane < CH_3COCH_3 < CH_2Cl_2, $H_2C{=}CH_2$ < Benzene

0.23 1.43 2.17 5.30 5.33 7.37

13.37

Compound	*Number of peaks*	*Peak Assignment*	*Splitting Pattern*	
(a) 1 2 $(CH_3)_3CH$	2	1	doublet	(9H)
		2	multiplet (dectet)	(1H)
(b) 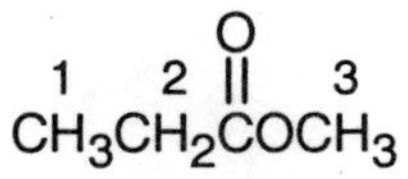 1 2 3 $CH_3CH_2\overset{O}{\overset{\Vert}{C}}OCH_3$	3	1	triplet	(3H)
		2	quartet	(2H)
		3	singlet	(3H)
(c) 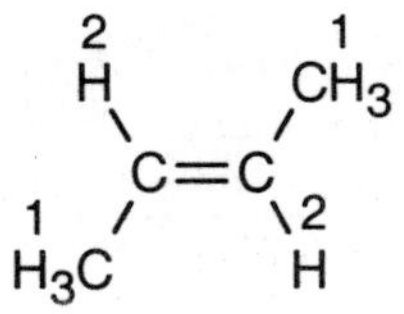 H(2), CH_3(1) C=C H_3C(1), H(2)	2	1	doublet	(6H)
		2	quartet	(2H)

13.38

$CH_3CH_2COCH(CH_3)_2$ (carbons numbered 1, 2, C=O, 3, 4)

	Peak Assignment	*Splitting Pattern*	
	1	triplet	(3H)
	2	quartet	(2H)
	3	septet	(1H)
	4	doublet	(6H)

13.39 Use of ^{13}C NMR to distinguish between the two isomers has been described in the text in Section 13.9. ^{1}H NMR can also be useful.

Isomer **A** has only four kinds of protons because of symmetry. Its vinylic proton absorption (4.5–6.5 δ) represents two hydrogens. Isomer **B** contains six different kinds of protons. Its ^{1}H NMR shows an unsplit methyl group signal and one vinylic proton signal of relative area 1. These differences make it possible to distinguish between **A** and **B**.

13.40 First, examine each isomer for structural differences that are obviously recognizable in the ^{1}H NMR spectrum. If it is not possible to pick out distinguishing features immediately, it may be necessary to sketch an approximate spectrum of each isomer for comparison.

(a) $CH_3CH{=}CHCH_2CH_3$ has two vinylic protons with chemical shifts at 5.4 – 5.5 δ. Because ethylcyclopropane shows no signal in this region, it should be easy to distinguish one isomer from the other.

(b) $CH_3CH_2OCH_2CH_3$ has two kinds of protons, and its ^{1}H NMR spectrum consists of two peaks — a triplet and a quartet. $CH_3OCH_2CH_2CH_3$ has four different types of protons, and its spectrum is more complex. In particular, the methyl group bonded to oxygen shows an unsplit singlet absorption.

(c) Each compound shows three peaks in its ^{1}H NMR spectrum. The ester, however, shows a downfield absorption due to the $-CH_2-$ hydrogens next to oxygen. No comparable peak shows in the spectrum of the ketone.

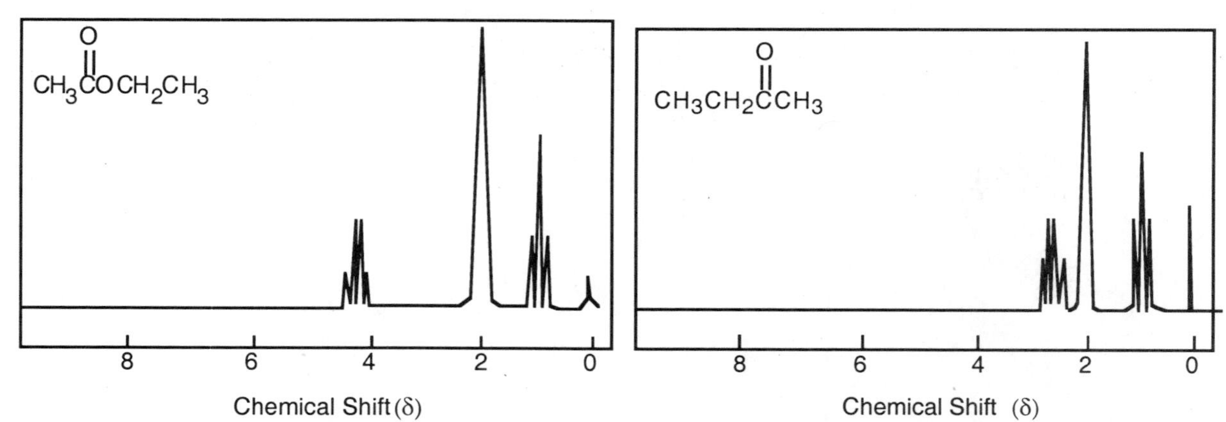

d) Each isomer contains four different kinds of protons — two kinds of methyl protons and two kinds of vinylic protons. For the first isomer, the methyl peaks are both singlets, whereas for the second isomer, one peak is a singlet and one is a doublet.

13.41

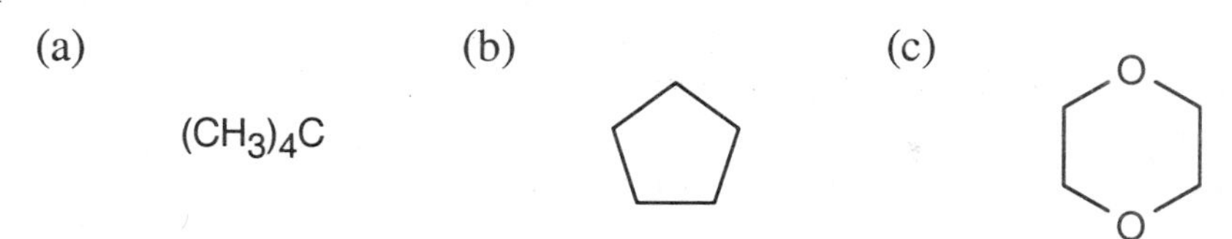

13.42

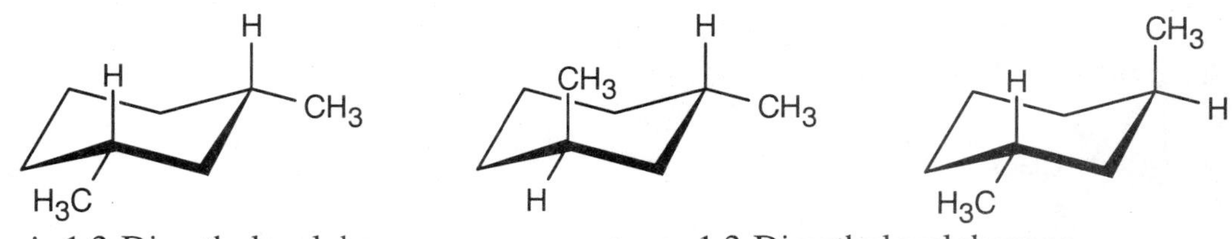

cis-1,3-Dimethylcyclohexane *trans*-1,3-Dimethylcyclohexane

cis-1,3-Dimethylcyclohexane is a meso compound. Because of symmetry, it shows 5 absorptions in its ^{13}C NMR spectrum. *trans*-1,3-Dimethylcyclohexane exists as a pair of enantiomers. Both enantiomers undergo ring-flips at a rate that is faster than the time frame of an NMR spectrum and that averages absorptions due to nonequivalent carbons. Like the cis isomer, the racemic mixture of trans enantiomers shows 5 absorptions in its ^{13}C spectrum.

13.43 (a),(b) C_3H_6O contains one double bond or ring. Possible structures for C_3H_6O include:

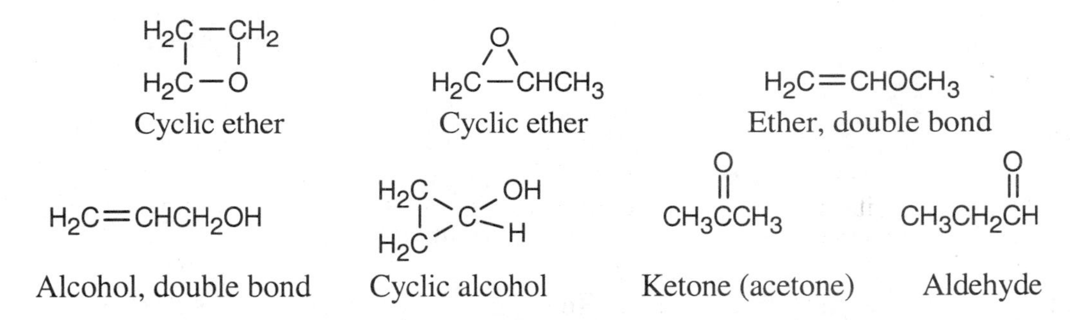

(c) Carbonyl functional groups (usually ketones) absorb at 1715 cm^{-1} in the infrared. Only the last two compounds above show an infrared absorption in this region.

(d) Because the aldehyde from part (b) has three different kinds of protons, its ^{1}H NMR spectrum shows three peaks. The ketone, however, shows only one peak. Since the unknown compound of this problem shows only one ^{1}H NMR absorption (in the methyl ketone region), it must be acetone.

13.44 Either ^{1}H NMR or ^{13}C NMR can be used to distinguish among these isomers. In either case, it is first necessary to find the number of different kinds of protons or carbon atoms.

Compound	*Kinds of Protons*	*Kinds of Carbon atoms*	*Number of ^{1}H NMR peaks*	*Number of ^{13}C NMR peaks*
H_2C-CH_2 / H_2C-CH_2 (cyclobutane)	1	1	1	1
$H_2C=CHCH_2CH_3$	5	4	5	4
$CH_3CH=CHCH_3$	2	2	2	2
$(CH_3)_2C=CH_2$	2	3	2	3

^{13}C NMR is the simplest method for identifying these compounds because each isomer differs in the number of absorptions in its ^{13}C NMR spectrum. ^{1}H NMR can also be used to distinguish among the isomers because the two isomers that show two ^{1}H NMR peaks differ in their splitting patterns.

13.45

Compound	*Number of Peaks*		*Distinguishing Absorptions*
3-methyl-2-cyclohexenone (CH_3)	^{13}C	7	Two vinylic peaks
	^{1}H	5	Unsplit vinylic peak, relative area 1
3-cyclopentenyl methyl ketone ($C(=O)CH_3$)	^{13}C	4	One vinylic peak
	^{1}H	5	Split vinylic peak, relative area 2

The two isomers have different numbers of peaks in both ^{1}H NMR and ^{13}C NMR. In addition, the distinguishing absorptions in the vinylic region of both the ^{1}H and ^{13}C spectra make it possible to identify each isomer by its NMR spectrum.

The ketone IR absorption of 3-methyl-2-cyclohexenone occurs near 1690 cm^{-1} because the double bond is next to the ketone group. The ketone IR absorption of 3-cyclopentenyl methyl ketone occurs near 1715 cm^{-1}, the usual position for ketone absorption.

13.46 The unknown compound has no degrees of unsaturation and has two different kinds of hydrogens. The unknown compound is $BrCH_2CH_2CH_2Br$.

13.47

(a)

$(CH_3)_2CHCCH_3$ (C=O; labels: 1 on $(CH_3)_2$, 3 on CH, 2 on CH_3)

1 = 0.95 δ (isopropyl group)
2 = 2.10 δ (methyl ketone)
3 = 2.43 δ (isopropyl group)

(b)

H_3C(1)(Br)C=C(H)(H) — H's: 2, 3

1 = 2.32 δ (methyl group attached to double bond)
2,3 = 5.35 δ, 5.54 δ (vinylic H)

13.48 Possible structures for $C_4H_7ClO_2$ are $CH_3CH_2CO_2CH_2Cl$ and $ClCH_2CO_2CH_2CH_3$. Chemical-shift data can distinguish between them.

$CH_3CH_2C(=O)OCH_2Cl$ **A** $ClCH_2C(=O)OCH_2CH_3$ **B**

In **A**, the protons attached to the carbon bonded to both oxygen and chlorine ($-OCH_2Cl$) absorb far downfield (5.0 – 6.0 δ). Because no signal is present in this region of the ^1H NMR spectrum given, the unknown must be **B**.

13.49

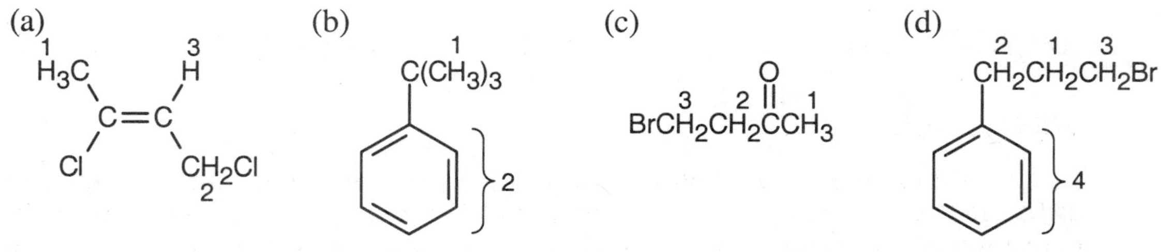

(a)	(b)	(c)	(d)
1 = 2.18 δ (allylic)	1 = 1.30 δ (saturated)	1 = 2.11 δ (next to C=O)	1 = 2.15 δ
2 = 4.16 δ (H,Cl bonded to same C)	2 = 7.30 δ (aromatic)	2 = 3.52 δ (next to C=O, Br)	2 = 2.75 δ (benzylic)
3 = 5.71 δ (vinylic)		3 = 4.40 δ (H,Br bonded to same C)	3 = 3.38 δ (H,Br bonded to same C)
The *E* isomer is also a correct answer			4 = 7.22 δ (aromatic)

13.50

(a) $(CH_3)_2CHCH_2Br$

(b) $CH_3CH(Cl)CH_2CH_2Cl$

13.51

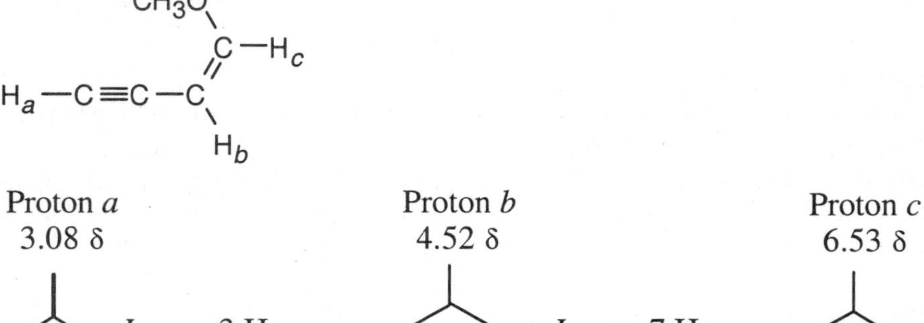

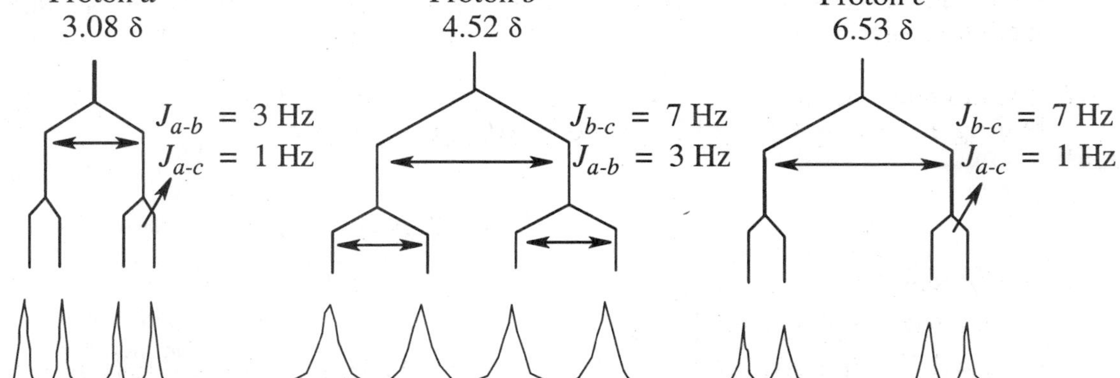

The absorptions in the ^1H NMR spectrum can be identified by comparison with the tree diagrams. H_a absorbs at 3.08 δ, H_b absorbs at 4.52 δ, and H_c absorbs at 6.35 δ.

13.52

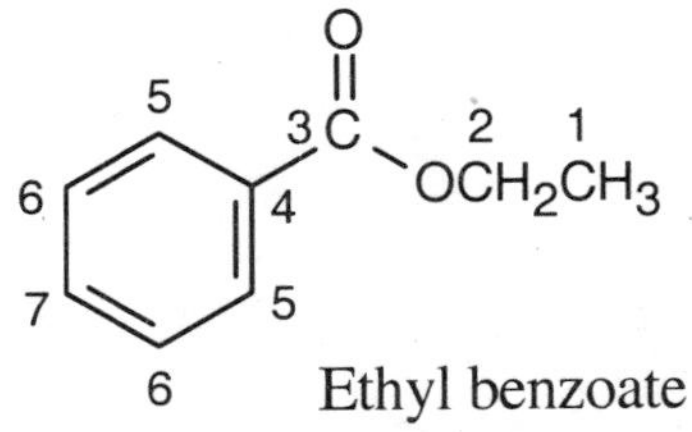

Ethyl benzoate ($C_6H_5COOCH_2CH_3$; carbons numbered: 1 = CH_3, 2 = OCH_2, 3 = C=O, 4 = ring carbon bonded to C=O, 5, 6, 7 = remaining ring carbons)

Carbon	*δ (ppm)*
1	14
2	61
3	166
4, 5, 6, 7	127–133 (4 peaks)

13.53 Compound **A** (4 multiple bonds and/or rings) must be symmetrical because it exhibits only six peaks in its ^{13}C NMR spectrum. Saturated carbons account for two of these peaks (δ = 15, 28 ppm), and unsaturated carbons account for the other four (δ = 119, 129, 131, 143 ppm).

^1H NMR shows a triplet (3 H at 1.1 δ), and a quartet (2 H at 2.5 δ), indicating the presence of an ethyl group. The other signals (4 H at 6.9 – 7.3 δ are due to aromatic protons.

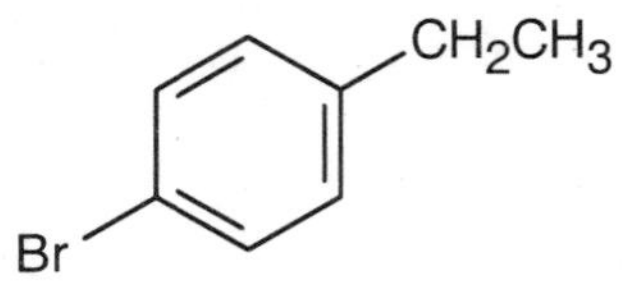

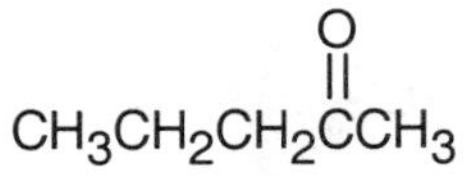

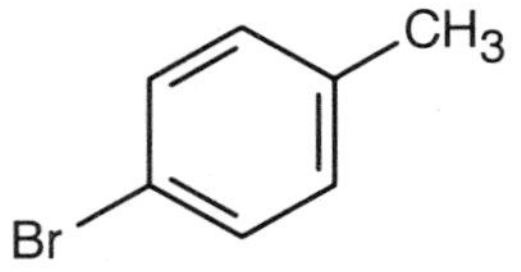

Br–C_6H_4–CH_2CH_3 (para)

A

13.54

(a) $CH_3CH_2CH_2CCH_3$ (C=O)

(b) Br–C_6H_4–CH_3 (para)

(c) C_6H_5–CH_2CH_2Br

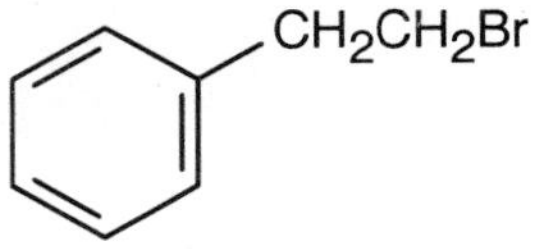

13.55 The peak in the mass spectrum at m/z = 84 is probably the molecular ion of the unknown compound and corresponds to a molecular weight of 84 (C_6H_{12} — one double bond or ring).

^{13}C NMR shows three different kinds of carbons and indicates a symmetrical hydrocarbon. The absorption at 132 δ is due to a vinylic carbon atom. A reasonable structure for the unknown is 3-hexene. Data do not distinguish between cis and trans isomers.

$CH_3CH_2CH{=}CHCH_2CH_3$ 3-Hexene

13.56 Compound **A**, a hydrocarbon having M^+ = 96, has the formula C_7H_{12}, indicating two degrees of unsaturation. Because it reacts with BH_3, Compound **A** contains a double bond. From the broadband decoupled ^{13}C NMR spectrum, we can see that C_7H_{12} is symmetrical, since it shows only five peaks.

The DEPT-135 spectrum of Compound **A** indicates three different CH_2 carbons, one $=CH_2$ carbon and one –C= carbon; the last two carbons are shown to be sp^2 hybridized by their chemical shifts. In the DEPT-135 spectrum of Compound **B**, the absorptions due to double bond carbons have been replaced by a CH carbon and a CH_2 carbon bonded to an electronegative group.

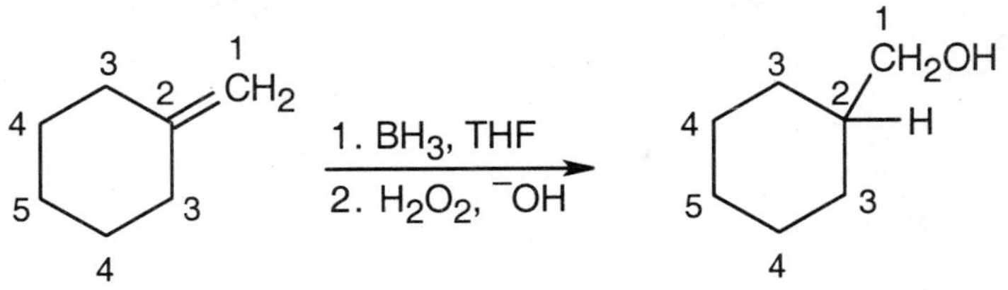

Compound **A** →(1. BH_3, THF; 2. H_2O_2, ^{-}OH) Compound **B**

Compound **A**		Compound **B**	
1	106.9 δ	1	68.2 δ
2	149.7 δ	2	40.5 δ
3	35.7 δ	3	29.9 δ
4, 5	26.8 δ, 28.7 δ	4, 5	26.1 δ, 26.9 δ

13.57 The IR absorption indicates that **C** is an alcohol. From M^+, we can arrive at a molecular formula of $C_5H_{10}O$, which indicates one degree of unsaturation. The broadband-decoupled spectrum shows five peaks; two are due to a double bond, and one is due to a carbon bonded to an electronegative atom (O).

The DEPT spectra show that **C** contains 4 CH_2 carbons and one CH carbon, and that **C** has a monosubstituted double bond. $HOCH_2CH_2CH_2CH{=}CH_2$ is a likely structure for **C**.

13.58 Compound **D** is very similar to Compound **C**. The DEPT spectra make it possible to distinguish between the isomers. **D** has 2 CH carbons, one CH_3 carbon, and 2 CH_2 carbons, and, like **C**, has a monosubstituted double bond. The peak at 74.4 δ is due to a secondary alcohol. The second structure shown is also a satisfactory answer.

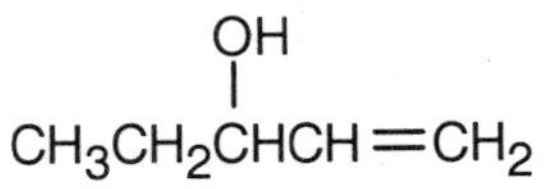

$CH_3CH_2CH(OH)CH{=}CH_2$ Compound **D** $CH_3CH(OH)CH_2CH{=}CH_2$

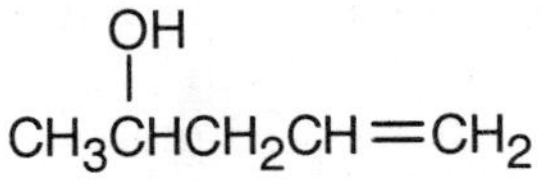

13.59 Compound **E** has two degrees of unsaturation and has two equivalent carbons (because its broadband-decoupled spectrum shows only 6 peaks). Two carbons absorb in the vinylic region of the spectrum; because one is a CH carbon and the other is a CH_2 carbon, **E** contains a monosubstituted double bond. The peak at 165.8 δ (not seen in the DEPT spectra) is due to a carbonyl group.

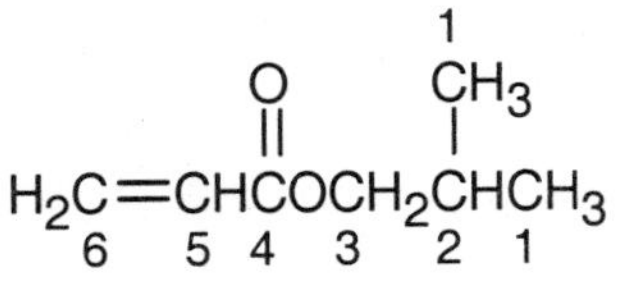

$H_2C{=}CHC(O)OCH_2CH(CH_3)CH_3$ (carbons numbered 6 5 4 3 2 1; 1 = CH_3)

Compound **E**

Carbon	*δ (ppm)*
1	19.1
2	28.0
3	70.5
4	165.8
5 }	129.8
6 }	129.0

13.60

$\xrightarrow{HBr}$

Compound **F**		Compound **G**	
Carbon	*δ (ppm)*	*Carbon*	*δ (ppm)*
1	132.4	1	56.0
2	32.2	2	39.9
3	29.3	3	27.7
4	27.6	4	25.1

13.61 Make a model of one enantiomer of 3-methyl-2-butanol and orient it as a staggered Newman projection along the C2-C3 bond. The *S* enantiomer is pictured.

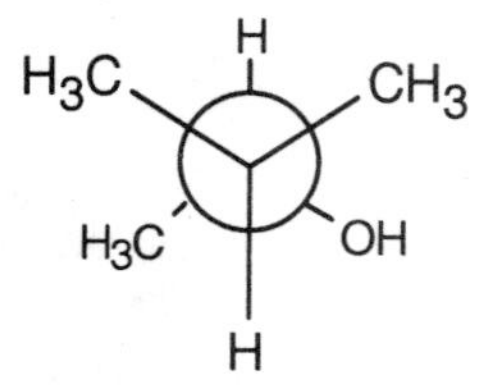

Because of the chirality center at C2, the two methyl groups at the front of the projection aren't equivalent. Since the methyl groups aren't equivalent, their carbons show slightly different signals in the ^{13}C NMR.

13.62 Commercial 2,4-pentanediol is a mixture of three stereoisomers - (*R*,*R*), (*S*,*S*), and (*R*,*S*). The meso isomer shows three signals in its ^{13}C NMR spectrum. Its diastereomers, the *R*,*R* and *S*,*S* enantiomeric pair, also show three signals, but two of these signals occur at different δ values from the meso isomer. This is expected, because diastereomers differ in physical and chemical properties.

A Look Ahead

13.63 The product (M^+= 88) has the formula $C_4H_8O_2$. the IR absorption indicates that the product is an ester. The ^{1}H NMR shows an ethyl group and an $-OCH_3$ group.

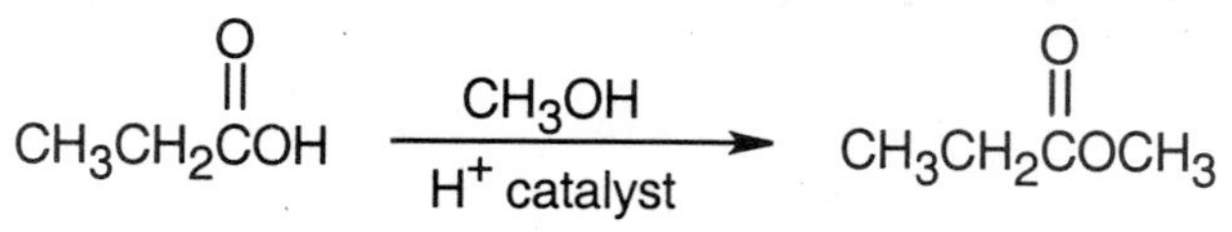

$$CH_3CH_2\overset{O}{\overset{\|}{C}}OH \xrightarrow[H^+ \text{ catalyst}]{CH_3OH} CH_3CH_2\overset{O}{\overset{\|}{C}}OCH_3$$

13.64 The product is a methyl ketone.

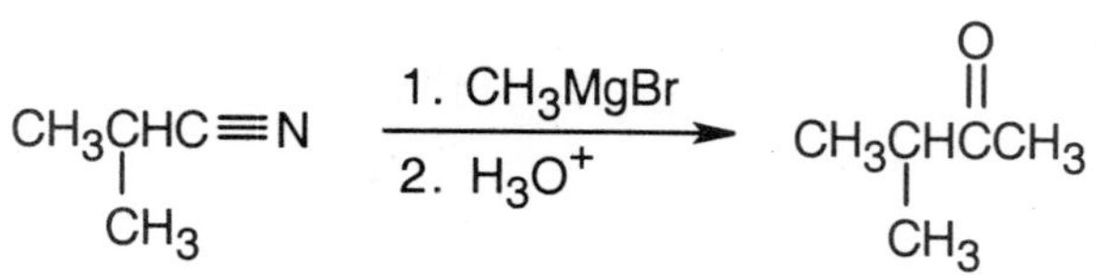

$$CH_3\underset{CH_3}{\underset{|}{C}}HC{\equiv}N \xrightarrow[2.\ H_3O^+]{1.\ CH_3MgBr} CH_3\underset{CH_3}{\underset{|}{C}}H\overset{O}{\overset{\|}{C}}CH_3$$

Molecular Modeling

13.65

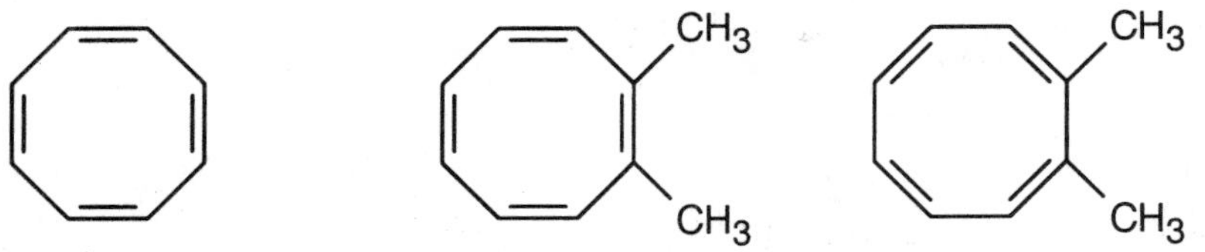

COT 1,2-DimethylCOT 2,3-DimethylCOT

COT has a non-planar tub shape, in which short and long C–C bonds alternate around the ring. Because the tub shape prevents orbital overlap between adjacent π orbitals, COT is not a resonance hybrid. However, all of the carbons in COT are chemically identical and have identical chemical shifts. The dimethylCOT derivatives have the same tub-shaped skeleton, but the methyl groups of each isomer have different chemical environments relative to each other and to the tub. Thus, each dimethyl derivative is expected to have a different ^{13}C NMR spectrum, although both spectra should show 5 peaks.

13.66

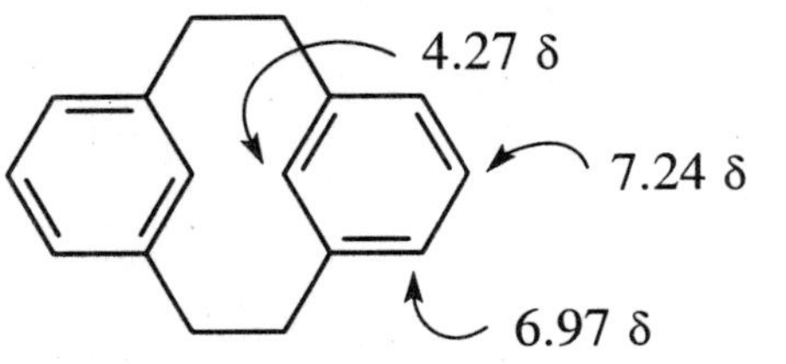

4.27 δ 7.24 δ 6.97 δ

13.67 2-Chloronorcamphor shows "W" coupling between H_a and H_b. None of the protons in 2-chlorocamphor have a "W" relationship with H_a.

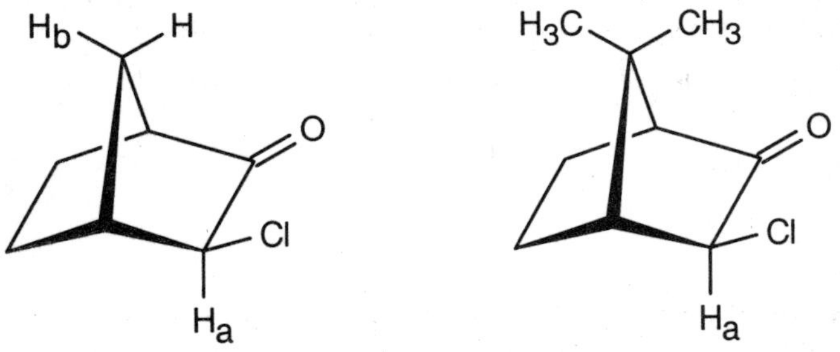

2-Chloronorcamphor 2-Chlorocamphor

Review Unit 5: Spectroscopy

Major Topics Covered (with vocabulary):

Mass Spectrometry:
cation radical mass spectrum base peak molecular ion alpha cleavage McLafferty rearrangement dehydration

The Electromagnetic Spectrum:
electromagnetic radiation wavelength frequency hertz amplitude quanta absorption spectrum

Infrared Spectroscopy:
wavenumber fingerprint region

Nuclear Magnetic Resonance Spectroscopy:
nuclear magnetic resonance rf energy effective magnetic field shielding downfield upfield chemical shift delta scale FT-NMR DEPT ^{13}C NMR integration multiplet spin-spin splitting coupling $n + 1$ rule coupling constant tree diagram

Types of Problems:

After studying these chapters, you should be able to:

- Write molecular formulas corresponding to a given molecular ion.
- Use mass spectra to determine molecular weights and base peaks, to distinguish between hydrocarbons, and to identify selected functional groups by their fragmentation patterns.
- Calculate the energy of electromagnetic radiation, and convert from wavelength to wavenumber and *vice versa.*
- Identify functional groups by their infrared absorptions.
- Use IR and MS to monitor reaction progress.

- Calculate the relationship between delta value, chemical shift, and spectrometer operating frequency.
- Identify nonequivalent carbons and hydrogens, and predict the number of signals appearing in the ^{1}H NMR and ^{13}C NMR spectra of compounds.
- Assign resonances to specific carbons or hydrogens of a given structure.
- Propose structures for compounds, given their NMR spectra.
- Predict splitting patterns, using tree diagrams if necessary.
- Use NMR to distinguish between isomers and to identify reaction products.

Points to Remember:

* In mass spectrometry, the molecular ion is a cation radical. Further fragmentations of the molecular ion can be of two types – those that produce a cation plus a radical, and those that produce a different cation radical plus a neutral atom. In all cases, the fragment bearing the charge – whether cation or cation radical – is the one that is detected.

* Although mass spectrometry has many uses in research, we are interested in it for only a limited amount of data. The most important piece of information it provides for us is the molecular weight of an unknown. A mass spectrum can also show if an unknown is branched

or straight-chain (branched hydrocarbons have more complex spectra than their straight-chain isomers). Finally, if we know if certain groups are present, we can obtain structural information about an unknown compound. For example if we know that a ketone is present, we can look for peaks that correspond to alpha cleavage and/or McLafferty rearrangement fragments.

* The position of an IR absorption is related to both the strength of the bond and to the nature of the two atoms that form the bond. For example, a carbon-carbon triple bond absorbs a higher frequency than a carbon-carbon double bond, which absorbs at a higher frequency than a carbon-carbon single bond. Bonds between two atoms of significantly different mass absorb at higher frequencies than bonds between two atoms of similar mass.

* Not all IR absorptions are due to bond stretches. Many of the absorptions in the fingerprint region of an IR spectrum are due to bending and out-of-plane motions.

* It is confusing, but true, that larger δ values in an NMR spectrum are associated with nuclei that are less shielded, and that these nuclei require a lower field strength for resonance. Nuclei with small values of δ are more shielded and require a higher field strength for resonance.

* Both ^{13}C NMR and ^{1}H NMR are indispensable for establishing the structure of an organic compound. ^{13}C NMR indicates if a molecule is symmetrical and shows the types of carbons in a molecule (by DEPT NMR). ^{1}H NMR shows how the carbons are connected (by spin-spin splitting) and how many protons are in the molecule (by integration). Both types of spectra show (by chemical shift) the electronic environment of the magnetic nuclei.

Self-Test:

Compound **A** is a hydrocarbon with $M^+ = 78$. What is its molecular formula? What is its degree of unsaturation? Draw three possible formulas for **A**. The ^{13}C NMR spectrum of **A** shows 3 peaks – at 18.5 δ, 69.4 δ and 82.4 δ, and the ^{1}H NMR spectrum shows two peaks. What is the structure of **A**? What significant absorptions would you see in the IR spectrum of **A**?

Compound **B** has the molecular formula C_8H_{14}, and shows 3 peaks in its ^{1}H NMR spectrum – at 1.7 δ (6H), 2.1 δ (4H) and 4.7 δ (4H). All 3 peaks are singlets. **B** also shows an IR absorption at 890 cm^{-1}. What is a possible structure for **B**? If you're still not sure, the following peaks were observed in the ^{13}C NMR spectrum of **B**: 22 δ, 36 δ, 110 δ, 146 δ. The peaks at 36 δ and 110 δ were negative signals in the DEPT-135 spectrum, and the peak at 22 δ was a positive signal.

Compound **C** is a hydrocarbon with $M^+ = 112$. What are possible molecular formulas for **C**? The five peaks in the ^{1}H NMR spectrum of **C** are all singlets and occur at the following δ values: 0.9 δ (9 H), 1.8 δ (3 H), 1.9 δ (2 H), 4.6 δ (1 H) and 4.8 δ (1 H). An IR absorption at 890 cm^{-1} is also present What is the structure of **C**?

$$CH_3CH_2\overset{\overset{\displaystyle O}{\|}}{C}CH_2OH$$

D

Describe the ^{13}C NMR and ^{1}H NMR spectra of **D**. For the ^{1}H NMR spectrum, include the spin-spin splitting patterns, peak areas, and positions of the chemical shifts. Give two significant absorptions that you might see in the IR spectrum. Would you expect to see products of McLafferty rearrangement in the mass spectrum of **D**? Of alpha cleavage?

Multiple choice:

1. Which of the following formulas could not arise from a compound with $M^+ = 142$ that contains C, H, and possibly O?
(a) $C_{11}H_{10}$ (b) $C_{10}H_8O$ (c) $C_9H_{18}O$ (d) $C_8H_{16}O_2$

2. Which of the following mass spectrum fragments is a cation, rather than a cation radical?
(a) base peak (b) molecular ion (c) product of McLafferty rearrangement (d) product of dehydration of an alcohol

3. Which element contributes significantly to $(M+1)^+$?
(a) N (b) H (c) C (d) O

4. In which type of spectroscopy is the wavelength of absorption the longest?
(a) NMR spectroscopy (b) infrared spectroscopy (c) ultraviolet spectroscopy
(d) X-ray spectroscopy

5. Which functional group is hard to detect in an IR spectrum?
(a) aldehyde (b) –C≡C– (c) alcohol (d) ether

6. IR spectroscopy is especially good for:
(a) determining if an alkyne triple bond is at the end of a carbon chain or is in the middle
(b) predicting the type of carbonyl group that is present in a compound
(c) deciding if a double bond is cis-disubstituted or trans-disubstituted
(d) all of these situations

7. If a nucleus is strongly shielded:
(a) The effective field is smaller than the applied field, and the absorption is shifted downfield. (b) The effective field is larger than the applied field, and the absorption is shifted upfield. (c) The effective field is smaller than the applied field, and the absorption is shifted upfield. (d) The effective field is larger than the applied field, and the absorption is shifted downfield.

8. When the operating frequency of an ^1H NMR spectrometer is changed:
(a) The value of chemical shift in δ and of the coupling constant remain the same. (b) The values of chemical shift in Hz and of the coupling constant change. (c) The value of chemical shift in Hz remains the same, but the coupling constant changes. (d) The values of chemical shift in δ and of the coupling constant change.

9. ^{13}C NMR can provide all of the following data except:
(a) the presence or absence of symmetry in a molecule (b) the connectivity of the carbons in a molecule (c) the chemical environment of a carbon (d) the number of hydrogens bonded to a carbon

10. Which kind of carbon is detected in DEPT-90 ^{13}C NMR spectroscopy?
(a) primary carbon (b) secondary carbon (c) tertiary carbon (d) quaternary carbon

Chapter 14 – Conjugated Dienes and Ultraviolet Spectroscopy

Chapter Outline

I. Conjugated Dienes (Sections 14.1 – 14.9).
  A. Preparation of conjugated dienes (Section 14.1).
    1. Base-induced elimination of allylic halides is the most common method.
    2. In industry, thermal cracking is sometimes used.
  B. Properties of conjugated dienes (Section 14.2 – 14.4).
    1. Stability of conjugated dienes (Section 14.2).
      a. Heats of hydrogenation show that conjugated dienes are somewhat more stable than nonconjugated dienes.
      b. Because conjugated dienes are more stable and contain less energy, they release less heat on hydrogenation.
    2. Molecular orbital description of 1,3-butadiene (Section 14.3).
      a. The stability of 1,3-butadiene may be due to the greater amount of *s* character of the C–C single bond between the double bonds.
      b. Molecular orbital theory offers another explanation.
        i. If we combine 4 adjacent *p* orbitals, we generate a set of 4 molecular orbitals.
        ii. Bonding electrons go into the lower two MOs.
        iii. The lowest MO has a bonding interaction between C2 and C3 that gives that bond partial double-bond character.
      c. The π electrons of butadiene are delocalized over the entire π framework.
    3. Bond lengths in 1,3-butadiene (Section 14.4).
      a. The C2–C3 bond length is 6 pm shorter than a C–C single bond.
      b. Both explanations (above) can account for this bond shortening.
  C Reactions of conjugated dienes (Sections 14.5 – 14.9).
    1. Electrophilic addition to conjugated dienes (Sections 14.5 – 14.6).
      a. Conjugated dienes react in electrophilic addition reactions to give products of both 1,2 addition and 1,4 addition (Section 14.5).
        i. Addition of an electrophile gives an allylic carbocation intermediate that is resonance-stabilized.
        ii. Addition can occur at either end of the allylic carbocation to yield two products.
      b. The ratio of products can vary if the reaction is carried out under conditions of kinetic control or of thermodynamic control (Section 14.6).
        i. Under conditions of thermodynamic control (high temperature), the product whose formation has a larger negative value of $\Delta G°$ forms in greater amounts.
        ii. Under conditions of kinetic control (lower temperature), the product whose formation has the lower energy of activation forms in greater amounts.
        iii. In electrophilic addition reactions of conjugated dienes, the 1,2 adduct forms preferentially at low T, and the 1,4 adduct forms preferentially at high T.
    2. Diene polymers (Section 14.7).
      a. Like simple alkenes, conjugated dienes can polymerize.
        i. Because double bonds remain in the polymer, cis-trans isomerism is possible.
        ii. Polymerization can be initiated by either a radical or by acid.
        iii. Polymerization occurs by 1,4 addition.

b. Natural rubber is a polymer of isoprene with *Z* double-bond stereochemistry, and *gutta-percha* is a polymer of isoprene with *E* double-bond stereochemistry.
c. Synthetic rubber and isoprene (a polymer of chloroprene) are also diene polymers.
d. Rubber needs to be hardened by vulcanization.
Heating rubber with sulfur forms cross-links that lock the chains together.

3. The Diels-Alder reaction (Sections 14.8 – 14.9).
   a. How the reaction occurs (Section 14.8).
      i. A diene can react with certain alkenes to form a cyclic product.
      ii. This reaction, the Diels-Alder reaction, forms two C–C bonds in a single step.
      iii. The reaction occurs by a pericyclic mechanism, which takes place in a single step by a cyclic redistribution of electrons.
      iv. In the reaction, σ overlap occurs between the two alkene *p* orbitals with the two *p* orbitals on carbons 1 and 4 of the diene.
      v. The two alkene carbons and C1 and C2 of the diene rehybridize from sp^2 to sp^3, and C2 and C3 of the diene remain sp^2 hybridized.
   b. The dienophile (Section 14.9).
      i. The dienophile must have an electron-withdrawing group and may contain a triple bond.
      ii. The stereochemistry of the dienophile is maintained during the reaction.
      iii. Only *endo* product is formed because orbital overlap is greater in the transition state than for *exo* product.
      A substituent in a bicyclic ring system is *endo* if it is cis to the larger of the other two bridges.
   c. The diene.
      i. A diene must adopt an *s-cis* conformation in order to undergo the Diels-Alder reaction.
      ii. Some dienes can rotate to achieve an *s-cis* conformation; those that are rigid can't react.
      iii. Dienes that have fixed *s-cis* geometry are very reactive.

II. Ultraviolet spectroscopy (Sections 14.10 – 14.13).
   A. Principles of ultraviolet spectroscopy (Section 14.10).
      1. The ultraviolet region of interest is between the wavelengths 200 nm and 400 nm.
      2. The energy absorbed is used to promote a π electron in a conjugated system from one orbital to another.
   B. Ultraviolet spectrum of 1,3-butadiene (Section 14.11).
      1. When 1,3-butadiene is irradiated with ultraviolet light, a π electron is promoted from the highest occupied molecular orbital (HOMO) to the lowest unoccupied molecular orbital.
      2. UV radiation of 217 nm is necessary to promote this transition.
      3. This transition is known as a $\pi \rightarrow \pi^*$ transition.
   C. The ultraviolet spectrum.
      1. A UV spectrum is a plot of absorbance vs. wavelength.
         a. The absorbance is $A = \log [I_0/I]$.
         b. I_0 = intensity of incident light.
         c. I = intensity of transmitted light.
         d. The baseline is zero absorbance.
      2. For a specific substance, *A* is related to the molar absorptivity (ε).
         a. Molar absorptivity is the absorbance of a sample whose concentration is 1 mol/L with a pathlength of 1 cm..
         b. $A = \varepsilon \times c \times l$
         c. The range of ε is 10,000 – 25,000.
      3. UV spectra usually consist of a single broad peak, whose maximum is λ_{max}.

D. Interpreting UV spectra (Section 14.12).
1. The wavelength necessary for a $\pi \rightarrow \pi^*$ transition depends on the energy difference between HOMO and LUMO.
2. By measuring this difference, it is possible to learn about the extent of conjugation in a molecule.
3. As the extent of conjugation increases, λ_{max} increases.
4. Different types of conjugated systems have characteristic values of λ_{max}.

F. Colored organic compounds (Section 14.13).
Compounds with extensive systems of conjugated bonds absorb in the visible range of the electromagnetic spectrum (400 – 800 nm).

Solutions to Problems

14.1 To find conjugation, look for alternating double bonds (or double and triple bonds).

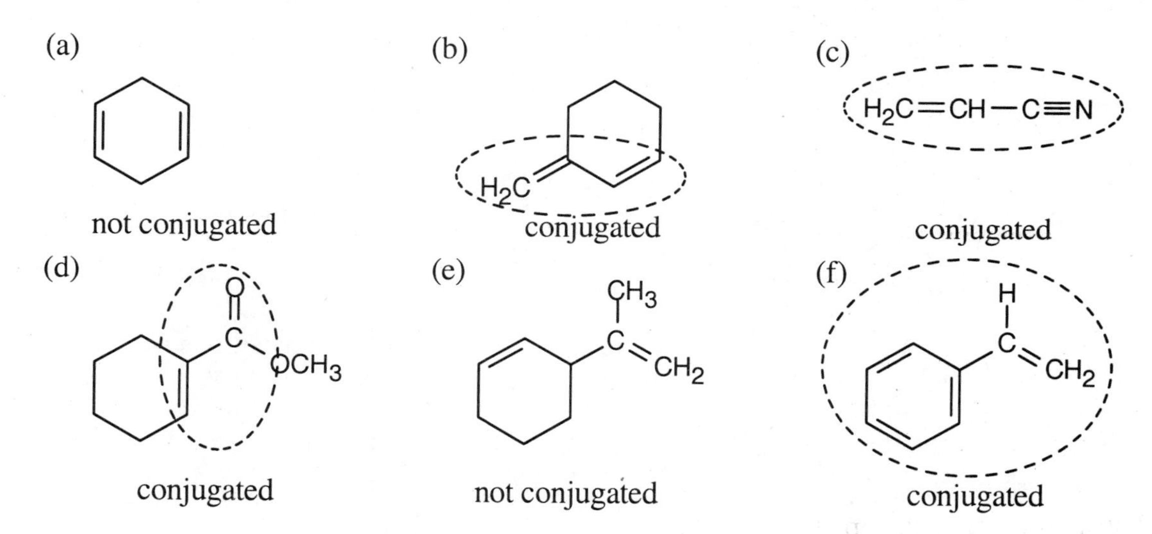

14.2 We would expect $\Delta H_{hydrog} = -126 + (-126) = -252$ kJ/mol for allene if the heat of hydrogenation for each double bond were the same as that for an isolated double bond. The measured ΔH_{hydrog}, –298 kJ/mol, is 46 kJ/mol more negative than the expected value. Thus, allene is higher in energy (less stable) than a nonconjugated diene, which in turn is less stable than a conjugated diene.

14.3

$CH_3CH{=}CHCH{=}CH_2$ 1,3-Pentadiene

Product	*Name*	*Results from:*
$CH_3CH{=}CHCH(Cl)CH_3$	4-Chloro-2-pentene	1,2 addition 1,4 addition
$CH_3CH_2CH(Cl)CH{=}CH_2$	3-Chloro-1-pentene	1,2 addition
$CH_3CH_2CH{=}CHCH_2Cl$	1-Chloro-2-pentene	1,4 addition

14.4

$$\underset{\textbf{D}}{\overset{\delta+}{CH_3CH_2CH}{=\!=}CH{=\!=}\overset{\delta+}{CH_2}} \xleftarrow[\text{protonation on carbon 4}]{H^+} CH_3CH{=}CHCH{=}CH_2 \xrightarrow[\text{protonation on carbon 1}]{H^+} \underset{\textbf{A}}{\overset{\delta+}{CH_3CH}{=\!=}CH{=\!=}\overset{\delta+}{CHCH_3}}$$

$$\underset{\textbf{C}}{CH_3\overset{+}{C}HCH_2CH{=}CH_2} \xleftarrow[\text{protonation on carbon 3}]{H^+} CH_3CH{=}CHCH{=}CH_2 \xrightarrow[\text{protonation on carbon 2}]{H^+} \underset{\textbf{B}}{CH_3CH{=}CHCH_2\overset{+}{C}H_2}$$

A and **D**, which are resonance-stabilized, are formed over **B** and **C**, which are not. The positive charge of allylic carbocation **A** is delocalized over two secondary carbons, while the positive charge of carbocation **D** is delocalized over one secondary and one primary carbon. We therefore predict that carbocation **A** is the major intermediate formed, and that 4-chloro-2-pentene predominates. Note that this product results from both 1,2– and 1,4–addition.

14.5

$$\underset{\text{3-Bromo-1-butene}}{CH_3\overset{Br}{\overset{|}{C}}HCH{=}CH_2} \rightleftharpoons \left[CH_3\overset{+}{C}HCH{=}CH_2 \longleftrightarrow CH_3CH{=}CH\overset{+}{C}H_2\right] + Br^- \rightleftharpoons \underset{\text{1-Bromo-2-butene}}{CH_3CH{=}CHCH_2Br}$$

Allylic halides can undergo slow dissociation to form stabilized carbocations. Both 3-bromo-1-butene and 1-bromo-2-butene form the same allylic carbocation, pictured above, on dissociation. Addition of bromide ion to the allylic carbocation then occurs to form a mixture of bromobutenes. Since the reaction is run under equilibrium conditions, the thermodynamically more stable 1-bromo-2-butene predominates.

14.6

$$H_2C{=}CHCH{=}CH_2 + X_2 \xrightarrow{1,2} H_2\overset{X}{\overset{|}{C}}\underset{X}{\underset{|}{C}}HCH{=}CH_2 \quad \text{monosubstituted double bond}$$

$$H_2C{=}CHCH{=}CH_2 + X_2 \xrightarrow{1,4} H_2\overset{X}{\overset{|}{C}}CH{=}CH\underset{X}{\underset{|}{C}}H_2 \quad \text{disubstituted double bond}$$

1,4–adducts are more stable than 1,2–adducts because disubstituted double bonds are more stable than monosubstituted double bonds (see Chapter 6).

14.7 The initiator may be either a radical or a cation. Diene polymerization is a 1,4 addition process that forms a polymer whose monomer units have 4 carbons and that contains a double bond every 4 bonds.

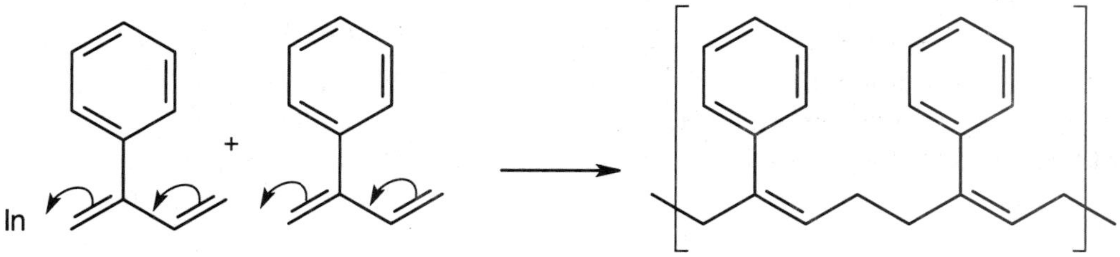

In

14.8

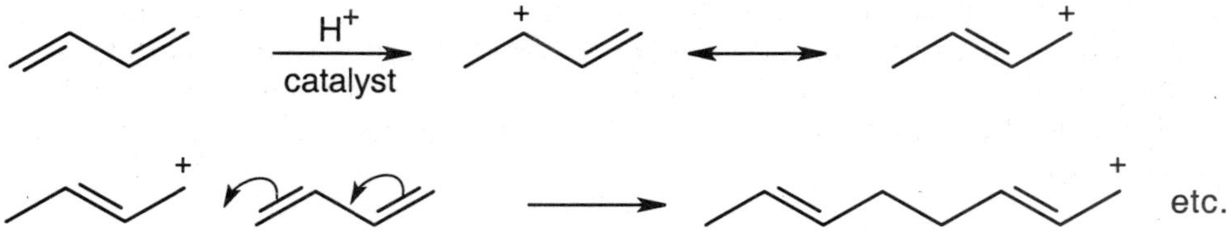

H^+
catalyst

etc.

14.9 Draw the reactants in an orientation that shows where the new bonds will form. Form the new bonds by connecting the two reactants, removing two double bonds, and relocating the remaining double bond so that it lies between carbon 2 and carbon 3 of the diene. The substituents on the dienophile retain their trans relationship in the product.

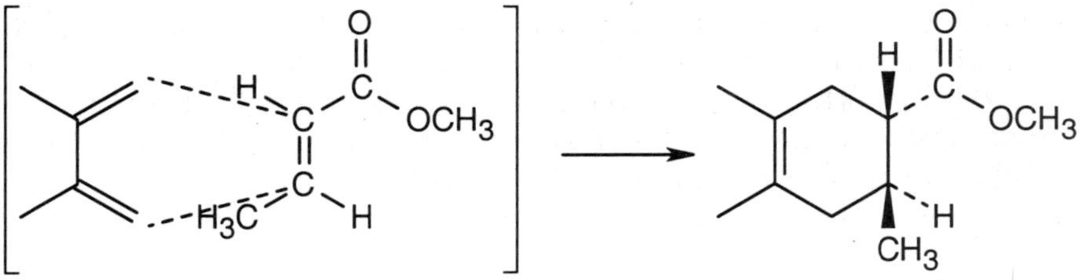

OCH_3 H_3C H CH_3

trans product

14.10

Good dienophiles: (a) $H_2C{=}CHCCl$ (with C=O)   (d)

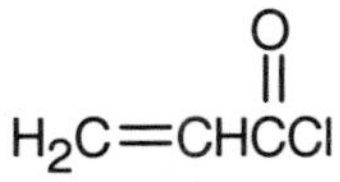

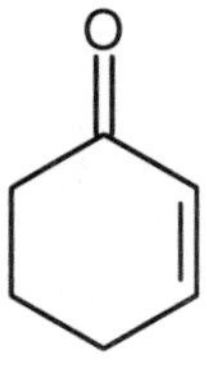

Poor dienophiles: (b) $H_2C{=}CHCH_2CH_2COCH_3$ (with C=O)   (c)    (e)

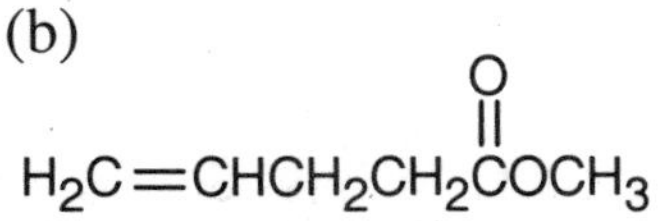

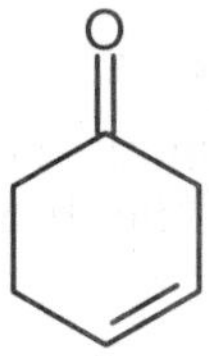

Compound (a) and (d) are good dienophiles because they have electron-withdrawing groups conjugated with a carbon-carbon double bond. Alkene (c) is a poor dienophile because it has no electron-withdrawing functional group. Compounds (b) and (e) are poor dienophiles because their electron-withdrawing groups are not conjugated with the double bond.

14.11 (a) This diene has an *s*-cis conformation and should undergo Diels-Alder cycloaddition.

(b) This diene has an *s*-trans conformation. Because the double bonds are in a fused ring system, it is not possible for them to rotate to an *s*-cis conformation.

(c) Rotation can occur about the single bond of this *s*-trans diene. The resulting *s*-cis conformation, however, has an unfavorable steric interaction of the interior methyl group with a hydrogen at carbon 1. Rotation to the *s*-cis conformation is therefore not favored energetically.

s-trans
(more stable)

s-cis
(less stable)

14.12 Rotation of the diene to the *s*-cis conformation must occur in order for reaction to take place.

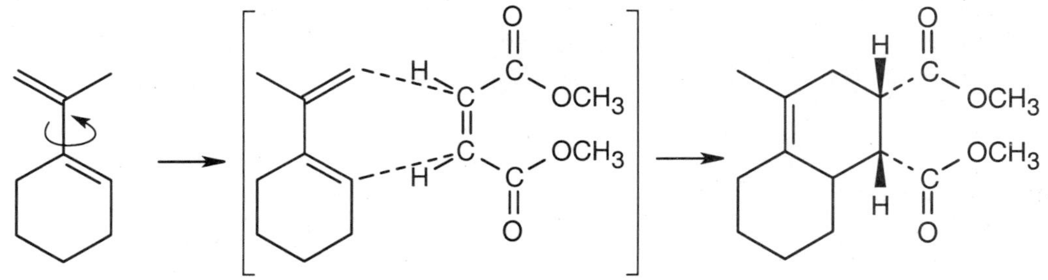

s-trans

s-cis

14.13

$$200 \text{ nm} = 200 \times 10^{-9} \text{ m} = 2 \times 10^{-7} \text{ m}$$

$$400 \text{ nm} = 400 \times 10^{-9} \text{ m} = 4 \times 10^{-7} \text{ m}$$

for $\lambda = 2 \times 10^{-7}$ m:

$$E = \frac{1.20 \times 10^{-4} \text{ kJ/mol}}{\lambda \text{ (in m)}} = \frac{1.20 \times 10^{-4} \text{ kJ/mol}}{2.0 \times 10^{-7}} = 6.0 \times 10^{2} \text{ kJ/mol}$$

for $\lambda = 4 \times 10^{-7}$ m:

$$E = \frac{1.20 \times 10^{-4} \text{ kJ/mol}}{\lambda \text{ (in m)}} = \frac{1.20 \times 10^{-4} \text{ kJ/mol}}{4.0 \times 10^{-7}} = 3.0 \times 10^{2} \text{ kJ/mol}$$

The energy of electromagnetic energy in the region of the spectrum from 200 nm to 400 nm is 300 – 600 kJ/mol.

14.14

	UV	IR	^{1}H NMR (at 60 MHz)
Energy (in kJ/mol)	300 – 600	4.7 – 47	2.4×10^{-5}

The energy required for UV transitions is greater than the energy required for IR or ^{1}H NMR transitions.

14.15

$$\varepsilon = \frac{A}{C \times l}$$

Where ε = molar absorptivity
A = absorbance
l = sample pathlength (in cm)
C = concentration (in M)

In this problem:

$\varepsilon = 50{,}100 = 5.01 \times 10^4$

$l = 1.00$ cm

$A = 0.735$

$$C = \frac{A}{\varepsilon \times l} = \frac{0.735}{5.01 \times 10^4 \times 1.00} = 1.47 \times 10^{-5}\ \text{M}$$

14.16 All compounds having alternating single and multiple bonds should show ultraviolet absorption in the range 200–400 nm. Only compound (a) is not UV-active. All of the compounds pictured below are UV active.

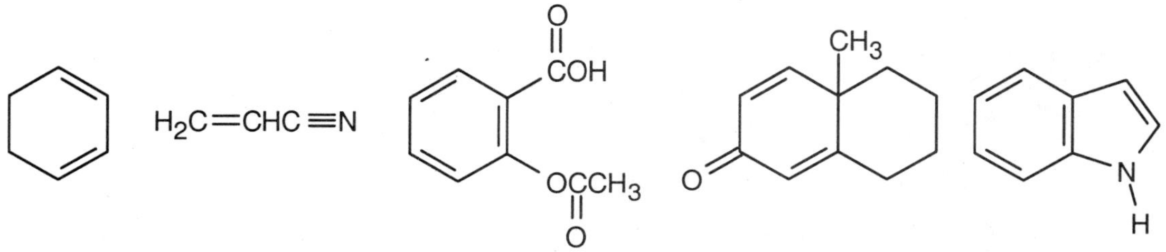

$H_2C{=}CHC{\equiv}N$

COH

OCCH$_3$

CH$_3$

O

N

H

Visualizing Chemistry

14.17

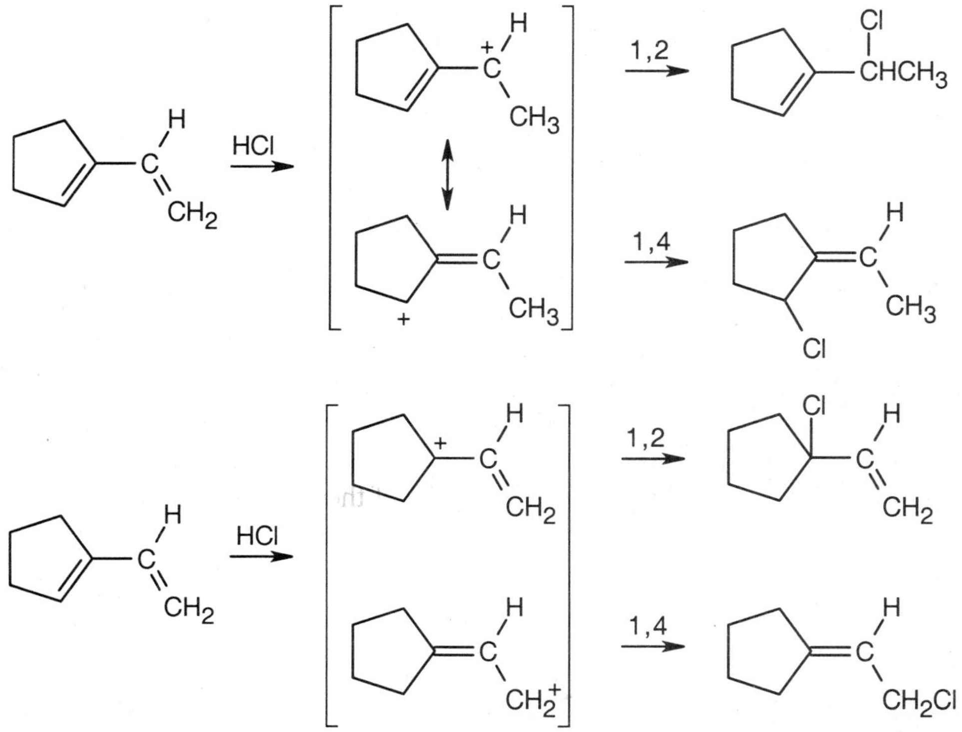

H C CH$_2$ HCl → [H C CH$_3$ ↕ H C CH$_3$] 1,2 → Cl CHCH$_3$; 1,4 → H C CH$_3$ Cl

H C CH$_2$ HCl → [H C CH$_2$; H C CH$_2^+$] 1,2 → Cl H C CH$_2$; 1,4 → H C CH$_2$Cl

14.18

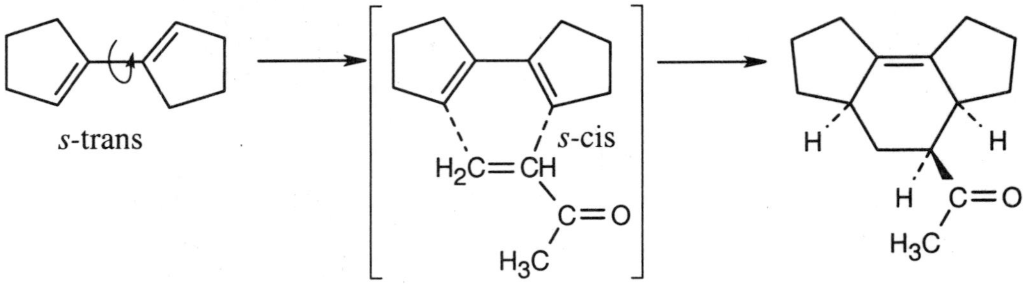

s-trans → [*s*-cis $H_2C{=}CH$ $C{=}O$ H_3C] → H H H $C{=}O$ H_3C

14.19 In the most stable conformation of conjugated dienes, the four carbons that form the conjugated system of π bonds must lie in a plane in order to have the overlap needed for stabilization. In the conformation pictured, the two double bonds don't lie in a plane. The stability from conjugation, which leads to lower energy, is lost, and the molecule behaves as if its double bonds were isolated – a higher energy conformation.

Additional Problems

14.20

(a) $CH_3CH{=}C(CH_3)CH{=}CHCH_3$
3-Methyl-2,4-hexadiene

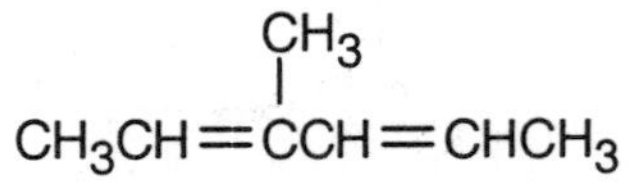

(b) $H_2C{=}CHCH{=}CHCH{=}CHCH_3$
1,3,5-Heptatriene

(c) $CH_3CH{=}C{=}CHCH{=}CHCH_3$
2,3,5-Heptatriene

(d) $CH_3CH{=}C(CH_2CH_2CH_3)CH{=}CH_2$
3-Propyl-1,3-pentadiene

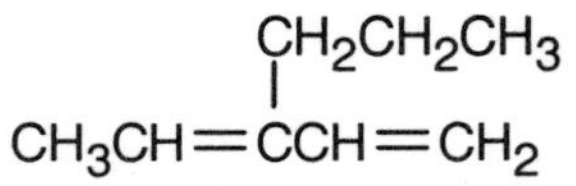

14.21

(a) 1 mol Br_2 → Br Br + Br Br

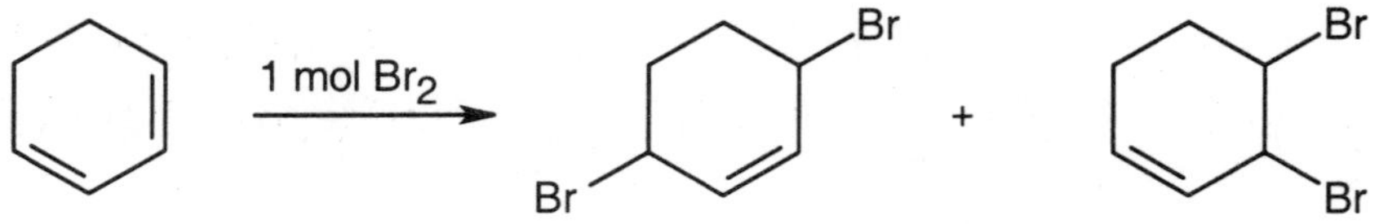

(b) 1. O_3 2. Zn, H_3O^+ → H C O H C O + O C H C O H

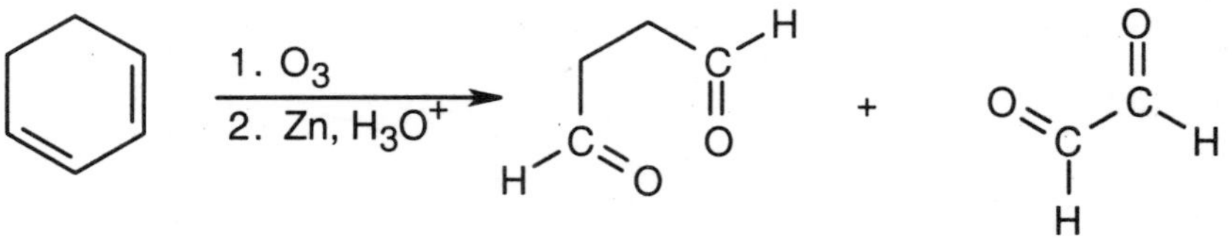

(c) 1 mol HCl, Ether → Cl

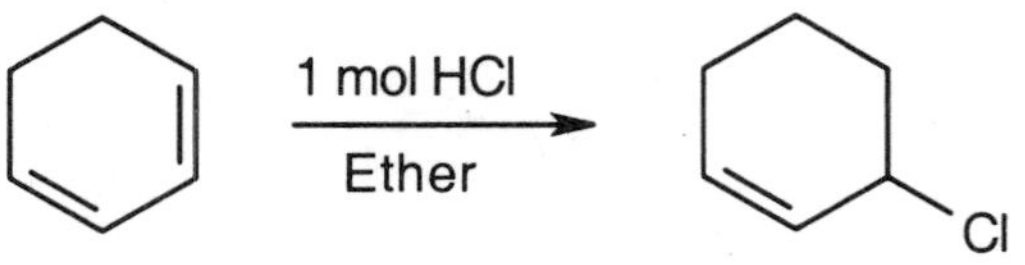

(d)

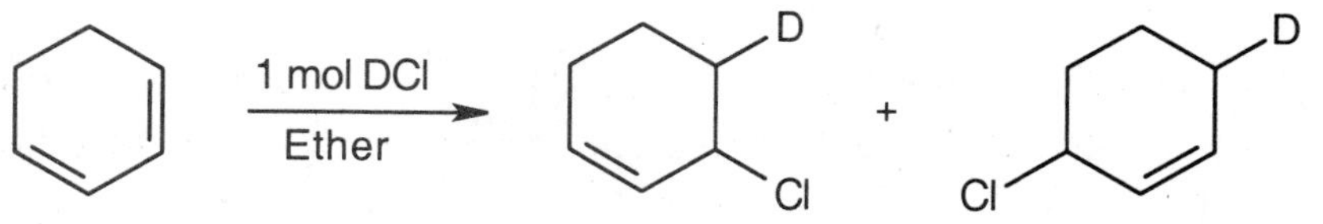

1 mol DCl / Ether

D, Cl + D, Cl

(e)

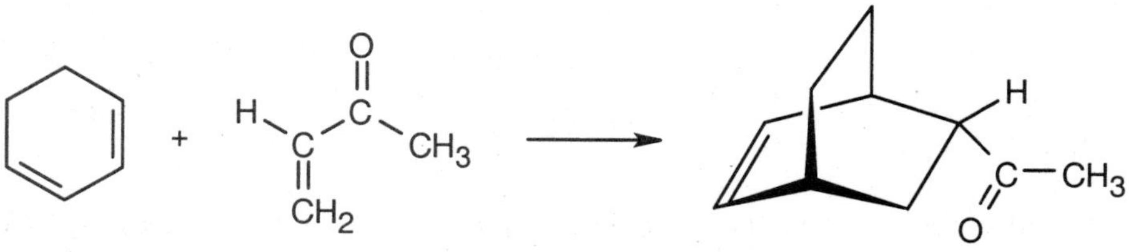

$+$ H–C(=CH$_2$)–C(=O)–CH$_3$ ⟶ H, C(=O)–CH$_3$

(f)

1. OsO_4
2. $NaHSO_3$

HO HO OH OH H H H H

meso

+ HO HO H H H OH H OH + H H OH OH HO H HO H

enantiomers

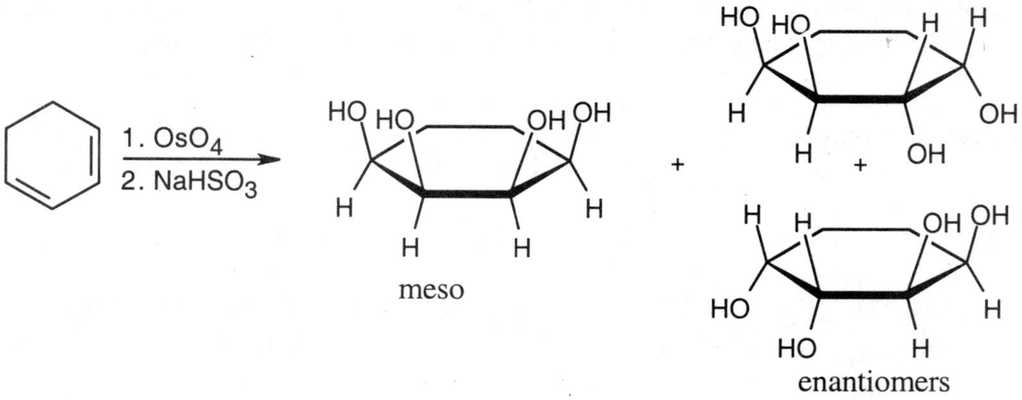

14.22

Conjugated dienes:

$CH_3CH{=}CHCH{=}CH_2$
1,3-Pentadiene

$H_2C{=}CH\overset{CH_3}{\overset{|}{C}}{=}CH_2$
2-Methyl-1,3-butadiene

Cumulated dienes:

$CH_3CH_2CH{=}C{=}CH_2$
1,2-Pentadiene

$CH_3CH{=}C{=}CHCH_3$
2,3-Pentadiene

$H_2C{=}C{=}C(CH_3)_2$
3-Methyl-1,2-butadiene

Nonconjugated diene:

$H_2C{=}CHCH_2CH{=}CH_2$
1,4-Pentadiene

14.23

	$CH_3CH_2C{\equiv}CCH_2CH_3$ 3-Hexyne	$CH_3CH{=}CHCH{=}CHCH_3$ 2,4-Hexadiene	$CH_3CH_2CH{=}C{=}CHCH_3$ 2,3-Hexadiene
^1H NMR:	2 peaks (triplet, quartet) below 2.0 δ	3 peaks, two in region 4.5–6.5 δ	5 peaks, two in region 4.5–6.5 δ
^{13}C NMR:	3 peaks, 8–55 δ (2) 65–85 δ (1)	3 peaks, 8–30 δ (1) 100–150 δ (2)	6 peaks, 8–55 δ (3) 100–150 δ (2) ~200 δ (1)(*sp* carbon)
UV absorption?	no	yes	no

2,4–Hexadiene can easily be distinguished from the other two isomers because it is the only isomer that absorbs in the UV region. The other two isomers show significant differences in their ^1H and ^{13}C NMR spectra and can be identified by either technique.

14.24

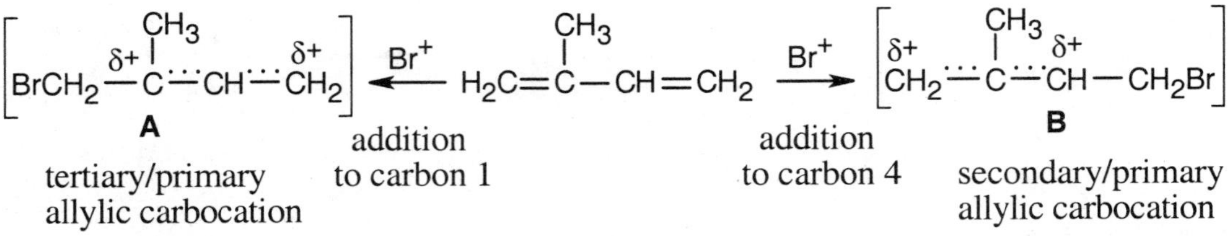

$$\left[BrCH_2\overset{\delta+}{-}\underset{}{\overset{CH_3}{\overset{|}{C}}}{\cdots}CH{\cdots}\overset{\delta+}{CH_2}\right]_{\mathbf{A}} \xleftarrow{Br^+} H_2C{=}\overset{CH_3}{\overset{|}{C}}{-}CH{=}CH_2 \xrightarrow{Br^+} \left[\overset{\delta+}{CH_2}{\cdots}\overset{CH_3}{\overset{|}{C}}{\cdots}\overset{\delta+}{CH}{-}CH_2Br\right]_{\mathbf{B}}$$

addition to carbon 1 (left); addition to carbon 4 (right)

A: tertiary/primary allylic carbocation

B: secondary/primary allylic carbocation

Tertiary/primary allylic carbocation **A** is more stable than secondary/primary allylic carbocation **B**. Since the products formed from the more stable intermediate predominate, 3,4–dibromo–3–methyl– 1–butene is the major product of 1,2 addition of bromine to isoprene.

14.25 Among the possible products:

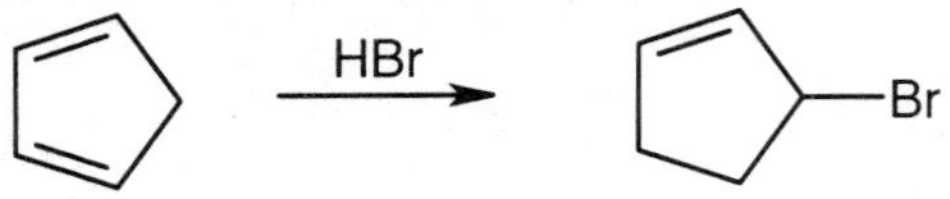

$$\text{cyclopentadiene} \xrightarrow{HBr} \text{3-bromocyclopentene (ring–Br)}$$

14.26

1 2 3 4
$CH{=}CHCH{=}CH_2$

1-Phenyl-1,3-butadiene

Protonation of carbon 1:

$C_6H_5CH_2\overset{\delta+}{C}H\text{---}CH\text{---}\overset{\delta+}{C}H_2$

A allylic

$\xrightarrow{Cl^-}$ $C_6H_5CH_2CH(Cl)CH{=}CH_2$ 3-Chloro-4-phenyl-1-butene

$\xrightarrow{Cl^-}$ $C_6H_5CH_2CH{=}CHCH_2Cl$ 1-Chloro-4-phenyl-2-butene

Protonation of carbon 2:

$C_6H_5\overset{+}{C}HCH_2CH{=}CH_2$

B

$\xrightarrow{Cl^-}$ $C_6H_5CH(Cl)CH_2CH{=}CH_2$ 4-Chloro-4-phenyl-1-butene

Protonation of carbon 3:

$C_6H_5CH{=}CHCH_2\overset{+}{C}H_2$

C

$\xrightarrow{Cl^-}$ $C_6H_5CH{=}CHCH_2CH_2Cl$ 4-Chloro-1-phenyl-1-butene

Protonation of carbon 4:

$C_6H_5\overset{\delta+}{C}H\text{---}CH\text{---}\overset{\delta+}{C}HCH_3$

D allylic

$\xrightarrow{Cl^-}$ $C_6H_5CH{=}CHCH(Cl)CH_3$ 3-Chloro-1-phenyl-1-butene

$\xrightarrow{Cl^-}$ $C_6H_5CH(Cl)CH{=}CHCH_3$ 1-Chloro-1-phenyl-2-butene

Carbocation **D** is most stable because it can use the π systems of both the benzene ring and the side chain to further delocalize positive charge. 3-Chloro-1-phenyl-1-butene is the major product because it results from cation **D** and because its double bond can conjugate with the benzene ring to provide extra stability.

14.27 A vinyl branch in a diene polymer is the result of an occasional 1,2–double bond addition to the polymer chain. Branching can also occur in cationic polymerization for the same reason.

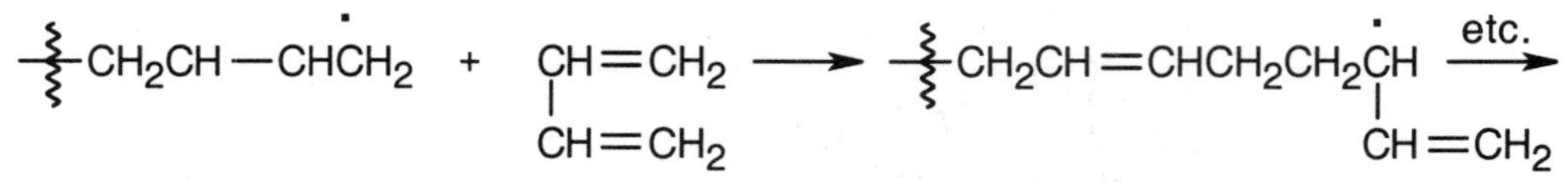

$$\text{-}CH_2CH{-}\dot{C}HCH_2 + \underset{CH=CH_2}{CH=CH_2} \longrightarrow \text{-}CH_2CH{=}CHCH_2CH_2\underset{CH=CH_2}{\dot{C}H} \xrightarrow{\text{etc.}}$$

14.28

$$\text{-}CH_2(H_3C)C{=}C(H)CH_2\text{-} \xrightarrow{O_3} \text{-}CH_2(H_3C)C{=}O + O{=}C(H)CH_2\text{-}$$

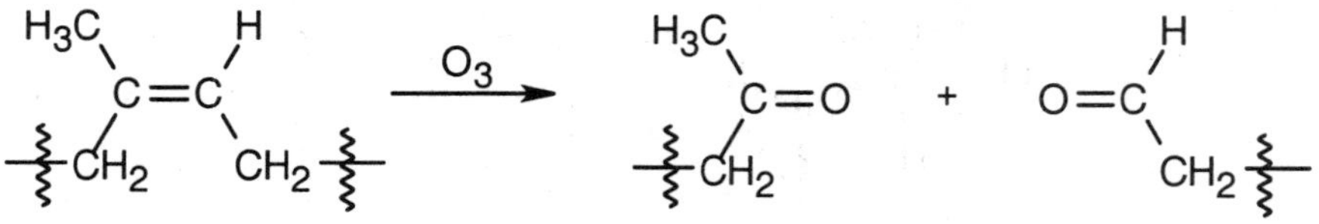

Ozone causes oxidative cleavage of the double bonds in rubber and breaks the polymer chain.

14.29 To absorb in the 200–400 nm range, an alkene must be conjugated. Since the double bonds of allene aren't conjugated, allene doesn't absorb light in the UV region.

14.30 Only compounds having alternating multiple bonds show $\pi \rightarrow \pi^*$ ultraviolet absorptions in the 200–400 nm range. Of the compounds shown, only pyridine (b) absorbs in this range.

14.31

(a)

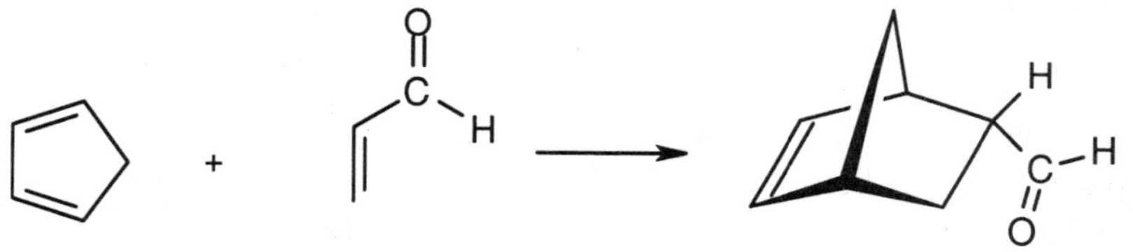

(b)

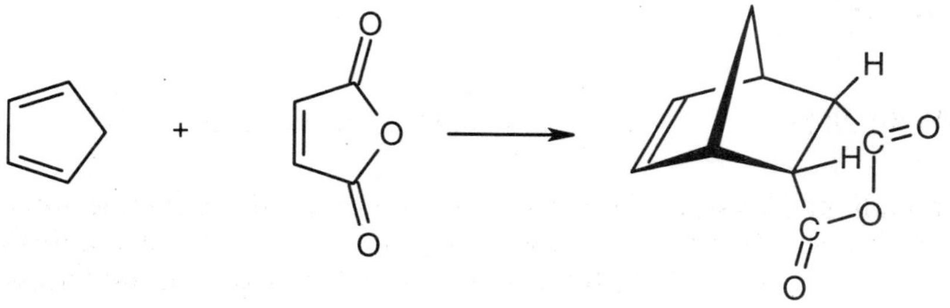

(c)

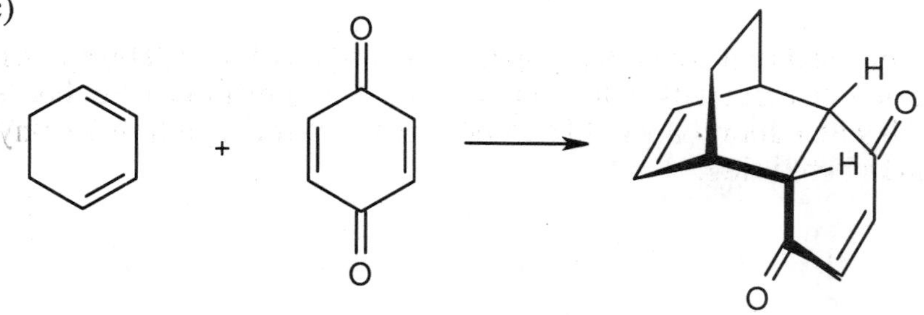

If two equivalents of cyclohexadiene are present for each equivalent of dienophile, you can also obtain a second product:

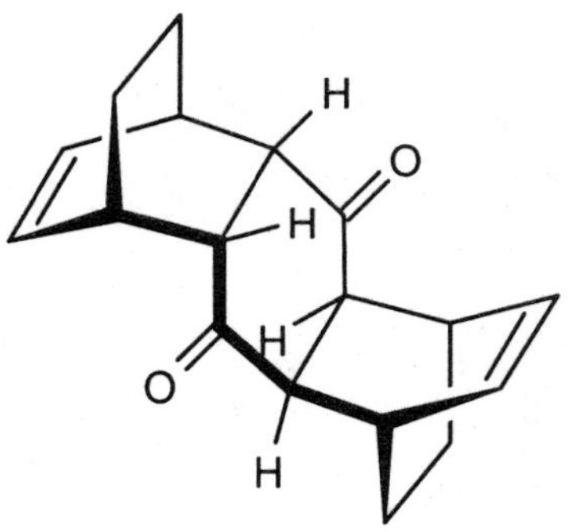

14.32

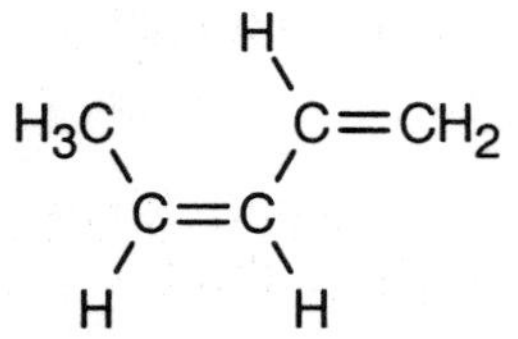

cis-1,3-Pentadiene 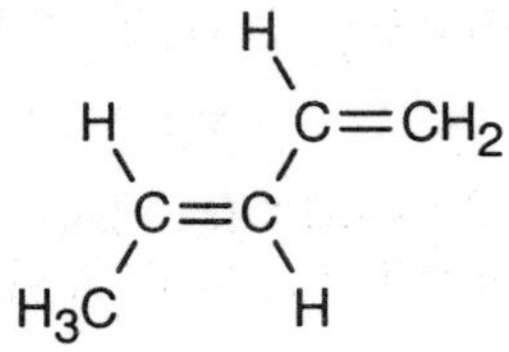 *trans*-1,3-Pentadiene

Both pentadienes are more stable in *s*-trans conformations. To undergo Diels-Alder reactions, however, they must rotate about the single bond between the double bonds to assume *s*-cis conformations.

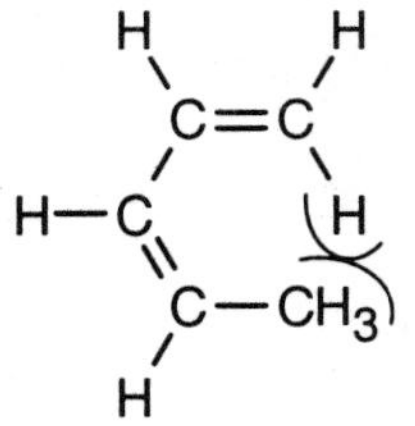

cis-1,3-Pentadiene 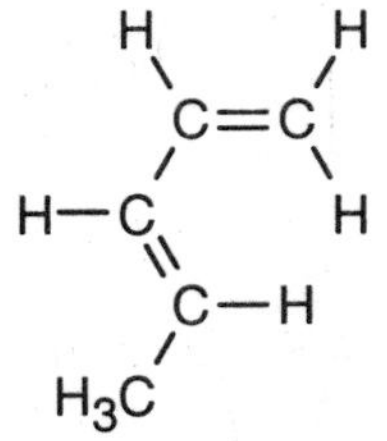 *trans*-1,3-Pentadiene

When *cis*-1,3-pentadiene rotates to the *s*-cis conformation, a steric interaction occurs between the methyl-group protons and a hydrogen on C1. Since it's more difficult for *cis*-1,3-pentadiene to assume the *s*-cis conformation, it is less reactive in the Diels-Alder reaction.

14.33 HC≡CC≡CH can't be used as a Diels-Alder diene because it is linear. The end carbons are too far apart to be able to react with a dienophile in a cyclic transition state. Furthermore, the product of Diels-alder addition would be impossibly strained, with two *sp*-hybridized carbons in a six-membered ring.

14.34

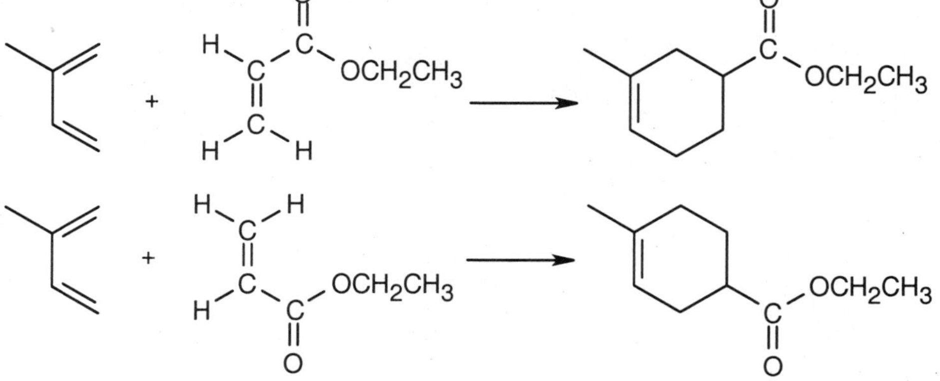

Two different orientations of the dienophile ester group are possible in the cyclic transition state, and two different products can form.

14.35 The most reactive dienophiles contain electron-withdrawing groups.

Most reactive ⟶ *Least reactive*

$(NC)_2C{=}C(CN)_2$	>	$H_2C{=}CHCHO$	>	$H_2C{=}CHCH_3$	>	$(CH_3)_2C{=}C(CH_3)_2$
Four electron-withdrawing groups		One electron-withdrawing group				Four electron-*donating* groups

The methyl groups of 2,3-dimethyl-2-butene also decrease reactivity for steric reasons.

14.36 The difference in reactivity of the three cyclic dienes is due to steric factors. As the "non-diene" part of the molecule becomes larger, the carbon atoms at the end of the diene portion of the ring are forced farther apart. Overlap with the π system of the dienophile in the pericyclic transition state is poorer, and reaction is slower.

14.37 First, find the cyclohexene ring formed by the Diels-Alder reaction. After you locate the new bonds, you should then be able to identify the diene and the dienophile.

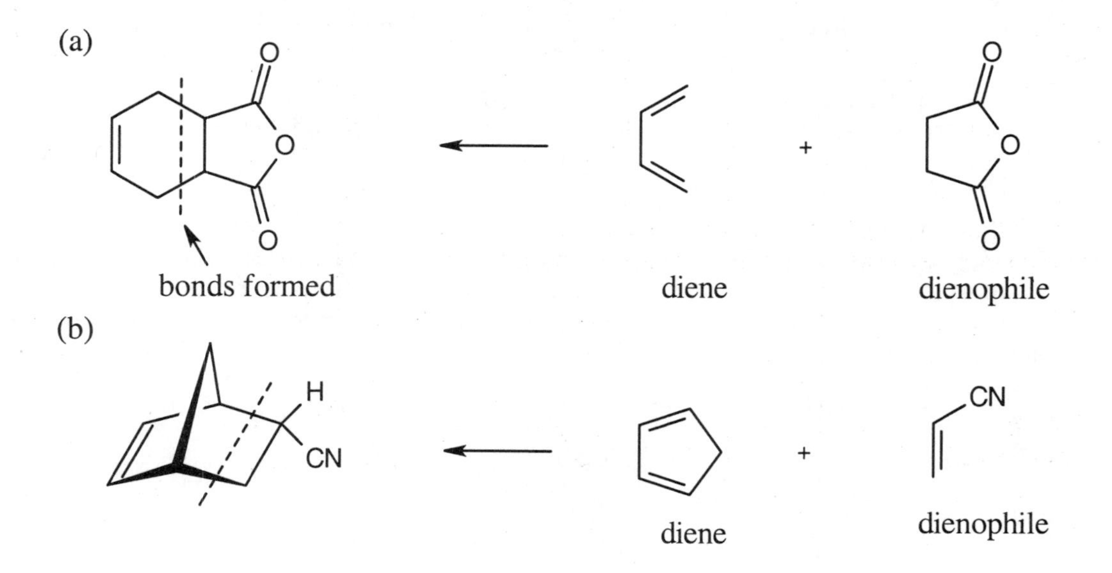

(c)

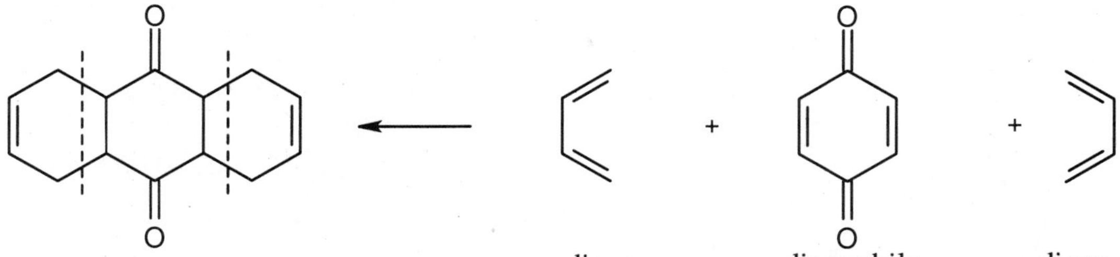

diene dienophile diene

(d)

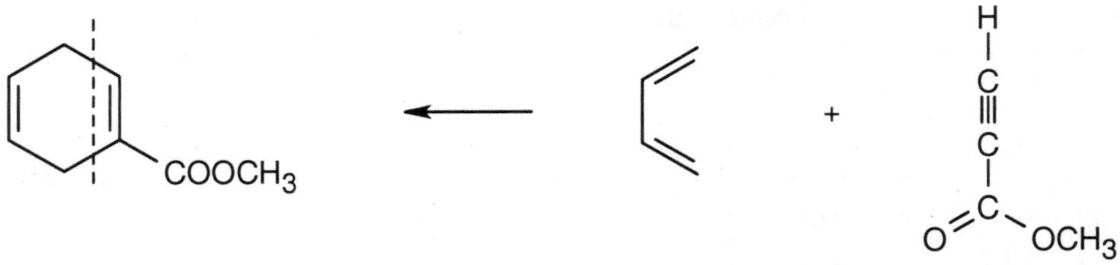

$COOCH_3$

diene dienophile

14.38

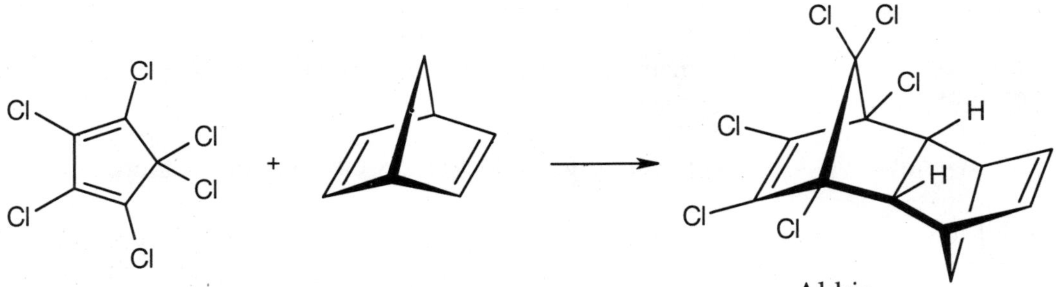

Aldrin

14.39

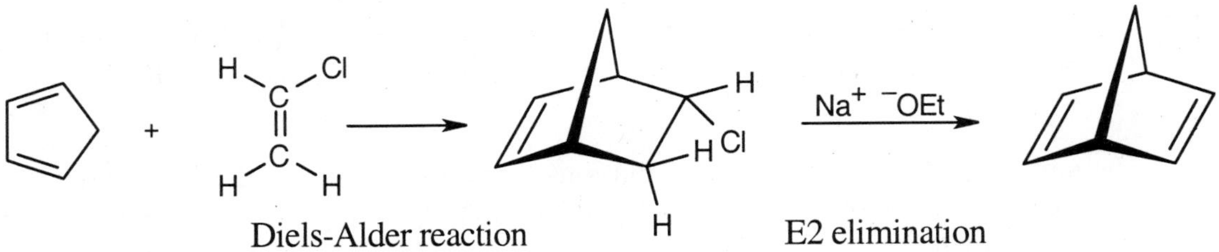

Diels-Alder reaction

$Na^+ \ {}^-OEt$

E2 elimination

14.40 Diels-Alder reaction are reversible when the products are much more stable (of lower energy) than the reactants. In this case, the reactant is a nonconjugated diene, and the products are benzene (a stable, conjugated molecule) and ethylene.

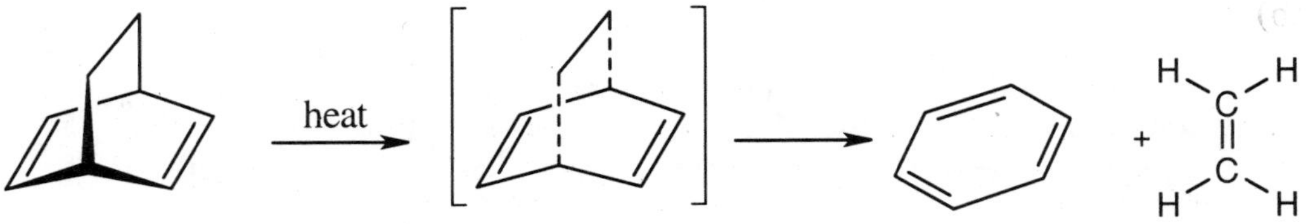

heat

14.41 A Diels-Alder reaction between α–pyrone (diene) and the alkyne dienophile yields the following product.

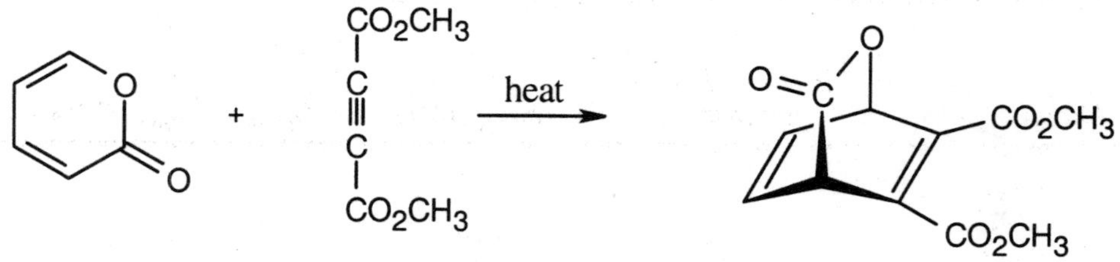

The double bonds in this product are not conjugated, and a more stable product can be formed by loss of CO_2.

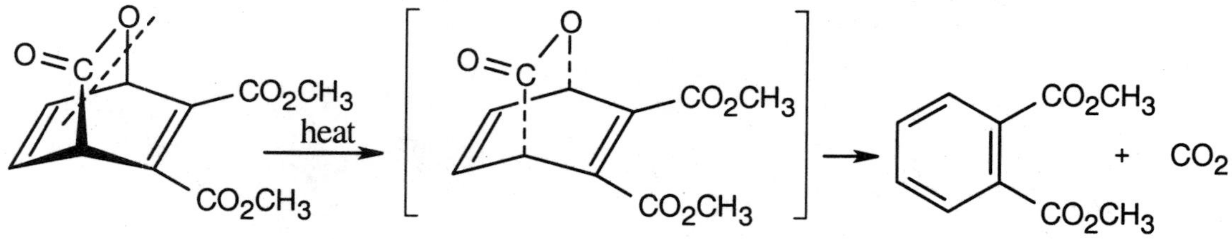

This process can occur in a manner similar to the reverse Diels-Alder reaction of the previous problem.

14.42 The first equivalent of maleic anhydride adds to the *s*-cis bond of the diene.

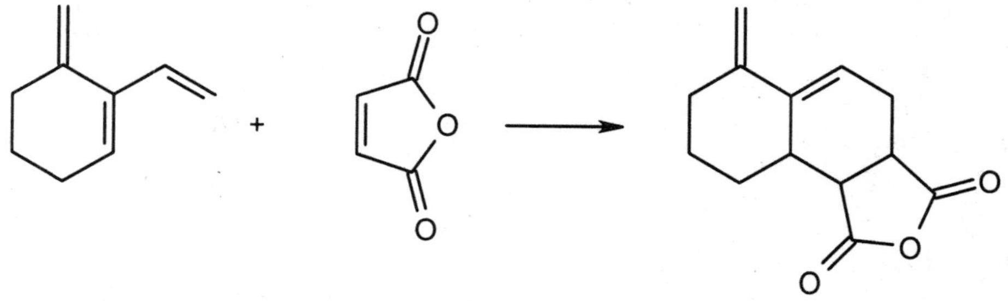

The new double bond has an *s*-cis relationship to the remaining double bond of the diene starting material. A second equivalent of maleic anhydride adds to the diene to form the product shown.

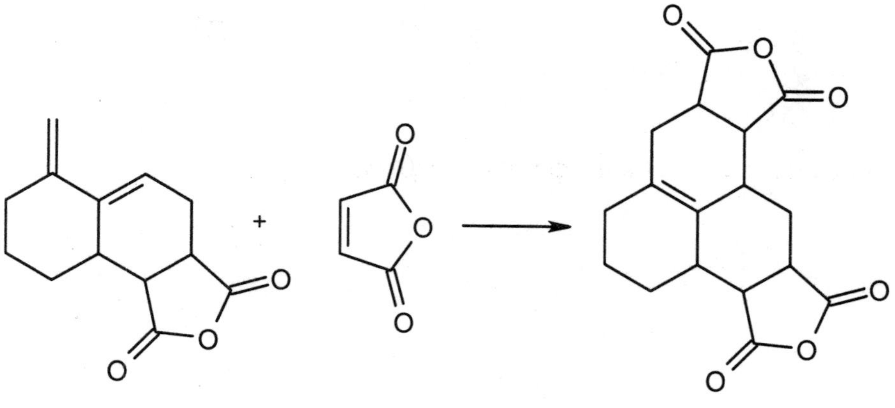

14.43 The value of λ_{max} in the ultraviolet spectrum of dienes becomes larger with increasing alkyl substitution. Since energy is inversely related to λ_{max}, the energy needed to produce ultraviolet absorption decreases with increasing substitution.

Diene	*# of $-CH_3$ groups*	λ_{max}(nm)	$\lambda_{max} - \lambda_{max}$ (butadiene)
$H_2C=CH-CH=CH_2$	0	217	0
$H_2C=CH-C(CH_3)=CH_2$	1	220	3
$H_3C-CH=CH-CH=CH_2$	1	223	6
$H_2C=C(CH_3)-C(CH_3)=CH_2$	2	226	9
$H_3C-CH=CH-CH=CH-CH_3$	2	227	10
$H_3C-C(CH_3)=CH-C(CH_3)=CH_2$	3	232	15
$H_3C-C(CH_3)=CH-CH=C(CH_3)-CH_3$	4	240	23

The average increase in λ_{max} is 5 nm per methyl group.

14.44

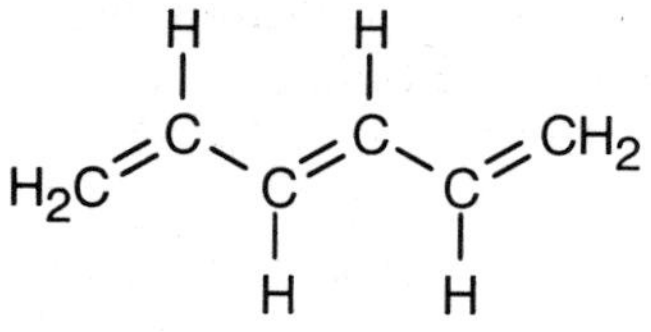

$$H_2C{=}CH{-}CH{=}CH{-}CH{=}CH_2$$

1,3,5-Hexatriene
λ_{max} = 258 nm

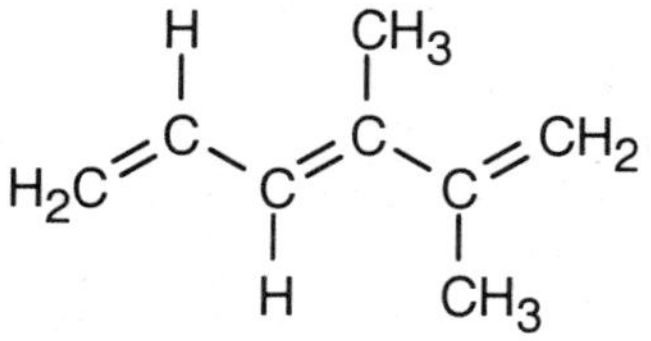

$$H_2C{=}CH{-}CH{=}C(CH_3){-}C(CH_3){=}CH_2$$

2,3-Dimethyl-1,3,5-hexatriene
$\lambda_{max} \approx$ 265 nm

In Problem 14.43, we concluded that one alkyl group increases λ_{max} of a conjugated diene by 5 nm. Since 2,3-dimethyl-1,3,5-hexatriene has two methyl substituents, its UV λ_{max} should be about 10 nm longer than the λ_{max} of 1,3,5-hexatriene.

14.45 (a) ß-Ocimene, $C_{10}H_{16}$, has three degrees of unsaturation. Catalytic hydrogenation yields a hydrocarbon of formula $C_{10}H_{22}$. ß-Ocimene thus contains three double bonds and no rings.
(b) The ultraviolet absorption at 232 nm indicates that ß-ocimene is conjugated.
(c) The carbon skeleton, as determined from hydrogenation, is:

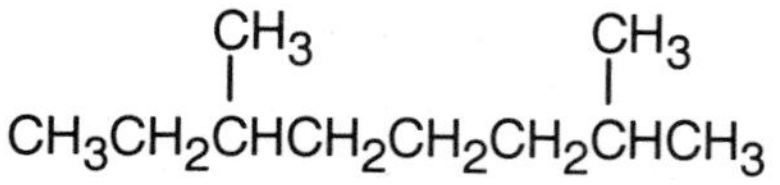

$$CH_3CH_2CH(CH_3)CH_2CH_2CH_2CH(CH_3)CH_3$$

2,6-Dimethyloctane

Ozonolysis data are used to determine the location of the double bonds. The acetone fragment, which comes from carbon atoms 1 and 2 of 2,6-dimethyloctane, fixes the position of one double bond. Formaldehyde results from ozonolysis of a double bond at the other end of ß-ocimene. Placement of the other fragments to conform to the carbon skeleton yields the following structural formula for ß-ocimene.

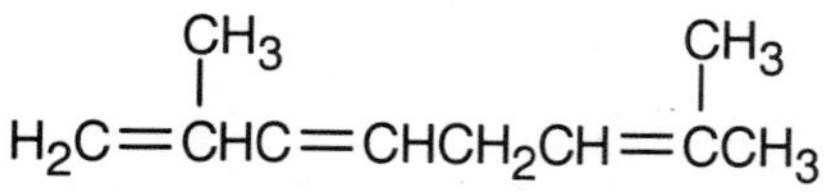

$$H_2C{=}CHC(CH_3){=}CHCH_2CH{=}C(CH_3)CH_3$$

β-Ocimene

(d)

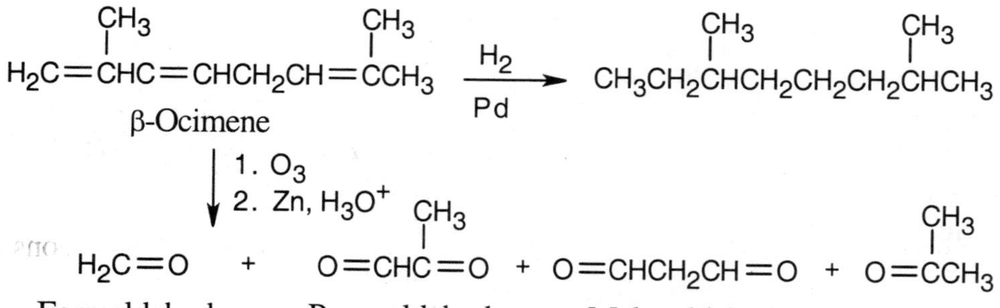

$$\underset{\beta\text{-Ocimene}}{H_2C{=}CHC(CH_3){=}CHCH_2CH{=}C(CH_3)CH_3} \xrightarrow[Pd]{H_2} CH_3CH_2CH(CH_3)CH_2CH_2CH_2CH(CH_3)CH_3$$

$$\downarrow \text{1. } O_3 \quad \text{2. Zn, } H_3O^+$$

$$\underset{\text{Formaldehyde}}{H_2C{=}O} + \underset{\text{Pyruvaldehyde}}{O{=}CHC(CH_3){=}O} + \underset{\text{Malonaldehyde}}{O{=}CHCH_2CH{=}O} + \underset{\text{Acetone}}{O{=}C(CH_3)CH_3}$$

14.46 Much of what was proven for ß–ocimene is also true for myrcene, since both hydrocarbons have the same carbon skeleton and contain conjugated double bonds. The difference between the two isomers is in the placement of double bonds.

The ozonolysis fragments from myrcene are 2-oxopentanedial (five carbon atoms), acetone (three carbon atoms), and two equivalents of formaldehyde (one carbon atom each). Putting these fragments together in a manner consistent with the data gives the following structural formula for myrcene:

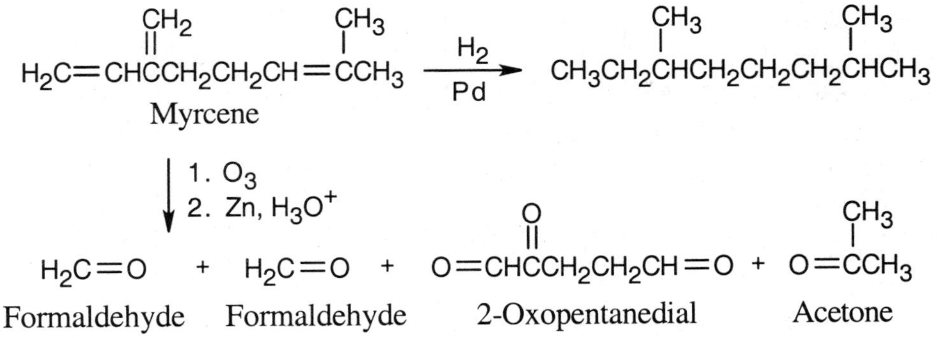

$$H_2C{=}CHC({=}CH_2)CH_2CH_2CH{=}C(CH_3)CH_3 \xrightarrow[Pd]{H_2} CH_3CH_2CH(CH_3)CH_2CH_2CH_2CH(CH_3)CH_3$$

Myrcene

1. O_3
2. Zn, H_3O^+

$$H_2C{=}O + H_2C{=}O + O{=}CHC({=}O)CH_2CH_2CH{=}O + O{=}C(CH_3)CH_3$$

Formaldehyde Formaldehyde 2-Oxopentanedial Acetone

14.47

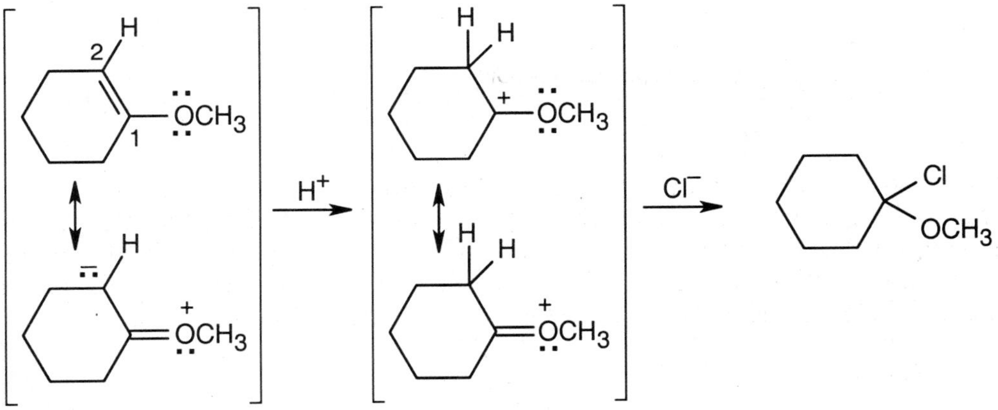

H+ → Cl− →

Conjugation with the oxygen non-bonding electrons makes the double bond more nucleophilic.

Reaction with HCl yields a cation intermediate that can be stabilized by the oxygen electrons.

Addition of Cl^- leads to the observed product.

There are two reasons why the other regioisomer is not formed: (1) Carbon 1 is less nucleophilic than carbon 2; (2) The cation intermediate that would result from protonation at carbon 1 can't be stabilized by the oxygen electrons.

14.48 (a) Hydrocarbon **A** must have two double bonds and two rings, since no carbons are lost on ozonolysis and a diketone–dialdehyde is formed.

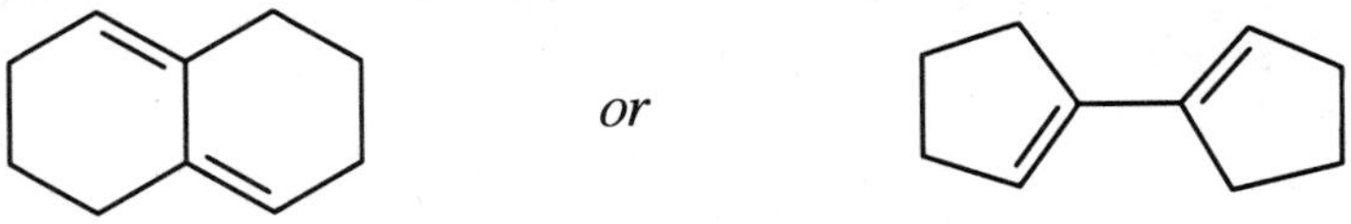

I *or* II

(b) Rotation about the central single bond of II allows the double bond to assume the *s-cis* conformation necessary for a Diels-Alder reaction. Rotation is not possible for I.

(c)

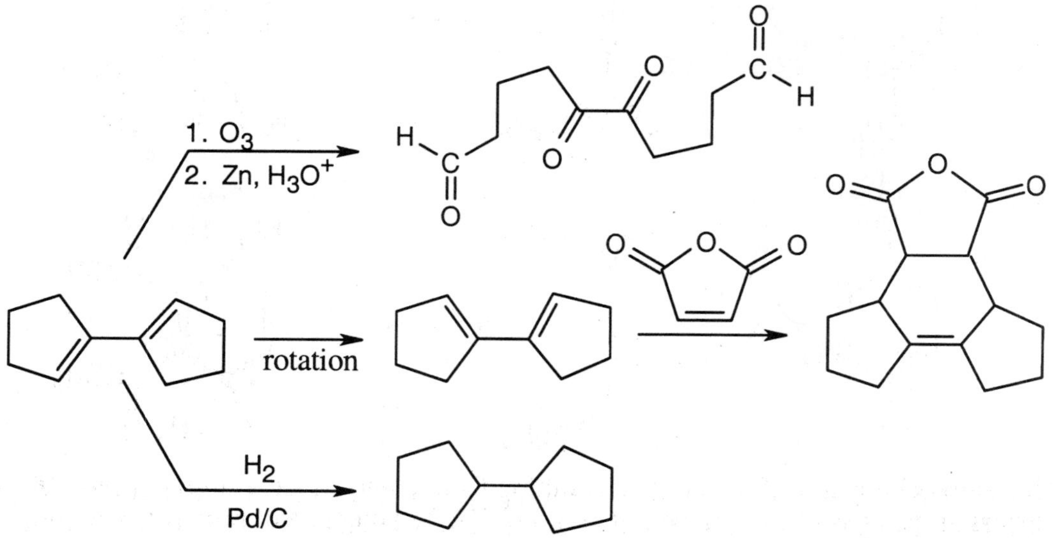

1. O_3
2. Zn, H_3O^+

rotation

H_2

Pd/C

14.49

$$H_2C{=}CHCH{=}CH_2 \xrightarrow[\text{1,4-Addition}]{Br_2} BrCH_2CH{=}CHCH_2Br \xrightarrow[\text{S}_N2\text{ Substitution}]{2\ NaCN} NCCH_2CH{=}CHCH_2CN$$

$$\downarrow H_2,\ Pd$$

$$NCCH_2CH_2CH_2CH_2CN$$

Adiponitrile

14.50

$$C = \frac{A}{\varepsilon \times l} = \frac{0.065}{11{,}900 \times 1.00\ \text{cm}} = \frac{6.5 \times 10^{-2}}{1.19 \times 10^{4}} = 5.5 \times 10^{-6}\text{M}$$

14.51

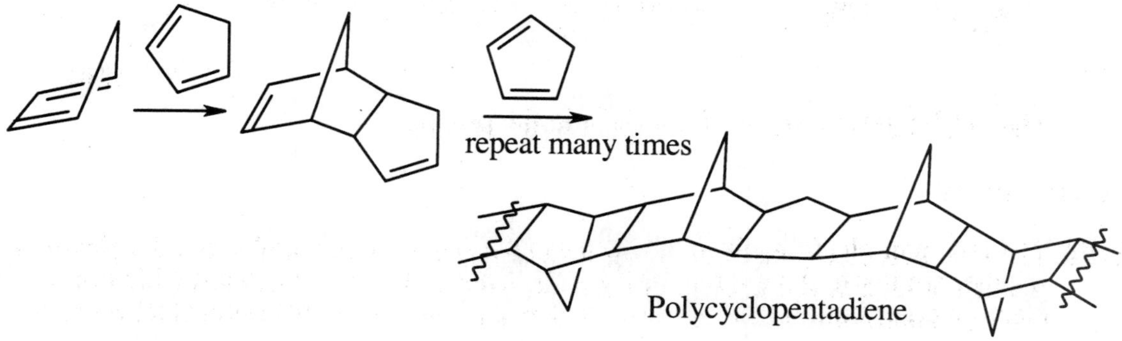

repeat many times

Polycyclopentadiene

Polycyclopentadiene is the product of successive Diels-Alder additions of cyclopentadiene to a growing polymer chain. Strong heat causes depolymerization of the chain and reversion to cyclopentadiene monomer units.

14.52

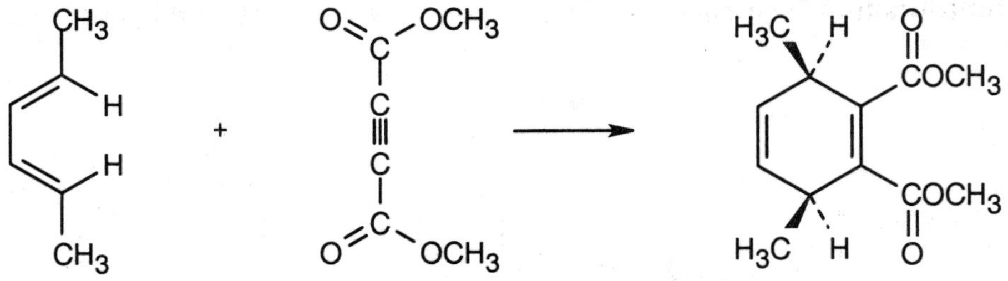

CH_3 H H CH_3 + O=C(–OCH_3)–C≡C–C(=O)–OCH_3 → H_3C H $COCH_3$ (C=O) $COCH_3$ (C=O) H_3C H

14.53

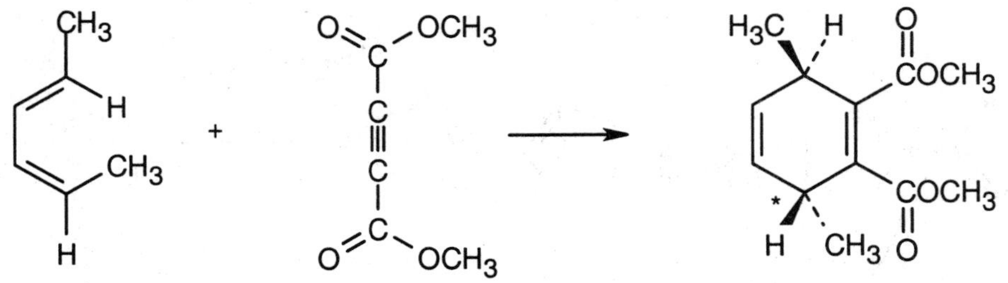

CH_3 H CH_3 H + O=C(–OCH_3)–C≡C–C(=O)–OCH_3 → H_3C H $COCH_3$ (C=O) $COCH_3$ (C=O) * H CH_3

The stereochemistry of the product resulting from Diels-Alder reaction of the (2*E*,4*Z*) diene differs at the starred carbon from that of the (2*E*,4*E*) diene. Not only is the stereochemistry of the dienophile maintained during the Diels-Alder reaction, the stereochemistry of the diene is also maintained.

14.54 Although it is usually best to work backwards in a synthesis problem, it sometimes helps to work *both* forwards and backwards. In this problem, we know that the starting materials are a diene and a dienophile. This suggests that the synthesis involves a Diels-Alder reaction. The product is a dialdehyde in which the two aldehyde groups have a cis relationship, indicating that they are the products of ozonolysis of a bridgehead double bond. These two pieces of information allow us to propose the following synthesis:

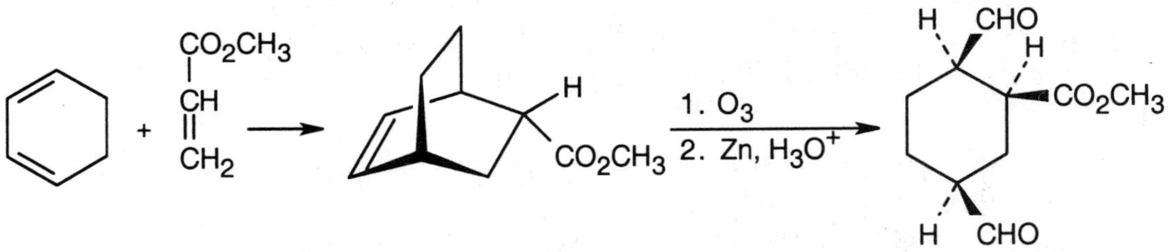

+ CO_2CH_3–CH=CH_2 → H CO_2CH_3 → (1. O_3; 2. Zn, H_3O^+) H CHO H CO_2CH_3 H CHO

The –CHO groups are cis to the ester in the product.

A Look Ahead

14.55 The lone pair electrons from nitrogen can overlap with the double bond π electrons in a manner similar to the overlap of the π electrons of two conjugated double bonds. This electron contribution from nitrogen makes an enamine double bond electron-rich.

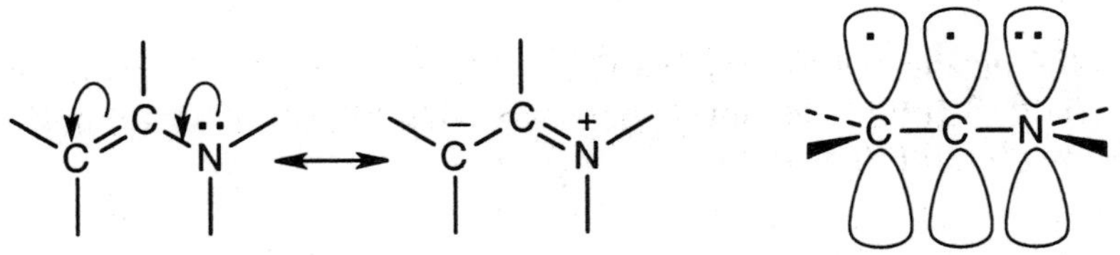

C=C–N: ⟷ $\bar{C}$–C=N^+ C–C–N

The orbital picture of an enamine shows a 4 *p*-electron system that resembles the system of a conjugated diene.

14.56 Double bonds can be conjugated not only with other multiple bonds but also with the lone-pair electrons of atoms such as oxygen and nitrogen. *p*–Toluidine has the same number of double bonds as benzene, yet its λ_{max} is 31 nm greater. The electron pair of the nitrogen atom can conjugate with the π electrons of the three double bonds of the ring, extending the π system and increasing λ_{max}.

14.57 Dilute NaOH removes the proton from the –OH group, leaving the phenoxide anion.

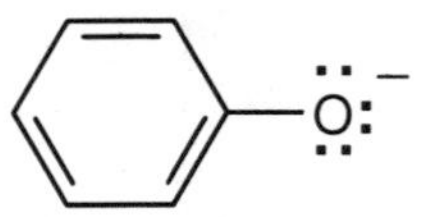

The increased electron density at oxygen increases conjugation of the oxygen lone pair electrons with the π electrons of the ring double bonds. The extended conjugation increases λ_{max} in a manner similar to *p*–toluidine (Problem 14.56).

Molecular Modeling

14.58 UV excitation shortens and strengthens the C2–C3 bond. This is consistent with promotion of an electron from the HOMO (which is π antibonding between C2–C3) to the LUMO (which is π bonding between C2–C3). Excitation also lengthens and weakens the C1–C2 and C3–C4 bonds. The HOMO is π bonding for these two bonds, and the LUMO is π antibonding.

14.59 The lowest and highest energy conformations for 1-butene have C–C–C–C dihedral angles of 120° and 45°, respectively. The barrier to rotation equals the energy difference between these conformations and is 9.4 kJ/mol. For 1,3-butadiene, the lowest and highest energy conformations have dihedral angles of 180° and 105°, respectively. The barrier to rotation is substantially higher (23.6 kJ/mol). 1,3-Butadiene has two minimum-energy conformations – planar *s*-trans and non-planar *s*-cis, which has a C–C–C–C dihedral angle of 30°–45° and is 11.8 kJ/mol higher in energy. Only *s*-cis geometry permits cycloaddition, but it is not the preferred geometry.

14.60 Electrostatic potential maps show that the alkene carbons in benzoquinone and 3,3,3-trifluoropropene are much more positive than those in ethylene. Benzoquinone and 3,3,3-trifluoropropene should be reactive dienophiles, and ethylene should be relatively unreactive.

14.61 Transition state B is lower in energy by 5.3 kJ/mol and leads to the kinetic (endo) product. Product A is lower in energy by 5.8 kJ/mol and is the thermodynamic (exo) product. The kinetic and thermodynamic products are different for this reaction.

Chapter 15 – Benzene and Aromaticity

Chapter Outline

I. Introduction to aromatic compounds (Sections 15.1 – 15.4).
   A. Sources of aromatic hydrocarbons (Section 15.1).
      1. Some aromatic hydrocarbons are obtained from distillation of coal tar.
      2. Other aromatic hydrocarbons are formed when petroleum is passed over a catalyst during refining.
   B. Naming aromatic compounds (Section 15.2).
      1. Many aromatic compounds have nonsystematic names.
      2. Monosubstituted benzenes are named in the same way as other hydrocarbons, with -benzene as the parent name.
         a. Alkyl-substituted benzenes are named in two ways:
            i. If the alkyl substituent has six or fewer carbons, the hydrocarbon is named as an alkyl-substituted benzene.
            ii. If the alkyl substituent has more than six carbons, the compound is named as a phenyl-substituted alkane.
         b. The $C_6H_5CH_2$– group is a benzyl group.
      3. Disubstituted benzenes are named by the ortho(*o*), meta(*m*),para(*p*) system.
         a. A benzene ring with two substituents in a 1,2 relationship is *o*-disubstituted.
         b. A benzene ring with two substituents in a 1,3 relationship is *m*-disubstituted.
         c. A benzene ring with two substituents in a 1,4 relationship is *p*-disubstituted.
         d. The *o, m, p*–system of nomenclature is also used in describing reactions.
      4. Benzenes with more than two substituents are named by numbering the position of each substituent.
         a. Number so that the lowest possible combination of numbers is used.
         b. Substituents are listed alphabetically.
      5. Any of the nonsystematic names can be used as a parent name.
   C. Stability and structure of benzene (Section 15.3).
      1. Stability of benzene.
         a. Benzene doesn't undergo typical alkene reactions.
            Benzene reacts slowly with Br_2 to give substitution, not addition, product.
         b. $\Delta H°_{hydrog}$ of benzene is 150 kJ/mol less than that predicted for 3 x $\Delta H°_{hydrog}$ of cyclohexene, indicating that benzene has extra stability.
         c. All six bonds of benzene are of equal length.
      2. Structure of benzene.
         a. Resonance theory explains that benzene is a resonance hybrid of two forms.
         b. Benzene is represented in this book as one line-bond structure, rather than as a hexagon with a circle to represent the double bonds.
   D. Molecular orbital picture of benzene (Section 15.4).
      1. Benzene is a planar molecule with 120° bond angles.
      2. All carbons are sp^2-hybridized and identical, and each carbon has an electron in a *p* orbital perpendicular to the plane of the ring.
      3. It is impossible to define 3 localized π bonds; the electrons are delocalized over the ring.
      4. Six molecular orbitals (MOs) can be constructed for benzene.
         a. The 3 lower-energy MOs are bonding MOs.
         b. The 3 higher energy MOs are antibonding.
         c. One pair of bonding orbitals is degenerate, as is one pair of antibonding orbitals.

d. The 6 bonding electrons of benzene occupy the 3 bonding orbitals and are delocalized over the ring.

II. Aromaticity (Sections 15.5 – 15.8).
- A. The Hückel $4n + 2$ rule (Section 15.5).
  1. For a compound to be aromatic, it must possess the qualities we have already mentioned.
     In addition, it must fulfill Hückel's Rule.
  2. Hückel's Rule: A molecule is aromatic only if it has a planar, monocyclic system of conjugation with a total of $4n + 2$ π electrons (where n is an integer).
  3. Molecules with 4, 8, 12 ... π electrons are antiaromatic.
  4. Examples:
     a. Cyclobutadiene ($n = 4$) is antiaromatic.
     b. Benzene ($n = 6$) is aromatic.
     c. Cyclooctatetraene ($n = 8$) is antiaromatic.
        i. Cyclooctatetraene is stable, but its chemical behavior is like an alkene, rather than an aromatic compound.
        ii. Cyclooctatetraene is tub-shaped, and its bonds have two different lengths.
- B. Aromatic ions (Section 15.6).
  1. Any cyclic conjugated molecule with $4n + 2$ electrons can be aromatic, even if it is an ion.
  2. The cyclopentadienyl anion.
     a. Although cyclopentadiene isn't aromatic, removal of H^+ produces a six-π-electron cyclic anion that is aromatic.
     b. This anion has a $pK_a = 16$, indicating that a stable anion is formed on dissociation of H^+.
     c. Both the cyclopentadienyl cation (4 π electrons) and the cyclopentadienyl radical (5 π electrons) are unstable.
  3. The cycloheptatrienyl cation.
     a. Removal of H^- from cycloheptatriene produces the cycloheptatrienyl cation, which has 6 π electrons and is stable.
     b. The cycloheptatrienyl radical and anion are unstable.
- C. Two aromatic heterocycles (Section 15.7).
  1. A heterocycle (a cyclic compound containing one or more elements in addition to carbon) can also be aromatic.
  2. Pyridine.
     a. The nitrogen atom of pyridine contributes one π electron to the π system of the ring, making pyridine aromatic.
     b. The nitrogen lone pair is not involved with the ring π system.
  3. Pyrrole.
     a. The nitrogen of pyrrole contributes both lone-pair electrons to the ring π system making pyrrole aromatic.
     b. The nitrogen atom makes a different contribution to the π ring system in pyrrole and in pyridine.
- D. Why $4n + 2$? (Section 15.8).
  1. For aromatic compounds, there is a single lowest-energy MO that can accept two π electrons.
  2. The next highest levels occur in degenerate pairs that can accept 4 π electrons.
  3. For all aromatic compounds and ions, a stable species occurs only when $4n + 2$ π electrons are available to completely fill the bonding MOs.

III. Polycyclic aromatic compounds (Section 15.9).
- A. Although Hückel's Rule strictly applies only to monocyclic compounds, some polycyclic compounds show aromatic behavior.
- B. Naphthalene and anthracene are two common aromatic polycyclic compounds.
  1. Both can be represented by several resonance forms.

2. Both show chemical and physical properties common to aromatic compounds.
3. Both have a Hückel number of π electrons.

IV. Spectroscopy of aromatic compounds (Section 15.10).
   A. IR spectroscopy.
      1. A C–H stretch occurs at 3030 cm^{-1}.
      2. As many as 4 absorptions occur in the region 1450–1600 cm^{-1}.
      3. Weak absorptions are visible in the range 1660–2000 cm^{-1}.
      4. Strong absorptions in the region 690–900 cm^{-1} can be used to determine the substitution pattern of an aromatic ring.
   B. UV spectroscopy.
      The conjugated π system of an aromatic ring gives rise to an intense absorption at 205 nm and weaker absorptions in the range 255–275 nm.
   C. NMR spectroscopy.
      1. ^1H NMR.
         a. Hydrogens directly bonded to an aromatic ring absorb in the region 6.5–8.0 δ.
            i. Spin-spin coupling can give information about the substitution pattern.
            ii. Aromatic protons are deshielded because the applied magnetic field sets up a ring-current, which produces a small magnetic field that opposes the applied field and deshields the aromatic protons.
            iii. If protons reside on the inside of an aromatic ring system, they are strongly shielded and absorb far upfield.
            iv. The presence of a ring-current is a test of aromaticity.
         b. Benzylic protons absorb at 2.3–3.0 δ.
      2. ^{13}C NMR.
         a. Aromatic carbons absorb in the range 110–140 δ.
         b. Since alkene protons also absorb in this region, ^{13}C NMR is not uniquely useful in identifying an aromatic ring.

Solutions to Problems

15.1 An ortho disubstituted benzene has two substituents in a 1,2 relationship. A meta disubstituted benzene has two substituents in a 1,3 relationship. A para disubstituted benzene has two substituents in a 1,4 relationship.

(a) Cl, CH_3

meta disubstituted

(b) NO_2, Br

para disubstituted

(c) SO_3H, OH

ortho disubstituted

15.2 Remember to give the lowest possible numbers to substituents on trisubstituted rings.

(a)

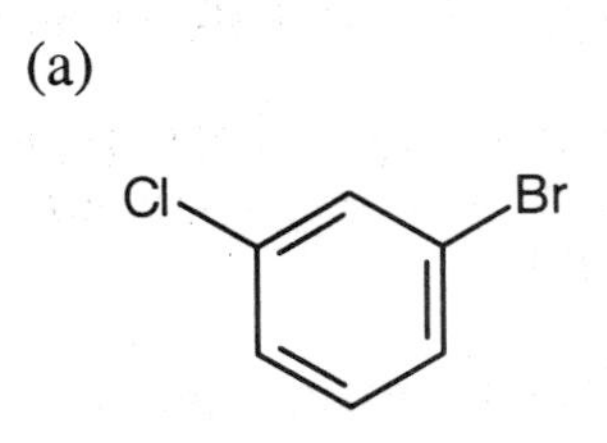

m-Bromochlorobenzene

(b)

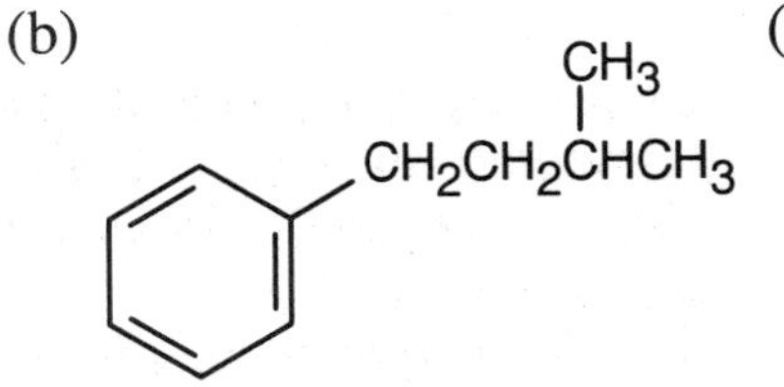

(3-Methylbutyl)benzene

(c) NH_2, Br

p-Bromoaniline

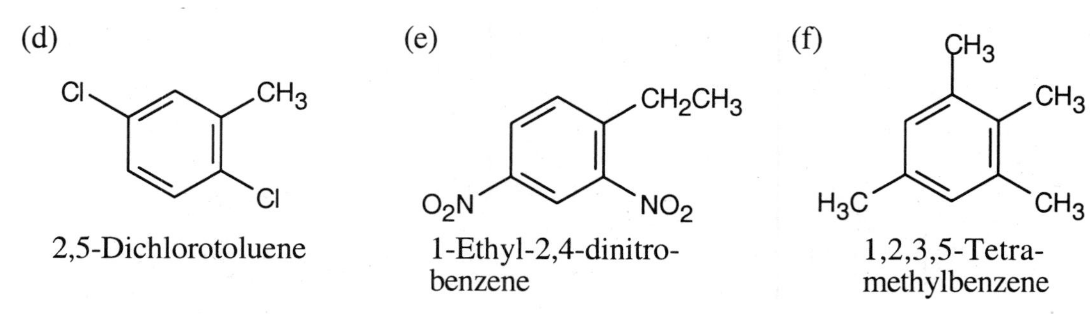

2,5-Dichlorotoluene

1-Ethyl-2,4-dinitro-benzene

1,2,3,5-Tetra-methylbenzene

15.3

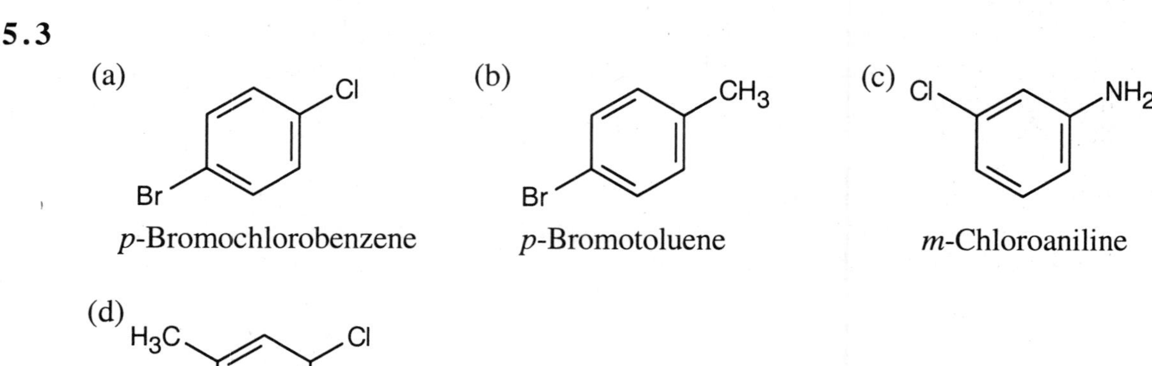

p-Bromochlorobenzene

p-Bromotoluene

m-Chloroaniline

(d) H_3C Cl CH_3

1-Chloro-3,5-dimethylbenzene

15.4

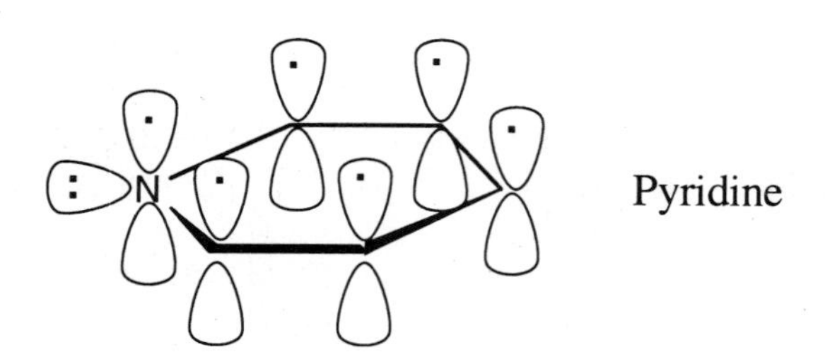

Pyridine

The electronic descriptions of pyridine and benzene are very similar. The pyridine ring is formed by the σ overlap of carbon and nitrogen sp^2 orbitals. In addition, six *p* orbitals, perpendicular to the plane of the ring, hold six electrons. These six *p* orbitals form six π molecular orbitals that allow electrons to be delocalized over the π system of the pyridine ring. The lone pair of nitrogen electrons occupies an sp^2 orbital that lies in the plane of the ring.

15.5

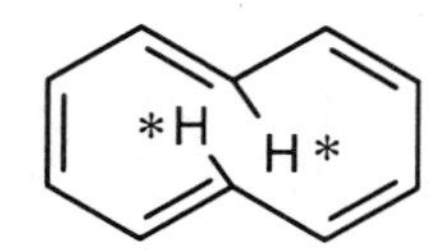

Cyclodecapentaene has $4n + 2$ π electrons (n = 2), but it is not flat. If cyclodecapentaene were flat, the hydrogen atoms starred would crowd each other across the ring. To avoid this interaction, this cyclodecaene has two trans double bonds that keep the molecule from being flat and aromatic.

15.6

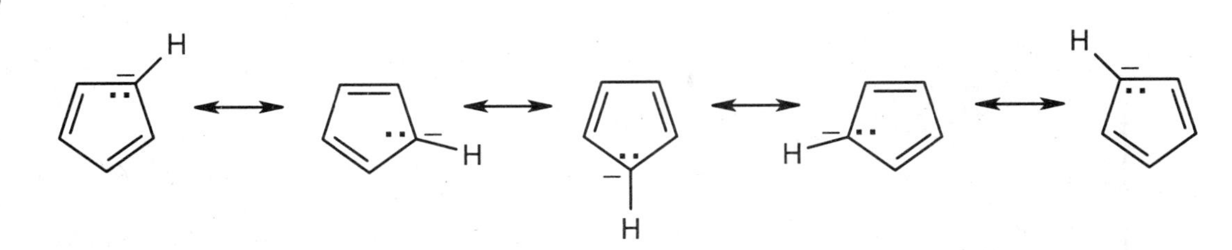

A compound that can be described by several resonance forms has a structure that can be represented by no one form. The structure of the cyclopentadienyl anion is a combination of all of the above structures and contains only one kind of carbon atom and one kind of hydrogen atom. All carbon-carbon bond lengths are equivalent, as are all carbon-hydrogen bonds lengths. Both the ^{1}H NMR and ^{13}C NMR spectra show only one absorption.

15.7 When cyclooctatetraene accepts two electrons, it becomes a $(4n + 2)$ π electron aromatic ion. Cyclooctatetraenyl dianion is planar with a carbon-carbon bond angle of 135° (a regular octagon).

15.8

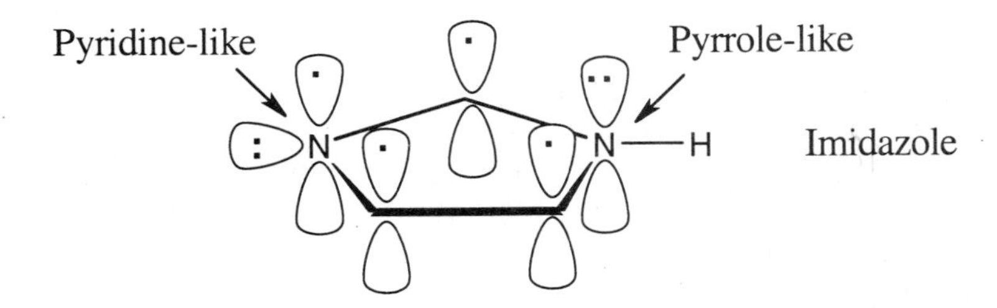

The aromatic heterocycle imidazole contains six π electrons. Each carbon contributes one electron, the nitrogen bonded to hydrogen (pyrrole-like) contributes two electrons, and the remaining nitrogen (pyridine-like) contributes one electron. Both nitrogens are sp^2 hybridized.

15.9 Furan is the oxygen analog of pyrrole. Furan is aromatic because it has 6 π electrons in a cyclic, conjugated system. Oxygen contributes two lone-pair electrons from a *p* orbital perpendicular to the plane of the ring.

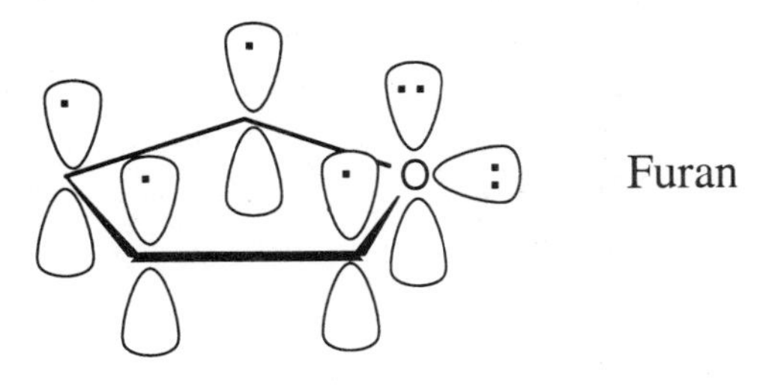

15.10

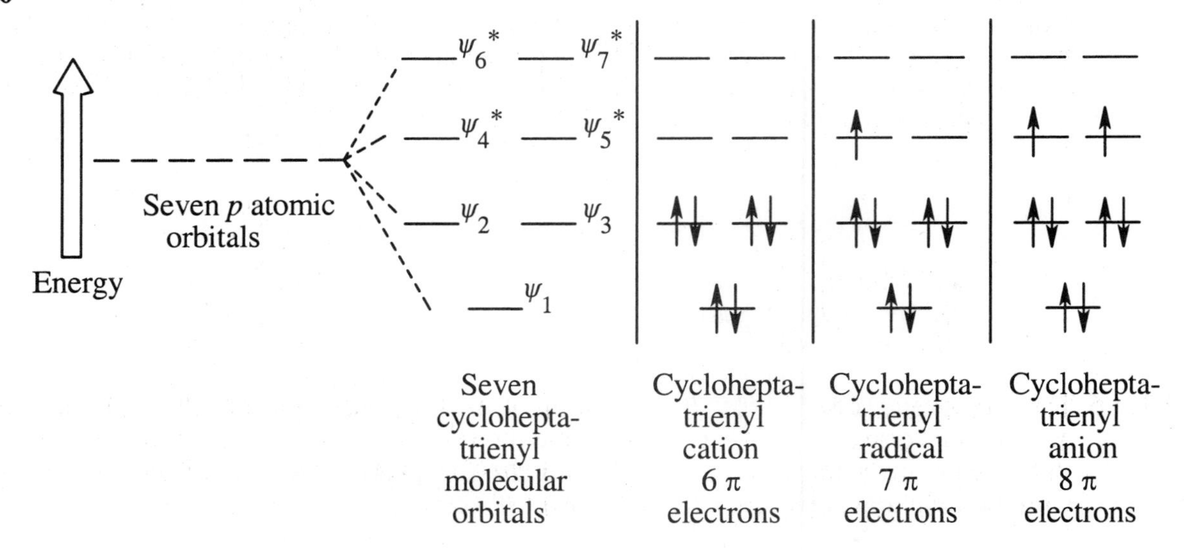

The cycloheptatrienyl cation has six π electrons (a Hückel number) and is aromatic.

15.11

Azulene

Azulene is aromatic because it has a conjugated cyclic π electron system containing ten π electrons (a Hückel number).

15.12 Naphthalene is a ten π electron compound; the circle in each ring represents five electrons.

Visualizing Chemistry

15.13

(a)

m-Isopropylphenol

(b)

o-Nitrobenzoic acid

15.14 The all-cis decapentaene shown here is not aromatic because it is not planar. All hydrogens, however, are equivalent and show one absorption in the vinylic region of the molecule's ^{1}H NMR spectrum. If the molecule were aromatic, the absorption would appear between 6.5–8.0 δ.

15.15 1,6-Methanonaphthalene has ten π electrons and is sufficiently planar to behave as an aromatic molecule. The perimeter hydrogens absorb in the aromatic region of the ^1H NMR spectrum (6.9–7.3 δ). Interaction of the applied magnetic field with the perimeter π electrons sets up a ring current (see Section 15.10) that strongly shields the CH_2 protons and causes them to absorb far upfield (–0.5 δ).

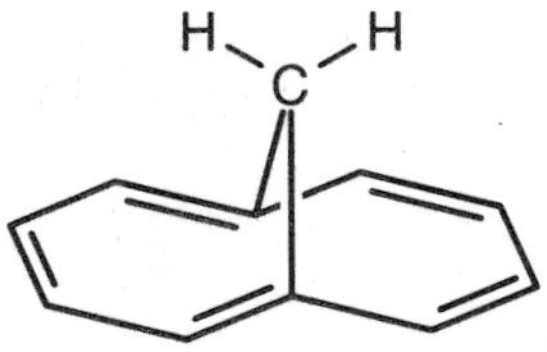

15.16 Three resonance forms for the carbocation of the formula $C_{13}H_9$ are shown below, and more can be drawn. These forms show that the positive charge of the carbocation can be stabilized in the same way as an allylic or benzylic carbocation is stabilized – by overlap with the neighboring π electrons of the ring system.

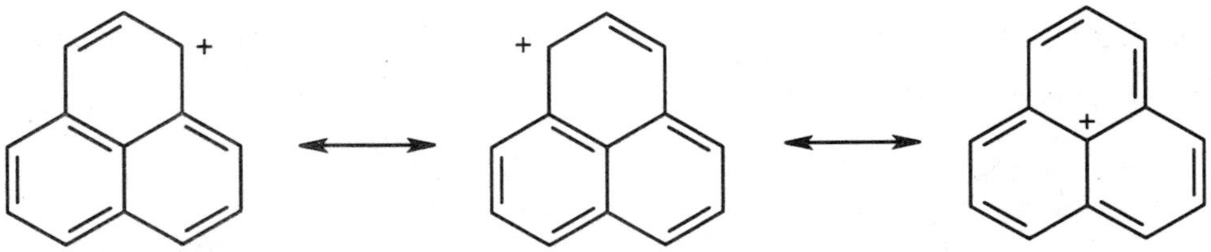

Additional Problems

15.17

(a) 2-Methyl-5-phenylhexane

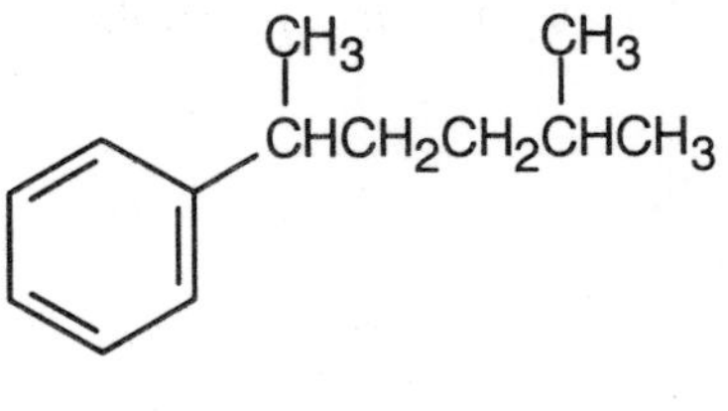

(b) *m*-Bromobenzoic acid

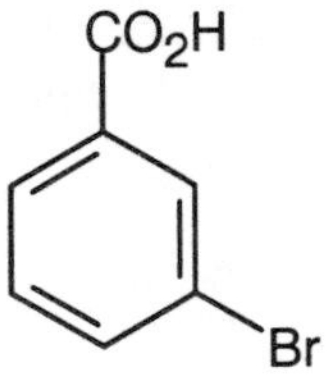

(c) 1-Bromo-3,5-dimethylbenzene

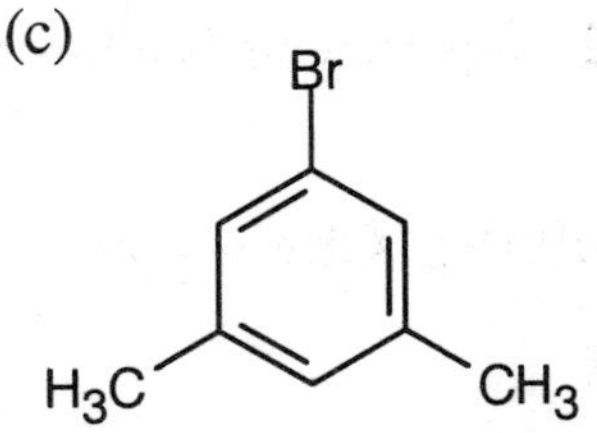

(d) *o*-Bromopropylbenzene

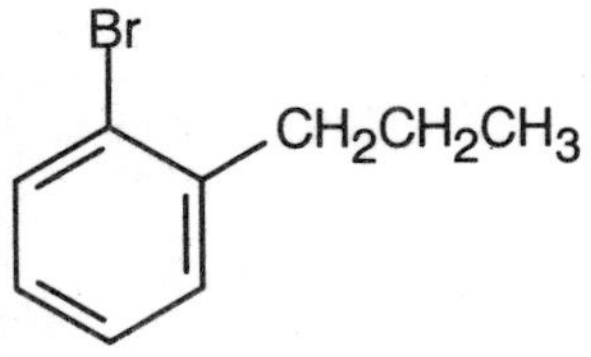

(e) 1-Fluoro-2,4-dinitrobenzene

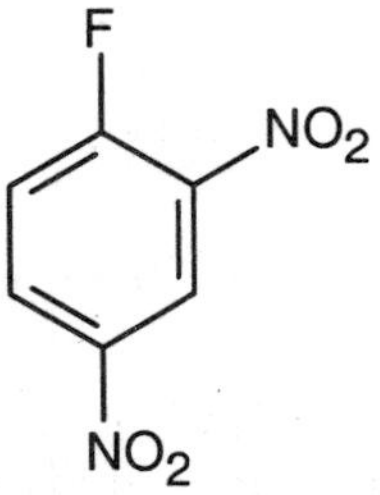

(f) *p*-Chloroaniline

15.18

(a) 3-Methyl-1,2-benzenediamine

(b) 1,3,5-Benzenetriol

(c) 3-Methyl-2-phenylhexane

(d) *o*-Aminobenzoic acid

(e) *m*-Bromophenol

(f) 2,4,6-Trinitrophenol

(g) *p*-Iodonitrobenzene

15.19

(a)

o-Dinitrobenzene

m-Dinitrobenzene

p-Dinitrobenzene

(b)

1-Bromo-2,3-dimethylbenzene

2-Bromo-1,3-dimethylbenzene

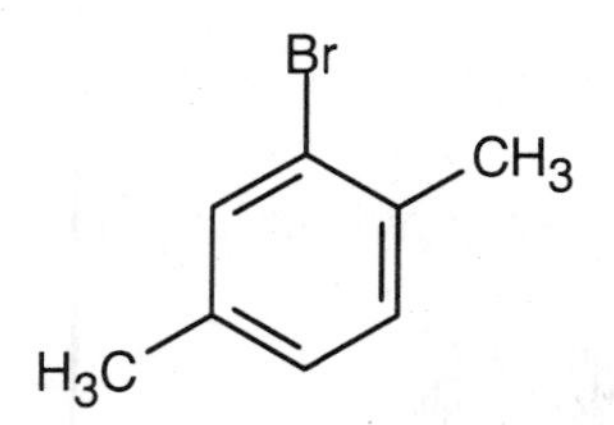

2-Bromo-1,4-dimethylbenzene

1-Bromo-2,4-dimethylbenzene

4-Bromo-1,2-dimethylbenzene

1-Bromo-3,5-dimethylbenzene

(c)

2,3,4-Trinitrophenol

2,3,5-Trinitrophenol

2,3,6-Trinitrophenol

2,4,5-Trinitrophenol

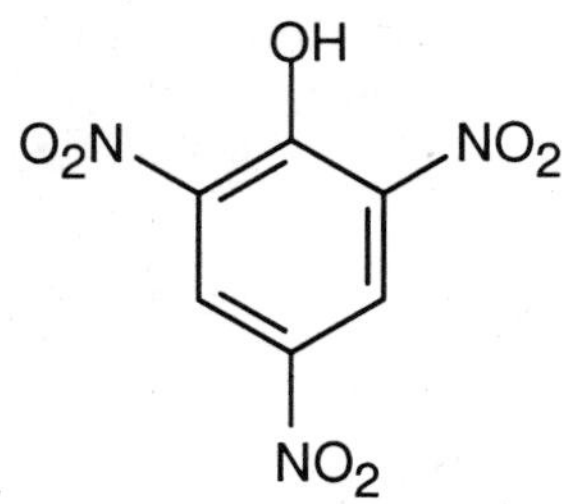

2,4,6-Trinitrophenol

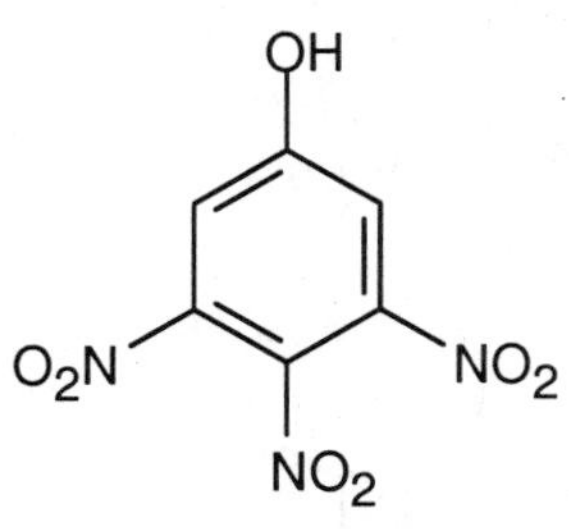

3,4,5-Trinitrophenol

15.20 All aromatic compounds of formula C_7H_7Cl have one ring and three double bonds.

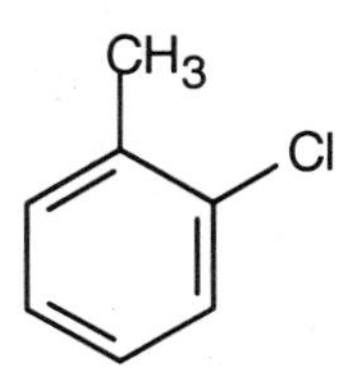

o-Chlorotoluene

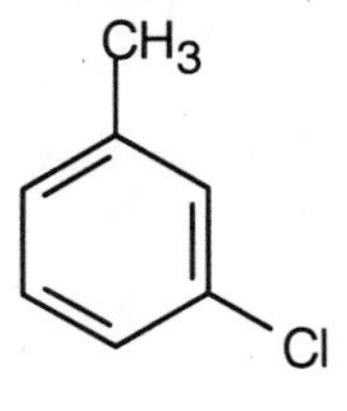

m-Chlorotoluene

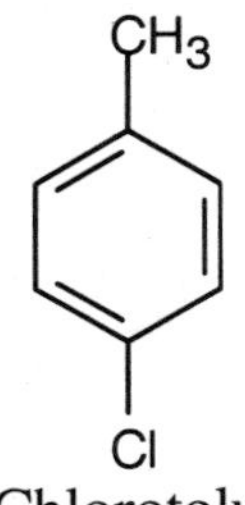

p-Chlorotoluene

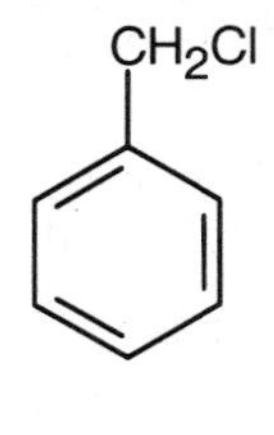

Benzyl chloride
or
(Chloromethyl)-
benzene

15.21 Six of these compounds are illustrated and named in Problem 15.19(b).The other eight are:

o-(Bromomethyl)-
toluene

m-(Bromomethyl)-
toluene

p-(Bromomethyl)-
toluene

(1-Bromoethyl)benzene

(2-Bromoethyl)benzene

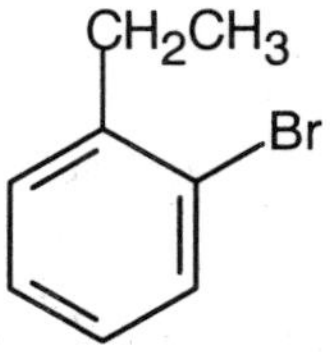

o-Bromoethylbenzene 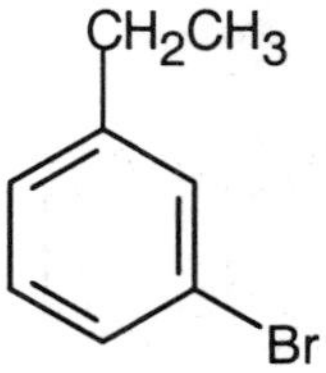 *m*-Bromoethylbenzene 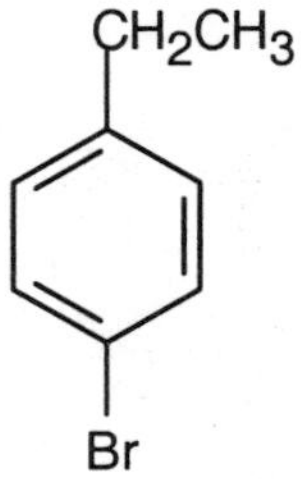 *p*-Bromoethylbenzene

15.22 All compounds in this problem have four double bonds and/or rings and must be substituted benzenes, if they are to be aromatic. They may be substituted by methyl, ethyl, propyl, or butyl groups.

(a)

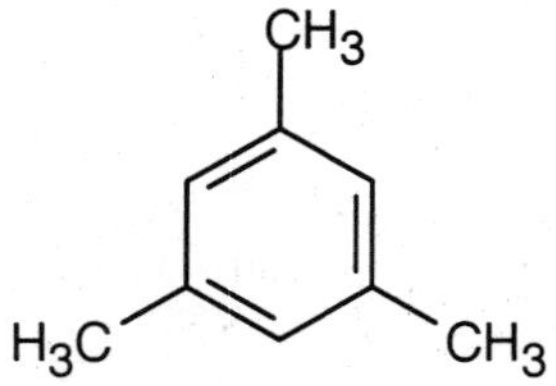

(b)

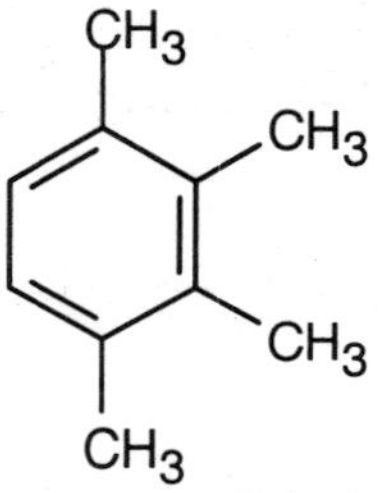

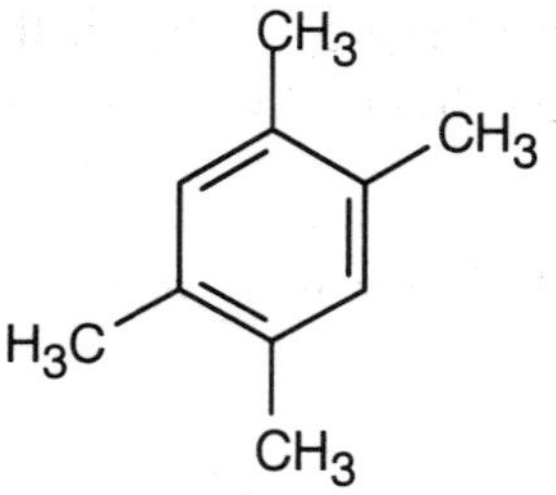

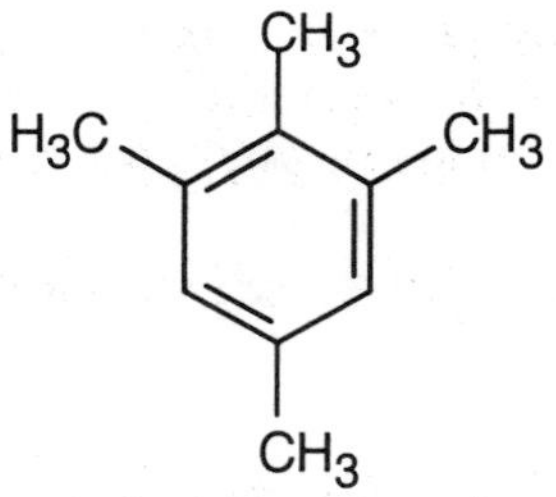

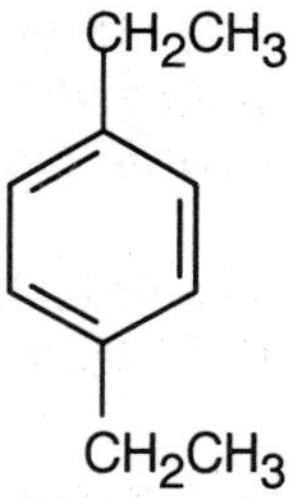

(c)

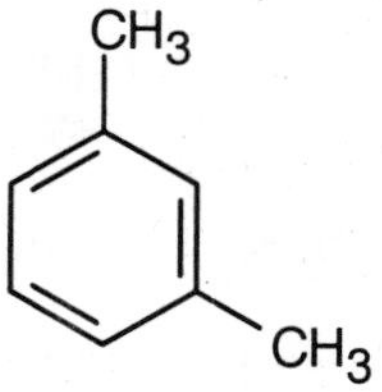

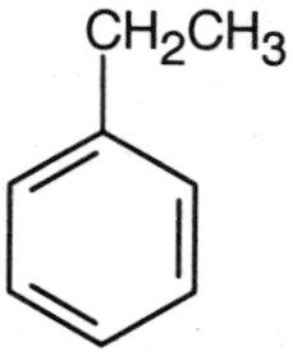

(d)

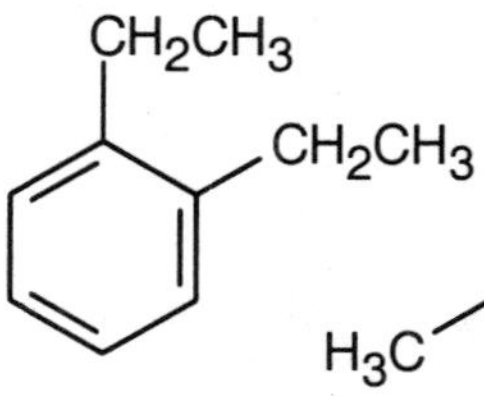

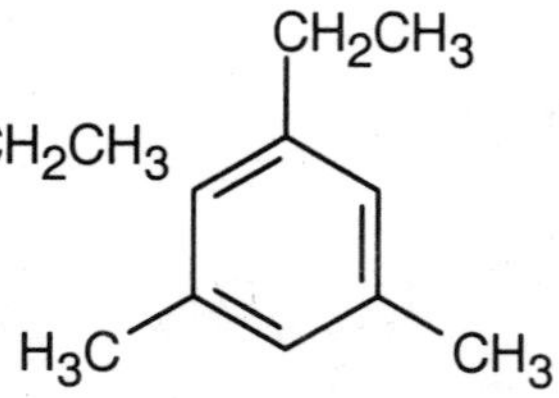

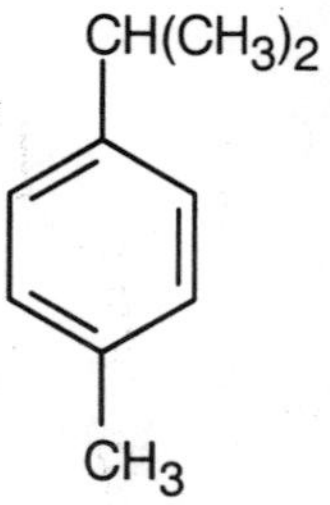

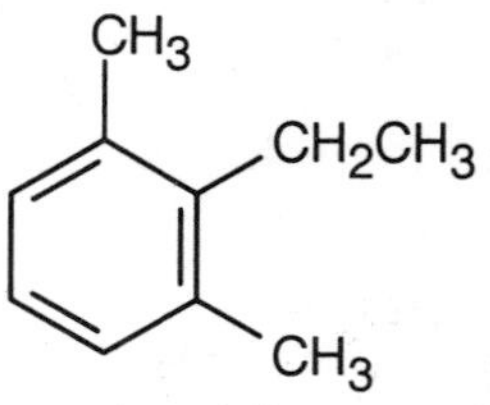

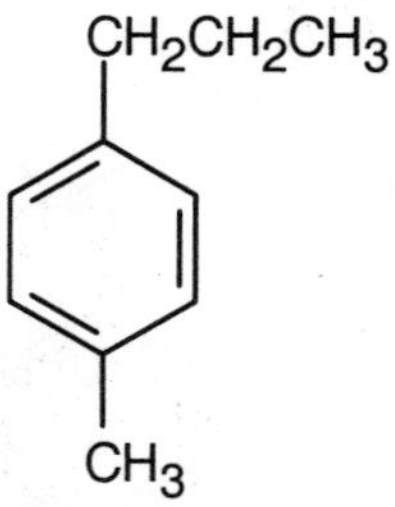

15.23

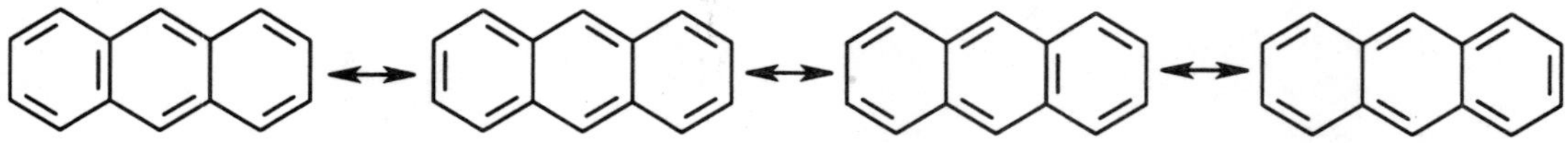

15.24

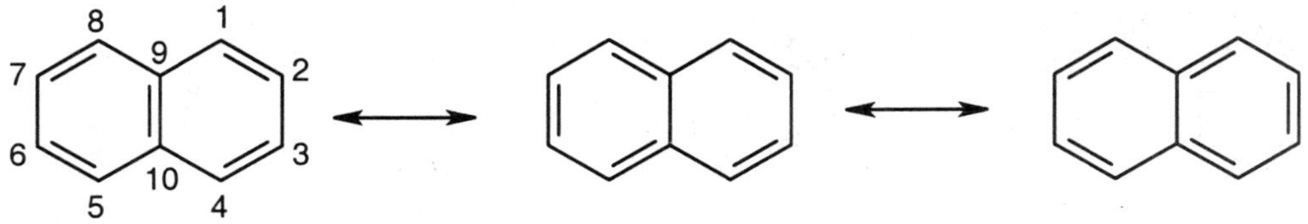

The bond between carbons 1 and 2 is represented as a double bond in two of the three resonance structures, but the bond between carbons 2 and 3 is represented as a double bond in only one resonance structure. The C1-C2 bond thus has more double bond character in the resonance hybrid, and it is shorter than the C2-C3 bond. The C3-C4, C5-C6, and C7-C8 bonds also have more double bond character than the remaining bonds.

15.25, 15.26

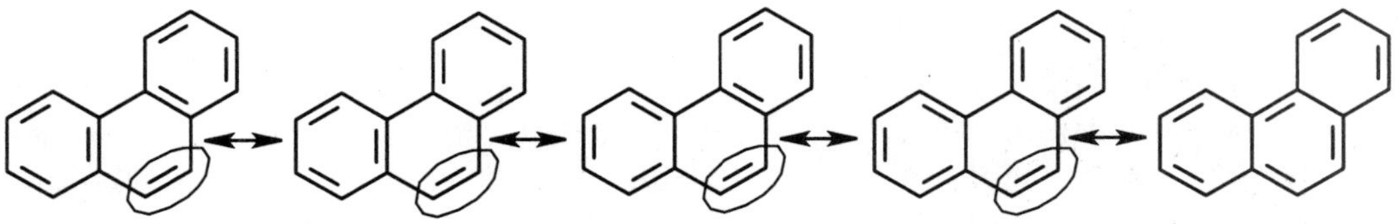

The circled bond is represented as a double bond in four of the five resonance forms of phenanthrene. This bond has more double-bond character and thus is shorter than the other carbon-carbon bonds of phenanthrene.

15.27 The heat of hydrogenation is the amount of heat liberated when a compound reacts with hydrogen.

(1) Benzene + 3 $H_2 \rightarrow$ Cyclohexane $\Delta H°_{hydrog} = -206$ kJ/mol

(2) 1,3–Cyclohexadiene + 2 $H_2 \rightarrow$ Cyclohexane $\Delta H°_{hydrog} = -230$ kJ/mol

---

Benzene + $H_2 \rightarrow$ 1,3–Cyclohexadiene $\Delta H°_{hydrog} = -206$ kJ/mol – (–230 kJ/mol) $= +24$ kJ/mol

The heat of hydrogenation for this reaction is positive, and the reaction is endothermic.

15.28

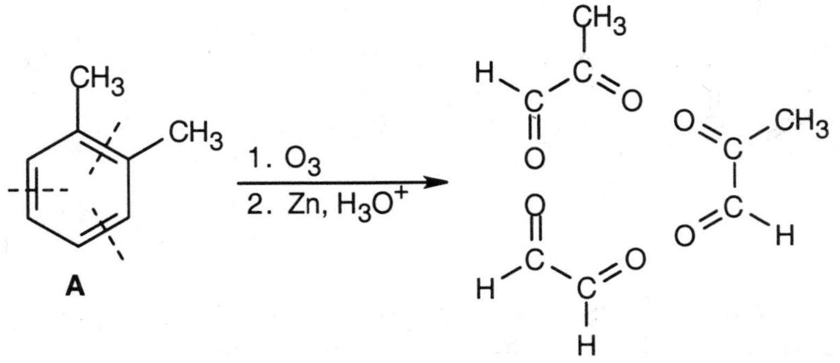

A

1. O_3

2. Zn, H_3O^+

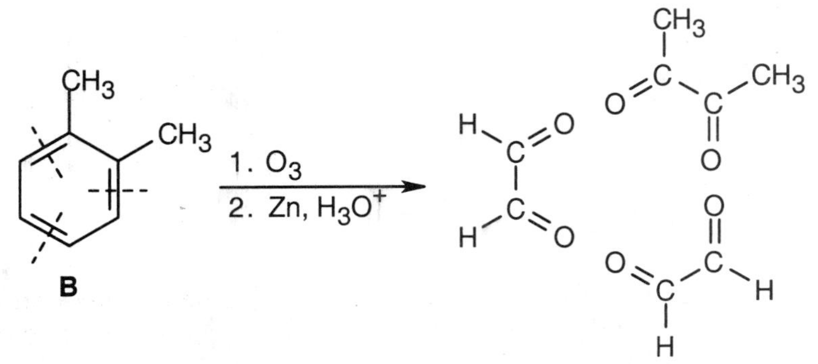

$$\text{B (}o\text{-xylene)} \xrightarrow[\text{2. Zn, } H_3O^+]{\text{1. } O_3} \text{OHC–CHO} + CH_3COCOCH_3 + \text{OHC–CHO}$$

If *o*-xylene exists only as a structure **A**, ozonolysis would cause cleavage at the bonds indicated and would yield two equivalents of pyruvaldehyde and one equivalent of glyoxal for each equivalent of **A** consumed. If *o*-xylene exists only as structure **B**, ozonolysis would yield one equivalent of 2,3-butanedione and two equivalents of glyoxal. If *o*–xylene exists as a resonance hybrid of **A** and **B**, the ratio of ozonolysis products would be glyoxal : pyruvaldehyde : 2,3-butanedione = 3:2:1. Since this ratio is identical to the experimentally determined ratio, we know that **A** and **B** contribute equally to the structure of *o*-xylene. Note that these data don't distinguish between the resonance hybrid structure and the alternate, equilibrium between two isomeric *o*-xylenes.

15.29

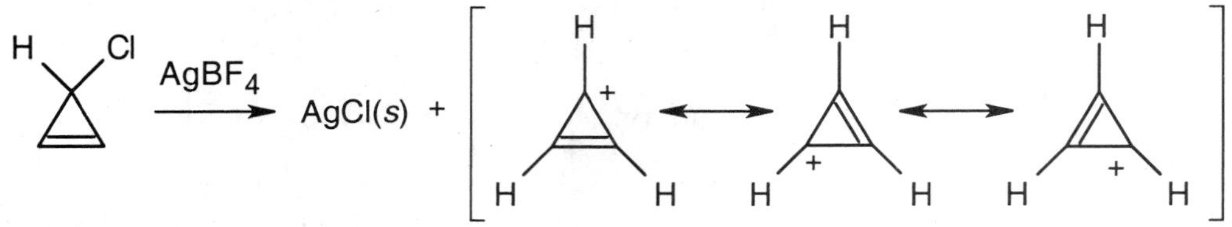

$$C_3H_3Cl \xrightarrow{AgBF_4} AgCl(s) + [C_3H_3^+ \longleftrightarrow C_3H_3^+ \longleftrightarrow C_3H_3^+]$$

The product of the reaction of 3-chlorocyclopropene with $AgBF_4$ is the cyclopropenyl cation $C_3H_3^+$. The resonance structures of the cation indicate that all hydrogen atoms are equivalent, and the ^1H NMR spectrum, which shows only one type of hydrogen atom, confirms this equivalence. The cyclopropenyl cation contains two π electrons and is aromatic according to Hückel's rule. (Here, $n = 0$.)

15.30

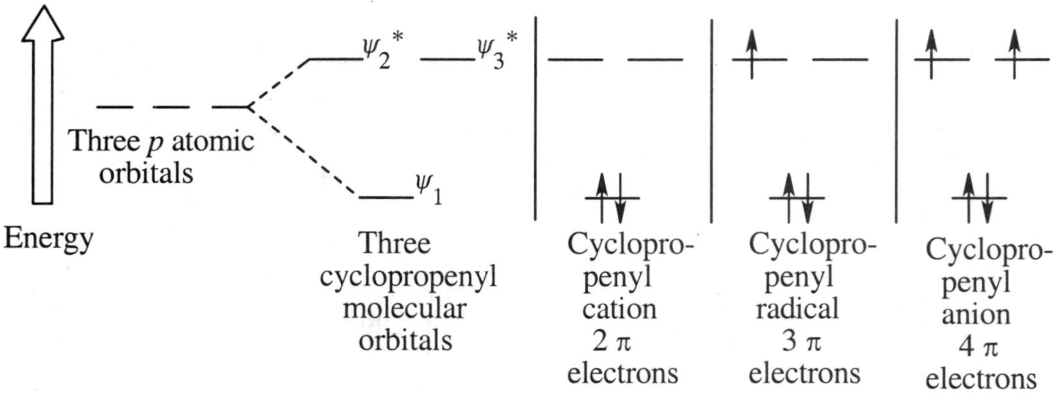

Energy

Three *p* atomic orbitals

ψ_2^* ψ_3^* ψ_1

Three cyclopropenyl molecular orbitals

Cyclopropenyl cation 2 π electrons

Cyclopropenyl radical 3 π electrons

Cyclopropenyl anion 4 π electrons

The cyclopropenyl cation is aromatic, according to Hückel's rule.

15.31

In resonance structure **A**, methylcyclopropenone is a cyclic conjugated compound with three π electrons in its ring. Because the electronegative oxygen attracts the π electrons of the carbon-oxygen π bond, however, a second resonance structure **B** can be drawn in which both carbonyl π electrons are located on oxygen, leaving only two π electrons in the ring. Since 2 is a Hückel number, the methylcyclopropenone ring fulfills the criteria of aromaticity and can be expected to be stable.

15.32

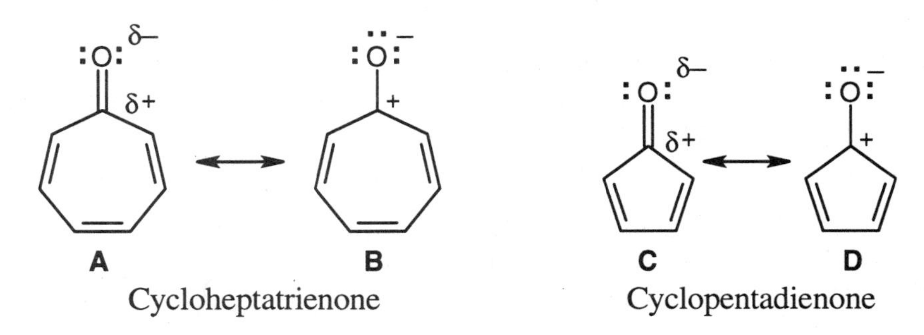

As in the previous problem, we can draw resonance forms in which both carbonyl π electrons are located on oxygen. The cycloheptatrienone ring in **B** contains six π electrons and is aromatic according to Hückel's rule. The cyclopentadienone ring in **D** contains four π electrons and is antiaromatic.

15.33 Check the number of electrons in the π system of each compound. The species with a Hückel ($4n + 2$) number of π electrons is the most stable.

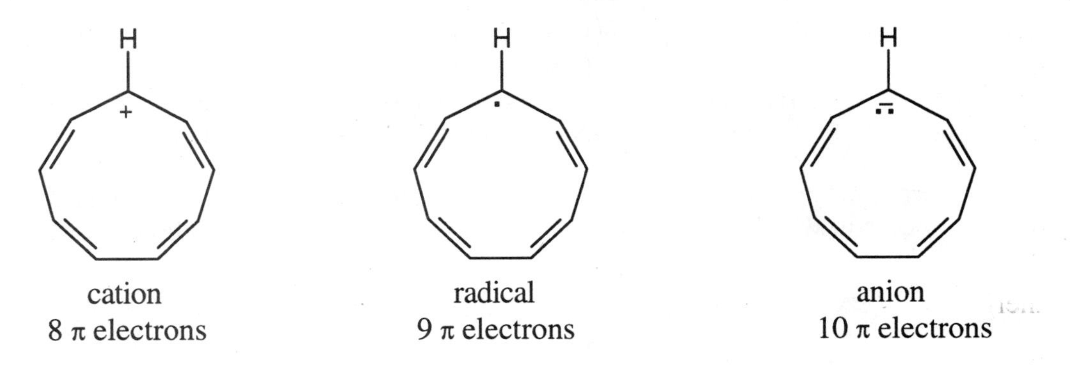

The 10 π electron anion is the most stable.

15.34 Treat 1,3,5,7-cyclononatetraene with a strong base to remove a proton.

15.35 Compound **A** has four multiple bonds and/or rings. Possible structures that yield three monobromo substitution products are:

Only structure I shows a six-proton singlet at 2.30 δ, because it contains two identical benzylic methyl groups unsplit by other protons. The presence of four protons in the aromatic region of the ^{1}H NMR spectrum confirms that I is the correct structure.

15.36

Molecules with dipole moments are polar because electron density is drawn from one part of the molecule to another. In azulene, electron density is drawn from the seven-membered ring to the five-membered ring, satisfying Hückel's rule for both rings and producing a dipole moment. The five-membered ring resembles the cyclopentadienyl anion in having six π electrons, while the seven-membered ring resembles the cycloheptatrienyl cation.

15.37 As in the previous problem, redistribution of the π electrons of calicene produces a resonance form in which both rings are aromatic and which has a dipole moment.

15.38 Pentalene has eight π electrons and is antiaromatic. Pentalene dianion, however, has ten π electrons and is a stable, aromatic ion.

15.39

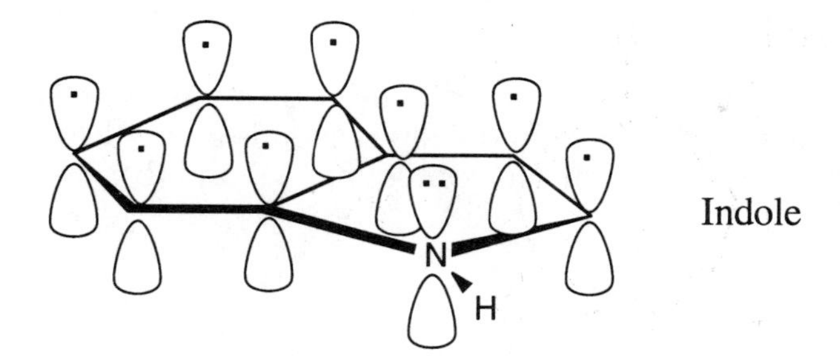

Indole

Indole, like naphthalene, has ten π electrons in two rings and is aromatic. Two π electrons come from the nitrogen atom.

15.40

$\xrightarrow{H^+}$ A $\longleftrightarrow$ B $\longleftrightarrow$ C $\longleftrightarrow$ D

Protonation of 4-pyrone gives structure **A**, which has resonance forms **B**, **C** and **D**. In **C** and **D**, a lone pair of electrons of the ring oxygen is delocalized into the ring to produce a six π electron system, which should be aromatic, according to Hückel's rule.

15.41

$C_6H_5CH_2CH{=}CHCH_3 + H{-}X \rightleftharpoons [C_6H_5CH(H){-}\overset{+}{C}H{-}CH_2CH_3 \;\; X^-] \rightleftharpoons C_6H_5CH{=}CHCH_2CH_3 + HX$

1-Phenyl-2-butene 1-Phenyl-1-butene

The alkene double bond is protonated to yield an intermediate carbocation, which loses a proton to give a product in which the double bond is conjugated with the aromatic ring, as shown by the increased value of λ_{max}.

15.42 The molecular weight of the hydrocarbon (120) corresponds to the molecular formula C_9H_{12}, which indicates four double bonds and/or rings. The ^1H NMR singlet at 7.25 δ indicates five aromatic ring protons. The septet at 2.90 δ is due to a benzylic proton that has six neighboring protons.

$C_6H_5{-}CH(CH_3)_2$ Isopropylbenzene

15.43 Both hydrocarbons are disubstituted benzenes. The compound in (a) has two ethyl groups, and the compound in (b) has an isopropyl group and a methyl group. The IR data must be used to find the pattern of substitution.

In Section 15.10, note that an IR absorption of 745 cm^{-1} corresponds to an *o*-disubstituted benzene, and an absorption of 825 cm^{-1} corresponds to a *p*-disubstituted benzene. It is now possible to assign structures to compounds (a) and (b).

(a) CH_2CH_3, CH_2CH_3

o-Diethylbenzene

(b) $CH(CH_3)_2$, H_3C

p-Isopropyltoluene

15.44 All of these compounds have 4 degrees of unsaturation and are substituted benzenes. The benzylic absorptions (2.3–3.0 δ) show the groups next to the aromatic ring. Remember that the data from the IR spectrum can be used to assign the substitution pattern of the ring.

(a) CH_2CH_3, Br

p-Bromoethylbenzene

(b) CH_2CH_3, CH_3

o-Ethyltoluene

(c) $C(CH_3)_3$, H_3C

p-tert-Butyltoluene

15.45 The compound has nine degrees of unsaturation. The ^1H NMR spectrum shows that the compound is symmetrical and that the only absorptions occur in the vinylic and aromatic regions of the spectrum. The IR spectrum shows peaks due to a monosubstituted benzene ring and to $R_2C{=}CH_2$ (890 cm^{-1}).

CH_2
||
C

A Look Ahead

15.46

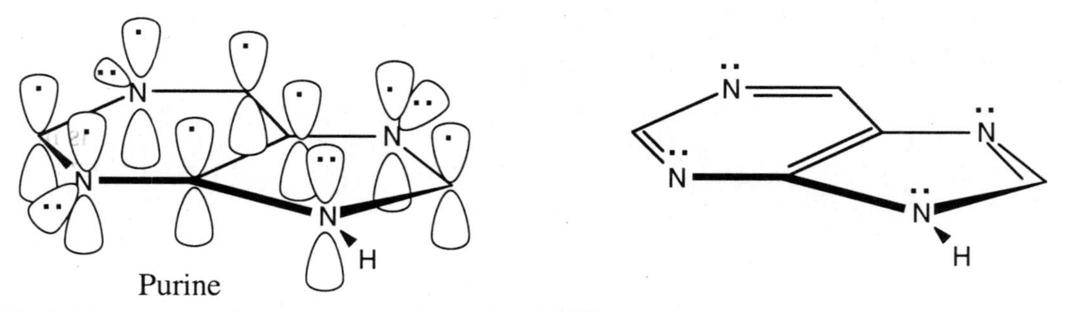

Purine

Purine is a ten-π-electron aromatic molecule. The N–H nitrogen atom in the five-membered ring donates both electrons of its lone pair to the π electron system, and each of the other three nitrogens donates one electron to the π electron system.

15.47

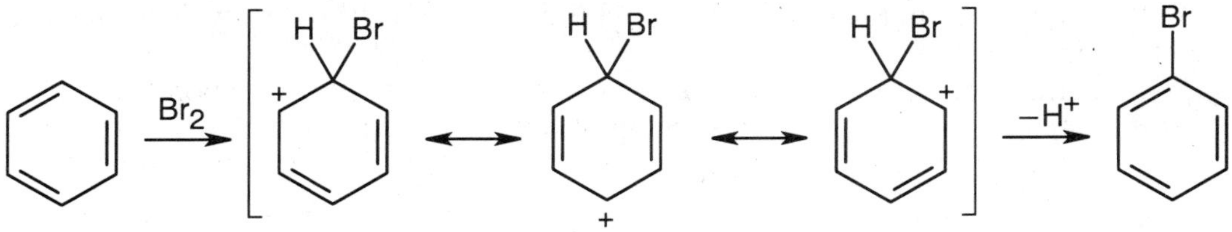

15.48

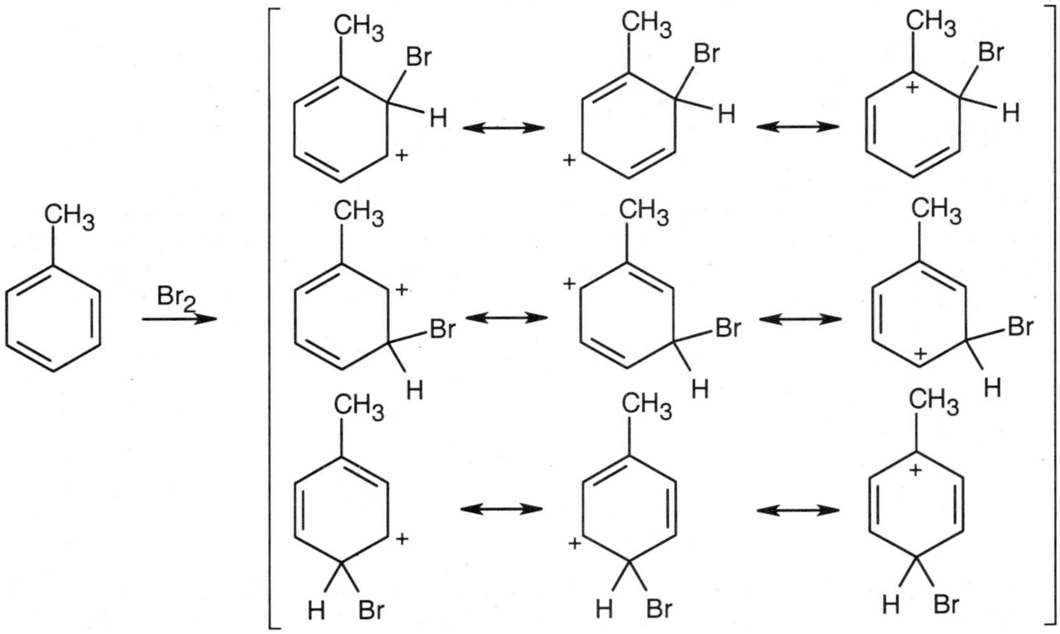

The ortho and para products predominate because the intermediate carbocation is more stabilized. The third resonance form drawn for ortho-para attack places the positive charge at the methyl-substituted carbon, which can form a more stable tertiary carbocation.

Molecular Modeling

15.49 The sulfur atoms in tetrahydrothiophene and 2,3-dihydrothiophene show negative regions that are due to lone pair electrons. The sulfur atom in thiophene is much less negative, indicating that the lone pair electrons are delocalized. Delocalization is favored in thiophene because the molecule contains 6 π electrons and is aromatic.

15.50 The hydrogens of the four-membered ring occupy the same positions relative to two of the three benzene rings in both isomers; they lie "outside" the two closest rings and are deshielded by these rings. The hydrogens also lie "outside" the more distant, third ring in the cis isomer and are more deshielded ($\delta = 5.2$) in this isomer. The hydrogens in the trans isomer approach more closely the middle "shielding" region of the distant benzene ring and are thus more shielded ($\delta = 4.4$).

15.51 The $=CH_2$ group in fulvene is less negative than the $=CH_2$ group in 2-methylpropene, and the $=CH_2$ group in methylenecyclopropene is more negative. These variations can be explained by resonance structures that show how the unequal distribution of electrons can yield a structure that satisfies Hückel's rule and is aromatic.

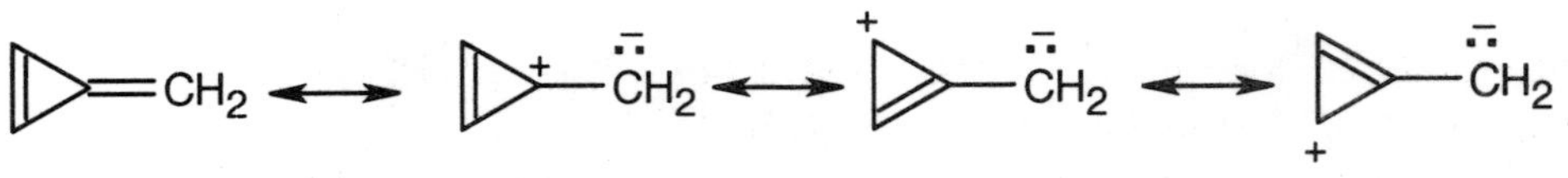

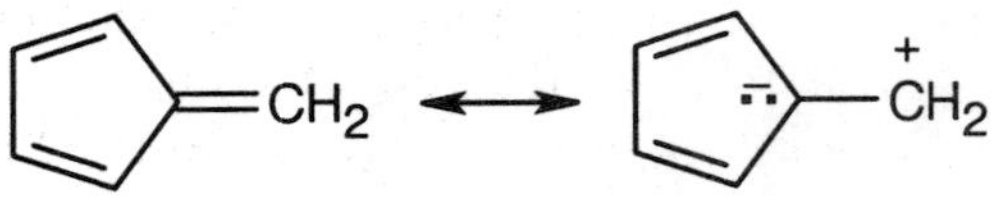

Chapter 16 – Chemistry of Benzene Electrophilic Aromatic Substitution

Chapter Outline

I. Electrophilic aromatic substitution reactions (Sections 16.1 – 16.4).
  A. Bromination of aromatic rings (Section 16.1).
    1. Characteristics of electrophilic aromatic substitution reactions.
      a. The accessibility of the π electrons of an aromatic ring make it a nucleophile.
      b. Aromatic rings are less reactive to electrophiles than are alkenes.
        A catalyst is needed to make the reacting molecule more electrophilic.
    2. Mechanism of bromination.
      a. Br_2 complexes with $FeBr_3$.
      b. The polarized electrophile is attacked by the π electrons of the ring in a slow, rate-limiting step.
      c. The cation intermediate is doubly allylic but is much less stable than the starting aromatic compound.
      d. The carbocation intermediate loses H^+ from the bromine-bearing carbon in a fast step to regenerate an aromatic ring.
  B. Other aromatic substitution reactions (Section 16.2).
    1. Chlorination and iodination.
      a. Chlorine reacts in the presence of $FeCl_3$ to yield chlorinated rings.
      b. Iodination occurs only in the presence of an oxidizing agent.
      c. Fluorine is too reactive to be useful.
    2. Nitration.
      a. A mixture of HNO_3 and H_2SO_4 is used for nitration.
      b. The reactive electrophile is NO_2^+.
      c. Products of nitration can be reduced with Fe or $SnCl_2$ to yield an arylamine.
    3. Sulfonation
      a. Rings can be sulfonated by a mixture of SO_3 and H_2SO_4 to yield sulfonic acids.
      b. The reactive electrophile is either SO_3 or HSO_3^+.
      c. Sulfonation is reversible.
      d. Sulfonic acids can be heated with NaOH to form phenols.
  C. Alkylation of aromatic rings (Section 16.3).
    1. The Friedel-Crafts reaction introduces an alkyl group onto an aromatic ring.
    2. An alkyl chloride, plus an $AlCl_3$ catalyst, produces an electrophilic carbocation.
    3. There are several limitations to using the Friedel-Crafts reaction.
      a. Only alkyl halides – not aryl or vinylic halides – can be used.
      b. Friedel-Crafts reactions don't succeed on rings that have amino substituents or deactivating groups.
      c. Polyalkylation is often seen.
      d. Rearrangements of the alkyl carbocation often occur.
        Rearrangements may occur by hydride shifts or by alkyl shifts.
  D. Acylation of aromatic rings (Section 16.4).
    1. Friedel-Crafts acylation occurs when an aromatic ring reacts with a carboxylic acid chloride.
    2. The reactive electrophile is an acyl cation, which doesn't rearrange.
    3. Polyacylation never occurs in acylation reactions.

II. Substituent effects in substituted aromatic rings (Sections 16.5 – 16.7).
  A. Types of substituent effects (Section 16.5).
    1. Substituents affect the reactivity of an aromatic ring.
    2. Substituents affect the orientation of further substitution.

3. Substituents can be grouped into three groups:
   a. Ortho/para directing activators.
   b. Ortho/para-directing deactivators.
   c. Meta-directing deactivators.
4. Two kinds of effects are responsible for reactivity and orientation.
   a. Inductive effects are due to differences in bond polarity.
   b. Resonance effects are due to overlap of a *p* orbital of a substituent with a *p* orbital on an aromatic ring.
      i. Carbonyl, cyano and nitro substituents withdraw electrons. These substituents have the structure –Y=Z.
      ii. Halogen, hydroxyl, alkoxyl and amino substituents donate electrons. These substituents have the structure –Y**:**.
      iii. Resonance effects are greatest at the ortho and para positions.
   c. Resonance and inductive effects don't always act in the same direction.

B. Explanation of substituent effects (Section 16.6).
1. All activating groups donate electrons to an aromatic ring.
2. All deactivating groups withdraw electrons from a ring.
3. Alkyl groups – ortho-and para-directing activators.
   a. Alkyl groups inductively donate electrons to a ring.
   b. Alkyl groups are *o,p*-directors because the carbocation intermediates from attack are best stabilized when attack occurs at the ortho and para positions.
4. OH, NH_2 groups – ortho-and para-directing activators.
   a. OH, NH_2 donate electrons by resonance involving the ring and the group.
   b. The intermediates of *o,p*-attack are more stabilized by resonance than are intermediates of *m*-attack.
5. Halogens – ortho-and para-directing deactivators.
   a. The electron-withdrawing inductive effect of halogen outweighs its electron-donating resonance effect.
   b. The resonance effect orients substitution to the *o,p* positions.
   c. The inductive effect deactivates the ring.
6. Meta-directing deactivators.
   a. Meta-directing deactivators act through both inductive and resonance effects.
   b. Because resonance effects destabilize ortho and para positions the most, substitution ion occurs at the meta position.

C. Trisubstituted benzenes: additivity of effects (Section 16.7).
1. If the effects of both groups are additive, the product of substitution is easy to predict.
2. If the directing effects of the groups are opposed, the more powerful activating group determines the product, although mixtures sometimes result.
3. Substitution rarely occurs between two groups that are meta to each other for steric reasons.

III. Other reactions of aromatic rings (Sections 16.8 – 16.11).

A. Nucleophilic aromatic substitution (Section 16.8).
1. An aryl halide with electron-withdrawing groups can undergo nucleophilic aromatic substitution.
2. This reaction occurs through an addition/elimination mechanism.
3. Addition of the nucleophile proceeds through an intermediate Meisenheimer complex that is stabilized by *o,p* electron-withdrawing substituents on the ring.
4. The halide is eliminated to yield product.

B. Benzyne (Section 16.9).
1. At high temperatures and with strong base, aryl halides without electron-withdrawing substituents can be converted to phenols .
2. This reaction occurs by an elimination/addition reaction that involves a benzyne intermediate.

a. Strong base cause elimination of HX from the halide to generate benzyne.
b. A nucleophile adds to benzyne to give the product.
3. The benzyne intermediate can be trapped in a Diels-Alder reaction.
4. Benzyne has the electronic structure of a distorted alkyne and has one very weak π bond.

C. Oxidation of aromatic compounds (Section 16.10).
1. Oxidation of alkylbenzene side chains.
a. Strong oxidizing agents cause the oxidation of alkyl side chains with benzylic hydrogens.
b. The products of side-chain oxidation are benzoic acids.
c. Reaction proceeds by a complex radical mechanism.
2. Bromination of alkylbenzene side chains.
a. NBS brominates alkylbenzene side chains at the benzylic position.
b. Bromination occurs by the mechanism described for allylic bromination and requires a radical initiator.
c. The intermediate benzylic radical is stabilized by resonance.

D. Reduction of aromatic compounds (Section 16.11).
1. Catalytic hydrogenation of aromatic rings.
a. It is possible to selectively reduce alkene bonds in the presence of aromatic rings because rings are relatively inert to catalytic hydrogenation.
b. With a stronger catalyst, aromatic rings can be reduced.
2. Reduction of aryl alkyl ketones.
a. Aryl alkyl ketones can undergo catalytic hydrogenation to form alkylbenzenes.
b. Acylation plus reduction is a route to alkyl substitution without rearrangement.
c. This reaction only occurs with aryl alkyl ketones and also reduces nitro groups to amino groups.

IV. Synthesis of substituted benzenes (Section 16.12).
A. To synthesize substituted benzenes, it is important to introduce groups so that they have the proper orienting effects.
B. It is best to use retrosynthetic analysis to plan a synthesis.

Solutions to Problems

16.1

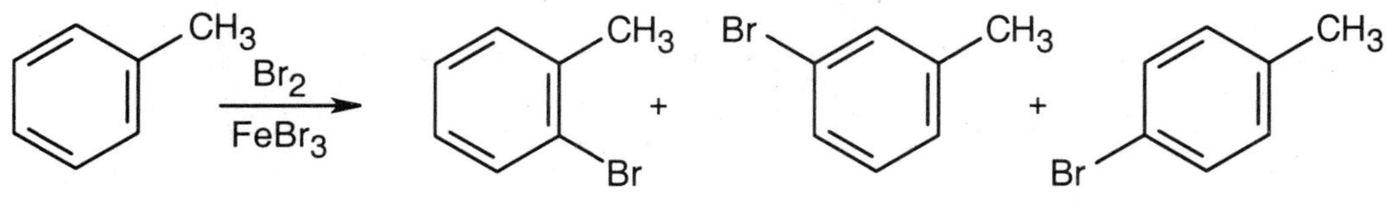

CH_3 $\xrightarrow[FeBr_3]{Br_2}$ CH_3, Br + Br, CH_3 + Br, CH_3

o-Bromotoluene *m*-Bromotoluene *p*-Bromotoluene

16.2

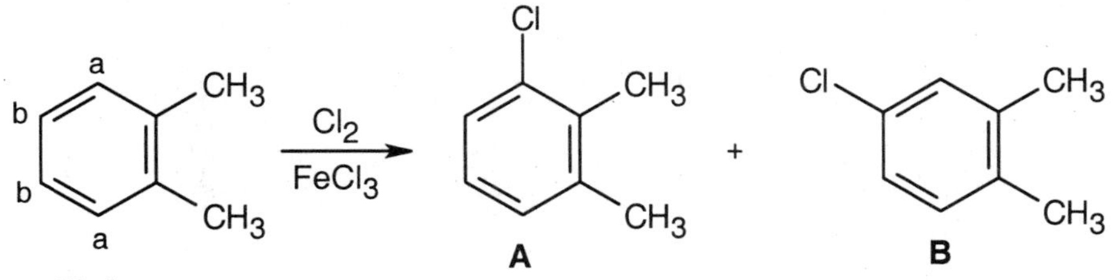

a, b, b, a, CH_3, CH_3 $\xrightarrow[FeCl_3]{Cl_2}$ Cl, CH_3, CH_3 + Cl, CH_3, CH_3

o-Xylene **A** **B**

Chlorination at position "a" of *o*–xylene yields product **A**, and chlorination at position "b" yields product **B**.

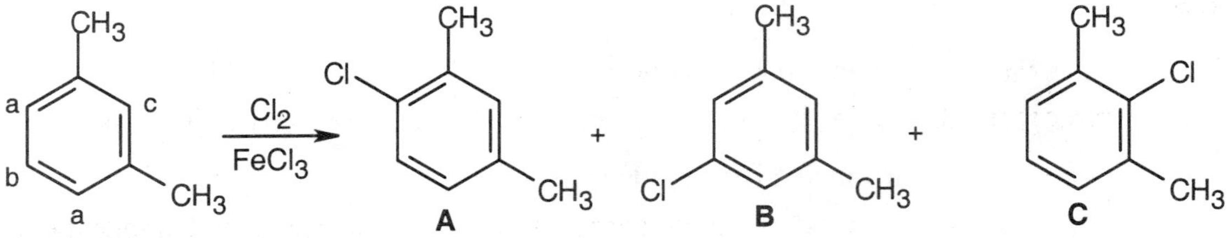

CH_3 a c b a CH_3 *m*-Xylene $\xrightarrow[FeCl_3]{Cl_2}$ A (CH_3, Cl, CH_3) + B (CH_3, Cl, CH_3) + C (CH_3, Cl, CH_3)

Three products might be expected to form on chlorination of *m*–xylene. Product **C** is unlikely to form because substitution rarely occurs between two meta substituents. Product **B** is also unlikely to form for reasons to be explained later in the chapter.

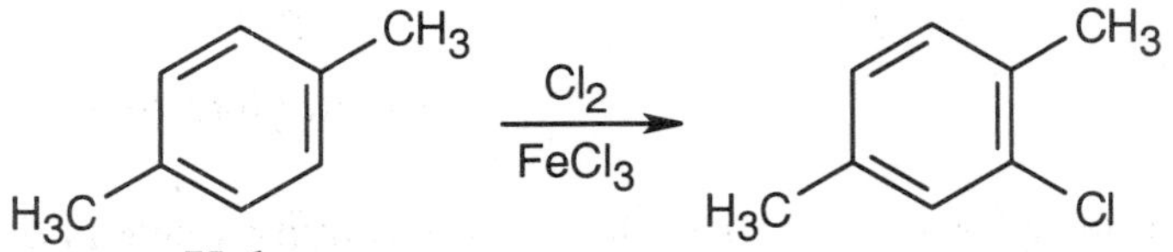

CH_3 H_3C *p*-Xylene $\xrightarrow[FeCl_3]{Cl_2}$ CH_3 H_3C Cl

Only one product results from chlorination of *p*–xylene because all sites for chlorination are equivalent.

16.3

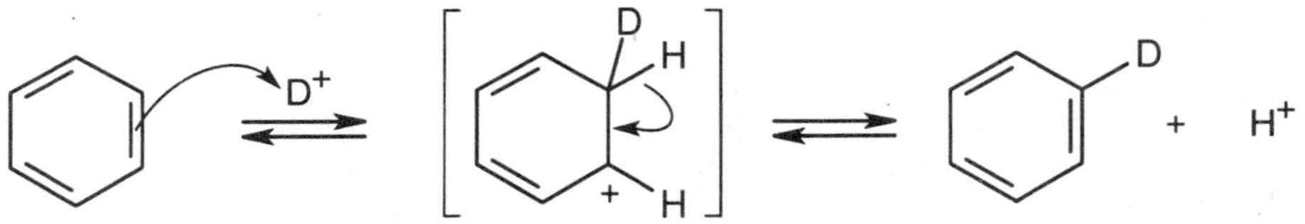

D^+ ⇌ [D H + H] ⇌ D + H^+

Benzene can be protonated by strong acids. The resulting intermediate can lose either deuterium or hydrogen. If H^+ is lost, deuterated benzene is produced. Attack by D^+ can occur at all positions of the ring and leads to eventual replacement of all hydrogens by deuterium.

16.4 Carbocation rearrangements of alkyl halides occur (1) if the initial carbocation is primary or secondary, and (2) if it is possible for the initial carbocation to rearrange to a more stable secondary or tertiary cation.

(a) Although $CH_3\overset{+}{C}H_2$ is a primary carbocation, it can't rearrange to a more stable cation.

(b) $CH_3CH_2CHClCH_3$ forms a secondary carbocation that doesn't rearrange.

(c) $CH_3CH_2\overset{+}{C}H_2$ rearranges to the more stable $CH_3\overset{+}{C}HCH_3$.

(d) $(CH_3)_3C\overset{+}{C}H_2$ (primary) undergoes an alkyl shift to yield $(CH_3)_2\overset{+}{C}CH_2CH_3$ (tertiary).

(e) The cyclohexyl carbocation doesn't rearrange.

In summary:
No rearrangement: CH_3CH_2Cl, $CH_3CH_2CHClCH_3$, chlorocyclohexane
Rearrangement: $CH_3CH_2CH_2Cl$, $(CH_3)_3CCH_2Cl$

16.5

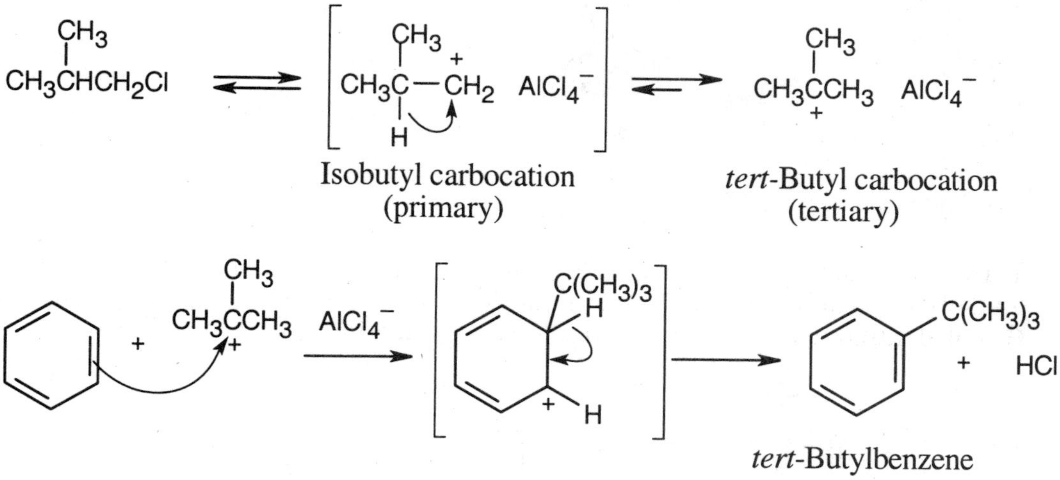

The isobutyl carbocation is initially formed when 1–chloro–2–methylpropane and $AlCl_3$ react. This carbocation rearranges via a hydride shift to give the more stable *tert*–butyl carbocation, which can then alkylate benzene to form *tert*–butylbenzene.

16.6 To identify the carboxylic acid chloride used in the Friedel-Crafts acylation of benzene, break the bond between benzene and the ketone carbon and replace it with a –Cl.

(a)

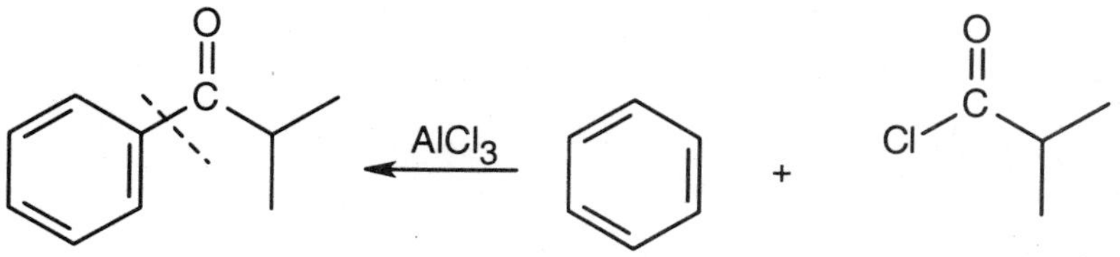

(b)

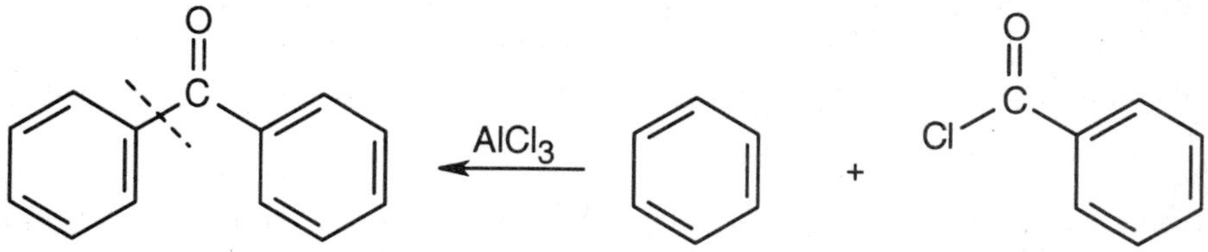

16.7 Refer to Figure 16.10 in the text for the directing effects of substituents. You should memorize the effects of the most important groups. As in Practice Problem 16.2, identify the directing effect of the substituent, and draw the product.

(a)

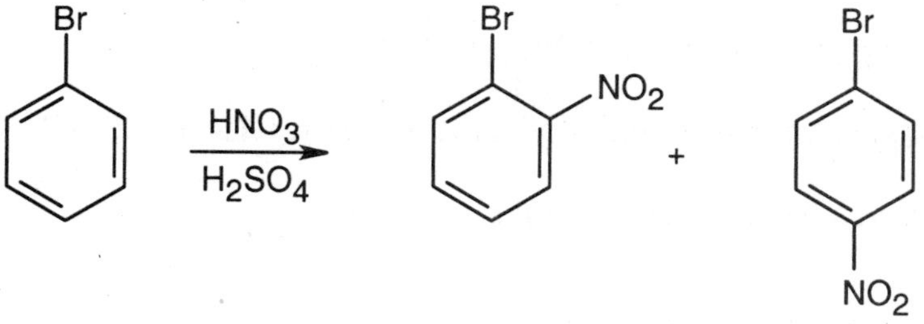

Even though bromine is a deactivator, it is an ortho–para director.

(b)

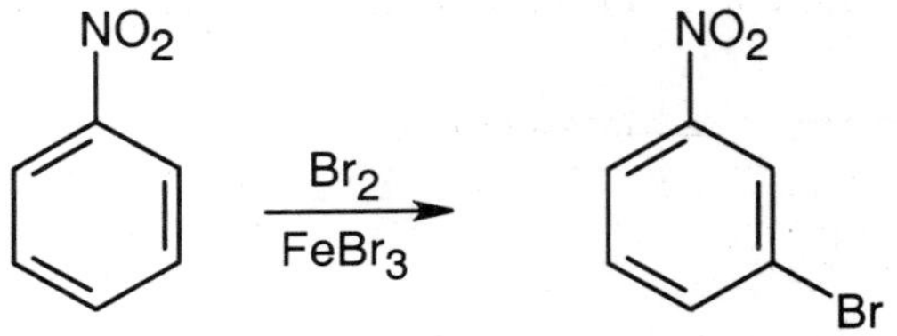

NO_2 $\xrightarrow[\text{FeBr}_3]{\text{Br}_2}$ NO_2, Br

The $-NO_2$ group is a meta–director.

(c)

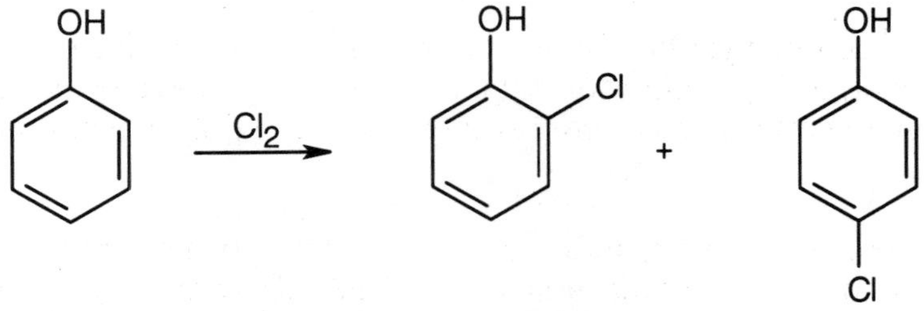

OH $\xrightarrow{\text{Cl}_2}$ OH, Cl + OH, Cl

(d)

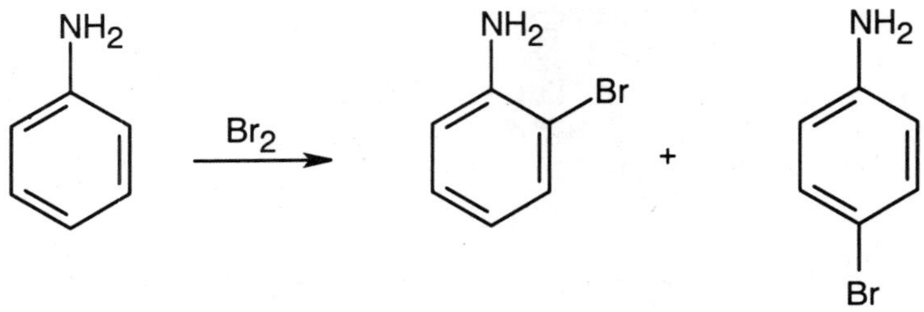

NH_2 $\xrightarrow{\text{Br}_2}$ NH_2, Br + NH_2, Br

No catalyst is necessary because aniline is highly activated.

16.8

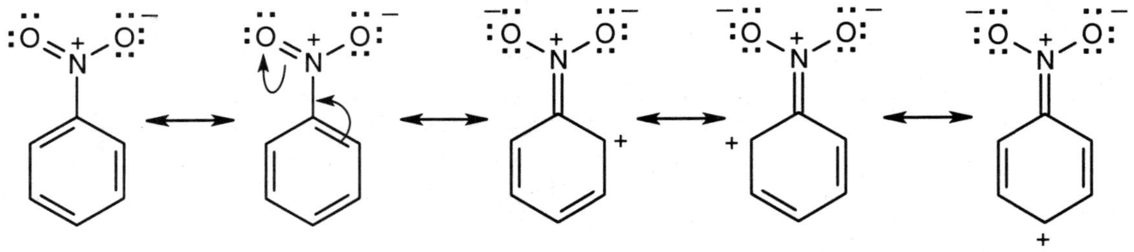

:Ö=N⁺–Ö:⁻ ⟷ :Ö=N⁺–Ö:⁻ ⟷ ⁻:Ö–N⁺–Ö:⁻ (ring +) ⟷ ⁻:Ö–N⁺–Ö:⁻ (ring +) ⟷ ⁻:Ö–N⁺–Ö:⁻ (ring +)

16.9

:C̈l: ⟷ :C̈l: ⟷ :Cl:⁺ (ring :⁻) ⟷ :Cl:⁺ (ring :⁻) ⟷ :Cl:⁺ (ring :⁻)

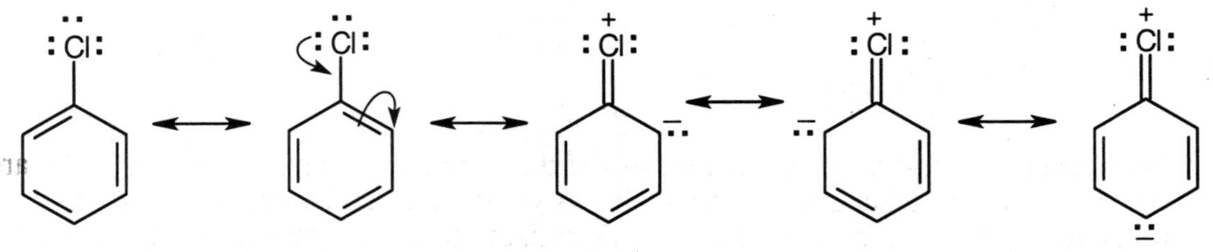

16.10 Use Figure 16.10 to find the activating and deactivating effects of groups.

Most Reactive ⟶ *Least Reactive*

(a) Phenol > toluene > benzene > nitrobenzene

(b) Phenol > benzene > chlorobenzene > benzoic acid

(c) Aniline > benzene > bromobenzene > benzaldehyde

16.11 An acyl substituent is deactivating. Once an aromatic ring has been acylated, it is much less reactive to further substitution. An alkyl substituent is activating, however, so an alkyl-substituted ring is more reactive than an unsubstituted ring, and polysubstitution occurs readily.

16.12 Toluene is more reactive toward electrophilic substitution than (trifluoromethyl)benzene. The electronegativity of the three fluorine atoms causes the trifluoromethyl group to be electron-withdrawing and deactivating toward electrophilic substitution.

16.13

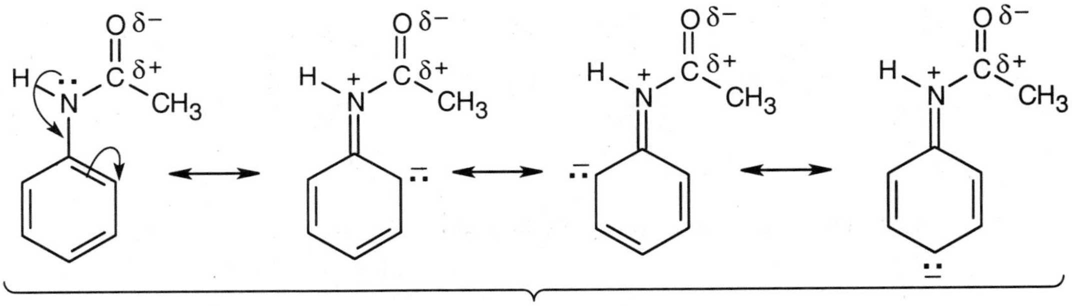

O δ− C δ+ H N CH3 ⟷ O δ− H + C δ+ N CH3 ⟷ O δ− H + C δ+ N CH3 ⟷ O δ− H + C δ+ N CH3

less favored

:O:δ+ H C δ− N CH3 ⟷ :O: − H + C N CH3

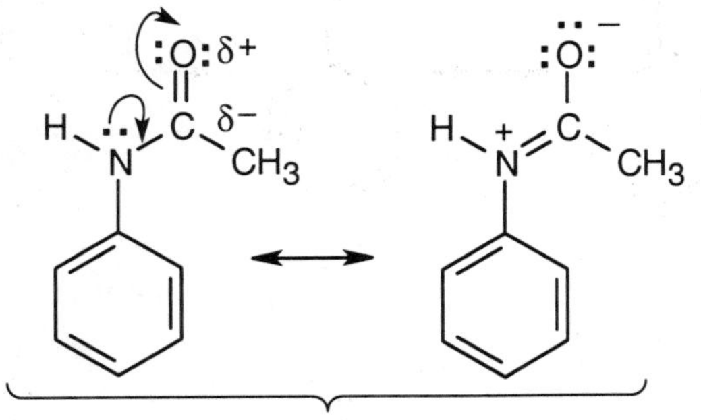

more favored

For acetanilide, resonance delocalization of the nitrogen lone pair electrons to the aromatic ring is less favored because the positive charge on nitrogen is next to the positively polarized carbonyl group. Resonance delocalization to the carbonyl oxygen *is* favored because of the electronegativity of oxygen. Since the nitrogen lone pair electrons are less available to the ring, the reactivity of the ring toward electrophilic substitution is decreased, and acetanilide is less reactive than aniline toward electrophilic substitution.

16.14

Ortho attack:

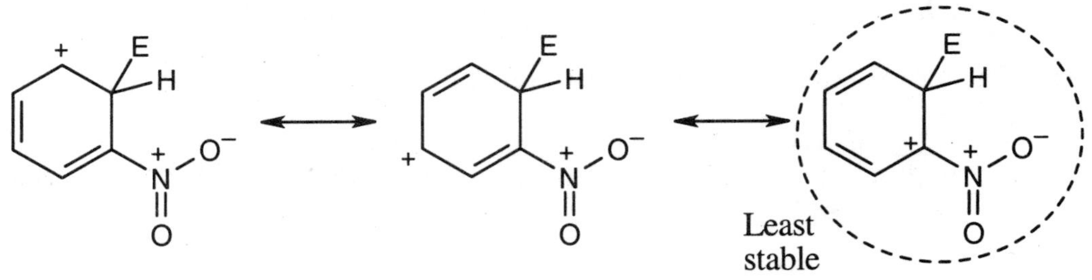

Least stable

Meta attack:

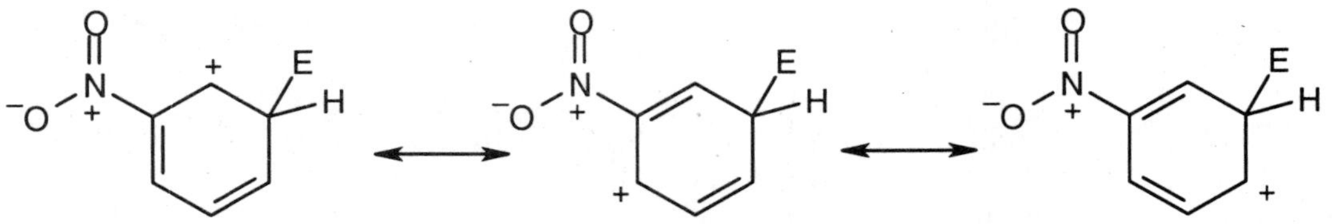

Para attack:

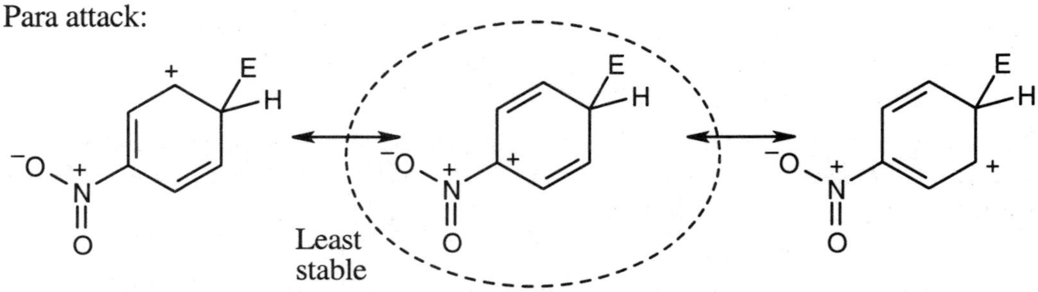

Least stable

The circled resonance forms are unfavorable, because they place two positive charges adjacent to each other. The intermediate from *meta* attack is thus favored.

16.15

(a)

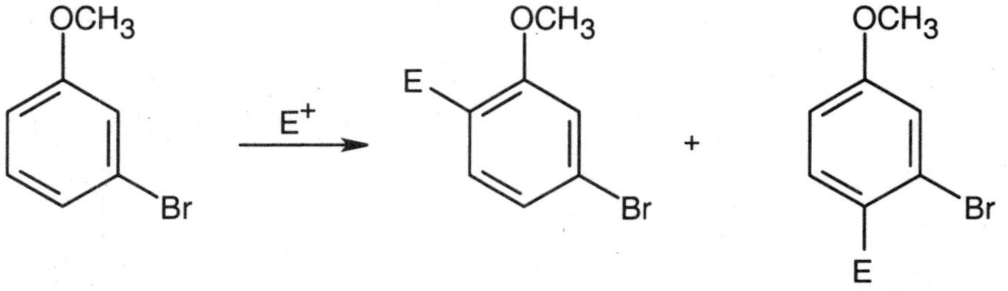

$\xrightarrow{E^+}$ +

Both groups are ortho-para directors and direct substitution to the same positions. Attack doesn't occur between the two groups for steric reasons.

(b)

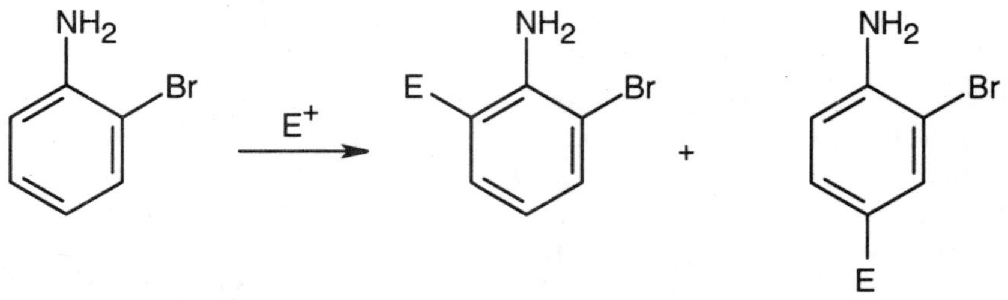

$\xrightarrow{E^+}$ +

Both groups are ortho-para directors, but direct to different positions. Because $-NH_2$ group is a more powerful activator, substitution occurs ortho and para to it.

(c)

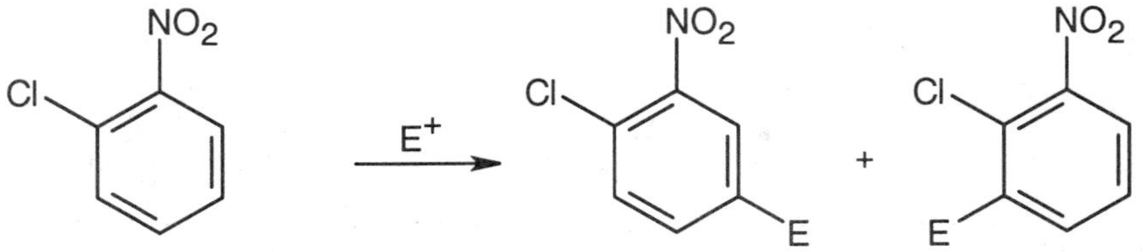

Both groups are deactivating, but they orient substitution toward the same positions.

16.16

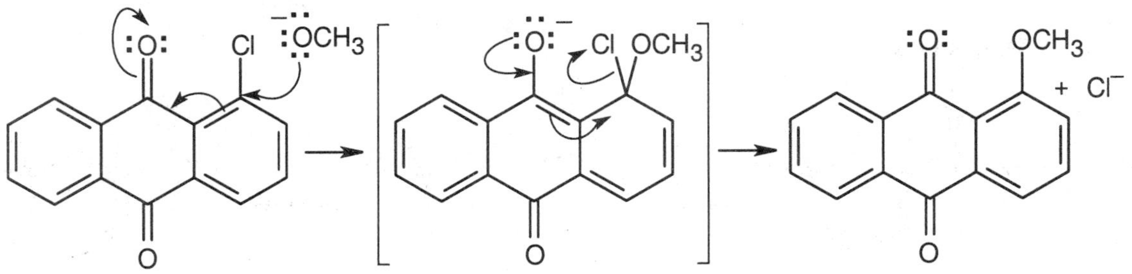

Addition of the nucleophile $^{-}OCH_3$ — Meisenheimer complex — Elimination of Cl^-

The carbonyl oxygens make the chlorine–containing ring electron-poor and vulnerable to attack by the nucleophile $^{-}OCH_3$. They also stabilize the negatively charged Meisenheimer complex.

16.17

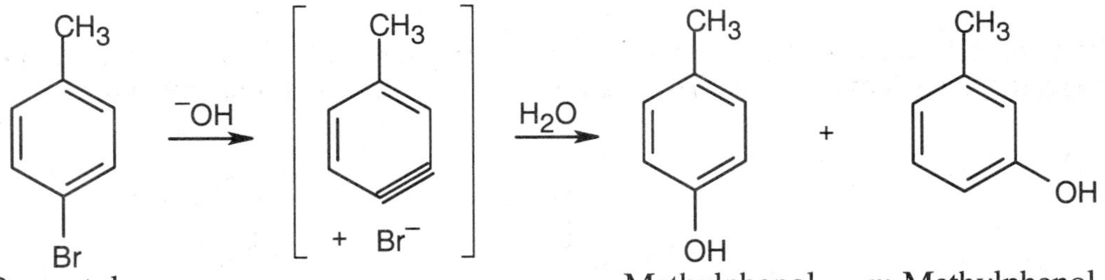

p-Bromotoluene → *p*-Methylphenol + *m*-Methylphenol

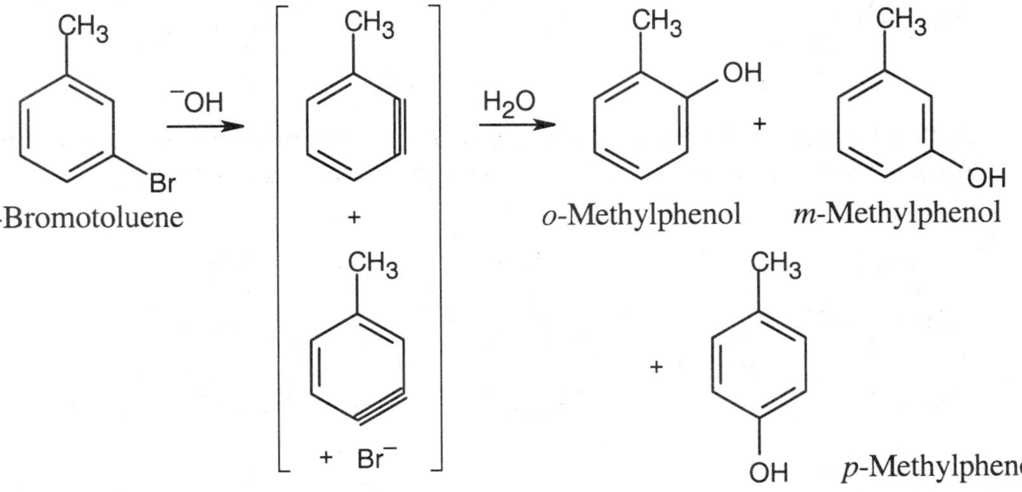

m-Bromotoluene → *o*-Methylphenol + *m*-Methylphenol + *p*-Methylphenol

Treatment of *m*–bromotoluene with NaOH leads to two benzyne intermediates, which react with water to yield three methylphenol products.

16.18

(a)

O_2N–C$_6$H$_4$–$CH(CH_3)_2$ $\xrightarrow[H_2O]{KMnO_4}$ O_2N–C$_6$H$_4$–COOH

m-Nitrobenzoic acid

(b)

H_3C–C$_6$H$_4$–$C(CH_3)_3$ $\xrightarrow[H_2O]{KMnO_4}$ HOOC–C$_6$H$_4$–$C(CH_3)_3$

p-tert-Butylbenzoic acid

Treatment with $KMnO_4$ oxidizes the methyl group but leaves the *tert*-butyl group untouched.

16.19

Benzene $\xrightarrow[AlCl_3]{CH_3CH_2Cl}$ C$_6$H$_5$–CH_2CH_3 $\xrightarrow[(PhCO_2)_2]{NBS}$ C$_6$H$_5$–CH(Br)CH_3 $\xrightarrow[CH_3CH_2OH]{KOH}$ C$_6$H$_5$–$CH{=}CH_2$

Styrene

16.20

Bond	$CH_3CH_2{-}H$	C$_6$H$_5$–$CH_2{-}H$	$H_2C{=}CHCH_2{-}H$
Bond dissociation energy	420 kJ/mol	368 kJ/mol	361 kJ/mol

Bond dissociation energies measure the amount of energy that must be supplied to cleave a bond into two radical fragments. A radical is thus higher in energy and less stable than the compound it came from. Since the C–H bond dissociation energy is 420 kJ/mol for ethane and 368 kJ/mol for a methyl group C–H bond of toluene, less energy is required to form a benzyl radical than to form an ethyl radical. A benzyl radical is thus more stable than a primary alkyl radical by 52 kJ/mol. The bond dissociation energy of an allyl C–H bond is 361 kJ/mol, indicating that a benzyl radical is nearly as stable as an allyl radical.

16.21

C$_6$H$_5$–C(=O)–Cl + Benzene $\xrightarrow{AlCl_3}$ C$_6$H$_5$–C(=O)–C$_6$H$_5$ $\xrightarrow[Pd]{H_2}$ C$_6$H$_5$–C(H)(H)–C$_6$H$_5$

16.22 (a) In order to synthesize the product with the correct orientation of substituents, benzene must be nitrated before it is chlorinated.

Benzene $\xrightarrow[H_2SO_4]{HNO_3}$ nitrobenzene (NO_2) $\xrightarrow[FeCl_3]{Cl_2}$ *m*-Chloronitrobenzene (NO_2, Cl)

(b) Chlorine can be introduced into the correct position if benzene is first acylated. The chlorination product can then be reduced.

Benzene $\xrightarrow[AlCl_3]{CH_3CCl\ (C{=}O)}$ acetophenone ($O{=}C{-}CH_3$) $\xrightarrow[FeCl_3]{Cl_2}$ *m*-chloroacetophenone (Cl) $\xrightarrow[Pd]{H_2}$ *m*-Chloroethylbenzene (CH_2CH_3, Cl)

(c) Two routes are possible:

Benzene $\xrightarrow[AlCl_3]{CH_3CH_2CCl\ (C{=}O)}$ propiophenone ($O{=}C{-}CH_2CH_3$) $\xrightarrow[Pd]{H_2}$ propylbenzene ($CH_2CH_2CH_3$) $\xrightarrow[FeCl_3]{Cl_2}$ *p*-Chloropropylbenzene ($CH_2CH_2CH_3$, Cl)

Benzene $\xrightarrow[FeCl_3]{Cl_2}$ chlorobenzene (Cl) $\xrightarrow[AlCl_3]{CH_3CH_2CCl\ (C{=}O)}$ *p*-chloropropiophenone ($O{=}C{-}CH_2CH_3$, Cl) $\xrightarrow[Pd]{H_2}$ *p*-Chloropropylbenzene

16.23 (a) Friedel-Crafts acylation, like Friedel-Crafts alkylation, does not occur at an aromatic ring carrying a meta-directing group. We will learn in a later chapter how to synthesize this compound by another route.

(b) There are two problems with this synthesis as it is written:

1. Rearrangement often occurs during Friedel-Crafts alkylations using primary halides.
2. Even if *p*–chloropropylbenzene could be synthesized, introduction of the second –Cl group would occur ortho to the alkyl group.

A possible route to this compound:

$$\text{chlorobenzene} \xrightarrow[AlCl_3]{CH_3CH_2CCl\ (O)} p\text{-}ClC_6H_4COCH_2CH_3 \xrightarrow[FeCl_3]{Cl_2} 3,4\text{-}Cl_2C_6H_3COCH_2CH_3 \xrightarrow[Pd]{H_2} \text{product}$$

(Final product substituents: $CH_2CH_2CH_3$, CH_3, Cl)

Visualizing Chemistry

16.24 (a) The methoxyl group is an *o,p*-director.

(i) $C_6H_5OCH_3 \xrightarrow[FeBr_3]{Br_2}$ *p*-Bromomethoxybenzene + *o*-Bromomethoxybenzene

(ii) $C_6H_5OCH_3 \xrightarrow[AlCl_3]{CH_3CCl\ (O)}$ *p*-Methoxyacetophenone + *o*-Methoxyacetophenone

(b) Both functional groups direct substituents to the same position.

(i) $p\text{-}CH_3C_6H_4CHO \xrightarrow[FeBr_3]{Br_2}$ 3-Bromo-4-methylbenzaldehyde

(ii) $p\text{-}CH_3C_6H_4CHO \xrightarrow[AlCl_3]{CH_3CCl\ (O)}$ 3-Acetyl-4-methylbenzaldehyde

16.25 In the most stable form of acetophenone, the carbon and oxygen of the carbonyl group lie in the same plane as the ring in order for the π electrons of the C=O bond to overlap with the π electrons of the ring. In the conformation pictured, oxygen is perpendicular to the ring, and overlap can't occur.

16.26 In the lowest-energy conformation of the biphenyl, the aromatic rings are perpendicular. If the rings were planar, steric strain between the methyl groups and the ring hydrogens would occur. Complete rotation around the single bond doesn't take place because the repulsive interaction between the methyl groups causes a barrier to rotation.

Additional Problems

16.27

(a) Br $\xrightarrow[H_2SO_4]{HNO_3}$ Br, NO_2 + Br, O_2N

o-Bromonitrobenzene *p*-Bromonitrobenzene

(b) CN $\xrightarrow[H_2SO_4]{HNO_3}$ O_2N, CN

m-Nitrobenzonitrile

(c) COOH $\xrightarrow[H_2SO_4]{HNO_3}$ O_2N, COOH

m-Nitrobenzoic acid

(d) NO_2 $\xrightarrow[H_2SO_4]{HNO_3}$ O_2N, NO_2

m-Dinitrobenzene

(e) SO_3H $\xrightarrow[H_2SO_4]{HNO_3}$ O_2N, SO_3H

m-Nitrobenzenesulfonic acid

(f) OCH_3 $\xrightarrow[H_2SO_4]{HNO_3}$ OCH_3, NO_2 + OCH_3, O_2N

o-Methoxynitrobenzene *p*-Methoxynitrobenzene

Only methoxybenzene reacts faster than benzene.

16.28 *Most reactive* ————————————————> *Least reactive*

(a) Benzene > Chlorobenzene > *o*-Dichlorobenzene
(b) Phenol > Nitrobenzene > *p*-Bromonitrobenzene
(c) *o*-Xylene > Fluorobenzene > Benzaldehyde
(d) *p*-Methoxybenzonitrile > *p*-Methylbenzonitrile > Benzonitrile

16.29

(a) Bromobenzene $\xrightarrow[AlCl_3]{CH_3Cl}$ *o*-Bromotoluene + *p*-Bromotoluene

(b) 3-Bromophenol $\xrightarrow[AlCl_3]{CH_3Cl}$ 5-Bromo-2-methylphenol + 3-Bromo-4-methylphenol

Both groups direct substitution to the same position.

(c) 4-Chloroaniline $\xrightarrow[AlCl_3]{CH_3Cl}$ No reaction. $AlCl_3$ combines with $-\ddot{N}H_2$ to form a complex that deactivates the ring toward Friedel-Crafts alkylation.

(d) 2,4-Dichloronitrobenzene $\xrightarrow[AlCl_3]{CH_3Cl}$ No reaction. The ring is deactivated.

(e) 2,4-Dichlorophenol $\xrightarrow[AlCl_3]{CH_3Cl}$ 2,4-Dichloro-6-methylphenol

(f) Benzoic acid (COOH) $\xrightarrow[AlCl_3]{CH_3Cl}$ No reaction.The ring is deactivated.

(g) *p*-Toluenesulfonic acid (SO_3H, H_3C) $\xrightarrow[AlCl_3]{CH_3Cl}$ No reaction.The ring is deactivated.

(h) Br, CH_3, Br benzene $\xrightarrow[AlCl_3]{CH_3Cl}$ Br, CH_3, H_3C, Br benzene

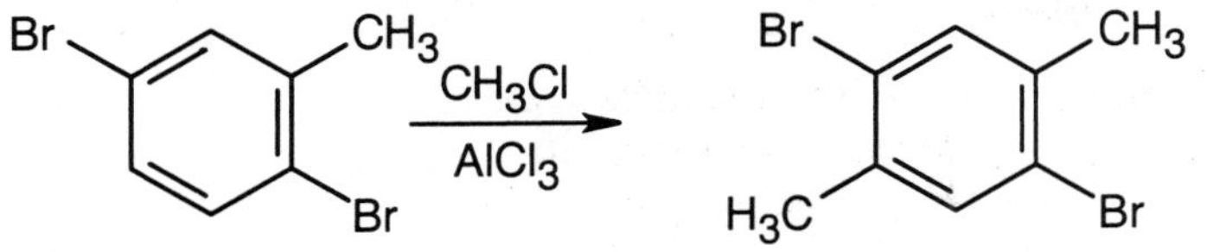

1,4-Dibromo-2,5-dimethylbenzene

Alkylation occurs in the indicated position because the methyl group is more activating than bromine, and because substitution rarely takes place between two groups.

16.30

(a) O_2N, OH benzene $\xrightarrow[FeCl_3]{Cl_2}$ O_2N, OH, Cl benzene + O_2N, OH, Cl benzene

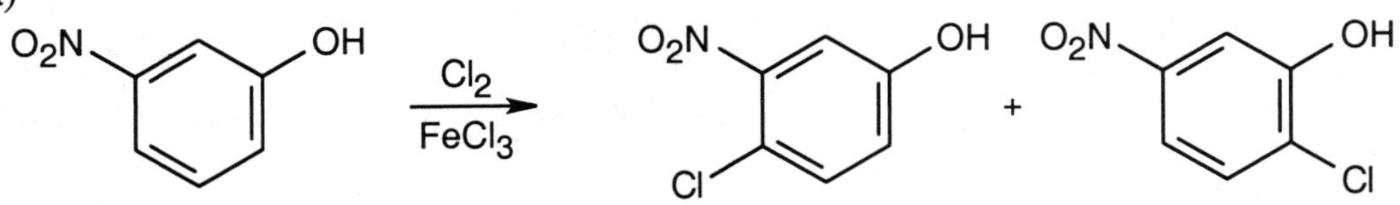

4-Chloro-3-nitrophenol 2-Chloro-5-nitrophenol

The –OH group directs the orientation of substitution.

(b) CH_3, CH_3 benzene $\xrightarrow[FeCl_3]{Cl_2}$ Cl, CH_3, CH_3 benzene + Cl, CH_3, CH_3 benzene

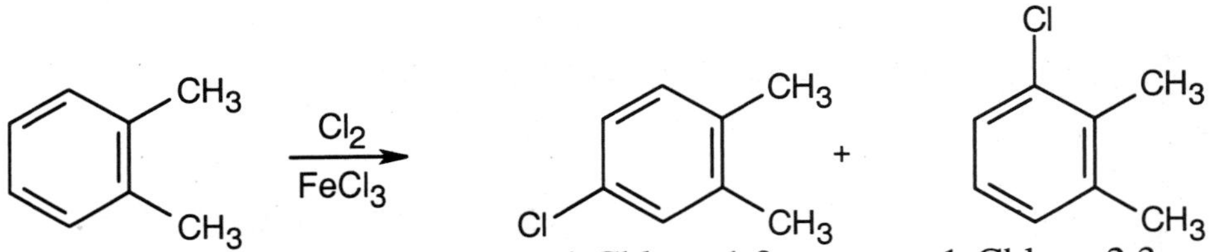

4-Chloro-1,2-dimethylbenzene 1-Chloro-2,3-dimethylbenzene

(c) COOH, O_2N benzene $\xrightarrow[FeCl_3]{Cl_2}$ Cl, COOH, O_2N benzene + COOH, Cl, O_2N benzene

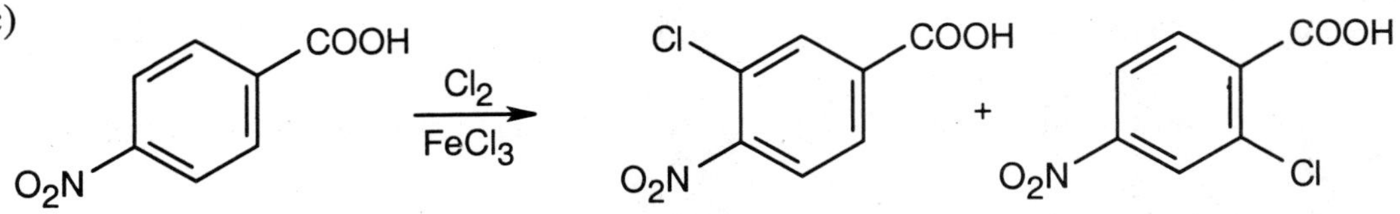

3-Chloro-4-nitrobenzoic acid 2-Chloro-4-nitrobenzoic acid

Both groups are deactivating to the same extent, and both possible products form.

(d) SO_3H, Br benzene $\xrightarrow[FeCl_3]{Cl_2}$ Cl, SO_3H, Br benzene

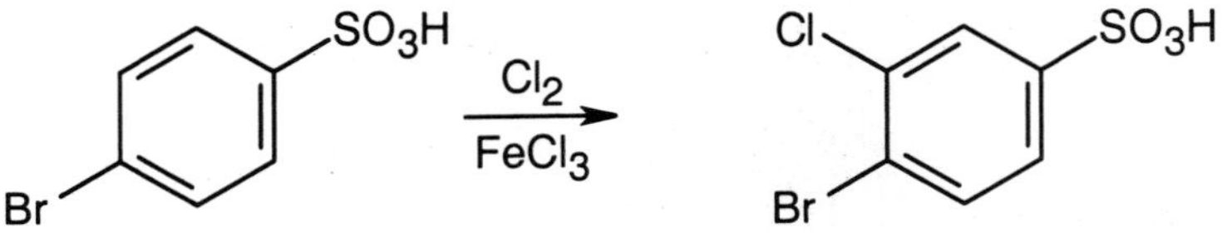

4-Bromo-3-chlorobenzenesulfonic acid

16.31

(a) F benzene $\xrightarrow[H_2SO_4]{SO_3}$ F, HO_3S benzene + F, SO_3H benzene

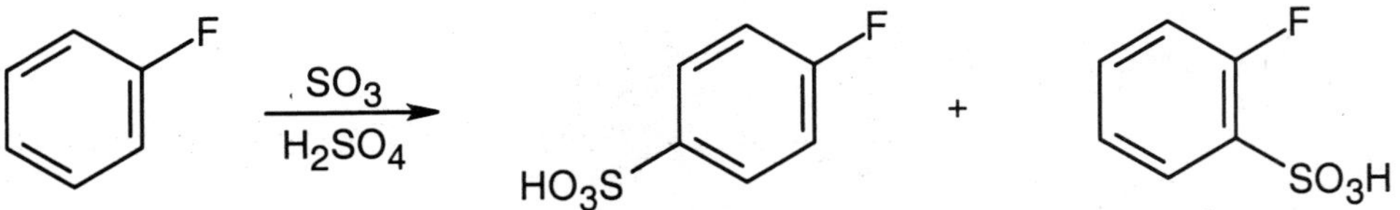

p-Fluorobenzenesulfonic acid *o*-Fluorobenzenesulfonic acid

(b)

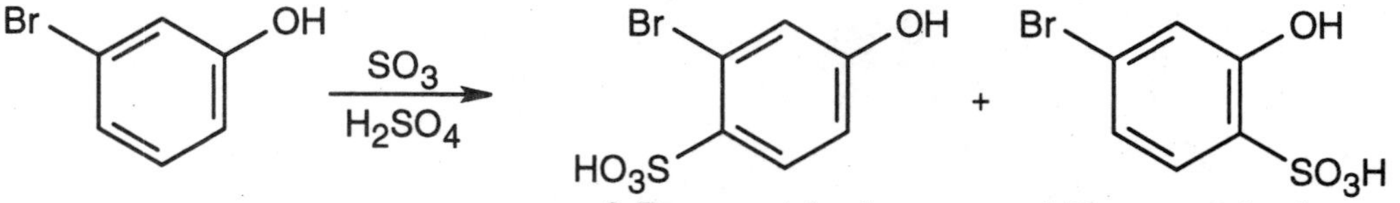

$$\text{3-bromophenol} \xrightarrow[H_2SO_4]{SO_3} \text{2-bromo-4-hydroxybenzenesulfonic acid} + \text{4-bromo-2-hydroxybenzenesulfonic acid}$$

2-Bromo-4-hydroxy-benzenesulfonic acid

4-Bromo-2-hydroxy-benzenesulfonic acid

(c)

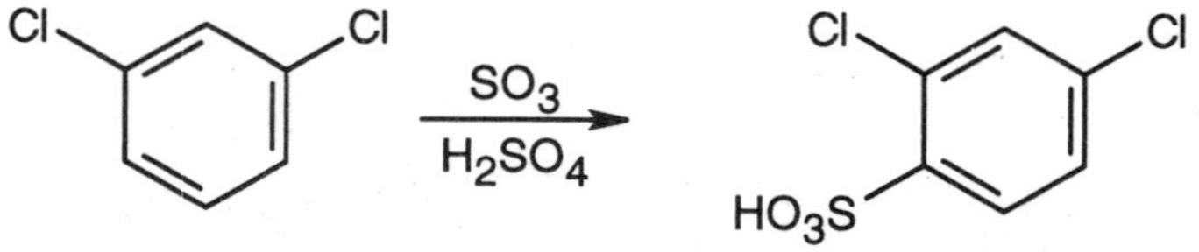

$$\text{1,3-dichlorobenzene} \xrightarrow[H_2SO_4]{SO_3} \text{2,4-dichlorobenzenesulfonic acid}$$

2,4-Dichlorobenzene-sulfonic acid

(d)

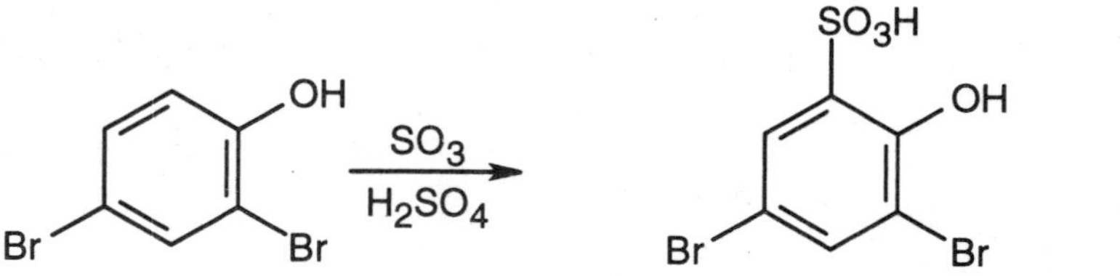

$$\text{2,4-dibromophenol} \xrightarrow[H_2SO_4]{SO_3} \text{3,5-dibromo-2-hydroxybenzenesulfonic acid}$$

3,5-Dibromo-2-hydroxy-benzenesulfonic acid

16.32 *Most reactive* ————————————————> *Least reactive*

Phenol > Toluene > *p*-Bromotoluene > Bromobenzene

Aniline and nitrobenzene don't undergo Friedel-Crafts alkylations.

16.33

(a)

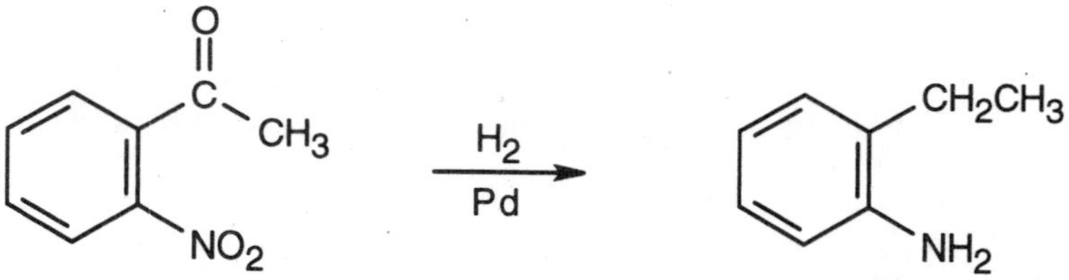

$$\text{o-nitroacetophenone } (C_6H_4(COCH_3)(NO_2)) \xrightarrow[Pd]{H_2} C_6H_4(CH_2CH_3)(NH_2)$$

o-Ethylaniline

Catalytic hydrogenation reduces both the aryl ketone and the nitro group.

(b)

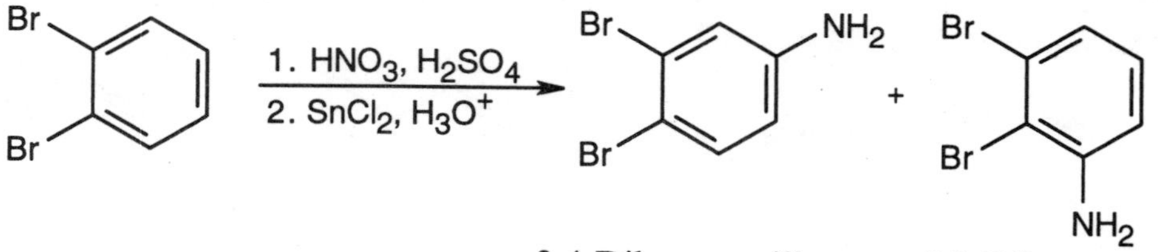

$$\text{1,2-dibromobenzene} \xrightarrow[2.\ SnCl_2,\ H_3O^+]{1.\ HNO_3,\ H_2SO_4} \text{3,4-dibromoaniline} + \text{2,3-dibromoaniline}$$

3,4-Dibromoaniline

2,3-Dibromoaniline

Nitration, followed by reduction with tin(II) chloride, produces substituted anilines.

(c)

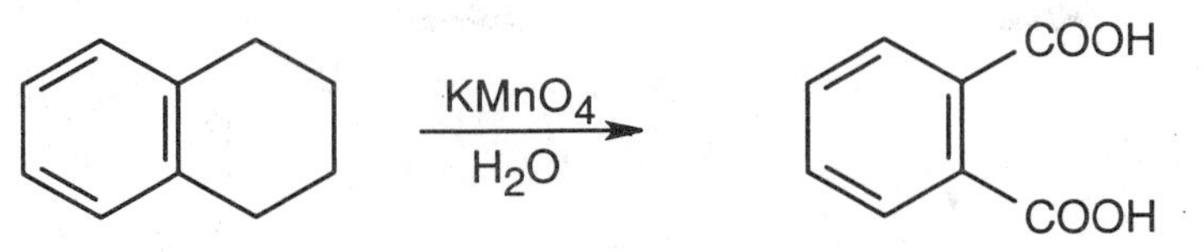

KMnO$_4$, H$_2$O → COOH, COOH

o-Benzenedicarboxylic acid
(Phthalic acid)

Aqueous KMnO$_4$ oxidizes alkyl side chains to benzoic acids.

(d)

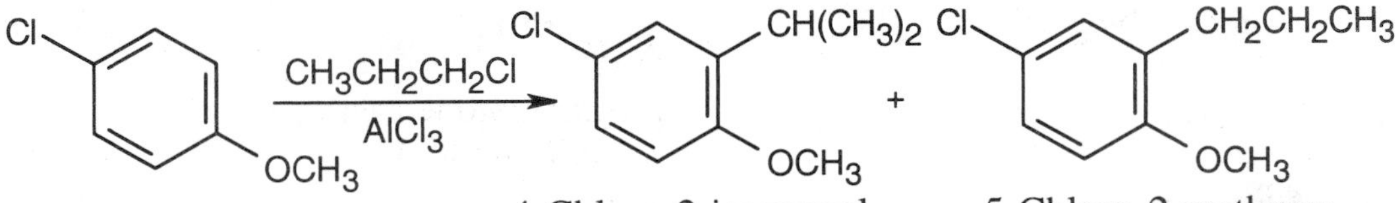

Cl, OCH$_3$ — CH$_3$CH$_2$CH$_2$Cl, AlCl$_3$ → Cl, CH(CH$_3$)$_2$, OCH$_3$ + Cl, CH$_2$CH$_2$CH$_3$, OCH$_3$

4-Chloro-2-isopropyl-methoxybenzene   5-Chloro-2-methoxy-propylbenzene

The methoxy group directs substitution because it is a more powerful activating group. Rearranged and unrearranged products are formed.

16.34

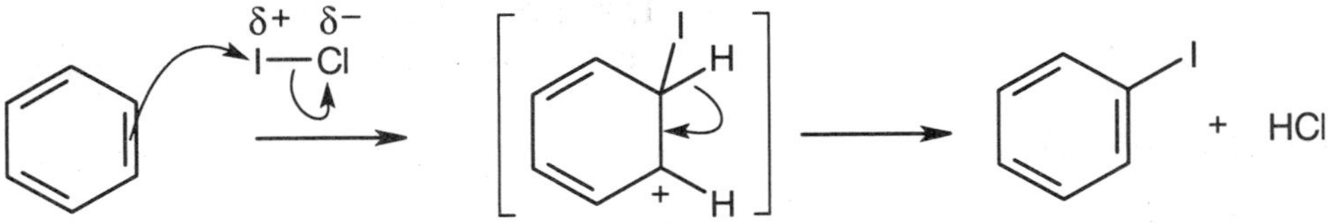

δ+ δ− I—Cl → [I, H, +, H] → I + HCl

ICl can be represented as $\overset{\delta+}{\text{I}}\text{—}\overset{\delta-}{\text{Cl}}$ because chlorine is a more electronegative element than iodine. The iodine atom can act as an electrophile in electrophilic aromatic substitution reactions.

16.35

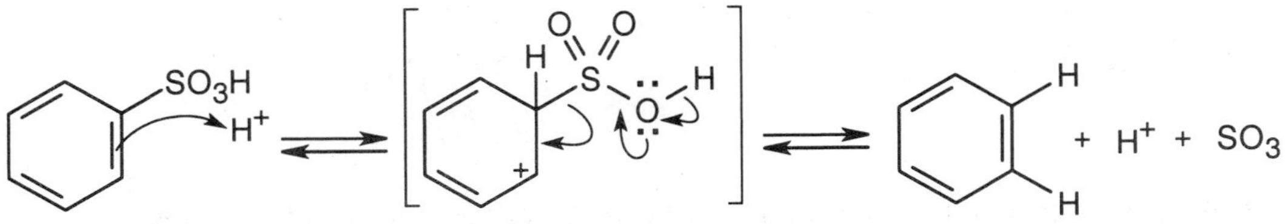

SO$_3$H, H$^+$ ⇌ [H, O, O, S, O, H, +] ⇌ H, H + H$^+$ + SO$_3$

This mechanism is the reverse of the sulfonation mechanism illustrated in the text. H$^+$ is the electrophile in this reaction.

16.36

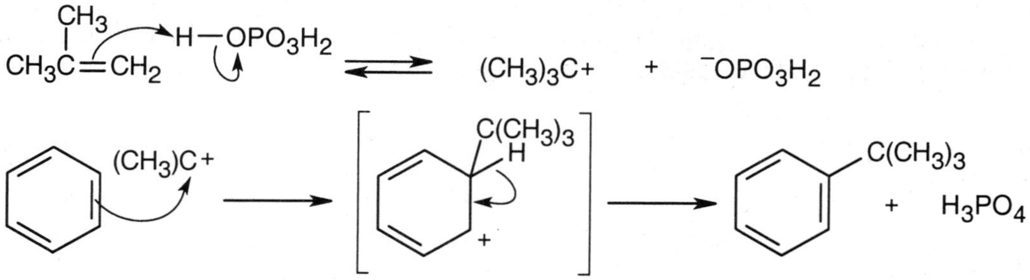

$CH_3C(CH_3){=}CH_2$ + H–OPO_3H_2 ⇌ $(CH_3)_3C^+$ + $^-OPO_3H_2$

$(CH_3)C^+$ + benzene → [carbocation intermediate with C(CH$_3$)$_3$, H] → C(CH$_3$)$_3$-benzene + H_3PO_4

Phosphoric acid protonates 2-methylpropene, forming a *tert*-butyl carbocation. The carbocation acts as an electrophile in a Friedel-Crafts reaction to yield *tert*-butylbenzene.

16.37 When an electrophile reacts with an aromatic ring bearing a $(CH_3)_3N^+$– group:

Ortho attack:

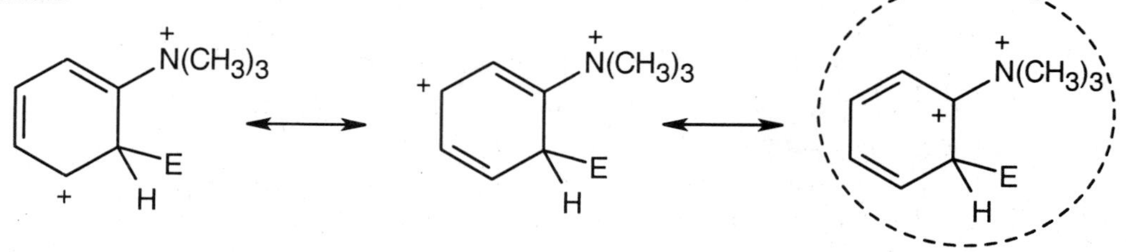

$\overset{+}{N}(CH_3)_3$ ... E, H

This is a destabilizing resonance form because two positive charges are next to each other.

Meta attack:

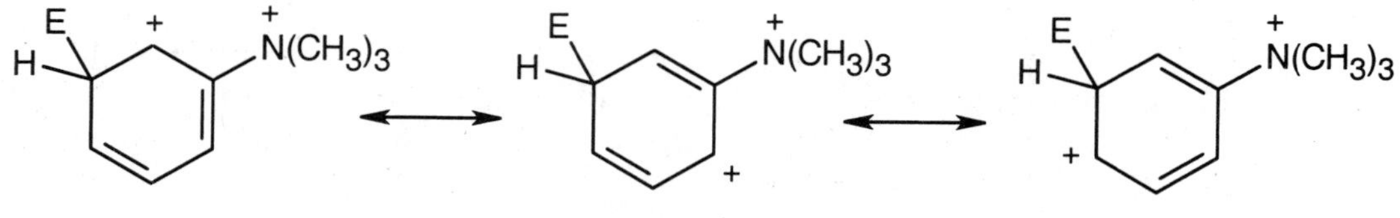

Para attack:

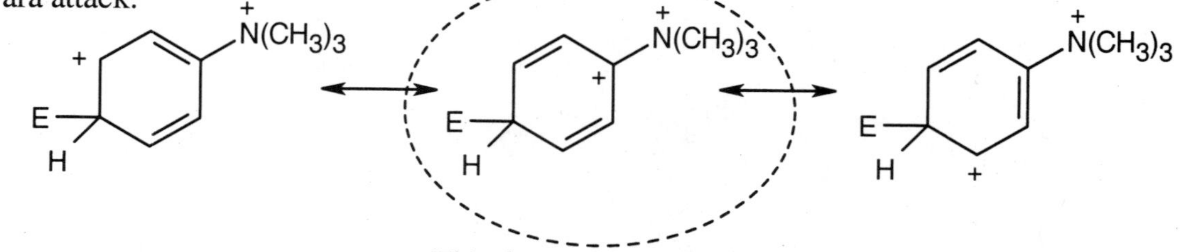

This form is destabilizing.

The *N,N,N*-trimethylammonium group has no electron-withdrawing resonance effect because it has no vacant *p* orbitals to overlap with the π orbital system of the aromatic ring. The $(CH_3)_3N^+$– group is inductively deactivating, however, because it is positively charged. It is meta-directing because the cationic intermediate resulting from meta attack is somewhat more stable than those resulting from ortho or para attack.

16.38 The aromatic ring is deactivated toward electrophilic aromatic substitution by the combined electron-withdrawing inductive effect of electronegative nitrogen and oxygen. The lone pair of electrons of nitrogen can, however, stabilize by resonance the ortho and para substituted intermediates but not the meta intermediate.

Ortho attack:

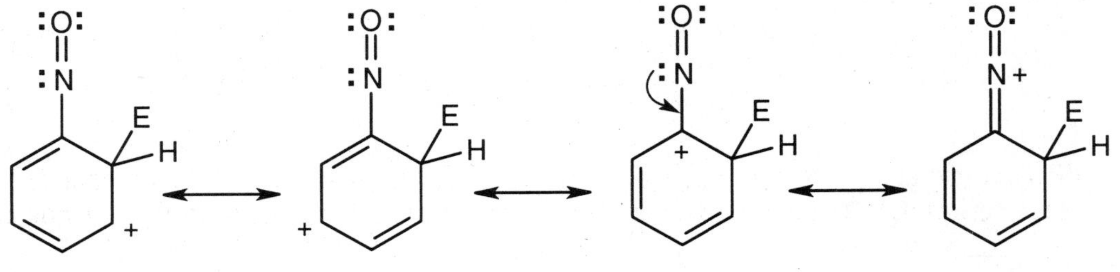

:O: || :N E H + ⟷ :O: || :N E H + ⟷ :O: || :N + E H ⟷ :O: || N+ E H

Meta attack:

:O: || :N + E H ⟷ :O: || :N + E H ⟷ :O: || :N E + H

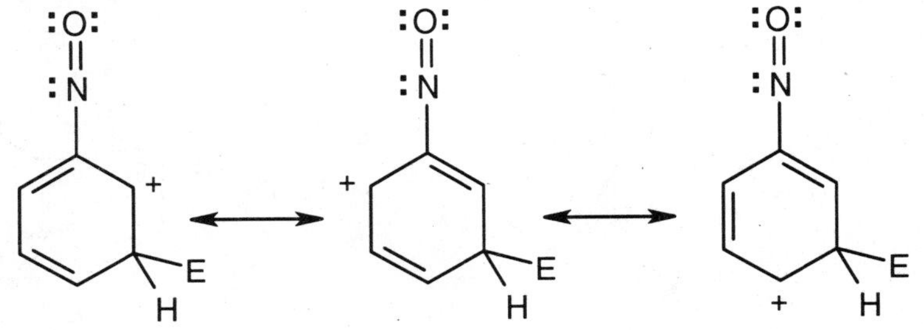

Para attack:

:O: || :N + H E ⟷ :O: || :N + H E ⟷ :O: || :N + H E ⟷ :O: || N+ H E

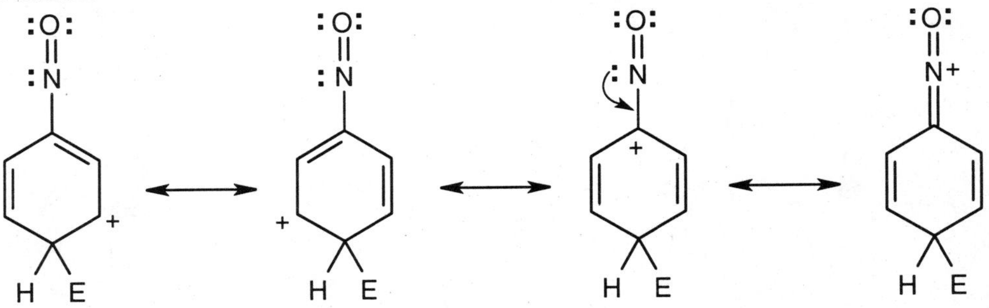

16.39

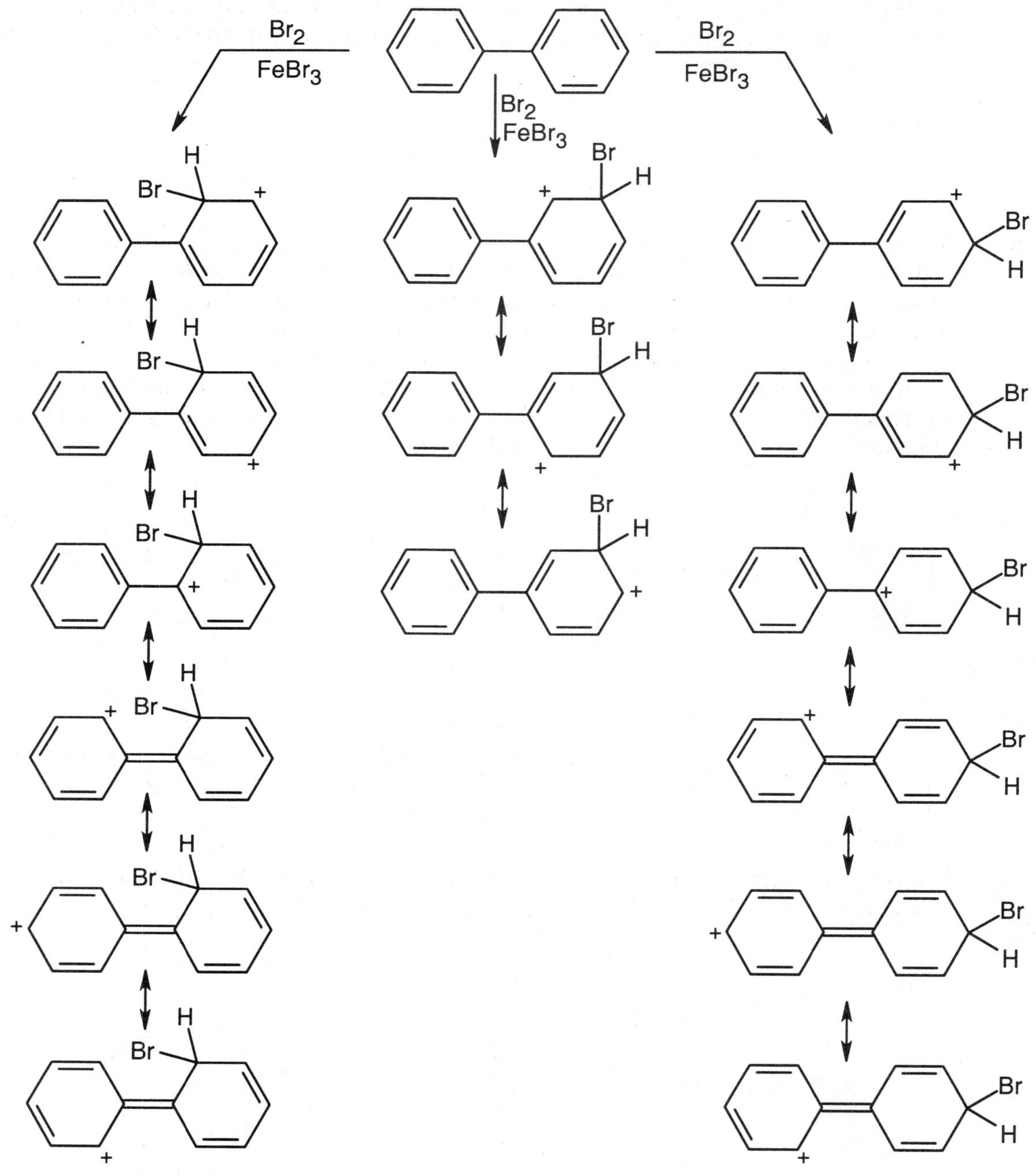

Br2
FeBr3
Br2
FeBr3
Br2
FeBr3
H
Br
+
Br
H
+
+
Br
H

Resonance structures show that bromination occurs in the ortho and para positions of the rings. The positively charged intermediate formed from ortho or para attack can be stabilized by resonance contributions from the second ring of biphenyl, but this stabilization is not possible for meta attack.

16.40 Attack occurs on the unsubstituted ring because bromine is a deactivating group. Attack occurs at the ortho and para positions of the ring because the positively charged intermediate can be stabilized by resonance contributions from bromine and from the second ring (Problem 16.39).

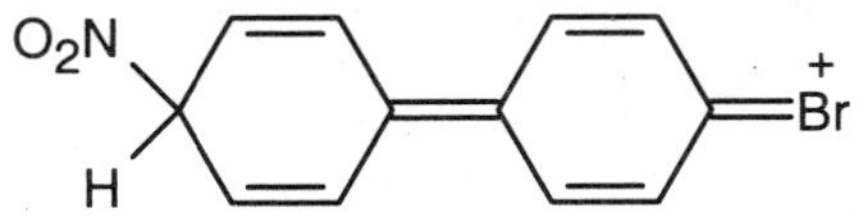

O_2N H $\overset{+}{Br}$

16.41 When directly attached to a ring, the –CN group is a meta-directing deactivator for both inductive and resonance reasons. In 3-phenylpropanenitrile, however, the saturated side chain does not allow resonance interactions of –CN with the aromatic ring, and the –CN group is too far from the ring for its inductive effect to be strongly felt. The side chain acts as an alkyl substituent, and ortho–para substitution is observed.

In 3-phenylpropenenitrile, the –CN group interacts with the ring through the π electrons of the side chain. Resonance forms show that –CN deactivates the ring toward electrophilic substitution, and substitution occurs at the meta position.

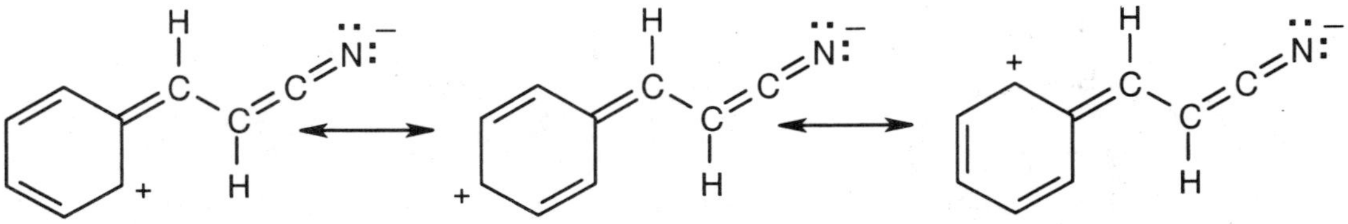

16.42

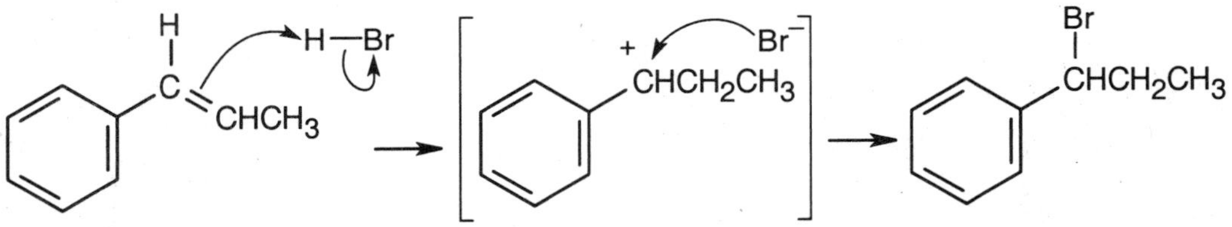

H–Br, $CHCH_3$ → [$\overset{+}{C}HCH_2CH_3$ Br^-] → Br, $CHCH_2CH_3$

Protonation of the double bond at carbon 2 of 1-phenylpropene leads to an intermediate that can be stabilized by resonance involving the benzene ring.

16.43

$$CHCl_3 + AlCl_3 \rightleftharpoons {}^+CHCl_2\ AlCl_4^-$$

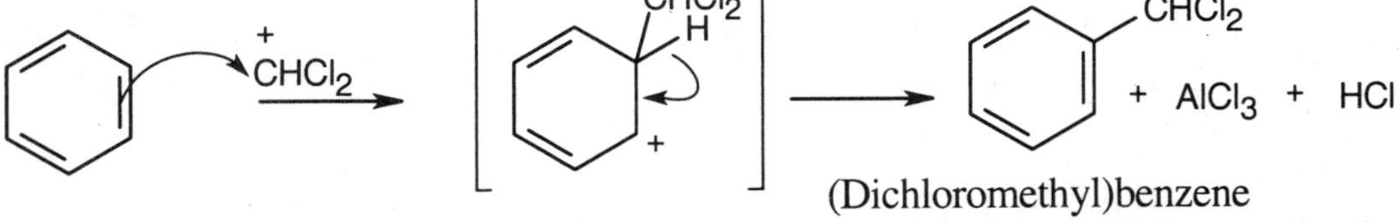

$\overset{+}{C}HCl_2$ → [$CHCl_2$, H, +] → $CHCl_2$ + $AlCl_3$ + HCl

(Dichloromethyl)benzene

(Dichloromethyl)benzene can react with two additional equivalents of benzene by the same mechanism to produce triphenylmethane.

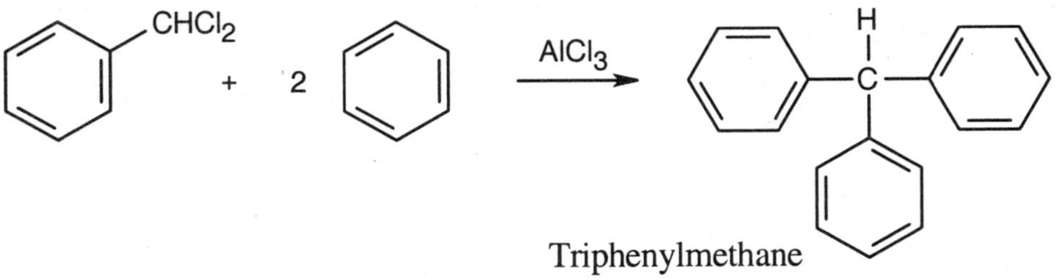

$CHCl_2$ + 2 (benzene) $\xrightarrow{AlCl_3}$ H–C

Triphenylmethane

16.44

(a)

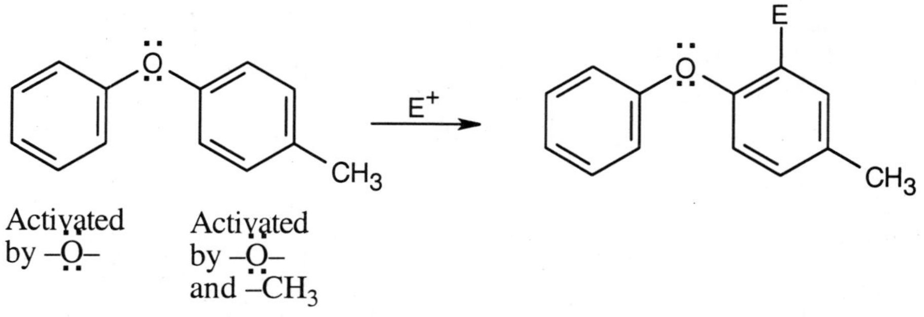

Activated by –O–   Activated by –O– and $-CH_3$

Substitution occurs in the more activated ring. The position of substitution is determined by the more powerful activating group – in this case, the ether oxygen.

(b)

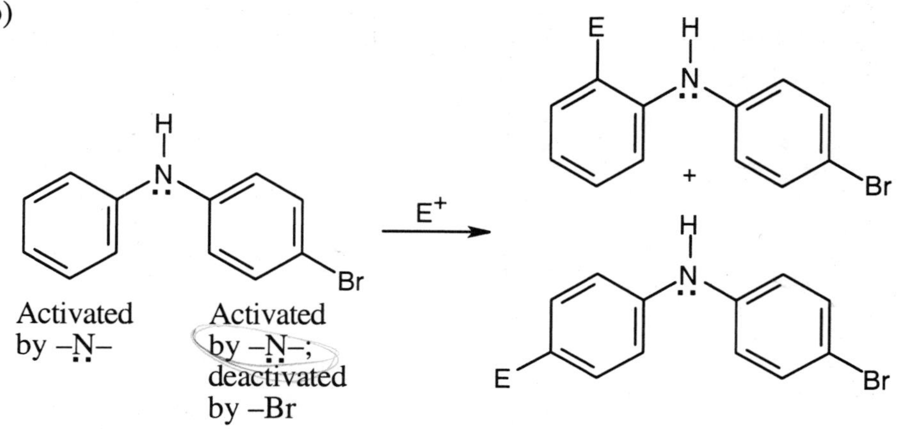

Activated by –N–   Activated by –N–; deactivated by –Br

The left ring is more activated than the right ring. –NH– is an *ortho-para* director.

(c)

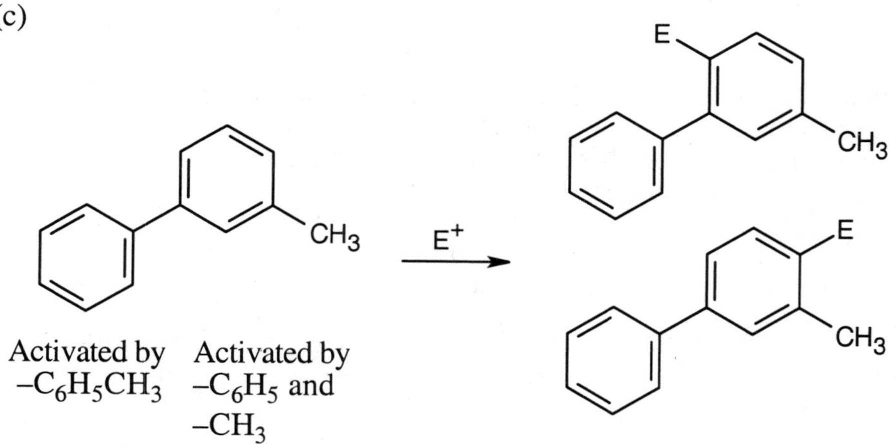

Activated by $-C_6H_5CH_3$ Activated by $-C_6H_5$ and $-CH_3$

Substitution occurs in the ortho and para positions of the more activated ring. Substitution doesn't occur between $-C_6H_5$ and $-CH_3$ for steric reasons.

16.45

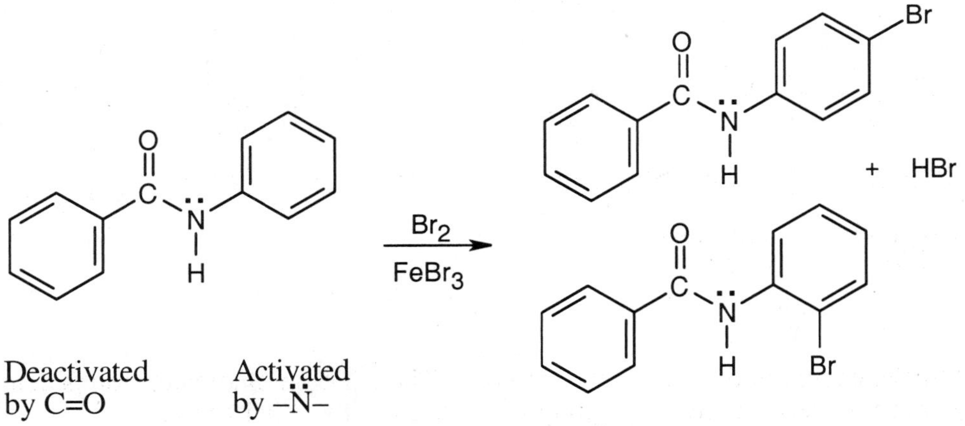

Br2 / FeBr3 → + HBr

Deactivated by C=O

Activated by –N̈–

Attack occurs in the activated ring and yields ortho and para bromination products. The intermediate is resonance-stabilized by overlap of the nitrogen lone pair electrons with the π electrons of the substituted ring.

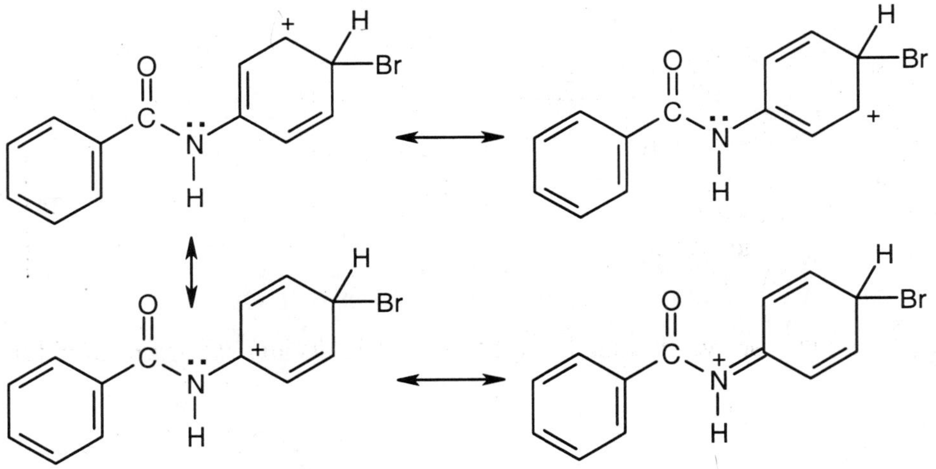

Similar drawings can be made of the resonance forms of the intermediate resulting from ortho attack. Even though the nitrogen lone-pair electrons are less available for delocalization than the lone-pair electrons of aniline (Problem 16.13), the –NH– group is nevertheless more activating than the C=O group.

16.46 Reaction of (*R*)-2-chlorobutane with $AlCl_3$ produces an ion pair [$CH_3{}^+CHCH_2CH_3$ $^-AlCl_4$]. The planar, sp^2-hybridized carbocation is achiral, and its reaction with benzene yields racemic product.

16.47

(a)

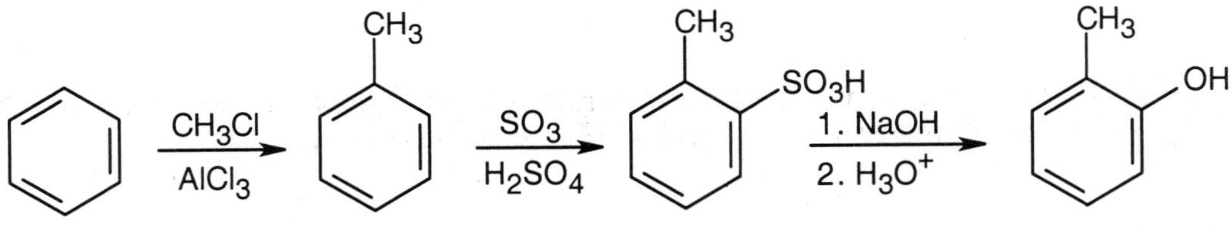

CH_3Cl / $AlCl_3$ → SO_3 / H_2SO_4 → 1. NaOH, 2. H_3O^+ →

o-Methylphenol

(b)

$$\text{Benzene} \xrightarrow[H_2SO_4]{SO_3} \text{C}_6\text{H}_5\text{SO}_3\text{H} \xrightarrow[2.\ H_3O^+]{1.\ NaOH} \text{C}_6\text{H}_5\text{OH} \xrightarrow[H_2SO_4]{3\ HNO_3} \text{2,4,6-}(O_2N)_3C_6H_2OH$$

2,4,6-Trinitrophenol

(c)

$$\text{Benzene} \xrightarrow[AlCl_3]{CH_3Cl} \text{C}_6\text{H}_5\text{CH}_3 \xrightarrow[H_2SO_4]{3\ HNO_3} \text{2,4,6-}(O_2N)_3C_6H_2CH_3 \xrightarrow[H_2O]{KMnO_4} \text{2,4,6-}(O_2N)_3C_6H_2COOH$$

2,4,6-Trinitrobenzoic acid

(d)

$$\text{Benzene} \xrightarrow[H_2SO_4]{HNO_3} \text{C}_6\text{H}_5\text{NO}_2 \xrightarrow[FeBr_3]{Br_2} m\text{-BrC}_6\text{H}_4\text{NO}_2 \xrightarrow[2.\ ^-OH]{1.\ SnCl_2,\ H_3O^+} m\text{-BrC}_6\text{H}_4\text{NH}_2$$

m-Bromoaniline

16.48 When synthesizing substituted aromatic rings, it is necessary to introduce substituents in the proper order. A group that is introduced out of order will not have the proper directing effect. Remember that in many of these reactions a mixture of ortho and para isomers may be formed.

(a)

$$\text{Benzene} \xrightarrow[H_2SO_4]{SO_3} \text{C}_6\text{H}_5\text{SO}_3\text{H} \xrightarrow[2.\ H_3O^+]{1.\ NaOH} \text{C}_6\text{H}_5\text{OH} \xrightarrow[FeCl_3]{Cl_2} p\text{-ClC}_6\text{H}_4\text{OH}$$

p-Chlorophenol

(b)

$$\text{Benzene} \xrightarrow[H_2SO_4]{HNO_3} \text{C}_6\text{H}_5\text{NO}_2 \xrightarrow[FeBr_3]{Br_2} m\text{-BrC}_6\text{H}_4\text{NO}_2$$

m-Bromonitrobenzene

(c)

Benzene $\xrightarrow[FeBr_3]{Br_2}$ Bromobenzene $\xrightarrow[H_2SO_4]{SO_3}$ *o*-Bromobenzenesulfonic acid

o-Bromobenzene-
sulfonic acid

(d)

Benzene $\xrightarrow[H_2SO_4]{SO_3}$ Benzenesulfonic acid $\xrightarrow[FeCl_3]{Cl_2}$ *m*-Chlorobenzenesulfonic acid

m-Chlorobenzene-
sulfonic acid

16.49

(a)

Toluene $\xrightarrow[H_2SO_4]{HNO_3}$ *p*-Nitrotoluene $\xrightarrow[FeBr_3]{Br_2}$ 2-Bromo-4-nitrotoluene

2-Bromo-4-nitrotoluene

(b)

Benzene $\xrightarrow[H_2SO_4]{3\ HNO_3}$ 1,3,5-Trinitrobenzene

1,3,5-Trinitrobenzene

(c)

Benzene $\xrightarrow[H_2SO_4]{HNO_3}$ Nitrobenzene $\xrightarrow[2.\ ^{-}OH]{1.\ SnCl_2,\ H_3O^+}$ Aniline $\xrightarrow{3\ Br_2}$ 2,4,6-Tribromoaniline

2,4,6-Tribromo-
aniline

No catalyst is needed for bromination because aniline is very activated toward substitution.

(d) CH_3 → (SO_3, H_2SO_4) → CH_3, SO_3H → (1. NaOH; 2. H_3O^+) → CH_3, OH → (Cl_2, $FeCl_3$) → CH_3, Cl, OH

2-Chloro-4-methylphenol

The –OH group is a stronger director than $–CH_3$.

16.50 (a) Chlorination of toluene occurs at the ortho and para positions. To synthesize the given product, first oxidize toluene to benzoic acid and then chlorinate.

(b) *p*-Chloronitrobenzene is inert to Friedel-Crafts alkylation because the ring is deactivated.

(c) The first two steps in the sequence are correct, but H_2/Pd reduces the nitro group as well as the ketone.

16.51 All of these syntheses involve NBS bromination of the benzylic position of a side chain.

(a) → (Cl_2, $FeCl_3$) → Cl → (CH_3CH_2Cl, $AlCl_3$) → CH_2CH_3, Cl → (NBS, $(PhCO_2)_2$) → $CHBrCH_3$, Cl → (KOH, ethanol) → $CH{=}CH_2$, Cl

(b) → (CH_3Cl, $AlCl_3$) → CH_3 → (CH_3Cl, $AlCl_3$) → CH_3, CH_3 → (2 NBS, $(PhCO_2)_2$) → CH_2Br, CH_2Br → (NaOH) → CH_2OH, CH_2OH

(c) → (CH_3CH_2Cl, $AlCl_3$) → CH_2CH_3 → (NBS, $(PhCO_2)_2$) → $CHBrCH_3$ → (KOH, ethanol) → $CH{=}CH_2$ → (1. BH_3, THF; 2. H_2O_2, ^-OH) → CH_2CH_2OH

16.52

(a)

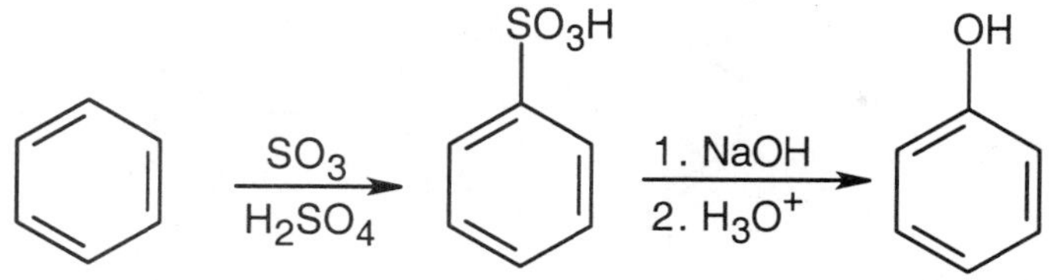

SO3 / H2SO4 → (benzenesulfonic acid, SO3H) → 1. NaOH, 2. H3O+ → OH

Phenol

(b)

(CH3)2CHCl / AlCl3 → (cumene) → NBS / (PhCO2)2 → C(CH3)2Br → Phenol / AlCl3 → C(CH3)2–C6H4–OH → 2 (CH3)3CCl / AlCl3 →

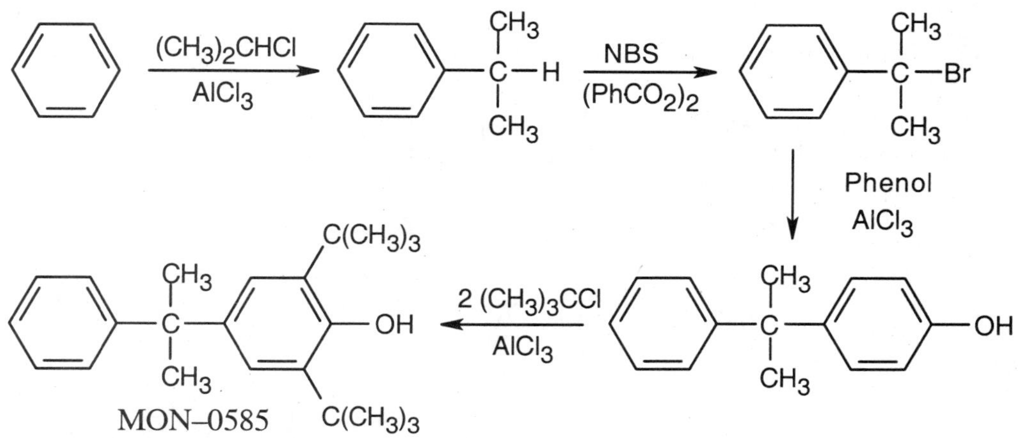

MON–0585

Synthetic analysis: The product is a substituted phenol, whose –OH group directs the orientation of the $-C(CH_3)_3$ groups. The precursor to MON–0585 is synthesized by a Friedel-Crafts alkylation of phenol by the appropriate hydrocarbon halide. This compound is synthesized by NBS bromination of the product of alkylation of benzene with 2-chloropropane.

16.53

(1)

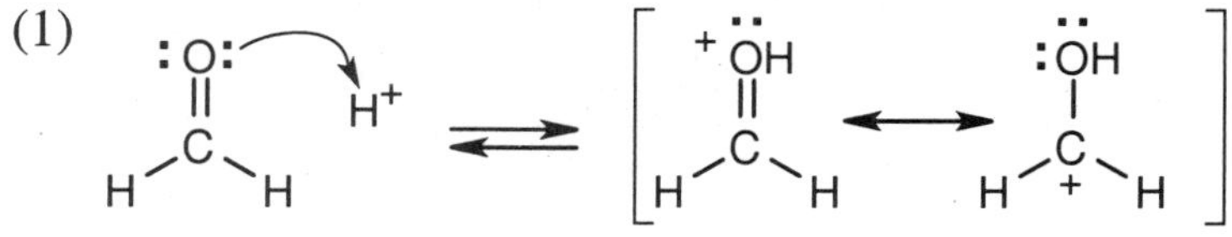

Formaldehyde is protonated to form a carbocation.

(2)

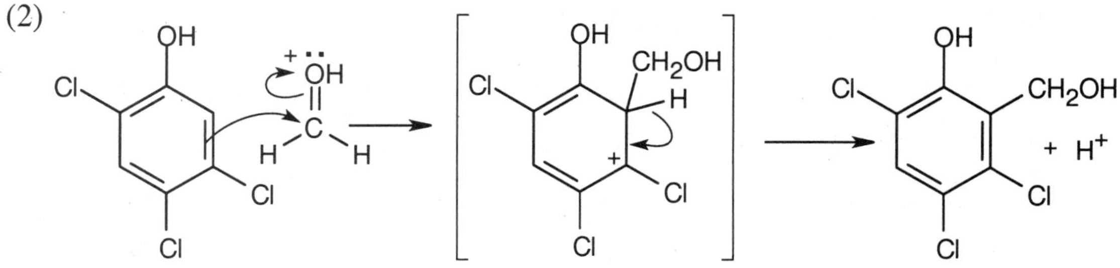

The formaldehyde cation acts as the electrophile in a substitution reaction at the 6 position of 2,4,5–trichlorophenol.

(3)

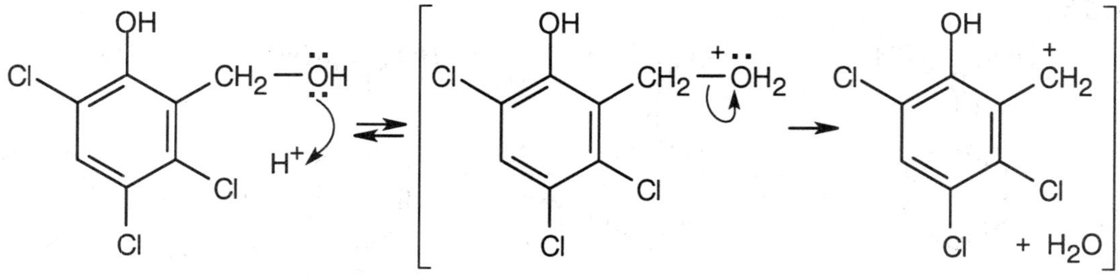

The product from step 2 is protonated by strong acid to produce a carbocation.

(4)

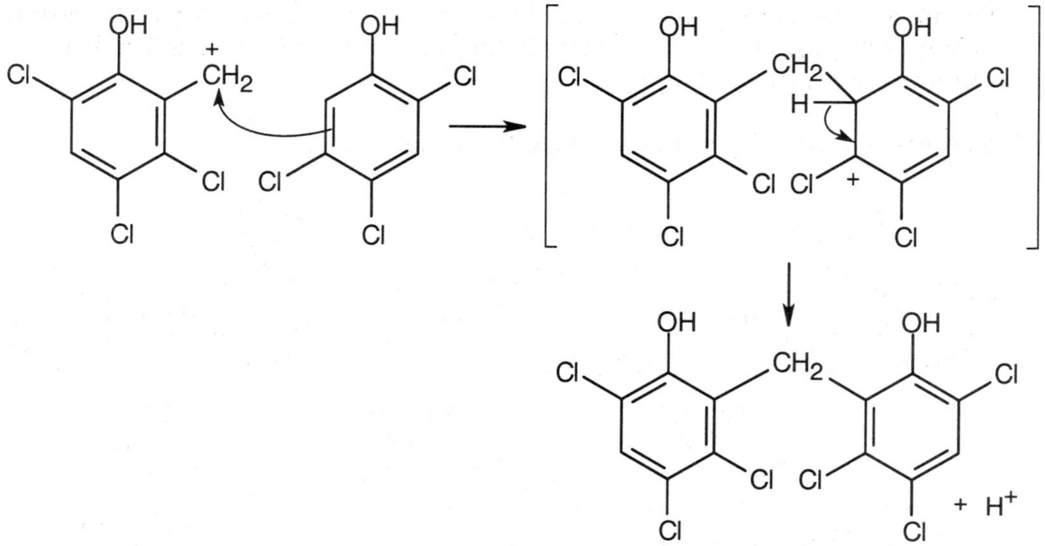

Hexachlorophene

This carbocation is attacked by a second molecule of 2,4,5–trichlorophenol to produce hexachlorophene.

16.54

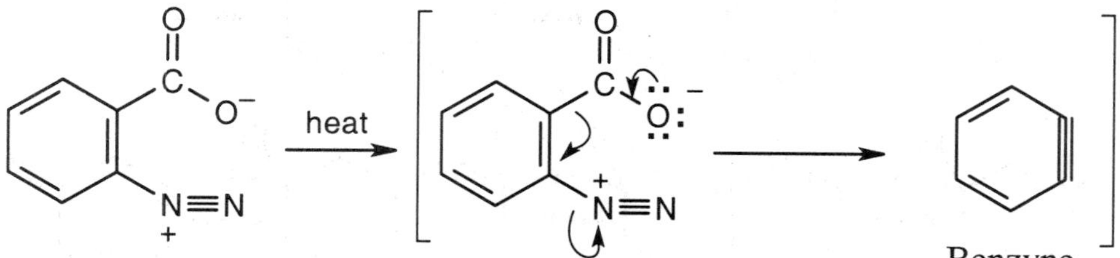

Benzyne
$+ CO_2 + N_2$

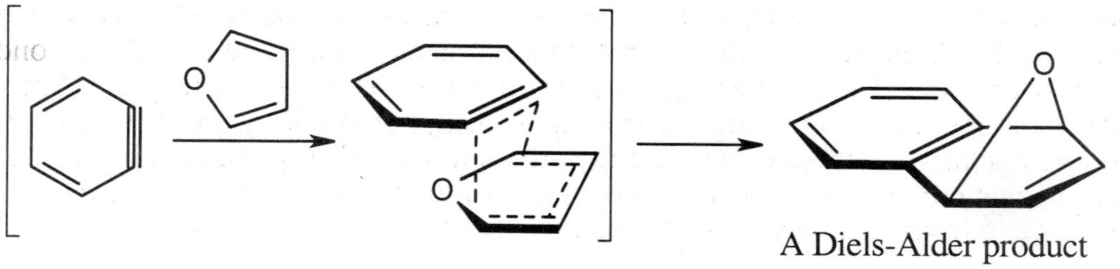

A Diels-Alder product

16.55

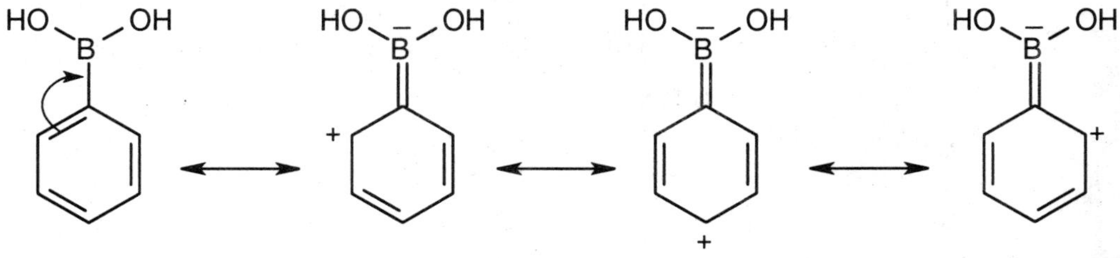

HO B OH

The trivalent boron in phenylboronic acid is a Lewis acid (electron pair acceptor). It is possible to write resonance forms for phenylboronic acid in which an electron pair from the phenyl ring is delocalized onto boron. In these resonance forms, the ortho and para positions of phenylboronic acid are the most electron-deficient, and substitutions occur primarily at the meta position.

16.56 Resonance forms for the intermediate from attack at C1:

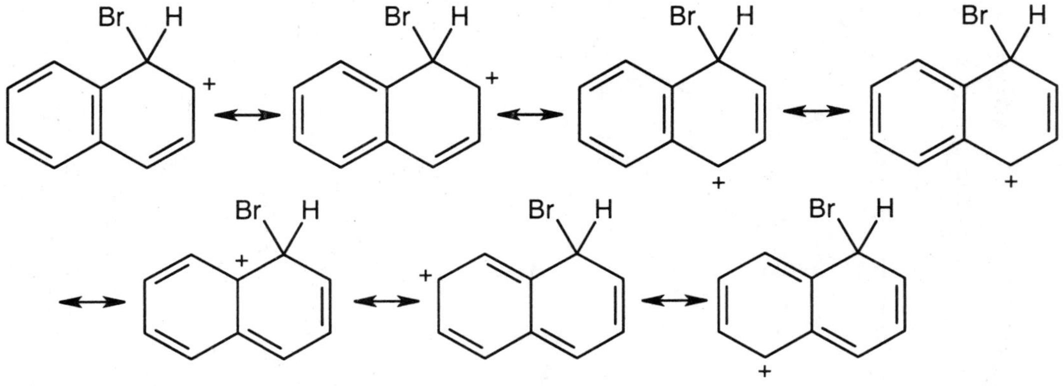

Resonance forms for the intermediate from attack at C2:

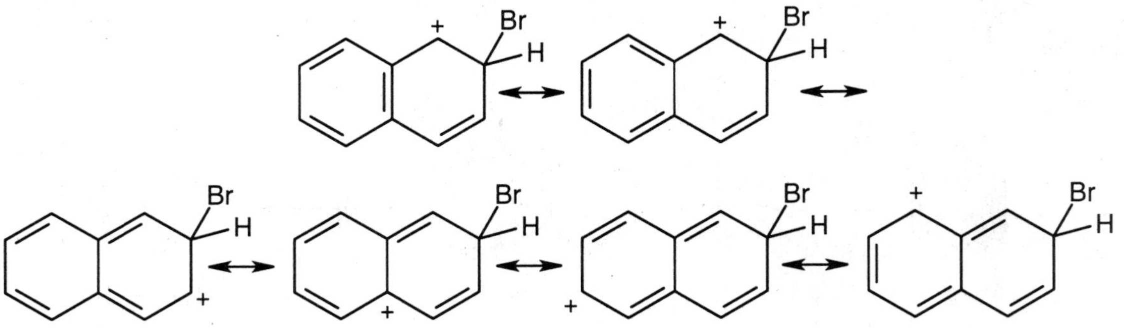

There are seven resonance forms for attack at C1 and six for attack at C2. Look carefully at the forms, however. In the first four resonance structures for C1 attack, the second ring is still fully aromatic. In the other three forms, however, the positive charge has been delocalized into the second ring, destroying the ring's aromaticity. For C2 attack, only the first two resonance structures have a fully aromatic second ring. Since stabilization is lost when aromaticity is disrupted, the intermediate from C2 attack is less stable than the intermediate from C1 attack, and C1 attack is favored.

16.57

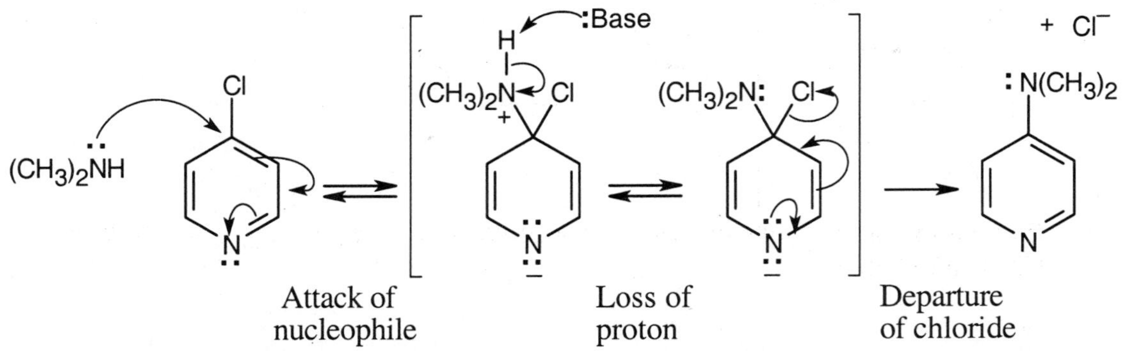

Attack of nucleophile — Loss of proton — Departure of chloride

This reaction is an example of nucleophilic aromatic substitution. Dimethylamine is a nucleophile, and the pyridine nitrogen acts as an electron-withdrawing group that can stabilize the negatively-charged intermediate.

16.58

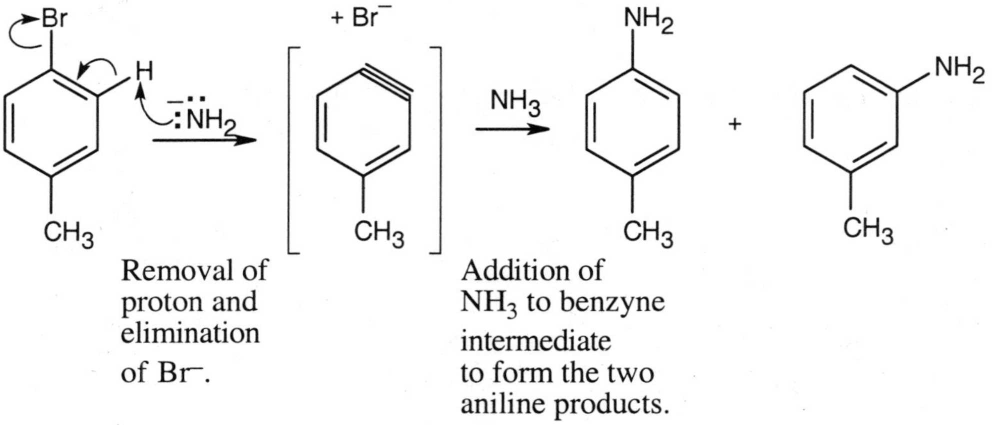

Removal of proton and elimination of Br^-. — Addition of NH_3 to benzyne intermediate to form the two aniline products.

The reaction of an aryl halide with potassium amide proceeds through a benzyne intermediate. Ammonia can then add to either end of the triple bond to produce the two methylanilines observed.

16.59

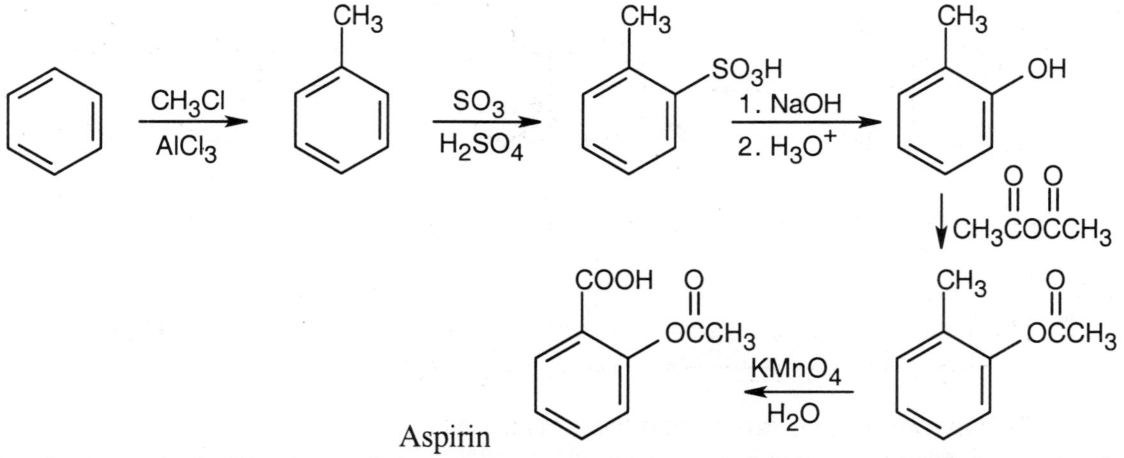

Aspirin

Synthetic analysis: The immediate precursor to aspirin probably has a methyl group that is oxidized to the carboxylic acid. This precursor is formed by acetylating *o*-methylphenol. *o*-Methylphenol is produced by treating *o*-methylbenzenesulfonic acid with NaOH.

16.60

(a)

H^+ O: ⇌ [OH + → + OH]

Protonation of the cyclic ether creates a carbocation intermediate that can react in a Friedel-Crafts alkylation.

(b)

+ OH → [H + OH] → OH + H^+

The intermediate alkylates benzene, forming an alcohol product.

(c)

H^+ OH ⇌ [$\overset{+}{O}H_2$ → +] + H_2O

Protonation of the alcohol, followed by loss of water, generates a second carbocation.

(d)

[+ → + H] → + H^+

This carbocation undergoes internal alkylation to yield the observed product.

16.61

(1)

$$^{-}:C \equiv \ddot{O}^{+} \xrightleftharpoons{HCl,\ AlCl_3} H-C \equiv \ddot{O}^{+} \ AlCl_4^{-}$$

Carbon monoxide is protonated to form an acyl cation.

(2)

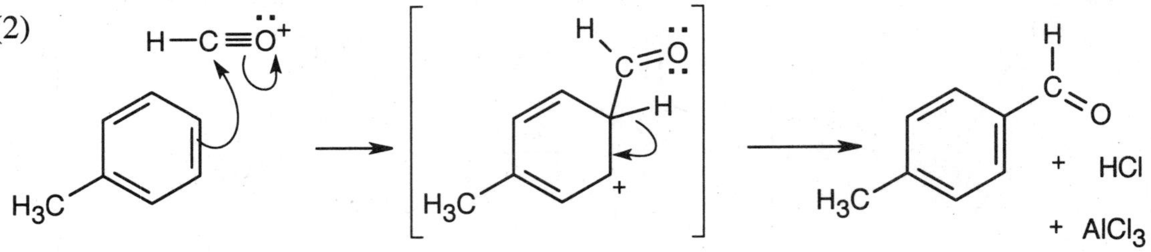

H—C≡O⁺ ; H_3C (toluene) → [intermediate: H, C=O, H, +, H_3C] → H_3C–C(H)=O + HCl + $AlCl_3$

The acyl cation reacts with benzene by a Friedel-Crafts acylation mechanism.

16.62

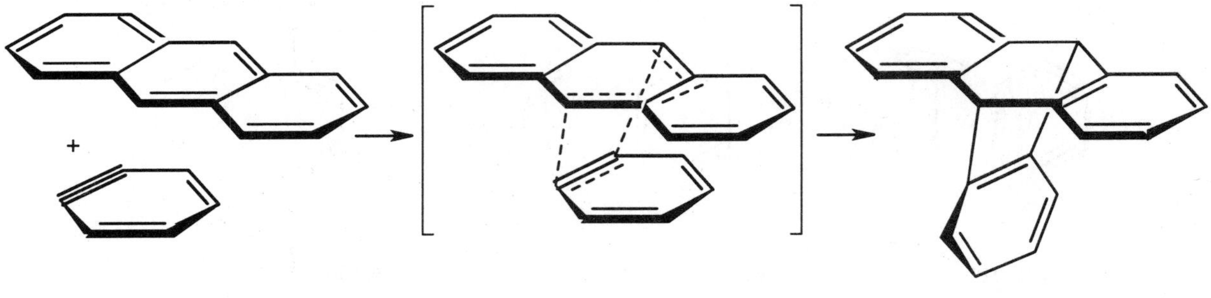

anthracene + benzyne → [transition state] → Trypticene

Trypticene

The reaction between benzyne and anthracene is a Diels-Alder reaction.

16.63

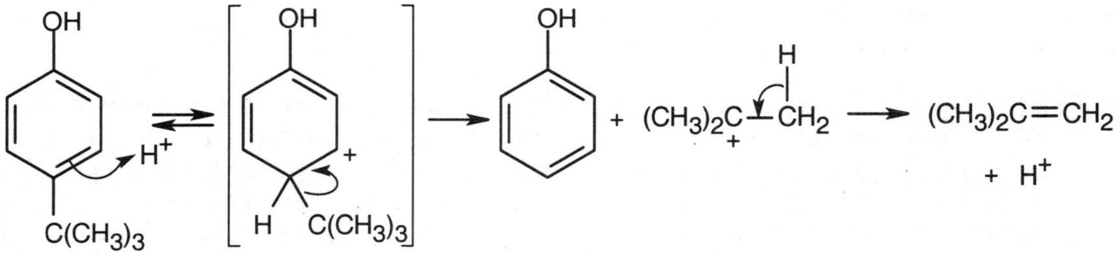

OH, C(CH_3)$_3$, H^+ ⇌ [OH, +, H, C(CH_3)$_3$] → OH (phenol) + $(CH_3)_2\overset{+}{C}$–CH_2–H → $(CH_3)_2C{=}CH_2$ + H^+

Protonation of aromatic ring

Loss of *tert*-butyl carbocation

Loss of proton

16.64

$$HO{-}OH \xrightarrow[\text{catalyst}]{\text{acid}} HO{-}\overset{+}{O}H_2$$ reactive electrophile

benzene + $HO{-}\overset{+}{O}H_2$ → [OH, H, +] → OH (phenol) + H^+

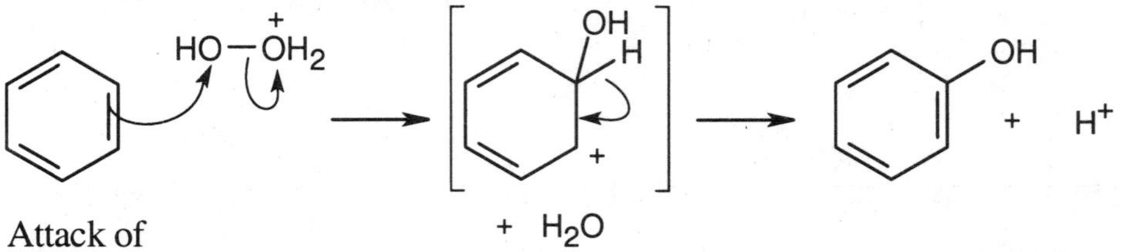

Attack of π electrons on reactive electrophile

$+ H_2O$

Loss of proton

The reactive electrophile (protonated H_2O_2) is equivalent to ^{+}OH.

16.65 Both of these syntheses test your ability to execute steps in the correct order.

(a)

Benzene $\xrightarrow[AlCl_3]{CH_3Cl}$ toluene $\xrightarrow[H_2SO_4]{HNO_3}$ p-nitrotoluene (CH_3, NO_2) + ortho isomer $\xrightarrow[FeBr_3]{Br_2}$ 2-bromo-4-nitrotoluene (CH_3, Br, NO_2)

(b)

Benzene $\xrightarrow[AlCl_3]{(CH_3)_2CHCCl \text{ (C=O)}}$ $C_6H_5C(=O)CH(CH_3)_2$ $\xrightarrow[FeCl_3]{Cl_2}$ m-chloro ketone (Cl, $C(=O)CH(CH_3)_2$) $\xrightarrow{H_2,\ Pd}$ m-chloro $CH_2CH(CH_3)_2$ benzene $\xrightarrow[H_2SO_4]{SO_3}$ product ($CH_2CH(CH_3)_2$, SO_3H, Cl)

16.66 Problem 16.42 shows the mechanism of the addition of HBr to 1–phenylpropene and shows how the aromatic ring stabilizes the carbocation intermediate. For the methoxy-substituted styrene, an additional resonance form can be drawn in which the cation is stabilized by the electron-donating resonance effect of the oxygen atom. For the nitro-substituted styrene, the cation is destabilized by the electron-withdrawing effect of the nitro group.

$CH_3\ddot{O}$–(ring$^+$)=$CHCH_3$ ⟷ $CH_3\overset{+}{O}$=(ring)=$CHCH_3$ — favorable

$O=\overset{+}{N}(O^-)$–(ring$^+$)=$CHCH_3$ — unfavorable

Thus, the intermediate resulting from addition of HBr to the methoxy-substituted styrene is more stable, and reaction of *p*–methoxystyrene is faster.

16.67

CH_3CH_2–C(H)(*R*)(Ph)–CH_3 $\xrightarrow[(PhCO_2)_2]{NBS}$ CH_3CH_2–C(CH_3)(*R*)(Ph)–Br + Br–C(CH_3)(*S*)(Ph)–CH_2CH_3

The planar, achiral benzyl radical reacts with bromine to produce a racemic mixture.

16.68

$$H_3C-\overset{+}{S}(-\ddot{O}:^-)-CH_3 + C_6H_5CH_2-Br \longrightarrow \left[(CH_3)_2\overset{+}{S}-\ddot{O}-CH(H)C_6H_5 \quad :Base \right] + Br^- \longrightarrow (CH_3)_2S + C_6H_5CHO$$

S_N2 displacement occurs when the negatively charged oxygen of dimethyl sulfoxide attacks the benzylic carbon of benzyl bromide, displacing Br^-

Base abstracts a benzylic proton, and dimethyl sulfide is eliminated in an E2 reaction.

16.69

C_6H_5Br	$C_6H_5NH_2$	$p\text{-}BrC_6H_4NH_2$
μ = 1.53 D	μ = 1.52 D	μ = 2.91 D
–Br has a strong electron-withdrawing inductive effect.	$-NH_2$ has a strong electron-donating resonance effect.	The polarities of the two groups are additive and produce a net dipole moment almost equal to the sum of the individual moments.

16.70

$$C_6H_6 \xrightarrow{a} C_6H_5COCH_2CH_3 \xrightarrow{b} C_6H_5CH_2CH_2CH_3 \xrightarrow{c} p\text{-}BrC_6H_4CH_2CH_2CH_3 \xrightarrow{d} p\text{-}BrC_6H_4CHBrCH_2CH_3 \xrightarrow{e} p\text{-}BrC_6H_4CH{=}CHCH_3$$

(a) CH_3CH_2COCl, $AlCl_3$; (b) H_2, Pd; (c) Br_2, $FeBr_3$; (d) NBS, $(PhCO_2)_2$; (e) KOH, ethanol

A Look Ahead

16.71

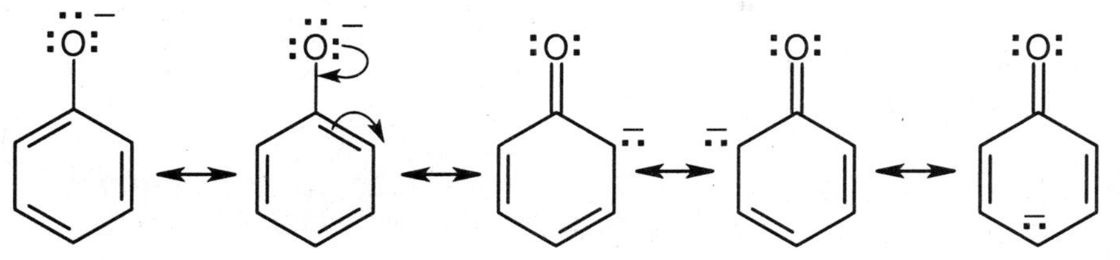

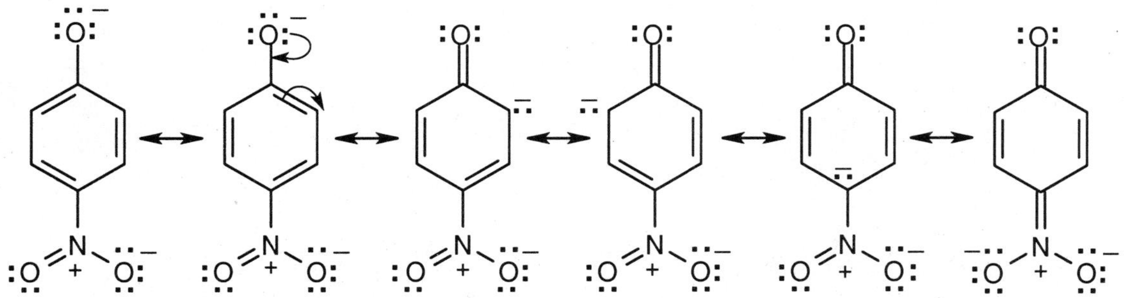

An electron-withdrawing substituent *destabilizes* a positively charged intermediate (as in electrophilic aromatic substitution) but *stabilizes* a negatively charged intermediate. In the case of the dissociation of a phenol, an $-NO_2$ group stabilizes the phenoxide anion by resonance, thus lowering ΔG^o and pK_a. In the last resonance form for *p*-nitrophenol, the negative charge has been delocalized onto the oxygens of the nitro group.

16.72 For the same reason as in the previous problem, a methyl group destabilizes the negatively charged intermediate, thus raising ΔG^o and pK_a.

16.73

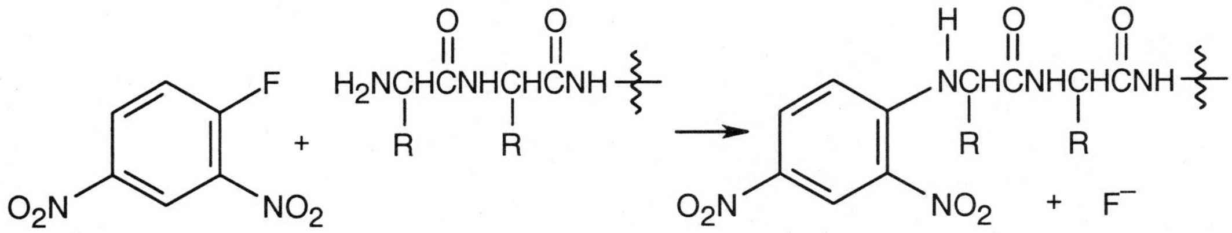

The reaction proceeds by a nucleophilic aromatic substitution mechanism, and the product is a protein with a 2,4-dinitrophenyl group bonded to its terminal amino group.

Molecular Modeling

16.74 The strain energy difference between isomers depends on ring conformation. Using lowest energy conformations (you may have built different ones), you can see that the ortho isomer is much lower in energy than the meta and para isomers (by 202 kJ/mol and 281 kJ/mol, respectively). The meta and para isomers both show strain by adopting a nonplanar benzene ring geometry and by pushing the two carbon atoms bonded to the benzene ring far outside of the ring plane.

16.75 The ring face is most negative in toluene, less negative in 4-nitrotoluene and least negative in 2,4-dinitrotoluene. The ring acts as a nucleophile during the nitration reaction, and more negative rings are better nucleophiles. Therefore, each nitration should be more difficult than the previous one.

16.76 1-Bromo-4-methoxy-3-methylbenzenium ion is the lower energy ion by 61 kJ/mol, and 5-bromo-2-methoxytoluene should be the major product.

16.77 The $-NO_2$ oxygen atoms are the most negatively charged atoms in the ortho complex. The CH_3O- oxygen and the fluorine atom are the most negatively charged atoms in the meta complex. Delocalization of charge from the reactive atoms to the nitro group stabilizes the ortho complex. Substitution of *o*-fluoronitrobenzene is more rapid.

Review Unit 6: Conjugation and Aromaticity

Major Topics Covered (with vocabulary):

Conjugated dienes:
delocalization 1,4-addition allylic position thermodynamic control kinetic control vulcanization Diels-Alder cycloaddition dienophile *endo* product *exo* product *s*-cis conformation

Ultraviolet spectroscopy:
highest occupied molecular orbital (HOMO) lowest unoccupied molecular orbital (LUMO) molar absorptivity

Aromaticity:
aromatic arene phenyl group ortho, meta, para substitution degenerate Hückel $4n + 2$ rule antiaromatic heterocycle polycyclic aromatic compound ring current

Chemistry of aromatic compounds:
electrophilic aromatic substitution sulfonation alkali fusion Friedel-Crafts alkylation polyalkylation Friedel-Crafts acylation ortho- and para-directing activator ortho- and para-directing deactivator meta-directing deactivator inductive effect resonance effect nucleophilic aromatic substitution Meisenheimer complex benzyne benzylic position

Types of Problems:

After studying these chapters, you should be able to:

- Predict the products of electrophilic addition to conjugated molecules.
- Understand the concept of kinetic *vs.* thermodynamic control of reactions.
- Recognize diene polymers, and draw a representative segment of a diene polymer.
- Predict the products of Diels-Alder reactions, and identify compounds that are good dienophiles and good dienes.
- Calculate the energy required for UV absorption, and use molar absorptivity to calculate concentration.
- Predict if and where a compound absorbs in the ultraviolet region.

- Name and draw substituted benzenes.
- Draw resonance structures and molecular orbital diagrams for benzene and other cyclic conjugated molecules.
- Use Hückel's rule to predict aromaticity.
- Draw orbital pictures of cyclic conjugated molecules.
- Use NMR, IR and UV data to deduce the structures of aromatic compounds.

- Predict the products of electrophilic aromatic substitution reactions.
- Formulate the mechanisms of electrophilic aromatic substitution reactions.
- Understand the activating and directing effects of substituents on aromatic rings, and use inductive and resonance arguments to predict orientation and reactivity.
- Predict the products of other reactions of aromatic compounds.
- Synthesize substituted benzenes.

Points to Remember:

* It's not always easy to recognize Diels-Alder products, especially if the carbon-carbon double bond of the initial product has been hydrogenated. If no hydrogenation has taken place, look for a double bond in a six-membered ring and at least one electron-withdrawing group across the ring from the double bond. When a bicyclic product has been formed, it has probably resulted from a Diels-Alder reaction in which the diene is cyclic.

* To be aromatic, a molecule must be planar, cyclic, conjugated, and it must have $4n + 2$ electrons in its π system.

* The carbocation intermediate of electrophilic aromatic substitution loses a proton to yield the aromatic product. In all cases, a base is involved with proton removal, but the nature of the base varies with the type of substitution reaction. Although this book shows the loss of the proton, it often doesn't show the base responsible for proton removal. This doesn't imply that the proton flies off, unassisted; it just means that the base involved has not been identified in the problem.

* Nucleophilic aromatic substitution reactions and substitution reactions proceeding through benzyne intermediates take place by different routes. In the first reaction, the substitution takes place by an addition, followed by an elimination. In the second case, the substitution involves an elimination, followed by an addition. Virtually all substitutions are equivalent to an addition and an elimination (in either order).

* Activating groups achieve their effects by making an aromatic ring more electron-rich and reactive toward electrophiles. Ortho-para directing groups achieve their effects by stabilizing the positive charge that results from ortho or para addition of an electrophile to the aromatic ring. The intermediate resulting from addition to a ring with an ortho-para director usually has one resonance form that is especially stable. The intermediate resulting from addition to a ring with a meta-director usually has a resonance form that is especially unfavorable when addition occurs ortho or para to the functional group. Meta-substitution results because it is less unfavorable than ortho-para substitution.

Self-test:

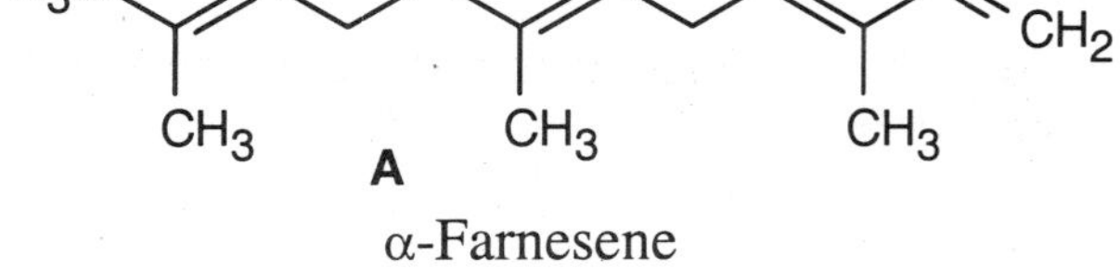

H_3C CH_2 CH_3 CH_3 CH_3

A

α-Farnesene

α-Farnesene (**A**), an important biological intermediate in the synthesis of many natural products, has double bonds that are both conjugated and unconjugated. Show the products you would expect from conjugate addition of HBr; of Br_2. What products would you expect from ozonolysis of **A**? Give one or more distinctive absorptions that you might see in the IR spectrum of **A** and distinguishing features of the ^{1}H NMR of **A**. Would you expect **A** to be UV-active?

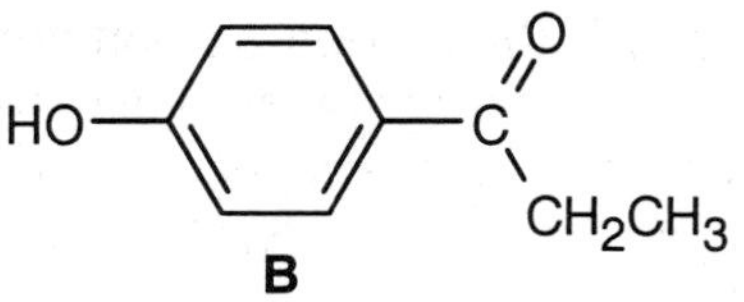

B
Paroxypropione

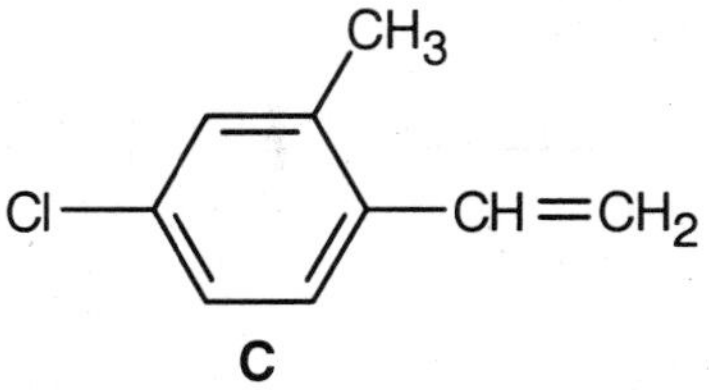

C

Paroxypropione (**B**) is a hormone inhibitor. Predict the products of reaction of **B** with: (a) Br_2, $FeBr_3$; (b) CH_3Cl, $AlCl_3$; (c) $KMnO_4$, H_3O^+; (d) H_2, Pd/C. If the product of (d) is treated with the reagents in (a) or (b), does the orientation of substitution change? What significant information can you obtain from the IR spectrum of **B**?

Name **C**. Plan a synthesis of **C** from benzene. Describe the ^1H NMR of **C** (include spin-spin splitting). Where might **C** show an absorption in a UV spectrum?

Multiple choice:

1. What are the hybridizations of the carbons in 1,2-butadiene, starting with C1?
(a) sp^2, sp^2, sp^2, sp^2 (b) sp^2, sp^2, sp^2, sp^3 (c) sp^2, sp, sp^2, sp^3 (d) sp, sp, sp^2, sp^3

2. In a reaction in which the less stable (ls) product is formed at lower temperature, and the more stable product (ms) is formed at higher temperature:
(a) $\Delta G_{ms}° > \Delta G_{ls}°$ and $\Delta G_{ms}^‡ > \Delta G_{ls}^‡$ (b) $\Delta G_{ms}° > \Delta G_{ls}°$ and $\Delta G_{ls}^‡ > \Delta G_{ms}^‡$
(c) $\Delta G_{ms}° < \Delta G_{ls}°$ and $\Delta G_{ms}^‡ > \Delta G_{ls}^‡$ (d) $\Delta G_{ms}° < \Delta G_{ls}°$ and $\Delta G_{ls}^‡ > \Delta G_{ms}^‡$
Note: In this problem, a large value for $\Delta G°$ means a large negative value.

3. Which of the following combinations is most likely to undergo a successful Diels-Alder reaction?

(a) (b) (c) (d)

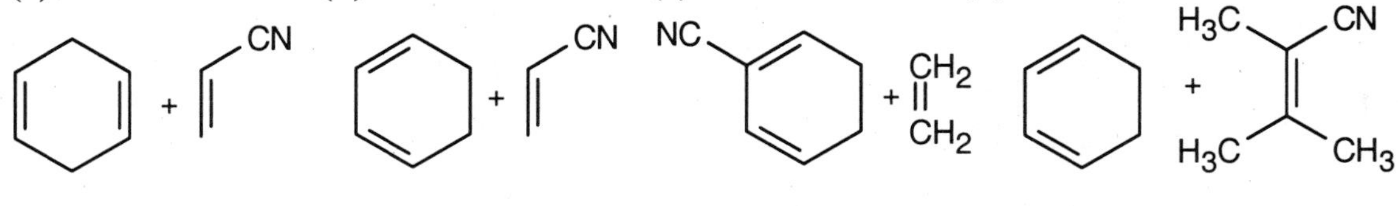

4. Which of the following groups, when bonded to the terminal carbon of a conjugated π system, probably affects the value of λ_{max} the least?
(a) $-NH_2$ (b) $-Cl$ (c) $-OH$ (d) $-CH_3$

5. If the value of λ_{max} for an unsubstituted diene is approximately 220 nm, and each additional double bond increases the value of λ_{max} by 30 nm, what is the minimum number of double bonds present in a compound that absorbs in the visible range of the electromagnetic spectrum?
(a) 6 (b) 7 (c) 8 (d) 9

6. Which of the following compounds is aromatic?

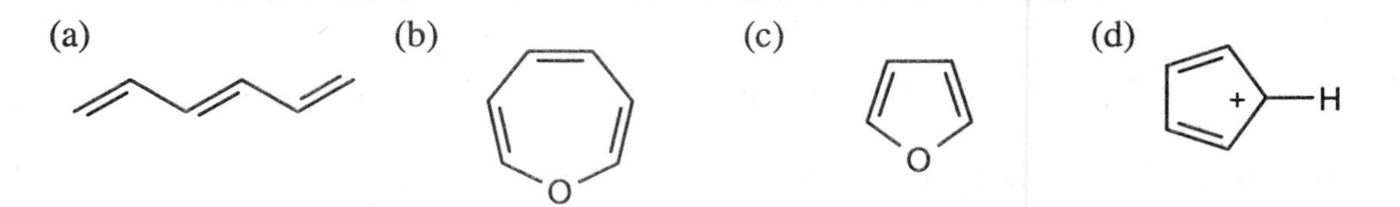

7. How many benzene isomers of $C_7H_6Br_2$ can be drawn?
(a) 10 (b) 11 (c) 12 (d) 14

8. Which of the following functional groups isn't a meta-directing deactivator?
(a) $-NO_2$ (b) $-N(CH_3)_2$ (c) $-N(CH_3)_3^+$ (d) $-NHCOCH_3$

9. Which of the following compounds can't be synthesized by an electrophilic aromatic substitution reaction that we have studied?
(a) *m*-Cresol (b) *p*-Chloroaniline (c) 2,4-Toluenedisulfonic acid (d) *m*-Bromotoluene

10. In only one of the following compounds can you reduce the aromatic ring without also reducing the side chain. Which compound is it?
(a) *p*-Bromoanisole (b) Acetophenone (methyl phenyl ketone) (c) Styrene
(d) Phenylacetylene

Chapter 17 – Alcohols and Phenols

Chapter Outline

I. Introduction to alcohols and phenols (Sections 17.1 – 17.3).
  A. Naming alcohols and phenols (Section 17.1).
    1. Alcohols are classified as primary, secondary or tertiary, depending on the number of organic groups bonded to the –OH carbon.
    2. Rules for naming alcohols.
      a. The longest chain containing the –OH group is the parent chain, and the parent name replaces *-e* with *-ol*.
      b. Numbering begins at the end of the chain nearer the –OH group.
      c. The substituents are numbered according to their position on the chain and cited in alphabetical order.
    3. Phenols are named according to rules discussed in Section 15.2.
  B. Hydrogen bonding of alcohols and phenols (Section 17.2).
    1. Alcohols have sp^3 hybridization and a nearly tetrahedral bond angle.
    2. Alcohols and phenols have elevated boiling points, relative to hydrocarbons, due to hydrogen bonding.
      a. In hydrogen bonding, an –OH hydrogen is attracted to a lone pair of electrons on another molecule, resulting in a weak electrostatic force that holds the molecules together.
      b. These weak forces must be overcome in boiling.
  C Acidity and basicity of alcohols and phenols (Section 17.3).
    1. Alcohols and phenols are both weakly acidic and weakly basic.
    2. Alcohols and phenols dissociate to a slight extent to form alkoxide ions and phenoxide ions.
    3. Acidity of alcohols.
      a. Alcohols are similar in acidity to water.
      b. Alkyl substituents decrease acidity by preventing solvation of the alkoxide ion.
      c. Electron-withdrawing substituents increase acidity by delocalizing negative charge.
      d. Alcohols don't react with weak bases, but they do react with alkali metals and strong bases.
    4. Acidity of phenols.
      a. Phenols are a million times more acidic than alcohols and are soluble in dilute NaOH.
      b. Phenol acidity is due to resonance stabilization of the phenoxide anion.
      c. Electron-withdrawing substituents increase phenol acidity, and electron-donating substituents decrease phenol acidity.

II. Alcohols (Sections 17.4 – 17.9).
  A. Preparation of alcohols (Sections 17.4 – 17.6).
    1. Familiar methods (Section 17.4).
      a. Hydration of alkenes.
        i. Hydroboration/oxidation yields non-Markovnikov products.
        ii. Oxymercuration/reduction yields Markovnikov products.
      b. 1,2-diols can be prepared by OsO_4 hydroxylation, followed by reduction.
        i. This reaction occurs with syn stereochemistry.
        ii. Ring-opening of epoxides produces 1,2-diols with anti stereochemistry.
    2. Reduction of carbonyl compounds (Section 17.5).
      a. Aldehydes are reduced to primary alcohols.

b. Ketones are reduced to secondary alcohols.
   Either $NaBH_4$(milder) or $LiAlH_4$(more reactive) can be used to reduce aldehydes and ketones.
c. Carboxylic acids and esters are reduced to primary alcohols with $LiAlH_4$.
   i. These reactions occur by addition of hydride to the positively polarized carbon of a carbonyl group.
   ii. Water adds to the alkoxide intermediate to yield alcohol product.

3. Reaction of carbonyl compounds with Grignard reagents (Section 17.6).
   a. RMgX adds to carbonyl compounds to give alcohol products.
      i. Reaction of RMgX with formaldehyde yields primary alcohols.
      ii. Reaction of RMgX with aldehydes yields secondary alcohols.
      iii. Reaction of RMgX with ketones yields tertiary alcohols.
      iv. Reaction of RMgX with esters yields tertiary alcohols with two –R groups bonded to the alcohol carbon.
      v. No reaction occurs with carboxylic acids because the acidic hydrogen quenches the Grignard reagent.
   b. Limitations of the Grignard reaction.
      i. Grignard reagents can't be prepared from reagents containing other reactive functional groups.
      ii. Grignard reagents can't be prepared from compounds having acidic hydrogens.
   c. Grignard reagents behave as carbon anions and add to the carbonyl carbon.
      A proton from water is added to the alkoxide intermediate to produce the alcohol.

B. Reactions of alcohols (Sections 17.7 – 17.9).
1. Dehydration to yield alkenes (Section 17.7).
   a. Tertiary alcohols can undergo acid-catalyzed dehydration with warm aqueous H_2SO_4.
      i. Zaitsev products are usually formed.
      ii. The severe conditions needed for dehydration of secondary and primary alcohols restrict this method to tertiary alcohols.
      iii. Tertiary alcohols react fastest because the intermediate carbocation formed in this E1 reaction is most stable.
   b. Secondary and primary alcohols are dehydrated with $POCl_3$ in pyridine.
      i. This reaction occurs by an E2 mechanism.
      ii. Pyridine serves as a base and as a solvent.
2. Conversion to alkyl halides.
   a. Tertiary alcohols (ROH) are converted to RX by treatment with HX.
      The reaction occurs by an S_N1 mechanism.
   b. Primary alcohols are converted by the reagents PBr_3 and $SOCl_2$.
      The reaction occurs by an S_N2 mechanism.
3. Conversion into tosylates.
   a. Reaction with *p*-toluenesulfonyl chloride converts alcohols to tosylates.
   b. Only the O–H bond is broken.
   c. Tosylates behave as halides in substitution reactions.
   d. S_N2 reactions involving tosylates proceed with inversion of configuration.
4. Oxidation of alcohols (Section 17.8).
   a. Primary alcohols can be oxidized to aldehydes or carboxylic acids.
   b. Secondary alcohols can be oxidized to ketones.
   c. Tertiary alcohols aren't oxidized.
   d. Oxidation to ketones and carboxylic acids can be carried out with $KMnO_4$, CrO_3, or $Na_2Cr_2O_7$.
   e. Oxidation of a primary alcohol to an aldehyde is achieved with PCC.
      PCC is also used on sensitive alcohols.

f. Oxidation occurs by a mechanism closely related to an E2 mechanism.
   The reaction involves a chromate intermediate.

5. Protection of alcohols (Section 17.9).
   a. It is sometimes necessary to protect an alcohol when it interferes with a reaction involving a functional group in another part of a molecule.
   b. The following reaction sequence may be applied:
      i. Protect the alcohol.
      ii. Carry out the reaction.
      iii. Remove the protecting group.
   c. A trimethylsilyl (TMS) ether can be used for protection.
      i. TMS ether formation occurs by an S_N2 route.
      ii. TMS ethers are quite unreactive.
      iii. TMS ethers can be cleaved by aqueous acid or by F^-.

III. Phenols (Sections 17.10 – 17.11).
   A. Preparation and uses of phenols (Section 17.10).
      1. Phenols can be prepared by treating chlorobenzene with NaOH.
      2. Phenols can also be prepared from isopropylbenzene (cumene).
         a. Cumene reacts with O_2 by a radical mechanism to form cumene hydroperoxide.
         b. Treatment of the hydroperoxide with acid gives phenol and acetone.
            The mechanism involves protonation, rearrangement, loss of water, readdition of water to form a hemiacetal, and breakdown to acetone and phenol.
      3. In the laboratory, phenols can be formed by treatment of sulfonic acids with NaOH.
      4. Chlorinated phenols, such as 2,4-D, are formed by chlorinating phenol.
      5. BHT is prepared by Friedel-Crafts alkylation of *p*-cresol with 2-methylpropene.
   B. Reactions of phenols (Section 17.11).
      1. Phenols undergo electrophilic aromatic substitution reactions (Chapter 16).
         The –OH group is a *o,p*-director.
      2. Strong oxidizing agents convert phenols to quinones.
         a. Reaction with Fremy's salt to form a quinone occurs by a radical mechanism.
         b. The redox reaction quinone → hydroquinone occurs readily.
         c. Ubiquinones are an important class of biochemical oxidizing agents that function as a quinone/hydroquinone redox system.

IV. Spectroscopy of alcohols and phenols (Section 17.12).
   A. UV spectroscopy.
      1. Both alcohols and phenols show –OH stretches in the region 3300–3600 cm^{-1}.
         a. Unassociated alcohols show a peak at 3600 cm^{-1}.
         b. Associated alcohols show a broader peak at 3300–3400 cm^{-1}.
      2. Alcohols show a C–O stretch near 1500 cm^{-1}.
      3. Phenols show aromatic bands at 1500–1600 cm^{-1}.
      4. Phenol shows monosubstituted aromatic bands at 690 and 760 cm^{-1}.
   B. NMR spectroscopy.
      1. In ^{13}C NMR spectroscopy, carbons bonded to –OH groups absorb in the range 50–80 δ.
      2. ^1H NMR.
         a. Hydrogens on carbons bearing –OH groups absorb in the range 3.5–4.5 δ.
            The hydroxyl hydrogen doesn't split these signals.
         b. D_2O exchange can be used to locate the O–H signal.
         c. Spin-spin splitting occurs between the oxygen-bearing carbon and neighboring –H.
         d. Phenols show aromatic ring absorptions, as well as an O–H absorption in the range 3–8 δ.

3. Mass Spectrometry.
   a. Alcohols undergo alpha cleavage to give a neutral radical and an oxygen-containing cation.
   b. Alcohols also undergo dehydration to give an alkene radical cation.

Solutions to Problems

17.1 The parent chain must contain the hydroxyl group, and the hydroxyl group(s) should receive the lowest possible number.

(a) $CH_3CH(OH)CH_2CH(OH)CH(CH_3)CH_3$

5-Methyl-2,4-hexanediol

(b) $C_6H_5CH_2CH_2C(OH)(CH_3)CH_3$

2-Methyl-4-phenyl-2-butanol

(c) HO–cyclohexane–$(CH_3)_2$

4,4-Dimethylcyclohexanol

(d) H, Br, H, OH

trans-2-Bromocyclopentanol

(e) H_3C, OH, Br

4-Bromo-3-methylphenol

17.2

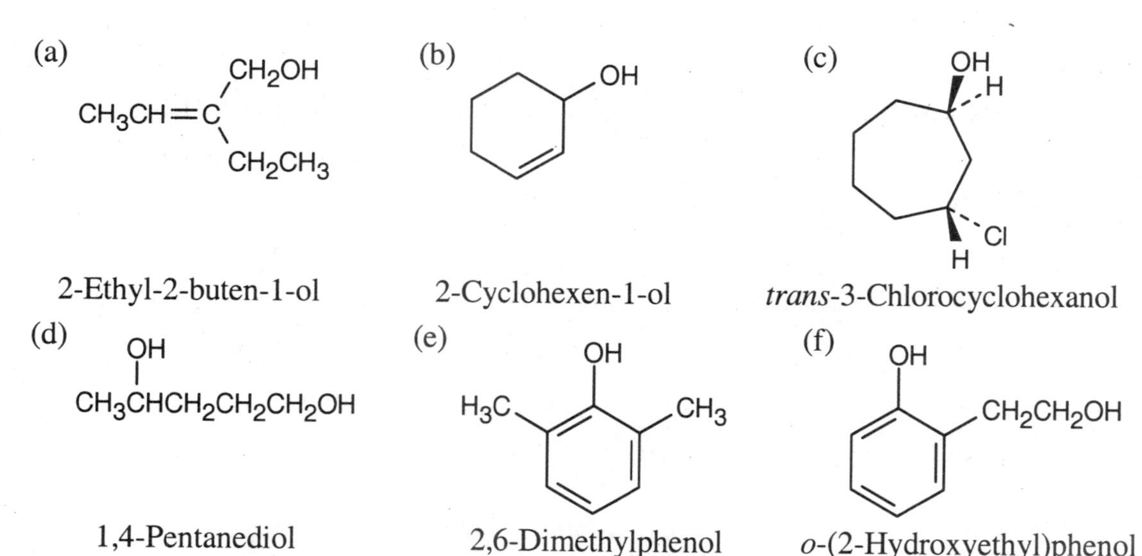

17.3 In general, the boiling points of a series of isomers decrease with branching. The more nearly spherical a compound becomes, the less surface area it has relative to a straight chain compound of the same molecular weight and functional group type. A smaller surface area allows fewer van der Waals interactions, the weak forces that cause covalent molecules to be attracted to each other.

In addition, branching in alcohols makes it more difficult for hydroxyl groups to approach each other to form hydrogen bonds. A given volume of 2–methyl–2–propanol therefore contains fewer hydrogen bonds than the same volume of 1–butanol, so less energy is needed to break them in boiling.

17.4

Least acidic ⟶ *Most acidic*

(a) $HC{\equiv}CH$ < $(CH_3)_2CHOH$ < CH_3OH < $(CF_3)_2CHOH$
alkyne; hindered alcohol; alcohol; alcohol with electron-withdrawing groups

(b) *p*-Methylphenol < Phenol < *p*-(Trifluoromethyl)phenol
phenol with electron-donating groups; phenol with electron-withdrawing groups

(c) Benzyl alcohol < Phenol < *p*-Hydroxybenzoic acid
alcohol; phenol; carboxylic acid

17.5 We saw in Chapter 16 that a nitro group is electron-withdrawing. Since electron-withdrawing groups stabilize phenoxide anions, *p*–nitrobenzyl alcohol is more acidic than benzyl alcohol. The methoxy group, which is electron-donating, makes *p*-methoxybenzyl alcohol less acidic than benzyl alcohol.

17.6

(a)

$$C_6H_5CH_2CH_2C(CH_3){=}CH_2 \xrightarrow[2.\ H_2O_2,\ ^-OH]{1.\ BH_3,\ THF} C_6H_5CH_2CH_2CH(CH_3)CH_2OH$$

2-Methyl-4-phenyl-1-butanol

Remember that the hydroxyl group is bonded to the less substituted carbon after hydroboration/oxidation.

(b)

$$CH_3CH_2CH{=}C(CH_3)_2 \xrightarrow[2.\ NaBH_4]{1.\ Hg(OAc)_2,\ H_2O} CH_3CH_2CH_2C(OH)(CH_3)_2$$

2-Methyl-2-pentanol

Markovnikov product results from oxymercuration/reduction.

(c)

$$\text{(H)(C}_4\text{H}_9\text{)C}{=}\text{C(H)(C}_4\text{H}_9\text{)} \xrightarrow[2.\ NaHSO_3,\ H_2O]{1.\ OsO_4} \text{H--C(OH)(C}_4\text{H}_9\text{)--C(OH)(C}_4\text{H}_9\text{)--H}$$

meso-5,6-decanediol

Hydroxylation results in a diol with syn stereochemistry.

17.7

(a)

$$CH_3COCH_2CH_2COOCH_3 \xrightarrow[2.\ H_3O^+]{1.\ NaBH_4} CH_3CH(OH)CH_2CH_2COOCH_3$$

$NaBH_4$ reduces aldehydes and ketones without disturbing other functional groups.

(b)

$$CH_3COCH_2CH_2COOCH_3 \xrightarrow[2.\ H_3O^+]{1.\ LiAlH_4} CH_3CH(OH)CH_2CH_2CH_2OH$$

$LiAlH_4$ reduces both ketones and esters.

(c)

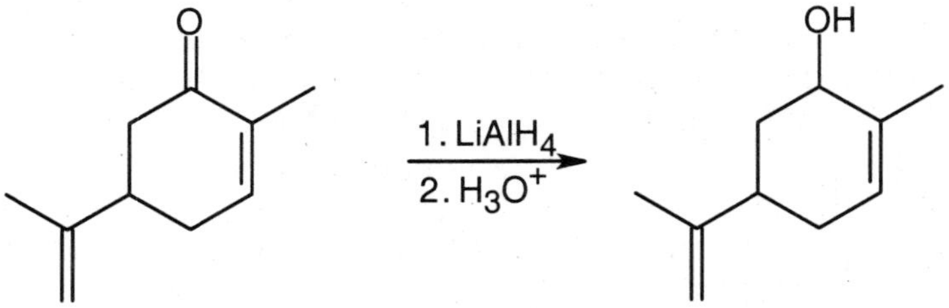

$$\xrightarrow[\text{2. } H_3O^+]{\text{1. } LiAlH_4}$$

$LiAlH_4$ reduces carbonyl functional groups without reducing double bonds.

17.8

(a)

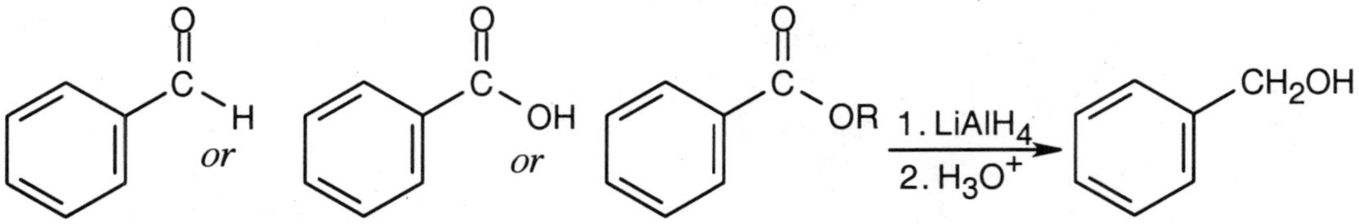

C_6H_5CHO *or* C_6H_5COOH *or* C_6H_5COOR $\xrightarrow[\text{2. } H_3O^+]{\text{1. } LiAlH_4}$ $C_6H_5CH_2OH$

Benzyl alcohol is a reduction product of an aldehyde or a carboxylic acid or an ester.

(b)

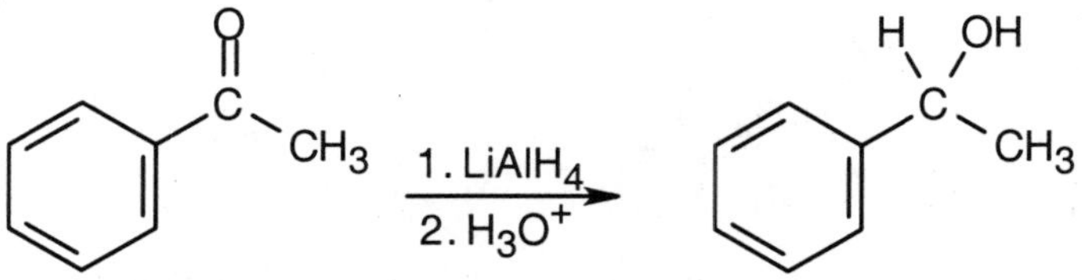

$C_6H_5COCH_3$ $\xrightarrow[\text{2. } H_3O^+]{\text{1. } LiAlH_4}$ $C_6H_5CH(OH)CH_3$

Reduction of a ketone yields this secondary alcohol.

(c)

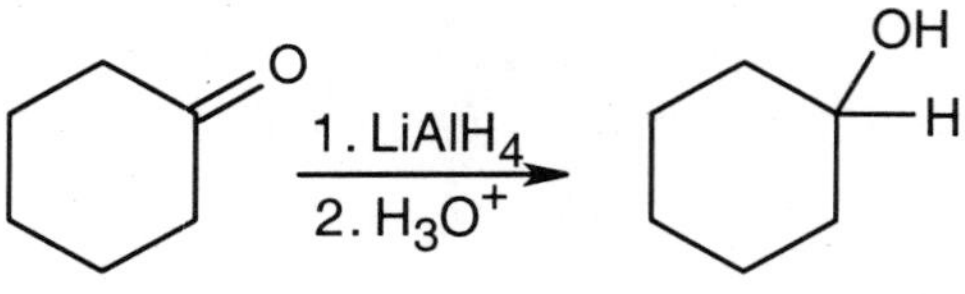

$$\xrightarrow[\text{2. } H_3O^+]{\text{1. } LiAlH_4}$$

(d)

$(CH_3)_2CHCHO$ *or* $(CH_3)_2CHCOOH$ *or* $(CH_3)_2CHCOOR$ $\xrightarrow[\text{2. } H_3O^+]{\text{1. } LiAlH_4}$ $(CH_3)_2CHCH_2OH$

17.9 All of the products have an –OH and a methyl group bonded to what was formerly a ketone carbon.

(a)

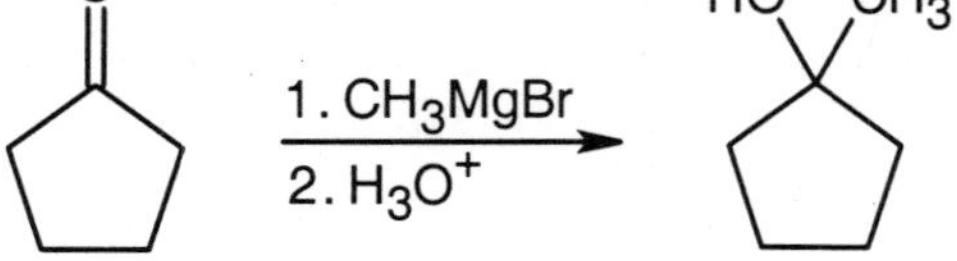

$$\xrightarrow[\text{2. } H_3O^+]{\text{1. } CH_3MgBr}$$

(b)

$C_6H_5COC_6H_5$ $\xrightarrow[\text{2. } H_3O^+]{\text{1. } CH_3MgBr}$ $(C_6H_5)_2C(OH)CH_3$

(c)

$CH_3CH_2CH_2COCH_2CH_3$ $\xrightarrow[\text{2. } H_3O^+]{\text{1. } CH_3MgBr}$ $CH_3CH_2CH_2C(OH)(CH_3)CH_2CH_3$

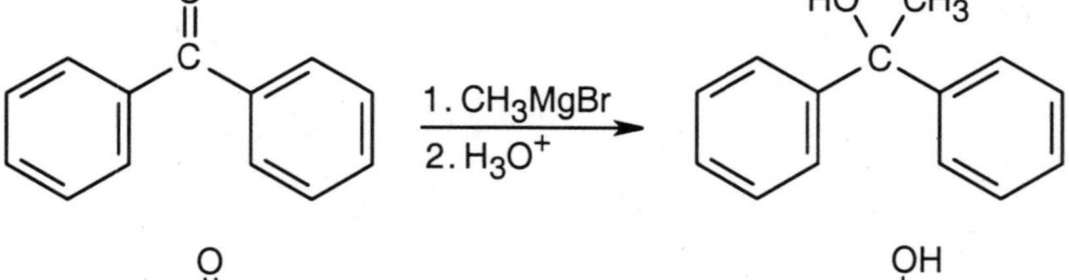

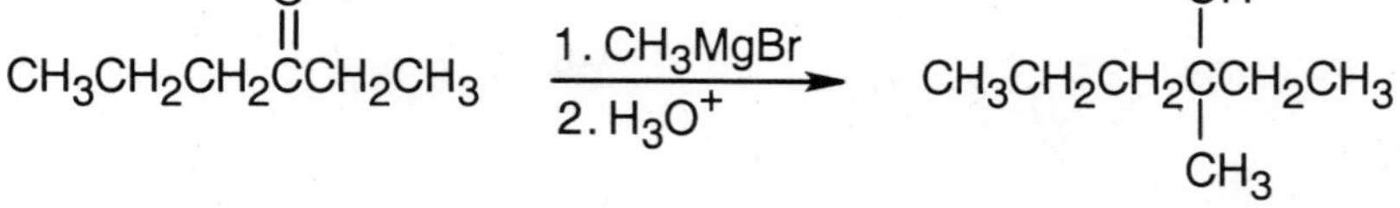

17.10 First, identify the type of alcohol. If the alcohol is primary, it can only be synthesized from formaldehyde plus the appropriate Grignard reagent. If the alcohol is secondary, it is synthesized from an aldehyde and a Grignard reagent. (Usually, there are two combinations of aldehyde and Grignard reagent). A tertiary alcohol is synthesized from a ketone and a Grignard reagent. If all three groups on the tertiary alcohol are different, there are often three different combinations of ketone and Grignard reagent. If two of the groups on the alcohol carbon are the same, the alcohol may also be synthesized from an ester and two equivalents of Grignard reagent.

(a) 2-Methyl-2-propanol is a tertiary alcohol. To synthesize a tertiary alcohol, start with a ketone.

$$CH_3C(=O)CH_3 \xrightarrow[2.\ H_3O^+]{1.\ CH_3MgBr} CH_3C(OH)(CH_3)CH_3$$

If two or more alkyl groups bonded to the carbon bearing the –OH group are the same, an alcohol can be synthesized from an ester and a Grignard reagent.

$$CH_3C(=O)OR \xrightarrow[2.\ H_3O^+]{1.\ 2\ CH_3MgBr} CH_3C(OH)(CH_3)CH_3$$

2-Methyl-2-propanol

(b) Since 1-methylcyclohexanol is a tertiary alcohol, start with a ketone.

$$\text{Cyclohexanone} \xrightarrow[2.\ H_3O^+]{1.\ CH_3MgBr} \text{1-Methylcyclohexanol (OH, } CH_3\text{)}$$

1-Methylcyclohexanol

(c) 3-Methyl-3-pentanol is a tertiary alcohol. When two of the three groups bonded to the alcohol carbon are the same, either a ketone or an ester can be used as a starting material.

$$CH_3CH_2C(=O)CH_2CH_3 \xrightarrow[2.\ H_3O^+]{1.\ CH_3MgBr}$$

or

$$CH_3CH_2C(=O)CH_3 \xrightarrow[2.\ H_3O^+]{1.\ CH_3CH_2MgBr}$$

or

$$CH_3C(=O)OR \xrightarrow[2.\ H_3O^+]{1.\ 2\ CH_3CH_2MgBr}$$

$$CH_3CH_2C(OH)(CH_3)CH_2CH_3$$

3-Methyl-3-pentanol

(d) Three possible combinations of ketone plus Grignard reagent can be used to synthesize this tertiary alcohol.

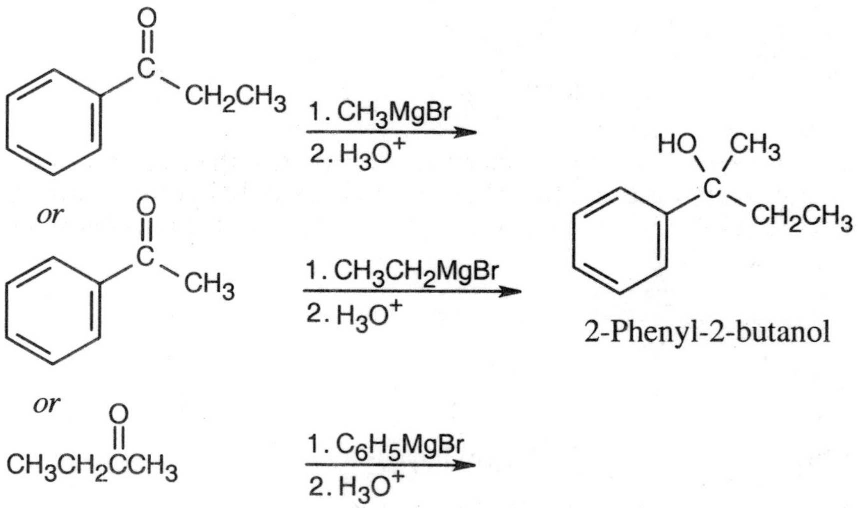

$C_6H_5COCH_2CH_3$ $\xrightarrow[2.\ H_3O^+]{1.\ CH_3MgBr}$

or

$C_6H_5COCH_3$ $\xrightarrow[2.\ H_3O^+]{1.\ CH_3CH_2MgBr}$

or

$CH_3CH_2COCH_3$ $\xrightarrow[2.\ H_3O^+]{1.\ C_6H_5MgBr}$

$C_6H_5C(OH)(CH_3)CH_2CH_3$

2-Phenyl-2-butanol

(e) Formaldehyde must be used to synthesize this primary alcohol.

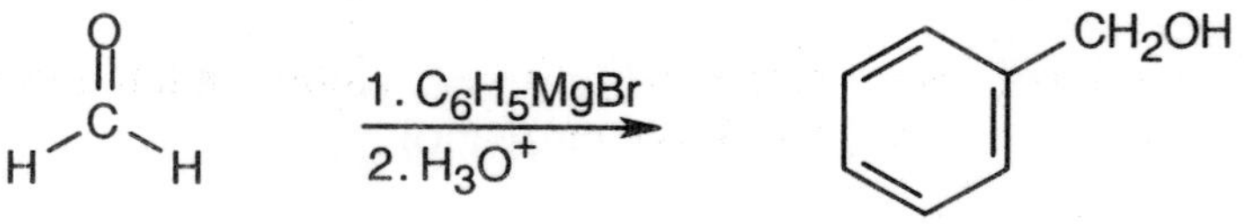

$HCHO$ $\xrightarrow[2.\ H_3O^+]{1.\ C_6H_5MgBr}$ $C_6H_5CH_2OH$

Benzyl alcohol

17.11 First, interpret the structure of the alcohol. This alcohol, 1-ethylcyclohexanol, is a tertiary alcohol that can be synthesized from a ketone. Only one combination of ketone and Grignard reagent is possible.

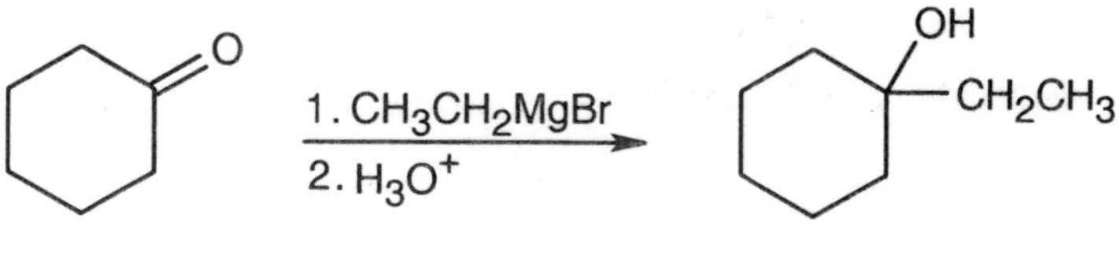

Cyclohexanone $\xrightarrow[2.\ H_3O^+]{1.\ CH_3CH_2MgBr}$ 1-ethylcyclohexanol (OH, CH_2CH_3)

17.12

(a)

$CH_3CH_2CH(OH)CH(CH_3)_2$ $\xrightarrow[pyridine]{POCl_3}$ $(CH_3CH_2)HC{=}C(CH_3)_2$ (major) + $(H_3C)HC{=}CH(CH(CH_3)_2)$ (minor, H and $CH(CH_3)_2$ trans) + $(H_3C)HC{=}CH(CH(CH_3)_2)$ (minor, H_3C and $CH(CH_3)_2$ cis)

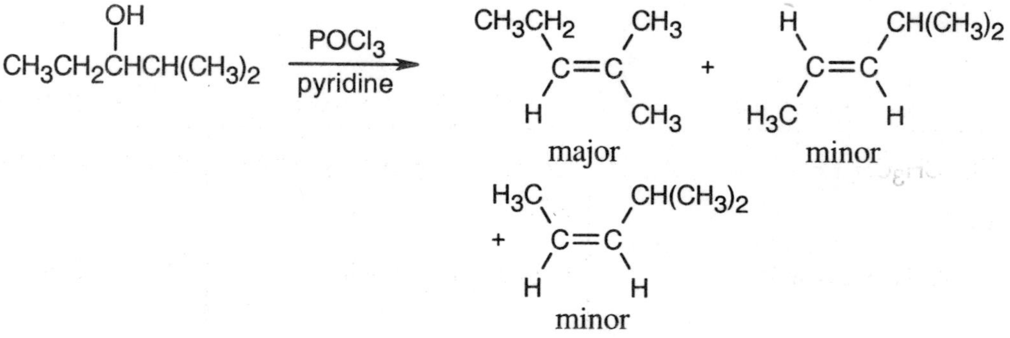

The major product has the more substituted double bond.

(b) $\xrightarrow[\text{pyridine}]{POCl_3}$ 3-Methylcyclohexene

In E2 elimination, dehydration proceeds most readily when the two groups to be eliminated have a trans–diaxial relationship. In this compound, the only hydrogen with the proper stereochemical relationship to the –OH group is at C6. Thus the non-Zaitsev product, 3-methylcyclohexene, is formed.

(c) $\xrightarrow[\text{pyridine}]{POCl_3}$ 1-Methylcyclohexene

Here, the hydrogen at C2 is trans to the hydroxyl, and dehydration yields the Zaitsev product, 1-methylcyclohexene.

17.13 Aldehydes are synthesized from oxidation of primary alcohols, and ketones are synthesized from oxidation of secondary alcohols.

(a) $C_6H_5CH(OH)CH_3 \xrightarrow[H_3O^+]{CrO_3} C_6H_5COCH_3$

(b) $CH_3CH(CH_3)CH_2OH \xrightarrow[CH_2Cl_2]{PCC} CH_3CH(CH_3)CHO$

(c) cyclopentanol $\xrightarrow[H_3O^+]{CrO_3}$ cyclopentanone

17.14

Starting material	*CrO_3, H_3O^+ Product*	*PCC Product*
(a) $CH_3CH_2CH_2CH_2CH_2CH_2OH$	$CH_3CH_2CH_2CH_2CH_2COOH$	$CH_3CH_2CH_2CH_2CH_2CHO$
(b) $CH_3CH_2CH_2CH_2CH(OH)CH_3$	$CH_3CH_2CH_2CH_2COCH_3$	$CH_3CH_2CH_2CH_2COCH_3$
(c) $CH_3CH_2CH_2CH_2CH_2CHO$	$CH_3CH_2CH_2CH_2CH_2COOH$	no reaction

17.15

$:\ddot{F}:^- + (CH_3)_3Si–O–C_6H_{11} \longrightarrow FSi(CH_3)_3 + Li^+\ ^-:\ddot{O}–C_6H_{11}$

This is an S_N2 reaction in which the nucleophile F^- attacks silicon and displaces an alkoxide ion as leaving group.

17.16

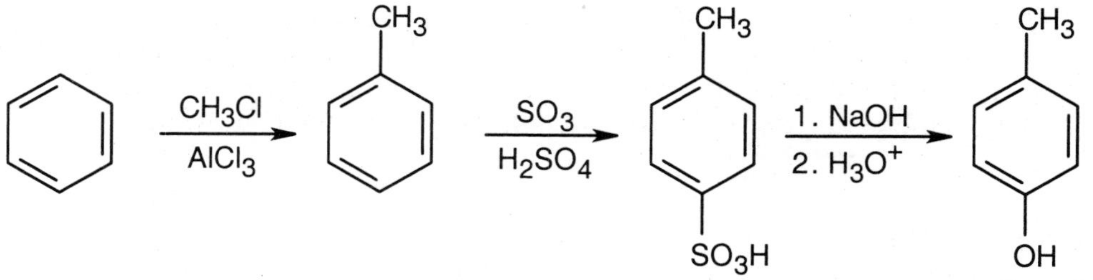

Benzene $\xrightarrow[AlCl_3]{CH_3Cl}$ toluene $\xrightarrow[H_2SO_4]{SO_3}$ p-toluenesulfonic acid (CH_3, SO_3H) $\xrightarrow[2.\ H_3O^+]{1.\ NaOH}$ (CH_3, OH)

p-Cresol

17.17

$CH_3C(CH_3){=}CH_2 + H–OPO_3H_2 \rightleftharpoons (CH_3)_3C+ \ + \ ^-OPO_3H_2$

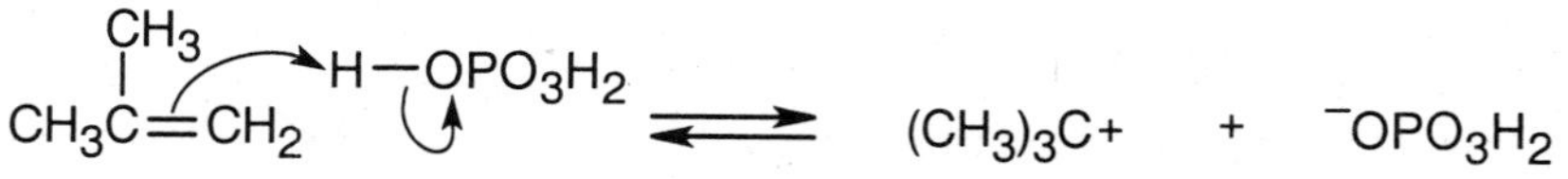

Phosphoric acid protonates 2-methylpropene, forming a *tert*-butyl carbocation.

p-cresol (OH, CH_3) + $(CH_3)C+$ ⟶ [intermediate: OH, $C(CH_3)_3$, H, +, CH_3] ⟶ (OH, $C(CH_3)_3$, CH_3) + H_3PO_4

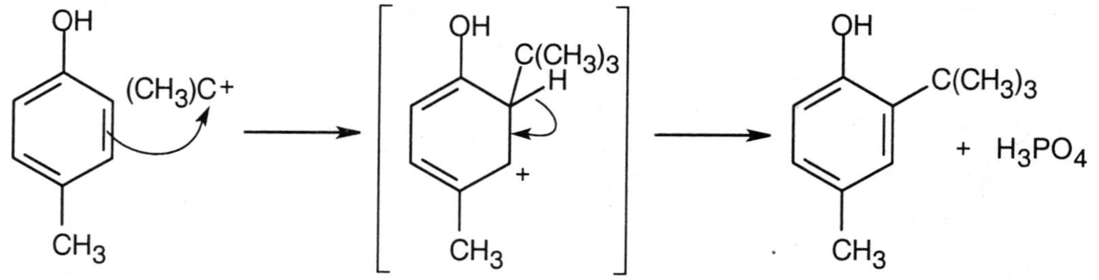

The *tert*-butyl carbocation acts as an electrophile and alkylates *p*-cresol. Alkylation occurs ortho to the –OH group for both steric and electronic reasons.

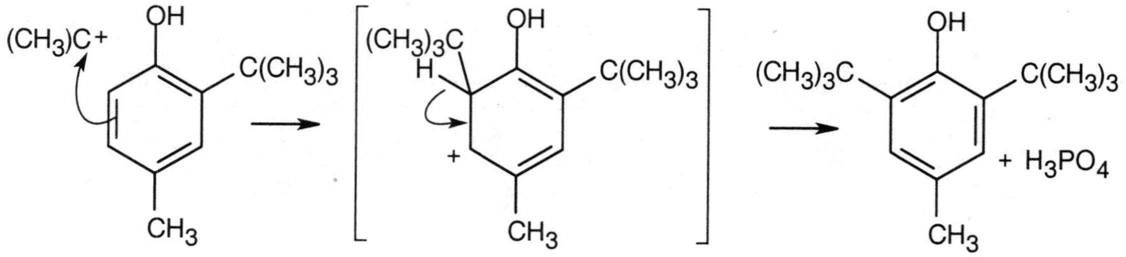

$(CH_3)C+$ + (OH, $C(CH_3)_3$, CH_3) ⟶ [intermediate: OH, $(CH_3)_3C$, H, +, $C(CH_3)_3$, CH_3] ⟶ (OH, $(CH_3)_3C$, $C(CH_3)_3$, CH_3) + H_3PO_4

BHT

A second *tert*-butyl carbocation alkylation forms BHT.

17.18 The infrared spectra of cholesterol and 5-cholestene-3-one each exhibit a unique absorption that makes it easy to distinguish between them. Cholesterol shows an –OH stretch at 3300-3600 cm^{-1}, and 5-cholestene-3-one shows a C=O stretch at 1715 cm^{-1}. In the oxidation of cholesterol to 5-cholestene-3-one, the –OH band will disappear and will be replaced by a C=O band. When oxidation is complete, no –OH absorption should be visible.

17.19 Under conditions of slow exchange, the –OH signal of a tertiary alcohol (R_3COH) is unsplit, the signal of a secondary alcohol (R_2CHOH) is split into a doublet, and the signal of a primary alcohol (RCH_2OH) is split into a triplet.

(a) 2-Methyl-2-propanol is a tertiary alcohol; its –OH signal is unsplit.
(b) Cyclohexanol is a secondary alcohol; its –OH absorption is a doublet.
(c) Ethanol is a primary alcohol; its –OH signal appears as a triplet.
(d) 2-Propanol is a secondary alcohol; its –OH absorption is split into a doublet.
(e) Cholesterol is a secondary alcohol; its –OH absorption is split into a doublet.
(f) 1-Methylcyclohexanol is a tertiary alcohol; its –OH signal is unsplit.

Visualizing Chemistry

17.20

(a) $CH_3CH(CH_3)CH_2$–C(H)(OH)–CH_2CH_3

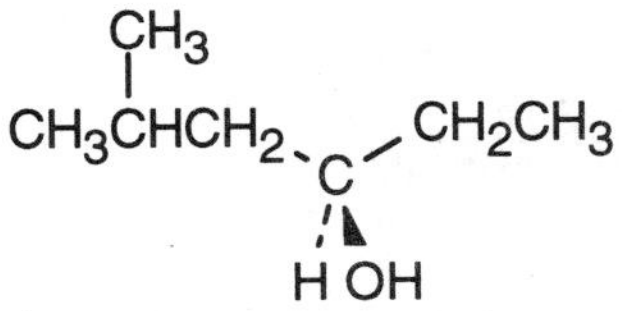

(*R*)-5-Methyl-3-hexanol

(b) H_3C, HO (cyclohexane chair)

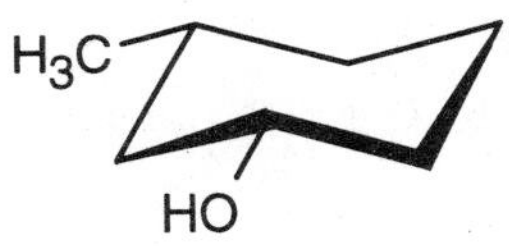

cis-3-Methylcyclohexanol

(c) cyclopentyl–$CHCH_3$(OH)

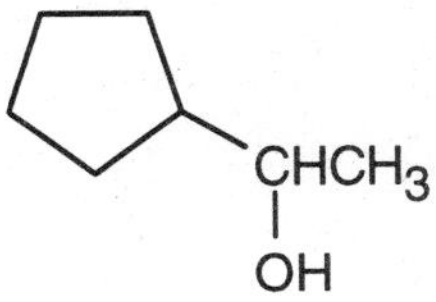

(*S*)-1-Cyclopentylethanol

(d) OH, H_3C, NO_2 (benzene ring)

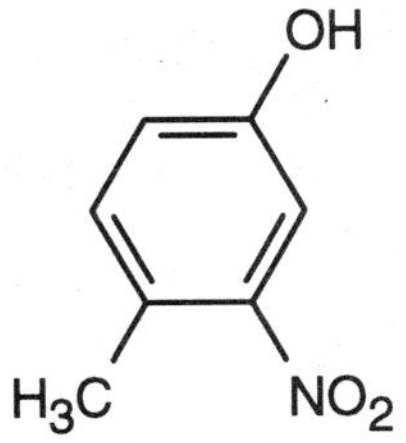

4-Methyl-3-nitrophenol

17.21

(a) H_3C–C(H)(OH)–C(H)(CH_3)–(3-bromophenyl) $\xrightarrow{PBr_3}$ H_3C–C(Br)(H)–C(H)(CH_3)–(3-bromophenyl)

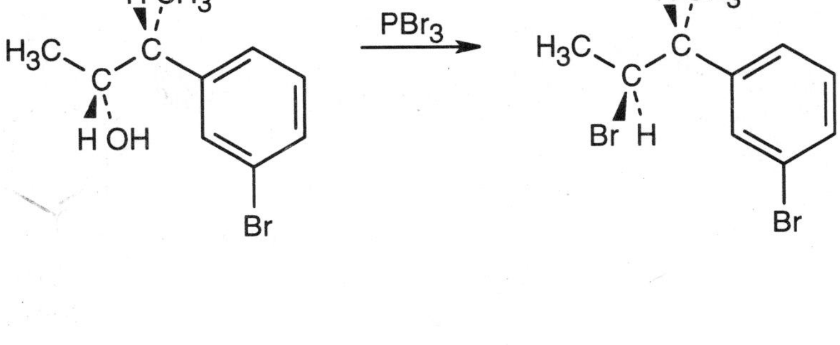

(b)

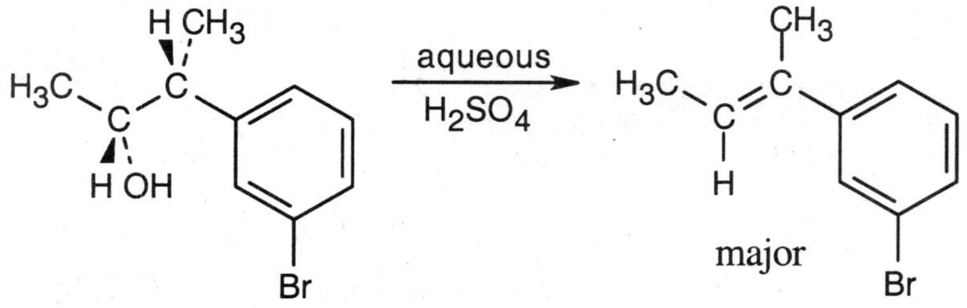

H CH3 / H3C C C / H OH / Br —— aqueous H2SO4 ——> CH3 / H3C C=C / H / major / Br

$$H_3C-CH(OH)-CH(CH_3)-C_6H_4Br \xrightarrow[H_2SO_4]{\text{aqueous}} H_3C-CH{=}C(CH_3)-C_6H_4Br$$

major

(c)

H CH3 / H3C C C / H OH / Br —— SOCl2 ——> H CH3 / H3C C C / Cl H / Br

$$H_3C-CH(OH)-CH(CH_3)-C_6H_4Br \xrightarrow{SOCl_2} H_3C-CH(Cl)-CH(CH_3)-C_6H_4Br$$

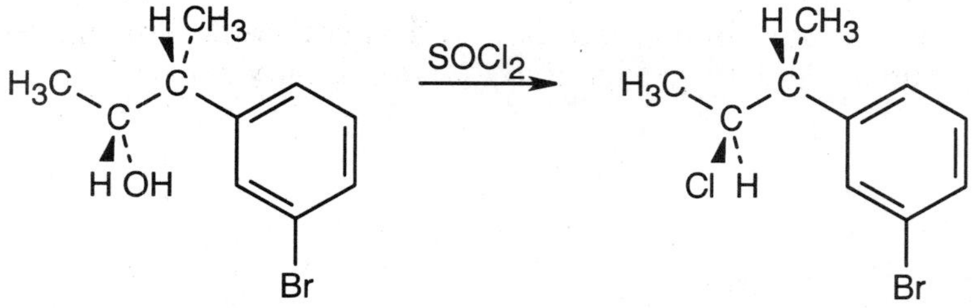

(d)

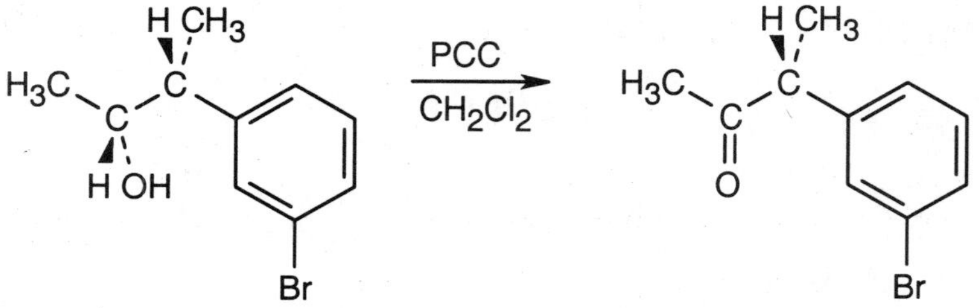

H CH3 / H3C C C / H OH / Br —— PCC CH2Cl2 ——> H CH3 / H3C C C / O / Br

$$H_3C-CH(OH)-CH(CH_3)-C_6H_4Br \xrightarrow[CH_2Cl_2]{PCC} H_3C-C(=O)-CH(CH_3)-C_6H_4Br$$

(e)

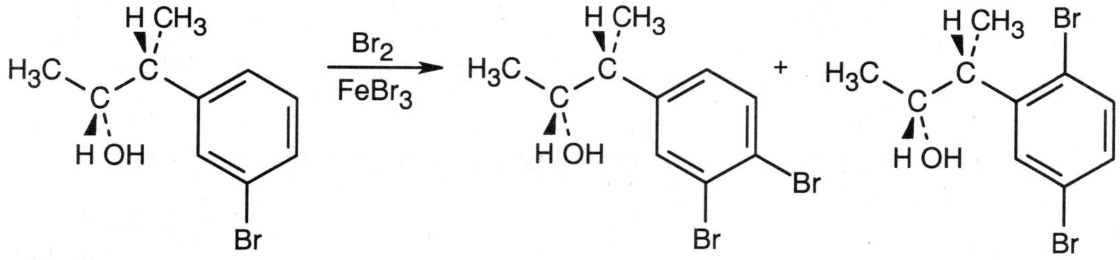

H CH3 / H3C C C / H OH / Br —— Br2 FeBr3 ——> H CH3 / H3C C C / H OH / Br / Br + H CH3 Br / H3C C C / H OH / Br

$$H_3C-CH(OH)-CH(CH_3)-C_6H_4Br \xrightarrow[FeBr_3]{Br_2} H_3C-CH(OH)-CH(CH_3)-C_6H_3Br_2 + H_3C-CH(OH)-CH(CH_3)-C_6H_3Br_2$$

17.22

(a)

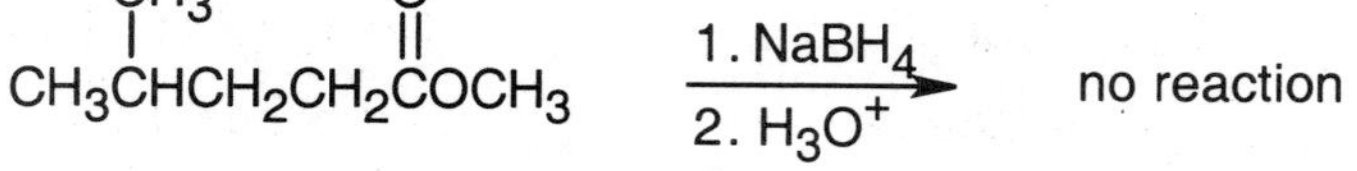

$$CH_3CH(CH_3)CH_2CH_2C(=O)OCH_3 \xrightarrow[2.\ H_3O^+]{1.\ NaBH_4} \text{no reaction}$$

(b)

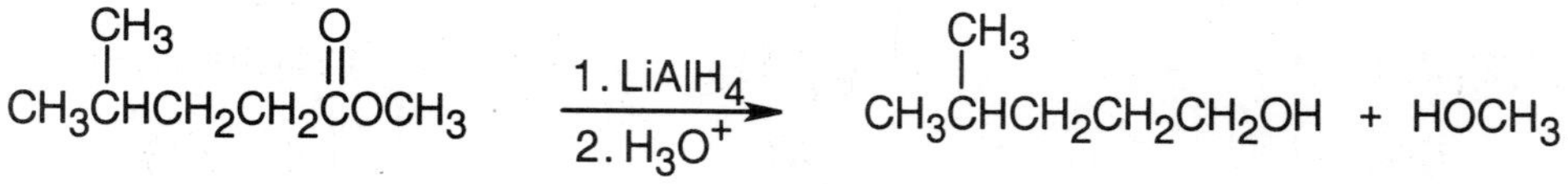

$$CH_3CH(CH_3)CH_2CH_2C(=O)OCH_3 \xrightarrow[2.\ H_3O^+]{1.\ LiAlH_4} CH_3CH(CH_3)CH_2CH_2CH_2OH + HOCH_3$$

(c)

$$CH_3CH(CH_3)CH_2CH_2C(=O)OCH_3 \xrightarrow[2.\ H_3O^+]{1.\ 2\ CH_3MgBr} CH_3CH(CH_3)CH_2CH_2C(OH)(CH_3)CH_3 + HOCH_3$$

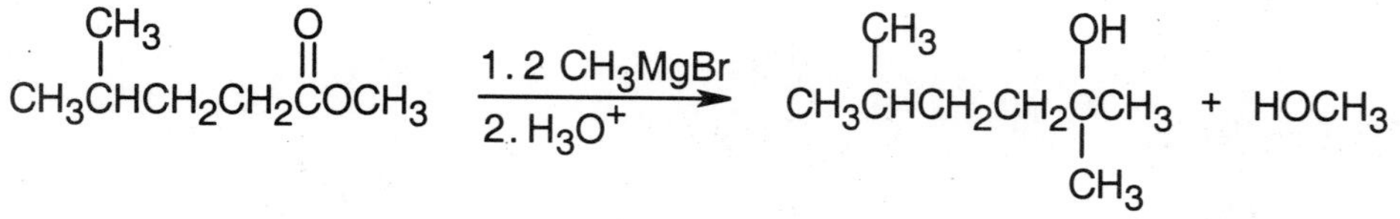

17.23

$CH_3CH_2C(S)H(CH_3)C(=O)CH_3$ $\xrightarrow[2.\ H_3O^+]{1.\ CH_3CH_2MgBr}$ $CH_3CH_2C(S)H(CH_3)C(R)(OH)(CH_2CH_3)CH_3$ + $CH_3CH_2C(S)H(CH_3)C(S)(OH)(CH_2CH_3)CH_3$

(3*R*,4*S*)-3,4-Dimethyl-3-hexanol (3*S*,4*S*)-3,4-Dimethyl-3-hexanol

The product is a mixture of the (3*R*,4*S*) and (3*S*,4*S*) diastereomers. The diastereomers are formed in unequal amounts, and the product mixture is optically active.

Additional Problems

17.24

(a) $HOCH_2CH_2CH(CH_3)CH_2OH$
2-Methyl-1,4-butanediol

(b) $CH_3CH(OH)CH(CH_2CH_2CH_3)CH_2CH_3$
3-Ethyl-2-hexanol

(c) *cis*-1,3-Cyclobutanediol

(d) *cis*-2-Methyl-4-cyclohepten-1-ol

(e) *cis*-3-Phenylcyclopentanol

(f) 2-Bromo-4-cyanophenol *or* 3-Bromo-4-hydroxybenzonitrile

17.25 None of these alcohols has multiple bonds or rings.

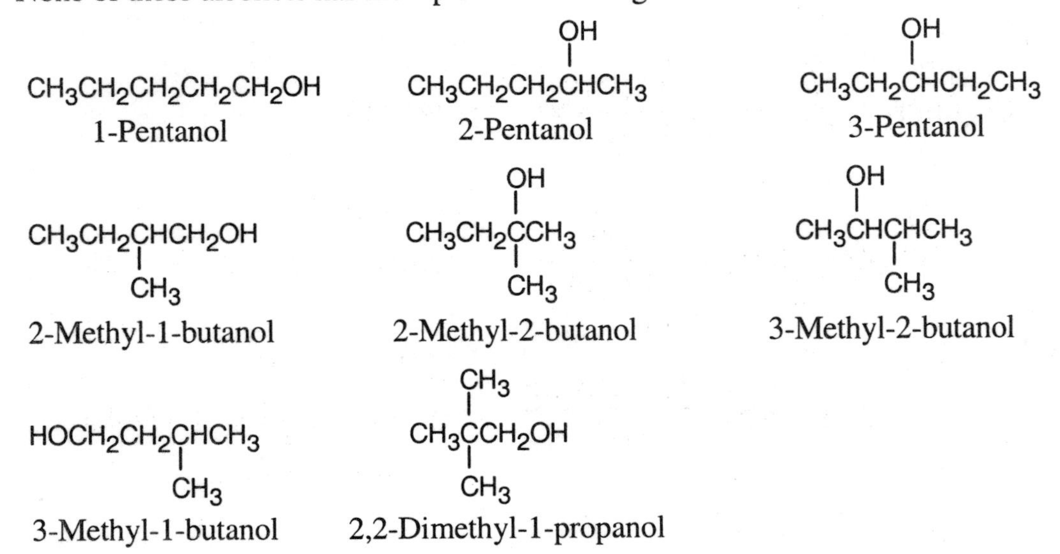

2-Pentanol, 2-methyl-1-butanol and 3-methyl-2-butanol have chiral carbons.

17.26 Primary alcohols react with CrO_3 in aqueous acid to form carboxylic acids, secondary alcohols yield ketones, and tertiary alcohols are unreactive to oxidation. Of the eight alcohols in the previous problem, only 2-methyl-2-butanol is unreactive to CrO_3 oxidation.

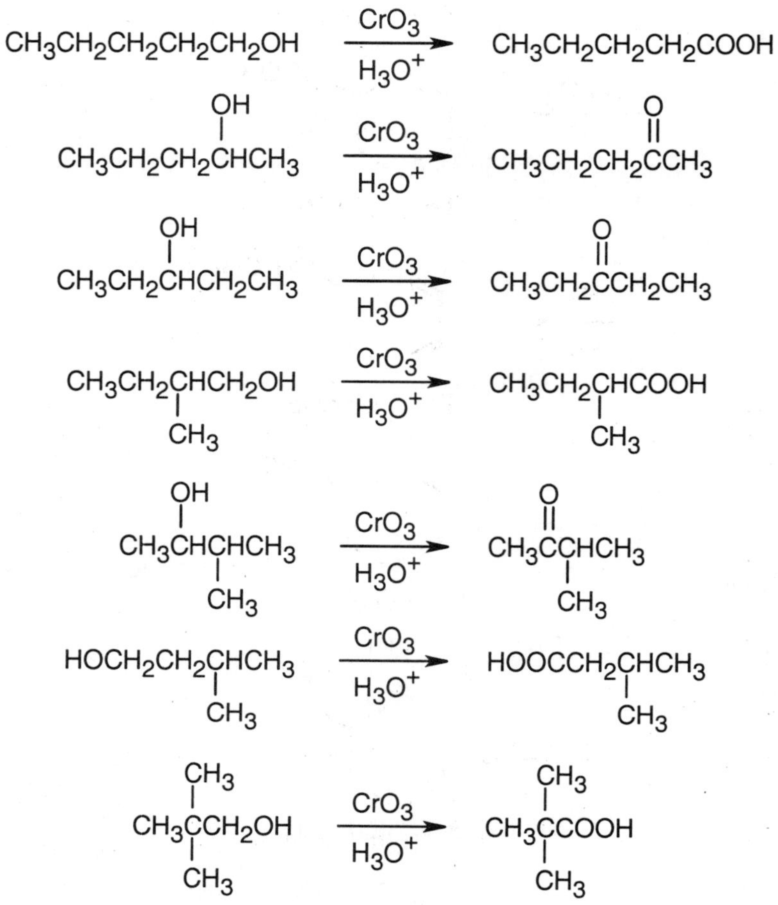

$$CH_3CH_2CH_2CH_2CH_2OH \xrightarrow[H_3O^+]{CrO_3} CH_3CH_2CH_2CH_2COOH$$

$$CH_3CH_2CH_2CH(OH)CH_3 \xrightarrow[H_3O^+]{CrO_3} CH_3CH_2CH_2C(=O)CH_3$$

$$CH_3CH_2CH(OH)CH_2CH_3 \xrightarrow[H_3O^+]{CrO_3} CH_3CH_2C(=O)CH_2CH_3$$

$$CH_3CH_2CH(CH_3)CH_2OH \xrightarrow[H_3O^+]{CrO_3} CH_3CH_2CH(CH_3)COOH$$

$$CH_3CH(OH)CH(CH_3)CH_3 \xrightarrow[H_3O^+]{CrO_3} CH_3C(=O)CH(CH_3)CH_3$$

$$HOCH_2CH_2CH(CH_3)CH_3 \xrightarrow[H_3O^+]{CrO_3} HOOCCH_2CH(CH_3)CH_3$$

$$CH_3C(CH_3)_2CH_2OH \xrightarrow[H_3O^+]{CrO_3} CH_3C(CH_3)_2COOH$$

17.27

(a)

$$C_6H_5CH_2CH_2OH \xrightarrow[pyridine]{POCl_3} C_6H_5CH{=}CH_2$$

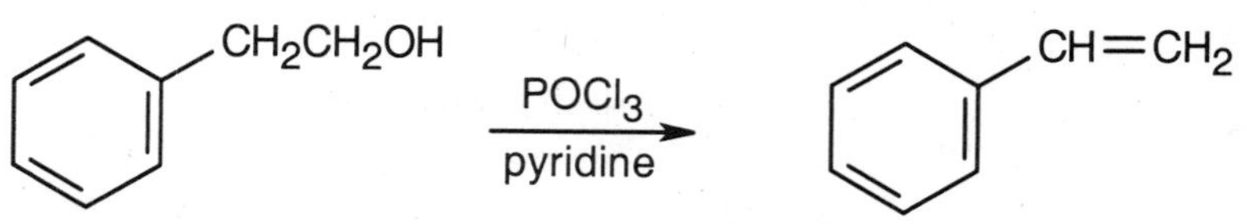

2-Phenylethanol → Styrene

(b)

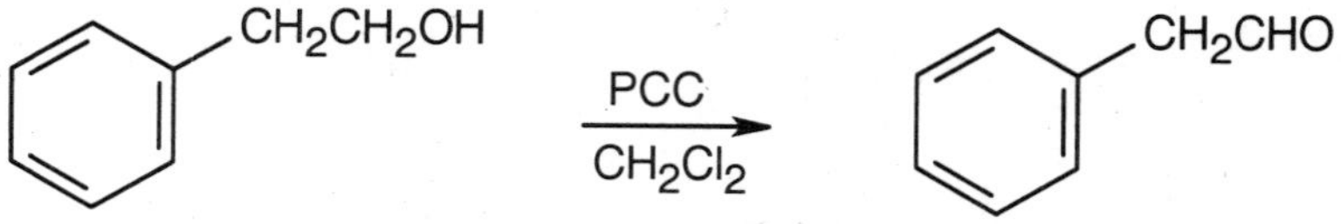

$$C_6H_5CH_2CH_2OH \xrightarrow[CH_2Cl_2]{PCC} C_6H_5CH_2CHO$$

Phenylacetaldehyde

PCC = $C_5H_6NCrO_3Cl$ (pyridinium chlorochromate)

(c)

$$C_6H_5CH_2CH_2OH \xrightarrow[H_3O^+]{CrO_3} C_6H_5CH_2COOH$$

Phenylacetic acid

(d)

$$C_6H_5CH_2CH_2OH \xrightarrow[H_2O]{KMnO_4} C_6H_5COOH$$

Benzoic acid

(e)

$$C_6H_5CH{=}CH_2 \xrightarrow[Pd]{H_2} C_6H_5CH_2CH_3$$

from (a) Ethylbenzene

(f)

$$C_6H_5CH{=}CH_2 \xrightarrow[2.\ Zn,\ H_3O^+]{1.\ O_3} C_6H_5CHO$$

from (a) Benzaldehyde

(g)

$$C_6H_5CH{=}CH_2 \xrightarrow[2.\ NaBH_4]{1.\ Hg(OAc)_2,\ H_2O} C_6H_5CH(OH)CH_3$$

from (a) 1-Phenylethanol

(h)

$$C_6H_5CH_2CH_2OH \xrightarrow{PBr_3} C_6H_5CH_2CH_2Br$$

1-Bromo-2-phenylethane

17.28

(a)

$$C_6H_5CH(OH)CH_3 \xrightarrow[CH_2Cl_2]{PCC} C_6H_5COCH_3$$

1-Phenylethanol Acetophenone

(b)

$$C_6H_5CH(OH)CH_3 \xrightarrow[H_2O]{KMnO_4} C_6H_5COOH \xrightarrow[2.\ H_3O^+]{1.\ LiAlH_4} C_6H_5CH_2OH$$

Benzyl alcohol

(c)

$C_6H_5COOH \xrightarrow[FeBr_3]{Br_2}$ m-BrC_6H_4COOH

from (b) — *m*-Bromobenzoic acid

(d)

$C_6H_5CCH_3$ (C=O) $\xrightarrow[2.\ H_3O^+]{1.\ CH_3MgBr}$ $C_6H_5C(OH)(CH_3)CH_3$

from (a) — 2-Phenyl-2-propanol

17.29 In some of these problems, different combinations of Grignard reagent and carbonyl compound are possible. Remember that aqueous acid is added to the initial Grignard adduct to yield the alcohol.

(a)

$CH_3CHO + CH_3CH_2MgBr$

or

$CH_3CH_2CHO + CH_3MgBr$

$\longrightarrow CH_3CH(OH)CH_2CH_3$

2-Butanol

(b)

$C_6H_5C(=O)CH_3 + CH_3MgBr$

or

$C_6H_5C(=O)OR + 2\ CH_3MgBr$

or

$C_6H_5MgBr + CH_3C(=O)CH_3$

$\longrightarrow C_6H_5C(OH)(CH_3)CH_3$

2-Phenyl-2-propanol

(c)

$H_2C{=}C(CH_3){-}MgBr + CH_2O \longrightarrow H_2C{=}C(CH_3)CH_2OH$

2-Methyl-2-propen-1-ol

(d)

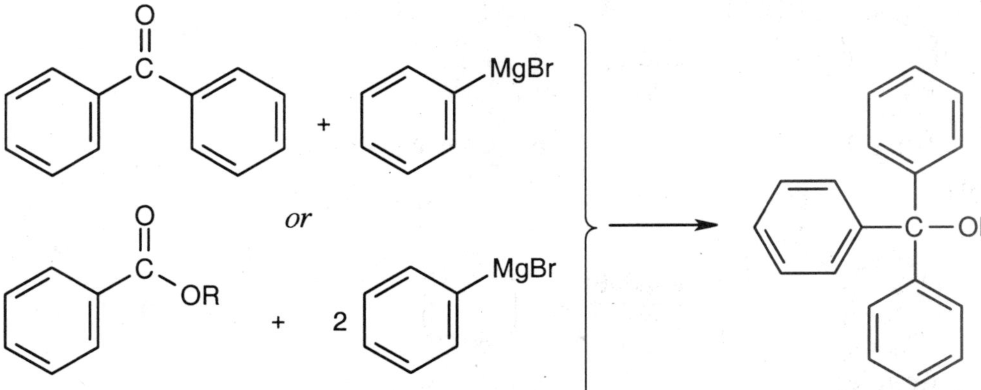

Benzophenone + Phenylmagnesium bromide (C_6H_5MgBr)

or

C_6H_5COOR + 2 C_6H_5MgBr

$\longrightarrow$ $(C_6H_5)_3C{-}OH$

Triphenylmethanol

(e)

$$CH_3MgBr + HCCH_2CH_2CH_2Br \text{ (with C=O on the first C)} \longrightarrow CH_3CH(OH)CH_2CH_2CH_2Br$$

5-Bromo-2-pentanol

17.30

$$HOCH_2CH_2CH_2CH_2Cl \xrightarrow{^-H:} [\,^-\!:\ddot{O}CH_2CH_2CH_2CH_2Cl\,] + H_2 \longrightarrow \text{tetrahydrofuran} + Cl^-$$

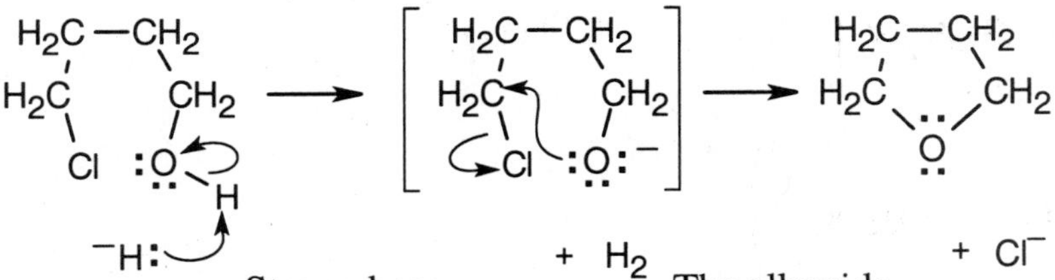

Strong base deprotonates the alcohol hydrogen

The alkoxide displaces Cl^- to form tetrahydrofuran

17.31

Alcohol	*Carbonyl precursor(s)*
(a) $CH_3CH_2CH_2CH_2C(CH_3)_2CH_2OH$	$CH_3CH_2CH_2CH_2C(CH_3)_2CHO$
	$CH_3CH_2CH_2CH_2C(CH_3)_2COOH$
	$CH_3CH_2CH_2CH_2C(CH_3)_2COOR$
(b) $(CH_3)CCH(OH)CH_3$	$(CH_3)CC(=O)CH_3$

(c) cyclohexyl–CH(OH)CH$_2$CH$_3$ cyclohexyl–C(=O)CH$_2$CH$_3$

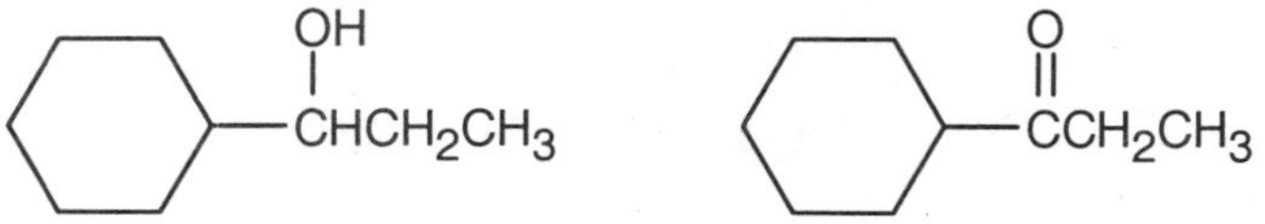

(d) cyclopentyl–CH$_2$OH cyclopentyl–CHO cyclopentyl–COOH cyclopentyl–COOR

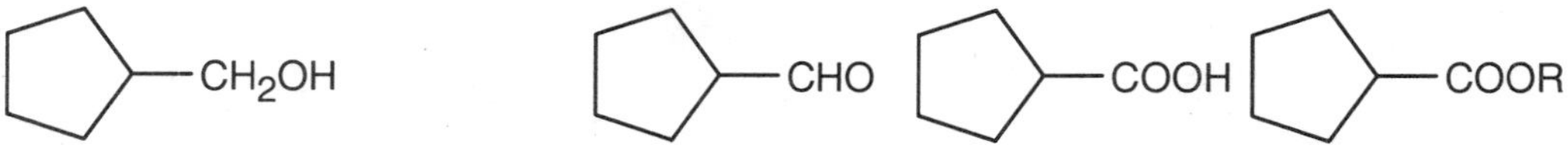

17.32 In these compounds you want to reduce some, but not all, of the functional groups present. To do this, you must select the correct reducing agent.

(a)

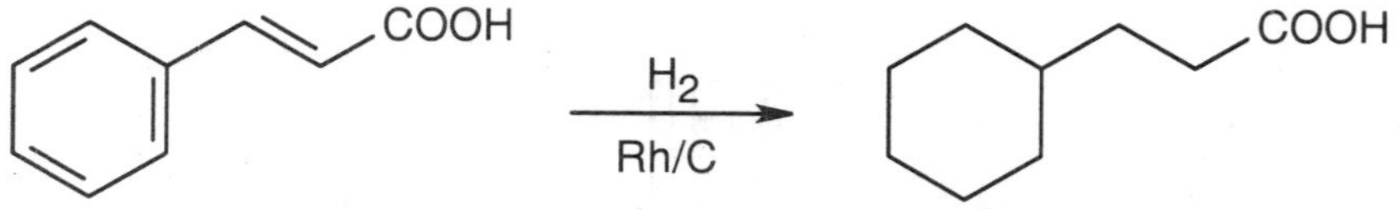

$$C_6H_5CH{=}CHCOOH \xrightarrow[Rh/C]{H_2} C_6H_{11}CH_2CH_2COOH$$

H_2, plus rhodium catalyst, hydrogenates all hydrocarbon double bonds without affecting carbonyl double bonds.

(b)

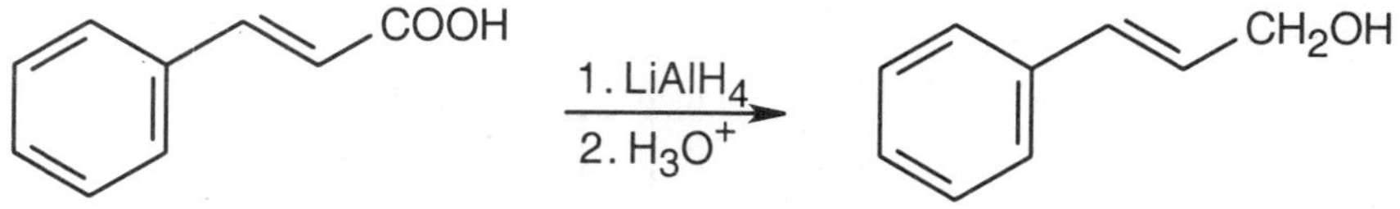

$$C_6H_5CH{=}CHCOOH \xrightarrow[2.\ H_3O^+]{1.\ LiAlH_4} C_6H_5CH{=}CHCH_2OH$$

$LiAlH_4$ reduces carbonyl groups without affecting carbon–carbon double bonds.

17.33

Grignard Reagent + *Carbonyl Compound* ⟶ *Product*

(a)

CH_3MgBr + $CH_3\overset{O}{\overset{\|}{C}}CH_3$

or

2 CH_3MgBr + $CH_3\overset{O}{\overset{\|}{C}}OR$

⟶ $(CH_3)COH$

(b)

CH_3CH_2MgBr + cyclohexanone ⟶ 1-ethylcyclohexanol (OH, CH_2CH_3)

(c)

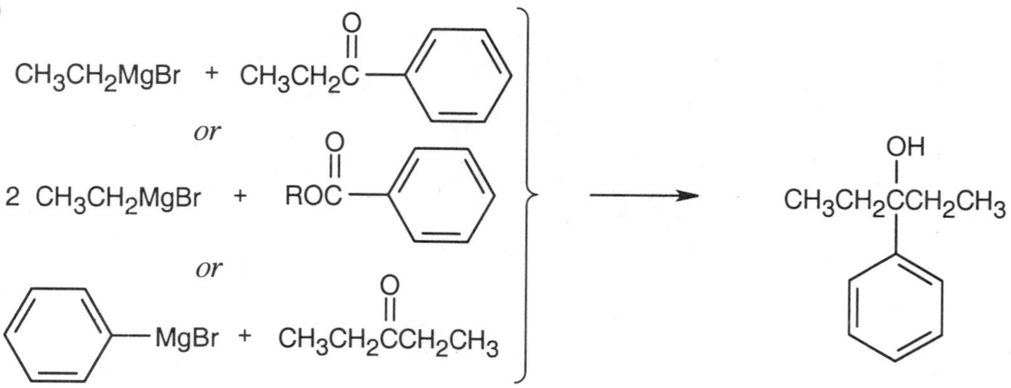

CH_3CH_2MgBr + $CH_3CH_2\overset{O}{\overset{\|}{C}}$–$C_6H_5$

or

2 CH_3CH_2MgBr + $RO\overset{O}{\overset{\|}{C}}$–$C_6H_5$

or

C_6H_5–MgBr + $CH_3CH_2\overset{O}{\overset{\|}{C}}CH_2CH_3$

⟶ $CH_3CH_2\overset{OH}{\overset{|}{C}}CH_2CH_3$ (with C_6H_5 on the central carbon)

(d)

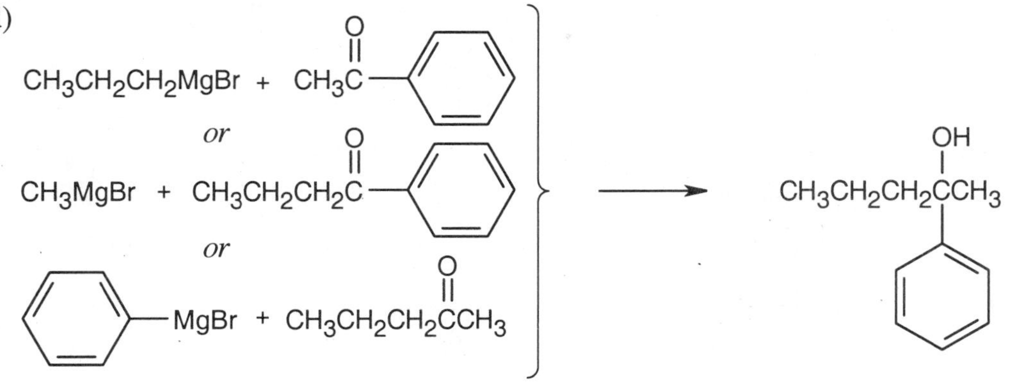

$$CH_3CH_2CH_2MgBr + CH_3\overset{O}{\overset{\|}{C}}-C_6H_5$$

or

$$CH_3MgBr + CH_3CH_2CH_2\overset{O}{\overset{\|}{C}}-C_6H_5$$

or

$$C_6H_5-MgBr + CH_3CH_2CH_2\overset{O}{\overset{\|}{C}}CH_3$$

$$\longrightarrow CH_3CH_2CH_2\overset{OH}{\overset{|}{C}}(C_6H_5)CH_3$$

(e)

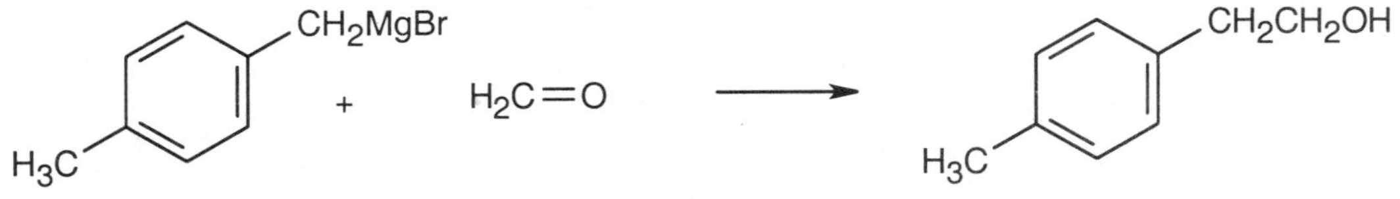

$$p\text{-}H_3C-C_6H_4-CH_2MgBr + H_2C{=}O \longrightarrow p\text{-}H_3C-C_6H_4-CH_2CH_2OH$$

(f)

$$CH_3MgBr + C_5H_9-CH_2\overset{O}{\overset{\|}{C}}CH_3$$

or

$$2\ CH_3MgBr + C_5H_9-CH_2\overset{O}{\overset{\|}{C}}OR$$

or

$$C_5H_9-CH_2MgBr + CH_3\overset{O}{\overset{\|}{C}}CH_3$$

$$\longrightarrow C_5H_9-CH_2\overset{OH}{\overset{|}{C}}(CH_3)_2$$

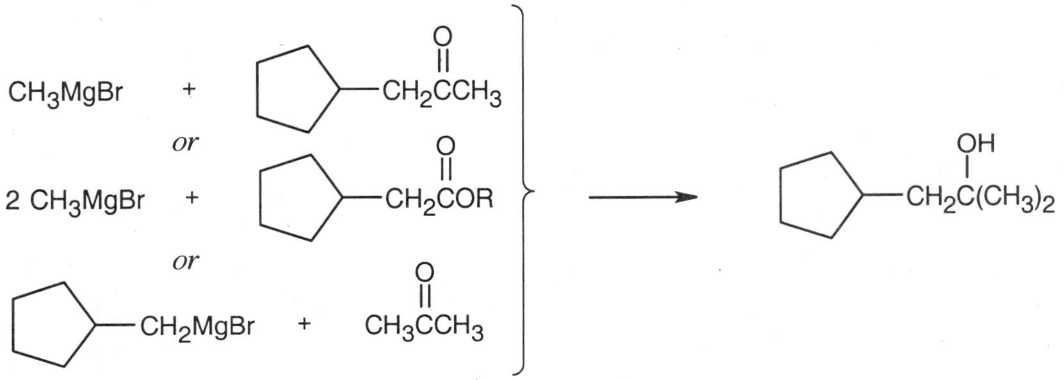

17.34

(a) $$CH_3CH_2CH_2CH_2CH_2OH \xrightarrow{PBr_3} CH_3CH_2CH_2CH_2CH_2Br$$

(b) $$CH_3CH_2CH_2CH_2CH_2OH \xrightarrow{SOCl_2} CH_3CH_2CH_2CH_2CH_2Cl$$

(c) $$CH_3CH_2CH_2CH_2CH_2OH \xrightarrow[H_3O^+]{CrO_3} CH_3CH_2CH_2CH_2COOH$$

(d) $$CH_3CH_2CH_2CH_2CH_2OH \xrightarrow[CH_2Cl_2]{PCC} CH_3CH_2CH_2CH_2CHO$$

17.35

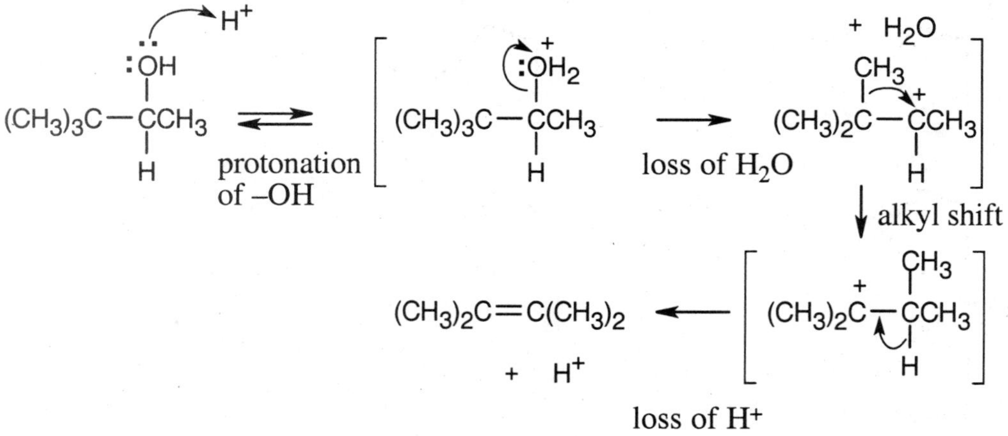

H^+

$:\ddot{O}H$

$(CH_3)_3C-CCH_3$

H

protonation of –OH

$:\ddot{O}H_2^+$

$(CH_3)_3C-CCH_3$

H

loss of H_2O

+ H_2O

CH_3

$(CH_3)_2C-\overset{+}{C}CH_3$

H

alkyl shift

CH_3

$(CH_3)_2\overset{+}{C}-CCH_3$

H

$(CH_3)_2C{=}C(CH_3)_2$

+ H^+

loss of H^+

17.36 This mechanism consists of the same steps as are seen in Problem 17.35. Two different alkyl shifts result in two different cycloalkenes.

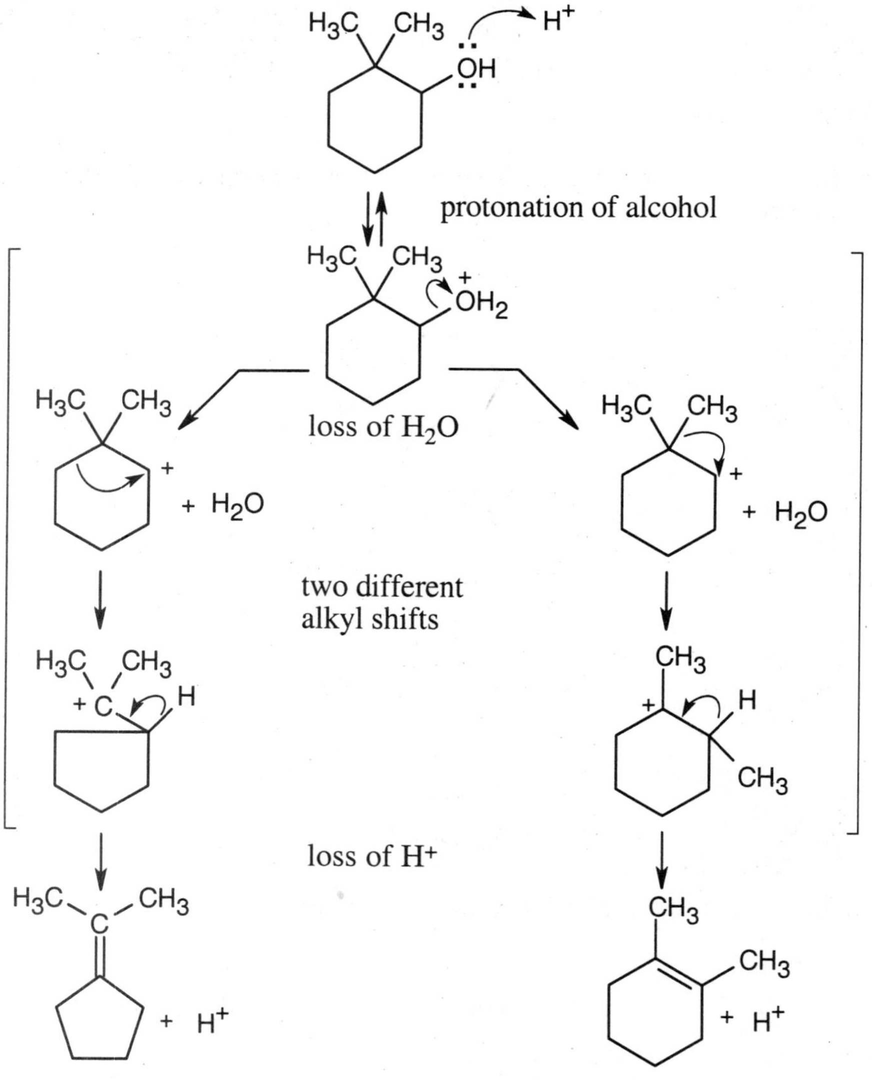

H_3C CH_3 H^+

$\ddot{O}H$

protonation of alcohol

H_3C CH_3

$\overset{+}{O}H_2$

loss of H_2O

H_3C CH_3

+

+ H_2O

H_3C CH_3

+

+ H_2O

two different alkyl shifts

H_3C CH_3

+ C H

CH_3

+ H

CH_3

loss of H^+

H_3C C CH_3

+ H^+

CH_3

CH_3

+ H^+

Isopropylidenecyclopentane

1,2-Dimethylcyclohexene

17.37

(a)

Cyclopentanol $\xrightarrow[H_3O^+]{CrO_3}$ cyclopentanone

(b)

Cyclopentanol $\xrightarrow[\text{pyridine}]{POCl_3}$ cyclopentene

(c)

Cyclopentanone (from (a)) $\xrightarrow[2.\ H_3O^+]{1.\ CH_3MgBr}$ 1-methylcyclopentanol

(d)

1-Methylcyclopentanol (from (c)) $\xrightarrow{H_3O^+}$ 1-methylcyclopentene $\xrightarrow[2.\ H_2O_2,\ ^-OH]{1.\ BH_3,\ THF}$ trans-2-methylcyclopentanol (H, OH, H, CH_3)

Remember that hydroboration proceeds with syn stereochemistry, and the –H and –OH added have a cis relationship.

17.38

(a)

1-Methylcyclohexanol (CH_3, OH) $\xrightarrow{HBr}$ 1-bromo-1-methylcyclohexane (CH_3, Br)

(b)

1-Methylcyclohexanol (CH_3, OH) $\xrightarrow{NaH}$ (CH_3, $O^-\ Na^+$) + H_2

(c)

1-Methylcyclohexanol (CH_3, OH) $\xrightarrow{H_2SO_4}$ 1-methylcyclohexene (CH_3)

(d)

1-Methylcyclohexanol (CH_3, OH) $\xrightarrow{Na_2Cr_2O_7}$ no reaction

Tertiary alcohols aren't oxidized by sodium dichromate.

17.39

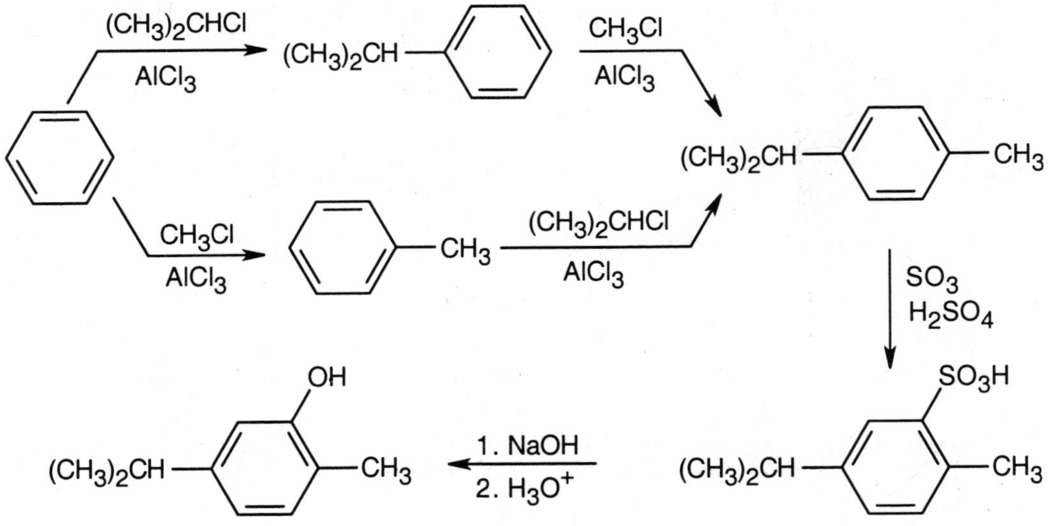

$(CH_3)_2CHCl$, $AlCl_3$ → $(CH_3)_2CH$– ; CH_3Cl, $AlCl_3$ →

CH_3Cl, $AlCl_3$ → –CH_3 ; $(CH_3)_2CHCl$, $AlCl_3$ →

$(CH_3)_2CH$– –CH_3

SO_3, H_2SO_4 ↓

SO_3H

$(CH_3)_2CH$– –CH_3

1. NaOH
2. H_3O^+

OH

$(CH_3)_2CH$– –CH_3

Carvacrol

Sulfonation occurs ortho to the methyl group for steric reasons.

17.40

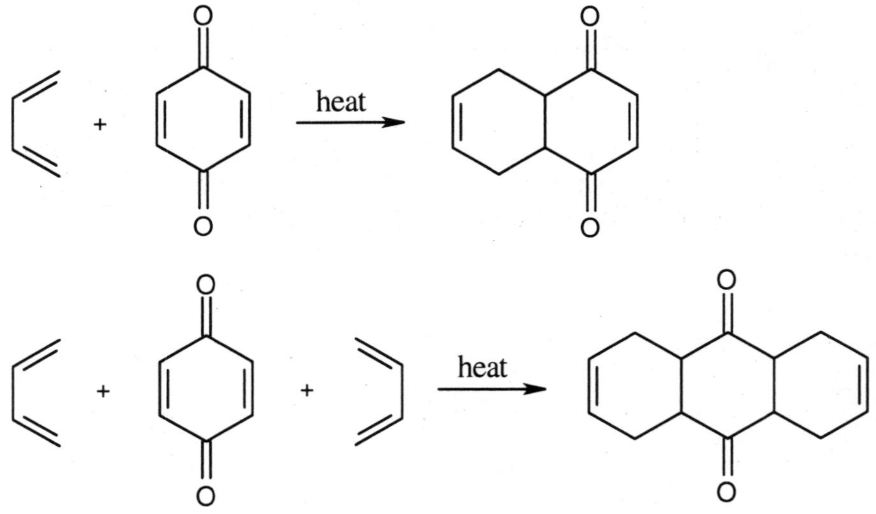

O, O, + , heat, O, O

O, + , + , heat, O, O

17.41 Remember that electron-withdrawing groups stabilize phenoxide anions and increase acidity. Electron-donating groups decrease phenol acidity.

Least acidic ⟶ *Most acidic*

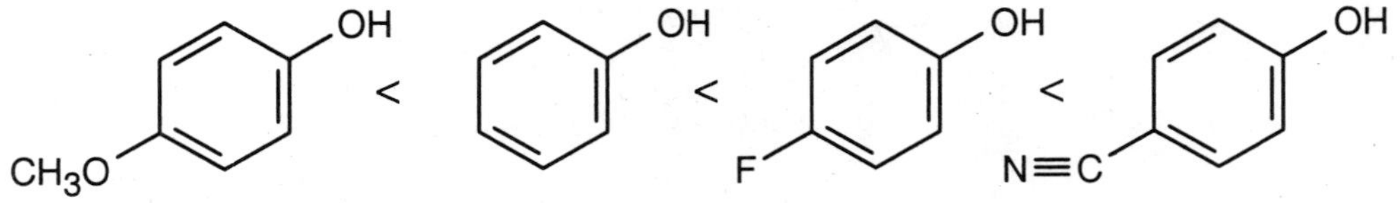

OH, CH_3O < OH < OH, F < OH, N≡C

electron-donating group

electron-withdrawing by inductive effect

electron-withdrawing by resonance

17.42

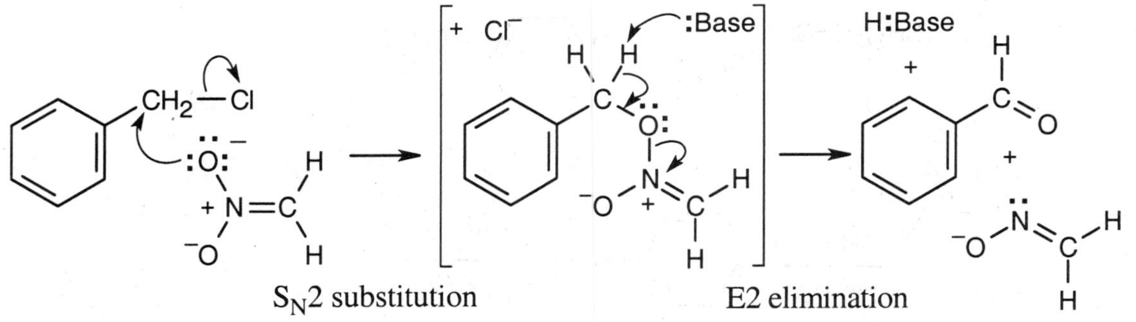
:Base
H:Base
S_N2 substitution
E2 elimination

17.43

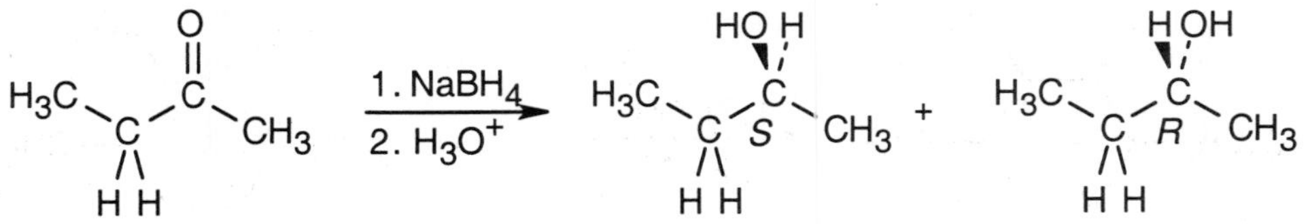
1. $NaBH_4$
2. H_3O^+

Reaction of 2-butanone with $NaBH_4$ produces a racemic mixture of (*R*)-2-butanol and (*S*)-2-butanol

17.44

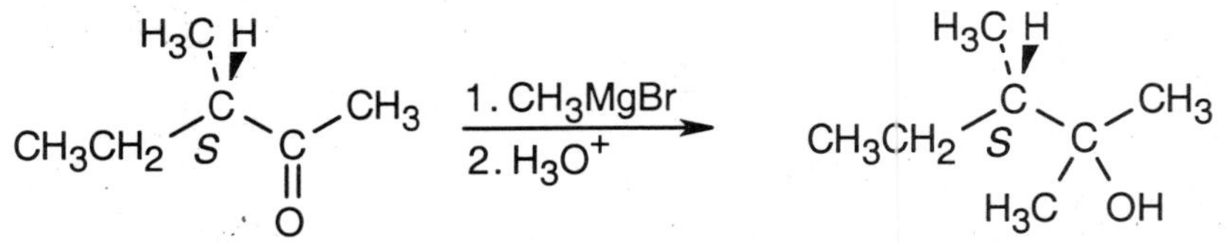
1. CH_3MgBr
2. H_3O^+

(*S*)-2,3-Dimethyl-2-pentanol

Despite this problem's resemblance to Problem 17.23, the stereochemical outcome is different. Addition of methylmagnesium bromide to the carbonyl group doesn't produce a new chirality center and doesn't affect the chirality center already present. The product is pure (*S*)-2,3-dimethyl-2-pentanol.

17.45

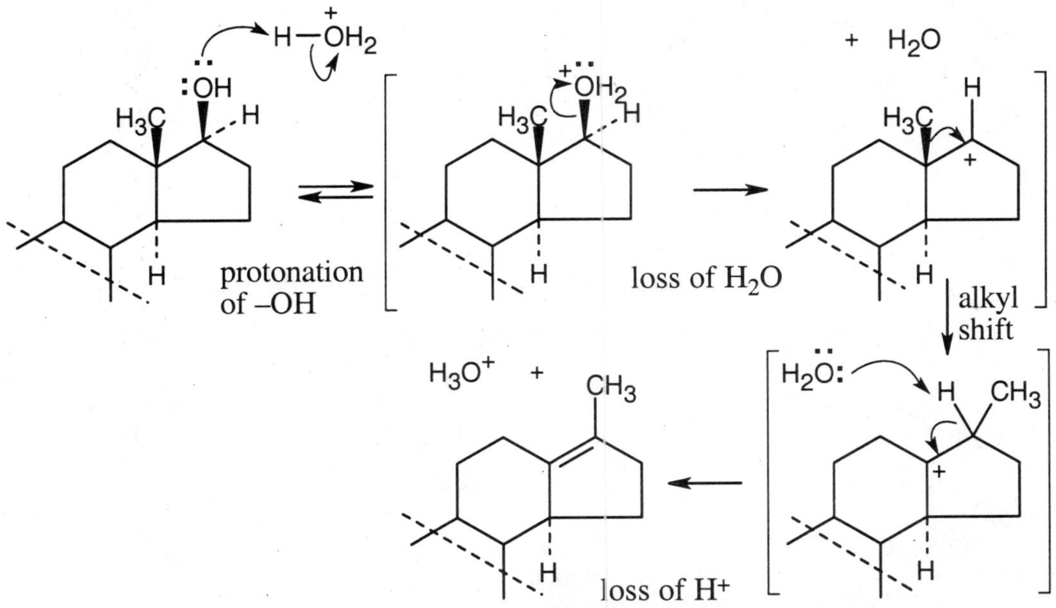
protonation of –OH
loss of H_2O
alkyl shift
loss of H^+

This is a carbocation rearrangement involving the shift of an alkyl group. The sequence of steps is the same as those seen in Problems 17.35 and 17.36.

17.46

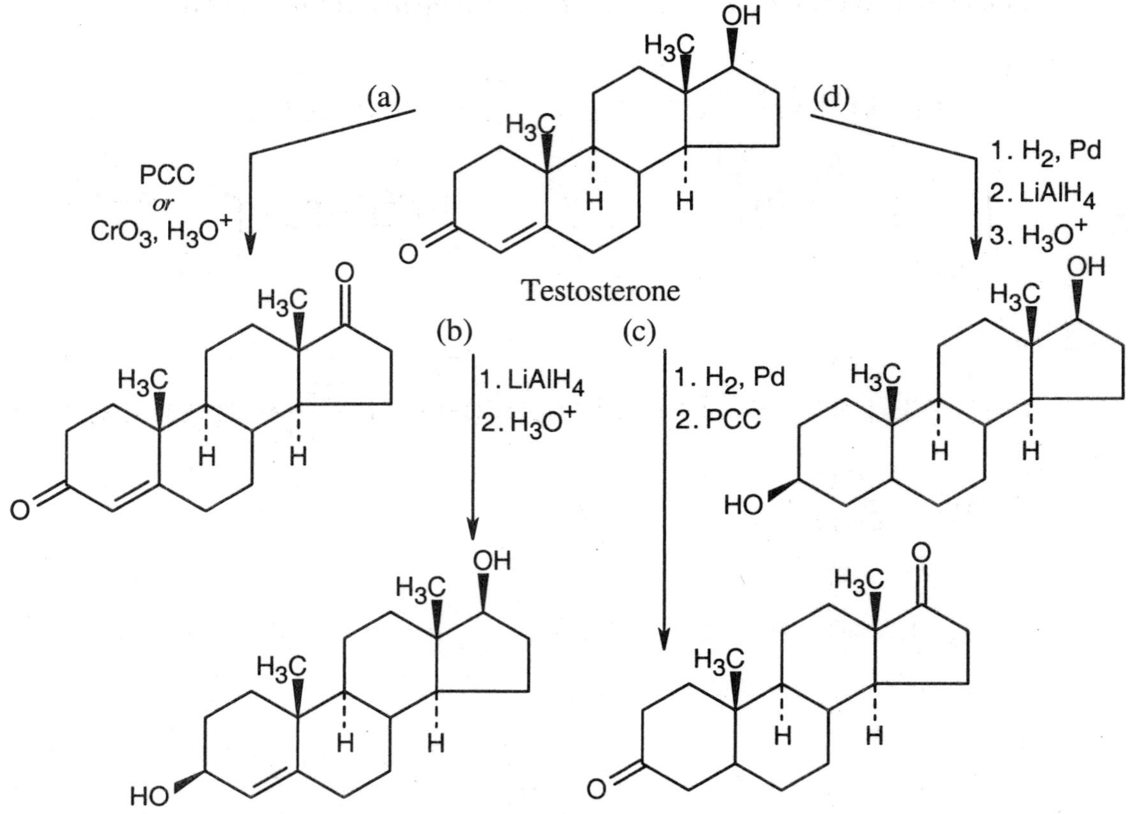

(a) PCC *or* CrO_3, H_3O^+

(b) 1. $LiAlH_4$ 2. H_3O^+

(c) 1. H_2, Pd 2. PCC

(d) 1. H_2, Pd 2. $LiAlH_4$ 3. H_3O^+

Testosterone

All of these transformations require the proper sequence of oxidations and reductions. In (d), $NaBH_4$ can also be used for reduction.

17.47

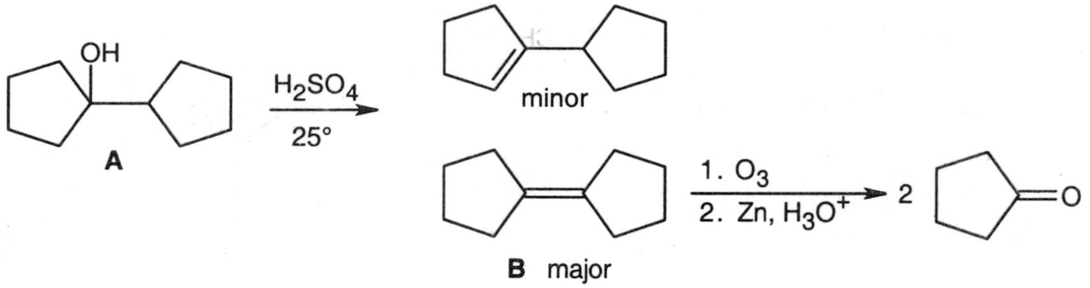

A $\xrightarrow[25°]{H_2SO_4}$ minor + **B** major

B $\xrightarrow[\text{2. Zn, } H_3O^+]{\text{1. } O_3}$ 2 (cyclopentanone)

17.48

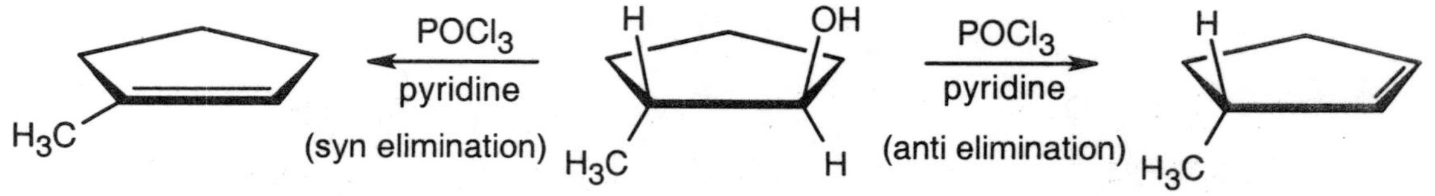

1-Methylcyclopentene $\xleftarrow[\text{pyridine}]{POCl_3}$ (syn elimination) *trans*-2-Methylcyclopentanol (anti elimination) $\xrightarrow[\text{pyridine}]{POCl_3}$ 3-Methylcyclopentene

The more stable dehydration product is 1-methylcyclopentene, which can be formed only via syn periplanar elimination. The product of anti periplanar elimination is 3-methylcyclopentene. Since this product predominates, the requirement of anti periplanar geometry must be more important than formation of the more stable product.

17.49 All of these syntheses involve a Grignard reaction at some step. Both the carbonyl compound and the Grignard reagent must be prepared from alcohols.

(a)

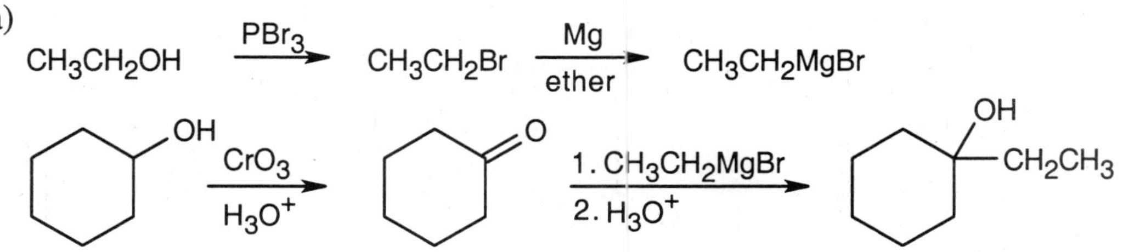

$$CH_3CH_2OH \xrightarrow{PBr_3} CH_3CH_2Br \xrightarrow[\text{ether}]{Mg} CH_3CH_2MgBr$$

$$\text{Cyclohexanol} \xrightarrow[H_3O^+]{CrO_3} \text{Cyclohexanone} \xrightarrow[2.\ H_3O^+]{1.\ CH_3CH_2MgBr} \text{1-ethylcyclohexanol (OH, } CH_2CH_3\text{)}$$

(b)

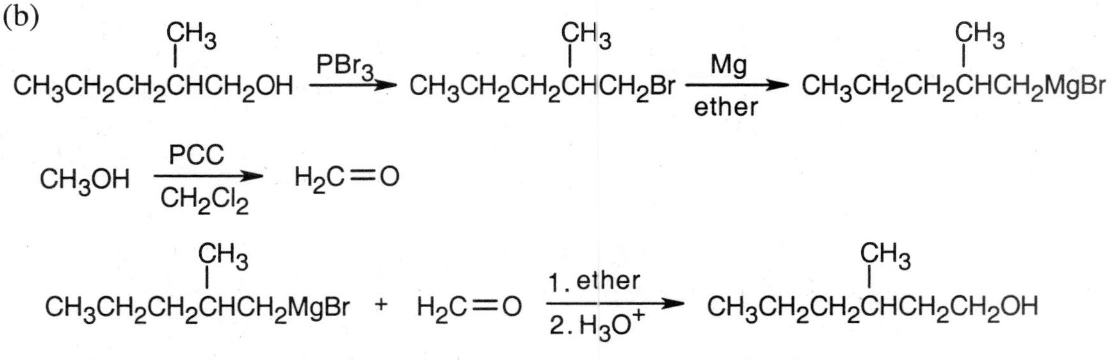

$$CH_3CH_2CH_2CH(CH_3)CH_2OH \xrightarrow{PBr_3} CH_3CH_2CH_2CH(CH_3)CH_2Br \xrightarrow[\text{ether}]{Mg} CH_3CH_2CH_2CH(CH_3)CH_2MgBr$$

$$CH_3OH \xrightarrow[CH_2Cl_2]{PCC} H_2C{=}O$$

$$CH_3CH_2CH_2CH(CH_3)CH_2MgBr + H_2C{=}O \xrightarrow[2.\ H_3O^+]{1.\ \text{ether}} CH_3CH_2CH_2CH(CH_3)CH_2CH_2OH$$

(c)

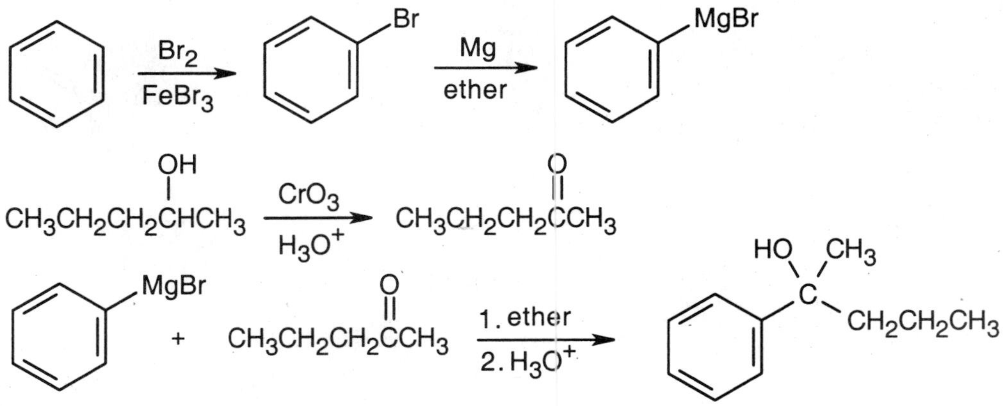

$$\text{Benzene} \xrightarrow[FeBr_3]{Br_2} C_6H_5Br \xrightarrow[\text{ether}]{Mg} C_6H_5MgBr$$

$$CH_3CH_2CH_2CH(OH)CH_3 \xrightarrow[H_3O^+]{CrO_3} CH_3CH_2CH_2C(O)CH_3$$

$$C_6H_5MgBr + CH_3CH_2CH_2C(O)CH_3 \xrightarrow[2.\ H_3O^+]{1.\ \text{ether}} C_6H_5C(OH)(CH_3)CH_2CH_2CH_3$$

(d)

$$CH_3CH(CH_3)CH_2CH_2OH \xrightarrow[CH_2Cl_2]{PCC} CH_3CH(CH_3)CH_2CHO$$

$$CH_3CH(CH_3)CH_2CHO + \underset{\text{from (a)}}{CH_3CH_2MgBr} \xrightarrow[2.\ H_3O^+]{1.\ \text{ether}} CH_3CH(CH_3)CH_2CH(OH)CH_2CH_3$$

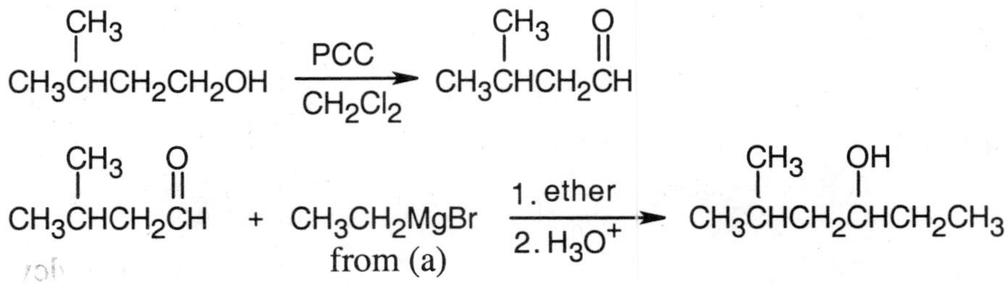

17.50 The pinacol rearrangement follows a sequence of steps similar to other rearrangements we have studied in this chapter. The second hydroxyl group assists in the alkyl shift.

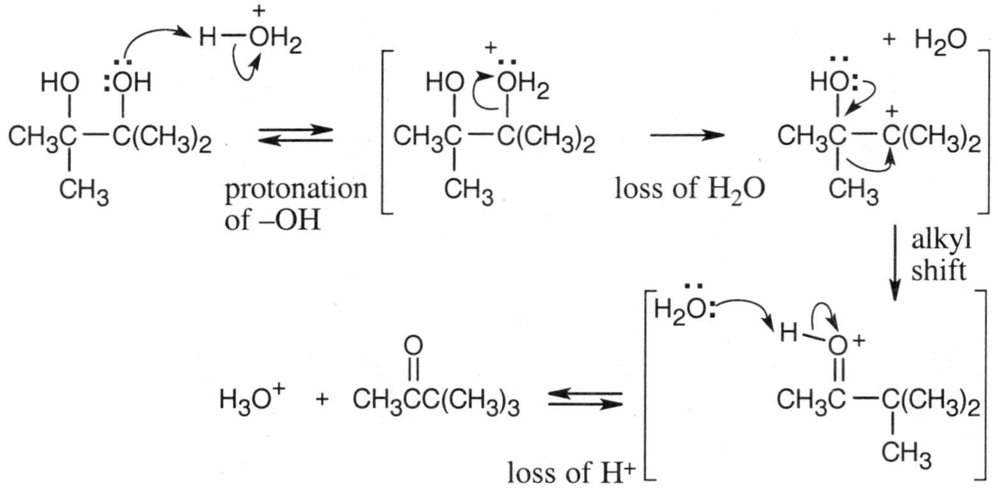

HO :OH, $CH_3C-C(CH_3)_2$, CH_3 ⇌ (protonation of –OH; $H-\overset{+}{O}H_2$) [HO $\overset{+}{O}H_2$, $CH_3C-C(CH_3)_2$, CH_3] → (loss of H_2O) [HO:, $CH_3C-\overset{+}{C}(CH_3)_2$, CH_3 + H_2O] ↓ (alkyl shift) [H_2O: $H-\overset{+}{O}$, $CH_3C-C(CH_3)_2$, CH_3] ⇌ (loss of H^+) H_3O^+ + $CH_3CC(CH_3)_3$ (C=O)

17.51 The hydroxyl group is axial in the cis isomer, which is expected to oxidize faster than the trans isomer. (Remember that the bulky *tert*-butyl group is always equatorial in the more stable isomer.)

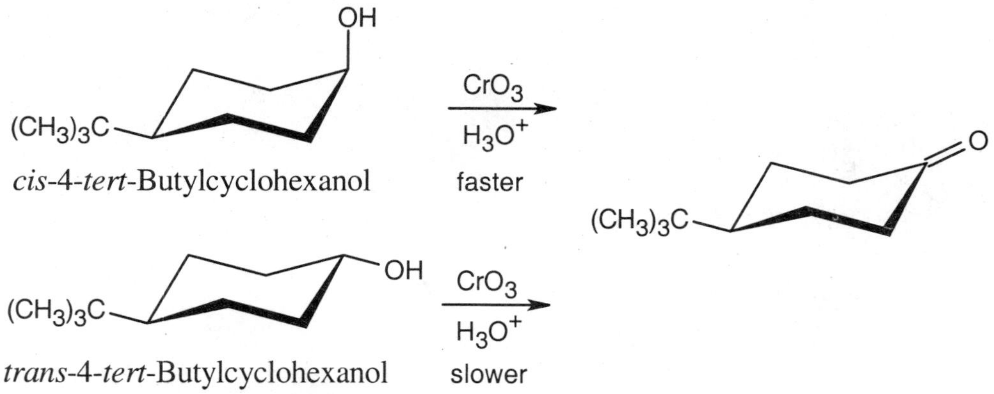

cis-4-*tert*-Butylcyclohexanol $\xrightarrow[H_3O^+]{CrO_3}$ (faster) 4-*tert*-butylcyclohexanone

trans-4-*tert*-Butylcyclohexanol $\xrightarrow[H_3O^+]{CrO_3}$ (slower) 4-*tert*-butylcyclohexanone

17.52

Cyclohexanone $\xrightarrow[2.\ H_3O^+]{1.\ LiAlH_4}$ cyclohexanol (OH) $\xrightarrow{PBr_3}$ bromocyclohexane (Br) $\xrightarrow{Mg,\ ether}$ cyclohexylmagnesium bromide (MgBr) $\xrightarrow[2.\ H_3O^+]{1.\ Cyclohexanone}$ 1-cyclohexylcyclohexanol (OH) $\xrightarrow{H_2SO_4}$ Bicyclohexylidene

Bicyclohexylidene

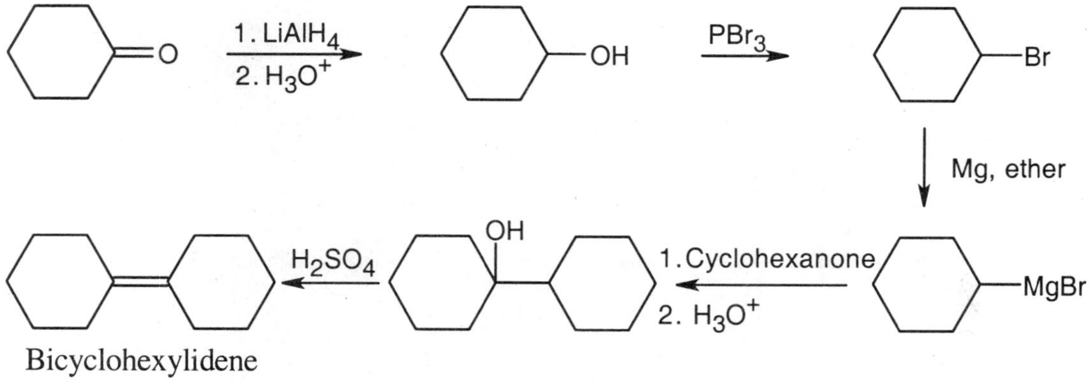

17.53 An alcohol adds to an aldehyde by a mechanism that we will study in a later chapter. The hydroxyl group of the addition intermediate undergoes oxidation (as shown in Section 17.8), and an ester is formed.

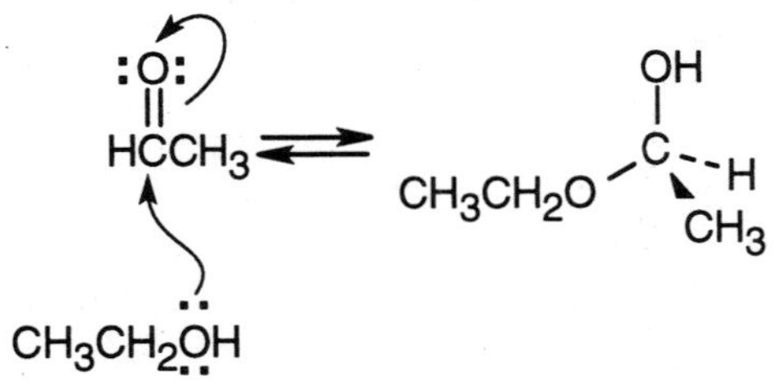

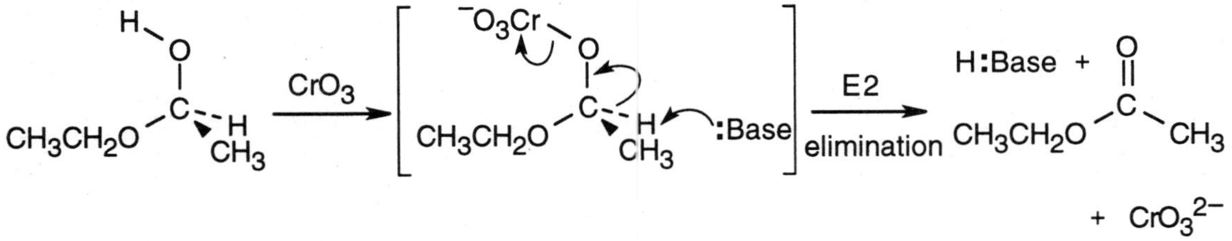

:Ö: ‖ HCCH$_3$ + CH$_3$CH$_2$ÖH ⇌ CH$_3$CH$_2$O–C(OH)(H)CH$_3$

CH$_3$CH$_2$O–C(OH)(H)CH$_3$ —CrO_3→ [CH$_3$CH$_2$O–C(OCrO$_3^-$)(H)CH$_3$:Base] —E2 elimination→ H:Base + CH$_3$CH$_2$O–C(=O)–CH$_3$ + CrO_3^{2-}

17.54

Cyclohexanone —a→ cyclohexanol —b→ bromocyclohexane —c→ cyclohexylmethanol (CH$_2$OH) —d→ cyclohexanecarbaldehyde (CHO) —e→ 1-cyclohexyl-2-phenylethanol (OH) —f→ 1-cyclohexyl-2-phenylethene

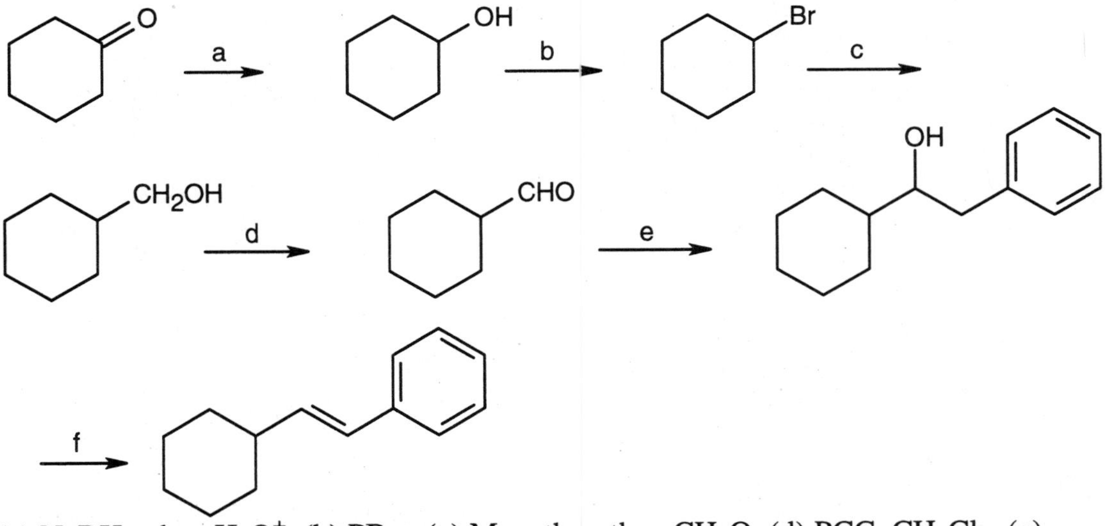

(a) NaBH$_4$, then H_3O^+ (b) PBr$_3$ (c) Mg, ether, then CH$_2$O (d) PCC, CH$_2$Cl$_2$ (e) C$_6$H$_5$CH$_2$MgBr, then H_3O^+ (f) POCl$_3$, pyridine

17.55
1. $C_8H_{18}O_2$ has *no* double bonds or rings.
2. The IR band at 3350 cm^{-1} shows the presence of a hydroxyl group.
3. The compound is symmetrical (simple NMR).
4. There is no splitting.

A structure that meets all these criteria:

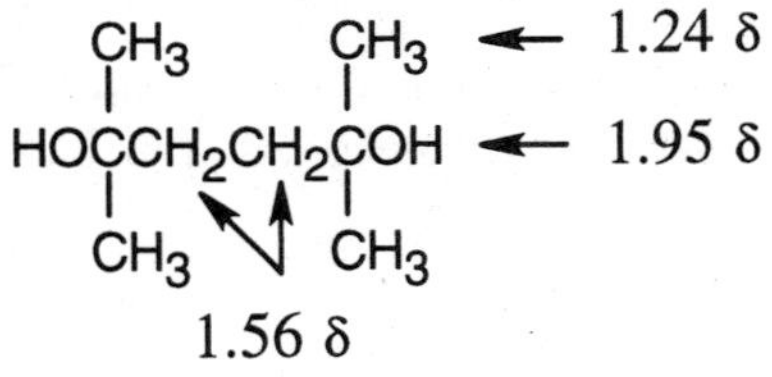

HOC(CH$_3$)$_2$CH$_2$CH$_2$C(CH$_3$)$_2$OH

CH$_3$ ← 1.24 δ; OH ← 1.95 δ; CH$_2$CH$_2$ ← 1.56 δ

2,5-Dimethyl-2,5-hexanediol

17.56

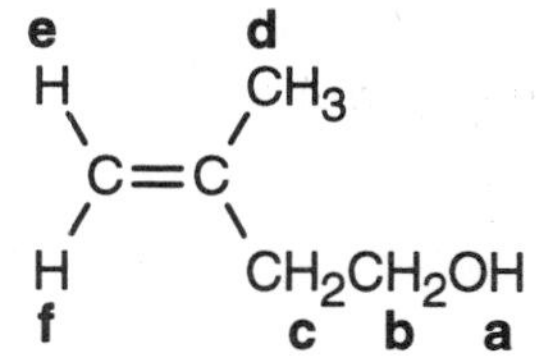

```
 e         d
 H        CH3
  \       /
   C=C
  /       \
 H        CH2CH2OH
 f         c  b  a
```

3-Methyl-3-buten-3-ol

The peak absorbing at 1.75 δ (3 H) is due to the **d** protons. This peak, which occurs in the allylic region of the spectrum, is unsplit.

The peak absorbing at 2.13 δ (1 H) is due to the –OH proton **a**.

The peak absorbing at 2.30 δ (2 H) is due to protons **c**. The peak is a triplet because of splitting by the adjacent b protons.

The peak absorbing at 3.70 δ (2 H) is due to the **b** protons. The adjacent oxygen causes the peak to be downfield, and the adjacent $–CH_2–$ group splits the peak into a triplet.

The peaks at 4.78 δ and 4.85 δ (2 H) are due to protons **e** and **f**.

17.57 (a) Compound **A** has one double bond or ring.

(b) The infrared absorption at 3400 cm^{-1} indicates the presence of an alcohol. (The weak absorption at 1640 cm^{-1} is due to a C=C stretch.)

(c) (1) The absorptions at 1.63 δ and 1.70 δ are due to unsplit methyl protons. Because the absorptions are shifted slightly downfield, the protons are adjacent to an unsaturated center.

(2) The broad singlet at 3.83 δ is due to an alcohol proton.

(3) The doublet at 4.15 δ is due to two protons bonded to a carbon bearing an electronegative atom (oxygen, in this case).

(4) The proton absorbing at 5.70 δ is a vinylic proton.

(d)

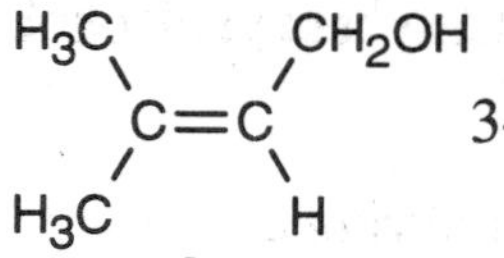

```
H3C        CH2OH
   \       /
    C=C
   /       \
H3C         H
     A
```

3-Methyl-2-buten-1-ol

17.58 (a) $C_5H_{12}O$, $C_4H_8O_2$, $C_3H_4O_3$

(b) The ^1H NMR data show that the compound has twelve protons.

(c) The IR absorption at 3600 cm^{-1} shows that the compound is an alcohol.

(d) The compound contains five carbons, two of which are identical.

(e) $C_5H_{12}O$ is the molecular formula of the compound.

(f)

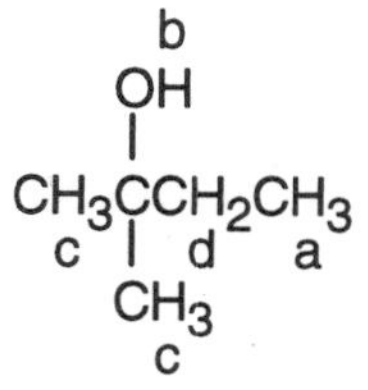

```
       b
      OH
       |
  CH3CCH2CH3
   c | d   a
     CH3
      c
```

a = 0.9 δ

b = 1.0 δ

c = 1.2 δ

d = 1.4 δ

2-Methyl-2-butanol

17.59

2.3 δ → H_3C–(C₆H₄)–CH_2–OH ← 2.5 δ

ring: 7.1 δ; CH_2: 4.5 δ

p-Methylbenzyl alcohol

17.60

(a)

$CH_3CH_2CH(OH)CH_2CH_3$ (a b c b a; OH = d)

3-Pentanol

a = 0.9 δ
b = 1.5 δ
c = 1.9 δ
d = 3.4 δ

(b)

C_6H_5–$CH(OH)CH_3$ (OH = b; CH = c; CH_3 = a; ring = d)

1-Phenylethanol

a = 1.5 δ
b = 2.4 δ
c = 4.8 δ
d = 7.3 δ

17.61

(a)

C_6H_5–$CH(OH)CH_2CH_3$ (OH = c; CH = d; CH_2 = b; CH_3 = a; ring = e)

1-Phenyl-1-propanol

a = 0.9 δ
b = 1.8 δ
c = 2.3 δ
d = 4.5 δ
e = 7.3 δ

(b)

CH_3O–(C₆H₄)–CH_2OH (CH_3O = b; CH_2 = c; OH = a; ring = d)

p-Methoxybenzyl alcohol

a = 2.6 δ
b = 3.8 δ
c = 4.5 δ
d = 7.0 δ

17.62

Structural formula: $C_8H_{10}O$ contains 4 multiple bonds and/or rings.

Infrared: The broad band at 3500 cm^{-1} indicates a hydroxyl group. The absorptions at 1500 cm^{-1} and 1600 cm^{-1} are due to an aromatic ring. The absorption at 830 cm^{-1} shows that the ring is disubstituted. Compound **A** is probably a phenol.

^1H NMR: The triplet at 1.18 δ (3 H) is coupled with the quartet at 2.56 δ (2 H). These two absorptions are due to an ethyl group.

The peaks at 6.75 δ-7.05 δ (4 H) are due to an aromatic ring. The symmetrical splitting pattern of these peaks indicate that the aromatic ring is *p*–disubstituted.

The absorption at 5.50 δ (1 H) is due to an –OH proton.

Compound **A**

HO–(C₆H₄)–CH_2CH_3

p-Ethylphenol

17.63

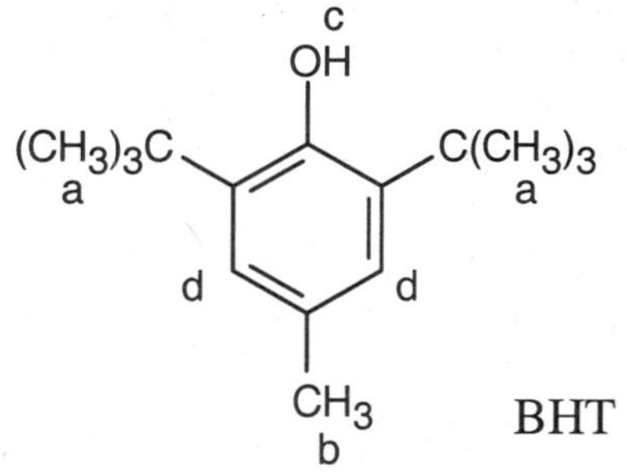

(BHT structure: 2,6-di-tert-butyl-4-methylphenol with labels a = $(CH_3)_3C$, b = CH_3, c = OH, d = ring H)

BHT

a = 1.4 δ
b = 2.2 δ
c = 5.0 δ
d = 7.0 δ

A Look Ahead

17.64

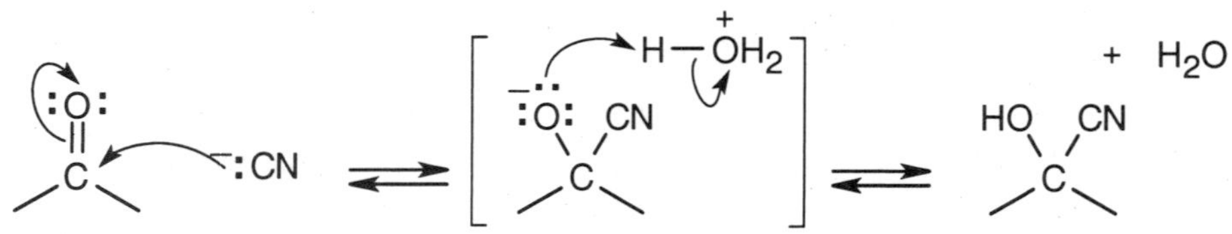

The nucleophile ^-CN adds to the positively polar-ized carbonyl carbon.

The tetrahedral intermediate is protonated to give the addition product.

17.65

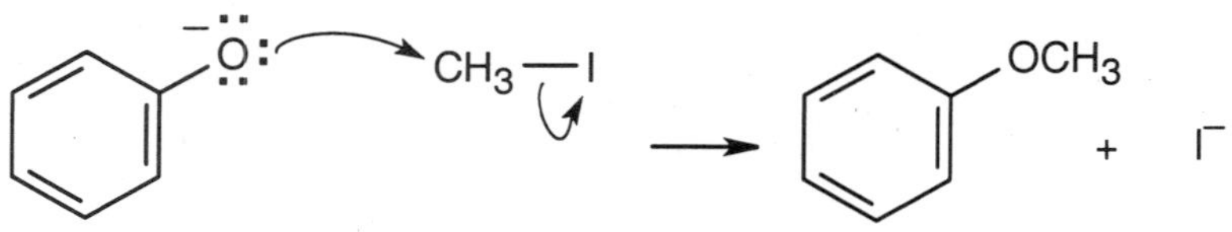

$C_6H_5O^- + CH_3{-}I \longrightarrow C_6H_5OCH_3 + I^-$

The reaction is an S_N2 displacement of iodide by phenoxide ion.

Molecular Modeling

17.66 Butane and 1,2-dimethoxyethane prefer the anti conformation by 3.2 kJ/mol and 1.3 kJ/mol, respectively. 1,2-Ethanediol prefers the gauche conformation by 11 kJ/mol. Electrostatic potential maps of 1,2-ethanediol's two conformations show that the gauche conformation is stabilized by an intramolecular hydrogen bond between the two hydroxyl groups.

17.67 *Most negative oxygen* $PhO^- > NCCH_2PhO^- > NCPhO^-$ *Least negative oxygen*

4-Cyanophenol is the most acidic phenol because the negative charge of the conjugate base is not only delocalized onto the ring, it is also delocalized onto the cyano substituent.

17.68 The O–H bond distances are 96.7 pm in *tert*-butyl alcohol and 96.8 and 97.6 pm in the dimer. Calculated O–H stretching frequencies are 3853 cm^{-1} in *tert*-butyl alcohol and 3861 cm^{-1} and 3681 cm^{-1} in the dimer. Dimerization makes the hydrogen-bonded O–H distance slightly longer and makes the O–H stretching frequency lower. Dimerization doesn't affect the other O–H bond distance or stretching frequency.

17.69 The C–O bond distance is longer in the intermediate (148.0 pm vs. 144.1 pm), indicating a weaker C–O bond. The electrostatic potential map shows that the $–CH_3$ group is much more positive (electrophilic) in the intermediate. These changes and the fact that oxygen is much less negative in the intermediate than in CH_3OH suggest that –SOCl is an electron-withdrawing group.

Chapter 18 – Ethers and Epoxides; Thiols and Sulfides

Chapter Outline

I. Acyclic ethers (Sections 18.1 – 18.6)
- A. Naming ethers (Section 18.1).
  1. Ethers with no other functional groups are named by citing the two organic substituents and adding the word "ether".
  2. When other functional groups are present, the ether is an alkoxy substituent.
- B. Properties of ethers (Section 18.2).
  1. Ethers have the same geometry as water and alcohols.
  2. Ethers have a small dipole moment that causes a slight boiling point elevation.
- C. Preparation of ethers (Sections 18.3 – 18.4).
  1. Symmetrical ethers can be synthesized by acid-catalyzed dehydration of alcohols.
  2. Williamson ether synthesis (Section 18.3).
     - a. Metal alkoxides react with primary alkyl halides and tosylates to form ethers.
     - b. The alkoxides are prepared by reacting an alcohol with a strong base, such as NaH.
       Reaction of the free alcohol with the halide can be achieved with Ag_2O.
     - c. The reaction occurs via an S_N2 mechanism.
       - i. The halide component must be primary.
       - ii. In cases where one ether component is hindered, reaction should occur between the alkoxide of the more hindered reagent and the halide of the less hindered reagent.
  3. Alkoxymercuration of alkenes (Section 18.4).
     - a. Ethers can be formed from the reaction of alcohols with alkenes.
     - b. The reaction is catalyzed by mercuric trifluoroacetate.
     - c. The mechanism is similar to that for hydration of alkenes.
       $NaBH_4$ is used for demercuration of the intermediate.
     - d. Many different types of ethers can be prepared by this method.
- D. Reactions of ethers (Sections 18.5 – 18.6).
  1. Acidic cleavage (Section 18.5).
     - a. Strong acids can be used to cleave ethers.
     - b. Cleavage can occur by S_N2 or S_N1 routes.
       - i. Primary and secondary alcohols react by an S_N2 mechanism, in which the halide attacks the ether at the less hindered site.
         This route selectively produces one halide and one alcohol.
       - ii. Tertiary, benzylic and allylic ethers react by either an S_N1 or an E1 route.
  2. Claisen rearrangement (Section 18.6).
     - a. The Claisen rearrangement is specific to allyl aryl ethers.
     - b. The result of Claisen rearrangement is an *o*-allyl phenol.
     - c. The reaction takes place by a pericyclic mechanism.
       Inversion of the allyl group is evidence for this mechanism.

II. Cyclic ethers (Sections 18.7 – 18.9).
- A. Epoxides (Sections 18.7 – 18.8).
  1. The three-membered ring of epoxides gives them unique chemical reactivity (Section 18.7).
  2. The nonsystematic name *–ene oxide* describes the method of formation.
  3. The systematic prefix *epoxy-* describes the location of the epoxide ring.

4. Preparation of epoxides.
   a. Epoxides can be prepared by reaction of an alkene with a peroxyacid RCO_3H.
      The reaction occurs in one step with syn stereochemistry.
   b. Epoxides are formed when halohydrins are treated with base.
      This reaction is an intramolecular Williamson ether synthesis.
5. Ring-opening reactions of epoxides (Section 18.8).
   a. Acid-catalyzed ring opening.
      i. Acid-catalyzed ring opening produces 1,2 diols.
      ii. Ring opening takes place by back-side attack of a nucleophile on the protonated epoxide ring.
         A *trans*-1,2-diol is formed from an epoxycycloalkane.
      iii. When both epoxide carbons are primary or secondary, attack occurs primarily at the less hindered site.
      iv. When one epoxide carbon is tertiary, attack occurs at the more highly substituted site.
      v. The mechanism is midway between S_N2 and S_N1 routes.
         The reaction occurs by back-side attack (S_N2), but positive charge is stabilized by a tertiary carbocation-like transition state (S_N1).
   b. Base-catalyzed ring-opening.
      i. Base-catalyzed ring opening occurs because of the reactivity of the strained epoxide ring.
      ii. Ring-opening also occurs when epoxides react with Grignard reagents, forming a product with two more carbons than the starting alkyl halide.
      iii. Ring-opening takes place by an S_N2 mechanism, in which the nucleophile attacks the less hindered epoxide carbon.

B. Crown ethers (Section 18.9).
   1. Crown ethers are large cyclic ethers.
   2. Crown ethers are named as *x*-crown-*y*, where x = the ring size and y = # of oxygens.
   3. Crown ethers are able to solvate metal cations.
      a. Different sized crown ethers solvate different cations.
      b. Complexes of crown ethers with ionic salts are soluble in organic solvents.
      c. This solubility allows many reactions to be carried out under aprotic conditions.
      d. The reactivity of many anions in S_N2 reactions is enhanced by crown ethers.

III. Spectroscopy of ethers (Section 18.10).
   A. IR spectroscopy.
      1. Ethers are difficult to identify in IR spectra because many other absorptions occur at 1050–1150 cm^{-1}, where ethers absorb.
   B. NMR spectroscopy.
      1. ^1H NMR spectroscopy.
         a. Hydrogens on a carbon next to an ether oxygen absorb downfield (3.4–4.5 δ).
         b. Hydrogens on a carbon next to an epoxide oxygen absorb at a slightly higher field (2.5–3.5 δ).
      2. ^{13}C NMR spectroscopy.
         Ether carbons absorb downfield (50–80 δ).

IV. Thiols and sulfides (Section 18.11).
   A. Naming thiols and sulfides.
      1. Thiols (sulfur analogs of alcohols) are named by the same system as alcohols, with the suffix *-thiol* replacing *-ol.*
         The –SH group is a mercapto- group.
      2. Sulfides (sulfur analogs of ethers) are named by the same system as ethers, with *sulfide* replacing *ether*.
         The –SR group is an alkylthio- group.

B. Thiols.
1. Thiols stink!
2. Thiols may be prepared by S_N2 displacement with a sulfur nucleophile.
   a. The reaction may proceed to form sulfides.
   b. Better yields occur when thiourea is used.
3. Thiols can be oxidized by Br_2 or I_2 to yield disulfides, RSSR.
   The reaction can be reversed by treatment with zinc and acid.

C. Sulfides.
1. Treatment of a thiol with base yields a thiolate anion, which can react with an alkyl halide to form a sulfide.
2. Thiolate anions are excellent nucleophiles.
3. Dialkyl sulfides can react with alkyl halides to form trialkylsulfonium salts, which are also good alkylating agents.
   Many biochemical reactions use trialkylsulfonium groups as alkylating agents.
4. Sulfides are easily oxidized to sulfoxides (R_2SO) and sulfones (R_2SO_2).
   Dimethyl sulfoxide is used as a polar aprotic solvent.

Solutions to Problems

18.1 Ethers can be named either as alkoxy-substituted compounds or by citing the two groups bonded to oxygen, followed by the word "ether".

(a) $CH_3CH(CH_3)OCH(CH_3)CH_3$

2-Isopropoxypropane *or* Diisopropyl ether

(b) Cyclopentyl–$OCH_2CH_2CH_3$

Propoxycyclopentane *or* Cyclopentyl propyl ether

(c) Benzene ring with OCH_3 and Br (para)

p-Bromoanisole *or* *p*-Bromomethoxybenzene

(d) Cyclohexene ring with OCH_3

1-Methoxycyclohexene

(e) $CH_3CH(CH_3)CH_2OCH_2CH_3$

1-Ethoxy-2-methylpropane *or* Ethyl isobutyl ether

(f) $H_2C{=}CHCH_2OCH{=}CH_2$

Allyl vinyl ether

18.2 The first step of the dehydration procedure is protonation of an alcohol. Water is then displaced by another molecule of alcohol to form an ether. If two different alcohols are present, either one can be protonated and either one can displace water, yielding a mixture of products.

If this procedure were used with ethanol and 1–propanol, the products would be diethyl ether, ethyl propyl ether, and dipropyl ether. If there were equimolar amounts of the alcohols, and if they were of equal reactivity, the product ratio would be diethyl ether : ethyl propyl ether : dipropyl ether = 1:2:1.

18.3 Remember that the halide in the Williamson ether synthesis should be primary or methyl, in order to avoid competing elimination reactions. The alkoxide anions shown are formed by treating the corresponding alcohols with NaH.

(a)

$$\left.\begin{array}{l} CH_3CH_2CH_2O^- + CH_3Br \\ \quad\quad \textit{or} \\ CH_3CH_2CH_2Br + CH_3O^- \end{array}\right\} \longrightarrow CH_3CH_2CH_2OCH_3 + Br^-$$

Methyl propyl ether

(b)

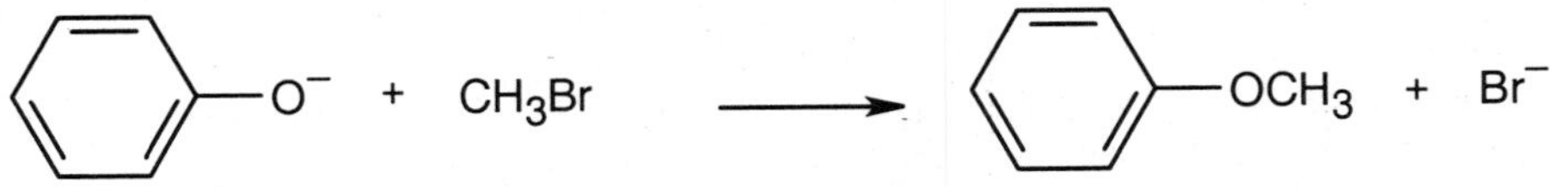

$$C_6H_5O^- + CH_3Br \longrightarrow C_6H_5OCH_3 + Br^-$$

Methyl phenyl ether
(Anisole)

(c)

$$(CH_3)_2CHO^- + C_6H_5CH_2Br \longrightarrow (CH_3)_2CHOCH_2C_6H_5 + Br^-$$

Benzyl isopropyl ether

(d)

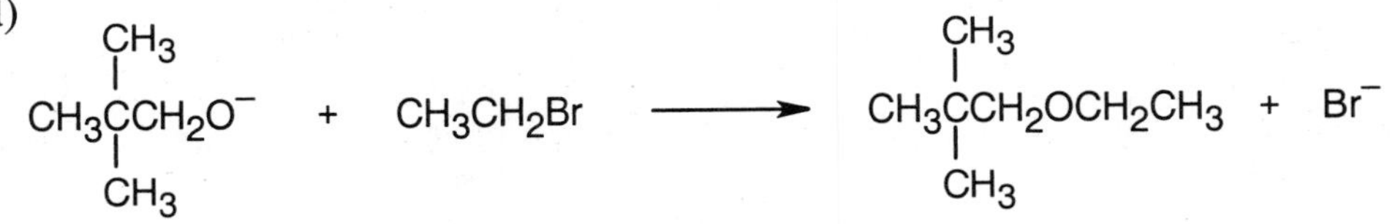

$$(CH_3)_3CCH_2O^- + CH_3CH_2Br \longrightarrow (CH_3)_3CCH_2OCH_2CH_3 + Br^-$$

Ethyl 2,2-dimethylpropyl ether

18.4 The compounds most reactive in the Williamson ether synthesis are also most reactive in any S_N2 process (review Chapter 11 if necessary).

Most reactive ⟶ *Least reactive*

(a)

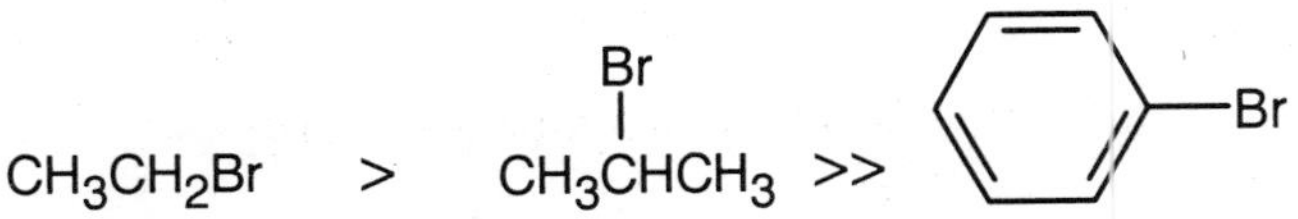

$$CH_3CH_2Br \; > \; CH_3CH(Br)CH_3 \; >> \; C_6H_5Br$$

primary halide	secondary halide	aryl halide (not reactive)

(b)

$$CH_3CH_2Br \; > \; CH_3CH_2Cl \; >> \; CH_3CH{=}CHI$$

better leaving group	poorer leaving group	vinylic (not reactive)

18.5

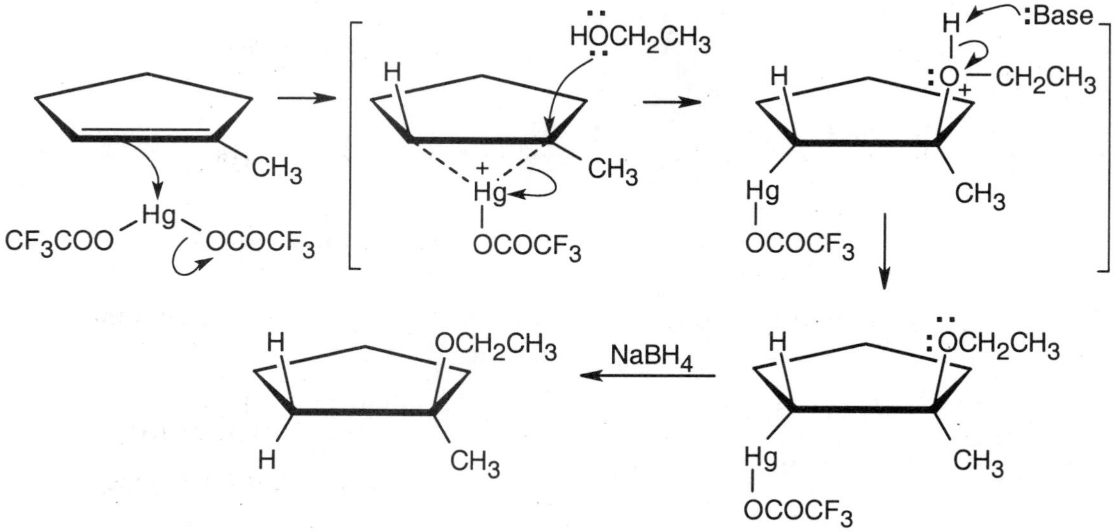

CF$_3$COO–Hg–OCOCF$_3$ → [H, Hg$^+$, OCOCF$_3$, CH$_3$, HÖCH$_2$CH$_3$ → H, Hg, OCOCF$_3$, H–Base, :O$^+$–CH$_2$CH$_3$, CH$_3$] → H, Hg, OCOCF$_3$, :ÖCH$_2$CH$_3$, CH$_3$ —NaBH$_4$→ H, H, OCH$_2$CH$_3$, CH$_3$

The reaction mechanism of alkoxymercuration/demercuration of an alkene is similar to other electrophilic additions we have studied. First, the cyclopentene π electrons attack Hg^{2+} with formation of a mercurinium ion. Next, the nucleophilic alcohol displaces mercury. Markovnikov addition occurs because the carbon bearing the methyl group is better able to stabilize the partial positive charge arising from cleavage of the carbon-mercury bond. The ethoxyl and the mercuric groups are trans to each other. Finally, removal of mercury by $NaBH_4$ by a mechanism that is not fully understood results in the formation of 1-ethoxy-1-methylcyclopentane.

18.6 (a) Either method of synthesis is appropriate.
Williamson:

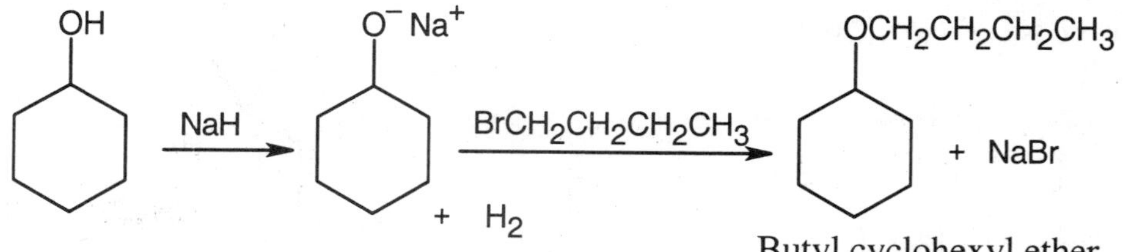

Cyclohexanol (OH) —NaH→ cyclohexyl O^-Na^+ + H_2 —$BrCH_2CH_2CH_2CH_3$→ cyclohexyl $OCH_2CH_2CH_2CH_3$ + NaBr

Butyl cyclohexyl ether

Alkoxymercuration:

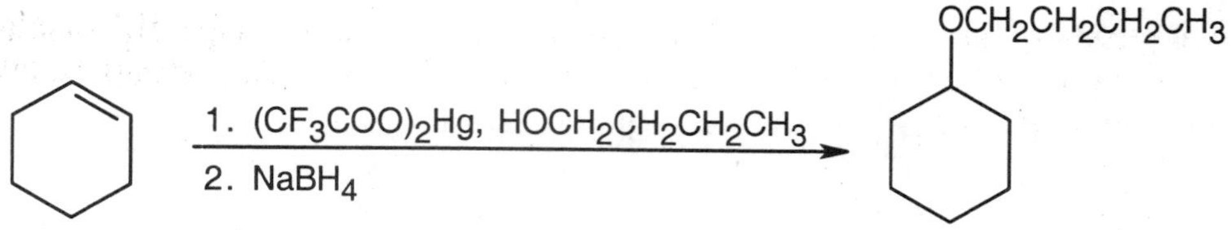

Cyclohexene —1. $(CF_3COO)_2Hg$, $HOCH_2CH_2CH_2CH_3$; 2. $NaBH_4$→ cyclohexyl $OCH_2CH_2CH_2CH_3$

(b) Either method is possible, but the Williamson synthesis is simpler.

$$C_6H_5CH_2OH \xrightarrow{NaH} C_6H_5CH_2O^-Na^+ + H_2 \xrightarrow{CH_3CH_2Br} C_6H_5CH_2OCH_2CH_3 + NaBr$$

Benzyl ethyl ether

(c) Because both parts of the ether have branching, use alkoxymercuration.

$$CH_3C(CH_3){=}CH_2 \xrightarrow[2.\ NaBH_4]{1.\ (CF_3COO)_2Hg,\ CH_3CH(OH)CH_2CH_3} (CH_3)_3COCH(CH_3)CH_2CH_3$$

sec-Butyl-*tert*-butyl ether

(d) The Williamson synthesis must be used.

$$BrCH_2CH_2CH_2CH_2OH \xrightarrow{NaH} BrCH_2CH_2CH_2CH_2O^-Na^+ + H_2 \longrightarrow \text{tetrahydrofuran} + NaBr$$

Tetrahydrofuran

18.7 (a) First, notice the substitution pattern of the ether. Bonded to the ether oxygen are a primary alkyl group and a tertiary alkyl group. When one group is tertiary, cleavage occurs by an S_N1 or E1 route to give either an alkene or a tertiary halide and a primary alcohol.

$$C_6H_5C(CH_3)_2{-}O{-}CH_3 \xrightarrow{HBr} C_6H_5C(CH_3){=}CH_2 + C_6H_5C(CH_3)_2Br + CH_3OH$$

tertiary primary

(b) In this problem, the groups are primary and secondary alkyl groups. Br^- displaces the less hindered primary group, and oxygen remains with the secondary group, to give a secondary alcohol.

$$CH_3CH_2CH(CH_3){-}O{-}CH_2CH_2CH_3 \xrightarrow{HBr} CH_3CH_2CH(CH_3)OH + Br{-}CH_2CH_2CH_3$$

secondary primary

18.8

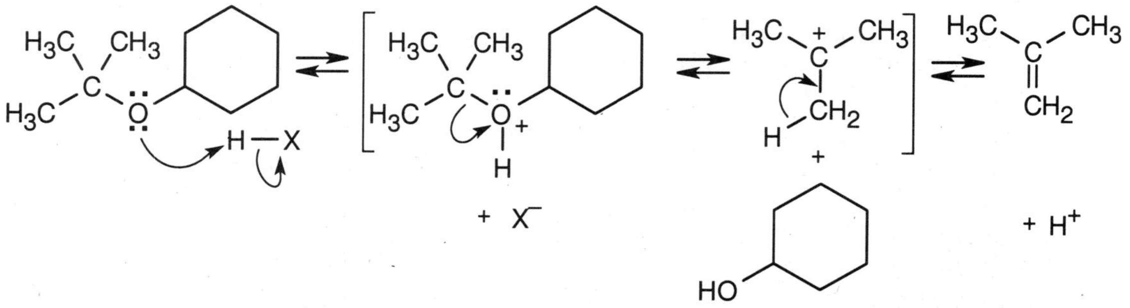

The first step of acid-catalyzed ether cleavage is protonation of the ether oxygen to give an intermediate, which dissociates to form an alcohol and a tertiary carbocation. The carbocation then loses a proton to form an alkene, 2-methylpropene. This is an example of E_1 elimination. The acid used is often trifluoroacetic acid.

18.9

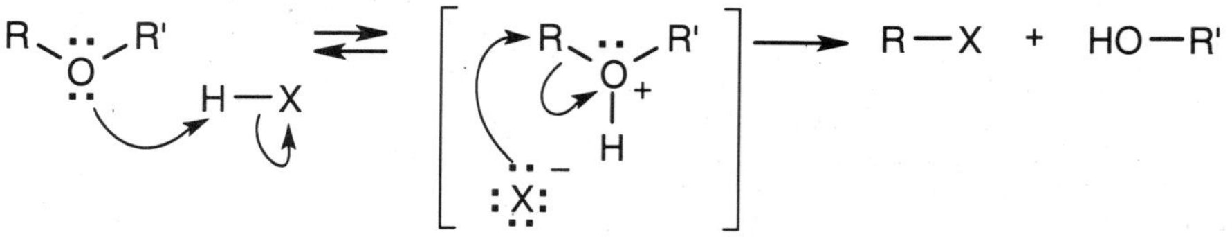

HX first protonates the oxygen atom, and halide then effects nucleophilic displacement to form an alcohol and an organic halide. The better the nucleophile, the more effective the displacement. Since I^- and Br^- are more nucleophilic than Cl^-, ether cleavage proceeds more smoothly with HI or HBr than with HCl.

18.10 Draw the ether with the groups involved in the rearrangement positioned as they will appear in the product. Six bonds will either be broken or formed in the product ; they are shown as dashed lines in the transition state. Redraw the bonds to arrive at the intermediate enone, which rearranges to the more stable phenol.

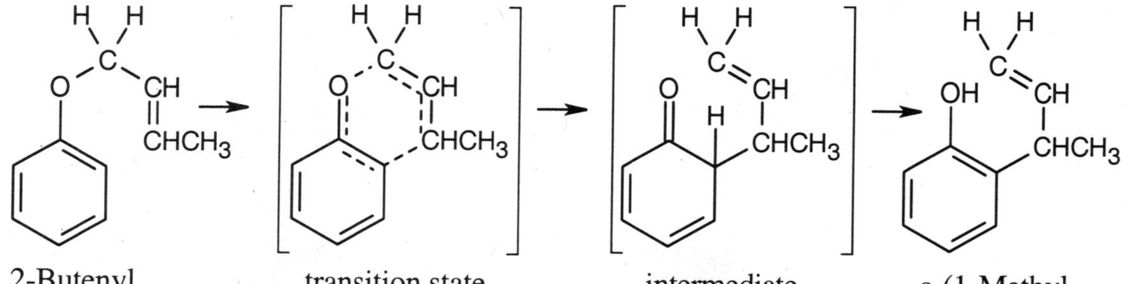

2-Butenyl phenyl ether — transition state — intermediate — *o*-(1-Methylallyl) phenol

18.11 Epoxidation by use of *m*-chloroperoxybenzoic acid (RCO_3H) is a syn addition of oxygen to a double bond. The original bond stereochemistry is retained.

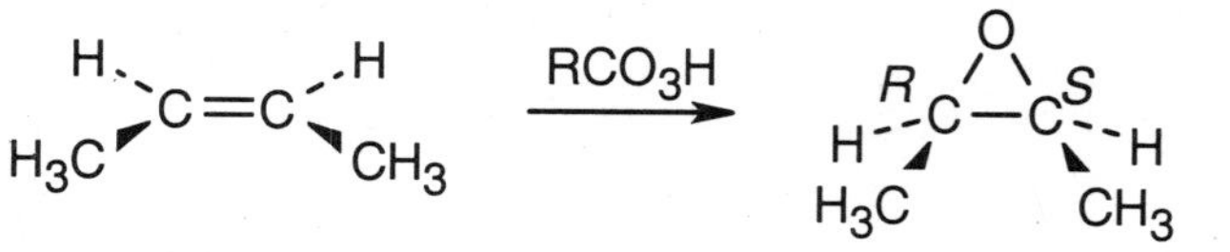

cis-2-Butene $\xrightarrow{RCO_3H}$ *cis*-2,3-Epoxybutane

In the epoxide product, as in the alkene starting material, the methyl groups are cis.

18.12

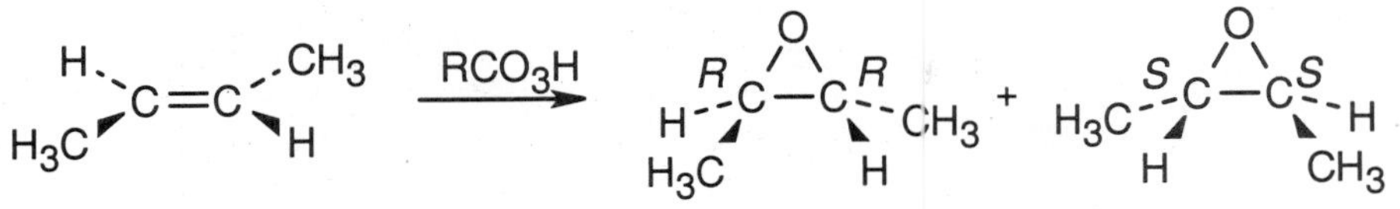

H, CH_3, C=C, H_3C, H —RCO_3H→ O, R, R, H–C–C–CH_3, H_3C, H + O, S, S, H_3C–C–C–H, H, CH_3

trans-2-Butene *trans*-2,3-Epoxybutane

The argument in the previous problem can be used to show that reaction of *trans*–2–butene with *m*-chloroperoxybenzoic acid yields *trans*-2,3-epoxybutane. A mixture of enantiomers is formed because the peroxyacid can attack either the top or bottom of the double bond.

18.13 As discussed in this section, epoxide ring opening occurs more often at the more hindered carbon when one of the epoxide carbons is tertiary. In both parts of this problem, one epoxide carbon is tertiary.

(a)

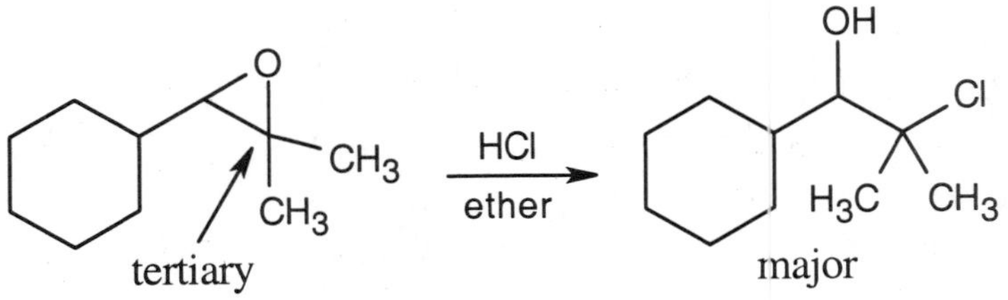

O, CH_3, CH_3, tertiary —HCl, ether→ OH, Cl, H_3C, CH_3, major

(b)

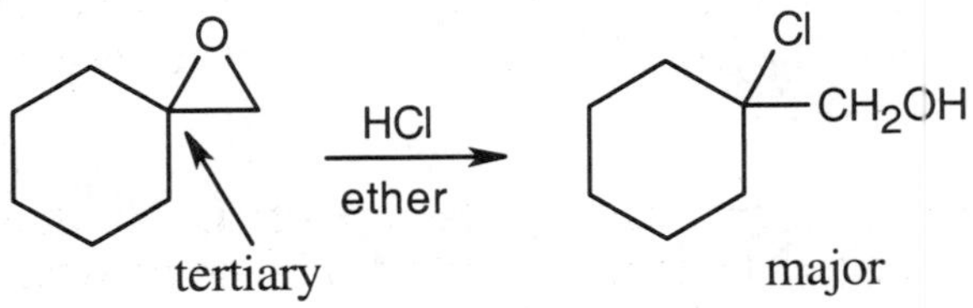

O, tertiary —HCl, ether→ Cl, CH_2OH, major

18.14

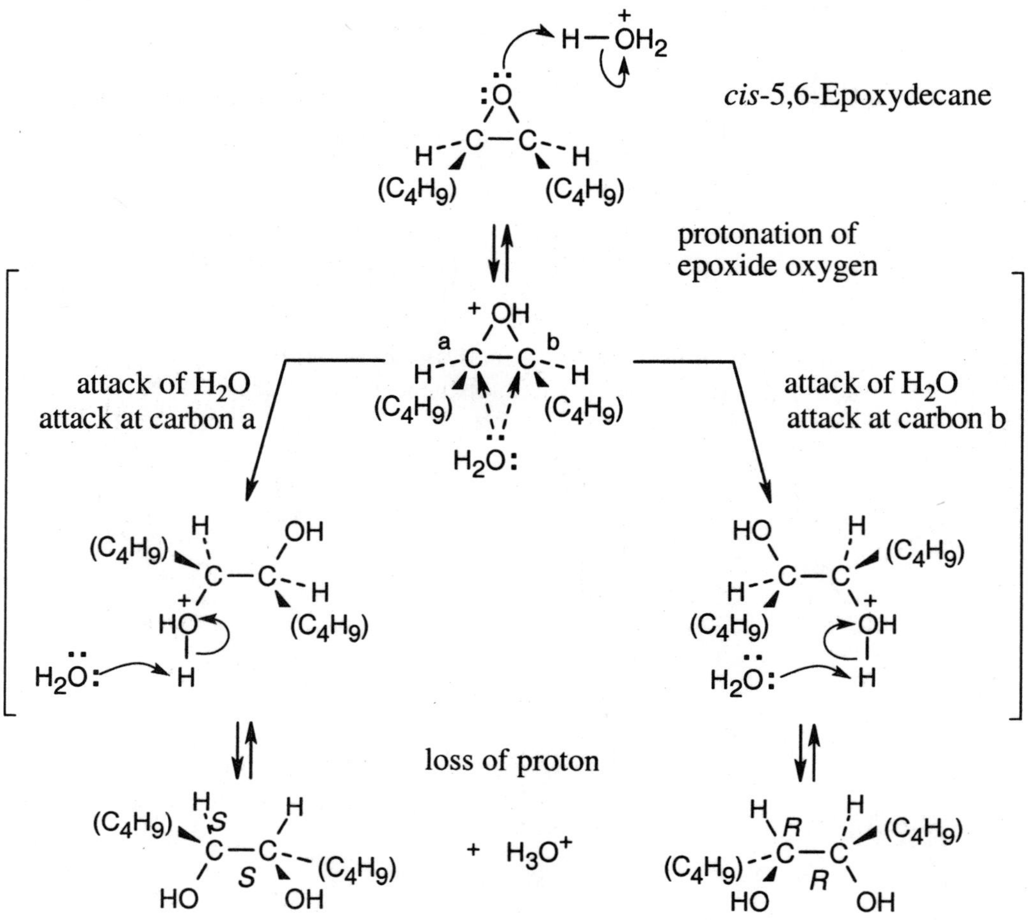

cis-5,6-Epoxydecane

protonation of epoxide oxygen

attack of H_2O
attack at carbon a

attack of H_2O
attack at carbon b

loss of proton

$+ H_3O^+$

The product of acid hydrolysis of *cis*-5,6-epoxydecane is a racemic mixture.

18.15

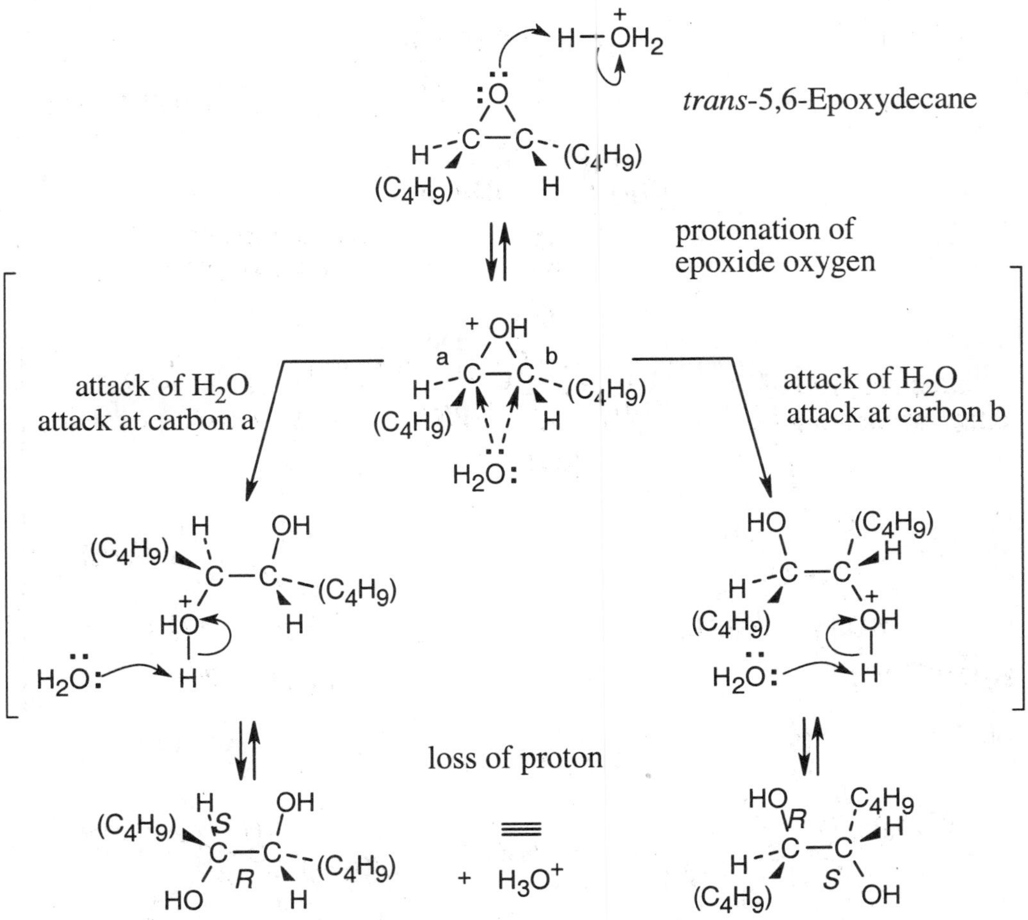

trans-5,6-Epoxydecane

protonation of epoxide oxygen

attack of H_2O attack at carbon a

attack of H_2O attack at carbon b

loss of proton

$+ H_3O^+$

The product of acid hydrolysis of *trans*-5,6-epoxydecane is a meso compound that is a diastereomer of the one formed in the previous problem.

18.16 (a) Attack of the basic nucleophile occurs at the less substituted epoxide carbon.

$$CH_3CH_2C(CH_3)(-O-)CH_2 \xrightarrow[H_2{}^{18}O]{NaOH} \left[CH_3CH_2C(O^-)(CH_3)-CH_2{}^{18}OH\right] \xrightarrow{H_2O} CH_3CH_2C(OH)(CH_3)-CH_2{}^{18}OH$$

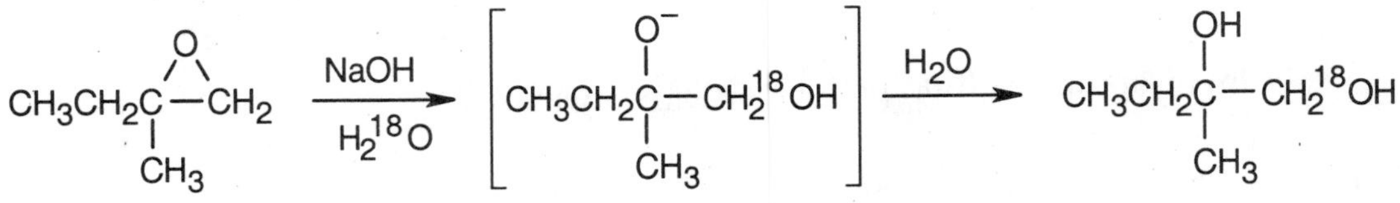

(b) Under acidic conditions, ring-opening occurs at the more substituted epoxide carbon when one of the carbons is tertiary.

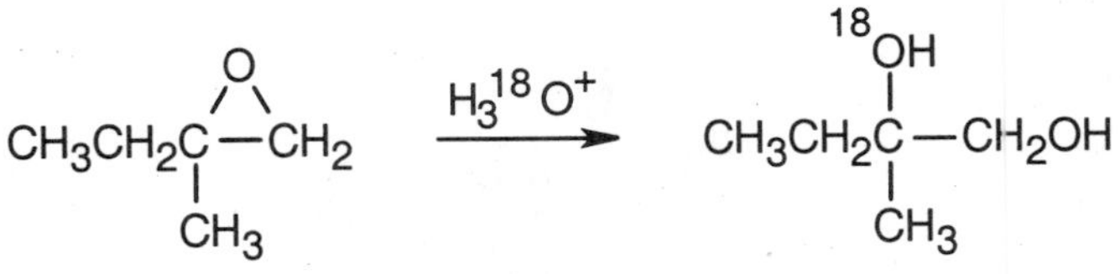

$$CH_3CH_2C(CH_3)(-O-)CH_2 \xrightarrow{H_3{}^{18}O^+} CH_3CH_2C({}^{18}OH)(CH_3)-CH_2OH$$

18.17

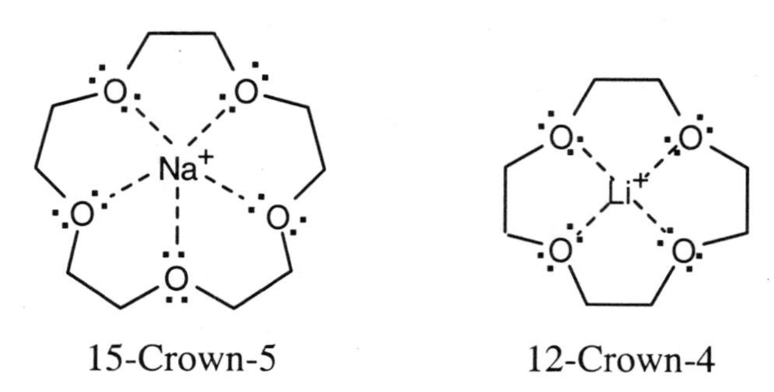

15-Crown-5 12-Crown-4

The ion-to-oxygen distance in 15-crown-5 is about 40% longer than the ion-to-oxygen distance in 12-crown-4.

18.18

CH3CH2C(H)—C(H)(H) epoxide (O bridging); labels: a = CH_3, b = CH_2, c = ring H on C2, d,e = the two H on C1

1,2-Epoxybutane

a = 1.0 δ
b = 1.5 δ
c = 2.9 δ
d,e = 2.5 δ, 2.7 δ

18.19 Thiols are named by the same rules as alcohols, with the suffix -ol replaced by the suffix -thiol. Sulfides are named by the same rules as ethers, with "sulfide" replacing "ether".

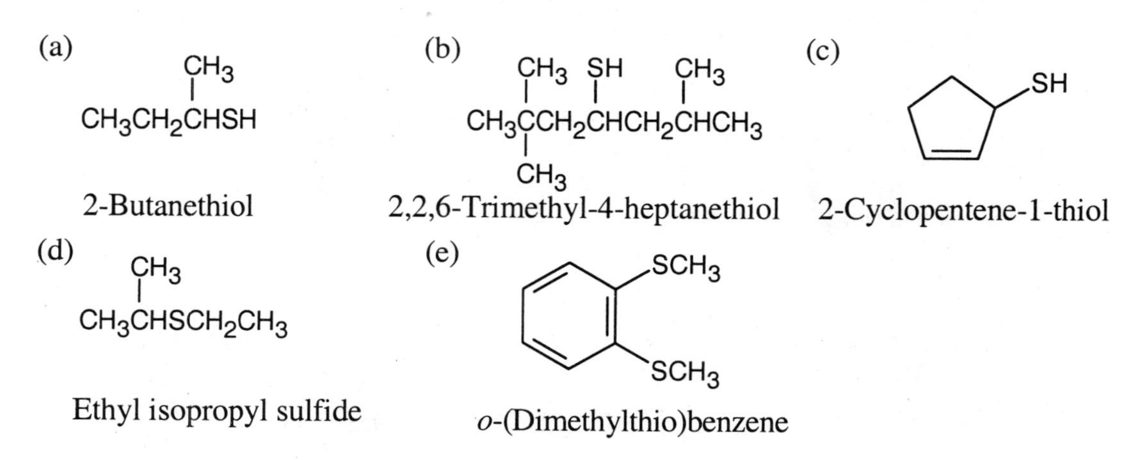

(a) 2-Butanethiol (b) 2,2,6-Trimethyl-4-heptanethiol (c) 2-Cyclopentene-1-thiol

(d) Ethyl isopropyl sulfide (e) *o*-(Dimethylthio)benzene

18.20 Thiourea is used to prepare thiols from haloalkanes.

$$CH_3CH{=}CHCOCH_3\ (\text{C=O}) \xrightarrow[2.\ H_3O^+]{1.\ LiAlH_4} CH_3CH{=}CHCH_2OH \xrightarrow{PBr_3} CH_3CH{=}CHCH_2Br$$

Methyl 2-butenoate

$$CH_3CH{=}CHCH_2Br \xrightarrow[2.\ ^-OH,\ H_2O]{1.\ (H_2N)_2C{=}S} CH_3CH{=}CHCH_2SH$$

$$H_2C{=}CHCH{=}CH_2 \xrightarrow{HBr} CH_3CH{=}CHCH_2Br \xrightarrow[2.\ ^-OH,\ H_2O]{1.\ (H_2N)_2C{=}S} CH_3CH{=}CHCH_2SH$$

1,3-Butadiene 2-Butene-1-thiol

18.21

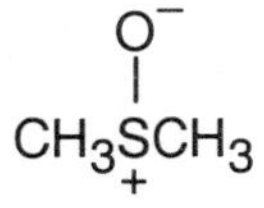

$CH_3\overset{+}{S}(O^-)CH_3$ — Dimethyl sulfoxide

CH_3SCH_3 — Dimethyl sulfide

The boiling point of dimethyl sulfoxide is high because it is a dipolar compound. Dimethyl sulfoxide is miscible with water because it can hydrogen-bond with water.

Visualizing Chemistry

18.22

(a) *cis*-1-Ethoxy-3-methylcyclohexane

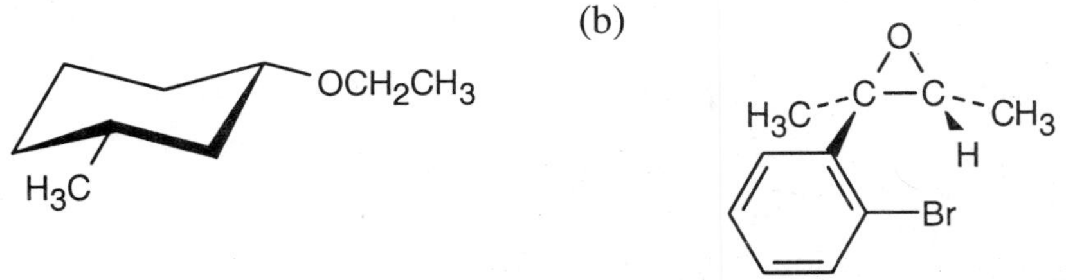

(b) *E*-2-(*o*-Bromophenyl)-2,3-epoxybutane

18.23 Ring-opening occurs at the tertiary carbon to give the transition state carbocation-like stability. Bromine approaches 180° from the C–OH bond, as it would in an S_N2 reaction.

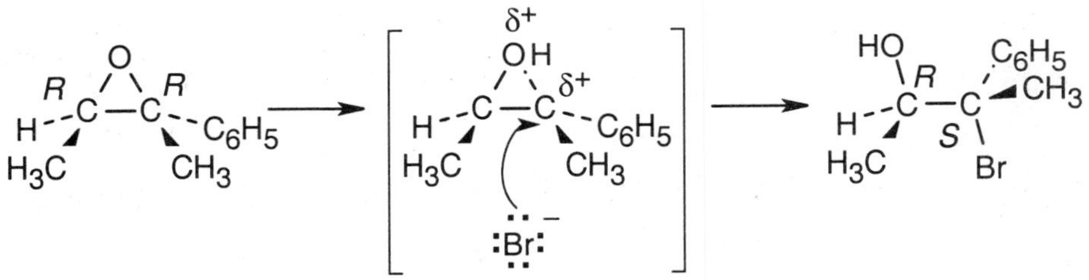

18.24 A molecular model of bornene shows that approach to the upper face of the double bond is hindered by a methyl group. Reaction with RCO_3H occurs at the lower face of the double bond to produce epoxide **A**.

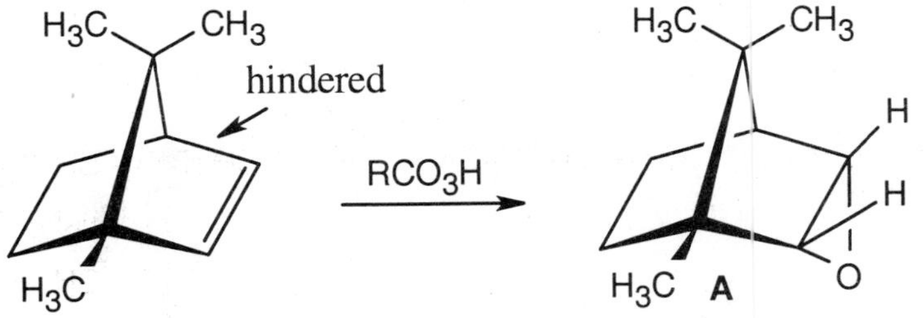

hindered

RCO_3H

A

In the reaction of Br_2 and H_2O with bornene, the intermediate bromonium ion also forms at the lower face. Reaction with water yields a bromohydrin which, when treated with base, forms epoxide **B**.

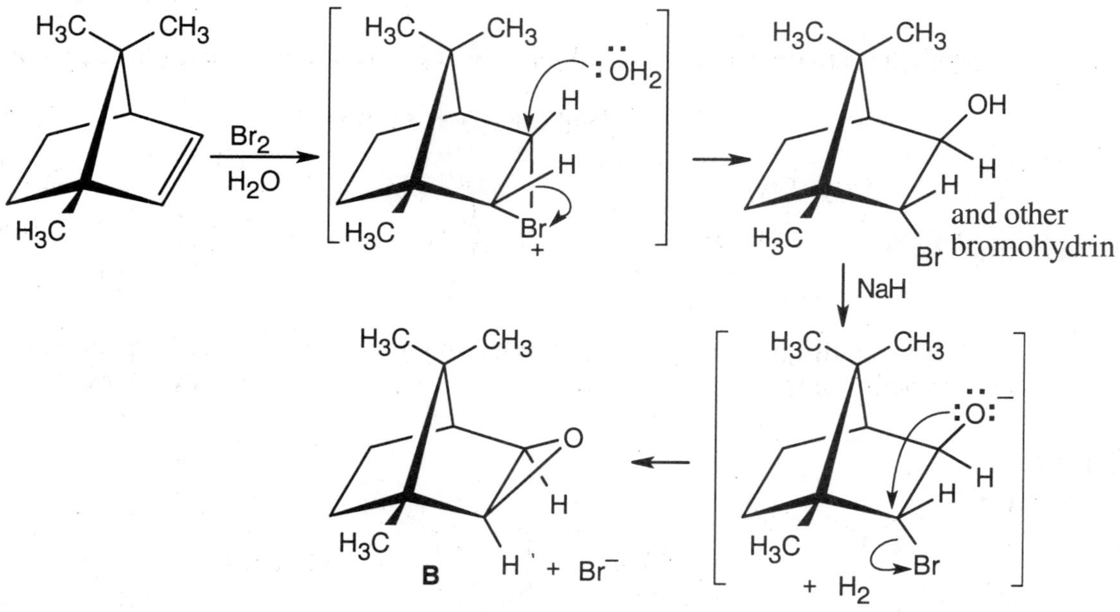

and other bromohydrin

B

Additional Problems

18.25

(a) Ethyl 1-ethylpropyl ether

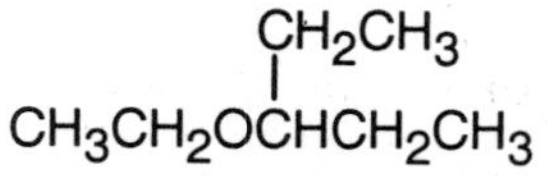

(b) Di(*p*-chlorophenyl) ether

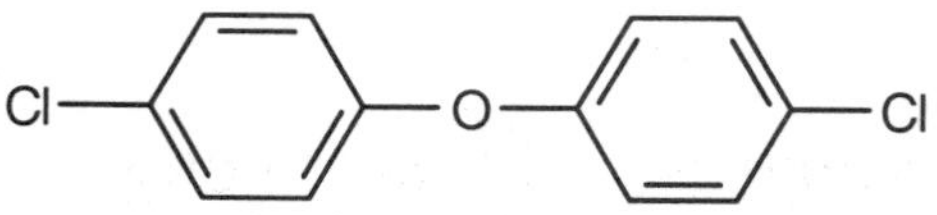

(c) 3,4-Dimethoxybenzoic acid

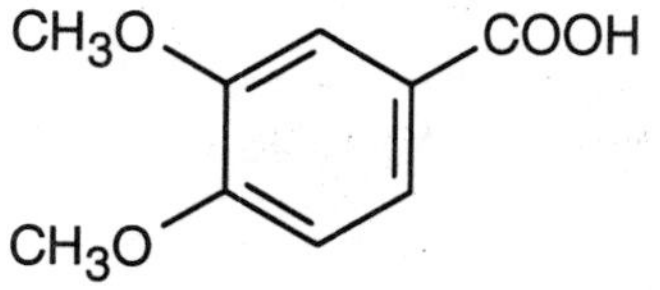

(d) Cyclopentyloxycyclohexane

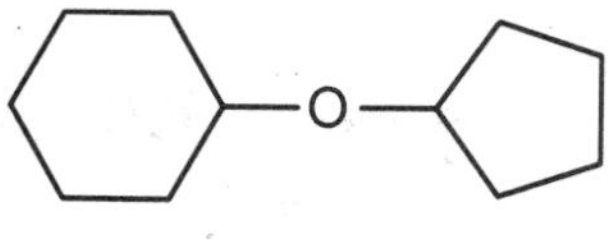

(e) 4-Allyl-2-methoxyphenol

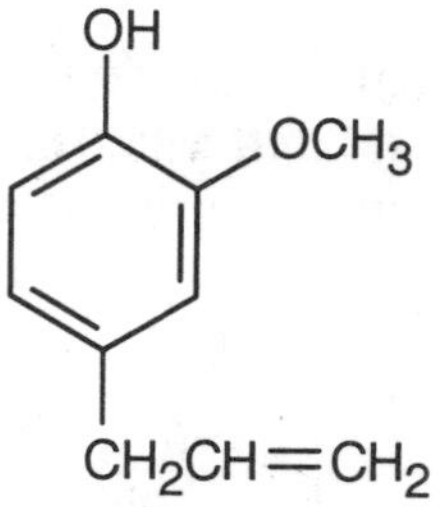

18.26

(a) Cyclohexyl isopropyl sulfide

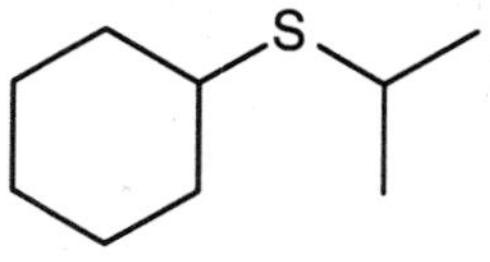

(b) *o*-Dimethoxybenzene

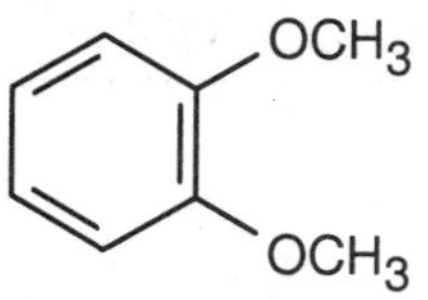

(c) 1,2-Epoxycyclopentane

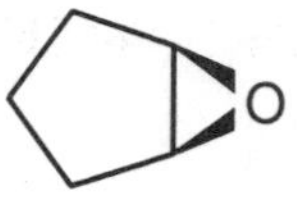

(d) 2-Methyltetrahydrofuran

(e) $CH_3CH(CH_3)-O-$cyclopropyl

Cyclopropyl isopropyl ether *or* Isopropoxycyclopropane

(f) o-Nitrobenzenethiol

o-Nitrobenzenethiol

(g) $CH_3CH_2CH(CH_3)CH(CH_3)CHSCHCH_3$ with CH_3 on each marked carbon

2-(Isopropylthio)-3,4-dimethylhexane

(h) $CH_3C(OCH_3)_2CH_3$

2,2-Dimethoxypropane

(i) cyclohexane with SCH_3, SCH_3 on one carbon

1,1-(Dimethylthio)-cyclohexane

18.27

(a) Cyclohexyl$-OCH_2CH_3$ $\xrightarrow[H_2O]{HI}$ Cyclohexyl$-OH$ + CH_3CH_2I

(b) $C_6H_5-OC(CH_3)_3$ $\xrightarrow{CF_3COOH}$ C_6H_5-OH + $H_3C-C(CH_3)=CH_2$

(c) $H_2C=CHOCH_2CH_3$ $\xrightarrow[H_2O]{HI}$ CH_3CH_2I + $[H_2C=CH(OH)]$ (enol) $\longrightarrow$ $CH_3CH(=O)$

The enol tautomerizes to an aldehyde.

(d) $(CH_3)_3CCH_2OCH_2CH_3$ $\xrightarrow[H_2O]{HI}$ $(CH_3)_3CCH_2OH$ + CH_3CH_2I

18.28

(a) C_6H_5-OH $\xrightarrow{NaH}$ $C_6H_5-O^- Na^+$ $\xrightarrow{CH_3CH_2Br}$ $C_6H_5-OCH_2CH_3$

(b) $CH_3CH=CH_2$ $\xrightarrow[2.\ NaBH_4]{1.\ C_6H_5OH,\ Hg(OCOCF_3)_2}$ $C_6H_5-OCH(CH_3)_2$

(c)

H_3C / H trans-alkene $C=C$ with H and CH_3 $\xrightarrow{RCO_3H}$ epoxide: H_3C, H, O, H, CH_3

(d)

Cyclopentanol (—OH) $\xrightarrow[\text{2. } NaBH_4]{\text{1. } Hg(OCOCF_3)_2,\ (CH_3)_2C{=}CH_2}$ cyclopentyl —$OC(CH_3)_3$

(e)

Cyclohexene $\xrightarrow{RCO_3H}$ cyclohexene oxide (O) $\xrightarrow[HOCH_3]{^-OCH_3}$ trans-2-methoxycyclohexanol (H, OH, H, OCH_3) $\xrightarrow[\text{2. } CH_3I]{\text{1. NaH}}$ trans-1,2-dimethoxycyclohexane (H, OCH_3, H, OCH_3)

(f)

Cyclohexene $\xrightarrow[\text{2. } H_2O_2,\ ^-OH]{\text{1. } BD_3}$ (H, OH, D, H) $\xrightarrow[\text{2. } CH_3I]{\text{1. NaH}}$ (H, OCH_3, D, H)

18.29

(a)

1-Phenylethanol ($C_6H_5CH(OH)CH_3$) $\xrightarrow{NaH}$ $C_6H_5CH(O^-\ Na^+)CH_3$ + H_2 $\xrightarrow{CH_3Br}$ $C_6H_5CH(OCH_3)CH_3$ + NaBr

Methyl 1-phenylethyl ether

(b)

$C_6H_5CH(OH)CH_3$ $\xrightarrow[\text{pyridine}]{POCl_3}$ $C_6H_5CH{=}CH_2$ $\xrightarrow{RCO_3H}$ C_6H_5—CH—CH_2 (O epoxide)

Styrene

Phenylepoxyethane

(c)

$C_6H_5CH(OH)CH_3$ $\xrightarrow[\text{2. } NaBH_4]{\text{1. } Hg(OCOCF_3)_2,\ (CH_3)_2C{=}CH_2}$ $C_6H_5CH(OC(CH_3)_3)CH_3$

tert-Butyl 1-phenylethyl ether

(d)

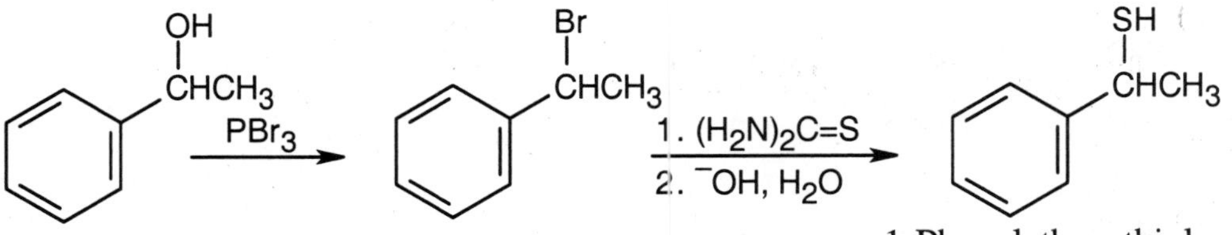

OH
CHCH$_3$
$\xrightarrow{PBr_3}$
Br
CHCH$_3$
$\xrightarrow[2.\ ^-OH,\ H_2O]{1.\ (H_2N)_2C{=}S}$
SH
CHCH$_3$

1-Phenylethanethiol

18.30

(a)

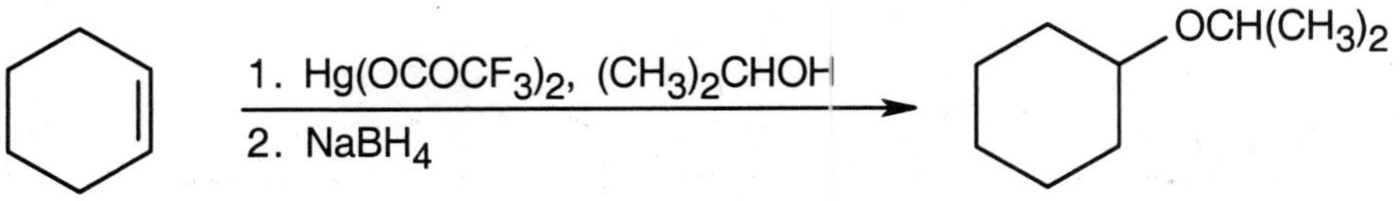

$\xrightarrow[2.\ NaBH_4]{1.\ Hg(OCOCF_3)_2,\ (CH_3)_2CHOH}$
OCH(CH$_3$)$_2$

(b)

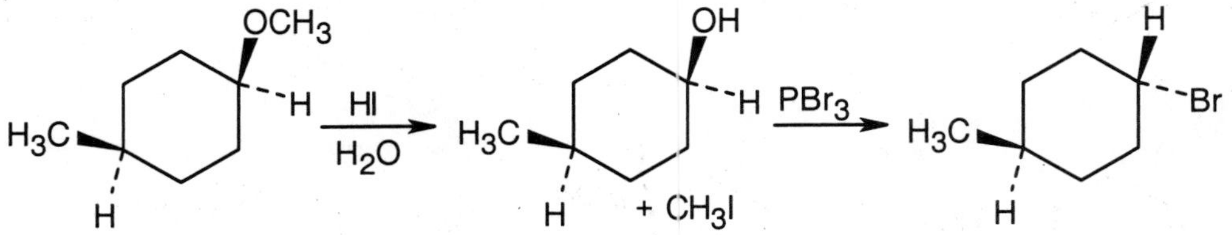

OCH$_3$
H$_3$C
H
H
$\xrightarrow[H_2O]{HI}$
OH
H$_3$C
H
H
+ CH$_3$I
$\xrightarrow{PBr_3}$
H
Br
H$_3$C
H

(c)

C(CH$_3$)$_3$
$\xrightarrow{RCO_3H}$
O
C(CH$_3$)$_3$
+
O
C(CH$_3$)$_3$
$\xrightarrow[H_2O]{^-OH}$
H OH
H
OH
C(CH$_3$)$_3$

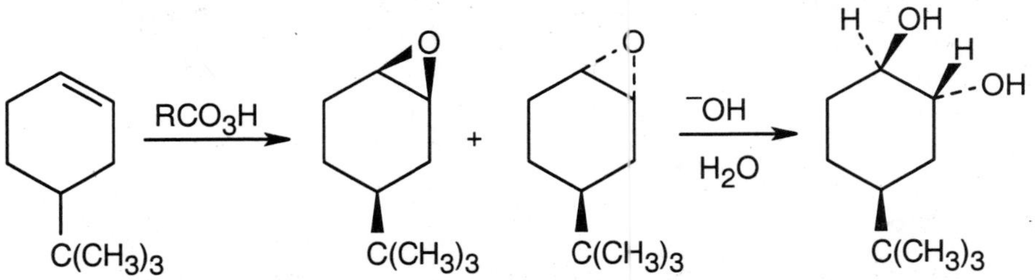

(d)

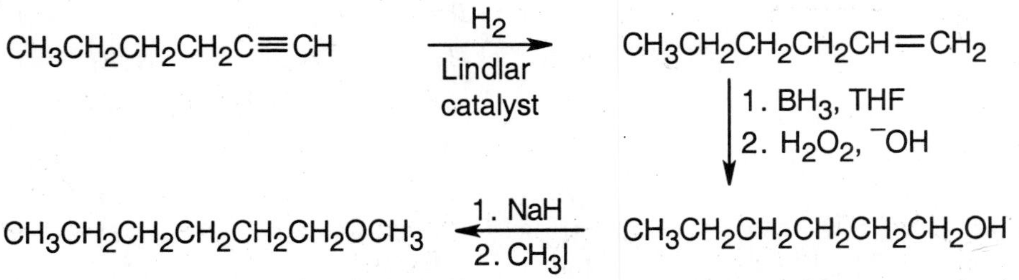

$$CH_3CH_2CH_2CH_2C{\equiv}CH \xrightarrow[\text{Lindlar catalyst}]{H_2} CH_3CH_2CH_2CH_2CH{=}CH_2$$

$$\downarrow 1.\ BH_3,\ THF;\ 2.\ H_2O_2,\ ^-OH$$

$$CH_3CH_2CH_2CH_2CH_2CH_2OCH_3 \xleftarrow[2.\ CH_3I]{1.\ NaH} CH_3CH_2CH_2CH_2CH_2CH_2OH$$

(e)

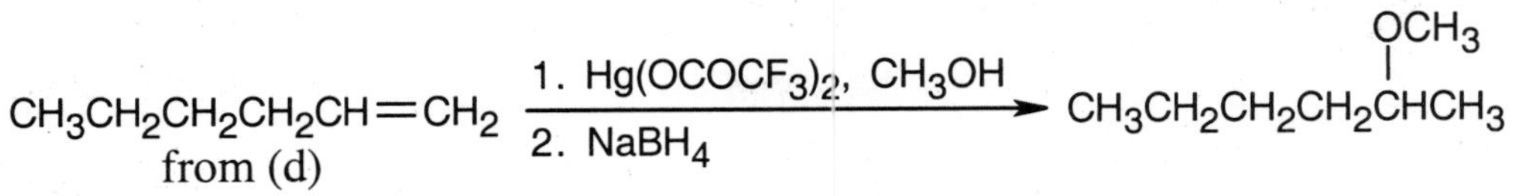

$$CH_3CH_2CH_2CH_2CH{=}CH_2 \xrightarrow[2.\ NaBH_4]{1.\ Hg(OCOCF_3)_2,\ CH_3OH} CH_3CH_2CH_2CH_2\overset{OCH_3}{CH}CH_3$$

from (d)

18.31

O
$\xrightarrow[H_2O]{HI}$ $ICH_2CH_2CH_2CH_2OH$;
O
$\xrightarrow[H_2O]{2\ HI}$ $ICH_2CH_2CH_2CH_2I$

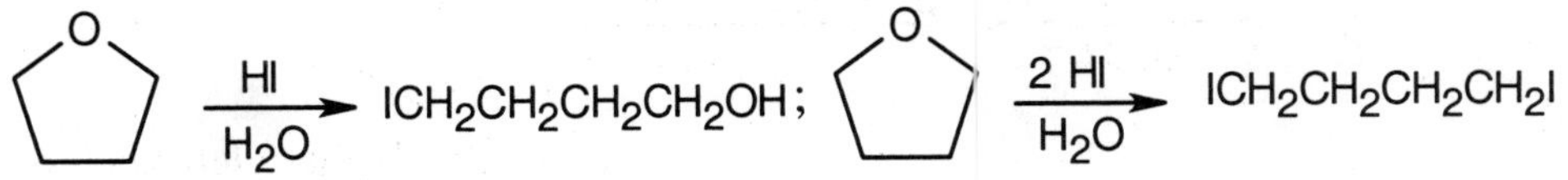

18.32

(a)

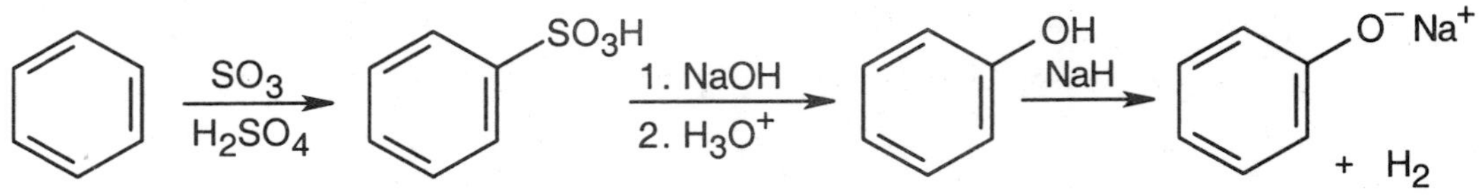

SO_3, H_2SO_4 → SO_3H ; 1. NaOH, 2. H_3O^+ → OH ; NaH → $O^- Na^+$ + H_2

(b)

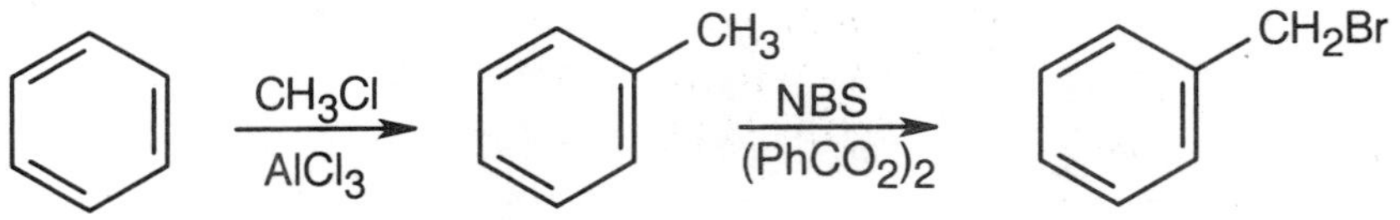

CH_3Cl, $AlCl_3$ → CH_3 ; NBS, $(PhCO_2)_2$ → CH_2Br

(c)

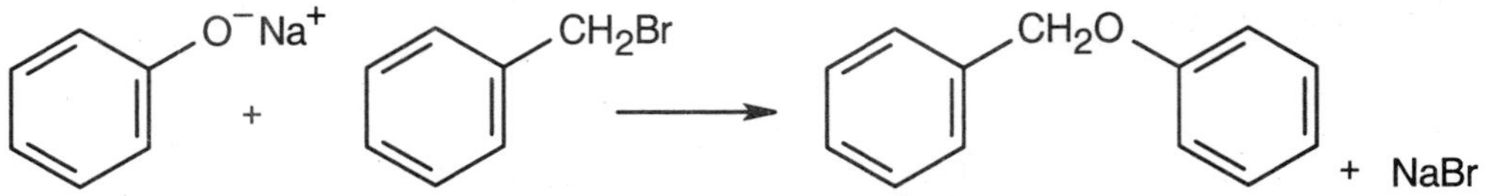

$O^- Na^+$ (from (a)) + CH_2Br (from (b)) → CH_2O (Benzyl phenyl ether) + NaBr

18.33

H_2C—CH_2, CH_3, CH_3, H_2C, C, HO, HO:, H^+ ⇌ (1.) [H_2C—CH_2, CH_3, H_2C, C, HO, H_2O^+, CH_3] ⇌ (2.) [H_2C—CH_2, CH_3, H_2C, C^+, HO:, CH_3 + H_2O] ⇌ (3.) [H_2C—CH_2, CH_3, H_2C, C, O^+, H, CH_3] ⇌ H_2C—CH_2, CH_3, H_2C, C, O, CH_3 + H^+

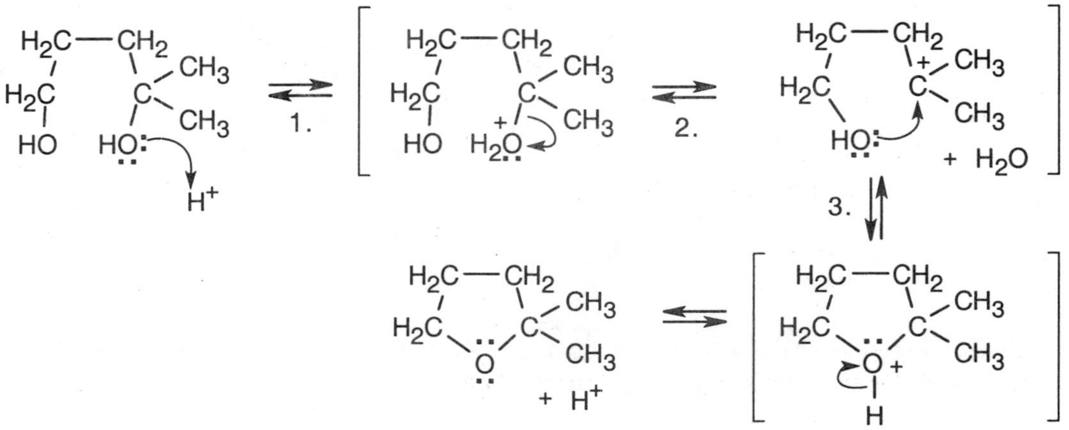

The reaction involves: (1) protonation of the tertiary hydroxyl group; (2) loss of water to form a tertiary carbocation; (3) nucleophilic attack on the carbocation by the second hydroxyl group. The tertiary hydroxyl group is more likely to be eliminated because the resulting carbocation is more stable.

18.34

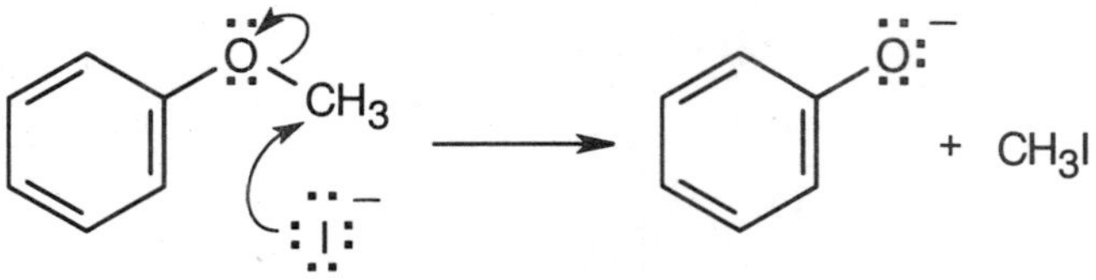

O, CH_3, I^- → O^- + CH_3I

This reaction is an S_N2 displacement and can't occur at an aryl carbon. DMF is a polar aprotic solvent that increases the rate of an S_N2 reaction by making anions more nucleophilic.

18.35

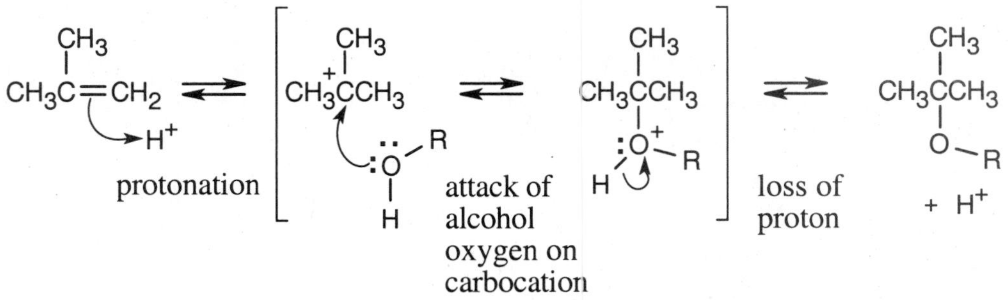

$$CH_3C(CH_3){=}CH_2 \rightleftharpoons \left[CH_3\overset{+}{C}(CH_3)CH_3 \rightleftharpoons CH_3C(CH_3)_2\overset{+}{O}(H)R \right] \rightleftharpoons CH_3C(CH_3)_2O{-}R + H^+$$

protonation (by H^+); attack of alcohol oxygen on carbocation (R–O–H); loss of proton

Notice that this reaction is the reverse of acid-catalyzed cleavage of a tertiary ether (Problem 18.8).

18.36

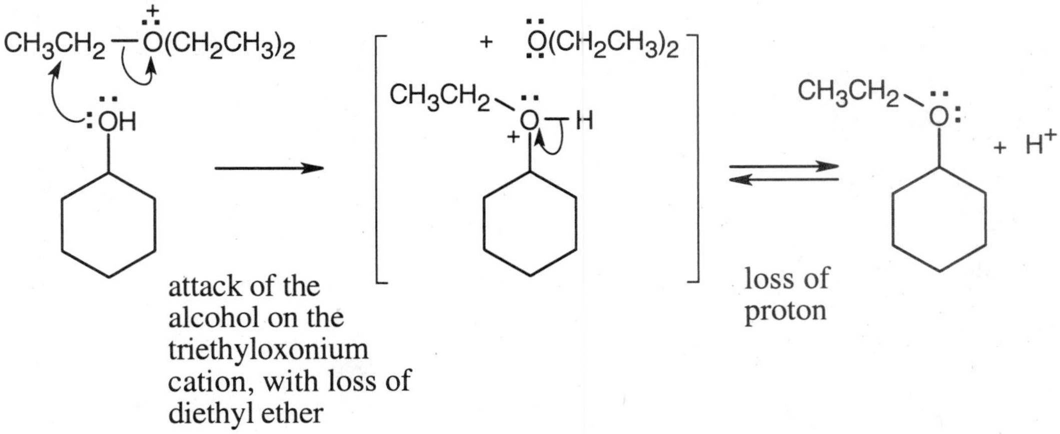

$$CH_3CH_2{-}\overset{+}{O}(CH_2CH_3)_2 + \text{cyclohexanol} \longrightarrow \left[CH_3CH_2{-}\overset{+}{O}(H){-}C_6H_{11} + O(CH_2CH_3)_2 \right] \rightleftharpoons CH_3CH_2{-}O{-}C_6H_{11} + H^+$$

attack of the alcohol on the triethyloxonium cation, with loss of diethyl ether

loss of proton

Trialkyloxonium salts are more reactive alkylating agents than alkyl iodides because a neutral ether is a better leaving group than an iodide ion.

18.37

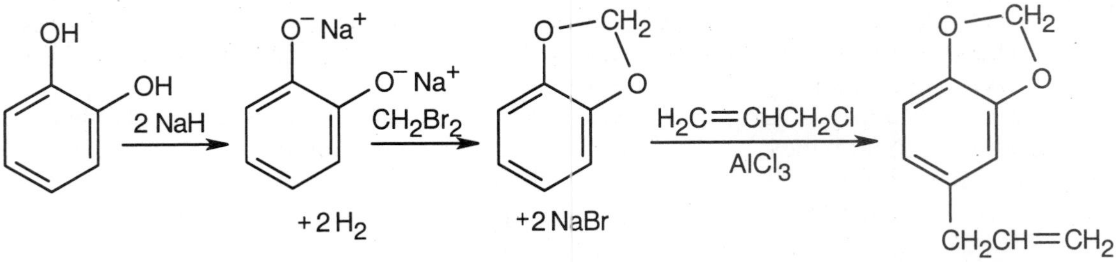

$$\text{catechol (OH, OH)} \xrightarrow{2\ NaH} \text{(O}^-\text{Na}^+\text{, O}^-\text{Na}^+) + 2H_2 \xrightarrow{CH_2Br_2} \text{(O–CH}_2\text{–O)} + 2\,NaBr \xrightarrow[AlCl_3]{H_2C{=}CHCH_2Cl} \text{(O–CH}_2\text{–O, } CH_2CH{=}CH_2)$$

Safrole

18.38

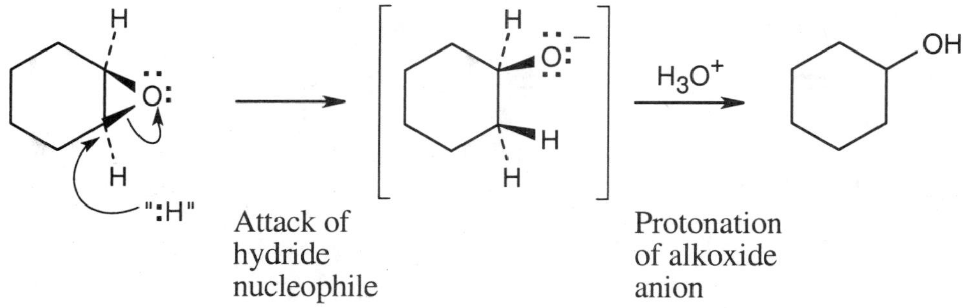

H O: H ":H" → [H O:⁻ H H] H_3O^+ → OH

Attack of hydride nucleophile

Protonation of alkoxide anion

The reaction is an S_N2 epoxide cleavage by attack a hydride nucleophile. The exact nature of the attacking nucleophile is not clear.

18.39

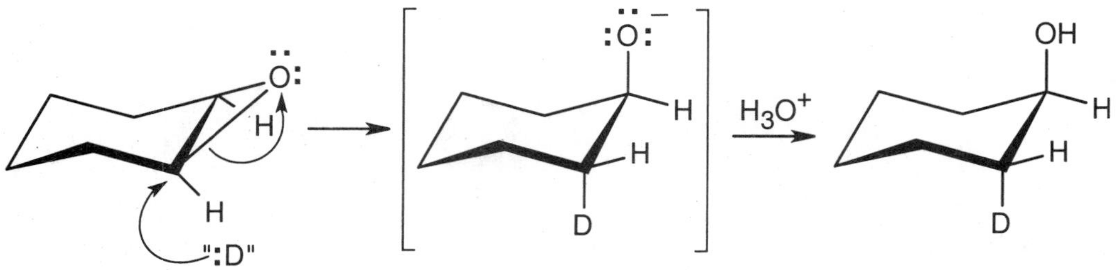

O: H H ":D" → [:O:⁻ H H D] H_3O^+ → OH H H D

Deuterium and –OH have a trans diaxial relationship in the product.

18.40

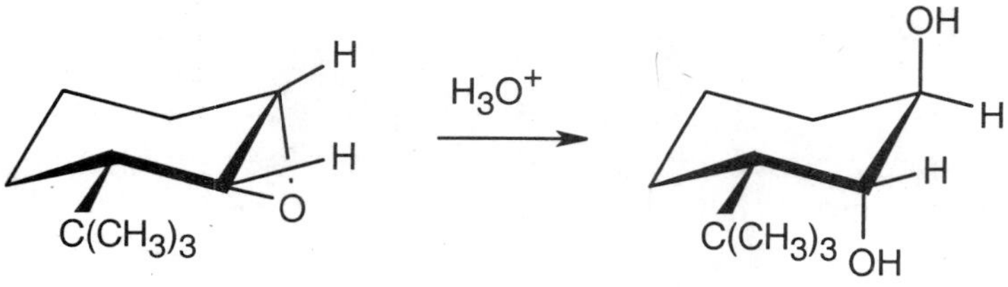

H H O C(CH3)3 H_3O^+ → OH H H C(CH3)3 OH

cis-3-*tert*-Butyl-1,2-epoxycyclohexane

The hydroxyl groups in the product have a trans diaxial relationship.

18.41 The mechanism of Grignard addition to oxetane is the same as the mechanism of Grignard addition to epoxides, described in Section 18.8. The reaction proceeds at a reduced rate because oxetane is less reactive than ethylene oxide. The four-membered ring oxetane is less strained, and therefore more stable, than the three-membered ethylene oxide ring.

18.42

ring flip

trans-2-Chlorocyclohexanol

Cl^- +

1,2-Epoxycyclohexane

+ H_2O + Cl^-

cis-2-Chlorocyclohexanol

enol

Cyclohexanone

In the trans isomer, the –OH and –Cl are in the trans orientation that allows epoxide formation to occur as described in Section 18.7. Epoxidation can't occur for the cis isomer, however. Instead, the base ^-OH brings about E2 elimination, producing an enol, which rearranges to cyclohexanone.

18.43

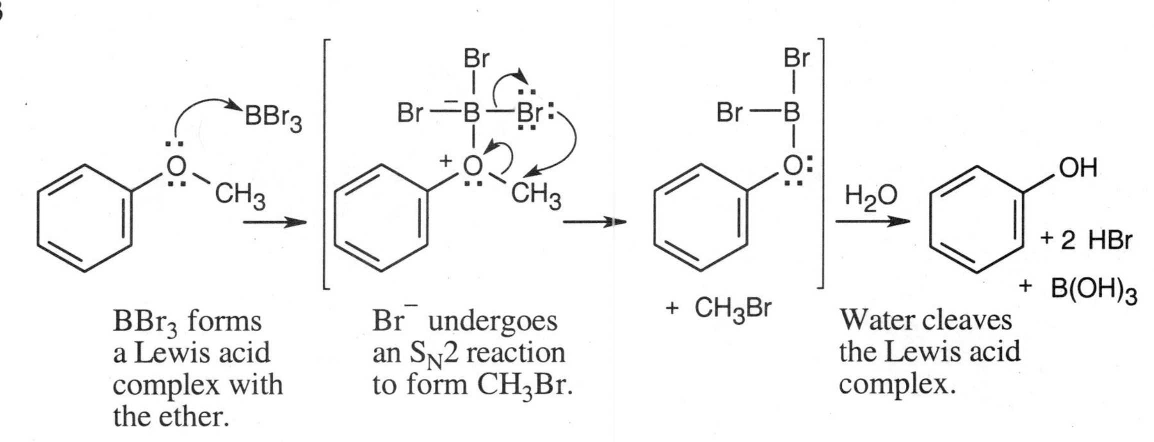
BBr3
CH3
Br
Br
CH3
Br
Br
H2O
OH
+ 2 HBr
+ B(OH)3
+ CH3Br
BBr3 forms a Lewis acid complex with the ether.
Br⁻ undergoes an SN2 reaction to form CH3Br.
Water cleaves the Lewis acid complex.

18.44

$$\frac{1.06 \text{ g vanillin}}{152 \text{ g/mol}} = 6.97 \times 10^{-3} \text{ mol vanillin}$$

$$\frac{1.60 \text{ g AgI}}{234.8 \text{ g/mol}} = 6.81 \times 10^{-3} \text{ mol AgI}$$

$$\underset{\text{AgI}}{6.81 \times 10^{-3} \text{ mol}} \longrightarrow \underset{I^-}{6.81 \times 10^{-3} \text{ mol}} \longrightarrow \underset{CH_3I}{6.81 \times 10^{-3} \text{ mol}} \longrightarrow \underset{-OCH_3}{6.81 \times 10^{-3} \text{ mol}}$$

Thus, 6.97×10^{-3} mol of vanillin contain 6.81×10^{-3} mol of methoxyl groups. Since the ratio of moles vanillin to moles methoxyl is approximately 1:1, each vanillin contains one methoxyl group.

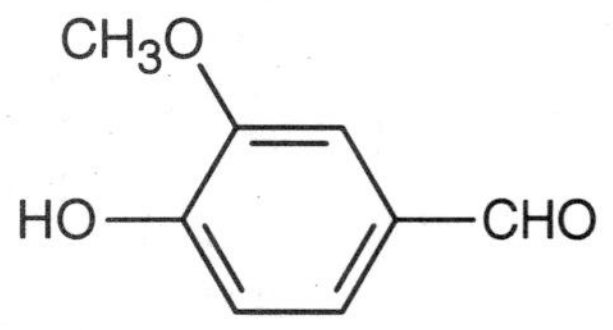

CH_3O HO CHO Vanillin

18.45 Disparlure, $C_{19}H_{38}O$, contains one degree of unsaturation, which the ^1H NMR absorption at 2.8 δ identifies as an epoxide ring.

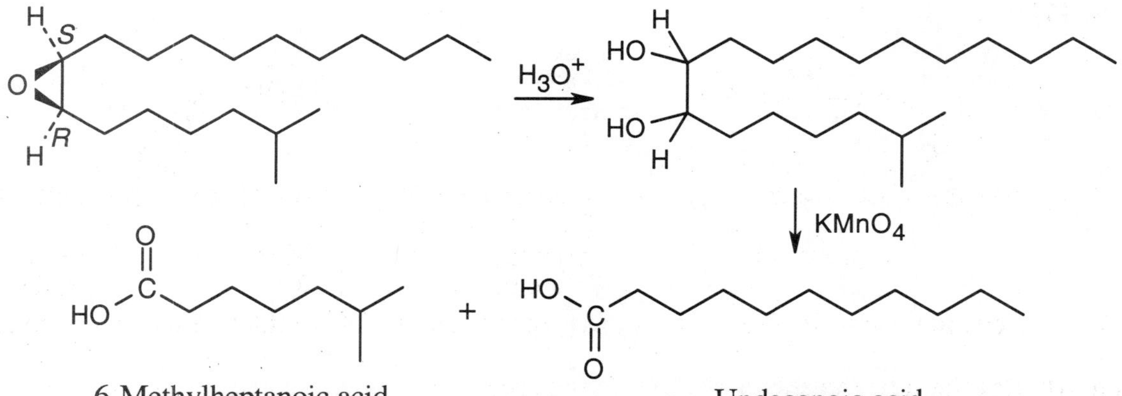

H S O R H $\xrightarrow{H_3O^+}$ H HO HO H

$\downarrow$ $KMnO_4$

HO C O + HO C O

6-Methylheptanoic acid Undecanoic acid

18.46

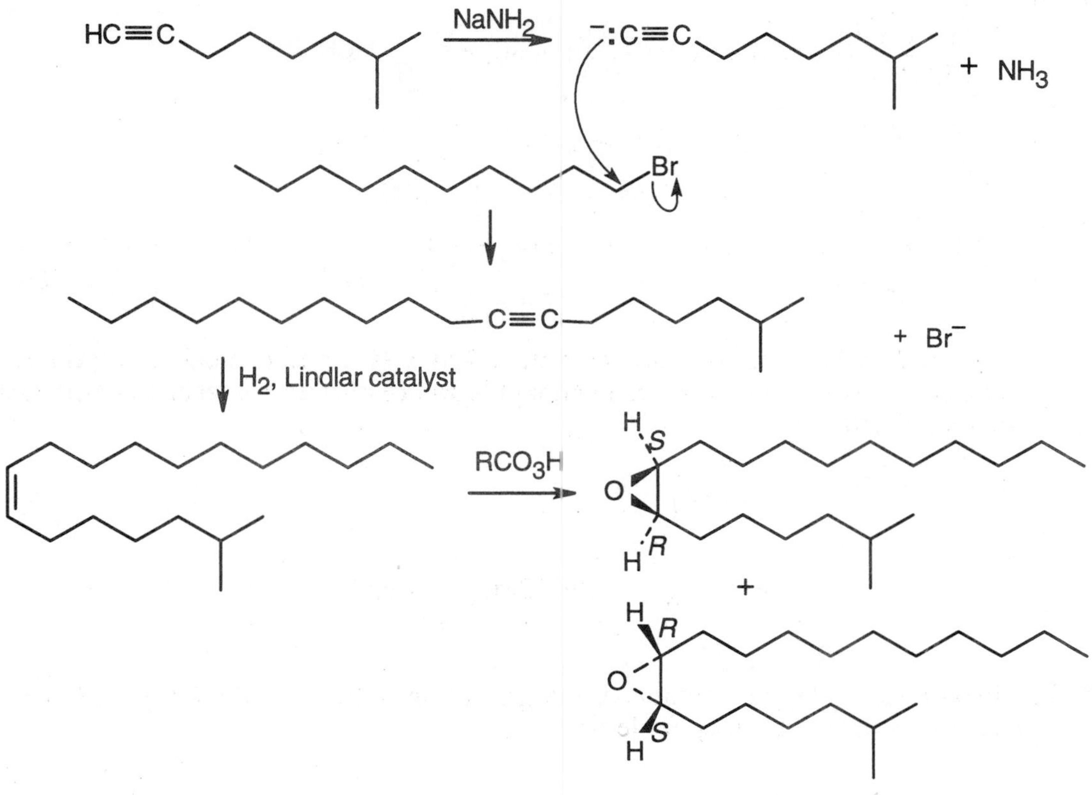

$HC{\equiv}C$–R $\xrightarrow{NaNH_2}$ $^{-}{:}C{\equiv}C$–R + NH_3

(R–Br) → R–$C{\equiv}C$–R' + Br^-

↓ H_2, Lindlar catalyst

$\xrightarrow{RCO_3H}$ (S,R epoxide) + (R,S epoxide)

18.47

$Ph_2C{-}CH_2$ (epoxide) ⇌ [$Ph_2C{-}CH_2$ (protonated epoxide, $\overset{+}{O}H$) ⇌ $Ph_2\overset{+}{C}{-}CH(OH)(H)$ ⟶ $Ph_2CH{-}CH{=}\overset{+}{O}H$] ⇌ $Ph_2CHCH{=}O$ + H_3O^+

protonation — epoxide opening — hydride shift — loss of proton

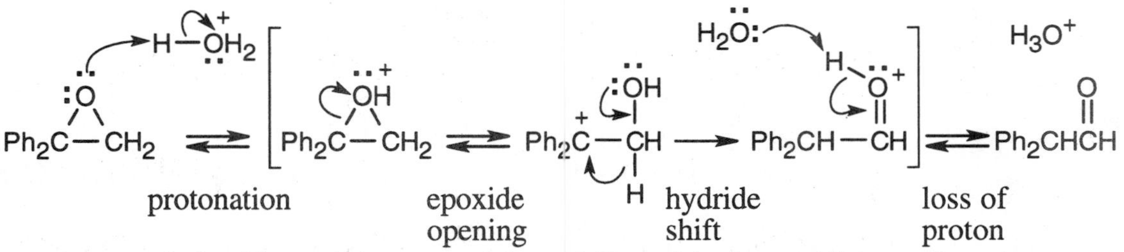

Reaction occurs by this route because of the stability of the intermediate carbocation.

18.48 Use the reaction shown in the previous problem.

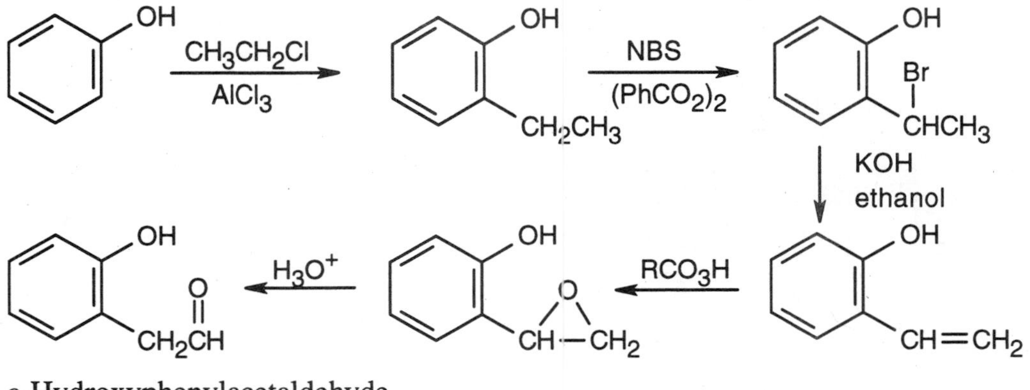

Phenol $\xrightarrow[AlCl_3]{CH_3CH_2Cl}$ o-ethylphenol (CH_2CH_3) $\xrightarrow[(PhCO_2)_2]{NBS}$ o-($CHBrCH_3$)phenol $\xrightarrow{KOH\text{, ethanol}}$ o-($CH{=}CH_2$)phenol $\xrightarrow{RCO_3H}$ o-(epoxide $CH{-}CH_2$)phenol $\xrightarrow{H_3O^+}$ o-(CH_2CHO)phenol

o-Hydroxyphenylacetaldehyde

18.49

(a) (b)

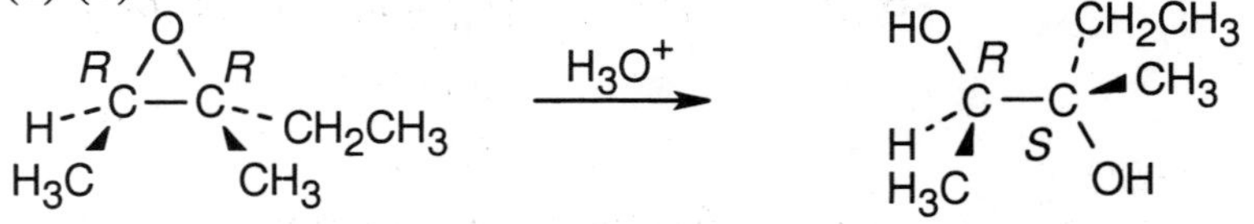

(2*R*,3*R*)-2,3-Epoxy-3-methylpentane (2*R*,3*S*)-3-Methyl-2,3-pentanediol

Reaction with aqueous acid causes ring opening to occur at C3 because the positive charge of the cationic intermediate is more stabilized at the tertiary carbon.

(c) If ring opening occurs exclusively at C3, the product is the 2*R*,3*S* isomer and is chiral. (If ring opening occurred equally at either carbon, the product would be a mixture of chiral enantiomers).

(d) The product is optically active because one chiral enantiomer is produced.

18.50

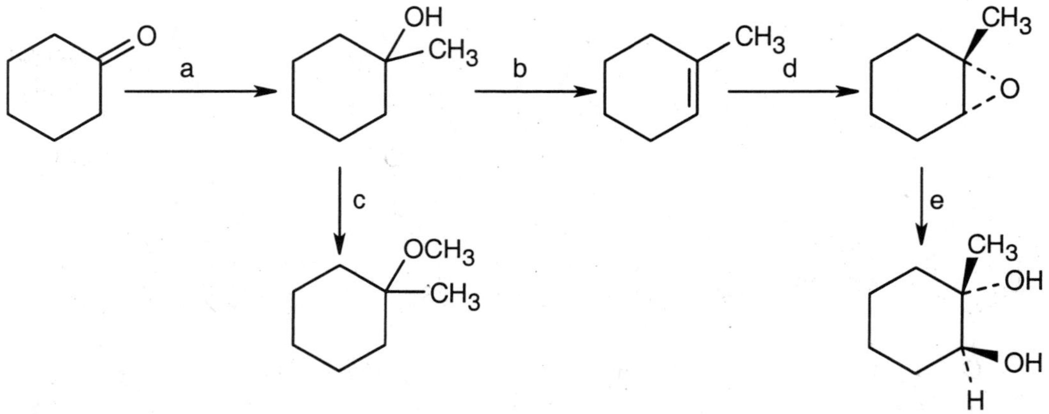

(a) CH_3MgBr, ether; (b) H_2SO_4, H_2O; (c) NaH, then CH_3I; (d) *m*-$ClC_6H_4CO_3H$; (e) H_3O^+

18.51 $M^+ = 116$ corresponds to a sulfide of molecular formula $C_6H_{12}S$, indicating one degree of unsaturation. The IR absorption at 890 cm^{-1} is due to a $R_2{=}CH_2$ group.

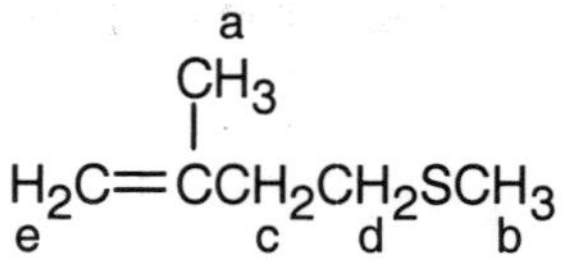

2-Methyl-4(methylthio)-1-butene

a = 1.74 δ
b = 2.11 δ
c = 2.27 δ
d = 2.57 δ
e = 4.73 δ

18.52

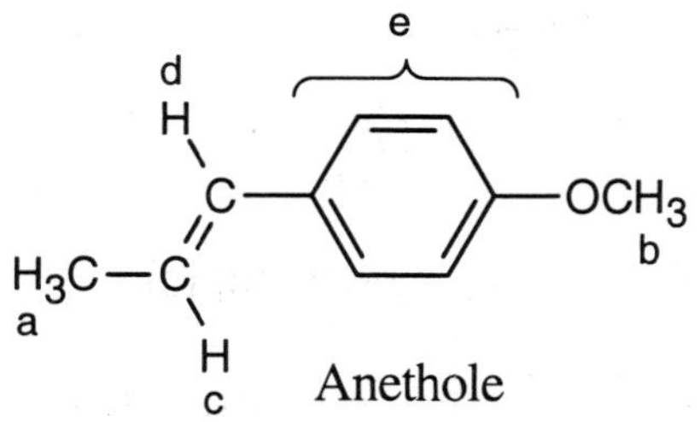

Peak	*Chemical shift*	*Multiplicity*	*Split by:*
a	1.83 δ	doublet	c
b	3.75 δ	singlet	
c	6.08 δ	two quartets	a,d
d	6.28 δ	doublet	c
e	6.80 δ, 7.23 δ	multiplet	

18.53

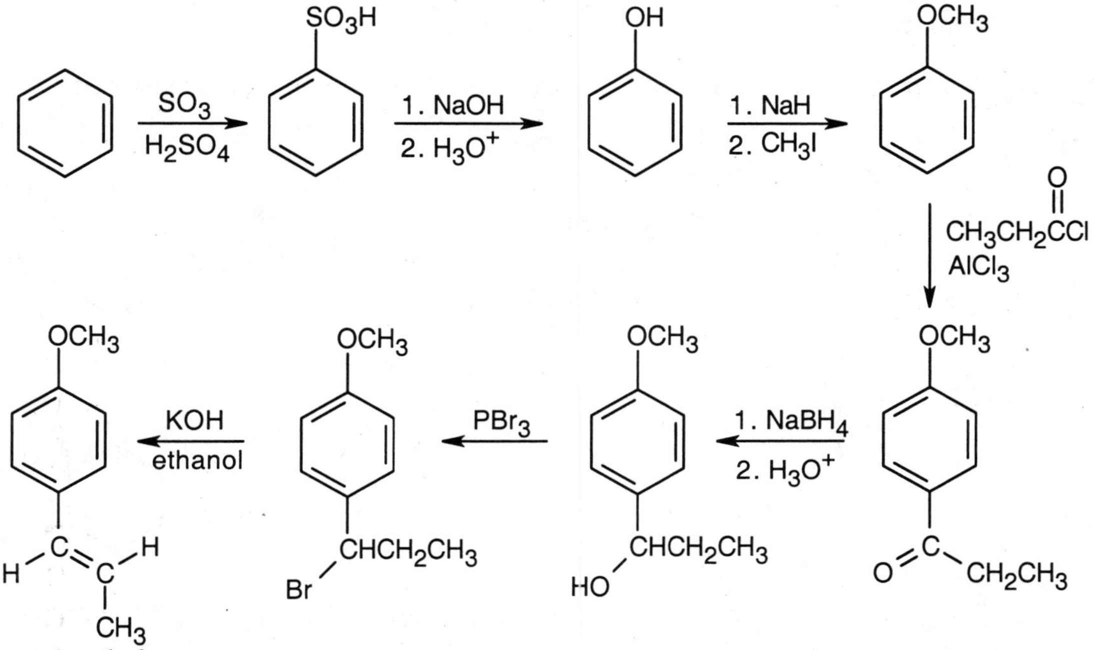

Synthetic analysis: The anethole ring has two functional groups – an ether and a hydrocarbon side chain with a double bond. The ether is synthesized first – by a Williamson ether synthesis from phenol and CH_3I. The hydrocarbon side chain results from a Friedel-Crafts acylation of the ether. Reduction of the ketone, bromination and dehydrohalogenation are used to introduce the double bond.

18.54

(a) $CH_3CH_2C(CH_3)_2SH$ (a = CH_3, c = CH_2, b = CH_3 groups)

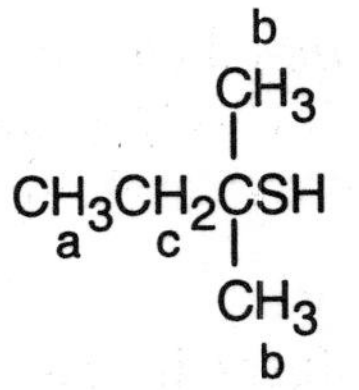

a = 1.0 δ
b = 1.3 δ
c = 1.7 δ

(b) C_6H_5–$OCH_2CH_2CH_2Br$ (c, a, b; d = aromatic ring)

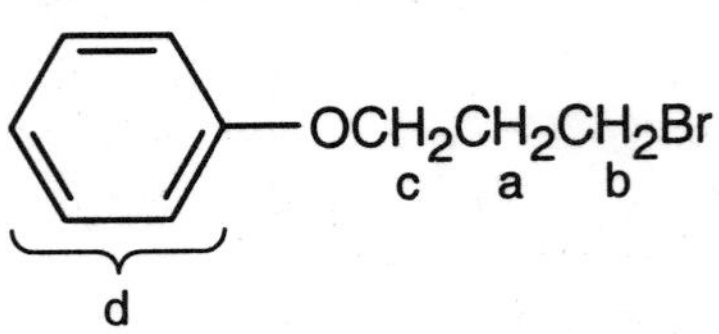

a = 2.3 δ
b = 3.6 δ
c = 4.1 δ
d = 6.9-7.3 δ

(c) $CH_3CH(OCH_3)_2$ (a = CH_3, c = CH, b = OCH_3)

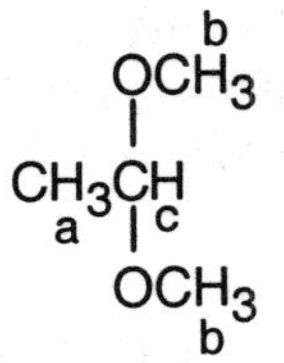

a = 1.3 δ
b = 3.3 δ
c = 4.6 δ

(d) C_6H_5–CH=CHOCH$_3$ (b, c, a; d = aromatic ring)

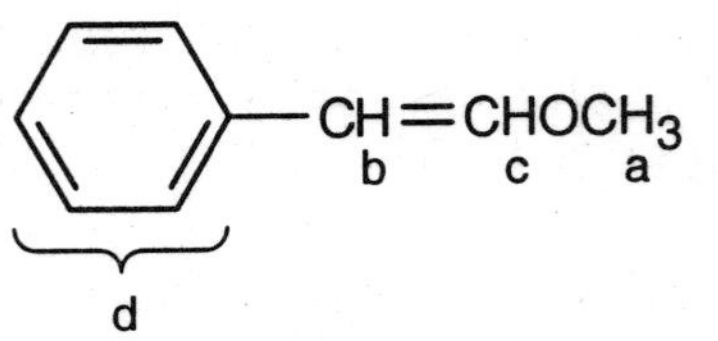

a = 3.7 δ
b = 5.2 δ
c = 6.1 δ
d = 7.1-7.6 δ

18.55

(a)

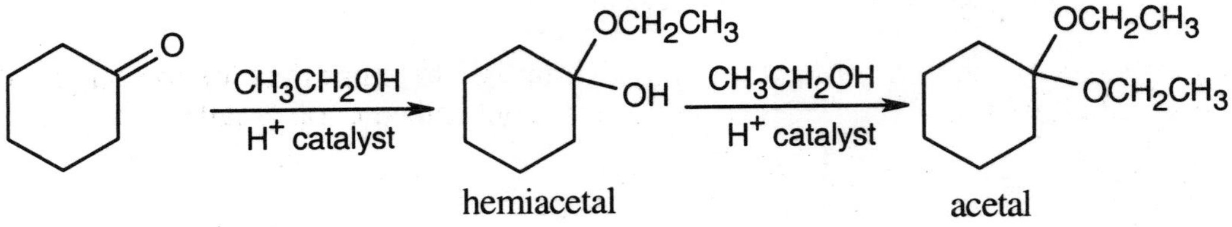

Cyclohexanone → (CH_3CH_2OH, H^+ catalyst) → hemiacetal (1-ethoxycyclohexanol) → (CH_3CH_2OH, H^+ catalyst) → acetal (1,1-diethoxycyclohexane)

hemiacetal acetal

(b)

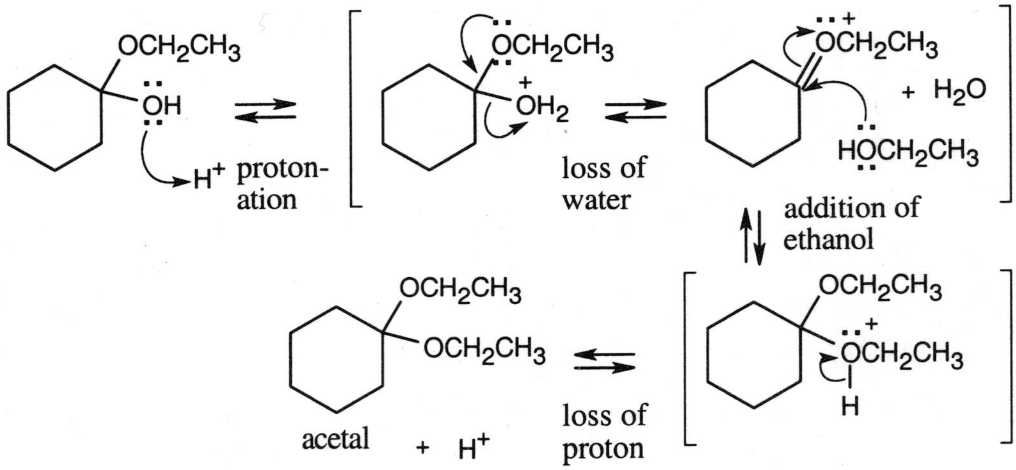

hemiacetal ⇌ (protonation, H^+) [protonated OH_2^+] ⇌ (loss of water) [oxonium ion + H_2O; $HOCH_2CH_3$] ⇌ (addition of ethanol) [protonated acetal] ⇌ (loss of proton) acetal + H^+

proton-ation

loss of water

addition of ethanol

loss of proton

acetal + H^+

18.56

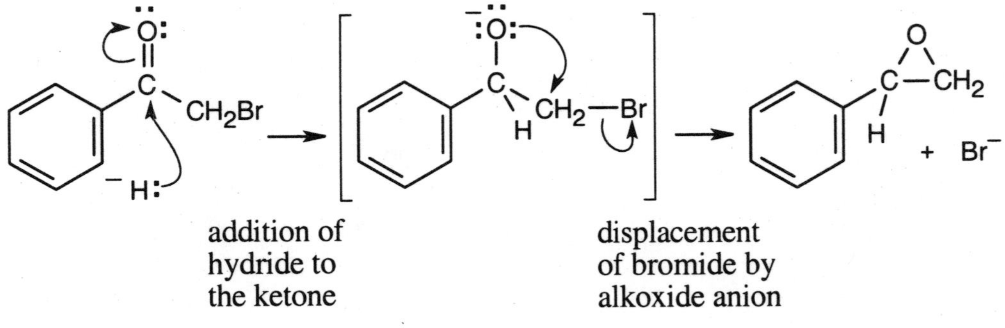

$C_6H_5COCH_2Br$ + $^-$H: → [$C_6H_5CH(O^-)CH_2$–Br] → C_6H_5–epoxide (C(H)–CH_2 bridged by O) + Br^-

addition of hydride to the ketone

displacement of bromide by alkoxide anion

The intermediate resulting from addition of H:$^-$ is similar to the intermediate in a Williamson ether synthesis. Intramolecular reaction occurs to form the epoxide.

Molecular Modeling

18.57 Conformer A adopts the "crown" shape that is useful for binding metal cations, but conformer B is lower in energy; the energy difference is 82 kJ/mol. Electrostatic interactions between all of the negative oxygens in conformer A destabilize this structure. Conformer B has a structure that minimizes these interactions.

18.58 $\Delta H°$ is unfavorable for both reactions, but formation of tetrahydrofuran is less unfavorable (45 kJ/mol) than formation of ethylene oxide (185 kJ/mol) because three-membered rings are more strained. $\Delta S°$ is very favorable for both reactions, but slightly less favorable for formation of tetrahydrofuran. More conformational flexibility is lost when 1,4-butanediol is converted to tetrahydrofuran.

18.59 Diethyl ether is the only solvent that can be used with strongly basic reagents. The electrostatic potential map of diethyl ether doesn't reveal any positive (acidic) hydrogens, whereas the maps of ethanol and acetone both reveal positive (acidic) hydrogens that would react with strongly basic reagents.

18.60 12-Crown-4 should bind lithium most strongly (the other ions are too large). 18-Crown-6 should bind potassium most strongly (the other ions are too small).

Review Unit 7: Alcohols, Ethers, and Related Compounds

Major Topics Covered (with vocabulary):

The –OH group:
alcohol phenol glycol wood alcohol hydrogen bonding alkoxide ion phenoxide ion acidity constant

Alcohols:
Grignard Reagent pyridinium chlorochromate tosylate protecting group TMS ether

Phenols:
cumene hydroperoxide quinone hydroquinone ubiquinone

Acyclic ethers:
Williamson ether synthesis Claisen rearrangement

Cyclic ethers:
epoxide oxirane peroxyacid crown ether 18-crown-6

Thiols and sulfides:
Thiol sulfide mercapto group alkylthio group disulfide thiolate ion trialkylsulfonium salt sulfoxide sulfone

Types of Problems:

After studying these chapters, you should be able to:

- Name and draw structures of alcohols, phenols, ethers, thiols and sulfides.
- Explain the properties and acidity of alcohols and phenols.
- Prepare all of the types of compounds studied.
- Predict the products of reactions involving alcohols, phenols and ethers.
- Formulate mechanisms of reactions involving alcohols, phenols and ethers.
- Identify alcohols, phenols and ethers by spectroscopic techniques.

Points to Remember:

* The great biochemical importance of hydroxyl groups is due to two factors:(1) Hydroxyl groups make biomolecules more soluble because they can hydrogen-bond with water. (2) Hydroxyl groups can be oxidized to aldehydes, ketones and carboxylic acids. The presence of a hydroxyl group in a biological molecule means that all functional groups derived from alcohols can be easily introduced.

* Carbon-carbon bond-forming reactions are always more difficult to learn than functional group transformations because it is often difficult to recognize the components that form a carbon skeleton. The product of a Grignard reaction contains a hydroxyl group bonded to at least one alkyl group (usually two or three). When looking at a product that might have been formed by a Grignard reaction, remember that a tertiary alcohol results from the addition of a Grignard reagent to either a ketone or an ester (the alcohol formed from the ester has two identical –R groups), a secondary alcohol results from addition of a Grignard reagent to an aldehyde, and a

primary alcohol results from addition of a Grignard reagent to formaldehyde or to ethylene oxide. Remember that any molecule taking part in a Grignard reaction must not contain functional groups that might also react with the Grignard reagent.

* Ethers are quite unreactive, relative to many other functional groups we study, and are often used as solvents for that reason. Concentrated halogen acids can cleave ethers to alcohols and halides. Remember that the halide is bonded to the less substituted alkyl group of the ether.

* Epoxide rings can be opened by both acid and base. In basic ring-opening of an unsymmetrical epoxide (and in ring-opening using a Grignard reagent), attack occurs at the less substituted carbon of the epoxide ring. In acidic ring opening, the position of attack depends on the substitution pattern of the epoxide. When one of the epoxide carbons is tertiary, attack occurs at the more substituted carbon, but when the epoxide carbons are both primary or secondary, attack occurs at the less substituted carbon.

* The most useful spectroscopic data for these compounds: (1) A broad IR absorption in the range 3300 cm^{-1}–3600 cm^{-1} shows the presence of the –OH group of an alcohol or a phenol. (2) Hydrogens bonded to the –O–C– carbon of an alcohol or ether absorb in the range 3.5–4.5 δ in an ^1H NMR spectrum or in the range 50–80 δ in a ^{13}C NMR spectrum.

Self-Test:

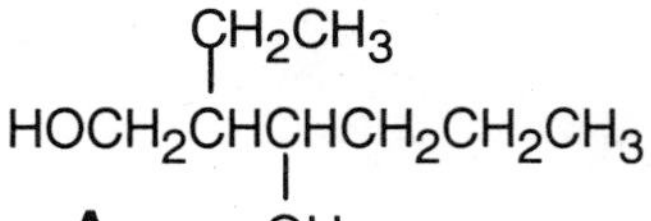

$HOCH_2CH(OH)CH(CH_2CH_3)CH_2CH_2CH_3$
A
an insecticide

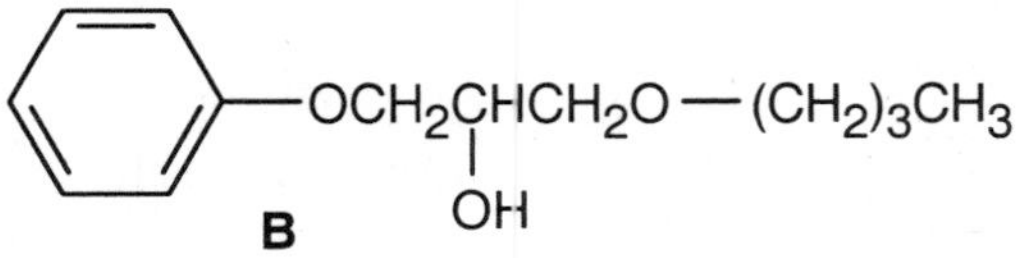

C_6H_5–$OCH_2CH(OH)CH_2O$–$(CH_2)_3CH_3$
B
Febuprol
(increases bile flow)

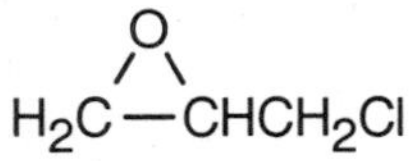

H_2C–$CHCH_2Cl$ (epoxide, O bridging)
Epichlorohydrin

Provide a IUPAC name for **A**. Would you expect **A** to be water-soluble? Label the hydroxyl groups of **A** as primary, secondary or tertiary. What products are formed when **A** reacts with: (a) CrO_3, H_3O^+ ; (b) PBr_3; (c) $(CH_3)_3SiCl$, Et_3N.

Name **B** by IUPAC rules. Show the three components that comprise **B**. The synthesis of **B** involves a ring-opening reaction of the epoxide epichlorohydrin. Use this information to propose a synthesis of **B** from epichlorohydrin and any alcohol or phenol.

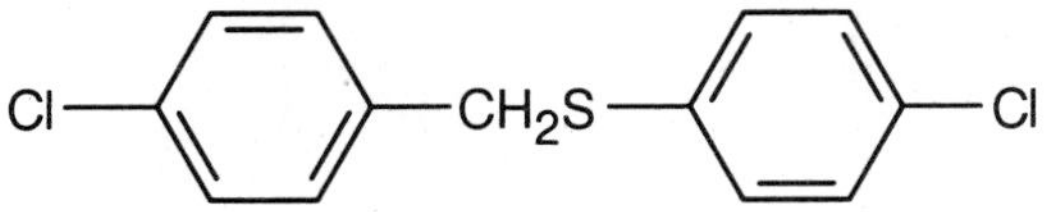

Cl–C_6H_4–CH_2S–C_6H_4–Cl
C
Chlorbenside
(larvicide)

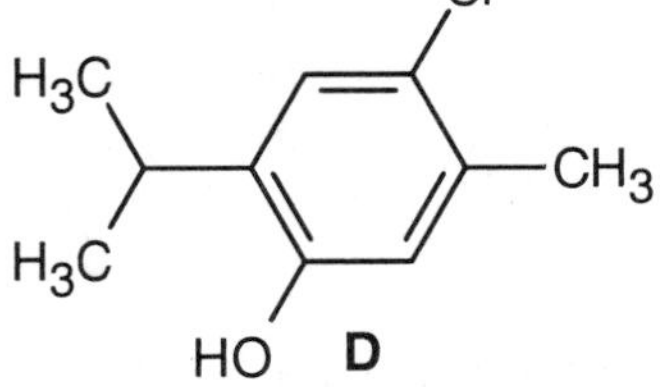

Cl, CH_3, HO, $(H_3C)_2CH$ substituted benzene
D
Chlorothymol

What type of compound is **C**? Name **C**. Synthesize **C** from benzenethiol and benzene. What products are formed when **C** is treated with: (a) CH_3I; (b) H_2O_2, H_2O; (c) product of (b) + CH_3CO_3H.

Synthesize **D** from toluene; assume that isomeric product mixtures can be separated. Describe the IR and ^1H NMR spectrum of **D**.

Multiple Choice:

1. Hydrogen bonding affects all of the following except:
   (a) boiling point (b) solubility (c) position of –OH absorption in IR spectrum (d) chemical shift of –C–O– carbon in ^{13}C NMR.

2. Which of the following alcohols can't be synthesized by a Grignard reaction?
   (a) Benzyl alcohol (b) Triphenylmethanol (c) 3–Bromo-1-hexanol (d) 1-Hexanol

3. Which of the following reactions of a chiral alcohol occurs with inversion of configuration?
   (a) reaction with NaH (b) reaction with PBr_3 (c) reaction with tosyl chloride (d) reaction with $(CH_3)_3SiCl$

4. How many diols of the formula $C_4H_{10}O_2$ are chiral?
   (a) 2 (b) 3 (c) 4 (d) 5

5. Which alcohol is the least acidic?
   (a) 2-Propanol (b) Methanol (c) Ethanol (d) 2-Chloroethanol

6. Which of the following compounds can't be reduced to form $C_6H_5CH_2OH$?
   (a) C_6H_5COOH (b) C_6H_5CHO (c) $C_6H_5COOCH_3$ (d) $C_6H_5OCH_3$

7. The reagent used for dehydration of an alcohol is:
   (a) PCl_3 (b) $POCl_3$ (c) $SOCl_2$ (d) PCC

8. All of the following are products of oxidation of a thiol except:
   (a) a sulfide (b) a disulfide (c) a sulfoxide (d) a sulfone

9. In which of the following epoxide ring-opening reactions does attack of the nucleophile occur at the more substituted carbon of the epoxide ring?

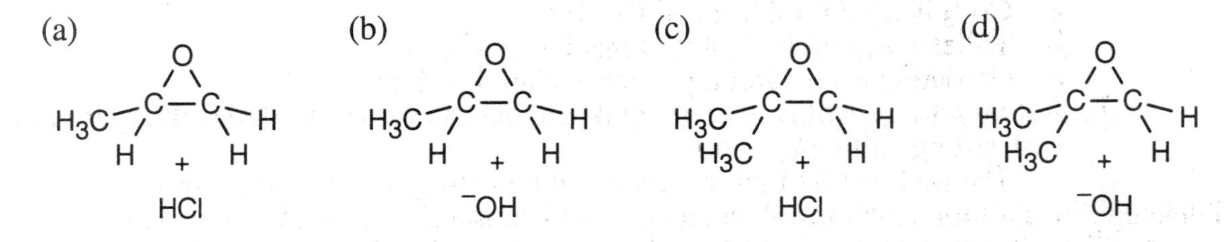

10. Ethers are stable to all of the following reagents except:
    (a) nucleophiles (b) bases (c) strong acids (d) dilute acids

Chapter 19 – Aldehydes and Ketones
Nucleophilic Addition Reactions

Chapter Outline

I. General information about aldehydes and ketones (Sections 19.1 – 19.3).
  A. Naming aldehydes and ketones (Section 19.1).
    1. Naming aldehydes.
      a. Aldehydes are named by replacing the *-e* of the corresponding alkane with *-al*.
      b. The parent chain must contain the –CHO group.
      c. The aldehyde carbon is always carbon 1.
      d. When the –CHO group is attached to a ring, the suffix *-carbaldehyde* is used.
    2. Naming ketones.
      a. Ketones are named by replacing the *-e* of the corresponding alkane with *-one*.
      b. Numbering starts at the end of the carbon chain nearer to the carbonyl carbon.
      c. The word *acyl* is used when a RCO– group is a substituent.
  B. Preparation of aldehydes and ketones (Section 19.2).
    1. Preparation of aldehydes.
      a. Oxidation of primary alcohols with PCC.
      b. Oxidative cleavage of alkenes with at least one vinylic hydrogen.
      c. Reduction of carboxylic acid derivatives.
    2. Preparation of ketones.
      a. Oxidation of secondary alcohols.
      b. Ozonolysis of alkenes with at least one disubstituted unsaturated carbon.
      c. Friedel-Crafts acylation of aromatic compounds.
      d. Hydration of terminal alkynes to produce methyl ketones.
      e. Preparation from carboxylic acid derivatives.
  C. Oxidation of aldehydes and ketones (Section 19.3).
    1. Aldehydes can be oxidized to carboxylic acids by many reagents.
      a. CrO_3 is used for normal aldehydes.
      b. Tollens reagent is used for sensitive aldehydes.
      c. Oxidation occurs through intermediate 1,1-diols.
    2. Ketones are generally inert to oxidation, but can be oxidized to carboxylic acids with strong oxidizing agents.
      The usefulness of this reaction is limited to symmetrical ketones.

II. Nucleophilic addition reactions of aldehydes and ketones (Sections 19.4 – 19.15).
  A. Characteristics of nucleophilic addition reactions (Sections 19.4 – 19.5).
    1. Mechanism of nucleophilic addition reactions (Section 19.4).
      a. A nucleophile attacks the electrophilic carbonyl carbon from a direction 45° from the plane of the carbonyl group.
      b. The carbonyl group rehybridizes from sp^2 to sp^3, and a tetrahedral alkoxide intermediate is produced.
      c. The attacking nucleophile may be neutral or negatively charged.
        Neutral nucleophiles usually have a hydrogen atom that can be eliminated.
      d. The tetrahedral intermediate has two fates:
        i. The intermediate can be protonated to give an alcohol.
        ii. The carbonyl oxygen can be eliminated as –OH to give a product with a C=Nu double bond.
      e. Possible reversibility and acid/base catalysis are important features of nucleophilic addition reactions.

2. Relative reactivity of aldehydes and ketones (Section 19.5).
   a. Aldehydes are usually more reactive than ketones in nucleophilic addition reactions for two reasons:
      i A nucleophile can approach the carbonyl group of an aldehyde with more ease because only one alkyl group is in the way.
      ii. Aldehyde carbonyl groups are more strongly polarized and electrophilic because of the electron-donating properties of alkyl groups.
   b. Aromatic aldehydes are less reactive than aliphatic aldehydes because the electron-donating aromatic ring makes the carbonyl carbon less electrophilic.

B. Nucleophilic addition reactions (Section 19.6 – 19.15).
1. Hydration (Section 19.6).
   a. Water adds to aldehydes and ketones to give 1,1-diols (often referred to as gem-diols).
   b. The reaction is reversible, but generally, the equilibrium favors the carbonyl compound.
   c. Reaction is slow in pure water, but is catalyzed by both aqueous acid and base.
      i. The base-catalyzed reaction is an addition of –OH, followed by protonation of the tetrahedral intermediate by water.
      ii. In the acid-catalyzed reaction, the carbonyl oxygen is protonated, and neutral water adds to the carbonyl carbon.
   d. The catalysts have different effects.
      i. Base catalysis makes water a better nucleophile.
      ii. Acid catalysis makes the carbonyl carbon a better electrophile.
   e. Reactions of carbonyl groups with H-X, where X is electronegative, are reversible; the equilibrium favors the aldehyde or ketone.
2. Cyanohydrin formation (Section 19.7).
   a. HCN adds to aldehydes and ketones to give cyanohydrins.
      i. The reaction is base-catalyzed and proceeds through a tetrahedral intermediate.
      ii. Equilibrium favors the cyanohydrin adduct.
   b. Cyanohydrin formation is useful for the transformations that the –CN group can undergo.
      i. The –CN group can be reduced, to form an amine.
      ii. The –CN group can be hydrolyzed, to produce a carboxylic acid.
3. Addition of Grignard and hydride reagents (Section 19.8).
   a. Addition of Grignard reagents.
      i. Mg^{2+} complexes with oxygen, making the carbonyl group more electrophilic.
      ii. $R:^-$ adds to the carbonyl carbon to form a tetrahedral intermediate.
      iii. Water is added in a separate step to protonate the intermediate, yielding an alcohol.
      iv. Grignard reactions are irreversible because $R:^-$ is not a leaving group.
   b. Hydride addition.
      i. $LiAlH_4$ and $NaBH_4$ act as if they are $H:^-$ donors and add to carbonyl compounds to form tetrahedral alkoxide intermediates.
      ii. In a separate step, water is added to protonate the intermediate, yielding an alcohol.
4. Addition of amines (Section 19.9).
   a. Amines add to aldehydes and ketones to form imines and enamines.
   b. Imines are formed when a primary amine adds to an aldehyde or ketone.
      i. The process is acid-catalyzed.
      ii. A proton transfer converts the initial adduct to a carbinolamine.
      iii. Acid-catalyzed elimination of water yields an imine.

iv. The reaction rate maximum occurs at pH = 4.5. At this pH, [H] is high enough to catalyze elimination of water, but low enough so that the amine is nucleophilic.
v. Some imine derivatives are useful for characterizing aldehydes and ketones.

c. Enamines are produced when aldehydes and ketones react with secondary alcohols.
The mechanism is similar to that of imine formation, except a proton from the α carbon is lost in the dehydration step.

5. Addition of hydrazine: the Wolff-Kishner reaction (Section 19.10).
   a. Hydrazine reacts with aldehydes and ketones in the presence of KOH to form alkanes.
   The intermediate hydrazone undergoes base-catalyzed bond migration, loss of N_2 and protonation to form the alkane.
   b. The Wolff-Kishner reduction can also be used to convert an acylbenzene to an alkylbenzene.
6. Addition of alcohols: acetal formation (Section 19.11).
   a. In the presence of an acid catalyst, two equivalents of an alcohol can add to an aldehyde or ketone to produce an acetal.
      i. The initial intermediate is a hemiacetal (a hydroxy ether).
      ii. Protonation of –OH, loss of water, with formation of the oxonium ion, and addition of a second molecule of ROH yields the acetal.
   b. The reaction is reversible, but changing the reaction conditions can drive the reaction in either direction.
   c. Because acetals are inert to many reagents, they can be used as protecting groups in syntheses.
   Diols are often used as protecting groups, forming cyclic acetals.
7. The Wittig reaction (Section 19.12).
   a. The Wittig reaction converts an aldehyde or ketone to an alkene.
   b. Steps in the Wittig reaction:
      i. An alkyl halide reacts with triphenylphosphine to form an alkyltriphenylphosphonium salt.
      ii. Butyllithium converts the salt to an ylide.
      iii. The ylide adds to an aldehyde or ketone to from a dipolar betaine.
      iv. The betaine forms a four-membered ring intermediate, which decomposes to form the alkene and triphenylphosphine oxide.
   c. Uses of the Wittig reaction.
      i. The Wittig reaction can be used to produce mono-, di-, and trisubstituted alkenes, but steric hindrance keeps tetrasubstituted alkenes from forming.
      ii. The Wittig reaction produces pure alkenes of known stereochemistry (excluding *E,Z* isomers).
8. The Cannizzaro reaction (Section 19.13).
   a. The Cannizzaro reaction is unique in that the tetrahedral intermediate of addition of a nucleophile can expel a leaving group.
   b. Steps in the Cannizzaro reaction.
      i. HO^- adds to an aldehyde with no α hydrogens to form a tetrahedral intermediate.
      ii. H^- is expelled and adds to another molecule of aldehyde.
      iii. The result is a disproportionation reaction, in which one molecule of aldehyde is oxidized and a second molecule is reduced.
   c. The Cannizzaro reaction isn't synthetically useful, but it resembles the mode of action of the enzyme cofactor NADH.

9. Conjugate addition to α,β-unsaturated aldehydes and ketones (Section 19.14).
   a. Steps in conjugate addition.
      i. Because the double bond of an α,β-unsaturated aldehyde/ketone is conjugated with the carbonyl group, addition can occur at the β position, which is an electrophilic site.
      ii. Protonation of the α carbon of the enolate intermediate results in a product having a carbonyl group and a nucleophile with a 1,3 relationship.
   b. Conjugate addition of amines.
      i. Primary and secondary amines add to α,β-unsaturated aldehydes and ketones.
      ii. The conjugate addition product is often formed exclusively.
   c. Conjugate addition of organocopper reagents.
      i. Conjugate addition of organocopper reagents alkylates the double bond of α,β-unsaturated aldehydes and ketones.
      ii. This type of addition doesn't occur with other organometallic reagents.
      iii. The mechanism involves radicals and isn't a typical nucleophilic addition.
10. Biochemical nucleophilic addition reactions (Section 19.15).
   a. Some amino acids are formed by nucleophilic addition reactions.
   b. The defense mechanism of the millipede involves the reverse of cyanohydrin formation and results in the formation of HCN.

III. Spectroscopy of aldehydes and ketones (Section 19.16).
   A. IR spectroscopy.
      1. The C=O absorption of aldehydes and ketones occurs in the range 1660–1770 cm^{-1}.
         a. The exact position of absorption can be used to distinguish between an aldehyde and a ketone.
         b. The position of absorption also gives information about other structural features, such as unsaturation and angle strain.
         c. The absorption values are constant from one compound to another.
      2. Aldehydes also show absorptions in the range 2720–2820 cm^{-1}.
   B. NMR spectroscopy.
      1. ^1H NMR spectroscopy.
         a. Aldehyde protons absorb near 10 δ, and show spin-spin coupling with protons on the adjacent carbon.
         b. Hydrogens on the carbon next to a carbonyl group absorb near 2.0–2.3 δ. Methyl ketone protons absorb at 2.1 δ.
      2. ^{13}C NMR spectroscopy.
         a. The carbonyl-group carbons absorb in the range 190–215 δ.
         b. These absorptions characterize aldehydes and ketones.
         c. Unsaturation lowers the value of δ.
   C. Mass spectrometry.
         a. Some aliphatic aldehydes and ketones undergo McLafferty rearrangement.
            i. A hydrogen on the γ carbon is transferred to the carbonyl oxygen, the bond between the α carbon and the β carbon is broken, and a neutral alkene fragment is produced.
            ii. The remaining cation radical is detected.
         b. Alpha cleavage.
            i. The bond between the carbonyl group and the α carbon is cleaved.
            ii. The products are a neutral radical and an acyl cation, which is detected.

Solutions to Problems

19.1 Remember that the principal chain must contain the aldehyde or ketone group and that an aldehyde group occurs only at the end of a chain. The aldehyde carbon is carbon 1 in an acyclic compound, and the suffix *-carbaldehyde* is used when the aldehyde group is attached to a ring.

(a) $CH_3CH_2\overset{O}{\overset{\|}{C}}CH(CH_3)_2$

2-Methyl-3-pentanone

(b) $C_6H_5CH_2CH_2CHO$

3-Phenylpropanal

(c) $CH_3\overset{O}{\overset{\|}{C}}CH_2CH_2CH_2\overset{O}{\overset{\|}{C}}CH_2CH_3$

2,6-Octanedione

(d) [cyclohexane ring with CH_3, H and H, CHO substituents]

trans-2-Methylcyclohexanecarbaldehyde

(e) $OHCCH_2CH_2CH_2CHO$

Pentanedial

(f) [cyclohexanone ring with H_3C, H and H, CH_3 substituents]

cis-2,5-Dimethylcyclohexanone

(g) $CH_3CH_2CH(CH_3)CH(CH_2CH_2CH_3)\overset{O}{\overset{\|}{C}}CH_3$

4-Methyl-3-propyl-2-hexanone

(h) $CH_3CH{=}CHCH_2CH_2CHO$

4-Hexenal

19.2 The grouping –CHO is used to represent an aldehyde.

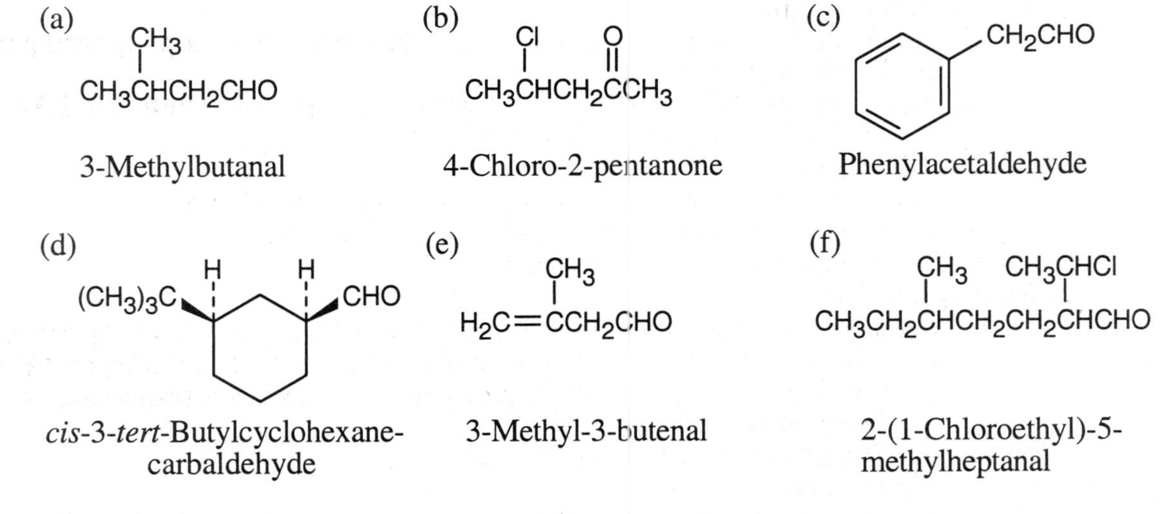

19.3 We have seen the first two methods of aldehyde preparation in earlier chapters.

(a)

$$CH_3CH_2CH_2CH_2CH_2OH \xrightarrow[CH_2Cl_2]{PCC} CH_3CH_2CH_2CH_2CHO$$

1-Pentanol

(b)

$$CH_3CH_2CH_2CH_2CH{=}CH_2 \xrightarrow[\text{2. Zn, } H_3O^+]{\text{1. } O_3} CH_3CH_2CH_2CH_2CHO + CH_2O$$

1-Hexene

(c)

$$CH_3CH_2CH_2CH_2COOCH_3 \xrightarrow[\text{2. } H_3O^+]{\text{1. DIBAH}} CH_3CH_2CH_2CH_2CHO$$

19.4 All of these methods are familiar.

(a)

$$CH_3CH_2C{\equiv}CCH_2CH_3 \xrightarrow[Hg(OAc)_2]{H_3O^+} CH_3CH_2CH_2\overset{O}{\overset{\|}{C}}CH_2CH_3$$

(b)

Benzene $\xrightarrow[AlCl_3]{CH_3COCl}$ acetophenone ($C_6H_5COCH_3$) $\xrightarrow[FeBr_3]{Br_2}$ m-bromoacetophenone (Br, $COCH_3$)

(c)

Bromobenzene (Br) $\xrightarrow{Mg}$ C_6H_5MgBr $\xrightarrow[\text{2. } H_3O^+]{\text{1. } CH_3CHO}$ $C_6H_5CH(OH)CH_3$ (HO–CHCH$_3$) $\xrightarrow[H_3O^+]{CrO_3}$ $C_6H_5COCH_3$ (O=C–CH$_3$)

(d)

1-Methylcyclohexene (CH_3) $\xrightarrow[\text{2. } H_2O_2, \ ^-OH]{\text{1. } BH_3, \text{ THF}}$ trans-2-methylcyclohexanol (H, CH_3, H, OH) $\xrightarrow[H_3O^+]{CrO_3}$ 2-methylcyclohexanone (CH_3, O)

19.5

$$(CH_3)_2C{=}O \ (\text{:O: } \delta-, \ C \ \delta+) + {}^-\text{:CN} \rightleftharpoons \left[(H_3C)_2C(\text{:}\ddot{O}\text{:}^-)CN \right] \underset{}{\overset{H_2O}{\rightleftharpoons}} (H_3C)_2C(OH)CN + HO^-$$

nucleophilic addition — protonation of tetrahedral intermediate — cyanohydrin

Cyanide anion adds to the positively polarized carbonyl carbon to form a tetrahedral intermediate. This intermediate is protonated to yield acetone cyanohydrin.

19.6

CH_3O–C₆H₄–CHO O_2N–C₆H₄–CHO

electron-donating electron-withdrawing

The electron-withdrawing nitro group makes the aldehyde carbon of *p*-nitrobenzaldehyde more electron-poor (more electrophilic) and more reactive toward nucleophiles than the aldehyde carbon of *p*–methoxybenzaldehyde.

19.7

Cl_3C–CHO (δ– on O, δ+ on C) + $:OH_2$ ⇌ [Cl_3C–CH(O^-)–OH_2^+] ⇌ Cl_3C–CH(OH)–OH

Chloral hydrate

19.8

$R_2C{=}O$ + $:\overset{*}{O}H_2$ ⇌ [$R_2C(O^-)(\overset{*}{O}H_2^+)$] ⇌ $R_2C(OH)(\overset{*}{O}H)$

The above mechanism is similar to other nucleophilic addition mechanisms we have studied. Since all steps are reversible, we can write the above mechanism in reverse to show how labeled oxygen is incorporated into an aldehyde or ketone.

$R_2C(OH)(\overset{*}{O}H)$ ⇌ [$R_2C(OH_2^+)(\overset{*}{O}^-)$] ⇌ $R_2C{=}O^*$ + H_2O

This exchange is very slow in water but proceeds more rapidly when either acid or base is present.

19.9

2,2,6-Trimethylcyclohexanone + ^-CN ⇌ [alkoxide: NC, O^-, H_3C, CH_3, CH_3] $\overset{HCN}{\rightleftharpoons}$ cyanohydrin (NC, OH, H_3C, CH_3, CH_3) + ^-CN

2,2,6-Trimethylcyclohexanone

Cyanohydrin formation is an equilibrium process. Because formation of the product of addition of HCN to 2,2,6-trimethylcyclohexanone is sterically hindered by the three methyl groups, the equilibrium lies toward the side of the unreacted ketone.

19.10

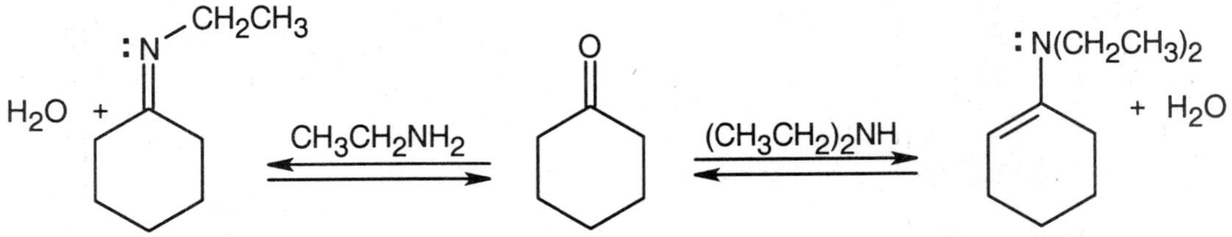

H_2O + imine (=N–CH_2CH_3) ⇌ ($CH_3CH_2NH_2$) cyclohexanone ⇌ ($(CH_3CH_2)_2NH$) enamine (N$(CH_2CH_3)_2$) + H_2O

imine enamine

Reaction of a ketone or aldehyde with a primary amine yields an imine, in which C=O has been replaced by C=NR. Reaction of a ketone or aldehyde with a secondary amine yields an enamine, in which C=O has been replaced by C–NR_2, and the double bond has moved.

19.11

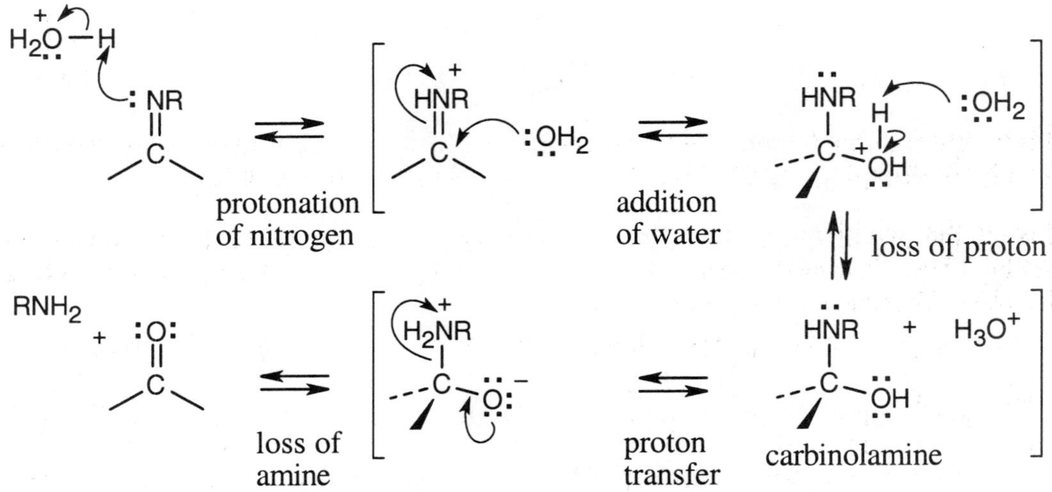

protonation of nitrogen

addition of water

loss of proton

RNH_2 + C=O

loss of amine

proton transfer

carbinolamine

H_3O^+

19.12 The structure is an enamine, which is prepared from a ketone and a secondary amine. Find the components, and draw the reaction.

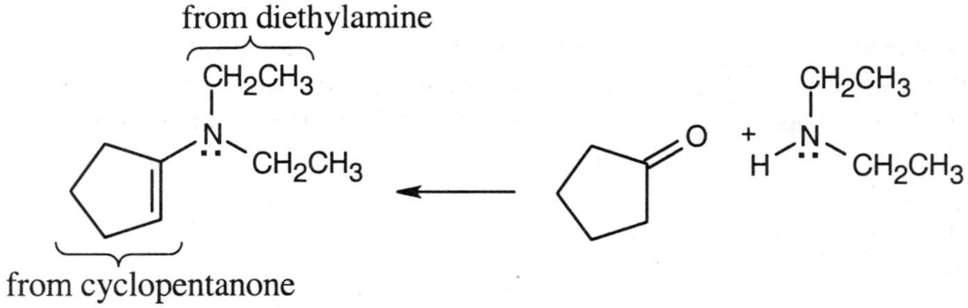

from diethylamine

from cyclopentanone

19.13

(a)

protonation of oxygen

addition of –OH

loss of proton

hemiacetal

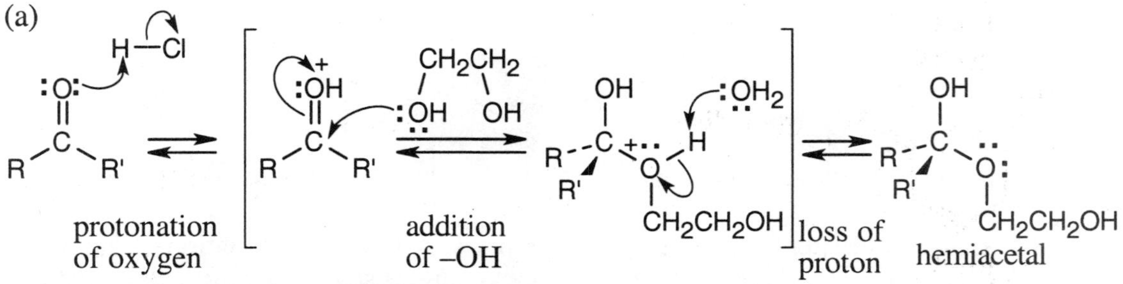

Formation of the hemiacetal is the first step.

(b)

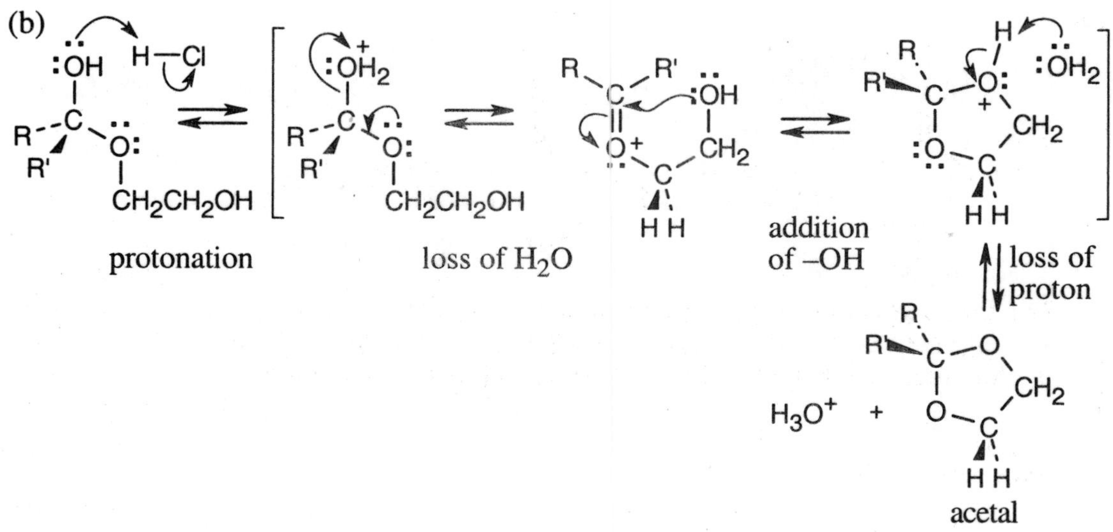

:ÖH H–Cl

R–C–Ö: R'

CH_2CH_2OH

⇌ [:ÖH₂ (+) R–C–Ö: R' CH_2CH_2OH ⇌ R, R' C=O+ :ÖH CH_2 C H H ⇌ R, R' C –O: + H :ÖH₂ CH_2 :O–C H H]

protonation

loss of H_2O

addition of –OH

loss of proton

R, R' C–O CH_2 O–C H H

H_3O^+ +

acetal

Protonation of the hemiacetal hydroxyl group is followed by loss of water. Attack by the second hydroxyl group of ethylene glycol forms the cyclic acetal ring.

19.14 Locate the two identical –OR groups to identify the alcohol that was used to form the acetal. (The illustrated acetal was formed from methanol.) Replace these two –OR groups by =O to find the carbonyl compound.

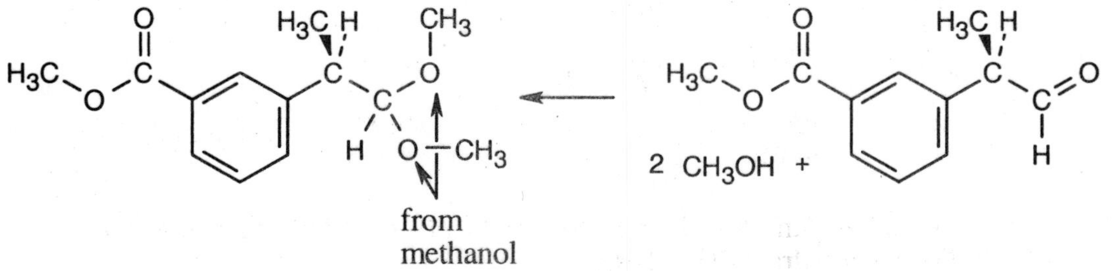

H_3C–O–C(=O)– (benzene ring) –C(H_3C, H)–C(H)(O–CH_3)(O–CH_3) ⟵ $2\ CH_3OH$ + H_3C–O–C(=O)– (benzene ring) –C(H_3C, H)–C(=O)–H

from methanol

19.15 Locate the double bond that is formed by the Wittig reaction. The simpler or less substituted component comes from the ylide, and the more substituted component comes from the aldehyde or ketone. Triphenylphosphine oxide is a byproduct of all these reactions.

(a)

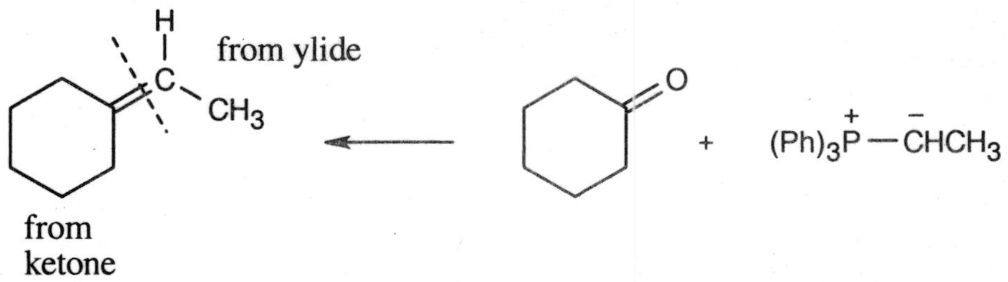

H

C

CH_3

from ylide

from ketone

⟵ (cyclohexanone) + $(Ph)_3\overset{+}{P}-\overset{-}{C}HCH_3$

(b)

CH_2 from ylide

from ketone

⟵ (cyclohexenone) + $(Ph)_3\overset{+}{P}-\overset{-}{C}H_2$

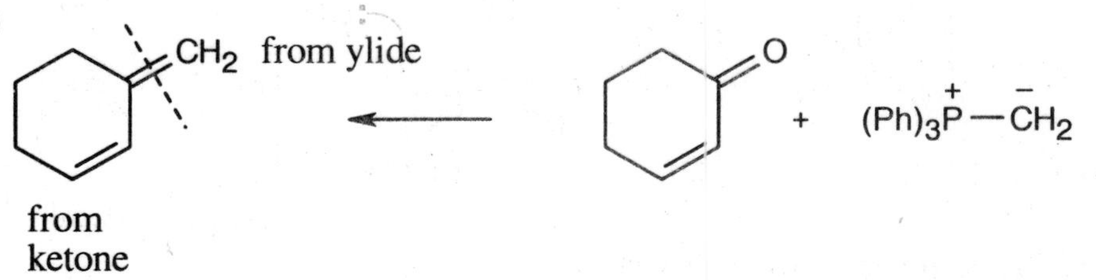

(c)

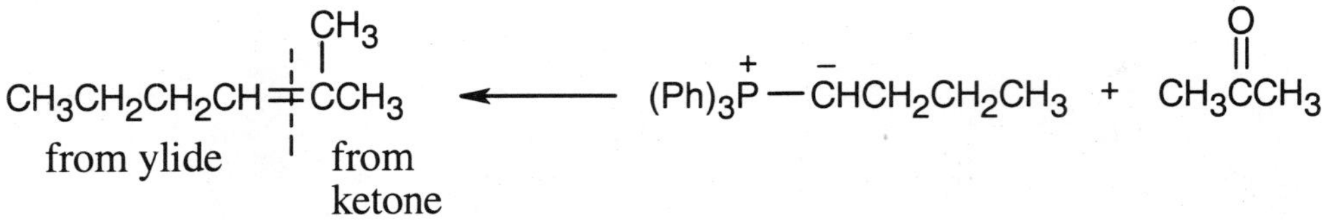

$$CH_3CH_2CH_2CH{=}C(CH_3)CH_3 \longleftarrow (Ph)_3\overset{+}{P}-\overset{-}{C}HCH_2CH_2CH_3 + CH_3COCH_3$$

from ylide | from ketone

(d)

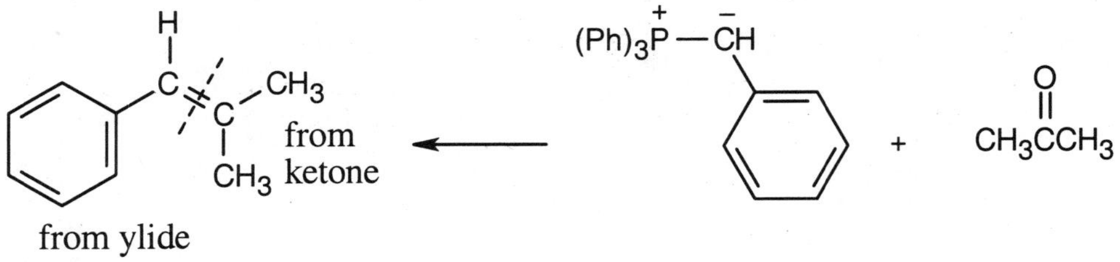

$$C_6H_5CH{=}C(CH_3)_2 \longleftarrow (Ph)_3\overset{+}{P}-\overset{-}{C}HC_6H_5 + CH_3COCH_3$$

from ylide | from ketone

(e)

$$C_6H_5CH{=}CHC_6H_5 \longleftarrow (Ph)_3\overset{+}{P}-\overset{-}{C}HC_6H_5 + C_6H_5CHO$$

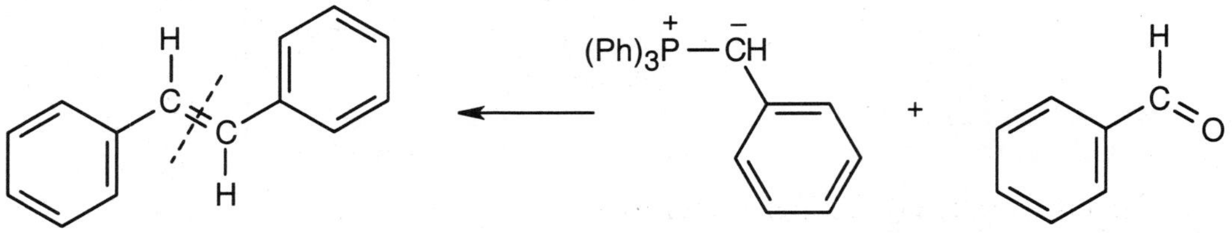

The *Z* isomer is also produced.

19.16

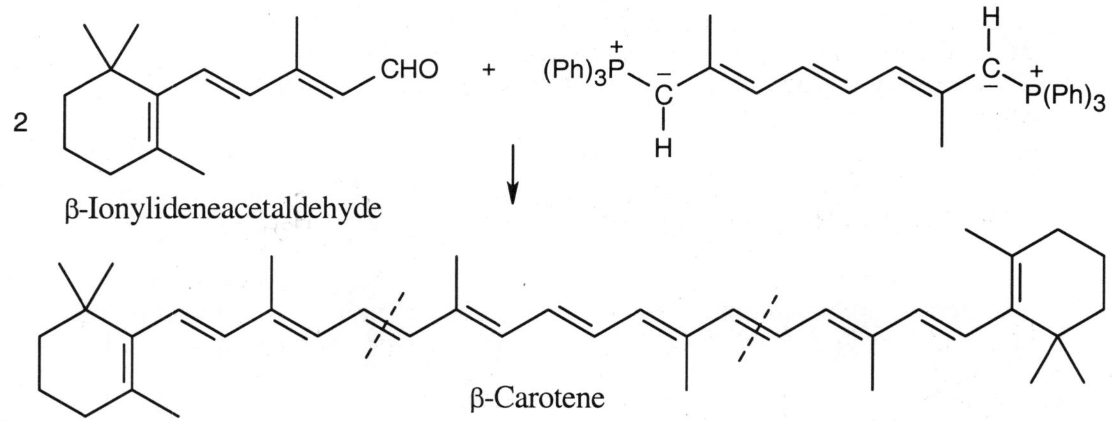

2 β-Ionylideneacetaldehyde (CHO) + $(Ph)_3\overset{+}{P}-\overset{-}{C}H-\cdots-\overset{-}{C}H-\overset{+}{P}(Ph)_3$

↓

β-Carotene

19.17

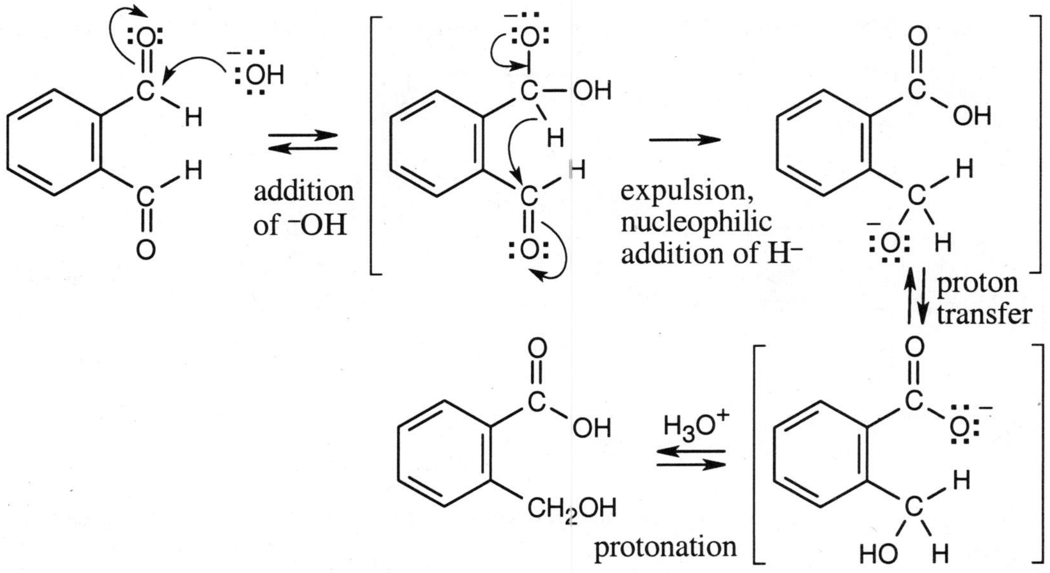

addition of ^{-}OH

expulsion, nucleophilic addition of H^{-}

proton transfer

H_3O^+

protonation

This is an internal Cannizzaro reaction.

19.18 To determine the reactants that form a conjugate addition product, follow these steps:

(1) Give to the aldehyde or ketone carbon the number "1", and count two carbons away from the carbonyl carbon. The double bond in the α, β-unsaturated starting material connected the carbons numbered "2" and "3".
(2) The grouping bonded to the "3" carbon (circled here) came from the alkyllithium reagent.

(a)

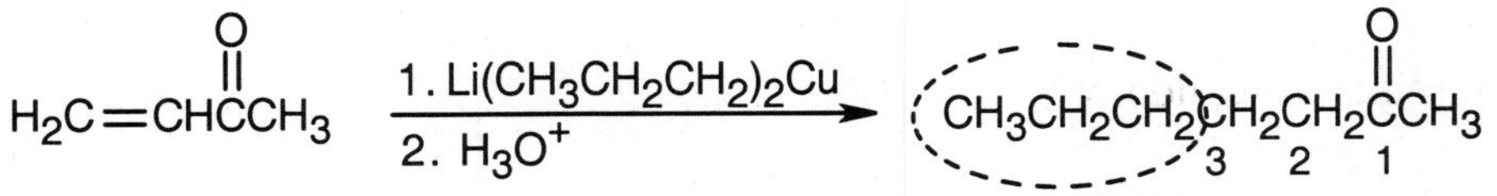

$$H_2C{=}CHCCH_3 \xrightarrow[\text{2. } H_3O^+]{\text{1. } Li(CH_3CH_2CH_2)_2Cu} CH_3CH_2CH_2CH_2CH_2CCH_3$$

2-Heptanone

(b)

1. $Li(CH_3)_2Cu$
2. H_3O^+

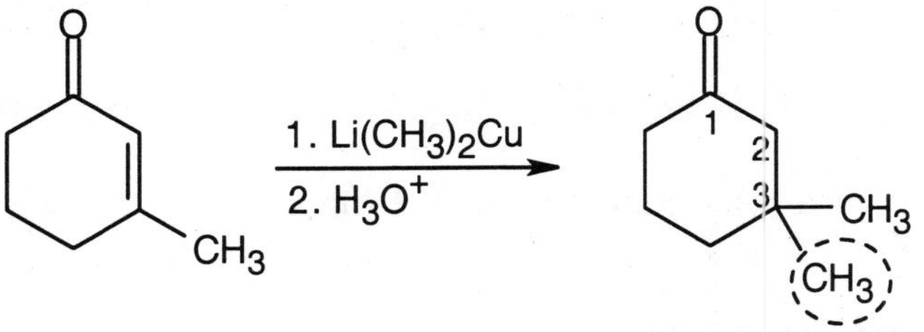

3,3-Dimethylcyclohexanone

(c)

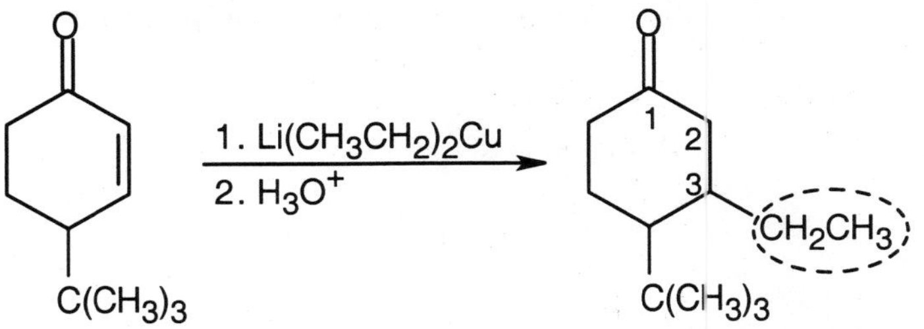

1. $Li(CH_3CH_2)_2Cu$
2. H_3O^+

$C(CH_3)_3$

4-*tert*-Butyl-3-ethylcyclohexanone

(d)

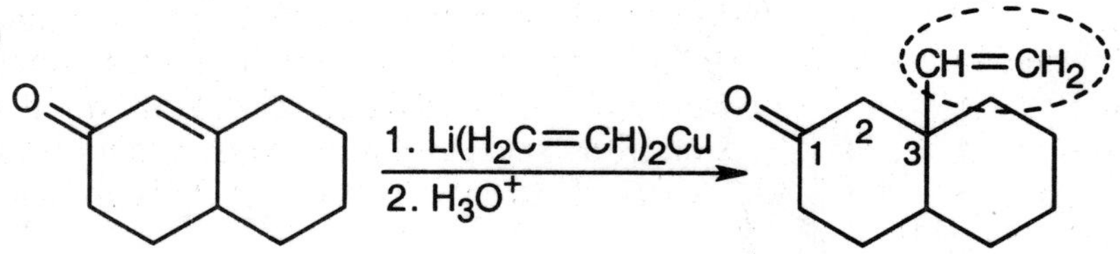

1. $Li(H_2C{=}CH)_2Cu$
2. H_3O^+

$CH{=}CH_2$

1 2 3

19.19

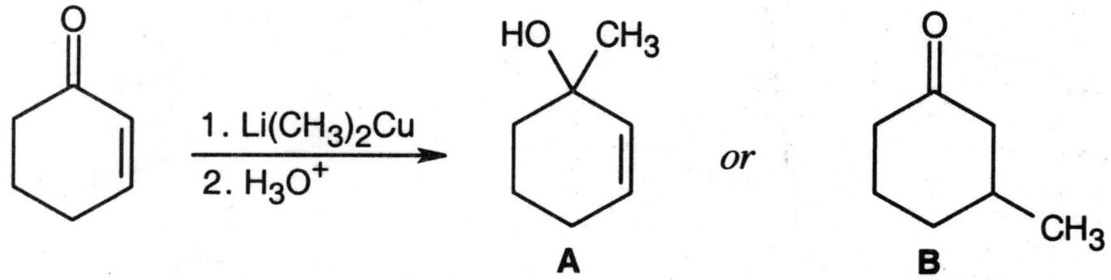

1. $Li(CH_3)_2Cu$
2. H_3O^+

HO CH_3 or CH_3

2-Cyclohexenone

A
1-Methyl-2-cyclohexen-1-ol

B
3-Methylcyclohexanone

2-Cyclohexenone is a cyclic α,ß-unsaturated ketone whose carbonyl IR absorption occurs at 1685 cm^{-1}. If direct addition product **A** is formed, the carbonyl absorption will vanish and a hydroxyl absorption will appear at 3300 cm^{-1}. If conjugate addition produces **B**, the carbonyl absorption will shift to 1715 cm^{-1}, where 6-membered-ring saturated ketones absorb.

19.20 (a) $H_2C{=}CHCH_2COCH_3$ absorbs at 1715 cm^{-1}. (4-Penten-2-one is not an α,ß–unsaturated ketone.)

(b) $CH_3CH{=}CHCOCH_3$ is an α,ß–unsaturated ketone and absorbs at 1685 cm^{-1}.

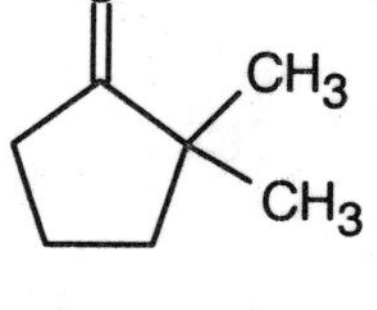

CH_3 CH_3 Cl C H

(c) 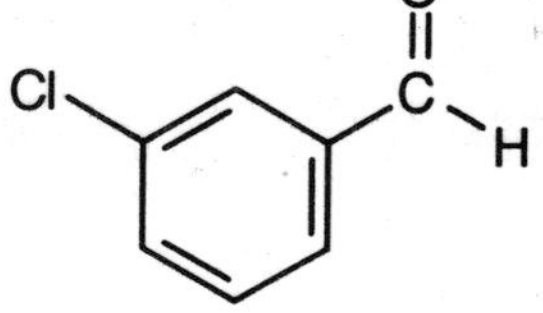 (d) 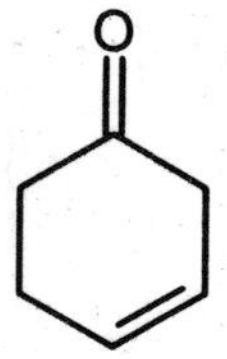 (e)

(c) 2,2-Dimethylcyclopentanone, a five-membered-ring ketone, absorbs at 1750 cm^{-1}.

(d) *m*-Chlorobenzaldehyde shows a singlet absorption at 1705 cm^{-1} and a doublet at 2720 cm^{-1} and 2820 cm^{-1}.

(e) 3-Cyclohexenone absorbs at 1715 cm^{-1}.

(f) $CH_3CH_2CH_2CH{=}CHCHO$ is an α,ß–unsaturated aldehyde and absorbs at 1705 cm^{-1}.

19.21 In mass spectra, only charged particles are detected. The McLafferty rearrangement produces an uncharged alkene (not detected) and an oxygen-containing fragment, which is a cation radical and is detected. Alpha cleavage produces a neutral radical (not detected) and an oxygen-containing cation, which is detected. Since alpha cleavage occurs primarily on the more substituted side of the aldehyde or ketone, only this cleavage is shown.

(a)

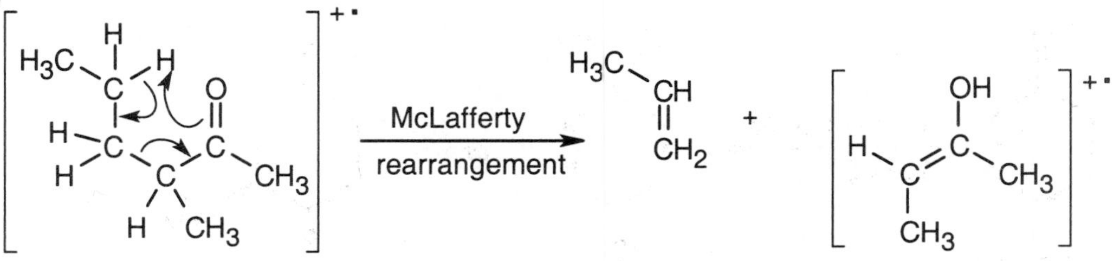

McLafferty rearrangement

$H_3C{-}CH{=}CH_2$ + $[H(CH_3)C{=}C(OH)CH_3]^{+\cdot}$

m/z = 72

Alpha cleavage

$CH_3CH_2CH_2\dot{C}HCH_3$ + $[CH_3C{\equiv}O]^{+}$

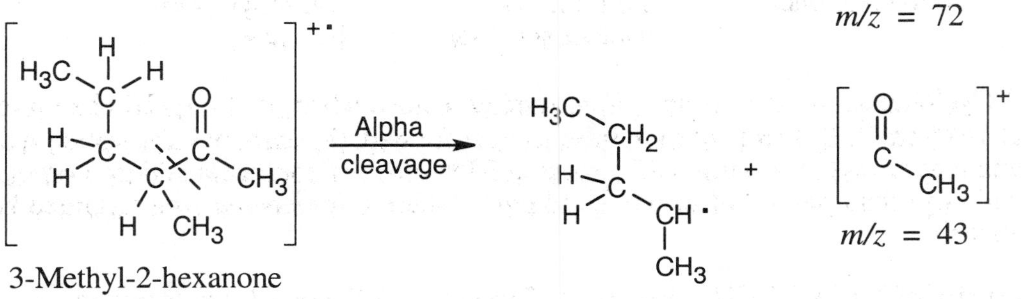

m/z = 43

3-Methyl-2-hexanone

m/z = 114

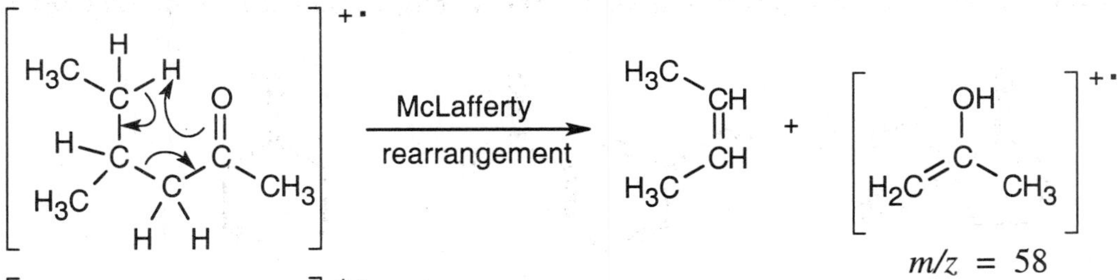

McLafferty rearrangement

$H_3C{-}CH{=}CH{-}CH_3$ + $[H_2C{=}C(OH)CH_3]^{+\cdot}$

m/z = 58

Alpha cleavage

$CH_3CH_2CH(CH_3)\dot{C}H_2$ + $[CH_3C{\equiv}O]^{+}$

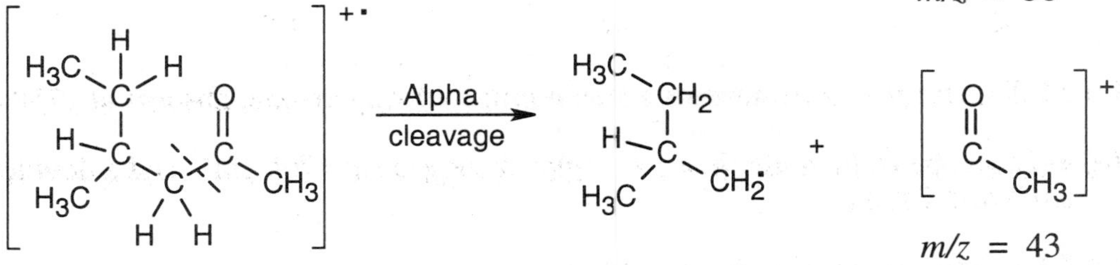

m/z = 43

4-Methyl-2-hexanone

m/z = 114

Both isomers exhibit peaks at $m/z = 43$ due to α-cleavage. The products of McLafferty rearrangement, however, occur at different values of m/z and can be used to identify each isomer.

(b)

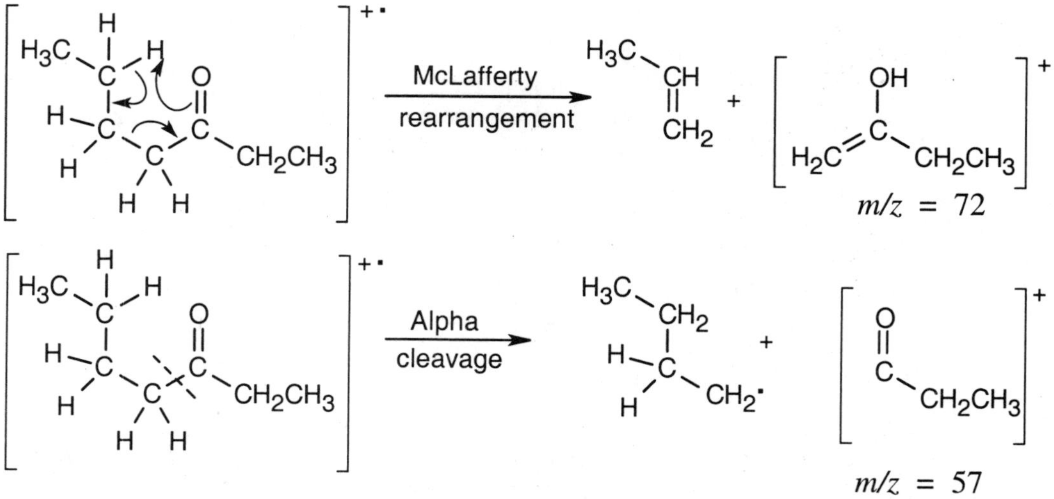

McLafferty rearrangement: $[CH_3CH_2CH_2CH_2COCH_2CH_3]^{+\cdot} \rightarrow H_3C{-}CH{=}CH_2 + [H_2C{=}C(OH){-}CH_2CH_3]^{+\cdot}$ m/z = 72

Alpha cleavage: $[CH_3CH_2CH_2CH_2COCH_2CH_3]^{+\cdot} \rightarrow H_3C{-}CH_2{-}CH_2{-}CH_2\cdot + [O{=}C{-}CH_2CH_3]^{+}$ m/z = 57

3-Heptanone m/z = 114

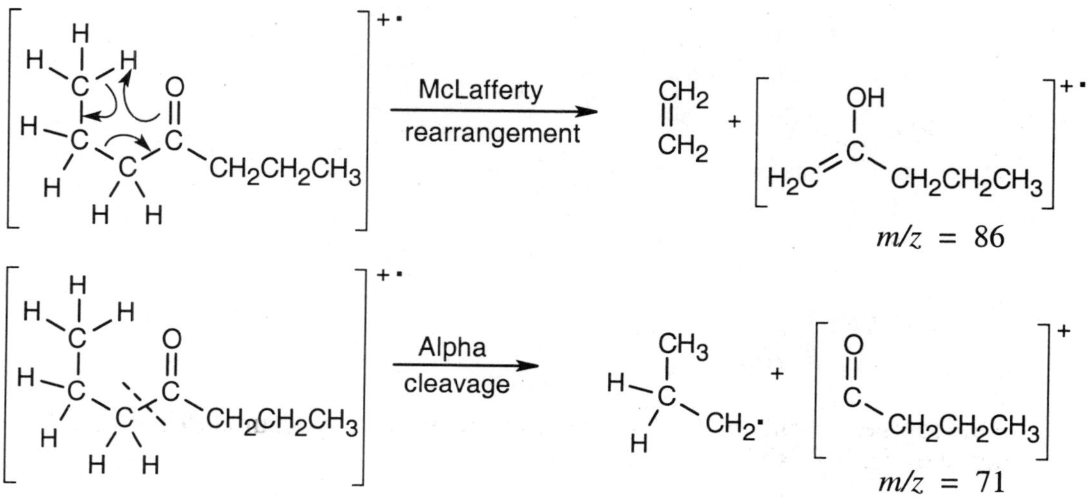

McLafferty rearrangement: $[CH_3CH_2CH_2COCH_2CH_2CH_3]^{+\cdot} \rightarrow CH_2{=}CH_2 + [H_2C{=}C(OH){-}CH_2CH_2CH_3]^{+\cdot}$ m/z = 86

Alpha cleavage: $[CH_3CH_2CH_2COCH_2CH_2CH_3]^{+\cdot} \rightarrow CH_3{-}CH_2{-}CH_2\cdot + [O{=}C{-}CH_2CH_2CH_3]^{+}$ m/z = 71

4-Heptanone m/z = 114

The isomers can be distinguished on the basis of both α–cleavage products (m/z = 57 *vs* m/z = 71) and McLafferty rearrangement products (m/z = 72 *vs* m/z = 86).

(c)

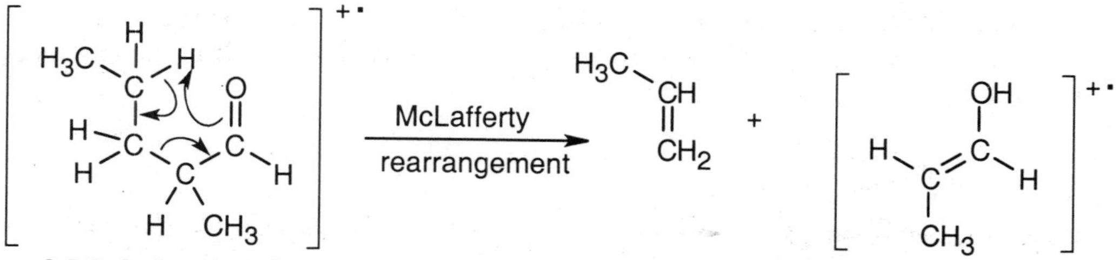

McLafferty rearrangement: $[CH_3CH_2CH_2CH(CH_3)CHO]^{+\cdot} \rightarrow H_3C{-}CH{=}CH_2 + [H(CH_3)C{=}C(OH){-}H]^{+\cdot}$ m/z = 58

2-Methylpentanal
m/z = 100

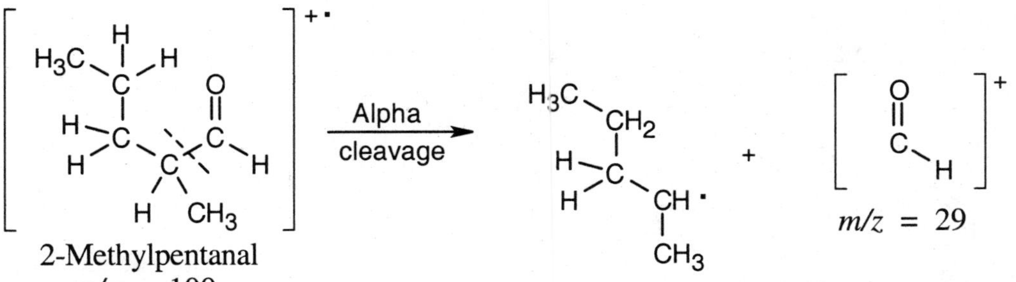

2-Methylpentanal
$m/z = 100$

Alpha cleavage

$m/z = 29$

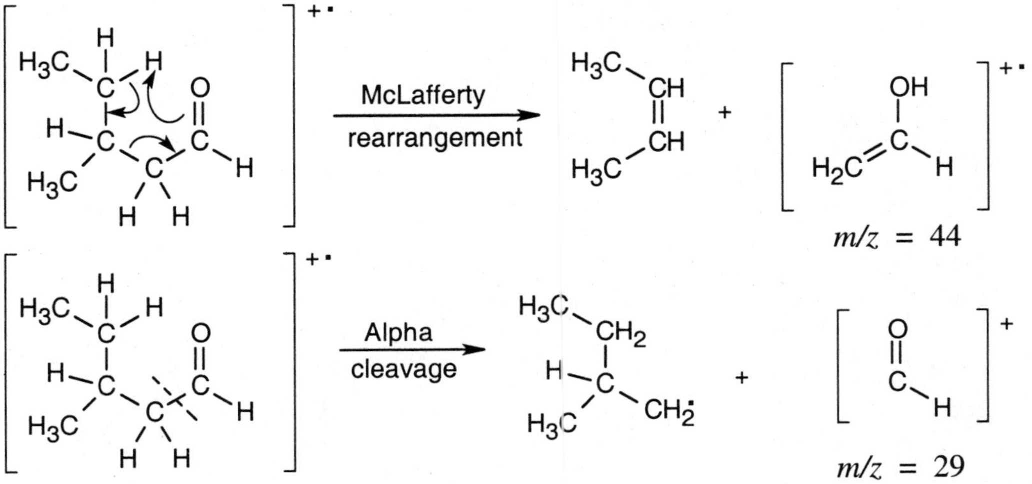

McLafferty rearrangement

$m/z = 44$

Alpha cleavage

$m/z = 29$

3-Methylpentanal
$m/z = 100$

The fragments from McLafferty rearrangement, which occur at different values of m/z, serve to distinguish the two isomers.

Visualizing Chemistry

19.22 It helps to know that all of these substances were prepared from aldehydes or ketones. Look for familiar groupings of atoms to recognize the starting materials.

(a) Notice that the substance pictured is a cyclic acetal. The starting materials were a diol (because cyclic acetals are prepared from diols) and an aldehyde (because an –H is bonded to the acetal carbon). Replace the two –OR groups with =O to find the aldehyde starting material.

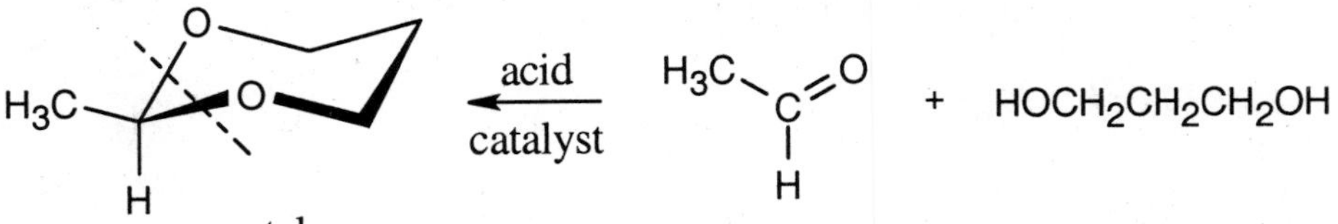

acetal $\xleftarrow[\text{catalyst}]{\text{acid}}$ H_3CCHO + $HOCH_2CH_2CH_2OH$

(b) We know that the product is an imine because it contains a carbon-nitrogen double bond. The carbon that is part of the C=N bond came from a ketone, and the nitrogen came from a primary amine.

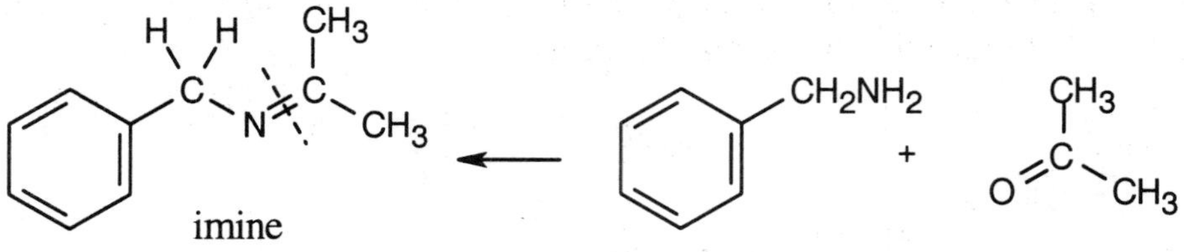

H H CH_3 C N C CH_3 ← CH_2NH_2 + CH_3 O=C CH_3

imine

(c) The product is an enamine, formed from a ketone and a secondary amine. Nitrogen is bonded to the carbon that once bore the carbonyl oxygen.

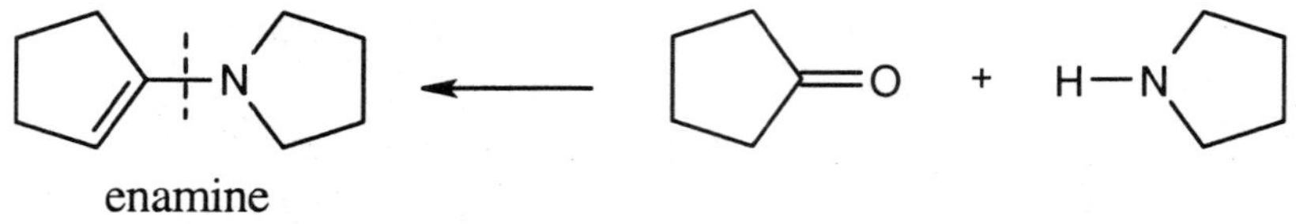

N ← =O + H—N

enamine

(d) The secondary alcohol product might have been formed by either of two routes – by reduction of a ketone or by Grignard addition to an aldehyde.

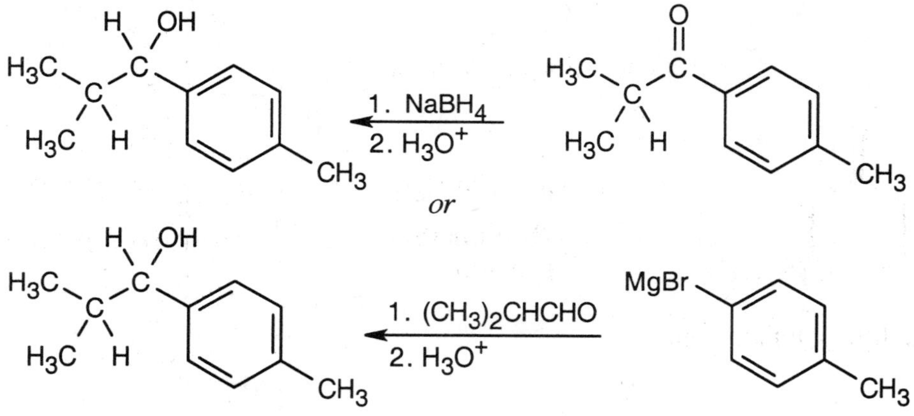

H OH H_3C C C H_3C H CH_3 ← 1. $NaBH_4$ 2. H_3O^+ — O H_3C C C H_3C H CH_3

or

H OH H_3C C C H_3C H CH_3 ← 1. $(CH_3)_2CHCHO$ 2. H_3O^+ — MgBr CH_3

19.23 The intermediate is a carbinolamine, resulting from the addition of an amine to a ketone. The product is an enamine because the amine nitrogen in the carbinolamine intermediate comes from a secondary amine.

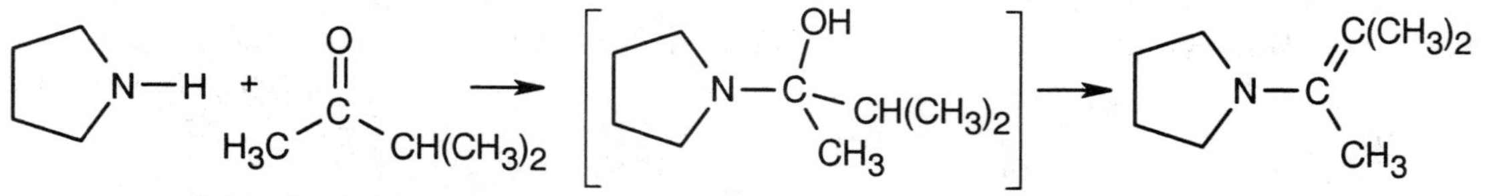

N—H + O C H_3C $CH(CH_3)_2$ → [OH N—C $CH(CH_3)_2$ CH_3] → N—C $C(CH_3)_2$ CH_3

3-Methyl-2-butanone

19.24 (a) The nitrogen atom is sp^2-hybridized, and nitrogen and the carbons bonded to it lie in a plane.
(b) A *p* orbital holds the lone-pair electrons of nitrogen.
(c) The *p* orbital holding the lone-pair electrons of nitrogen is aligned for overlap with the π electrons of the enamine double bond. With this geometry, the nitrogen lone-pair electrons can be conjugated with the double bond.

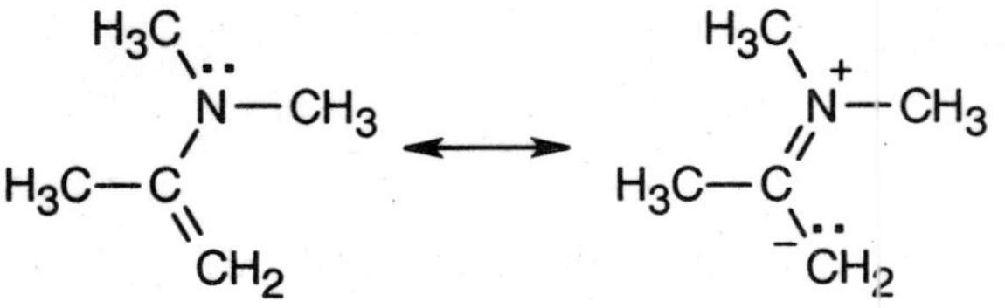

H_3C, $N{-}CH_3$, $H_3C{-}C$, CH_2 ⟷ H_3C, $\overset{+}{N}{-}CH_3$, $H_3C{-}C$, $^{-}CH_2$

Additional Problems

19.25

(a) $CH_3\overset{O}{\overset{\|}{C}}CH_2Br$
Bromoacetone

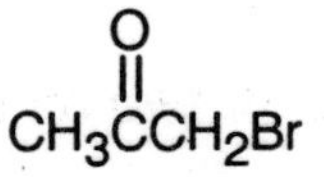

(b) O_2N, CHO, NO_2
3,5-Dinitrobenzenecarbaldehyde

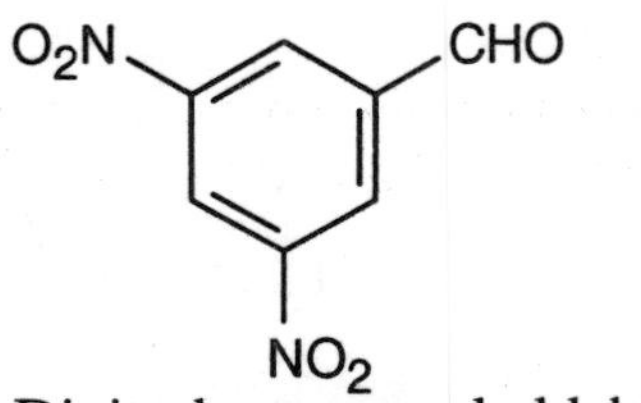

(c) $CH_3CH_2CH_2CH_2\overset{O}{\overset{\|}{C}}CH(CH_3)_2$
2-Methyl-3-heptanone

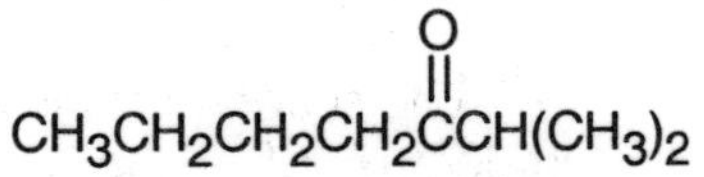

(d) H_3C, CH_3
3,5-Dimethylcyclohexanone

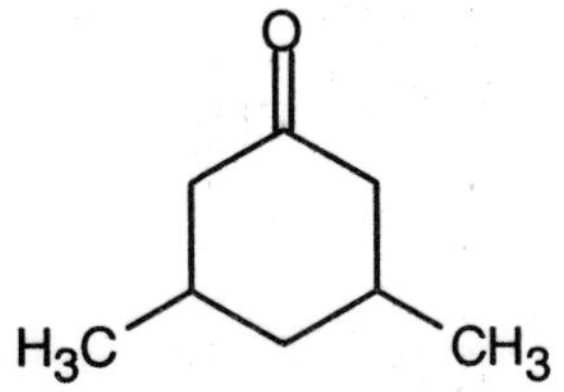

(e) $(CH_3)_3C\overset{O}{\overset{\|}{C}}C(CH_3)_3$
2,2,4,4-Tetramethyl-3-pentanone

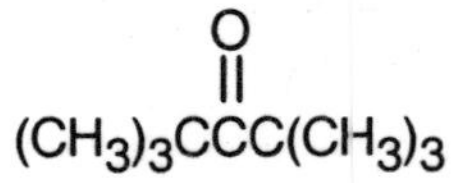

(f) $CH_3\overset{CH_3}{\overset{|}{C}}{=}CH\overset{O}{\overset{\|}{C}}CH_3$
4-Methyl-3-penten-2-one

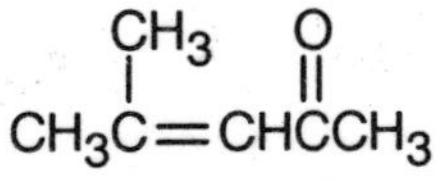

(g) $OHCCH_2CH_2CHO$
Butanedial

(h) $-CH{=}CHCHO$
3-Phenyl-2-propenal

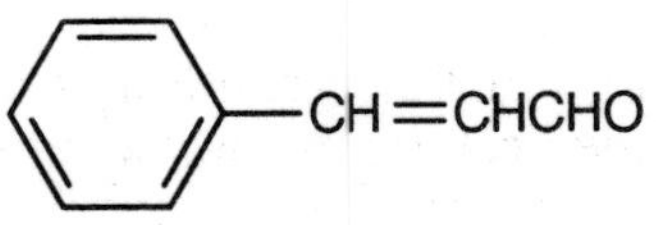

(i) H_3C, H_3C, O
6,6-Dimethyl-2,4-cyclohexadienone

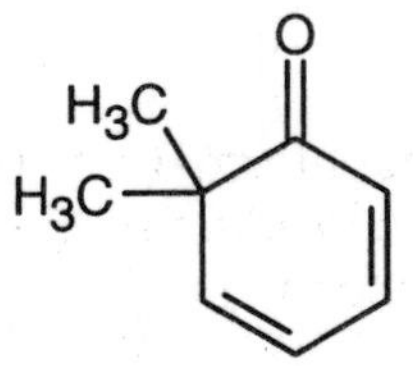

(j) O, C, CH_3, O_2N
p-Nitroacetophenone

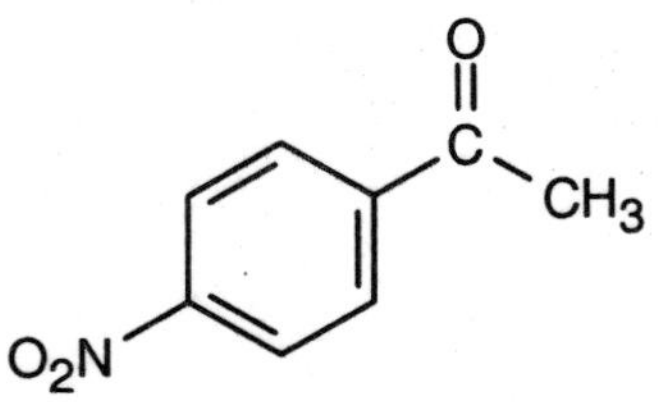

(k) CHO, *S*, H, H_3C, OH
(*S*)-2-Hydroxypropanal

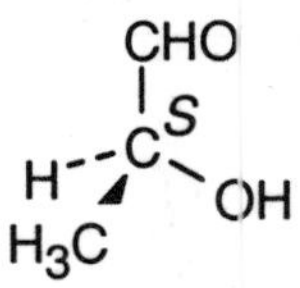

(l) CHO, HO, H, *S*, *R*, H, OH, CH_2OH
(2*S*,3*R*)-2,3,4-Trihydroxybutanal

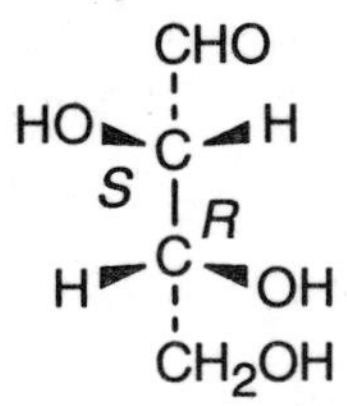

19.26

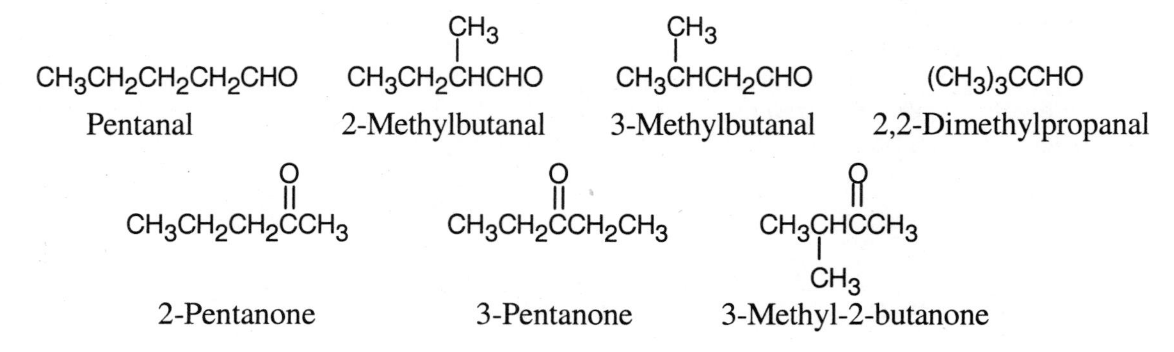

19.27

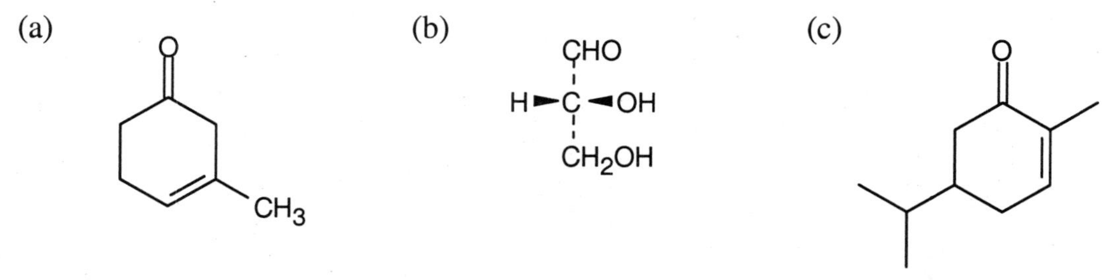

3-Methyl-3-cyclohexenone (*R*)-2,3-Dihydroxypropanal (D-Glyceraldehyde) 5-Isopropyl-2-methyl-2-cyclohexenone

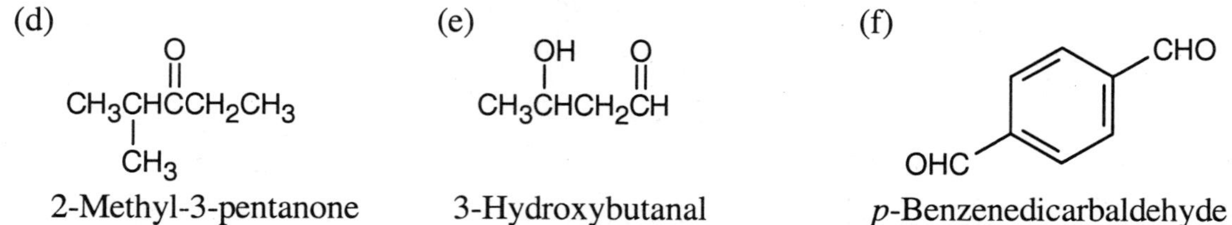

2-Methyl-3-pentanone 3-Hydroxybutanal *p*-Benzenedicarbaldehyde

19.28 (a) The α,ß–unsaturated ketone C_6H_8O contains one ring. Possible structures include:

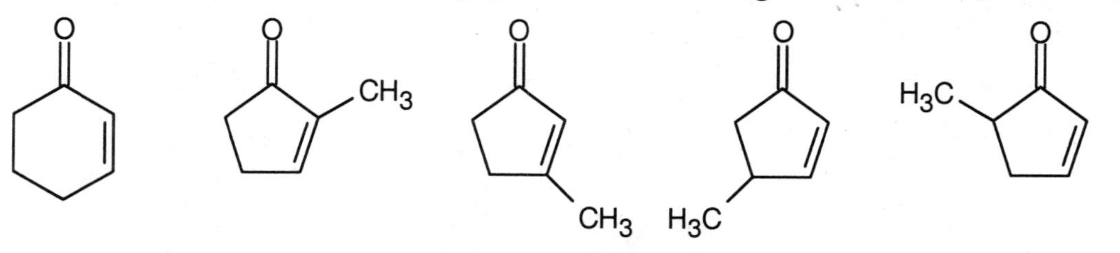

Cyclobutenones and cyclopropenones are also possible.

(b)

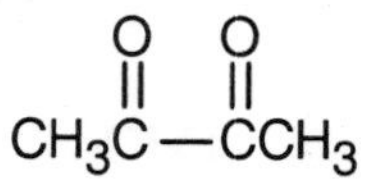

$CH_3\overset{O}{\overset{||}{C}}—\overset{O}{\overset{||}{C}}CH_3$ and many other structures.

(c)

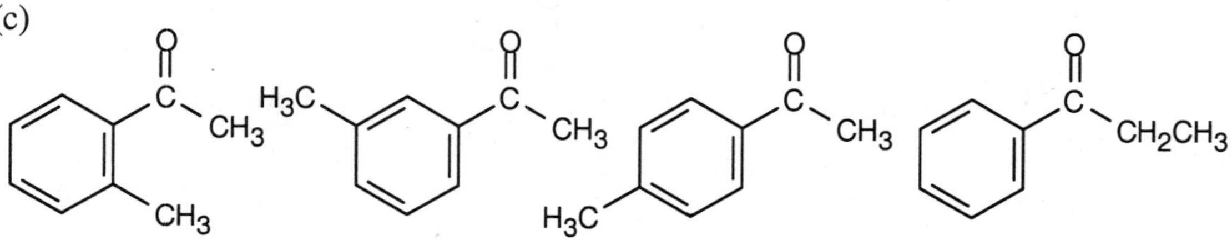

C, CH3, CH3 H3C, C, CH3 C, CH3, H3C C, CH2CH3

(d) cyclohexene-CHO (1-cyclohexenecarbaldehyde) and many other structures.

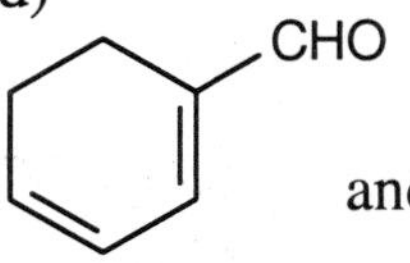

19.29

(a) $C_6H_5CH_2CH_2OH$

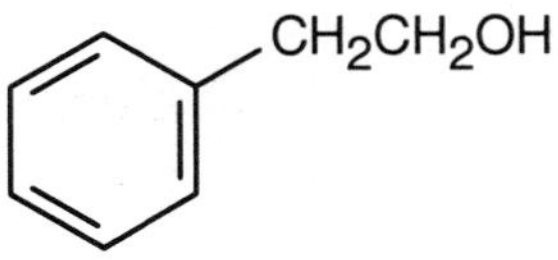

(b) $C_6H_5CH_2COOH$

(c) $C_6H_5CH_2CH{=}NOH$

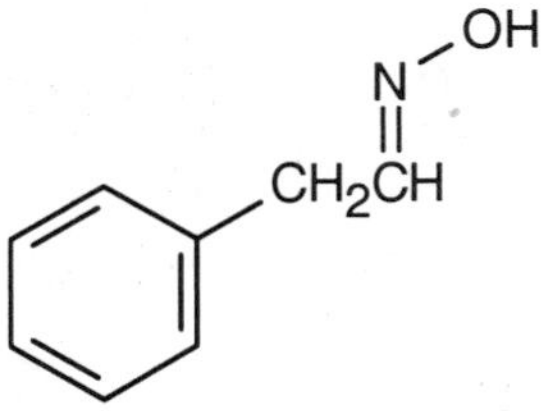

(d) $C_6H_5CH_2CH(OH)CH_3$

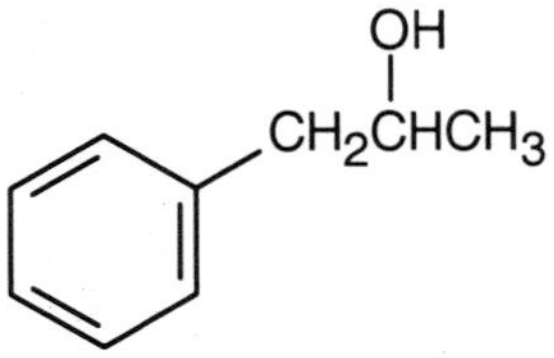

(e) $C_6H_5CH_2CH(OCH_3)OCH_3$

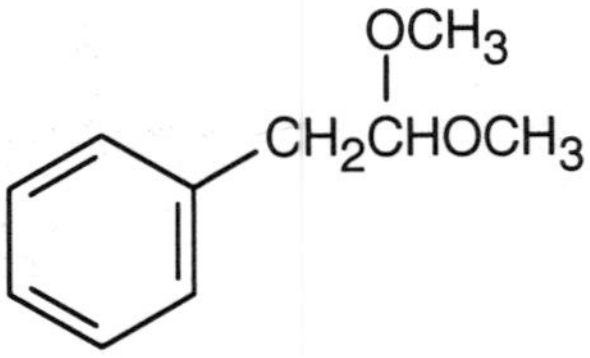

(f) $C_6H_5CH_2CH_3$

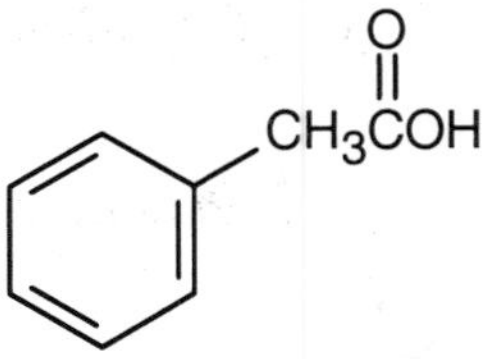

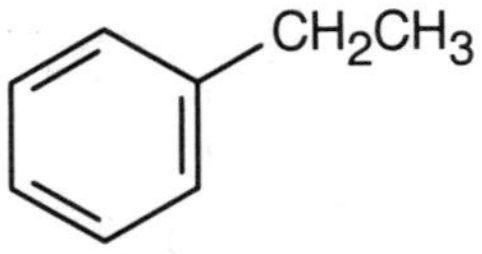

(g) $C_6H_5CH_2CH{=}CH_2$

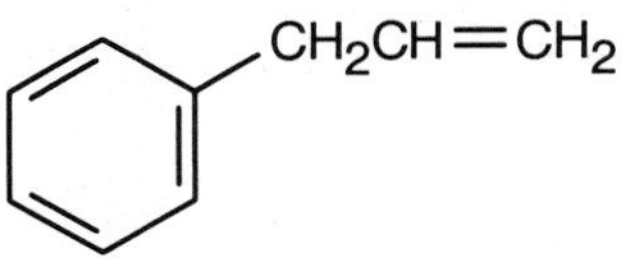

(h) $C_6H_5CH_2CH(OH)CN$

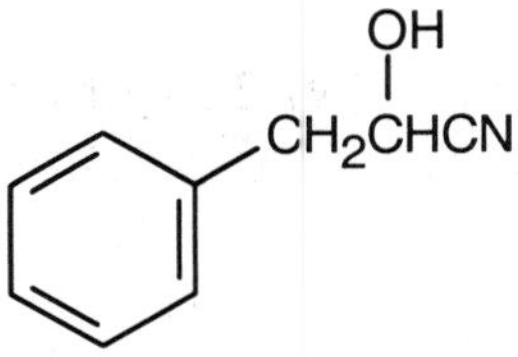

19.30

(a) $C_6H_5CH(OH)CH_3$

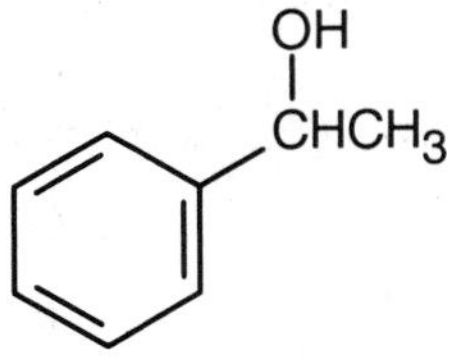

(b) no reaction

(c) $C_6H_5C(CH_3){=}NOH$

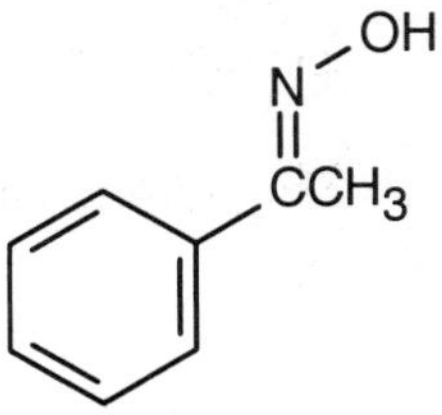

(d) $C_6H_5C(OH)(CH_3)_2$

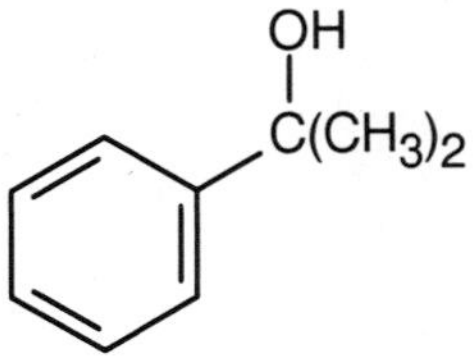

(e) $C_6H_5C(CH_3)(OCH_3)OCH_3$

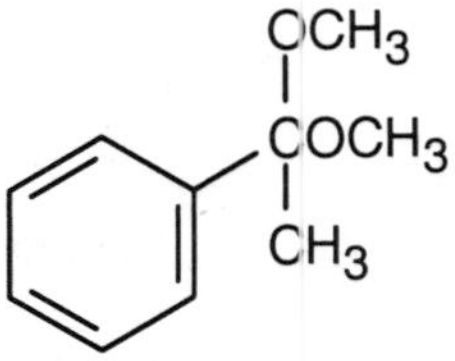

(f) $C_6H_5CH_2CH_3$

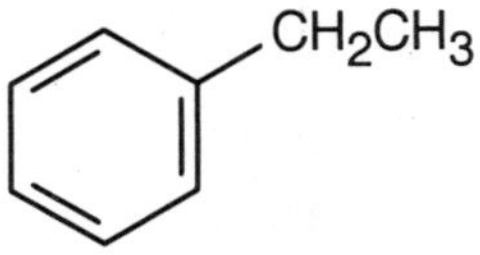

(g) $C_6H_5C(CH_3){=}CH_2$

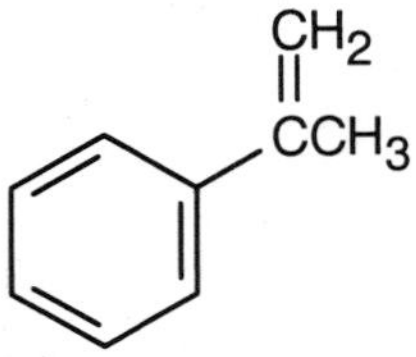

(h) $C_6H_5C(OH)(CH_3)CN$

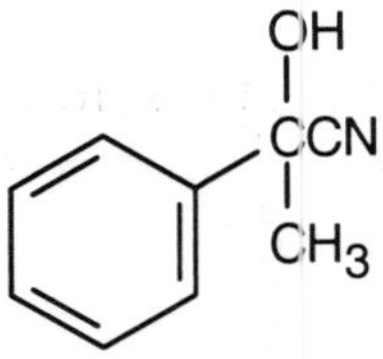

19.31

(a)

$\xrightarrow[\text{KOH}]{H_2NNH_2}$

(b)

$\xrightarrow[\text{2. } H_3O^+]{\text{1. } Li(C_6H_5)_2Cu}$

C_6H_5

(c)

$\xrightarrow[\text{2. } H_3O^+]{\text{1. } Li(H_2C{=}CH)_2Cu}$

$CH{=}CH_2$ $\xrightarrow[H_3O^+]{KMnO_4}$

COOH

(d)

$\xrightarrow[\text{2. } H_3O^+]{\text{1. } Li(CH_3)_2Cu}$

CH_3 $\xrightarrow[\text{KOH}]{H_2NNH_2}$

CH_3

or

$\xrightarrow{(C_6H_5)_3\overset{+}{P}-\overset{-}{C}H_2}$

CH_2 $\xrightarrow[\text{Pd/C}]{H_2}$

CH_3

19.32 Remember:

RCH_2-X + $(C_6H_5)_3P:$ $\longrightarrow$ $(C_6H_5)_3\overset{+}{P}CH_2R\ X^-$

alkyl halide; Triphenyl phosphine; phosphonium salt

$(C_6H_5)_3\overset{+}{P}CH_2R\ X^-$ + $CH_3CH_2CH_2CH_2^-\ Li^+$ $\longrightarrow$ $(C_6H_5)_3\overset{+}{P}-\overset{-}{C}HR$

phosphonium salt; Butyllithium; ylide

$(C_6H_5)_3\overset{+}{P}-\overset{-}{C}HR$ + $O{=}C\langle$ $\longrightarrow$ $(R)(H)C{=}C\langle$

ylide; aldehyde or ketone; alkene

	Alkyl halide	*Aldehyde/ketone*	*Product*
(a)	$C_6H_5CH_2Br$	$HCCH{=}CH{-}C_6H_5$ (with C=O)	$C_6H_5{-}CH{=}CHCH{=}CH{-}C_6H_5$
(b)	$C_6H_5CH_2Br$	cyclohexanone	$C_6H_5CH{=}C_6H_{10}$
(c)	CH_3Br	2-cyclohexenone	3-methylenecyclohexene ($=CH_2$)
(d)	CH_3Br	1-cyclohexenecarbaldehyde ($H{-}C{=}O$)	1-vinylcyclohexene ($H{-}C{=}CH_2$)

19.33 Suppose that tri*methyl*phosphine were to react with alkyl halide.

$$\underset{\text{alkyl halide}}{RCH_2CH_2{-}X} + \underset{\text{Trimethyl phosphine}}{(CH_3)_3P:} \longrightarrow \underset{\text{phosphonium salt}}{(CH_3)_3\overset{+}{P}CH_2CH_2R\ X^-}$$

Treatment of the phosphonium salt with strong base would yield two different ylides.

$$(CH_3)_3\overset{+}{P}CH_2CH_2R\ X^- \xrightarrow{BuLi} (CH_3)_3\overset{+}{P}{-}\overset{-}{C}HCH_2R \text{ and } (CH_3)_2\overset{+}{P}(\overset{-}{C}H_2)CH_2CH_2R$$

Reaction of the ylides with a carbonyl compound would produce two different alkenes, a problem that can't occur when triphenylphosphine is used.

$$R_2C{=}O + (CH_3)_3\overset{+}{P}{-}\overset{-}{C}HCH_2R \longrightarrow R_2C{=}C(H)CH_2R$$

or

$$R_2C{=}O + (CH_3)_2\overset{+}{P}(\overset{-}{C}H_2)CH_2CH_2R \longrightarrow R_2C{=}CH_2$$

19.34 Remember from Chapter 17:
Primary alcohols are formed from formaldehyde + Grignard reagent.
Secondary alcohols are formed from an aldehyde + Grignard reagent.
Tertiary alcohols are formed from a ketone (or an ester) + Grignard reagent.

	Aldehyde/ Ketone	*Grignard reagent*	*Product*
(a)	CH_3CHO	$CH_3CH_2CH_2MgBr$	$CH_3CH_2CH_2CH(OH)CH_3$
	or $CH_3CH_2CH_2CHO$	CH_3MgBr	
(b)	CH_2O	$CH_3CH_2CH_2MgBr$	$CH_3CH_2CH_2CH_2OH$
(c)	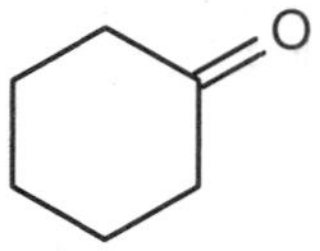 cyclohexanone	C_6H_5MgBr	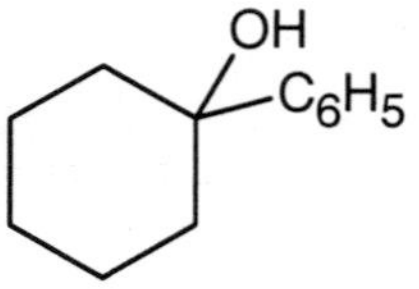 1-phenylcyclohexanol (OH, C_6H_5)
(d)	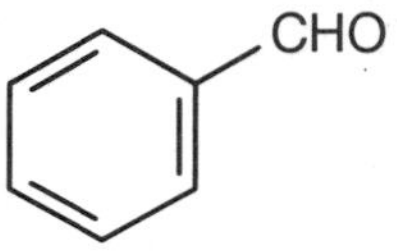 benzaldehyde (CHO)	C_6H_5MgBr	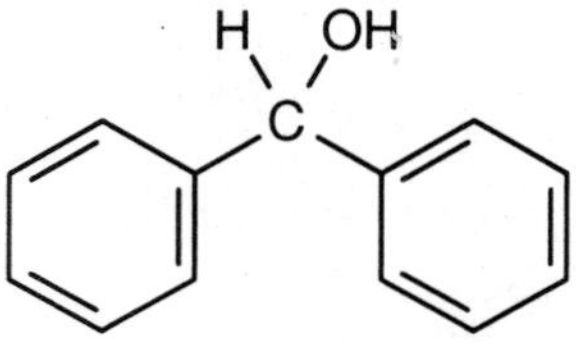 diphenylmethanol (H, OH, C)

19.35

(a)

$$(Ph)_3P: + ClCH_2OCH_3 \longrightarrow (Ph)_3\overset{+}{P}CH_2OCH_3\ Cl^- \xrightarrow{BuLi} (Ph)_3\overset{+}{P}-\overset{-}{C}HOCH_3 + LiCl$$

(b)

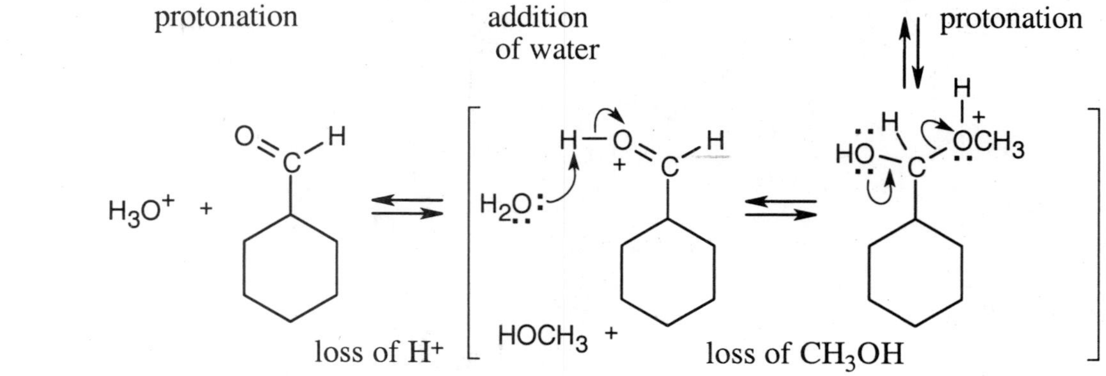

protonation; addition of water; $-H^+$; protonation; loss of CH_3OH; loss of H+

H_3O^+ + cyclohexanecarboxaldehyde ($O{=}CH$) ⇌ [H_2O: + protonated aldehyde; $HOCH_3$ +] ⇌ ...

19.36 4-Hydroxybutanal forms a cyclic acetal when the hydroxyl oxygen adds to the aldehyde group.

protonation ⇌ addition of –OH ⇌ loss of proton ($-H^+$) → hemiacetal

Methanol reacts with the cyclic hemiacetal to form 2-methoxytetrahydrofuran, which is a cyclic acetal.

protonation ⇌ loss of water ($+ H_2O$) ⇌ addition of methanol ($HOCH_3$) ⇌ loss of proton ($-H^+$) → a cyclic acetal

2–Methoxytetrahydrofuran is a cyclic acetal. The hydroxyl oxygen of 4–hydroxybutanal reacts with the aldehyde to form the cyclic ether linkage.

19.37 In general, ketones are less reactive than aldehydes for both steric (excess crowding) and electronic reasons. If the keto aldehyde in this problem were reduced with one equivalent of $NaBH_4$, the aldehyde functional group would be reduced in preference to the ketone.

For the same reason, reaction of the keto aldehyde with one equivalent of ethylene glycol selectively forms the acetal of the aldehyde functional group. The ketone can then be reduced with $NaBH_4$ and the acetal protecting group can be removed.

$$CH_3C(=O)CH_2CH_2CH_2CHO \xrightarrow[2.\ H_3O^+]{1.1\ \text{equiv}\ NaBH_4} CH_3C(=O)CH_2CH_2CH_2CH_2OH$$

$$CH_3C(=O)CH_2CH_2CH_2CHO \xrightarrow[\text{acid catalyst}]{HOCH_2CH_2OH} CH_3C(=O)CH_2CH_2CH_2CH(OCH_2CH_2O)$$

$$\xrightarrow[2.\ H_3O^+]{1.\ NaBH_4} CH_3CH(OH)CH_2CH_2CH_2CHO$$

19.38

(a)

C_6H_5CHO —(1. $NaBH_4$; 2. H_3O^+)→ $C_6H_5CH_2OH$ —(PBr_3)→ $C_6H_5CH_2Br$ —(Mg, ether)→ $C_6H_5CH_2MgBr$ —(1. CH_2O; 2. H_3O^+)→ $C_6H_5CH_2CH_2OH$ —(PCC)→ $C_6H_5CH_2CHO$

Synthetic analysis: The product resembles the starting material in having an aldehyde group, but a –CH_2– group lies between the aldehyde and the aromatic ring. The aldehyde results from oxidation of an alcohol that is the product of a Grignard reaction between formaldehyde and benzylmagnesium bromide. The Grignard reagent is formed from benzyl bromide, which results from treatment of benzyl alcohol with PBr_3. Reduction of benzaldehyde yields the alcohol.

(b)

C_6H_5CHO —(1. CH_3MgBr; 2. H_3O^+)→ $C_6H_5CH(OH)CH_3$ —(CrO_3, H_3O^+)→ $C_6H_5COCH_3$

$C_6H_5COCH_3$ + pyrrolidine (N–H) —(H_3O^+)→ $C_6H_5C(=CH_2)$–N(pyrrolidinyl)

Synthetic analysis: When you see a secondary amine and a double bond, you should recognize an enamine. The enamine is formed from the amine and benzophenone. Benzophenone, in turn, results from reaction of benzaldehyde with methylmagnesium bromide, followed by oxidation.

(c)

$C_6H_5CH_2Br$ (from (a)) —(1. $(Ph)_3P$:; 2. BuLi)→ $C_6H_5\overset{-}{C}H-\overset{+}{P}(Ph)_3$ —(Cyclopentanone)→ $C_6H_5CH=C_5H_8$ (benzylidenecyclopentane)

Synthetic analysis: The trisubstituted double bond suggests a Wittig reaction. Reaction of cyclopentanone with the Wittig reagent formed from benzaldehyde yields the desired product.

19.39

(a) (b) 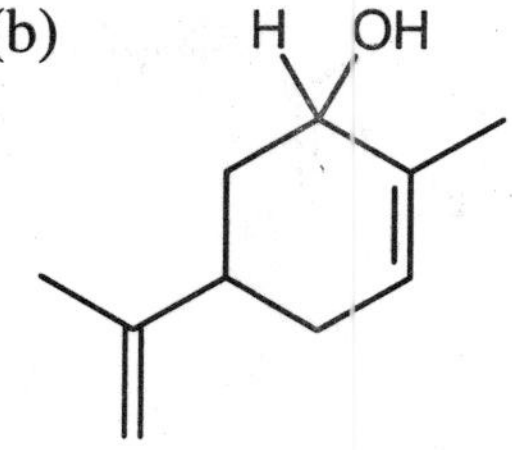 (c)

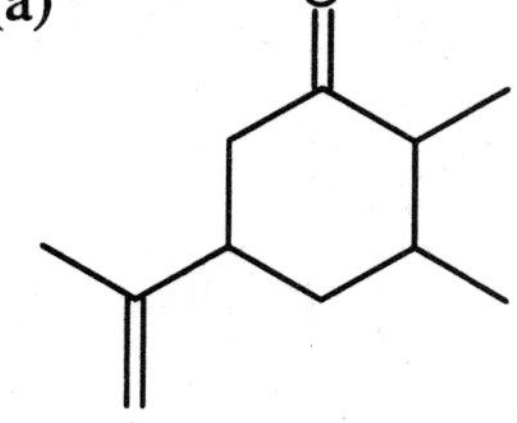

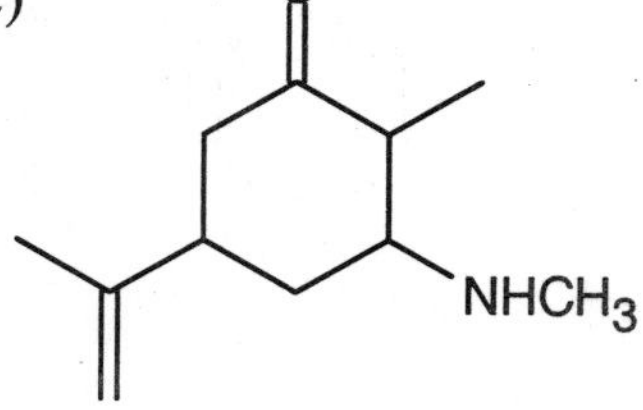

(c) $NHCH_3$

(d) HO C_6H_5 (e) 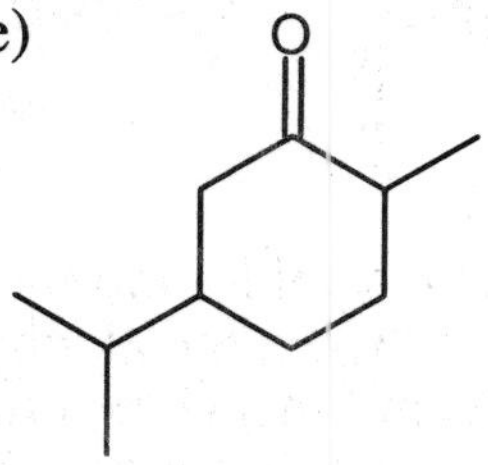 (f) no reaction

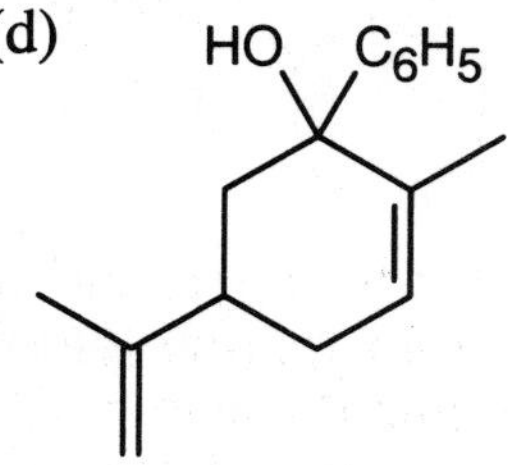

(g) H CH$_3$ + H_3C H (h) O O

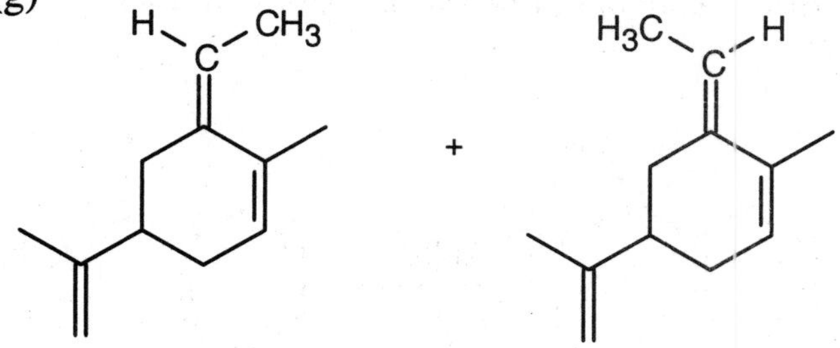

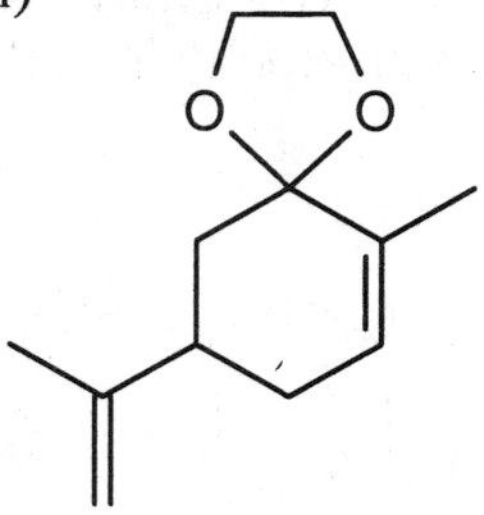

19.40

HO^{-} + $C_6H_5CHBr_2$ ⟶ [HO–CHBr(C_6H_5) + Br^-] ⟶ O=CH(C_6H_5) + HBr

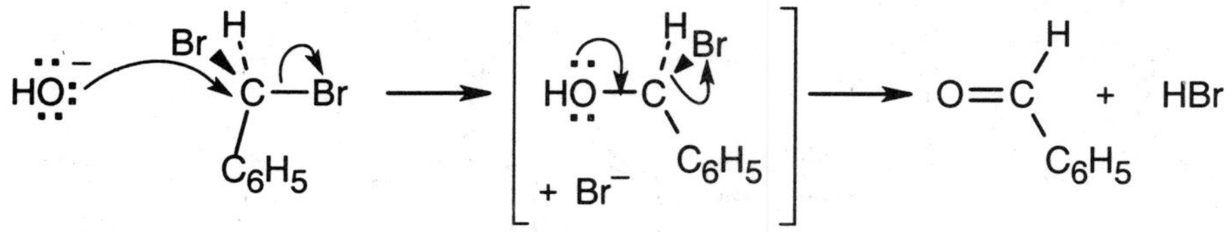

S_N2 addition of hydroxide to $C_6H_5CHBr_2$ yields an unstable bromoalcohol intermediate, which loses HBr to yield benzaldehyde.

19.41

(a)

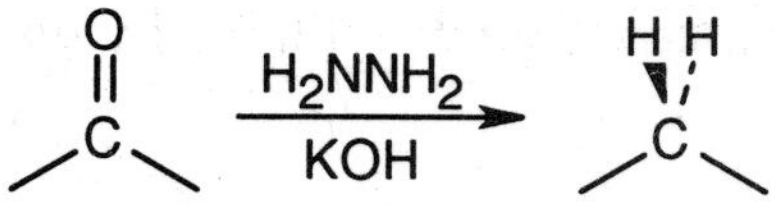

C=O $\xrightarrow[\text{KOH}]{H_2NNH_2}$ CH_2

Advantage: reduction is one-step

Disadvantage: can't be used when base-sensitive functional groups are present

(b)

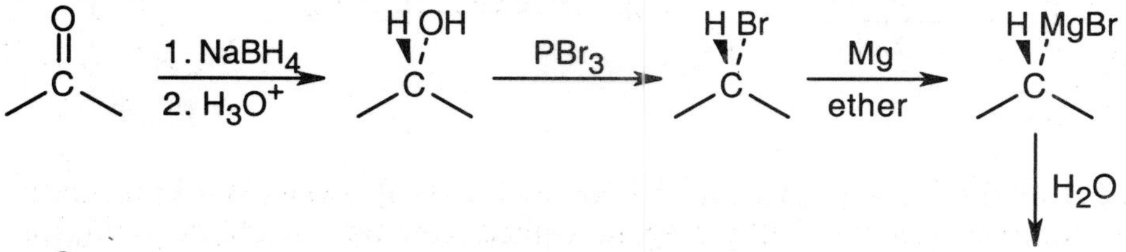

C=O $\xrightarrow[2.\ H_3O^+]{1.\ NaBH_4}$ CH(OH) $\xrightarrow{PBr_3}$ CH(Br) $\xrightarrow[\text{ether}]{Mg}$ CH(MgBr) $\xrightarrow{H_2O}$ CH_2

(c)

C=O $\xrightarrow[2.\ H_3O^+]{1.\ NaBH_4}$ CH(OH) $\xrightarrow[\text{pyridine}]{POCl_3}$ CH=C $\xrightarrow[\text{Pd/C}]{H_2}$ CH_2

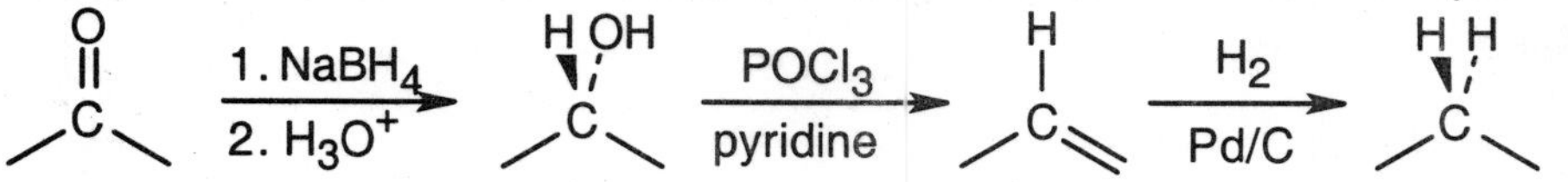

Disadvantage: these two methods require several steps.

19.42

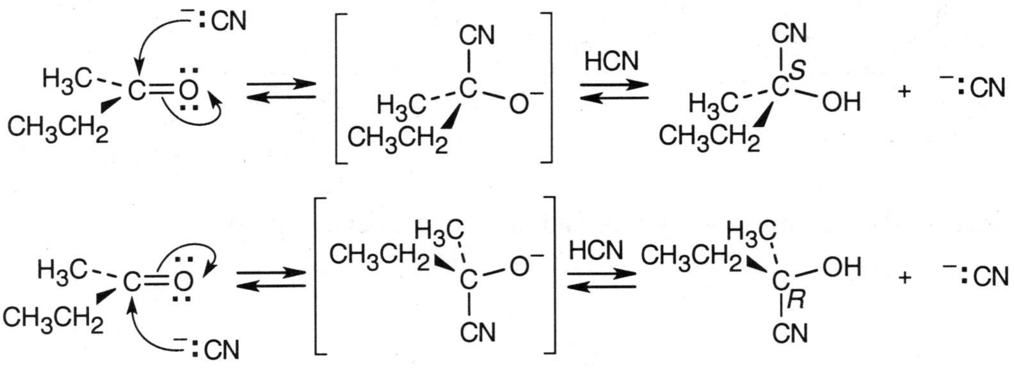

Attack can occur with equal probability on either side of the planar carbonyl group to yield a racemic product mixture that is optically inactive.

19.43

(a)

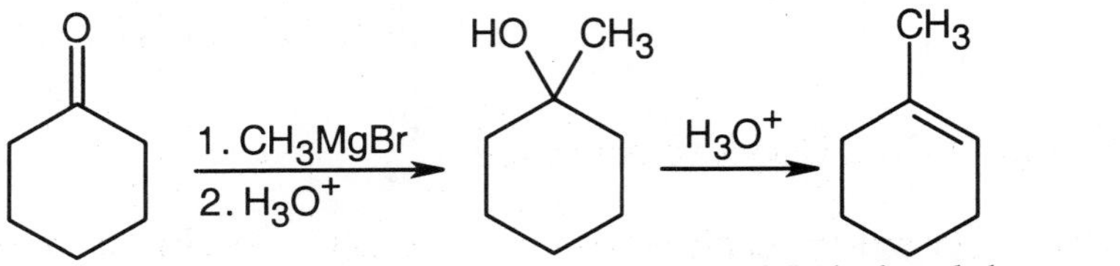

1-Methylcyclohexene

Synthetic analysis: The methyl group is introduced by a Grignard reaction with methylmagnesium bromide. Dehydration of the resulting tertiary alcohol produces 1-methylcyclohexene.

(b)

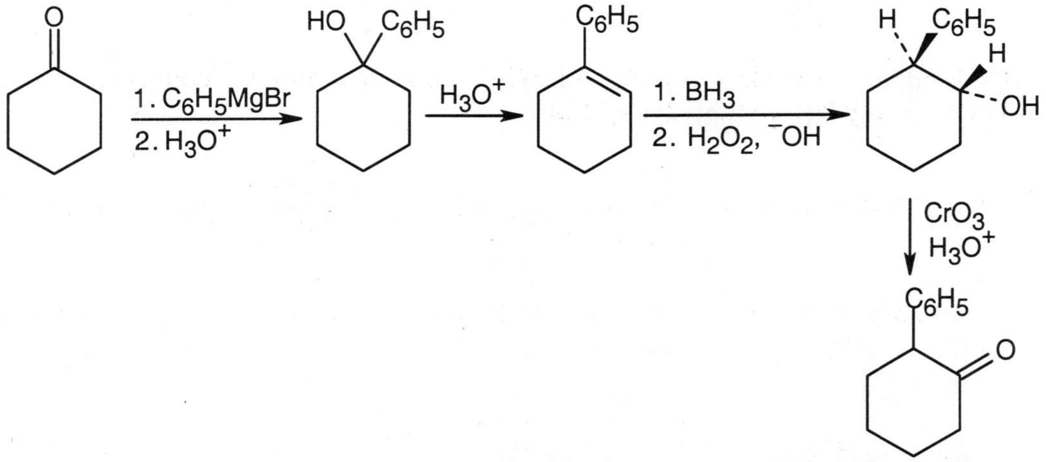

2-Phenylcyclohexanone

Reaction with phenylmagnesium bromide yields a tertiary alcohol that can be dehydrated. The resulting double bond can be treated with BH_3 to give an alcohol that can be oxidized to produce the desired ketone.

(c)

$$\text{Cyclohexanone} \xrightarrow[2.\ H_3O^+]{1.\ NaBH_4} \text{Cyclohexanol} \xrightarrow[\text{pyridine}]{POCl_3} \text{Cyclohexene} \xrightarrow[2.\ NaHSO_3,\ H_2O]{1.\ OsO_4} \textit{cis}\text{-1,2-Cyclohexanediol}$$

Reduction, dehydration and hydroxylation yield the desired product.

(d)

$$\text{Cyclohexanol (from (c))} \xrightarrow[2.\ Mg,\ ether]{1.\ PBr_3} \text{Cyclohexyl-MgBr} \xrightarrow[2.\ H_3O^+]{1.\ Cyclohexanone} \text{1-Cyclohexylcyclohexanol}$$

A Grignard reaction forms 1-cyclohexylcyclohexanol.

19.44 (a) Basic silver ion does not oxidize secondary alcohols to ketones. Grignard addition to a conjugated ketone yields the 1,2 product, not the 1,4 product. The correct scheme:

$$\text{HO-octalin} \xrightarrow[CH_2Cl_2]{PCC} \text{octalone} \xrightarrow[2.\ H_3O^+]{1.\ Li(CH_3)_2Cu} \text{CH}_3\text{-decalone}$$

(b) Reaction of an alcohol with acidic CrO_3 converts primary alcohols to carboxylic acids, not to aldehydes. The correct scheme:

$$C_6H_5CH{=}CHCH_2OH \xrightarrow[CH_2Cl_2]{PCC} C_6H_5CH{=}CHCHO \xrightarrow[H^+]{CH_3OH} C_6H_5CH{=}CHCH(OCH_3)_2$$

(c) Treatment of a cyanohydrin with H_3O^+ produces a carboxylic acid, not an amine. The correct scheme:

$$CH_3COCH_3 \xrightarrow{HCN} CH_3C(OH)(CN)CH_3 \xrightarrow[2.\ H_2O]{1.\ LiAlH_4} CH_3C(OH)(CH_2NH_2)CH_3$$

19.45

$$CH_3\overset{O}{\overset{\|}{C}}CH_2CH_2\overset{O}{\overset{\|}{C}}OCH_3 \xrightarrow[H^+]{HOCH_2CH_2OH} \text{(ethylene ketal)}\ CH_3CCH_2CH_2\overset{O}{\overset{\|}{C}}OCH_3$$

1. LiAlH$_4$ 2. H_3O^+ → ketal-protected $CH_3CCH_2CH_2CH_2OH$ —PCC, CH_2Cl_2→ ketal-protected $CH_3CCH_2CH_2CHO$

1. DIBAH 2. H_3O^+ (from the ketal ester) → ketal-protected $CH_3CCH_2CH_2CHO$

$\xrightarrow{(C_6H_5)_3\overset{+}{P}-\overset{-}{C}(CH_3)_2}$ ketal-protected $CH_3CCH_2CH_2CH{=}C(CH_3)_2$

$\xrightarrow{H_3O^+}$ $CH_3\overset{O}{\overset{\|}{C}}CH_2CH_2CH{=}C(CH_3)_2$

6-Methyl-5-hepten-2-one

The reaction sequence involves protecting the ketone, converting the ester to an aldehyde, using a Wittig reaction to introduce a substituted double bond, and deprotecting the ketone.

19.46 The same series of steps used to form an acetal is followed in this mechanism.

Cyclopentanone + H^+ ⇌ (protonation) [protonated cyclopentanone ($\overset{+}{O}H$) + $HSCH_2CH_3$ ⇌ (addition of RSH) HO, $\overset{+}{S}(H)CH_2CH_3$ on ring carbon] ⇌ ($-H^+$, loss of proton) HO / SCH_2CH_3 hemithioacetal

HO / SCH_2CH_3 + H^+ ⇌ (protonation) [$H_2\overset{+}{O}$ / SCH_2CH_3] ⇌ (loss of H_2O) [$CH_3CH_2\overset{+}{S}{=}$ring + H_2O, + $HSCH_2CH_3$]

⇅ addition of RSH

[CH_3CH_2S / $\overset{+}{S}(H)CH_2CH_3$] ⇌ ($-H^+$, loss of proton) CH_3CH_2S / SCH_2CH_3 thioacetal

19.47 Even though the product looks unusual, this reaction is made up of steps with which you are familiar.

:O: $\overset{-}{:}CH_2\overset{+}{S}(CH_3)_2$ ⇌ [$^{-}$:O: CH_2—$\overset{+}{S}(CH_3)_2$] ⟶ O—CH_2 + $S(CH_3)_2$

addition of ylide

S_N2 displacement of dimethyl sulfide by O$^-$

19.48

:O—H $^-$:OH, H_3C, H_3C, C, CN ⇌ [:O:$^-$ H_3C, H_3C, C, CN] ⇌ O, C, H_3C, CH_3 + $^-$:CN

removal of proton by $^-$OH

elimination of $^-$CN

The above steps are the reverse of Problem 19.5.

:O: $^-$:CN ⇌ [$^-$:O: CN] $\underset{}{\overset{H_2O}{\rightleftharpoons}}$ HO CN + HO^-

nucleophilic addition

protonation of tetrahedral intermediate

This step is a nucleophilic addition of cyanide.

19.49

O, H^+ ⇌ [$\overset{+}{O}$, HOR ⇌ O, $\overset{+}{O}$R, H] $\overset{-H^+}{\rightleftharpoons}$ O, OR

protonation of double bond

addition of alcohol

loss of proton

The steps of this reaction are similar to some of the steps in the formation of an acetal from a hemiacetal.

19.50

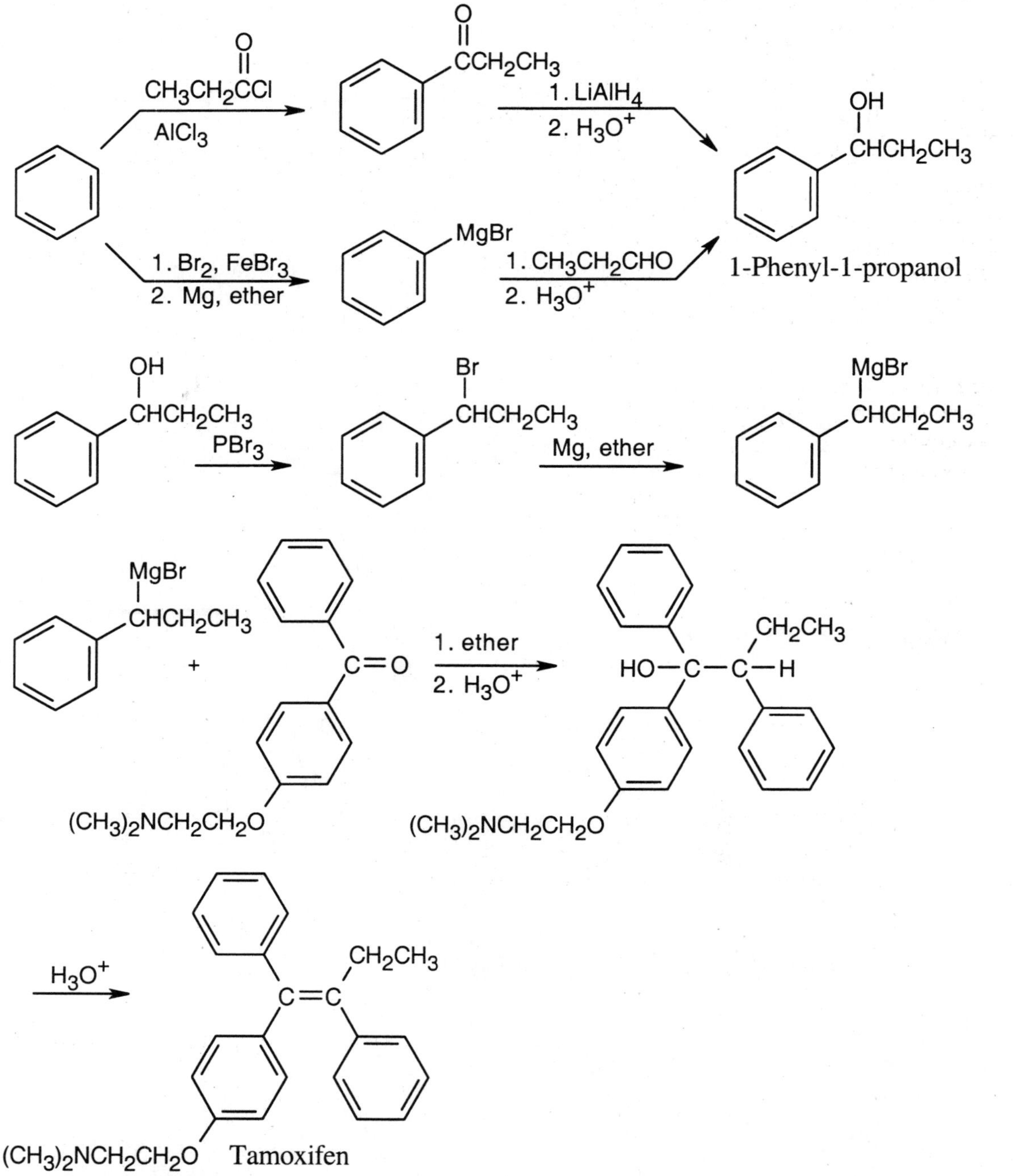

Benzene + CH_3CH_2COCl, $AlCl_3$ → $C_6H_5COCH_2CH_3$; 1. $LiAlH_4$ 2. H_3O^+ → $C_6H_5CH(OH)CH_2CH_3$

Benzene, 1. Br_2, $FeBr_3$ 2. Mg, ether → C_6H_5MgBr; 1. CH_3CH_2CHO 2. H_3O^+ → $C_6H_5CH(OH)CH_2CH_3$

1-Phenyl-1-propanol

$C_6H_5CH(OH)CH_2CH_3$ —PBr_3→ $C_6H_5CH(Br)CH_2CH_3$ —Mg, ether→ $C_6H_5CH(MgBr)CH_2CH_3$

$C_6H_5CH(MgBr)CH_2CH_3$ + $(CH_3)_2NCH_2CH_2O$–C_6H_4–C(=O)–C_6H_5 —1. ether 2. H_3O^+→ HO–C(C_6H_5)($C_6H_4OCH_2CH_2N(CH_3)_2$)–CH(CH_2CH_3)(C_6H_5)

—H_3O^+→ (C_6H_5)($(CH_3)_2NCH_2CH_2O$–C_6H_4)C=C(CH_2CH_3)(C_6H_5) Tamoxifen

Synthetic analysis: When you see a product that contains a double bond, and you also know that one of the starting materials is a ketone, it is tempting to use a Wittig reaction for synthesis. In this case, however, the tetrasubstituted double bond can't be formed by a Wittig reaction because of steric hindrance. The coupling step is achieved by a Grignard reaction between the illustrated ketone and a Grignard reagent, followed by dehydration. The Grignard reagent is synthesized from 1-phenyl-1-propanol. which can be prepared from benzene by either of two routes.

19.51

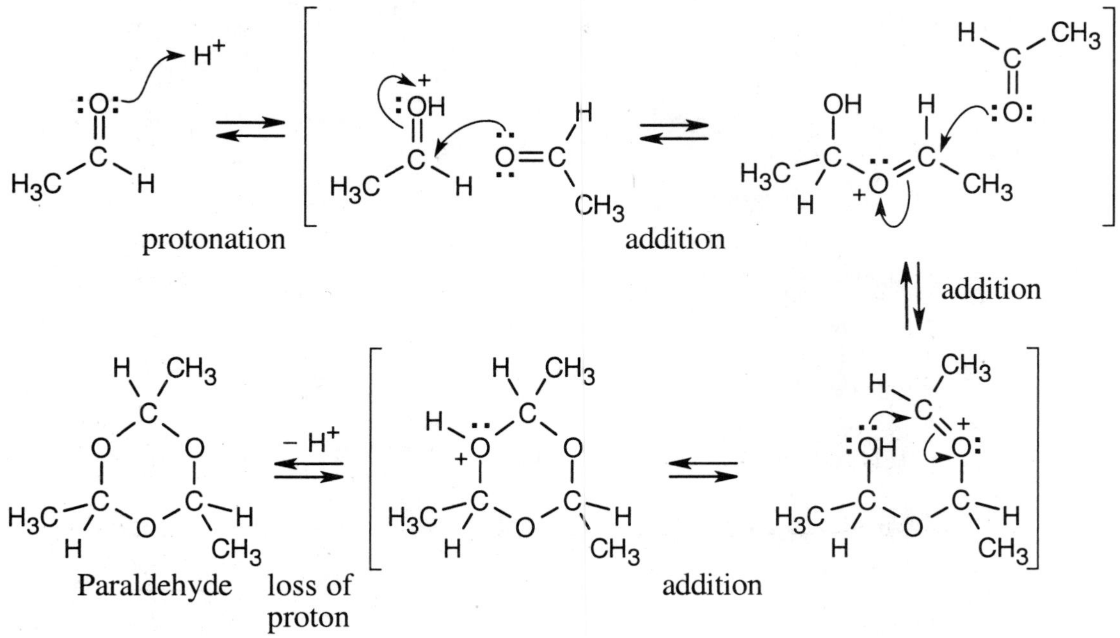

protonation

addition

addition

Paraldehyde loss of proton

addition

Protonation makes the carbonyl carbon more electrophilic. Three successive additions of the carbonyl oxygen of acetaldehyde to the electrophilic carbonyl carbon, followed by loss of a proton, give the cyclic product.

19.52

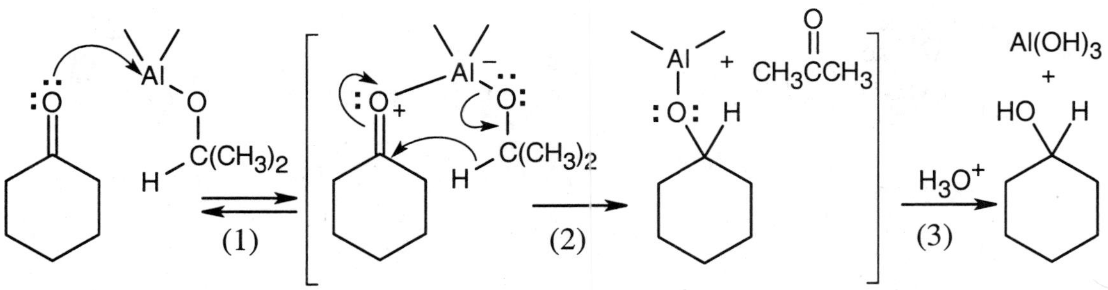

(1) (2) H_3O^+ (3)

(1) Aluminum, a Lewis acid, complexes with the carbonyl oxygen.
(2) Complexation with aluminum makes the carbonyl group electrophilic and facilitates hydride transfer from isopropoxide.
(3) Treatment of the reaction mixture with aqueous acid cleaves the aluminum-oxygen bond and produces cyclohexanol.

Both the MPV reaction and the Cannizzaro reaction are hydride transfers in which a carbonyl group is reduced by an alkoxide group, which is oxidized. Note that each aluminum triisopropoxide molecule is capable of reducing three ketone molecules.

19.53 (a) Nucleophilic addition of one nitrogen of hydrazine to one of the carbonyl groups, followed by elimination of water, produces a hydrazone.

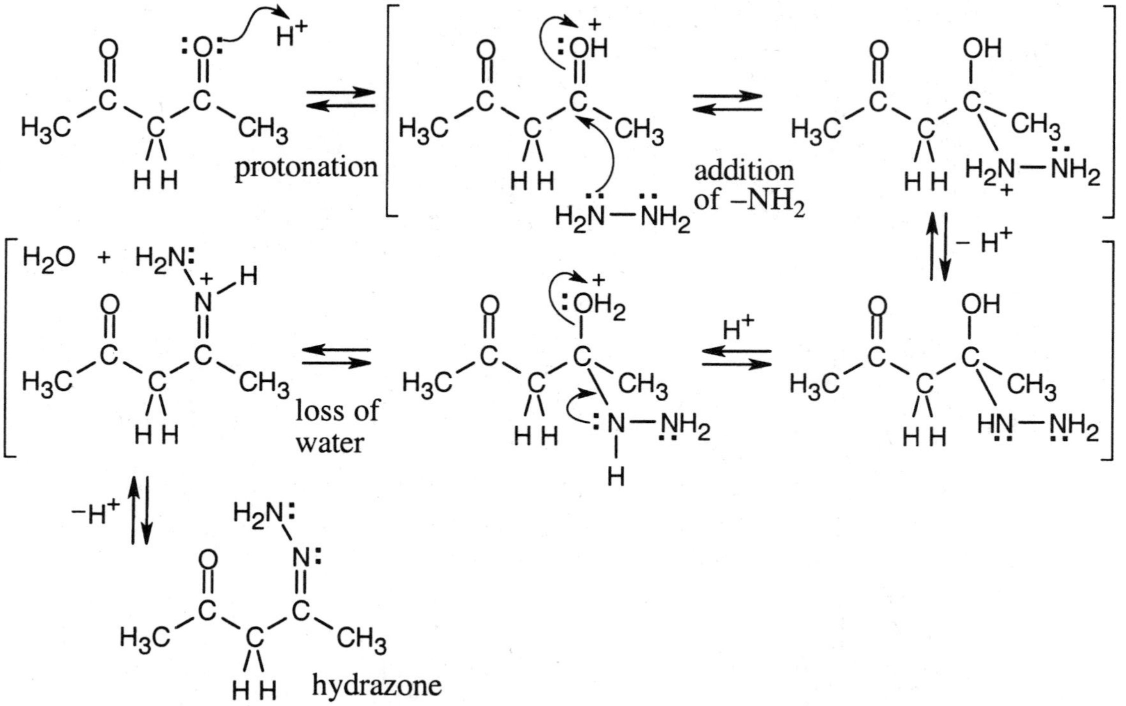

protonation

addition of $-NH_2$

$- H^+$

H^+

loss of water

$-H^+$

hydrazone

(b) In a similar manner, the *other* nitrogen of hydrazine can add to the *other* carbonyl group of 2,4-pentanedione to form the pyrazole.

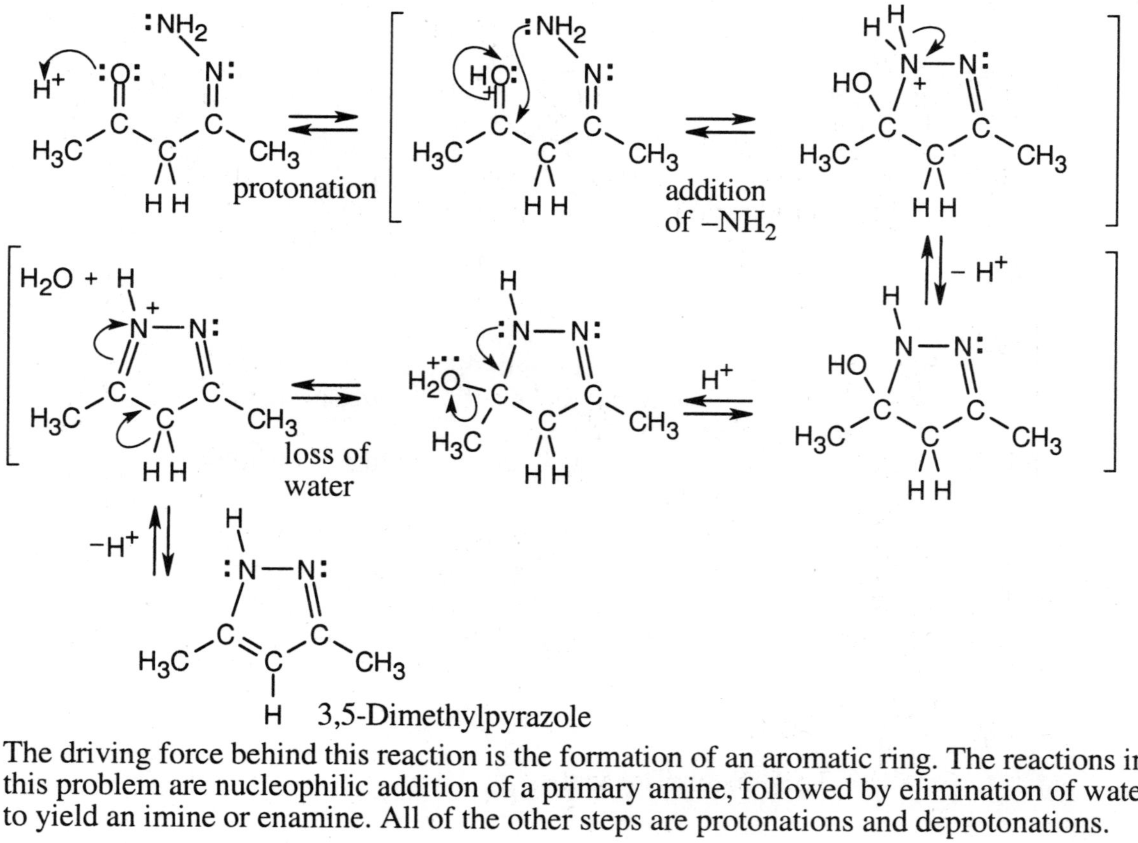

protonation

addition of $-NH_2$

$- H^+$

H^+

loss of water

$-H^+$

3,5-Dimethylpyrazole

The driving force behind this reaction is the formation of an aromatic ring. The reactions in this problem are nucleophilic addition of a primary amine, followed by elimination of water to yield an imine or enamine. All of the other steps are protonations and deprotonations.

19.54 The same sequence of steps used in the previous problem leads to the formation of 3,5–dimethylisoxazole when hydroxylamine is the reagent. The proton lost in the last step of (b) results in a ring that is aromatic.

(a)

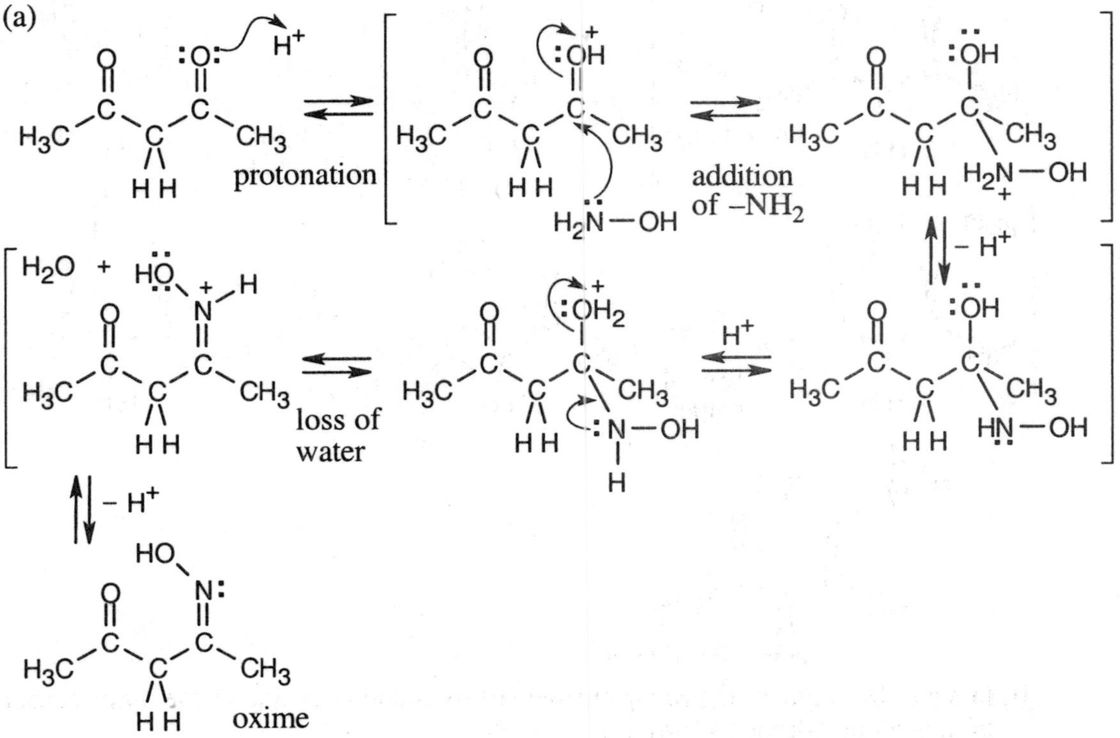

protonation

addition of $-NH_2$

$- H^+$

loss of water

H^+

$- H^+$

oxime

(b)

protonation

addition of –OH

$- H^+$

loss of water

H^+

$- H^+$

3,5-Dimethylisoxazole

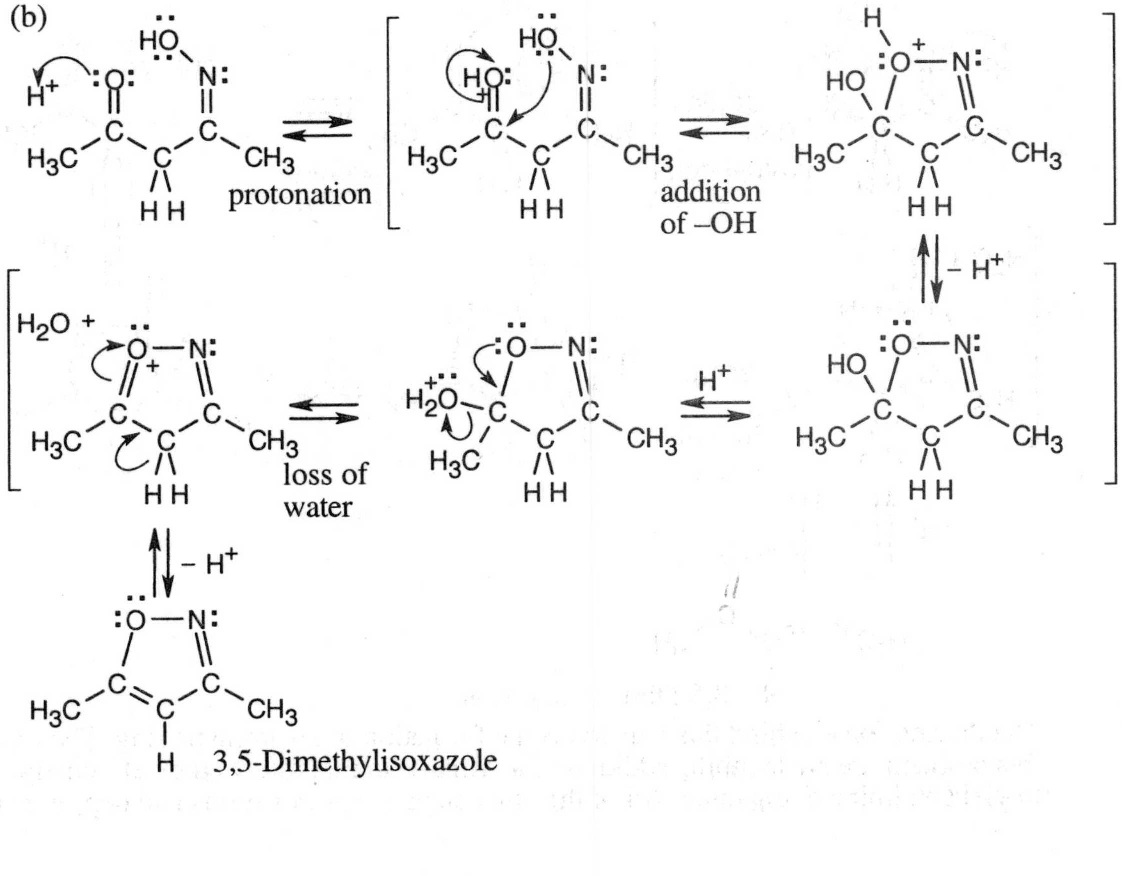

19.55

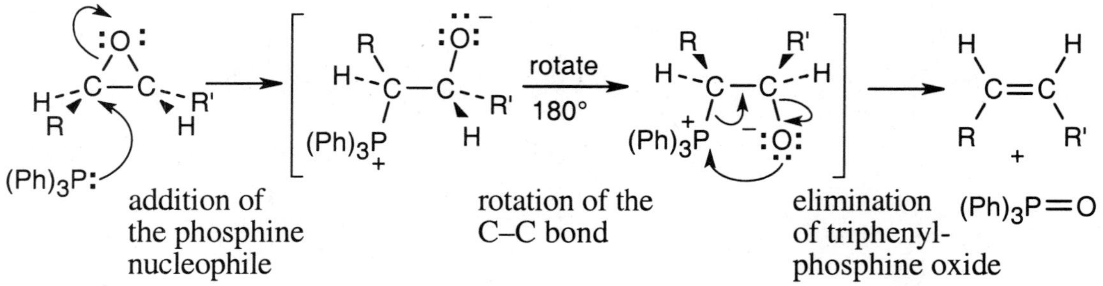

(Ph)$_3$P: addition of the phosphine nucleophile

rotate 180°

rotation of the C–C bond

elimination of triphenylphosphine oxide

(Ph)$_3$P=O

The final step is the same as the last step in a Wittig reaction.

19.56

$$HO{-}OH + {}^{-}OH \rightleftharpoons {}^{-}O{-}OH + H_2O$$

Hydrogen peroxide and hydroxide react to form water and peroxide anion.

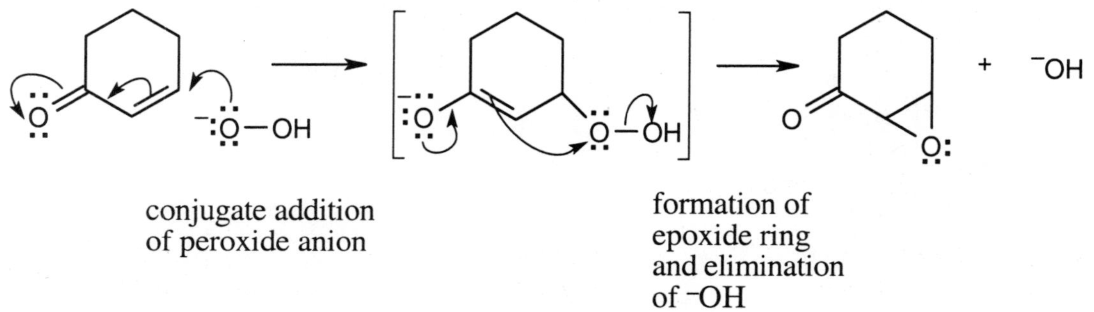

+ $^{-}$OH

conjugate addition of peroxide anion

formation of epoxide ring and elimination of $^{-}$OH

Conjugate addition of peroxide anion is followed by elimination of hydroxide ion, with formation of the epoxide ring.

19.57 Use Table 19.2 if you need help.

Absorption:	*Due to:*
(a) 1750 cm^{-1}	5–membered ring ketone
1685 cm^{-1}	α,ß–unsaturated ketone
(b) 1720 cm^{-1}	5–membered ring *and* aromatic ketone
(c) 1750 cm^{-1}	5–membered ring ketone
(d) 1705 cm^{-1}, 2720 cm^{-1}, 2820 cm^{-1}	aromatic aldehyde
1715 cm^{-1}	aliphatic ketone

Compounds in parts b-d also show aromatic ring IR absorptions in the range 1450 cm^{-1} – 1600 cm^{-1} and in the range 690 – 900 cm^{-1}.

19.58

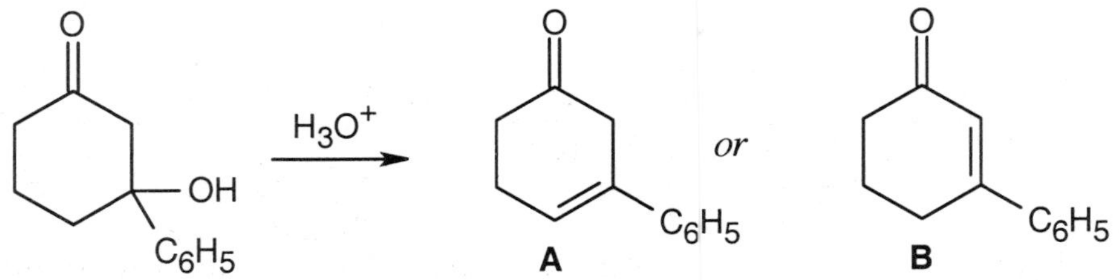

H_3O^+

3-Hydroxy-3-phenyl-cyclohexanone → **A** *or* **B**

C_6H_5 — OH

Compound **A** is a cyclic, nonconjugated keto alkene whose carbonyl infrared absorption should occur at 1715 cm^{-1}. Compound **B** is an α,ß–unsaturated, cyclic ketone; additional conjugation with the phenyl ring should lower its IR absorption below 1685 cm^{-1}. Because the actual IR absorption occurs at 1670 cm^{-1}, **B** is the correct structure.

19.59 The molecular weight of Compound A shows that the molecular formula of A is $C_5H_{10}O$ (one degree of unsaturation), and the IR absorption shows that A is an aldehyde. The uncomplicated ^1H NMR is that of 2,2-dimethylpropanal.

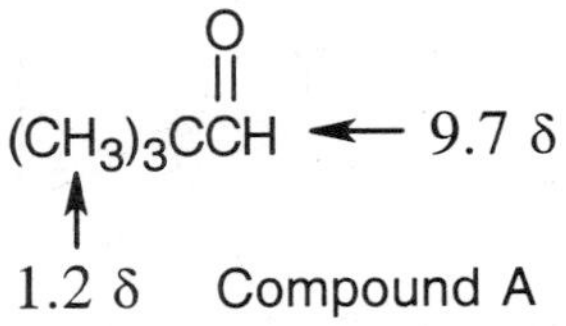

$(CH_3)_3CCH$ (C=O) ← 9.7 δ

1.2 δ Compound A

19.60 The IR of Compound B shows a ketone absorption. The splitting pattern of the ^1H NMR spectrum indicates an isopropyl group and indicates that the compound is a methyl ketone.

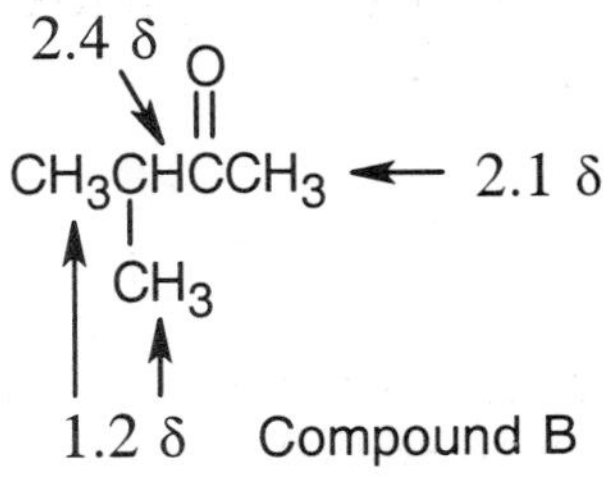

2.4 δ

CH_3CHCCH_3 (C=O; CH_3 branch) ← 2.1 δ

1.2 δ Compound B

19.61 Before looking at the ^1H NMR spectrum, we know that the compound of formula $C_9H_{10}O$ has 5 degrees of unsaturation, and we know from the IR spectrum that the unknown is an aromatic ketone. The splitting pattern in the ^1H NMR spectrum shows an ethyl group, which chemical shift data shows is next to a ketone.

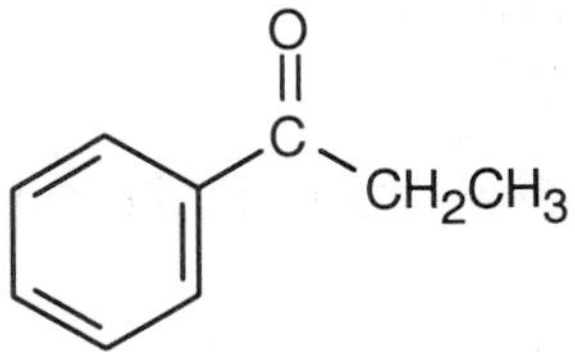

$C_6H_5C(=O)CH_2CH_3$

19.62 The IR absorption is that of an aldehyde that isn't conjugated with the aromatic ring. The two triplets in the ^1H NMR spectrum are due to two adjacent methylene groups.

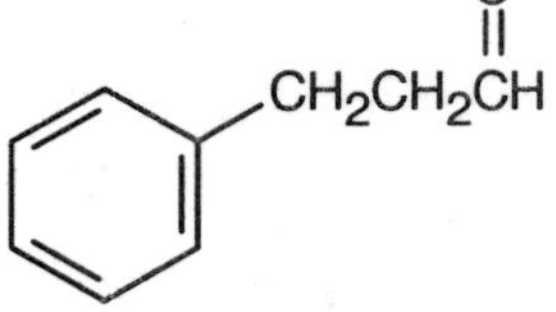

$C_6H_5CH_2CH_2CHO$

19.63 (a) The unknown is a ketone (from IR) whose carbonyl group is flanked by a secondary carbon and a tertiary carbon (from ^{13}C NMR).
(b) This aldehyde has an isopropyl group.
(c) The IR absorption shows that this compound is an α,β-unsaturated ketone, and the molecular formula shows 3 degrees of unsaturation. The ^{13}C NMR spectrum indicates 3 sp^2-hybridized carbons and 3 secondary carbons.

(a) $(CH_3)_2CHCOCH_2CH_3$

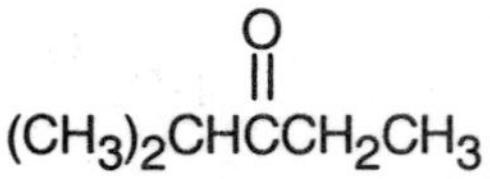

(b) $(CH_3)_2CHCH_2CHO$

(c) 2-cyclohexenone

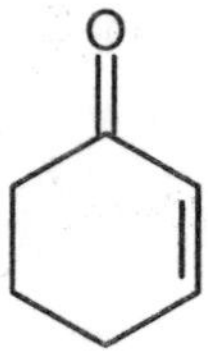

19.64 Compound A has 4 degrees of unsaturation and is a five-membered ring ketone. The ^{13}C NMR spectrum has only three peaks and indicates that A is very symmetrical.

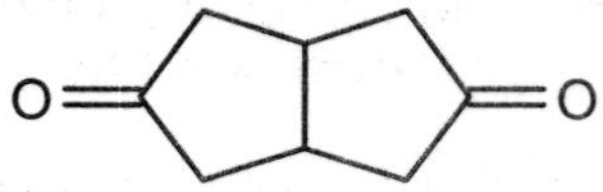

Compound A

19.65 As always, calculate the degree of unsaturation first, then use the available IR data to assign the principal functional groups.

(a) $CH_3CH(Cl)COCH_3$ (a = CH_3, c = CH, b = $COCH_3$)
a = 1.7 δ
b = 2.3 δ
c = 4.3 δ

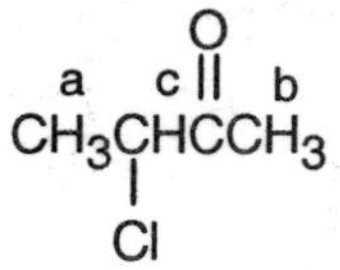

(b) $(CH_3)_3CCH_2COCH_3$ (a = $(CH_3)_3$, c = CH_2, b = $COCH_3$)
a = 1.0 δ
b = 2.1 δ
c = 2.3 δ

(c) p-ethoxybenzaldehyde (e = CHO; c,d = ring H; b = OCH_2; a = CH_3)
a = 1.5 δ
b = 4.1 δ
c,d = 7.0 δ, 7.8 δ
e = 9.9 δ

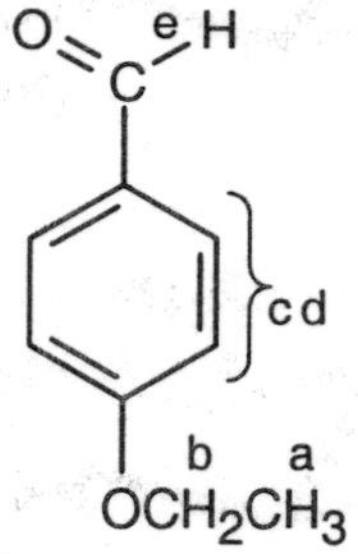

19.66

(a) $C_6H_5CH_2COCH_2CH_3$ (d = C_6H_5, c = CH_2, b = CH_2, a = CH_3)
a = 1.0 δ
b = 2.5 δ
c = 3.7 δ
d = 7.3 δ

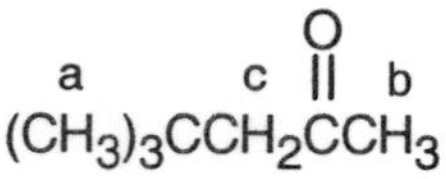

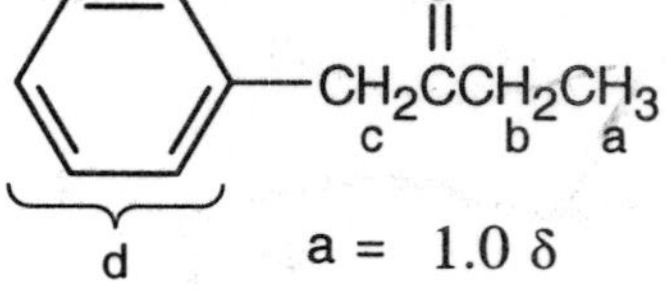

(b) $(CH_3O)_2CHCH_2COCH_3$ (c = CH_3O, d = CH, b = CH_2, a = CH_3)
a = 2.2 δ
b = 2.7 δ
c = 3.4 δ
d = 4.8 δ

(c) $H_2C{=}C(CH_3)CHO$ (b,c = $=CH_2$, a = CH_3, d = CHO)
a = 1.8 δ
b,c = 6.0 δ, 6.3 δ
d = 9.6 δ

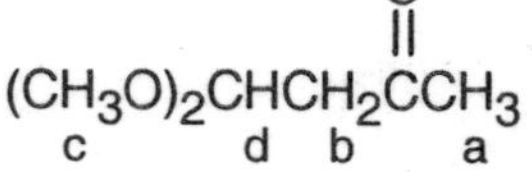

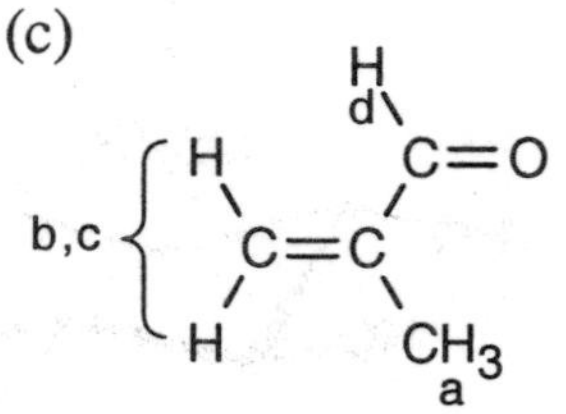

A Look Ahead

19.67

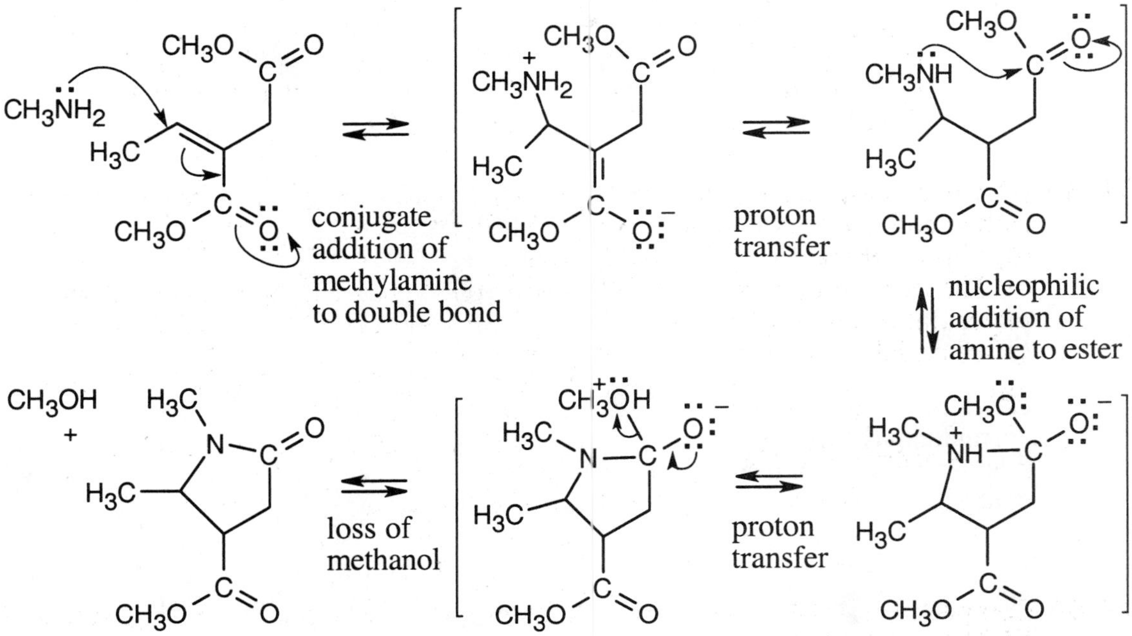

conjugate addition of methylamine to double bond

proton transfer

nucleophilic addition of amine to ester

loss of methanol

proton transfer

In his series of reactions, conjugate addition of an amine to an α,β-unsaturated ester is followed by nucleophilic addition of the amine to a second ester, with loss of methanol.

19.68 In order to make the drawing less cluttered, many of the hydrogens have been omitted.

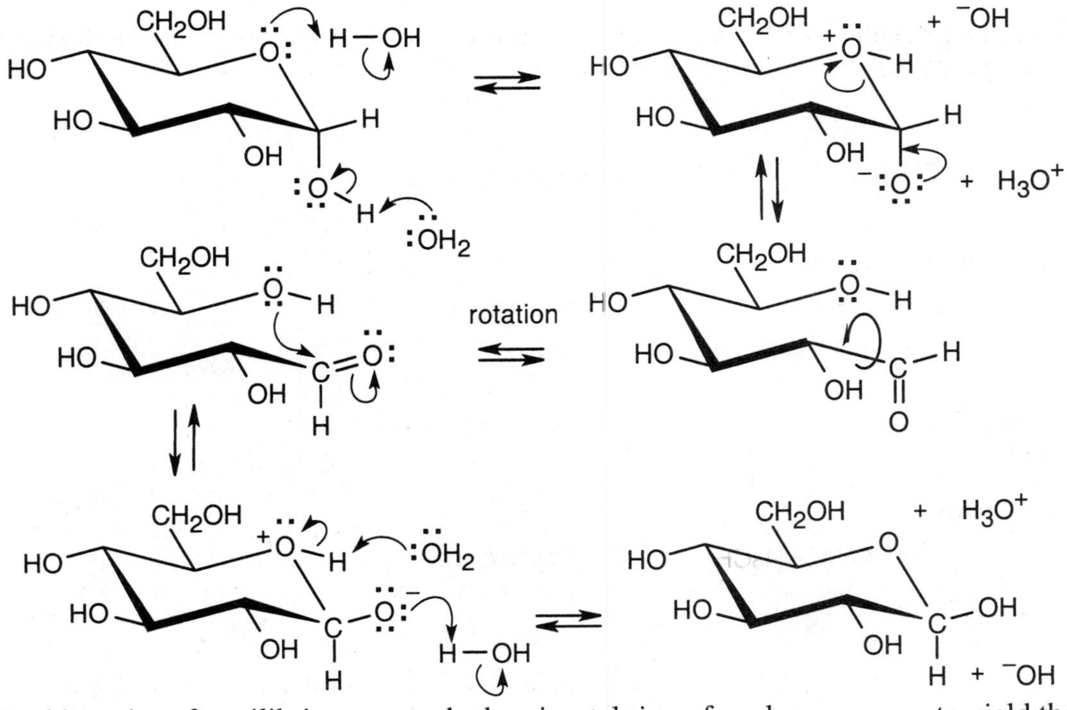

rotation

In this series of equilibrium steps, the hemiacetal ring of α–glucose opens to yield the free aldehyde. Bond rotation is followed by formation of the cyclic hemiacetal of ß–glucose. The reaction is catalyzed by both acid and base.

19.69 The free aldehyde form of glucose (Problem 19.68) is reduced in the same manner described in the text for other aldehydes to produce the polyalcohol sorbitol.

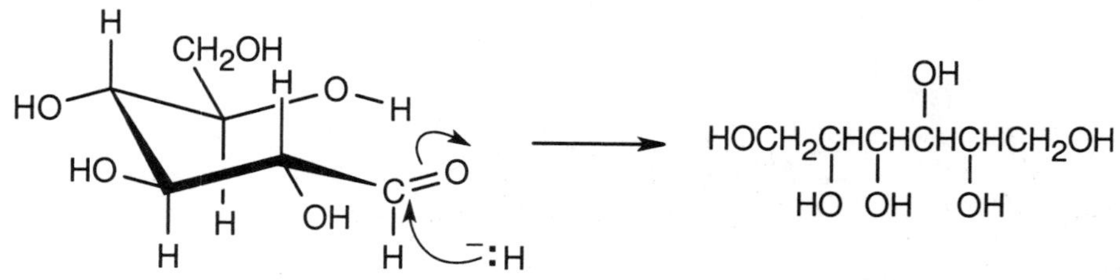

Glucose → Sorbitol ($HOCH_2CH(OH)CH(OH)CH(OH)CH(OH)CH_2OH$)

Molecular Modeling

19.70 The bond length of the C–C bond that is forming increases from 180.7 pm (formaldehyde), to 182.7 pm (acetone), to 220.9 pm (benzophenone). Transition state strain increases in the same way. This increase is reasonable because formaldehyde has the smallest substituents and benzophenone has the largest substituents.

19.71 Monoacetal B is lower in energy by 22 kJ/mol and forms selectively under thermodynamic control. This molecule is stabilized by conjugation between the second carbonyl group and the carbon-carbon double bond. The electrostatic potential map of the of the starting material shows that the conjugated carbonyl carbon is less positive than the isolated carbonyl group.

19.72 The calculated C=O stretching frequency for cyclohexanone is 1938 cm^{-1}. A fluorine on the α carbon increases the frequency regardless of fluorine orientation – 1957 cm^{-1} for axial 2-fluorocyclohexanone, and 1958 cm^{-1} for equatorial 2-fluorocyclohexanone. Two fluorines produce an even larger effect (1975 cm^{-1} for 2,2-difluorocyclohexanone). A fluorine on the β carbon has a similar, but smaller, effect (1945 cm^{-1}).

19.73

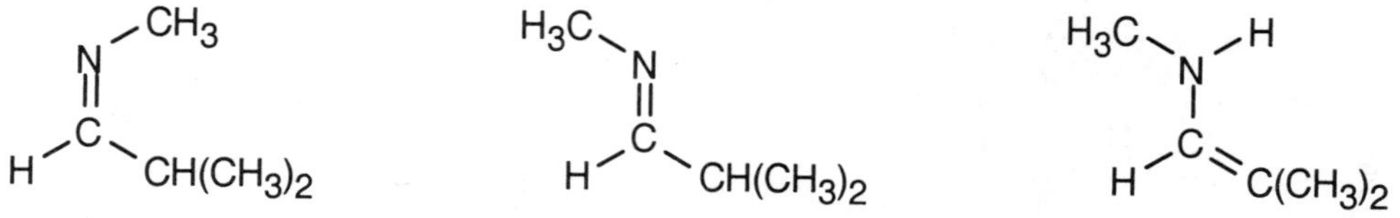

imine A: $H(CH(CH_3)_2)C{=}N{-}CH_3$ (imine A)

imine B: $H(CH(CH_3)_2)C{=}N{-}CH_3$ (imine B)

enamine: $H_3C{-}NH{-}CH{=}C(CH_3)_2$ (enamine)

Imine B is the most stable of the three possible products and should be the major product. The enamine is 11 kJ/mol higher in energy, and imine A is 15 kJ/mol higher in energy. The bonds of the enamine (N–H and C=C) differ from those of the imine (N=C and C–H), and the combination of enamine bonds is apparently less stable than the imine bonds. Although imines A and B contain the same bonds, the methyl group in imine A is cis to the isopropyl group, and steric repulsion between these two groups may destabilize this imine. Evidence for this destabilization can be seen in the C–C=N bond angle, which is 131° in imine A and 123° in imine B.

19.74 The more accessible (less hindered) lobe of the LUMO is exo in 2-norbornanone and endo in camphor. Addition of hydride to the more accessible lobe yields the following alcohols:

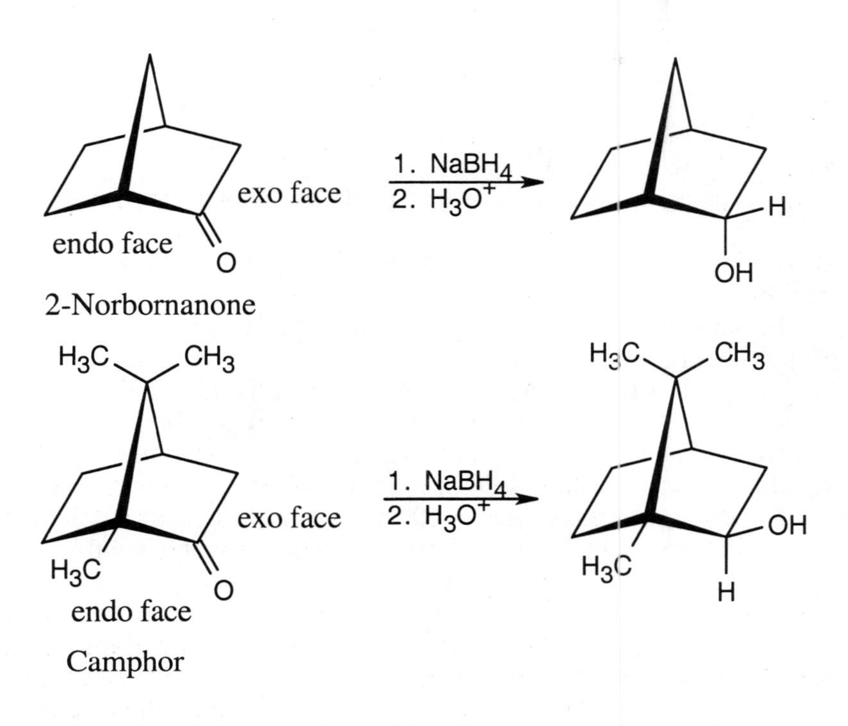

Chapter 20 – Carboxylic Acids

Chapter Outline

I. General information about carboxylic acids (Sections 20.1 – 20.5).
  A. Naming carboxylic acids (Section 20.1).
    1. Noncyclic carboxylic acids are named by replacing the *-e* of the corresponding alkane by *-oic acid.*
    2. Compounds that have a carboxylic acid bonded to a ring are named by using the suffix *-carboxylic acid.*
    3. Many carboxylic acids have historical, nonsystematic names.
  B. Structure and properties of carboxylic acids (Section 20.2).
    1. The carbonyl group of carboxylic acids is sp^2-hybridized and planar.
    2. Carboxylic acids are strongly associated because of hydrogen-bonding, and their boiling points are elevated.
  C. Carboxylic acid acidity (Sections 20.3 – 20.5).
    1. Dissociation of carboxylic acids (Section 20.3).
      a. Carboxylic acids react with bases to form salts that are water-soluble.
      b. Carboxylic acids dissociate slightly in dilute aqueous solution to give H_3O^+ and carboxylate anions.
        The K_a values for carboxylic acids are near 10^{-5}, making them weaker than mineral acids but stronger than alcohols.
      c. The relative strength of carboxylic acids is due to resonance stabilization of the carboxylate anion.
        i. Both carbon-oxygen bonds of carboxylic acids are the same length.
        ii. The bond length is intermediate between single and double bonds.
    2. Substituent effects on acidity (Section 20.4).
      a. Carboxylic acids differ in acid strength.
        i. Electron-withdrawing groups stabilize carboxylate anions and increase acidity.
        ii. Electron-donating groups decrease acidity.
      b. These inductive effects decrease with increasing distance from the carboxyl group.
    3. Substituent effects in substituted benzoic acids (Section 20.5).
      a. Groups that are deactivating in electrophilic aromatic substitution reactions increase the acidity of substituted benzoic acids.
      b. The acidity of benzoic acids can be used to predict electrophilic reactivity.

II. Preparation of carboxylic acids (Section 20.6).
  A Methods already studied.
    1. Oxidation of substituted alkylbenzenes.
    2. Oxidative cleavage of alkenes and alkynes.
    3. Oxidation of primary alcohols and aldehydes.
  B. Nitrile hydrolysis.
    1. Nitriles can be hydrolyzed by strong aqueous acids or bases to yield carboxylic acids.
    2. The sequence nitrile formation → nitrile hydrolysis can be used to prepare a carboxylic acid from a halide.
    3. This method is generally limited to compounds that can undergo S_N2 reactions.
  C. Carboxylation of Grignard reagents.
    1. A Grignard reagent can be treated with CO_2 (and protonated) to form a carboxylic acid.

2. This method is limited to compounds that don't have other functional groups that interfere with Grignard reagent formation.

III. Reactions of carboxylic acids (Sections 20.7 – 20.8).
   A. Carboxylic acid can undergo reactions typical of alcohols and ketones (Section 20.7).
   B. Other types of reactions of carboxylic acids:
      1. Alpha substitution.
      2. Decarboxylation.
      3. Nucleophilic acyl substitution.
      4. Reduction (Section 20.8).
         a. Carboxylic acids can be reduced to alcohols with either $LiAlH_4$ or BH_3.
         b. BH_3 is a more selective reagent.

IV. Spectroscopy of carboxylic acids.
   A. Infrared spectroscopy.
      1. The O–H absorption occurs at 2500–3300 cm^{-1} and is easy to identify.
      2. The C=O absorption occurs at 1710–1760 cm^{-1}
         The position of this absorption depends on whether the acid is free (1760 cm^{-1}) or associated (1710 cm^{-1}).
   B. NMR spectroscopy.
      1. ^{13}C NMR spectroscopy.
         a. Carboxylic acids absorb between 165–185 δ.
         b. Saturated acids absorb downfield from α,β-unsaturated acids.
      2. ^{1}H NMR spectroscopy.
         The carboxylic acid proton absorbs at 12 δ.

Solutions to Problems

20.1 Carboxylic acids are named by replacing *-e* of the corresponding alkane with *-oic acid.* The carboxylic acid carbon is C1.
When-COOH is a substituent of a ring, the suffix *-carboxylic acid* is used; the carboxyl carbon is not C1 in this system.

(a)

$(CH_3)_2CHCH_2COOH$

3-Methylbutanoic acid

(b)

$CH_3CH(Br)CH_2CH_2COOH$

4-Bromopentanoic acid

(c)

$CH_3CH{=}CHCH{=}CHCOOH$

2,4-Hexadienoic acid

(d)

COOH
|
$CH_3CH_2CHCH_2CH_2CH_3$

2-Ethylpentanoic acid

(e)

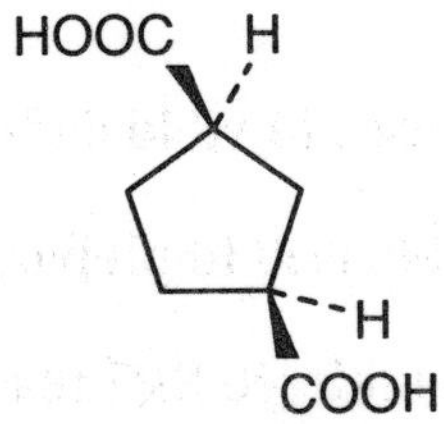

HOOC H

H

COOH

cis-1,3-Cyclopentane-dicarboxylic acid

(f)

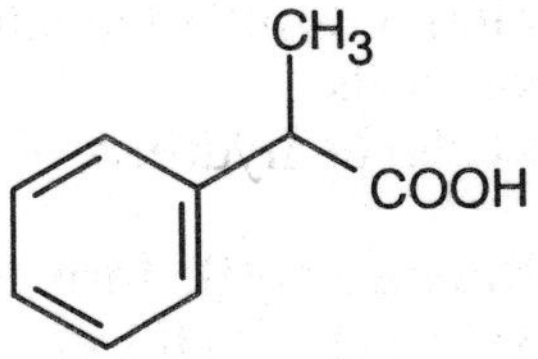

CH_3

COOH

2-Phenylpropanoic acid

20.2

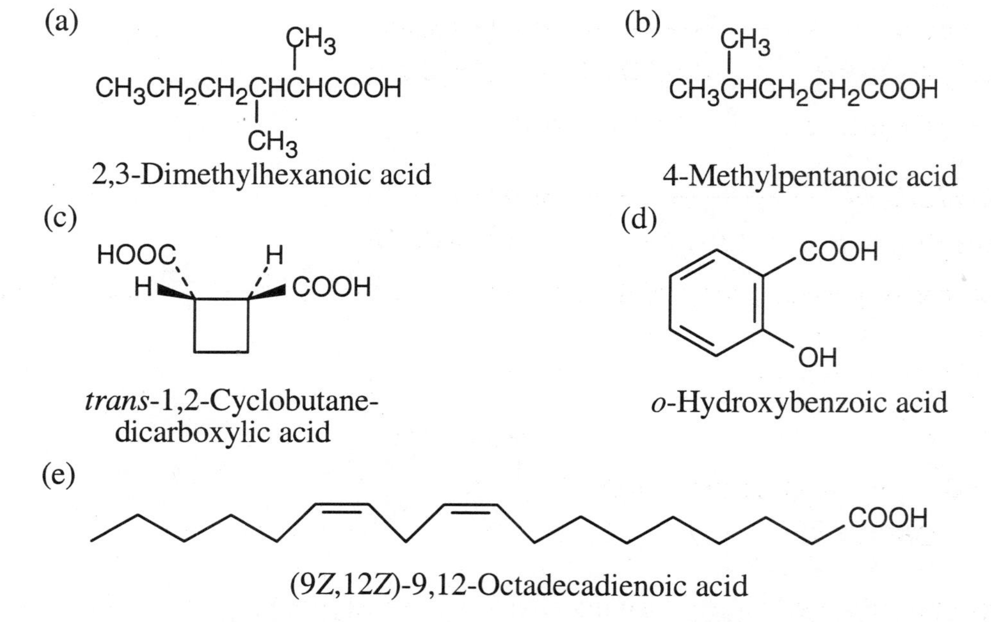

20.3 Naphthalene is insoluble in water and benzoic acid is only slightly soluble. The *salt* of benzoic acid is very soluble in water, however, and we can take advantage of this solubility in separating naphthalene from benzoic acid.

Dissolve the mixture in an organic solvent, and extract with a dilute aqueous solution of sodium hydroxide or sodium bicarbonate, which will neutralize benzoic acid. Naphthalene will remain in the organic layer, and all benzoic acid, now converted to the benzoate salt, will be in the aqueous layer. To recover benzoic acid, remove the aqueous layer, acidify it with dilute mineral acid, and extract with an organic solvent.

20.4

$$Cl_2CHCOOH + H_2O \overset{K_a}{\rightleftharpoons} Cl_2CHCOO^- + H_3O^+$$

$$K_a = \frac{[Cl_2CHCOO^-][H_3O^+]}{[Cl_2CHCOOH]} = 3.32 \times 10^{-2}$$

	Initial molarity	*Molarity after dissociation*
$Cl_2CHCOOH$	0.10 M	0.10 M – y
Cl_2CHCOO^-	0	y
H_3O^+	0	y

$$K_a = \frac{y \cdot y}{0.10 - y} = 3.32 \times 10^{-2}$$

Using the quadratic formula to solve for y, we find that y = 0.0434

$$\text{Percent dissociation} = \frac{0.0434}{0.1000} \times 100\% = 43.4\%$$

20.5

Weaker acid ⟶ *Stronger acid*

(a) CH_3CH_2COOH < $BrCH_2COOH$ < FCH_2COOH

Fluoride is the most electronegative group and can stabilize the carboxylate anion the best.

Weaker acid ⟶ *Stronger acid*

(c) $CH_3CH_2NH_2$ < CH_3CH_2OH < CH_3CH_2COOH

Alcohols are very weak acids, and the electron lone pair of amines makes them basic, not acidic.

20.6

$$HO{-}C({=}O){-}C({=}O){-}OH + H_2O \rightleftharpoons H_3O^+ + HO{-}C({=}O){-}C({=}O){-}O^- \longleftrightarrow HO{-}C({=}O){-}C({=}O){-}O^-$$

The pK_1 of oxalic acid is lower than that of a monocarboxylic acid because the carboxylate anion is stabilized both by resonance and by the inductive effect of the nearby second carboxylic acid group.

$$HO{-}C({=}O){-}C({=}O){-}O^- + H_2O \rightleftharpoons H_3O^+ + {}^-O{-}C({=}O){-}C({=}O){-}O^- \longleftrightarrow {}^-O{-}C({=}O){-}C({=}O){-}O^-$$

The pK_2 of oxalic acid is higher than pK_1 for two reasons: (1) The first carboxylate group inductively destabilizes the negative charge resulting from dissociation of the second proton. (2) Electrostatic repulsion between the two adjacent negative charges destabilizes the dianion.

20.7 A pK_a of 4.45 indicates that *p*–cyclopropylbenzoic acid is a weaker acid than benzoic acid. This, in turn, indicates that a cyclopropyl group must be electron-donating. Since electron-donating groups increase reactivity in electrophilic substitution reactions, *p*–cyclopropylbenzene should be more reactive than benzene toward electrophilic bromination.

20.8 Remember that electron-withdrawing groups increase carboxylic acid acidity, and electron donating groups decrease carboxylic acid acidity. Benzoic acid is more acidic than acetic acid.

Least acidic ⟶ *Most acidic*

(a) $H_3C{-}C_6H_4{-}COOH$ < $C_6H_5{-}COOH$ < $Cl{-}C_6H_4{-}COOH$

Least acidic ——————————> *Most acidic*

(b)

CH_3COOH < C_6H_5COOH (benzoic acid) < p-$O_2NC_6H_4COOH$

20.9 In parts (a) and (b), Grignard carboxylation must be used because the starting materials can't undergo S_N2 reactions. In (c), either method can be used.

(a)

$$C_6H_5Br \xrightarrow[\text{3. } H_3O^+]{\text{1. Mg, ether; 2. } CO_2\text{, ether}} C_6H_5COOH$$

(b)

$$(CH_3)_3CCl \xrightarrow[\text{3. } H_3O^+]{\text{1. Mg, ether; 2. } CO_2\text{, ether}} (CH_3)_3CCOOH$$

(c)

$$CH_3CH_2CH_2Br \xrightarrow[\text{3. } H_3O^+]{\text{1. Mg, ether; 2. } CO_2\text{, ether}} \textit{or} \xrightarrow[\text{2. } H_3O^+]{\text{1. NaCN}} CH_3CH_2CH_2COOH$$

20.10

$$C_6H_5CH_2Br \xrightarrow[\text{3. } H_3O^+]{\text{1. Mg, ether; 2. } CO_2\text{, ether}} \textit{or} \xrightarrow[\text{2. } H_3O^+]{\text{1. NaCN}} C_6H_5CH_2COOH \xrightarrow[\text{2. } H_3O^+]{\text{1. } LiAlH_4} \textit{or} \xrightarrow[\text{2. } H_3O^+]{\text{1. } BH_3\text{, THF}} C_6H_5CH_2CH_2OH$$

Synthetic analysis: The alcohol product can be formed by reduction of a carboxylic acid with either $LiAlH_4$ or BH_3. The carboxylic acid can be synthesized either by Grignard carboxylation or by nitrile hydrolysis.

20.11

$$C_5H_9CH_2OH \xrightarrow{PBr_3} C_5H_9CH_2Br \xrightarrow[\text{3. } H_3O^+]{\text{1. Mg, ether; 2. } CO_2\text{, ether}} \textit{or} \xrightarrow[\text{2. } H_3O^+]{\text{1. NaCN}} C_5H_9CH_2COOH \xrightarrow[\text{2. } H_3O^+]{\text{1. } LiAlH_4} \textit{or} \xrightarrow[\text{2. } H_3O^+]{\text{1. } BH_3\text{, THF}} C_5H_9CH_2CH_2OH$$

Synthetic analysis: After treating the initial alcohol with PBr_3, the same steps as used in the previous problem can be followed.

20.12

O 1715 cm^{-1}

3300–3400 cm^{-1} HO — =O 1710 cm^{-1}

OH 2500–3300 cm^{-1}

The positions of the carbonyl absorptions are too similar to be useful. The –OH absorptions, however, are sufficiently different for distinguishing between the compounds; the broad band of the carboxylic acid hydroxyl group is especially noticeable.

20.13 ^{1}H NMR:

O — 2.0–2.3 δ

3.5–4.5 δ → H — =O

OH ← 12 δ HO

The distinctive peak at 12 δ serves to identify the carboxylic acid. For the hydroxyketone, the absorption of the hydrogen on the oxygen-bearing carbon (3.5–4.5 δ) is significant. The position of absorption of the hydroxyl hydrogen is unpredictable, but addition of D_2O to the sample can be used to identify this peak.

^{13}C NMR:

O 50–60 δ 210 δ

C ← 180 δ HO — =O

OH

The positions of the carbonyl carbon absorptions can be used to distinguish between these two compounds. The hydroxy ketone also shows an absorption in the range 50–60 δ due to the hydroxyl group carbon.

Visualizing Chemistry

20.14

(a) HOOC, OCH3, Br

3-Bromo-4-methoxy-benzoic acid

(b) H, H3C, C=C, COOH, CH3

3-Methyl-2-butenoic acid

(c) COOH

1,3-Cyclopentadiene-carboxylic acid

20.15

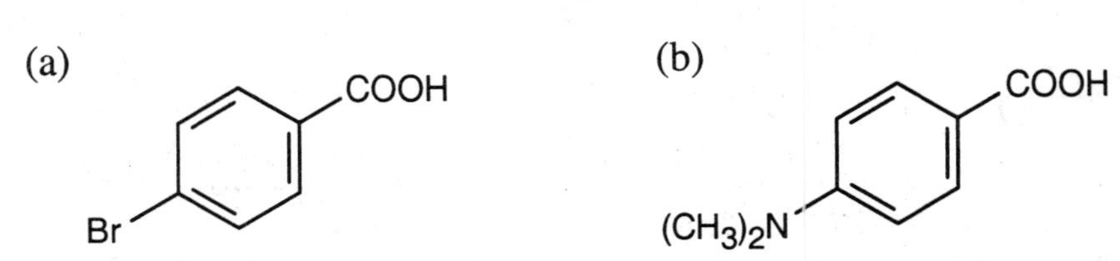

(a) *p*-Bromobenzoic acid is more acidic than benzoic acid because the electron-withdrawing bromine stabilizes the carboxylate anion.

(b) This *p*-substituted aminobenzoic acid is less acidic than benzoic acid because the electron-donating group destabilizes the carboxylate anion.

20.16

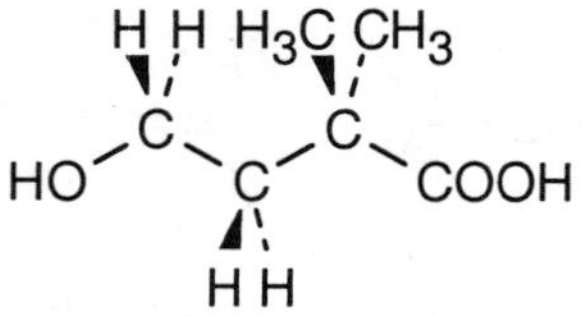

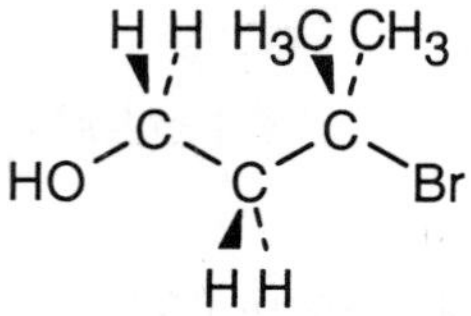

[Structures: HO–CH₂–CH₂–C(CH₃)₂–COOH and HO–CH₂–CH₂–C(CH₃)₂–Br, drawn with wedge bonds]

Nitrile reduction can't be used to synthesize the above carboxylic acid because the tertiary halide precursor (shown on the right) doesn't undergo S_N2 substitution with cyanide. Reaction occurs instead at the hydroxyl group. Grignard carboxylation also can't be used because the hydroxyl group interferes with formation of the Grignard reagent. If the hydroxyl group is protected, however, Grignard carboxylation can take place.

Additional Problems

20.17

(a) $CH_3CH(COOH)CH_2CH_2CH(COOH)CH_3$
2,5-Dimethylhexanedioic acid

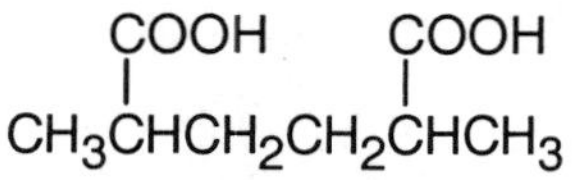

(b) $(CH_3)_3CCOOH$
2,2-Dimethylpropanoic acid

(c) $CH_3CH_2CH_2CH(CH_2CH_2CH_3)CH_2COOH$
3-Propylhexanoic acid

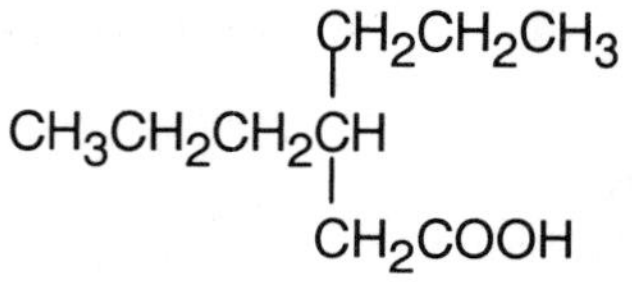

(d) O_2N–C₆H₄–COOH (para)
p-Nitrobenzoic acid

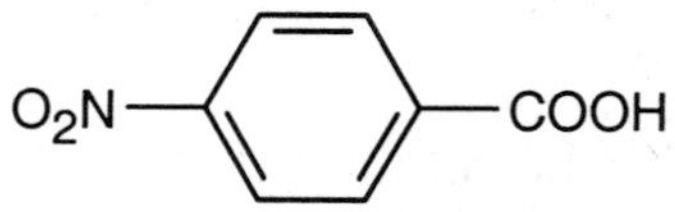

(e) [cyclodecene ring bearing COOH on the double-bond carbon]
1-Cyclodecenecarboxylic acid

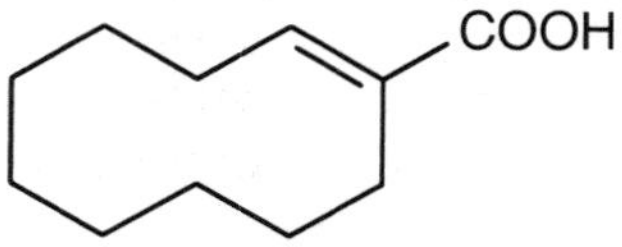

(f) $BrCH_2CH(Br)CH_2CH_2COOH$
4,5-Dibromopentanoic acid

20.18

(a) [cyclohexane ring with COOH groups on adjacent carbons, both wedged, H atoms dashed]
cis-1,2-Cyclohexane-dicarboxylic acid

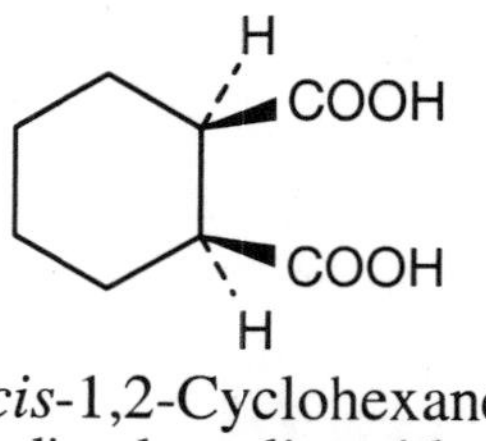

(b) $HOOCCH_2CH_2CH_2CH_2CH_2COOH$
Heptanedioic acid

(c) $CH_3C{\equiv}CCH{=}CHCOOH$
2-Hexen-4-ynoic acid

(d) $CH_3CH_2CH_2CH_2CH(CH_2CH_3)CH_2CH(CH_2CH_2CH_3)COOH$
4-Ethyl-2-propyloctanoic acid

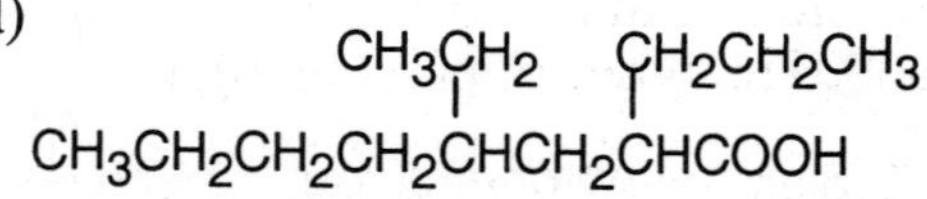

(e) 3-Chlorophthalic acid (benzene ring bearing two COOH groups and Cl)

(f) $(C_6H_5)_3CCOOH$
Triphenylacetic acid

20.19

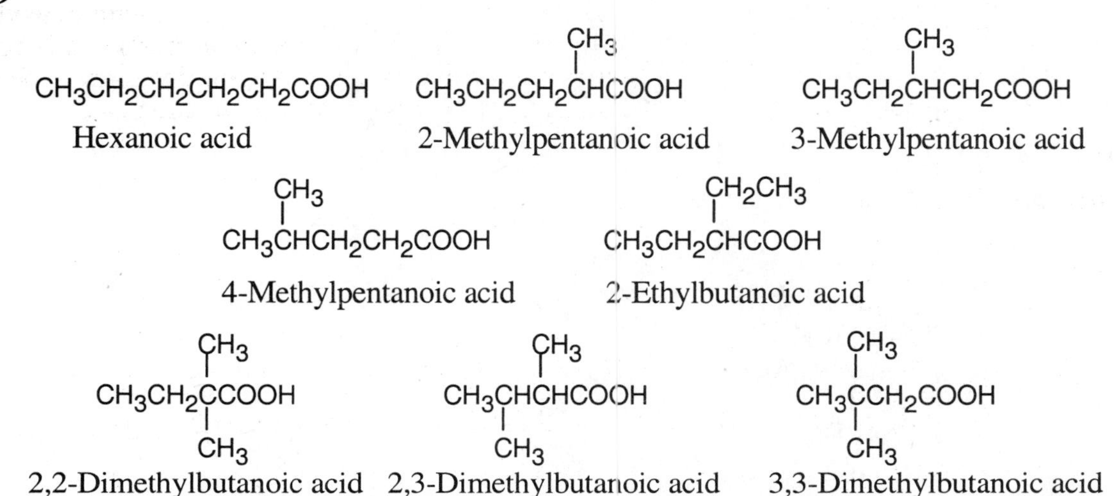

20.20

Less acidic ⟶ *More acidic*

(a) CH_3COOH < $HCOOH$ < $HOOC—COOH$
Acetic acid, Formic acid, Oxalic acid

(b) *p*-Bromobenzoic acid (weakly electron-withdrawing substituent) < *p*-Nitrobenzoic acid (strongly electron-withdrawing substituent) < 2,4-Dinitrobenzoic acid (two strongly electron-withdrawing substituents)

(c) FCH_2CH_2COOH < ICH_2COOH < FCH_2COOH

In (c), the strongest acid has the most electronegative atom next to the carboxylic acid group. The next strongest acid has a somewhat less electronegative atom next to the carboxylic acid group. The weakest acid has an electronegative atom two carbons away from the carboxylic acid group.

20.21 Remember that the conjugate base of a weak acid is a strong base. In other words, the stronger the acid, the weaker the base derived from that acid.

Less basic ⟶ *More basic*

(a) $Mg(OAc)_2$ < $Mg(OH)_2$ < $H_3C^- MgBr^+$

Acetic acid is a much stronger acid than water, which is a much, much stronger acid than methane. The order of base strength is just the reverse.

(b) Sodium *p*-nitrobenzoate < Sodium benzoate < $HC{\equiv}C^- Na^+$

p-Nitrobenzoic acid is stronger than benzoic acid, which is much stronger than acetylene.

(c) $HCOO^- Li^+$ < $HO^- Li^+$ < $CH_3CH_2O^- Li^+$

20.22

(a)

$$CH_3CH_2CH_2COOH \xrightarrow[2.\ H_3O^+]{1.\ BH_3 \text{ or } LiAlH_4} CH_3CH_2CH_2CH_2OH$$

1-Butanol

(b)

$$CH_3CH_2CH_2CH_2OH \xrightarrow{PBr_3} CH_3CH_2CH_2CH_2Br$$

from (a) — 1-Bromobutane

(c)

$$CH_3CH_2CH_2CH_2Br \xrightarrow[2.\ H_3O^+]{1.\ NaCN} CH_3CH_2CH_2CH_2COOH$$

from (b) — Pentanoic acid

Grignard carboxylation can also be used.

(d)

$$CH_3CH_2CH_2CH_2Br \xrightarrow{K^+\ ^-OC(CH_3)_3} CH_3CH_2CH{=}CH_2$$

from (b) — 1-Butene

(e)

$$CH_3CH_2CH_2CH_2Br \xrightarrow{2\ Li} 2\ CH_3CH_2CH_2CH_2^-Li^+ \xrightarrow{CuI} (CH_3CH_2CH_2CH_2)_2Cu^-Li^+$$

from (b)

$$(CH_3CH_2CH_2CH_2)_2Cu^-Li^+ + CH_3CH_2CH_2CH_2Br \longrightarrow CH_2(CH_2)_6CH_3$$

Octane

20.23

(a)

$$CH_3CH_2CH_2CH_2OH \xrightarrow[H_3O^+]{CrO_3} CH_3CH_2CH_2COOH$$

(b)

$$CH_3CH_2CH_2CH_2Br \xrightarrow{NaOH} CH_3CH_2CH_2CH_2OH \xrightarrow[H_3O^+]{CrO_3} CH_3CH_2CH_2COOH$$

(c)

$$CH_3CH_2CH{=}CH_2 \xrightarrow[2.\ H_2O_2,\ ^-OH]{1.\ BH_3,\ THF} CH_3CH_2CH_2CH_2OH \xrightarrow[H_3O^+]{CrO_3} CH_3CH_2CH_2COOH$$

(d)

$$CH_3CH_2CH_2Br \xrightarrow[2.\ H_3O^+]{1.\ NaCN} CH_3CH_2CH_2COOH$$

or

$$CH_3CH_2CH_2Br \xrightarrow[3.\ H_3O^+]{1.\ Mg,\ ether;\ 2.\ CO_2,\ ether} CH_3CH_2CH_2COOH$$

(e)

$$CH_3CH_2CH_2CH{=}CHCH_2CH_2CH_3 \xrightarrow[H_3O^+]{KMnO_4} 2\ CH_3CH_2CH_2COOH$$

20.24

(a)

Benzene $\xrightarrow[AlCl_3]{CH_3Cl}$ Toluene (CH_3) $\xrightarrow[H_2O]{KMnO_4}$ Benzoic acid (COOH) $\xrightarrow[FeCl_3]{Cl_2}$ (COOH, Cl)

m-Chlorobenzoic acid

(b)

Benzene $\xrightarrow[AlCl_3]{CH_3Cl}$ (CH_3) $\xrightarrow[FeBr_3]{Br_2}$ (CH_3, Br) $\xrightarrow[H_2O]{KMnO_4}$ (COOH, Br)

p-Bromoobenzoic acid

(c)

Benzene $\xrightarrow[AlCl_3]{CH_3Cl}$ (CH_3) $\xrightarrow[(PhCO_2)_2]{NBS}$ (CH_2Br) $\xrightarrow[2.\ H_3O^+]{1.\ NaCN}$ (CH_2COOH)

Phenylacetic acid

Alternatively, benzyl bromide can be converted to a Grignard reagent, poured over CO_2, and the resulting mixture can be treated with aqueous acid.

20.25 (a) $K_a = 8.4 \times 10^{-4}$ for lactic acid
$pK_a = -\log (8.4 \times 10^{-4}) = 3.08$

(b) $K_a = 5.6 \times 10^{-6}$ for acrylic acid
$pK_a = -\log (5.6 \times 10^{-6}) = 5.25$

20.26 (a) $pK_a = 3.14$ for citric acid
$K_a = 10^{-3.14} = 7.2 \times 10^{-4}$

(b) $pK_a = 2.98$ for tartaric acid
$K_a = 10^{-2.98} = 1.0 \times 10^{-3}$

20.27 $\Delta G° = -RT \ln K_a = -2.303\ RT \log K_a$. Here, R = 8.315 J/mol • K; T = 300 K
$\Delta G° = (-2.303)(8.315\ \text{J/mol} \cdot \text{K})(300\ \text{K})(\log K_a)$
$= (-5.745 \times 10^3\ \text{J/mol})(\log K_a)$
$= (-5.745\ \text{kJ/mol})(\log K_a)$

For ethanol: $pK_a = 16$; $\log K_a = -16$
$\Delta G° = (-5.745\ \text{kJ/mol})(-16) = +92\ \text{kJ/mol}$

For acetic acid: $pK_a = 4.75$; $\log K_a = -4.75$
$\Delta G° = (-5.745\ \text{kJ/mol})(-4.75) = +27\ \text{kJ/mol}$

Dissociation of acetic acid is more favorable. Since $\Delta G°$ for acetic acid is a smaller number, less energy is required for dissociation of acetic acid than for dissociation of ethanol.

20.28 Inductive effects of functional groups are transmitted through σ bonds. For oxalic acid, the electron-withdrawing inductive effect of one carboxyl group on the ionization of the second group is significant. However, as the length of the carbon chain increases, the effect of one functional group on another decreases. In this example, the influence of the second carboxyl group on the ionization of the first is barely felt by succinic and adipic acids.

20.29

p-Methylbenzoic acid ($H_3C-C_6H_4-COOH$):

(a) 1. BH_3, THF; 2. H_3O^+ → $H_3C-C_6H_4-CH_2OH$

(b) NBS, $(PhCO_2)_2$ → $BrH_2C-C_6H_4-COOH$

(c) 1. CH_3MgBr; 2. H_3O^+ → $H_3C-C_6H_4-COOH$ + CH_4

(d) $KMnO_4$, H_2O → $HOOC-C_6H_4-COOH$

(e) 1. $LiAlH_4$; 2. H_3O^+ → $H_3C-C_6H_4-CH_2OH$

In (c), the acidic proton reacts with the Grignard reagent to form methane.

20.30

(a)

$$CH_3CH_2Br \xrightarrow{\text{Mg, ether}} CH_3CH_2MgBr \xrightarrow[\text{2. } H_3O^+]{\text{1. } ^{13}CO_2\text{, ether}} CH_3CH_2{}^{13}COOH$$

(b)

$$CH_3Br \xrightarrow[\text{ether}]{\text{Mg}} CH_3MgBr \xrightarrow[\text{2. } H_3O^+]{\text{1. } ^{13}CO_2\text{, ether}} CH_3{}^{13}COOH \xrightarrow[\text{2. } H_3O^+]{\text{1. } BH_3\text{, THF}} CH_3{}^{13}CH_2OH$$

$$CH_3{}^{13}CH_2OH \xrightarrow{PBr_3} CH_3{}^{13}CH_2Br \xrightarrow[\text{ether}]{\text{Mg}} CH_3{}^{13}CH_2MgBr \xrightarrow[\text{2. } H_3O^+]{\text{1. } CO_2\text{, ether}} CH_3{}^{13}CH_2COOH$$

20.31

(a)

$$\text{methylenecyclohexane} \xrightarrow[\text{2. } H_2O_2,\ ^-OH]{\text{1. } BH_3,\ THF} \text{cyclohexyl(H)}CH_2OH \xrightarrow{PBr_3} \text{cyclohexyl}CH_2Br \xrightarrow[\text{2. } H_3O^+]{\text{1. NaCN}} \text{cyclohexyl}CH_2COOH$$

Grignard carboxylation can also be used to form the carboxylic acid.

(b)

$$\text{methylenecyclohexane} \xrightarrow{HBr} \text{1-bromo-1-methylcyclohexane } (H_3C,\ Br) \xrightarrow[\text{ether}]{Mg} (H_3C,\ MgBr) \xrightarrow[\text{2. } H_3O^+]{\text{1. } CO_2\text{, ether}} (H_3C,\ COOH)$$

Only Grignard carboxylation can be used because ^-CN brings about elimination of the tertiary bromide to form a double bond.

20.32 (a) Grignard carboxylation can't be used to prepare the carboxylic acid because of the acidic hydroxyl group. Use nitrile hydrolysis.

(b) Either method produces the carboxylic acid. Grignard carboxylation is a better reaction for preparing a carboxylic acid from a secondary bromide. Nitrile hydrolysis produces an optically active carboxylic acid from an optically active bromide.

c) Neither method of acid synthesis yields the desired product. Any Grignard reagent formed will react with the carbonyl functional group present in the starting material. Reaction with cyanide occurs at the carbonyl functional group, producing a cyanohydrin, as well as at halogen. However, if the ketone is first protected by forming an acetal, either method can be used.

d) Since the hydroxyl proton interferes with formation of the Grignard reagent, nitrile hydrolysis must be used to form the carboxylic acid.

20.33 2-Chloro-2-methylpentane is a tertiary alkyl halide and ^-CN is a base. Instead of the desired S_N2 reaction of cyanide with a halide, E2 elimination occurs and yields 2-methyl-2-pentene.

$$CH_3CH_2CH_2C(CH_3)(Cl)CH_3 \xrightarrow[S_N2]{^-CN} \!\!\!\!\!\!\times\ CH_3CH_2CH_2C(CH_3)(CN)CH_3$$

$$CH_3CH_2CH_2C(CH_3)(Cl)CH_3 \xrightarrow[E2]{^-CN} (CH_3CH_2)(H)C{=}C(CH_3)_2$$

20.34 (a) BH_3 is a reducing agent, not an oxidizing agent. To obtain benzoic acid from toluene, use $KMnO_4$.

(b) Use CO_2 instead of NaCN to form the carboxylic acid, or eliminate Mg from this reaction scheme and form the acid by nitrile hydrolysis.

(c) Reduction of a carboxylic acid with $LiAlH_4$ yields an alcohol, not an alkyl group.

(d) Acidic hydrolysis of the nitrile will also dehydrate the tertiary alcohol. Use basic hydrolysis to form the carboxylic acid.

20.35

CH_3 (toluene) $\xrightarrow[H_2SO_4]{HNO_3}$ CH_3 / NO_2 $\xrightarrow[H_2O]{KMnO_4}$ COOH / NO_2 $\xrightarrow[2.\ ^-OH]{1.\ SnCl_2,\ H_3O^+}$ COOH / NH_2 PABA

Notice that the order of the reactions is very important. If toluene is oxidized first, the nitro group will be introduced in the meta position. If the nitro group is reduced first, oxidation to the carboxylic acid will reoxidize the $-NH_2$ group.

20.36 Before starting this type of problem, identify the functional groups present in the starting material. Lithocholic acid contains only alcohol and carboxylic acid functional groups. The given reagents can react with one, both, or neither functional group. Remember to keep track of stereochemistry.

no reaction

(a) CrO_3, H_3O^+

(b) Tollens reagent

(c) 1. BH_3 2. H_3O^+

(d) $(CH_3)_3SiCl$, $(CH_3CH_2)_3N$

(e) 1. 2 CH_3MgBr 2. H_3O^+ → Lithocholic acid + 2 CH_4

(f) 1. $LiAlH_4$ 2. H_3O^+

Lithocholic acid

20.37

Br_2, $FeBr_3$ → Br

Mg, ether → MgBr

1. CH_2O 2. H_3O^+ → CH_2OH

PCC → CHO

Cl_2, $FeCl_3$ → Cl, CHO

1. NaCN 2. H_3O^+ → Cl, OH, CHCOOH

1. $SOCl_2$, pyr 2. H_3O^+ → Cl, Cl, CHCOOH

Fenclorac

Other routes to this compound are possible. The illustrated route was chosen because it introduced the potential benzylic functional group and the potential carboxylic acid in one step. Notice that the aldehyde functional group and the cyclohexyl group both serve to direct the aromatic chlorination to the correct position. Also, reaction of the hydroxy acid with $SOCl_2$ converts –OH to –Cl and –COOH to –COCl. Treatment with H_3O^+ regenerates the carboxylic acid.

20.38

Substituent	pK_a	*Acidity*	**E.A.S. reactivity*
$—PCl_2$	3.59	Most acidic	Least reactive (most deactivating)
$—OSO_2CH_3$	3.84	↓	↓
—CH=CHCN	4.03		
$—HgCH_3$	4.10		
—H	4.19		
$—Si(CH_3)_3$	4.27	Least acidic	Most reactive (least deactivating)

*Electrophilic aromatic substitution

Recall from Section 20.5 that substituents that increase acidity also decrease reactivity in electrophilic aromatic substitution reactions. Of the above substituents, only $–Si(CH_3)_3$ is an activator.

20.39

(a)

Benzene $\xrightarrow[\text{AlCl}_3]{\text{CH}_3\text{CHClCH}_3}$ $C_6H_5CH(CH_3)_2$ $\xrightarrow[(\text{PhCO}_2)_2]{\text{NBS}}$ $C_6H_5CBr(CH_3)_2$ $\xrightarrow{\text{Mg, ether}}$ $C_6H_5C(CH_3)_2MgBr$ $\xrightarrow[\text{2. H}_3\text{O}^+]{\text{1. CO}_2\text{, ether}}$ $C_6H_5C(CH_3)_2COOH$

Again, other routes to this compound are possible. The above route was chosen because it had relatively few steps and because the Grignard reagent could be prepared without competing reactions. Notice that nitrile hydrolysis is not a possible route to this compound because the necessary halide is tertiary.

(b)

Cyclohexanone $\xrightarrow[\text{2. H}_3\text{O}^+]{\text{1. PhMgBr, ether}}$ 1-phenylcyclohexanol (C_6H_5, OH) $\xrightarrow{\text{HBr}}$ 1-bromo-1-phenylcyclohexane (C_6H_5, Br) $\xrightarrow{\text{Mg, ether}}$ (C_6H_5, MgBr) $\xrightarrow[\text{2. H}_3\text{O}^+]{\text{1. CO}_2\text{, ether}}$ 1-phenylcyclohexanecarboxylic acid (C_6H_5, COOH)

Synthetic analysis: The product appears to have been formed from a Grignard adduct. Bromination of this adduct, followed by conversion to a Grignard reagent and carboxylation of this reagent, gives the desired product. Again, nitrile hydrolysis is not a route to this compound.

20.40 As we have seen throughout this book, the influence of substituents on reactions can be due to inductive effects or to resonance effects. For *m*-hydroxybenzoic acid, the negative charge of the carboxylate anion is stabilized by the electron-withdrawing *inductive* effect of –OH, making this isomer more acidic. For *p*-hydroxybenzoic acid, the negative charge of the anion is destabilized by the electron-donating *resonance* effect of –OH that acts over the π electron system of the ring, but is not important for *m*-substituents.

$m\text{-HOC}_6H_4COOH \rightleftharpoons m\text{-HOC}_6H_4CO_2^- + H^+$

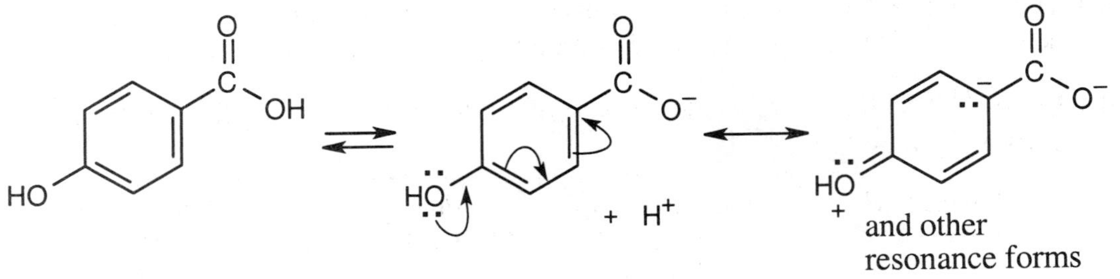

and other resonance forms

20.41

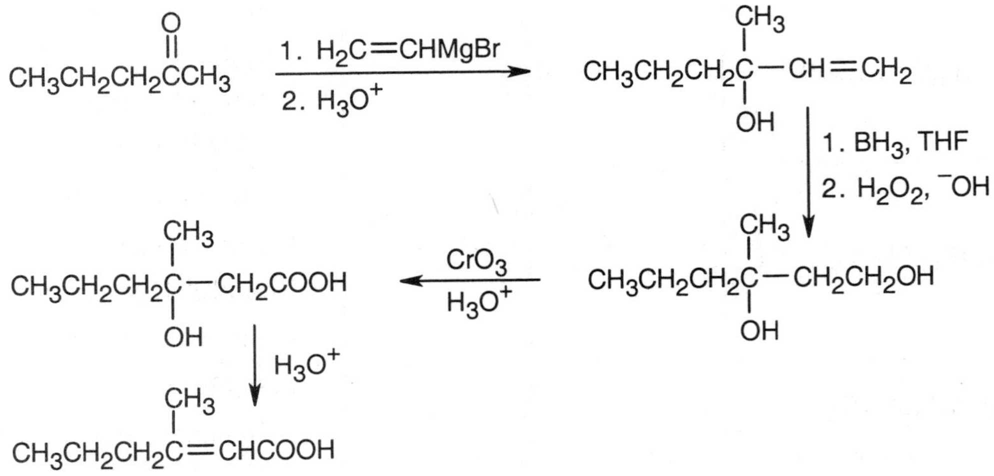

$$CH_3CH_2CH_2CCH_3 \xrightarrow[2.\ H_3O^+]{1.\ H_2C{=}CHMgBr} CH_3CH_2CH_2C(CH_3)(OH){-}CH{=}CH_2$$

$$\xrightarrow[2.\ H_2O_2,\ ^-OH]{1.\ BH_3,\ THF} CH_3CH_2CH_2C(CH_3)(OH){-}CH_2CH_2OH \xrightarrow[H_3O^+]{CrO_3} CH_3CH_2CH_2C(CH_3)(OH){-}CH_2COOH$$

$$\xrightarrow{H_3O^+} CH_3CH_2CH_2C(CH_3){=}CHCOOH$$

3-Methyl-2-hexenoic acid

Synthetic analysis: As in all of these more complex syntheses, other routes to the target compound are possible. This route was chosen because the Grignard reaction introduces a double bond without removing functionality at carbon 3. Dehydration occurs in the desired direction to produce a double bond conjugated with the carboxylic acid carbonyl group.

20.42

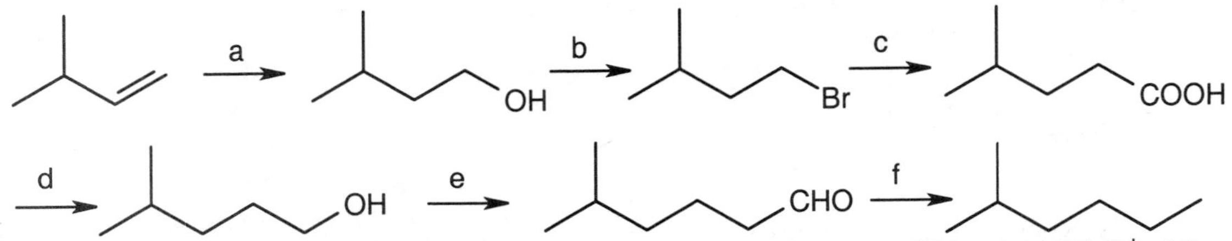

(a) BH_3, THF, then H_2O_2, OH^-; (b) PBr_3; (c) Mg, then CO_2, or CN^-, then H_3O^+; (d) $LiAlH_4$, then H_3O^+; (e) PCC; (f) N_2H_4, KOH

20.43

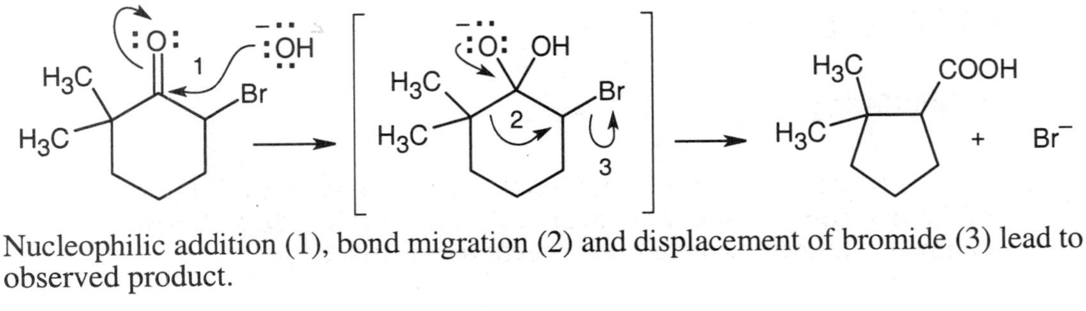

Nucleophilic addition (1), bond migration (2) and displacement of bromide (3) lead to the observed product.

20.44 The peak at 1.08 δ is due to a *tert*-butyl group, and the peak at 11.2 δ is due to a carboxylic acid group. The compound is 3,3-dimethylbutanoic acid, $(CH_3)_3CCH_2COOH$.

20.45 Either ^{13}C NMR or ^{1}H NMR can be used to distinguish among these three isomeric carboxylic acids.

Compound	*Number of ^{13}C NMR absorptions*	*Number of ^{1}H NMR absorptions*	*Splitting of ^{1}H NMR signals*
$CH_3(CH_2)_3COOH$	5	5	1 triplet, peak area 3, 1.0 δ 1 triplet, peak area 2, 2.4 δ 2 multiplets, peak area 4, 1.5 δ 1 singlet, peak area 1, 12.0 δ
$(CH_3)_2CHCH_2COOH$	4	4	1 doublet, peak area 6, 1.0 δ 1 doublet, peak area 2, 2.4 δ 1 multiplet, peak area 1, 1.6 δ 1 singlet, peak area 1, 12.0 δ
$(CH_3)_3CCOOH$	3	2	1 singlet, peak area 9, 1.3 δ 1 singlet, peak area 1, 12.1 δ

20.46 In all of these pairs, different numbers of peaks occur in the spectra of each isomer. (a), (b) Use either ^{1}H NMR or ^{13}C NMR to distinguish between the isomers.

Compound	*Number of ^{13}C NMR absorptions*	*Number of ^{1}H NMR absorptions*
(a) HOOC, COOH	3	2
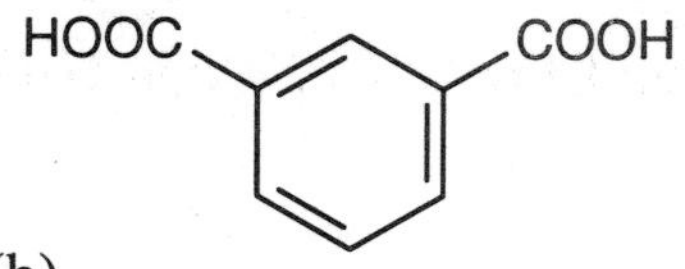 HOOC, COOH	5	4
(b) $HOOCCH_2CH_2COOH$	2	2
$CH_3CH(COOH)_2$	3	3

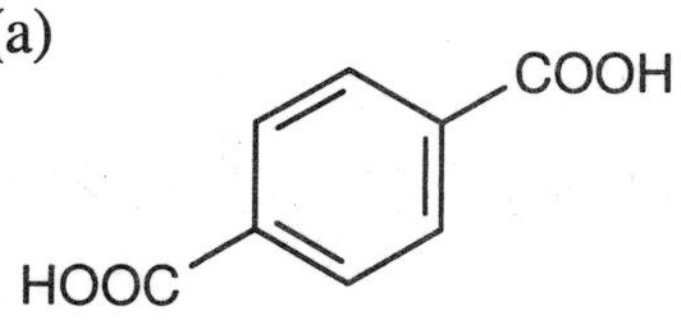

(c) Use ^{1}H NMR to distinguish between these two compounds. The carboxylic acid proton of the first compound absorbs near 12 δ, and the aldehyde proton of the second compound absorbs near 10 δ and is split into a triplet.

(d) The cyclic acid shows four absorptions in both its ^1H NMR and ^{13}C NMR spectra. The unsaturated acid shows six absorptions in its ^{13}C NMR and five in its ^1H NMR spectrum; one of the ^1H NMR signals occurs in the vinylic region (4.5 – 6.5 δ) of the spectrum. The ^{13}C NMR spectrum of the unsaturated acid also shows two absorptions in the C=C bond region (100–150 δ).

20.47 The compound has one degree of unsaturation, due to the carboxylic acid absorption seen in the IR spectrum.

$$\overset{a}{CH_3}\overset{b}{CH_2}O\overset{c}{CH_2}CO\overset{d}{O}H$$

a = 1.3 δ
b = 3.7 δ
c = 4.1 δ
d = 11.1 δ

20.48 Both compounds contain four different kinds of protons (remember that the $H_2C=$ protons are nonequivalent). The carboxylic acid proton absorptions are easy to identify; the other three absorptions in each spectrum are more complex.

It is possible to assign the spectra by studying the methyl group absorptions. The methyl group peak of crotonic acid is split into a doublet by the geminal ($CH_3CH=$) proton, while the methyl group absorption of methacrylic acid is a singlet. The first spectrum is that of crotonic acid, and the second spectrum is that of methacrylic acid.

20.49 (a) From the formula, we know that the compound has 2 degrees of unsaturation, one of which is due to the carboxylic acid group that absorbs at 183.0 δ. The ^{13}C NMR spectrum also shows that no other sp^2 carbons are present in the sample and indicates that the other degree of unsaturation is due to a ring, which is shown to be a cyclohexane ring by symmetry and by the types of carbons in the structure.

(b) The compound has 5 degrees of unsaturation, and is a methyl-substituted benzoic acid. The symmetry shown by the aromatic absorptions identifies the compound as *p*-methylbenzoic acid.

(a) COOH

(b) COOH

H_3C

A Look Ahead

20.50 This mechanism divides into two sequences of steps. In the first part of the mechanism, an acetal is hydrolyzed to acetone and a dihydroxycarboxylic acid.

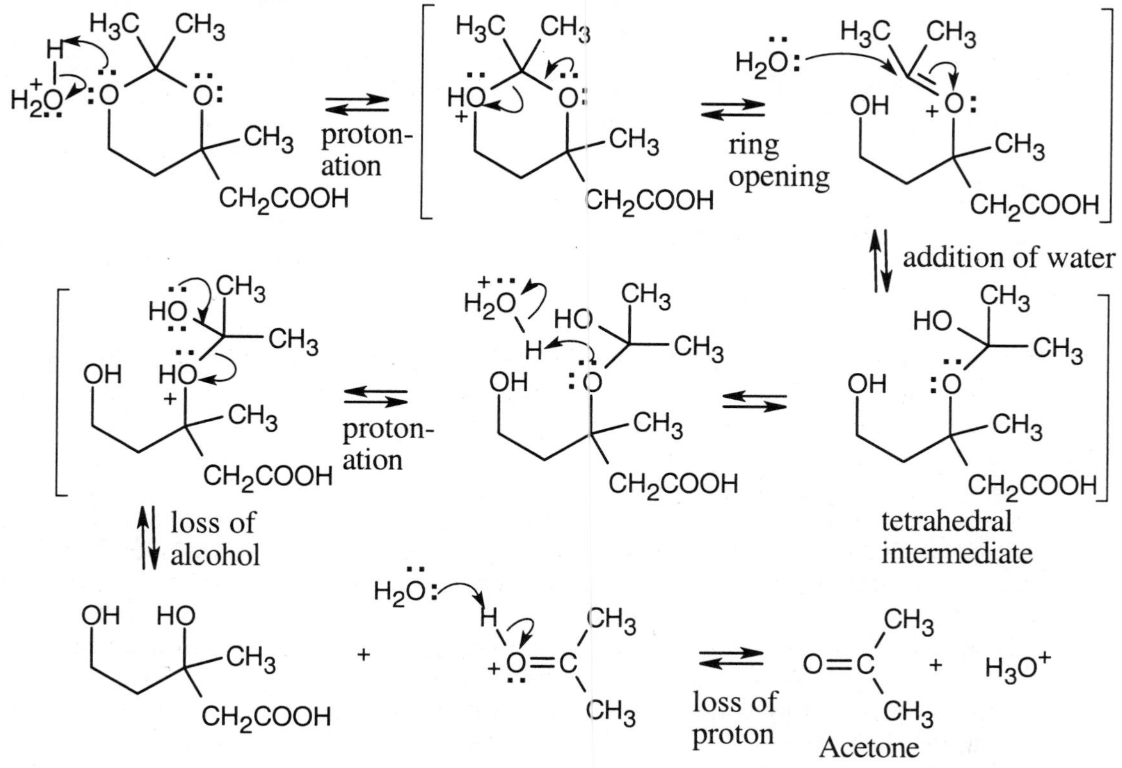

In the second set of steps, the dihydroxycarboxylic acid forms a cyclic ester (a lactone).

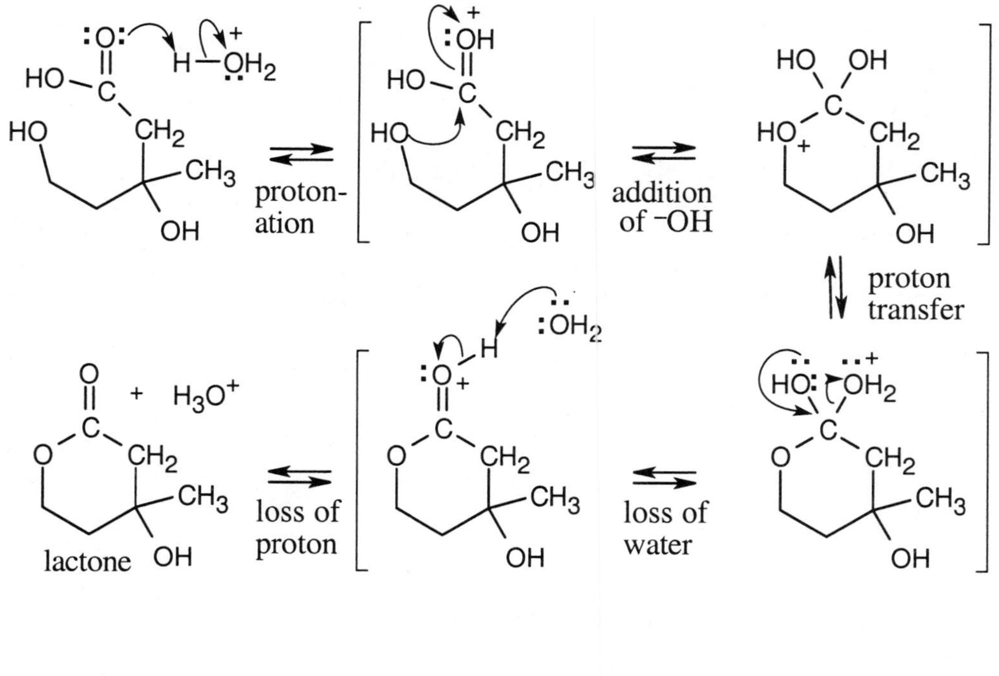

20.51

deprotonation

loss of CO_2

enolate

+ CO_2

+ ^{-}OH

This reaction proceeds because of the loss of CO_2 and the stability of the enolate anion.

Molecular Modeling

20.52 Acetic acid contains a more negative carbonyl group and a more positive –OH hydrogen (compared to –SH), and acetic acid molecules should associate more strongly. This observation is confirmed by the energy required to break the hydrogen bonds in acetic acid (113 kJ/mol), relative to the energy required to break the hydrogen bonds in thioacetic acid (41 kJ/mol).

20.53 The electrostatic map of pentanoic acid shows that the O–H bond is polar, and the bond density surface shows that the bond is covalent (the short O–H distance of 97 pm is additional confirmation). The hydrogen bonds to only one oxygen because covalent bonds require electron-sharing, and the carbonyl oxygen already has a full valence shell. The electrostatic potential map shows that sodium pentanoate exists as an ion pair, and the bond density surface confirms that sodium and oxygen do not share electrons. The map and the two nearly identical Na–O distances (212 pm) indicate that sodium forms ionic bonds to both oxygens. This is possible because ionic bonds do not require electron-sharing. Rather, the sodium ion occupies a position that brings it close to as many electron-rich atoms as possible.

20.54 Butanoate has the most negative oxygens, and 2-chlorobutanoate has the least negative oxygens. These data confirm chlorine's electron-withdrawing inductive effect, which is strongest over short distances. Anions with more negative oxygens should be stronger bases, and their conjugate acids should be weaker acids. Thus, butanoic acid should be the weakest acid, and 2-chlorobutanoic acid should be the strongest acid.

20.55 The nitro groups are electron-withdrawing, and they make the oxygens less negative in 4-nitrobenzoate anion (relative to benzoate anion) and in 4-nitrophenoxide anion (relative to phenoxide anion). The nitro group has a greater effect on the charge on oxygen in phenoxide anion because the charge can be delocalized into the ring and is closer to the nitro group, whereas the charge in benzoate anion is delocalized only over the carboxylate group.

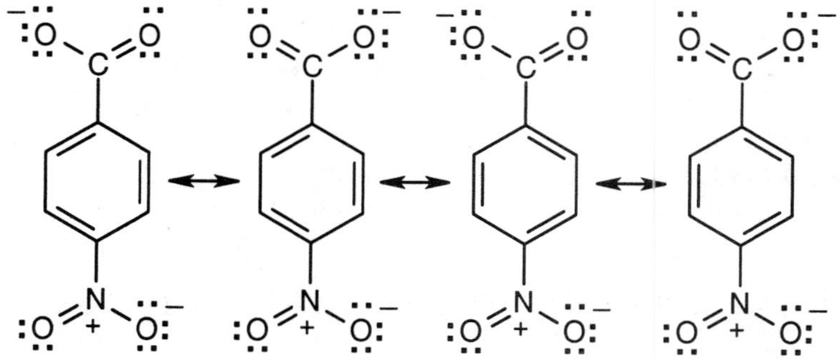

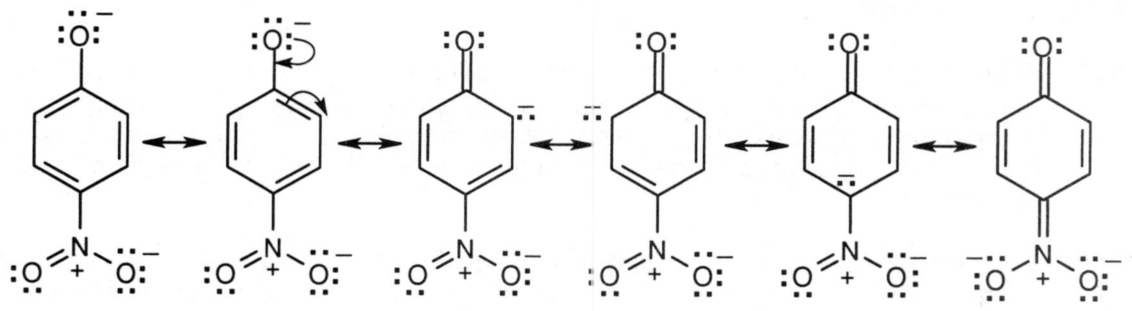

(Note: To save space, resonance forms involving the $-NO_2$ group aren't shown.)

Chapter 21 – Carboxylic Acid Derivatives. Nucleophilic Acyl Substitution Reactions

Chapter Outline

I. Introduction to carboxylic acid derivatives (Sections 21.1 – 21.2).
  A. Naming carboxylic acid derivatives (Section 21.1).
    1. Acid halides.
      a. The acyl group is named first, followed by the halide.
      b. For acyclic compounds, the *-ic acid* of the carboxylic acid name is replaced by *-yl,* followed by the name of the halide.
      c. For cyclic compounds, the *-carboxylic acid* ending is replaced by *-carbonyl,* followed by the name of the halide.
    2. Acid anhydrides.
      a. Symmetrical anhydrides are named by replacing *acid* by *anhydride.*
      b. For symmetrical substituted anhydrides, the prefix *bis-* is added to the acid name, followed by *anhydride.*
      c. Unsymmetrical anhydrides are named by citing the two acids alphabetically, followed by *anhydride.*
    3. Amides.
      a. Amides with an unsubstituted $-NH_2$ group are named by replacing *-oic acid* by *-amide* or by replacing *-carboxylic acid* with *-carboxamide.*
      b. If nitrogen is substituted, the nitrogen substituents are named, and an *N-* is put before each.
    4. Esters.
      Esters are named by first identifying the alkyl group and then the carboxylic acid group, replacing *-oic acid* by *-ate.*
    5. Nitriles.
      1. Simple nitriles are named by adding *-nitrile* to the alkane name.
      2. More complex nitriles are named as derivatives of carboxylic acids by replacing *-oic acid* by *-onitrile* or by replacing *-carboxylic acid* by *-carbonitrile.*
  B. Nucleophilic acyl substitution reactions (Section 21.2).
    1. Mechanism of nucleophilic acyl substitution reactions.
      a. A nucleophile adds to the polar carbonyl group.
      b. The tetrahedral intermediate eliminates one of the two substituents originally bonded to it, resulting in a net substitution reaction.
      c. Reactions of carboxylic acid derivatives take this course because one of the groups bonded to the carbonyl carbon is a good leaving group.
      d. The addition step is usually rate-limiting.
    2. Relative reactivity of carboxylic acid derivatives.
      a. Both steric and electronic factors determine relative reactivity.
        i. Steric hindrance in the acyl group decreases reactivity.
        ii. More polarized acid derivatives are more reactive than less polarized derivatives.
        iii. The effect of substituents on reactivity is similar to their effect on electrophilic aromatic substitution reactions.
      b. It is possible to convert more reactive derivatives into less reactive derivatives.
        i. In order of decreasing reactivity: acid chlorides > acid anhydrides > esters > amides.
        ii. Only esters, amides, and carboxylic acids are found in nature.
    3. Kinds of reactions of carboxylic acid derivatives:
      a. Hydrolysis: reaction with water to yield a carboxylic acid.
      b. Alcoholysis: reaction with an alcohol to yield an ester.

c. Aminolysis: reaction with ammonia or an amine to yield an amide.
d. Reduction.
i. Reaction with a hydride reducing agent yields an aldehyde or an alcohol.
ii Amides are reduced to yield amines.
e. Reaction with an organometallic reagent to yield a ketone or alcohol.

II. Reactions of carboxylic acids and their derivatives (Section 21.3 – 21.9).
A. Nucleophilic acyl substitution reactions of carboxylic acids (Section 21.3).
1. Carboxylic acids can be converted to acid chlorides by reaction with $SOCl_2$.
The reaction proceeds through a chlorosulfite intermediate.
2. Acid anhydrides are usually formed by heating the corresponding carboxylic acid to remove 1 equivalent of water.
3. Conversion to esters.
a. Conversion can be effected by the S_N2 reaction of a carboxylate and an alkyl halide.
b. Esters can be produced by the acid-catalyzed reaction of a carboxylic acid and an alcohol.
i. This reaction is known as a Fischer esterification.
ii. Mineral acid makes the acyl carbon more reactive toward the alcohol.
iii. All steps are reversible.
iv. The reaction can be driven to completion by removing water or by using a large excess of alcohol.
v. Isotopic labelling studies have confirmed the mechanism.
4. Amides are difficult to form from carboxylic acids because amines convert carboxylic acids to carboxylate salts that no longer have electrophilic carbons.

B. Chemistry of carboxylic acid halides (Section 21.4).
1. Carboxylic acid halides are prepared by reacting carboxylic acids with either $SOCl_2$ or PBr_3.
2. Acyl halides are very reactive.
Most reactions occur by nucleophilic acyl substitution mechanisms.
3. Hydrolysis.
a. Acyl halides react with water to form carboxylic acids.
b. The reaction mixture usually contains a base to scavenge the HCl produced.
4. Alcoholysis.
a. Acyl halides react with alcohols to form esters.
b. Base is usually added to scavenge the HCl produced.
c. Primary alcohols are more reactive than secondary or tertiary alcohols.
It's often possible to esterify a less hindered alcohol selectively.
5. Aminolysis.
a. Acid chlorides react with ammonia and amines to give amides.
b. Either two equivalents of ammonia/amine must be used, or NaOH must be present, in order to scavenge HCl.
6. Reduction.
a. $LiAlH_4$ reduces acid halides to alcohols.
The reaction is a substitution of H^- for Cl^- that proceeds through an intermediate aldehyde, which is then reduced.
b. Reduction with $LiAlH(OC(CH_3)_3)_3$ stops at the intermediate aldehyde.
7. Reaction with organometallic reagents.
a. Reaction with Grignard reagents yields tertiary alcohols and proceeds through an intermediate ketone.
b. Reaction with organocopper reagents yields ketones.
i. Reaction occurs by a radical mechanism.
ii. This reaction doesn't occur with other carboxylic acid derivatives.

C. Chemistry of carboxylic acid anhydrides (Section 21.5).
1. Acid anhydrides can be prepared by reaction of carboxylate anions with acid chlorides.
   Both symmetrical and unsymmetrical anhydrides can be prepared by this route.
2. Acid anhydrides react more slowly than acid chlorides.
   a. Acid anhydrides undergo most of the same reactions as acid chlorides.
   b. Acetic anhydride is often used to prepare acetate esters.
   c. In reactions of acid anhydrides, one-half of the molecule is "thrown away", making anhydrides inefficient to use.

D. Chemistry of esters (Section 21.6).
1. Esters can be prepared by:
   a. S_N2 reaction of a carboxylate anion with an alkyl halide.
   b. Fischer esterification.
   c. Reaction of an acid chloride with an alcohol, in the presence of base.
2. Esters are less reactive than acid halides and anhydrides but undergo the same types of reactions.
3. Hydrolysis.
   a. Basic hydrolysis (saponification) occurs through a nucleophilic acyl substitution mechanism.
      i. Loss of alkoxide ion yields a carboxylic acid which is deprotonated to give a carboxylate anion.
      ii. Isotope-labelling studies confirm this mechanism.
   b. Acidic hydrolysis can occur by more than one mechanism.
      The usual route is by the reverse of Fischer esterification.
4. Aminolysis.
   Esters can be converted to amides by heating with ammonia/amines, but it's easier to start with an acid chloride.
5. Reduction.
   a. $LiAlH_4$ reduces esters to primary alcohols by a route similar to that described for acid chlorides.
   b. If DIBAH at –78°C is used, reduction yields an aldehyde.
6. Reaction with Grignard reagents.
   Esters react twice with Grignard reagents to produce tertiary alcohols containing two identical substituents.

E. Chemistry of amides (Section 21.7).
1. Amides are prepared by the reaction of acid chlorides with ammonia/amines.
2. Hydrolysis.
   a. Hydrolysis occurs under more severe conditions than needed for hydrolysis of other acid derivatives.
   b. Acid hydrolysis occurs by addition of water to a protonated amide, followed by loss of ammonia or an amine.
   c. Basic hydrolysis occurs by attack of HO^-, followed by loss of $^-NH_2$.
3. Reduction.
   $LiAlH_4$ reduces amides to amines.

F. Chemistry of nitriles (Section 21.8).
1. Preparation of nitriles.
   a. Nitriles can be prepared by S_N2 reaction of ^-CN with a primary alkyl halide.
   b. They can also be prepared by $SOCl_2$ dehydration of primary amides.
2. Nitriles can react with nucleophiles via sp^2-hybridized imine intermediates.
3. Hydrolysis.
   a. Aqueous base hydrolyzes nitriles to carboxylates, plus an amine/ammonia.
      i. The reaction formation of a hydroxyimine that isomerizes to an amide, which is further hydrolyzed.
      ii. Milder conditions allow isolation of the amide.

b. Aqueous acid hydrolyzes nitriles to carboxylic acids and ammonium ions.
4. Reduction.
   a. $LiAlH_4$ reduces nitriles to primary amines.
   b. If DIBAH is used, an aldehyde is formed.
5. Reaction with Grignard reagents.
   Reaction of a nitrile with Grignard reagents yields a ketone.

G. Thiol esters (Section 21.9).
1. Nature uses thiol esters in nucleophilic acyl substitution reactions.
2. Acetyl CoA is used as an acylating agent.

III. Polyamides and polyesters (Section 21.10).

A. Formation of polyesters and polyamides.
1. When a diamine and a diacid chloride react, a polyamide is formed.
2. When a diacid and a diol react, a polyester is formed.
3. These polymers are called step-growth polymers because each bond is formed independently of the others.

B. Types of polymers.
1. Nylons are the most common polyamides.
2. The most common polyester, Dacron, is formed from dimethylterephthalate and ethylene glycol.

IV. Spectroscopy of carboxylic acid derivatives and nitriles (Section 21.11).

A. Infrared spectroscopy.
1. All of these compounds have characteristic carbonyl absorptions that help identify them; these are listed in Table 21.3.
2. Nitriles have an intense absorption at 2250 cm^{-1} that readily identifies them.

B. NMR spectroscopy is of limited usefulness.
1. Hydrogens next to carbonyl groups absorb at around 2.1 δ in a ^1H NMR spectrum, but this absorption can't be used to distinguish among carboxylic acid derivatives.
2. Carbonyl carbons absorb in the range 160–180 δ, but, again, this absorption can't be used to distinguish among carboxylic acid derivatives.

Solutions to Problems

21.1 Table 21.1 lists the suffixes for naming carboxylic acid derivatives. The suffixes used when the functional group is part of a ring are in parentheses.

(a) $CH_3CH(CH_3)CH_2CH_2C(=O)Cl$

4-Methylpentanoyl chloride

(b) Cyclohexyl–$CH_2C(=O)NH_2$

Cyclohexylacetamide

(c) $CH_3CH_2CH(CH_3)CN$

2-Methylbutanenitrile

(d) $(C_6H_5C(=O))_2O$

Benzoic anhydride

(e) Cyclopentyl–$C(=O)OCH(CH_3)CH_3$

Isopropyl cyclopentanecarboxylate

(f) Cyclopentyl–$O–C(=O)CH(CH_3)CH_3$

Cyclopentyl 2-methylpropanoate

(g) $H_2C{=}CHCH_2CH_2\overset{O}{\overset{\|}{C}}NH_2$

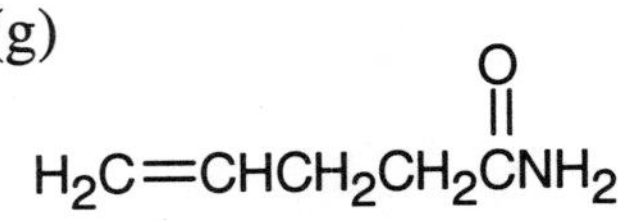

4-Pentenamide

(h) $CH_3CH_2\underset{}{\overset{CN}{\overset{|}{C}}}HCH_2CH_3$

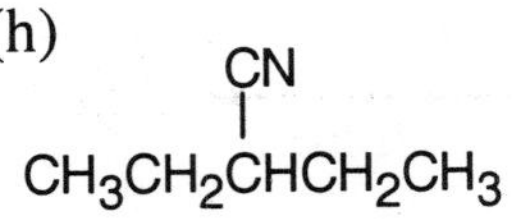

2-Ethylbutanenitrile

(i) $(H_3C)_2C{=}C(CH_3)COCl$

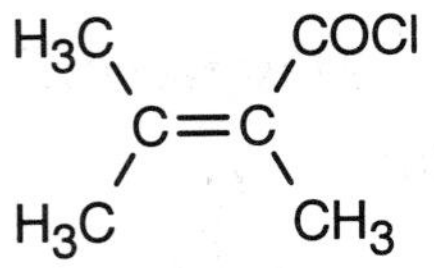

2.3-Dimethyl-2-butenoyl chloride

21.2

(a) $CH_3CH_2CH{=}CHCN$

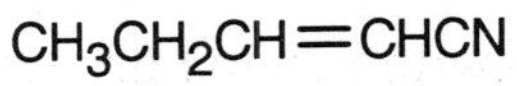

2-Pentenenitrile

(b) $CH_3CH_2CH_2\overset{O}{\overset{\|}{C}}N(CH_3)CH_2CH_3$

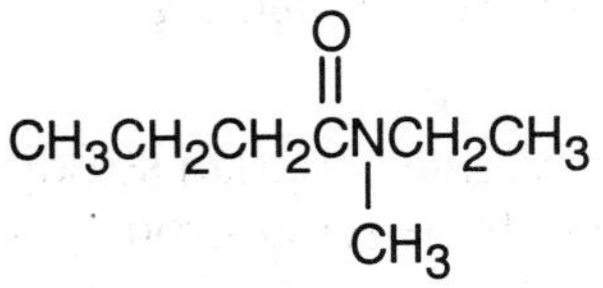

N-Ethyl-*N*-methylbutanamide

(c) $CH_3CH(CH_3)CH_2CH(CH_3)\overset{O}{\overset{\|}{C}}Cl$

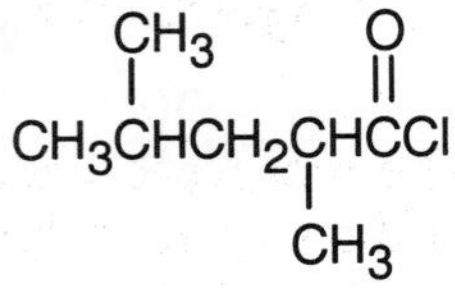

2,4-Dimethylpentanoyl chloride

(d) 1-methylcyclohexyl–$\overset{O}{\overset{\|}{C}}$–$OCH_3$

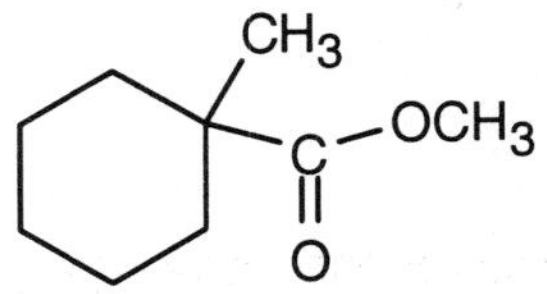

Methyl 1-methylcyclohexanecarboxylate

(e) $CH_3CH_2\overset{O}{\overset{\|}{C}}CH_2\overset{O}{\overset{\|}{C}}OCH_2CH_3$

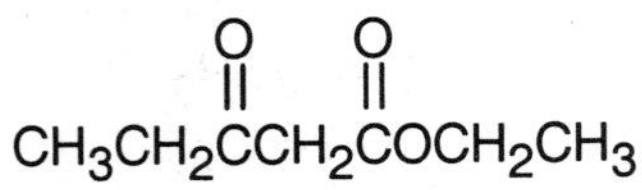

Ethyl-3-oxopenatanoate

(f) $(p\text{-}Br\text{-}C_6H_4\text{-}\overset{O}{\overset{\|}{C}})_2O$

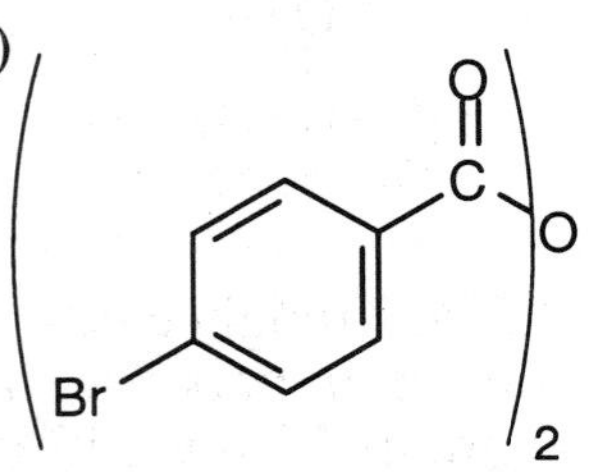

Bis(*p*-Bromobenzoic) anhydride

(g) $H\overset{O}{\overset{\|}{C}}\text{-}O\text{-}\overset{O}{\overset{\|}{C}}CH_2CH_3$

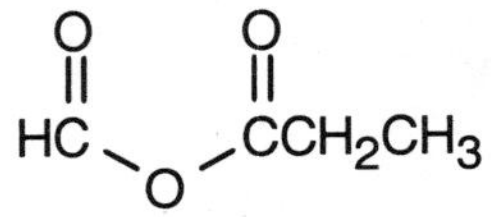

Formic propanoic anhydride

(h) cyclopentane with COBr and CH_3 cis (H, H)

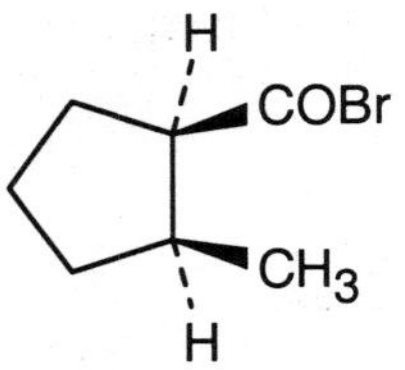

cis-2-Methylcyclohexanecarbonyl bromide

21.3

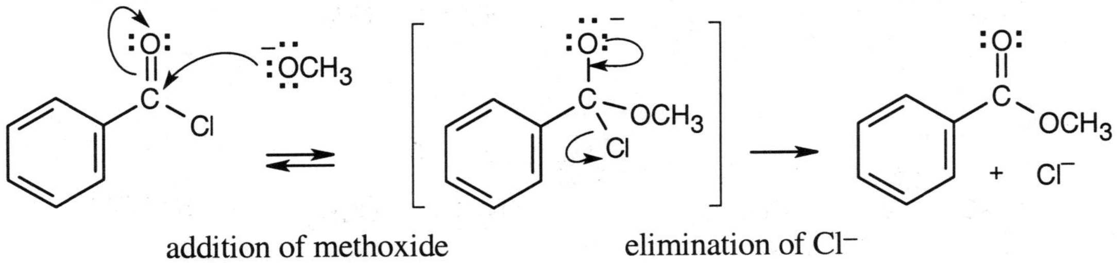

$C_6H_5COCl + {:}\ddot{O}CH_3^- \rightleftharpoons [C_6H_5C(O^-)(OCH_3)Cl] \rightarrow C_6H_5COOCH_3 + Cl^-$

addition of methoxide to form a tetrahedral intermediate

elimination of Cl^-

21.4

Most reactive ⟶ *Least reactive*

(a) $CH_3\overset{O}{\overset{||}{C}}Cl$ > $CH_3\overset{O}{\overset{||}{C}}OCH_3$ > $CH_3\overset{O}{\overset{||}{C}}NH_2$

(b) $CH_3\overset{O}{\overset{||}{C}}OCH(CF_3)_2$ > $CH_3\overset{O}{\overset{||}{C}}OCH_2CCl_3$ > $CH_3\overset{O}{\overset{||}{C}}OCH_3$

The most reactive acyl derivatives contain strongly electron-withdrawing groups in the part of the structure that is to be the leaving group.

21.5

$$\underset{\delta-}{F_3C}-\underset{\delta+}{\overset{O\ \delta-}{\overset{||}{C}}}-\underset{\delta-}{OCH_3} \qquad H_3C-\underset{\delta+}{\overset{O\ \delta-}{\overset{||}{C}}}-\underset{\delta-}{OCH_3}$$

The strongly electron-withdrawing trifluoromethyl group makes the carbonyl carbon more electron-poor and more reactive toward nucleophiles than the methyl acetate carbonyl group. Methyl trifluoroacetate is thus more reactive than methyl acetate in nucleophilic acyl substitution reactions.

21.6 Identify the nucleophile and the leaving group, and replace the leaving group by the nucleophile in the product. In this problem, the leaving groups are circled, and the nucleophiles are boxed.

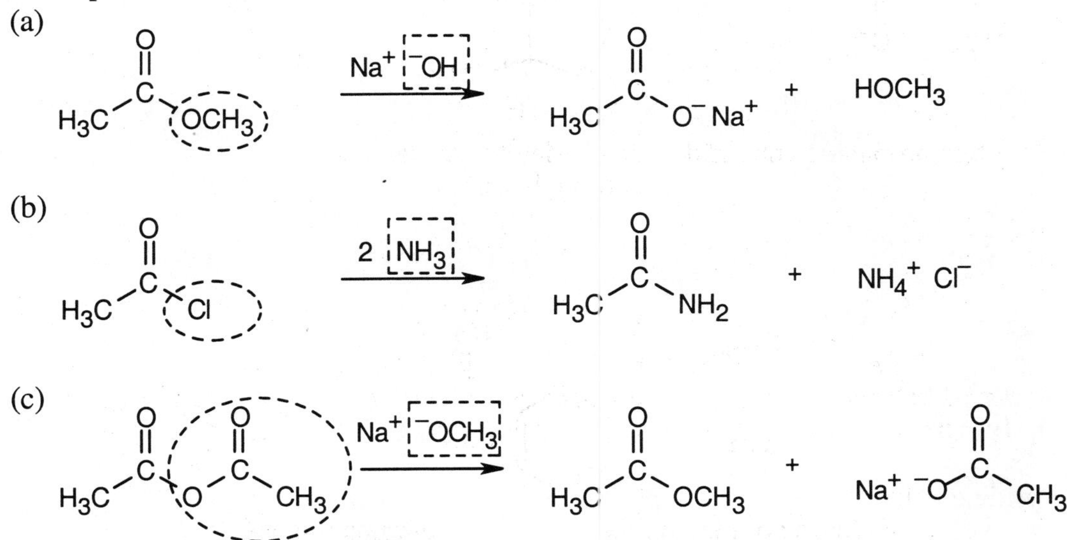

21.7 The structure represents the tetrahedral intermediate in the reaction of methyl cyclopentanecarboxylate with hydroxide, a nucleophile. The products are cyclopentanecarboxylate anion and methanol.

Methyl cyclopentanecarboxylate + ^{-}OH (Hydroxide) $\rightleftharpoons$ [tetrahedral intermediate: $C(O^{-})(OH)(OCH_3)$] $\longrightarrow$ Cyclopentanecarboxylate ($C(=O)O^{-}$) + $HOCH_3$ (Methanol)

21.8 In Fischer esterification, an alcohol undergoes a nucleophilic acyl substitution with a carboxylic acid to yield an ester. The mineral acid catalyst makes the carboxyl group of the acid more nucleophilic. Predicting the products is easier if the two reagents are positioned so that the reacting functional groups point towards each other.

(a)

$$H_3C\text{-}C(=O)\text{-}OH + HOCH_2CH_2CH_2CH_3 \xrightleftharpoons{HCl} H_3C\text{-}C(=O)\text{-}OCH_2CH_2CH_2CH_3 + H_2O$$

Acetic acid + Butanol → Butyl acetate

(b)

$$CH_3CH_2CH_2\text{-}C(=O)\text{-}OH + HOCH_3 \xrightleftharpoons{HCl} CH_3CH_2CH_2\text{-}C(=O)\text{-}OCH_3 + H_2O$$

Butanoic acid + Methanol → Methyl butanoate

21.9 Under Fischer esterification conditions, many hydroxycarboxylic acids can form intramolecular esters (lactones).

5-Hydroxypentanoic acid (H_2C–$C(=O)OH$, H_2C–CH_2–CH_2OH) $\xrightleftharpoons{HCl}$ a lactone (six-membered ring: $C(=O)$–O–CH_2–CH_2–CH_2–CH_2) + H_2O

21.10 Pyridine neutralizes the HCl byproduct by forming pyridinium chloride. This neutralization removes from the product mixture acid that might cause side reactions. As mentioned previously, positioning the reacting groups so that they face each other makes it easier to predict the products.

(a)

$$CH_3CH_2\text{-}C(=O)\text{-}Cl + HOCH_3 \xrightarrow{\text{Pyridine}} CH_3CH_2\text{-}C(=O)\text{-}OCH_3$$

Propanoyl chloride + Methanol → Methyl propanoate

(b)

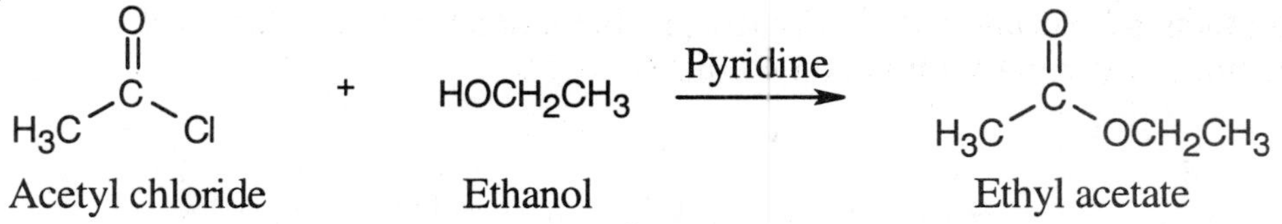

Acetyl chloride + Ethanol ($HOCH_2CH_3$) $\xrightarrow{\text{Pyridine}}$ Ethyl acetate ($H_3C-CO-OCH_2CH_3$)

(c)

Benzoyl chloride + Ethanol ($HOCH_2CH_3$) $\xrightarrow{\text{Pyridine}}$ Ethyl benzoate

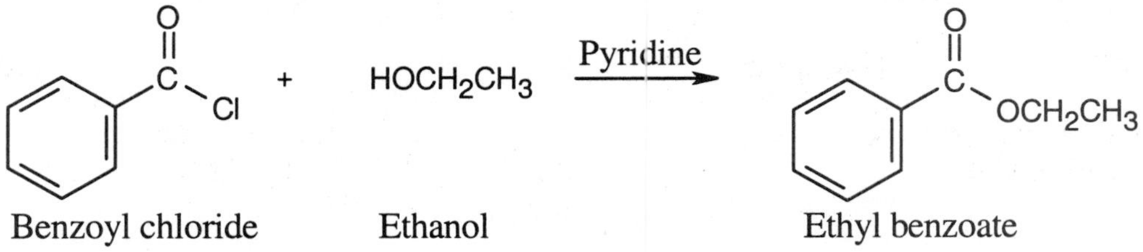

21.11 Cyclohexanol is a secondary alcohol, which for steric reasons is less reactive in the Fischer esterification reaction.Thus, reaction of cyclohexanol with benzoyl chloride is the preferred method for preparing cyclohexyl benzoate.

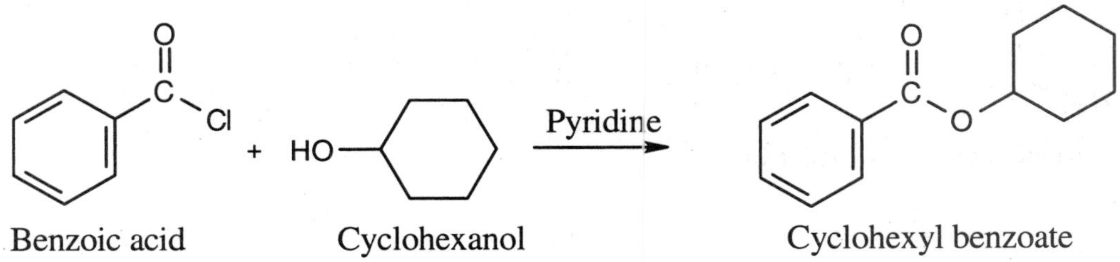

Benzoic acid + Cyclohexanol $\xrightarrow{\text{Pyridine}}$ Cyclohexyl benzoate

21.12

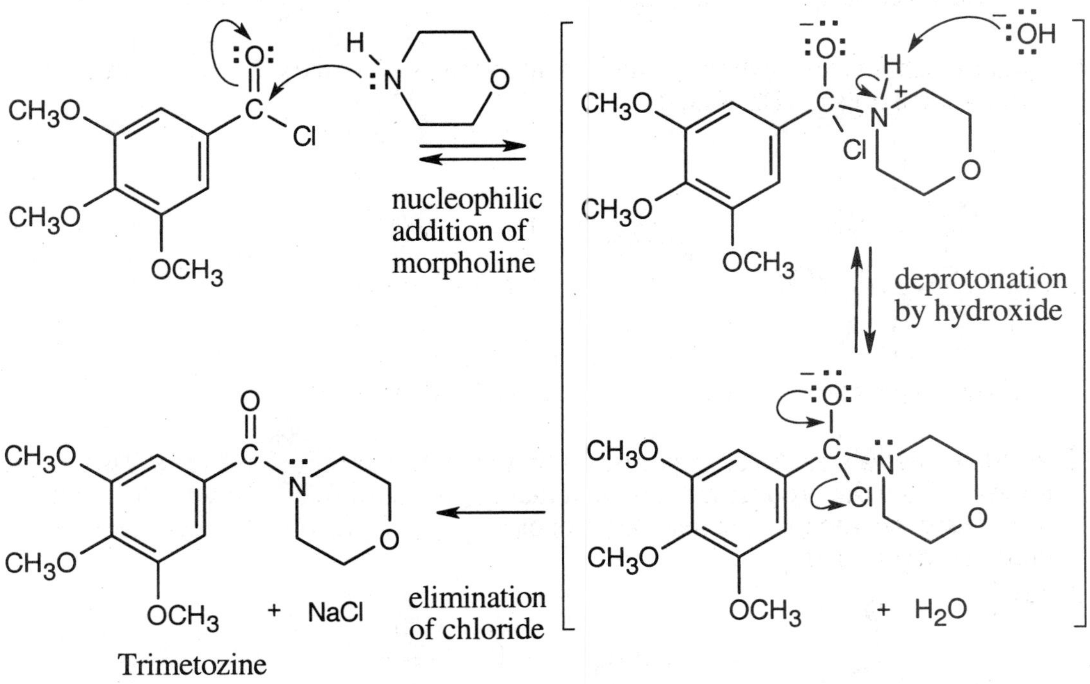

nucleophilic addition of morpholine

deprotonation by hydroxide

elimination of chloride

+ NaCl

+ H_2O

Trimetozine

21.13 In all of these reactions, an extra equivalent of base must be added to neutralize the acid produced. In (a) and (b), two equivalents of the amine may be used in place of NaOH.

(a)

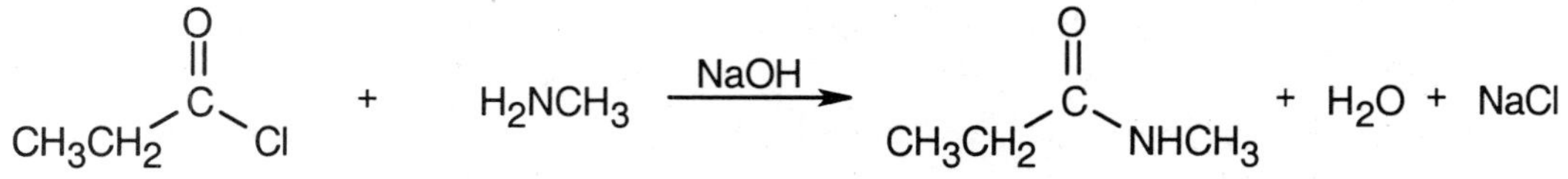

$$CH_3CH_2COCl + H_2NCH_3 \xrightarrow{NaOH} CH_3CH_2CONHCH_3 + H_2O + NaCl$$

Propanoyl chloride, Methylamine, *N*-Methylpropanamide

(b)

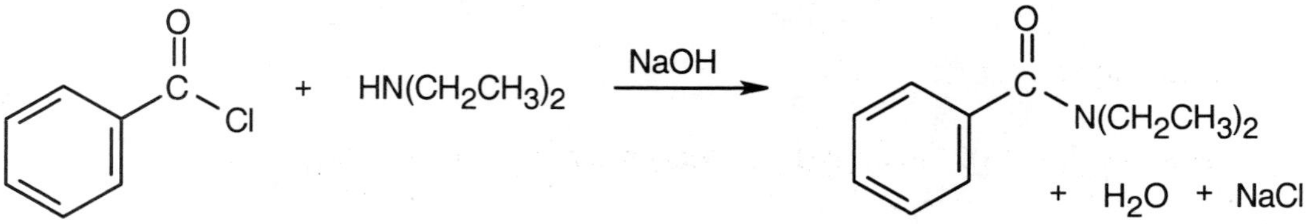

$$C_6H_5COCl + HN(CH_2CH_3)_2 \xrightarrow{NaOH} C_6H_5CON(CH_2CH_3)_2 + H_2O + NaCl$$

Benzoyl chloride, Diethylamine, *N,N*-Diethylbenzamide

(c)

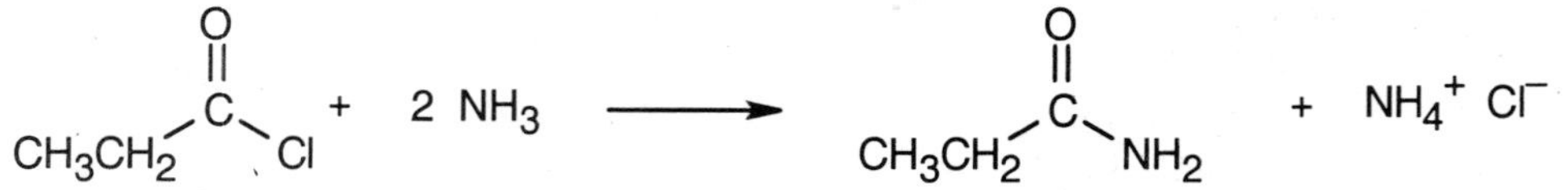

$$CH_3CH_2COCl + 2\ NH_3 \longrightarrow CH_3CH_2CONH_2 + NH_4^+\ Cl^-$$

Propanoyl chloride, Propanamide

21.14 To draw the product, replace the halide with an alkyl group from the organocopper reagent. Two combinations of acid chloride and organocopper reagent are possible.

(a)

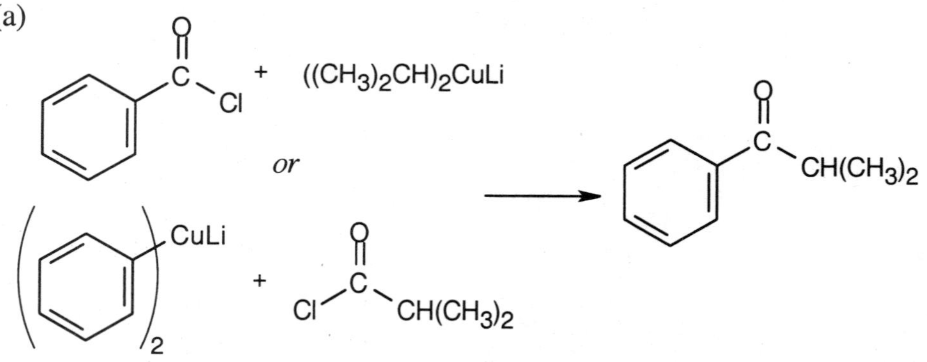

$$C_6H_5COCl + ((CH_3)_2CH)_2CuLi$$

or

$$(C_6H_5)_2CuLi + ClCOCH(CH_3)_2$$

$$\longrightarrow C_6H_5COCH(CH_3)_2$$

(b)

$$H_2C{=}CHCOCl + (CH_3CH_2CH_2)_2CuLi$$

or

$$(H_2C{=}CH)_2CuLi + ClCOCH_2CH_2CH_3$$

$$\longrightarrow H_2C{=}CHCOCH_2CH_2CH_3$$

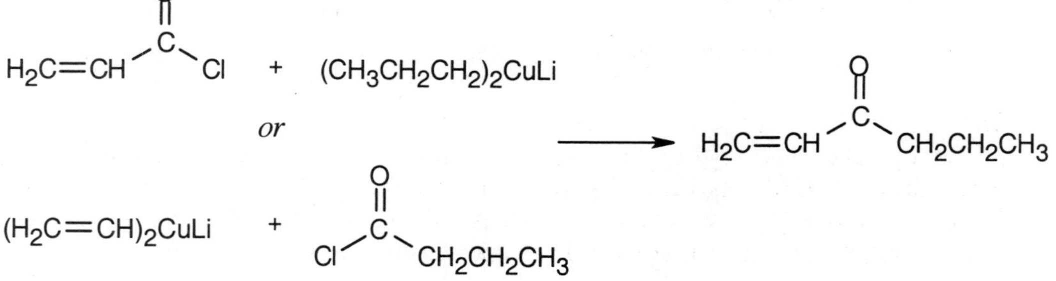

21.15

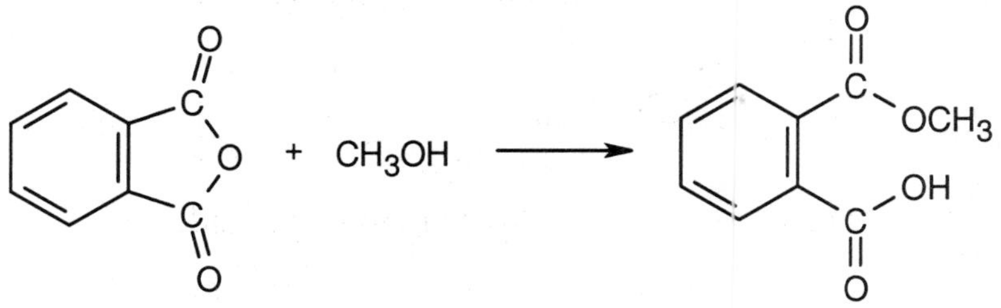

+ CH_3OH ⟶

Phthalic anhydride

The second half of the anhydride becomes a carboxylic acid.

21.16

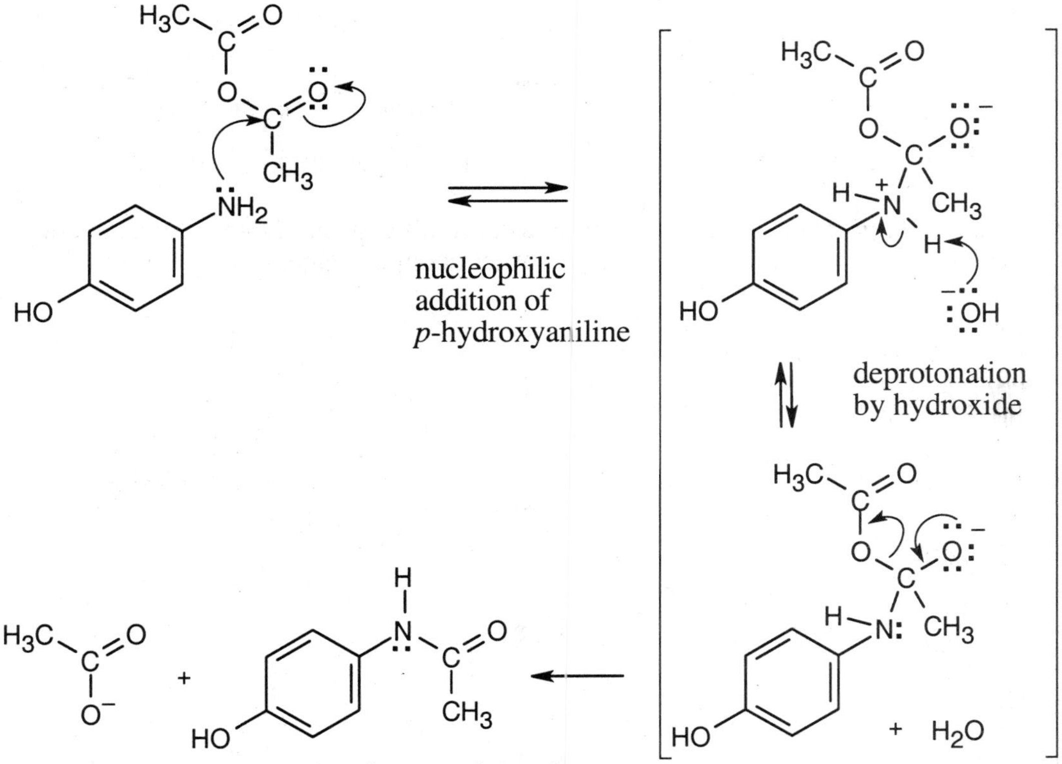

nucleophilic addition of *p*-hydroxyaniline

deprotonation by hydroxide

Acetaminophen

loss of acetate

21.17 One equivalent of base must be added to the reaction mixture when an amine reacts with an anhydride. This base removes a proton from nitrogen after the formation of the initial tetrahedral intermediate. If base were not added, the amine starting material would serve as the base, and the reaction would stop when half of the starting amine has been converted to amide. The rest of the amine would be protonated and would no longer be nucleophilic.

21.18 Acidic hydrolysis of an ester is a reversible reaction because the products are an alcohol and a carboxylic acid, which can reform the original ester. Basic hydrolysis of an ester is irreversible because the products are an alcohol and a carboxylate anion, which has a negative charge and is not attacked by nucleophiles.

21.19

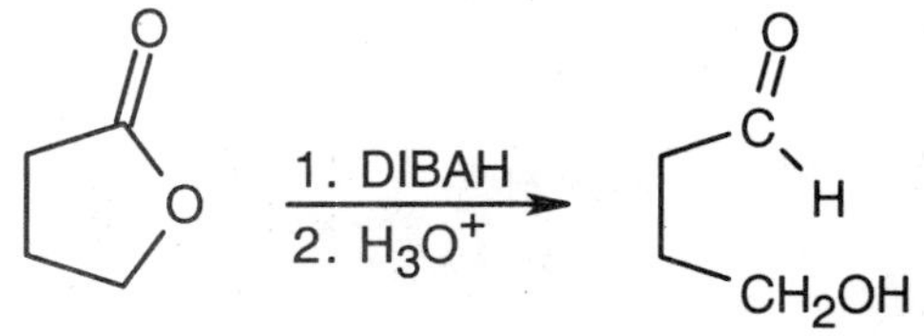

$$\text{butyrolactone} \xrightarrow[\text{2. } H_3O^+]{\text{1. DIBAH}} HOCH_2CH_2CH_2CHO$$

The product of reduction of butyrolactone is 4-hydroxybutanal.

21.20 Lithium aluminum hydride reduces an ester to form two alcohols.

(a)

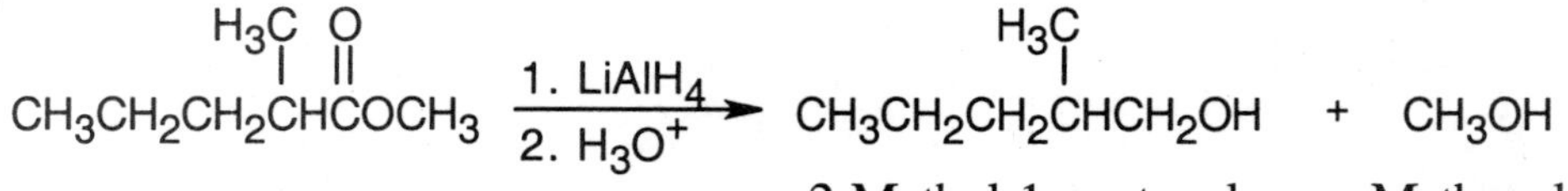

$$CH_3CH_2CH_2CH(CH_3)COCH_3 \xrightarrow[\text{2. } H_3O^+]{\text{1. } LiAlH_4} CH_3CH_2CH_2CH(CH_3)CH_2OH + CH_3OH$$

2-Methyl-1-pentanol   Methanol

(b)

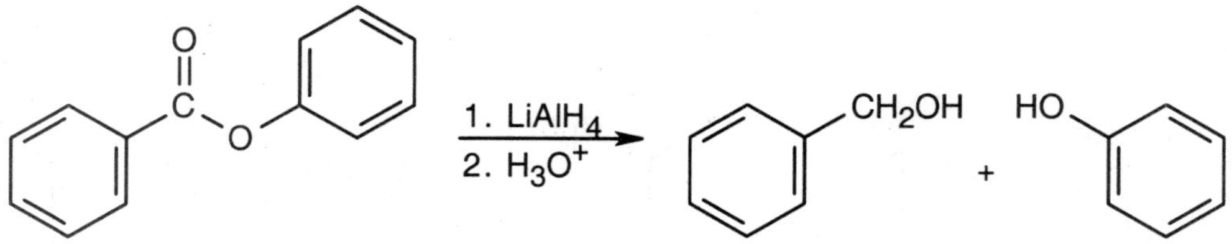

$$C_6H_5COOC_6H_5 \xrightarrow[\text{2. } H_3O^+]{\text{1. } LiAlH_4} C_6H_5CH_2OH + HOC_6H_5$$

Benzyl alcohol   Phenol

21.21 Remember that Grignard reagents can only be used with esters if it is desired to form a tertiary alcohol that has two identical substituents. Identify these two substituents, which come from the Grignard reagent, and work backward to select the ester (the alkyl group of the ester is unimportant).

Tertiary Alcohol ⟵ *Grignard Reagent* + *Ester*

(a) $C_6H_5C(CH_3)_2OH$ ⟵ $2\ CH_3MgBr + C_6H_5COOR$

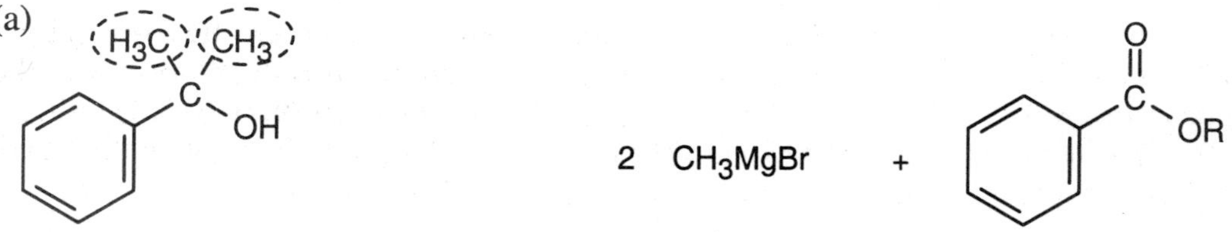

(b) $(C_6H_5)_2C(CH_3)OH$ ⟵ $2\ C_6H_5MgBr + H_3CCOOR$

(c) $CH_3CH_2CH_2CH_2C(CH_2CH_3)_2OH$ ⟵ $2\ CH_3CH_2MgBr + CH_3CH_2CH_2CH_2COOR$

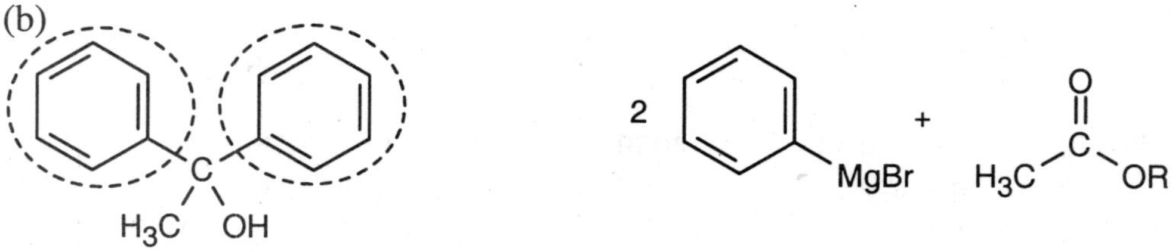

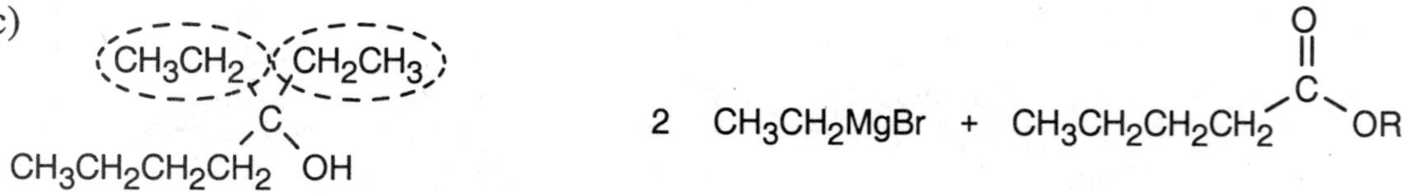

21.22

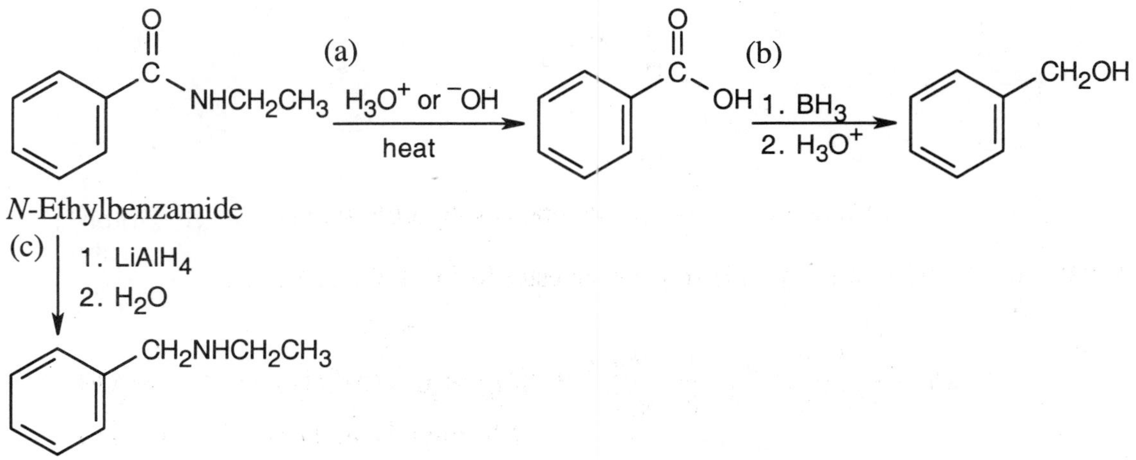

N-Ethylbenzamide ($C_6H_5CONHCH_2CH_3$) —(a) H_3O^+ or ^-OH, heat→ C_6H_5COOH —(b) 1. BH_3; 2. H_3O^+→ $C_6H_5CH_2OH$

N-Ethylbenzamide —(c) 1. $LiAlH_4$; 2. H_2O→ $C_6H_5CH_2NHCH_2CH_3$

21.23

Cyclohexyl–Br —Mg, ether→ Cyclohexyl–MgBr —1. CO_2, ether; 2. H_3O^+→ Cyclohexyl–COOH —$SOCl_2$→ Cyclohexyl–COCl —$(CH_3)_2NH$, NaOH→ Cyclohexyl–$CON(CH_3)_2$ —1. $LiAlH_4$; 2. H_2O→ Cyclohexyl–$CH_2N(CH_3)_2$

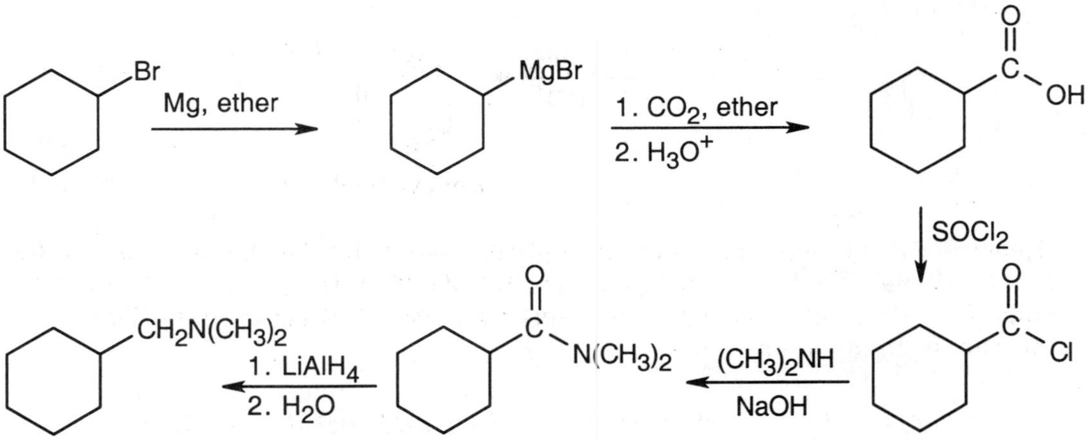

Synthetic analysis: The product is a *N*-substituted amine, which can be formed by reduction of an amide. The amide results from treatment of an acid chloride with the appropriate amine. The acid chloride is the product of the reaction of $SOCl_2$ with a carboxylic acid that is formed by carboxylation of the Grignard reagent synthesized from the starting material.

21.24

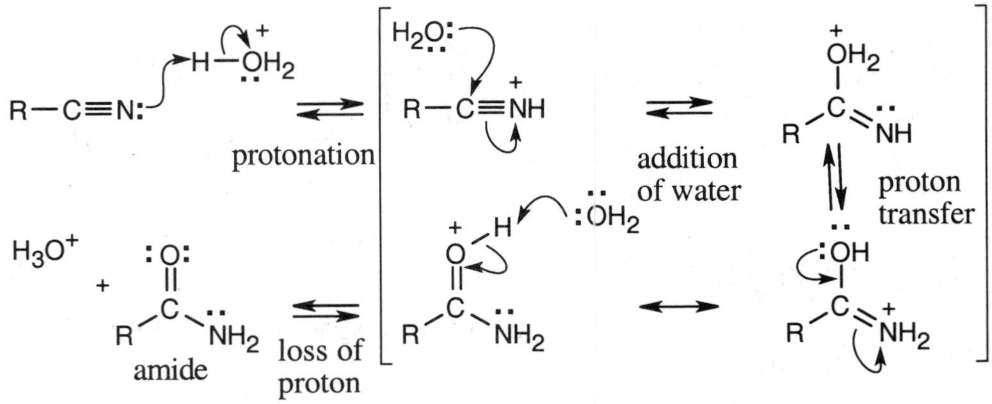

$R{-}C{\equiv}N{:}$ + $H{-}\overset{+}{O}H_2$ ⇌ (protonation) [$R{-}C{\equiv}\overset{+}{N}H$ + $H_2\ddot{O}{:}$ ⇌ (addition of water) $R{-}C(\overset{+}{O}H_2){=}\ddot{N}H$ ⇌ (proton transfer) $R{-}C(\ddot{O}H){=}\overset{+}{N}H_2$ ⟷ $R{-}C(={\overset{+}{O}}{-}H)\ddot{N}H_2$ + $:\ddot{O}H_2$] ⇌ (loss of proton) $R{-}C(={:}\ddot{O}{:})\ddot{N}H_2$ (amide) + H_3O^+

The first equivalent of water adds to a nitrile to produce an amide.

$$R-C(=\ddot{O}:)-\ddot{N}H_2 + H-\ddot{O}H_2^+ \underset{\text{protonation}}{\rightleftharpoons} \left[R-C(=\ddot{O}^+-H)-\ddot{N}H_2 \;(H_2\ddot{O}:) \underset{\text{addition of water}}{\rightleftharpoons} R-C(:\ddot{O}-H)(H_2O^+)-\ddot{N}H_2 \right.$$

$$\underset{\text{proton transfer}}{\rightleftharpoons} R-C(:\ddot{O}-H)(HO)-\overset{+}{N}H_3 \underset{\text{loss of ammonia}}{\rightleftharpoons} \left. R-C(=\overset{+}{\ddot{O}}-H)-OH \;(:NH_3) \right] \underset{\text{loss of proton}}{\rightleftharpoons} R-C(=O)-OH + NH_4^+$$

The second equivalent of water adds to the amide to yield a carboxylic acid, plus ammonium ion.

21.25

(a)

$$CH_3CH_2C{\equiv}N \xrightarrow[2.\ H_2O]{1.\ CH_3CH_2MgBr} CH_3CH_2\overset{O}{\overset{\|}{C}}CH_2CH_3$$

This symmetrical ketone can be synthesized by a Grignard reaction between propanenitrile and ethylmagnesium bromide.

(b)

$$(CH_3)_2CHC{\equiv}N \xrightarrow[2.\ H_2O]{1.\ DIBAH} (CH_3)_2CHCHO$$

DIBAH reduces nitriles to aldehydes.

(c)

$$C_6H_5C{\equiv}N \xrightarrow[2.\ H_2O]{1.\ CH_3MgBr} C_6H_5\overset{O}{\overset{\|}{C}}CH_3$$

or

$$CH_3C{\equiv}N \xrightarrow[2.\ H_2O]{1.\ C_6H_5MgBr} C_6H_5\overset{O}{\overset{\|}{C}}CH_3$$

Acetophenone can be synthesized by either of two Grignard routes.

21.26

$$C_6H_5CH_2Br \xrightarrow{NaCN} C_6H_5CH_2CN \xrightarrow[2.\ H_2O]{1.\ CH_3CH_2MgBr} C_6H_5CH_2\overset{O}{\overset{\|}{C}}CH_2CH_3$$

1-Phenyl-2-butanone

Once you realize that the product results from a Grignard reaction with a nitrile, this synthesis is easy.

21.27 In each example, if n molecules of one component react with n molecules of the other component, a polymer with n repeating units is formed, and $2n$ small molecules are formed as byproducts; these are shown in each reaction.

(a)

$$BrCH_2CH_2CH_2Br + HOCH_2CH_2CH_2OH \xrightarrow{\text{Base}} -(CH_2CH_2CH_2OCH_2CH_2CH_2O)- + HBr$$

(b)

$$HOCH_2CH_2OH + HOOC(CH_2)_6COOH \xrightarrow[\text{catalyst}]{H_2SO_4} -(OCH_2CH_2O-\overset{O}{\overset{\|}{C}}(CH_2)_6\overset{O}{\overset{\|}{C}})- + H_2O$$

(c)

$$H_2N(CH_2)_6NH_2 + Cl\overset{O}{\overset{\|}{C}}(CH_2)_4\overset{O}{\overset{\|}{C}}Cl \longrightarrow -(HN(CH_2)_6NH-\overset{O}{\overset{\|}{C}}(CH_2)_4\overset{O}{\overset{\|}{C}})- + HCl$$

21.28

1,4-Benzenedicarboxylic acid + 1,4-Benzenediamine → Kevlar

$$HO\overset{O}{\overset{\|}{C}}-C_6H_4-\overset{O}{\overset{\|}{C}}OH + H_2N-C_6H_4-NH_2 \longrightarrow -(\overset{O}{\overset{\|}{C}}-C_6H_4-\overset{O}{\overset{\|}{C}}NH-C_6H_4-NH\overset{O}{\overset{\|}{C}}-C_6H_4-\overset{O}{\overset{\|}{C}}NH-C_6H_4-NH)-$$

Kevlar

21.29

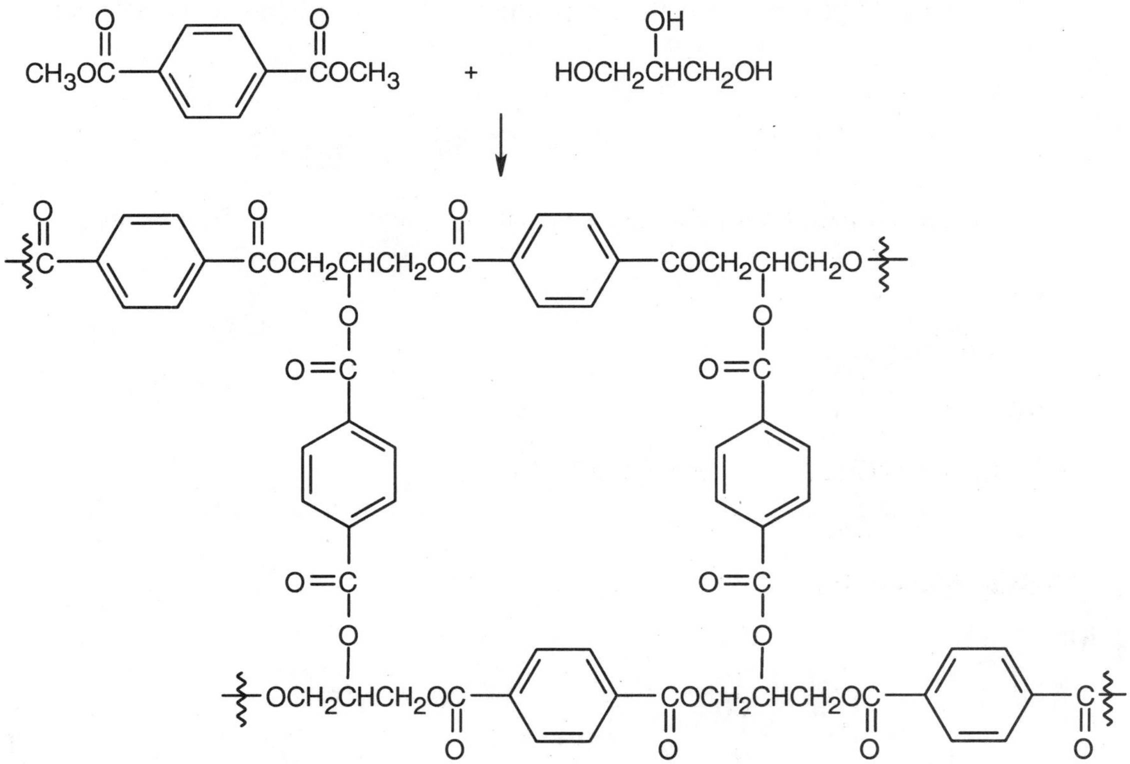

$CH_3OC(=O)$–(benzene ring)–$C(=O)OCH_3$ + $HOCH_2CH(OH)CH_2OH$

The product of the reaction of dimethyl terephthalate with glycerol has a high degree of cross-linking and is more rigid than Dacron.

21.30 Use Table 21.3 if you need help.

Absorption	*Functional group present*
(a) 1735 cm^{-1}	Aliphatic ester *or* 6-membered ring lactone
(b) 1810 cm^{-1}	Aliphatic acid chloride
(c) 2500 – 3300 cm^{-1} and 1710 cm^{-1}	Carboxylic acid
(d) 2250 cm^{-1}	Aliphatic nitrile
(e) 1715 cm^{-1}	Aliphatic ketone *or* 6-membered ring ketone

21.31 To solve this type of problem:

1. Use the IR absorption to determine the functional group(s) present.
2. Draw the functional group.
3. Use the remaining atoms to complete the structure.

(a) 1. IR 2250 cm^{-1} corresponds to a nitrile.
   2. –C≡N
   3. CH_3CH_2CN is the structure of the compound.

(b) 1. IR 1735 cm^{-1} corresponds to an aliphatic ester.
   2. —C(=O)—O—

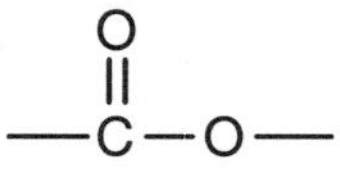

3. The remaining five carbons and twelve hydrogens can be arranged in a number of ways to produce a satisfactory structure for this compound. For example:

$$CH_3CH_2\overset{O}{\overset{\|}{C}}OCH_2CH_2CH_3 \quad or \quad H\overset{O}{\overset{\|}{C}}OCH_2CH_2CH_2CH_2CH_3$$

The structural formula indicates that this compound can't be a lactone.

(c)

$$CH_3\overset{O}{\overset{\|}{C}}N(CH_3)_2$$

(d)

$$CH_3CH{=}CH\overset{O}{\overset{\|}{C}}Cl \quad or \quad H_2C{=}C(CH_3)\overset{O}{\overset{\|}{C}}Cl$$

Visualizing Chemistry

21.32

(a) (b)

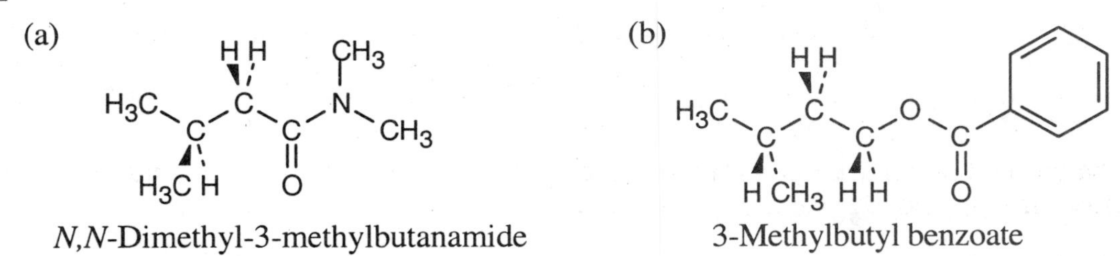

N,N-Dimethyl-3-methylbutanamide — 3-Methylbutyl benzoate

21.33

(a)

o-Bromobenzoic acid $\xrightarrow{SOCl_2}$ o-bromobenzoyl chloride $\xrightarrow[\text{pyridine}]{CH_3CH(OH)CH_3}$ Isopropyl *o*-bromobenzoate

o-Bromobenzoic acid; Isopropanol; Isopropyl *o*-bromobenzoate

This compound can also be synthesized by Fischer esterification of *o*-bromobenzoic acid with isopropanol and an acid catalyst.

(b)

$$C_5H_9{-}CH_2COOH \xrightarrow{2\ NH_3} C_5H_9{-}CH_2CONH_2$$

Cyclopentylacetic acid — Cyclopentylacetamide

21.34

$$H_2C{=}CHCH(CH_3)CH_2COCl \xrightarrow{2\ NH_3} \left[H_2C{=}CHCH(CH_3)CH_2C(O^-)(Cl)NH_2 \right] \longrightarrow H_2C{=}CHCH(CH_3)CH_2CONH_2 + NH_4^+\ Cl^-$$

3-Methyl-4-pentenoyl chloride — 3-Methyl-4-pentenamide

The starting material is 3-methyl-4-pentenoyl chloride. Ammonia adds to give the observed tetrahedral intermediate, which eliminates Cl^- to yield the above amide.

Additional Problems

21.35

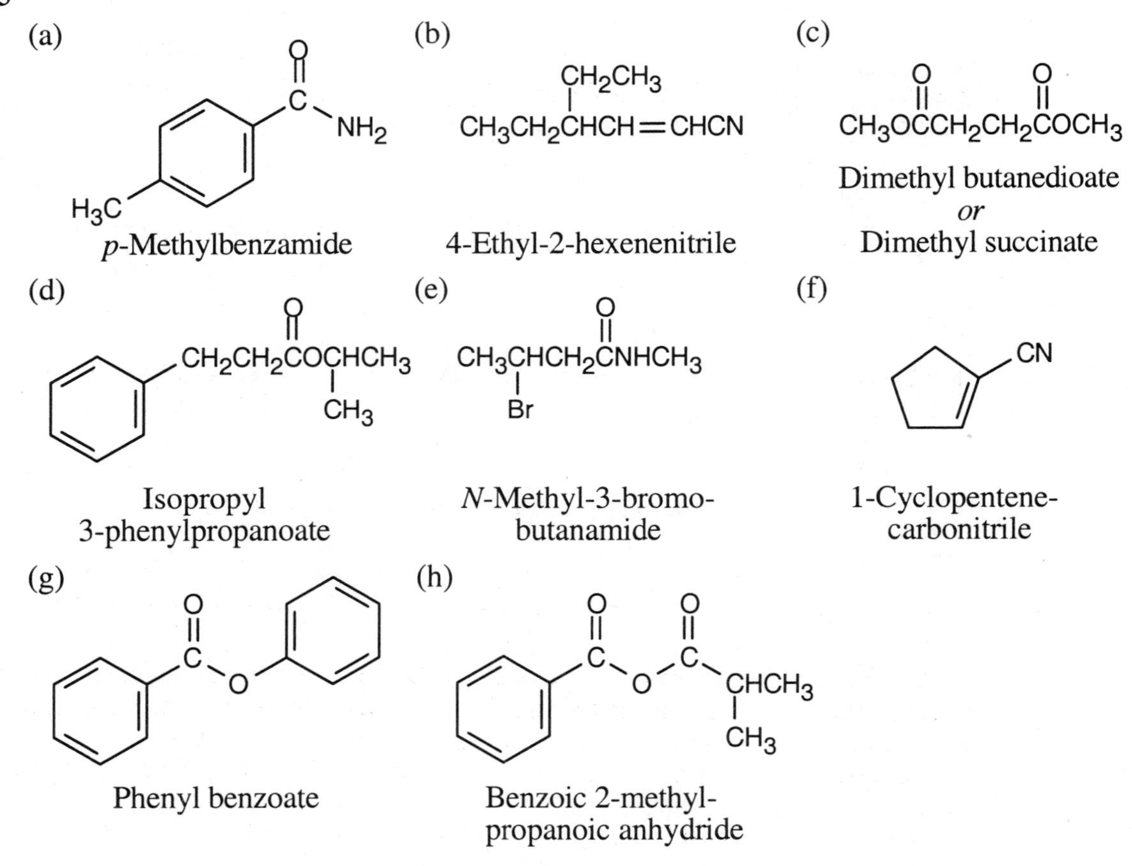

21.36

(a) $p\text{-}BrC_6H_4CH_2CONH_2$

p-Bromophenylacetamide

(b) $C_6H_5CO\text{-}C_6H_4\text{-}CN\ (meta)$

m-Benzoylbenzonitrile

(c) $CH_3CH_2CH_2CH_2C(CH_3)_2CONH_2$

2,2-Dimethylhexanamide

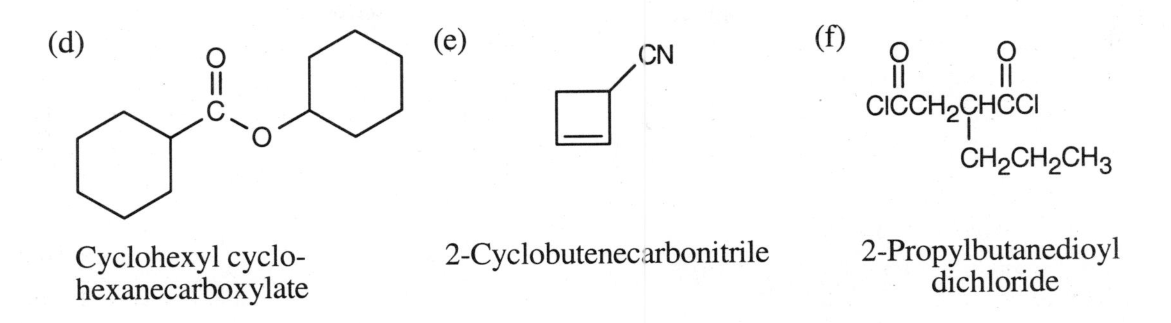

Cyclohexyl cyclohexanecarboxylate

2-Cyclobutenecarbonitrile

2-Propylbutanedioyl dichloride

21.37 Many structures can be drawn for each part of this problem.

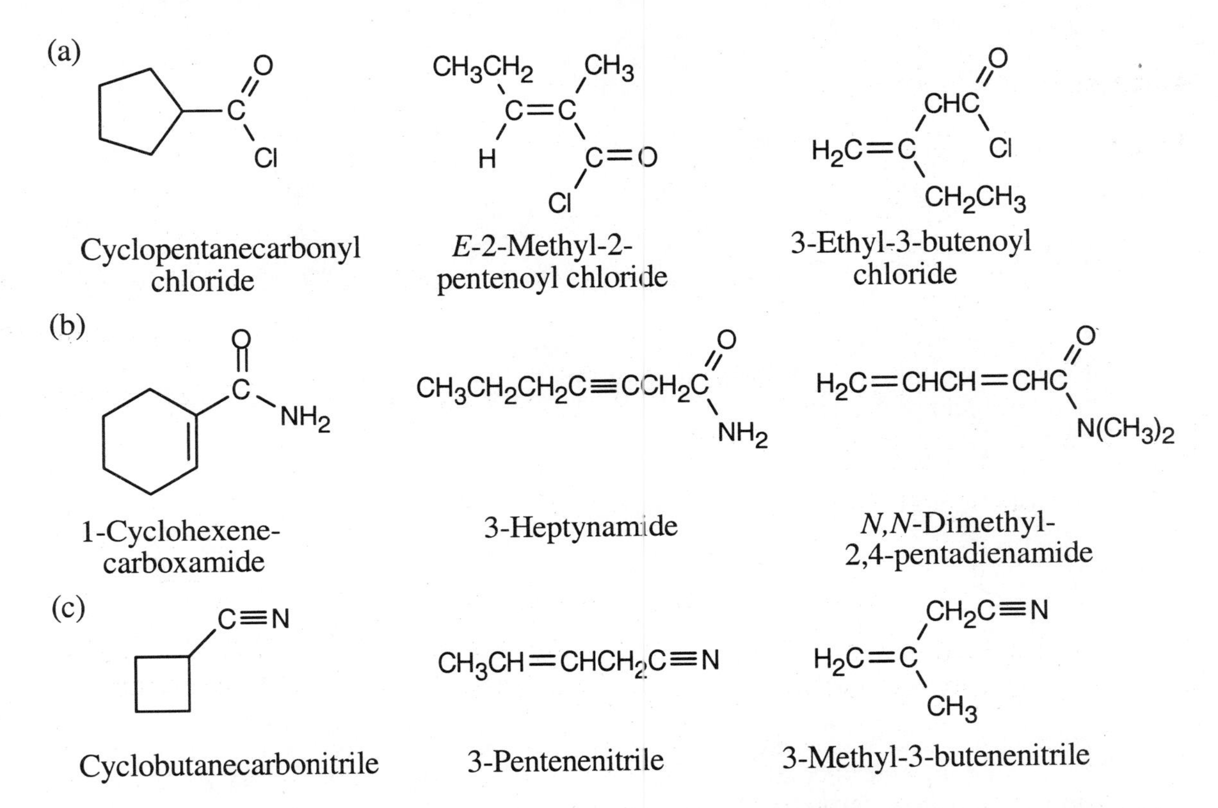

21.38 The reactivity of esters in saponification reactions is influenced by steric factors. Branching in both the acyl and alkyl portions of an ester hinders attack of the hydroxide nucleophile. This effect is less pronounced in the alkyl portion of the ester than in the acyl portion because alkyl branching is one atom farther away from the site of attack, but it is still significant.

Most reactive ⟶ *Least reactive*

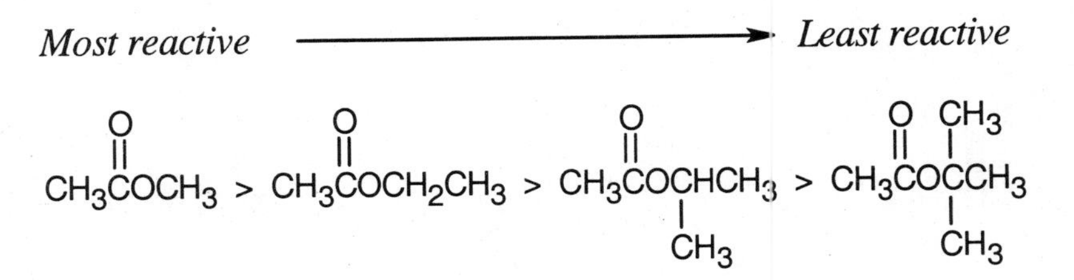

21.39

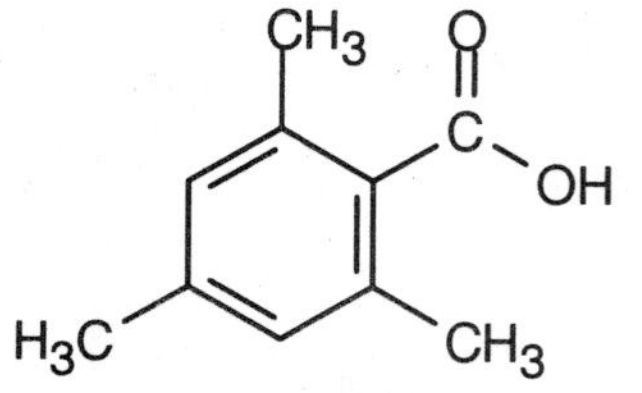

2,4,6-Trimethylbenzoic acid

2,4,6–Trimethylbenzoic acid has two methyl groups ortho to the carboxylic acid functional group. These bulky methyl groups block the approach of the alcohol and prevent esterification from occurring under Fischer esterification conditions. Another possible route to the methyl ester:

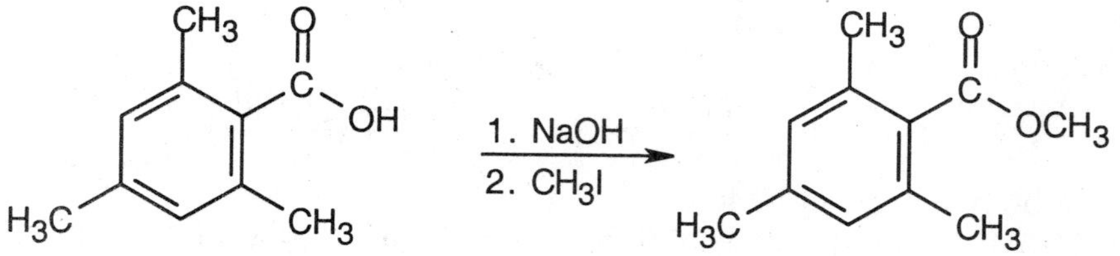

1. NaOH
2. CH_3I

This route succeeds because reaction occurs farther away from the site of steric hindrance. It is also possible to form the acid chloride of 2,4,6-trimethylbenzoic acid and react it with methanol and pyridine.

21.40

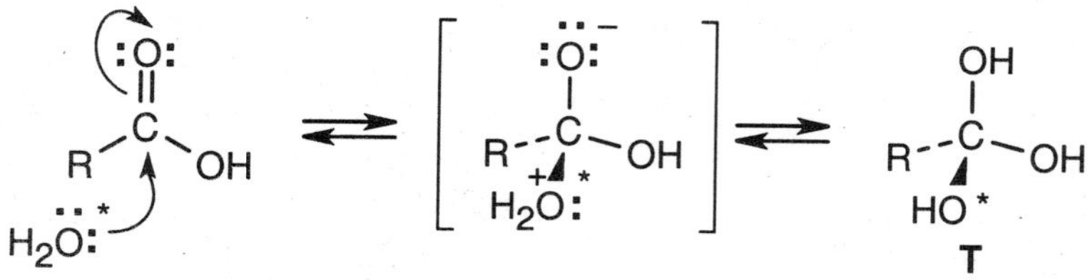

T

The tetrahedral intermediate **T** can eliminate any one of the three –OH groups to reform either the original carboxylic acid or labeled carboxylic acid. Further reaction of water with mono-labeled carboxylic acid leads to the doubly labeled product.

21.41

(a)

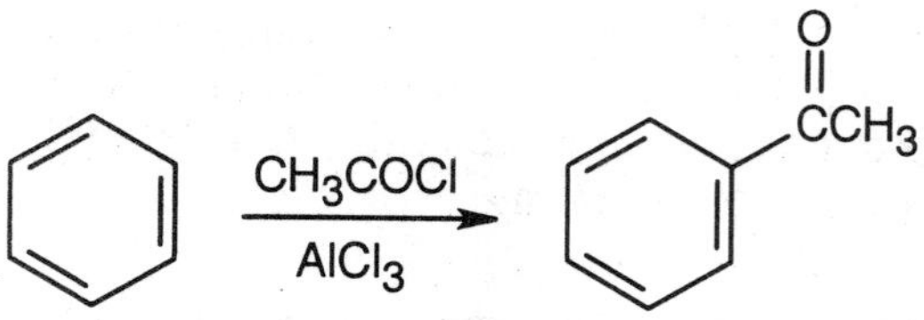

CH_3COCl
$AlCl_3$

(b)

Br; Mg →; MgBr; 1. CO_2 2. H_3O^+ →; COH (C=O); $SOCl_2$ →; CCl (C=O)

CH_3CHO ↓; OH CHCH$_3$; PCC, CH_2Cl_2 →; $(CH_3)_2CuLi$ ↓; CCH$_3$ (C=O)

(c)

COCH$_3$ (C=O); ^-OH, H_2O →; COH (C=O); $SOCl_2$ →; CCl (C=O); $(CH_3)_2CuLi$ →; CCH$_3$ (C=O)

Reaction of an ester with Grignard reagent produces a tertiary alcohol, not a ketone. Acid can also be used to catalyze hydrolysis of the ester.

(d)

CN; 1. CH_3MgBr 2. H_3O^+ →; CCH$_3$ (C=O)

(e)

CH=CH$_2$; 1. $Hg(OAc)_2$, H_2O 2. $NaBH_4$ →; OH CHCH$_3$; PCC, CH_2Cl_2 →; CCH$_3$ (C=O)

21.42 A negatively charged tetrahedral intermediate is formed when the nucleophile ^-OH attacks the carbonyl carbon of an ester. An electron-withdrawing substituent can stabilize this negatively charged tetrahedral intermediate and increase the rate of reaction. (Contrast this effect with substituent effects in electrophilic aromatic substitution, in which positive charge developed in the intermediate is stabilized by electron-*donating* substituents.) Substituents that are deactivating in electrophilic aromatic substitution are activating in ester hydrolysis, as the observed reactivity order shows. The substituents –CN and –CHO are electron-withdrawing; –NH_2 is strongly electron-donating.

Most reactive ——→ *Least reactive*

Y = –NO_2 > –C≡N > –CHO > –Br > –H > –CH_3 > –OCH_3 > –NH_2

21.43

(a)

$$CH_3CH_2CH_2COOH \xrightarrow[2.\ H_3O^+]{1.\ BH_3} CH_3CH_2CH_2CH_2OH$$

(b)

$$\underset{\text{from (a)}}{CH_3CH_2CH_2CH_2OH} \xrightarrow[CH_2Cl_2]{PCC} CH_3CH_2CH_2CHO$$

or

$$CH_3CH_2CH_2COOH \xrightarrow[3.\ H_3O^+]{1.\ SOCl_2;\ 2.\ Li[OC(CH_3)_3]_3AlH} CH_3CH_2CH_2CHO$$

(c)

$$\underset{\text{from (a)}}{CH_3CH_2CH_2CH_2OH} \xrightarrow{PBr_3} CH_3CH_2CH_2CH_2Br$$

(d)

$$\underset{\text{from (c)}}{CH_3CH_2CH_2CH_2Br} \xrightarrow{NaCN} CH_3CH_2CH_2CH_2CN$$

(e)

$$CH_3CH_2CH_2CH_2Br \xrightarrow{K^+\ ^-OC(CH_3)_3} CH_3CH_2CH{=}CH_2$$

(f)

$$\underset{\text{from (d)}}{CH_3CH_2CH_2CH_2CN} \xrightarrow{H_3O^+} CH_3CH_2CH_2CH_2COOH \xrightarrow{SOCl_2} CH_3CH_2CH_2CH_2COCl \xrightarrow{2\ CH_3NH_2} CH_3CH_2CH_2CH_2CONHCH_3$$

(g)

$$\underset{\text{from (d)}}{CH_3CH_2CH_2CH_2CN} \xrightarrow[2.\ H_3O^+]{1.\ CH_3MgBr} CH_3CH_2CH_2CH_2\overset{O}{\overset{\|}{C}}CH_3$$

(h)

$$CH_3CH_2CH_2COOH \xrightarrow{SOCl_2} CH_3CH_2CH_2COCl \xrightarrow{C_6H_6,\ AlCl_3} C_6H_5\overset{O}{\overset{\|}{C}}CH_2CH_2CH_3 \xrightarrow{H_2,\ Pd/C} C_6H_5CH_2CH_2CH_2CH_3$$

21.44 Dimethyl carbonate is a diester. Use your knowledge of the Grignard reaction to work your way through this problem.

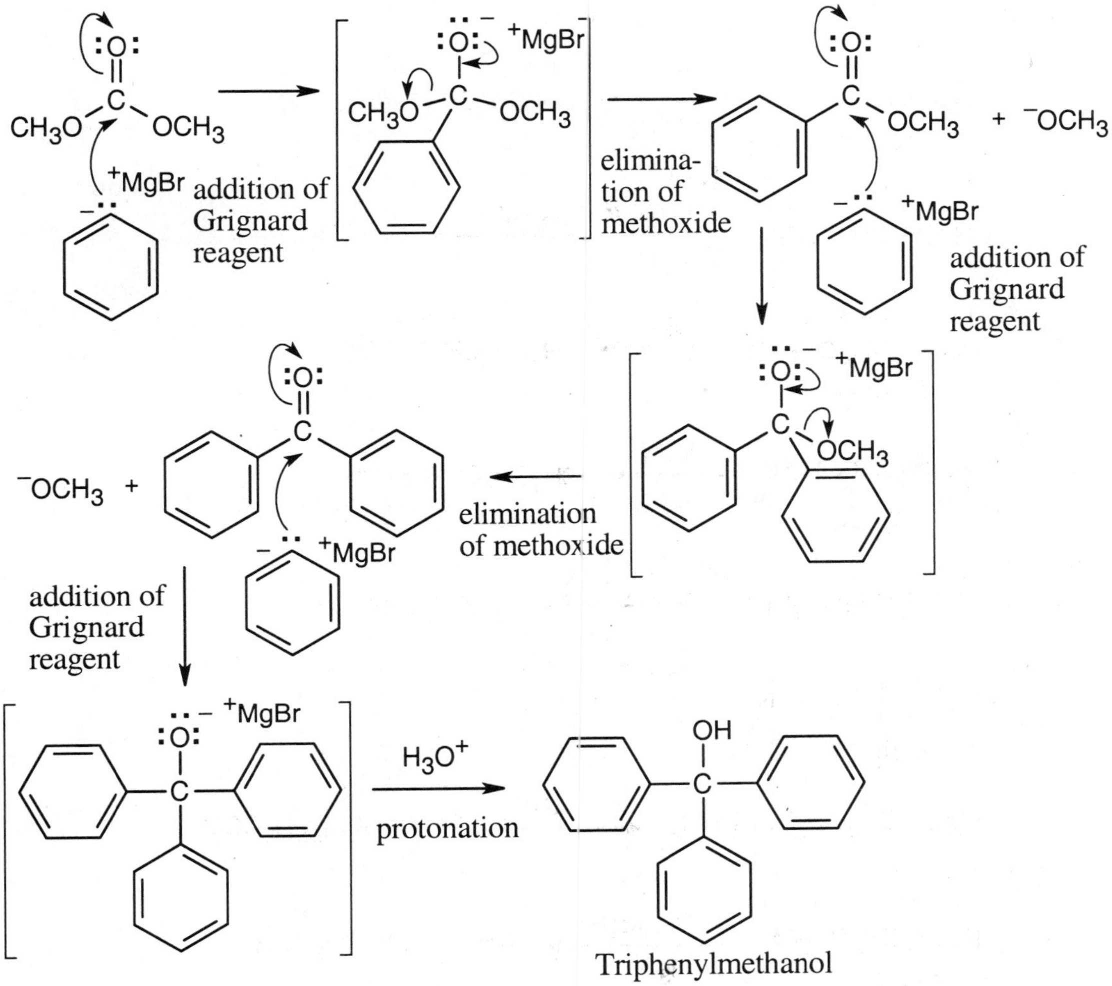

addition of Grignard reagent

elimina-tion of methoxide

addition of Grignard reagent

elimination of methoxide

addition of Grignard reagent

H_3O^+

protonation

Triphenylmethanol

The overall reaction consists of three additions of phenylmagnesium bromide, two eliminations of methoxide and one protonation.

21.45

$C_6H_5COOCH_2CH_3 + H^+ \rightleftharpoons$ (protonation) $[C_6H_5C(=\overset{+}{O}H)OCH_2CH_3] \rightleftharpoons$ (addition of methanol, $HOCH_3$) $[C_6H_5C(OH)(OCH_2CH_3)(\overset{+}{O}HCH_3)]$

$\rightleftharpoons$ (proton transfer) $[C_6H_5C(OH)(\overset{+}{O}HCH_2CH_3)(OCH_3)] \rightleftharpoons$ (loss of ethanol) $[C_6H_5C(=\overset{+}{O}H)OCH_3] + HOCH_2CH_3 \rightleftharpoons$ (loss of proton) $C_6H_5COOCH_3 + H^+$

In acidic methanol, the ethyl ester reacts by a nucleophilic acyl substitution reaction to yield a methyl ester. The equilibrium favors the methyl ester because of the large excess of methanol present.

21.46

$(CH_3)_3COC(=O)Cl + :N_3^- \rightleftharpoons$ (addition of azide) $[(CH_3)_3CO-C(O^-)(N_3)(Cl)] \longrightarrow$ (elimination of chloride) $(CH_3)_3COC(=O)N_3 + Cl^-$

This reaction is a typical nucleophilic acyl substitution reaction, with azide as the nucleophile and chloride as the leaving group.

21.47

(a)

$$CH_3CH_2\overset{O}{\overset{\|}{C}}Cl \xrightarrow[\text{ether}]{(Ph)_2CuLi} CH_3CH_2\overset{O}{\overset{\|}{C}}-C_6H_5$$

(b)

$$CH_3CH_2\overset{O}{\overset{\|}{C}}Cl \xrightarrow[2.\ H_3O^+]{1.\ LiAlH_4} CH_3CH_2CH_2OH$$

(c)

$$CH_3CH_2\overset{O}{\overset{\|}{C}}Cl \xrightarrow[2.\ H_3O^+]{1.\ 2\ CH_3MgBr} CH_3CH_2\overset{OH}{\overset{|}{C}}(CH_3)_2$$

(d)

$$CH_3CH_2\overset{O}{\overset{\|}{C}}Cl \xrightarrow[2.\ H_3O^+]{1.\ Li[OC(CH_3)_3]_3AlH} CH_3CH_2\overset{O}{\overset{\|}{C}}H$$

(e)

$$CH_3CH_2\overset{O}{\overset{\|}{C}}Cl \xrightarrow{H_3O^+} CH_3CH_2\overset{O}{\overset{\|}{C}}OH + HCl$$

(f)

$$CH_3CH_2\overset{O}{\overset{\|}{C}}Cl + HO\text{—}C_6H_{11} \xrightarrow{\text{Pyridine}} CH_3CH_2\overset{O}{\overset{\|}{C}}\text{—}O\text{—}C_6H_{11}$$

(g)

$$CH_3CH_2\overset{O}{\overset{\|}{C}}Cl + H_2N\text{—}C_6H_5 \xrightarrow{NaOH} CH_3CH_2\overset{O}{\overset{\|}{C}}\text{—}\underset{H}{\underset{|}{N}}\text{—}C_6H_5$$

(h)

$$CH_3CH_2\overset{O}{\overset{\|}{C}}Cl \xrightarrow{CH_3COO^-\ Na^+} CH_3CH_2\overset{O}{\overset{\|}{C}}\text{—}O\overset{O}{\overset{\|}{C}}CH_3$$

21.48 The reagents in parts (a), (d), (f), and (h) don't react with methyl propanoate.

(b)

$$CH_3CH_2\overset{O}{\overset{\|}{C}}OCH_3 \xrightarrow[2.\ H_3O^+]{1.\ LiAlH_4} CH_3CH_2CH_2OH + CH_3OH$$

(c)

$$CH_3CH_2\overset{O}{\overset{\|}{C}}OCH_3 \xrightarrow[2.\ H_3O^+]{1.\ 2\ CH_3MgBr} CH_3CH_2\overset{OH}{\overset{|}{C}}(CH_3)_2$$

(e)

$$CH_3CH_2\overset{O}{\overset{\|}{C}}OCH_3 \xrightarrow{H_3O^+} CH_3CH_2\overset{O}{\overset{\|}{C}}OH + CH_3OH$$

(g)

$$CH_3CH_2\overset{O}{\overset{\|}{C}}OCH_3 + H_2N\text{—}C_6H_5 \longrightarrow CH_3CH_2\overset{O}{\overset{\|}{C}}\text{—}\underset{H}{\underset{|}{N}}\text{—}C_6H_5$$

21.49 The reagents in parts (a), (d), (f), (g), and (h) don't react with propanamide.

(b)

$$CH_3CH_2\overset{O}{\overset{\|}{C}}NH_2 \xrightarrow[2.\ H_2O]{1.\ LiAlH_4} CH_3CH_2CH_2NH_2$$

(c)

$$CH_3CH_2\overset{O}{\overset{\|}{C}}NH_2 \xrightarrow[2.\ H_3O^+]{1.\ CH_3MgBr} CH_3CH_2\overset{O}{\overset{\|}{C}}NH_2 + CH_4$$

(e)

$$CH_3CH_2\overset{O}{\overset{\|}{C}}NH_2 \xrightarrow{H_3O^+} CH_3CH_2\overset{O}{\overset{\|}{C}}OH + NH_4^+$$

The reagent in parts (a), (d), (f), (g), and (h) don't react with propanenitrile.

(b)

$$CH_3CH_2C{\equiv}N \xrightarrow[2.\ H_2O]{1.\ LiAlH_4} CH_3CH_2CH_2NH_2$$

(c)

$$CH_3CH_2C{\equiv}N \xrightarrow[2.\ H_3O^+]{1.\ CH_3MgBr} CH_3CH_2\overset{O}{\overset{\|}{C}}CH_3$$

(e)

$$CH_3CH_2C{\equiv}N \xrightarrow{H_3O^+} CH_3CH_2\overset{O}{\overset{\|}{C}}OH + NH_4^+$$

21.50

$$CH_3\overset{^{18}O}{\overset{\|}{C}}-{}^{18}OH \xrightarrow[2.\ H_3O^+]{1.\ BH_3} CH_3CH_2{}^{18}OH \xrightarrow[\text{Pyridine}]{CH_3COCl} CH_3\overset{O}{\overset{\|}{C}}-{}^{18}OCH_2CH_3 + H_2O$$

Ethyl propanoate

Remember that the ^{18}O label appears in both oxygens of the acetic acid starting material.

21.51

(a)

$$F_3C\overset{:O:}{\overset{\|}{C}}OCOCF_3 + HO\overset{O}{\overset{\|}{C}}R \rightleftharpoons \left[F_3C\overset{:\ddot{O}:^-}{\overset{|}{C}}(OCOCF_3)-\overset{H}{O^+}-\overset{O}{\overset{\|}{C}}R \underset{\text{loss of proton}}{\overset{-H^+}{\rightleftharpoons}} F_3C\overset{:\ddot{O}:^-}{\overset{|}{C}}(OCOCF_3)-O-\overset{O}{\overset{\|}{C}}R \right]$$

addition of carboxylic acid

elimination of trifluoroacetate

$$\downarrow$$

$$^-OCOCF_3 + F_3C\overset{O}{\overset{\|}{C}}-O-\overset{O}{\overset{\|}{C}}R$$

(b) The electron-withdrawing fluorine atoms polarize the carbonyl group, making it more reactive toward nucleophiles.

(c) Because trifluoroacetate is a better leaving group than other carboxylate anions, the reaction proceeds as indicated.

21.52

(a)

$$CH_3CH_2CH_2CH_2C{\equiv}N \xrightarrow[2.\ H_2O]{1.\ LiAlH_4} CH_3CH_2CH_2CH_2CH_2NH_2$$

(b)

$$CH_3CH_2CH_2CH_2C{\equiv}N \xrightarrow{H_3O^+} CH_3CH_2CH_2CH_2\overset{O}{\overset{\|}{C}}OH \xrightarrow{SOCl_2} CH_3CH_2CH_2CH_2\overset{O}{\overset{\|}{C}}Cl$$

$$\downarrow 2\ (CH_3)_2NH$$

$$CH_3CH_2CH_2CH_2CH_2N(CH_3)_2 \xleftarrow[2.\ H_2O]{1.\ LiAlH_4} CH_3CH_2CH_2CH_2\overset{O}{\overset{\|}{C}}N(CH_3)_2$$

Reaction of pentylamine (a) with CH_3Br also yields the desired product, but other alkylation products are also formed.

(c)

$$CH_3CH_2CH_2CH_2C{\equiv}N \xrightarrow[2.\ H_3O^+]{1.\ CH_3MgBr} CH_3CH_2CH_2CH_2\overset{O}{\overset{\|}{C}}CH_3$$

$$\downarrow 1.\ CH_3MgBr;\ 2.\ H_3O^+$$

$$CH_3CH_2CH_2CH_2\overset{OH}{\overset{|}{C}}(CH_3)$$

(d)

$$\underset{\text{from (c)}}{CH_3CH_2CH_2CH_2\overset{O}{\overset{\|}{C}}CH_3} \xrightarrow[2.\ H_3O^+]{1.\ NaBH_4} CH_3CH_2CH_2CH_2\overset{OH}{\overset{|}{C}}HCH_3$$

(e)

$$CH_3CH_2CH_2CH_2C{\equiv}N \xrightarrow[2.\ H_2O]{1.\ DIBAH} CH_3CH_2CH_2CH_2CHO$$

21.53 The first two transformations start with bromocyclohexane, which can be prepared by reaction of cyclohexanol with PBr_3. Subsequent transformations start with cyclohexanone, which is produced by oxidation of cyclohexanol with either PCC or acidic CrO_3.

(a)

$$C_6H_{11}Br \xrightarrow[\text{ether}]{Mg} C_6H_{11}MgBr \xrightarrow[2.\ H_2O]{1.\ CO_2} C_6H_{11}COOH \xrightarrow[H^+\ \text{catalyst}]{CH_3OH} C_6H_{11}COOCH_3 \xrightarrow[2.\ H_3O^+]{1.\ DIBAH} C_6H_{11}CHO$$

(b)

$$C_6H_{11}Br \xrightarrow[\text{ether}]{Mg} C_6H_{11}MgBr \xrightarrow[2.\ H_3O^+]{1.\ CH_2O} C_6H_{11}CH_2OH \xrightarrow[CH_2Cl_2]{PCC} C_6H_{11}CHO$$

(c)

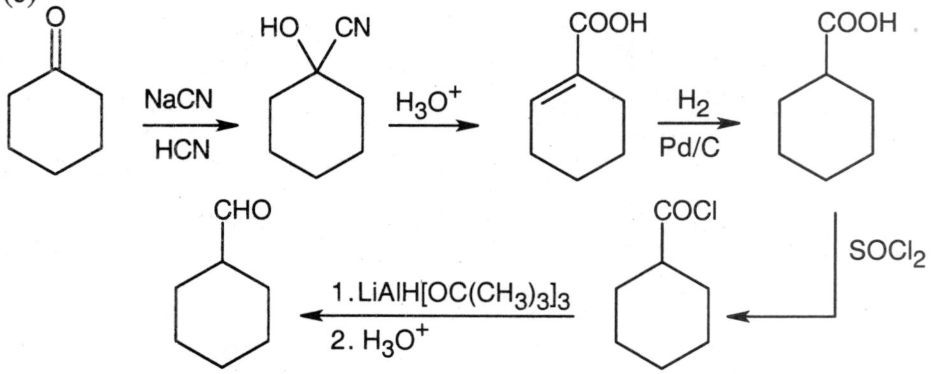

Cyclohexanone $\xrightarrow[HCN]{NaCN}$ 1-hydroxycyclohexanecarbonitrile (HO, CN) $\xrightarrow{H_3O^+}$ cyclohex-1-enecarboxylic acid (COOH) $\xrightarrow[Pd/C]{H_2}$ cyclohexanecarboxylic acid (COOH) $\xrightarrow{SOCl_2}$ cyclohexanecarbonyl chloride (COCl) $\xrightarrow[2.\ H_3O^+]{1.\ LiAlH[OC(CH_3)_3]_3}$ cyclohexanecarbaldehyde (CHO)

(d)

Cyclohexanone $\xrightarrow{(C_6H_5)_3\overset{+}{P}-\overset{-}{C}H_2}$ methylenecyclohexane (CH_2) $\xrightarrow[2.\ H_2O_2,\ ^-OH]{1.\ BH_3,\ THF}$ cyclohexylmethanol (CH_2OH) $\xrightarrow[CH_2Cl_2]{PCC}$ cyclohexanecarbaldehyde (CHO)

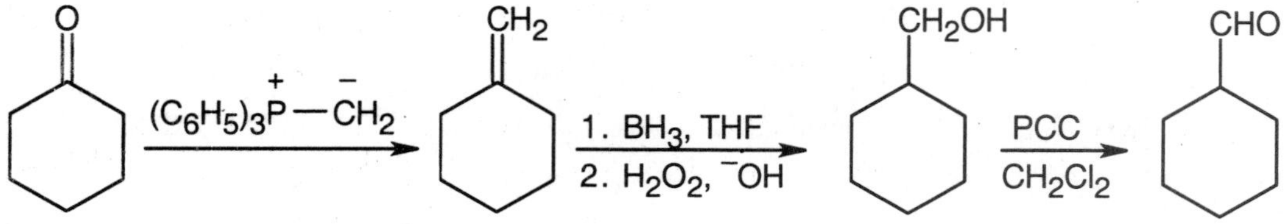

The following route is also possible, but E2 elimination caused by $^-$CN is a possibility.

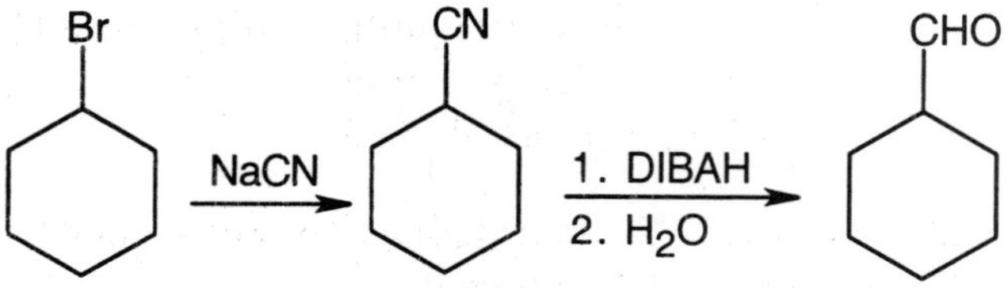

Bromocyclohexane (Br) $\xrightarrow{NaCN}$ cyclohexanecarbonitrile (CN) $\xrightarrow[2.\ H_2O]{1.\ DIBAH}$ cyclohexanecarbaldehyde (CHO)

21.54

$$H_2C{=}CHCH{=}CH_2 \xrightarrow[1,4\text{-addition}]{Br_2,\ CH_2CCl_2} BrCH_2CH{=}CHCH_2Br$$

$$\downarrow \ 2\ NaCN,\ HCN$$

$$H_2CCH_2CH_2CH_2CH_2NH_2 \xleftarrow[Pd/C]{H_2} NCCH_2CH{=}CHCH_2CN \ + \ 2\ NaBr$$

21.55

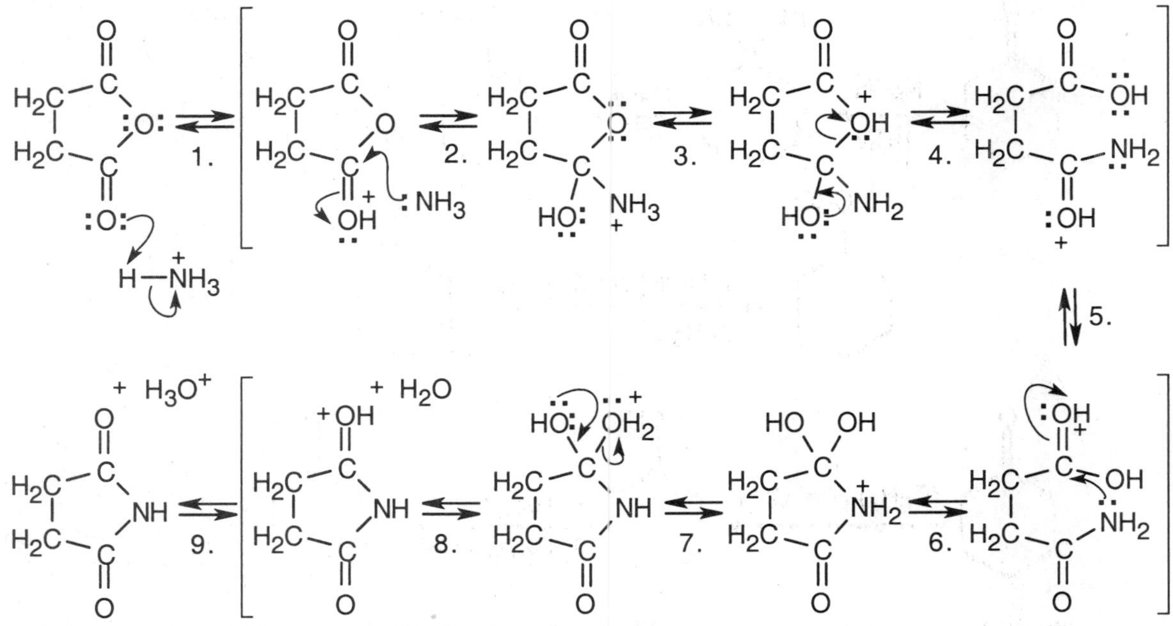

A summary of steps:

Step 1: Protonation

Step 2: Nucleophilic addition of NH_3

Steps 3,5,7: Proton transfers

Step 4: Ring opening

Step 6: Nucleophilic addition of $–NH_2$

Step 8: Loss of H_2O

Step 9: Loss of proton

This reaction requires high temperatures because the intermediate amide is a poor nucleophile and the carboxylic acid carbonyl group is unreactive.

21.56 This synthesis requires a nucleophilic aromatic substitution reaction, explained in Section 16.8.

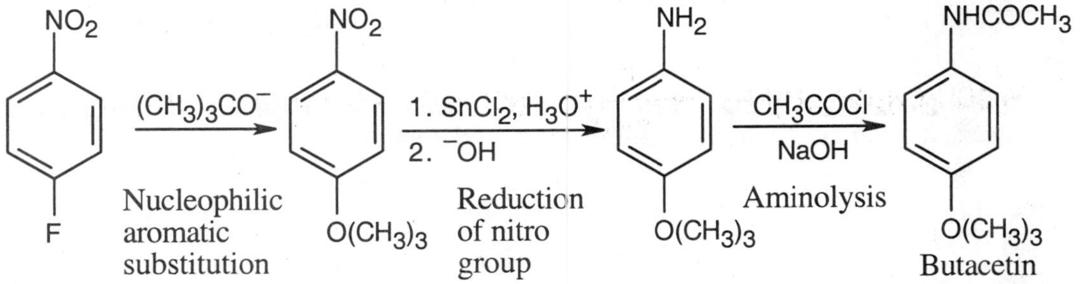

$(CH_3)_3CO^-$ — Nucleophilic aromatic substitution

1. $SnCl_2$, H_3O^+; 2. ^-OH — Reduction of nitro group

CH_3COCl, NaOH — Aminolysis

Butacetin

Synthetic analysis: The amide can be formed by the reaction of acetyl chloride with the appropriate amine, which is produced by reduction of the nitro group of the starting material. A nucleophilic aromatic substitution of –F by $–O(CH_3)_3$ can take place because the ring has an electron-withdrawing nitro group para to the site of substitution.

21.57

HO-C6H3(NO2)-COOH $\xrightarrow[\text{Formation of acid chloride}]{SOCl_2}$ HO-C6H3(NO2)-COCl $\xrightarrow[\text{Esterification}]{\text{Phenol}}$ HO-C6H3(NO2)-CO-O-C6H5 $\xrightarrow[\text{Reduction of nitro group}]{1.\ SnCl_2,\ H_3O^+;\ 2.\ ^-OH}$ HO-C6H3(NH2)-CO-O-C6H5

Phenyl 4-amino-salicylate

Synthetic analysis: The product results from esterification of an acid chloride, followed by reduction of the nitro group.

21.58

H_3C-C6H4-Br $\xrightarrow{\text{Mg, ether}}$ H_3C-C6H4-MgBr $\xrightarrow[\text{Grignard carboxylation}]{1.\ CO_2;\ 2.\ H_3O^+}$ H_3C-C6H4-COOH $\xrightarrow[\text{formation of acid chloride}]{SOCl_2}$ H_3C-C6H4-COCl $\xrightarrow[\text{NaOH}]{HN(CH_2CH_3)_2}$ H_3C-C6H4-C(=O)N$(CH_2CH_3)_2$

N,N-Diethyl-*m*-toluamide formation of amide

Synthetic analysis: Grignard carboxylation yields *m*-methylbenzoic acid, which can be converted to an acid chloride and treated with diethylamine.

21.59

NC-C6H4-CH_3 $\xrightarrow{H_3O^+}$ HOOC-C6H4-CH_3 $\xrightarrow[(PhCO_2)_2]{NBS}$ HOOC-C6H4-CH_2Br $\xrightarrow[NH_3]{\text{excess}}$ HOOC-C6H4-CH_2NH_2 $\xrightarrow[Rh/C]{H_2}$ trans-HOOC-C6H10-CH_2NH_2 (H, H)

Synthetic analysis: Using a rhodium catalyst, the aromatic ring is hydrogenated to form the cis-substituted cyclohexane, which is converted to the trans isomer by heating to 300°. The nitrile group is hydrolyzed to form a carboxylic acid, and the methyl group is brominated and treated with ammonia to form the amine.

Note: Another route to this compound is (1) oxidation of the methyl group (2) hydrogenation of both the nitrile and the aromatic ring.

21.60 (a)

$H_2\ddot{C}^{-}-\overset{+}{N}\equiv N: \longleftrightarrow H_2C=\overset{+}{N}=\ddot{N}:^{-}$

Resonance forms show that the carbon of diazomethane is basic, and reaction with an acid can occur to form a methyldiazonium ion.

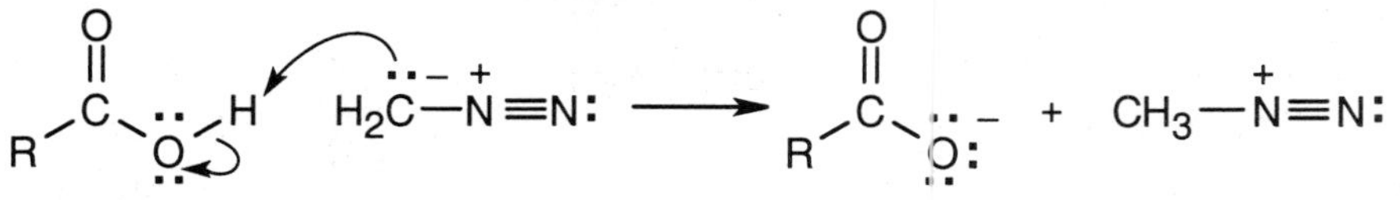

$RCO_2H + H_2\ddot{C}^{-}-\overset{+}{N}\equiv N: \longrightarrow RCO_2^{-} + CH_3-\overset{+}{N}\equiv N:$

(b)

$RCO_2^{-} + CH_3-\overset{+}{N}\equiv N: \longrightarrow RCOOCH_3 + N_2$

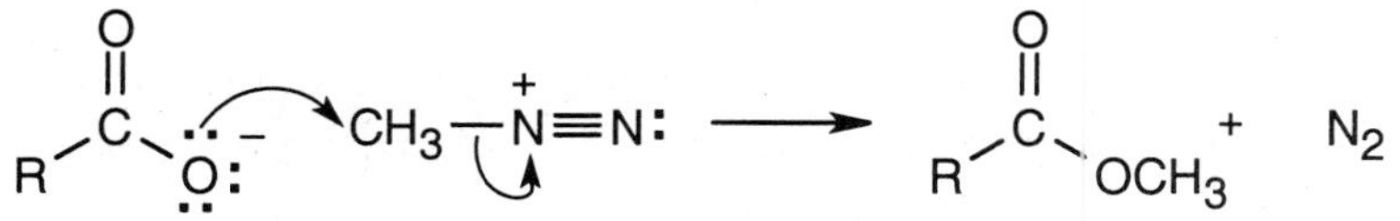

An S_N2 reaction then occurs in which the carboxylate ion displaces N_2 as the leaving group to form the methyl ester.

21.61

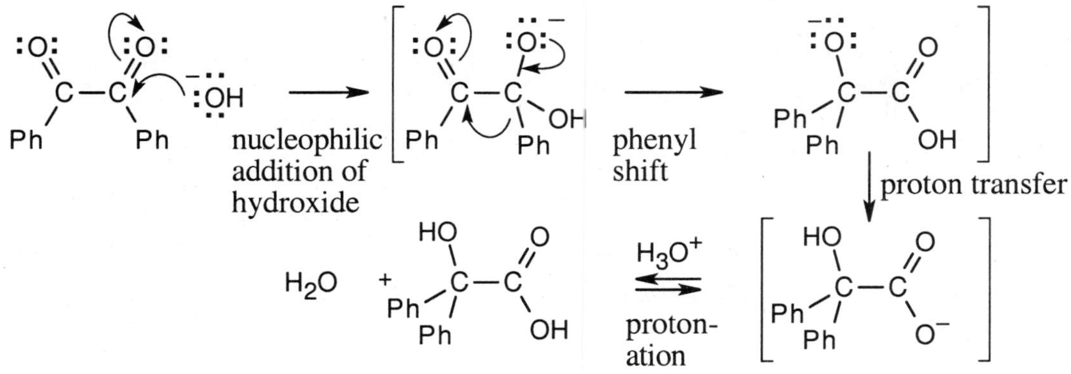

nucleophilic addition of hydroxide

phenyl shift

proton transfer

H_3O^+

protonation

H_2O + $Ph_2C(OH)COOH$

21.62

$HCOCOPh \xrightarrow[\text{2. } H_3O^+]{\text{1. NaOH, } H_2O} PhCH(OH)COOH$

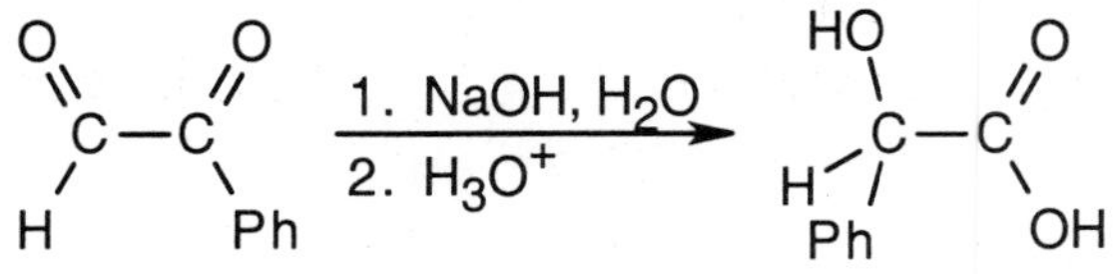

Use the mechanism shown in the previous problem to help predict the product.

21.63

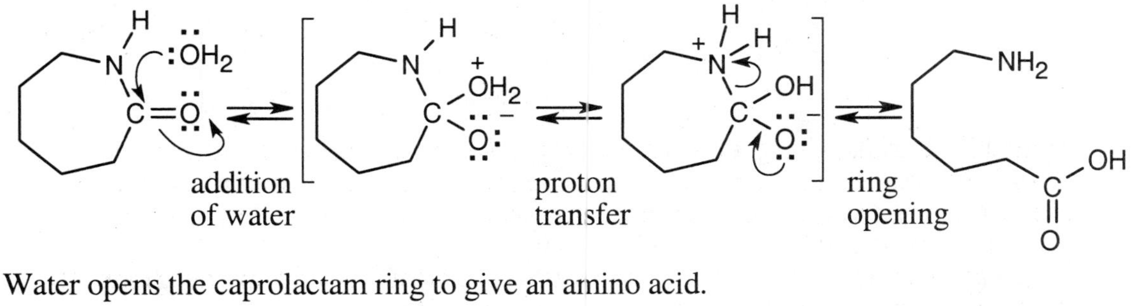

addition of water

proton transfer

ring opening

$H_2N(CH_2)_5COOH$

Water opens the caprolactam ring to give an amino acid.

$H_2N(CH_2)_5C(=O)OH + H_2N(CH_2)_5C(=O)OH \rightleftharpoons$ addition of amino group to carboxylic acid group $\rightarrow$ [$H_2N(CH_2)_5C(O^-)(OH)-N^+H_2(CH_2)_5C(=O)OH$]

proton transfer

[$H_2N(CH_2)_5C(O^-)(^+OH_2)-NH(CH_2)_5C(=O)OH$]

loss of water

$H_2N(CH_2)_5C(=O)-NH(CH_2)_5C(=O)OH + H_2O$

The amino group of one monomer adds to the carboxylic acid group of a second monomer to form a dimer. Further additions yield the nylon 6 polymer. Heat forces the equilibrium in the direction of product formation by driving off water.

$\text{-}(HN(CH_2)_5C(=O)-NH(CH_2)_5C(=O))\text{-}$ Nylon 6

21.64

$HOC(=O)(CH_2)_6C(=O)OH + H_2N-C_6H_{10}-CH_2-C_6H_{10}-NH_2$

$\downarrow$

$\text{-}(C(=O)(CH_2)_6C(=O)-NH-C_6H_{10}-CH_2-C_6H_{10}-NH)_n\text{-}$ Qiana

21.65

HO^- + β-propiolactone $\rightarrow$ [tetrahedral intermediate] $\rightarrow$ $HOC(=O)CH_2CH_2O^-$ + β-propiolactone

$\downarrow$

[$HOC(=O)CH_2CH_2O$–tetrahedral intermediate]

$HOC(=O)CH_2CH_2O-C(=O)CH_2CH_2O^-$

repeat many times

$\text{-}(OC(=O)CH_2CH_2O-C(=O)CH_2CH_2O)_n\text{-}$

The polymer is a polyester.

21.66 The polyimide pictured is a step-growth polymer of a benzene tetracarboxylic acid and an aromatic diamine.

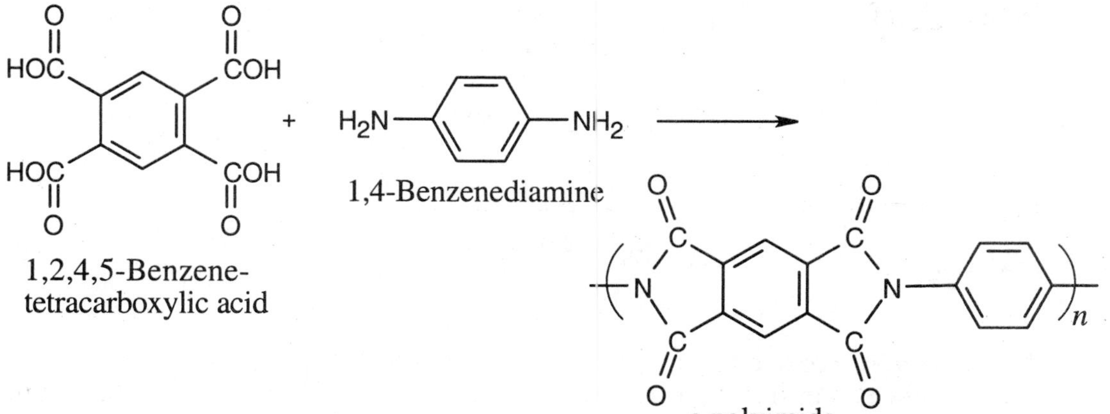

1,2,4,5-Benzene-tetracarboxylic acid + $H_2N-C_6H_4-NH_2$ (1,4-Benzenediamine) ⟶ a polyimide

21.67 In some of these pairs, IR spectroscopy is sufficient for differentiating between the isomers. For others, either ^{1}H NMR or a combination of ^{1}H NMR and IR data is necessary.

(a)

	$CH_3CH_2CNHCH_3$ (C=O)	$CH_3CN(CH_3)_2$ (C=O)
	N-Methylpropanamide	*N,N*-Dimethylacetamide
IR:	1680 cm^{-1} (*N*-substituted amide)	1650 cm^{-1} (*N,N*-disubstituted amide)
^{1}H NMR:	one methyl group one ethyl group	three methyl groups

(b)

	$HOCH_2CH_2CH_2CH_2C{\equiv}N$	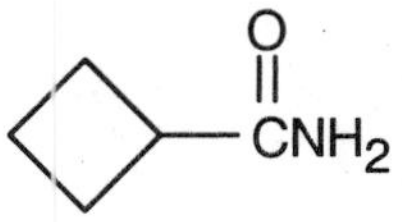 cyclobutyl–CNH_2 (C=O)
	5-Hydroxypentanenitrile	Cyclobutanecarboxamide
IR:	3300–3400 cm^{-1} (hydroxyl) 2250 cm^{-1} (nitrile)	1690 cm^{-1} (amide)

(c)

	$ClCH_2CH_2CH_2COH$ (C=O)	$CH_3OCH_2CH_2CCl$ (C=O)
	4-Chloropentanoic acid	3-Methoxypropanoyl chloride
IR:	2500–3300 cm^{-1} (hydroxyl) 1710 cm^{-1} (carboxylic acid)	1810 cm^{-1} (carboxylic acid chloride)

(d)

$CH_3CH_2\overset{O}{\overset{\|}{C}}OCH_2CH_3$
Ethyl propanoate

$CH_3\overset{O}{\overset{\|}{C}}OCH_2CH_2CH_3$
Propyl acetate

¹H NMR:	two triplets, two quartets	one singlet, one triplet, one quartet, one multiplet

21.68 The IR spectrum indicates that this compound has a carbonyl group.

$CH_3CHCl\overset{O}{\overset{\|}{C}}OCH_3$ (a = CH_3, c = CHCl, b = OCH_3)

a = 1.7 δ
b = 3.8 δ
c = 4.4 δ

21.69 The IR absorption at 2250 cm^{-1} identifies this compound as a nitrile.

$CH_3CH_2CH_2C{\equiv}N$ (a, b, c)

a = 1.1 δ
b = 1.7 δ
c = 2.3 δ

21.70

(a) $CH_3CH_2CH_2\overset{O}{\overset{\|}{C}}Cl$ (a, b, c)

a = 1.0 δ
b = 1.7 δ
c = 2.3 δ

(b) $N{\equiv}CCH_2\overset{O}{\overset{\|}{C}}OCH_2CH_3$ (b, c, a)

a = 1.3 δ
b = 3.6 δ
c = 4.3 δ

(c) $CH_3\overset{O}{\overset{\|}{C}}OCH(CH_3)_2$ (b, c, a)

a = 1.3 δ
b = 2.0 δ
c = 5.0 δ

21.71

(a) $ClCH_2CH_2\overset{O}{\overset{\|}{C}}OCH_2CH_3$ (d, b, c, a)

a = 1.3 δ
b = 2.8 δ
c = 3.8 δ
d = 4.2 δ

(b) $CH_3CH_2O\overset{O}{\overset{\|}{C}}CH_2\overset{O}{\overset{\|}{C}}OCH_2CH_3$ (a, c, b, c, a)

a = 1.3 δ
b = 3.4 δ
c = 4.2 δ

(c) $C_6H_5CH{=}CH\overset{O}{\overset{\|}{C}}OCH_2CH_3$ (phenyl = d; e, c, b, a)

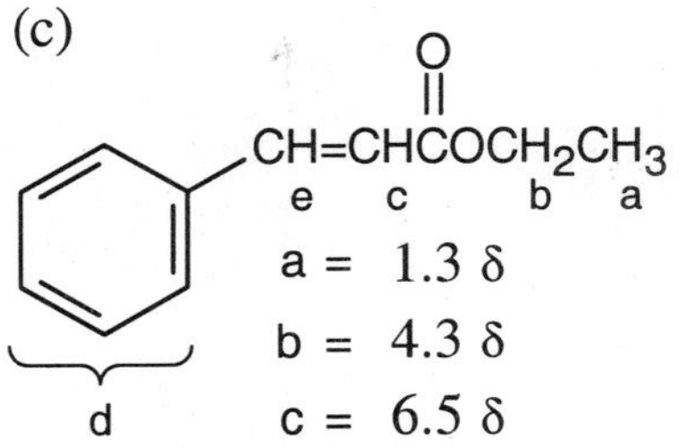

a = 1.3 δ
b = 4.3 δ
c = 6.5 δ
d = 7.4–7.6 δ
e = 7.7 δ

A Look Ahead

21.72 Addition of the triamine causes formation of cross-links between prepolymer chains.

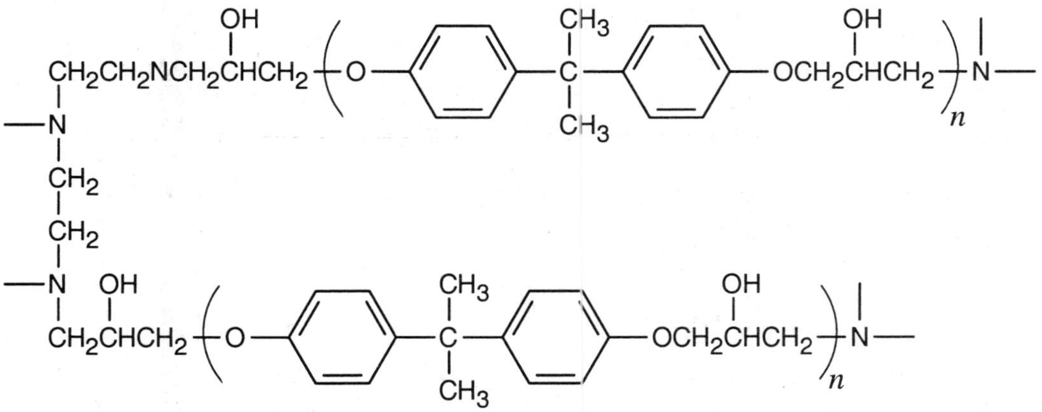

21.73 This a nucleophilic acyl substitution reaction whose mechanism is similar to others we have studied.

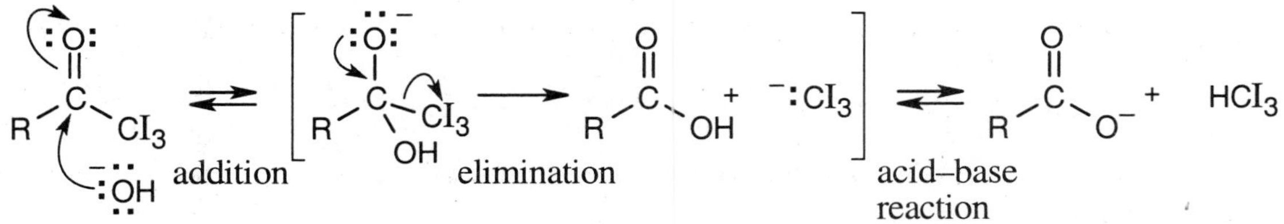

addition elimination acid–base reaction

$I_3C:^-$ can act as a leaving group because the electron-withdrawing iodine atoms stabilize the carbanion.

Molecular Modeling

21.74 The carbonyl oxygen is the most basic (negative) atom in both structures.

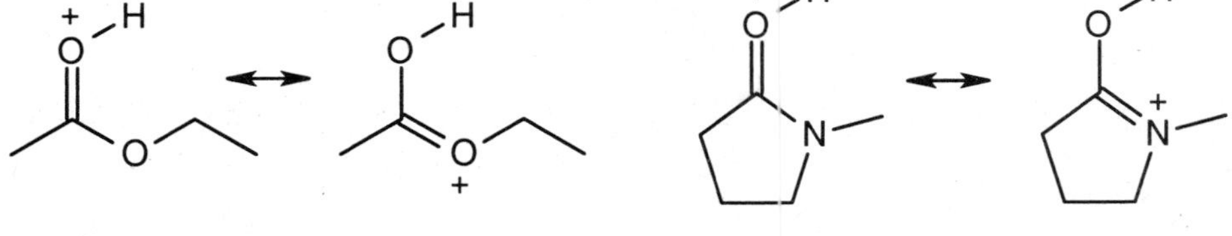

21.75 Methyl acetate has two preferred geometries, with C–O–C=O dihedral angles of 0° and 180°; planar geometries are stabilized by a resonance interaction between the alkyl oxygen and the carbonyl group. The 0° geometry is lower in energy by 40 kJ/mol. Conformers A and B of the cyclic ester-amide have C–O–C=O dihedral angles of 143° and 46°, respectively. Thus, conformer B's geometry is more like that of the lower energy conformer of methyl acetate, and conformer B is lower in energy by 25 kJ/mol.

21.76 Ring strain in the 4-membered ring of penicillin forces the amide nitrogen to be pyramidal. Electrostatic potential maps show that this localizes the electron lone pair on nitrogen and makes the carbonyl carbon more positive and reactive toward nucleophiles.

21.77 Recall that calculated frequencies are about 10% larger than observed frequencies. The calculated C=O stretching frequencies are 1941 cm^{-1} (acetone), 1943 cm^{-1} (ethyl acetate), and 1877 cm^{-1} (*N,N*-dimethylformamide). The stretching frequencies of the ketone and ester are similar, but the amide's stretching frequency is lower and is easily distinguished.

Review Unit 8: Carbonyl Compounds 1. Reaction at the Carbonyl Group

Major Topics Covered (with vocabulary);

Aldehydes and ketones:
-carbaldehyde acyl group acetyl group formyl group benzoyl group hydrate Tollens reagent

Reactions of aldehydes and ketones:
nucleophilic addition reaction gem diol cyanohydrin imine enamine carbinolamine 2,4-dinitrophenylhydrazone Wolff-Kishner reaction acetal hemiacetal Wittig reaction ylide betaine Cannizzaro reaction conjugate addition α,β-unsaturated carbonyl compound

Carboxylic acids and their derivatives:
carboxylation carboxylic acid derivative acid halide acid anhydride amide ester nitrile –carbonitrile

Reactions of carboxylic acids and their derivatives:
nucleophilic acyl substitution hydrolysis alcoholysis aminolysis Fischer esterification reaction lactone saponification DIBAH lactam thiol ester polyamide polyester step-growth polymer chain-growth polymer nylon

Types of Problems:

After studying these chapters you should be able to:
- Name and draw aldehydes, ketones, carboxylic acids and their derivatives.
- Prepare all of these compounds.
- Explain the reactivity difference between aldehydes and ketones and between carboxylic acids and all their derivatives.
- Calculate dissociation constants of carboxylic acids, and predict the relative acidities of substituted carboxylic acids.
- Formulate mechanisms for reactions related to the reactions we have studied.
- Predict the products of the reactions for all functional groups we have studied.
- Use spectroscopic techniques to identify these compounds.
- Draw representative segments of step-growth polymers.

Points to Remember:

* In all of these reactions, a nucleophile adds to a positively polarized carbonyl carbon to form a tetrahedral intermediate. There are three possible fates for the tetrahedral intermediate: (1) The intermediate can be protonated, as occurs in Grignard reactions, reductions, and cyanohydrin formation. (2) The intermediate can lose water (or ^{-}OH), as happens in imine and enamine formation. (3) The intermediate can lose a leaving group, as occurs in most reactions of carboxylic acid derivatives.

* Many of the reactions in these three chapters require acid or base catalysis. An acid catalyst, protonates the carbonyl oxygen, making the carbonyl carbon more reactive toward electrophiles, and/or protonates the tetrahedral intermediate, making loss of a leaving group easier. A base catalyst deprotonates the nucleophile, making it more nucleophilic. The pH optimum for these reactions is a compromise between the two needs.

* Here are a few reminders for drawing the mechanisms of nucleophilic addition and substitution reactions. (1) When a reaction is acid-catalyzed, none of the intermediates are negatively charged, although, occasionally, a few may be neutral. Check your mechanisms for charge balance. (2) Make sure you have drawn arrows correctly. The point of the arrow shows the new location of the electron pair at the base of the arrow. (3) In a polar reaction, two arrows never point at each other. If you find two arrows pointing at each other, redraw the mechanism.

* Reactions of acyl halides are almost always carried out with an equivalent of base present. The base is used to scavenge the hydrogen ions produced when a nucleophile adds to an acyl halide. If base were not present, hydrogen ions would protonate the nucleophile and make it unreactive.

* The products of acidic cleavage of an amide are a carboxylic acid and a protonated amine. The products of basic cleavage of an amide are a carboxylate anion and an amine.

* In some of the mechanisms shown in the answers, a series of protonations and deprotonations occur. These steps convert the initial tetrahedral intermediate into an intermediate that more easily loses a leaving group. These deprotonations may be brought about by the solvent, by the conjugate base of the catalyst, by other molecules of the carbonyl compound or may occur intramolecularly. When a "proton transfer" is shown as part of a mechanism, the base that removes the proton has often not been shown. However, it is implied that the proton transfer is assisted by a base: the proton doesn't fly off the intermediate unassisted.

* The most useful spectroscopic information for identifying carbonyl compounds comes from IR spectroscopy and ^{13}C NMR spectroscopy. Carbonyl groups have distinctive identifying absorptions in their infrared spectra. ^{13}C NMR is also useful for identifying aldehydes, ketones, and nitriles, although other groups are harder to distinguish. The ^{1}H NMR absorptions of aldehydes and carboxylic acids are also significant. Look at mass spectra for McLafferty rearrangements and alpha-cleavage reactions of aldehydes and ketones.

Self-Test:

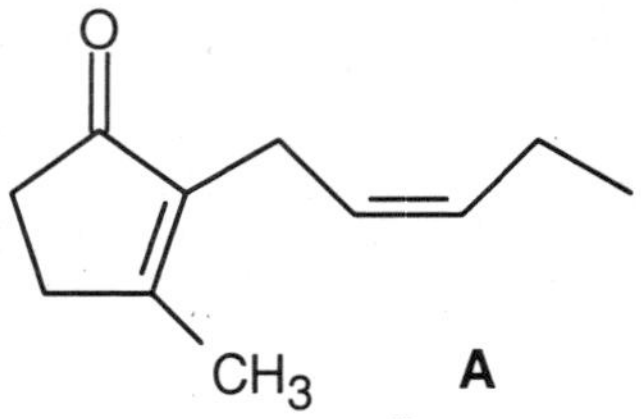

O CH3 **A**
Jasmone
(used in perfumery)

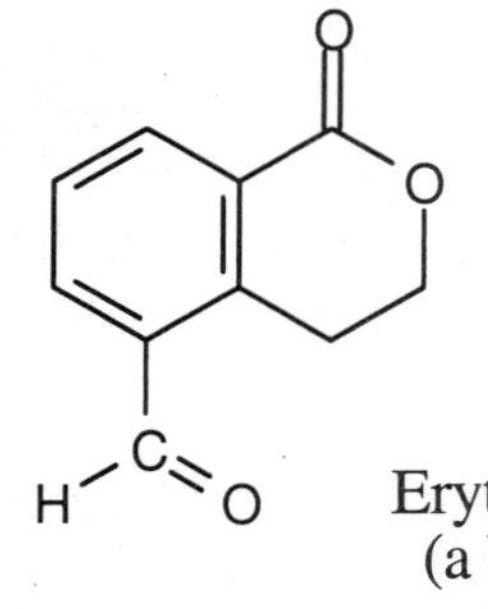

O O H C O **B**
Erythrocentaurin
(a bitter tonic)

Predict the products of the reaction of **A** with: (a) $LiAlH_4$, then H_3O^+; (b) C_6H_5MgBr, then H_3O^+; (c) $(CH_3)_2NH$, H_3O^+; (d) CH_3OH, H^+ catalyst (e) $(C_6H_5)_3P^+—CH_2^-$; (f) 1 equiv. $CH_3CH_2NH_2$, H_3O^+. How would you reduce **A** to yield a saturated hydrocarbon? Where would you expect the carbonyl absorption of **A** to occur in its IR spectrum?

Predict the products of **B** with the reagents (a) – (d) above. What product(s) would be formed if **B** was treated with Br_2, $FeBr_3$? Where do the carbonyl absorptions occur in the IR spectrum of **B**? Describe the ^{13}C NMR spectrum of **B**.

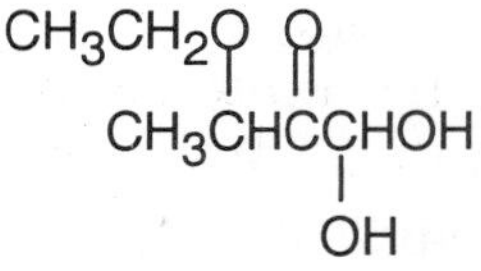

CH_3CH_2O O
$CH_3CHCCHOH$
OH

C
Kethoxal
(antiviral)

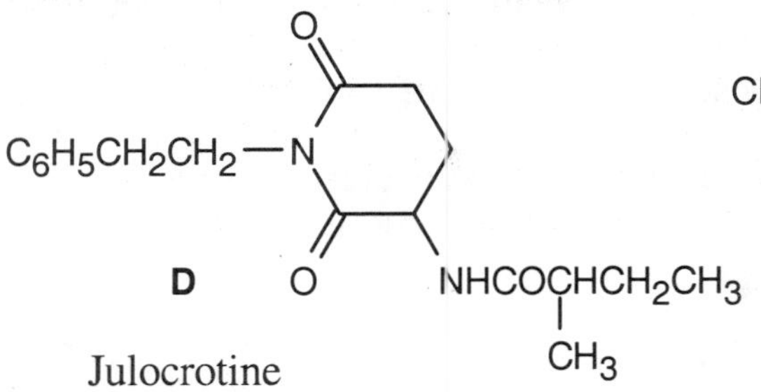

$C_6H_5CH_2CH_2$—N
O
O
NHCOCHCH$_2$CH$_3$
CH$_3$

D
Julocrotine

COCH$_3$
CH$_3$O OH
OCH$_3$

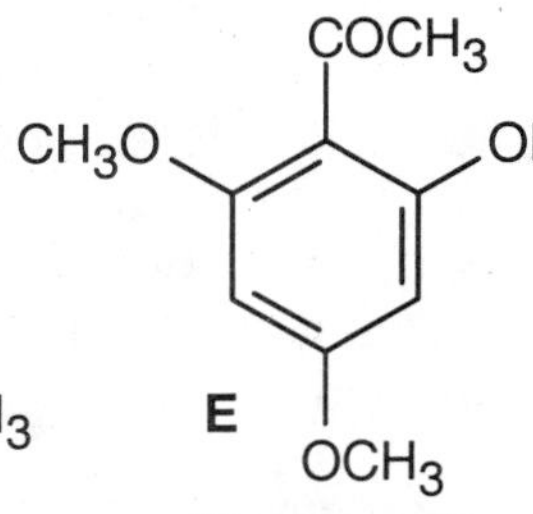

E
Xanthoxylin

Kethoxal (**C**) exists in solution as an equilibrium mixture. With what compound is it in equilibrium. Why does the equilibrium lie on the side of kethoxal?

Identify the carboxylic acid derivatives present in **D**. Show the products of treatment of **D** with (a) ^-OH, H_2O (b) $LiAlH_4$, then H_2O.

Name **E**. Describe the IR spectrum and ^1H NMR spectrum of **E**.

Multiple Choice:

1. In which of the following nucleophilic addition reactions does the equilibrium lie on the side of the products?
(a) Acetone + HCN (b) Propanal + H_2O (c) Acetaldehyde + HBr
(d) 2,2,4,4-Tetramethyl-3-pentanone + HCN

2. Which alcohol can be formed by three different combinations of carbonyl compound + Grignard reagent?
(a) 2-Butanol (b) 3-Methyl-3-hexanol (c) Triphenylmethanol (d) 1-Phenylethanol

3. A nitrile can be converted to all of the following except:
(a) an aldehyde (b) an amide (c) an amine (d) A nitrile can be converted to all of the above compounds.

4. Which of the following *p*- substituted benzoic acids is the least acidic?
(a) $CH_3COC_6H_5COOH$ (b) $CH_3OC_6H_5COOH$ (c) BrC_6H_5COOH (d) NCC_6H_5COOH

5. A carboxylic acid can be reduced by all of the following except:
(a) $LiAlH_4$, then H_3O^+ (b) BH_3, THF, then H_3O^+ (c) $NaBH_4$, then H_3O^+ (d) All of these reagents can reduce a carboxylic acid.

6. Which of the following carboxylic acids can be formed by both Grignard carboxylation and by nitrile hydrolysis?
(a) Phenylacetic acid (b) Benzoic acid (c) Trimethylacetic acid (d) 3-Butynoic acid

7. Acid anhydrides are used mainly for:
(a) synthesizing carboxylic acids (b) forming alcohols (c) introducing acetyl groups
(d) forming aldehydes

8. An aldehyde is formed from an acid halide by reduction with:
(a) DIBAH (b) $LiAlH_4$ (c) $(CH_3CH_2)CuLi$ (d) $LiAlH[OC(CH_3)_3]_3$

9. From which carboxylic acid derivative can you form a ketone as the product of a Grignard reaction?
(a) acid chloride (b) ester (c) nitrile (d) amide

10. An infrared absorption at 1650 cm^{-1} indicates the presence of:
(a) aromatic acid chloride (b) *N,N*-disubstituted amide (c) α,β-unsaturated ketone
(d) aromatic ester

Chapter 22 – Carbonyl α-Substitution Reactions

Chapter Outline

I. Keto-enol tautomerism (Section 22.1).
   A. Nature of tautomerism.
      1. Carbonyl compounds with hydrogens bonded to their α carbons equilibrate with their corresponding enols.
      2. This rapid equilibration is called tautomerism, and the individual isomers are tautomers.
      3. Unlike resonance forms, tautomers are different isomers.
      4. Despite the fact that very little of the enol isomer is present at room temperature, enols are very important because they are reactive.
   B. Mechanism of tautomerism.
      1. In acid catalysis, the carbonyl carbon is protonated to form an intermediate that can lose a hydrogen from its α carbon to yield a neutral enol.
      2. In base-catalyzed enol formation, an acid-base reaction occurs between a base and an α hydrogen.
         a. The resultant enolate is protonated to yield an enol.
         b. Protonation can occur either on carbon or on oxygen.
         c. Only hydrogens on the α positions of carbonyl compounds are acidic.

II. Enols (Sections 22.2 – 22.4).
   A. Reactivity of enols (Section 22.2).
      1. The electron-rich double bonds of enols cause them to behave as nucleophiles.
         The electron-donating enol –OH groups make enols more reactive than alkenes.
      2. When an enol reacts with an electrophile, the initial adduct loses –H from oxygen to give a substituted carbonyl compound.
   B. Reactions of enols (Sections 22.3 – 22.4).
      1. Alpha halogenation of aldehydes and ketones (Section 22.3).
         a. Aldehydes and ketones can be halogenated at their α positions by reaction of X_2 in acidic solution.
         b. The reaction proceeds by acid-catalyzed formation of an enol intermediate.
         c. Halogen isn't involved in the rate-limiting step: the rate doesn't depend on the identity of the halogen, but only on [ketone] and $[H^+]$.
         d. α-Bromo ketones are useful in syntheses because they can be dehydrobrominated by base treatment to form α,β-unsaturated ketones.
      2. Alpha-bromination of carboxylic acids (Section 22.4).
         a. In the Hell-Volhard-Zelinskii (HVZ) reaction, a mixture of Br_2 and PBr_3 can be used to brominate carboxylic acids in the α position.
         b. The initially formed acid bromide reacts with Br_2 to form an α-bromo acid bromide, which is hydrolyzed by water to give the α-bromo carboxylic acid.
         c. The reaction proceeds through an acid bromide enol.

III. Enolates (Sections 22.5 – 22.8).
   A. Enolate ion formation (Section 22.5).
      1. Hydrogens α to a carbonyl group are weakly acidic.
         a. This stability is due to overlap of a vacant *p* orbital with the carbonyl group *p* orbitals, allowing the carbonyl group to stabilize the negative charge by resonance.
         b. The two resonance forms aren't equivalent; the form with the negative charge on oxygen is of lower energy.
      2. Strong bases are needed for enolate ion formation.
         a. Alkoxide ions are too weak to use in enolate formation.

b. Lithium diisopropylamide (LDA) is used for forming enolates because it is a very strong base, it is soluble in THF, it is hindered and it can be used at low temperatures.
c. LDA can be used to form the enolates of many different carbonyl compounds.
3. When a hydrogen is flanked by two carbonyl groups, it is much more acidic. Both carbonyl groups can stabilize the negative charge.

B. Reactivity of enolate ions (Section 22.6).
1. Enolates are more useful than enols for two reasons:
a. Unlike enols, stable solutions of enolates are easily prepared.
b. Enolates are more reactive than enols because they are more nucleophilic.
2. Enolates can react either at carbon or at oxygen.
a. Reaction at carbon yields an α-substituted carbonyl compound.
b. Reaction at oxygen yields an enol derivative.

C Reactions of enolate ions (Sections 22.7 – 22.8).
1. The haloform reaction (Section 22.7).
a. Base-promoted halogenation of aldehydes and ketones proceeds readily because each halogen added makes the carbonyl compound more reactive.
b. Consequently, polyhalogenated compounds are usually produced.
c. This reaction is only useful with methyl ketones, which form HCX_3 when reacted with halogens.
i. The HCX_3 is a solid that can be identified.
ii. The last step of the reaction involves a carbanion leaving group.
2. Alkylation reactions of enolates (Section 22.8).
a. General features.
i. Alkylations are useful because they form a new C–C bond.
ii. Alkylations have the same limitations as S_N2 reactions; the alkyl groups must be methyl, primary, allylic or benzylic.
b. The malonic ester synthesis.
i. The malonic ester synthesis is used for preparing a carboxylic acid from a halide while lengthening the chain by two atoms.
ii. Diethyl malonate is useful because its enolate is easily prepared by reaction with sodium ethoxide.
iii. Since diethyl malonate has two acidic hydrogens, two alkylations can take place.
iv. Heating in aqueous HCl causes hydrolysis and decarboxylation of the alkylated malonate.
Decarboxylations are common only to β-keto acids and malonic acids.
v. Cycloalkanecarboxylic acids can also be prepared.
c. The acetoacetic ester synthesis.
i. The acetoacetic ester synthesis is used for converting an alkyl halide to a methyl ketone, while lengthening the carbon chain by 3 atoms.
ii. As with malonic ester, acetoacetic ester has two acidic hydrogens which are flanked by a ketone and an ester, and two alkylations can take place.
iii. Heating in aqueous HCl hydrolyzes the ester and decarboxylates the acid to yield the ketone.
iv. All β-keto esters can undergo this type of reaction.
d. Direct alkylation of ketones, esters, and nitriles.
i. LDA in a nonprotic solvent can be used to convert the above compounds to their enolates.
ii. Alkylation of an unsymmetrical ketone leads to a mixture of products, but the major product is alkylated at the less hindered position.

Solutions to Problems

22.1–22.2 Acidic hydrogens in the keto form of each of these compounds are bold.

	Keto Form	*Enol Form*	*Number of Acidic Hydrogens*
(a)	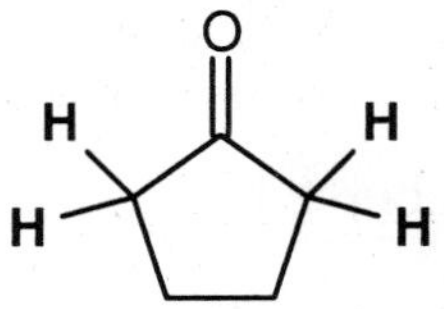	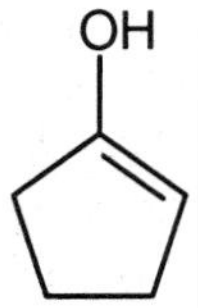	4
(b)	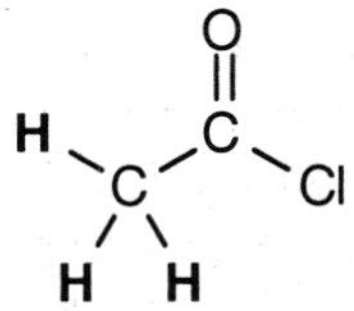	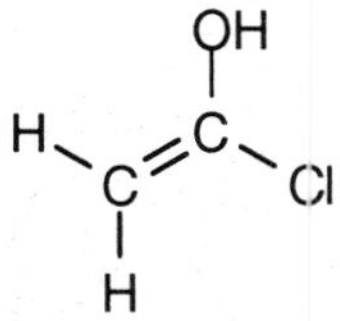	3
(c)	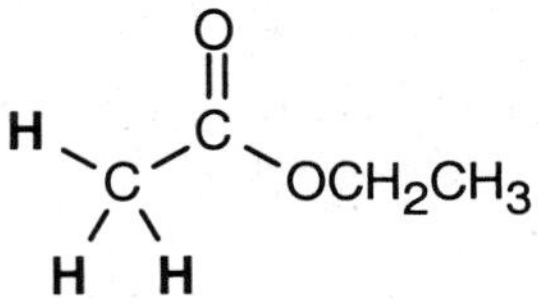	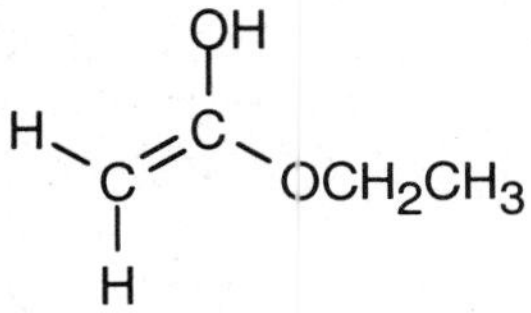	3
(d)	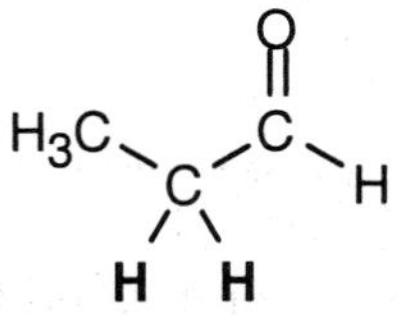	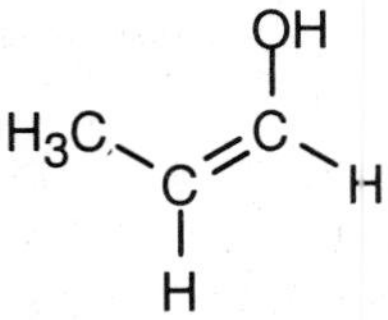	2
(e)	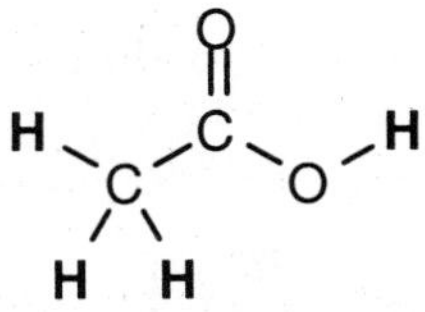	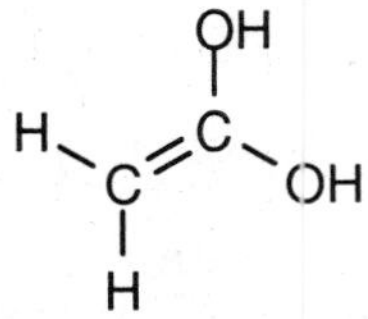	4
(f)	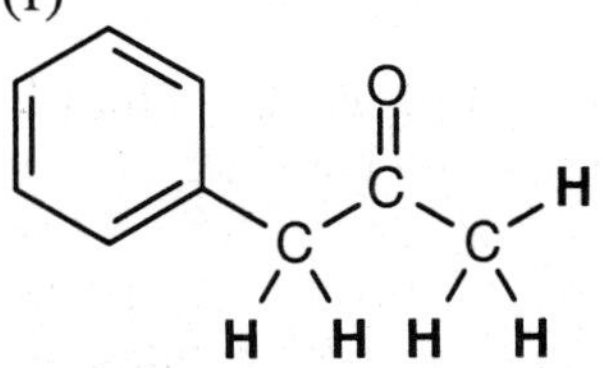	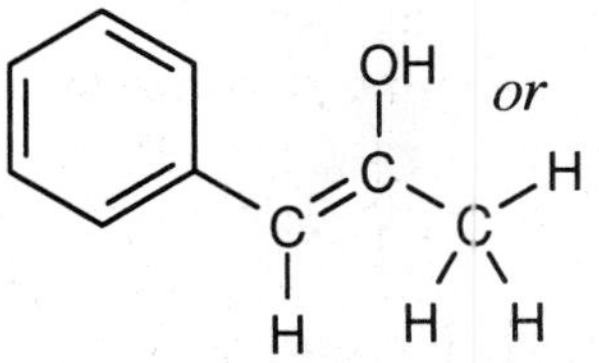 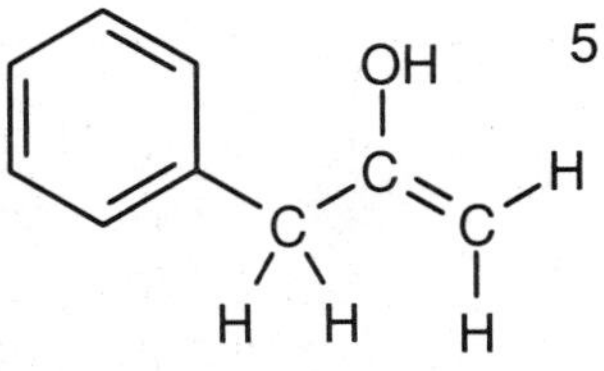	5
(g)	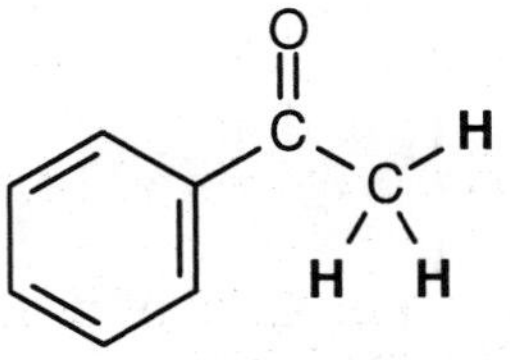	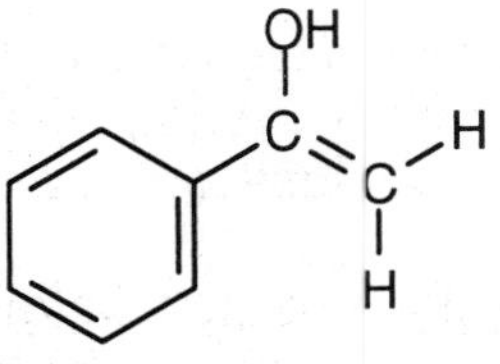	3

In (d) and (f), cis and trans enolates are possible.

22.3

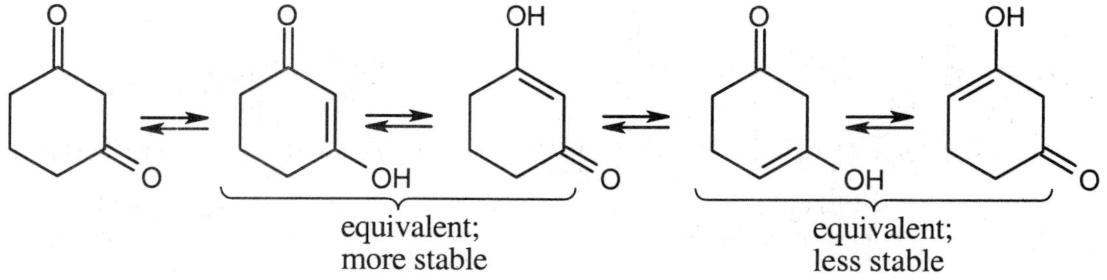

The first two monoenols are more stable because the enol double bond is conjugated with the carbonyl group.

22.4

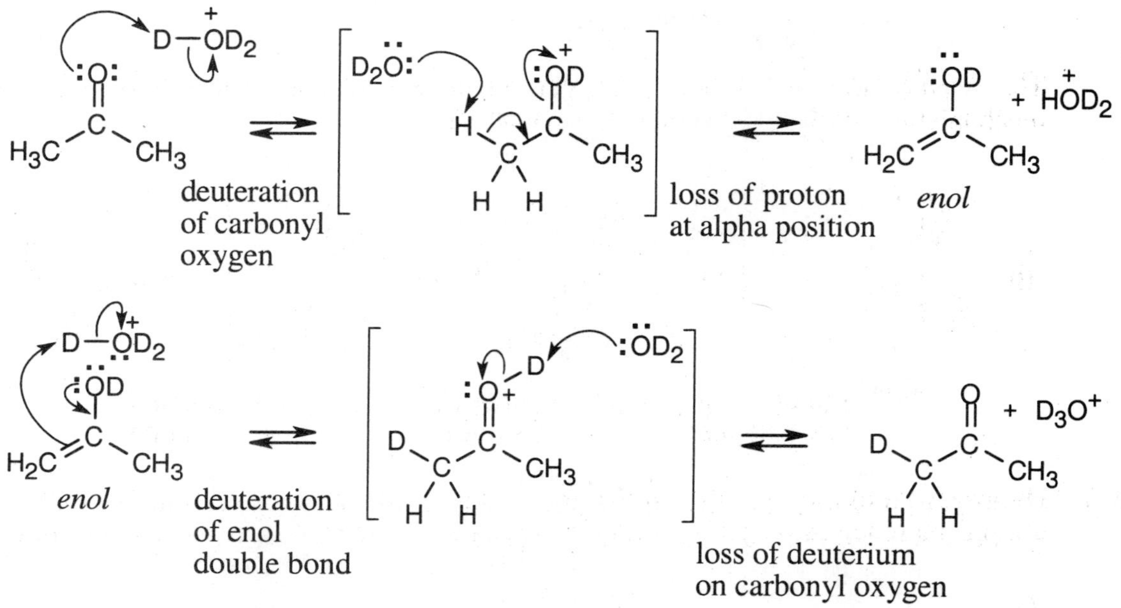

22.5 Alpha-bromination, followed by dehydration using pyridine, yields the enone pictured.

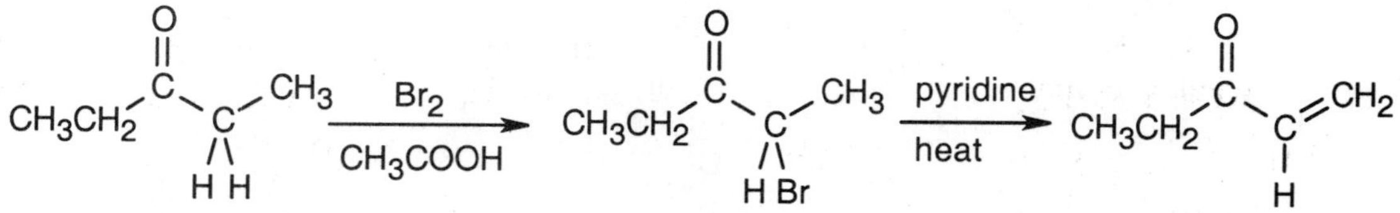

22.6

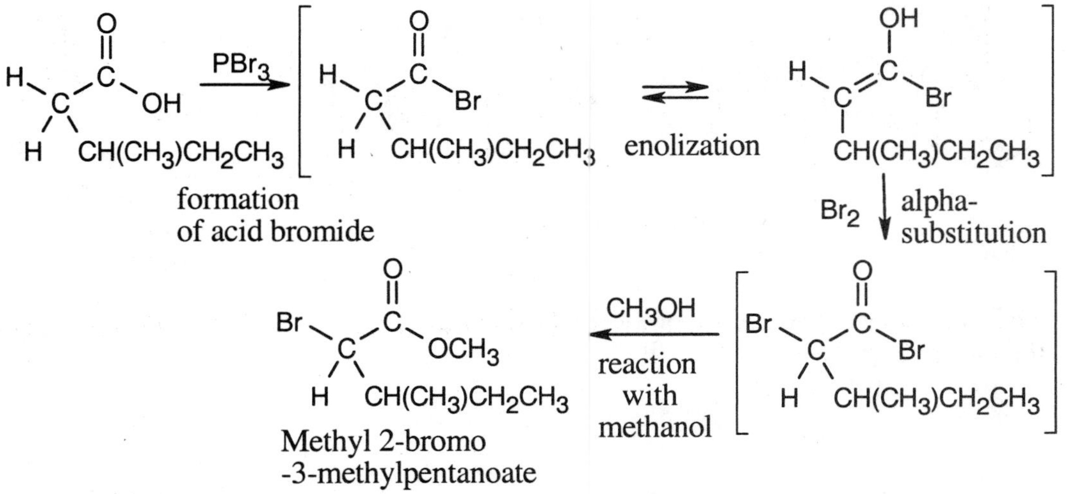

formation of acid bromide

enolization

Br_2 alpha-substitution

CH_3OH reaction with methanol

Methyl 2-bromo -3-methylpentanoate

The mechanism of the ester-forming step is a nucleophilic acyl substitution and resembles mechanisms that we have studied in Chapter 21.

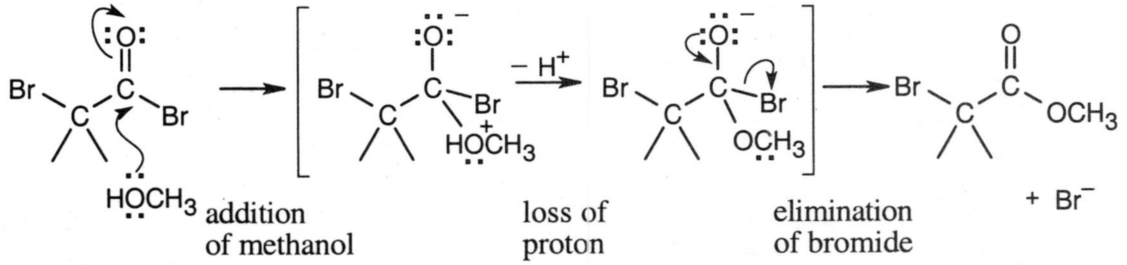

addition of methanol

loss of proton

elimination of bromide

$+ Br^-$

22.7 Hydrogens α to one carbonyl group are weakly acidic. Hydrogens α to two carbonyl groups are much more acidic, but they are not as acidic as carboxylic acid protons.

(a) weakly acidic

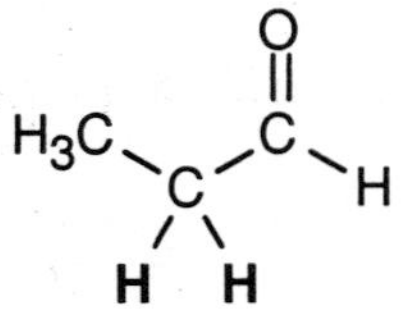

(b) weakly acidic

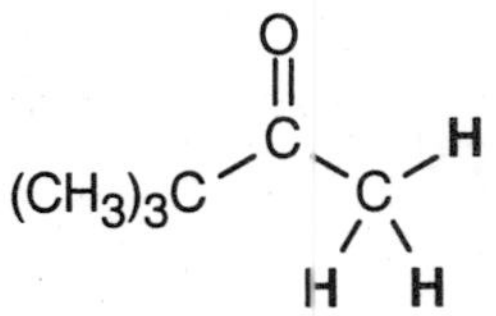

(c) weakly acidic; most acidic

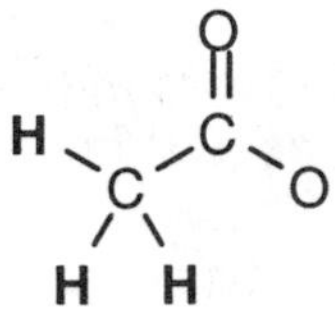

(d) weakly acidic

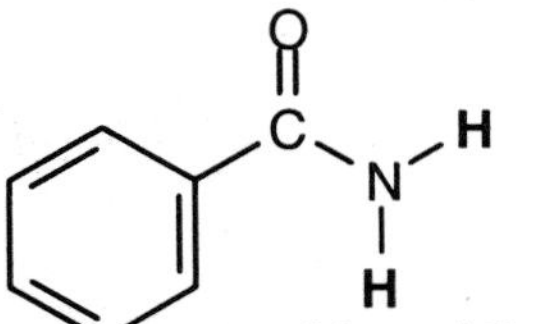

(e) weakly acidic

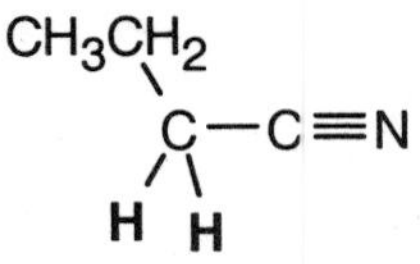

(f) weakly acidic

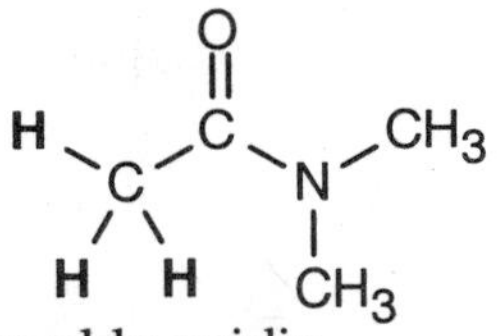

(g)

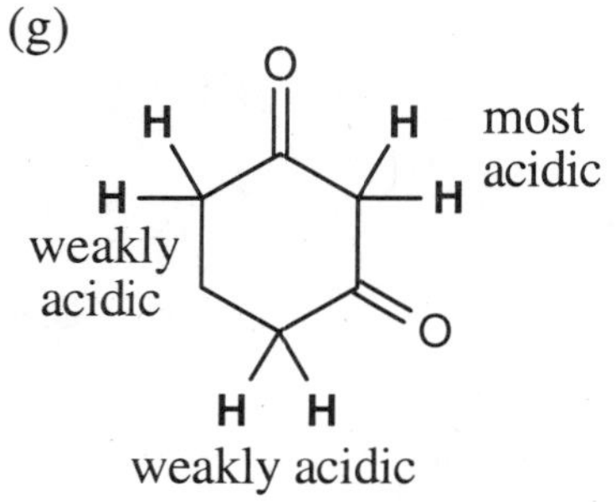

H H most acidic
H— —H
weakly acidic

H H
weakly acidic

22.8 Nitriles are weakly acidic because the nitrile anion can be stabilized by resonance involving a *p* orbital of the nitrile nitrogen.

$$H_2\ddot{C}^{-}-C\equiv N: \longleftrightarrow H_2C=C=\ddot{N}:^{-}$$

22.9 Halogenation in acid medium is acid-*catalyzed* because hydrogen ions are regenerated:

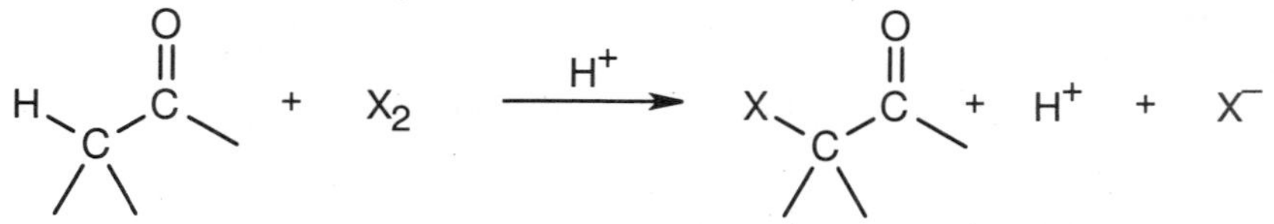

$$\text{H–C–C(=O)–} + X_2 \xrightarrow{H^+} \text{X–C–C(=O)–} + H^+ + X^-$$

Halogenation in basic medium is base-*promoted* because a stoichiometric amount of base is consumed:

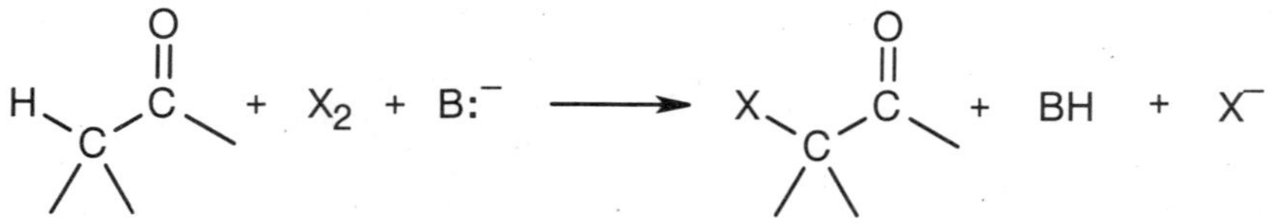

$$\text{H–C–C(=O)–} + X_2 + B:^- \longrightarrow \text{X–C–C(=O)–} + BH + X^-$$

22.10 In Problem 21.73 we formulated a mechanism for the haloform reaction in which the last step was the elimination of $^-:CX_3$. Despite the fact that this leaving group is a carbanion, its negative charge is stabilized by the three electron-withdrawing halogens.

22.11 The malonic ester synthesis converts an alkyl halide to a carboxylic acid with two more carbons (a substituted acetic acid). Identify the component that originates from malonic ester (the acid component). The rest of the molecule comes from the alkyl halide, which should be primary or methyl.

(a)

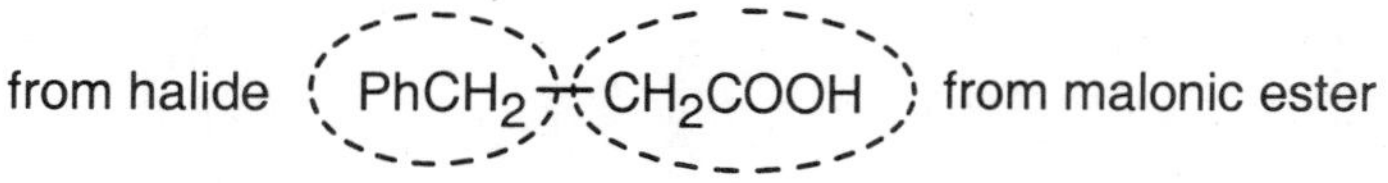

from halide $PhCH_2$ + CH_2COOH from malonic ester

$$CH_2(COOEt)_2 \xrightarrow[\text{2. } PhCH_2Br]{\text{1. } Na^+ \ ^-OEt} PhCH_2-CH(COOEt)_2 + NaBr$$

$$\downarrow H_3O^+, \text{heat}$$

$$PhCH_2-CH_2COOH + CO_2 + 2\ EtOH$$

3-Phenylpropanoic acid

(b)

$CH_3CH_2CH_2$ (from halide) — $CHCOOH$ (from malonic ester), with CH_3 (from halide)

$$CH_2(COOEt)_2 \xrightarrow[2.\ CH_3CH_2CH_2Br]{1.\ Na^+\ ^-OEt} CH_3CH_2CH_2-CH(COOEt)_2 + NaBr$$

$$\downarrow 1.\ Na^+\ ^-OEt;\ 2.\ CH_3Br$$

$$CH_3CH_2CH_2-C(CH_3)(COOEt)_2 + NaBr \xrightarrow[heat]{H_3O^+} CH_3CH_2CH_2-CH(CH_3)COOH + CO_2 + 2\ EtOH$$

2-Methylpentanoic acid

(c)

from halide $(CH_3)_2CHCH_2$ — CH_2COOH from malonic ester

$$CH_2(COOEt)_2 \xrightarrow[2.\ (CH_3)_2CHCH_2Br]{1.\ Na^+\ ^-OEt} (CH_3)_2CHCH_2-CH(COOEt)_2 + NaBr$$

$$\downarrow H_3O^+,\ heat$$

$$(CH_3)_2CHCH_2-CH_2COOH + CO_2 + 2\ EtOH$$

4-Methylpentanoic acid

(d)

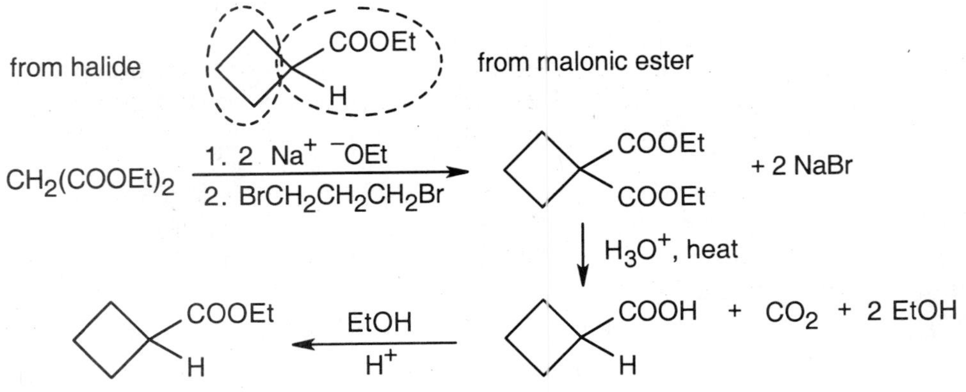

from halide (cyclobutane ring) — $C(COOEt)H$ from malonic ester

$$CH_2(COOEt)_2 \xrightarrow[2.\ BrCH_2CH_2CH_2Br]{1.\ 2\ Na^+\ ^-OEt} \text{cyclobutane-1,1-}(COOEt)_2 + 2\ NaBr$$

$$\downarrow H_3O^+,\ heat$$

$$\text{cyclobutane-}COOH\ (H) + CO_2 + 2\ EtOH \xrightarrow[H^+]{EtOH} \text{cyclobutane-}COOEt\ (H)$$

Ethyl cyclohexanecarboxylate

22.12 Since malonic ester has only two acidic hydrogen atoms, it can be alkylated only two times. Formation of trialkylated acetic acids is thus not possible.

22.13 As in the malonic acid synthesis, you should identify the structural fragments of the target compound. The acetoacetic acid synthesis converts an alkyl halide to a methyl ketone ("substituted acetone"). The methyl ketone component comes from acetoacetic ester; the other component comes from a halide.

(a)

from halide $(CH_3)_2CHCH_2$ | $CH_2\overset{O}{\overset{\|}{C}}CH_3$ from acetoacetic ester

$$\underset{COOEt}{CH_2\overset{O}{\overset{\|}{C}}CH_3} \xrightarrow[\text{2. }(CH_3)_2CHCH_2Br]{\text{1. }Na^+\ ^-OEt} (CH_3)_2CHCH_2{-}\underset{COOEt}{CH\overset{O}{\overset{\|}{C}}CH_3} + NaBr$$

$$\downarrow H_3O^+,\ heat$$

$$(CH_3)_2CHCH_2{-}CH_2\overset{O}{\overset{\|}{C}}CH_3 + CO_2 + EtOH$$

5-Methyl-2-hexanone

(b)

from halide $C_6H_5CH_2CH_2$ | $CH_2\overset{O}{\overset{\|}{C}}CH_3$ from acetoacetic ester

$$\underset{COOEt}{CH_2\overset{O}{\overset{\|}{C}}CH_3} \xrightarrow[\text{2. }C_6H_5CH_2CH_2Br]{\text{1. }Na^+\ ^-OEt} C_6H_5CH_2CH_2{-}\underset{COOEt}{CH\overset{O}{\overset{\|}{C}}CH_3} + NaBr$$

$$\downarrow H_3O^+,\ heat$$

$$C_6H_5CH_2CH_2{-}CH_2\overset{O}{\overset{\|}{C}}CH_3 + CO_2 + EtOH$$

5-Phenyl-2-pentanone

22.14

$$\underset{COOEt}{CH_2\overset{O}{\overset{\|}{C}}CH_3} \xrightarrow[\text{2. }BrCH_2CH_2CH_2CH_2Br]{\text{1. 2 }Na^+\ ^-OEt} \text{1-acetyl-1-(COOEt)cyclopentane } (H_3C{-}\overset{O}{\overset{\|}{C}}{-}C_5H_8{-}COOEt) + 2\ NaBr$$

$$\downarrow H_3O^+,\ heat$$

$$H_3C{-}\overset{O}{\overset{\|}{C}}{-}C_5H_9 \text{ (cyclopentyl)} + CO_2 + EtOH$$

22.15 The acetoacetic ester synthesis can only be used for certain products:

(1) Three carbons must originate from acetoacetic ester. In other words, compounds of the type $RCOCH_3$ can't be synthesized by the reaction of RX with acetoacetic ester.
(2) Alkyl halides must be primary or methyl.
(3) The acetoacetic ester synthesis can't be used to prepare compounds that are trisubstituted at the α position.

(a) Phenylacetone can't be produced by an acetoacetic ester synthesis because bromobenzene, the necessary halide, does not enter into S_N2 reactions. [See (2) above.]

(b) Acetophenone can't be produced by an acetoacetic ester synthesis. [See (1) above.]

(c) 3,3–Dimethyl–2–butanone can't be prepared because it is trisubstituted at the α position. [See (3) above.]

22.16 Direct alkylation is used to introduce substituents α to an ester, ketone or nitrile. Look at the target molecule to identify these substituents. Alkylation is achieved by treating the starting material with LDA, followed by a primary halide.

(a)

$$C_6H_5CH_2\overset{O}{\overset{\|}{C}}CH_3 \xrightarrow[2.\ CH_3I]{1.\ LDA} C_6H_5\underset{CH_3}{\underset{|}{C}}H\overset{O}{\overset{\|}{C}}CH_3$$

3-Phenyl-2-butanone

Alkylation occurs at the carbon next to the phenyl group because the phenyl group can help stabilize the enolate anion intermediate.

(b)

$$CH_3CH_2CH_2CH_2C{\equiv}N \xrightarrow[2.\ CH_3CH_2I]{1.\ LDA} CH_3CH_2CH_2\underset{CH_2CH_3}{\underset{|}{C}}HC{\equiv}N$$

2-Ethylpentanenitrile

(c)

$$\text{cyclohexanone} \xrightarrow[2.\ ICH_2CH{=}CH_2]{1.\ LDA} \text{2-}(CH_2CH{=}CH_2)\text{cyclohexanone}$$

2-Allylcyclohexanone

(d)

$$\text{cyclohexanone} \xrightarrow[2.\ \text{excess}\ CH_3I]{1.\ \text{excess}\ K^+\ {}^-OC(CH_3)_3} \text{2,6-}(H_3C)_2\text{,2,6-}(CH_3)_2\text{cyclohexanone}$$

2,2,6,6-Tetramethylcyclohexanone

This alkylation can also be carried out using LDA as the base.

Visualizing Chemistry

21.17 (a) Check to see if the desired product is a methyl ketone or a substituted carboxylic acid. (The target molecule is a methyl ketone.) Next, identify the halide or halides that react with acetoacetic ester. (The halide is 1-bromo-3-methyl-2-butene.) Formulate the reaction, remembering to include a decarboxylation step.

from halide $(CH_3)_2C{=}CHCH_2$ | $CH_2\overset{O}{\overset{\|}{C}}CH_3$ from acetoacetic ester

$$CH_3\overset{O}{\overset{\|}{C}}CH_2COOEt \xrightarrow[2.\ (CH_3)_2C=CHCH_2Br]{1.\ Na^+\ ^-OEt} (CH_3)_2C{=}CHCH_2{-}CH(COOEt)\overset{O}{\overset{\|}{C}}CH_3 + NaBr$$

$$\xrightarrow{H_3O^+,\ \text{heat}} CO_2 + EtOH + (CH_3)_2C{=}CHCH_2{-}CH_2\overset{O}{\overset{\|}{C}}CH_3$$

6-Methyl-5-hepten-2-one

(b) This product is formed from the reaction of malonic ester with both benzyl bromide and bromomethane.

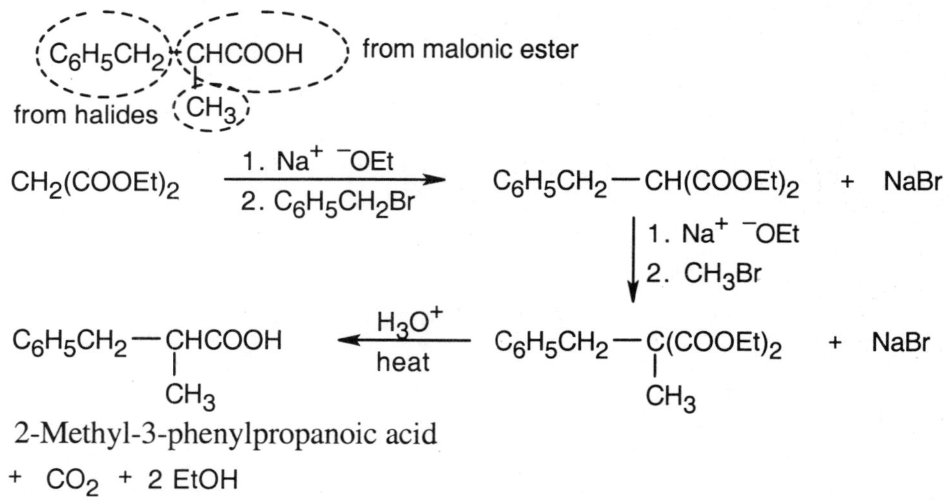

22.18

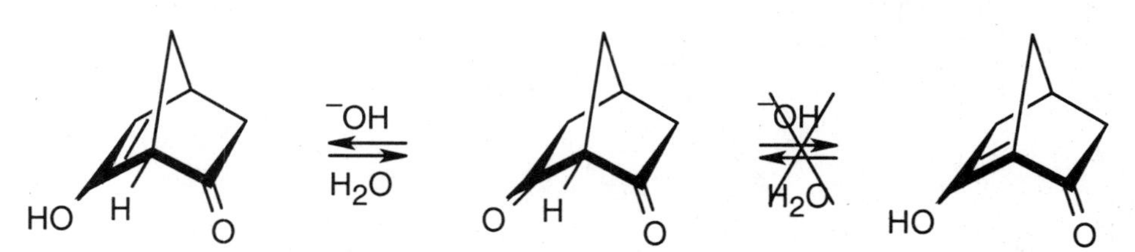

Ordinarily, β-diketones are acidic because of enolization to form an enolate that can be stabilized by delocalization over both carbonyl groups. In this case, loss of the proton at the bridgehead carbon doesn't occur because of the ring strain that would be produced by a bridgehead double bond. Instead, enolization takes place in the opposite direction, and the diketone resembles acetone, rather than a β-diketone, in it pK_a and degree of ionization.

22.19

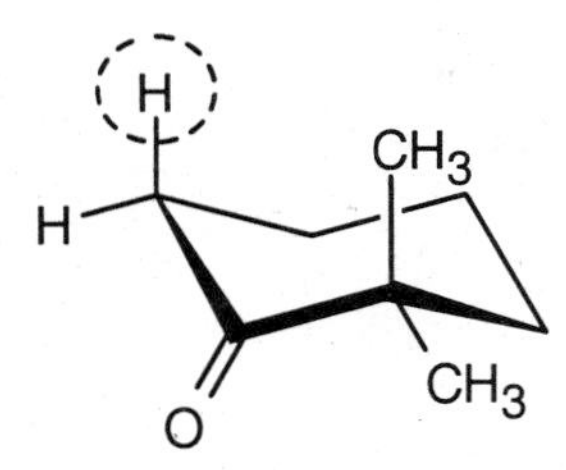

Enolization can occur on only one side of the carbonyl group because of the two methyl groups. The circled axial hydrogen is more acidic because the p orbital that remains after its removal is aligned for optimum overlap with the π electrons of the carbonyl oxygen.

Additional Problems

22.20 Acidic hydrogens are bold. The most acidic hydrogens are the two between the carbonyl groups in (b) and the hydroxyl hydrogen in (c). Structure (d) has no acidic hydrogens. The hydrogens in (c) that are bonded to the methyl group in (c) are acidic (draw resonance forms to prove it).

(a) $CH_3CH_2\mathbf{C}\mathbf{H}(CH_3)C(=O)CH_3$

(b) H H, O, O, H, H, H, H

(c) $HOCH_2CH_2C(=O)C{\equiv}CCH_3$

(d) $C_6H_4(COOCH_3)_2$

no acidic hydrogens

(e) O, C, Cl, **H**

(f) $CH_3CH_2C(=O)C(CH_3){=}CH_2$

22.21 Check your answer by using Table 22.1.

Least Acidic ⟶ *Most Acidic*

$(CH_3CH_2)_2NH < CH_3\overset{O}{\overset{\|}{C}}CH_3 < CH_3CH_2OH < CH_3\overset{O}{\overset{\|}{C}}CH_2\overset{O}{\overset{\|}{C}}CH_3 < CH_3CH_2\overset{O}{\overset{\|}{C}}OH < CCl_3\overset{O}{\overset{\|}{C}}OH$

22.22

(a) :O: :O: H3C C C C CH3 H ⟷ :O:⁻ :O: H3C C=C C CH3 H ⟷ :O: :O:⁻ H3C C C=C CH3 H

(b)

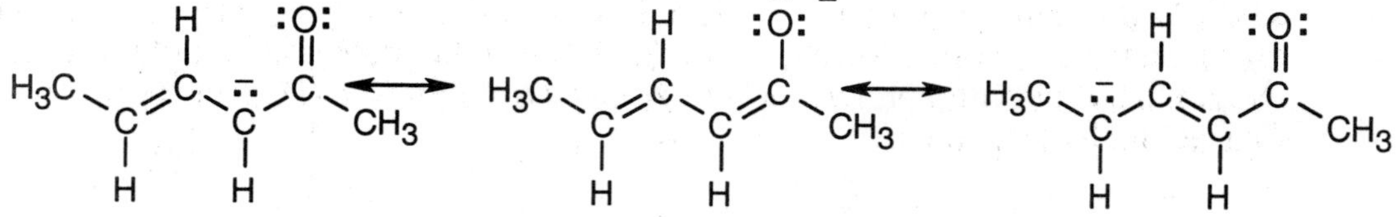

(c)

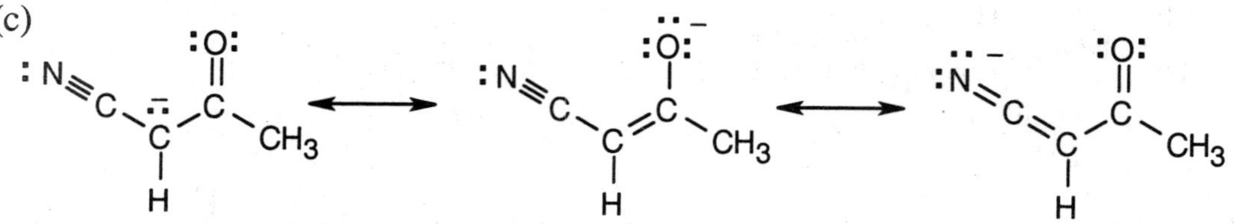

(d)

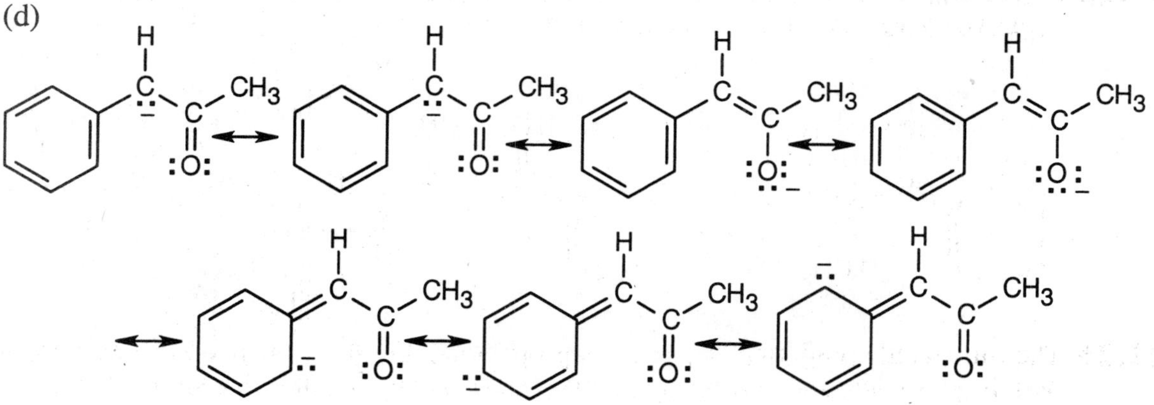

(e)

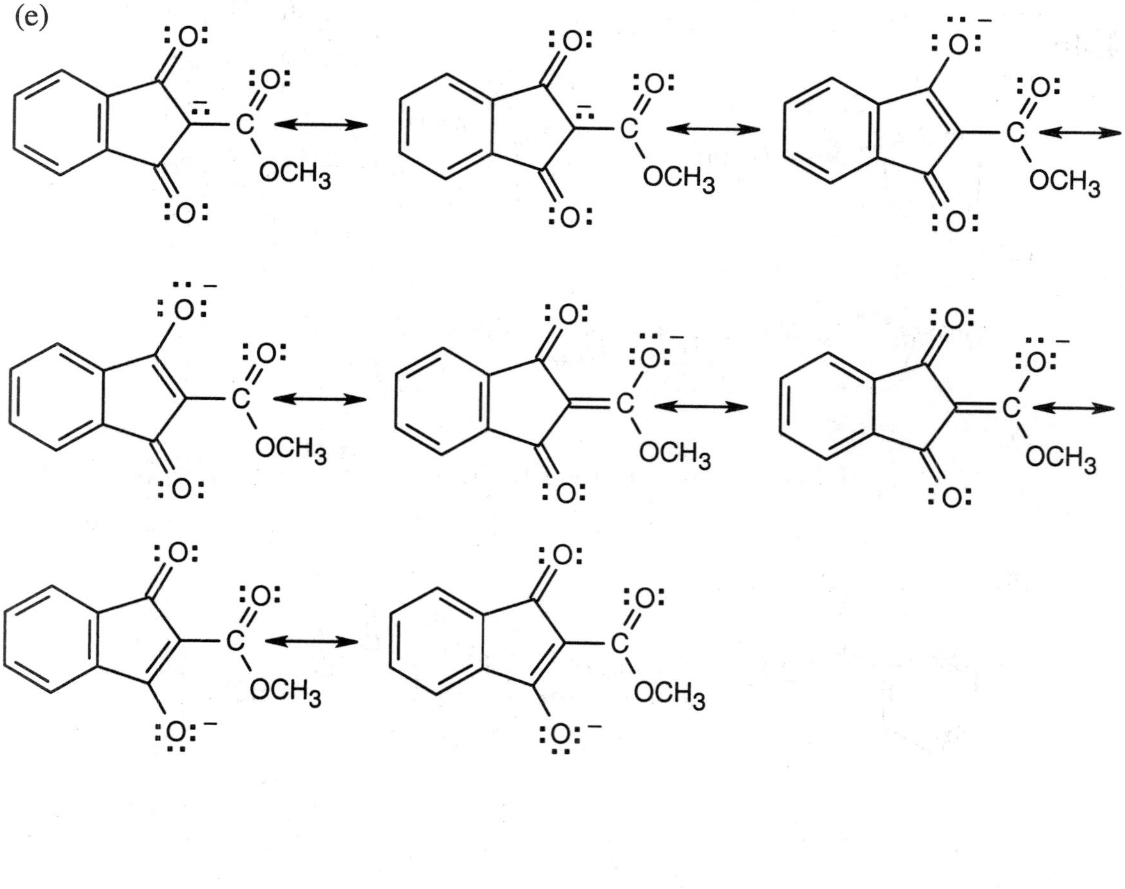

22.23 When a compound containing acidic hydrogen atoms is treated with NaOD in D_2O, all acidic hydrogens are gradually replaced by deuteriums. For each proton (atomic weight 1) lost, a deuteron (atomic weight 2) is added. Since the molecular weight of cyclohexanone increases by four after $NaOD/D_2O$ treatment (from 98 to 102), cyclohexanone contains four acidic hydrogen atoms.

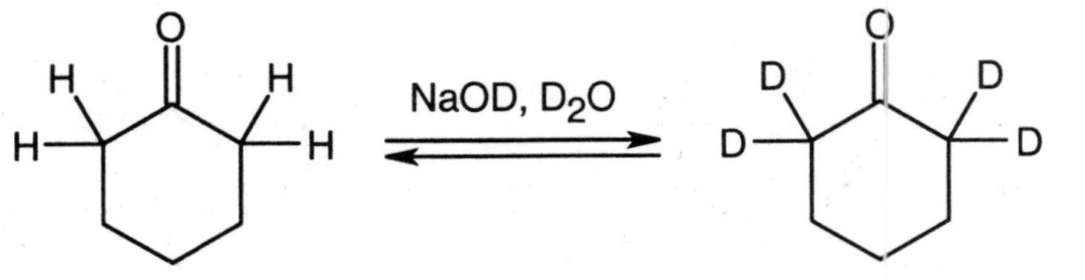

NaOD, D_2O

22.24 Enolization at the γ position produces an anion that is stabilized by delocalization of the negative charge over the π system of five atoms.

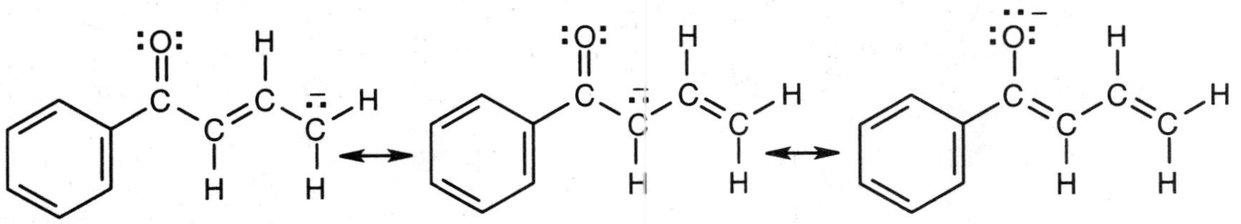

22.25 The illustrated compound, 1-phenyl-2-propenone, doesn't yield an anion when treated with base because the hydrogen on the α carbon is vinylic and isn't acidic (check Table 22.1 for acidity constants).

22.26

(a)

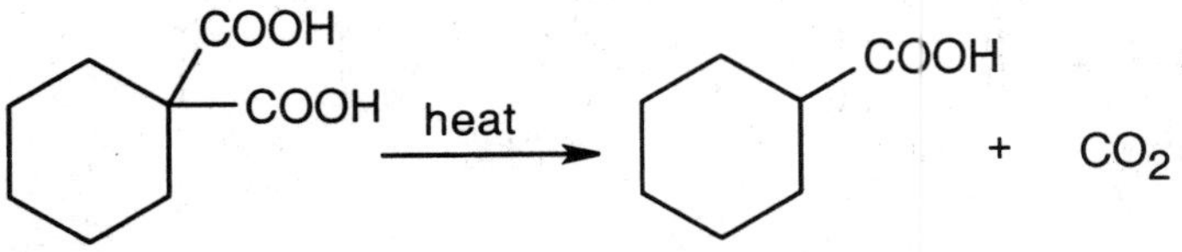

COOH, COOH → heat → COOH + CO_2

(b)

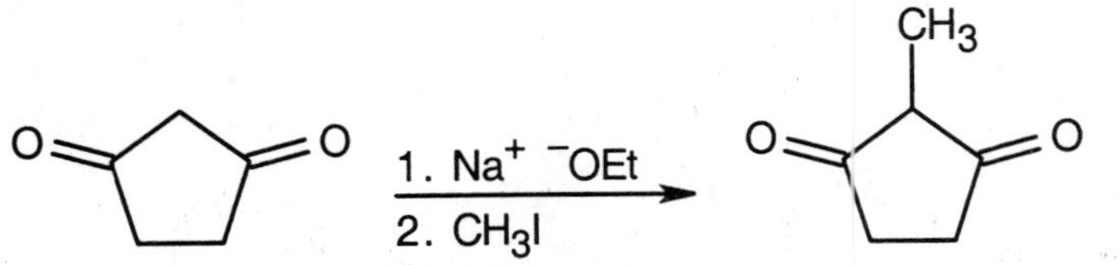

1. $Na^+ \ {}^-OEt$
2. CH_3I

CH_3

(c)

$CH_3CH_2CH_2COOH \xrightarrow[Br_2]{PBr_3} CH_3CH_2CHBrCOBr$ (**A**) $\xrightarrow{H_2O} CH_3CH_2CHBrCOOH$ (**B**)

(d)

$\xrightarrow[I_2]{^-OH,\ H_2O}$ + HCI_3

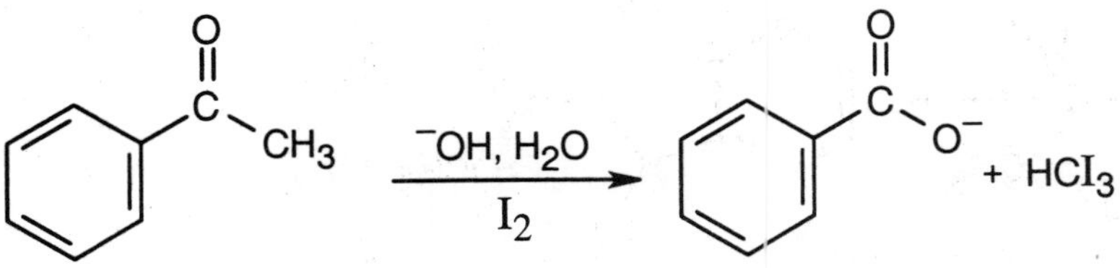

22.27 Since both base-promoted chlorination and base-promoted bromination occur at the same rate, the step involving halogen must come after the slow, or rate-limiting, step. Formation of the enolate ion is the rate-limiting step and is dependent only on the concentrations of ketone and base.

rate = *k* [base] [ketone]

22.28

(a)

$$CH_2(COOEt)_2 \xrightarrow[\text{2. } CH_3CH_2CH_2Br]{\text{1. } Na^+ \ ^-OEt} CH_3CH_2CH_2{-}CH(COOEt)_2 + NaBr$$

$$\downarrow H_3O^+, \text{heat}$$

$$\underset{\text{Ethyl pentanoate}}{CH_3CH_2CH_2CH_2COOEt} \xleftarrow[H^+]{EtOH} CH_3CH_2CH_2CH_2COOH + CO_2 + 2\ EtOH$$

(b)It would be difficult to prepare ethyl 3-methylpentanoate by a malonic ester synthesis. The halide needed (2-bromobutane) is a secondary halide that undergoes elimination, as well as substitution, when treated with strong base.

(c)

$$CH_2(COOEt)_2 \xrightarrow[\text{2. } CH_3CH_2Br]{\text{1. } Na^+ \ ^-OEt} CH_3CH_2CH(COOEt)_2 + NaBr \xrightarrow[\text{2. } CH_3Br]{\text{1. } Na^+ \ ^-OEt} CH_3CH_2C(CH_3)(COOEt)_2 + NaBr$$

$$\downarrow H_3O^+, \text{heat}$$

$$CH_3CH_2CH(CH_3)COOEt \xleftarrow[H^+]{EtOH} CH_3CH_2CH(CH_3)COOH + CO_2 + 2\ EtOH$$

Ethyl 2-methylbutanoate

(d) The malonic acid synthesis can't be used to synthesize carboxylic acids that are trisubstituted at the alpha position.

22.29

(a)

$$CH_3COCH_2COOEt \xrightarrow[\text{2. 2 } CH_3CH_2Br]{\text{1. 2 } Na^+ \ ^-OEt} (CH_3CH_2)_2C(COOEt)COCH_3 + 2\ NaBr \xrightarrow[\text{heat}]{H_3O^+} (CH_3CH_2)_2CHCOCH_3 + CO_2 + EtOH$$

3-Ethyl-2-pentanone

(b)

$$CH_3COCH_2COOEt \xrightarrow[\text{2. } CH_3CH_2CH_2Br]{\text{1. } Na^+ \ ^-OEt} CH_3CH_2CH_2CH(COOEt)COCH_3 + NaBr$$

$$\downarrow \text{1. } Na^+ \ ^-OEt; \text{ 2. } CH_3Br$$

$$CH_3CH_2CH_2CH(CH_3){-}COCH_3 \xleftarrow[\text{heat}]{H_3O^+} CH_3CH_2CH_2C(CH_3)(COOEt){-}COCH_3 + NaBr$$

3-Methyl-2-hexanone

$+ CO_2 + EtOH$

22.30 Use a malonic ester synthesis if the product you want is an α–substituted carboxylic acid or derivative. Use an acetoacetic acid synthesis if the product you want is an α–substituted methyl ketone.

(a)

$$CH_2(COOEt)_2 \xrightarrow[\text{2. 2 } CH_3Br]{\text{1. 2 } Na^+ \ ^-OEt} CH_3C(CH_3)(COOEt)_2 + 2\ NaBr$$

(b)

$$CH_3COCH_2COOEt \xrightarrow[\text{2. } BrCH_2(CH_2)_4CH_2Br]{\text{1. 2 } Na^+ \ ^-OEt} \text{1-acetylcycloheptane-1-carboxylic ester (}C(COCH_3)(COOEt)\text{ ring)} + NaBr$$

$$\downarrow H_3O^+, \text{heat}$$

$$\text{cycloheptyl-}CCH_3\text{ (C=O), H} + CO_2 + EtOH$$

(c)

$$CH_2(COOEt)_2 \xrightarrow[\text{2. } BrCH_2CH_2CH_2Br]{\text{1. 2 } Na^+ \ ^-OEt} \text{cyclobutane-1,1-(COOEt)}_2 + 2\ NaBr$$

$$\downarrow H_3O^+, \text{heat}$$

$$\text{cyclobutane-COOH, H} + CO_2 + 2\ EtOH$$

(d)

$$CH_3COCH_2COOEt \xrightarrow[\text{2. } H_2C{=}CHCH_2Br]{\text{1. } Na^+ \ ^-OEt} H_2C{=}CHCH_2CH(COOEt)CCH_3\ (C{=}O) + NaBr$$

$$\downarrow H_3O^+, \text{heat}$$

$$H_2C{=}CHCH_2CH_2CCH_3\ (C{=}O) + CO_2 + EtOH$$

H_3O^+, heat

22.31 The haloform reaction occurs only with methyl ketones.

Positive haloform reaction:

(a) CH_3COCH_3, (b) $C_6H_5COCH_3$

Negative haloform reaction:

(c) CH_3CH_2CHO, (d) CH_3COOH, (e) CH_3CN

22.32 Reaction of (*R*)-2-methylcyclohexanone with aqueous base is shown below. Reaction with aqueous acid proceeds by a related mechanism.

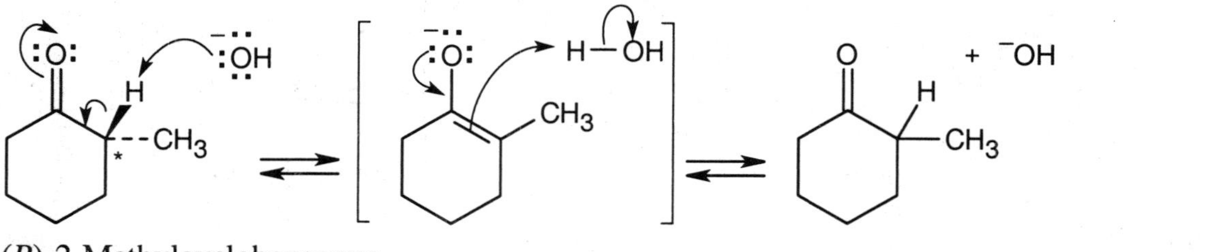

(*R*)-2-Methylcyclohexanone

Carbon 2 loses its chirality when the enolate ion double bond is formed. Protonation occurs with equal probability from either side of sp^2–hybridized carbon 2, resulting in racemic product.

22.33

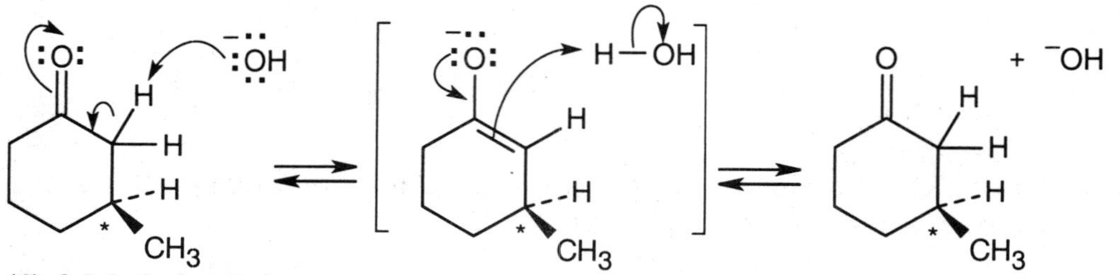

(*S*)-3-Methylcyclohexanone

(*S*)-3-Methylcyclohexanone isn't racemized by base because its chirality center is not involved in the enolization reaction.

22.34 The Hell-Volhard-Zelinskii reaction involves formation of an intermediate acid bromide enol, with loss of stereochemical configuration at the chirality center. Bromination of (*R*)-2-phenylpropanoic acid can occur from either face of the enol double bond, producing racemic 2-bromo-2-phenylpropanoic acid. If the molecule had a chirality center that didn't take part in enolization (Problem 22.33), the product would be optically active.

22.35

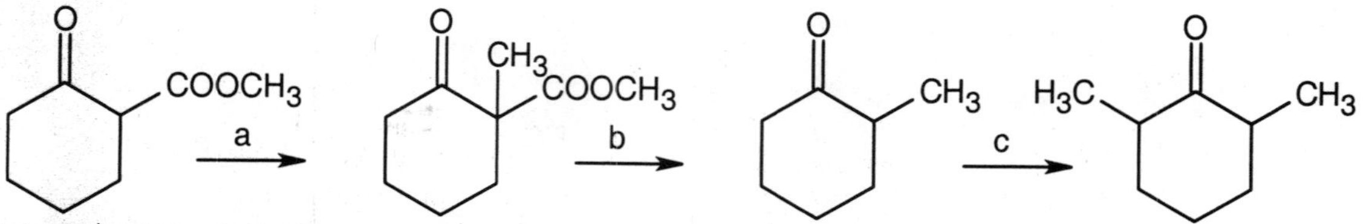

(a) $Na^+ \; ^-OEt$, then CH_3I; (b) H_3O^+, heat; (c) LDA, then CH_3I

22.36

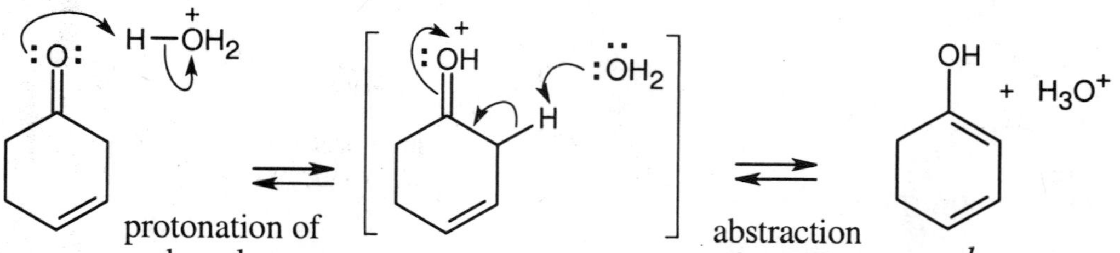

protonation of carbonyl oxygen

abstraction of α proton

enol

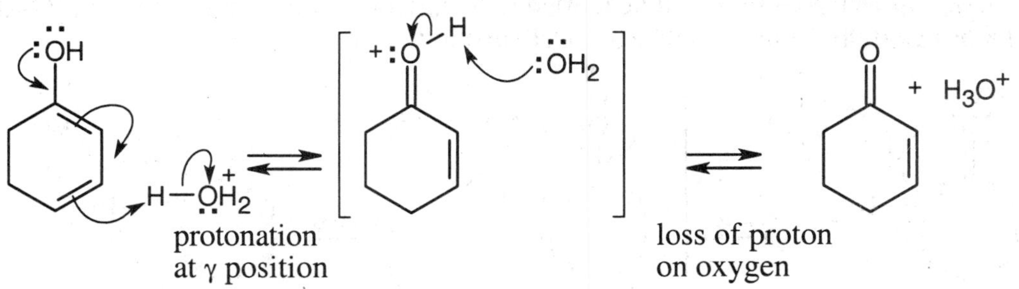

protonation at γ position

loss of proton on oxygen

22.37

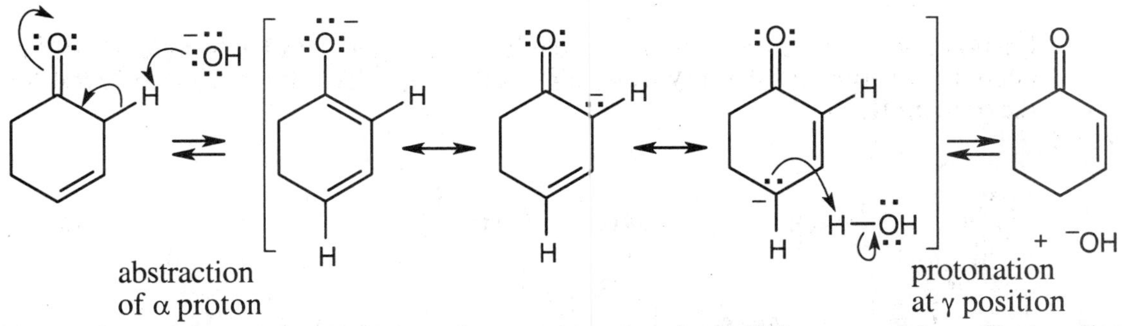

abstraction of α proton

protonation at γ position

The enolate of 3-cyclohexenone can be protonated at three different positions. Protonation at the γ position yields the α,ß-unsaturated ketone.

22.38

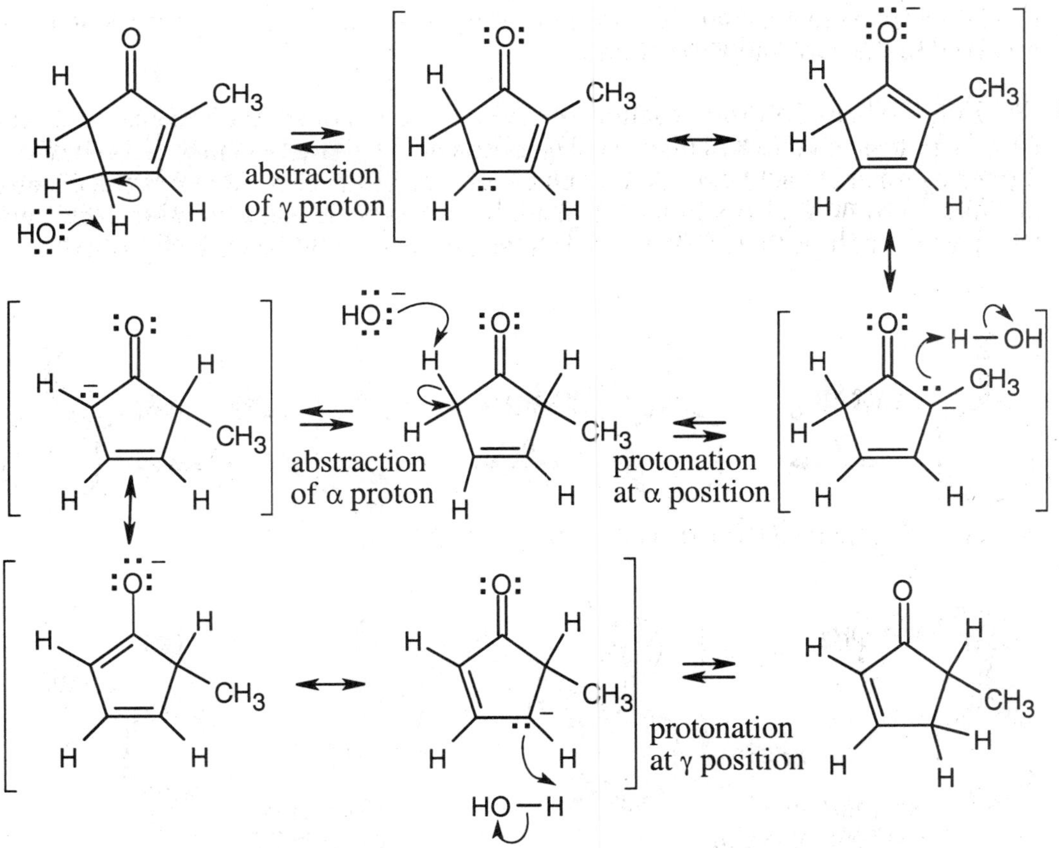

abstraction of γ proton

abstraction of α proton

protonation at α position

protonation at γ position

All protons in the five-membered ring can be exchanged by base treatment.

22.39 Protons α to a carbonyl group or γ to an enone carbonyl group are acidic (Problem 22.38). Thus for 2-methyl-2-cyclopentenone, protons at the starred positions are acidic.

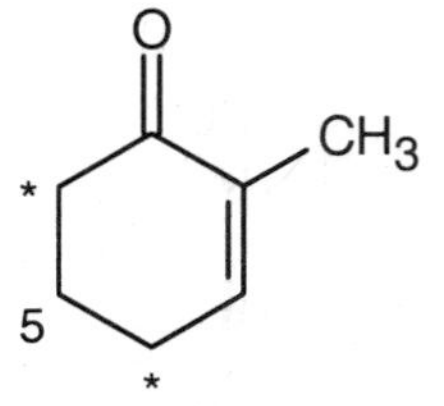

Isomerization of a 2–substituted 2–cyclohexenone to a 6–substituted 2–cyclohexenone requires removal of a proton from the 5–position of the 2–substituted isomer. Since protons in this position are not acidic, double bond isomerization does not occur.

22.40 A nitroso compound is analogous to a carbonyl compound. If there are hydrogens α to the nitroso group, enolization similar to that observed for carbonyl compounds can occur, leading to formation of an oxime. If no hydrogens are adjacent to the nitroso group, enolization to the oxime can't occur, and the nitroso compound is stable.

22.41 First treat geraniol with PBr_3 to form $(CH_3)_2C{=}CHCH_2CH_2C(CH_3){=}CHCH_2Br$ (geranyl bromide).

(a)

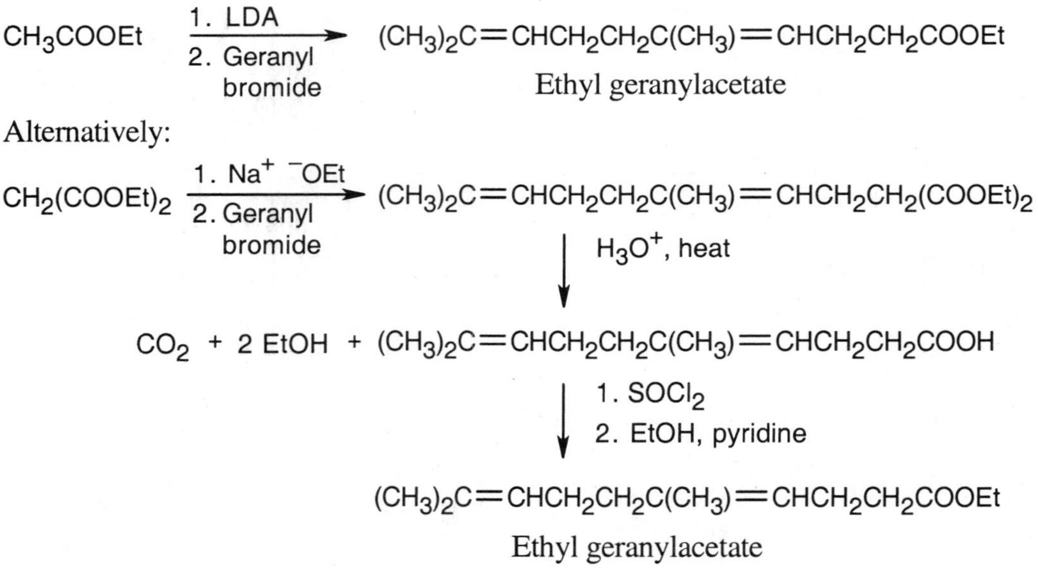

$$CH_3COOEt \xrightarrow[\text{2. Geranyl bromide}]{\text{1. LDA}} (CH_3)_2C{=}CHCH_2CH_2C(CH_3){=}CHCH_2CH_2COOEt$$

Ethyl geranylacetate

Alternatively:

$$CH_2(COOEt)_2 \xrightarrow[\text{2. Geranyl bromide}]{\text{1. } Na^+ \ ^-OEt} (CH_3)_2C{=}CHCH_2CH_2C(CH_3){=}CHCH_2CH_2(COOEt)_2$$

$\downarrow$ H_3O^+, heat

$$CO_2 + 2\ EtOH + (CH_3)_2C{=}CHCH_2CH_2C(CH_3){=}CHCH_2CH_2COOH$$

$\downarrow$ 1. $SOCl_2$ 2. EtOH, pyridine

$$(CH_3)_2C{=}CHCH_2CH_2C(CH_3){=}CHCH_2CH_2COOEt$$

Ethyl geranylacetate

(b)

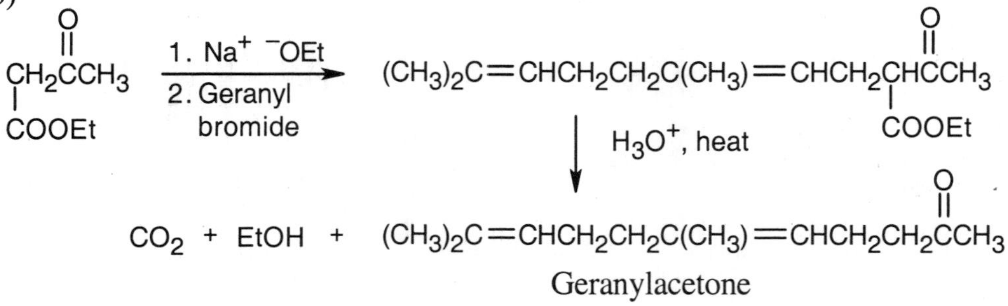

$$\underset{\displaystyle COOEt}{CH_2\overset{O}{\overset{\|}{C}}CH_3} \xrightarrow[\text{2. Geranyl bromide}]{\text{1. } Na^+ \ ^-OEt} (CH_3)_2C{=}CHCH_2CH_2C(CH_3){=}CHCH_2\underset{\displaystyle COOEt}{CH}\overset{O}{\overset{\|}{C}}CH_3$$

$\downarrow$ H_3O^+, heat

$$CO_2 + EtOH + (CH_3)_2C{=}CHCH_2CH_2C(CH_3){=}CHCH_2CH_2\overset{O}{\overset{\|}{C}}CH_3$$

Geranylacetone

22.42

(a)

Cyclohexanone $\xrightarrow[\text{THF}]{Ph_3\overset{+}{P}-\overset{-}{C}H_2}$ methylenecyclohexane ($=CH_2$)

(b)

product of (a) $\xrightarrow[\text{2. } H_2O_2,\ ^{-}OH]{\text{1. } BH_3\text{, THF}}$ cyclohexyl–CH_2OH $\xrightarrow{PBr_3}$ cyclohexyl–CH_2Br

(c)

Cyclohexanone $\xrightarrow[\text{2. } C_6H_5CH_2Br]{\text{1. LDA}}$ 2-($CH_2C_6H_5$)cyclohexanone

(d)

$CH_2(COOEt)_2$ $\xrightarrow[\text{2. product of (b)}]{\text{1. } Na^+\ ^{-}OEt}$ cyclohexyl–$CH_2CH(COOEt)_2$ $\xrightarrow{H_3O^+\text{, heat}}$ cyclohexyl–CH_2CH_2COOH + 2 EtOH + CO_2

(e)

Cyclohexanone $\xrightarrow[H_3O^+]{\text{NaCN}}$ 1-hydroxycyclohexane-1-carbonitrile (HO, CN) $\xrightarrow[\text{heat}]{H_3O^+}$ cyclohex-1-ene-1-COOH

Warm aqueous acid both hydrolyzes the nitrile and dehydrates the alcohol.

(f)

Cyclohexanone $\xrightarrow[CH_3COOH]{Br_2}$ 2-bromocyclohexanone (Br) $\xrightarrow[\text{heat}]{\text{Pyridine}}$ cyclohex-2-enone

(g)

cyclohex-1-ene-1-COOH (from (e)) $\xrightarrow[\text{Pd/C}]{H_2}$ cyclohexane-COOH

22.43 Treatment of either the cis or trans isomer with base causes enolization α to the carbonyl group and results in loss of configuration at the α–position. Reprotonation at carbon 2 produces either of the diastereomeric 4-*tert*-butyl-2-methylcyclohexanones. In both diastereomers the *tert*-butyl group of carbon 4 occupies the equatorial position for steric reasons. The methyl group of the cis isomer is also equatorial, but the methyl group of the trans isomer is axial. The trans isomer is less stable because of 1,3-diaxial interactions of the methyl group with the ring protons.

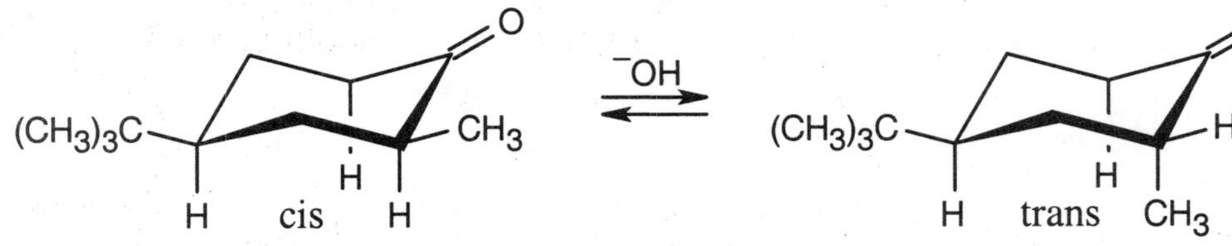

$(CH_3)_3C$ ... CH_3 (cis) $\underset{}{\overset{^-OH}{\rightleftharpoons}}$ $(CH_3)_3C$ ... H, CH_3 (trans)

22.44 (a) Reaction with Br_2 at the α position occurs only with aldehydes and ketones, not with esters.

(b) Aryl halides can't be used in malonic ester syntheses because they don't undergo S_N2 displacement reactions.

(c) The product of this reaction sequence, $H_2C{=}CHCH_2CH_2COCH_3$, is a methyl ketone, not a carboxylic acid.

22.45 The reaction of cyclohexanone and *tert*–butylmagnesium bromide gives the expected carbonyl addition product. The yield of the *tert*–butylmagnesium bromide addition product is very low, however, because of the difficulty of approach of the bulky *tert*- butyl Grignard reagent to the carbonyl carbon. More favorable is the acid-base reaction between the Grignard reagent and a carbonyl α proton.

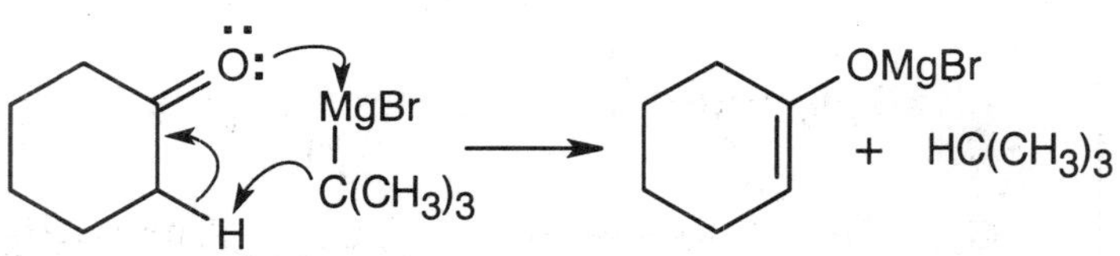

Cyclohexanone + MgBr–$C(CH_3)_3$ ⟶ 1-cyclohexenyl–OMgBr + $HC(CH_3)_3$

When D_3O^+ is added to the reaction mixture, the deuterated ketone is produced.

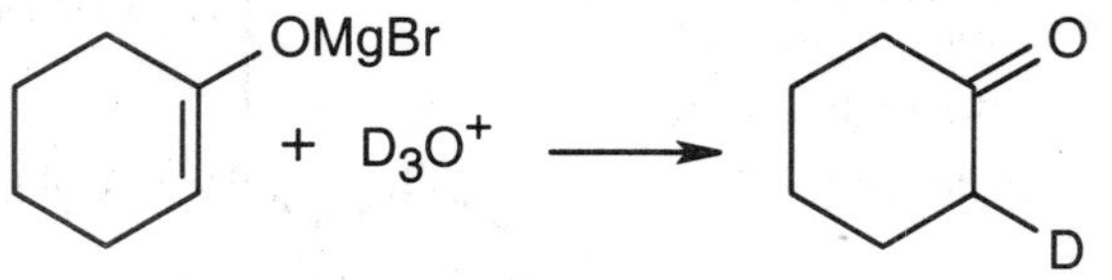

1-cyclohexenyl–OMgBr + D_3O^+ ⟶ 2-deuteriocyclohexanone

22.46

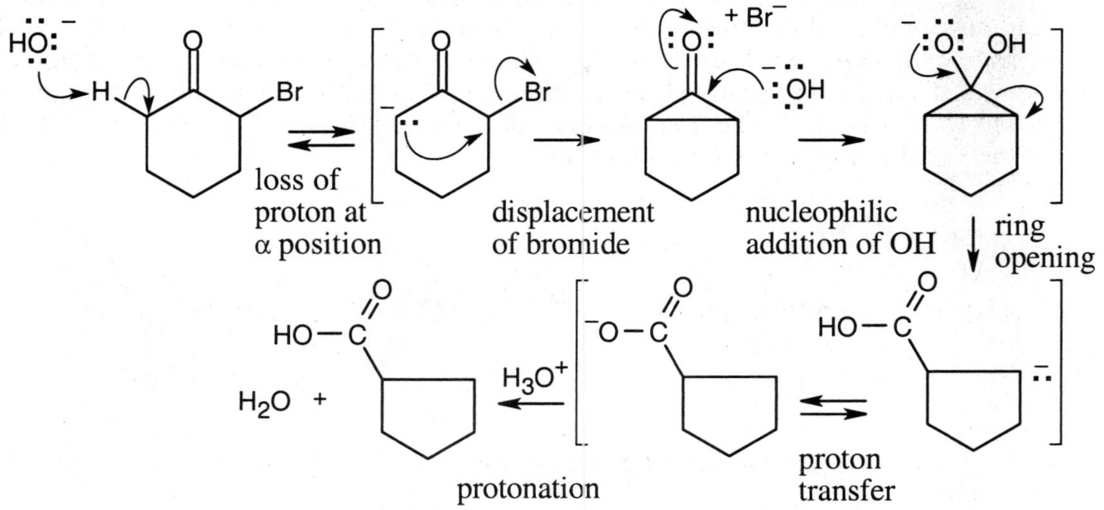

loss of proton at α position

displacement of bromide

nucleophilic addition of OH

ring opening

proton transfer

protonation

H_3O^+

H_2O +

22.47

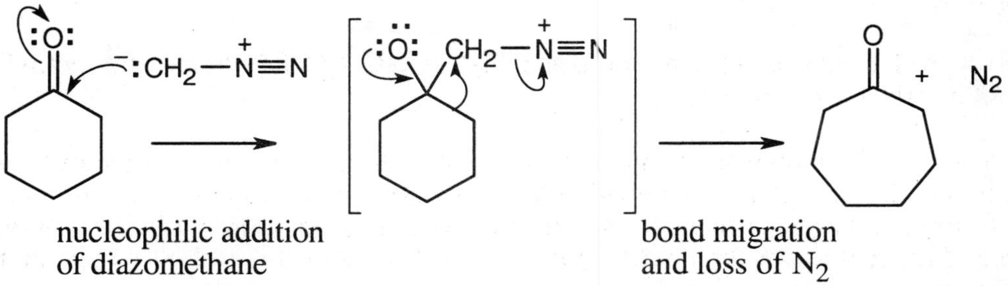

nucleophilic addition of diazomethane

bond migration and loss of N_2

+ N_2

22.48

HCl

acid-catalyzed enolization (Figure 22.1)

attack of enol π electrons on phenylselenyl chloride, with loss of Cl^-

loss of proton

+ H^+

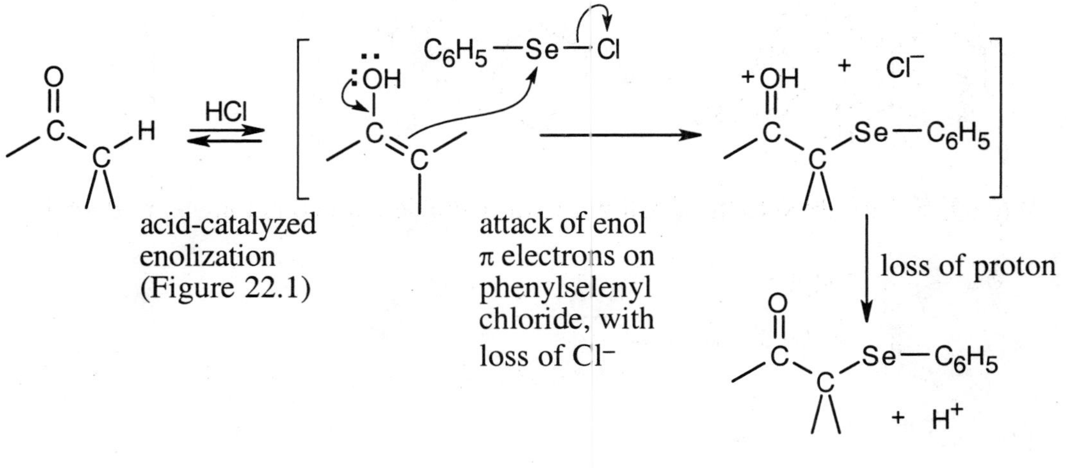

22.49 Start at the end of the sequence of reactions and work backwards. If necessary, cover up pieces of information you are not using at the moment to keep them from distracting you.

(a) Because the *keto acid* $C_9H_{13}NO_3$ loses CO_2 on heating, it must be a ß-keto acid. Neglecting stereoisomerism, we can draw the structure of the ß-keto acid as:

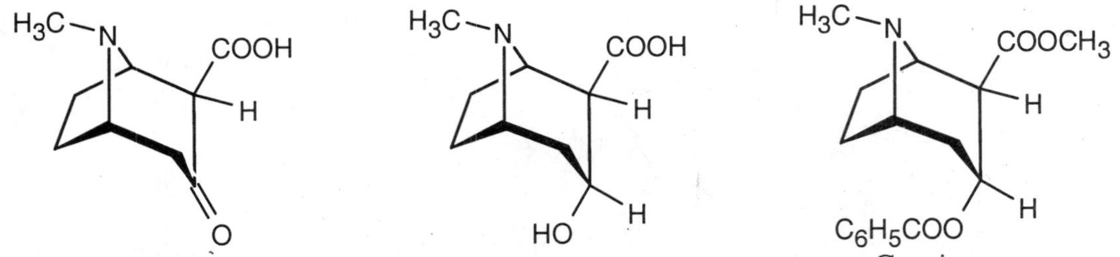

H_3C—N, COOH, H, O
Keto acid

H_3C—N, COOH, H, HO, H
Ecgonine

H_3C—N, $COOCH_3$, H, C_6H_5COO, H
Cocaine

(b) When ecgonine ($C_9H_{15}NO_3$) is treated with CrO_3, the keto acid $C_9H_{13}NO_3$ is produced. Since CrO_3 is used for oxidizing alcohols to carbonyl compounds, ecgonine has the following structure. Again, the stereochemistry is unspecified.

(c) Ecgonine contains carboxylic acid and alcohol functional groups. The other products of hydroxide treatment of cocaine are a carboxylic acid (benzoic acid) and an alcohol (methanol). Cocaine thus contains two ester functional groups, which are saponified on reaction with hydroxide.

The complete reaction sequence:

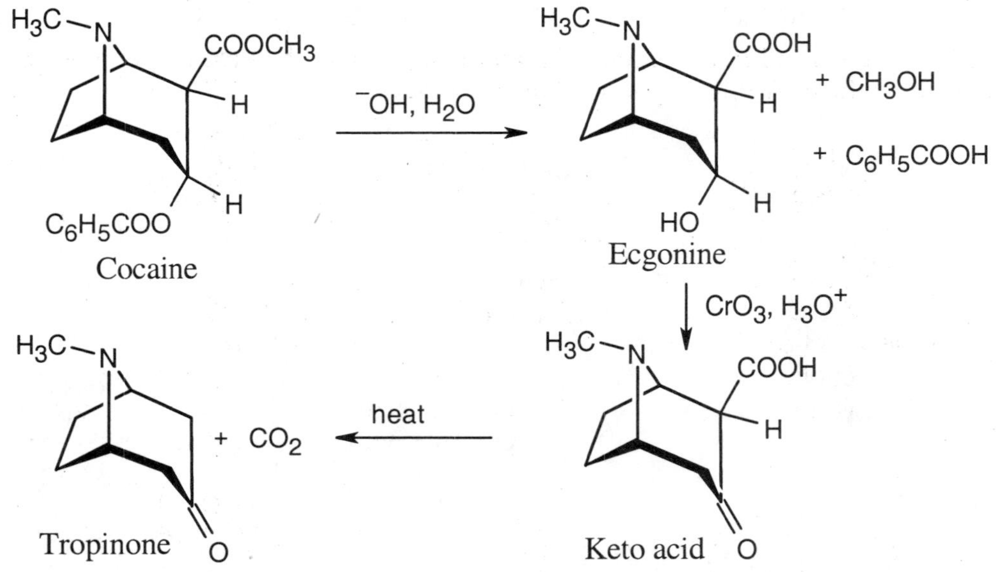

Cocaine (H_3C—N, $COOCH_3$, H, C_6H_5COO, H) $\xrightarrow{^-OH,\ H_2O}$ Ecgonine (H_3C—N, COOH, H, HO, H) + CH_3OH + C_6H_5COOH

Ecgonine $\xrightarrow{CrO_3,\ H_3O^+}$ Keto acid (H_3C—N, COOH, H, O)

Keto acid $\xrightarrow{heat}$ Tropinone (H_3C—N, O) + CO_2

22.50 Laurene differs in stereochemical configuration from the starting material at the carbon α to the methylene group. Since this position is α to the carbonyl group in the starting material, enolization and isomerization must have occurred during the reaction.

Isomerization of the starting ketone is brought about by a reversible reaction with the basic Wittig reagent, which yields an equilibrium mixture of two diastereomeric ketones. The isomeric ketone then reacts preferentially with the Wittig reagent to give only the observed product.

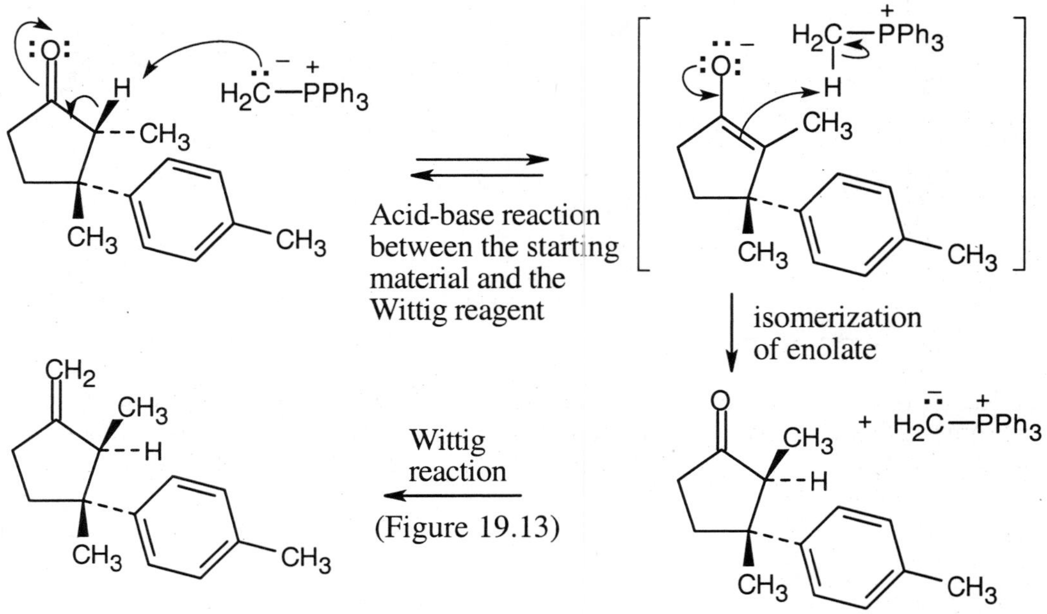

Acid-base reaction between the starting material and the Wittig reagent

isomerization of enolate

Wittig reaction (Figure 19.13)

22.51 The key step is an intramolecular alkylation reaction of the ketone α carbon, with the tosylate in the adjacent ring serving as the leaving group.

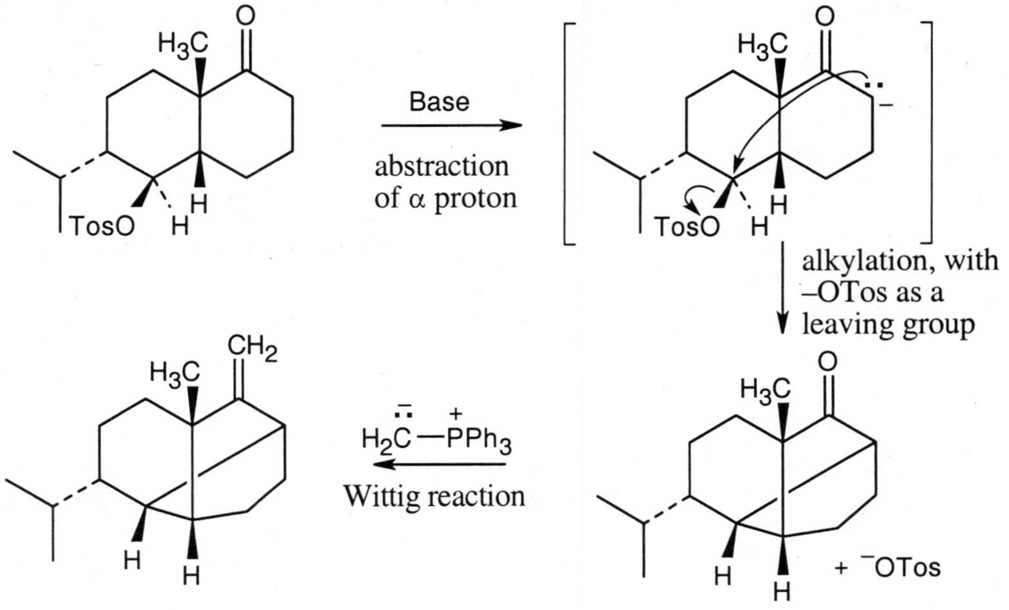

Base

abstraction of α proton

alkylation, with –OTos as a leaving group

$H_2\ddot{C}^{-}-\overset{+}{P}Ph_3$

Wittig reaction

A Look Ahead

22.52

$$CH_3\overset{O}{\overset{\|}{C}}NH\overset{H}{\overset{|}{C}}COOEt \text{ (with } COOEt \text{ on the } \alpha\text{-carbon)} \xrightarrow[\text{formation of enolate}]{Na^+\ ^-OEt} \left[CH_3\overset{O}{\overset{\|}{C}}NH\ddot{C}^{-}COOEt \text{ (with } COOEt) \right]$$

$$\downarrow CH_3I \quad \text{alkylation}$$

$$H_2N\underset{CH_3}{\underset{|}{C}}H\overset{O}{\overset{\|}{C}}OH \;\; \text{Alanine} \xleftarrow[\text{ester cleavage, amide cleavage, decarboxylation}]{H_3O^+,\ \text{heat}} CH_3\overset{O}{\overset{\|}{C}}\text{-}NH\overset{H_3C}{\overset{|}{C}}\text{—}\overset{O}{\overset{\|}{C}}\text{-}OEt \text{ (with } COOEt \text{ on the } \alpha\text{-carbon)}$$

$$+ 2\ EtOH + CO_2 + CH_3COOH$$

Acid cleaves both ester bonds, as well as the amide bond, by mechanisms that were shown in Figure 21.10 and Section 21.7. Subsequent decarboxylation of the β-keto acid produces alanine.

22.53

$$CH_2(COOEt)_2 \xrightarrow[2.\ (CH_3)_2CHCH_2Br]{1.\ Na^+\ ^-OEt} (CH_3)_2CHCH_2CH(COOEt)_2$$

$$\downarrow H_3O^+,\ \text{heat}$$

$$(CH_3)_2CHCH_2\overset{Br}{\overset{|}{C}}HCOOH \xleftarrow[2.\ H_2O]{1.\ Br_2,\ PBr_3} (CH_3)_2CHCH_2CH_2COOH + CO_2 + 2\ EtOH$$

$$\downarrow NH_3$$

$$(CH_3)_2CHCH_2\overset{NH_2}{\overset{|}{C}}HCOOH$$

Isoleucine

A malonic ester synthesis is used to form 4-methylpentanoic acid. HVZ bromination of the acid, followed by reaction with ammonia yields isoleucine. The last reaction is an S_N2 displacement of bromide by ammonia.

Molecular Modeling

22.54 2-Butanone is lower in energy than its enol by 80 kJ/mol, and 2,4-pentanedione is lower in energy than its enol, but only by 11 kJ/mol. The enol tautomer is stabilized by resonance (the enol is conjugated) and by intramolecular hydrogen bonding. Cyclohexadieneone is lower in energy by 51 kJ/mol, because it is stabilized by resonance (this molecule is actually phenol).

22.55 The α carbon is much more negative (nucleophilic) in the lithium enolate. The C–O–H bond angle, 112.7°, reflects the fact that sp^3 hybridization of oxygen leads to better orbital overlap and a stronger bond. The C–O–Li bond angle, 178.3°, is nearly linear and reflects the fact that the O–Li bond is ionic and that no orbital overlap is required. Li occupies the position that brings it closest to the most negative region of oxygen.

22.56 The observed lithium enolate (**A**) is 4.2 kJ/mol higher in energy than lithium enolate **B**. The electrostatic potential map of LDA shows that the nitrogen is sterically hindered. The observed enolate is formed by abstracting hydrogen from the more sterically accessible alkyl group. Thus, the observed stereochemistry is controlled by steric factors.

22.57 The two most important enolates (**A** and **C**) are formed by abstraction of a proton α to the carbonyl group and by abstraction of a proton γ to the carbonyl group that has an allylic relationship to the carbon-carbon double bond. Resonance structures show that 5 protons are acidic, and all can exchange with D_2O to produce the D_5 product.

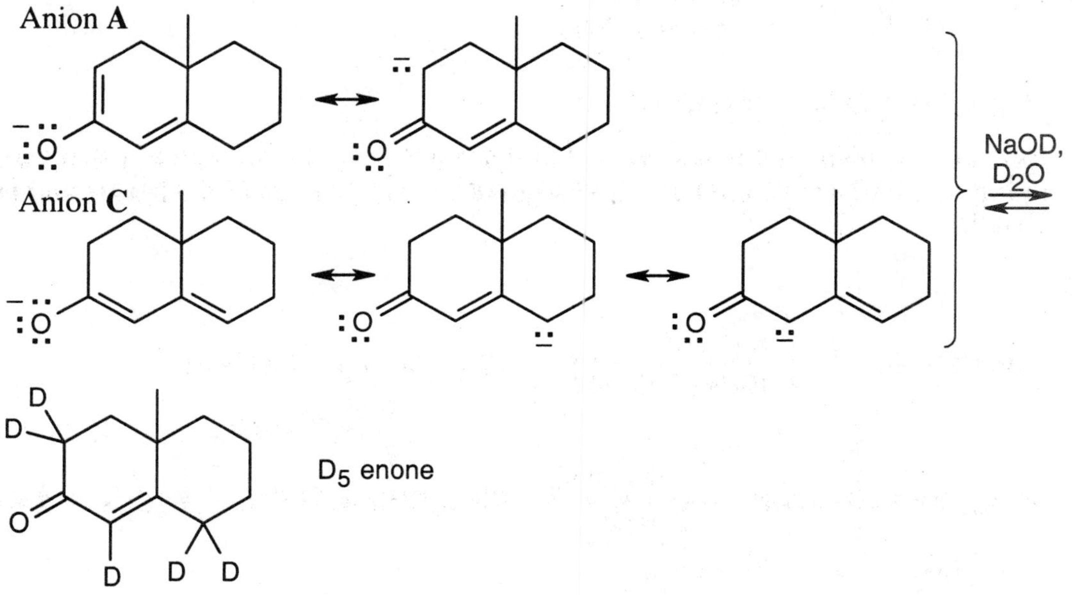

Anion **A**

Anion **C**

NaOD, D_2O

D D D D D

D_5 enone

Chapter 23 – Carbonyl Condensation Reactions

Chapter Outline

I. Mechanism of carbonyl condensation reactions (Section 23.1).
  A. Carbonyl condensation reactions take place between two carbonyl components.
    1. One component (the nucleophilic donor) is converted to its enolate and undergoes an α–substitution reaction.
    2. The other component (the electrophilic acceptor) undergoes nucleophilic addition.
  B. Many kinds of carbonyl compounds undergo carbonyl condensation reactions.
II. The aldol reaction (Sections 23.2 – 23.7).
  A. Characteristics of the aldol reaction (Sections 23.2 – 23.3).
    1. The aldol condensation is a base-catalyzed dimerization of two aldehydes or ketones (Section 23.2).
    2. The reaction can occur between two components that have α hydrogens.
    3. For monosubstituted aldehydes, the equilibrium favors products, but for other aldehydes and ketones, the equilibrium favors reactants.
    4. The reaction occurs by the mechanism described above.
    5. Carbonyl condensation reactions require only a catalytic amount of base (Section 23.3).
      Alpha-substitution reactions, on the other hand, use one equivalent of base.
  B. Dehydration of aldol products (Section 23.4).
    1. Aldol products are easily dehydrated to yield conjugated enones.
      a. Dehydration is catalyzed by both acid and base.
      b. Reaction conditions for dehydration are only slightly more severe than for condensation.
      c. Often, dehydration products are isolated directly from condensation reactions.
    2. Conjugated enones are more stable than nonconjugated enones.
    3. Removal of the water byproduct drives the aldol equilibrium towards product formation.
  C. Aldol products (Sections 23.5 – 23.6).
    1. Using aldol reactions in synthesis (Section 23.5).
      a. Obvious aldol products are:
        i. α,β-Unsaturated aldehydes/ketones.
        ii. β-Hydroxy aldehydes/ketones.
      b. Often, it's possible to work backwards from a product that doesn't resemble an aldol product and recognize aldol components.
    2. Mixed aldol reactions (23.6).
      a. If two similar aldehydes/ketones react under aldol conditions, 4 products may be formed.
      b. A single product can be formed from two different components :
        i. If one carbonyl component has no α-hydrogens.
        ii. If one carbonyl compound is much more acidic than the other.
  D. Intramolecular aldol condensations (Section 23.7).
    1. Treatment of certain dicarbonyl compounds with base can lead to cyclic products.
    2. A mixture of cyclic products may result, but the more strain-free ring is usually formed.
III. The Claisen condensation (Sections 23.8 – 23.10).
  A. Features of the Claisen condensation (Section 23.8).
    1. Treatment of an ester with 1 equivalent of base yields a β-keto ester.
    2. The reaction is reversible and has a mechanism similar to that of the aldol reaction.

3. A major difference from the aldol condensation is the expulsion of an alkoxide ion from the tetrahedral intermediate of the initial Claisen adduct.
4. Because the product is often acidic, one equivalent of base is needed; addition of base drives the reaction to completion.
5. Addition of acid yields the final product.

B Mixed Claisen condensations (Section 23.9).
1. Mixed Claisen condensations of two different esters can succeed if one component has no α hydrogens.
2. Mixed Claisen condensations between a ketone and an ester with no α hydrogens are also successful.

C. Intramolecular Claisen condensations: the Dieckmann cyclization (Section 23.10).
1. The Dieckmann cyclization is used to form cyclic β-keto esters.
   a. 1,6-Diesters form 5-membered rings.
   b. 1,7-Diesters form 6-membered rings.
2. The mechanism is similar to the Claisen condensation mechanism.
3. The product β-keto esters can be further alkylated.
   This is a good route to 2-substituted cyclopentanones and cyclohexanones.

IV. Other carbonyl condensation reactions (Sections 23.11 – 23.14).

A. The Michael reaction (Section 23.11).
1. The Michael reaction is the conjugate addition of an enolate to an α,β-unsaturated carbonyl compound.
   The highest-yielding reactions occur between stable enolates and unhindered α,β-unsaturated carbonyl compounds.
2. The mechanism is a conjugate addition of a nucleophilic enolate to the β carbon of a α,β-carbonyl acceptor.
3. Stable enolates are Michael donors, and α,β-unsaturated compounds are Michael acceptors.

B. The Stork enamine reaction (Section 23.12).
1. A ketone that has been converted to an enamine can act as a Michael donor in a reaction known as the Stork enamine reaction.
2. The sequence of reactions in the Stork enamine reaction:
   a. Enamine formation from a ketone.
   b. Michael-type addition to a α,β-unsaturated carbonyl compound.
   c. Enamine hydrolysis back to a ketone.
3. This sequence is equivalent to the Michael addition of a ketone to an α,β-unsaturated carbonyl compound.

C. The Robinson annulation reaction (Section 23.13).
1. The Robinson annulation reaction combines a Michael reaction with an intramolecular aldol condensation to synthesize substituted ring systems.
2. The components are a nucleophilic donor, such as a β-keto ester, and an α,β-unsaturated ketone acceptor.
3. The intermediate 1,5-diketone undergoes an intramolecular aldol condensation to yield a cyclohexenone.

D. Biological carbonyl condensation reactions (Section 23.14).
1. Many biomolecules are synthesized by carbonyl condensation reactions.
2. Acetyl CoA is the major building block for synthesis of biomolecules.
   a. Acetyl CoA can act as an electrophilic acceptor by being attacked at its carbonyl group.
   b. Acetyl CoA can act as a nucleophilic donor by loss of its acidic α hydrogen.

Solutions to Problems

23.1 When you are first learning the aldol condensation, write all the steps.
(1) Form the enolate of one molecule of the carbonyl compound.

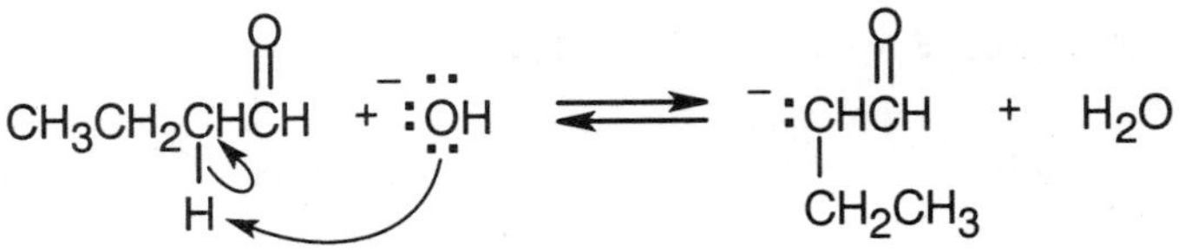

$$CH_3CH_2CH(H)CHO + {}^{-}:\ddot{O}H \rightleftharpoons {}^{-}:CH(CH_2CH_3)CHO + H_2O$$

(2) Have the enolate attack the electrophilic carbonyl of the second carbonyl molecule.

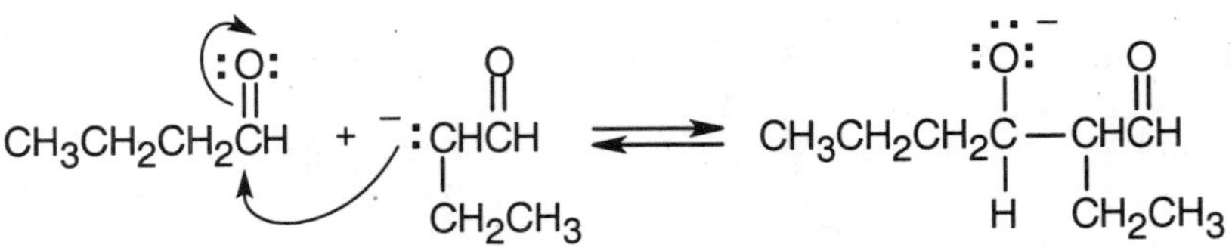

$$CH_3CH_2CH_2CHO + {}^{-}:CH(CH_2CH_3)CHO \rightleftharpoons CH_3CH_2CH_2CH(O^{-})-CH(CH_2CH_3)CHO$$

(3) Protonate the anionic oxygen.

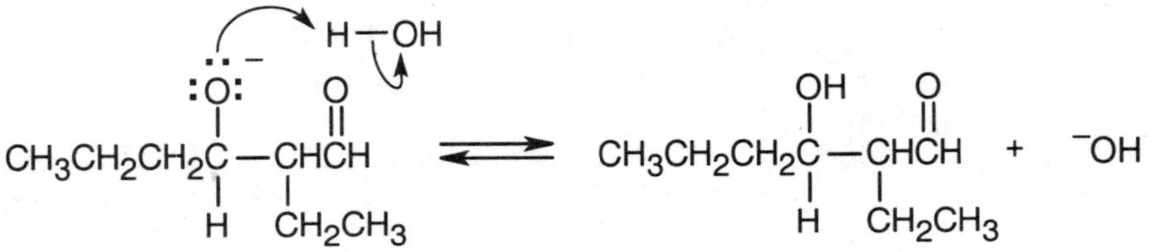

$$CH_3CH_2CH_2CH(O^{-})-CH(CH_2CH_3)CHO + H-OH \rightleftharpoons CH_3CH_2CH_2CH(OH)-CH(CH_2CH_3)CHO + {}^{-}OH$$

Practice writing out these steps for the other aldol condensations.

(b)

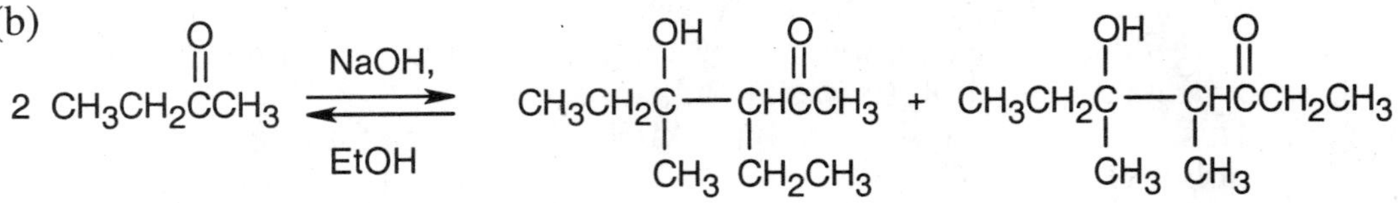

$$2\ CH_3CH_2COCH_3 \xrightarrow[\text{EtOH}]{\text{NaOH,}} CH_3CH_2C(OH)(CH_3)-CH(CH_2CH_3)COCH_3 + CH_3CH_2C(OH)(CH_3)-CH(CH_3)COCH_2CH_3$$

(c)

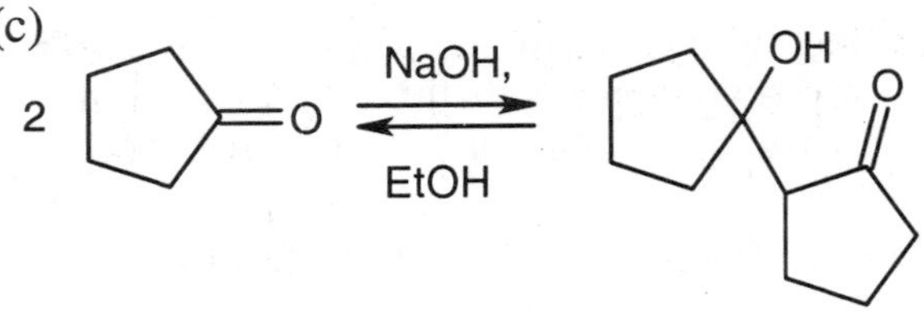

2 cyclopentanone $\xrightarrow[\text{EtOH}]{\text{NaOH,}}$ 2-(1-hydroxycyclopentyl)cyclopentanone (OH, O)

23.2

$$CH_3C(OH)(CH_3)-CH_2COCH_3$$

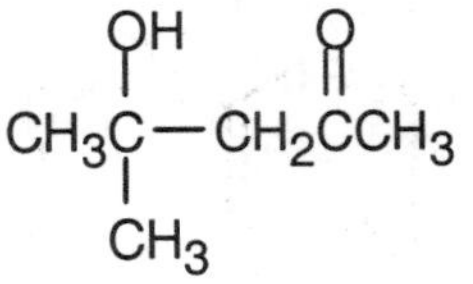

4-Hydroxy-4-methyl-2-pentanone

The steps for the reverse aldol are the reverse of those described in Problem 23.1.

(1) Deprotonate the alcohol oxygen.

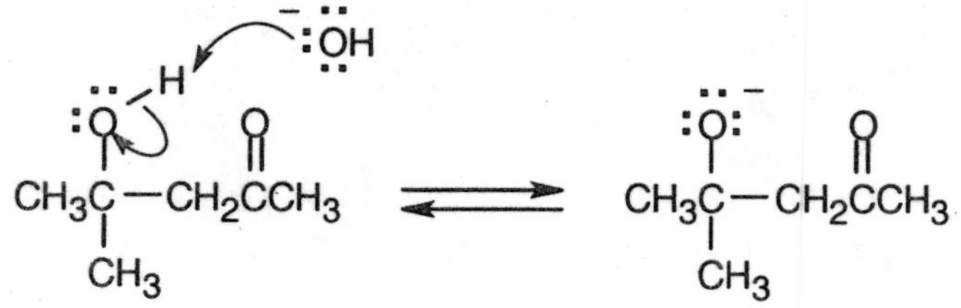

$$CH_3C(OH)(CH_3)-CH_2CCH_3(=O) + {}^-:OH \rightleftharpoons CH_3C(O^-)(CH_3)-CH_2CCH_3(=O)$$

(2) Eliminate the enolate anion.

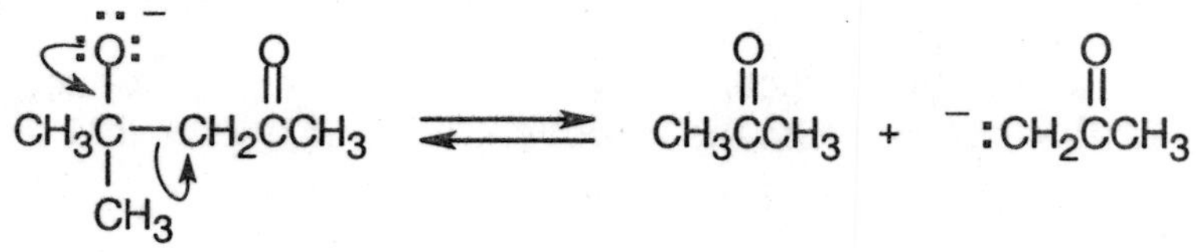

$$CH_3C(O^-)(CH_3)-CH_2CCH_3(=O) \rightleftharpoons CH_3CCH_3(=O) + {}^-:CH_2CCH_3(=O)$$

(3) Reprotonate the enolate anion.

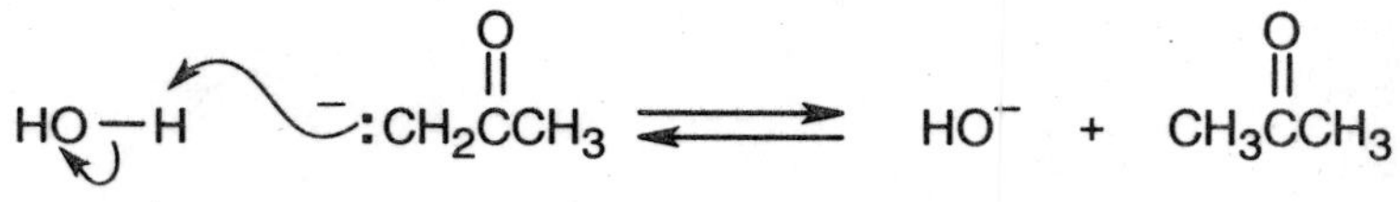

$$HO-H + {}^-:CH_2CCH_3(=O) \rightleftharpoons HO^- + CH_3CCH_3(=O)$$

23.3 As in Problem 23.1, align the two carbonyl compounds so that the location of the new bond is obvious. After drawing the product, eliminate water to form the enone.

(a) 2 cyclopentanone $\xrightleftharpoons[\text{EtOH}]{\text{NaOH,}}$ [aldol product: OH, O] $\rightleftharpoons$ enone + H_2O

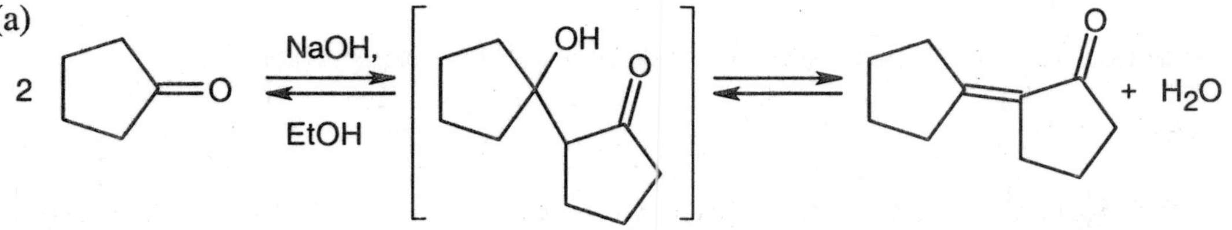

(b) $C_6H_5C(=O)CH_3$ + $H_3CC(=O)C_6H_5$ $\xrightleftharpoons[\text{EtOH}]{\text{NaOH,}}$ $C_6H_5C(H_3C)(OH)-CH_2-C(=O)C_6H_5$

heat

$C_6H_5C(CH_3)=CH-C(=O)C_6H_5$ + H_2O

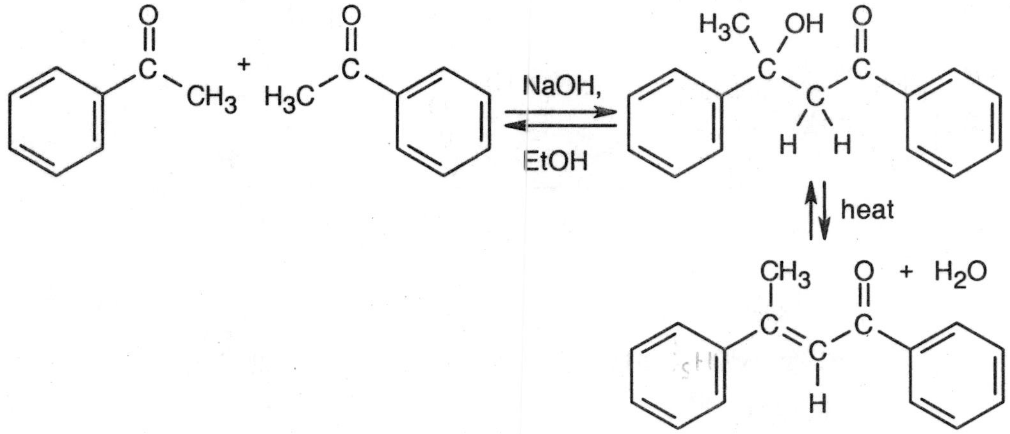

(c)

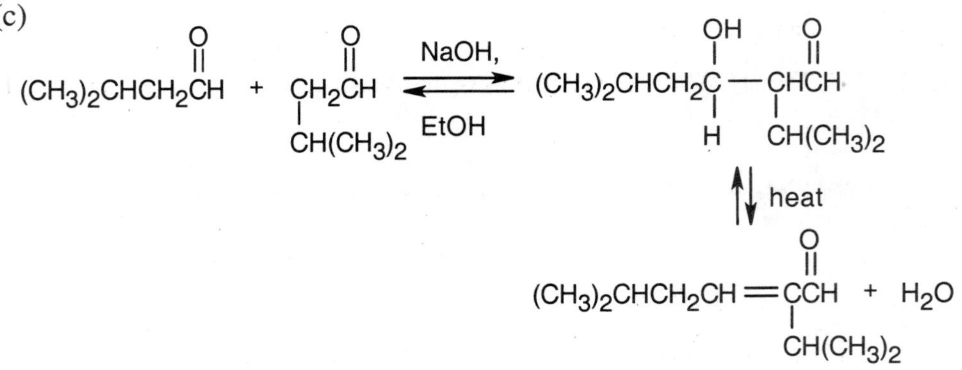

$$(CH_3)_2CHCH_2\overset{O}{\overset{\|}{C}}H + \underset{CH(CH_3)_2}{CH_2}\overset{O}{\overset{\|}{C}}H \underset{EtOH}{\overset{NaOH,}{\rightleftharpoons}} (CH_3)_2CHCH_2\underset{H}{\overset{OH}{C}}-\underset{CH(CH_3)_2}{CH}\overset{O}{\overset{\|}{C}}H$$

heat

$$(CH_3)_2CHCH_2CH=\underset{CH(CH_3)_2}{C}\overset{O}{\overset{\|}{C}}H + H_2O$$

23.4 Including double bond isomers, 4 products can be formed. The major product is formed by attack of the enolate that isn't between the carbonyl group and the methyl group.

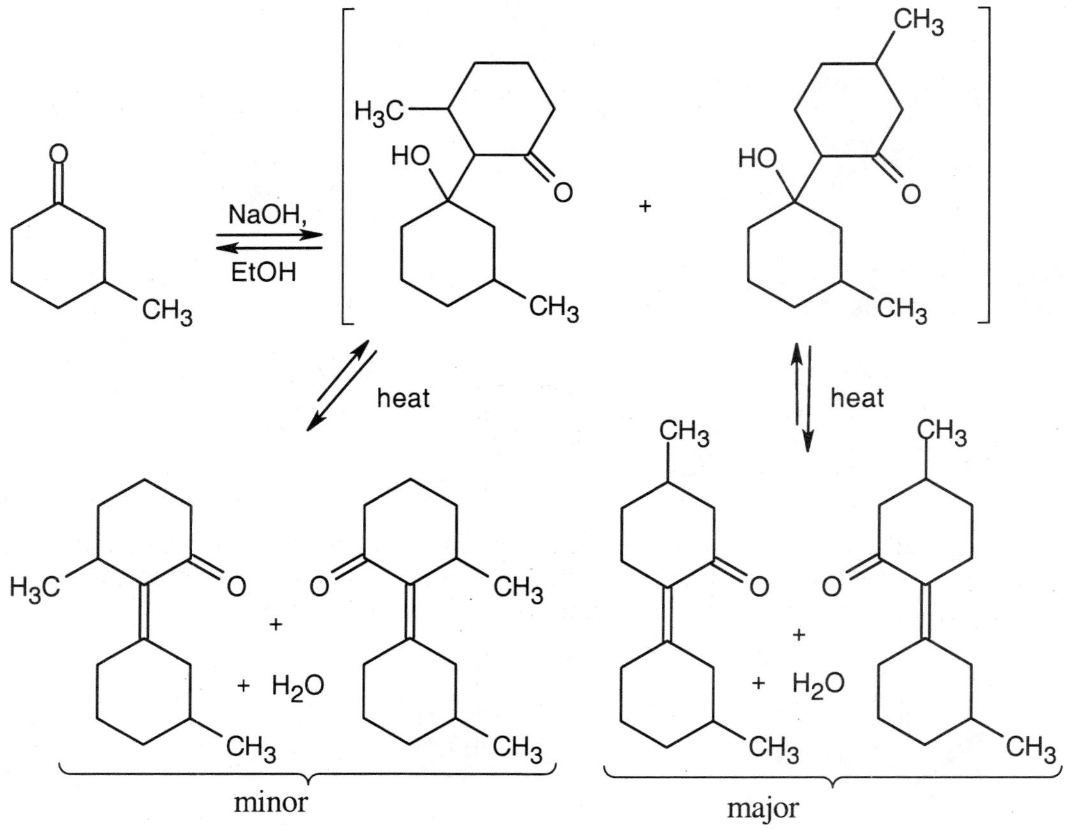

NaOH, EtOH

heat heat

minor major

23.5

(a)

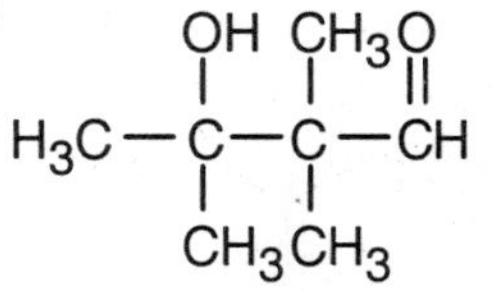

$$H_3C-\underset{CH_3}{\overset{OH}{C}}-\underset{CH_3}{\overset{CH_3}{C}}-\overset{O}{\overset{\|}{C}}H$$ 3-Hydroxy-2,2,3-trimethylbutanal

3-Hydroxy-2,2,3-trimethylbutanal is not an aldol self-condensation product. (Note that no aldol *self*-condensation can yield a product with an odd number of carbons.)

(b)

$$CH_3CH_2CH_2\overset{\displaystyle OH}{\underset{\displaystyle CH_3}{C}}-\overset{\displaystyle O}{\overset{\|}{C}}H$$

2-Hydroxy-2-methylpentanal

This is not an aldol product. The hydroxyl group in an aldol product must be ß, not α, to the carbonyl group.

(c)

$$CH_3CH_2\underset{\displaystyle CH_2CH_3}{C}=C(CH_3)\overset{\displaystyle O}{\overset{\|}{C}}CH_2CH_3$$

5-Ethyl-4-methyl-4-hepten-3-one

This product results from the aldol self-condensation of 3-pentanone, followed by dehydration.

23.6

$$2\ CH_3\overset{\displaystyle O}{\overset{\|}{C}}H \xrightarrow[\text{2. heat}]{\text{1. NaOH, EtOH}} CH_3CH=CH\overset{\displaystyle O}{\overset{\|}{C}}H \xrightarrow[\text{2. } H_3O^+]{\text{1. } NaBH_4} CH_3CH=CHCH_2OH$$

$$\downarrow H_2,\ Pd/C$$

1-Butanol $CH_3CH_2CH_2CH_2OH$

23.7

(a)

$$C_6H_5\overset{\displaystyle O}{\overset{\|}{C}}H + CH_3\overset{\displaystyle O}{\overset{\|}{C}}CH_3 \underset{\text{EtOH}}{\overset{\text{NaOH,}}{\rightleftharpoons}} C_6H_5\overset{\displaystyle OH}{C}HCH_2\overset{\displaystyle O}{\overset{\|}{C}}CH_3 \xrightarrow{\text{heat}} C_6H_5CH=CH\overset{\displaystyle O}{\overset{\|}{C}}CH_3 + H_2O$$

4-Phenyl-3-buten-2-one

This mixed aldol will succeed because one of the components, benzaldehyde, is a good acceptor yet has no α–hydrogen atoms. Although it is possible for acetone to undergo self-condensation, the mixed aldol reaction is much more favorable.

(b)

$$C_6H_5\overset{\displaystyle O}{\overset{\|}{C}}CH_3 + CH_3\overset{\displaystyle O}{\overset{\|}{C}}CH_3 \xrightarrow[\text{2. heat}]{\text{1. NaOH, EtOH}} C_6H_5\overset{\displaystyle CH_3}{C}=CH\overset{\displaystyle O}{\overset{\|}{C}}CH_3 + (CH_3)_2C=CH\overset{\displaystyle O}{\overset{\|}{C}}C_6H_5$$

$$+ (CH_3)_2C=CH\overset{\displaystyle O}{\overset{\|}{C}}CH_3 + C_6H_5\overset{\displaystyle CH_3}{C}=CH\overset{\displaystyle O}{\overset{\|}{C}}C_6H_5 + H_2O$$

Four products result from the aldol condensation of acetone and acetophenone. The two upper compounds are mixed aldol products, and the bottom two are self-condensation products.

(c)

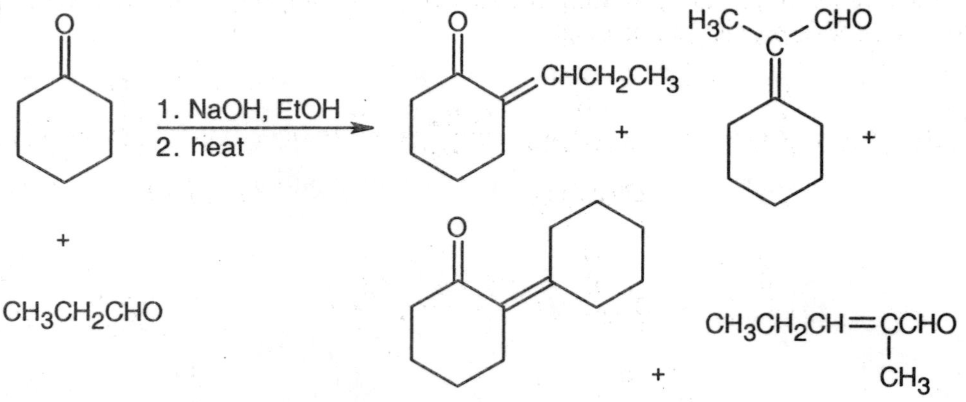

1. NaOH, EtOH
2. heat

CHCH$_2$CH$_3$ + H$_3$C–C–CHO +

CH$_3$CH$_2$CHO

+ CH$_3$CH$_2$CH=CCHO (CH$_3$)

As in (b), a mixture of products is formed because both carbonyl partners contain α–hydrogen atoms. The upper two products result from mixed aldol condensations; the lower two are self-condensation products.

23.8

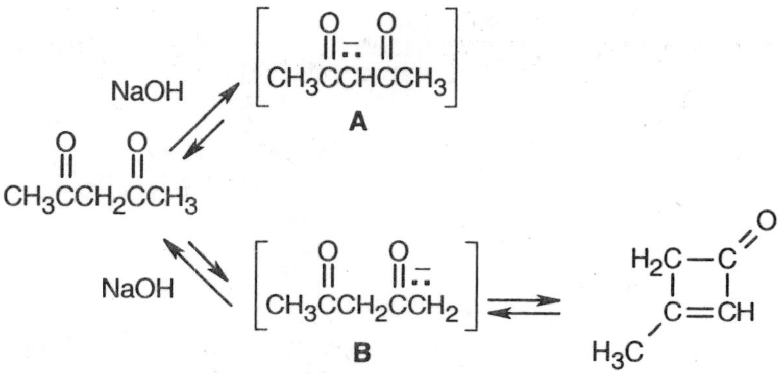

NaOH

CH$_3$CCHCCH$_3$ **A**

CH$_3$CCH$_2$CCH$_3$

NaOH

CH$_3$CCH$_2$CCH$_2$ **B**

H$_2$C—C=O, C=CH, H$_3$C

2,4–Pentanedione is in equilibrium with two enolate ions after treatment with base. Enolate **A** is stable and unreactive, while enolate **B** can undergo internal aldol condensation to form a cyclobutenone product. But because the aldol reaction is reversible and the cyclobutenone product is highly strained, there is little of this product present when equilibrium is reached. At equilibrium, only the stable, diketone enolate ion **A** is present.

23.9 This intramolecular aldol condensation gives a product with a seven-membered ring fused to a five-membered ring.

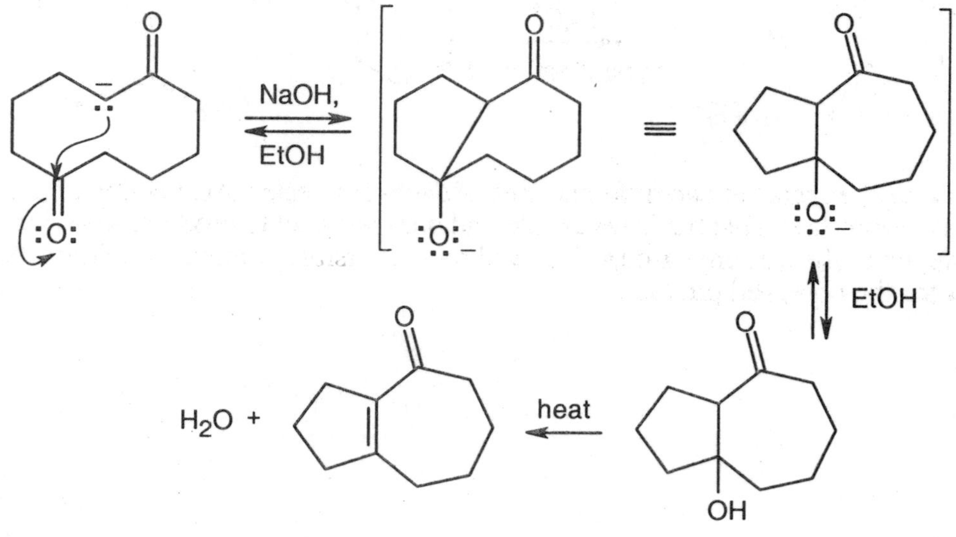

NaOH, EtOH

EtOH

heat

H_2O +

OH

23.10 As in the aldol condensation, writing the two Claisen components in the correct orientation makes it easier to predict the product.

(a)

$$(CH_3)_2CHCH_2\overset{O}{\overset{\|}{C}}OEt + \underset{CH(CH_3)_2}{\underset{|}{CH_2}}\overset{O}{\overset{\|}{C}}OEt \xrightarrow[2.\ H_3O^+]{1.\ Na^+\ ^-OEt} (CH_3)_2CHCH_2\overset{O}{\overset{\|}{C}}-\underset{CH(CH_3)_2}{\underset{|}{CH}}\overset{O}{\overset{\|}{C}}OEt + EtOH$$

(b)

$$C_6H_5CH_2\overset{O}{\overset{\|}{C}}OEt + \underset{C_6H_5}{\underset{|}{CH_2}}\overset{O}{\overset{\|}{C}}OEt \xrightarrow[2.\ H_3O^+]{1.\ Na^+\ ^-OEt} C_6H_5CH_2\overset{O}{\overset{\|}{C}}-\underset{C_6H_5}{\underset{|}{CH}}\overset{O}{\overset{\|}{C}}OEt + EtOH$$

(c)

$$C_6H_{11}CH_2\overset{O}{\overset{\|}{C}}OEt + \underset{C_6H_{11}}{\underset{|}{CH_2}}\overset{O}{\overset{\|}{C}}OEt \xrightarrow[2.\ H_3O^+]{1.\ Na^+\ ^-OEt} C_6H_{11}CH_2\overset{O}{\overset{\|}{C}}-\underset{C_6H_{11}}{\underset{|}{CH}}\overset{O}{\overset{\|}{C}}OEt + EtOH$$

23.11

$C_6H_5COCH_2COOEt$ + $^-:\ddot{O}H$ $\rightleftharpoons$ (addition of hydroxide) [$C_6H_5C(O^-)(OH)CH_2COOEt$ ↓ (elimination of acetate) C_6H_5COOH + $^-:CH_2COOEt$] $\xleftarrow{H_3O^+}$ (neutralization) C_6H_5COOH + $H_3CCOOEt$

Hydroxide ion can react at two different sites of the ß–keto ester. Abstraction of the acidic α–proton is more favorable but is reversible and does not lead to product. Addition of hydroxide ion to the carbonyl group, followed by irreversible elimination of ethyl acetate, accounts for the observed product.

23.12 As shown in Practice Problem 23.4, diethyl oxalate is a very effective reagent in mixed Claisen reactions.

EtOC(=O)—C(=O)OEt + 2,2-dimethylcyclohexanone $\xrightarrow[\text{2. } H_3O^+]{\text{1. } Na^+ \ ^-OEt}$ EtOC(=O)—C(=O)-(3,3-dimethyl-2-oxocyclohexyl) + EtOH

23.13

$\xrightarrow[\text{2. } H_3O^+]{\text{1. } Na^+ \ ^-OEt}$ + EtOH

Diethyl 4-methylcycloheptanedioate

23.14

H_3O^+ + EtOH

C1–C6 bond formation

$Na^+ \ ^-OEt$ EtOH

H_3O^+ + EtOH

C2–C7 bond formation

Unlike diethyl 4-methylheptanedioate, diethyl 3-methylheptanedioate is unsymmetrical. Two different enolates can form, and each can cyclize to a different product.

23.15 A Michael reaction takes place between a stable enolate (Michael donor) and an α,β-unsaturated carbonyl compound (Michael acceptor). The enolate adds to the double bond of the conjugated system. Predicting Michael products is easier when the donor and acceptor are positioned so that the product is evident.

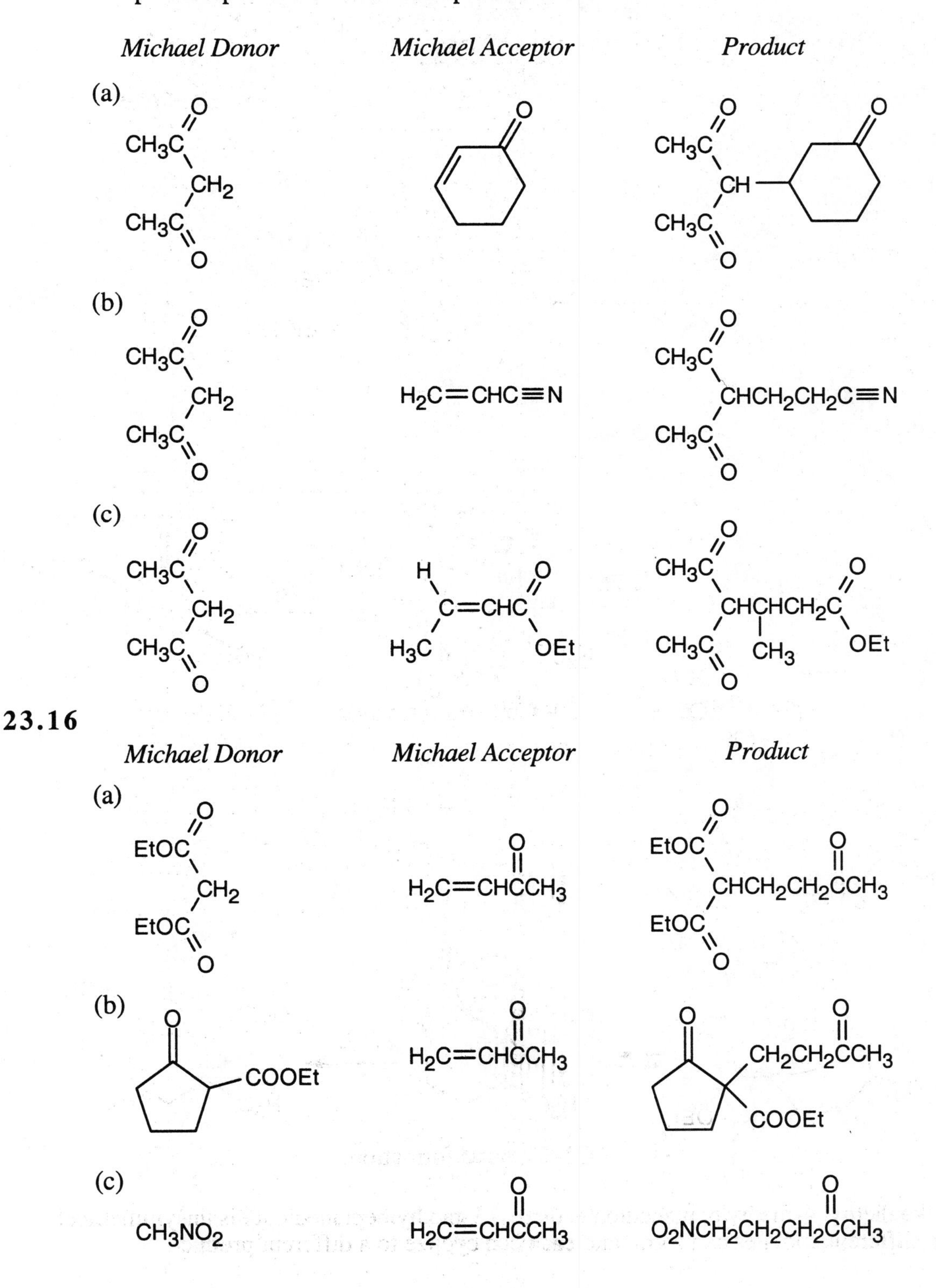

23.17 An enamine is formed from a ketone when it is necessary to synthesize a 1,5-diketone or a 1,5-dicarbonyl compound containing an aldehyde or ketone. The ketone starting material is converted to an enamine in order to increase the reactivity of the ketone and to direct the regiochemistry of addition. The process, as described in Section 23.12, is: (1) conversion of a ketone to its enamine; (2) Michael addition to an α,β-unsaturated carbonyl compound; (3) hydrolysis of the enamine to the starting ketone.

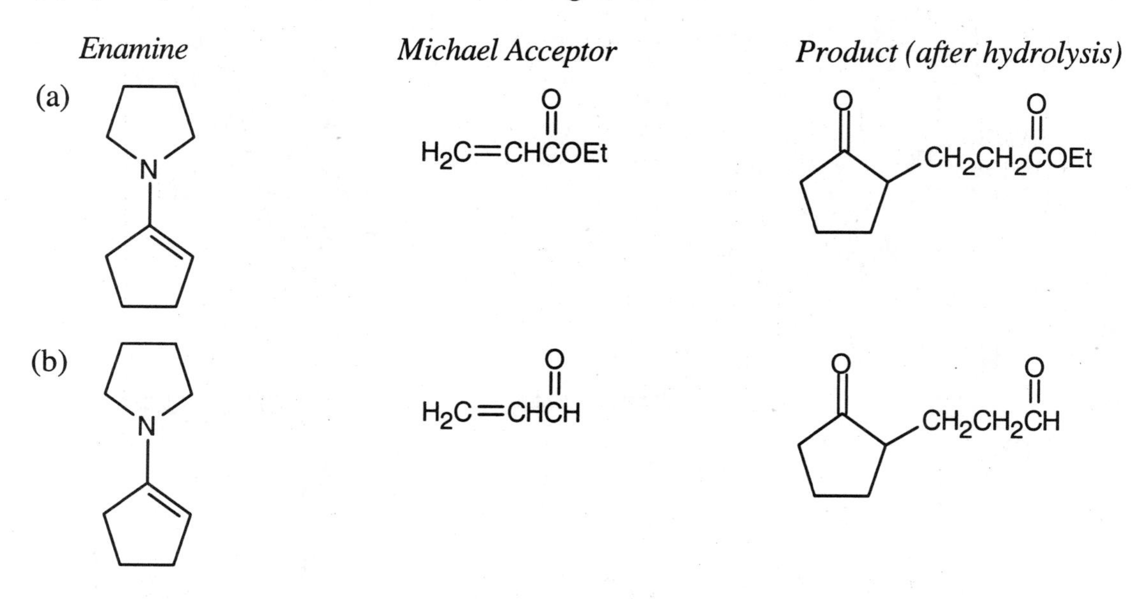

23.18 Analyze the product for the Michael acceptor and the ketone. In (a), the Michael acceptor is propenenitrile. The ketone is cyclopentanone, which is treated with pyrrolidine to form the enamine.

(a)

N–H + cyclopentanone ⇌ enamine + H_2O; enamine + $H_2C{=}CHC{\equiv}N$ → [iminium, $CH_2\ddot{C}HCN$] $\xrightarrow{H_3O^+}$ N–H + CH_2CHCN (2-substituted cyclopentanone)

(b)

N–H + cyclohexanone ⇌ enamine + H_2O; enamine + $H_2C{=}CHCOEt$ → [iminium, $CH_2\ddot{C}HCOEt$] $\xrightarrow{H_3O^+}$ N–H + CH_2CH_2COEt (2-substituted cyclohexanone)

23.19 The Robinson annulation is a combination of two reactions covered in this chapter. First, a Michael reaction takes place between a nucleophilic donor (the diketone in this problem) and an α,β-unsaturated carbonyl compound (the enone shown). The resulting product can cyclize in an aldol reaction. The base catalyzes both reactions.

Michael donor + $H_2C{=}CHCCH_3$ (Michael acceptor) $\xrightarrow{Na^+\ ^-OEt,\ EtOH}$ [Michael adduct] $\xrightarrow{Na^+\ ^-OEt,\ EtOH}$ [aldol product] $\xrightarrow{heat}$ H_2O + product

23.20

$CH_3CH_2CCH{=}CH_2$ (Michael acceptor) + Michael donor $\xrightarrow{Na^+\ ^-OEt,\ EtOH}$ [Michael adduct] $\xrightarrow{Na^+\ ^-OEt,\ EtOH}$ [aldol product] $\xrightarrow{heat}$ H_2O + product

Synthetic analysis: This is one of the more complicated-looking syntheses that we have seen. First, analyze the product for the two Michael components. The carbon-carbon double bond arises from dehydration of the aldol addition product, and is located where one of the two C=O groups of the original diketone used to be. The Michael addition takes place at the carbon between these ketone groups. The Michael acceptor is an enone that can also enter into the aldol condensation and furnishes the methyl group attached to the double bond.

Visualizing Chemistry

23.21

(a)

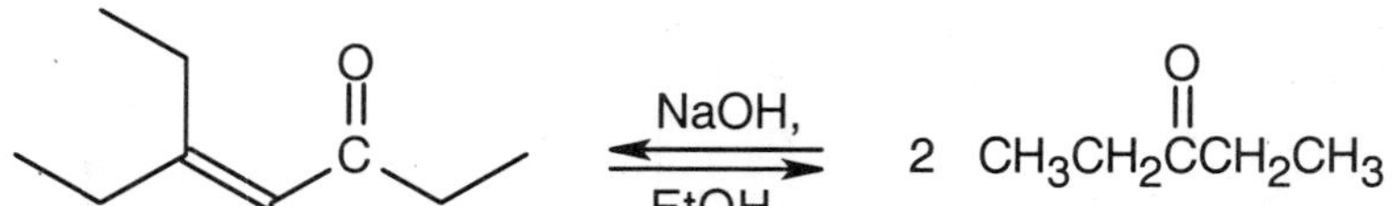

NaOH, EtOH

2 $CH_3CH_2\overset{O}{\overset{\|}{C}}CH_2CH_3$

3-Pentanone

(b)

NaOH, EtOH

2 $HC(=O)CH_2CH(CH_3)_2$

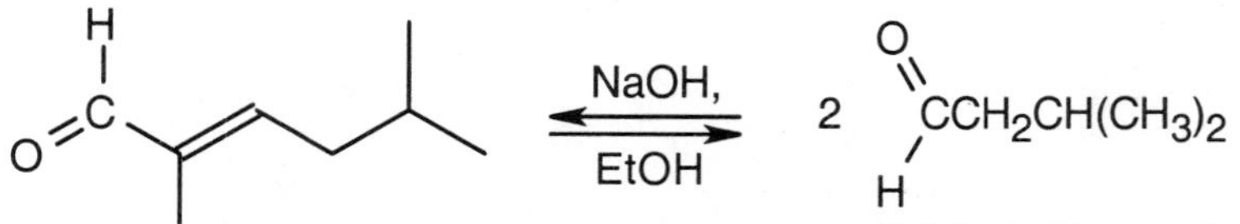

3-Methylbutanal

23.22 The enolate of methyl phenylacetate adds to a second molecule of methyl phenylacetate to form the intermediate that is pictured. Elimination of methoxide (circled) and acidification give the product shown.

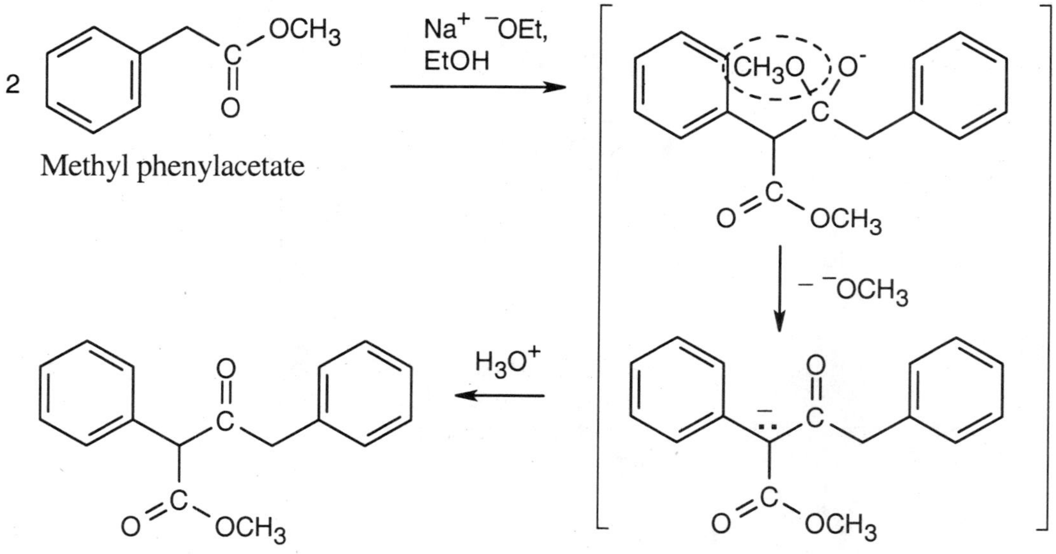

2 Methyl phenylacetate $\xrightarrow{Na^+\ ^-OEt,\ EtOH}$ [intermediate (CH_3O, O^-, OCH_3) $\xrightarrow{-\ ^-OCH_3}$ enolate (OCH_3)] $\xrightarrow{H_3O^+}$ product (OCH_3)

23.23

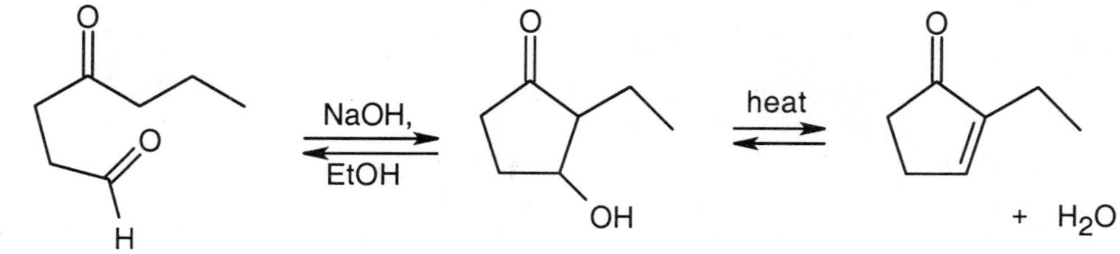

4-Oxoheptanal $\underset{EtOH}{\overset{NaOH,}{\rightleftharpoons}}$ (OH) $\overset{heat}{\rightleftharpoons}$ + H_2O

23.24 Remember that the carbon-carbon double bond in the product connects one of the carbonyl carbons of the Michael donor with an α carbon of the Michael acceptor.

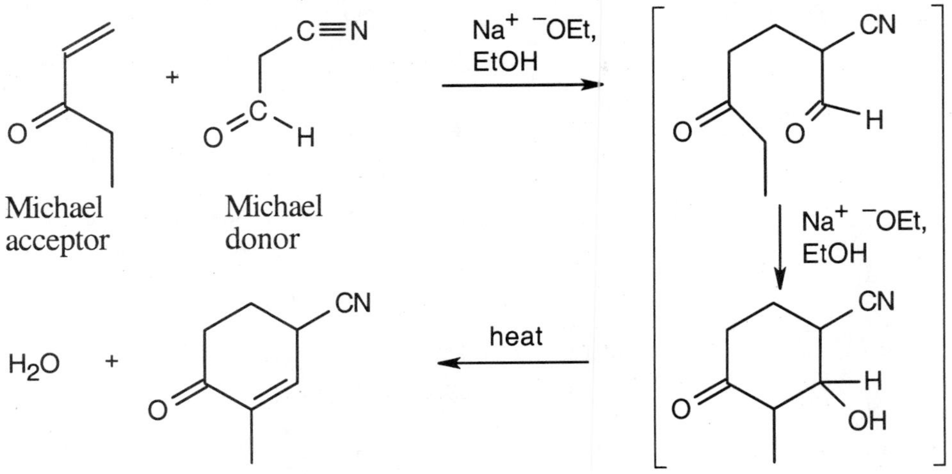

Michael acceptor + Michael donor $\xrightarrow{Na^+\ ^-OEt,\ EtOH}$ [ ] $\xrightarrow{Na^+\ ^-OEt,\ EtOH}$ [ ] $\xrightarrow{heat}$ H_2O +

Additional Problems

23.25 (a) $(CH_3)_3CCHO$ has no α hydrogens and does not undergo aldol self-condensation.

(b)

2 $\xrightarrow{NaOH,\ EtOH}$ [ ] $\xrightarrow{heat}$ + H_2O

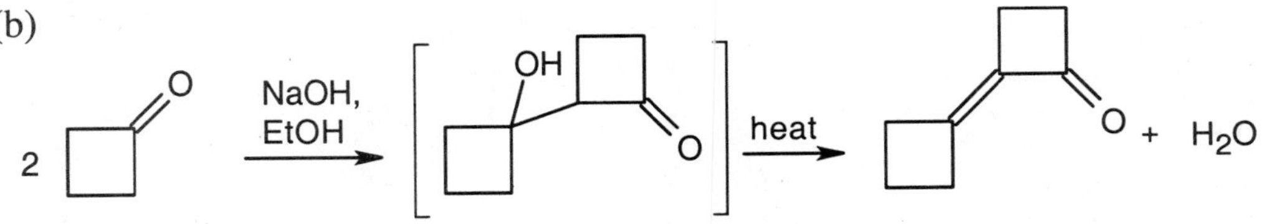

(c)

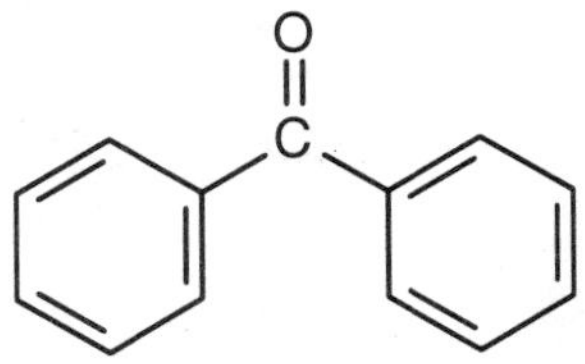

Benzophenone doesn't undergo aldol self condensation because it has no α hydrogens

(d)

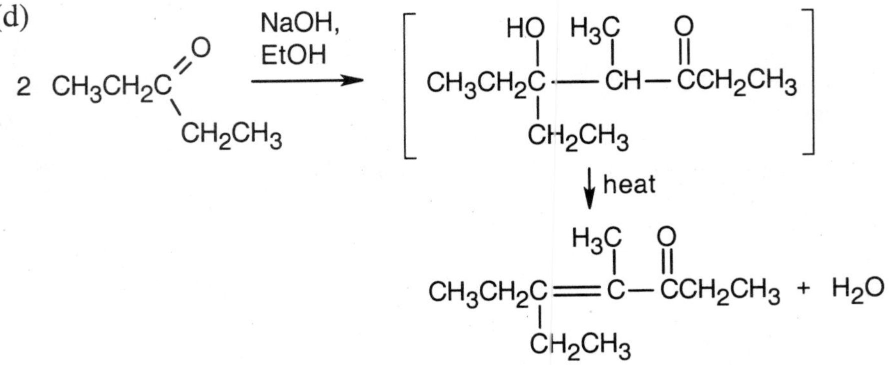

$$2\ CH_3CH_2C(=O)CH_2CH_3 \xrightarrow{NaOH,\ EtOH} [CH_3CH_2C(OH)(CH_2CH_3)-CH(CH_3)-C(=O)CH_2CH_3] \xrightarrow{heat} CH_3CH_2C(CH_2CH_3)=C(CH_3)-C(=O)CH_2CH_3 + H_2O$$

(e)

$$2\ CH_3(CH_2)_8CHO \xrightarrow{NaOH,\ EtOH} \left[CH_3(CH_2)_8CH(OH)\text{—}CH((CH_2)_7CH_3)CHO \right] \xrightarrow{heat} CH_3(CH_2)_8CH{=}C((CH_2)_7CH_3)CHO + H_2O$$

(f) $C_6H_5CH{=}CHCHO$ does not undergo aldol reactions because its sp^2-hybridized α proton isn't acidic.

23.26 As always, analyze the product for the carbon-carbon double bond that is formed by dehydration of the initial aldol adduct. Break the bond, and add a carbonyl oxygen to the appropriate carbon.

(a)

$$C_6H_5CHO + H_3CCOC_6H_5 \xrightarrow[2.\ heat]{1.\ NaOH,\ EtOH} C_6H_5CH{=}CHCOC_6H_5 + H_2O$$

(b)

$$CH_3CO(CH_2)_3CHO \xrightarrow[2.\ heat]{1.\ NaOH,\ EtOH} \text{2-cyclohexenone} + H_2O$$

(c)

$$o\text{-}C_6H_4(CH_2CHO)_2 \xrightarrow[2.\ heat]{1.\ NaOH,\ EtOH} \text{indene-3-carbaldehyde (CHO)} + H_2O$$

(d)

$$C_6H_5CH_2COCH_2C_6H_5 + C_6H_5COCOC_6H_5 \xrightarrow[2.\ heat]{1.\ NaOH,\ EtOH} \text{tetraphenylcyclopentadienone} + 2\ H_2O$$

23.27

CHO $\xrightarrow{^-OH}$ [CHO] $\xrightarrow{EtOH}$ CHO, HO H $\xrightarrow{heat}$ CHO + H_2O

23.28

NaOH, EtOH ⇌ [ ] $\xrightarrow{\text{1. EtOH, 2. heat}}$ **A** + H_2O

NaOH, EtOH ⇌ [ ] $\xrightarrow{\text{1. EtOH, 2. heat}}$ **B** + H_2O

23.29 Product **A**, which has two singlet methyl groups and no vinylic protons in its ^{1}H NMR, is the major product of the intramolecular cyclization of 2,5–heptanedione.

23.30

addition of ^-OH ⇌ proton transfer ⇌ ring opening ⇌ H_2O protonation ⇌ ^-OH deprotonation ⇌ H_2O + aldol condensation ⇌ 1. EtOH 2. heat, protonation, dehydration → **A** + H_2O

23.31 The reactive nucleophile in the acid-catalyzed aldol condensation is the *enol* of the carbonyl compound. The electrophile is the protonated carbonyl compound.

Step 1: Enol formation.

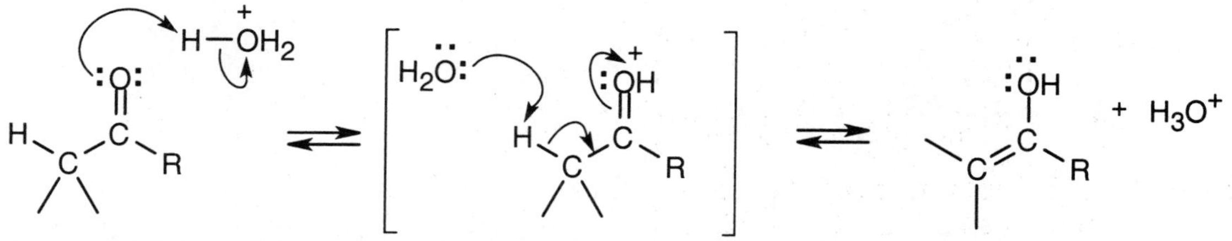

$H-\overset{+}{O}H_2$ $H_2\ddot{O}$: $\overset{+}{O}H$:OH + H_3O^+

Step 2: Addition of the enol nucleophile to the protonated carbonyl compound.

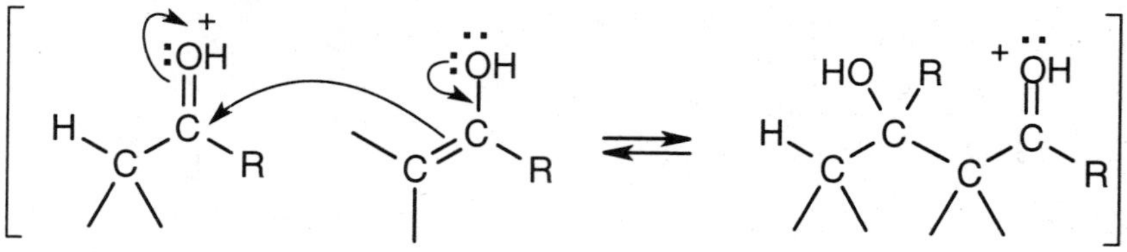

electrophile nucleophile

Step 3: Loss of proton from the carbonyl oxygen.

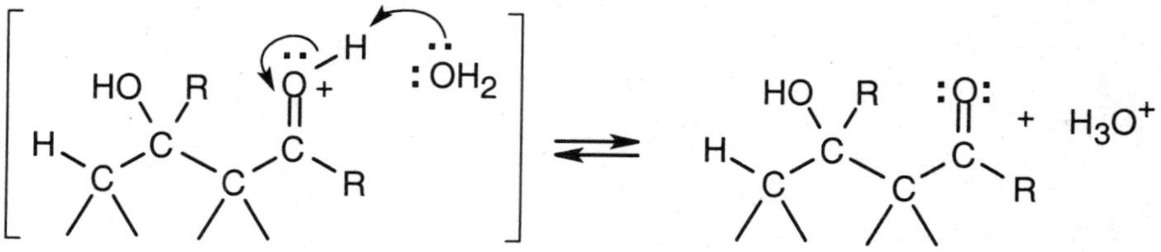

$:OH_2$ + H_3O^+

23.32

A NaOH, EtOH formation of enolate B

addition of enolate to carbonyl group

D H_2O protonation C

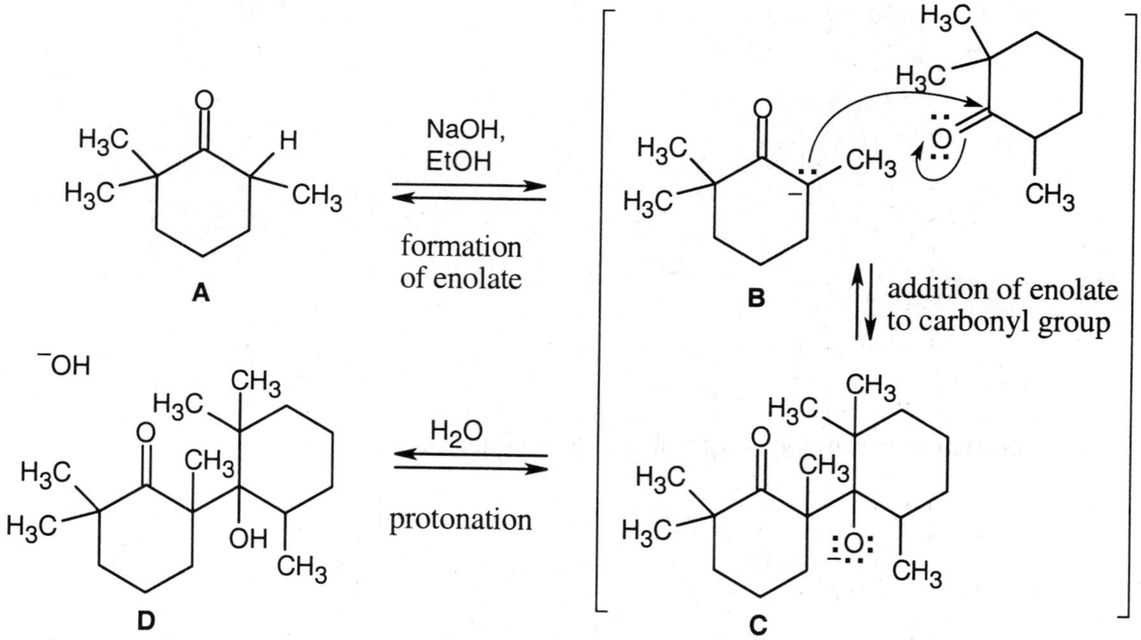

An aldol condensation involves a series of reversible equilibrium steps. In general, formation of product is favored by the dehydration of the ß-hydroxy ketone to form a conjugated enone. Here, dehydration to form conjugated product can't occur. In addition, the **B** $\rightleftharpoons$ **C** equilibrium favors **B** because of steric hindrance.

23.33

$$C_6H_5CHO + CH_3CHO \xrightarrow[\text{2. heat}]{\text{1. NaOH, EtOH}} C_6H_5CH{=}CHCHO + H_2O$$

Cinnamaldehyde

The mixed aldol product predominates.

23.34 The first step of an aldol condensation is enolate formation. The ketone shown here does not enolize because double bonds at the bridgehead of small bicyclic ring systems are too strained to form. Since the bicyclic ketone does not enolize, it doesn't undergo aldol condensation.

23.35

(a)

$$CH_3CH_2CH_2\overset{O}{\overset{\|}{C}}OEt + \underset{CH_2CH_3}{\underset{|}{CH_2}}\overset{O}{\overset{\|}{C}}OEt \xrightarrow[\text{2. } H_3O^+]{\text{1. NaOEt, EtOH}} CH_3CH_2CH_2\overset{O}{\overset{\|}{C}}{-}\underset{CH_2CH_3}{\underset{|}{CH}}\overset{O}{\overset{\|}{C}}OEt + HOEt$$

(b)

$$\text{cycloheptanone} + \text{cycloheptanone} \xrightarrow[\text{2. heat}]{\text{1. NaOH, EtOH}} \text{2-cycloheptylidenecycloheptanone} + H_2O$$

(c)

$$\text{diketone} \xrightarrow[\text{2. heat}]{\text{1. NaOH, EtOH}} \text{cyclohexenone product (major)} + H_2O$$

major

(d)

$$C_6H_5CH_2CH_2\overset{O}{\overset{\|}{C}}H + \underset{CH_2C_6H_5}{\underset{|}{CH_2}}\overset{O}{\overset{\|}{C}}H \xrightarrow[\text{2. heat}]{\text{1. NaOH, EtOH}} C_6H_5CH_2CH_2CH{=}\underset{CH_2C_6H_5}{\underset{|}{C}}\overset{O}{\overset{\|}{C}}H + H_2O$$

23.36 If cyclopentanone and base are mixed first, aldol self-condensation of cyclopentanone can occur before ethyl formate is added. If both carbonyl components are mixed together before adding base, the more favorable mixed Claisen condensation occurs with less competition from the aldol self-condensation reaction.

23.37

(a)

$$CH_3\overset{O}{\overset{\|}{C}}OEt + CH_3CH_2\overset{O}{\overset{\|}{C}}OEt \xrightarrow[2.\ H_3O^+]{1.\ NaOEt,\ EtOH} CH_3\overset{O}{\overset{\|}{C}}-CH_2\overset{O}{\overset{\|}{C}}OEt + CH_3CH_2\overset{O}{\overset{\|}{C}}-CH_2\overset{O}{\overset{\|}{C}}OEt + CH_3CH_2\overset{O}{\overset{\|}{C}}-\underset{CH_3}{\underset{|}{C}}H\overset{O}{\overset{\|}{C}}OEt + CH_3\overset{O}{\overset{\|}{C}}-\underset{CH_3}{\underset{|}{C}}H\overset{O}{\overset{\|}{C}}OEt + EtOH$$

self-condensation products   mixed condensation products

Approximately equal amounts of each product will form if the two esters are of similar reactivity.

(b)

$$C_6H_5\overset{O}{\overset{\|}{C}}OEt + C_6H_5CH_2\overset{O}{\overset{\|}{C}}OEt \xrightarrow[2.\ H_3O^+]{1.\ NaOEt,\ EtOH} C_6H_5CH_2\overset{O}{\overset{\|}{C}}-\underset{C_6H_5}{\underset{|}{C}}H\overset{O}{\overset{\|}{C}}OEt + C_6H_5\overset{O}{\overset{\|}{C}}-\underset{C_6H_5}{\underset{|}{C}}H\overset{O}{\overset{\|}{C}}OEt + EtOH$$

self-condensation product   mixed condensation product

The mixed condensation product predominates.

(c)

$$\text{cyclohexanone} + EtO\overset{O}{\overset{\|}{C}}OEt \xrightarrow[2.\ H_3O^+]{1.\ NaOEt,\ EtOH} \text{2-(}\overset{O}{\overset{\|}{C}}OEt\text{)cyclohexanone} + EtOH$$

This is the only *Claisen* monocondensation product (aldol self-condensation of cyclohexanone also occurs).

(d)

$$C_6H_5\overset{O}{\overset{\|}{C}}H + CH_3\overset{O}{\overset{\|}{C}}OEt \xrightarrow[2.\ H_3O^+]{1.\ NaOEt,\ EtOH} CH_3\overset{O}{\overset{\|}{C}}-CH_2\overset{O}{\overset{\|}{C}}OEt + C_6H_5\overset{O}{\overset{\|}{C}}-CH_2\overset{O}{\overset{\|}{C}}OEt + EtOH$$

self-condensation product   mixed condensation product

The mixed Claisen product is the major product.

23.38

$$CH_3\overset{O}{\overset{\|}{C}}-C(CH_3)_2\overset{O}{\overset{\|}{C}}OEt + {}^{-}:\ddot{O}Et \rightleftharpoons \left[CH_3\overset{:\ddot{O}:^-}{\underset{OEt}{\overset{|}{\underset{|}{C}}}}-C(CH_3)_2\overset{O}{\overset{\|}{C}}OEt\right] \longrightarrow \ddot{C}(CH_3)_2\overset{O}{\overset{\|}{C}}OEt + CH_3\overset{O}{\overset{\|}{C}}OEt$$

addition of ethoxide; elimination of ethyl dimethylacetate anion

$$\ddot{C}(CH_3)_2\overset{O}{\overset{\|}{C}}OEt + EtOH \rightleftharpoons (CH_3)_2CH\overset{O}{\overset{\|}{C}}OEt + {}^{-}OEt$$

This is a reverse Claisen reaction.

23.39 Two different reactions are possible when ethyl acetoacetate reacts with ethoxide anion. One possibility involves attack of ethoxide ion on the carbonyl carbon, followed by elimination of the anion of ethyl acetate—a reverse Claisen reaction similar to the one illustrated in 23.38. More likely, however, is the acid-base reaction of ethoxide ion and a doubly activated α–hydrogen of ethyl acetoacetate.

$$CH_3\overset{O}{\overset{\|}{C}}-\underset{H}{\underset{|}{C}}H\overset{O}{\overset{\|}{C}}OEt + {}^{-}:\ddot{O}Et \rightleftharpoons CH_3\overset{O}{\overset{\|}{C}}-\overset{..}{C}H\overset{O}{\overset{\|}{C}}OEt + HOEt$$

The resonance-stabilized acetoacetate anion is no longer reactive toward nucleophiles, and no further reaction occurs at room temperature. Elevated temperatures are required to make the cleavage reaction proceed. This complication doesn't occur with ethyl dimethylacetoacetate because it has no acidic hydrogens between its two carbonyl groups.

23.40 Michael reactions occur between stabilized enolate anions and α,ß-unsaturated carbonyl compounds. Learn to locate these components in possible Michael products. Usually, it is easier to recognize the enolate nucleophile; in (a), the nucleophile is the ethyl acetoacetate anion.The rest of the compound is the Michael acceptor. Draw a double bond in conjugation with an electron-withdrawing group in this part of the molecule.

Michael donor: $CH_3\overset{O}{\overset{\|}{C}}\underset{COOEt}{\underset{|}{C}H_2}$ — $CH_3\overset{O}{\overset{\|}{C}}\underset{COOEt}{\underset{|}{C}H}-CH_2CH_2\overset{O}{\overset{\|}{C}}C_6H_5$ — Michael acceptor: $H_2C=CH\overset{O}{\overset{\|}{C}}C_6H_5$

$$\underset{\text{Michael donor}}{CH_3\overset{O}{\overset{\|}{C}}\underset{COOEt}{\underset{|}{C}H_2}} + \underset{\text{Michael acceptor}}{H_2C=CH\overset{O}{\overset{\|}{C}}C_6H_5} \xrightarrow[2.\ H_3O^+]{1.\ NaOEt,\ EtOH} CH_3\overset{O}{\overset{\|}{C}}\underset{COOEt}{\underset{|}{C}H}-CH_2CH_2\overset{O}{\overset{\|}{C}}C_6H_5$$

(b) When the Michael product has been decarboxylated after the addition reaction, it is more difficult to recognize the original enolate anion.

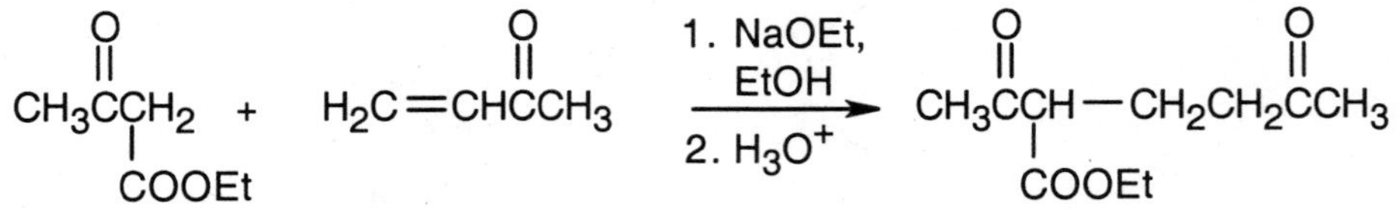

$$CH_3C(=O)CH_2(COOEt) + H_2C{=}CHC(=O)CH_3 \xrightarrow[2.\ H_3O^+]{1.\ NaOEt,\ EtOH} CH_3C(=O)CH(COOEt){-}CH_2CH_2C(=O)CH_3$$

Michael donor Michael acceptor

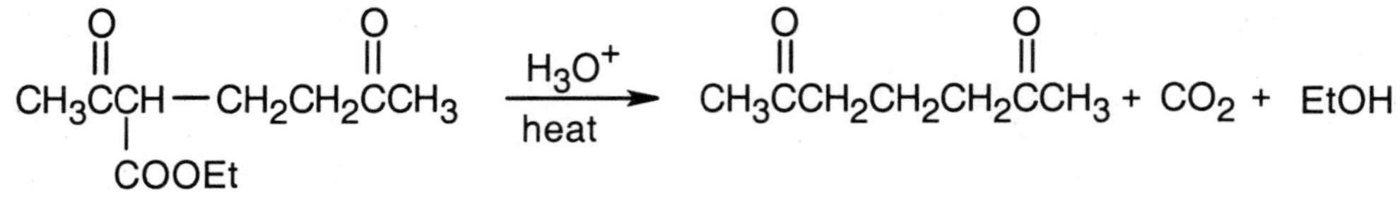

$$CH_3C(=O)CH(COOEt){-}CH_2CH_2C(=O)CH_3 \xrightarrow[heat]{H_3O^+} CH_3C(=O)CH_2CH_2CH_2C(=O)CH_3 + CO_2 + EtOH$$

(c)

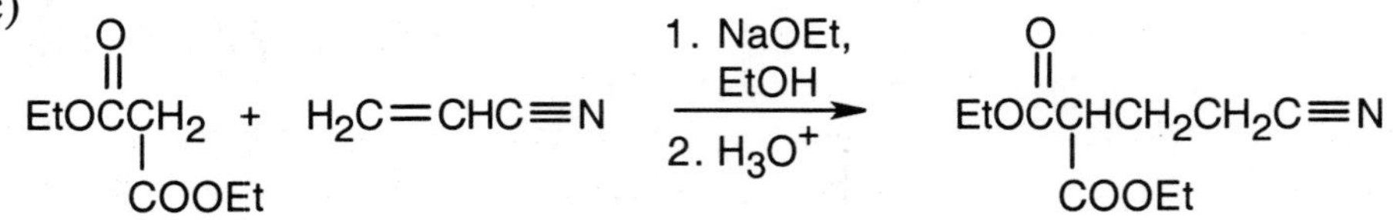

$$EtOC(=O)CH_2(COOEt) + H_2C{=}CHC{\equiv}N \xrightarrow[2.\ H_3O^+]{1.\ NaOEt,\ EtOH} EtOC(=O)CH(COOEt)CH_2CH_2C{\equiv}N$$

Michael donor Michael acceptor

(d)

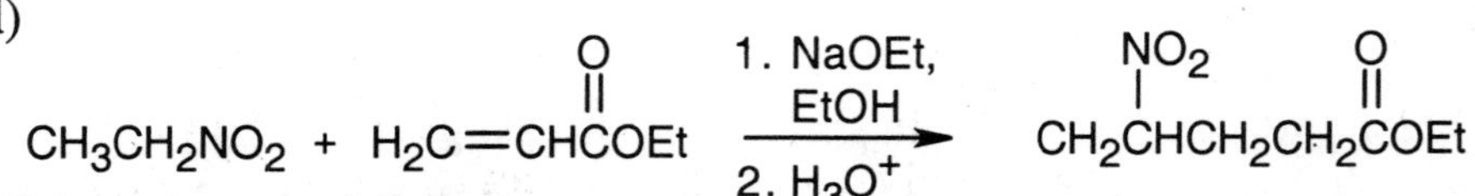

$$CH_3CH_2NO_2 + H_2C{=}CHC(=O)OEt \xrightarrow[2.\ H_3O^+]{1.\ NaOEt,\ EtOH} CH_3CH(NO_2)CH_2CH_2C(=O)OEt$$

Michael donor Michael acceptor

(e)

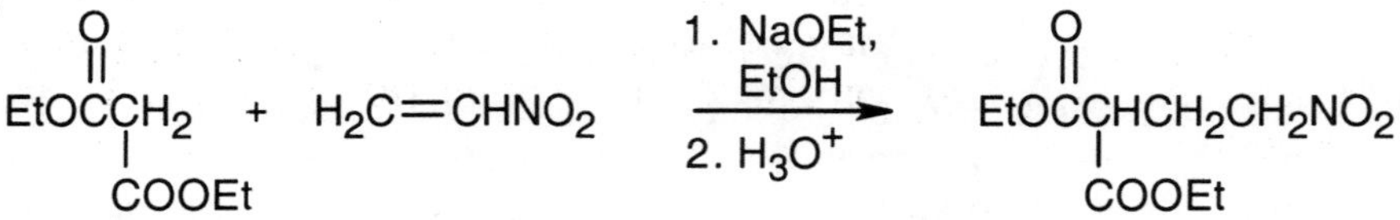

$$EtOC(=O)CH_2(COOEt) + H_2C{=}CHNO_2 \xrightarrow[2.\ H_3O^+]{1.\ NaOEt,\ EtOH} EtOC(=O)CH(COOEt)CH_2CH_2NO_2$$

Michael donor Michael acceptor

(f)

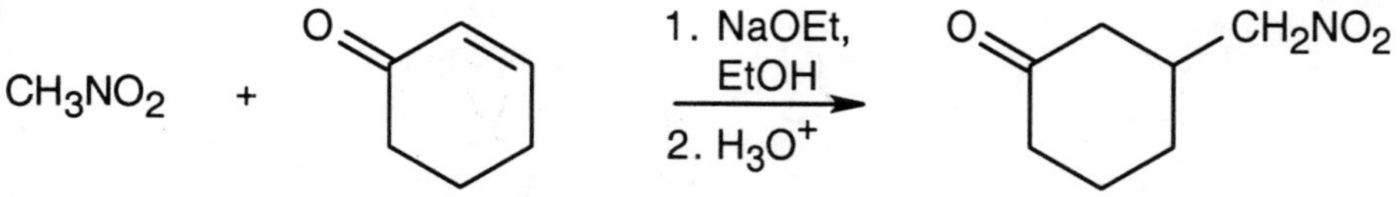

$$CH_3NO_2 + \text{2-cyclohexenone} \xrightarrow[2.\ H_3O^+]{1.\ NaOEt,\ EtOH} \text{3-(nitromethyl)cyclohexanone } (CH_2NO_2)$$

Michael donor Michael acceptor

23.41

1,3-Cyclohexanedione $\xrightarrow[\text{2. } CH_3I]{\text{1. } Na^+ \, ^-OEt, \text{ EtOH}}$ 2-methyl-1,3-cyclohexanedione $\xrightarrow[\text{3. } H_3O^+]{\text{1. } Na^+ \, ^-OEt, \text{ EtOH; 2. } H_2C{=}CHCOCH_3}$ [2-methyl-2-(3-oxobutyl)-1,3-cyclohexanedione] $\xrightarrow[\text{2. heat}]{\text{1. } Na^+ \, ^-OEt, \text{ EtOH}}$ bicyclic enedione product

Synthetic analysis: This sequence of reactions consists of an alkylation of a 1,3-diketone, followed by a Robinson annulation. The carbon-carbon double bond appears where the second carbonyl group of the diketone used to be and is the site of the ring-forming aldol reaction. A Michael reaction between the diketone and the Michael acceptor 3-buten-2-one adds the carbon atoms used to form the second ring, and an alkylation with CH_3I adds the methyl group.

23.42 (a) Several other products are formed in addition to the one pictured. Self-condensation of acetaldehyde and acetone (less likely) can occur, and an additional mixed product is formed.

(b) There are two problems with this reaction. (1) Michael reactions occur in low yield with mono-ketones. Formation of the enamine, followed by the Michael reaction, gives a higher yield of product. (2) Addition can occur on either side of the ketone to give a mixture of products.

(c) Internal aldol condensation of 2,6-heptanedione can product a four-membered ring or a six-membered ring. The six-membered ring is more likely to form because it is less strained.

$$\left[{}^-:CH_2\overset{O}{\overset{\|}{C}}CH_2CH_2CH_2\overset{:O:}{\overset{\|}{C}}CH_3 \rightleftharpoons \text{(cyclic alkoxide: } H_3C, O^-) \xrightleftharpoons{\text{EtOH}} \text{(cyclic alcohol: } H_3C, OH) \right] \xrightleftharpoons{\text{heat}} \text{3-methyl-2-cyclohexenone} + H_2O$$

23.43

1. a, 2. b → 1. c, 2. d → CO_2Et / CO_2Et

1. e, 2. f → CO_2Et → 1. g, 2. h → CH_3

(a) $LiAlH_4$, then H_3O^+; (b) $POCl_3$, pyridine (c) $KMnO_4$, H_3O^+; (d) CH_3OH, H^+; (e) Na^+ ^-OEt; (f) H_3O^+; (g) Na^+ ^-OEt, then CH_3Br; (h) H_3O^+

23.44

(a)

+ 2 C_6H_5CHO → 1. NaOH, EtOH; 2. heat → C_6H_5CH= =CHC_6H_5 + 2 H_2O

(b)

+ pyrrolidine (N–H) → [enamine] → 1. $H_2C{=}CHCN$; 2. H_3O^+ → CH_2CH_2CN + pyrrolidine (N–H)

(c)

1. LDA; 2. $BrCH_2CH{=}CH_2$ → $CH_2CH{=}CH_2$

23.45

(a)

$H_2C(COOEt)COCH_3$ ⇌ (Na^+ ^-OEt, EtOH; formation of enolate) ⇌ [$H\ddot{C}^-(COOEt)COCH_3$ + $H_2C{=}O$ ⇌ (aldol condensation) ⇌ $HOCH_2\ddot{C}^-(COOEt)COCH_3$] → loss of ^-OH → $H_2C{=}C(COOEt)COCH_3$ + ^-OH

(b)

Michael addition

proton transfer

23.46

(c)

internal aldol conden-sation

1. H_2O
2. heat

dehy-dration

(d)

H_3O^+, heat

ester cleavage and decarboxylation of a β-keto acid

+ EtOH

+ CO_2

Hagemann's ester

23.47

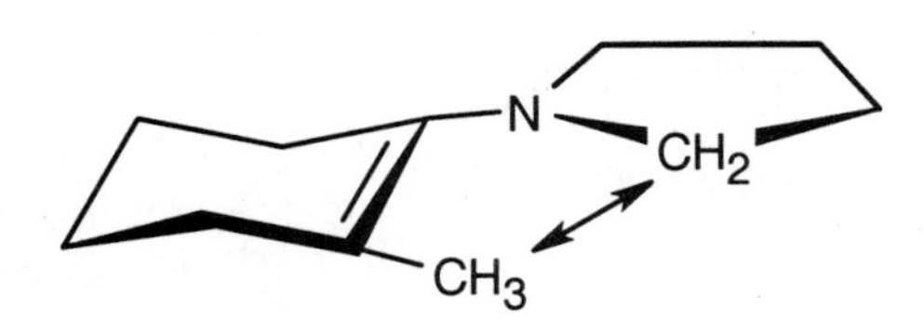

Crowding between the methyl group and the pyrrolidine ring disfavors this enamine.

ring flip

The crowding in this enamine can be relieved by a ring-flip, which puts the methyl group in an axial position. This enamine is the only one formed.

23.48

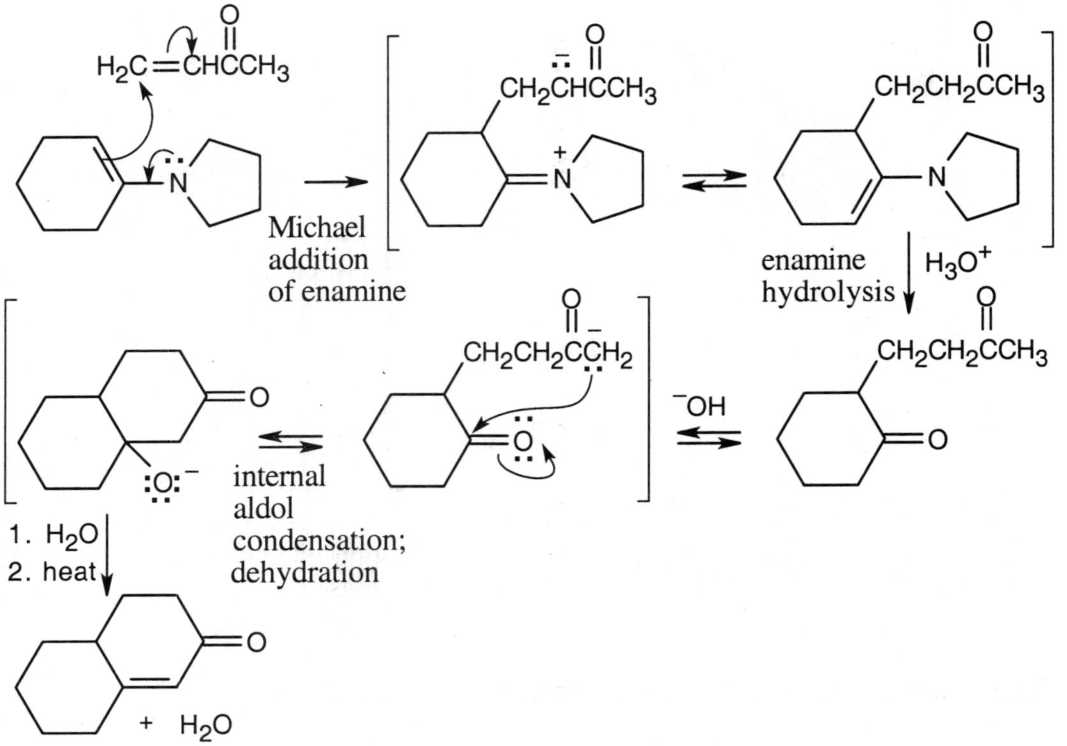

$H_2C{=}CHCCH_3$ (C=O)

Michael addition of enamine

$CH_2\ddot{C}HCCH_3$ (C=O); $\overset{+}{N}$

$CH_2CH_2CCH_3$ (C=O)

enamine hydrolysis H_3O^+

$CH_2CH_2CCH_3$ (C=O)

^{-}OH

$CH_2CH_2C\ddot{C}H_2^{-}$ (C=O)

internal aldol condensation; dehydration

$:\ddot{O}:^{-}$

1. H_2O
2. heat

+ H_2O

23.49

(a)

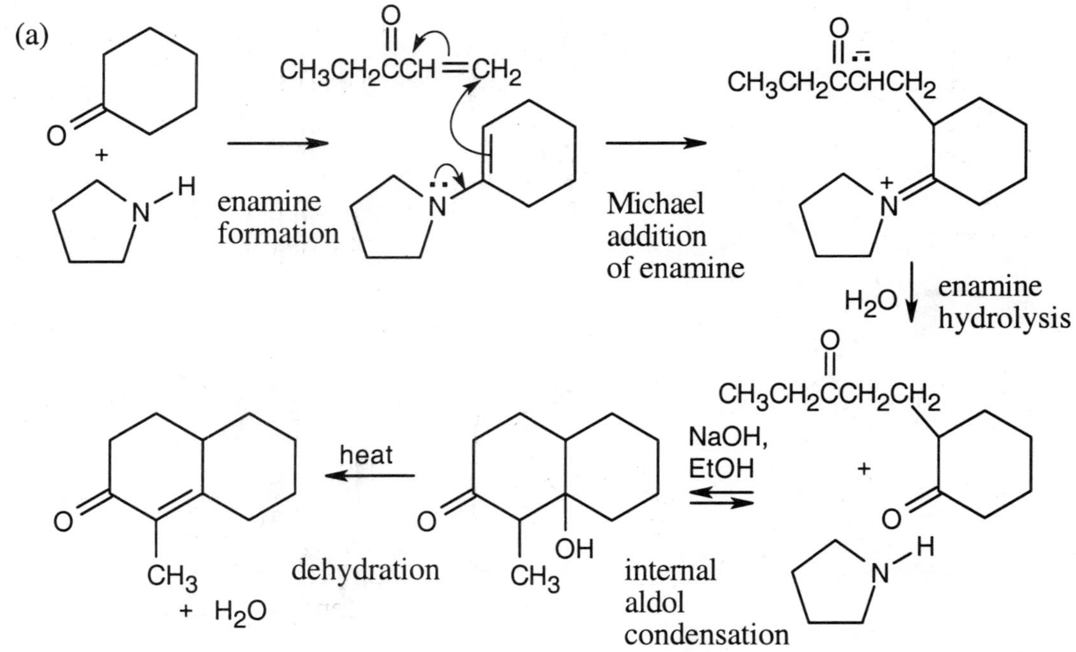

+

enamine formation

$CH_3CH_2CCH{=}CH_2$ (C=O)

Michael addition of enamine

$CH_3CH_2C\ddot{C}HCH_2$ (C=O); $\overset{+}{N}$

H_2O enamine hydrolysis

$CH_3CH_2CCH_2CH_2$ (C=O)

+

NaOH, EtOH

internal aldol condensation

OH

CH_3

heat

dehydration

CH_3

+ H_2O

(b)

$$\text{2-methylcyclohexanone} + \text{pyrrolidine} \xrightarrow{\text{enamine formation}} \text{enamine} + CH_3CCH{=}CH_2 \xrightarrow{\text{Michael addition of enamine}} \text{iminium ion} \xrightarrow[\text{enamine hydrolysis}]{H_3O^+} \text{2-(3-oxobutyl)-6-methylcyclohexanone} + \text{pyrrolidine}$$

$$\xrightleftharpoons[\text{internal aldol condensation}]{\text{NaOH, EtOH}} \text{hydroxy ketone} \xrightarrow[\text{dehydration}]{\text{heat}} \text{bicyclic enone} + H_2O$$

Notice that the correct enamine is formed (see Problem 23.47).

(c)

$$\beta\text{-tetralone} + \text{pyrrolidine} \xrightarrow{\text{enamine formation}} \text{enamine} + CH_3CCH{=}CH_2 \xrightarrow{\text{Michael addition of enamine}} \text{iminium ion} \xrightarrow[\text{enamine hydrolysis}]{H_3O^+} \text{1-(3-oxobutyl)-2-tetralone} + \text{pyrrolidine}$$

$$\xrightleftharpoons[\text{internal aldol condensation}]{\text{NaOH, EtOH}} \text{hydroxy ketone} \xrightarrow[\text{dehydration}]{\text{heat}} \text{tricyclic enone} + H_2O$$

The enamine forms so that the enamine double bond is conjugated with the aromatic ring.

23.50

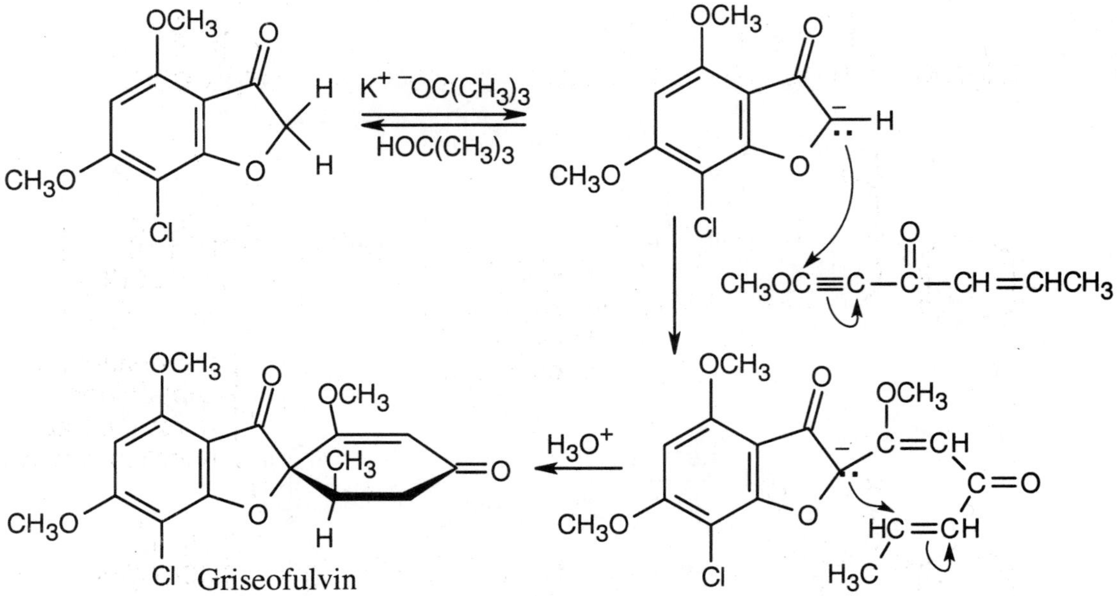

OCH_3 O H $K^+\ {}^{-}OC(CH_3)_3$ / $HOC(CH_3)_3$ H CH_3O Cl

OCH_3 O H CH_3O Cl

$CH_3OC{\equiv}C-\overset{O}{\overset{\|}{C}}-CH{=}CHCH_3$

OCH_3 O OCH_3 CH_3 =O CH_3O H Cl Griseofulvin

H_3O^+

OCH_3 O OCH_3 C=CH C=O CH_3O HC=CH H_3C Cl

Two Michael reactions are involved in the key step that forms the cyclohexenone ring.

23.51 Formation of the enolate of diethyl malonate is the first step:

$$CH_2(COOEt)_2 + {}^{-}OEt \rightleftharpoons {}^{-}\!:CH(COOEt)_2 + HOEt$$

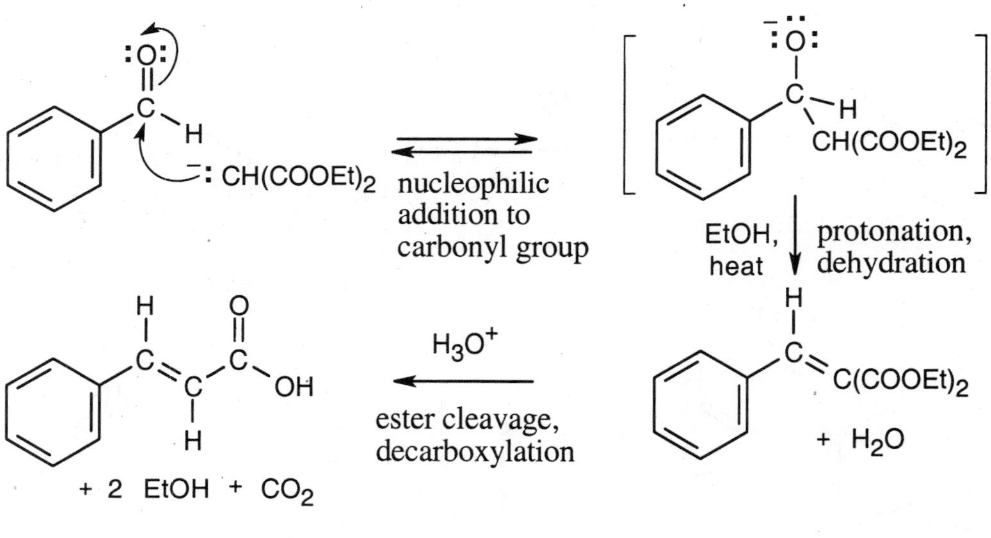

:O: C H ${}^{-}$: $CH(COOEt)_2$ — nucleophilic addition to carbonyl group → [$^{-}$:Ö: C H $CH(COOEt)_2$]

EtOH, heat | protonation, dehydration

H C= $C(COOEt)_2$ + H_2O

H_3O^+ — ester cleavage, decarboxylation

H C=C(H)–C(=O)–OH

$+ 2\ EtOH + CO_2$

23.52 Formation of enolate:

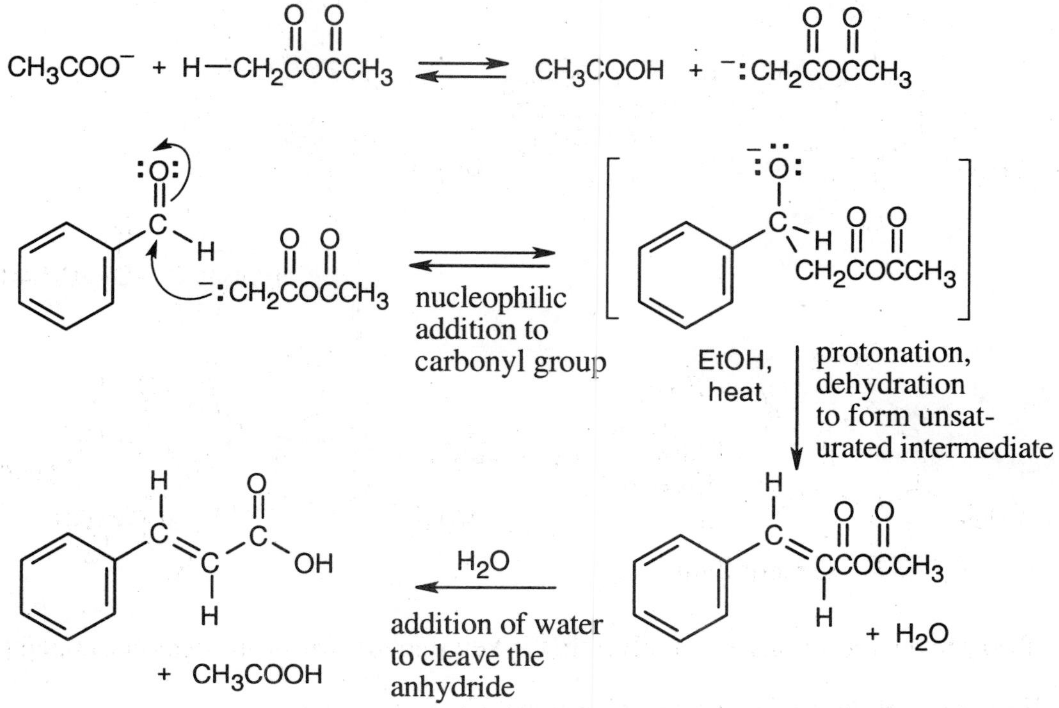

$$CH_3COO^- + H{-}CH_2COCCH_3 \rightleftharpoons CH_3COOH + {}^-{:}CH_2COCCH_3$$

nucleophilic addition to carbonyl group

EtOH, heat

protonation, dehydration to form unsaturated intermediate

H₂O

addition of water to cleave the anhydride

$+ H_2O$

$+ CH_3COOH$

23.53

$$ClCH_2COOEt + Na^+ \ {}^-OEt \rightleftharpoons Cl\ddot{C}HCOOEt + EtOH$$

ClCHCOOEt

CHCOOEt

Cl

CHCOOEt

$+ Cl^-$

nucleophilic addition

S_N2 displacement

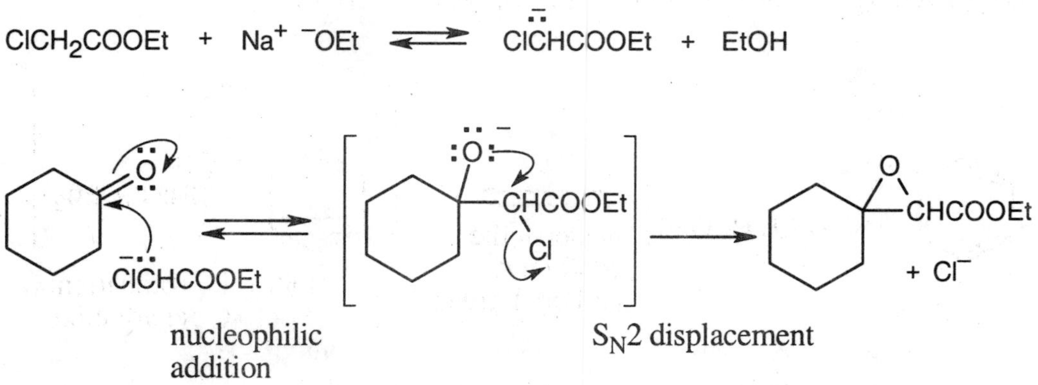

23.54 This problem becomes easier if you draw the starting material so that it resembles the product.

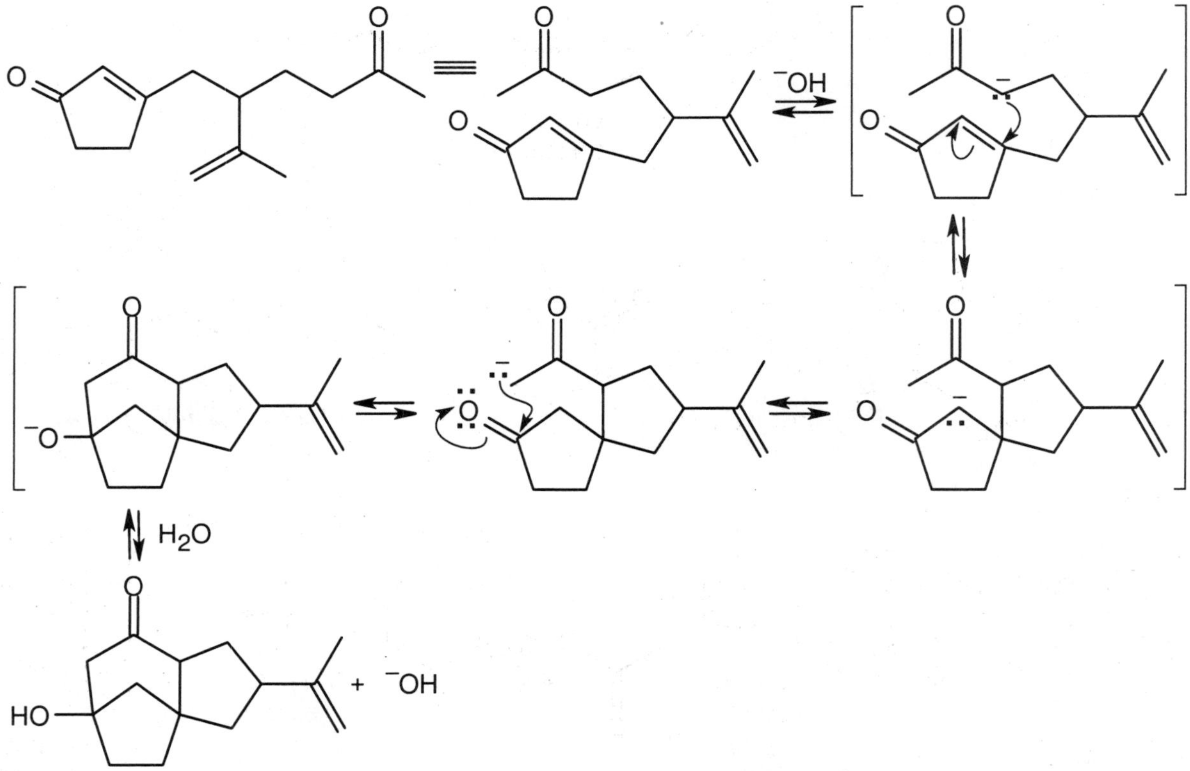

$\overset{-}{O}H$

H_2O

+ ^{-}OH

HO

23.55

COCH3

C≡CCOOCH3

$H_3C:^-$

conjugate addition of $H_3C:^-$

COCH3

C—COOCH3

CH3

nucleophilic addition

:O: :OCH3

COOCH3

CH3

loss of methoxide

COOCH3

+ $^{-}OCH_3$

CH3

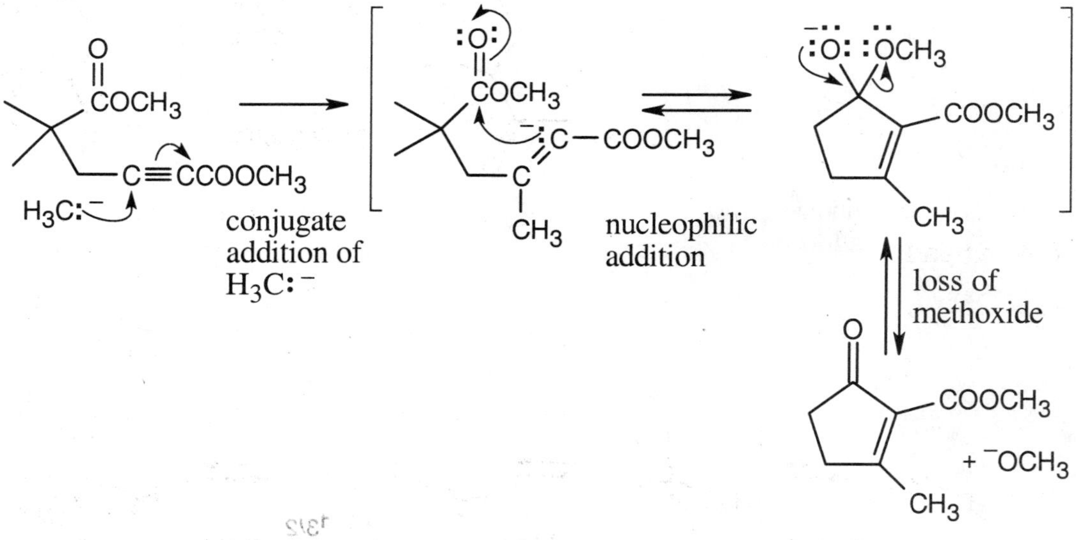

23.56

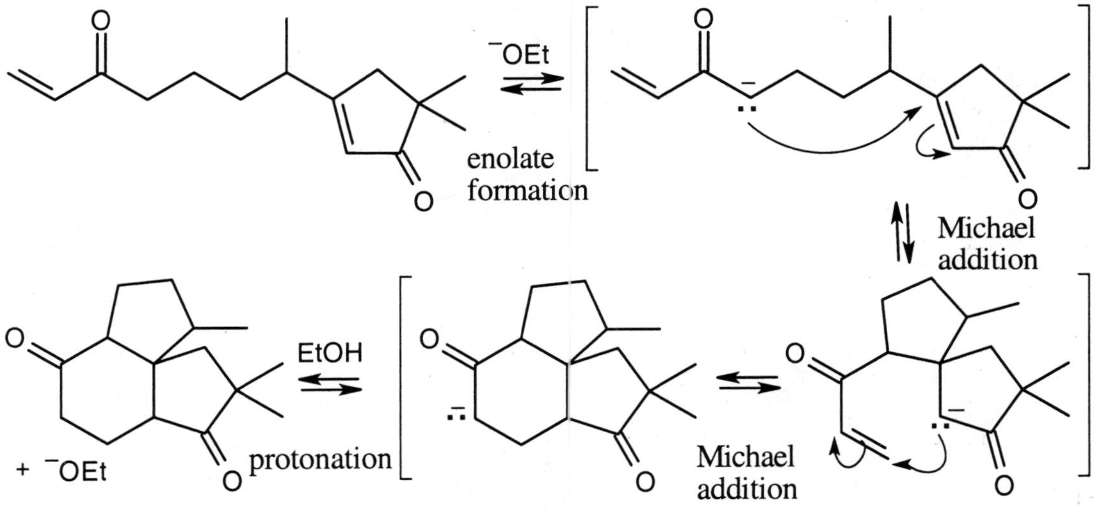

O

$^{-}$OEt

O

enolate formation

O

Michael addition

O

EtOH

O

+ $^{-}$OEt

O protonation

O

Michael addition

O

23.57

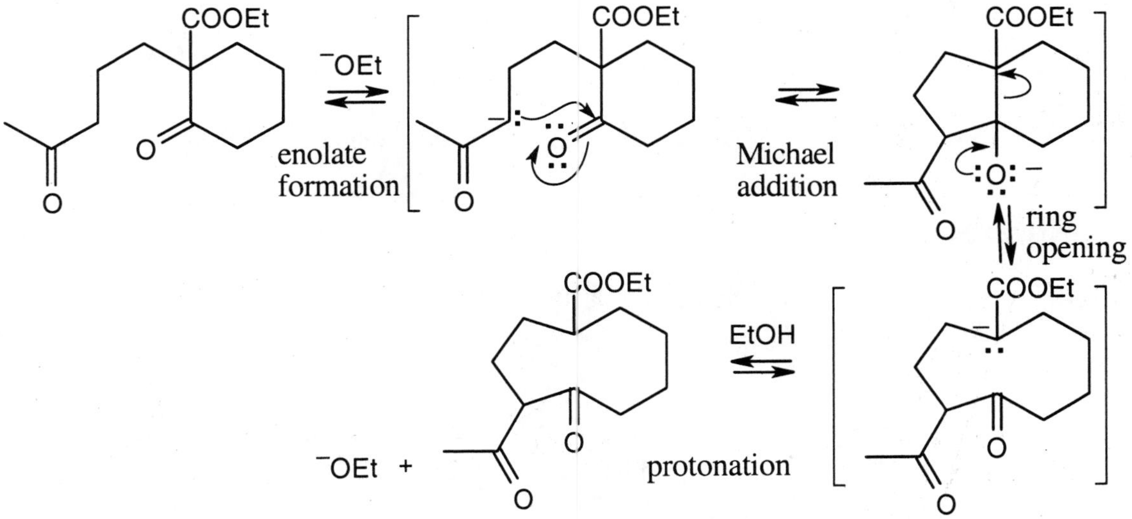

COOEt

O

O

$^{-}$OEt

enolate formation

COOEt

O

O

Michael addition

COOEt

:O: $^{-}$

O

ring opening

COOEt

EtOH

protonation

COOEt

O

O

$^{-}$OEt +

O

O

A Look Ahead

23.58

H^+

:O:

H_3C C H

proton-ation

:$\overset{+}{O}$H

H_3C C H

$\ddot{H}N(CH_3)_2$

addition of amine

:$\ddot{O}$H

H_3C C H

$H\overset{+}{N}(CH_3)_2$

proton transfer

$\ddot{O}H_2^+$

H_3C C H

$:N(CH_3)_2$

elimination of water

H_2O + $CH_3CH{=}\overset{+}{N}(CH_3)_2$

iminium ion

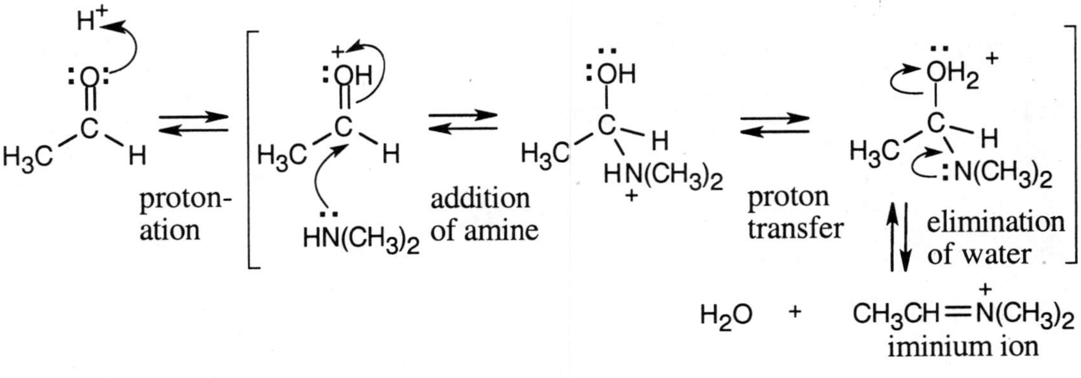

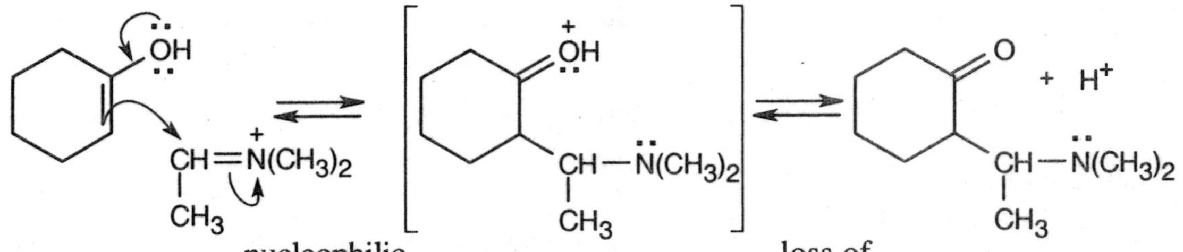

nucleophilic attack on iminium ion

loss of proton

23.59

Cocaine

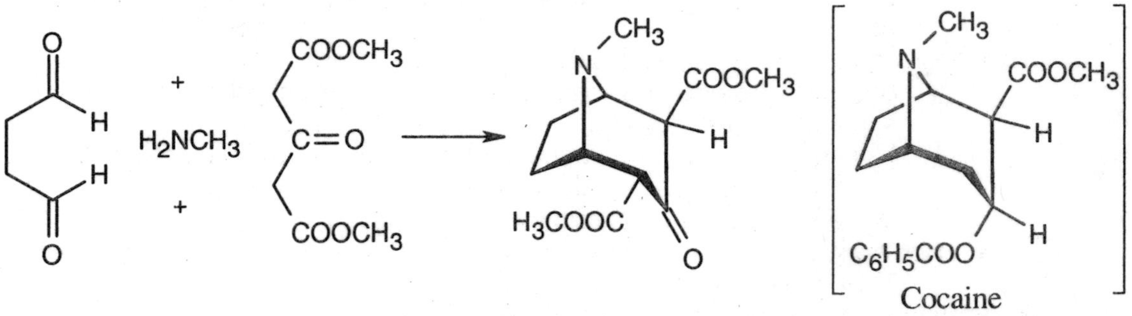

The Mannich reaction occurs between the dicarboxylate, butanedial and methylamine.

Molecular Modeling

23.60 Transition State B, which leads to a five-membered ring, is lower in energy by 31 kJ/mol. The kinetic products are:

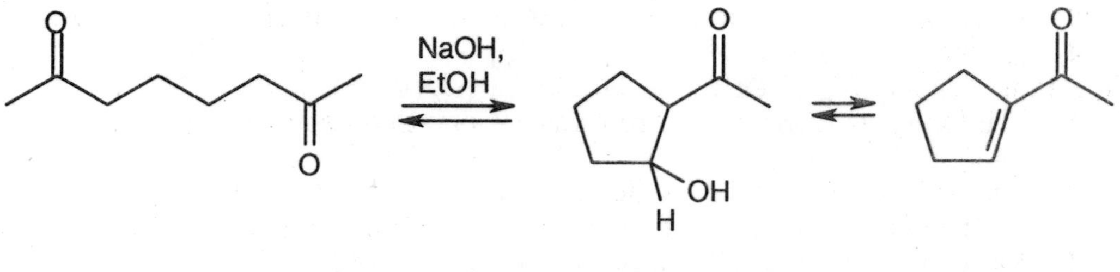

NaOH, EtOH

23.61 The first equilibrium, resulting in formation of the keto-ester, is endergonic by 113 kJ/mol. The second equilibrium, formation of the keto ester enolate, is exergonic by 223 kJ/mol. Formation of the keto ester enolate is necessary to drive the condensation to completion.

23.62 The imine anion is a stronger nucleophile.

Chapter 24 – Amines

Chapter Outline

I. Facts about amines (Section 24.1 –24.5).
  A. Naming amines (Section 24.1).
    1. Amines are classified as primary (RNH_2), secondary (R_2NH), tertiary (R_3N) or quaternary ammonium salts (R_4N^+).
    2. Primary amines are named two ways:
      a. For simple amines, the suffix *-amine* is added to the name of the alkyl substituent.
      b. For more complicated amines, the $-NH_2$ group is an amino substituent on the parent molecule.
    3. Secondary and tertiary amines:
      a. Symmetrical amines are named by using the prefixes *di-* and *tri-* before the name of the alkyl group.
      b. Unsymmetrical amines are named as *N*-substituted primary amines.
        The largest group is the parent.
    4. The simplest arylamine is aniline.
    5. Heterocyclic amines (nitrogen is part of a ring) have specific parent names.
        Nitrogen receives the lowest possible numbers.
  B. Structure and bonding in amines (Section 24.2).
    1. The three amine bonds and the lone pair occupy the corners of a tetrahedron.
    2. An amine with three different substituents is chiral.
      a. The two amine enantiomers interconvert by pyramidal inversion.
      b. This process is rapid at room temperature.
  C. Sources and properties of amines (Section 24.3).
    1. Simple amines are made by the alumina-catalyzed reaction of ammonia and methanol.
    2. Amines with fewer than 5 carbons are water-soluble.
    3. Amines have higher boiling points than alkanes of similar molecular weight.
    4. Amines smell really bad.
  D. Amine basicity (Sections 24.4 – 24.5).
    1. The lone pair of amine electrons makes amines both nucleophilic and basic (Section 24.4).
    2. The basicity constant K_b is the measure of the equilibrium of an amine with water.
        The larger the value of K_b (smaller pK_b), the stronger the base.
    3. More often, K_a is used to describe amine basicity.
      a. K_a is the dissociation constant of the conjugate acid of an amine.
      b. $pK_a + pK_b = 14$
      c. The smaller the value of K_a (larger pK_a), the stronger the base.
    4. Base strength.
      a. Alkylamines have similar basicities.
      b. Arylamines and heterocyclic amines are less basic than alkylamines.
        i. The sp^2 electrons of the pyridine lone pair are less available for bonding.
        ii. The pyrrole lone pair electrons are part of the ring π system.
      c. Amides are nonbasic.
      d. Amine basicity can be used as a means of separating amines.
        An amine can be converted to its salt, extracted from a solution with water, neutralized, and re-extracted with an organic solvent.
      e. Some amines are very weak acids.
        LDA is formed from diisopropylamine and acts as a strong base.

5. Basicity of substituted arylamines (Section 24.5).
    a. Arylamines are less basic than alkylamines for two reasons:
        i. Arylamine lone-pair electrons are delocalized over the aromatic ring and are less available for bonding.
        ii. Arylamines lose resonance stabilization when they are protonated.
    b. Electron-donating substituents increase arylamine basicity.

II. Synthesis of amines (Section 24.6).
  A. Reduction of amides, nitriles and nitro groups.
    1. S_N2 displacement with ^-CN, followed by reduction, turns a primary alkyl halide into an amine with one more carbon atom.
    2. Amide reduction converts an amide or nitrile into an amine with the same number of carbons.
    3. Arylamines can be prepared by reducing nitro compounds.
        a. Catalytic hydrogenation can be uses if no other interfering groups are present.
        b. $SnCl_2$ can also be used.
  B. S_N2 reactions of alkyl halides.
    1. It is possible to alkylate ammonia or an amine with RX.
        Unfortunately, it is difficult to avoid overalkylation.
    2. An alternative is displacement of ^-X by azide, followed by hydrogenation.
    3. Also, reaction of an alkyl halide with phthalimide anion, followed by hydrolysis, gives a primary amine.
  C. Reductive amination of aldehydes and ketones.
    Treatment of an aldehyde or ketone with ammonia or an amine in the presence of a reducing agent yields an amine
        a. The reaction proceeds through an imine, which is reduced.
        b. $NaBH_3CN$ is the reducing agent most commonly used.
  D. Rearrangements.
    1. Hofmann rearrangement.
        a. When a primary amide is treated with Br_2 and base, CO_2 is eliminated, and an amine with one less carbon is produced.
        b. The mechanism is lengthy and proceeds through an isocyanate intermediate.
        c. In the rearrangement step, the –R group migrates at the same time as the Br^- ion leaves.
    2. The Curtius rearrangement starts with an acyl azide and occurs by a mechanism very similar to that of the Hofmann rearrangement.

III. Reactions of amines (Sections 24.7 – 24.9).
  A. Alkylation and acylation (Section 24.7).
    1. Alkylation of primary and secondary amines is hard to control.
    2. Primary and secondary amines can also be acylated.
  B. Hofmann elimination.
    1. Alkylamines can be converted to alkenes by the Hofmann elimination reaction.
        a. The amine is treated with an excess of methyl iodide to form a quaternary ammonium salt.
        b. Treatment of the quaternary salt with Ag_2O, followed by heat, gives the alkene.
    2. The elimination is an S_N2 reaction.
    3. The less substituted double bond is formed because of the bulk of the leaving group.
    4. The reaction was formerly used for structure determination and is rarely used today.
  C. Reactions of arylamines (Section 24.8).
    1. Electrophilic aromatic substitution.
        a. Electrophilic aromatic substitutions are usually carried out on *N*-acetylated amines, rather than on unprotected amines.
            i. Amino groups are *o,p*-activators, and polysubstitution sometimes occurs.
            ii. Friedel-Crafts reactions don't take place with unprotected amines.

b. Aromatic amines are acetylated by treatment with acetic anhydride.
c. The *N*-acetylated amines are *o,p*-directing activators, but are less reactive than unprotected amines.
d. Synthesis of sulfa drugs was achieved by electrophilic aromatic substitution reactions on *N*-protected aromatic compounds.

2. The Sandmeyer reaction.
   a. When a primary arylamine is treated with HNO_2, an arenediazonium salt is formed.
   b. The diazonio group of arenediazonium salts can be replaced by many types of nucleophiles in radical substitution reactions.
      i. Aryl halides are formed by treatment with CuCl, CuBr or NaI.
      ii. Aryl nitriles are formed by treatment with CuCN.
      iii. Phenols are formed by treatment with Cu_2O and $Cu(NO_3)_2$.
      iv H_3PO_2 converts a diazonium salt to an arene, and is used when a substituent must be introduced and then removed.
3. Diazonium coupling reactions.
   a. Diazonium salts can react with activated aromatic rings to form colored azo compounds.
   b. The reaction is an electrophilic aromatic substitution that usually occurs at the *p*-position of the activated ring.
   c. The extended π system of the azo ring system makes these compounds brightly colored.
4. Tetraalkylammonium salts as phase-transfer catalysts (Section 24.9).
   a. If one reactant is water-soluble and the other is soluble in organic solvents, reaction can't occur because the two reactants can't encounter each other.
   b. In many cases, adding a tetraalkylammonium salt can bring about reaction. The cation is soluble in organic solvents, and can bring an anion from the aqueous layer into the organic layer, where it can take part in a reaction.

IV. Spectroscopy of amines (Section 24.10).
   A. IR spectroscopy.
      1. Primary and secondary amines absorb in the region 3300–3500 cm^{-1}.
         a. Primary amines show a pair of bands at 3350 cm^{-1} and 3450 cm^{-1}.
         b. Secondary amines show a single band at 3350 cm^{-1}.
         c. These absorptions are sharper than alcohol absorptions, which also occur in this range.
      2. Adding a small amount of HCl causes a broad band in the range 2200–3000 cm^{-1} that is due to ammonium ion.
   B. NMR spectroscopy.
      1. ^{1}H NMR.
         a. Amine protons are hard to identify because they appear as broad signals.
         b. Exchange with D_2O causes the amine signal to disappear and allows identification.
         c. Hydrogens on the carbon next to nitrogen are somewhat deshielded.
      2. ^{13}C NMR.
         Carbons next to nitrogen are slightly deshielded.
   C. Mass spectrometry.
      1. The nitrogen rule: A compound with an odd number of nitrogens has an odd-numbered molecular weight (and molecular ion).
      2. Alkylamines undergo α-cleavage and show peaks that correspond to both possible modes of cleavage.

Solutions to Problems

24.1 Facts to remember about naming amines:

(1) Primary amines are named by adding the suffix *-amine* to the name of the alkyl substituent.

(2) The prefix *di-* or *tri-* is added to the names of symmetrical secondary and tertiary amines.

(3) Unsymmetrical secondary and tertiary amines are named as *N*-substituted primary amines. The parent amine has the largest alkyl group.

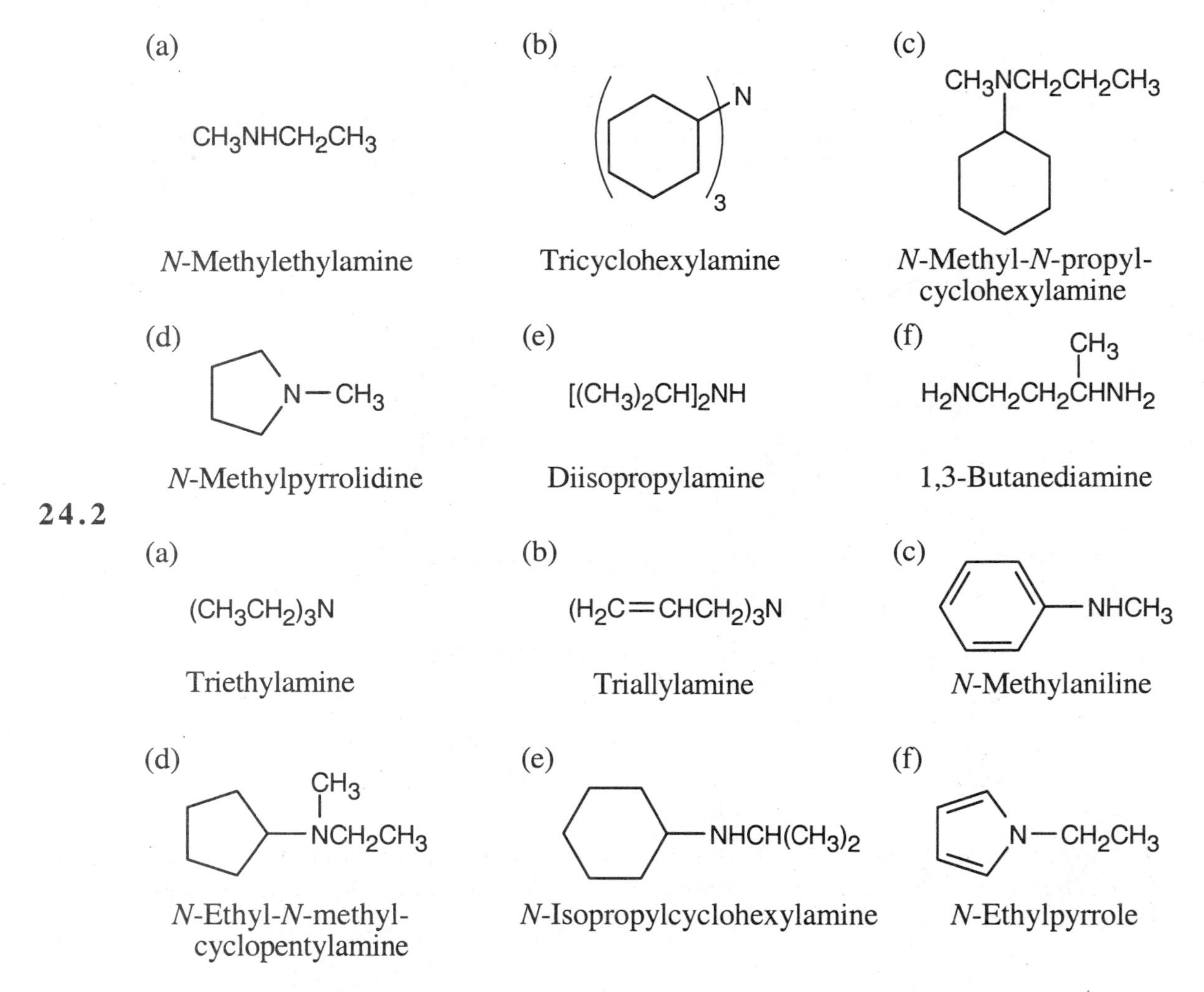

24.3 The numbering of heterocyclic rings is described in Section 24.1.

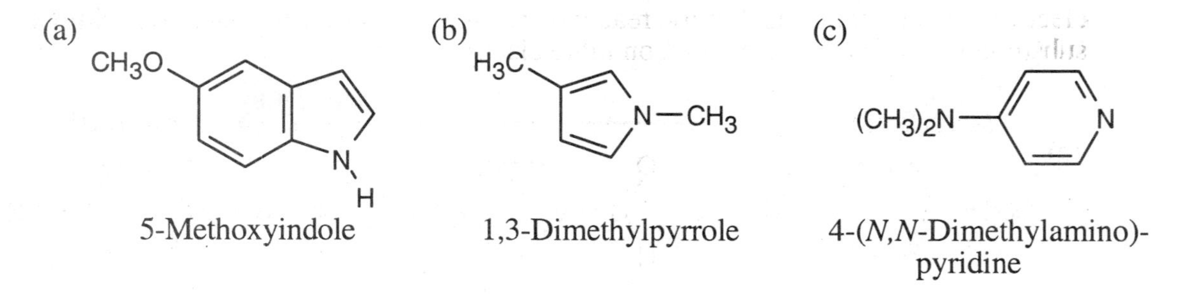

(d)

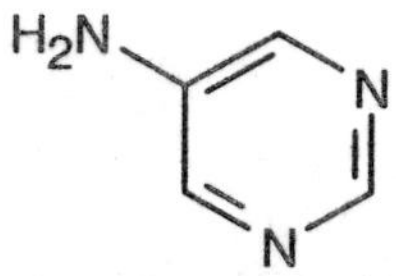

H_2N– (pyrimidine ring)

5-Aminopyrimidine

24.4 Amines are less basic than hydroxides but more basic than amides. The pK_a values of the conjugate acids of the amines in (c) are shown. The larger the pK_a, the stronger the base.

	More Basic	*Less Basic*
(a)	$CH_3CH_2NH_2$	$CH_3CH_2CONH_2$
(b)	NaOH	CH_3NH_2
(c)	$(CH_3)_2NH$ pK_a = 10.73	pyridine pK_a = 5.25

24.5

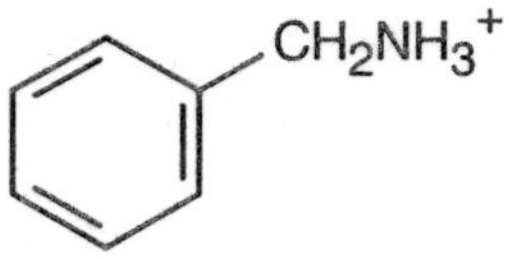

$C_6H_5CH_2NH_3^+$	$CH_3CH_2CH_2NH_3^+$
pK_a = 9.33 stronger acid (smaller pK_a)	pK_a = 10.71 weaker acid (larger pK_a)
$C_6H_5CH_2NH_2$	$CH_3CH_2CH_2NH_2$
pK_b = 14 − 9.33 = 4.67 weaker base	pK_b = 14 − 10.71 = 3.29 stronger base

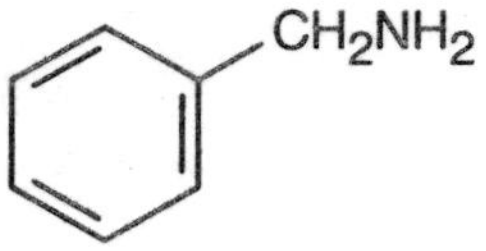

The stronger base (propylamine) holds onto a proton more tightly than the weaker base (benzylamine). Thus, the propylammonium ion is less acidic (larger pK_a) than the benzylammonium ion (smaller pK_a).

To calculate pK_b: $K_a \cdot K_b = 10^{-14}$, pK_a + pK_b = 14 and pK_b = 14 – pK_a.

24.6 The basicity order of substituted arylamines is the same as their reactivity order in electrophilic aromatic substitution reactions because, in both cases, electron-withdrawing substituents make the site of reaction more electron-poor.

Least Basic ⟶ *Most Basic*

(a) O_2N–C_6H_4–NH_2 < $HC(=O)$–C_6H_4–NH_2 < Br–C_6H_4–NH_2

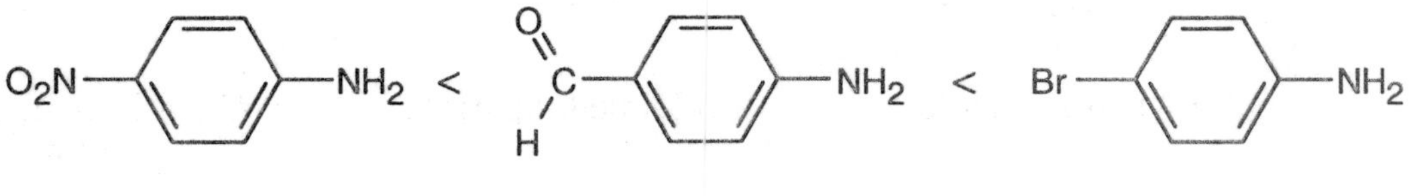

Least Basic ————————————————> *Most Basic*

(b)

$CH_3C(=O)-C_6H_4-NH_2$ < $Cl-C_6H_4-NH_2$ < $H_3C-C_6H_4-NH_2$

(c)

$F_3C-C_6H_4-NH_2$ < $H_2FC-C_6H_4-NH_2$ < $H_3C-C_6H_4-NH_2$

24.7 Amide reduction can be used to synthesize most amines, but nitrile reduction can be used to synthesize only primary amines. Thus, the compounds in (b) and (d) can be synthesized only by amide reduction.

	Amine	*Nitrile Precursor*	*Amide Precursor*
(a)	$CH_3CH_2NH_2$	$CH_3C{\equiv}N$	$CH_3C(=O)NH_2$
(b)	$(CH_3CH_2)_2NH$		$CH_3C(=O)NHCH_2CH_3$
(c)	$C_6H_5-CH_2NH_2$	$C_6H_5-C{\equiv}N$	$C_6H_5-C(=O)NH_2$
(d)	$C_6H_5-NHCH_2CH_3$		$C_6H_5-NHC(=O)CH_3$

The compounds in parts b and d can't be prepared by reduction of a nitrile.

24.8

addition of hydroxide

ring opening

proton transfer

addition of hydroxide

elimination of amide

proton transfer

$\ddot{N}H_2R$ +

24.9

HO, OH-substituted benzene with CH_2CH_2Br →(NaN_3)→ $CH_2CH_2N_3$ →(1. $LiAlH_4$; 2. H_2O)→ $CH_2CH_2NH_2$ Dopamine

potassium phthalimide → phthalimide N—CH_2CH_2—(OH, OH-benzene) →(^-OH, H_2O)→ Dopamine

24.10 Look at the target molecule to find the groups bonded to nitrogen. One group comes from the aldehyde/ketone precursor, and the other group comes from the amine precursor. In most cases, two combinations of amine and aldehyde/ketone are possible.

	Amine	*Amine Precursor*		*Carbonyl Precursor*
(a)	$CH_3CH_2NHCH(CH_3)_2$	$CH_3CH_2NH_2$		$CH_3\overset{O}{\overset{\parallel}{C}}CH_3$
			or	
		$H_2NCH(CH_3)_2$		CH_3CHO
(b)	C_6H_5—$NHCH_2CH_3$	C_6H_5—NH_2		$OHCCH_3$
(c)	cyclopentyl—$NHCH_3$	cyclopentyl—NH_2		$OHCCH_3$
			or	
		H_2NCH_3		cyclopentanone (=O)

24.11

$R-C(-\ddot{O}:^-)=\ddot{N}-\overset{+}{N}\equiv N:$ → $\ddot{O}=C=\ddot{N}-R$ + N_2

alkyl group migration and loss of N_2

$\ddot{O}=C=\ddot{N}-R$ + $H_2\ddot{O}:$ ⇌ [$H_2\overset{+}{O}-C(=\ddot{O}:)-\ddot{N}^-R$] ⇌ $H\ddot{O}-C(=\ddot{O}:)-\ddot{N}HR$ ⇌ [$^-:\ddot{O}-C(=\ddot{O}:)-\overset{+}{N}H_2R$] → CO_2 + RNH_2

attack of H_2O — proton transfer — **carbamic acid** — proton transfer — loss of CO_2

24.12 In both of these reactions, the product amine is formed from a carboxylic acid derivative precursor that has one more carbon than the amine. In the Hofmann rearrangement, the precursor is an amide, which is treated with ammonia, then with Br_2, NaOH and H_2O. In the Curtius rearrangement, the precursor is an acid chloride, which is treated with NaN_3, then with H_2O and heat.

(a)

$$(CH_3)_3CCH_2CH_2COOH \xrightarrow{SOCl_2} (CH_3)_3CCH_2CH_2COCl \xrightarrow{NH_3} (CH_3)_3CCH_2CH_2CONH_2$$

$$(CH_3)_3CCH_2CH_2COCl \xrightarrow{NaN_3} (CH_3)_3CCH_2CH_2CON_3$$

$$(CH_3)_3CCH_2CH_2CONH_2 \xrightarrow{Br_2,\ NaOH,\ H_2O} (CH_3)_3CCH_2CH_2NH_2$$

$$(CH_3)_3CCH_2CH_2CON_3 \xrightarrow[heat]{H_2O} (CH_3)_3CCH_2CH_2NH_2$$

(b)

$$H_3C\text{-}C_6H_4\text{-}COOH \xrightarrow{SOCl_2} H_3C\text{-}C_6H_4\text{-}COCl \xrightarrow{NH_3} H_3C\text{-}C_6H_4\text{-}COCl$$

$$H_3C\text{-}C_6H_4\text{-}COCl \xrightarrow{NaN_3} H_3C\text{-}C_6H_4\text{-}CON_3$$

$$H_3C\text{-}C_6H_4\text{-}COCl \xrightarrow{Br_2,\ NaOH,\ H_2O} H_3C\text{-}C_6H_4\text{-}NH_2$$

$$H_3C\text{-}C_6H_4\text{-}CON_3 \xrightarrow[heat]{H_2O} H_3C\text{-}C_6H_4\text{-}NH_2$$

24.13 The Hofmann elimination yields alkenes and amines from larger amines. The major alkene product has the less substituted double bond, but all possible products may be formed. The hydrogens that can be eliminated are starred. When possible, cis and trans double bond isomers are both formed.

	Amine	*Alkene Products*	*Amine products*
(a)	$CH_3CH_2CH_2CH(NH_2)CH_2CH_2CH_2CH_3$ (H's starred on C3 and C5)	$CH_3CH_2CH{=}CHCH_2CH_2CH_2CH_3$	$(CH_3)_3N$
		or	
		$CH_3CH_2CH_2CH{=}CHCH_2CH_2CH_3$	$(CH_3)_3N$

Both hydrogens that might be eliminated are secondary, and both possible products should form in approximately equal amounts.

	Amine	*Alkene Products*	*Amine products*
(b)	cyclohexylamine (NH_2; both adjacent ring H's starred)	cyclohexene	$(CH_3)_3N$
(c)	$CH_3CH_2CH_2CH(NH_2)CH_2CH_2CH_3$ (H's starred on C3 and C5)	$CH_3CH_2CH{=}CHCH_2CH_2CH_3$	$(CH_3)_3N$

In each of the above reactions, only one product can form.

(d) [cyclohexyl–$NHCH_2CH_3$] → $H_2C{=}CH_2$ *(major)* + [cyclohexyl–$N(CH_3)_2$]; *or* [cyclohexene] *(minor)* + $(CH_3)_2NCH_2CH_3$

The first pair of products in (d) result from elimination of a primary hydrogen and are the major products. the second pair of products result from elimination of a secondary hydrogen.

24.14

$$\text{piperidine (N–H)} \xrightarrow[\text{(excess)}]{CH_3I} \text{N,N-dimethylpiperidinium } I^- \xrightarrow[H_2O]{Ag_2O} \text{N,N-dimethylpiperidinium } ^-OH + AgI \xrightarrow{\text{heat}} CH_2{=}CHCH_2CH_2CH_2N(CH_3)_2 + H_2O$$

The product contains both the double bond and the tertiary amine in an ring-opened structure.

24.15 This reaction sequence is similar to the sequence used to synthesize sulfanilamide. Key steps are: (1) treatment of aniline with acetic anhydride (to modulate reactivity), (2) reaction of acetanilide with chlorosulfonic acid, (3) treatment of the chlorosulfonate with the heterocyclic base, and (4) removal of the acetyl group.

$$C_6H_6 \xrightarrow[H_2SO_4]{HNO_3} O_2N\text{–}C_6H_5 \xrightarrow[EtOH]{H_2,\ Pt} H_2N\text{–}C_6H_5$$

$$H_2N\text{–}C_6H_5 \xrightarrow{(1)\ (CH_3CO)_2O} CH_3\overset{O}{\overset{\|}{C}}NH\text{–}C_6H_5 + CH_3COOH$$

$$CH_3\overset{O}{\overset{\|}{C}}NH\text{–}C_6H_5 \xrightarrow[(2)]{HOSO_2Cl} CH_3\overset{O}{\overset{\|}{C}}NH\text{–}C_6H_4\text{–}SO_2Cl$$

$$CH_3\overset{O}{\overset{\|}{C}}NH\text{–}C_6H_4\text{–}SO_2Cl \xrightarrow{(3)\ \text{2-aminothiazole } (H_2N\text{–}C_3H_2NS)} CH_3\overset{O}{\overset{\|}{C}}NH\text{–}C_6H_4\text{–}SO_2\text{-}NH\text{–(2-thiazolyl)}$$

$$\xrightarrow[H_2O,\ (4)]{NaOH} H_2N\text{–}C_6H_4\text{–}SO_2\text{-}NH\text{–(2-thiazolyl)}$$

Sulfathiazole

24.16 In all of these reactions, benzene is nitrated and the nitro group is ultimately reduced, but the timing of the reduction step is important in arriving at the correct product. In (a), nitrobenzene is immediately reduced and alkylated. In (c), chlorination occurs before reduction so that chlorine can be introduced in the *m*-position. In (b) and (d), nitrobenzene is reduced and then acetylated, in order to overcome amine basicity and to control reactivity. In both cases, the acetyl group is removed in the last step.

Either method of nitro group reduction can be used in all parts of this problem; both methods are shown.

(a)

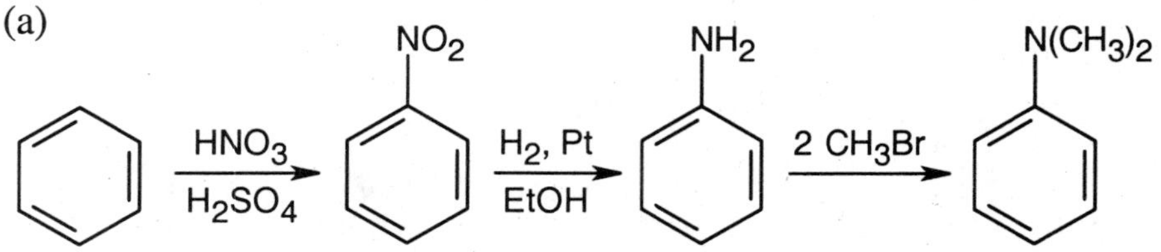

benzene $\xrightarrow[H_2SO_4]{HNO_3}$ NO_2-benzene $\xrightarrow[EtOH]{H_2,\ Pt}$ NH_2-benzene $\xrightarrow{2\ CH_3Br}$ $N(CH_3)_2$-benzene

Mono- and trialkylated anilines are also formed.

(b)

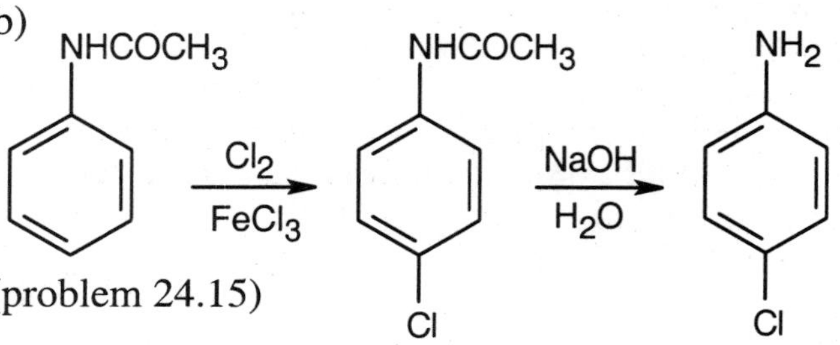

$NHCOCH_3$-benzene (problem 24.15) $\xrightarrow[FeCl_3]{Cl_2}$ $NHCOCH_3$, Cl (para) $\xrightarrow[H_2O]{NaOH}$ NH_2, Cl (para)

(c)

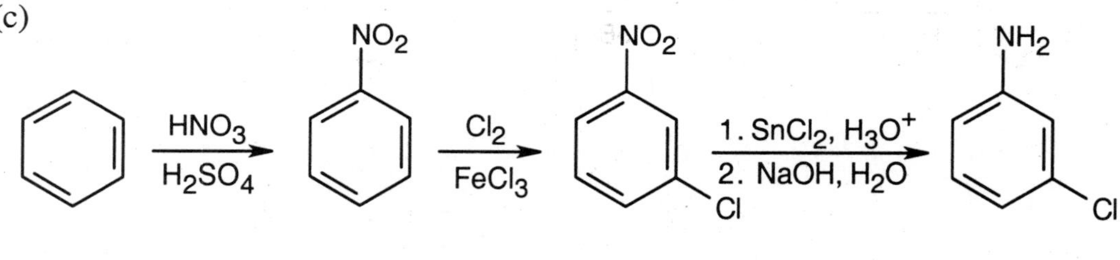

benzene $\xrightarrow[H_2SO_4]{HNO_3}$ NO_2-benzene $\xrightarrow[FeCl_3]{Cl_2}$ NO_2, Cl (meta) $\xrightarrow[2.\ NaOH,\ H_2O]{1.\ SnCl_2,\ H_3O^+}$ NH_2, Cl (meta)

(d)

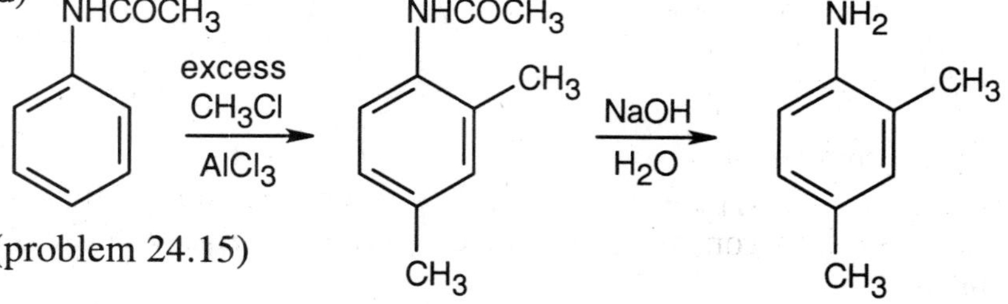

$NHCOCH_3$-benzene (problem 24.15) $\xrightarrow[AlCl_3]{\text{excess } CH_3Cl}$ $NHCOCH_3$, CH_3 (ortho), CH_3 (para) $\xrightarrow[H_2O]{NaOH}$ NH_2, CH_3 (ortho), CH_3 (para)

24.17

(a)

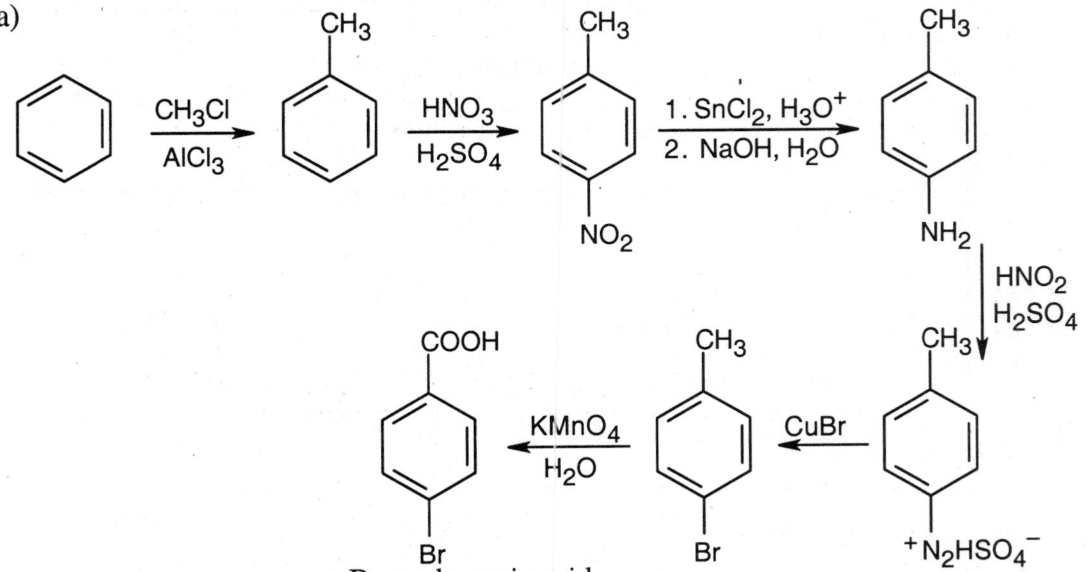

CH_3Cl / $AlCl_3$; HNO_3 / H_2SO_4; 1. $SnCl_2$, H_3O^+ 2. NaOH, H_2O; HNO_2 / H_2SO_4; CuBr; $KMnO_4$ / H_2O

CH_3 CH_3 CH_3 NO_2 NH_2 CH_3 $^+N_2HSO_4^-$ CH_3 Br   COOH   Br

p-Bromobenzoic acid

Synthetic analysis: The route shown above is one of several ways to synthesize *p*-bromobenzoic acid and is definitely not the simplest way. (The simplest route is Friedel-Crafts alkylation → bromination → oxidation). The illustrated synthesis shows the use of the diazonium replacement reaction that substitutes bromine for a nitro group. Oxidation of the methyl group yields the substituted benzoic acid.

(b)

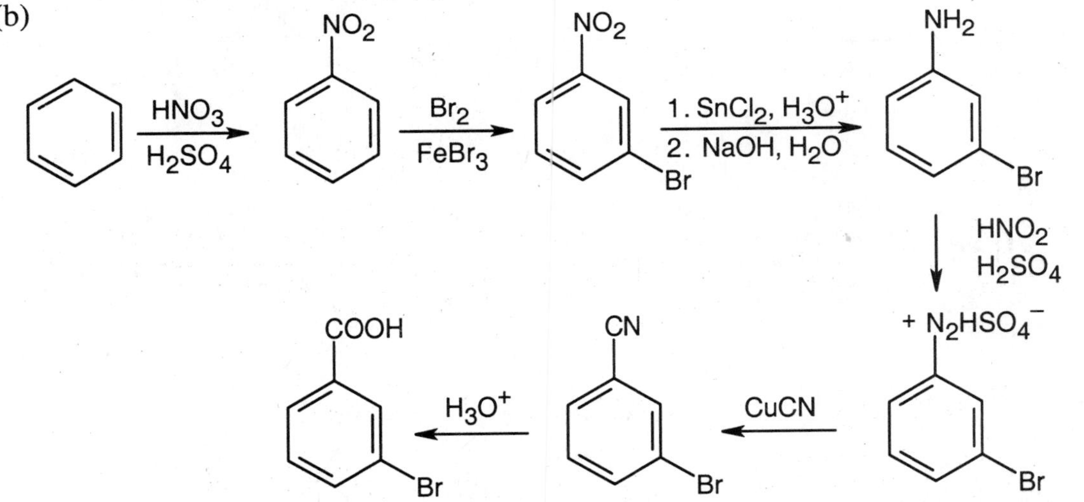

HNO_3 / H_2SO_4; Br_2 / $FeBr_3$; 1. $SnCl_2$, H_3O^+ 2. NaOH, H_2O; HNO_2 / H_2SO_4; CuCN; H_3O^+

NO_2 NO_2, Br NH_2, Br $^+N_2HSO_4^-$, Br   CN, Br   COOH, Br

m-Bromobenzoic acid

Synthetic analysis: Again, this isn't the easiest route to this compound. In this case, nitration is followed by bromination, then by diazotization, treatment with CuCN, and hydrolysis of the nitrile.

(c)

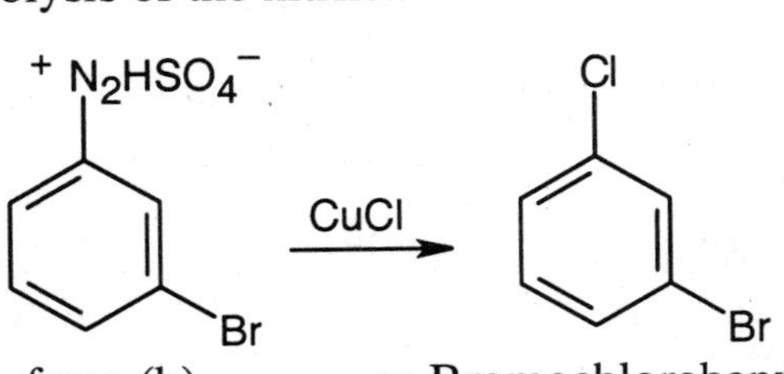

$^+N_2HSO_4^-$, Br   CuCl   Cl, Br

from (b)   *m*-Bromochlorobenzene

(d)

$$\text{p-}CH_3C_6H_4N_2^+HSO_4^- \text{ (from (a))} \xrightarrow{CuCN} \text{p-}CH_3C_6H_4CN \xrightarrow{H_3O^+} \text{p-}CH_3C_6H_4COOH$$

from (a)

p-Methylbenzoic acid

(e)

$$C_6H_5NHCOCH_3 \xrightarrow{2\ Br_2} \text{2,4-dibromoacetanilide} \xrightarrow[H_2O]{NaOH} \text{2,4-dibromoaniline} \xrightarrow[H_2SO_4]{HNO_2} \text{2,4-}Br_2C_6H_3N_2^+HSO_4^- \xrightarrow{CuBr} \text{1,2,4-}C_6H_3Br_3$$

(Problem 24.15)

1,2,4-Tribromobenzene

24.18

$$C_6H_6 \xrightarrow[H_2SO_4]{HNO_3} C_6H_5NO_2 \xrightarrow[2.\ NaOH,\ H_2O]{1.\ SnCl_2,\ H_3O^+} C_6H_5NH_2 \xrightarrow[H_2SO_4]{HNO_2} C_6H_5N_2HSO_4^-$$

$$C_6H_5\overset{+}{N}{\equiv}N: + C_6H_5\ddot{N}(CH_3)_2 \longrightarrow \left[C_6H_5\text{-}\ddot{N}{=}\ddot{N}\text{-}C_6H_5(H){=}\overset{+}{N}(CH_3)_2 \right] \xrightarrow{-\ H^+} C_6H_5\text{-}\ddot{N}{=}\ddot{N}\text{-}C_6H_4\text{-}\ddot{N}(CH_3)_2$$

Problem 24.16(a)

p-(*N,N*-Dimethylamino)azobenzene

Coupling takes place between *N,N*-dimethylaniline and a benzenediazonium salt to yield the desired product.

24.19

$$\underset{\mathbf{A}}{(CH_3)_3C\overset{O}{\overset{\|}{C}}CH_3} \xrightarrow[NaBH_3CN]{NH_3} \underset{\mathbf{B}}{(CH_3)_3C\overset{NH_2}{\overset{|}{C}}HCH_3}$$

The IR spectrum shows that **B** is a primary amine, and the ^1H NMR spectrum shows a 9-proton singlet, a one-proton quartet, and a 3-proton doublet. An absorption due to the amine protons is not visible.

Visualizing Chemistry

24.20

(a) *N*-Methylisopropylamine
secondary amine

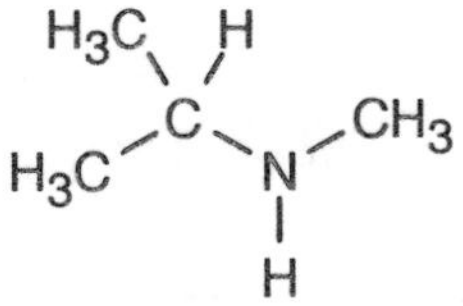

(b) *trans*-(2-Methylcyclopentyl)amine
primary amine

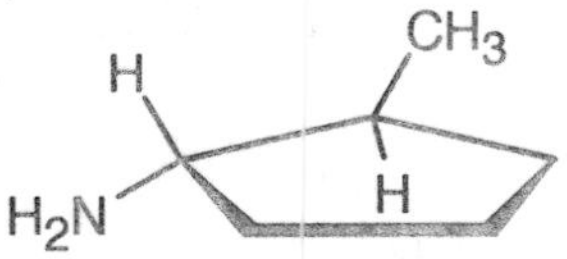

(c) *N*-Isopropylaniline
secondary amine

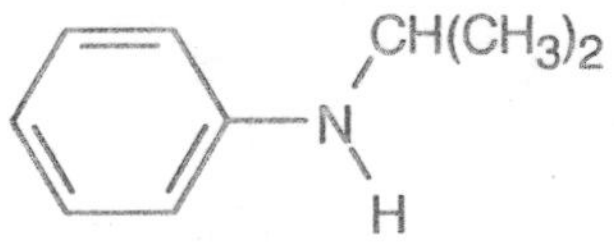

24.21

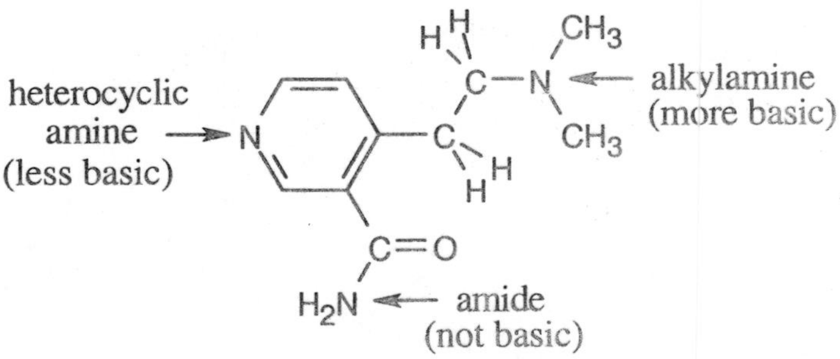

heterocyclic amine (less basic)

alkylamine (more basic)

amide (not basic)

24.22

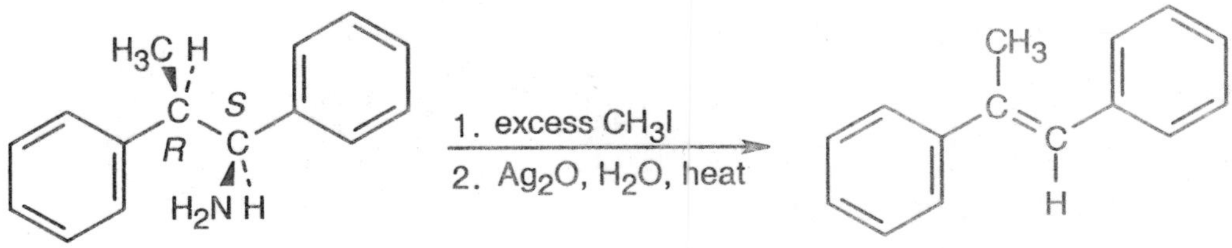

$$\text{(1}S\text{,2}R\text{)-(1,2-Diphenylpropyl)amine} \xrightarrow[\text{2. } Ag_2O,\ H_2O,\ \text{heat}]{\text{1. excess } CH_3I} (E)\text{-1,2-Diphenyl-1-propene}$$

(1*S*,2*R*)-(1,2-Diphenylpropyl)amine

(*E*)-1,2-Diphenyl-1-propene

Hofmann elimination is an E2 elimination, in which the two groups to be eliminated are 180° apart. The product that results from this elimination geometry is the *E* isomer.

Additional Problems

24.23

(a) secondary amine

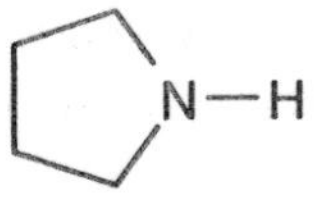

(b) secondary amine

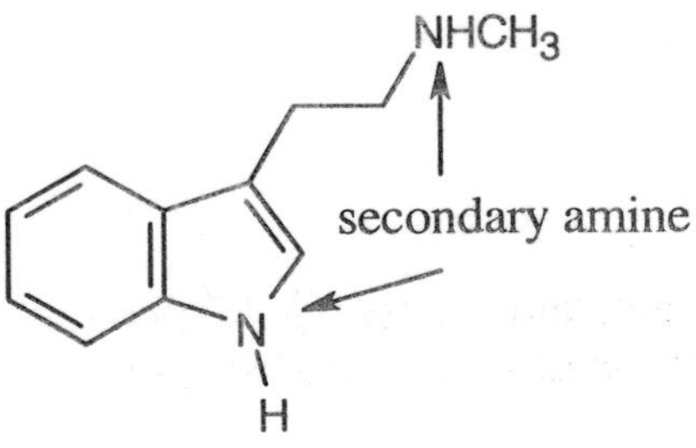

(c) tertiary amide; tertiary amine; secondary amine

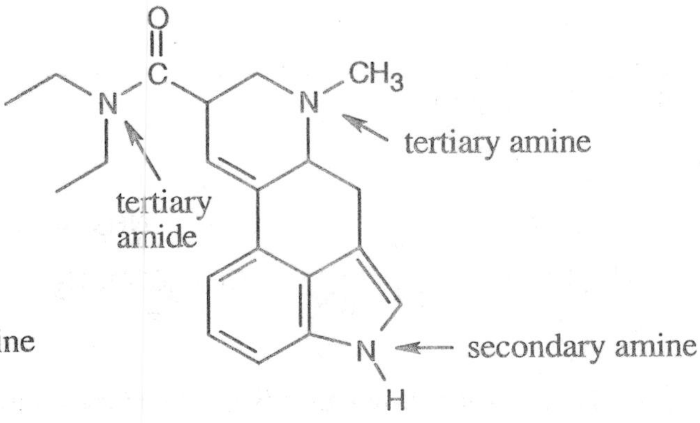

Lysergic acid diethylamide

24.24

(a) $N(CH_3)_3$ — *N,N*-Dimethylaniline

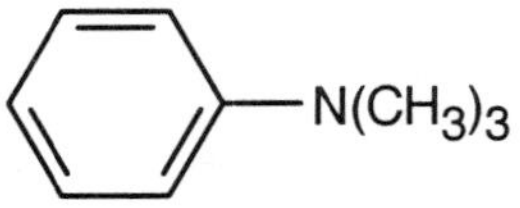

(b) CH_2NH_2 — (Cyclohexylmethyl)amine

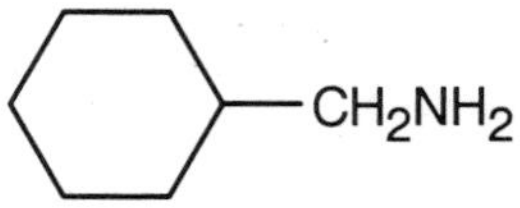

(c) $NHCH_3$ — *N*-Methylcyclohexylamine

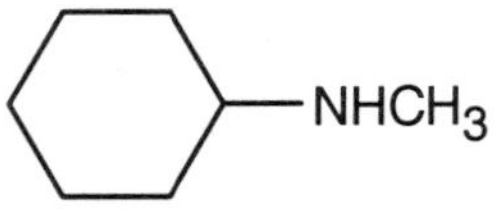

(d) CH_3, NH_2 — (2-Methylcyclohexyl)amine

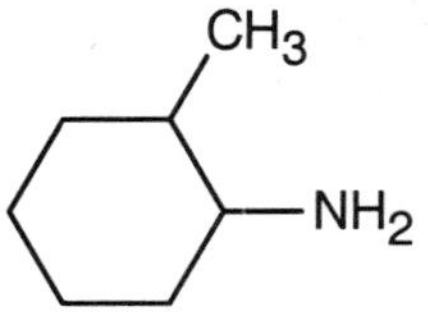

(e) $(H_3C)_2NCH_2CH_2COOH$ — 3-(*N,N*-dimethylamino)-propanoic acid

(f) CH_3, N, $CH(CH_3)_2$ — *N*-Isopropyl-*N*-methyl-cyclohexylamine

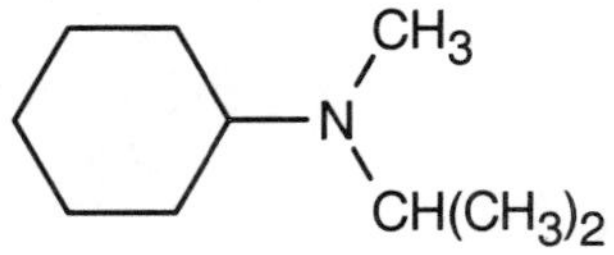

24.25

(a) NH_2, Br, Br — 2,4-Dibromoaniline

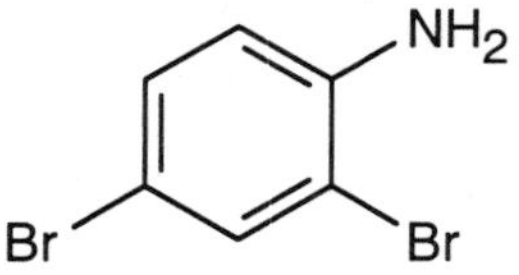

(b) $CH_2CH_2NH_2$ — (2-Cyclopentylethyl)amine

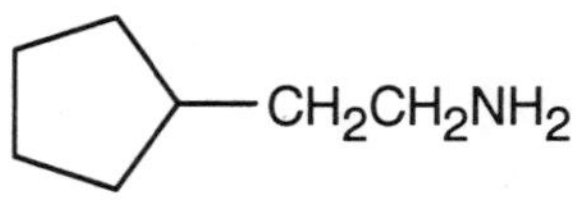

(c) $NHCH_2CH_3$ — *N*-Ethylcyclopentylamine

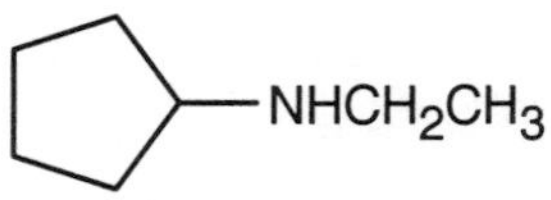

(d) CH_3, N, CH_3 — *N,N*-Dimethylcyclopentylamine

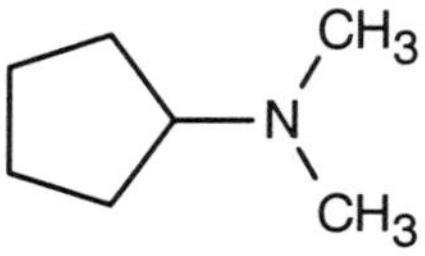

(e) $N-CH_2CH_2CH_3$ — *N*-Propylpyrrolidine

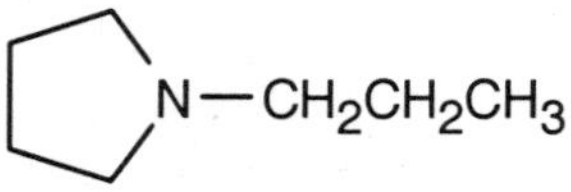

(f) $H_2NCH_2CH_2CH_2CN$ — 4-Aminobutanenitrile

24.26

(a) CH_2CH_3, H_3C, N^+, Br^-, $CH_2CH_2CH_3$, $C(CH_3)_3$ — (*S*)-*tert*-Butylethylmethyl-propylammonium bromide

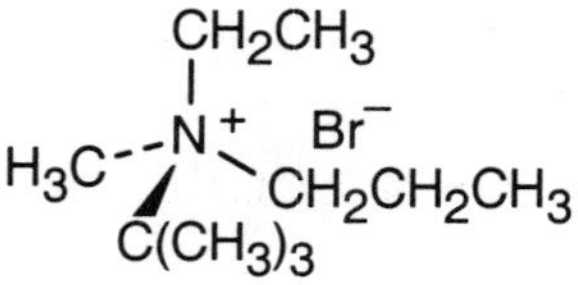

(b) N—H — Pyrrole

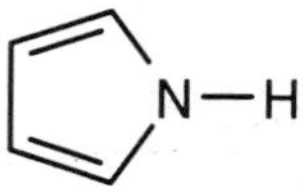

(c) H, $H_2C{=}CHCH_2NCH_2CH{=}CH_2$ — Diallylamine

24.27

(a) H_3C, NH_2 (*m*-Toluidine) $\xrightarrow[\text{1 mol}]{Br_2}$ H_3C, NH_2, Br + H_3C, NH_2, Br

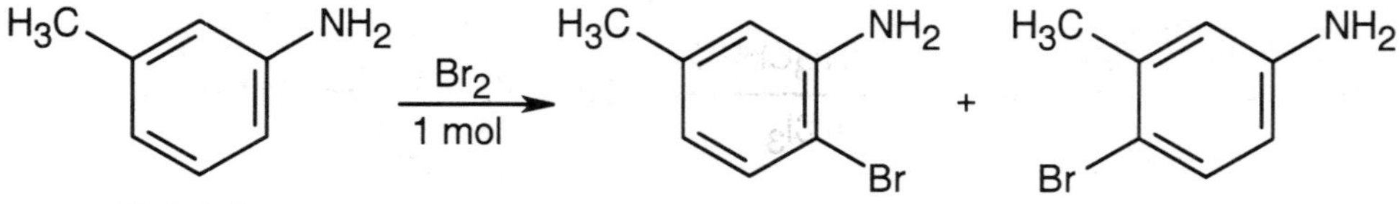

(b) H_3C, NH_2 $\xrightarrow[\text{excess}]{CH_3I}$ H_3C, $\overset{+}{N}(CH_3)_3I^-$ plus other *N*-methylated compounds

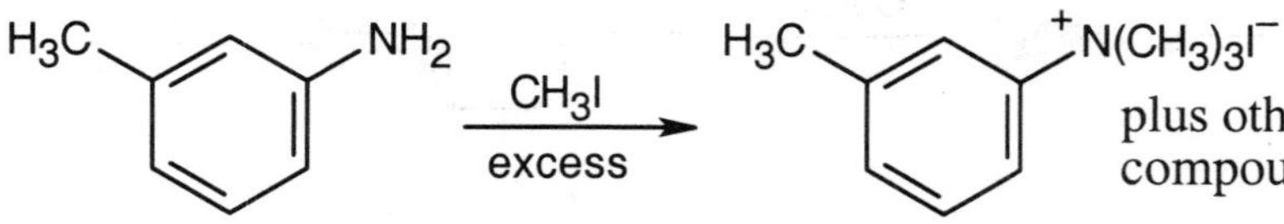

(c)

$$m\text{-}H_3C\text{-}C_6H_4\text{-}NH_2 \xrightarrow[AlCl_3]{CH_3Cl} m\text{-}H_3C\text{-}C_6H_4\text{-}NH_2\text{:}AlCl_3$$

(d)

$$m\text{-}H_3C\text{-}C_6H_4\text{-}NH_2 \xrightarrow[\text{pyridine}]{CH_3COCl} m\text{-}H_3C\text{-}C_6H_4\text{-}NHCOCH_3$$

(e)

$$m\text{-}H_3C\text{-}C_6H_4\text{-}NHCOCH_3 \xrightarrow{HOSO_2Cl} \text{4-}ClO_2S\text{-3-}H_3C\text{-}C_6H_3\text{-}NHCOCH_3$$

24.28

(a)

$$Br\text{-}C_6H_4\text{-}NH_2 \xrightarrow[\text{excess}]{CH_3I} Br\text{-}C_6H_4\text{-}\overset{+}{N}(CH_3)_3I^-$$

(b)

$$Br\text{-}C_6H_4\text{-}NH_2 \xrightarrow{HCl} Br\text{-}C_6H_4\text{-}\overset{+}{N}H_3\ Cl^-$$

(c)

$$Br\text{-}C_6H_4\text{-}NH_2 \xrightarrow[H_2SO_4]{HNO_2} Br\text{-}C_6H_4\text{-}\overset{+}{N}_2\ HSO_4^-$$

(d)

$$Br\text{-}C_6H_4\text{-}NH_2 \xrightarrow{CH_3COCl} Br\text{-}C_6H_4\text{-}NHCOCH_3$$

(e)

$$Br\text{-}C_6H_4\text{-}NH_2 \xrightarrow{CH_3MgBr} Br\text{-}C_6H_4\text{-}NH\ MgBr + CH_4$$

(f)

$$Br\text{-}C_6H_4\text{-}NH_2 \xrightarrow[AlCl_3]{CH_3CH_2Cl} Br\text{-}C_6H_4\text{-}NH_2\text{:}AlCl_3$$

(g)

$$Br\text{-}C_6H_4\text{-}\overset{+}{N}_2\ HSO_4^- \xrightarrow{CuCl} Br\text{-}C_6H_4\text{-}Cl$$

(h)

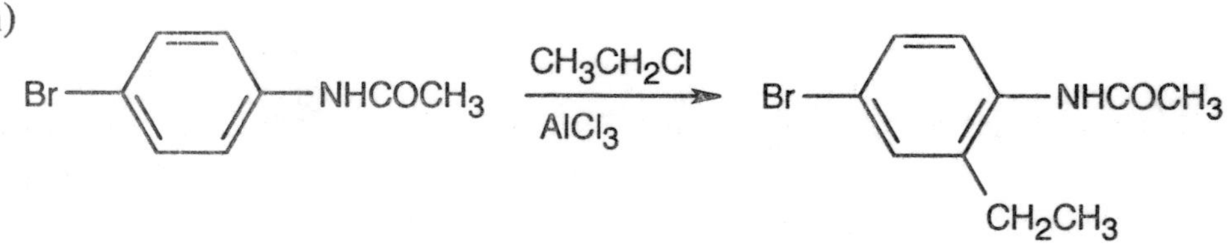

$$Br-C_6H_4-NHCOCH_3 \xrightarrow[AlCl_3]{CH_3CH_2Cl} Br-C_6H_3(CH_2CH_3)-NHCOCH_3$$

24.29

(a)

$$CH_2CH_2CH_2CH_2OH \xrightarrow{PBr_3} CH_2CH_2CH_2CH_2Br \xrightarrow{NaN_3} CH_2CH_2CH_2CH_2N_3 \xrightarrow[2.\ H_2O]{1.\ LiAlH_4} CH_2CH_2CH_2CH_2NH_2$$

Butylamine

(b)

$$CH_2CH_2CH_2CH_2OH \xrightarrow[H_3O^+]{CrO_3} CH_2CH_2CH_2COOH \xrightarrow{SOCl_2} CH_2CH_2CH_2COCl$$

$$CH_2CH_2CH_2COCl \xrightarrow[NaOH]{CH_2CH_2CH_2CH_2NH_2 \text{ from (a)}} CH_3CH_2CH_2\overset{O}{\overset{\|}{C}}NHCH_2CH_2CH_2CH_3 \xrightarrow[2.\ H_2O]{1.\ LiAlH_4} CH_3CH_2CH_2CH_2NHCH_2CH_2CH_2CH_3$$

Dibutylamine

(c)

$$CH_2CH_2CH_2CH_2OH \xrightarrow{CrO_3,\ H_3O^+} CH_2CH_2CH_2COOH \xrightarrow{SOCl_2} CH_2CH_2CH_2COCl$$

$$CH_2CH_2CH_2COCl \xrightarrow{2\ NH_3} CH_2CH_2CH_2CONH_2 \xrightarrow[H_2O]{Br_2,\ NaOH} CH_2CH_2CH_2NH_2$$

$$CH_2CH_2CH_2COCl \xrightarrow{NaN_3} CH_2CH_2CH_2CON_3 \xrightarrow{H_2O,\ heat} CH_2CH_2CH_2NH_2$$

Propylamine

(d)

$$CH_2CH_2CH_2CH_2OH \xrightarrow{PBr_3} CH_2CH_2CH_2CH_2Br \xrightarrow{NaCN} CH_2CH_2CH_2CH_2CN \xrightarrow[2.\ H_2O]{1.\ LiAlH_4} CH_2CH_2CH_2CH_2CH_2NH_2$$

Pentylamine

(e)

$$CH_2CH_2CH_2CH_2OH \xrightarrow{PCC} CH_2CH_2CH_2CHO \xrightarrow[(CH_3)_2NH]{NaBH_3CN} CH_2CH_2CH_2CH_2N(CH_3)_2$$

N,N-Dimethylbutylamine

(f)

$$\underset{\text{from (c)}}{CH_2CH_2CH_2NH_2} \xrightarrow[CH_3I]{excess} CH_2CH_2CH_2\overset{+}{N}(CH_3)I^- \xrightarrow[2.\ heat]{1.\ Ag_2O,\ H_2O} CH_3CH{=}CH_2 + (CH_3)_3N$$

Propene + $(CH_3)_3N$

24.30

(a)

$$CH_3CH_2CH_2CH_2COOH \xrightarrow{SOCl_2} CH_3CH_2CH_2CH_2COCl \xrightarrow{2\ NH_3} CH_3CH_2CH_2CH_2CONH_2$$

Pentanamide

(b)

$$\underset{\text{from (a)}}{CH_3CH_2CH_2CH_2CONH_2} \xrightarrow[H_2O]{Br_2,\ NaOH} \underset{\text{Butylamine}}{CH_3CH_2CH_2CH_2NH_2}$$

(c)

$$\underset{\text{from (a)}}{CH_3CH_2CH_2CH_2CONH_2} \xrightarrow[2.\ H_2O]{1.\ LiAlH_4} \underset{\text{Pentylamine}}{CH_3CH_2CH_2CH_2CH_2NH_2}$$

(d)

$$CH_3CH_2CH_2CH_2COOH \xrightarrow[2.\ H_2O]{1.\ Br_2,\ PBr_3} CH_3CH_2CH_2CH(Br)COOH$$

2-Bromopentanoic acid

(e)

$$CH_3CH_2CH_2CH_2COOH \xrightarrow[2.\ H_3O^+]{1.\ BH_3} CH_3CH_2CH_2CH_2CH_2OH \xrightarrow{PBr_3} CH_3CH_2CH_2CH_2CH_2Br \xrightarrow{NaCN} CH_3CH_2CH_2CH_2CH_2CN$$

Hexanenitrile

(f)

$$\underset{\text{from (e)}}{CH_3CH_2CH_2CH_2CH_2CN} \xrightarrow[2.\ H_2O]{1.\ LiAlH_4} \underset{\text{Hexylamine}}{CH_3CH_2CH_2CH_2CH_2CH_2NH_2}$$

24.31

(a)

$$C_6H_6 \xrightarrow[H_2SO_4]{HNO_3} C_6H_5NO_2 \xrightarrow[2.\ NaOH,\ H_2O]{1.\ SnCl_2,\ H_3O^+} C_6H_5NH_2$$

(b)

$$C_6H_5CONH_2 \xrightarrow[H_2O]{Br_2,\ NaOH} C_6H_5NH_2$$

(c)

$$C_6H_5CH_3 \xrightarrow[H_2O]{KMnO_4} C_6H_5COOH \xrightarrow{SOCl_2} C_6H_5COCl \xrightarrow{2\ NH_3} C_6H_5CONH_2 \xrightarrow[H_2O]{Br_2,\ NaOH} C_6H_5NH_2$$

24.32

$$C_6H_5NH_2 \xrightarrow[H_2SO_4]{HNO_2} C_6H_5\overset{+}{N}_2\ HSO_4^- \xrightarrow{H_3PO_2} \underset{(a)}{C_6H_6} \xrightarrow[AlCl_3]{CH_3Cl} \underset{(c)}{C_6H_5CH_3}$$

$$C_6H_5\overset{+}{N}_2\ HSO_4^- \xrightarrow{CuCN} C_6H_5CN \xrightarrow[H_2O]{NaOH} \underset{(b)}{C_6H_5CONH_2}$$

24.33

(a)

$$CH_3CH_2CH_2CH_2CONH_2 \xrightarrow[2.\ H_2O]{1.\ LiAlH_4} CH_3CH_2CH_2CH_2CH_2NH_2$$

(b)

$$CH_3CH_2CH_2CH_2CN \xrightarrow[2.\ H_2O]{1.\ LiAlH_4} CH_3CH_2CH_2CH_2CH_2NH_2$$

(c)

$$CH_3CH_2CH{=}CH_2 \xrightarrow[2.\ H_2O_2,\ ^-OH]{1.\ BH_3,\ THF} CH_3CH_2CH_2CH_2OH \xrightarrow{PBr_3} CH_3CH_2CH_2CH_2Br$$

$$CH_3CH_2CH_2CH_2Br \xrightarrow{NaCN} CH_3CH_2CH_2CH_2CN \xrightarrow[2.\ H_2O]{1.\ LiAlH_4} CH_3CH_2CH_2CH_2CH_2NH_2$$

(d)

$$CH_3CH_2CH_2CH_2CH_2CONH_2 \xrightarrow[H_2O]{Br_2,\ NaOH} CH_3CH_2CH_2CH_2CH_2NH_2$$

(e)

$$CH_3CH_2CH_2CH_2OH \xrightarrow{PBr_3} CH_3CH_2CH_2CH_2Br \xrightarrow{NaCN} CH_3CH_2CH_2CH_2CN$$

$$CH_3CH_2CH_2CH_2CN \xrightarrow[2.\ H_2O]{1.\ LiAlH_4} CH_3CH_2CH_2CH_2CH_2NH_2$$

(f)

$$CH_3CH_2CH_2CH_2CH{=}CHCH_2CH_2CH_2CH_3 \xrightarrow[2.\ Zn,\ H_3O^+]{1.\ O_3} 2\ CH_3CH_2CH_2CH_2CHO$$

$$CH_3CH_2CH_2CH_2CHO \xrightarrow[NaBH_3CN]{NH_3} CH_3CH_2CH_2CH_2CH_2NH_2$$

(g)

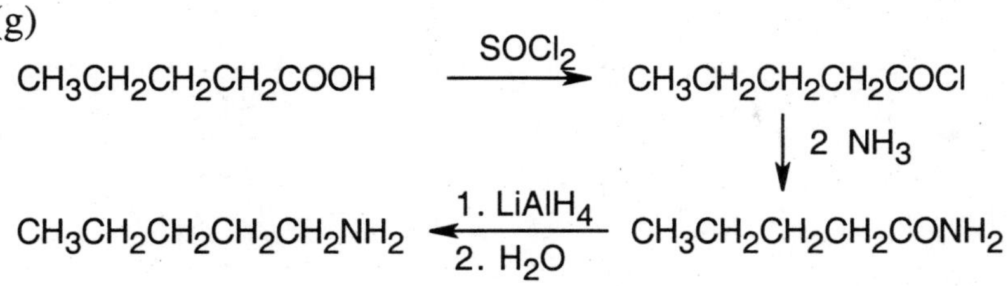

$$CH_3CH_2CH_2CH_2COOH \xrightarrow{SOCl_2} CH_3CH_2CH_2CH_2COCl$$

$$\downarrow 2\ NH_3$$

$$CH_3CH_2CH_2CH_2CH_2NH_2 \xleftarrow[2.\ H_2O]{1.\ LiAlH_4} CH_3CH_2CH_2CH_2CONH_2$$

24.34 Hydrogens that can be eliminated are starred. In cases where more than one alkene can form, the alkene with the less substituted double bond is the major product..

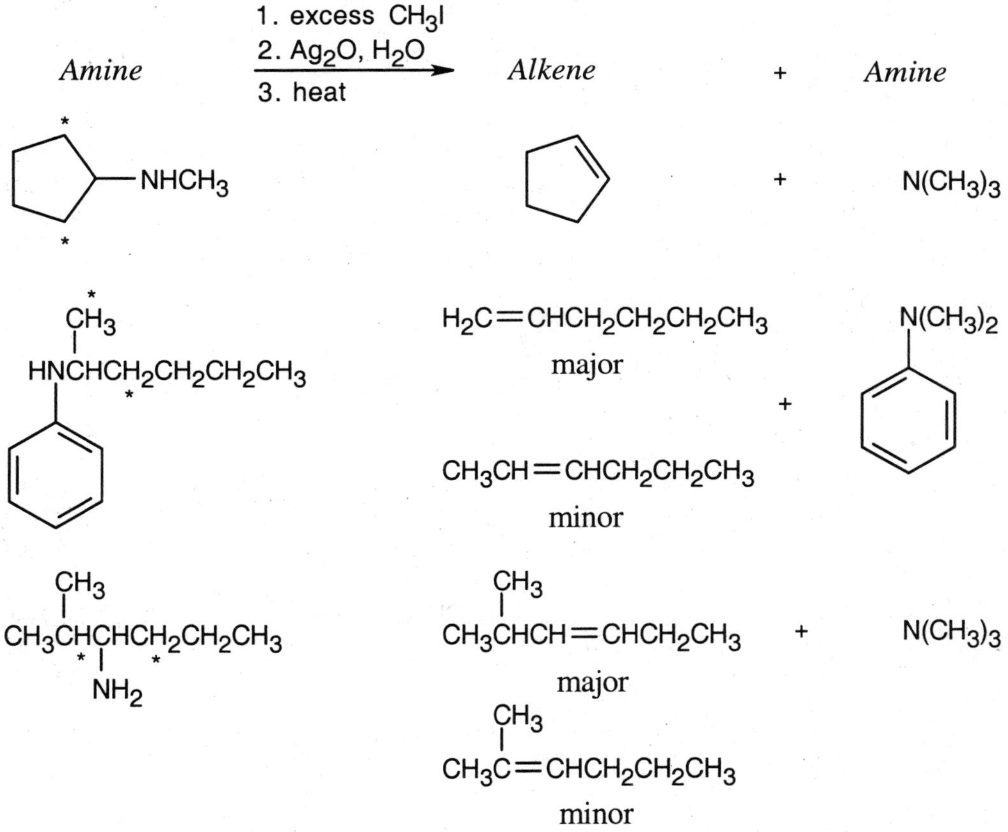

Amine $\xrightarrow[3.\ heat]{1.\ excess\ CH_3I;\ 2.\ Ag_2O,\ H_2O}$ *Alkene* + *Amine*

(a) Cyclopentyl–$NHCH_3$ (starred ring hydrogens) → cyclopentene + $N(CH_3)_3$

(b) $HN(C_6H_5)CH(CH_3^*)CH^*_2CH_2CH_2CH_3$ → $H_2C{=}CHCH_2CH_2CH_2CH_3$ (major) + $CH_3CH{=}CHCH_2CH_2CH_3$ (minor) + $C_6H_5N(CH_3)_2$

major

minor

(c) $CH_3CH^*(CH_3)CH(NH_2)CH^*_2CH_2CH_3$ → $CH_3CH(CH_3)CH{=}CHCH_2CH_3$ (major) + $CH_3C(CH_3){=}CHCH_2CH_2CH_3$ (minor) + $N(CH_3)_3$

major

minor

24.35

(a) Decahydroquinoline (N–H) $\xrightarrow[excess]{CH_3I}$ **A** (N,N-dimethyl quaternary ammonium, I^-) $\xrightarrow{Ag_2O,\ H_2O}$ **B** (quaternary ammonium, ^-OH) $\xrightarrow{heat}$

Cyclohexene (ring-fused double bond) with $(CH_2)_3N(CH_3)_2$ side chain — minor + **C** (2-allylcyclohexyl $N(CH_3)_2$) — major + cyclohexene with $(CH_2)_3N(CH_3)_2$ side chain — minor

minor major minor

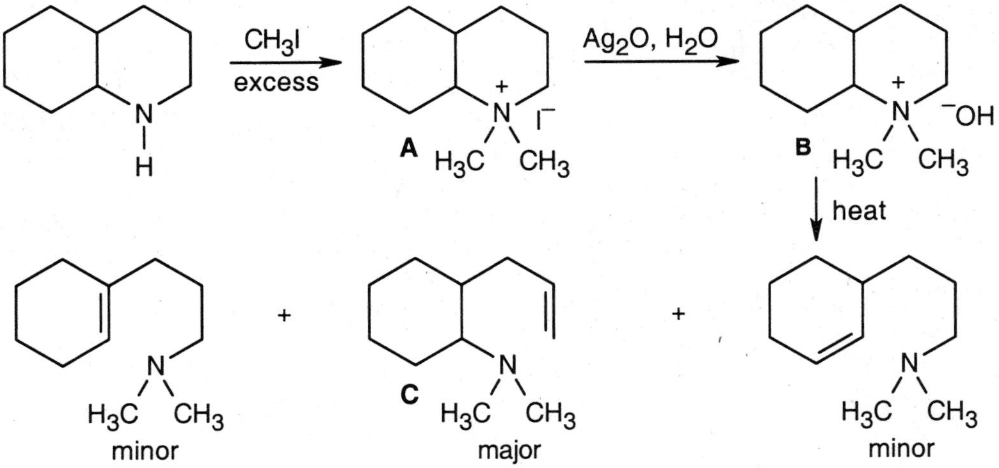

(b)

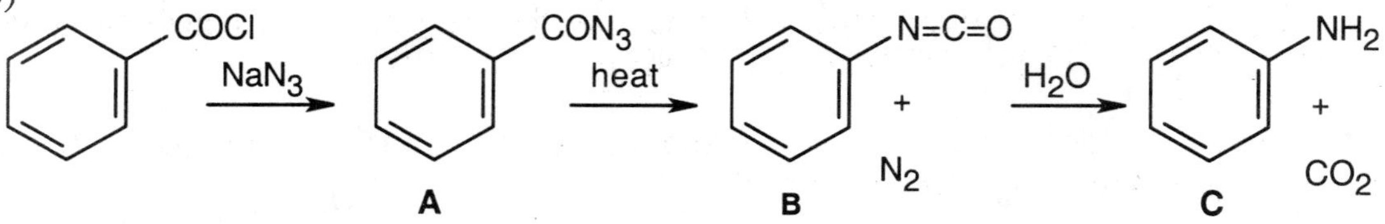

$C_6H_5COCl \xrightarrow{NaN_3} C_6H_5CON_3$ (**A**) $\xrightarrow{heat} C_6H_5N{=}C{=}O$ (**B**) $+ N_2 \xrightarrow{H_2O} C_6H_5NH_2$ (**C**) $+ CO_2$

(c)

Phthalimide (NH) $\xrightarrow{KOH}$ phthalimide anion $N^- K^+$ (**A**) $+ H_2O \xrightarrow{C_6H_5CH_2Br}$ N$-CH_2C_6H_5$ phthalimide (**B**) $+ KBr$

B $\xrightarrow{KOH, H_2O}$ phthalate (COO^-, COO^-) $+ C_6H_5CH_2NH_2$ (**C**)

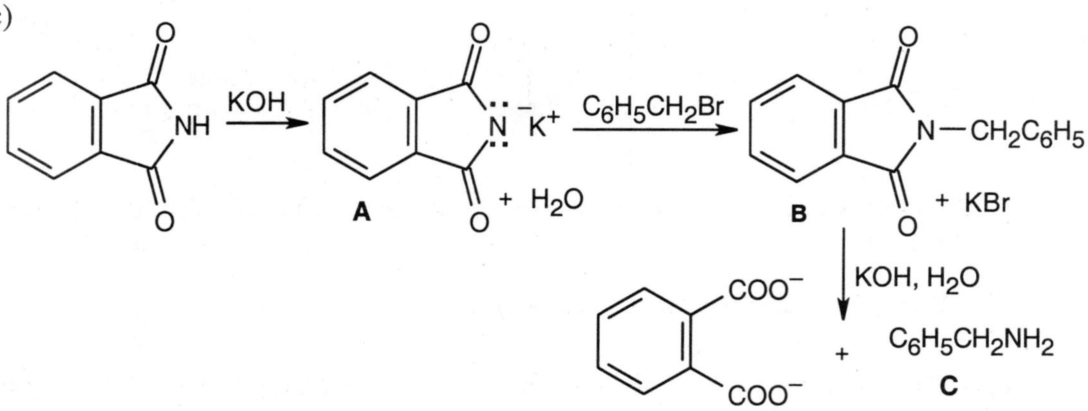

(d)

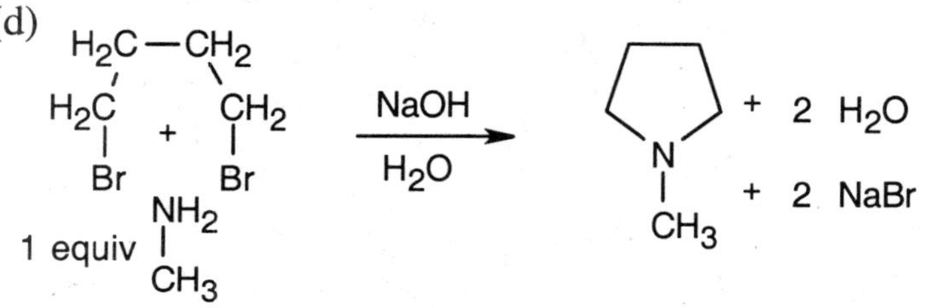

$BrH_2C-CH_2-CH_2-CH_2Br$ + 1 equiv CH_3NH_2 $\xrightarrow[H_2O]{NaOH}$ N-methylpyrrolidine ($N-CH_3$) $+ 2\ H_2O + 2\ NaBr$

24.36

$C_6H_5COCH_3 \xrightarrow{a} C_6H_5CH(NH_2)CH_3 \xrightarrow{b,\ c} C_6H_5CH{=}CH_2 \xrightarrow{d} C_6H_5CH(O)CH_2$ (epoxide) $\xrightarrow{e} C_6H_5CH(OH)CH_2N(CH_3)CH_3$

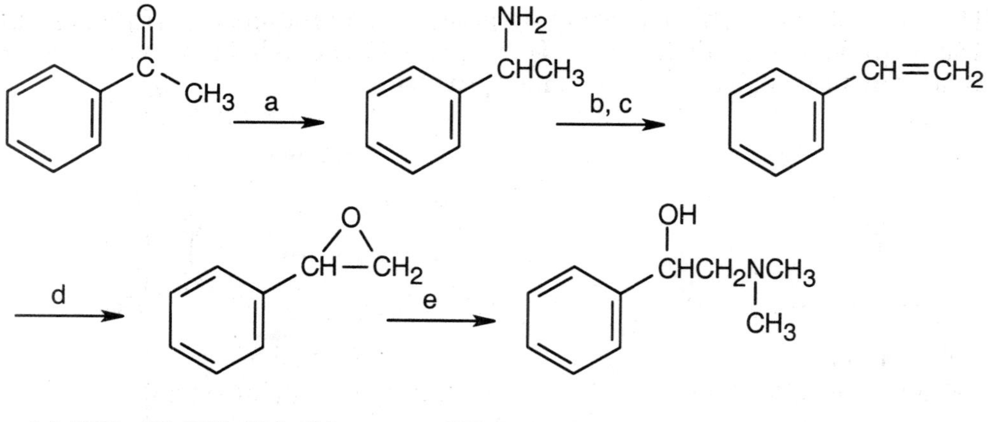

(a) NH_3, $NaBH_3CN$; (b) excess CH_3I; (c) Ag_2O, H_2O, heat; (d) RCO_3H
(e) $(CH_3)_2NH$

24.37 (a) The Hofmann rearrangement of amides yields an amine containing one less carbon atom than the starting amide. In this case, the product of Hofmann rearrangement is $CH_3CH_2NH_2$, not $CH_3CH_2CH_2NH_2$.

(b) No reaction occurs between a tertiary amine and a carbonyl group. To obtain the product shown, use $(CH_3)_2NH$.

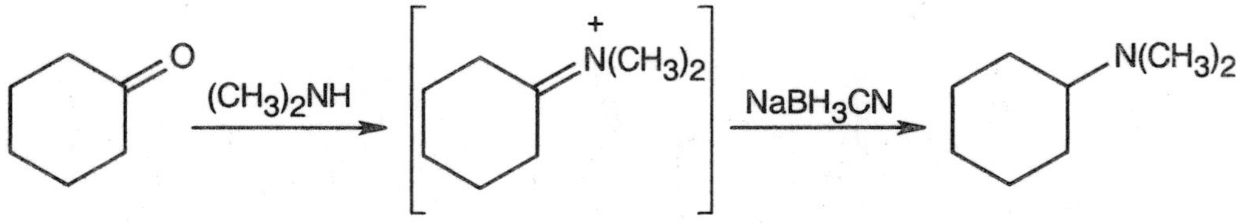

$(CH_3)_2NH$ → [$\overset{+}{N}(CH_3)_2$] $NaBH_3CN$ → $N(CH_3)_2$

(c) An elimination product is obtained when the tertiary bromide $(CH_3)_3CBr$ reacts with ammonia.

$$(CH_3)_3C{-}Br \xrightarrow{NH_3} (CH_2)_2C{=}CH_2$$

(d) An isocyanate intermediate in the Hofmann rearrangement results from treatment of an amide with Br_2 and ^-OH. Heat is used to decarboxylate the carbamic acid intermediate.

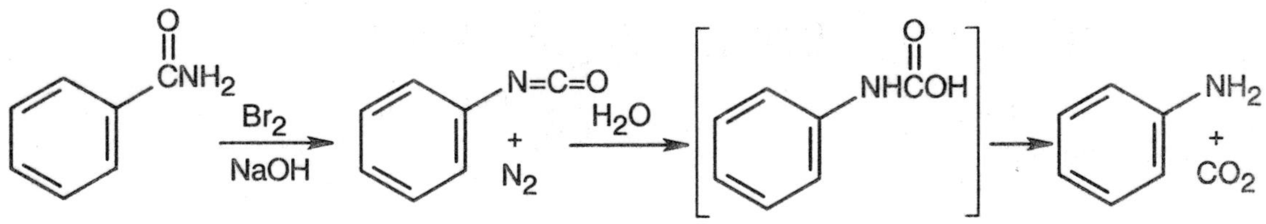

CNH_2 (O) $\xrightarrow[NaOH]{Br_2}$ $N{=}C{=}O$ + N_2 $\xrightarrow{H_2O}$ [$NHCOH$ (O)] → NH_2 + CO_2

carbamic acid

(e) The amine in this problem is being subjected to the conditions of Hofmann elimination. The major alkene product, $CH_3CH_2CH_2CH{=}CH_2$, contains the *less* substituted double bond. The product shown is the minor product.

24.38

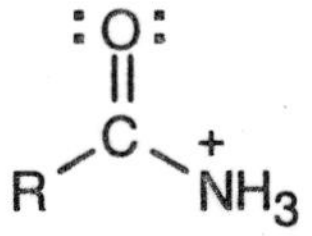

:O: R C $\overset{+}{N}H_3$ — *N*-Protonation (no resonance stabilization)

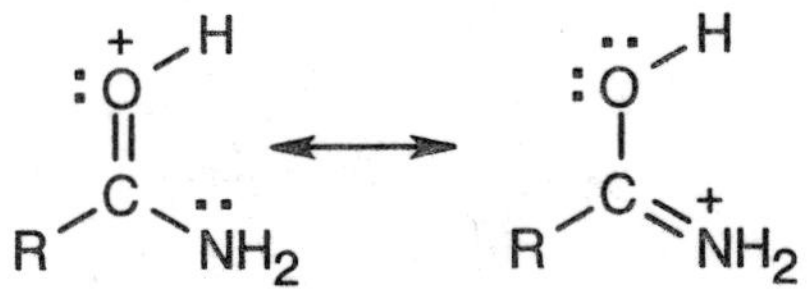

$\overset{+}{O}$–H R C $\ddot{N}H_2$ ⟷ $\ddot{O}$–H R C $\overset{+}{N}H_2$ — *O*-Protonation (resonance stabilization)

Protonation occurs on oxygen because an *O*–protonated amide is stabilized by resonance.

25.39 The nitrogen lone pair of electrons of diphenylamine can overlap the π electron system of either ring. Electron delocalization occurs to an even greater extent for diphenylamine than for aniline. Because the energy difference between non-protonated and protonated amine is much greater for diphenylamine than for aniline, diphenylamine is non-basic.

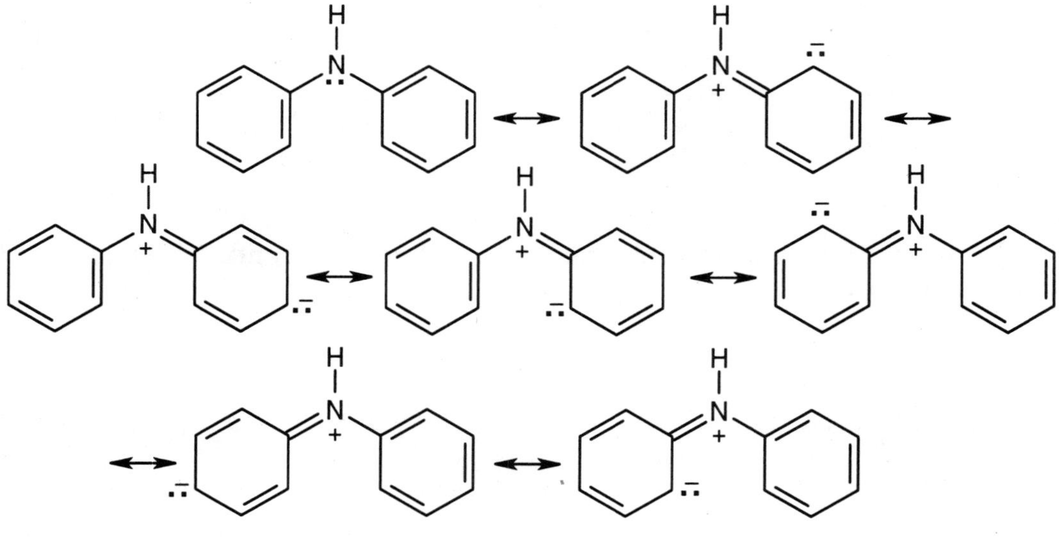

24.40

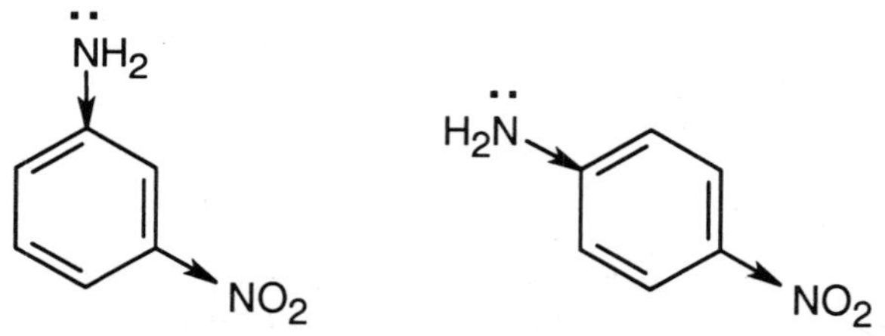

The inductive effect of the electron-withdrawing nitro group makes the amine nitrogens of both *m*-nitroaniline and *p*-nitroaniline less electron-rich and less basic than aniline.

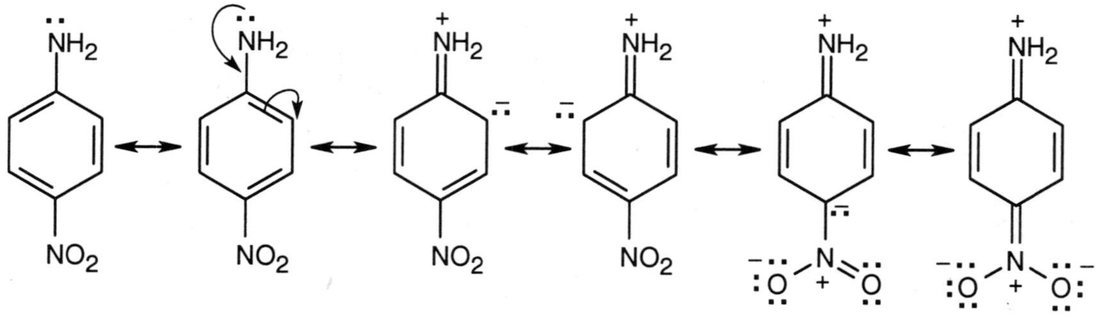

When the nitro group is para to the amino group, conjugation of the amino group with the nitro group can also occur. *p*–Nitroaniline is thus even less basic than *m*–nitroaniline.

24.41

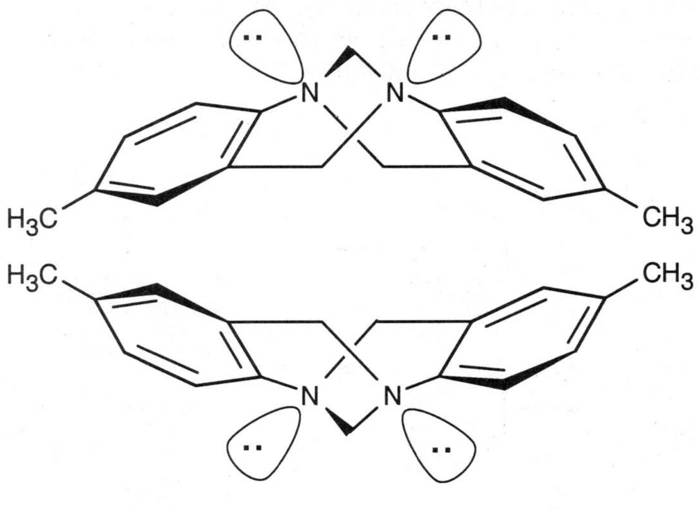

N
N
H3C
CH3
H3C
CH3
N
N

Tröger's base

The enantiomers of Tröger's base do not interconvert. Because of the rigid ring system, the substituents bonded to nitrogen can't be forced into the planar sp^2 geometry necessary for inversion at nitrogen to occur. Since inversion is not possible, the enantiomers are resolvable.

24.42

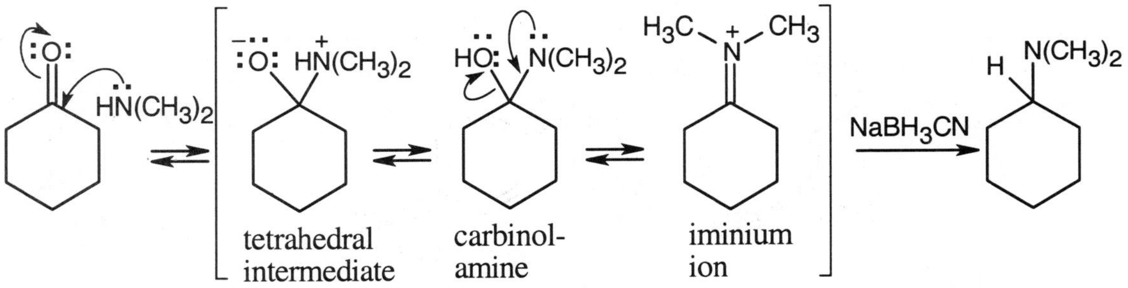

:O:
HN(CH3)2
:O: HN(CH3)2
HO: N(CH3)2
H3C CH3
N
NaBH3CN
H N(CH3)2
tetrahedral
intermediate
carbinol-
amine
iminium
ion

24.43

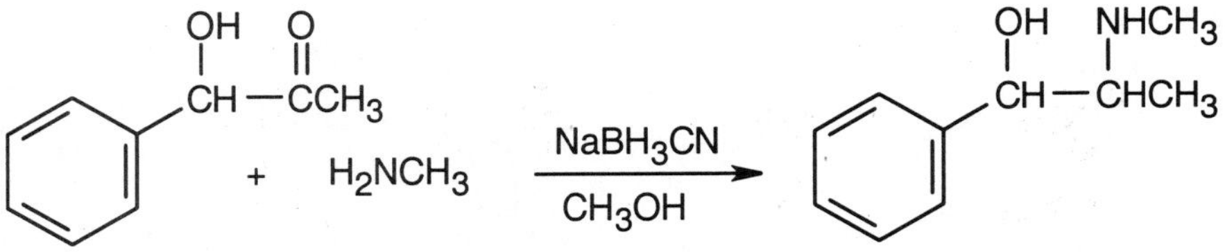

OH O
CH—CCH3
+ H2NCH3
NaBH3CN
CH3OH
OH NHCH3
CH—CHCH3

Ephedrine

24.44

Benzaldehyde first reacts with methylamine and $NaBH_3CN$ in the usual way to give the reductive amination product *N*-methylbenzylamine. This product then reacts further with benzaldehyde in a second reductive amination to give *N*-methyldibenzylamine.

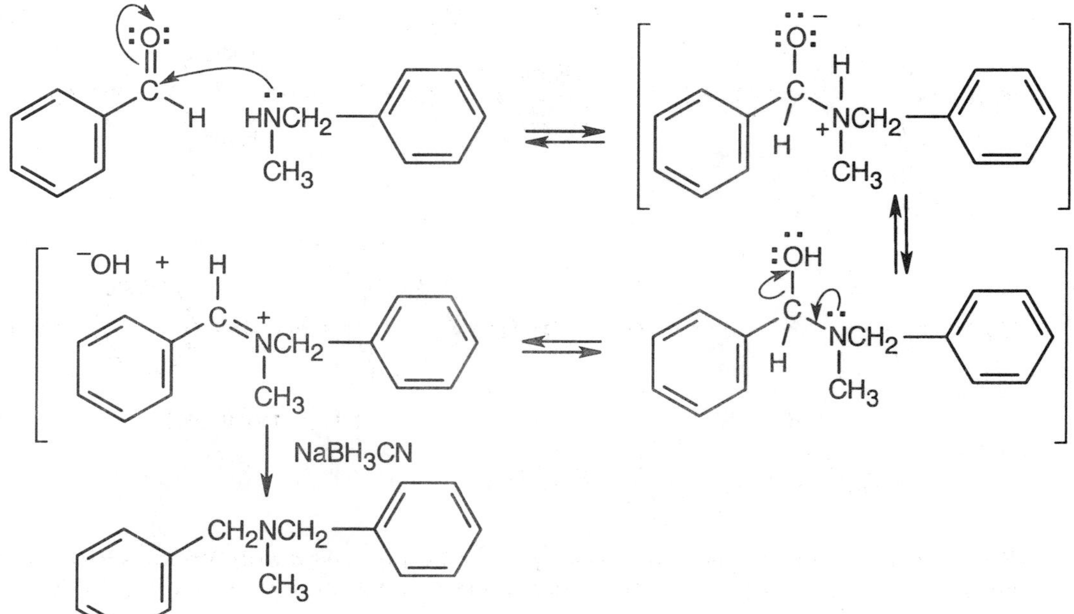

N-Methyldibenzylamine

24.45

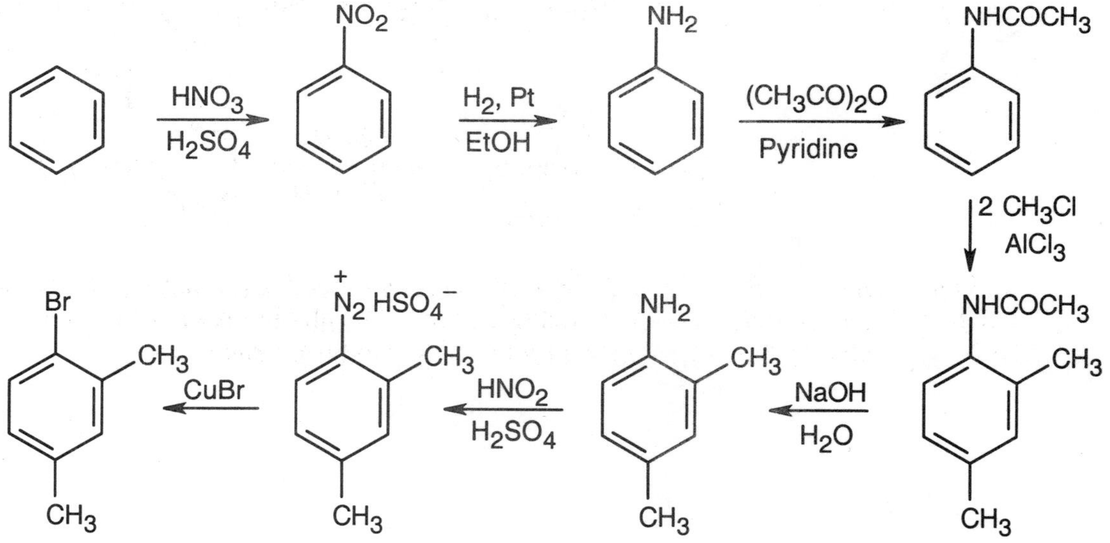

1-Bromo-2,4-dimethylbenzene

Synthetic analysis: We know that a diazotization is involved in this synthesis. The last step of this sequence involves the reaction of CuBr with a diazonium salt, which is formed from 2,4-dimethylaniline. In order to introduce the methyl groups, it is necessary to acetylate aniline and remove the acetyl group after methylation.

24.46

$$\text{C}_6\text{H}_6 \xrightarrow[H_2SO_4]{2\ HNO_3} m\text{-}O_2N\text{-}C_6H_4\text{-}NO_2 \xrightarrow[2.\ NaOH,\ H_2O]{1.\ SnCl_2,\ H_3O^+} m\text{-}H_2N\text{-}C_6H_4\text{-}NH_2$$

$$p\text{-}H_2N\text{-}C_6H_4\text{-}SO_2NH_2 \xrightarrow[H_2SO_4]{HNO_2} HSO_4^-\ \overset{+}{N_2}\text{-}C_6H_4\text{-}SO_2NH_2$$

Sulfanilamide (Section 24.8)

$$m\text{-}H_2N\text{-}C_6H_4\text{-}NH_2 + HSO_4^-\ \overset{+}{N_2}\text{-}C_6H_4\text{-}SO_2NH_2 \xrightarrow{-\ H^+} H_2N\text{-}C_6H_3(NH_2)\text{-}N{=}N\text{-}C_6H_4\text{-}SO_2NH_2$$

Prontosil

Prontosil is formed by the diazo coupling reaction between *m*-diaminobenzene and the diazonium salt of sulfanilamide, whose synthesis was shown in Section 24.8.

24.47

$$\text{C}_5\text{H}_9\text{-Br} \xrightarrow{Mg} \text{C}_5\text{H}_9\text{-MgBr} \xrightarrow[2.\ H_3O^+]{1.\ H_2C\text{—}CHCH_3\ (\text{epoxide, O})} \text{C}_5\text{H}_9\text{-}CH_2CH(OH)CH_3$$

$$\text{C}_5\text{H}_9\text{-}CH_2CH(OH)CH_3 \xrightarrow{PCC} \text{C}_5\text{H}_9\text{-}CH_2C(=O)CH_3$$

$$\text{C}_5\text{H}_9\text{-}CH_2C(=O)CH_3 \xrightarrow[NaBH_3CN]{H_2NCH_3} \text{C}_5\text{H}_9\text{-}CH_2CH(CH_3)NHCH_3$$

Cyclopentamine

Synthetic analysis: The last step of the synthesis is a reductive amination of a ketone that is formed by oxidation of the corresponding alcohol. The alcohol results from the Grignard reaction between cyclopentylmagnesium bromide and propylene oxide.

24.48

(a)

$CH_3(CH_2)_3NH-C_6H_4-COCH_2CH_2N(CH_3)_2$ (C=O) ← ($CH_3CH_2CH_2CHO$, $NaBH_3CN$)

Tetracaine

$O_2N-C_6H_4-COCH_2CH_2N(CH_3)_2$ (C=O) $\xrightarrow[EtOH]{H_2,\ Pt}$ $H_2N-C_6H_4-COCH_2CH_2N(CH_3)_2$ (C=O)

(b) ↑ 1. $SOCl_2$ 2. $HOCH_2CH_2N(CH_3)_2$, Pyridine

(c) Benzene $\xrightarrow{CH_3Cl,\ AlCl_3}$ $C_6H_5-CH_3$ $\xrightarrow[H_2SO_4]{HNO_3}$ $O_2N-C_6H_4-CH_3$ $\xrightarrow[H_2O]{KMnO_4}$ $O_2N-C_6H_4-COH$ (C=O)

Synthetic analysis: The synthesis in (a) is achieved by a reductive amination reaction. Reactions in (b) include formation of an acid chloride, esterification, and reduction of the nitro group.

24.49

Atropine (N–CH_3 bicyclic; $OC(=O)-CHC_6H_5$ with CH_2OH) $\xrightarrow[H_2O]{^-OH}$ Tropine (N–CH_3 bicyclic, OH) + $HOOCCHC_6H_5$ (CH_2OH) Tropic acid $\xrightarrow{H_2SO_4}$ Tropidene (N–CH_3) + H_2O

Synthetic analysis: We know the location of the –OH group of tropine because it is stated that tropine is an optically active alcohol. This hydroxyl group results from basic hydrolysis of the ester that is composed of tropine and tropic acid.

24.50

Tropidene (N–CH_3) $\xrightarrow{\text{1. } CH_3I \text{ 2. } Ag_2O,\ H_2O \text{ 3. heat}}$ $N(CH_3)_2$-cycloheptadiene $\xrightarrow{\text{1. } CH_3I \text{ 2. } Ag_2O,\ H_2O \text{ 3. heat}}$ Tropilidene + $N(CH_3)_3$

Tropilidene results from two cycles of Hofmann elimination on tropidene.

24.51

H_2, Pt — reduction of nitrile

nucleophilic addition

dehydration

H_2, Pt — reduction of double bond

+ H_2O

24.52 The molecular formula indicates that coniine has one double bond or ring, and the Hofmann elimination product shows that the nitrogen atom is part of a ring.

1. excess CH_3I
2. Ag_2O, H_2O
3. heat

$(CH_3)_2N$

Coniine

5-(*N*,*N*-Dimethylamino)-1-octene

24.53

$CH_3CH_2CH_2C(=O)CH_2COOEt$ + $CH_2=CHCN$

1. Na^+ ^-OEt, EtOH
2. H_3O^+

Michael addition

$CH_3CH_2CH_2C(=O)CH(COOEt)CH_2CH_2CN$

ester hydrolysis, decarboxylation — H_3O^+, heat

$CH_3CH_2CH_2C(=O)CH_2CH_2CH_2CN$ + CO_2 + EtOH

H_2, Pt — reduction

$CH_3CH_2CH_2C(=O)CH_2CH_2CH_2CH_2NH_2$

cyclization

H_2, Pt

Coniine

24.54

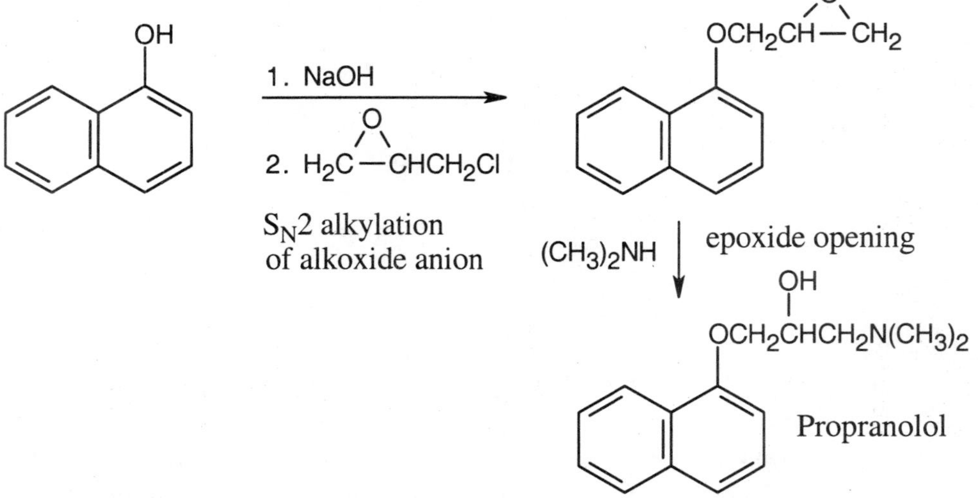

1-Naphthol (OH) $\xrightarrow[\text{2. } H_2C\text{—}CHCH_2Cl \text{ (epoxide)}]{\text{1. NaOH}}$ 1-naphthyl $OCH_2CH\text{—}CH_2$ (epoxide)

S_N2 alkylation
of alkoxide anion

$(CH_3)_2NH$ ↓ epoxide opening

OH
|
$OCH_2CHCH_2N(CH_3)_2$ (1-naphthyl) Propranolol

Remember from Chapter 18 that epoxide ring opening under basic conditions occurs by attack of the base at the less hindered carbon.

24.55

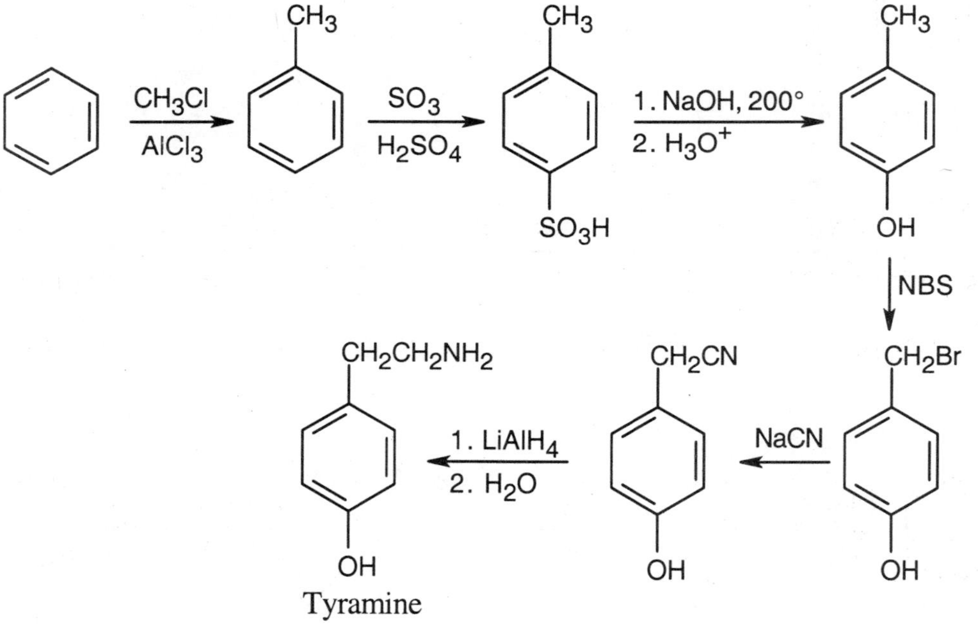

Benzene $\xrightarrow[AlCl_3]{CH_3Cl}$ toluene (CH_3) $\xrightarrow[H_2SO_4]{SO_3}$ p-$CH_3C_6H_4SO_3H$ $\xrightarrow[\text{2. } H_3O^+]{\text{1. NaOH, 200°}}$ p-$CH_3C_6H_4OH$

↓ NBS

p-$BrCH_2C_6H_4OH$ (CH_2Br, OH) $\xrightarrow{NaCN}$ p-$NCCH_2C_6H_4OH$ (CH_2CN, OH) $\xrightarrow[\text{2. } H_2O]{\text{1. } LiAlH_4}$ p-$H_2NCH_2CH_2C_6H_4OH$ ($CH_2CH_2NH_2$, OH)

Tyramine

Synthetic analysis: When you see $-CH_2NH_2$, think of the reduction of a nitrile. The nitrile comes from substitution of a benzylic bromide by ^-CN. We've had a lot of practice synthesizing compounds such as *p*-cresol.

24.56

(a)

CH_3 (toluene) $\xrightarrow[H_2SO_4]{HNO_3}$ NO_2, CH_3 $\xrightarrow[EtOH]{H_2, Pt}$ NH_2, CH_3

(b)

NH_2, CH_3 from (a) $\xrightarrow[H_2SO_4]{HNO_2}$ $^+N_2\ HSO_4^-$, CH_3 $\xrightarrow{CuCN}$ CN, CH_3 $\xrightarrow[2.\ H_2O]{1.\ LiAlH_4}$ CH_2NH_2, CH_3

(c)

CH_3 $\xrightarrow[H_2O]{KMnO_4}$ $COOH$ $\xrightarrow[H_2SO_4]{HNO_3}$ NO_2, $COOH$ $\xrightarrow[2.\ NaOH,\ H_2O]{1.\ SnCl_2,\ H_3O^+}$ NH_2, COO^-

NH_2, COO^- $\xrightarrow[H_2SO_4]{HNO_2}$ $^+N_2\ HSO_4^-$, $COOH$ $\xrightarrow{NaI}$ I, $COOH$ $\xrightarrow[H^+\ catalyst]{CH_3OH}$ I, $COOCH_3$

24.57

(benzene) $\xrightarrow[AlCl_3]{CH_3Cl}$ CH_3 $\xrightarrow[H_2SO_4]{SO_3}$ CH_3, SO_3H $\xrightarrow[2.\ H_3O^+]{1.\ NaOH,\ 200°}$ CH_3, OH

CH_3, OH $\xrightarrow[2.\ BrCH_2CH{=}CH_2]{1.\ NaOH}$ CH_3, $OCH_2CH{=}CH_2$ $\xrightarrow[2.\ NaHSO_3,\ H_2O]{1.\ OsO_4,\ pyridine}$ CH_3, $OCH_2CH(OH)CH_2OH$

Mephenesin

Synthetic analysis: A look at Mephenesin shows a diol that can be prepared by hydroxylation of a carbon-carbon double bond. The allyl group that is hydroxylated is introduced by a Williamson ether synthesis between 3-bromopropene and *o*-cresol, which we have synthesized before.

24.58

(a)

$$\text{2-aminobenzoic acid (COOH, }NH_2) \xrightarrow[H_2SO_4]{HNO_2} \text{(COOH, } \overset{+}{N_2}\ HSO_4^-) \xrightarrow{\text{Base}} \text{(COO}^-, \overset{+}{N}{\equiv}N\text{:)}$$

(b)

$$\text{(COO}^-, \overset{+}{N}{\equiv}N) \xrightarrow{\text{heat}} [\text{benzyne} + CO_2 + N_2 \longrightarrow \text{benzyne + cyclopentadiene}] \longrightarrow \text{product}$$

The reactive intermediate is benzyne, which undergoes a Diels Alder reaction with cyclopentadiene to yield the observed product.

24.59

$$\xrightarrow[\text{3. heat}]{\text{1. } CH_3I \quad \text{2. } Ag_2O,\ H_2O} \text{(} N(CH_3)_2 \text{)} \xrightarrow[\text{3. heat}]{\text{1. } CH_3I \quad \text{2. } Ag_2O,\ H_2O} \text{Cyclooctatriene} + N(CH_3)_3$$

Cyclooctatriene

Two successive cycles of Hofmann elimination lead to formation of cyclooctatriene.

$$\xrightarrow{\text{NBS}} \text{(Br)} \xrightarrow[\text{EtOH}]{\text{KOH}} \text{Cyclooctatetraene}$$

Cyclooctatetraene

Allylic bromination followed by elimination yield cyclooctatetraene.

24.60

$$\text{HO–CH(Ph)–C(=O)NH}_2 \xrightarrow[H_2O]{Br_2,\ NaOH} [\text{HO–CH(Ph)–NH}_2 \rightleftarrows \text{:O}^-\text{–CH(Ph)–}\overset{+}{N}H_3] \rightleftarrows \text{O=CH–Ph} + NH_3$$

Hofmann elimination of an α-hydroxy amide produces a carbinolamine intermediate that expels ammonia to give an aldehyde.

24.61

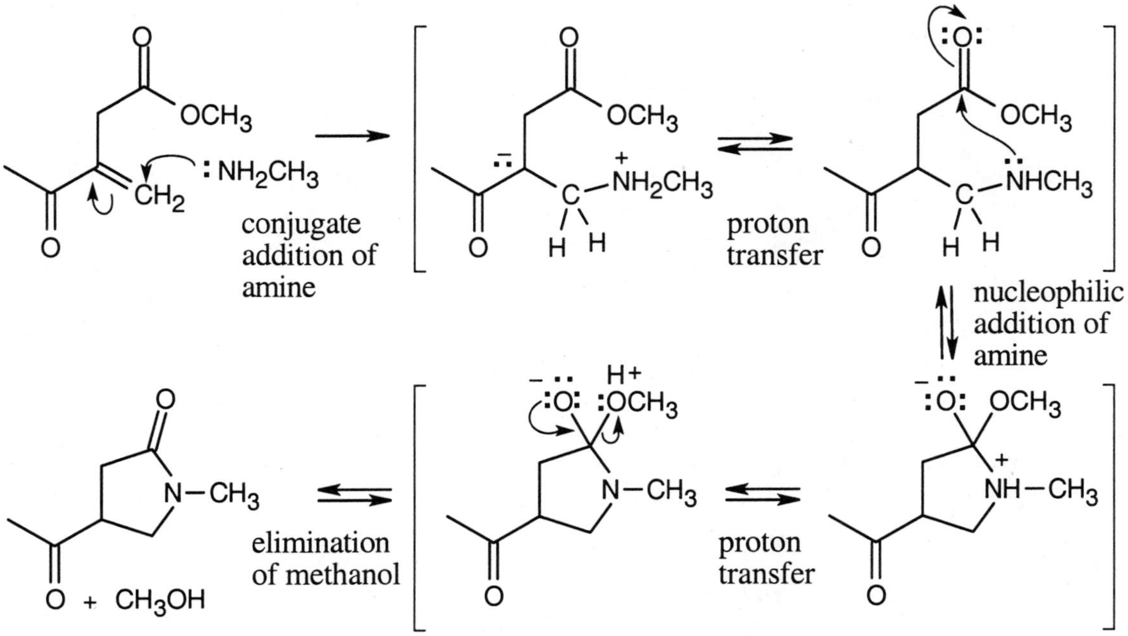

OCH3 : NH2CH3 CH2 O O

conjugate addition of amine

O OCH3 + NH2CH3 C H H O

proton transfer

:O: OCH3 NHCH3 C H H O

nucleophilic addition of amine

:O: OCH3 + NH–CH3 O

proton transfer

:O: H+ :OCH3 N–CH3 O

elimination of methanol

O N–CH3 O + CH3OH

24.62

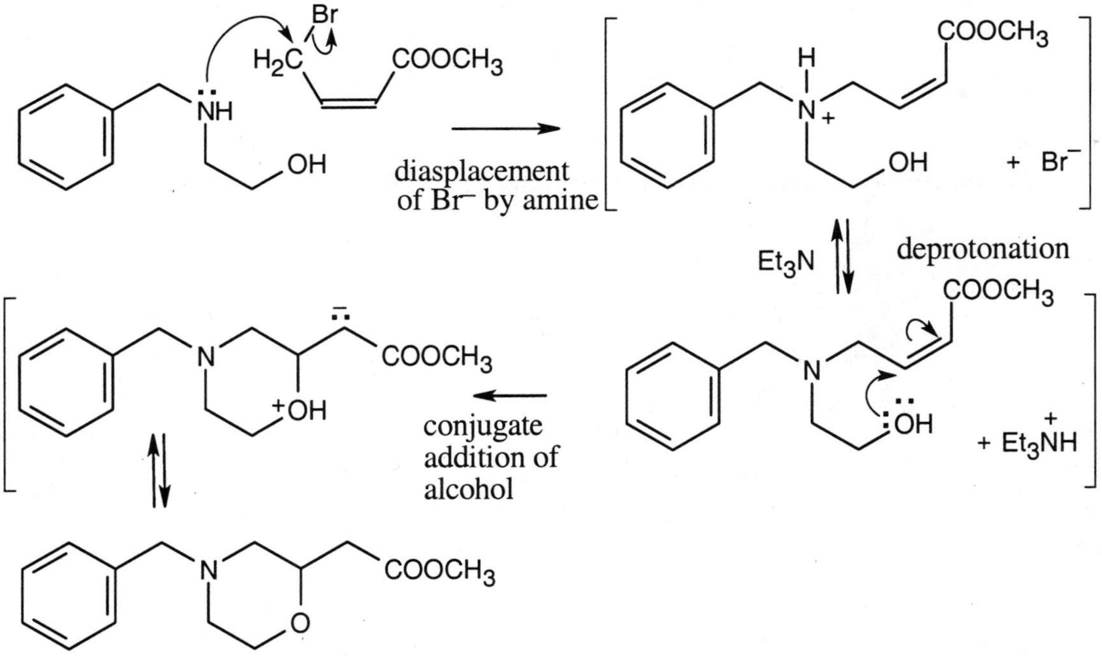

Br H2C COOCH3 NH OH

diasplacement of Br– by amine

COOCH3 H N+ OH + Br–

Et3N deprotonation

COOCH3 N :OH + Et3NH+

conjugate addition of alcohol

N COOCH3 +OH

N COOCH3 O

24.63 The ^{1}H NMR of the amine shows 5 peaks. Two are due to an ethyl group bonded to an electronegative element (oxygen), two are due to 4 aromatic ring hydrogens, and the peak at 3.4 δ is due to 2 amine hydrogens.

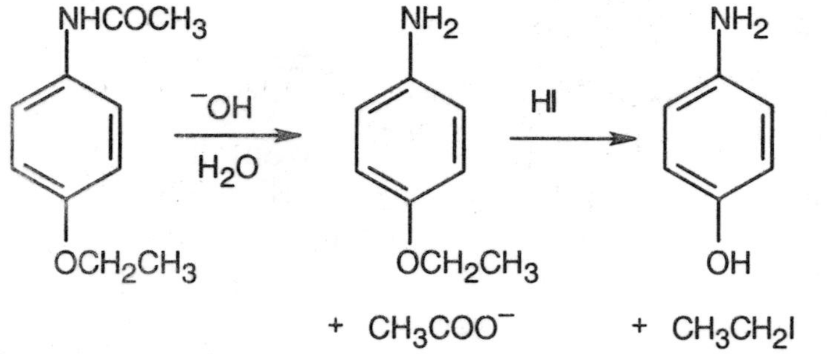

Phenacetin ($C_{10}H_{13}NO_2$; p-$NHCOCH_3$, OCH_2CH_3 benzene) $\xrightarrow[H_2O]{^-OH}$ p-Ethoxyaniline ($C_8H_{11}NO$; p-NH_2, OCH_2CH_3 benzene) + CH_3COO^- $\xrightarrow{HI}$ p-Aminophenol (C_6H_7NO; p-NH_2, OH benzene) + CH_3CH_2I

$C_{10}H_{13}NO_2$ — Phenacetin

$C_8H_{11}NO$ — *p*-Ethoxyaniline

C_6H_7NO — *p*-Aminophenol

24.64

(a) $HOCH_2CH_2CH_2NH_2$ (e d a c b)

a = 1.7 δ
b = 2.7 δ
c = 2.9 δ
d = 3.7 δ

(b) $(CH_3O)_2CHCH_2NH_2$ (c d b a)

a = 1.3 δ
b = 2.8 δ
c = 3.4 δ
d = 4.3 δ

24.65

(a) H_3C–C_6H_4–$N(CH_3)_2$ (H_3C = a; $N(CH_3)_2$ = b; ring hydrogens = c)

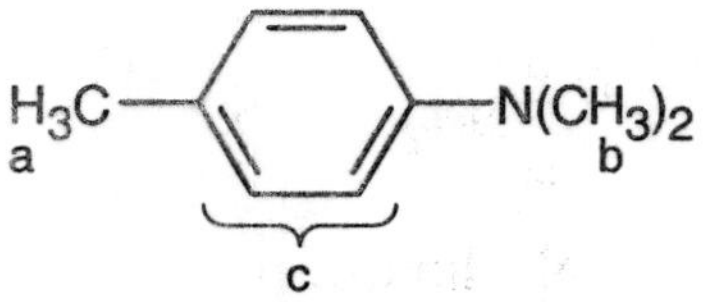

a = 2.3 δ
b = 2.9 δ
c = 6.7 δ, 7.0 δ

(b) C_6H_5–$N(CH_2CH_3)$–CH_2–C_6H_5 (CH_2 = c; CH_2CH_3 = b a; ring hydrogens = d or e)

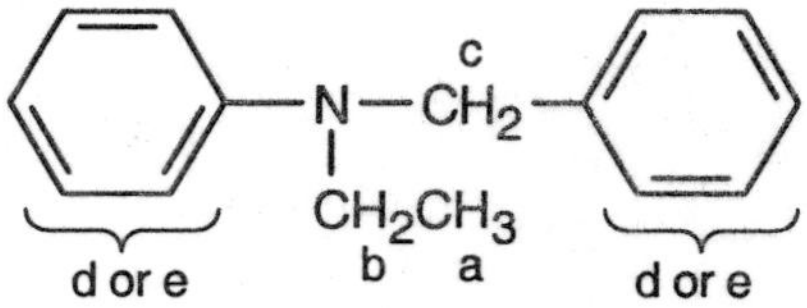

a = 1.2 δ
b = 3.4 δ
c = 4.5 δ
d = 6.7 δ
e = 7.2 δ

A Look Ahead

24.66

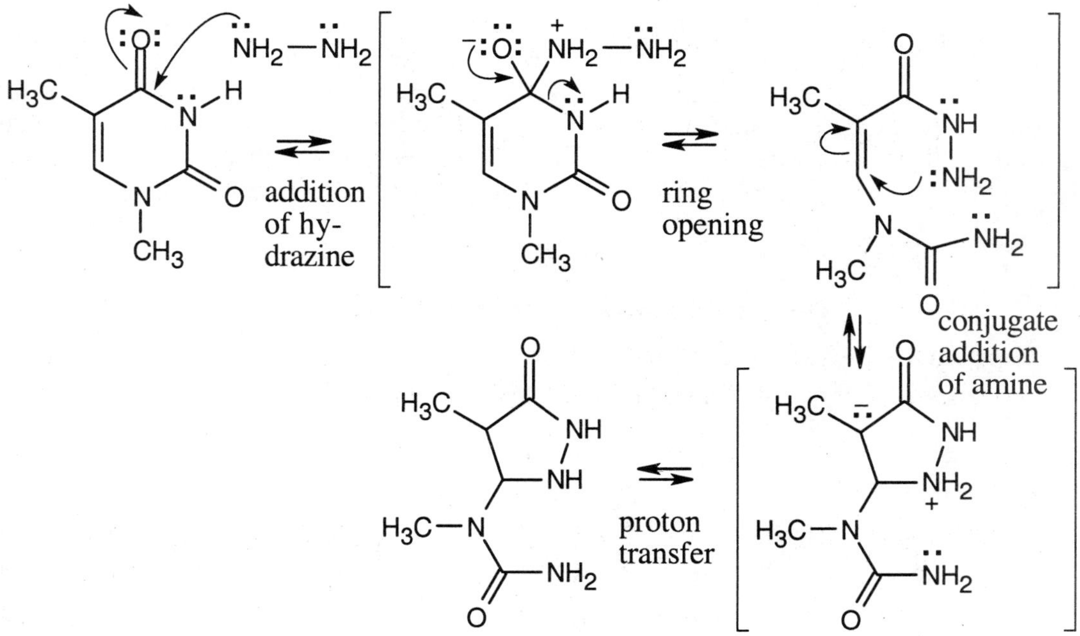

addition of hydrazine

ring opening

conjugate addition of amine

proton transfer

24.67

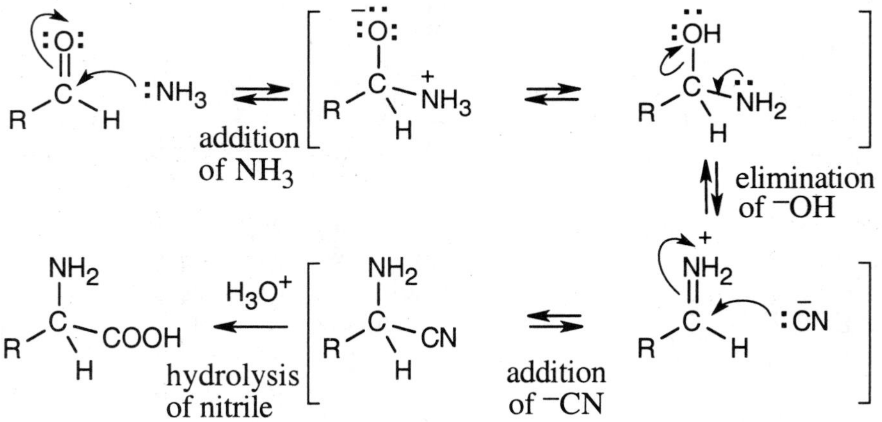

addition of NH_3

elimination of ^-OH

addition of ^-CN

H_3O^+ hydrolysis of nitrile

The mechanism of acid-catalyzed nitrile hydrolysis is shown in Problem 21.24.

Molecular Modeling

24.68 The dipole moment of 4-nitroaniline (7.5 D) is greater than the sum of the dipole moments of nitrobenzene and aniline (5.1 D + 1.5 D = 6.6 D), indicating a greater amount of charge separation in 4-nitroaniline. Electrostatic potential maps confirm this result. The amino group of 4-nitroaniline is less negative, relative to aniline, and its nitro group is more negative, relative to nitrobenzene, indicating charge transfer from the amino group to the nitro group. Charge transfer is consistent with the following type of resonance structure and is supported by the shorter C–N bond distances that are observed in 4-nitroaniline, 136 pm and 144 pm, relative to aniline (140 pm) and nitrobenzene (146 pm), respectively.

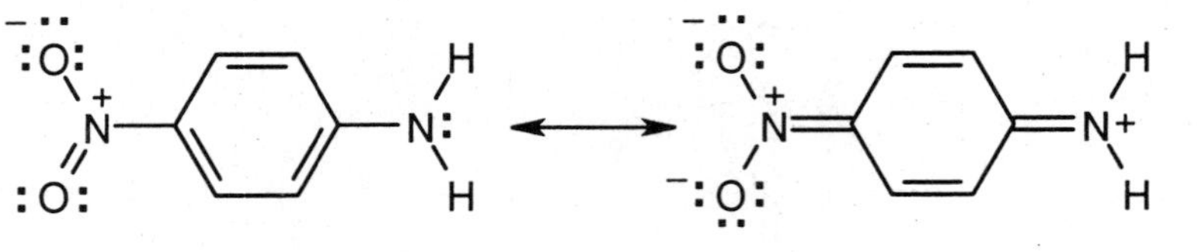

24.69

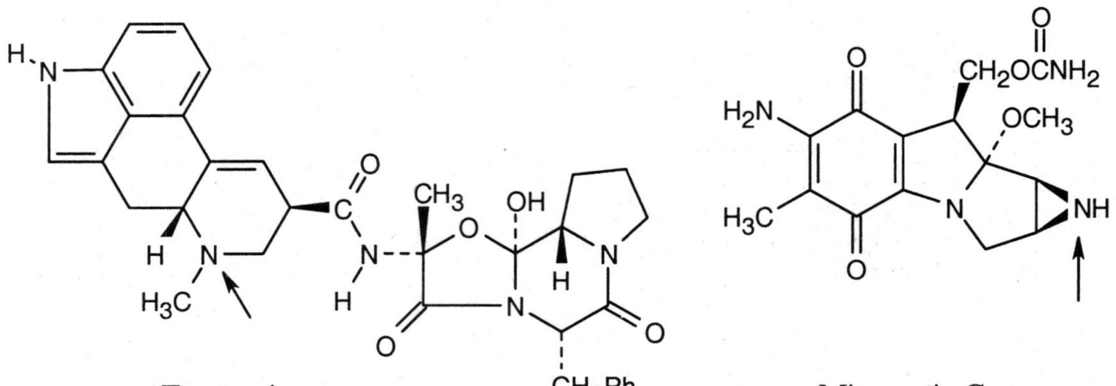

Ergotamine

Mitomycin C

24.70 The positive charge of these ions is buried, to varying degrees, and ion-solvent interactions should vary accordingly. Tetraethylammonium ion "looks" most positive and should be most water-soluble. Tetrabenzylammonium ion "looks" least positive and should be least water-soluble. Benzyltrimethylammonium ion contains both a positive, water soluble region and a nonpolar, organic-soluble region.

24.71 The transition state is more polar than the reactants. A polar, aprotic solvent like DMSO would interact weakly with the reactants and more strongly with the positive (NH_3) end of the transition state. A polar protic solvent like water would interact even more strongly with the transition state because it could hydrogen-bond to both positive and negative ends of the transition state. These interactions stabilize the transition state and should lead to a lower energy barrier and a faster reaction.

Review Unit 9: Carbonyl Compounds II
Reaction at the α Carbon; Amines

Major Topics Covered (with vocabulary):

Carbonyl α-substitution reactions:
α-substitution reaction   tautomerism   tautomer   enolate ion   Hell-Volhard-Zelinskii reaction   β-diketone   β-keto eater   haloform reaction   malonic ester synthesis   acetoacetic eater synthesis   LDA

Carbonyl condensation reactions:
carbonyl condensation reactions   aldol reaction   enone   Claisen condensation reaction   Dieckmann cyclization   Michael reaction   Michael acceptor   Michael donor   Stork enamine reaction   Robinson annulation reaction

Amines:
primary, secondary, tertiary amine   quaternary ammonium salt   arylamine   heterocyclic amine   pyramidal inversion K_b azide synthesis   Gabriel amine synthesis   reductive amination   Hofmann rearrangement   Curtius rearrangement   Hofmann elimination reaction   arenediazonium salt   diazotization   Sandmeyer reaction   azo compound   diazonium coupling reaction   phase-transfer catalyst   nitrogen rule

Types of Problems:

After studying these chapters, you should be able to:

- Draw keto-enol tautomers of carbonyl compounds, identify acidic hydrogens, and draw the resonance forms of enolates.
- Formulate the mechanisms of acid- and base-catalyzed enolization and of other α-substitution reactions.
- Predict the products of α-substitution reactions.
- Use α-substitution reactions in synthesis.

- Predict the products of carbonyl condensation reactions.
- Formulate the mechanisms of carbonyl condensation reactions.
- Use carbonyl condensation reactions in synthesis.

- Name and draw amines, and classify amines as primary, secondary, tertiary , quaternary, arylamines, or heterocyclic amines.
- Predict the basicity of alkylamines, arylamines and heterocyclic amines.
- Synthesize alkylamines and arylamines by several routes.
- Predict the products of reactions involving alkylamines and arylamines.
- Use diazonium salts in reactions involving arylamines, including diazo coupling reactions.
- Propose mechanisms for reactions involving alkylamines and arylamines.
- Identify amines by spectroscopic techniques.

Points to Remember:

* It is unusual to think of a carbonyl compound as an acid, but the protons α to a carbonyl group can be removed by a strong base. Protons α to two carbonyl groups are even more acidic: in some cases, acidity approaches that of phenols. This acidity is the basis for α-substitution reactions of compounds having carbonyl groups. Abstraction by base of an α proton produces a resonance-stabilized enolate anion that can be used in alkylations involving alkyl halides and tosylates.

* Alkylation of an unsymmetrical LDA-generated enolate generally occurs at the less hindered α carbon.

* When you need to synthesize a β-hydroxy ketone or aldehyde or an α,β-unsaturated ketone or aldehyde, use an aldol reaction. When you need to synthesize a β-diketone or β-keto ester, use a Claisen reaction. When you need to synthesize a 1,5-dicarbonyl compound, use a Michael reaction. The Robinson annulation is used to synthesize polycyclic molecules by a combination of a Michael reaction with an aldol condensation.

* In many of the mechanisms in this group of chapters, the steps involving proton transfer are not explicitly shown. The proton transfers occur between the proton and the conjugate base with the most favorable p*K* of those present in the solution. These steps have been omitted at times to simplify the mechanisms.

* In the Claisen condensation, the enolate of the β-dicarbonyl compound is treated with H_3O^+ to yield the neutral product.

* For an amine, the larger the value of pK_a of its ammonium ion, the stronger the base. The smaller the value of pK_b of the amine, the stronger the base.

* The Sandmeyer reaction allows the synthesis of substituted benzenes that can't be formed by electrophilic aromatic substitution reactions. These reactions succeed because N_2 is a very good leaving group.

Self-Test:

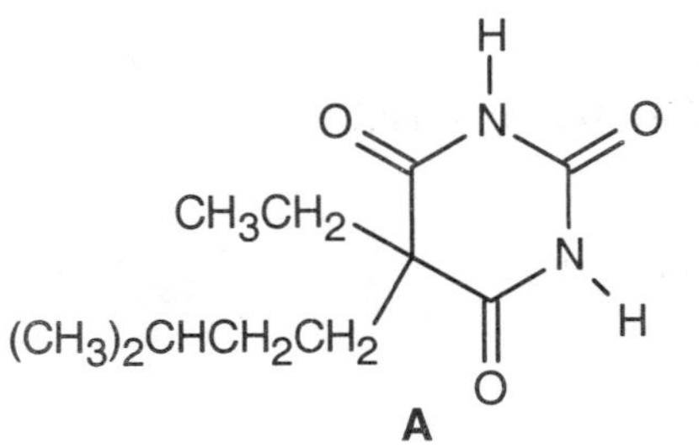

A
Pentymal
(a sedative)

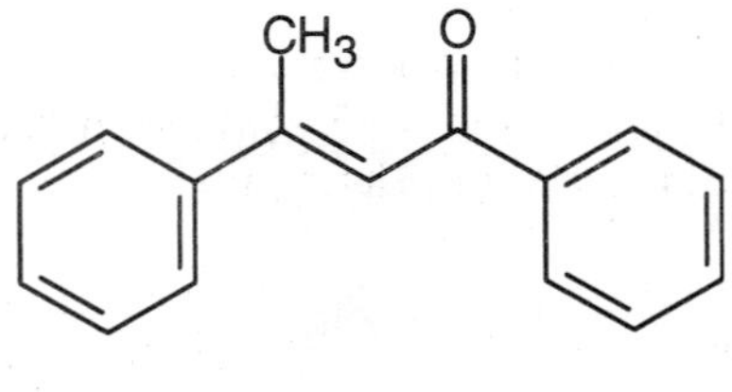

B
Dypnone
(sunscreen)

The six-membered ring in **A** is formed by the cyclization of two difunctional compounds. What are they? What type of reaction occurs to form the ring? The two alkyl groups are introduced into one of the difunctional compounds prior to cyclization. What type of reaction is occurring, and how is it carried out? What type of reaction occurs in the formation of Dypnone (**B**)? Why might **D** be effective as a sunscreen?

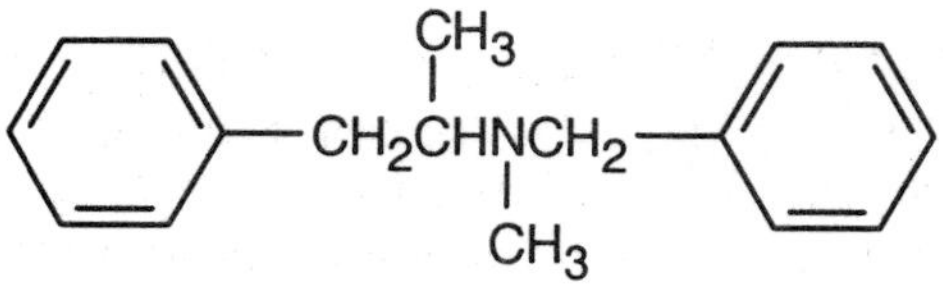

C

Benzphetamine
(an appetite suppressant)

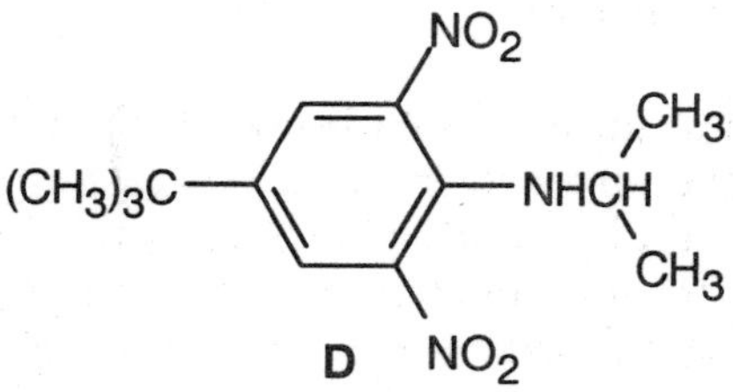

D

Butralin
(an herbicide)

What type of amine is **C**? Do you expect it to be more or less basic than ammonia? Than aniline? What product do you expect from Hofmann elimination of **C**? What significant absorptions might be seen in the IR spectrum of **C**? What information can be obtained from the mass spectrum? Plan a synthesis of **D** from benzene.

Multiple Choice:

1. Which of the following compounds has four acidic hydrogens?
(a) 2-Pentanone (b) 3-Pentanone (c) Acetophenone (d) Phenylacetone

2. In which of the following reactions is an enol, rather than an enolate, the reacting species?
(a) haloform reaction (b) malonic ester synthesis (c) LDA alkylation (d) Hell-Volhard-Zelinskii reaction

3. Cyclobutanecarboxylic acid is probably the product of a:
(a) malonic ester synthesis (b) acetoacetic ester synthesis (c) LDA alkylation
(d) Hell-Volhard-Zelinskii reaction

4. An LDA alkylation can be used to alkylate all of the following, except:
(a) aldehydes (b) ketones (c) esters (d) nitriles

5. If you want to carry out a carbonyl condensation, and you don't want to form α-substitution product, you should:
(a) lower the temperature (b) use one equivalent of base (c) use a catalytic amount of base
(d) use a polar aprotic solvent

6. Which reaction forms a cyclohexenone?
(a) Dieckmann cyclization (b) Michael reaction (c) Claisen condensation
(d) intramolecular aldol condensation

7. All of the following molecules are good Michael donors except:
(a) Ethyl acetoacetate (b) Nitroethylene (c) Malonic ester
(d) Ethyl 2-oxocyclohexanecarboxylate

8. The ammonium ion of which of the following amines has the smallest value of pK_a?
(a) Methylamine (b) Trimethylamine (c) Aniline (d) *p*-Bromoaniline

9. All of the following methods of amine synthesis are limited to primary amines, except:
(a) Curtius rearrangement (b) reductive amination (c) Hofmann rearrangement
(d) azide synthesis

10. To form an azo compound, an aryldiazonium salt should react with:
(a) CuCN (b) benzene (c) nitrobenzene (d) phenol

Chapter 25 – Biomolecules: Carbohydrates

Chapter Outline

I. Classification of carbohydrates(Section 25.1).
- A. Simple *vs.* complex:
  - 1. Simple carbohydrates can't be hydrolyzed to smaller units.
  - 2. Complex carbohydrates are made up of two or more simple sugars linked together.
    - a. A disaccharide is composed of two monosaccharides.
    - b. A polysaccharide is composed of three or more monosaccharides.
- B. Aldoses *vs.* ketoses:
  - 1. A monosaccharide with an aldehyde carbonyl group is an aldose.
  - 2. A monosaccharide with a ketone carbonyl group is a ketose.
- C. *Tri-*, *tetr-* , *pent-*, etc. indicate the number of carbons in the monosaccharide.

II. Monosaccharides (Sections 25.2 – 25.8).
- A. Configurations of monosaccharides (Section 25.2 – 25.4).
  - 1. Fischer projections (Section 25.2).
    - a. Each chirality center of a monosaccharide is represented by a pair of crossed lines.
    - b. The carbonyl carbon is placed at or near the top of the Fischer projection.
    - c. Fischer projections can be rotated by 180°, but not by 90° or 270°.
    - d. Carbohydrates with more than one chirality center are shown by stacking the centers on top of each other.
  - 2. D,L sugars (Section 25.3).
    - a. (*R*)-Glyceraldehyde is also known as D-glyceraldehyde.
    - b. In D sugars, the –OH group farthest from the carbonyl group points to the right. Most naturally-occurring sugars are D sugars.
    - c. In L sugars, the –OH group farthest from the carbonyl group points to the left.
    - d. D,L designations refer only to the configuration farthest from the carbonyl carbon and are unrelated to the direction of rotation of plane-polarized light.
  - 3. Configurations of the aldoses (Section 25.4).
    - a. There are 4 aldotetroses – D and L erythrose and threose.
    - b. There are 4 D,L pairs of aldopentoses: ribose, arabinose, xylose and lyxose.
    - c. There are 8 D,L pairs of aldohexoses : allose, altrose, glucose, mannose, gulose, idose, galactose, and talose.
    - d. A scheme for drawing and memorizing the D-aldohexoses:
      - i. Draw all –OH groups ar C5 pointing to the left.
      - ii. Draw the first four –OH groups at C4 pointing to the right and the second four pointing to the left.
      - iii. Alternate –OH groups at C3: two right, two left, two right, two left.
      - iv. Alternate –OH groups at C2: right, left, etc.
      - v. Use the mnemonic " All altruists gladly make gum in gallon tanks" to assign names.
- B. Cyclic structures of monosaccharides (Sections 25.5 – 25.6).
  - 1. Hemiacetal formation (Section 25.5).
    - a. Monosaccharides are in equilibrium with their internal hemiacetals.
      - i. Glucose exists primarily as a six-membered pyranose ring, formed by the –OH group at C5 and the aldehyde.
      - ii. Fructose exists primarily as a five-membered furanose ring.
    - b. Structure of pyranose rings.
      - i. Pyranose rings have a chair-like geometry.

ii. The hemiacetal oxygen is at the right rear for D-sugars.
iii. An –OH group on the right in a Fischer projection is on the bottom in a pyranose ring, and an –OH group on the left is on the top.
iv. For D sugars, the $-CH_2OH$ group is on the top.

2. Mutarotation (Section 25.6).
   a. When a monosaccharide cyclizes, a new chirality center is generated.
      i. The two diastereomers are anomers.
      ii. The form with the anomeric –OH group trans to the $-CH_2OH$ group is the α anomer (minor anomer).
      iii. The form with the anomeric –OH group cis to the $-CH_2OH$ group is the β anomer (major anomer).
   b. When a solution of either pure anomer is dissolved in water, the optical rotation of the solution reaches a constant value.
      i. This process is called mutarotation.
      ii. Mutarotation is due to the reversible opening and recyclizing of the hemiacetal ring and is catalyzed by both acid and base.

C. Reactions of monosaccharides (Section 25.7).
1. Ester and ether formation.
   a. Esterification occurs by treatment with an acid anhydride or acid chloride
   b. Ethers are formed by treatment with methyl iodide and Ag_2O.
   c. Ester and ether derivatives are crystalline and easy to purify.
2. Glycoside formation.
   a. Treatment of a hemiacetal with an alcohol and an acid catalyst yields an acetal.
      i. Acetals aren't in equilibrium with an open-chain form.
      ii. Aqueous acid reconverts the acetal to a monosaccharide.
   b. These acetals, called glycosides, occur in nature.
   c. The laboratory synthesis of glycosides is achieved by the Koenigs–Knorr reaction.
      i. Treatment of the acetylpyranose with HBr, followed by treatment with the appropriate alcohol and Ag_2O, gives the acetylglycoside.
      ii. Both anomers give the same product.
      iii. The reaction involves neighboring-group participation by acetate.
3. Reduction of monosaccharides.
   Reaction of a monosaccharide with $NaBH_4$ yields an alditol
4. Oxidation of monosaccharides.
   a. Several mild reagents can oxidize the carbonyl group to a carboxylic acid (aldonic acid).
      i. Tollens reagent, Fehling's reagent and Benedict's reagent all serve as tests for reducing sugars.
      ii. All aldoses and some ketoses are reducing sugars, but glycosides are nonreducing.
      iii. In the laboratory, aqueous Br_2 is used to oxidize aldoses (not ketoses).
   b. The more powerful oxidizing agent, dilute HNO_3, oxidizes aldoses to dicarboxylic acids (aldaric acids).
5. Chain-lengthening: the Kiliani–Fischer synthesis.
   a. In the Kiliani-Fischer synthesis, an aldehyde group becomes C2 of a chain-lengthened monosaccharide.
   b. The reaction involves cyanohydrin formation, reduction and hydrolysis.
   c. The products are two diastereomeric aldoses that differ in configuration at C2.
6. Chain-shortening: the Wohl degradation.
   a. The Wohl degradation shortens an aldose by one carbon.
   b. The reaction involves treatment of the aldose with hydroxylamine, dehydration and loss of HCN from the resulting cyanohydrin.

D. Stereochemistry of glucose: the Fischer proof.
  1. (+)-Glucose is an aldohexose.
    Fischer arbitrarily made what was later found to be the correct assignment for the stereochemistry at C5.
  2. Arabinose is converted to a mixture of glucose and mannose by chain extension.
    Glucose and mannose must have the same configuration at C3, C4, and C5.
  3. Arabinose is converted to an optically active aldaric acid.
    This fact excludes 4 of the 8 possible structures for glucose and mannose.
  4. Both glucose and mannose are oxidized to optically active aldaric acids.
    This fact excludes two of the remaining possible structures for glucose.
  5. One of the other 15 aldohexoses gives the same aldaric acid as that derived from glucose.
    This fact eliminates the structure of mannose, which is the only stereoisomer that can yield its aldaric acid.
  6. Fischer used this logic to elucidate the structures of 12 of the 16 aldohexoses.

III. Other carbohydrates (Sections 25.9 – 25.12).
  A. Disaccharides (Section 25.9).
    1. Cellobiose and maltose.
      a. Cellobiose and maltose contain a 1,4'-glycosidic acetal bond between two glucose monosaccharide units.
        The prime (') shows that the glycosidic bond is between two different sugars.
      b. Maltose consists of two glucopyranose units joined by a 1,4'-α-glycosidic bond.
      c. Cellobiose consists of two glucopyranose units joined by a 1,4'-β-glycosidic bond.
      d. Both maltose and cellobiose are reducing sugars and exhibit mutarotation.
      e. Humans can't digest cellobiose but can digest maltose.
    2. Lactose.
      a. Lactose consists of a unit of galactose joined by a β-glycosidic bond between C1 and C4 of a glucose unit.
      b. Lactose is a reducing sugar found in milk.
    3. Sucrose.
      a. Sucrose is a disaccharide that yields glucose and fructose on hydrolysis.
        a. Sucrose is called "invert sugar" because the sign of rotation changes when sucrose is hydrolyzed.
        b. Sucrose is one of the most abundant pure organic chemicals in the world.
      b. The two monosaccharides are joined by a glycosidic link between C1 of glucose and C2 of fructose.
      c. Sucrose isn't a reducing sugar and doesn't exhibit mutarotation.
  B. Polysaccharides and their synthesis (Section 25.10).
    1. Polysaccharides have a reducing end and undergo mutarotation, but aren't considered to be reducing sugars because of their size.
    2. Important polysaccharides.
      a. Cellulose.
        i. Cellulose consists of thousands of D-glucose units linked by 1,4'-β-glycosidic bonds.
        ii. In nature, cellulose is used as structural material.
      b. Starch.
        i. Starch consists of thousands of D-glucose units linked by 1,4'-α-glycosidic bonds.

ii. Starch can be separated into amylose (water-soluble) and amylopectin (water-insoluble) fractions.
Amylopectin contains 1,6'-α-glycosidic branches.
iii. Starch is digested in the mouth by glycosidase enzymes, which only cleave α-glycosidic bonds.
c. Glycogen.
i. Glycogen is an energy-storage polysaccharide.
ii. Glycogen contains both 1,4'- and 1,6'-links.
3. An outline of the glycan assembly method of polysaccharide synthesis.
a. A glycal (a monosaccharide with a C1–C2 double bond) is protected at C6 by formation of a silyl ether.
b. The protected glycal is epoxidized.
c. Treatment of the glycal epoxide (in the presence of $ZnCl_2$) with a second glycal with a free C6 hydroxyl group forms a disaccharide.
d. The process can be repeated.
C. Other important carbohydrates (Section 25.11).
1. Deoxy sugars have an –OH group missing and are components of nucleic acids.
2. In amino sugars, an –OH is replaced by a $-NH_2$.
Amino sugars are found in chitin and in antibiotics.
D. Cell surface carbohydrates and carbohydrate vaccines (Section 25.12).
1. Polysaccharides are involved in cell-surface recognition.
a. Polysaccharide markers on the surface of red blood cells are responsible for blood-group incompatibility.
b. Red blood cells have two types of markers (antigenic determinants) – A and B.
c. Unusual carbohydrates are components of these markers.
2. Possible anticancer vaccines have been synthesized from antibodies to cell-surface polysaccharides found on the surface of cancer cells.

Solutions to Problems

25.1

(a)
```
   H   O
    \ //
     C
     |
HO - C - H
     |
 H - C - OH
     |
    CH2OH
```
Threose
an aldotetrose

(b)
```
    CH2OH
     |
     C=O
     |
 H - C - OH
     |
 H - C - OH
     |
    CH2OH
```
Ribulose
a ketopentose

(c)
```
    CH2OH
     |
     C=O
     |
HO - C - H
     |
HO - C - H
     |
 H - C - OH
     |
    CH2OH
```
Tagatose
a ketohexose

(d)
```
   H   O
    \ //
     C
     |
 H - C - H
     |
 H - C - OH
     |
 H - C - OH
     |
    CH2OH
```
2-Deoxyribose
an aldopentose

25.2 Review Section 9.13 if you need to. To decide if two Fischer projections are identical, use the two allowable rotations to superimpose two groups of each projection. If the remaining groups are also superimposed after rotation, the projections represent the same enantiomer.
Projections A, B and C represent the *S* enantiomer of glyceraldehyde; projection D represents the *R* enantiomer.

25.3 Horizontal bonds of Fischer projections point out of the page, and vertical bonds point into the page.

(a)

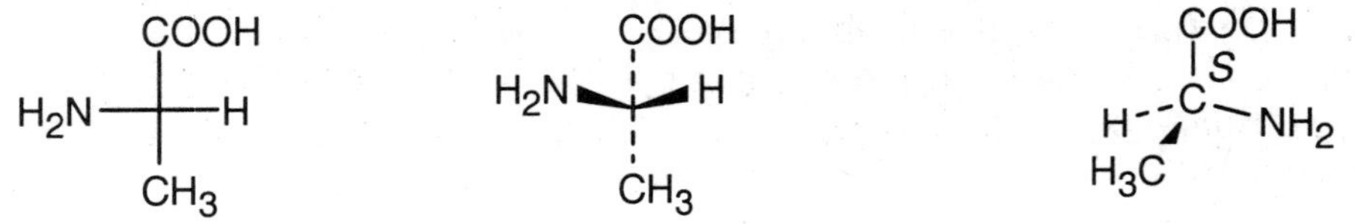

COOH; H_2N—C—H; CH_3 (Fischer projection) COOH; H_2N (wedge)—C—(wedge) H; CH_3 COOH; *S*; H- -C—NH_2; H_3C

(b)

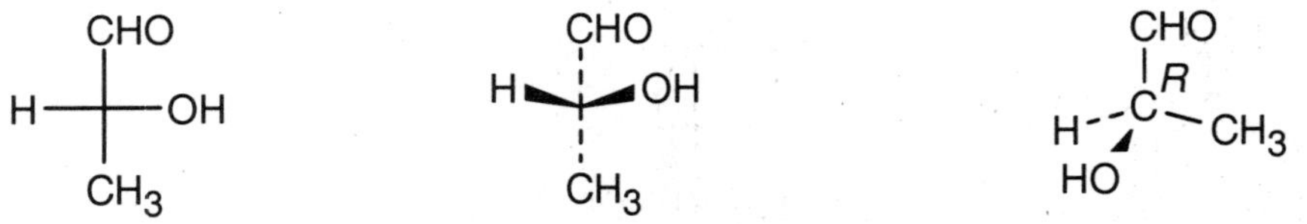

CHO; H—C—OH; CH_3 CHO; H (wedge)—C—(wedge) OH; CH_3 CHO; *R*; H- -C—CH_3; HO

(c)

CH_3; H—C—CHO; CH_2CH_3 CH_3; H (wedge)—C—(wedge) CHO; CH_2CH_3 CHO; *S*; H- -C—CH_3; CH_2CH_3

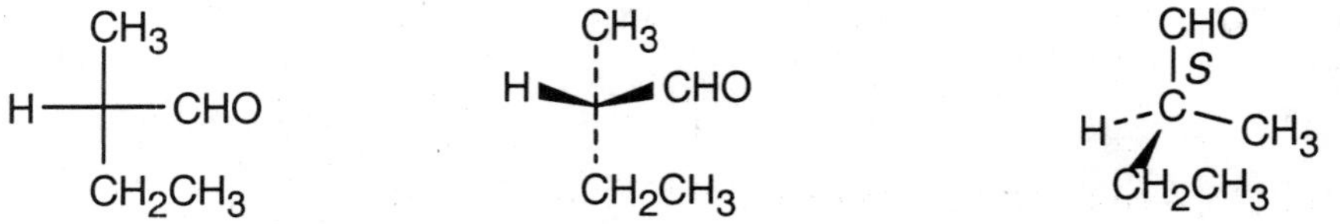

25.4 The hydroxyl group bonded to the chiral carbon farthest from the carbonyl group points to the right in a D sugar, and points to the left in an L sugar.

(a) CHO; HO—(*S*)—H; HO—(*S*)—H; CH_2OH

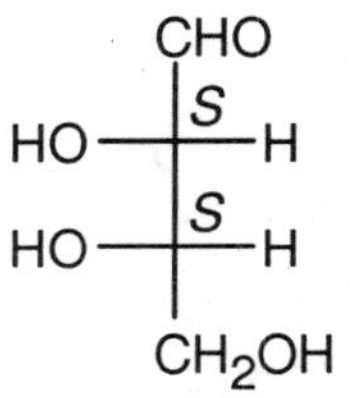

L-Erythrose

(b) CHO; H—(*R*)—OH; HO—(*S*)—H; H—(*R*)—OH; CH_2OH

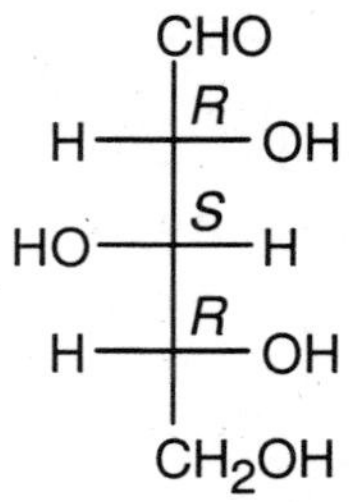

D-Xylose

(c) CH_2OH; C=O; HO—(*S*)—H; H—(*R*)—OH; CH_2OH

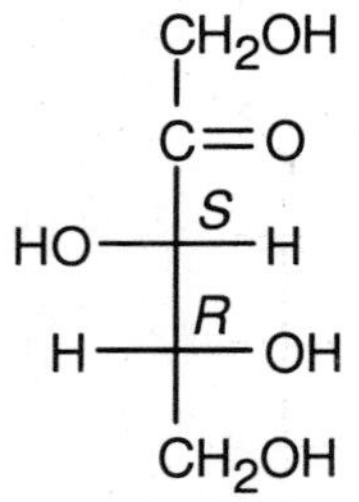

D-Xylulose

25.5

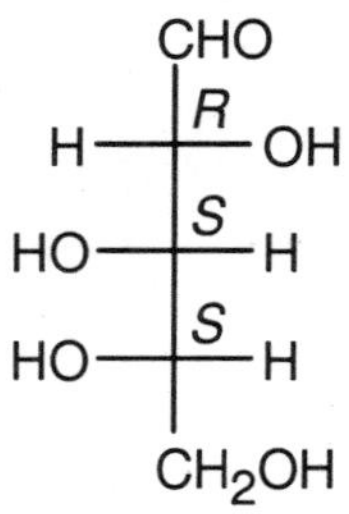

CHO; H—(*R*)—OH; HO—(*S*)—H; HO—(*S*)—H; CH_2OH

L-(+)-Arabinose

25.6

(a) CHO; HO—H; H—OH; HO—H; CH_2OH

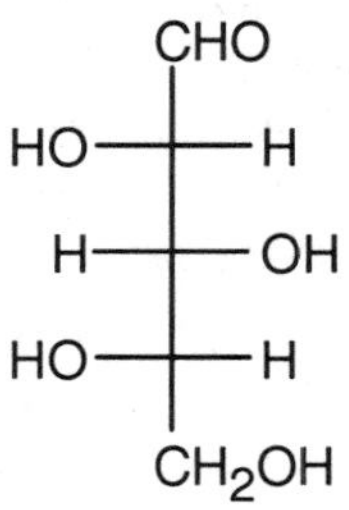

L-Xylose

(b) CHO; HO—H; H—OH; H—OH; HO—H; CH_2OH

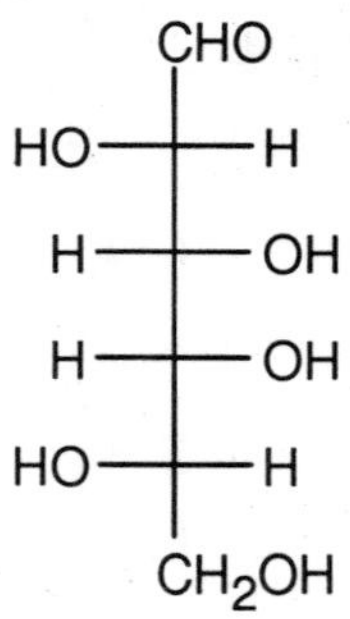

L-Galactose

(c) CHO; HO—H; HO—H; HO—H; HO—H; CH_2OH

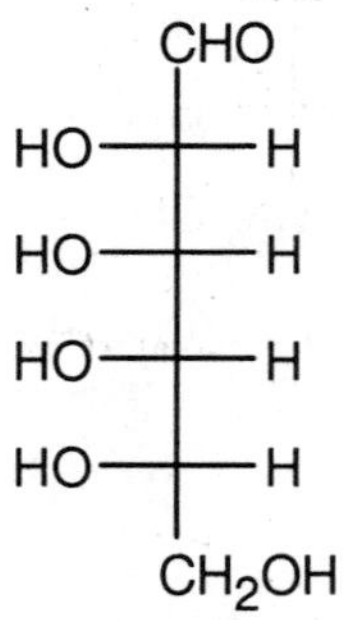

L-Allose

25.7 An aldoheptose has 5 chirality centers. Thus, there are $2^5 = 32$ aldoheptoses — 16 D aldoheptoses and 16 L aldoheptoses.

25.8 The best way to approach this problem is to build a model of the model (!) and copy your drawing as a Fischer projection, remembering that horizontal bonds point out of the page and vertical bonds point into the page. Use your model to assign configurations to the chirality centers.

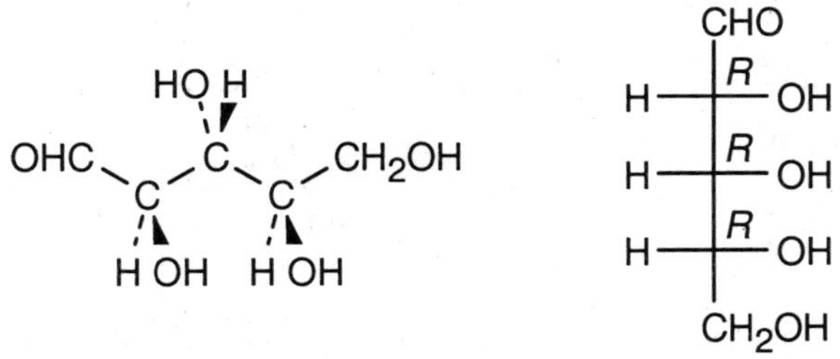

HO H
OHC C C C CH_2OH
H OH H OH

CHO
H — *R* — OH
H — *R* — OH
H — *R* — OH
CH_2OH

D-Ribose

25.9 Follow the steps in Practice Problem 25.2. First, draw the Fischer projection of D-allose. Lay the projection on its side, and curl it around so that the aldehyde is at the right front, and the CH_2OH is at the left rear. Connect the –OH at C5 to the C1 carbonyl group, raise the left carbon (C4), and lower the right carbon (C1) to form the chair. Note: For simplicity, hydrogens have been left off the ring structures.

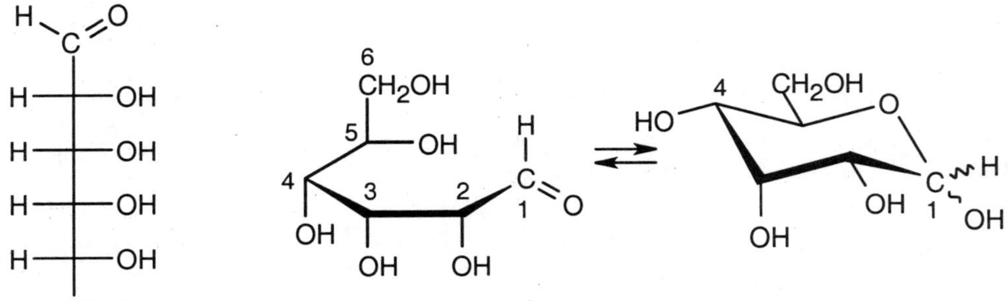

H C=O
H — OH
H — OH
H — OH
H — OH
CH_2OH

6
CH_2OH
5 OH
H
4 3 2 C=O
1
OH OH OH

D-Allose

4 CH_2OH
HO O
C H
OH 1 OH
OH

(Pyranose form)

25.10 The steps for drawing a furanose are similar to the steps for drawing a pyranose. Ring formation occurs between the –OH group at C4 and the carbonyl carbon.

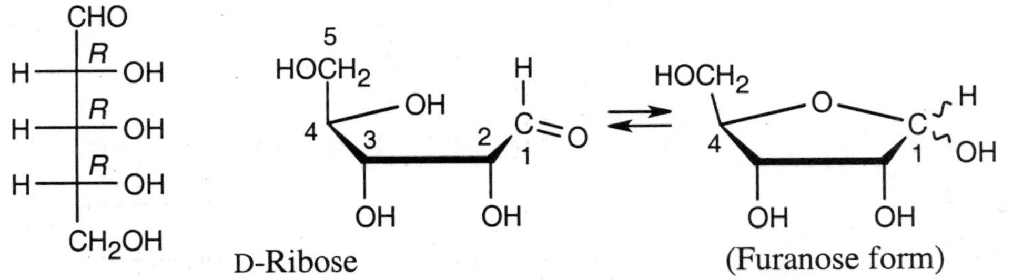

CHO
H — *R* — OH
H — *R* — OH
H — *R* — OH
CH_2OH

5
$HOCH_2$
OH
H
4 3 2 C=O
1
OH OH

D-Ribose

$HOCH_2$
O
C H
4 1 OH
OH OH

(Furanose form)

25.11 Recall from Section 25.5 that the furanose of fructose results from ring formation between the –OH group at C5 and the ketone at C2. In the α anomer, the anomeric –OH group is trans to the C6 –CH_2OH group, and in the β anomer the two groups are cis.

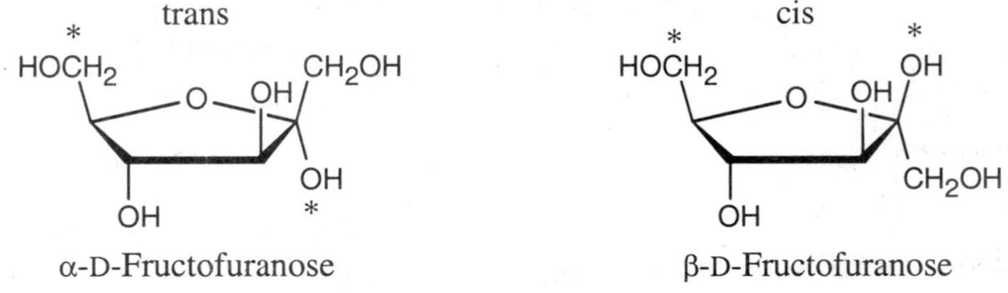

trans

$HOCH_2$ O OH CH_2OH OH OH

α-D-Fructofuranose

cis

$HOCH_2$ O OH OH CH_2OH OH

β-D-Fructofuranose

25.12 There are two ways to draw these anomers: (1) Following the steps in Problem 25.9, draw the Fischer projection, lay it on its side, form the pyranose ring, and convert it to a chair, remembering that the anomeric –OH group is cis to the C6 group. (2) Draw β-D-glucopyranose, and exchange the hydroxyl groups that differ between glucose and the other two hexoses.

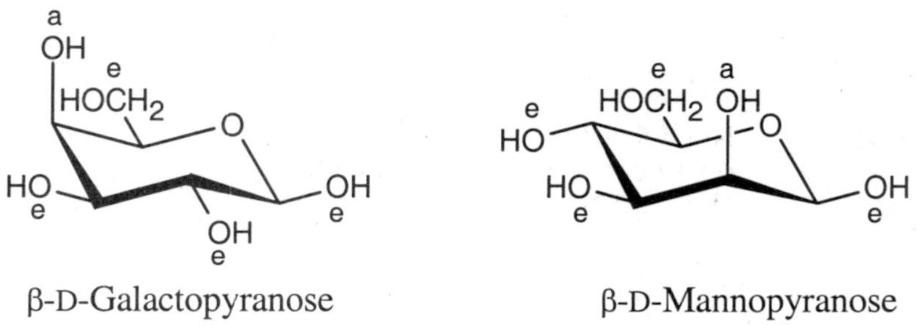

β-D-Galactopyranose

β-D-Mannopyranose

β-D-Galactopyranose and β-D-mannopyranose each have one hydroxyl group in the axial position and are therefore of similar stability.

25.13 In the previous problem we drew β-D-galactopyranose. In this problem, invert the configuration at each chirality center of the D enantiomer and perform a ring-flip to arrive at the structure of the L enantiomer.

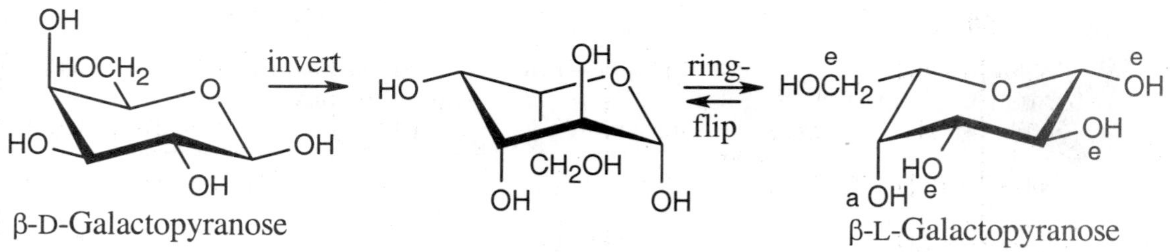

β-D-Galactopyranose —invert→ —ring-flip⇌ β-L-Galactopyranose

All substituents, except for the –OH at C4, are equatorial in the more stable conformation of β-L-galactopyranose.

25.14

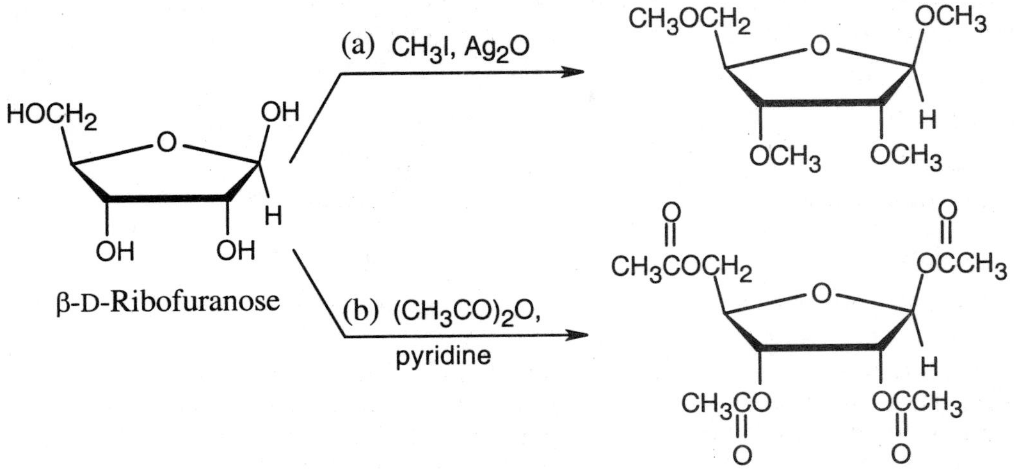
(a) CH3I, Ag2O
(b) (CH3CO)2O,
pyridine
HOCH2
OH
H
CH3OCH2
OCH3
CH3COCH2
OCCH3
CH3CO
β-D-Ribofuranose

25.15

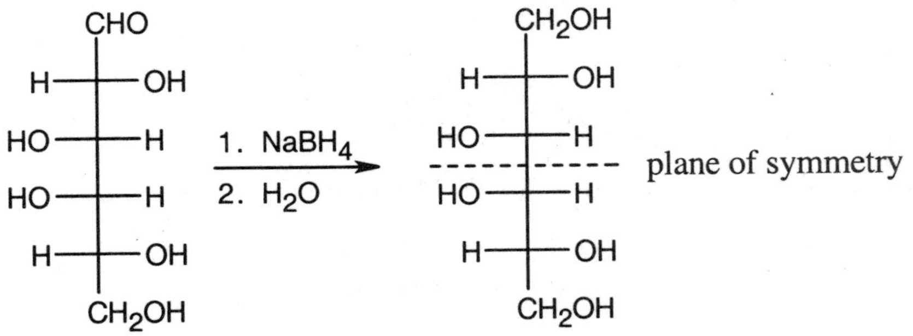
CHO
CH2OH
1. NaBH4
2. H2O
plane of symmetry

D-Galactose Galactitol

Reaction of D–galactose with $NaBH_4$ yields an alditol that has a plane of symmetry and is a meso compound.

25.16

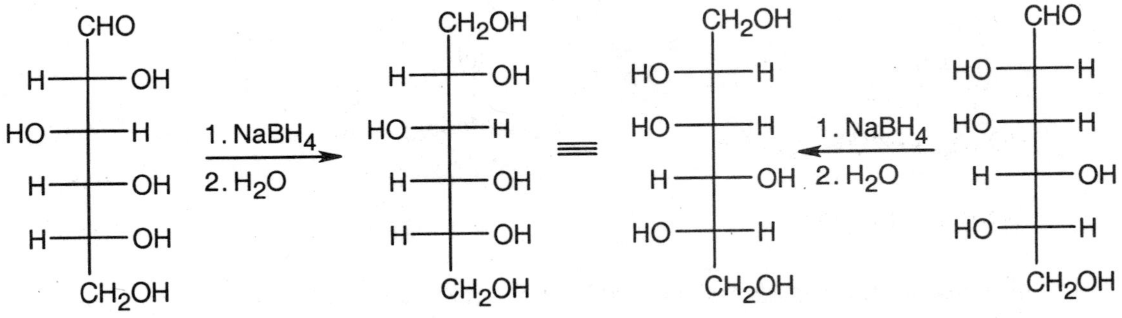
CHO
CH2OH
1. NaBH4
2. H2O

D-Glucose D-Glucitol L-Gulose

Reaction of an aldose with $NaBH_4$ produces a polyol (alditol). Because an alditol has the same functional group at both ends, two different aldoses can yield the same alditol. Here, L-gulose and D-glucose form the same alditol (rotate the Fischer projection of L-gulitol 180° to see the identity).

25.17

CHO / H–OH / HO–H / H–OH / H–OH / CH_2OH
D-Glucose

dil. HNO_3, heat →

COOH / H–OH / HO–H / H–OH / H–OH / COOH
Glucaric acid

CHO / H–OH / H–OH / H–OH / H–OH / CH_2OH
D-Allose

dil. HNO_3, heat →

COOH / H–OH / H–OH / H–OH / H–OH / COOH
Allaric acid

Allaric acid has a plane of symmetry and is an optically inactive meso compound. Glucaric acid has no symmetry plane.

25.18 D-Allose and D-galactose yield meso aldaric acids. All other D-hexoses produce optically active aldaric acids on oxidation.

25.19 The products of Kiliani-Fischer reaction of D-ribose have the same configuration at C3, C4 and C5 as D-ribose.

CHO / H–OH / H–OH / H–OH / CH_2OH
D-Ribose

1. HCN
2. H_2, Pd catalyst
3. H_3O^+
→

CHO / H–OH / H–OH / H–OH / H–OH / CH_2OH
D-Allose

+

CHO / HO–H / H–OH / H–OH / H–OH / CH_2OH
D-Altrose

25.20 The aldopentose, L-xylose has the same configuration as the configuration at C3, C4 and C5 of L-idose and L-gulose.

CHO / HO–H / H–OH / HO–H / CH_2OH
L-Xylose

1. HCN
2. H_2, Pd catalyst
3. H_3O^+
→

CHO / H–OH / HO–H / H–OH / HO–H / CH_2OH
L-Idose

+

CHO / HO–H / HO–H / H–OH / HO–H / CH_2OH
L-Gulose

25.21 The aldopentoses have the same configurations at C3 and C4 as D-threose.

CHO		CHO		CHO
H—OH		HO—H	1. H_2NOH	
HO—H	*or*	HO—H	2. $(CH_3CO)_2O$, $CH_3COO^-Na^+$ →	HO—H
H—OH		H—OH	3. $Na^+ {}^-OCH_3$	H—OH
CH_2OH		CH_2OH		CH_2OH
D-Xylose		D-Lyxose		D-Threose

25.22

CHO	CHO
HO—H	HO—H
H—OH	HO—H
H—OH	H—OH
CH_2OH	CH_2OH
D-Arabinose	D-Lyxose

D-Glucose has the same configuration at C3, C4, and C5 as D-arabinose. D-Lyxose is the only other aldopentose that yields an optically active aldaric acid on nitric acid oxidation.

25.23

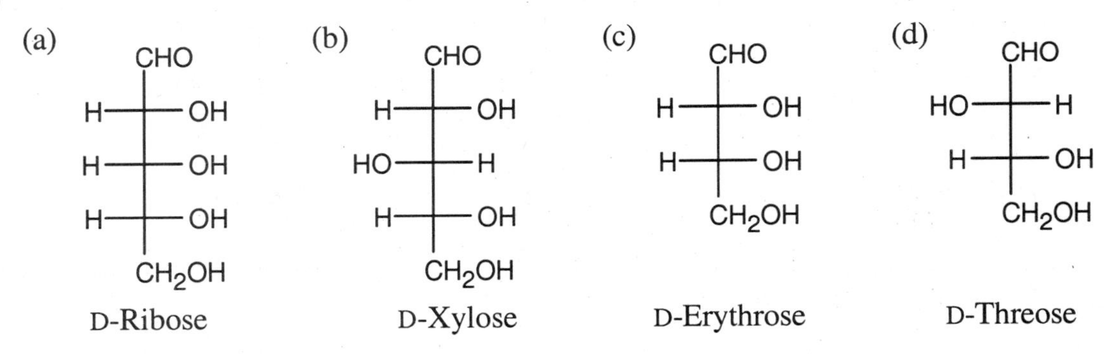

D-Ribose must have the above structure since it has the same configuration at C3 and C4 as D-arabinose. The structure of D-xylose is the remaining aldopentose isomer. D-Erythrose has the same configuration as C3 and C4 as D-ribose and D-arabinose.

25.24

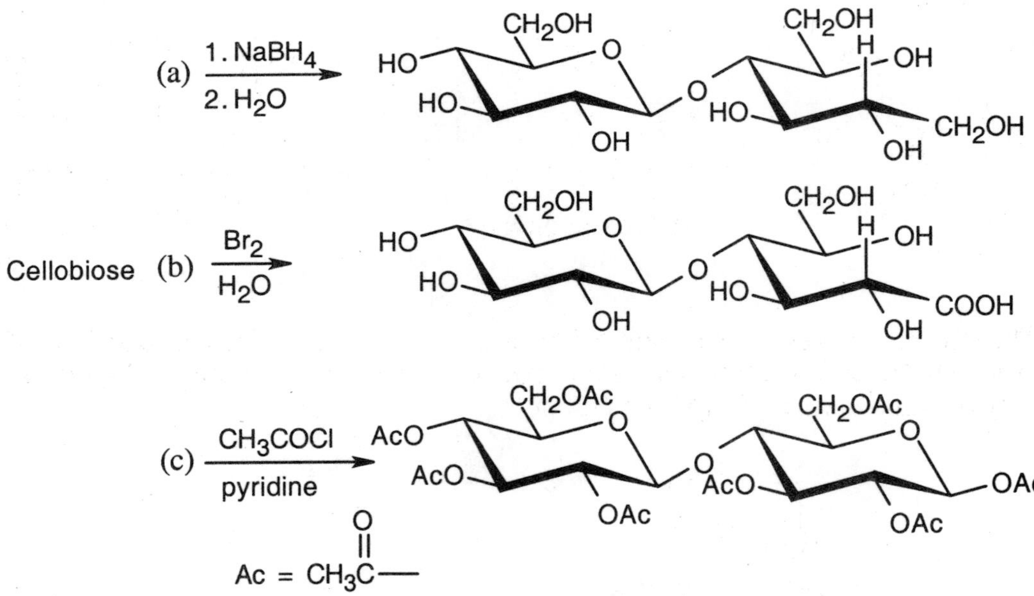

(a) Cellobiose $\xrightarrow[2.\ H_2O]{1.\ NaBH_4}$

(b) Cellobiose $\xrightarrow[H_2O]{Br_2}$

(c) Cellobiose $\xrightarrow[\text{pyridine}]{CH_3COCl}$

Ac = $CH_3C(=O)$—

Visualizing Chemistry

25.25 (a) Convert the model to a Fischer projection, remembering that the aldehyde group is on top, pointing into the page, and that the groups bonded to the carbons below point out of the page. The model represents a D-aldose because the –OH group at the chiral carbon farthest from the aldehyde points to the right.

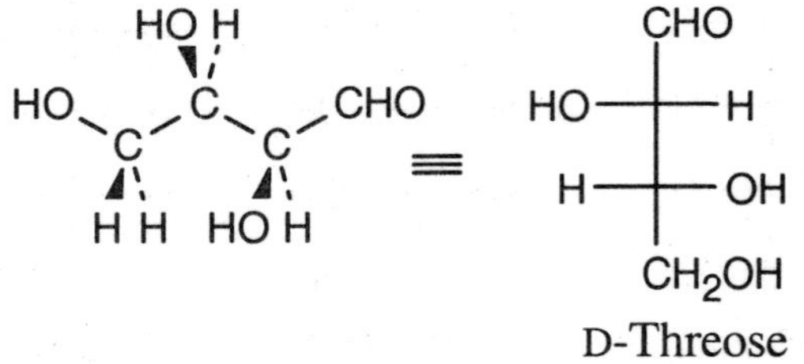

D-Threose

(b) Break the hemiacetal bond and uncoil the aldohexose. Notice that all hydroxyl groups point to the right in the Fischer projection. The model represents the β anomer of D-allopyranose.

(b)

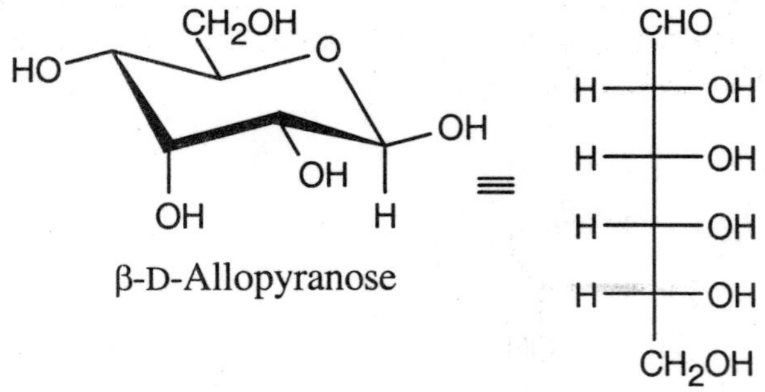

β-D-Allopyranose

25.26 The hints in the previous problem also apply here. Molecular models are also helpful.

(a)

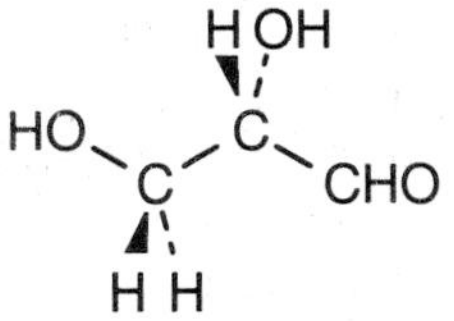

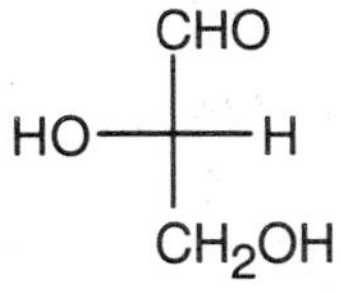

L-Glyceraldehyde

(b)

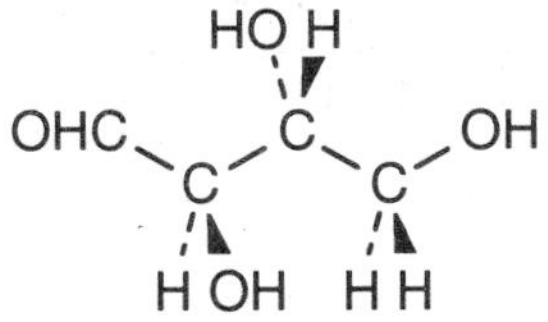

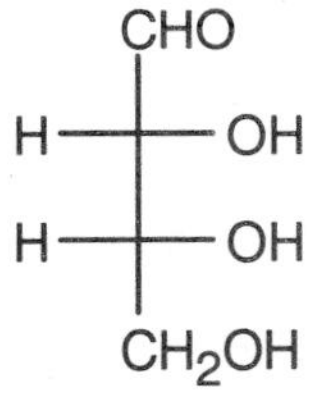

D-Erythrose

25.27 The structure represents an α anomer because the anomeric –OH group and the $-CH_2OH$ group are trans. The compound is α-L-mannopyranose because the –OH group at C2 is the only non-anomeric axial hydroxyl group.

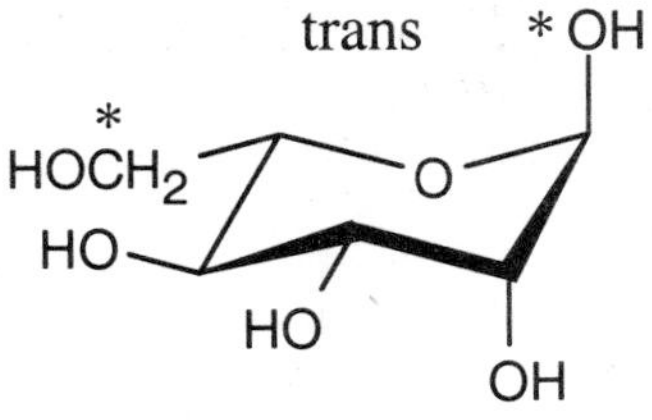

α-L-Mannopyranose

25.28

(a)

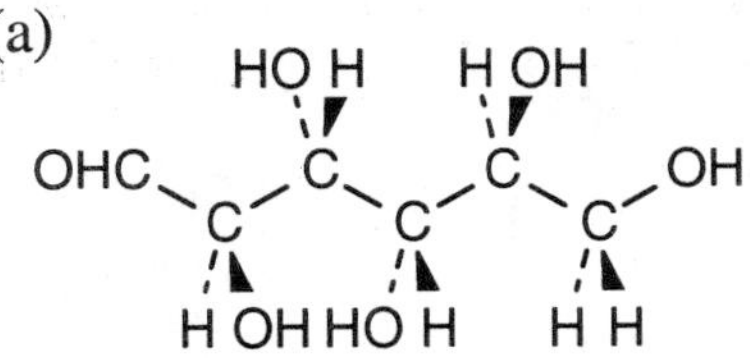

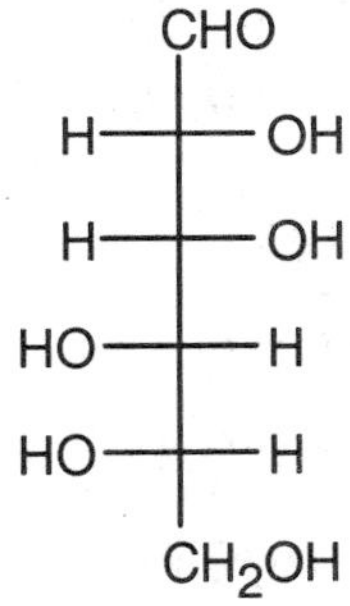

L-Mannose

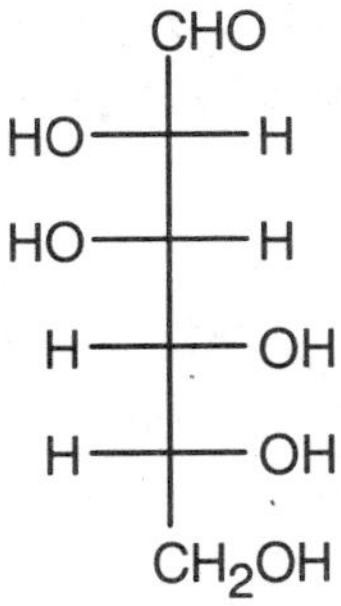

D-Mannose (enantiomer)

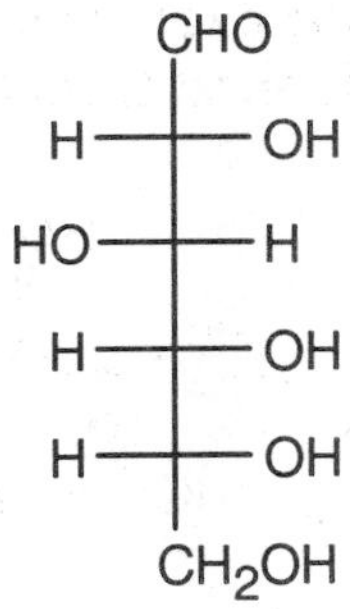

D-Glucose (diastereomer)

(b) The model represents an L-aldohexose because the hydroxyl group on the chiral carbon farthest from the aldehyde group points to the left.

(c) This is tricky! The furanose ring of an aldohexose is formed by connecting the –OH group at C4 to the aldehyde carbon. The best way to draw the anomer is to lie L-mannose on its side and form the ring. All substituents point down in the furanose, and the anomeric –OH and the $-CH(OH)CH_2OH$ group are cis.

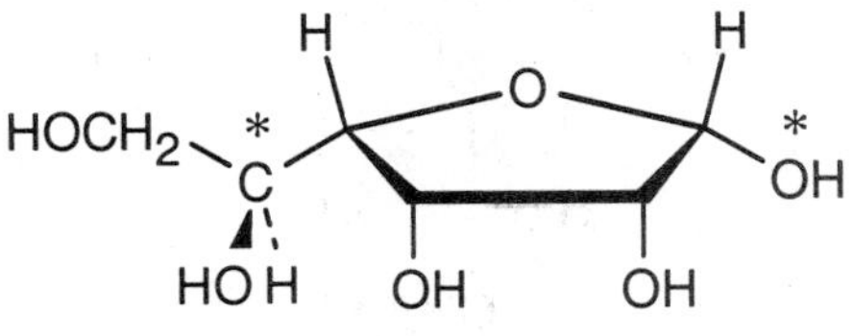

β-L-Mannofuranose

Additional Problems

25.29

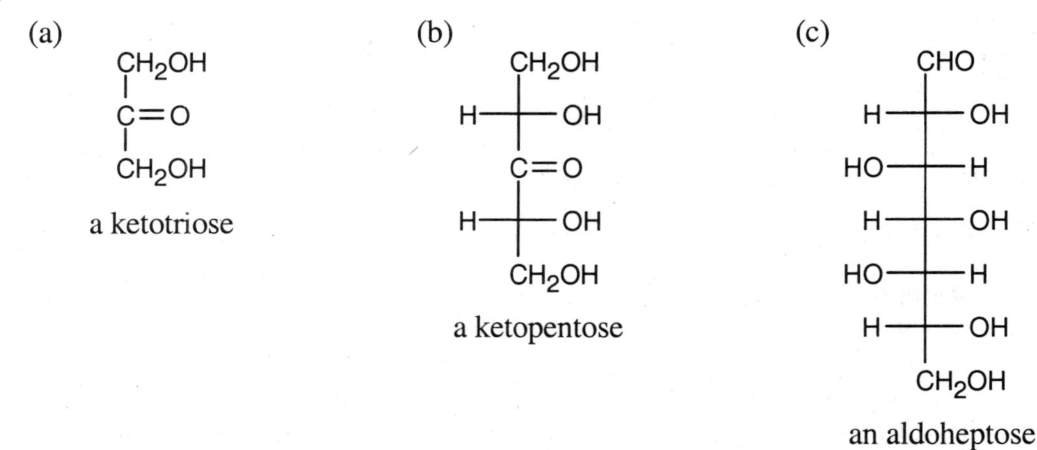

(a) a ketotriose (b) a ketopentose (c) an aldoheptose

25.30

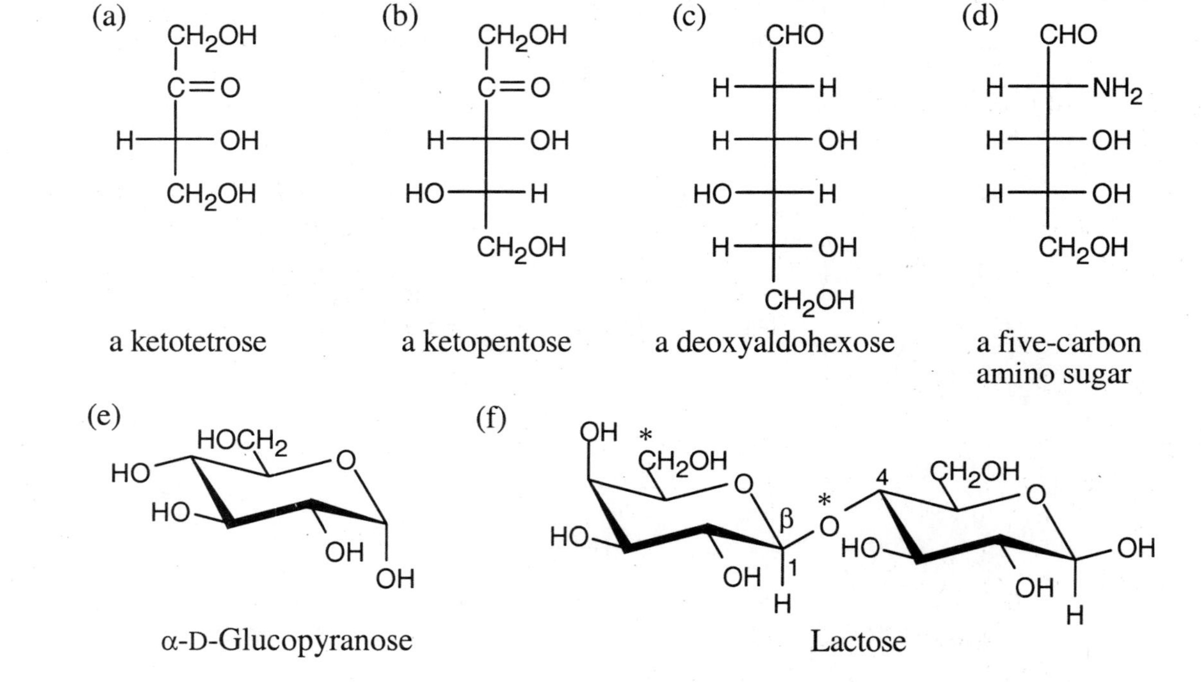

(a) a ketotetrose (b) a ketopentose (c) a deoxyaldohexose (d) a five-carbon amino sugar

(e) α-D-Glucopyranose (f) Lactose

25.31 – 25.32

Ascorbic acid has an L configuration because the hydroxyl group at the lowest chirality center points to the left.

L-Ascorbic acid

25.33

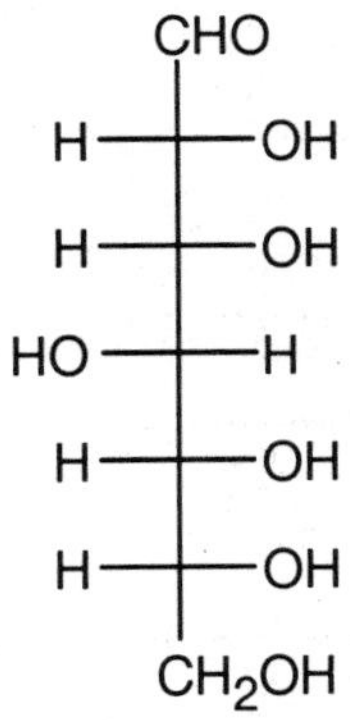
CHO
H OH
H OH
HO H
H OH
H OH
CH2OH

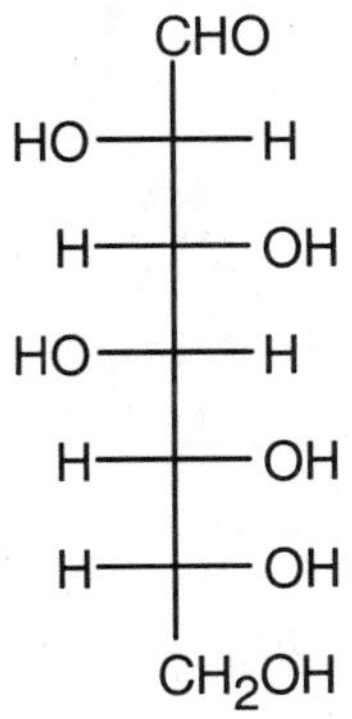
CHO
HO H
H OH
HO H
H OH
H OH
CH2OH

25.34

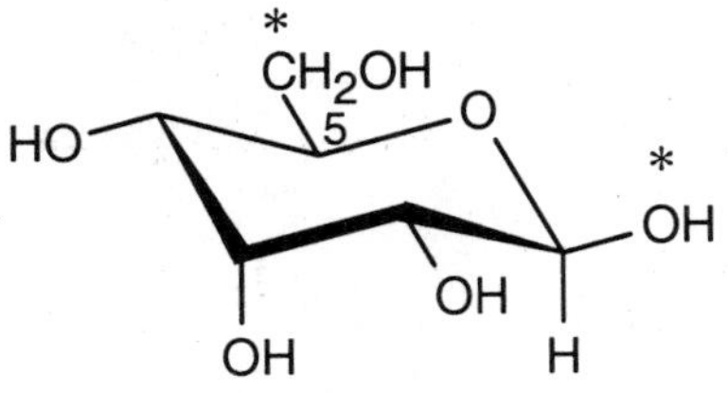
*
CH2OH
5
O
HO
*
OH
OH
OH
H

β-D-Allopyranose

This structure is a pyranose (6-membered ring) and is a β anomer (the C1 hydroxyl group and the $-CH_2OH$ groups are cis). It is a D sugar because the -O- at C5 is on the right in the uncoiled form.

25.35

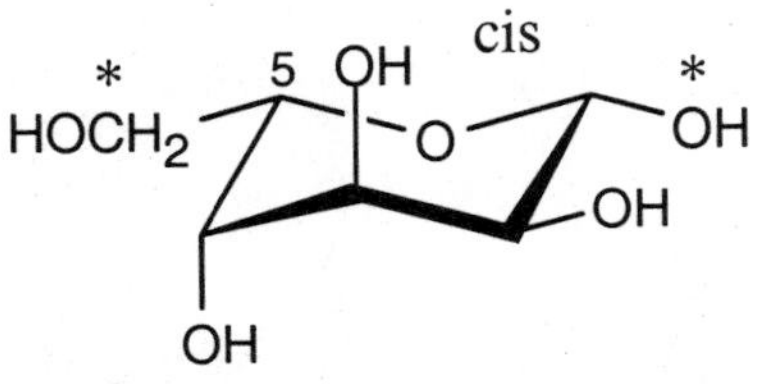
cis
*
5 OH
*
HOCH2
O
OH
OH
OH

β-L-Gulopyranose

This sugar is a β-pyranose. It is an L sugar because the -O- at C5 points to the left in the uncoiled form. It's also possible to recognize this as an L sugar by the fact that the configuration at C5 is *S*.

25.36

(a)

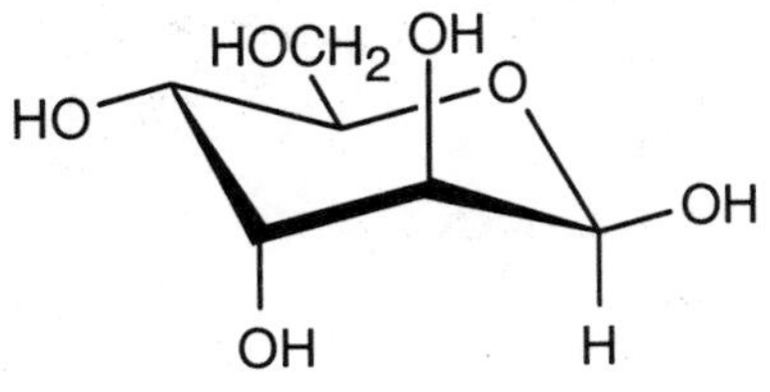
HOCH2 OH
O
HO
OH
OH
H

β-D-Altropyranose

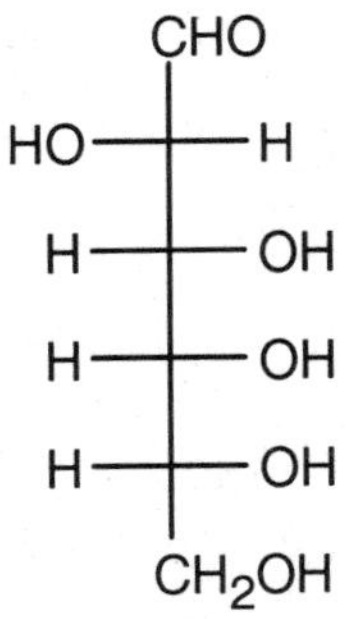
CHO
HO H
H OH
H OH
H OH
CH2OH

(b)

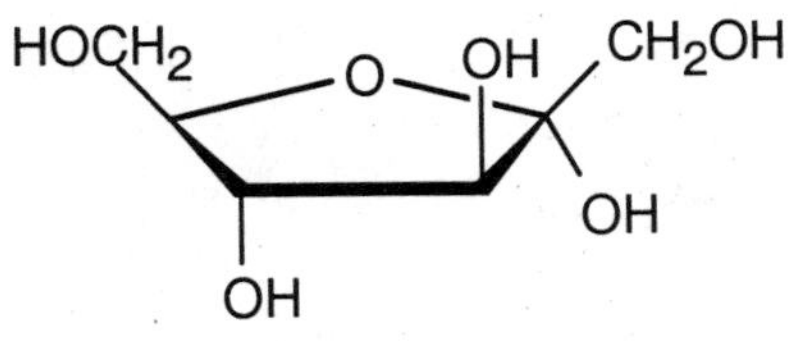
HOCH2
O
OH
CH2OH
OH
OH

α-D-Fructofuranose

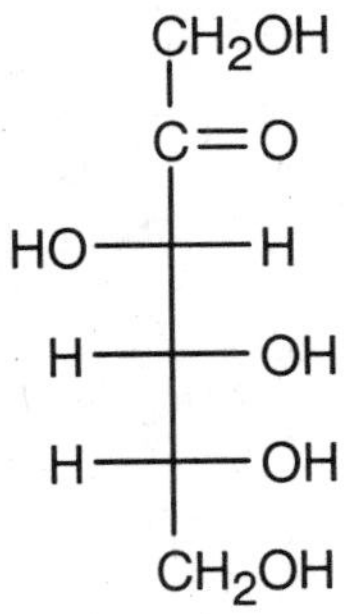
CH2OH
C=O
HO H
H OH
H OH
CH2OH

(c)

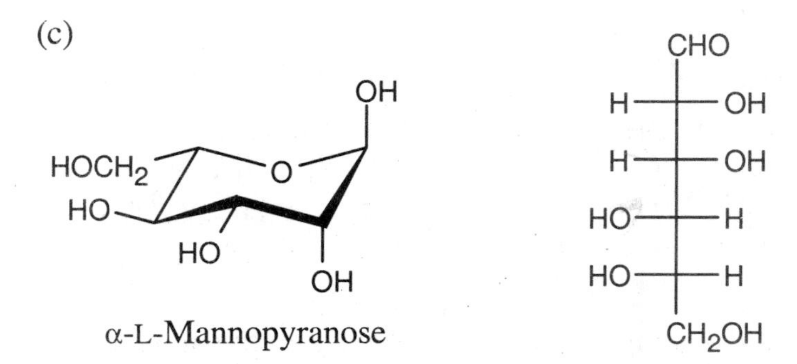

α-L-Mannopyranose

25.37

H O OH

H CH2OH

OH OH

β-D-Ribulofuranose

25.38–25.39

β-D-Talopyranose (a OH, e HOCH2, a OH, e HO, e OH)

(a) 1. $NaBH_4$ 2. H_2O → CH_2OH, HO–H, HO–H, HO–H, H–OH, CH_2OH

(b) dil. HNO_3 → COOH, HO–H, HO–H, HO–H, H–OH, COOH

(c) Br_2, H_2O → COOH, HO–H, HO–H, HO–H, H–OH, CH_2OH

(d) CH_3CH_2OH, HCl → OH, $HOCH_2$, OH, O, HO, OCH_2CH_3, H and α anomer

(e) CH_3I, Ag_2O → CH_3O, CH_2OCH_3, OCH_3, O, CH_3O, OCH_3, H

(f) $(CH_3CO)_2O$, pyridine → OAc, $AcOCH_2$, OAc, O, AcO, OAc, H

25.40

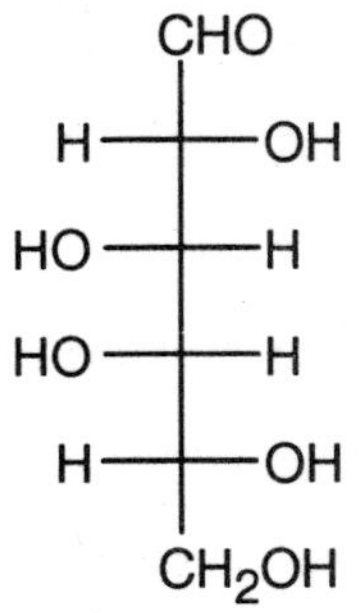

D-Galactose

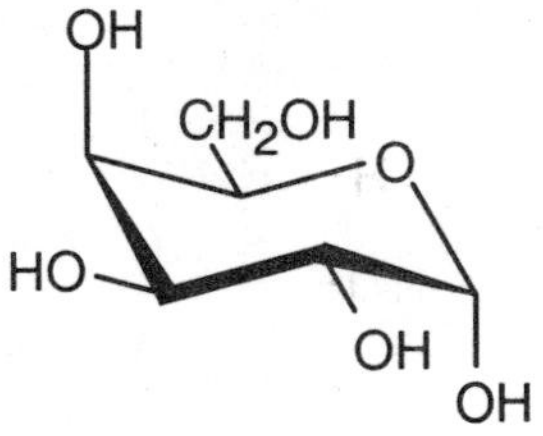

α-D-Galactopyranose
$[\alpha]_D = +150.7°$

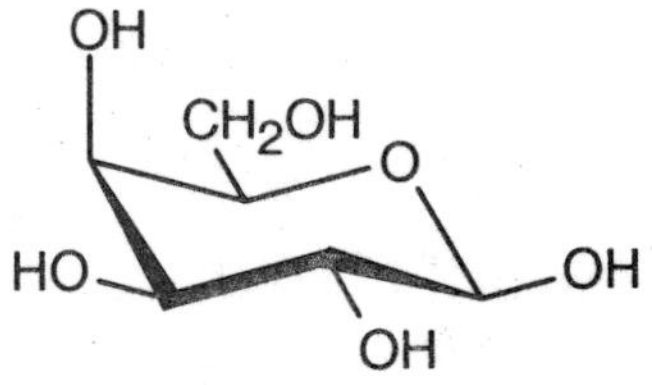

β-D-Galactopyranose
$[\alpha]_D = +52.8°$

Let x be the fraction of D–galactose present as the α anomer and y be the fraction of D–galactose present as the ß anomer.

$$150.7°x + 52.8°y = 80.2° \qquad x + y = 1; \quad y = 1 - x$$
$$150.7°x + 52.8°(1–x) = 80.2°$$
$$97.9°x = 27.4°$$
$$x = 0.280$$
$$y = 0.720$$

28.0% of D–galactose is present as the α anomer, and 72.0% is present as the β anomer.

25.41–25.43 Four D–2–ketohexoses are possible.

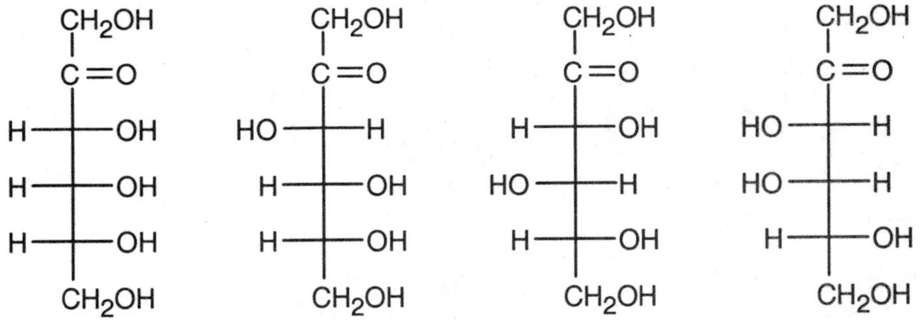

D-Psicose D-Fructose D-Sorbose D-Tagatose

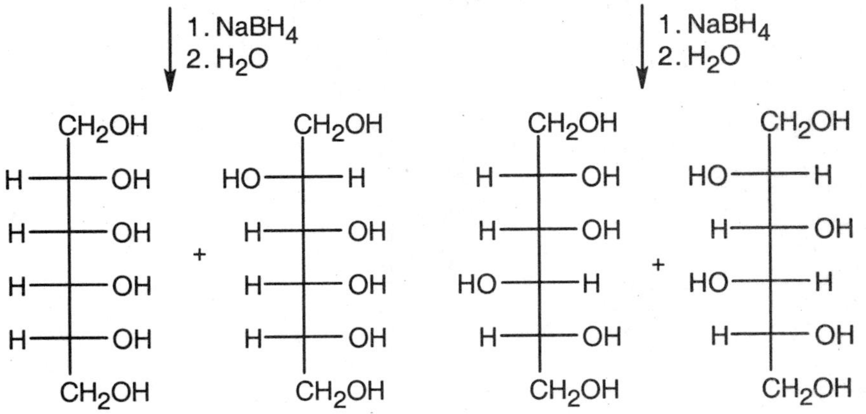

1. $NaBH_4$
2. H_2O

Allitol + Altritol Gulitol + Iditol

25.44 The two lactones are formed between a carboxylic acid and a hydroxyl group 4 carbons away. When the lactones are reduced with sodium amalgam, the resulting hexoses have an aldehyde at one end and a hydroxyl group at the other end.

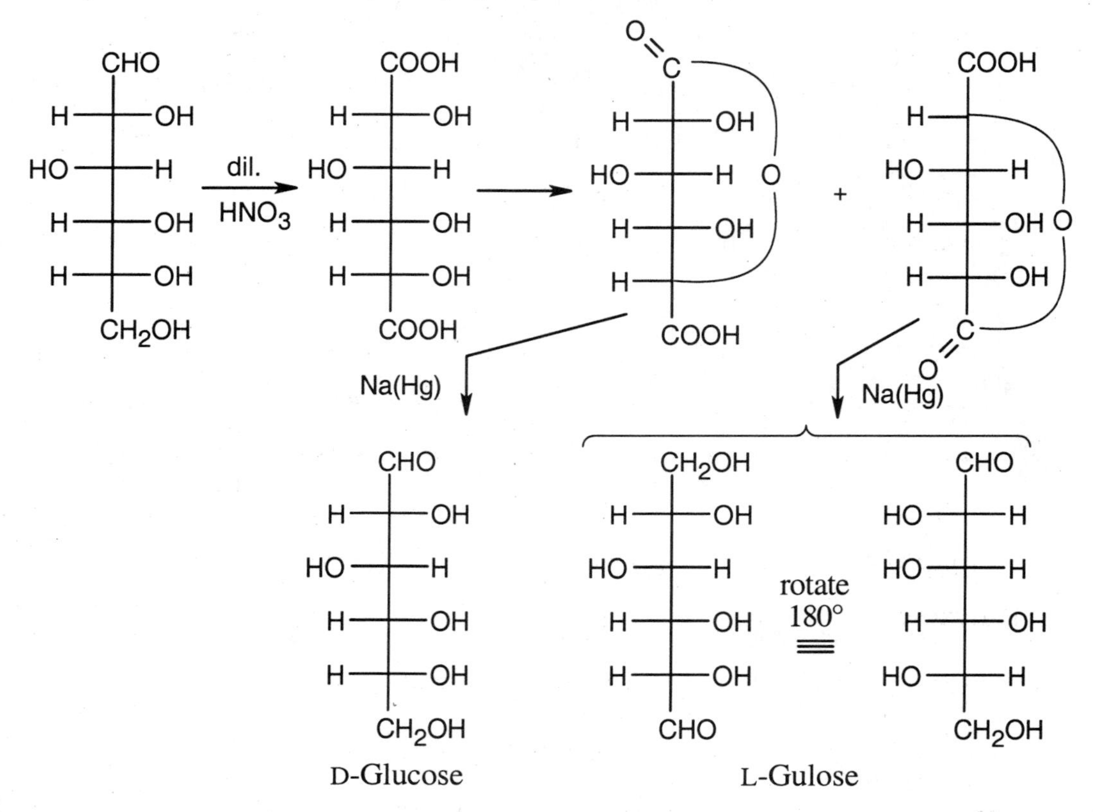

25.45

1. NaBH$_4$
2. H$_2$O

rotate 180°

1. NaBH$_4$
2. H$_2$O

D-Talose

D-Altrose

25.46

dil. HNO$_3$

rotate 180°

dil. HNO$_3$

D-Allose

L-Allose

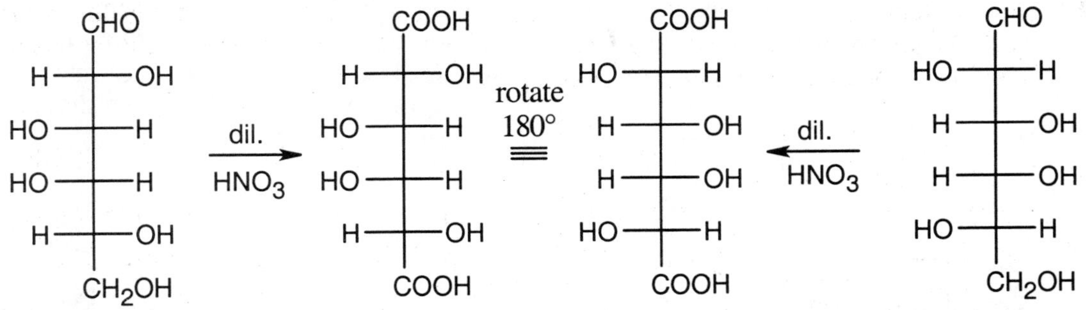

CHO / H—OH / HO—H / HO—H / H—OH / CH_2OH
D-Galactose

$\xrightarrow[\text{HNO}_3]{\text{dil.}}$

COOH / H—OH / HO—H / HO—H / H—OH / COOH

rotate 180° ≡

COOH / HO—H / H—OH / H—OH / HO—H / COOH

$\xleftarrow[\text{HNO}_3]{\text{dil.}}$

CHO / HO—H / H—OH / H—OH / HO—H / CH_2OH
L-Galactose

25.47

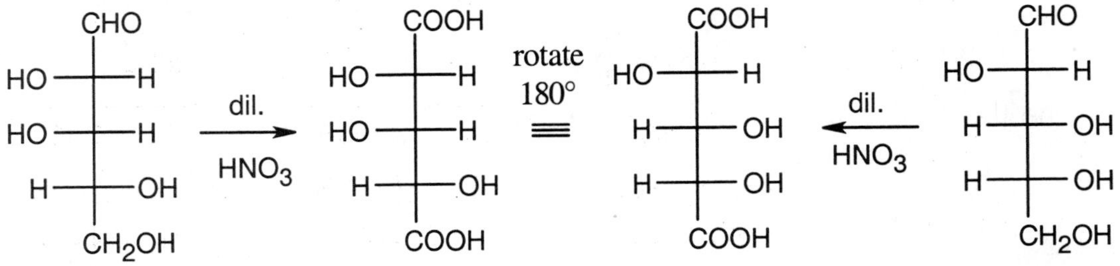

CHO / HO—H / HO—H / H—OH / CH_2OH
D-Lyxose

$\xrightarrow[\text{HNO}_3]{\text{dil.}}$

COOH / HO—H / HO—H / H—OH / COOH

rotate 180° ≡

COOH / HO—H / H—OH / H—OH / COOH

$\xleftarrow[\text{HNO}_3]{\text{dil.}}$

CHO / HO—H / H—OH / H—OH / CH_2OH
D-Arabinose

25.48 (a) D-Galactose gives the same aldaric acid as L-galactose.

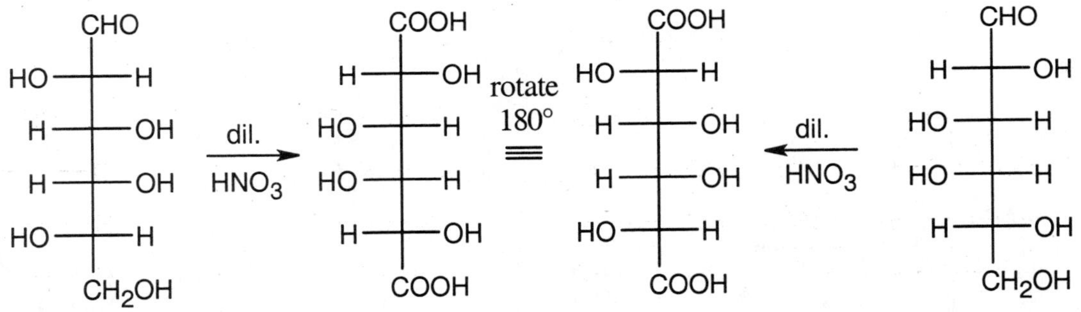

CHO / HO—H / H—OH / H—OH / HO—H / CH_2OH
L-Galactose

$\xrightarrow[\text{HNO}_3]{\text{dil.}}$

COOH / H—OH / HO—H / HO—H / H—OH / COOH

rotate 180° ≡

COOH / HO—H / H—OH / H—OH / HO—H / COOH

$\xleftarrow[\text{HNO}_3]{\text{dil.}}$

CHO / H—OH / HO—H / HO—H / H—OH / CH_2OH
D-Galactose

(b) The other aldohexose is a D-sugar.

(c)

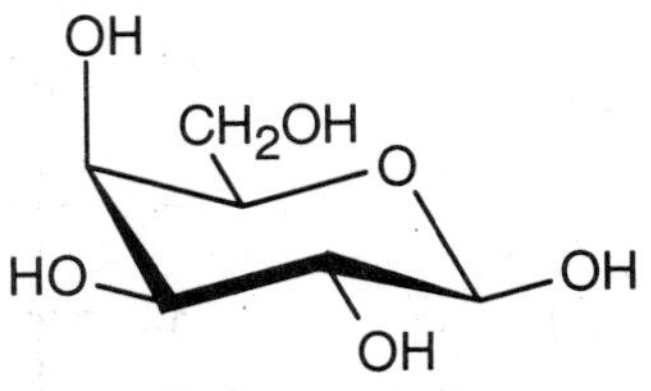

OH, CH_2OH, O, HO, OH, OH

β-D-Galactopyranose

25.49 The hard part of this problem is determining where the glycosidic bond occurs on the second glucopyranose ring. Treatment with iodomethane yields a tetra-*O*-methyl glucopyranose and a tri-*O*-methyl glucopyranose. The oxygen in the tri-*O*-methylated ring that is not part of the hemiacetal group and is not methylated is the oxygen that forms the acetal bond. In this problem, the C6 oxygen forms the glycosidic link.

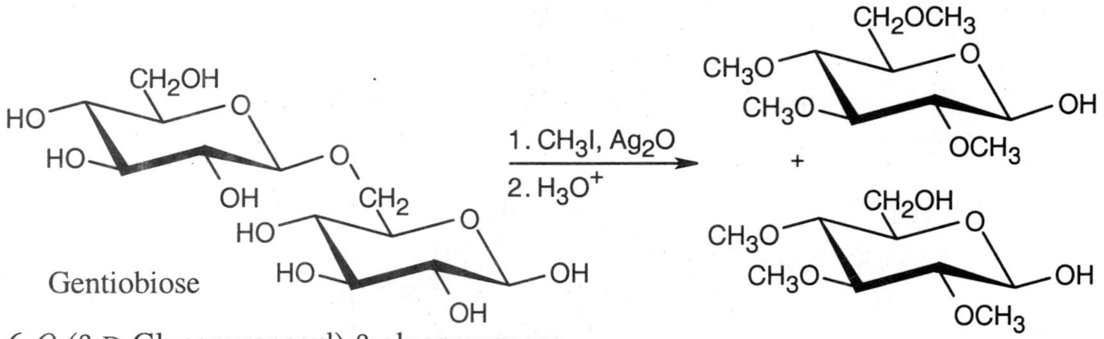

Gentiobiose

1. CH_3I, Ag_2O
2. H_3O^+

6-*O*-(β-D-Glucopyranosyl)-β-glucopyranose

25.50 Amygdalin has the same carbohydrate skeleton as gentiobiose. Draw the cyanohydrin of benzaldehyde, and form a bond between the hemiacetal oxygen and the carbonyl carbon of benzaldehyde, with elimination of water.

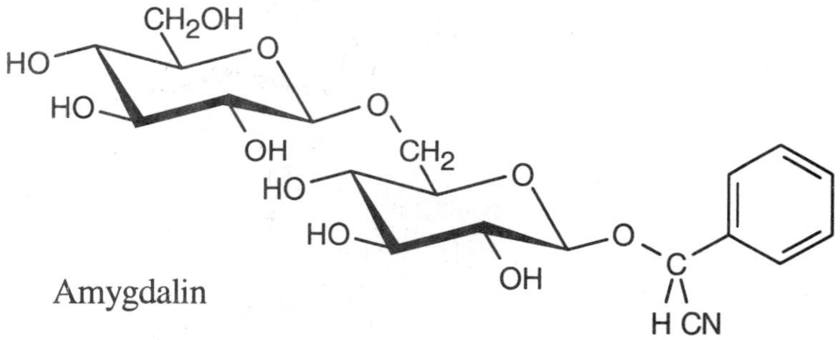

Amygdalin

25.51 Since trehalose is a nonreducing sugar, the two glucose units must be connected through an oxygen atom at the anomeric carbon of each glucose. There are three possible structures for trehalose: The two glucopyranose rings can be connected (α,α), (β,β), or (α,β).

25.52 Since trehalose is not cleaved by ß–glycosidases, it must have an α,α glycosidic linkage.

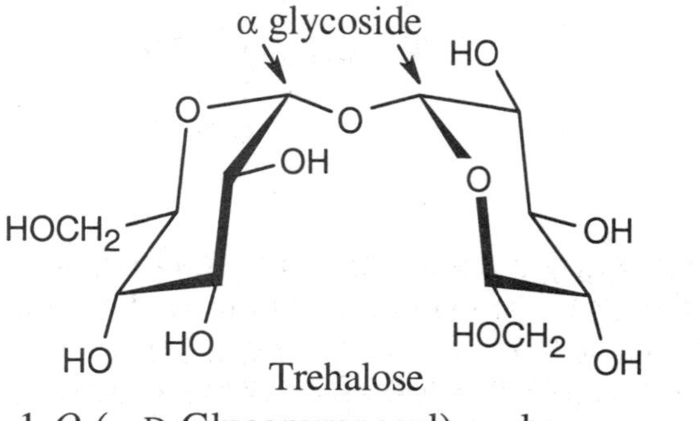

α glycoside

Trehalose

1-*O*-(α-D-Glucopyranosyl)-α-glucopyranose

25.53

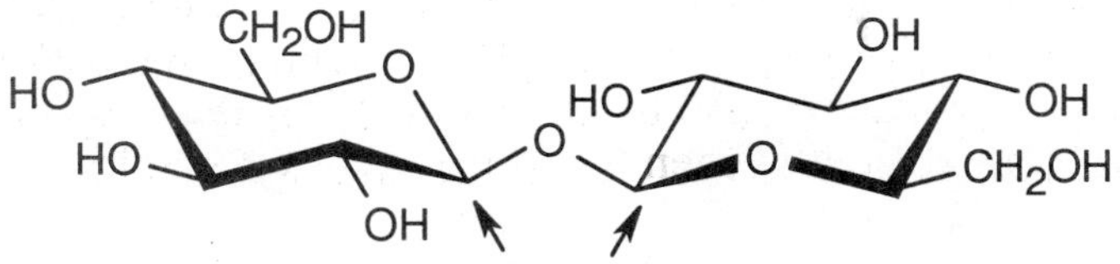

β glycoside

Neotrehalose

1-*O*-(β-D-Glucopyranosyl)-β-glucopyranose

α glycoside

β glycoside

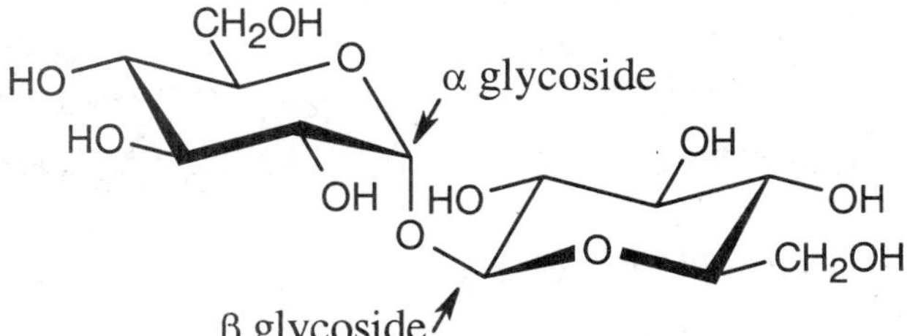

Isotrehalose

1-*O*-(α-D-Glucopyranosyl)-β-glucopyranose

25.54

Glucopyranose is in equilibrium with glucofuranose

Reaction with two equivalents of acetone occurs by the mechanism we learned for acetal formation (Sec 19.11)

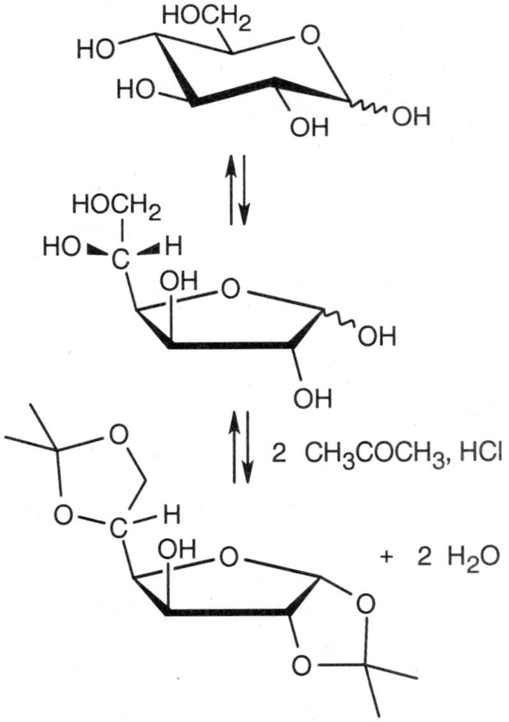

$2\ CH_3COCH_3$, HCl

$+\ 2\ H_2O$

A five-membered acetal ring forms much more readily when the hydroxyl groups are cis to one another. In glucofuranose, the C3 hydroxyl group is trans to the C2 hydroxyl group, and acetal formation occurs between acetone and the C1 and C2 hydroxyls of glucofuranose. Since the C1 hydroxyl group is part of the acetone acetal, the furanose is no longer in equilibrium with the free aldehyde, and the diacetone derivative is not a reducing sugar.

25.55

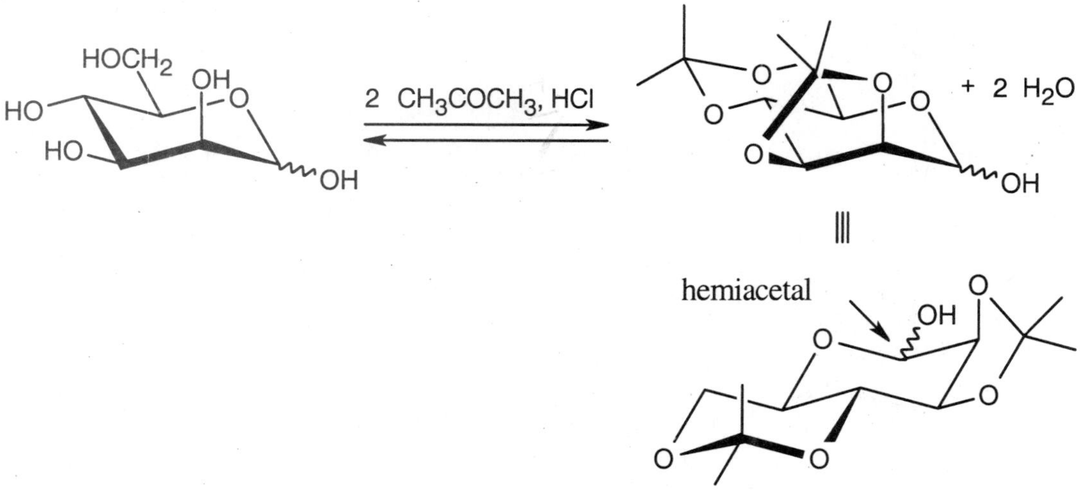

2 CH_3COCH_3, HCl

+ 2 H_2O

hemiacetal

2,3:4,6-Diacetone mannopyranoside

Acetone forms an acetal with the hydroxyl groups at C2 and C3 of D-mannopyranoside because the hydroxyl groups at these positions are cis to one another. The pyranoside ring is still a hemiacetal that is in equilibrium with free aldehyde, which is reducing toward Tollens' reagent.

25.56

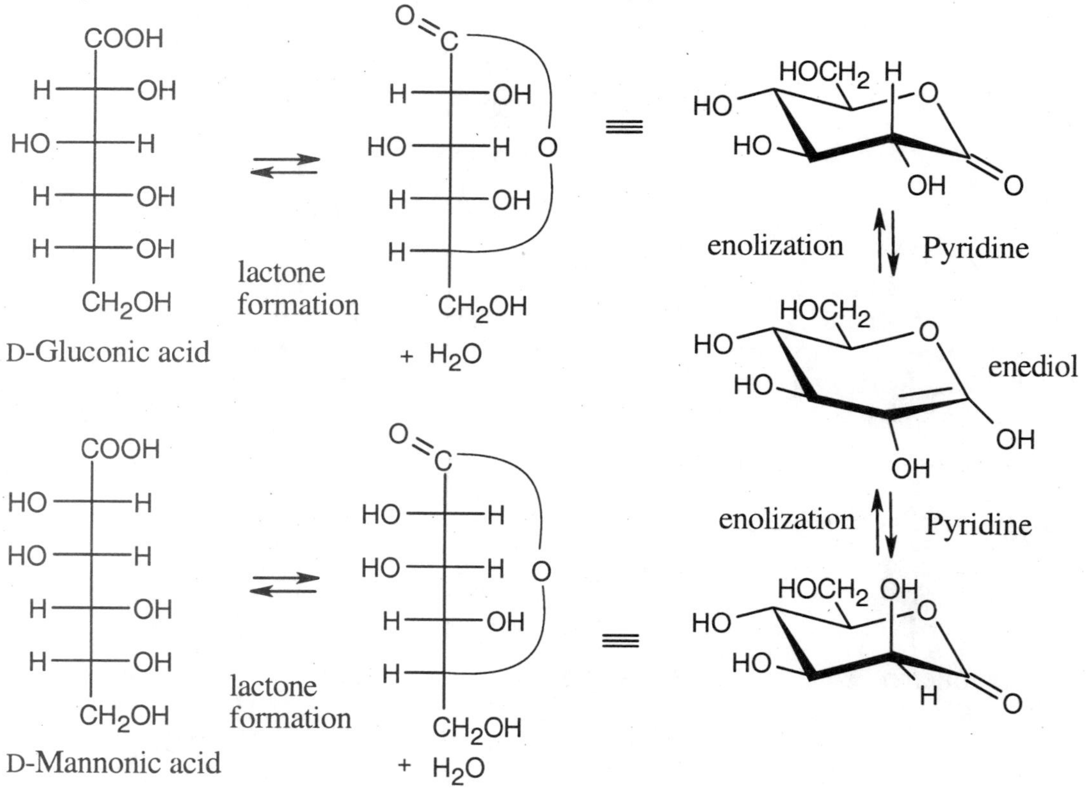

D-Gluconic acid ⇌ (lactone formation) + H_2O

enolization ⇅ Pyridine

enediol

enolization ⇅ Pyridine

D-Mannonic acid ⇌ (lactone formation) + H_2O

Isomerization at C2 occurs because the enediol can be reprotonated on either side of the double bond.

25.57 There are eight diastereomeric cyclitols.

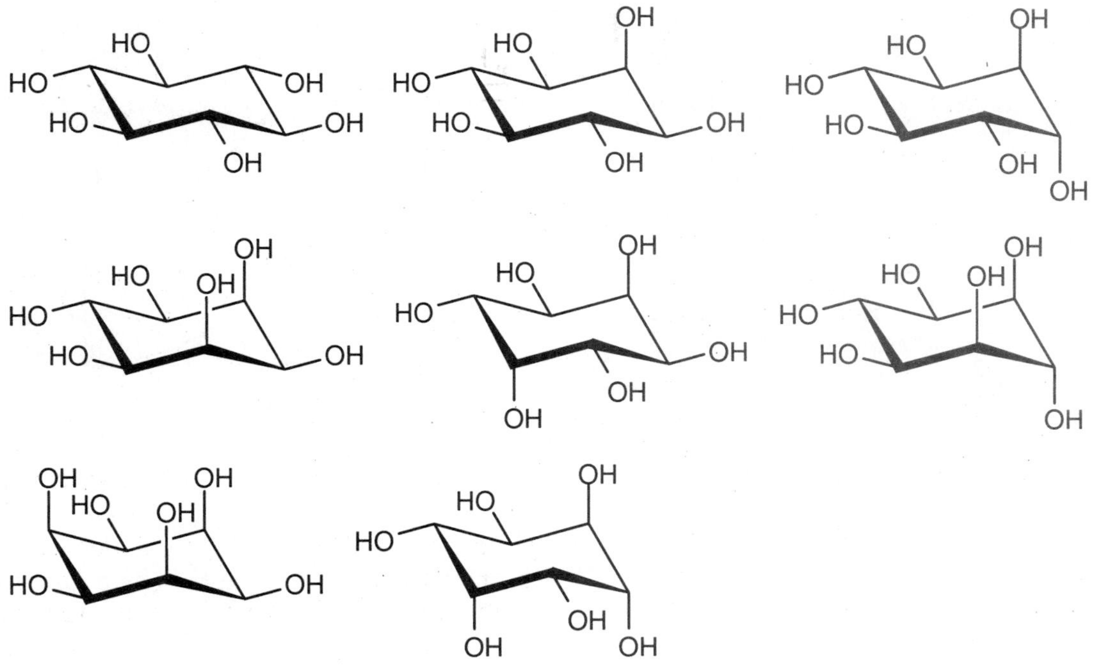

25.58

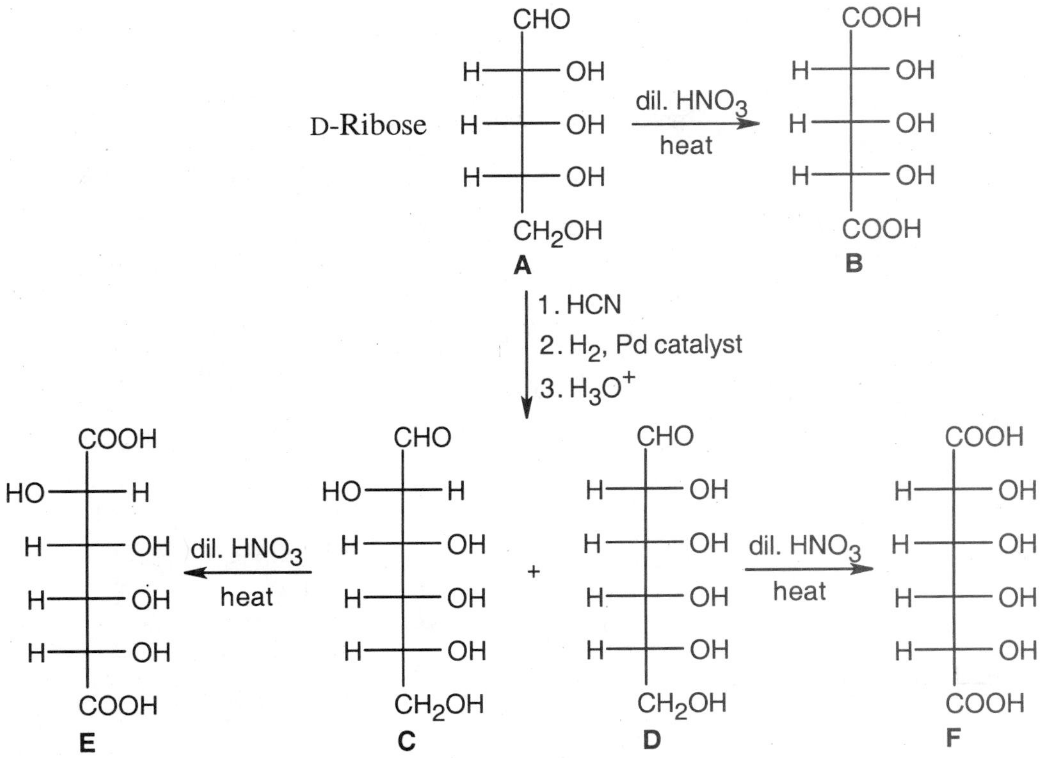

D-Ribose (A) $\xrightarrow[\text{heat}]{\text{dil. } HNO_3}$ B (COOH, H–OH, H–OH, H–OH, COOH)

A: CHO, H–OH, H–OH, H–OH, CH_2OH

A $\xrightarrow{\text{1. HCN; 2. } H_2\text{, Pd catalyst; 3. } H_3O^+}$ C + D

C (D-Altrose): CHO, HO–H, H–OH, H–OH, H–OH, CH_2OH $\xrightarrow[\text{heat}]{\text{dil. } HNO_3}$ E: COOH, HO–H, H–OH, H–OH, H–OH, COOH

D (D-Allose): CHO, H–OH, H–OH, H–OH, H–OH, CH_2OH $\xrightarrow[\text{heat}]{\text{dil. } HNO_3}$ F: COOH, H–OH, H–OH, H–OH, H–OH, COOH

Because **A** is oxidized to an optically inactive aldaric acid, the possible structures for **A** are D-ribose and D-xylose. Chain extension of D-xylose, however, produces two hexoses that, when oxidized, yield optically active aldaric acids.

25.59

(a) D-Glucose *or* D-Fructose *or* D-Mannose $\xrightarrow{\text{2 } H_2NNHPh}$ An osazone

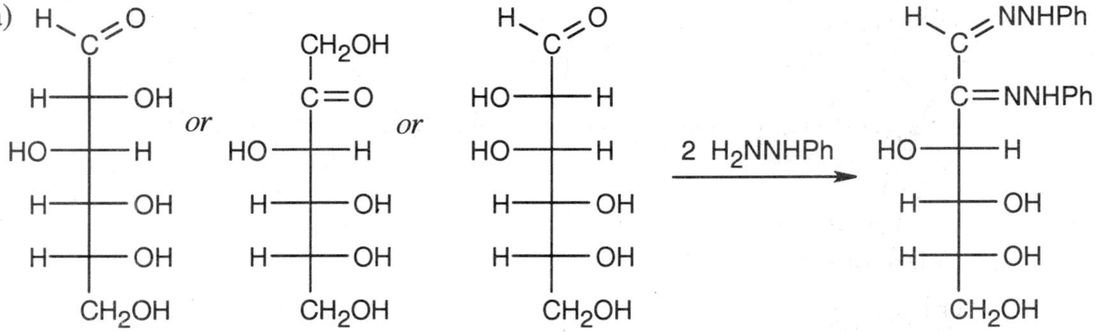

D-Glucose: HC=O; H–C–OH; HO–C–H; H–C–OH; H–C–OH; CH_2OH

D-Fructose: CH_2OH; C=O; HO–C–H; H–C–OH; H–C–OH; CH_2OH

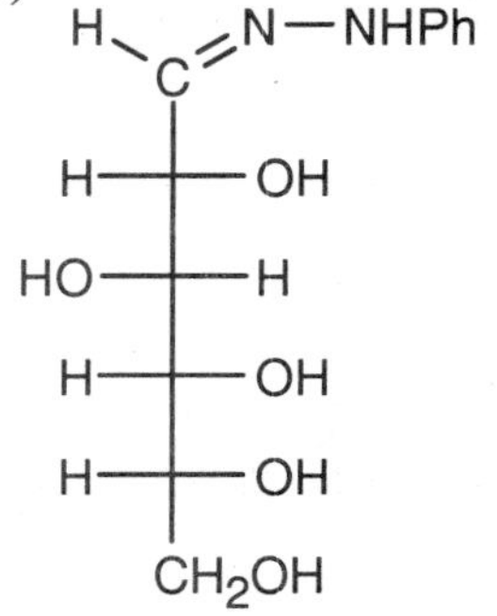

D-Mannose: HC=O; HO–C–H; HO–C–H; H–C–OH; H–C–OH; CH_2OH

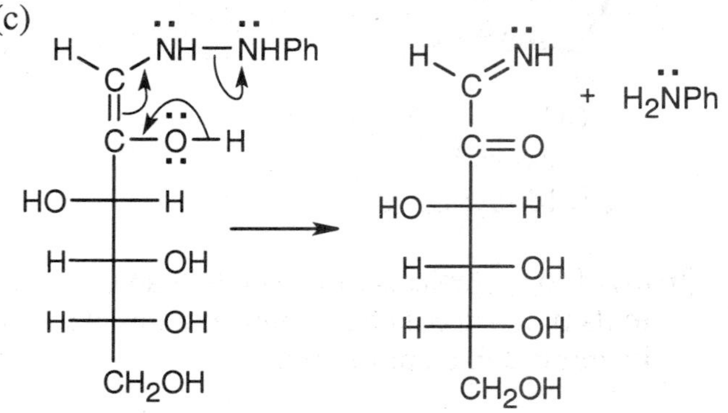

An osazone: HC=NNHPh; C=NNHPh; HO–C–H; H–C–OH; H–C–OH; CH_2OH

(b) phenylhydrazone: HC=N–NHPh; H–C–OH; HO–C–H; H–C–OH; H–C–OH; CH_2OH

(c) enol: HC(–NH–NHPh)=C(–OH); HO–C–H; H–C–OH; H–C–OH; CH_2OH → keto imine: HC=NH; C=O; HO–C–H; H–C–OH; H–C–OH; CH_2OH + H_2NPh

(d) $PhNH–NH_2$ + keto imine (HC=NH; C=O; HO–C–H; H–C–OH; H–C–OH; CH_2OH) ⇌ $PhNHN(H)–CH–NH_2$; C=O; ... ⇌ $PhNHN^{-}–CH–NH_3^{+}$; C=O; ... ⇌ PhNHN=CH; C=O; ... + NH_3

PhNHN=CH; C=O; ... + $PhNH–NH_2$ ⇌ PhNHN=CH; $PhNHNH_2^{+}–C–O^{-}$; ... ⇌ PhNHN=CH; PhNHNH–C–OH; ... ⇌ PhNHN=CH; PhNHN=C; HO–C–H; H–C–OH; H–C–OH; CH_2OH + H_2O

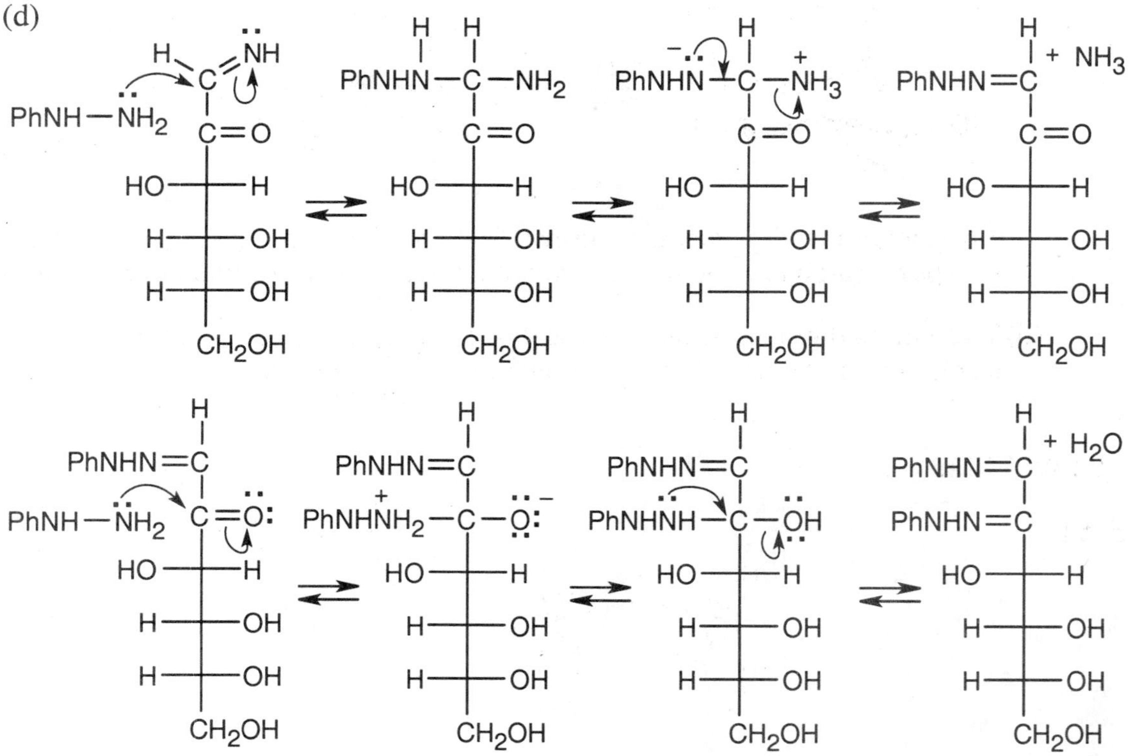

In these last steps, two nucleophilic addition reactions take place to yield imine products. The mechanism has been worked out in greater detail in Section 19.9, but the essential steps are additions of phenylhydrazine, first to the imine, then to the ketone. Proton transfers are followed by eliminations, first of ammonia, then of H_2O.

25.60 (a)

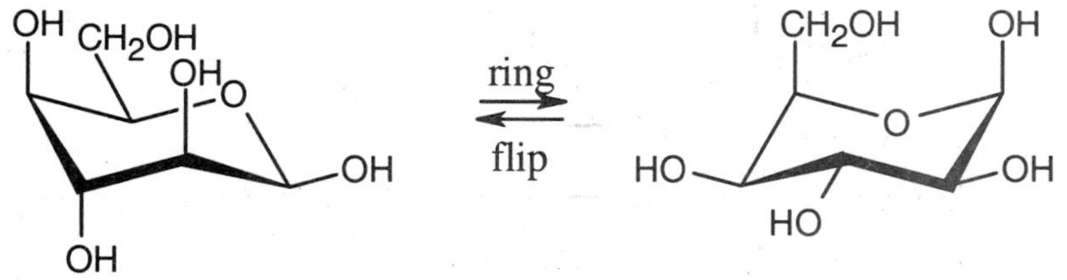

less stable
β-D-Idopyranose

more stable

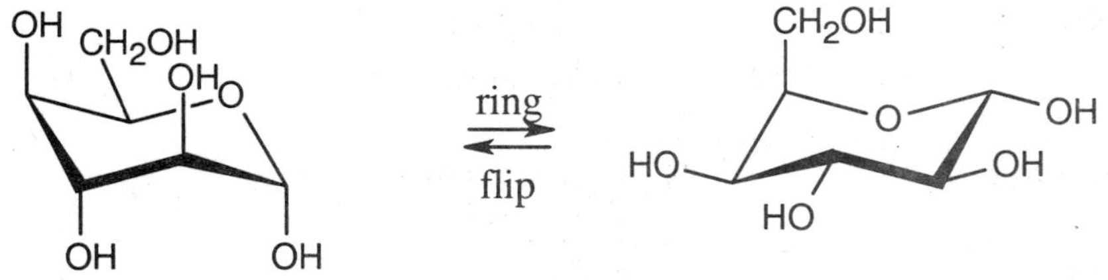

less stable
α-D-Idopyranose

more stable

(b) α-D-Idopyranose is more stable than β-D-idopyranose because only one group is axial in its more stable chair conformation, whereas β-D-idopyranose has two axial groups in its more stable conformation.

(c)

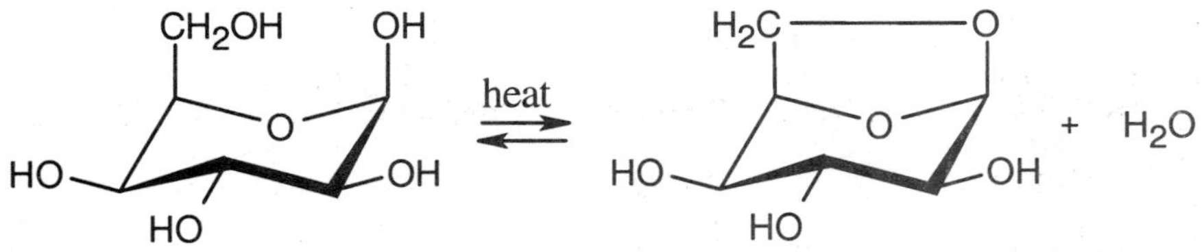

$$\xrightarrow{\text{heat}} \quad + \ H_2O$$

1,6-Anhydro-D-idopyranose is formed from the β anomer because the axial hydroxyl groups on carbons 1 and 6 are close enough for the five-membered ring to form.

(d) The hydroxyl groups at carbons 1 and 6 of D-glucopyranose are equatorial in the most stable conformation and are too far apart for a ring to form.

A Look Ahead

25.61

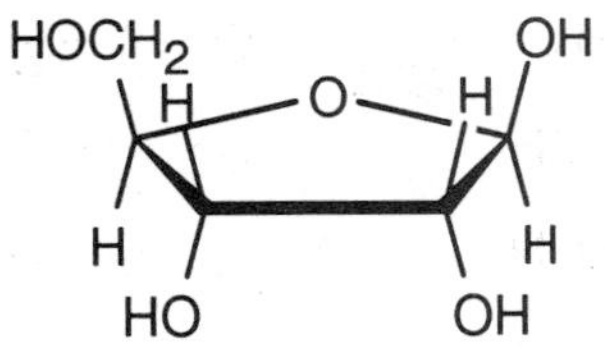

D-Ribofuranose is the sugar present in acetyl CoA.

Molecular Modeling

25.62

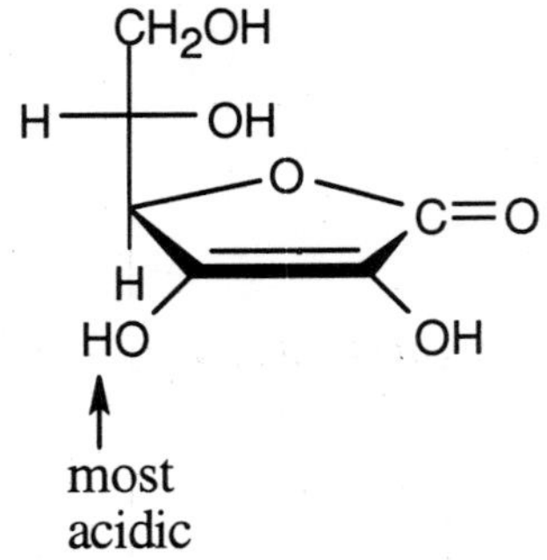

CH_2OH H OH O C=O H HO OH

most acidic

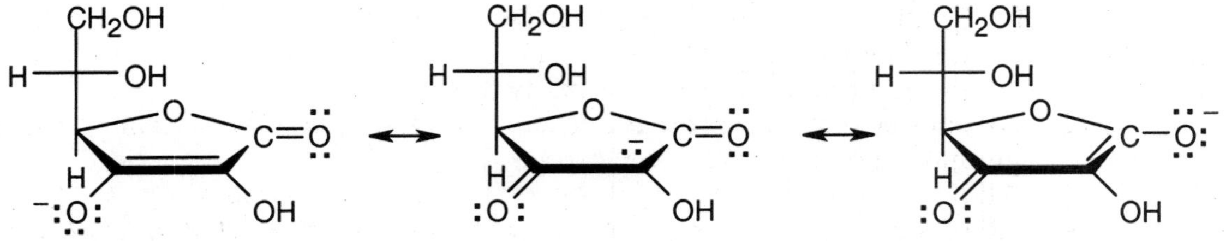

CH_2OH H OH O C=O: H :O: OH ↔ CH_2OH H OH O C=O H :O: OH ↔ CH_2OH H OH O C–O: H :O: OH

25.63 The pyranose is lower in energy by 17 kJ/mol and is thermodynamically favored.

Chapter 26 – Biomolecules: Amino Acids, Peptides and Proteins

Chapter Outline

I. Amino acids (Sections 26.1 – 26.4).
- A. Structure of amino acids (Section 26.1).
  1. Amino acids exist in solution as zwitterions.
     - a. Zwitterions are internal salts and have many of the properties associated with salts.
       - i. They have large dipole moments.
       - ii. They are soluble in water.
       - iii. They are crystalline and high-melting.
     - b. Zwitterions can act either as acids or as bases.
       - i. The $-COO^-$ group acts as a base.
       - ii. The ammonium group acts an acid.
  2. All natural amino acids are α-amino acids: the amino group and the carboxylic acid group are bonded to the same carbon.
  3. All but one (proline) of the 20 common amino acids are primary amines.
  4. All of the amino acids are represented by both a three-letter code and a one-letter code.
  5. All amino acids except glycine are chiral.
     - a. Only one enantiomer (L) of each pair is naturally-occurring.
     - b. In Fischer projections, the carboxylic acid is at the top, and the amino group points to the left.
     - c. α-Amino acids are referred to as L-amino acids.
  6. Side chains can be acidic or basic.
     - a. Fifteen of the amino acids are neutral.
     - b. Two (aspartic acid and glutamic acid) are acidic.
       At pH = 7.3, their side chains exist as carboxylate ions.
     - c. Three (lysine, arginine and histidine) are basic.
       - i. At pH = 7.3, the side chains of lysine and arginine exist as ammonium ions.
       - ii. Histidine is not quite basic enough to be protonated at pH = 7.3.
       - iii. The double-bonded nitrogen in the histidine ring is basic.
     - d. Cysteine and tyrosine are weakly acidic.
- B. Isoelectric points (Section 26.2).
  1. The isoelectric point (p*I*) is the pH at which an amino acid exists as a neutral, dipolar zwitterion.
     - a. p*I* is related to side chain structure.
       - i. The 15 amino acids that are neutral have p*I* near neutrality.
       - ii. The two acidic amino acids have p*I* at a lower pH.
       - iii. The 3 basic amino acids have p*I* at a higher pH.
     - b. For neutral amino acids, p*I* is the average of the two pK_a values.
       - i. For acidic amino acids, p*I* is the average of the two lowest pK_a values.
       - ii. For basic amino acids, p*I* is the average of the two highest pK_a values.
  2. Electrophoresis allows the separation of amino acids by differences in their p*I*.
     - a. A buffered solution of amino acids is placed on a paper or gel.
     - b. Electrodes are connected to the solution, and current is applied.
     - c. Negatively charged amino acids migrate to the positive electrode, and positively charged amino acids migrate to the negative electrode.
     - d. Amino acids can be separated because the extent of migration depends on p*I*.

3. The Henderson-Hasselbalch equation.
    a. If we know the values of pH and pK_a, we can calculate the percentages of protonated, neutral and deprotonated forms of an amino acid.
    b. If we do these calculations at several pH values, we can construct a titration curve for each amino acid.

C. Synthesis of α-amino acids (Section 26.3).
1. The Hell-Volhard-Zelinskii method and the phthalimide method.
    a. An α-bromo acid is produced from a carboxylic acid by α-bromination.
    b. Displacement of –Br gives the α-amino acid.
2. The Strecker synthesis.
    An aldehyde is treated with KCN and aqueous ammonia to give an intermediate α-amino nitrile, which is hydrolyzed.
3. The amidomalonate synthesis.
    a. An alkyl halide reacts with the anion of diethyl amidomalonate.
    b. Hydrolysis of the adduct yields the α-amino acid.
4. Reductive amination.
    a. Reductive amination of an α-keto carboxylic acid gives an α-amino acid.
    b. This method is related to the biosynthetic pathway for synthesis of amino acids.

D. Resolution of *R,S* amino acids (Section 26.4).
1. All of the methods listed above produce a racemic mixture of amino acids.
2. Resolution of the mixture:
    a. The mixture can react with a chiral reagent, followed by separation of the diastereomers and reconversion to amino acids.
    b. Enzymes selectively catalyze reactions that form one of the enantiomers, but not the other.

II. Peptides (Sections 26.5 – 26.11).

A. Peptide structure (Sections 26.5 – 26.6).
1. Peptide bonds (Section 26.5).
    a. A peptide is an amino acid polymer in which the amine group of one amino acid forms an amide bond with the carboxylic acid group of a second amino acid.
    b. This amino acid sequence is known as the backbone of the peptide or protein.
    c. Rotation about the amide bond is restricted.
2. The N-terminal amino acid of the polypeptide is always drawn on the left.
3. The C-terminal amino acid of the polypeptide is always drawn on the right.
4. Peptide structure is described by using the three-letter codes for the individual amino acids, stating with the N-terminal amino acid.
5. Disulfide bonds (Section 26.6).
    a. Two cysteines can form a disulfide bond (–S–S–).
    b. Disulfide bonds can link two polypeptides or introduce a loop in a polypeptide chain.

B. Structure determination of peptides (Sections 26.7 – 26.9).
1. Amino acid analysis (Section 26.7).
    a. Amino acid analysis provides the amount of each amino acid present in a protein or peptide.
    b. First, all disulfide bonds are broken and all peptide bonds are hydrolyzed.
    c. The mixture is placed on a chromatography column, and the residues are eluted.
    d. As each amino acid elutes, it undergoes reaction with ninhydrin, which produces a purple color that is detected and measured spectrophotometrically.
    e. Amino acid analysis is reproducible on properly maintained equipment; residues always elute at the same time, and only small sample sizes are needed.
2. The Edman degradation (Section 26.8).
    a. The Edman degradation removes one amino acid at a time from the $–NH_2$ end of a peptide.

i. The peptide is treated with phenylisothiocyanate, which reacts with the amino-terminal residue.
ii. The PITC derivative is split from the peptide.
iii. The residue undergoes rearrangement to a PTH, which is identified chromatographically.
iv. The shortened chain undergoes another round of Edman degradation.

b. Since the Edman degradation can only be used on peptides containing fewer than 50 amino acids, a protein must be cleaved into smaller fragments.
i. Partial acid hydrolysis is unselective.
ii. The enzyme trypsin cleaves proteins at the carboxyl side of arg and lys residues.
iii. The enzyme chymotrypsin cleaves proteins at the carboxyl side of Phe, Tyr and Trp residues.

c. The complete sequence of a protein results from determining the individual sequences of peptides and overlapping them.

3. C-Terminal residue determination (Section 26.9).
a. The enzyme carboxypeptidase is used to cleave the C-terminal residue in a peptide.
b. The peptide is incubated with carboxypeptidase, and the first free amino acid is detected.

C. Synthesis of peptides (Sections 26.10 – 26.11).
1. Laboratory synthesis of peptides (Section 26.10).
a. Groups that are not involved in peptide bond formation are protected.
i. Carboxyl groups are often protected as methyl or benzyl esters.
ii. Amino groups are protected as BOC derivatives.
b. The peptide bond is formed by coupling with DCC.
c. The protecting group are removed.
i. BOC groups are removed by brief treatment with trifluoroacetic acid.
ii. Esters are removed by mild hydrolysis or by hydrogenolysis (benzyl).
2. Automated peptide synthesis (Section 26.11).
a. The carboxyl group of a BOC-protected amino acid is attached to a polystyrene resin.
b. The resin is washed, and the BOC group is removed.
c. A second BOC-protected amino acid is coupled to the first, and the resin is washed.
d. The cycle is repeated as many times as needed.
e. Finally, treatment with anhydrous HF removes the final BOC group and frees the polypeptide.

III. Proteins (Sections 26.12 – 26.16).
A. Classification of proteins (Section 26.12).
1. Proteins can be classified by structure.
a. Simple proteins yield only amino acids on hydrolysis.
b. Conjugated proteins yield other non-protein components on hydrolysis.
2. Proteins can be classified by shape.
a. Fibrous proteins consist of long, filamentous polypeptide chains.
b. Globular proteins are compact and roughly spherical.
B. Protein structure (Section 26.13).
1. Levels of protein structure.
a. Primary structure refers to the amino acid sequence of a protein.
b. Secondary structure refers to the organization of segments of the peptide backbone into a regular pattern, such as a helix or sheet..
c. Tertiary structure describes the overall three-dimensional shape of a protein.
d. Quaternary structure describes how polypeptide subunits aggregate into a larger structure.

2. Examples of specific proteins.
   a. α-Keratin.
      i. α-Keratin is a fibrous protein found in wool, hair and nails.
      ii. Segments of α-keratin are coiled into a right-handed helix (α-helix).
   b. Fibroin.
      i. Fibroin is found in silk.
      ii. Fibroin has a β-pleated sheet secondary structure.
   c. Myoglobin.
      i. Myoglobin is a small globular protein.
      ii. The nonpolar amino acid side chains congregate in the center of myoglobin to avoid water.
      iii. The polar side chain residues are on the surface, where they can take part in hydrogen bonding and salt bridge formation.
      iv. Myoglobin also contains a covalently-bonded heme group.

C. Enzymes (Sections 26.14 – 26.15).
1. Description of enzymes (Section 26.14).
   a. An enzyme is a substance (usually protein) that catalyzes a biochemical reaction.
   b. An enzyme is specific and usually catalyzes the reaction of only one substrate. Some enzymes, such as papain, can operate on a range of substrates.
   c. Most enzymes are globular proteins, and many consist of a protein portion (apoenzyme) and a cofactor.
      i. Cofactors may be small organic molecules (coenzymes) or inorganic ions.
      ii. Many coenzymes are vitamins.
   d. Enzymes are grouped into 6 classes according to the reactions they catalyze.
      i. Hydrolases catalyze hydrolysis reactions.
      ii. Isomerases catalyze isomerizations.
      iii. Ligases catalyze bond formation between two molecules.
      iv. Lyases catalyze the loss of a small molecule from a substrate.
      v. Oxidoreductases catalyze oxidations and reductions.
      vi. Transferases catalyze the transfer of a group from one substrate to another.
   e. The name of an enzyme has two parts, ending with *-ase*.
      i. The first part identifies the substrate.
      ii. The second part identifies the enzyme's class.
2. How enzymes work – citrate synthase (Section 26.15).
   a. Citrate synthase catalyzes the aldol-like addition of acetyl CoA to oxaloacetate to produce citrate.
   b. Functional groups in a cleft of the enzyme bind oxaloacetate.
   c. Functional groups in a second cleft bind acetyl CoA. The two reactants are now in close proximity.
   d Two enzyme amino acid residues generate the enol of acetyl CoA.
   e. The enol undergoes nucleophilic addition to the ketone carbonyl group of oxaloacetate.
   f. Two enzyme amino acid residues deprotonate the enol and protonate the carbonyl oxygen.
   g. Water hydrolyzes the thiol ester, releasing citrate and CoA.

D. Denaturation of proteins (Section 26.26).
1. Modest changes in temperature and pH can disrupt a protein's tertiary structure.
   a. This process is known as denaturation.
   b. Denaturation doesn't affect protein primary structure.
2. Denaturation affects both physical and catalytic properties of proteins.
3. Occasionally, spontaneous renaturation can occur.

Solutions to Problems

26.1 Amino Acids with aromatic rings: Phe, Tyr, Trp, His.
Amino acids containing sulfur: Cys, Met.
Amino acids that are alcohols: Ser, Thr. (Tyr is a phenol.)
Amino acids having hydrocarbon side chains: Ala, Ile, Leu, Val, Phe.

26.2

COOH
H_2N—+—H = HOOC—+—NH_2 (H above, R below)
R

A Fischer projection of the α-carbon of an L–amino acid is pictured above.

For most L–amino acids:

Group	*Priority*
$-NH_2$	1
$-COOH$	2
$-R$	3
$-H$	4

For cysteine:

Group	*Priority*
$-NH_2$	1
$-CH_2SH$	2
$-COOH$	3
$-H$	4

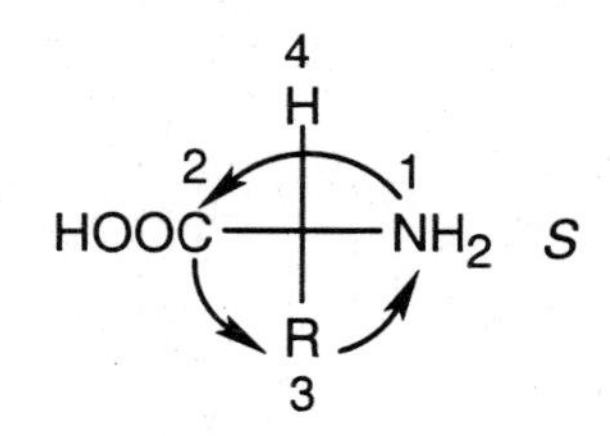

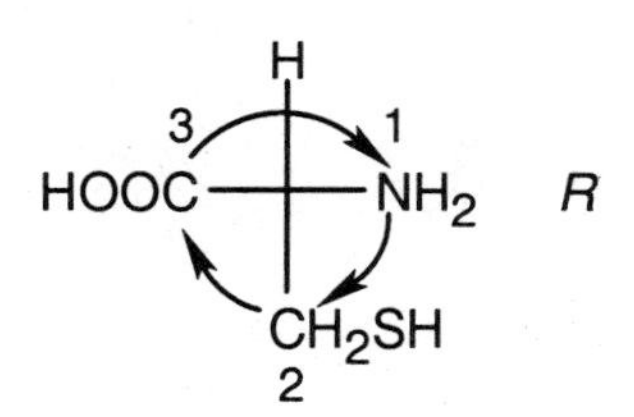

26.3–26.4

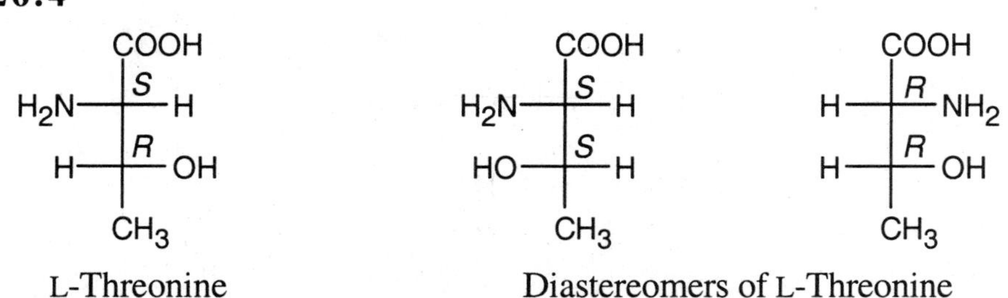

L-Threonine

Diastereomers of L-Threonine

26.5 Remember that an amino acid is positively charged at a pH lower than its pI and migrates to the negative electrode. At a pH higher than pI, the amino acid is negatively charged and migrates to the positive electrode.

(a)

Amino acid	*Isoelectric point*
Val	6.0
Glu	3.2
His	7.6

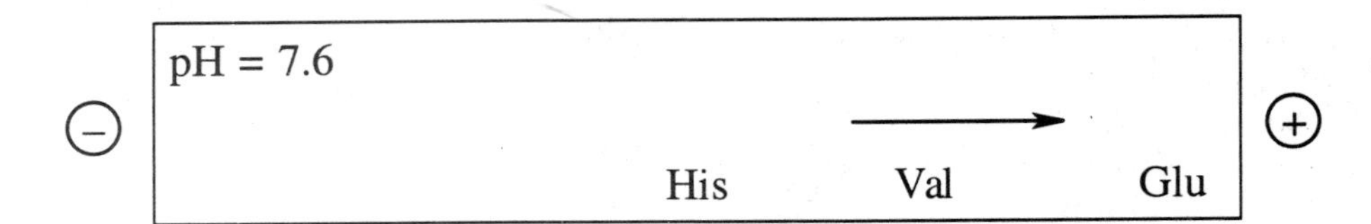

His has no net charge and doesn't migrate. The other two amino acids are negatively charged and migrate to the positive electrode.

(b)

Amino Acid	*Isoelectric point*
Gly	6.0
Phe	5.5
Ser	5.7

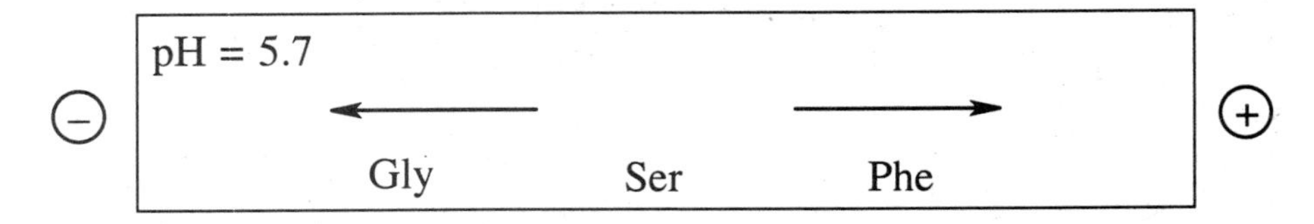

At pH = 5.7, Ser has no net charge. Phe is negatively charged and migrates to the positive electrode, and Gly is negatively charged and migrates to the positive electrode.

(c)

pH = 5.5

Gly Ser Phe

(d)

pH = 6.0

Gly Ser Phe

26.6 The Henderson-Hasselbalch equation states that:

$$\log \frac{[A^-]}{[HA]} = pH - pK_a$$

Follow these steps to find the ratios:

(1) Substitute the pH and pK_a into the expression. For threonine at pH = 1.50, use the value of pK_{a1}.

$$\log \frac{[A^-]}{[HA]} = 1.50 - 2.09 = -0.59; \ \log \frac{[HA]}{[A^-]} = 0.59$$

(2) Find the antilog of the expression.

$$\log\frac{[HA]}{[A^-]} = 0.59\ ;\ \frac{[HA]}{[A^-]} = 3.89$$

The ratio of [HA] / [A⁻] is 3.89.

(3) To find the percent of each form, remember that the fraction of each form adds to 1.

$$[HA] + [A^-] = 1$$

At pH = 1.50:

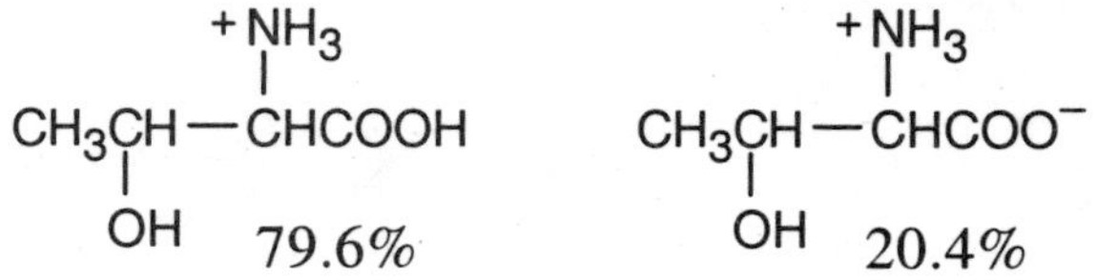

$CH_3CH(OH){-}CH(^+NH_3)COOH$ 79.6% $\quad$ $CH_3CH(OH){-}CH(^+NH_3)COO^-$ 20.4%

At pH = 10.00:

$$\log\frac{[A^-]}{[HA]} = 10.00 - 9.10 = 0.90;\ \frac{[A^-]}{[HA]} = 7.94$$

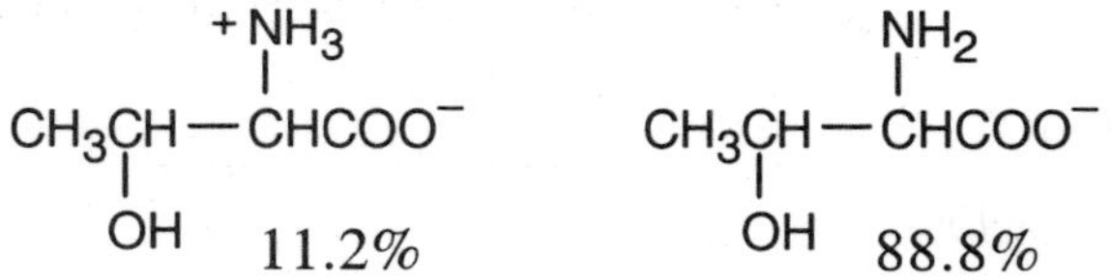

$CH_3CH(OH){-}CH(^+NH_3)COO^-$ 11.2% $\quad$ $CH_3CH(OH){-}CH(NH_2)COO^-$ 88.8%

26.7 This method of amino acid synthesis is simple and uses methods we have already studied. The phthalimide synthesis can also be used to introduce the amino group. Remember that only racemic amino acids are produced by this method and by the method in the next problem.

(a)

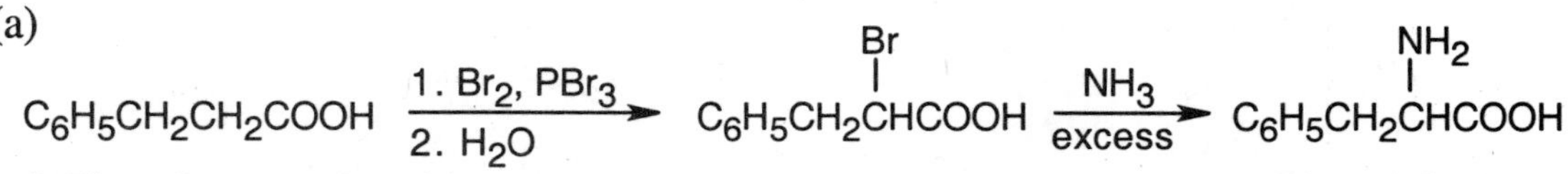

$$C_6H_5CH_2CH_2COOH \xrightarrow[2.\ H_2O]{1.\ Br_2,\ PBr_3} C_6H_5CH_2CH(Br)COOH \xrightarrow[excess]{NH_3} C_6H_5CH_2CH(NH_2)COOH$$

3-Phenylpropanoic acid → Phenylalanine

(b)

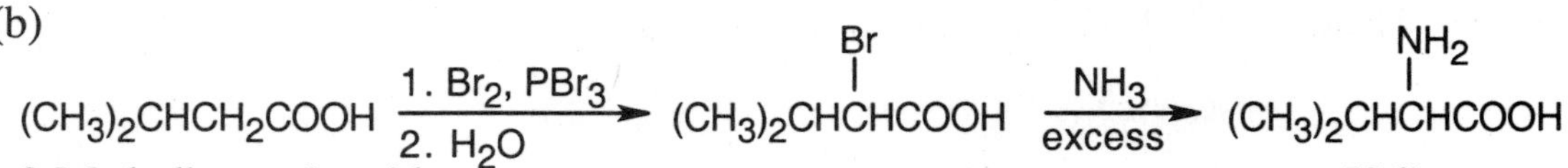

$$(CH_3)_2CHCH_2COOH \xrightarrow[2.\ H_2O]{1.\ Br_2,\ PBr_3} (CH_3)_2CHCH(Br)COOH \xrightarrow[excess]{NH_3} (CH_3)_2CHCH(NH_2)COOH$$

3-Methylbutanoic acid → Valine

26.8

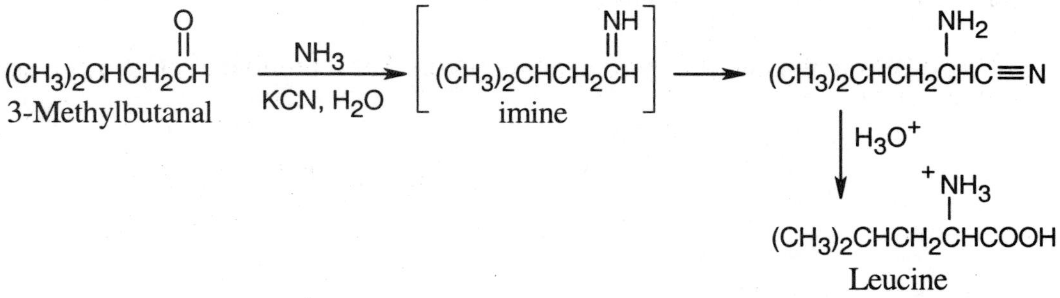

$$(CH_3)_2CHCH_2CH{=}O \xrightarrow[KCN,\ H_2O]{NH_3} \left[(CH_3)_2CHCH_2CH{=}NH\right] \longrightarrow (CH_3)_2CHCH_2CH(NH_2)C{\equiv}N \xrightarrow{H_3O^+} (CH_3)_2CHCH_2CH(^+NH_3)COOH$$

3-Methylbutanal → imine → Leucine

26.9

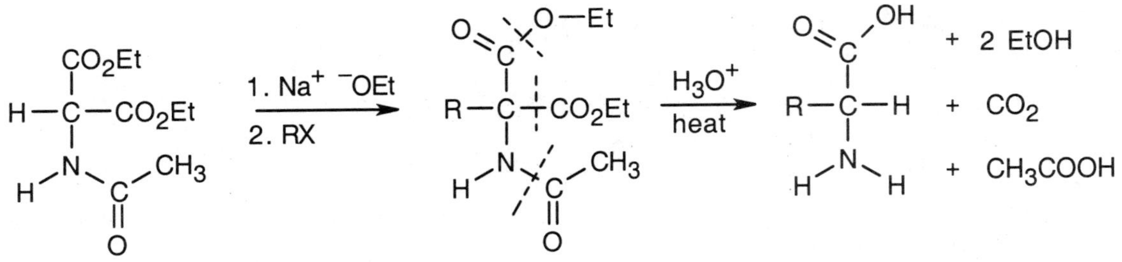

$$\mathrm{H{-}C(CO_2Et)_2{-}NHCOCH_3} \xrightarrow[\text{2. RX}]{\text{1. Na}^+\ ^-\text{OEt}} \mathrm{R{-}C(CO_2Et)_2{-}NHCOCH_3} \xrightarrow[\text{heat}]{\mathrm{H_3O^+}} \mathrm{R{-}CH(NH_2){-}COOH} + 2\ \mathrm{EtOH} + \mathrm{CO_2} + \mathrm{CH_3COOH}$$

In the amidomalonate synthesis, shown above, an alkyl halide RX is converted to $RCH(NH_2)COOH$. Choose an alkyl halide that completes the structure of the desired amino acid.

	Amino Acid	*Halide*
(a)	$(CH_3)_2CHCH_2CH(NH_2)COOH$ Leucine	$(CH_3)_2CHCH_2Br$
(b)	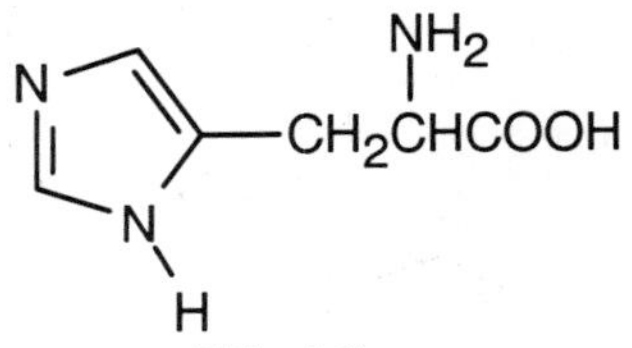 (imidazol-4-yl)$-CH_2CH(NH_2)COOH$ Histidine	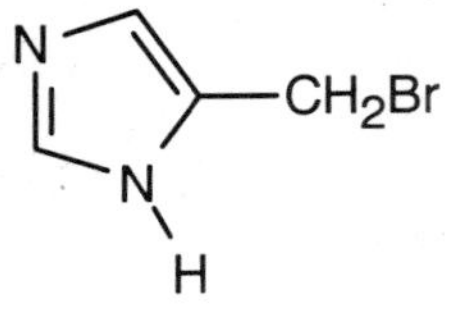 (imidazol-4-yl)$-CH_2Br$
(c)	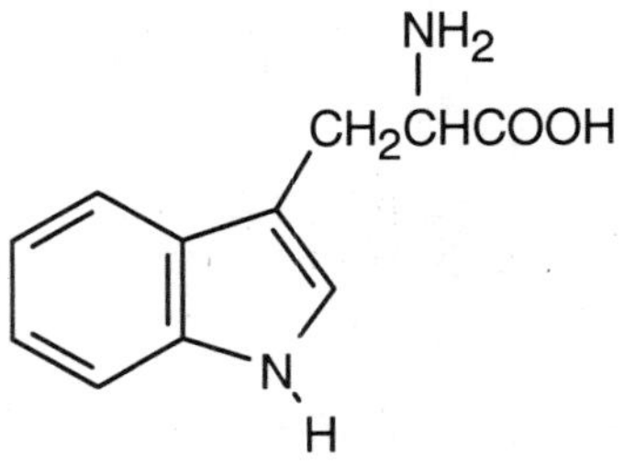 (indol-3-yl)$-CH_2CH(NH_2)COOH$ Tryptophan	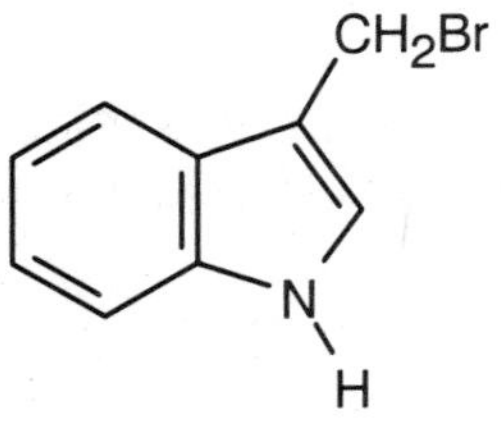 (indol-3-yl)$-CH_2Br$
(d)	$CH_3SCH_2CH_2CH(NH_2)COOH$ Methionine	$CH_3SCH_2CH_2Br$

26.10 Val–Tyr–Gly (VYG) Tyr–Gly–Val (YGV) Gly–Val–Tyr (GVY)
Val–Gly–Tyr (VGY) Tyr–Val–Gly (YVG) Gly–Tyr–Val (GYV)

26.11

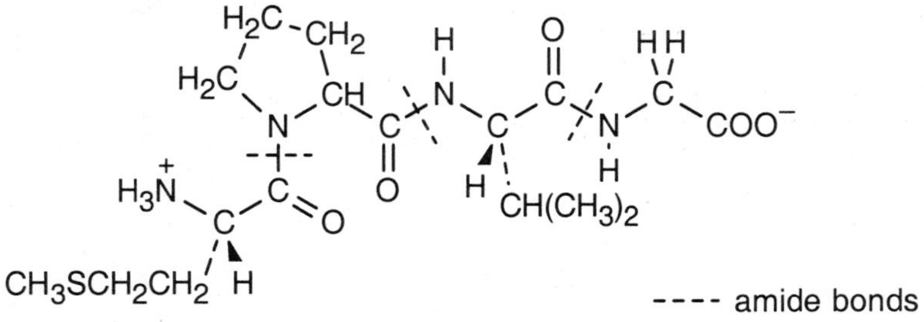

Met—Pro—Val—Gly

26.12

$H_3\overset{+}{N}CH(COO^-)CH_2SH \;+\; ICH_2COOH \longrightarrow H_3\overset{+}{N}CH(COOH)CH_2S{-}CH_2COOH \;+\; I^-$

The cysteine sulfur is a good nucleophile, and iodide is a good leaving group.

26.13 One product of the reaction of an amino acid with ninhydrin is the extensively conjugated purple ninhydrin product. The other major product is the aldehyde that results from reaction of the amino acid with ninhydrin. When valine reacts, the resulting aldehyde is 2-methylpropanal. The other products are carbon dioxide and water.

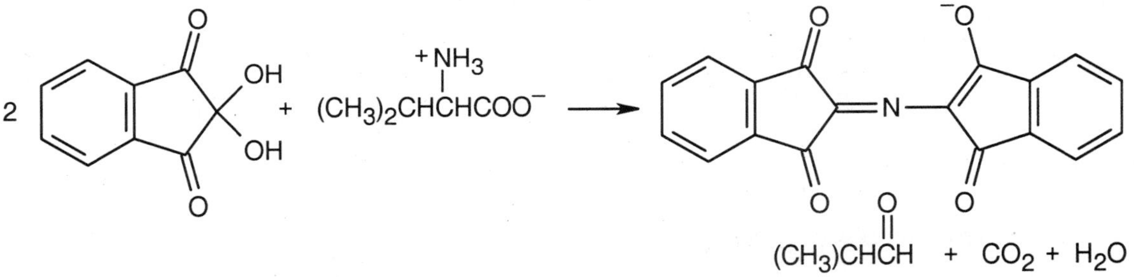

26.14 Trypsin cleaves peptide bonds at the carboxyl side of lysine and arginine. Chymotrypsin cleaves peptide bonds at the carboxyl side of phenylalanine, tyrosine and tryptophan.

Asp–Arg–Val–Tyr–Ile–His–Pro–Phe

Trypsin → Asp–Arg + Val–Tyr–Ile–His–Pro–Phe

Chymotrypsin → Asp–Arg–Val–Tyr + Ile–His–Pro–Phe

26.15 The part of the PTH derivative that lies to the right of the indicated dotted lines comes from the N-terminal residue. Complete the structure to identify the amino acid, which in this problem is methionine.

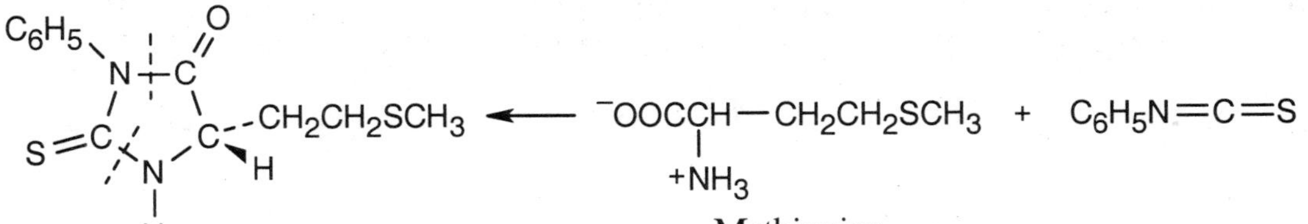

Methionine

26.16 The N-terminal residue of angiotensin II is aspartic acid. Replace the –R group of the PTH derivative in Figure 26.4 with $-CH_2COOH$ to arrive at the correct structure.

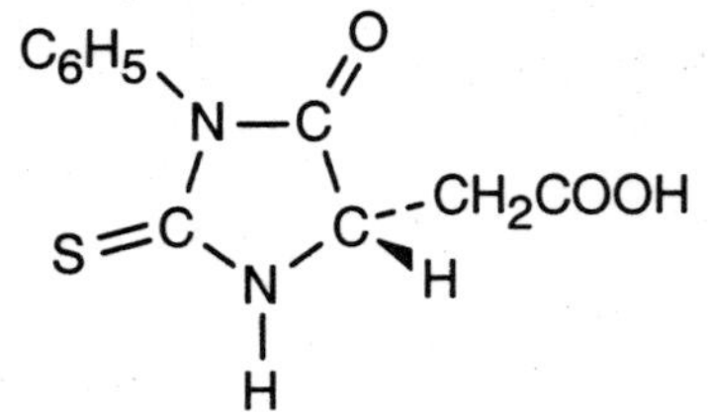

C_6H_5, O, N—C, S=C, N, H, C, CH_2COOH, H

26.17 Line up the fragments so that the amino acids overlap.

(a) Arg–Pro
Pro–Leu–Gly
Gly–Ile–Val

The complete sequence:

Arg–Pro–Leu–Gly–Ile–Val

(b) Val–Met–Trp
Trp–Asp-Val
Val–Leu

The complete sequence:

Val–Met–Trp-Asp–Val–Leu

26.18 Pro–Leu–Gly
Gly–Pro–Arg
Arg–Pro

The complete sequence:

Pro–Leu–Gly–Pro–Arg–Pro

26.19

The tripeptide is cyclic.

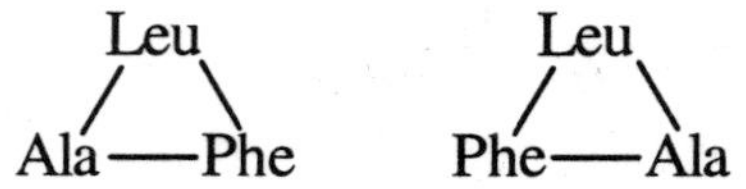

Leu, Ala—Phe   Leu, Phe—Ala

26.20

$(CH_3)_3CO-C(=O)-O-C(=O)-OC(CH_3)_3$ + $H_2NCHRCOOH$ ⇌ (nucleophilic addition of amino acid nitrogen) $[(CH_3)_3CO-C(O^-)(\overset{+}{N}H_2CHRCOOH)-O-C(=O)-OC(CH_3)_3]$

$Et_3N:$ ⇅ deprotonation

$[(CH_3)_3CO-C(O^-)(NHCHRCOOH)-O-C(=O)-OC(CH_3)_3$ + $Et_3NH^+]$

→ (loss of CO_2 and *tert*-butoxide)

$(CH_3)_3CO-C(=O)-NHCHRCOO^-$ + CO_2 + $HOC(CH_3)_3$

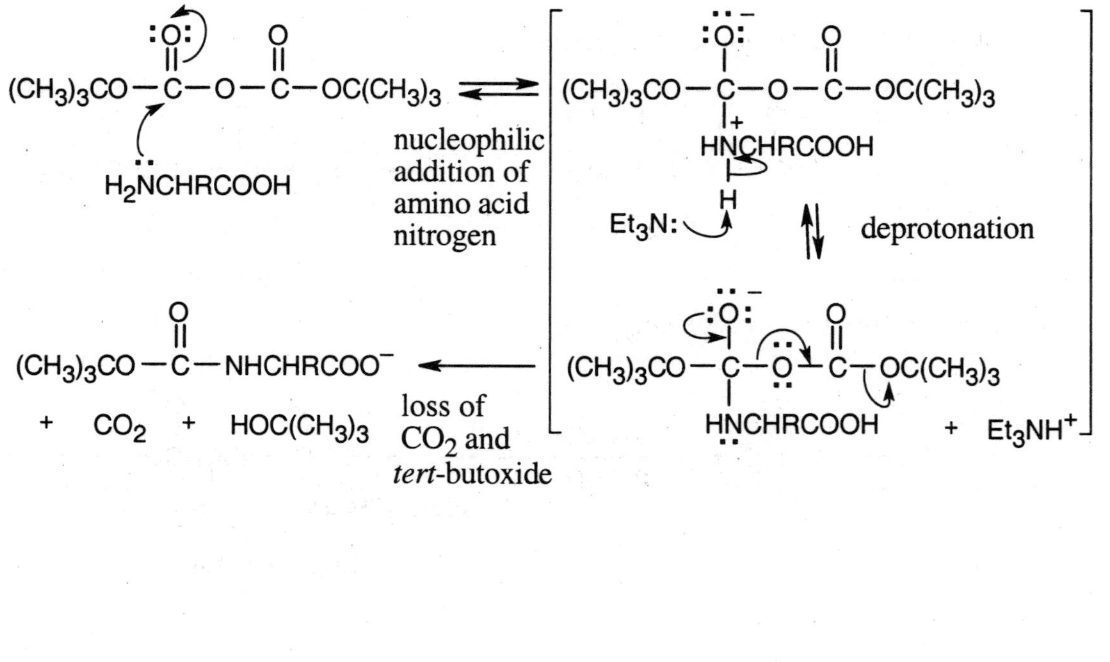

26.21

$$\text{Leu} = \underset{\displaystyle H_2N\overset{|}{C}HCOOH}{CH_2CH(CH_3)_2} = \underset{\displaystyle H_2N\overset{|}{C}HCOOH}{R} \qquad R = CH_2CH(CH_3)_2$$

1. Protect the amino group of leucine.

$$(CH_3)_3CO\overset{O}{\overset{\|}{C}}O\overset{O}{\overset{\|}{C}}OC(CH_3)_3 + \underset{\text{Leu}}{H_2N\overset{R}{\overset{|}{C}}HCOOH} \xrightarrow{Et_3N} (CH_3)_3CO\overset{O}{\overset{\|}{C}}NH\overset{R}{\overset{|}{C}}HCOOH + CO_2 + HOC(CH_3)_3$$

2. Protect the carboxylic acid group of alanine.

$$\underset{\text{Ala}}{H_2N\overset{CH_3}{\overset{|}{C}}HCOOH} + CH_3OH \xrightarrow[\text{catalyst}]{H^+} H_2N\overset{CH_3}{\overset{|}{C}}HCOOCH_3$$

3. Couple the protected acids with DCC.

$$(CH_3)_3CO\overset{O}{\overset{\|}{C}}NH\underset{R}{\underset{|}{C}}HCOOH + H_2N\underset{CH_3}{\underset{|}{C}}H\overset{O}{\overset{\|}{C}}OCH_3 + C_6H_{11}-N{=}C{=}N-C_6H_{11}$$

$$\downarrow$$

$$(CH_3)_3CO\overset{O}{\overset{\|}{C}}NH\underset{R}{\underset{|}{C}}H\overset{O}{\overset{\|}{C}}-NH\underset{CH_3}{\underset{|}{C}}H\overset{O}{\overset{\|}{C}}OCH_3 + C_6H_{11}-NH-\overset{O}{\overset{\|}{C}}-NH-C_6H_{11}$$

4. Remove the leucine protecting group.

$$(CH_3)_3CO\overset{O}{\overset{\|}{C}}NH\underset{R}{\underset{|}{C}}H\overset{O}{\overset{\|}{C}}-NH\underset{CH_3}{\underset{|}{C}}H\overset{O}{\overset{\|}{C}}OCH_3 \xrightarrow{CF_3COOH} H_3\overset{+}{N}\underset{R}{\underset{|}{C}}H\overset{O}{\overset{\|}{C}}-NH\underset{CH_3}{\underset{|}{C}}H\overset{O}{\overset{\|}{C}}OCH_3 + (CH_3)_2C{=}CH_2 + CO_2$$

5. Remove the alanine protecting group.

$$H_3\overset{+}{N}\underset{R}{\underset{|}{C}}H\overset{O}{\overset{\|}{C}}-NH\underset{CH_3}{\underset{|}{C}}H\overset{O}{\overset{\|}{C}}OCH_3 \xrightarrow[2.\ H_3O^+]{1.\ NaOH,\ H_2O} H_3\overset{+}{N}\underset{(CH_3)_2CHCH_2}{\underset{|}{C}}H\overset{O}{\overset{\|}{C}}-NH\underset{CH_3}{\underset{|}{C}}H\overset{O}{\overset{\|}{C}}OH + CH_3OH$$

Leu——Ala

26.22 (a) Pyruvate decarboxylase is a lyase.
(b) Chymotrypsin is a hydrolase.
(c) Alcohol dehydrogenase is an oxidoreductase.

Visualizing Chemistry

26.23

(a) Isoleucine

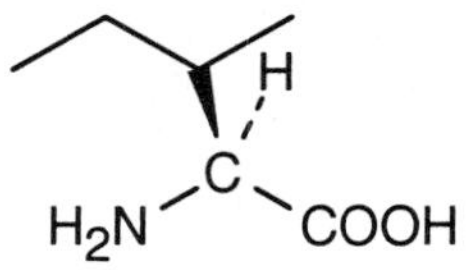

(b) Histidine

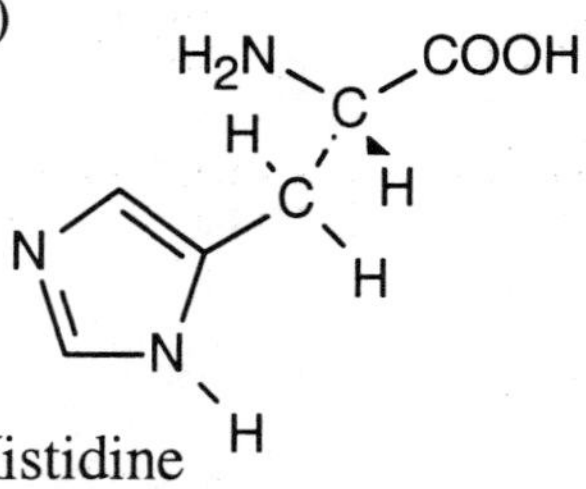

(c) Glutamine

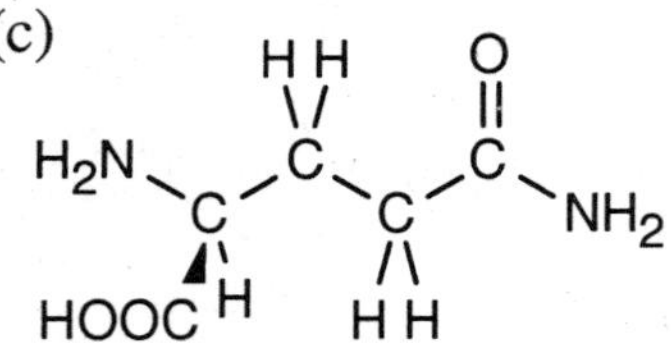

26.24

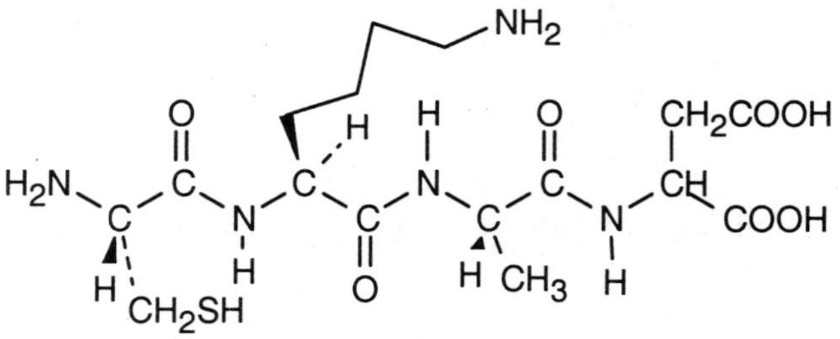

Cys——Lys——Ala——Asp

26.25

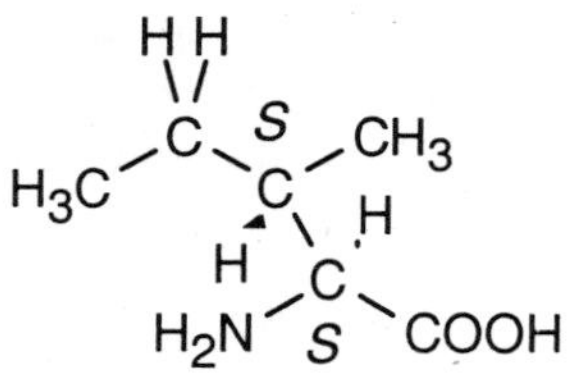

26.26 It's possible to identify this representation of valine as the D enantiomer by drawing its Fischer projection. The amino group is on the right, just as a hydroxyl group is on the right in the Fischer projection of D-glyceraldehyde.

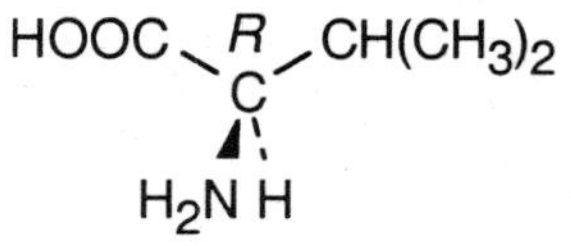

D-Valine
(*R*)-Valine

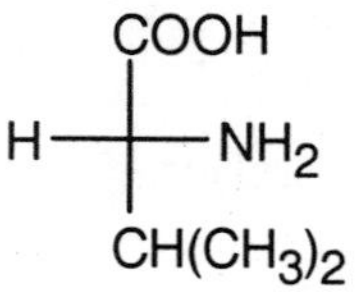

Additional Problems

26.27

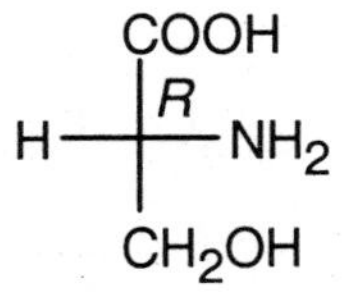

COOH
H — R — NH_2
CH_2OH

(*R*)-Serine

COOH
H — R — NH_2
CH_3

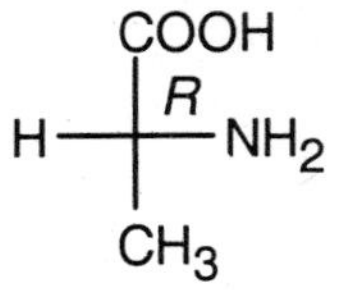

(*R*)-Alanine

Both (*R*)-serine and (*R*)-alanine are D-amino acids.

26.28

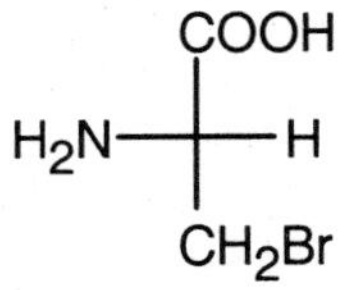

COOH
H_2N — — H
CH_2Br

L-Bromoalanine
(*R*)-Bromoalanine

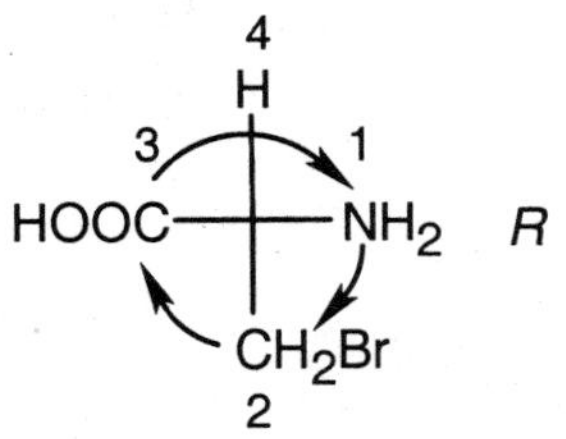

4 H
3 HOOC — — NH_2 1 *R*
CH_2Br 2

This L "amino acid" also has an *R* configuration because the –CH_2Br “side chain” is higher in priority than the –COOH group.

26.29

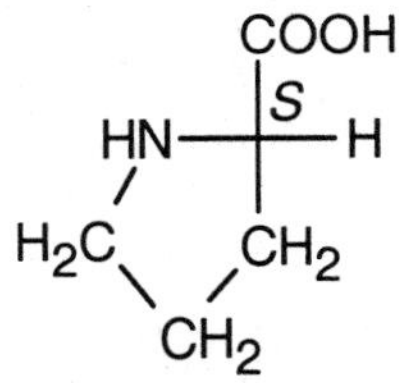

COOH
HN — S — H
H_2C CH_2
CH_2

(*S*)-Proline

26.30

(a)

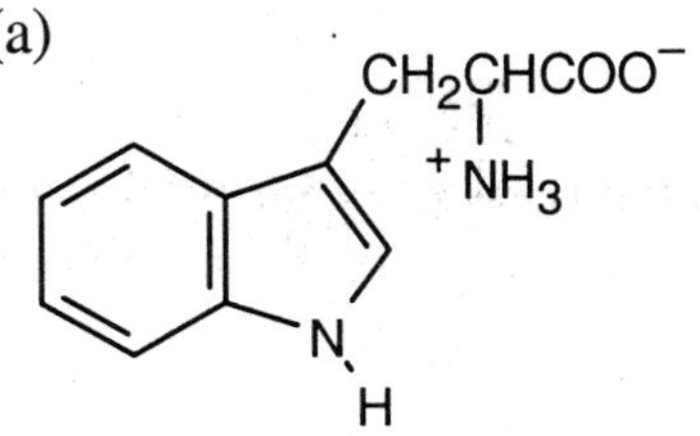

CH_2CHCOO^-
$^+NH_3$
N
H

Tryptophan (Trp)

(b)

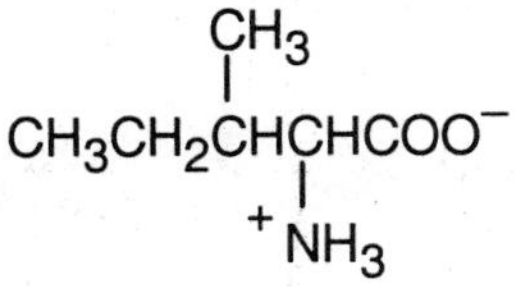

CH_3
$CH_3CH_2CHCHCOO^-$
$^+NH_3$

Isoleucine (Ile)

(c)

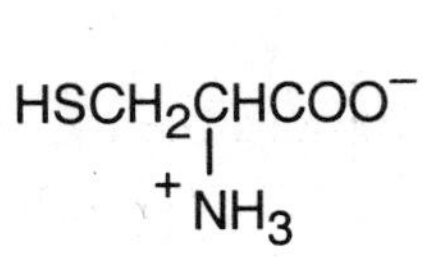

$HSCH_2CHCOO^-$
$^+NH_3$

Cysteine (Cys)

(d)

N
—CH_2CHCOO^-
N $^+NH_3$
H

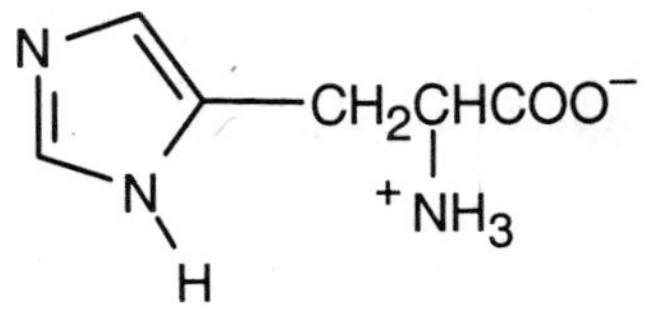

Histidine (His)

26.31 Water is a polar solvent, and chloroform is nonpolar. Since charged species are less stable in nonpolar solvents than in polar solvents, amino acids exist as the non-ionic amino carboxylic acid form in chloroform.

26.32

Amino Acid	*Isoelectric point*
Histidine	7.59
Serine	5.68
Glutamic acid	3.22

The optimum pH for the electrophoresis of three amino acid occurs at the isoelectric point of the amino acid intermediate in acidity. At this pH, one amino acid migrates toward the negative electrode (the least acidic), one migrates toward the positive electrode (the most acidic), and the amino acid intermediate in acidity does not migrate. In this example, electrophoresis at pH = 5.68 allows the maximum separation of the three amino acids.

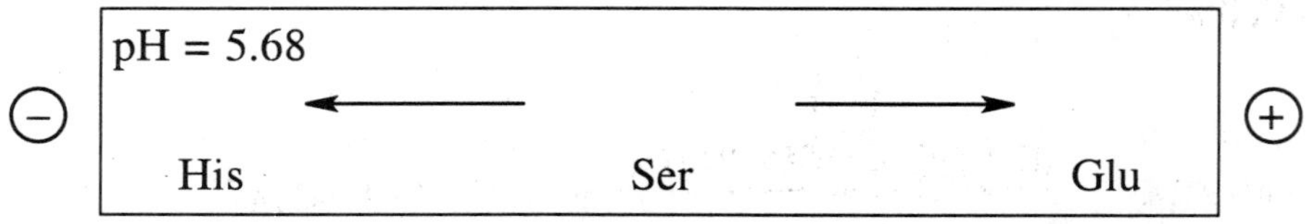

pH = 5.68

⊖ His ← Ser → Glu ⊕

26.33

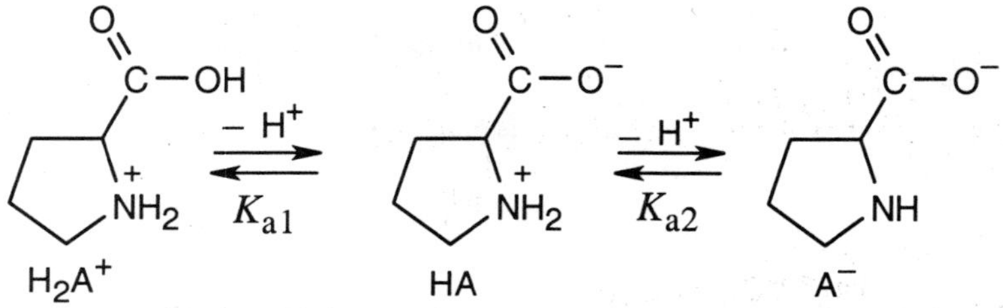

H_2A^+ (protonated) $\underset{K_{a1}}{\overset{-H^+}{\rightleftharpoons}}$ HA (neutral) $\underset{K_{a2}}{\overset{-H^+}{\rightleftharpoons}}$ A^- (deprotonated)

At pH = 2.50:

$$\log\frac{[HA]}{[H_2A^+]} = pH - pK_{a1} = 2.50 - 1.99 = 0.51; \quad \frac{[HA]}{[H_2A^+]} = 3.24$$

At pH = 2.50, approximately three times as many proline molecules exist in the neutral form as exist in the protonated form.

At pH = 9.70:

$$\log\frac{[A^-]}{[HA]} = pH - pK_{a2} = 9.70 - 10.60 = -0.90; \quad \frac{[A^-]}{[HA]} = 0.126$$

At pH = 9.70, the ratio of deprotonated proline to neutral proline is approximately 1:8.

26.34 (a) Val–Leu–Ser   Ser–Val–Leu
Val–Ser–Leu   Leu–Val–Ser
Ser–Leu–Val   Leu–Ser–Val

(b) Ser–Leu–Leu–Pro   Leu–Leu–Ser–Pro
Ser–Leu–Pro–Leu   Leu–Leu–Pro–Ser
Ser–Pro–Leu–Leu   Leu–Ser–Leu–Pro
Pro–Leu–Leu–Ser   Leu–Ser–Pro–Leu
Pro–Leu–Ser–Leu   Leu–Pro–Leu–Ser
Pro–Ser–Leu–Leu   Leu–Pro–Ser–Leu

26.35

(a)

$$\underset{\text{L-Valine}}{(CH_3)_2CHCH(\overset{+}{N}H_3)COO^-} \xrightarrow[\text{H}^+ \text{ catalyst}]{CH_3CH_2OH} (CH_3)_2CHCH(\overset{+}{N}H_3)COOCH_2CH_3$$

(b)

$$(CH_3)_2CHCH(\overset{+}{N}H_3)COO^- \xrightarrow[Et_3N]{(CH_3)_3COC(=O)OC(=O)OC(CH_3)_3} (CH_3)_2CHCH(NHC(=O)OC(CH_3)_3)COO^-$$

(c)

$$(CH_3)_2CHCH(\overset{+}{N}H_3)COO^- \xrightarrow{KOH,\ H_2O} (CH_3)_2CHCH(NH_2)COO^-$$

(d)

$$(CH_3)_2CHCH(\overset{+}{N}H_3)COO^- \xrightarrow[\text{2. } H_2O]{\text{1. } CH_3COCl\text{, pyridine}} (CH_3)_2CHCH(NHC(=O)CH_3)COO^-$$

26.36

(a)

$$\underset{\text{Formaldehyde}}{HC(=O)H} \xrightarrow[KCN,\ H_2O]{NH_3} HCH(NH_2)C{\equiv}N \xrightarrow{H_3O^+} \underset{\text{Glycine}}{H_2C(\overset{+}{N}H_3)-COOH}$$

(b)

$$\underset{\text{2-Methylpropanal}}{(CH_3)_2CHC(=O)H} \xrightarrow[KCN,\ H_2O]{NH_3} (CH_3)_2CHCH(NH_2)C{\equiv}N \xrightarrow{H_3O^+} \underset{\text{Valine}}{(CH_3)_2CHCH(\overset{+}{N}H_3)COOH}$$

26.37

(a)

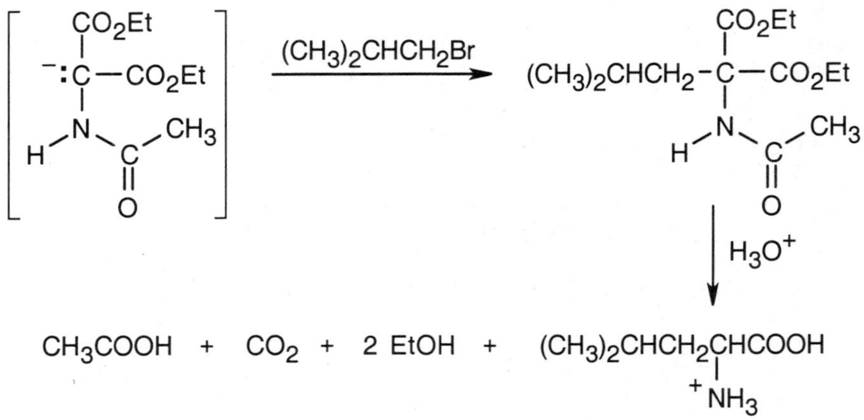

$$\left[{}^{-}\!:\!C(CO_2Et)_2(NHCOCH_3) \right] \xrightarrow{(CH_3)_2CHCH_2Br} (CH_3)_2CHCH_2\text{–}C(CO_2Et)_2(NHCOCH_3)$$

$$\xrightarrow{H_3O^+} CH_3COOH + CO_2 + 2\ EtOH + (CH_3)_2CHCH_2CH(NH_3^+)COOH$$

Leucine

(b)

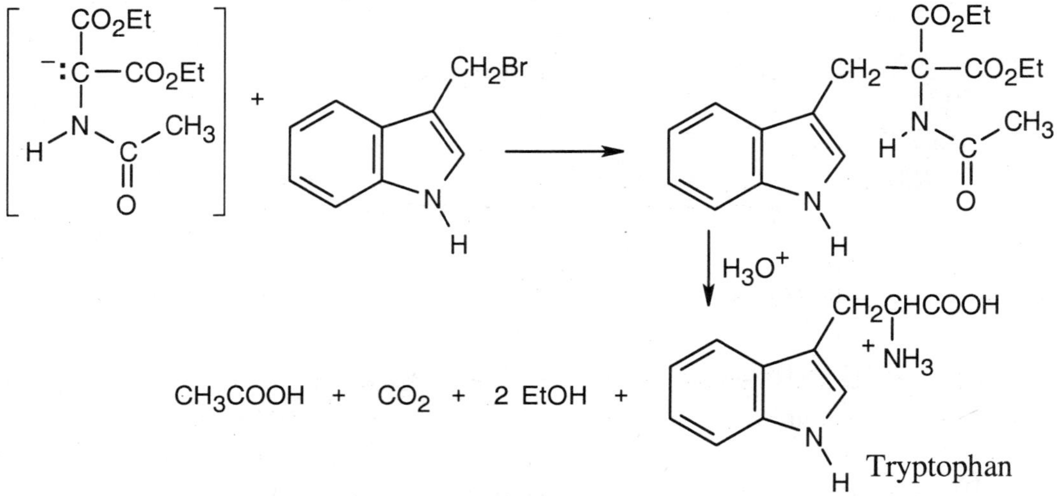

$$\left[{}^{-}\!:\!C(CO_2Et)_2(NHCOCH_3) \right] + \text{3-(}CH_2Br\text{)indole} \longrightarrow \text{indol-3-yl–}CH_2\text{–}C(CO_2Et)_2(NHCOCH_3)$$

$$\xrightarrow{H_3O^+} CH_3COOH + CO_2 + 2\ EtOH + \text{indol-3-yl–}CH_2CH(NH_3^+)COOH$$

Tryptophan

26.38

(a)

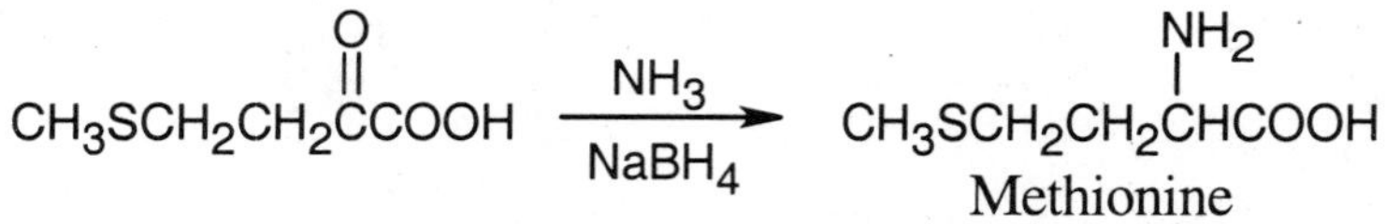

$$CH_3SCH_2CH_2C(=O)COOH \xrightarrow[NaBH_4]{NH_3} CH_3SCH_2CH_2CH(NH_2)COOH$$

Methionine

(b)

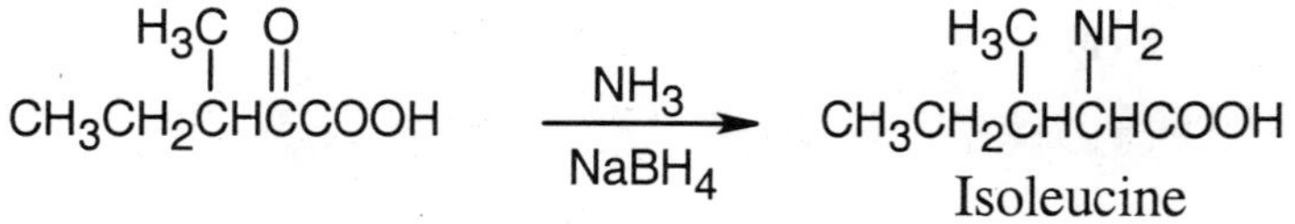

$$CH_3CH_2CH(CH_3)C(=O)COOH \xrightarrow[NaBH_4]{NH_3} CH_3CH_2CH(CH_3)CH(NH_2)COOH$$

Isoleucine

26.39

$$H_2C{=}O + {}^{-}\!:\!C(CO_2Et)_2(NHCOCH_3) \longrightarrow \left[{}^{-}OCH_2\text{–}C(CO_2Et)_2(NHCOCH_3) \right] \xrightarrow{H_3O^+} HOCH_2CH(NH_3^+)COOH$$

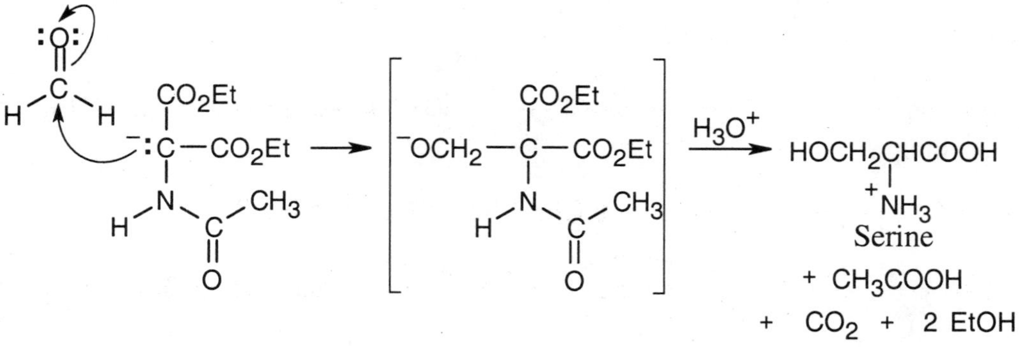

Serine

$+ CH_3COOH$

$+ CO_2 + 2\ EtOH$

26.40

(a)

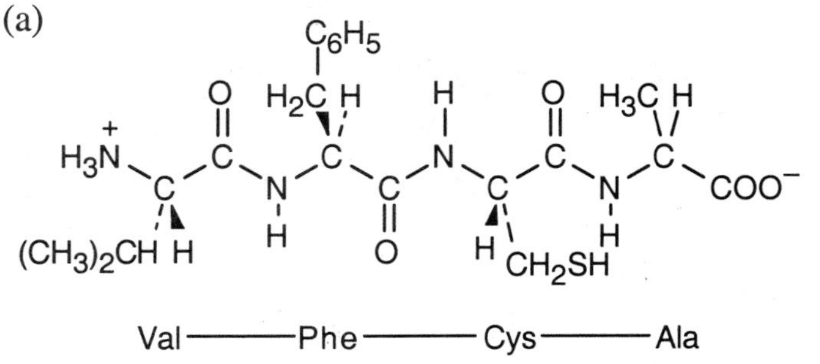

$H_3\overset{+}{N}CH(CH(CH_3)_2)C(=O)NHCH(CH_2C_6H_5)C(=O)NHCH(CH_2SH)C(=O)NHCH(CH_3)COO^-$

Val——Phe——Cys——Ala

(b)

$H_3\overset{+}{N}CH(CH_2CH_2COO^-)C(=O)$–Pro–$C(=O)NHCH(CH(CH_3)CH_2CH_3)C(=O)NHCH(CH_2CH(CH_3)_2)COO^-$

Glu——Pro——Ile——Leu

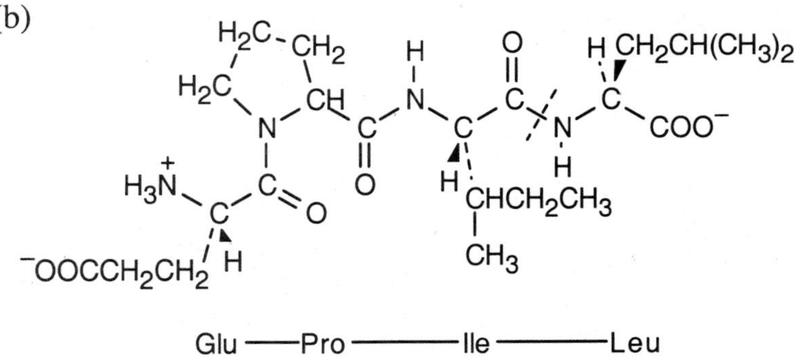

26.41

Step 1: Valine is protected as its BOC derivative.

$$(CH_3)_3COC(=O)OC(=O)OC(CH_3)_3 + \text{Val} \xrightarrow{Et_3N} (CH_3)_3COC(=O)\text{—Val—OH}$$

(BOC—Val—OH)

Step 2: BOC–Val bonds to the polymer in an S_N2 reaction.

$$\text{BOC—Val—OH} + ClCH_2\text{—(Polymer)} \xrightarrow{\text{Base}} \text{BOC—Val—}OCH_2\text{—(Polymer)}$$

Step 3: The polymer is first washed, then is treated with CF_3COOH to cleave the BOC group.

$$\text{BOC—Val—}OCH_2\text{—(Polymer)} \xrightarrow[\text{2. } CF_3COOH]{\text{1. wash}} \text{Val—}OCH_2\text{—(Polymer)}$$

Step 4: A BOC-protected Ala is coupled to the polymer-bound valine by reaction with DCC. The polymer is washed.

$$\text{BOC—Ala} + \text{Val—}OCH_2\text{—(Polymer)} \xrightarrow[\text{2. wash}]{\text{1. DCC}} \text{BOC—Ala—Val—}OCH_2\text{—(Polymer)}$$

Step 5: The polymer is treated with CF_3COOH to remove BOC.

$$\text{BOC—Ala—Val—}OCH_2\text{—(Polymer)} \xrightarrow{CF_3COOH} \text{Ala—Val—}OCH_2\text{—(Polymer)}$$

Step 6: A BOC-protected Phe is coupled to the polymer by reaction with DCC. The polymer is washed.

BOC—Phe + Ala—Val—OCH_2—(Polymer) $\xrightarrow[\text{2. wash}]{\text{1. DCC}}$ BOC—Phe—Ala—Val—OCH_2—(Polymer)

Step 7: Treatment with anhydrous HF removes the BOC group and cleaves the ester bond between the peptide and the polymer.

BOC—Phe—Ala—Val—OCH_2—(Polymer) $\xrightarrow{\text{HF}}$

Phe—Ala—Val + $(CH_3)_2C{=}CH_2$ + CO_2 + $HOCH_2$—(Polymer)

26.42

Peptide $\xrightarrow{\text{PITC}}$ *Phenylthiohydantoin* + *Shortened Peptide*

(a)

Ile–Leu–Pro–Phe → phenylthiohydantoin (C_6H_5–N, C=O, C=S, N–H; C bearing H and $CH(CH_3)CH_2CH_3$) + Leu–Pro–Phe

(b)

Asp–Ser–Thr–Gly–Ala → phenylthiohydantoin (C_6H_5–N, C=O, C=S, N–H; C bearing H and CH_2COOH) + Ser–Thr–Gly–Ala

26.43

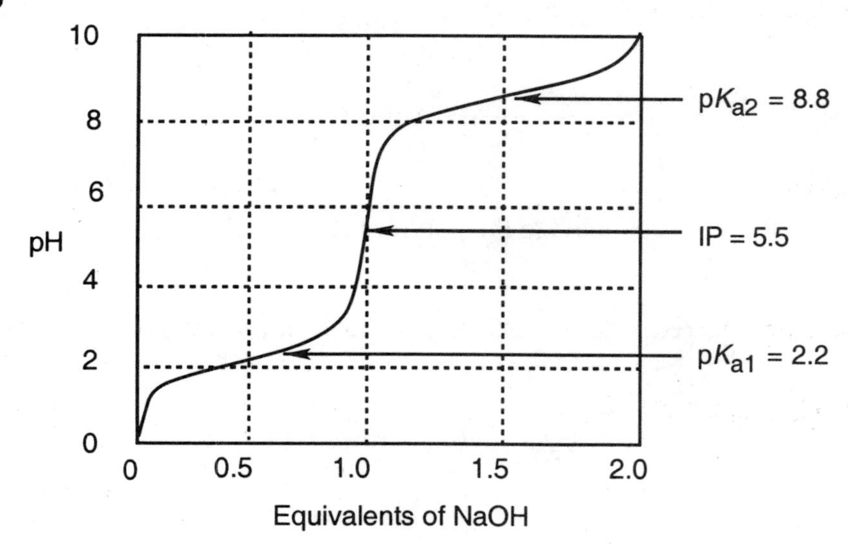

(a) The pK_a of an acid is the pH at which half of a compound is in its protonated form and half is in its deprotonated form. The values of pK_{a1} and pK_{a2} can be found from the graph by noting the pH values at which 0.5 equivalents and 1.5 equivalents of NaOH have been added. Thus, $pK_{a1} = 2.2$, and $pK_{a2} = 8.8$.

The isoelectric point is the average of the two pK_a values. IP = 1/2 ($pK_{a1} + pK_{a2}$) = 1/2 (2.2 + 8.8) = 5.5

(b) This is a neutral amino acid because its IP is similar to the values for IP of other neutral amino acids.

26.44 A proline residue in a polypeptide chain interrupts α–helix formation because the amide nitrogen of proline has no hydrogen that can contribute to the hydrogen-bonded structure of an α–helix.

26.45

Phe~~Leu—Met—Lys--Tyr~~Asp—Gly—Gly—Arg--Val—Ile—Pro—Tyr

Cleaved by trypsin = ----

Cleaved by chymotrypsin = ∿∿∿

26.46 (a) Hydrolases catalyze the cleavage of bonds by addition of water (hydrolysis).
(b) Lyases catalyze the elimination of a small molecule (H_2O, CO_2) from a molecule.
(c) Transferases catalyze the transfer of a functional group between substrates.

26.47 Amino acids with polar side chains are likely to be found on the outside of a globular protein, where they can form hydrogen bonds with water and with each other. Amino acids with nonpolar side chains are found on the inside of a globular protein, where the can avoid water. Thus, aspartic acid (b) and lysine (d) are found on the outside of a globular protein, and valine (a) and phenylalanine (c) are likely to be found on the inside.

26.48

Formation of cation:

$$H_3C-\ddot{O}-CH_2-Cl \quad SnCl_4 \rightleftharpoons H_3C-\overset{+}{O}=CH_2 \; SnCl_5^-$$

Electrophilic aromatic substitution:

$$H_2C=\overset{+}{O}CH_3 \quad SnCl_5^- \rightarrow \left[H \quad CH_2OCH_3 \right] \rightarrow CH_2OCH_3 + H^+$$

Protonation of the ether oxygen, followed by displacement of methanol by Cl^-.

$$CH_2OCH_3 \overset{H^+}{\rightleftharpoons} \left[CH_2-\overset{+}{O}(H)CH_3 \quad :\ddot{Cl}:^- \right] \rightarrow CH_2Cl + HOCH_3$$

26.49

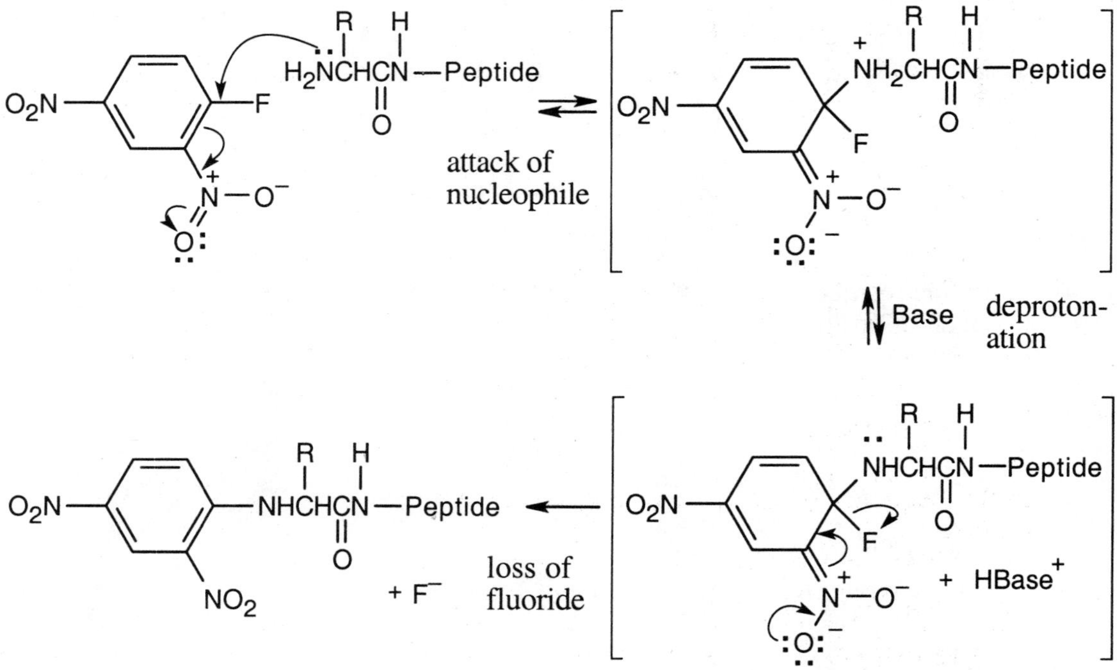

O_2N—C₆H₄(NO₂)—F + H_2NCHCN(R)(H)(=O)—Peptide ⇌ [attack of nucleophile] ⇌ [Base, deprotonation] → O_2N—C₆H₃(NO₂)—NHCHCN(R)(H)(=O)—Peptide + F^- (loss of fluoride)

Intermediate labels: $^+NH_2CHCN$—Peptide; F; $N^+—O^-$; $:O:^-$; NHCHCN—Peptide; + $HBase^+$

This reaction proceeds via a nucleophilic aromatic substitution mechanism.

26.50

(a)

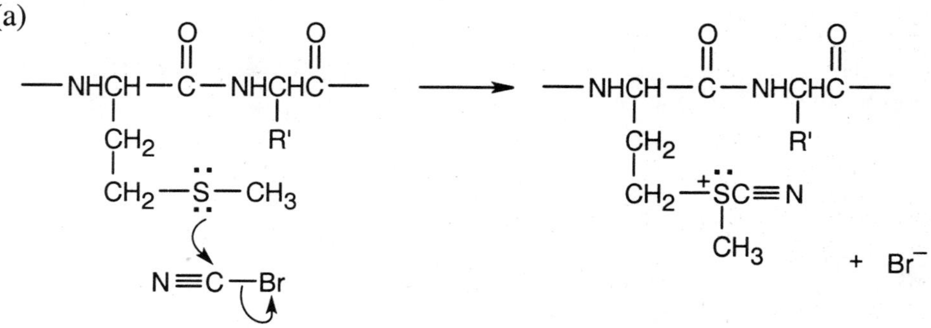

—NHCH(CH₂CH₂SCH₃)—C(=O)—NHCH(R')C(=O)— + N≡C—Br → —NHCH(CH₂CH₂$\overset{+}{S}$(CH₃)C≡N)—C(=O)—NHCH(R')C(=O)— + Br^-

The first step is an S_N2 displacement of bromide by sulfide.

(b)

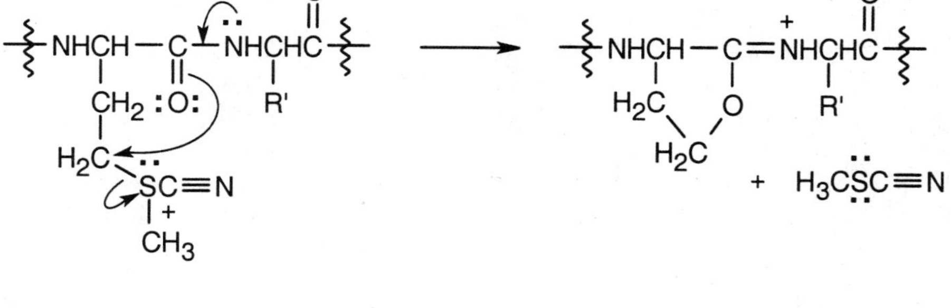

—NHCH(CH₂CH₂$\overset{+}{S}$(CH₃)C≡N)—C(=O)—NHCH(R')C(=O)— → —NHCH—C(=$\overset{+}{N}$HCH(R')C(=O)—)—O— (ring: H₂C, H₂C) + $H_3CSC{\equiv}N$

Internal S_N2 displacement of sulfide results in formation of a 5-membered ring containing an iminium group.

(c)

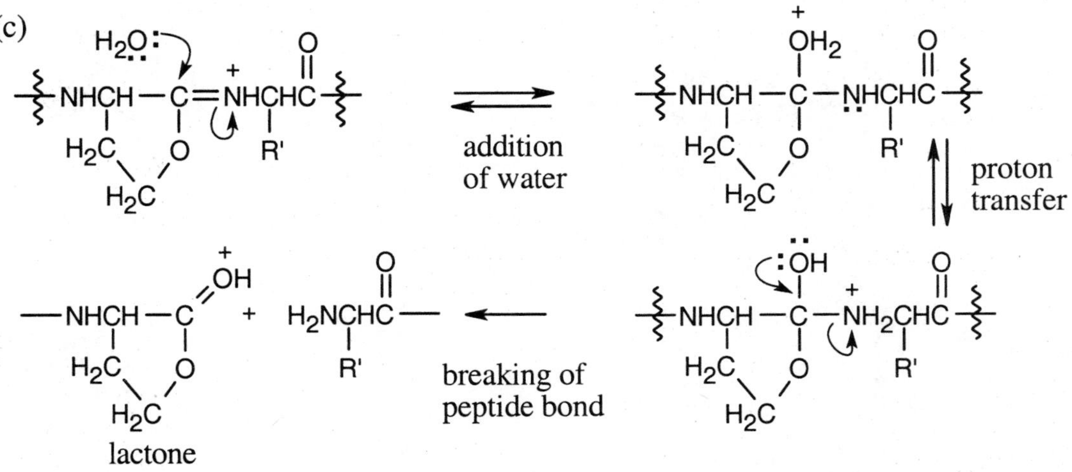

addition of water

proton transfer

breaking of peptide bond

lactone

In this sequence of steps, water adds to the imine double bond, and the peptide bond is cleaved.

(d)

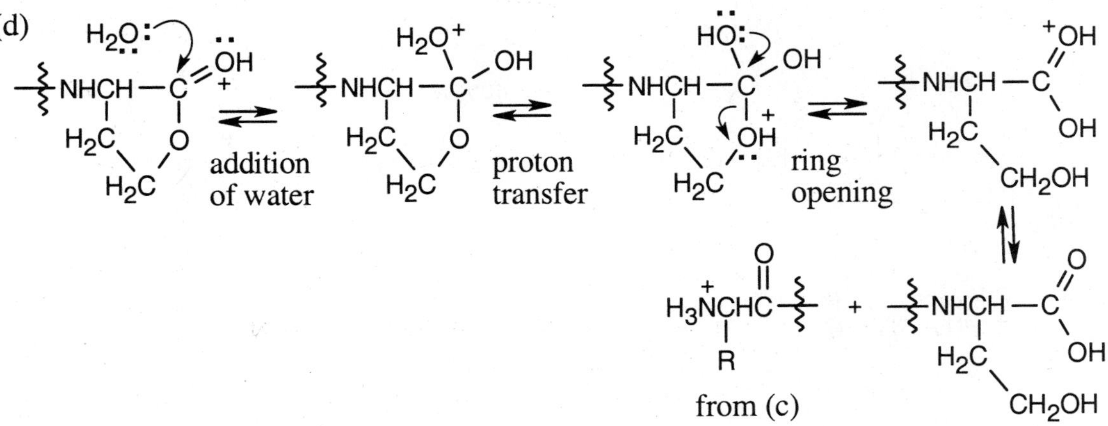

addition of water

proton transfer

ring opening

from (c)

Water opens the lactone ring to give the product shown.

26.51

(a)

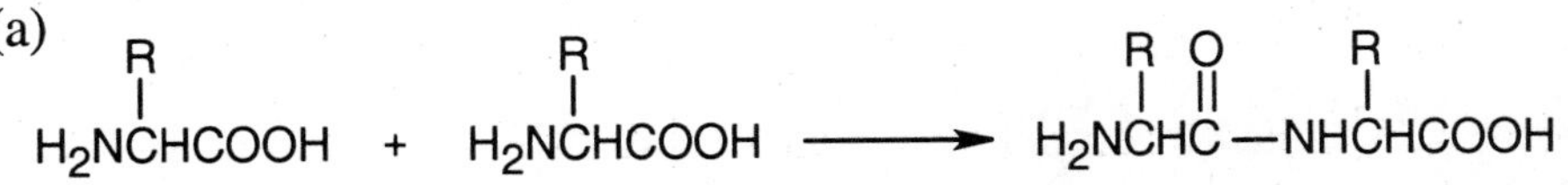

$H_2NCH(R)COOH + H_2NCH(R)COOH \longrightarrow H_2NCH(R)C(=O){-}NHCH(R)COOH$

A dipeptide is formed from two equivalents of the amino acid. The mechanism is shown in Figure 26.5.

(b)

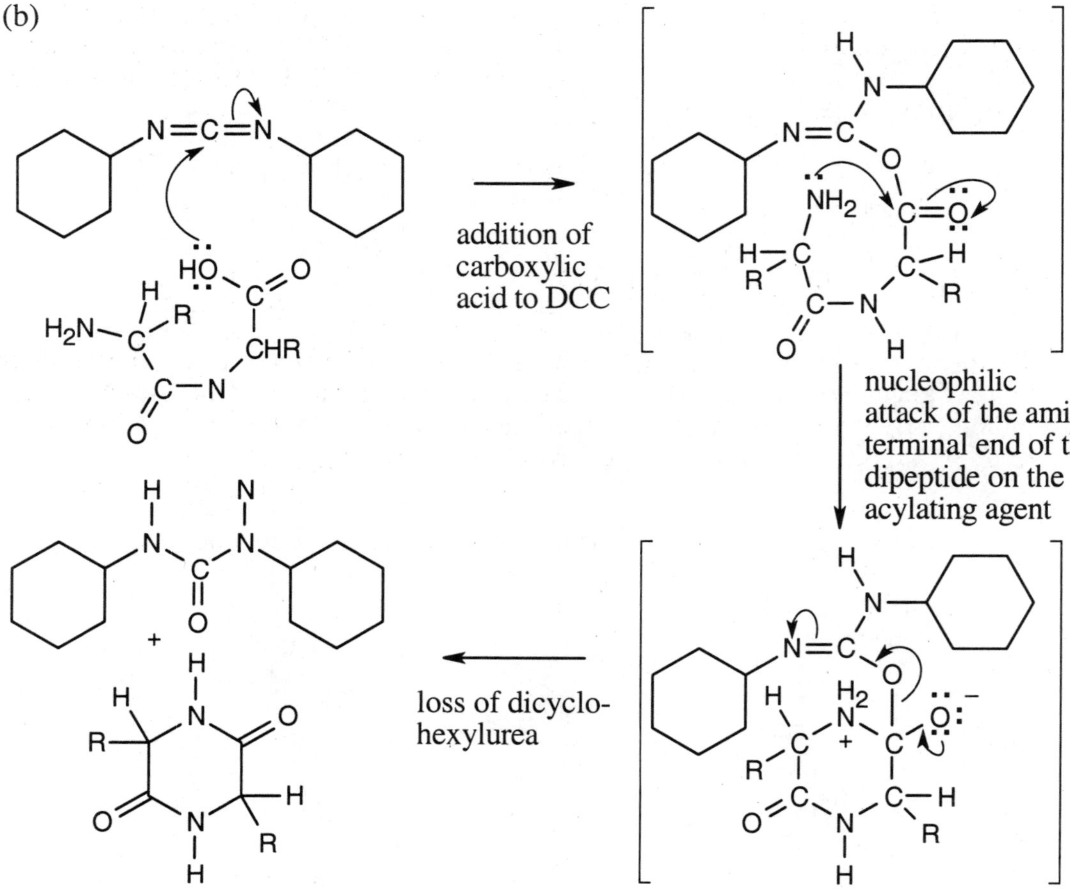

N=C=N (DCC, two cyclohexyl groups); HO–C(=O)–CHR–N–C(=O)–C(H)(R)–NH2 (dipeptide)

addition of carboxylic acid to DCC

[N=C(–O–C(=O)–CH(R)–N(H)–C(=O)–C(H)(R)–NH2)–N(H)–cyclohexyl]

nucleophilic attack of the amino-terminal end of the dipeptide on the acylating agent

[tetrahedral intermediate: H2N+ –C(O–)–O–C(=N)–N(H)]

loss of dicyclohexylurea

dicyclohexylurea (N–C(=O)–N) + 2,5-diketopiperazine (H, R, O, H, R, H)

DCC couples the carboxylic acid end of the dipeptide to the amino end to yield the 2,5–diketopiperazine.

26.52

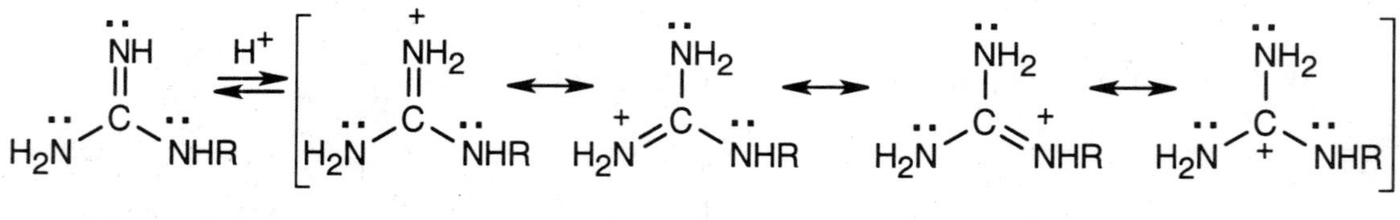

$H_2N-C(=NH)-NHR \overset{H^+}{\rightleftharpoons} [H_2N-C(=\overset{+}{N}H_2)-NHR \leftrightarrow H_2\overset{+}{N}=C(NH_2)-NHR \leftrightarrow H_2N-C(NH_2)=\overset{+}{N}HR \leftrightarrow H_2N-\overset{+}{C}(NH_2)-NHR]$

The protonated guanidino group can be stabilized by resonance.

26.53 100 g of cytochrome C contains 0.43 g iron, or 0.0077 mol Fe:

$$0.43 \text{ g Fe} \times \frac{1 \text{ mol Fe}}{55.8 \text{ g Fe}} = 0.0077 \text{ mol Fe}$$

Assuming that each mole of protein contains 1 mol Fe, then mol Fe = mol protein.

$$\frac{100 \text{ g Cytochrome C}}{0.0077 \text{ mol Cytochrome C}} = \frac{X \text{ g Cytochrome C}}{1 \text{ mol Cytochrome C}}$$

$$13{,}000 \text{ g/mol} = X$$

Cytochrome C has a minimum molecular weight of 13,000 g/ mol.

26.54 ^{1}H NMR shows that the two methyl groups of *N,N*–dimethylformamide are non-equivalent at room temperature. If rotation around the CO–N bond were unrestricted, the methyl groups would be interconvertible, and their ^{1}H NMR absorptions would coalesce into a single signal.

(a) H_3C / (b) H_3C –N–C(=O)H ⇌ rotate ⇌ (b) H_3C / (a) H_3C –N–C(=O)H

The presence of two absorptions shows that there is a barrier to rotation around the CO–N bond. This barrier is due to the partial double-bond character of the CO–N bond, as indicated by the two resonance forms. Heating to 180° supplies enough energy to allow rapid rotation and to cause the two NMR absorptions to merge.

$(H_3C)_2\ddot{N}-C(=\ddot{O}:)H \longleftrightarrow (H_3C)_2\overset{+}{N}=C(-:\ddot{O}:^-)H$

26.55 Gly

Gly–Asp–Phe–Pro
Phe–Pro–Val
Val–Pro–Leu

The complete sequence:

Gly–Gly–Asp–Phe–Pro–Val–Pro–Leu

26.56

(a)

Ninhydrin (OH, OH) ⇌ 1. (− H_2O) [indanetrione + $H_2\ddot{N}CHCOOH$ (R)] ⇌ 2. [OH, $NH_2^+CHCOO^-$ (R)] ⇌ 3. [$^+\ddot{O}H_2$, $\ddot{N}HCHCOO^-$ (R)] ⇌ 4. [$=\overset{+}{N}HCHCOO^-$ (R) + H_2O]

The steps involved in imine formation: (1) dehydration, (2) nucleophilic addition of the amino group of the amino acid, (3) proton transfer and (4) loss of water.

(b)

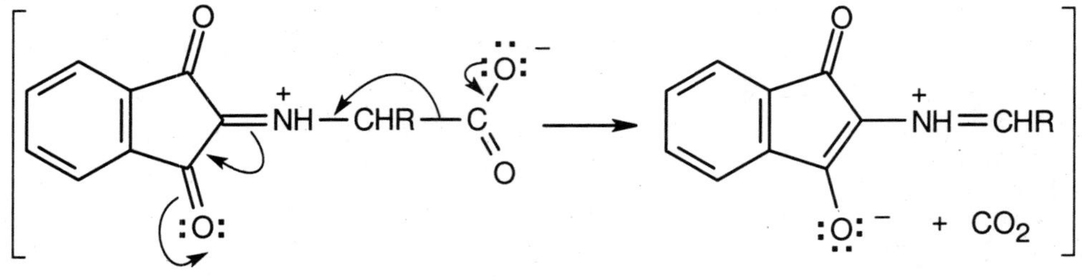

O
$\overset{+}{N}H$—CHR—C(=O)—$\ddot{O}$:$^-$
:O:
→
O
$\overset{+}{N}H$=CHR
:$\ddot{O}$:$^-$ + CO_2

Decarboxylation produces a second imine.

(c)

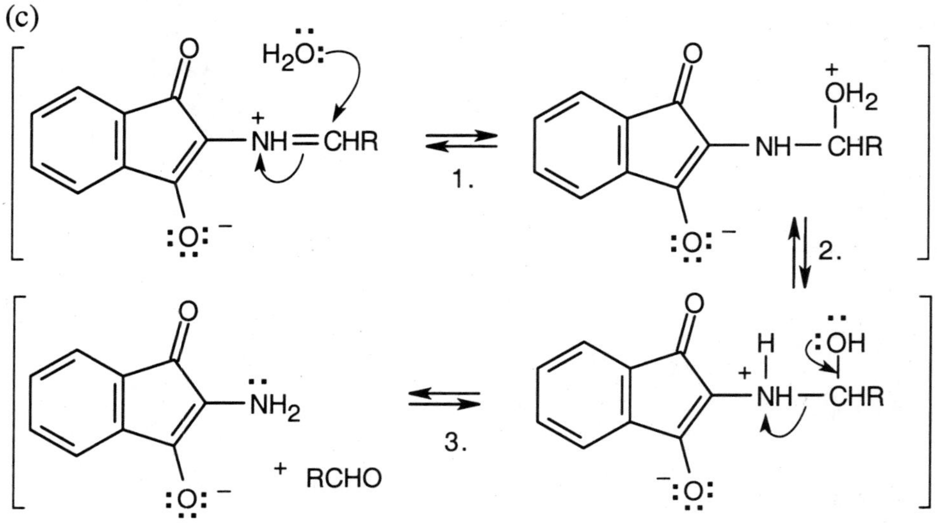

O
$H_2\ddot{O}$:
$\overset{+}{N}H$=CHR
:$\ddot{O}$:$^-$
⇌ 1.
O
$\overset{+}{O}H_2$
NH—CHR
:$\ddot{O}$:$^-$
⇅ 2.

O
$\ddot{N}H_2$
:$\ddot{O}$:$^-$
+ RCHO
⇌ 3.
O
H
+
NH—CHR
:$\ddot{O}$H
$^-$:$\ddot{O}$:

Hydrolysis of this imine occurs by addition of water (1), proton transfer (2) and bond cleavage (3) to yield an aldehyde and an amine.

(d)

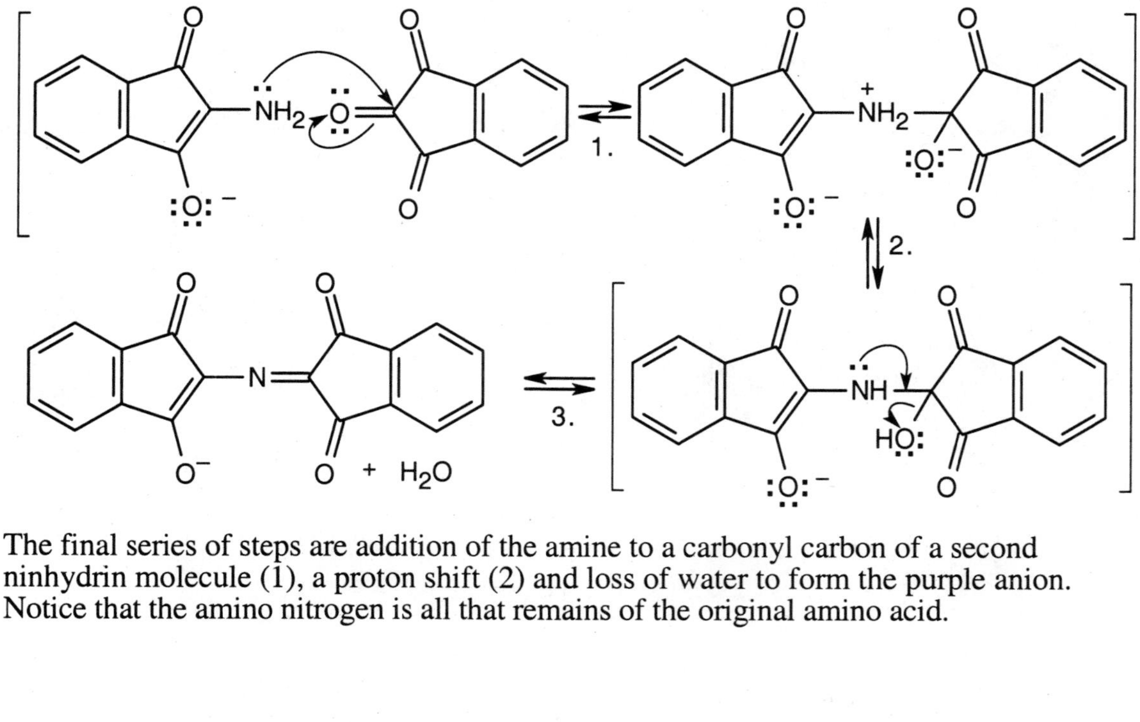

O
$\ddot{N}H_2$
:$\ddot{O}$:$^-$
O
$\ddot{O}$=
O
⇌ 1.
O
$\overset{+}{N}H_2$
:$\ddot{O}$:$^-$
O
:$\ddot{O}$:$^-$
O
⇅ 2.

O
N=
O^-
O
O + H_2O
⇌ 3.
O
$\ddot{N}H$
:$\ddot{O}$:$^-$
O
H$\ddot{O}$:
O

The final series of steps are addition of the amine to a carbonyl carbon of a second ninhydrin molecule (1), a proton shift (2) and loss of water to form the purple anion. Notice that the amino nitrogen is all that remains of the original amino acid.

26.57

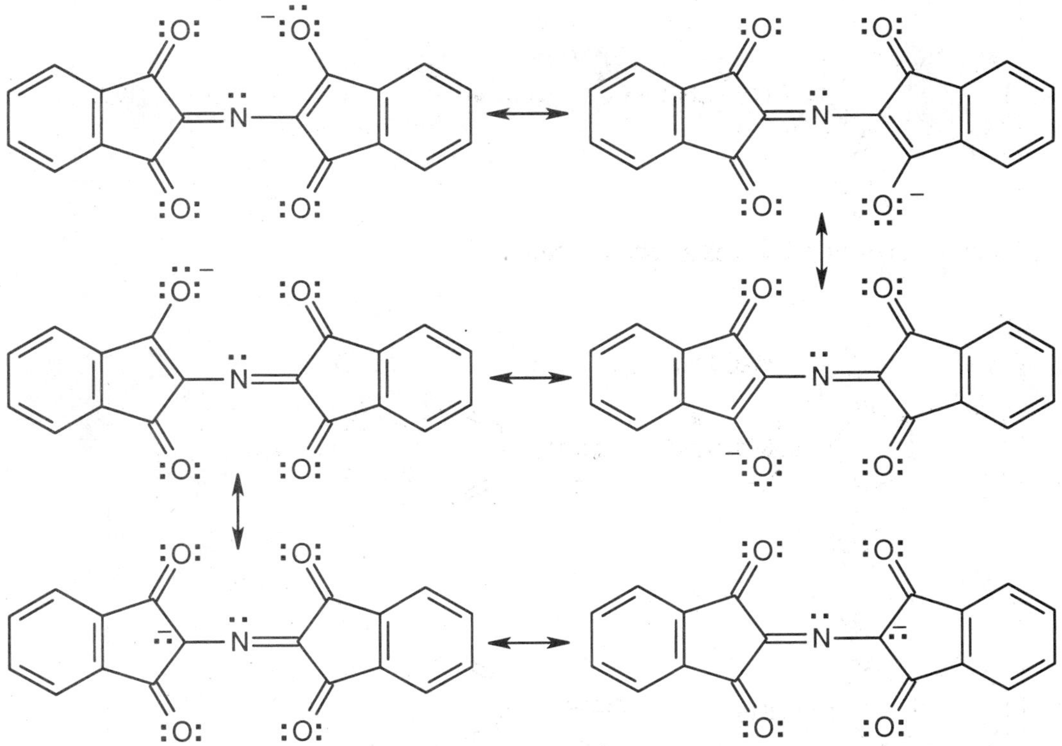

It is also possible to draw many other resonance forms that involve the π electrons of the aromatic 6-membered rings.

26.58

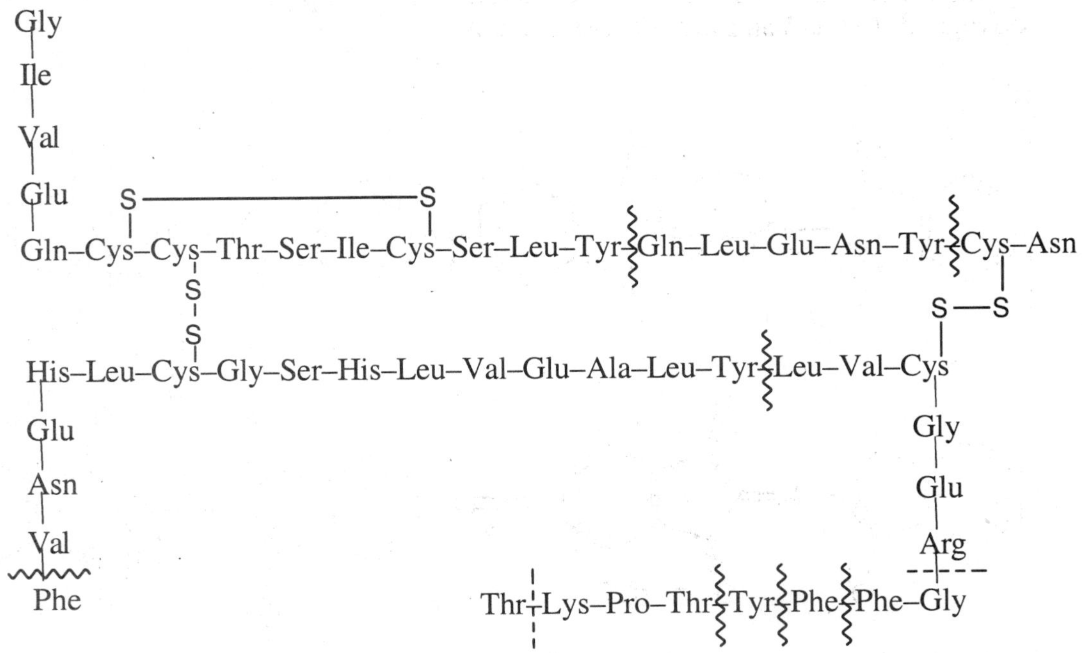
Gly
Ile
Val
Glu
Gln–Cys–Cys–Thr–Ser–Ile–Cys–Ser–Leu–Tyr–Gln–Leu–Glu–Asn–Tyr–Cys–Asn
His–Leu–Cys–Gly–Ser–His–Leu–Val–Glu–Ala–Leu–Tyr–Leu–Val–Cys
Glu
Asn
Val
Phe
Gly
Glu
Arg
Thr–Lys–Pro–Thr–Tyr–Phe–Phe–Gly

Cleaved by Trypsin = ----

Cleaved by Chymotrypsin = ∿∿∿

26.59 Ser–Ile–Arg–Val–Val–Pro–Tyr–Leu–Arg

26.60

Reduced oxytocin: Cys–Tyr–Ile–Gln–Asn–Cys–Pro–Leu–Gly–NH_2

Oxidized oxytocin: Cys–Tyr–Ile–Gln–Asn–Cys–Pro–Leu–Gly–NH_2 (S–S bond between the two Cys residues)

The C–terminal end of oxytocin is an amide, but this can't be determined from the information given.

26.61

(a)

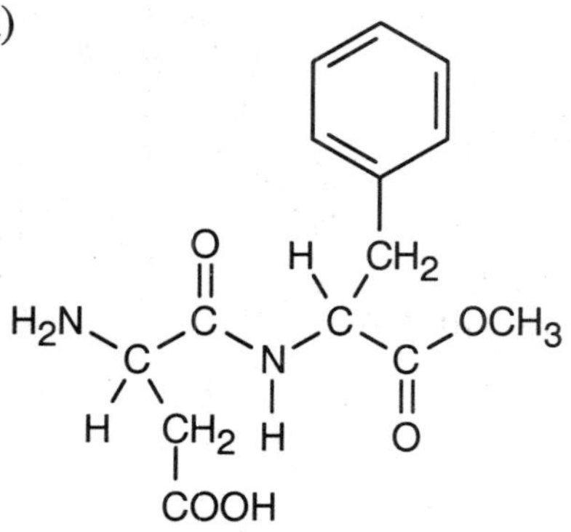

Aspartame (nonzwitterionic form)

(b)

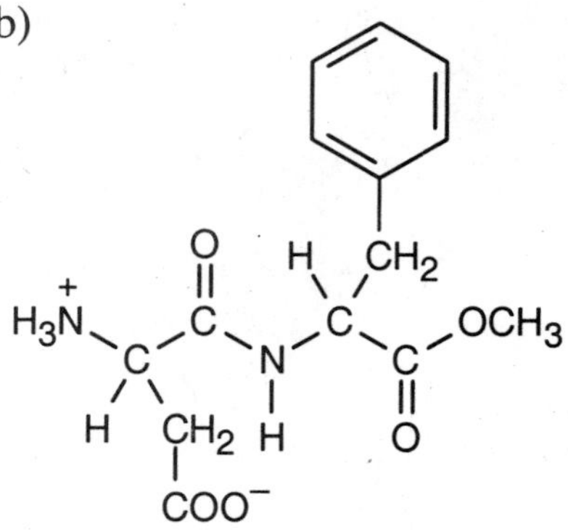

Aspartame at IP (pH = 5.9)

(c)

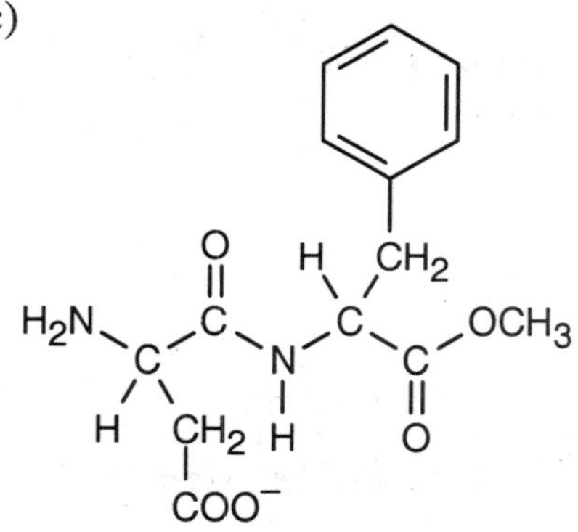

Aspartame at pH = 7.3

26.62

H_3O^+ protonation

addition of water

proton transfer

ring opening

bond rotation, addition of amine

proton transfer

loss of water

$- H^+$ deprotonation

A Look Ahead

26.63

$$CH_3CH_2CH(CH_3)CH(NH_2)COOH + \alpha\text{-keto acid} \xrightarrow[\text{amination}]{\text{reductive}} CH_3CH_2CH(CH_3)COCOOH + \text{amino acid}$$

Molecular Modeling

26.64 In the gas phase, neutral glycine is more stable than the zwitterion by 117 kJ/mol. Electrostatic potential maps show that the zwitterion contains separated charges, which are unstable in the gas phase but promote formation of strong hydrogen bonds with water.

26.65 Alpha-helix: Ala – Gly – Phe – Cys – Gly – Ser – Ala – Arg – Val – Trp
N-terminus C-terminus
Beta sheet: all glycine.
The helix contains no empty space.

26.66 Imidazolium ion **A** is more stable by 242 kJ/mol. The electrostatic potential map shows that the positive charge is more delocalized in this ion.

Review Unit 10: Biomolecules I – Carbohydrates, Amino Acids, Peptides

Major Topics Covered (with vocabulary):

Monosaccharides:
Carbohydrate monosaccharide aldose ketose D,L sugars pyranose furanose anomer anomeric center α anomer β anomer glycoside Koenigs-Knorr reaction aldonic acid alditol reducing sugar aldaric acid Kiliani-Fischer synthesis Wohl degradation

Other sugars:
disaccharide 1,4' link cellobiose maltose lactose sucrose polysaccharide cellulose amylose amylopectin glycogen glycal assembly method deoxy sugar amino sugar cell-surface carbohydrate

Amino acids:
amino acid zwitterion amphoteric α-amino acid side chain isoelectric point (p*I*) electrophoresis Henderson-Hasselbalch equation Strecker synthesis amidomalonate synthesis reductive amination resolution

Peptides:
residue backbone *N*-terminal amino acid C-terminal amino acid disulfide link amino acid analysis Edman degradation phenylthiohydantoin carboxypeptidase trypsin chymotrypsin peptide synthesis protection BOC derivative Merrifield solid-phase technique

Proteins:
primary structure secondary structure tertiary structure quaternary structure α-helix β-pleated sheet salt bridge prosthetic group enzyme cofactor apoenzyme holoenzyme coenzyme vitamin isomerase hydrolase ligase lyase oxidoreductase transferase denaturation

Types of Problems:

After studying these chapters, you should be able to:

- Classify carbohydrates as aldoses, ketoses, D or L sugars, monosaccharides, or polysaccharides.
- Draw monosaccharides as Fischer projections or chair conformations.
- Predict the products of reactions of monosaccharides and disaccharides.
- Deduce the structures of monosaccharides and disaccharides.
- Formulate mechanisms for reactions involving carbohydrates.

- Identify the common amino acids and draw them with correct stereochemistry in dipolar form.
- Explain the acid-base behavior of amino acids.
- Synthesize amino acids.
- Draw the structure of simple peptides.
- Deduce the structure of peptides and proteins.
- Outline the synthesis of peptides.
- Explain the classification of proteins and the levels of structure of proteins.
- Draw structures of reaction products of amino acids and peptides.

Points to Remember:

* Aldohexoses, ketohexoses and aldopentoses can all exist in both pyranose forms and furanose forms.

* A reaction that produces the same functional group at both ends of a monosaccharide halves the number of possible stereoisomers of the monosaccharide.

* The reaction conditions that form a glycoside are different from those that form a polyether, even though both reactions, technically, form –OR bonds.

* At physiological pH, the side chains of the amino acids aspartic acid and glutamic acid exist as anions, and the side chains of the amino acids lysine and arginine exist as cations. The imidazole ring of histidine exists as a mixture of protonated and neutral forms.

* Since the amide backbone of a protein is neutral and uncharged, the isoelectric point of a protein or peptide is determined by the relative numbers of acidic and basic amino acid residues present in the peptide.

* In the Merrifield technique of protein synthesis, a protecting group isn't needed for the carboxyl group because it is attached to the polymer support.

Self-Test:

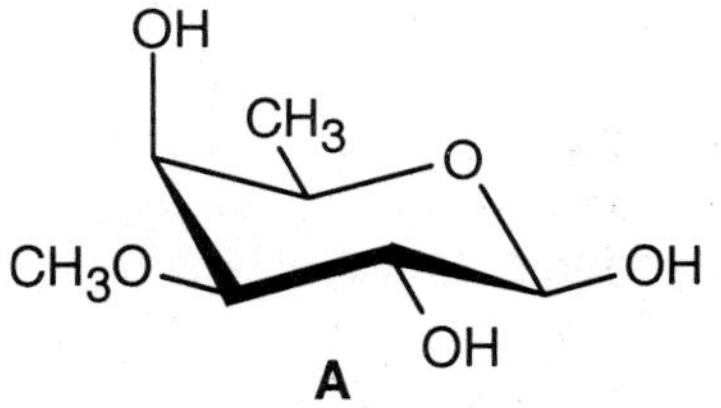

OH, CH_3, O, CH_3O, OH, OH

A

Digitalin (hydrolysis product of gitoxigenin, a heart medication)

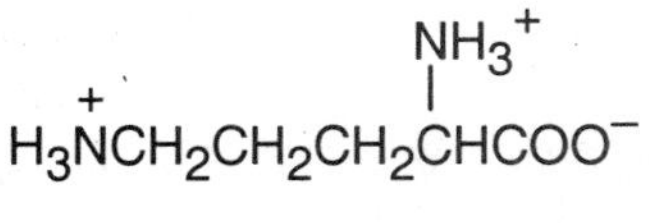

$$\overset{+}{H_3N}CH_2CH_2CH_2\overset{\overset{NH_3^+}{|}}{C}HCOO^-$$

C

Ornithine

Digitalin (**A**) is related to which D-aldohexose? Provide a name for **A**, including the configuration at the anomeric carbon. Predict the products of the reaction of **A** with: (a) CH_3OH, H^+ catalyst; (b) CH_3I, Ag_2O.

Vicianose (**B**) is a disaccharide associated with a natural product found in seeds. Treatment of **B** with CH_3I and Ag_2O, followed by hydrolysis, gives 2,3,4-tri-*O*-methyl-D-glucose and 2,3,4-tri-*O*-methyl-D-arabinose. What is the structure of **B**? Is **B** a reducing sugar?

Ornithine (**C**) is a nonstandard amino acid that occurs in metabolic processes. Which amino acid does it most closely resemble? Estimate pK_a values and pI for ornithine, and draw the major form present at pH = 2, pH = 6, and pH = 11. If ornithine were a component of proteins, how would it affect the tertiary structure of a protein?

Tyr–Gly–Gly–Phe–Leu–Arg–Arg–Ile–Arg–Pro–Lys–Leu–Lys–Trp–Asp–Asn–Gln
Porcine Dynorphin (**D**)

Dynorphin (**D**)is a neuropeptide. Indicate the *N*-terminal end and the C-terminal end. Show the products of cleavage with: (a) carboxypeptidase; (b) trypsin; (c) chymotrypsin. Show the *N*-phenylthiohydantoin that results from treatment of **D** with phenyl isothiocyanate. Do you expect **D** to be an acidic, a neutral or a basic peptide?

Kallidin (**E**) is a decapeptide that serves as a vasodilator. The composition of **E** is Arg_2 Gly Lys Phe_2 Pro_3 Ser. Treatment with carboxypeptidase shows that the C-terminal residue is Arg. Partial acid hydrolysis yields the following fragments:

Pro–Gly–Phe, Lys–Arg-Pro, Pro–Phe–Arg, Pro–Pro–Gly, Phe–Ser–Pro

What is the structure of **E**.

Multiple choice:

1. The enantiomer of α-D-glucopyranose is:
(a) β-D-Glucopyranose (b) α-L-Glucopyranose (c) β-L-Glucopyranose (d) none of these

2. All of the following reagents convert an aldose to an aldonic acid except:
(a) dilute HNO_3 (b) Fehling's reagent (c) Benedict's reagent (d) aqueous Br_2

3. Which two aldoses yield D-lyxose after Wohl degradation?
(a) D-Glucose and D-Mannose (b) D-Erythrose and D-Threose (c) D-Galactose and D-Altrose (d) D-Galactose and D-Talose

4. All of the following disaccharides are reducing sugars except:
(a) Cellobiose (b) Sucrose (c) Maltose (d) Lactose

5. Which of the following polysaccharides contains β-glycosidic bonds?
(a) Amylose (b) Amylopectin (c) Cellulose (d) Glycogen

6. To find the p*I* of an acidic amino acid:
(a) find the average of the two lowest pK_a values (b) find the average of the two highest pK_a values (c) find the average of all pK_a values (d) use the value of the pK_a of the side chain.

7. Which of the following techniques can synthesize a single enantiomer of an amino acid?
(a) Strecker synthesis (b) reductive amination (c) amidomalonate synthesis
(d) none of them

8. The purple product that results from the reaction of ninhydrin with an amino acid contains which group of the amino acid?
(a) the amino group (b) the amino nitrogen (c) the carboxylic acid group (d) the side chain

9. Which of the following reagents is not used in peptide synthesis?
(a) Phenylthiohydantoin (b) Di-*tert*-butyl dicarbonate (c) Benzyl alcohol
(d) Dicyclohexylcarbodiimide

10. Which structural element is not present in myoglobin?
(a) a prosthetic group (b) regions of α-helix (c) hydrophobic regions (d) quaternary structure

Chapter 27 – Biomolecules: Lipids

Chapter Outline

I. Esters (Sections 27.1 – 27.3).
   A. Waxes, fats and oils (Section 27.1).
      1. Waxes are esters of long-chain fatty acids with long-chain alcohols.
      2. Fats and oils are triacylglycerols.
         a. Hydrolysis of a fat yields glycerol and three fatty acids.
         b. The fatty acids need not be the same.
      3. Fatty acids.
         a. Fatty acids are even-numbered,unbranched long-chain (C_{12}–C_{20}) carboxylic acids.
         b. The most abundant saturated fatty acids are palmitic (C_{16}) and stearic(C_{18}) acids.
         c. The most abundant unsaturated fatty acids are oleic and linoleic acids (both C_{18}).
            Linoleic and arachidonic acids are polyunsaturated fatty acids.
         d. Unsaturated fatty acids are lower-melting than saturated fatty acids because the double bonds keep molecules from packing closely.
         e. The C=C bonds can be hydrogenated to produce higher-melting fats.
   B. Soap (Section 27.2).
      1. Soap is a mixture of the sodium and potassium salts of fatty acids produced by hydrolysis of animal fat.
      2. Soap acts as a cleanser because the two ends of a soap molecule are different.
         a. The hydrophilic carboxylate end dissolves in water.
         b. The hydrophobic hydrocarbon tails solubilize greasy dirt.
         c. In water, the hydrocarbon tails aggregate into micelles, where greasy dirt can accumulate.
      3. Soaps can form scum when they encounter Mg^{2+} and Ca^{2+} salts.
         This problem is circumvented by detergents, which don't form insoluble metal salts.
   3. Phospholipids (Section 27.3).
      1. Phosphoglycerides.
         a. Phosphoglycerides consist of glycerol, two fatty acids (at C1 and C2 of glycerol), and a phosphate group bonded to an amino alcohol at C3 of glycerol.
         b. The most important phospholipids are lecithins and cephalins.
         c. Phosphoglycerides comprise the major lipids in cell membranes.
            The phospholipid molecules are organized into a lipid bilayer, which has polar groups on the inside and outside, and nonpolar tails in the middle.
      2. Sphingolipids.
         a. Sphingolipids have sphingosine as their backbone.
         b. They are abundant in brain and nerve tissue as sphingomyelins.

II. Prostaglandins (Section 27.4).
   A. Prostaglandins are C_{20} carboxylic acids that contain a C_5 ring and two side chains.
   B. Prostaglandins are present in small amounts in all body tissues and fluids.
   C. Prostaglandins have many effects: they lower blood pressure, affect blood platelet aggregation, affect kidney function and stimulate uterine contractions.
   D. Prostaglandins are biosynthesized from arachidonic acid, which is synthesized from linoleic acid.
      1. The transformation from arachidonic acid is catalyzed by the cyclooxygenase (COX) enzyme.

2. One form of the COX enzyme catalyzes the usual functions, and a second form produces additional prostaglandin as a result of inflammation.

III. Terpenes (Section 27.5 – 27.6).
- A. Facts about terpenes (Section 27.5).
  1. Terpenes occur as essential oils in lipid extractions of plants.
  2. Terpenes are small organic molecules with diverse structures.
  3. All terpenes are structurally related.
     - a. Terpenes arise from head-to-tail bonding of isoprene units.
     - b. Carbon 1 is the head, and carbon 4 is the tail.
  4. Terpenes are classified by the number of isoprene units they contain.
     - a. Monoterpenes are synthesized from two isoprene units.
     - b. Sesquiterpenes are synthesized from three isoprene units.
     - c. Larger terpenes occur in both animals and plants.
- B. Biosynthesis of terpenes (Section 27.6).
  1. Nature uses two isoprene equivalents, isopentenyl pyrophosphate and dimethylallyl pyrophosphate, to synthesize terpenes.
     These units are created from 3 acetyl CoA molecules.
  2. The C=C bond of isopentenyl pyrophosphate displaces the pyrophosphate group of dimethallyl pyrophosphate, to form geranyl pyrophosphate, the precursor to all monoterpenes.
  3. Geranyl pyrophosphate reacts with isopentenyl pyrophosphate to yield farnesyl pyrophosphate, the precursor to sesquiterpenes.
  4. Further reactions give the 20- and 25-carbon units that lead to other terpenes.
  5. Triterpenes arise from tail-to-tail coupling of two farnesyl pyrophosphate units, to give squalene, the precursor to steroids.

IV. Steroids (Section 27.6 – 27.9).
- A. Facts about steroids (Section 27.7).
  1. Steroids are found in the lipid extracts of plants and animals.
  2. Steroids have a tetracyclic ring structure.
     The rings adopt chair conformations but are unable to undergo ring-flips.
  3 Steroids function as hormones in humans.
- B. Types of steroid hormones.
  1. Sex hormones.
     - a. Androgens (testosterone, androsterone) are male sex hormones.
     - b. Estrogens (estrone, estradiol) and progestins are female sex hormones.
  2. Adrenocortical hormones.
     - a. Mineralocorticoids (aldosterone) regulate cellular Na^+ and K^+ balance.
     - b. Glucocorticoids (hydrocortisone) regulate glucose metabolism and control inflammation.
  3. Synthetic steroids.
     Oral contraceptives and anabolic steroids are examples of synthetic steroids.
- C. Stereochemistry of steroids (Section 27.8).
  1. Two cyclohexane rings can be joined either cis or trans.
     - a. In a trans-fused ring, the groups at the ring junction are trans.
     - b. In cis-fused rings, the groups at the ring junction are cis.
     - c. Cis ring fusions usually occur between rings A and B.
  2. In both kinds of ring fusions, the angular methyl groups usually protrude above the rings.
  3. Steroids with A–B trans systems are more common.
  4. Substituents can be either axial or equatorial.
     Equatorial substituents are more favorable.
- D. Steroid biosynthesis (Section 27.9).
  1. All steroids are biosynthesized from squalene.
  2. Squalene is first epoxidized to form squalene oxide.

3. Ten additional steps are needed to form lanosterol.
   a. The first several steps are cyclization reactions.
   b. The last steps are carbocation rearrangements.
4. Other enzymes convert lanosterol to cholesterol.

Solutions to Problems

27.1

$$CH_3(CH_2)_{18}C(=O)O(CH_2)_{31}CH_3$$

$CH_3(CH_2)_{18}C(=O)$– from C_{20} acid; $O(CH_2)_{31}CH_3$ from C_{32} alcohol

27.2

$$\begin{array}{l} CH_2OC(=O)(CH_2)_{14}CH_3 \\ | \\ CHOC(=O)(CH_2)_{14}CH_3 \\ | \\ CH_2OC(=O)(CH_2)_{14}CH_3 \end{array}$$

Glyceryl tripalmitate

$$\begin{array}{ll} CH_2OC(=O)(CH_2)_7CH{=}CH(CH_2)_7CH_3 & \text{(cis)} \\ | & \\ CHOC(=O)(CH_2)_7CH{=}CH(CH_2)_7CH_3 & \text{(cis)} \\ | & \\ CH_2OC(=O)(CH_2)_7CH{=}CH(CH_2)_7CH_3 & \text{(cis)} \end{array}$$

Glyceryl trioleate

Glyceryl tripalmitate is higher melting because it is saturated.

27.3

$$CH_3(CH_2)_7CH{=}CH(CH_2)_7C(=O)O^- \ Mg^{2+} \ {}^-OC(=O)(CH_2)_7CH{=}CH(CH_2)_7CH_3$$

Magnesium oleate

The double bonds are cis.

27.4

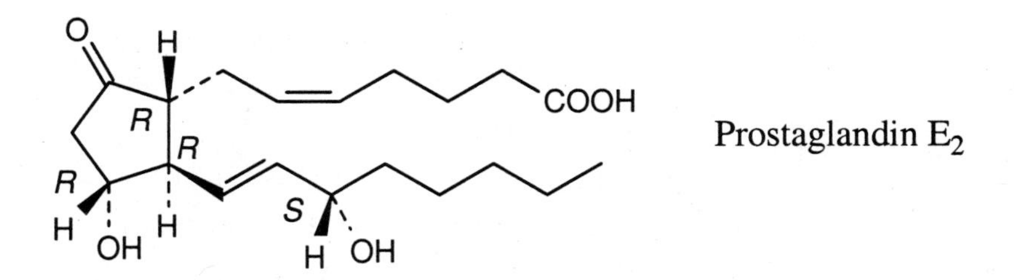

Prostaglandin E_2

27.5 To locate isoprene units:

(1) Determine the number of isoprene units you need to find by counting the carbons in the structures. Both (a) and (b) have 10 carbons and thus have two isoprene units. Caryophyllene has 3 units.

(2) Locate the first unit. The easiest part to find is a carbon with two methyl groups; sometimes this is an isopropyl group. Once you have found this piece, which contains 3 carbons (and is starred), count until you have reached 5 carbons. At this point, there may be more than one possible choice for the first isoprene unit.

(3) Look at the rest of the molecule to decide where the first isoprene unit ends. The second unit should be apparent. In this problem, the bonds formed by isoprene units are drawn as dashed lines

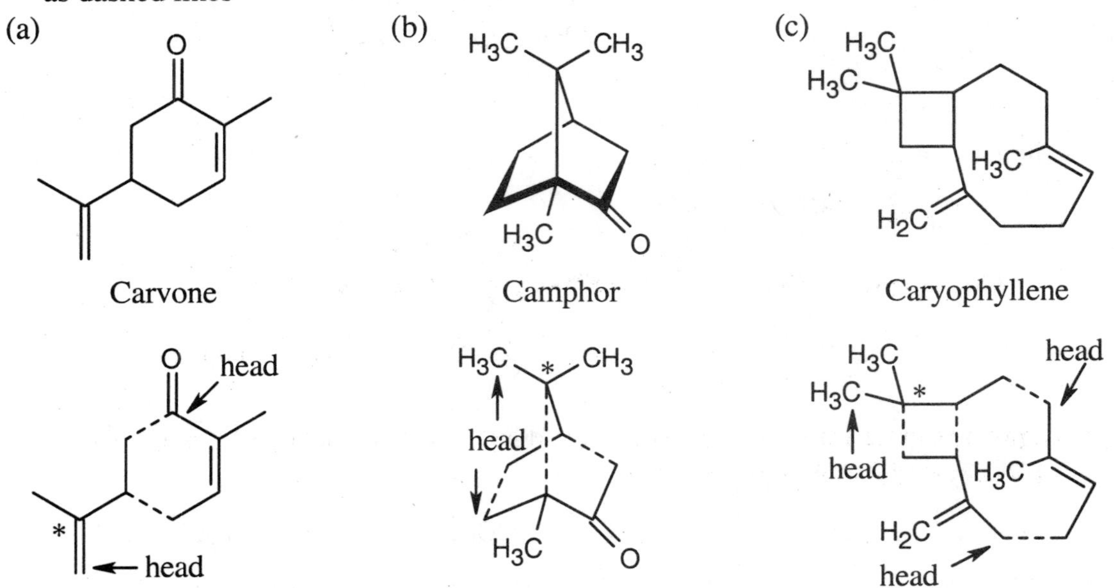

27.6 As described in Practice Problem 27.1, draw the pyrophosphate precursor so that it resembles the product. In (a), it's not easy to see the relationship, but once you've arrived at the product, rotate the structure.

(a)

PPO

+ $^{-}$OPP

H

$-H^+$

Neryl pyrophosphate

α-Pinene

(b)

CH_2—OPP

H

+ $^{-}$OPP

$-H^+$

γ-Bisabolene

27.7 Both ring systems are trans-fused, and both hydrogens at the ring junctions are axial. Refer back to Chapter 4 if you have trouble remembering the relationships of substituents on a cyclohexane ring.

(a)

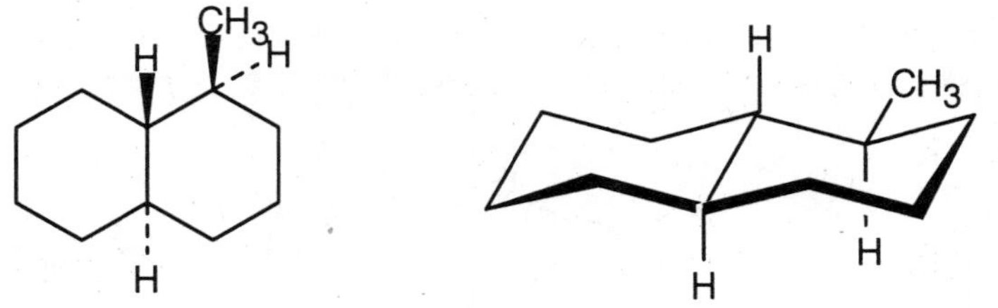

H CH3 H H H CH3 equatorial H H

(b)

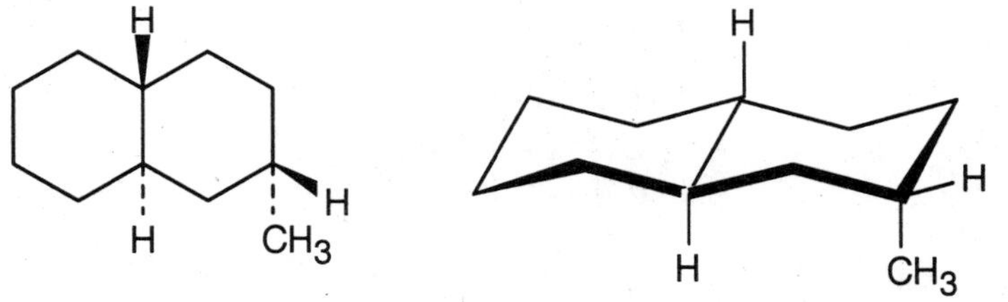

H H CH3 H H H CH3 axial

27.8 Draw the three-dimensional structure and note the relationship of the hydroxyl group to groups whose orientation is known.

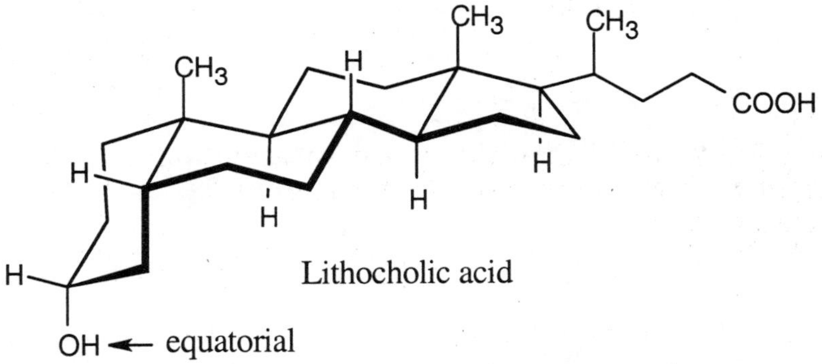

CH3 CH3 CH3 H COOH H H H H H OH ← equatorial

Lithocholic acid

27.9

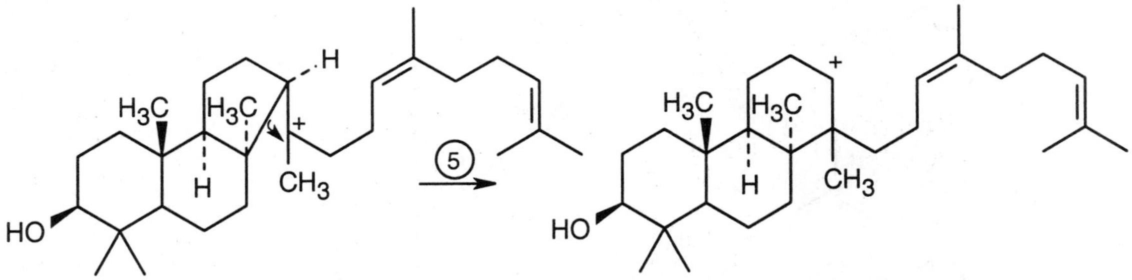

H3C H3C H H CH3 HO 5 H3C H3C H CH3 HO

In step 5, a tertiary carbocation is converted to a less stable secondary carbocation. Although this step seems energetically unfavorable, it must be necessary if the synthetic pathway is to lead to the desired product, lanosterol

27.10

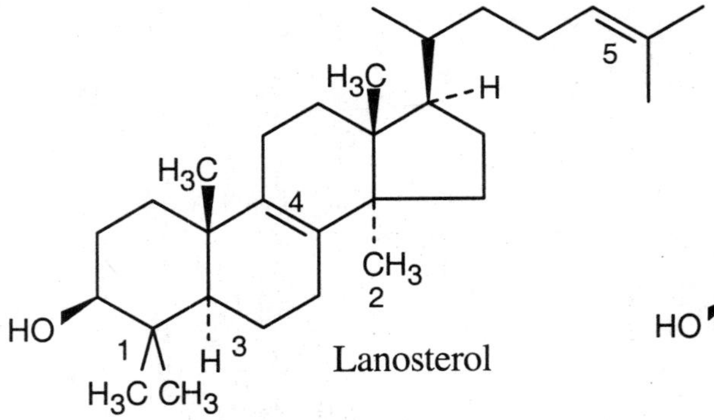

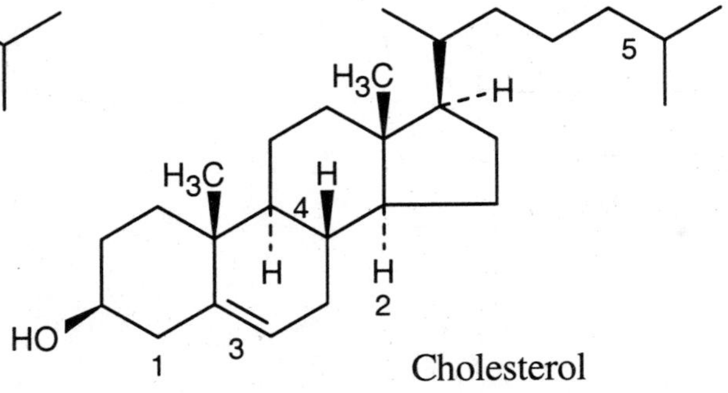

Lanosterol   Cholesterol

Lanosterol

1. Two methyl groups at C4.
2. One methyl group at C14.
3. C5–C6 single bond.
4. C8–C9 double bond.
5. Double bond in side chain

Cholesterol

1. Two hydrogens at C4.
2. One hydrogen at C14.
3. C5–C6 double bond
4. C8–C9 single bond.
5. Saturated side chain.

Visualizing Chemistry

27.11

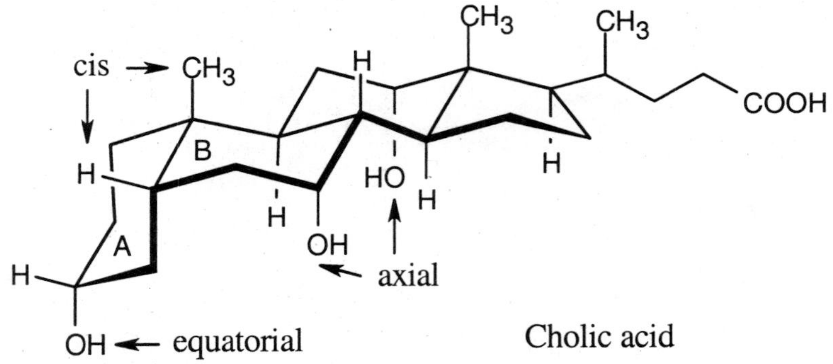

Cholic acid

Cholic acid is an A–B cis steroid because the groups at the fusion of ring A and ring B have a cis relationship.

27.12

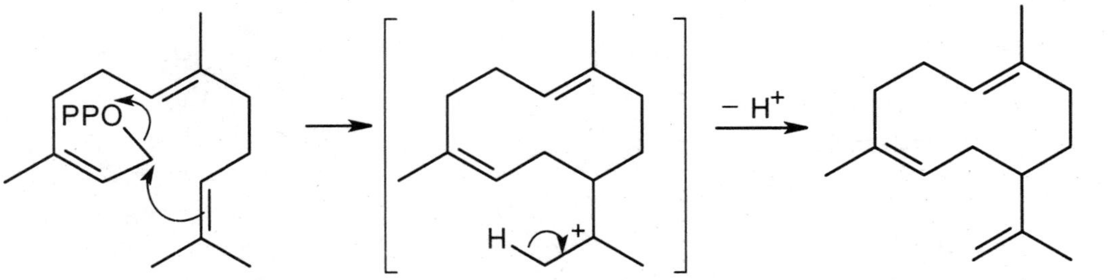

Farnesyl pyrophosphate   Helminthogermacrene

Draw farnesyl pyrophosphate in the configuration that resembles the product. In this reaction, a cyclization, followed by loss of a proton to form the double bond, gives helminthogermacrene.

Additional Problems

27.13

$$\begin{array}{l} CH_2OC(=O)(CH_2)_{16}CH_3 \\ |\\ {}^{*}CHOC(=O)(CH_2)_{16}CH_3 \\ | \\ CH_2OC(=O)(CH_2)_7CH{=}CH(CH_2)_7CH_3 \ \text{(cis)} \end{array}$$

optically active

or

$$\begin{array}{l} CH_2OC(=O)(CH_2)_{16}CH_3 \\ | \\ CHOC(=O)(CH_2)_7CH{=}CH(CH_2)_7CH_3 \ \text{(cis)} \\ | \\ CH_2OC(=O)(CH_2)_{16}CH_3 \end{array}$$

optically inactive

1. ^{-}OH, H_2O
2. H_3O^+

$$\begin{array}{l} CH_2OH \\ | \\ CHOH \\ | \\ CH_2OH \end{array} + HOC(=O)(CH_2)_7CH{=}CH(CH_2)_7CH_3 \ \text{(cis)} + 2\ HOC(=O)(CH_2)_{16}CH_3$$

Oleic acid; Stearic acid

Four different groups are bonded to the central glycerol carbon atom in the optically active fat.

27.14

$$CH_3(CH_2)_{14}C(=O)OCH_2(CH_2)_{14}CH_3$$

Cetyl palmitate

27.15 Fats, lecithins, cephalins and plasmalogens are all esters of a glycerol molecule that has carboxylic acid ester groups at C1 and C2. The third group bonded to glycerol, however, differs with the type of lipid.

Lipid	*Functional group at C3 of glycerol*
Fat	carboxylic acid ester
Cephalin	phosphate ester (also bonded to ethanolamine, an amino alcohol)
Lecithin	phosphate ester (also bonded to choline, an amino alcohol)
Plasmalogen	vinyl ether

27.16

$$\begin{array}{l} CH_2OH + HOC(=O)R'' \\ | \\ CHOH + HOC(=O)R' \\ | \\ CH_2OH + HC(=O)CH_2R \end{array} \xleftarrow{H_3O^+} \begin{array}{l} CH_2OC(=O)R'' \\ | \\ CHOC(=O)R' \\ | \\ CH_2OCH{=}CHR \end{array} \xrightarrow[H_2O]{NaOH} \begin{array}{l} CH_2OH + Na^+\ ^{-}OC(=O)R'' \\ | \\ CHOH + Na^+\ ^{-}OC(=O)R' \\ | \\ CH_2OCH{=}CHR \end{array}$$

Basic hydrolysis cleaves the carboxylic acid ester bonds but doesn't affect the ether bond. Acidic hydrolysis cleaves all three groups bonded to glycerol and produces an aldehyde from the vinyl ether group.

27.17

$$\text{a cardiolipin} \xrightarrow[H_2O]{NaOH} 3\ HOCH_2CH(OH)CH_2OH + 2\ Na_3PO_4 + RCO_2^-\,Na^+ + R'CO_2^-\,Na^+ + R''CO_2^-\,Na^+ + R'''CO_2^-\,Na^+$$

a cardiolipin

Saponification of a cardiolipin yields 4 different carboxylates, 3 equivalents of glycerol and two equivalents of phosphate.

27.18

$$\underset{\text{Stearolic acid}}{CH_3(CH_2)_7C{\equiv}C(CH_2)_7COOH} \xrightarrow[\text{2. Zn, }CH_3COOH]{\text{1. }O_3} \underset{\text{Nonanoic acid}}{CH_3(CH_2)_7COOH} + \underset{\text{Nonanedioic acid}}{HOOC(CH_2)_7COOH}$$

Stearolic acid contains a triple bond because the products of ozonolysis are carboxylic acids.

27.19

$$CH_3(CH_2)_7C{\equiv}CH \xrightarrow[NH_3]{NaNH_2} [CH_3(CH_2)_7C{\equiv}C{:}^-\,Na^+] + NH_3$$

$$[CH_3(CH_2)_7C{\equiv}C{:}^-\,Na^+] + I{-}CH_2(CH_2)_5CH_2{-}Cl \longrightarrow CH_3(CH_2)_7C{\equiv}C(CH_2)_6CH_2{-}Cl$$

$$CH_3(CH_2)_7C{\equiv}C(CH_2)_6CH_2{-}Cl \xrightarrow{NaCN} CH_3(CH_2)_7C{\equiv}C(CH_2)_6CH_2CN$$

$$CH_3(CH_2)_7C{\equiv}C(CH_2)_6CH_2CN \xrightarrow{H_3O^+} \underset{\text{Stearolic acid}}{CH_3(CH_2)_7C{\equiv}C(CH_2)_6CH_2COOH}$$

I^-, rather than Cl^-, is displaced by acetylide because iodide is a better leaving group than chloride.

27.20

$CH_2OC(O)(CH_2)_7CH{=}CH(CH_2)_7CH_3$ (cis)

$CHOC(O)(CH_2)_7CH{=}CH(CH_2)_7CH_3$ (cis)

$CH_2OC(O)(CH_2)_7CH{=}CH(CH_2)_7CH_3$ (cis)

Glyceryl trioleate

(a)

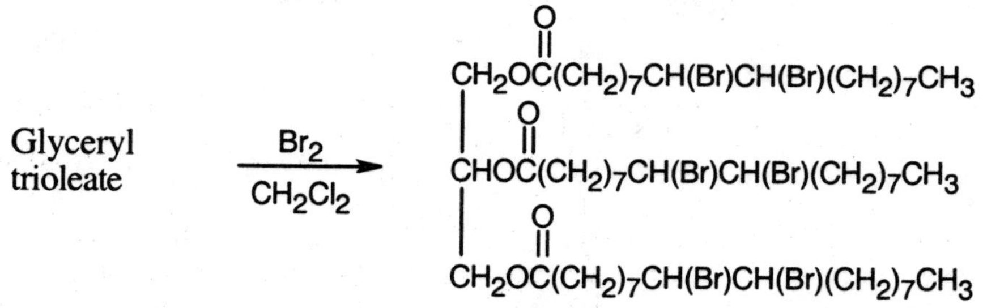

$$\text{Glyceryl trioleate} \xrightarrow[CH_2Cl_2]{Br_2} \begin{array}{l} CH_2OC(=O)(CH_2)_7CH(Br)CH(Br)(CH_2)_7CH_3 \\ | \\ CHOC(=O)(CH_2)_7CH(Br)CH(Br)(CH_2)_7CH_3 \\ | \\ CH_2OC(=O)(CH_2)_7CH(Br)CH(Br)(CH_2)_7CH_3 \end{array}$$

(b)

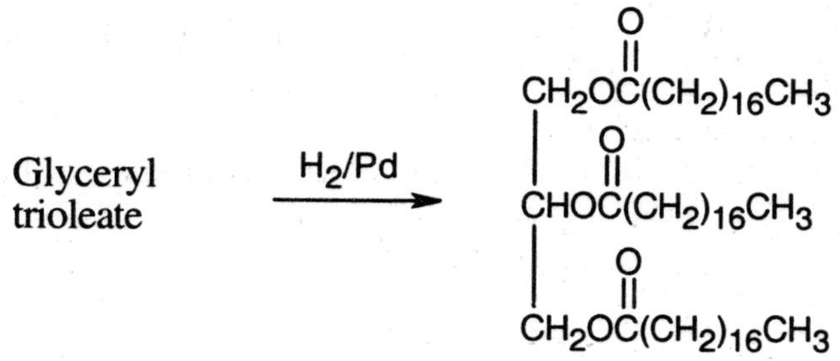

$$\text{Glyceryl trioleate} \xrightarrow{H_2/Pd} \begin{array}{l} CH_2OC(=O)(CH_2)_{16}CH_3 \\ | \\ CHOC(=O)(CH_2)_{16}CH_3 \\ | \\ CH_2OC(=O)(CH_2)_{16}CH_3 \end{array}$$

(c)

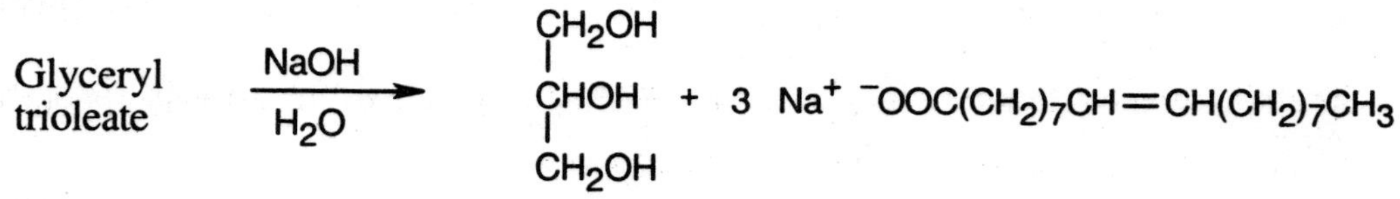

$$\text{Glyceryl trioleate} \xrightarrow[H_2O]{NaOH} \begin{array}{l} CH_2OH \\ | \\ CHOH \\ | \\ CH_2OH \end{array} + 3\ Na^+\ {}^-OOC(CH_2)_7CH{=}CH(CH_2)_7CH_3$$

(d)

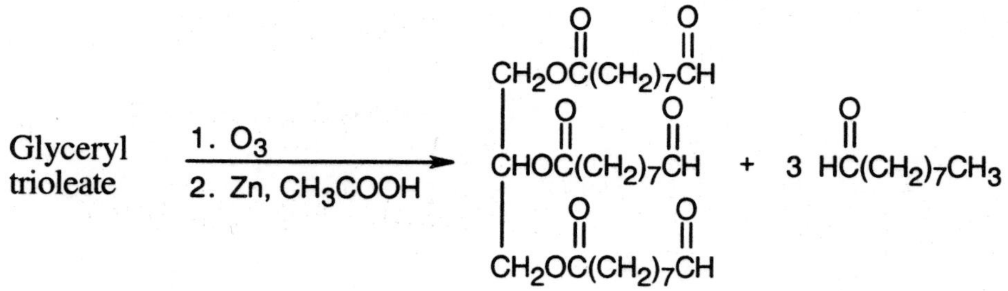

$$\text{Glyceryl trioleate} \xrightarrow[\text{2. Zn, } CH_3COOH]{\text{1. } O_3} \begin{array}{l} CH_2OC(=O)(CH_2)_7CH(=O) \\ | \\ CHOC(=O)(CH_2)_7CH(=O) \\ | \\ CH_2OC(=O)(CH_2)_7CH(=O) \end{array} + 3\ HC(=O)(CH_2)_7CH_3$$

(e)

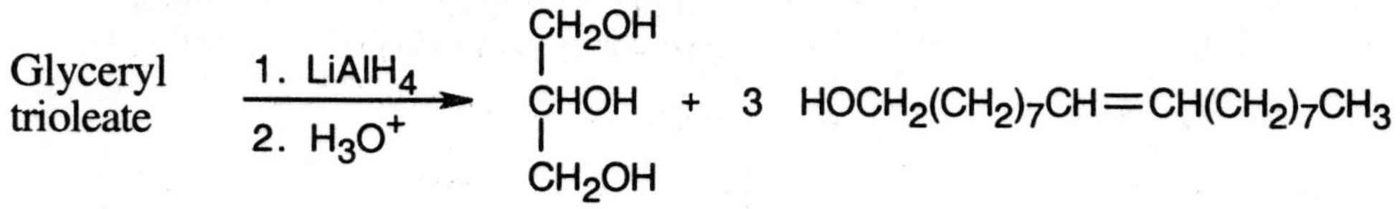

$$\text{Glyceryl trioleate} \xrightarrow[\text{2. } H_3O^+]{\text{1. } LiAlH_4} \begin{array}{l} CH_2OH \\ | \\ CHOH \\ | \\ CH_2OH \end{array} + 3\ HOCH_2(CH_2)_7CH{=}CH(CH_2)_7CH_3$$

(f)

$$\text{Glyceryl trioleate} \xrightarrow[\text{2. } H_3O^+]{\text{1. } CH_3MgBr} \begin{array}{l} CH_2OH \\ | \\ CHOH \\ | \\ CH_2OH \end{array} + 3\ HOC(CH_3)_2(CH_2)_7CH{=}CH(CH_2)_7CH_3$$

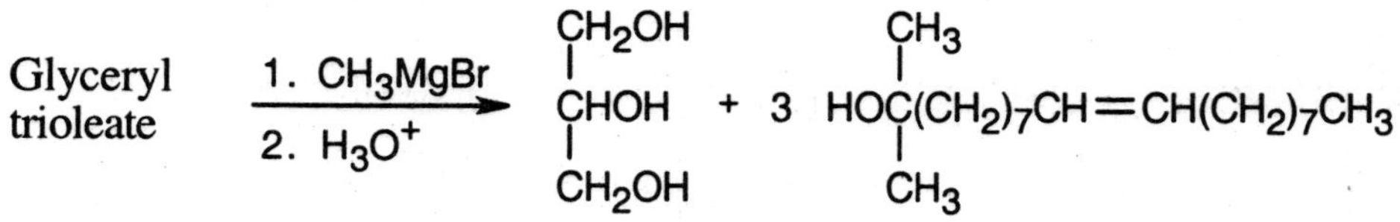

27.21

$$CH_3(CH_2)_7CH{=}CH(CH_2)_7COOH \text{ (cis)}$$
Oleic acid

(a)
$$\text{Oleic acid} \xrightarrow[HCl]{CH_3OH} CH_3(CH_2)_7CH{=}CH(CH_2)_7COOCH_3$$
Methyl oleate

(b)
$$\text{Methyl oleate from (a)} \xrightarrow{H_2/Pd} CH_3(CH_2)_{16}COOCH_3$$
Methyl stearate

(c)
$$\text{Oleic acid} \xrightarrow[\text{2. Zn, } CH_3COOH]{\text{1. } O_3} CH_3(CH_2)_7CHO + OHC(CH_2)_7COOH$$
Nonanal + 9-Oxononanoic acid

(d)
$$\text{9-Oxononanoic acid from (d)} \xrightarrow[NH_4OH]{Ag_2O} HOOC(CH_2)_7COOH$$
Nonanedioic acid

This is a Tollens oxidation.

(e)
$$\text{Oleic acid} \xrightarrow[CH_2Cl_2]{Br_2} CH_3(CH_2)_7CH(Br)CH(Br)(CH_2)_7COOH$$
$$\xrightarrow[\text{2. } H_3O^+]{\text{1. 3 } NaNH_2, NH_3} CH_3(CH_2)_7C{\equiv}C(CH_2)_7COOH$$
Stearolic acid

Three equivalents of the base are needed because one of them is neutralized by the carboxylic acid.

(f)
$$\text{Oleic acid} \xrightarrow{H_2/Pd} CH_3(CH_2)_{15}CH_2COOH \xrightarrow[\text{2. } H_2O]{\text{1. } Br_2, PBr_3} CH_3(CH_2)_{15}CH(Br)COOH$$
Stearic acid; 2-Bromostearic acid

(g)
$$2\ CH_3(CH_2)_{16}COOCH_3 \text{ from (b)} \xrightarrow[\text{2. } H_3O^+]{\text{1. } Na^+\ ^-OCH_3} CH_3(CH_2)_{16}C(=O)CH(COOCH_3)(CH_2)_{15}CH_3 + HOCH_3$$
$$\xrightarrow{H_3O^+, \text{ heat}} CH_3(CH_2)_{16}C(=O)CH_2(CH_2)_{15}CH_3 + CO_2 + HOCH_3$$
18-Pentatriacontanone

This synthesis uses a Claisen condensation, followed by a β-keto ester decarboxylation.

27.22–27.24

Remember that a compound with n chirality centers can have a maximum of 2^n stereoisomers. Not all the possible stereoisomers of these compounds are found in nature or can be synthesized. Some stereoisomers have highly strained ring fusions; others contain 1,3-diaxial interactions.

An (h) is next to the head of each isoprene unit. Each new bond formed is represented by a dashed line.

(a)

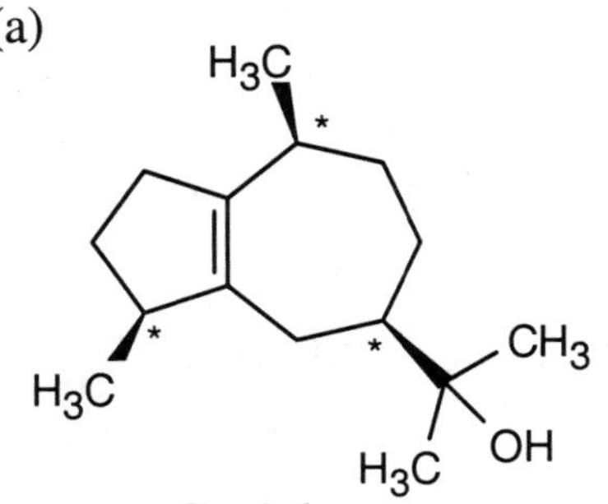

Guaiol
(8 possible stereoisomers)

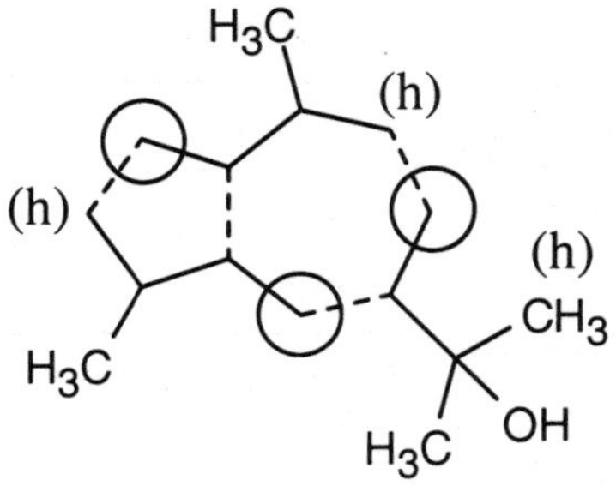

(b)

Sabinene
(4 possible stereoisomers)

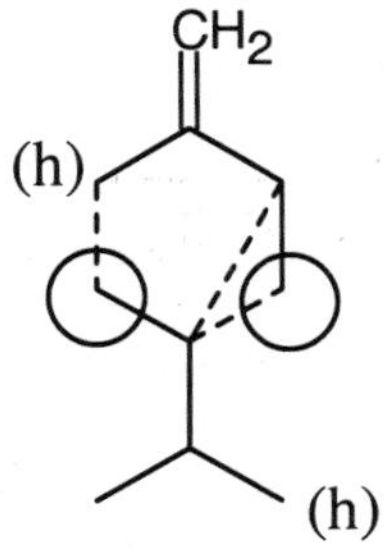

(c)

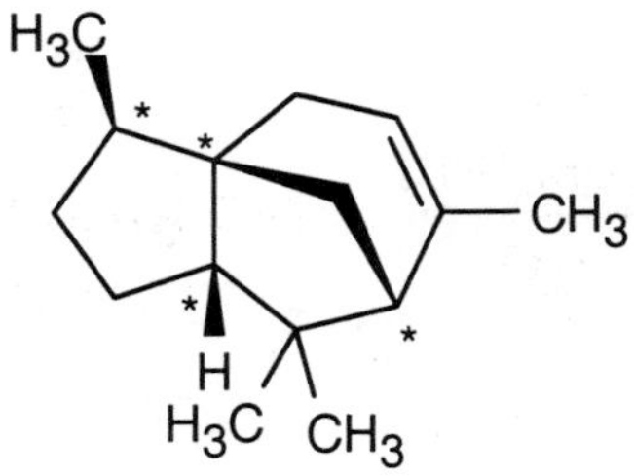

Cedrene
(16 possible stereoisomers)

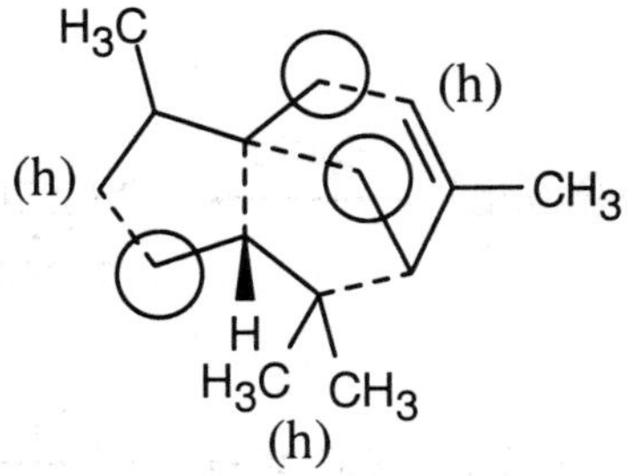

If carbon 1 of each pyrophosphate were isotopically labeled, the labels would appear at the circled positions of the terpenes.

27.25

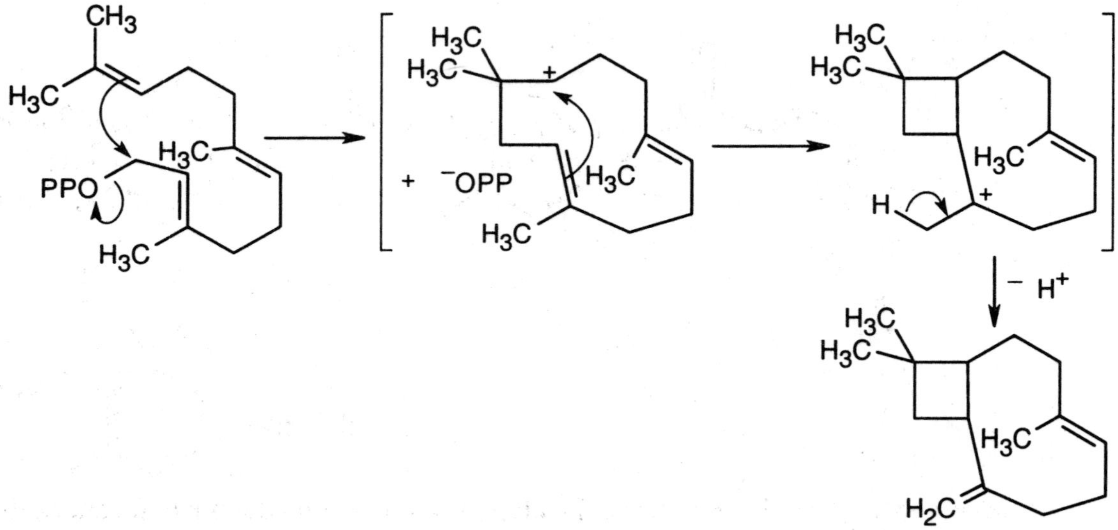

CH_3 H_3C H_3C PPO H_3C → [H_3C H_3C + + ^-OPP H_3C H_3C → H_3C H_3C H_3C H +]

$- H^+$

H_3C H_3C H_3C H_2C

Caryophyllene

Drawing farnesyl pyrophosphate in the correct orientation makes this problem much easier. Internal displacement of pyrophosphate by the electrons of one double bond is followed by attack of the electrons of the second double bond on the resulting carbocation. Loss of a proton from the carbon next to the resulting carbocation produces the double bond.

27.26

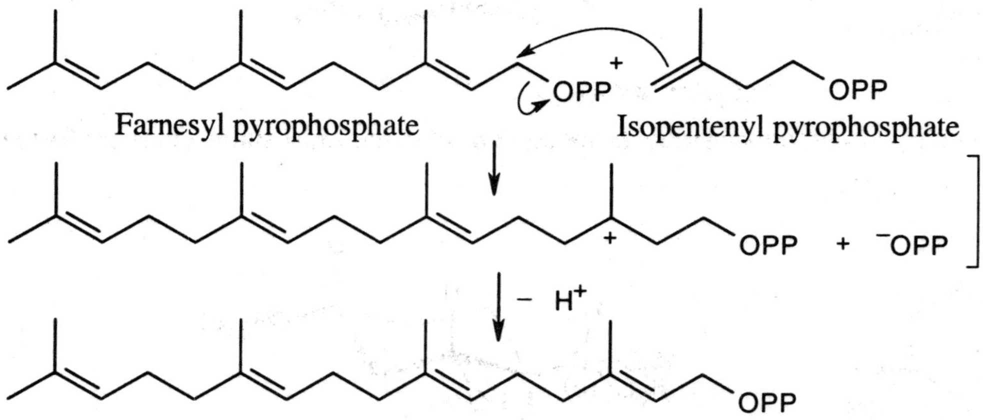

OPP + OPP

Farnesyl pyrophosphate Isopentenyl pyrophosphate

[+ OPP + ^-OPP]

$- H^+$

OPP

The precursor to flexibilene is formed from the reaction of farnesyl pyrophosphate and isopentenyl pyrophosphate.

OPP → [H + + ^-OPP] $-H^+$ →

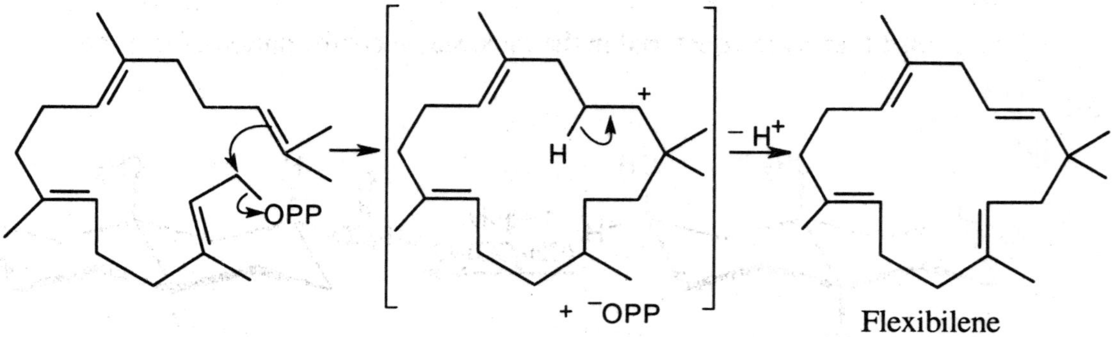

Flexibilene

The precursor cyclizes by the now-familiar mechanism to produce flexibilene.

27.27

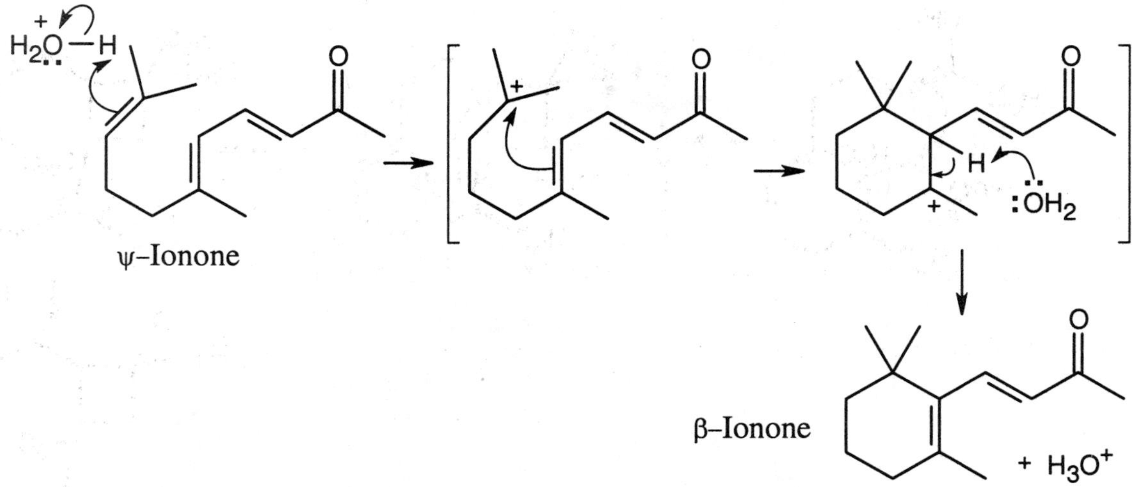

ψ–Ionone

β–Ionone

$+ \ H_3O^+$

Acid protonates a double bond, and the electrons of a second double bond attack the carbocation. Deprotonation yields β-ionone.

27.28

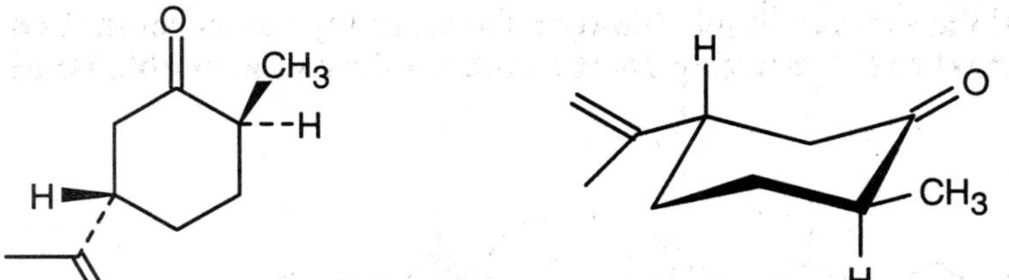

Dihydrocarvone

The two hydrocarbon substituents are equatorial in the most stable chair conformation.

27.29

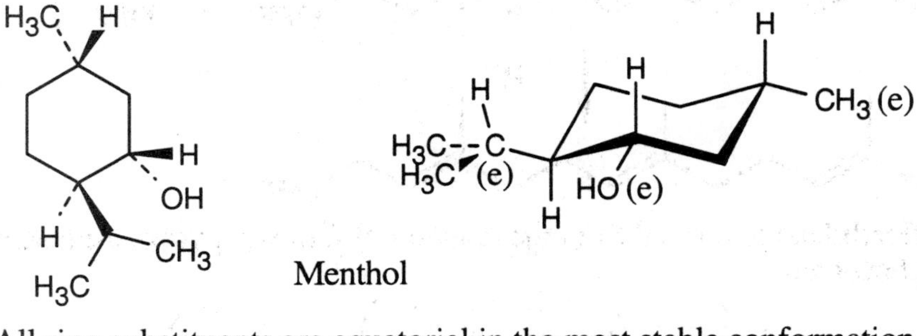

Menthol

All ring substituents are equatorial in the most stable conformation of menthol.

27.30

(a)

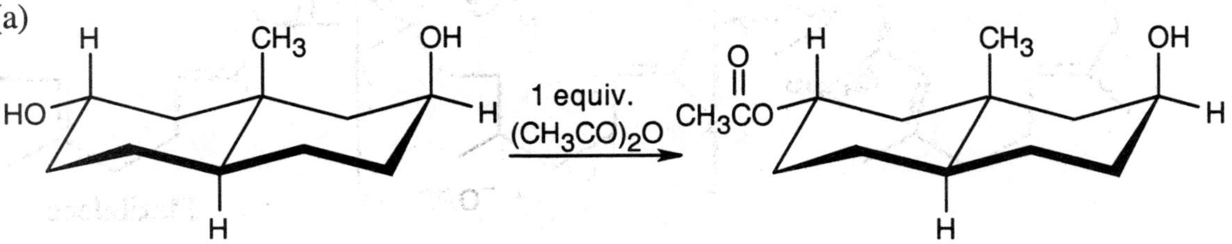

1 equiv. $(CH_3CO)_2O$

(b)

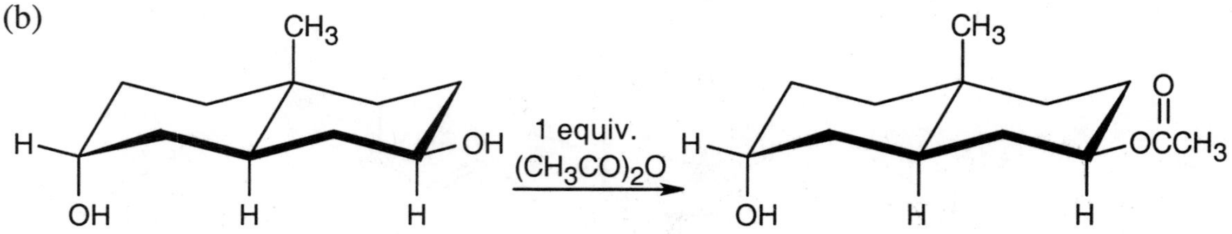

CH3
H
OH
OH
H
H
1 equiv.
(CH3CO)2O
CH3
O
OCCH3
H
OH
H
H

As always, use the stereochemistry of the groups at the ring junction to label the other substituents.

27.31

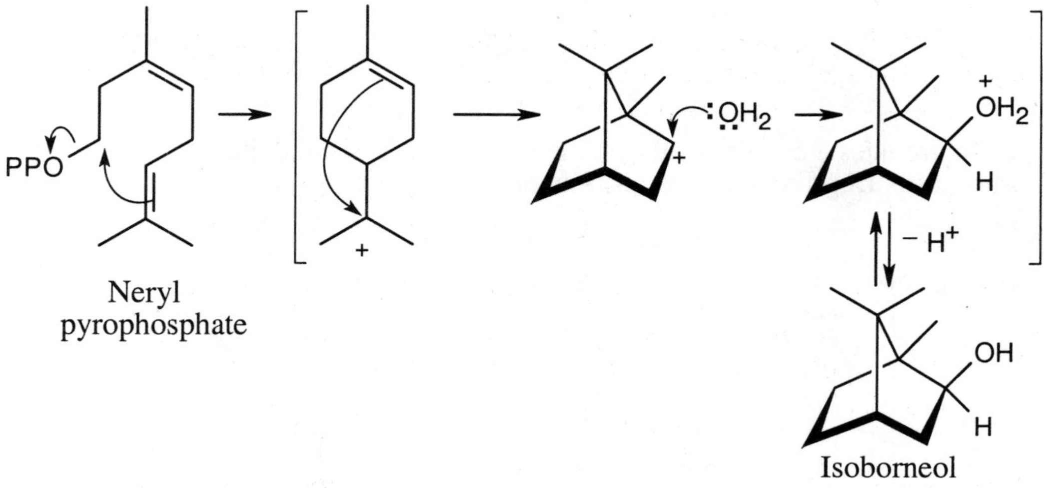

PPO
Neryl
pyrophosphate
+
:OH2
+
+
OH2
H
- H+
OH
H
Isoborneol

The cyclizations result in a secondary carbocation, which reacts with water to yield the secondary alcohol.

27.32

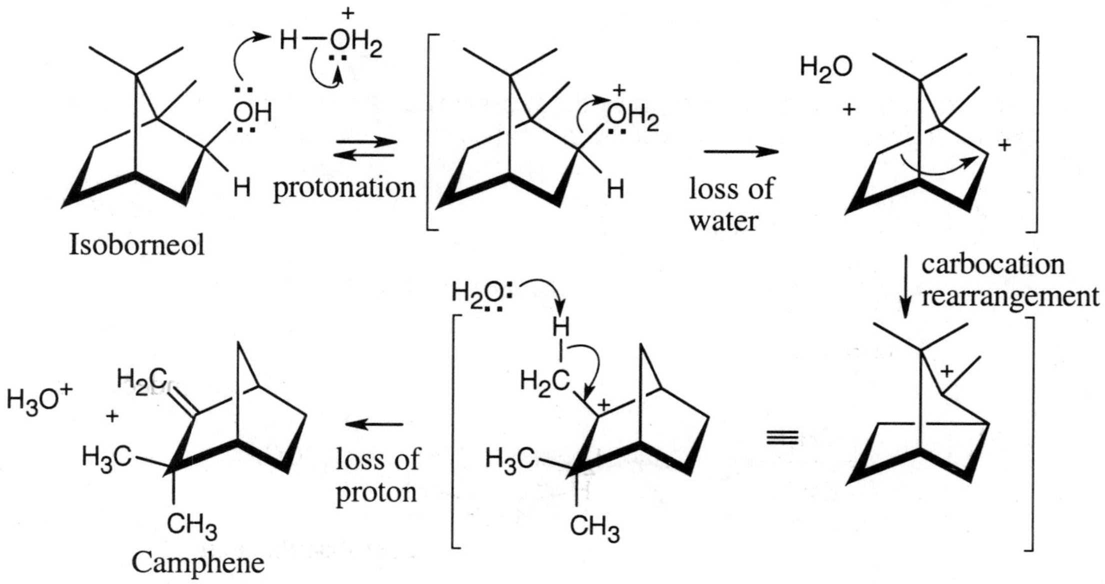

+
H—OH2
OH
H protonation
Isoborneol
+
OH2
H
loss of
water
H2O
+
+
carbocation
rearrangement
+
≡
H2O:
H
H2C
+
H3C
CH3
loss of
proton
H3O+
+
H2C
H3C
CH3
Camphene

The key step is the carbocation rearrangement, which occurs by the migration of one of the ring bonds.

27.33

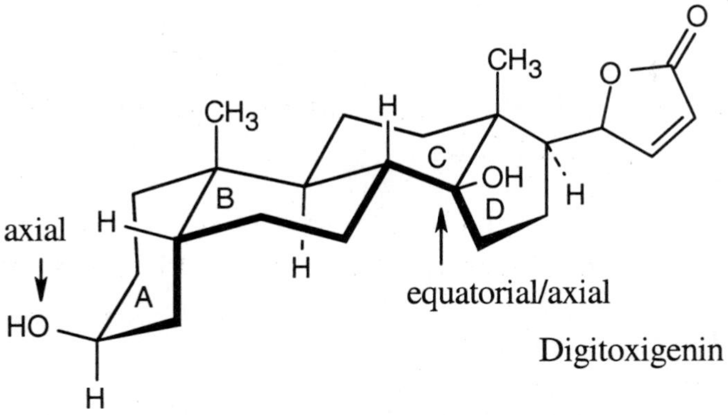

axial

equatorial/axial

Digitoxigenin

The hydroxyl group in ring A is axial, and the hydroxyl group at the ring C-D fusion is equatorial to ring C and axial to ring D. Notice that digitoxigenin has both an A–B cis ring fusion and a C–D cis ring fusion.

27.34

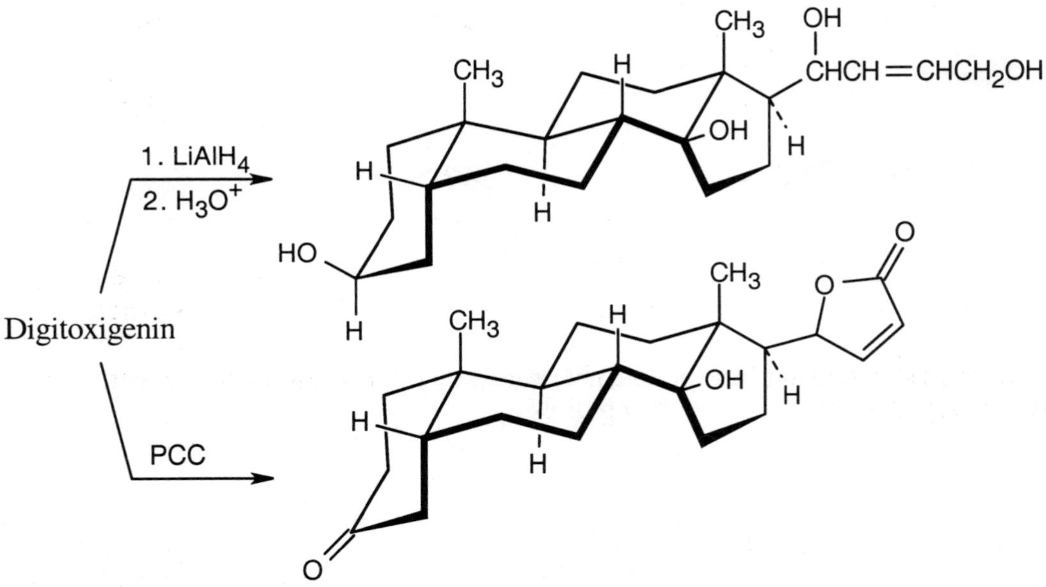

Digitoxigenin

1. $LiAlH_4$
2. H_3O^+

PCC

Lithium aluminum hydride reduces the lactone ring to a diol. PCC oxidizes only one hydroxyl group because the second group is tertiary and isn't oxidized.

27.35

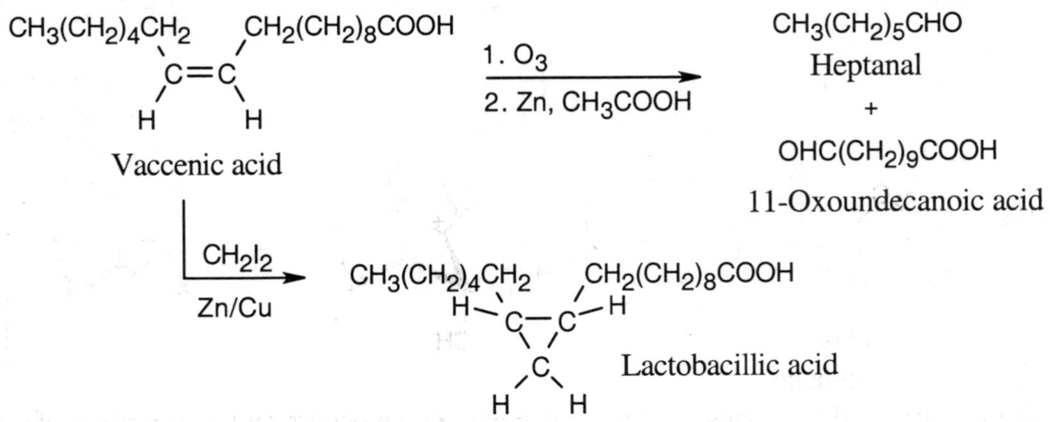

$CH_3(CH_2)_4CH_2CH=CHCH_2(CH_2)_8COOH$

Vaccenic acid

1. O_3
2. Zn, CH_3COOH

$CH_3(CH_2)_5CHO$
Heptanal
+
$OHC(CH_2)_9COOH$
11-Oxoundecanoic acid

CH_2I_2
Zn/Cu

$CH_3(CH_2)_4CH_2$ and $CH_2(CH_2)_8COOH$ on a cyclopropane ring (CH_2)

Lactobacillic acid

27.36

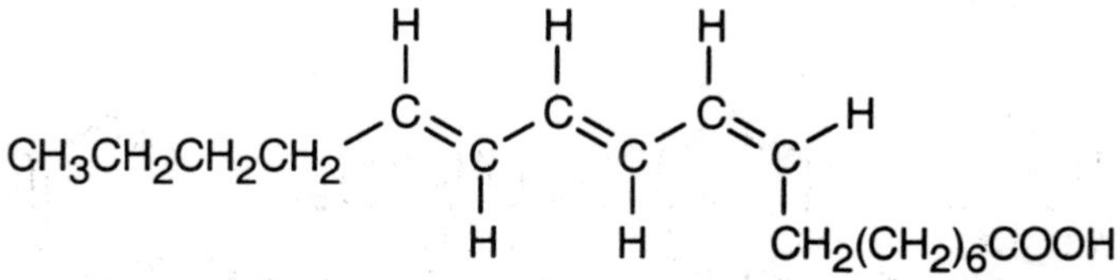

$CH_3CH_2CH_2CH_2$–CH=CH–CH=CH–CH=CH–$CH_2(CH_2)_6COOH$

(9*Z*,11*E*,13*E*)-9,11,13-Octadecatrienoic acid
(Eleostearic acid)

1. O_3
2. Zn, CH_3COOH

$CH_3CH_2CH_2CH_2CHO$ + OHC—CHO + OHC—CHO + $OHC(CH_2)_7COOH$

The stereochemistry of the double bonds can't be determined from the information given.

27.37

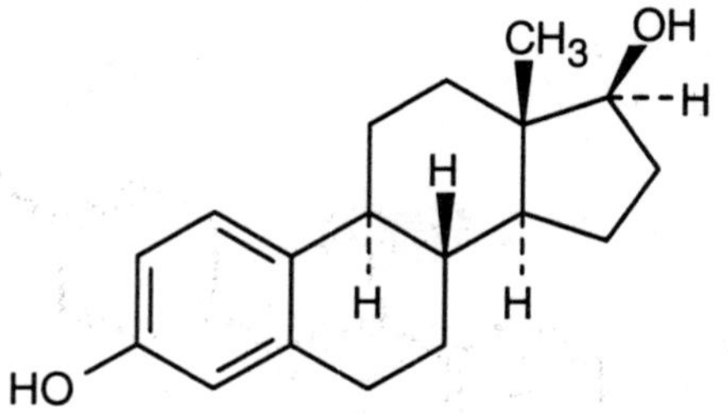

Estradiol

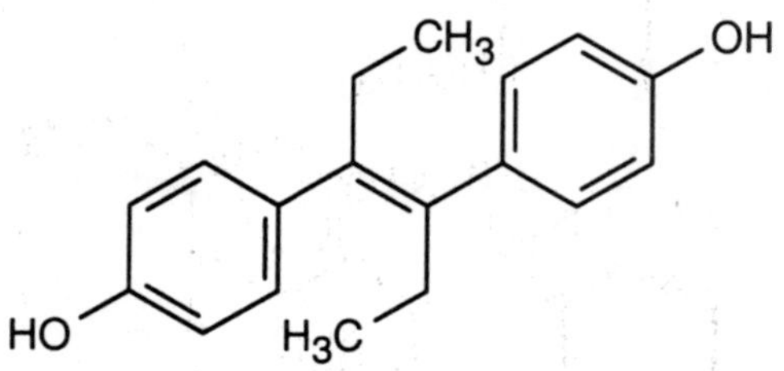

Diethylstilbestrol

Estradiol and diethylstilbestrol resemble each other in having similar carbon skeletons, in having a phenolic ring, and in being diols.

27.38

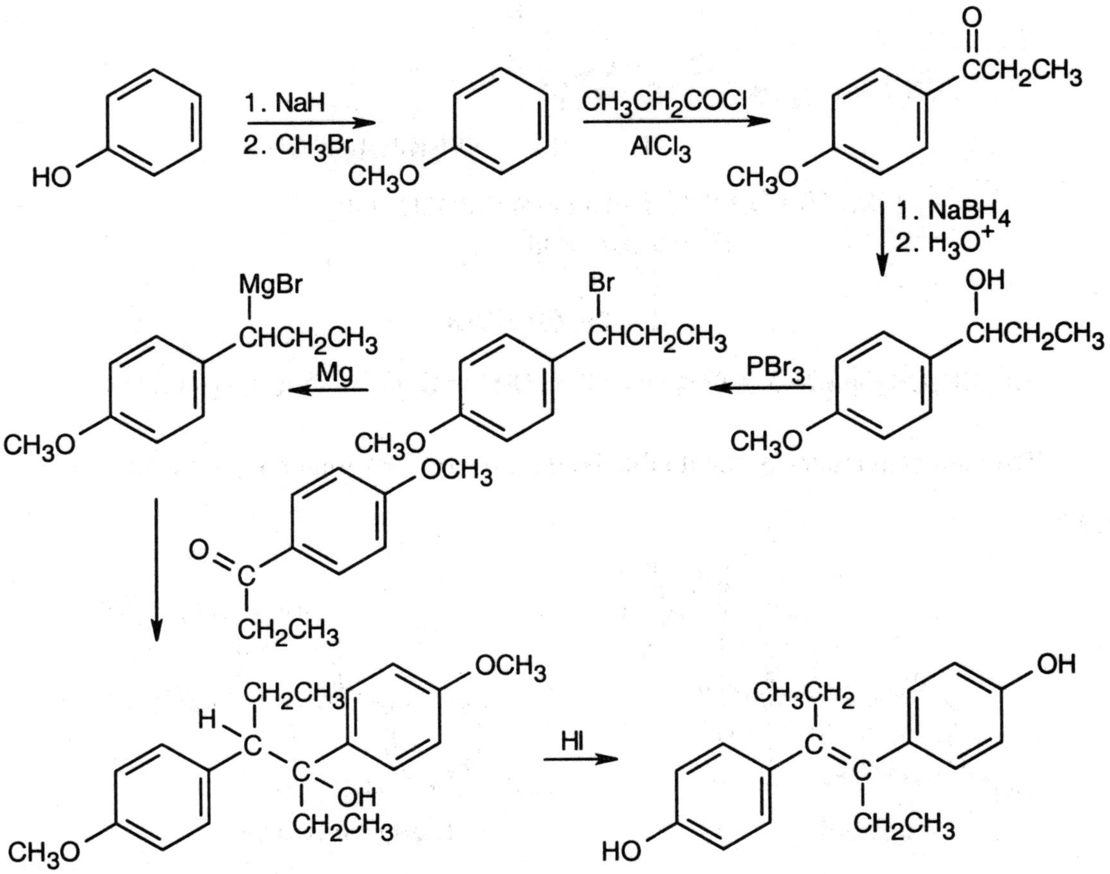

1. NaH
2. CH_3Br

CH_3CH_2COCl
$AlCl_3$

1. $NaBH_4$
2. H_3O^+

PBr_3

Mg

HI

Diethylstilbestrol

Synthetic analysis: The key reaction is a Grignard reaction between two molecules that are both synthesized from phenol. Phenol is first converted to anisole, in order to avoid problems with acidic hydrogens interfering with the Grignard reaction. Next, anisole undergoes Friedel-Crafts acylation with propanoyl chloride.The resulting ketone is one of the Grignard components. The other component is prepared by reduction, bromination and treatment with magnesium of a quantity of the ketone. After the Grignard reaction, HI serves to both dehydrate the alcohol and cleave the methyl ether groups.

27.39

(a)

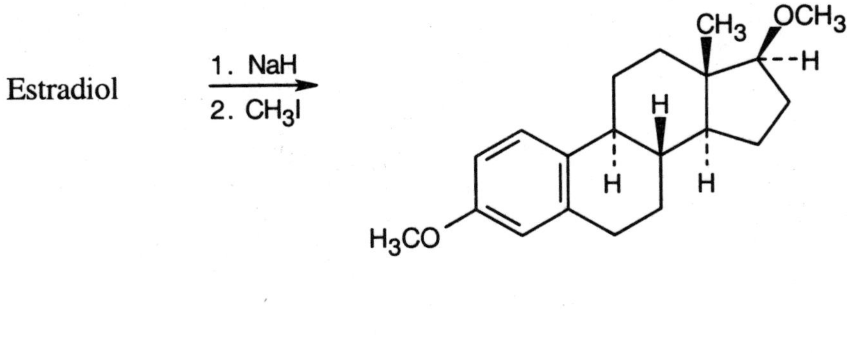

Estradiol

1. NaH
2. CH_3I

(b)

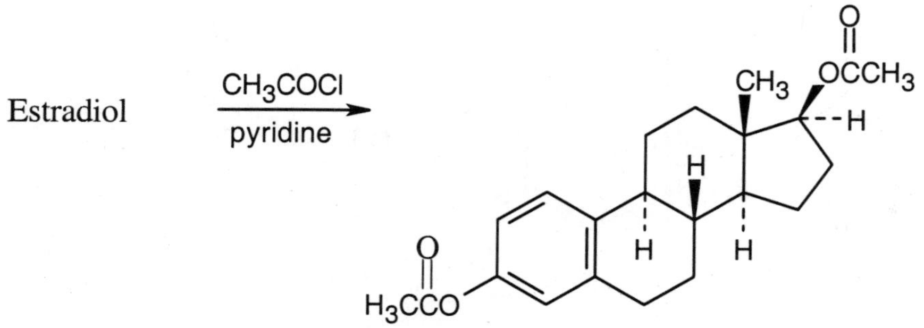

Estradiol $\xrightarrow[\text{pyridine}]{CH_3COCl}$

(c)

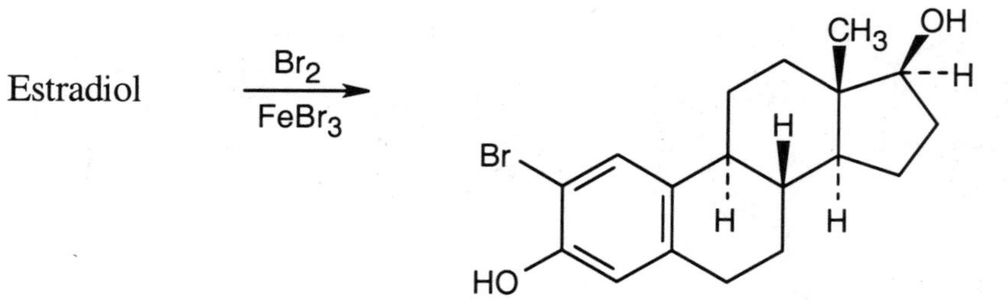

Estradiol $\xrightarrow[FeBr_3]{Br_2}$

(d)

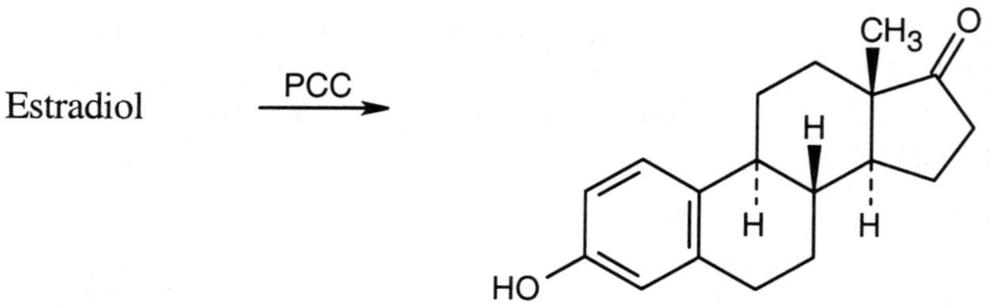

Estradiol $\xrightarrow{PCC}$

27.40

Cembrene $\xrightarrow[\text{2. Zn, } CH_3COOH]{\text{1. } O_3}$

1 equiv H_2

Dihydrocembrene

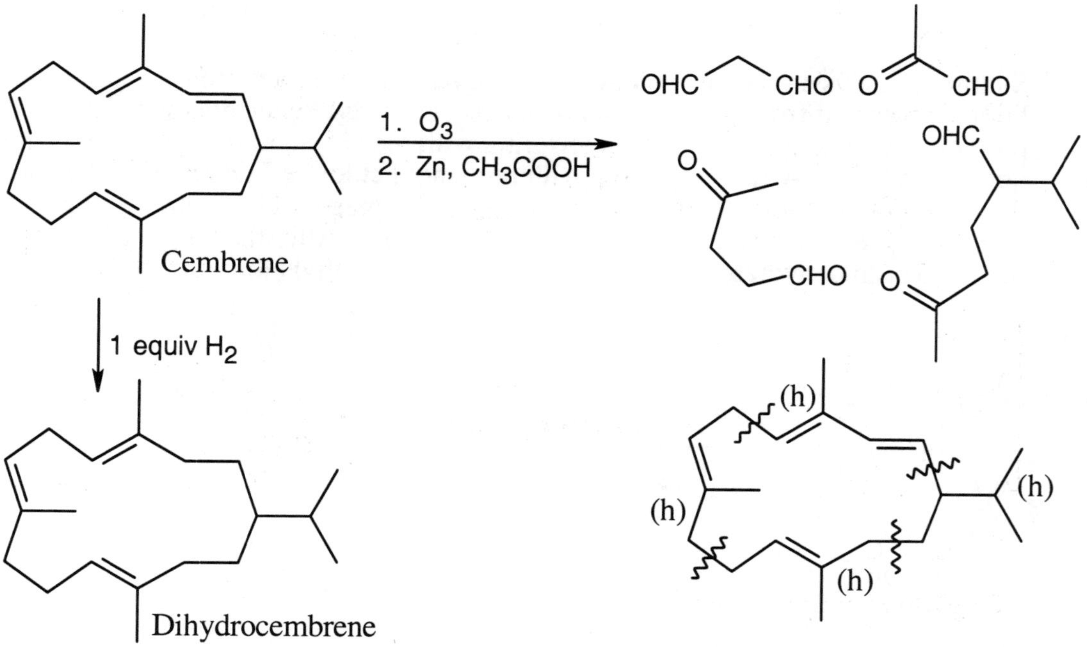

One equivalent of H_2 hydrogenates the least substituted double bond. Dihydrocembrene has no ultraviolet absorption because it is not conjugated. The isoprene units are shown above; "heads" are indicated by (h).

27.41

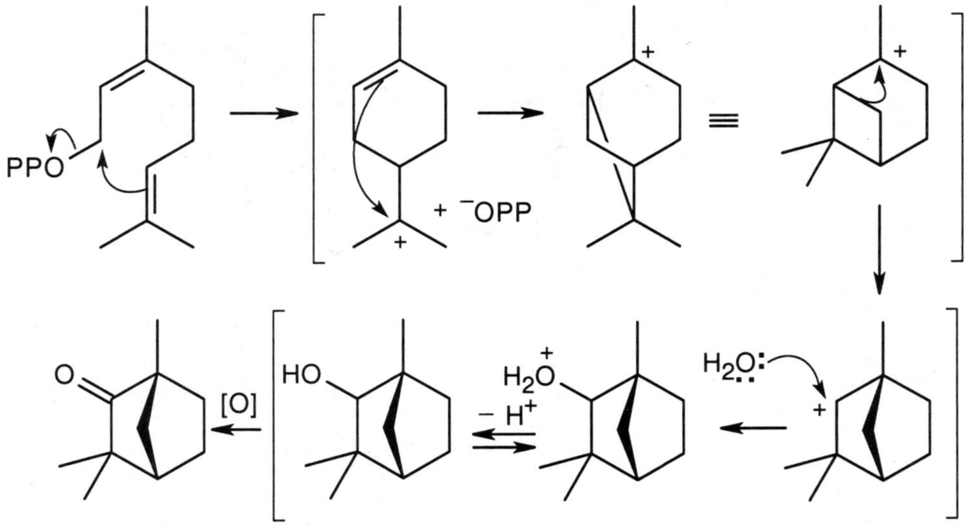
PPO
+ ⁻OPP
[O]
HO
H₂O
− H⁺
H₂O:
O

α-Fenchone

The mechanism follows the usual path: cyclization, followed by attack of the π electrons of the second double bond, produce an intermediate carbocation. A carbocation rearrangement occurs, and the resulting carbocation reacts with water to form an alcohol that is oxidized to give α-fenchone.

A Look Ahead

27.42

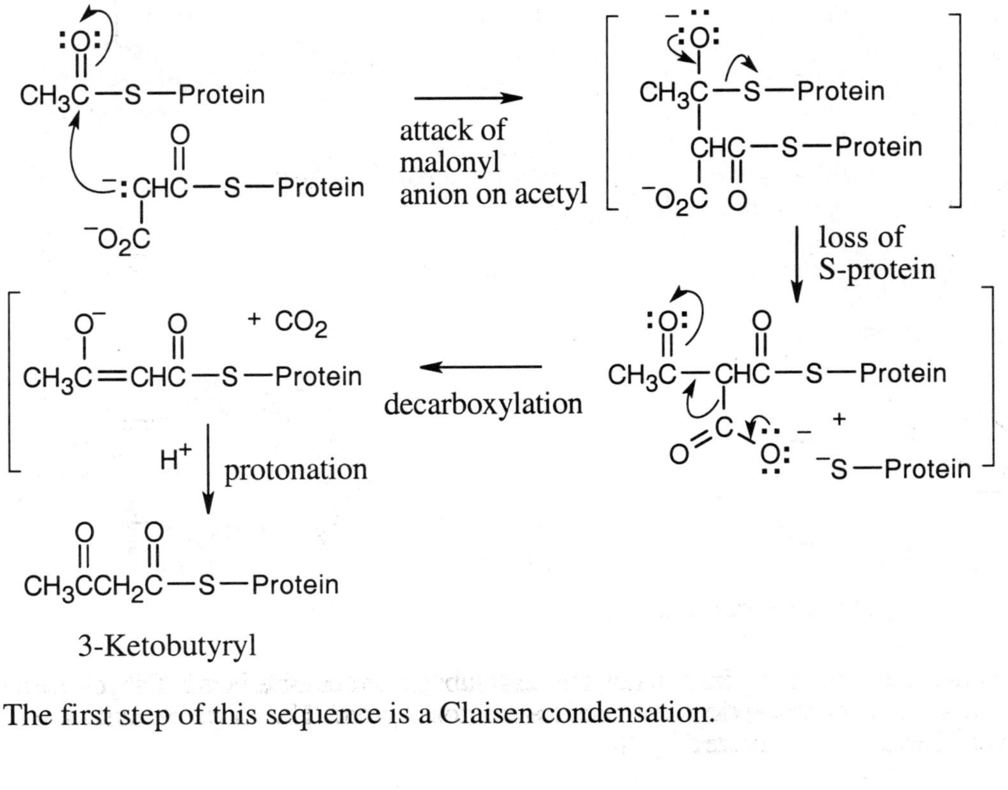
CH3C—S—Protein
:CHC—S—Protein
⁻O2C
attack of
malonyl
anion on acetyl
CH3C—S—Protein
CHC—S—Protein
⁻O2C O
loss of
S-protein
CH3C—CHC—S—Protein
⁻S—Protein
decarboxylation
+ CO2
CH3C=CHC—S—Protein
H⁺
protonation
CH3CCH2C—S—Protein

3-Ketobutyryl

The first step of this sequence is a Claisen condensation.

27.43

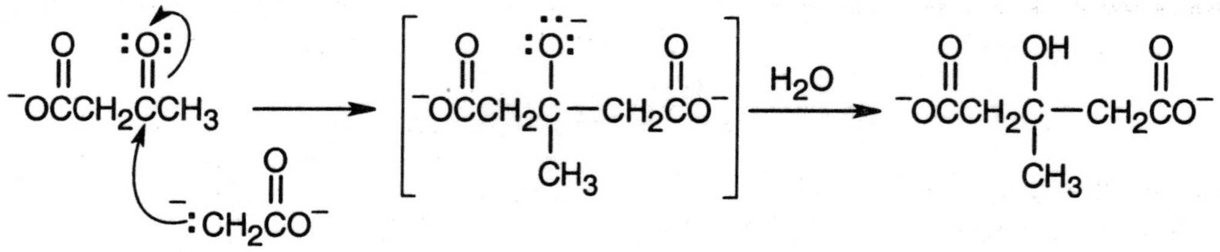

$^{-}OCCH_2CCH_3$ (with C=O groups) + $^{-}:CH_2CO^{-}$ ⟶ $[\,^{-}OCCH_2C(O^{-})(CH_3)-CH_2CO^{-}\,]$ $\xrightarrow{H_2O}$ $^{-}OCCH_2C(OH)(CH_3)-CH_2CO^{-}$

This reaction is an aldol condensation.

Molecular Modeling

27.44

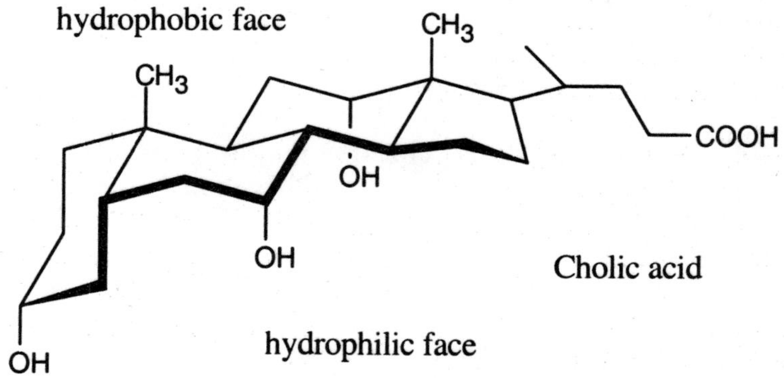

hydrophobic face

CH_3 CH_3 COOH OH OH OH

Cholic acid

hydrophilic face

The hydrophobic and hydrophilic regions correspond to the two "faces" of the ring system. The hydrophilic face is the convex region that includes the three polar hydroxyl groups (the side chain containing the polar carboxylic acid group can also rotate so that this group lies in the hydrophilic region). The other face includes only carbon-hydrogen bonds and is hydrophobic.

27.45 **A** = phospholipid, **B** = steroid, **C** = triacylglycerol, **D** = prostaglandin. **A** and **C** contain polar head groups; **B** and **D** do not.

27.46 The maps of fat-soluble vitamins contain few, and relatively small, polar regions, whereas the water-soluble vitamins are more polar. Vitamins B6 and C are water-soluble, and Vitamins A and E are fat-soluble.

Chapter 28 – Biomolecules: Heterocycles and Nucleic Acids

Chapter Outline

I. Heterocycles (Sections 28.1 – 28.7).
   A. Five-membered unsaturated heterocycles (Sections 28.1 – 28.3).
      1. Types of 5-membered unsaturated heterocycles (Section 28.1).
         a. Pyrrole, thiophene, and furan are the most common examples.
         b. None of these compounds behave as conjugated dienes
         c. All yield products of electrophilic aromatic substitution.
      2. Structures of pyrrole, furan and thiophene (Section 28.2).
         a. All are aromatic because they have six π electrons in a cyclic conjugated system.
         b. Pyrrole is nonbasic because all 5 of its electrons are used in bonding.
         c. The carbon atoms in pyrrole are electron-rich and are reactive toward electrophiles.
      3. Electrophilic substitution reactions (Section 28.3).
         a. All three compounds undergo electrophilic aromatic substitution reactions readily.
         b. Halogenation, nitration, sulfonation and Friedel-Crafts alkylation can take place if reaction conditions are modified.
         c. The reactivity order is furan > pyrrole > thiophene.
         d. Reaction occurs at the 2-position because the reaction intermediate from attack is more stable.
   B. Pyridine (Sections 28.4 – 28.6).
      1. Structure of pyridine (Section 28.4).
         a. Pyridine is the nitrogen-containing analog of benzene.
         b. The nitrogen lone pair isn't part of the π electron system .
         c. Pyridine is a stronger base than pyrrole but a weaker base than alkylamines.
      2. Electrophilic substitution of pyridine (Section 28.5).
         Electrophilic substitutions take place with great difficulty.
         i. The pyridine ring is electron-poor due to the electron-withdrawing inductive effect of nitrogen.
         ii. Acid-base complexation between nitrogen and an electrophile puts a positive charge on the ring.
      3. Nucleophilic substitution of pyridine (Section 28.6).
         a. Nucleophilic substitutions occur with 2- and 4-substituted halopyridines.
         b. These reactions occur by the nucleophilic aromatic substitution mechanism we studied in Chapter 16.
   C. Fused-ring heterocycles (Section 28.7).
      1. The reactivity of fused-ring heterocyclic compounds is related to the type of heteroatom and to the size of the ring.
      2. Purine and pyrimidine are two types of heterocyclic compounds that occur in nucleic acids.
         a. Pyrimidines have two pyridine-like nitrogens in a six-membered ring.
         b. Purines have 4 nitrogens (3 pyridine-like, and one pyrrole-like) in a fused-ring structure.

II. Nucleic acids (Sections 28.8 – 28.10).
   A. Nucleosides and nucleotides (Section 28.8).
      1. Nucleosides are composed of a heterocyclic purine or pyrimidine base plus an aldopentose.

a. In RNA, the purines are adenine and guanine, the pyrimidines are uracil and cytosine, and the sugar is ribose.
b. In DNA, thymine replaces uracil, and the sugar is 2'-deoxyribose.
2. Positions on the base receive non-prime superscripts, and positions on the sugar receive prime superscripts.
3. The heterocyclic base is bonded to C1' of the sugar.
4. In nucleotides, a phosphate group is bonded to C5' of the sugar.

B. Nucleic acids (Sections 28.9 – 28.10).
1. Structure of nucleic acids (Section 28.9).
a. Nucleic acids are composed of nucleotides connected by a phosphate ester bond between the 5' ester and the 3' hydroxyl group.
i. One end of the nucleic acid polymer has a free hydroxyl group and is called the 3' end.
ii. The other end has a free phosphate group and is called the 5' end.
b. The structure of a nucleic acid depends on the order of bases.
c. The sequence of bases is described by starting at the 5' end and listing the bases by their one-letter abbreviations.
d. DNA is vastly larger than RNA and is found in the cell nucleus.
2. Base-pairing in DNA (Section 28.10).
a. DNA consists of two polynucleotide strands coiled in a double helix.
Adenine and thymine hydrogen-bond with each other, and cytosine and guanine hydrogen-bond with each other.
b. Because the two DNA strands are complementary, the amount of A equals the amount of T, and the amount of C equals the amount of G.
c. The double helix is 2.0 nm wide, there are 10 bases in each turn, and each turn is 3.4 nm in height.
d. The double helix has a major groove and a minor groove into which polycyclic aromatic molecules can intercalate.

III. The transfer of genetic information (Sections 28.11 – 28.14).

A. The "central dogma" of molecular genetics (Section 28.11).
1. The function of DNA is to store genetic information and to pass it on to RNA, which uses it to make proteins.
2. Replication, transcription and translation are the three processes that are responsible for carrying out the central dogma.

B. Replication of DNA (Section 28.12).
1. Replication is the enzyme-catalyzed process whereby DNA makes a copy of itself.
2. Replication is semiconservative: each new strand of DNA consists of one old strand and one newly synthesized strand.
3. How replication occurs:
a. The DNA helix partially unwinds.
b. New nucleotides form base-pairs with their complementary partners.
c. Formation of new bonds is catalyzed by DNA polymerase and takes place in the 5' $\rightarrow$ 3' direction.
Bond formation occurs by attack of the 3' hydroxyl group on the 5' triphosphate, with loss of a diphosphate leaving group.
d. Both new chains are synthesized in the 5' $\rightarrow$ 3' direction.
i. One chain is synthesized continuously.
ii. The other strand is synthesized in small pieces, which are later joined by DNA ligase enzymes.

C. Transcription (Section 28.13).
1. There are 3 types of RNA:
a. Messenger RNA (mRNA) carries genetic information to ribosomes when protein synthesis takes place.

b. Ribosomal RNA (rRNA), complexed with protein, comprises the physical makeup of the ribosomes.
c. Transfer RNA (tRNA) brings amino acids to the ribosomes, where they are joined to make proteins.
2. mRNA is synthesized in the nucleus by transcription of DNA.
a. The DNA partially unwinds, forming a "bubble".
b. Ribonucleotides form base pairs with their complementary DNA bases.
c. Bond formation occurs in the 5' $\rightarrow$ 3' direction.
d. Only one strand of DNA (the template strand) is transcribed.
e. Thus, the synthesized mRNA is a copy of the coding strand (with U replacing T).
3. DNA contains "promoter sites", which indicate where mRNA synthesis is to begin, and base sequences that indicate where mRNA synthesis stops.
4. Synthesis of mRNA is not necessarily continuous.
a. Often, synthesis begins in a region of DNA called an exon and is interrupted by a seemingly nonsensical region of DNA called an intron.
b. In the final mRNA, the nonsense sections have been removed and the remaining pieces have been spliced together.

D. Translation (Section 28.14).
1. Translation is the process in which proteins are synthesized at the ribosomes by using mRNA as a template.
2. The message delivered by mRNA is contained in "codons" – 3-base groupings that are specific for an amino acid.
a. Amino acids are coded by 61 of the possible 64 codons.
b. The other 3 codons are "stop" codons.
3. Each tRNA is responsible for bringing an amino acid to the growing protein chain.
a. A tRNA has a cloverleaf-shaped secondary structure and consists of 70–100 ribonucleotides.
b. Each tRNA contains an anticodon complementary to the mRNA codon.
4. The protein chain is synthesized by enzyme-catalyzed peptide bond formation.
5. A 3-base "stop" codon on mRNA signals when synthesis is complete.

IV. DNA technology (Sections 28.15 – 28.17).
A. DNA sequencing (Section 28.15).
1. Maxam–Gilbert DNA sequencing.
a. Step 1: The DNA chain is cleaved at specific sites by restriction endonucleases.
i. The restriction endonuclease recognizes both a sequence on the coding strand and its complement on the template strand.
ii. The DNA strand is cleaved by several different restriction endonucleases, to produce fragments that overlap those from a different cleavage.
b. Step 2: Each restriction fragment is labelled at the 5' terminal end by incorporation of a ^{32}P-tagged phosphate group.
The labelled fragments are heated to separate the two strands of DNA.
c. Step 3: Each fragment is subjected to 4 sets of chemical reactions that cause:
i. Splitting of the DNA chain next to A, by treatment of the fragment with dimethyl sulfide, then piperidine.
ii. Splitting of the DNA chain next to G under conditions similar to those for splitting next to A.
iii. Splitting of the DNA chain next to both T and C, by treatment with hydrazine, then piperidine.
iv. Splitting of the DNA chain next to C, under similar conditions.
v. Reaction conditions are such that only a few of the many possible splittings occur.
vi. Only fragments with a ^{32}P label are important for sequencing.

d. Step 4: Fragments are separated by gel electrophoresis, and the products are located by autoradiography.
e. Step 5: The DNA sequence is read from the gel.
    i. The smallest fragments are farthest from the origin.
    ii. The sequence can be checked by determining the sequence of the complementary strand.
2. Sanger dideoxy DNA sequencing.
    a. The following mixture is assembled:
        i. The restriction fragment to be sequenced.
        ii. A primer (a small piece of DNA whose sequence is complementary to that on the 3' end of the fragment).
        iii. The 4 DNA nucleotide triphosphates.
        iv. Small amounts of the four dideoxynucleotide triphosphates, each of which is labeled with a different fluorescent dye.
    b. DNA polymerase is added to the mixture, and a strand begins to grow from the end of the primer.
    c. Whenever a dideoxynucleotide is incorporated, chain growth stops.
    d. When reaction is complete, the fragments are separated by gel electrophoresis.
    e. Because fragments of all possible lengths are represented, the sequence can be read by noting the color of fluorescence of each fragment.

B. DNA synthesis (Section 28.16).
1. DNA synthesis is based on principles similar to those for protein synthesis.
2. The following steps are needed:
    a. The nucleosides are protected and bound to a silica support.
        i. Adenine and cytosine bases are protected by benzoyl groups.
        ii. Guanine is protected by an isobutyryl group.
        iii. Thymine isn't protected.
        iv. The 5' –OH group is protected as a DMT ether.
    b. The DMT group is removed.
    c. The polymer-bound nucleoside is coupled with a protected nucleoside containing a phosphoramidite group.
        i. One of the phosphoramidite oxygens is protected as a β-cyano ether.
        ii. Tetrazole catalyzes the coupling.
    d. The phosphite is oxidized to a phosphate with I_2.
    e. Steps b – d are repeated until the desired chain is synthesized.
    f. All protecting groups are removed and the bond to the support is cleaved by treatment with aqueous ammonia.

C. The polymerase chain reaction (Section 28.17).
1. The polymerase chain reaction (PCR) can produce vast quantities a DNA fragment.
2. The key to PCR is *Taq* DNA polymerase, a heat-stable enzyme.
3. Steps in PCR:
    a. The following mixture is heated to 95°C (a temperature at which DNA becomes single-stranded);
        i. *Taq* polymerase.
        ii. Mg^{2+} ion.
        iii. The 4 deoxynucleotide triphosphates.
        iv. A large excess of two oligonucleotide primers, each of which is complementary to the ends of the fragment to be synthesized.
    b. The temperature is lowered to 37°C–50°C, causing the primers to hydrogen-bond to the single-stranded DNA.
    c. After raising the temperature to 72°C, *Taq* catalyzes the addition of further nucleotides, yielding two copies of the original DNA.
    d. The process is repeated until the desired quantity of DNA is produced.

Solutions to Problems

28.1

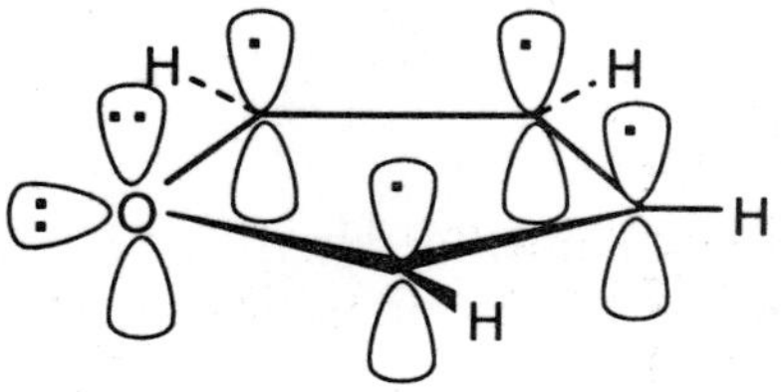

Furan

One oxygen lone pair is in a p orbital that is part of the π electron system of furan. The other oxygen lone pair is in an sp^2 orbital that lies in the plane of the furan ring.

28.2

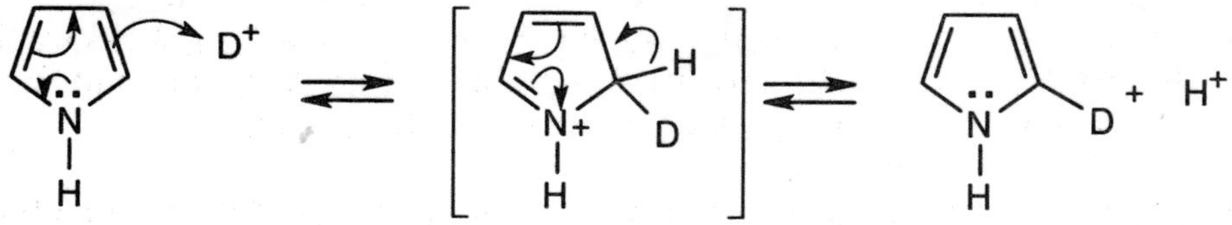

28.3

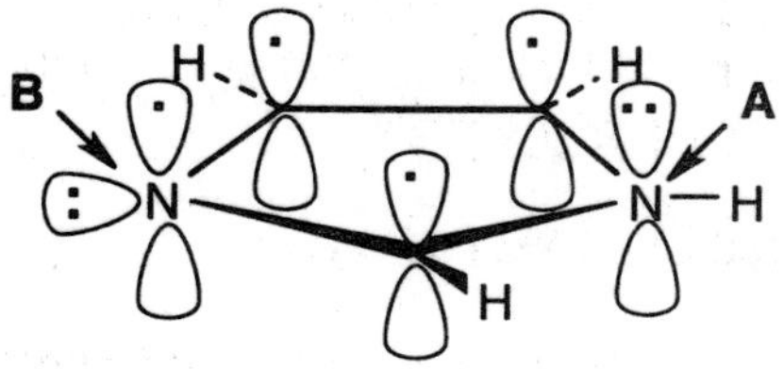

Imidazole

Nitrogen atom **B** is more basic than **A** because its lone pair of electrons lies in an sp^2 orbital and is more available for donation to a Lewis acid than the lone pair of electrons of nitrogen **A**, which is part of the ring π system.

28.4

Attack at C2:

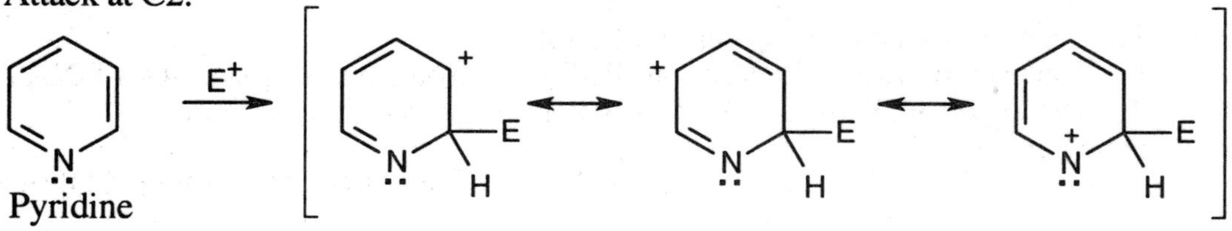

Attack at C3:

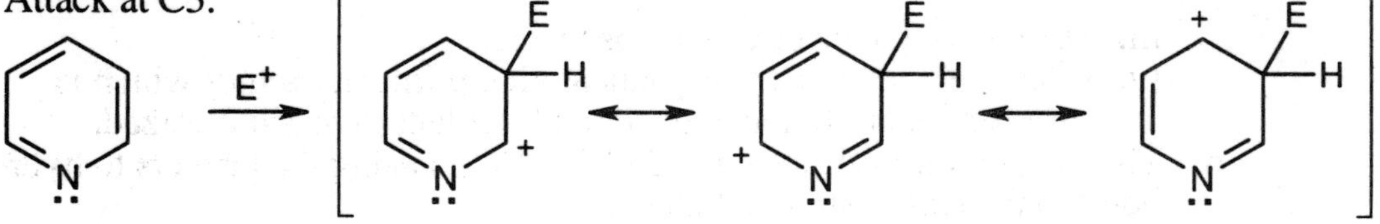

Attack at C4:

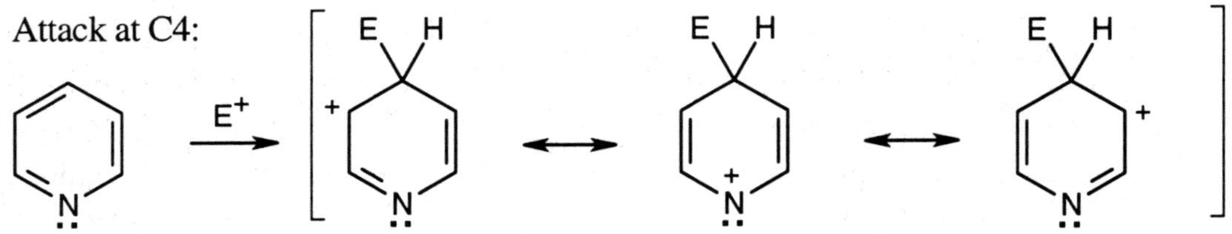

C3 attack is favored over C2 or C4 attack. The positive charge of the cationic intermediate of C3 attack is delocalized onto three carbon atoms, rather than onto two carbons and the electronegative pyridine nitrogen.

28.5

Attack at C3:

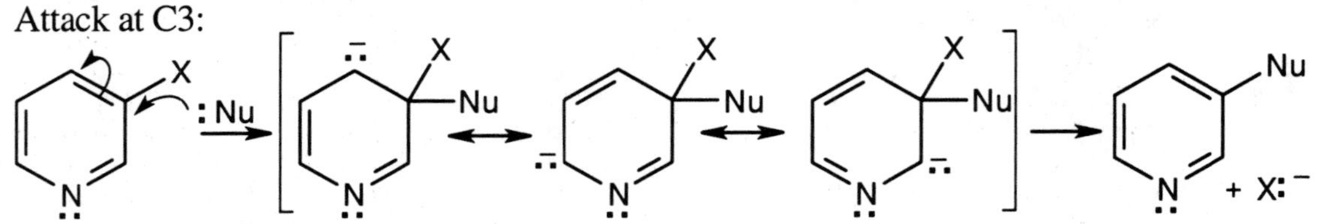

Attack at C4:

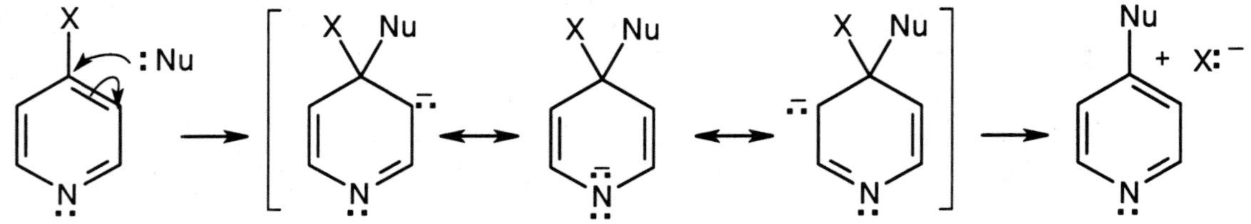

The negative charge resulting from C4 attack can be stabilized by nitrogen. Since no such stabilization is possible for C3 attack, reaction at C3 doesn't occur.

28.6

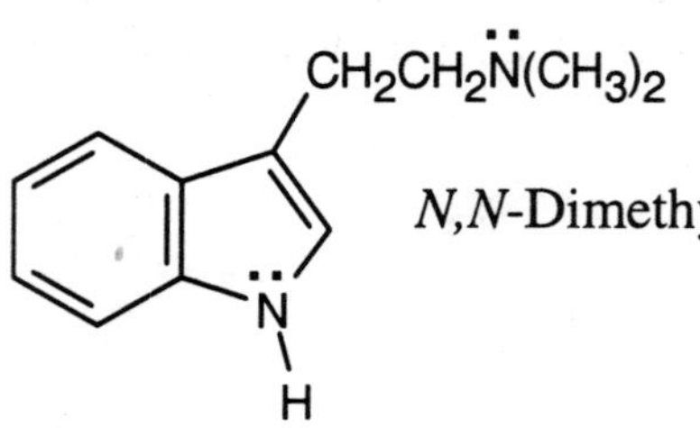

$CH_2CH_2\ddot{N}(CH_3)_2$

N,N-Dimethyltryptamine

The aliphatic nitrogen atom of *N,N*–dimethyltryptamine is more basic than the ring nitrogen atom because its lone electron pair is more available for donation to a Lewis acid. The aromatic nitrogen electron lone pair is part of the ring π electron system.

28.7

Attack at C2:

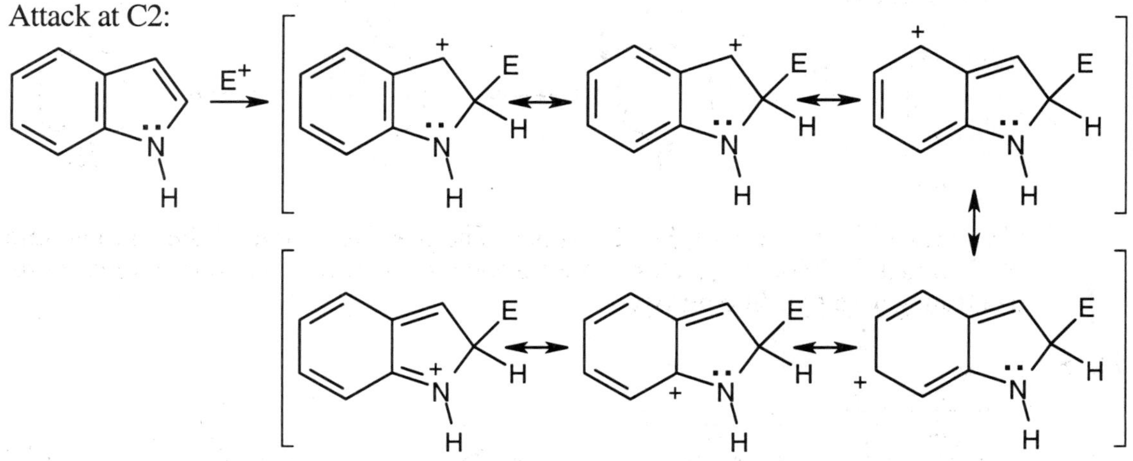

E+

Attack at C3:

E+

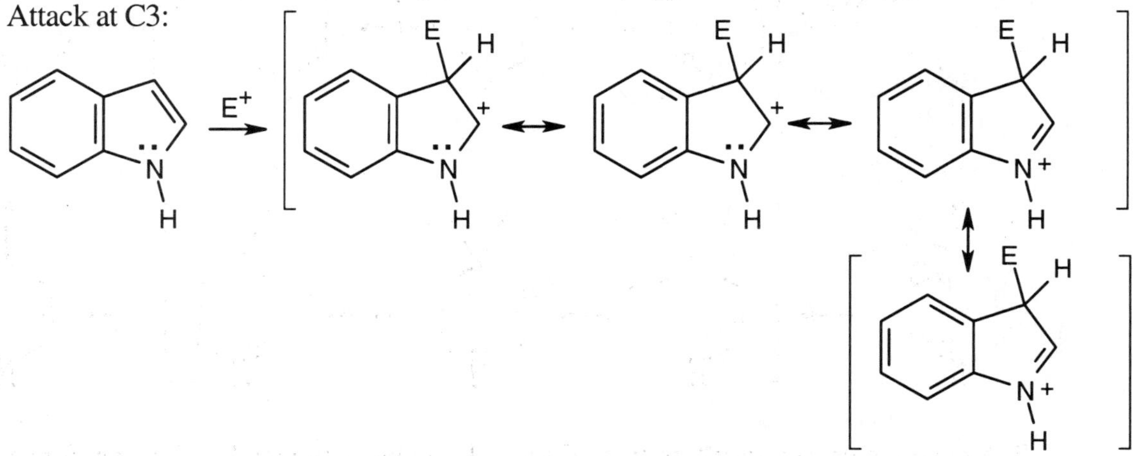

Positive charge can be stabilized by the nitrogen lone pair electrons in both C2 and C3 attack. In C2 attack, however, stabilization by nitrogen destroys the aromaticity of the fused benzene ring. Reaction at C3 is favored, even though the cationic intermediate has fewer resonance forms, because the aromaticity of the six-membered-ring is preserved in the most favored resonance form.

28.8

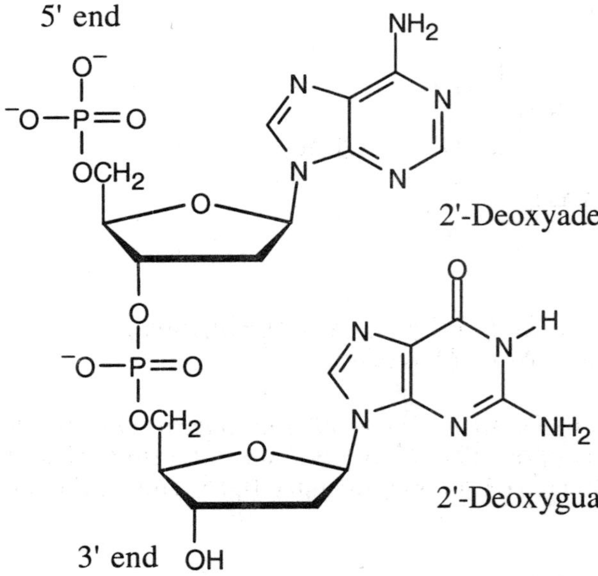

5' end

2'-Deoxyadenosine 5'-phosphate (A)

2'-Deoxyguanosine 5'-phosphate (G)

3' end

28.9

5' end

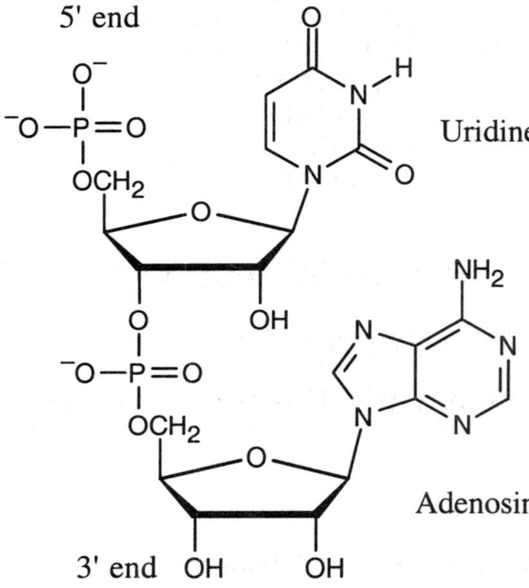

Uridine 5'-phosphate (U)

Adenosine 5'-phosphate (A)

3' end

28.10 DNA (5' end) G–G–C–T–A–A–T–C–C–G–T (3' end) is complementary to DNA (3' end) C–C–G–A–T–T–A–G–G–C–A (5' end)

Remember that the complementary strand has the 3' end on the left and the 5' end on the right.

28.11

Uracil 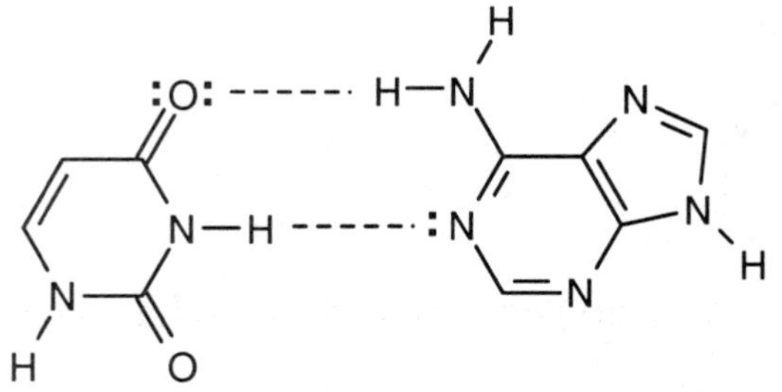 Adenine

28.12 DNA (5' end) G–A–T–T–A–C–C–G–T–A (3' end) is complementary to
RNA (3' end) C–U–A–A–U–G–G–C–A–U (5' end)

28.13–28.14 Several different codons can code for the same amino acid. The corresponding anticodon follows the slash mark after each codon. The mRNA codons are written with the 5' end on the left and the 3' end on the right, and the tRNA anticodons have the 3' end on the left and the 5' end on the right.

Amino acid:	Ala	Phe	Leu	Tyr
Codon sequence/ tRNA anticodon:	GCU/CGA	UUU/AAA	UUA/AAU	UAU/AUA
	GCC/CGG	UUC/AAG	UUG/AAC	UAC/AUG
	GCA/CGU		CUU/GAA	
	GCG/CGC		CUC/GAG	
			CUA/GAU	
			CUG/GAC	

28.15–28.16

The mRNA base sequence: (5' end) CUU–AUG–GCU–UGG–CCC–UAA (3' end)

The amino acid sequence: Leu—Met—Ala—Trp—Pro–(stop)

The DNA sequence: (template strand) (3' end) GAA–TAC–CGA–ACC–GGG–ATT (5' end)

28.17 Remember:

1. Only a few of the many possible splittings occur in each reaction.

2. Cleavage occurs at both sides of the reacting nucleotide.

(5' end) ^{32}P–A–A–C–A–T–G–G–C–G–C–T–T–A–T–G–A–C–G–A (3' end)

Reaction *Fragments*

(a) A
^{32}P
^{32}P–A
^{32}P–A–A–C
^{32}P–A–A–C–A–T–G–G–C–G–C–T–T
^{32}P–A–A–C–A–T–G–G–C–G–C–T–T–A–T–G
^{32}P–A–A–C–A–T–G–G–C–G–C–T–T–A–T–G–A–C–G
^{32}P–A–A–C–A–T–G–G–C–G–C–T–T–A–T–G–A–C–G–A

(b) G

^{32}P–A–A–C–A–T
^{32}P–A–A–C–A–T–G
^{32}P–A–A–C–A–T–G–G–C
^{32}P–A–A–C–A–T–G–G–C–G–C–T–T–A–T
^{32}P–A–A–C–A–T–G–G–C–G–C–T–T–A–T–G–A–C
^{32}P–A–A–C–A–T–G–G–C–G–C–T–T–A–T–G–A–C–G–A

(c) C

^{32}P–A–A
^{32}P–A–A–C–A–T–G–G
^{32}P–A–A–C–A–T–G–G–C–G
^{32}P–A–A–C–A–T–G–G–C–G–C–T–T–A–T–G–A
^{32}P–A–A–C–A–T–G–G–C–G–C–T–T–A–T–G–A–C–G–A

(d) C + T

^{32}P–A–A
^{32}P–A–A–C–A
^{32}P–A–A–C–A–T–G–G
^{32}P–A–A–C–A–T–G–G–C–G
^{32}P–A–A–C–A–T–G–G–C–G–C
^{32}P–A–A–C–A–T–G–G–C–G–C–T
^{32}P–A–A–C–A–T–G–G–C–G–C–T–T–A
^{32}P–A–A–C–A–T–G–G–C–G–C–T–T–A–T–G–A
^{32}P–A–A–C–A–T–G–G–C–G–C–T–T–A–T–G–A–C–G–A

28.18

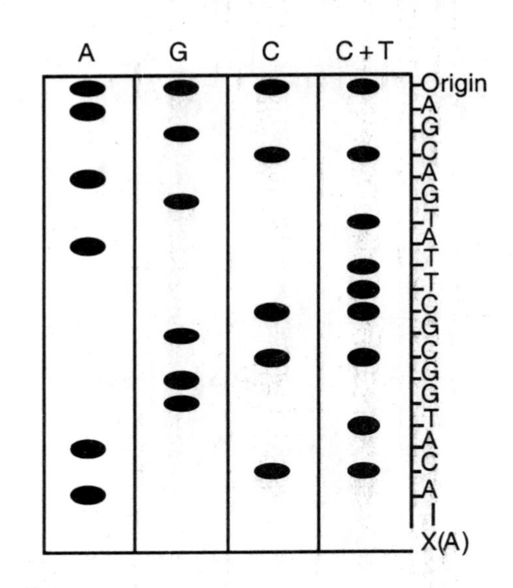

28.19 (5′ end) X–A–T–C–A–G–C–G–A–T–T–C–G–G–T–A–C (3′ end)

28.20

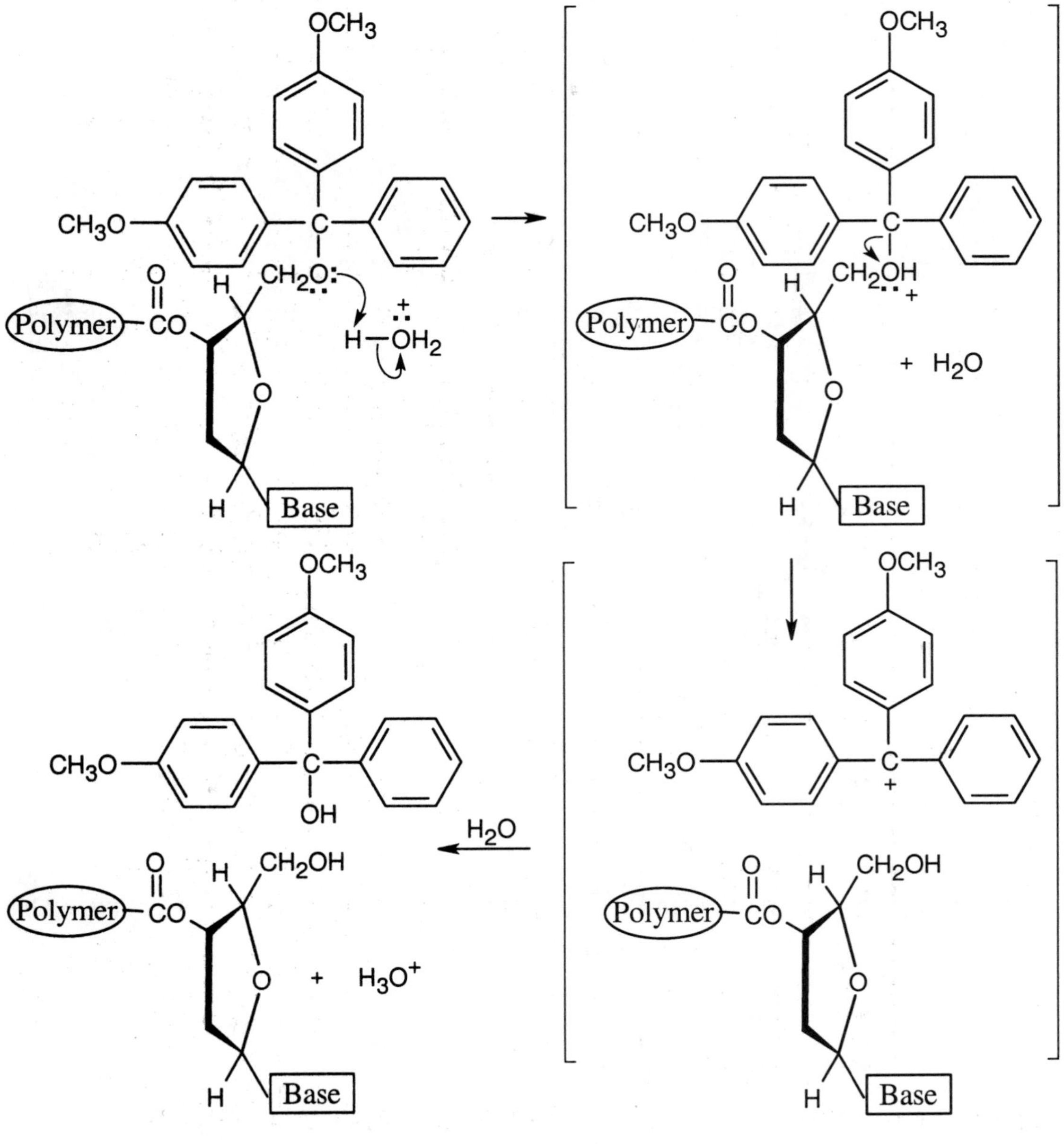

Cleavage of DMT ethers proceeds by an S_N1 mechanism and is rapid because the DMT cation is unusually stable.

28.21

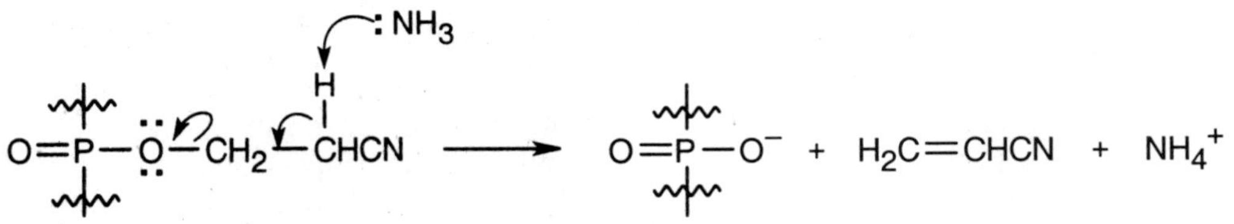

This is an E2 elimination reaction, which proceeds easily because the hydrogen α to the nitrile group is acidic.

Visualizing Chemistry

28.22

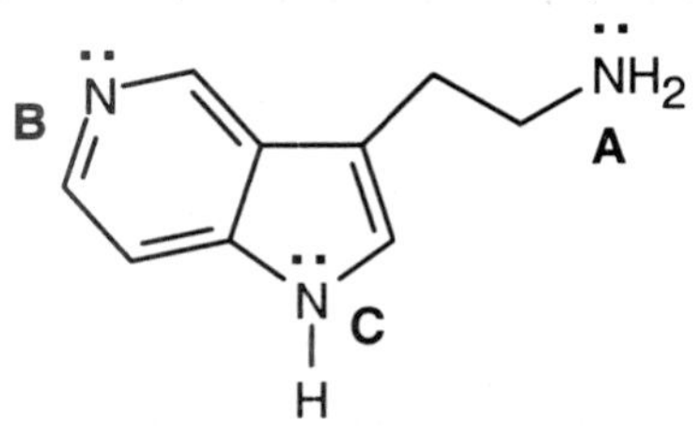

Nitrogen **A** is the most basic because it is an alkylamine nitrogen. Nitrogen **B** is less basic than **A** because its electron lone pair is held closer to the nucleus and is less available for bonding than the lone pair of **A**. Nitrogen **C** is nonbasic because its electron lone pair is part of the fused-ring aromatic system and is unavailable for bonding.

28.23

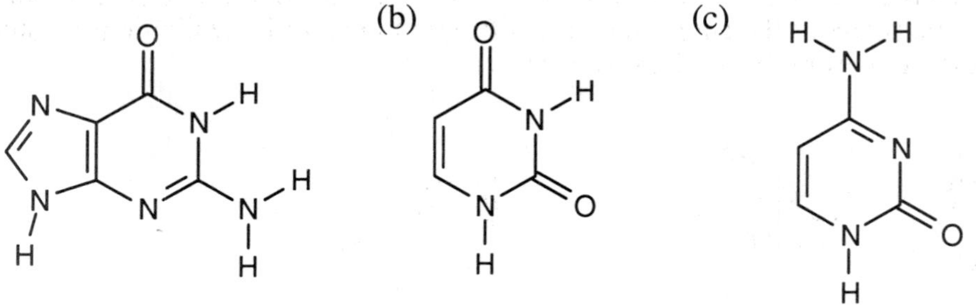

(a) Guanine (G)
DNA
RNA

(b) Uracil (U)
RNA

(c) Cytosine (C)
DNA
RNA

All three bases are found in RNA, but only guanine and cytosine are found in DNA.

28.24

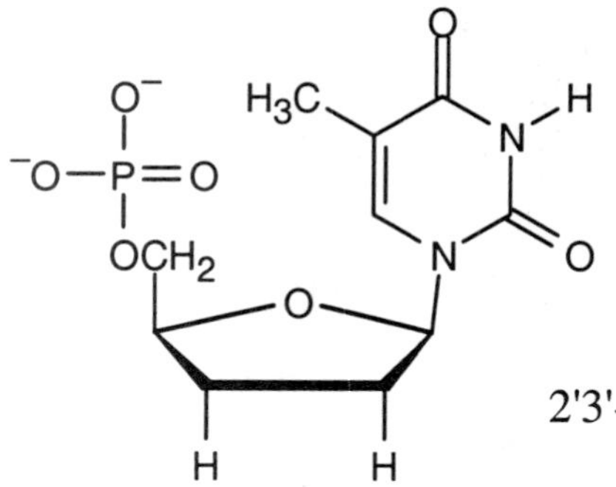

2'3'-Dideoxythymidine 5'-phosphate

The triphosphate made from 2'3'-dideoxythymidine 5' phosphate is labeled with a fluorescent dye and used in the Sanger method of DNA sequencing. Along with the restriction fragment to be sequenced, a DNA primer, and a mixture of the four dNTPs, small quantities of the four labeled dideoxyribonucleotide triphosphates are mixed together. DNA polymerase is added, and a strand of DNA complementary to the restriction fragment is synthesized. Whenever a dideoxyribonucleotide is incorporated into the DNA chain, chain growth stops. The fragments are separated by electrophoresis, and each terminal dideoxynucleotide can be identified by the color of its fluorescence. By identifying these terminal dideoxynucleotides, the sequence of the restriction fragment can be read.

Additional Problems

28.25 The pyrrole anion, $C_4H_4N:^-$, is a 6 π electron species that has the same electronic structure as the cyclopentadienyl anion. Both of these anions possess the aromatic stability of 6 π electron systems.

28.26

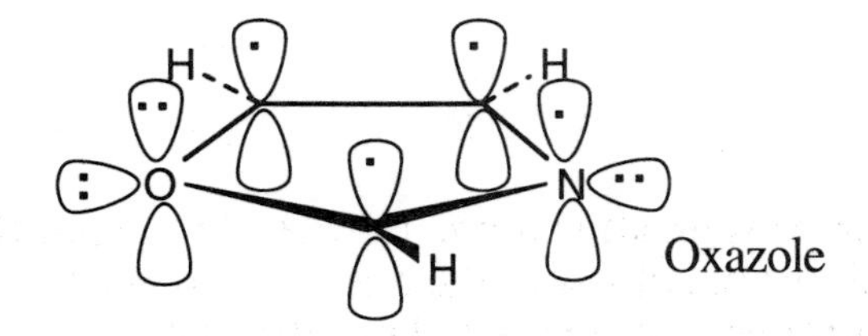

Oxazole is an aromatic 6 π electron heterocycle. Two oxygen electrons and one nitrogen electron are in *p* orbitals that are part of the π electron system of the ring, along with one electron from each carbon. An oxygen lone pair and a nitrogen lone pair are in sp^2 orbitals that lie in the plane of the ring. Since the nitrogen lone pair is available for donation to acids, oxazole is more basic than pyrrole.

28.27

(a) O $\xrightarrow[\text{dioxane}]{Br_2}$ O, Br

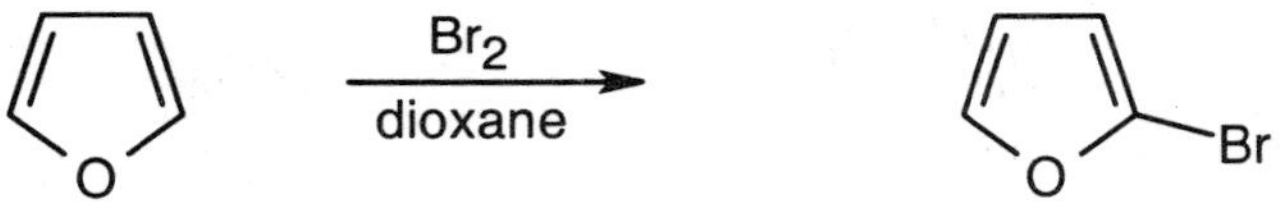

(b) O $\xrightarrow[\text{acetic anhydride}]{HNO_3}$ O, NO_2

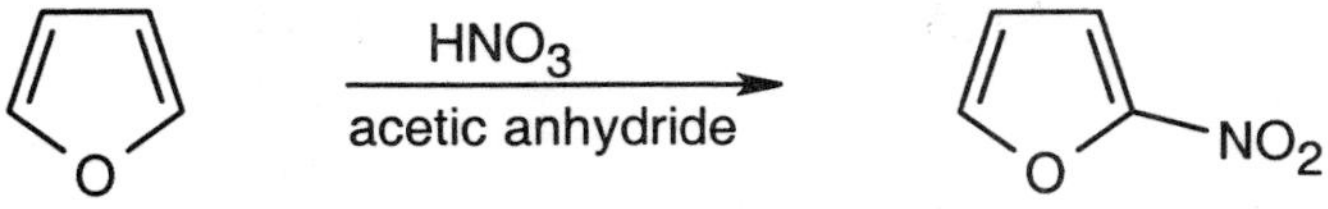

(c) O $\xrightarrow[SnCl_4]{CH_3COCl}$ O, C(=O)CH_3

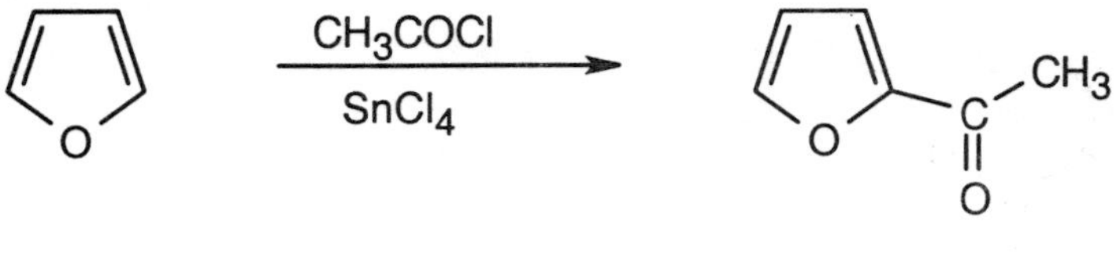

(d) O $\xrightarrow{H_2/Pd}$ O

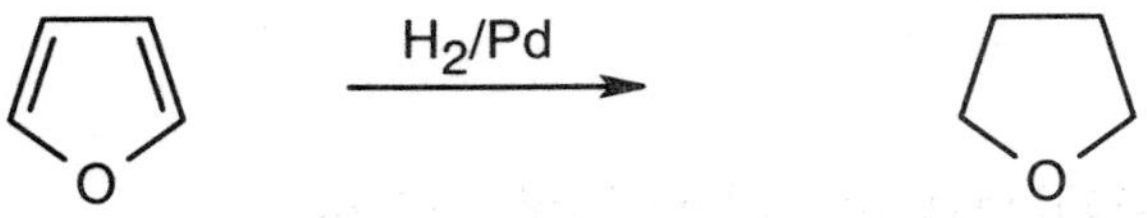

(e) O $\xrightarrow[\text{pyridine}]{SO_3}$ O, SO_3H

Notice that all of these electrophilic aromatic substitutions occur at C2 of the furan ring. Note also that the reaction conditions are much less severe than those used for substitutions on benzene.

28.28

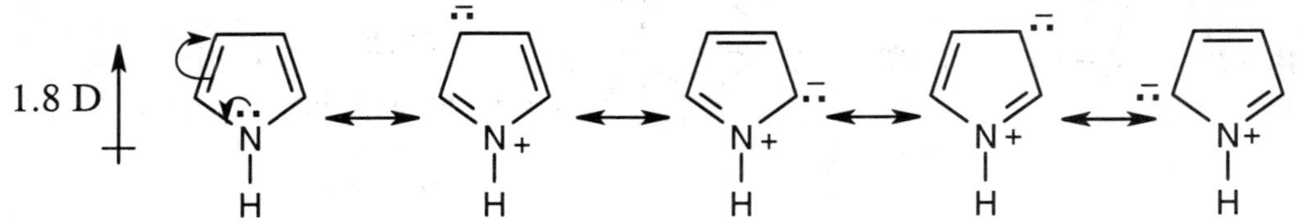

1.8 D

The dipole moment of pyrrole points in the direction indicated because resonance structures show that the nitrogen atom is electron-poor.

28.29

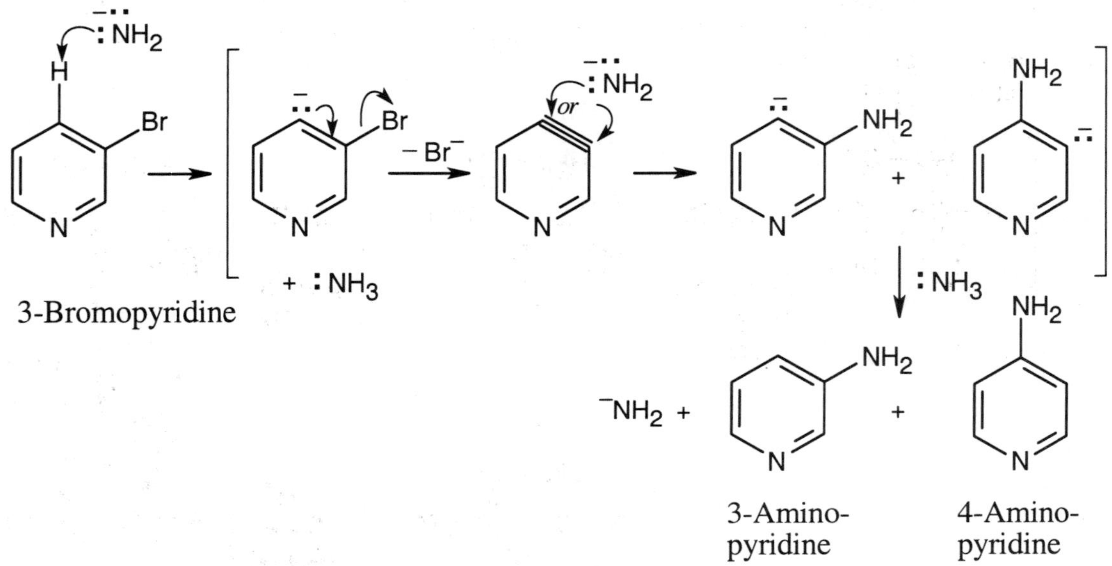

3-Bromopyridine

3-Aminopyridine

4-Aminopyridine

Reaction of 3-bromopyridine with $NaNH_2$ occurs by a benzyne mechanism. Since $^-:NH_2$ can add to either end of the triple bond of the benzyne intermediate, a mixture of products is formed.

28.30

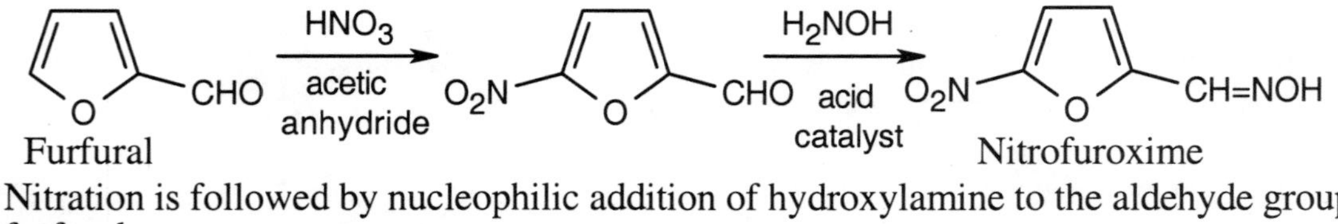

Furfural $\xrightarrow[\text{acetic anhydride}]{HNO_3}$ O_2N–furan–CHO $\xrightarrow[\text{acid catalyst}]{H_2NOH}$ O_2N–furan–CH=NOH

Nitrofuroxime

Nitration is followed by nucleophilic addition of hydroxylamine to the aldehyde group of furfural.

28.31

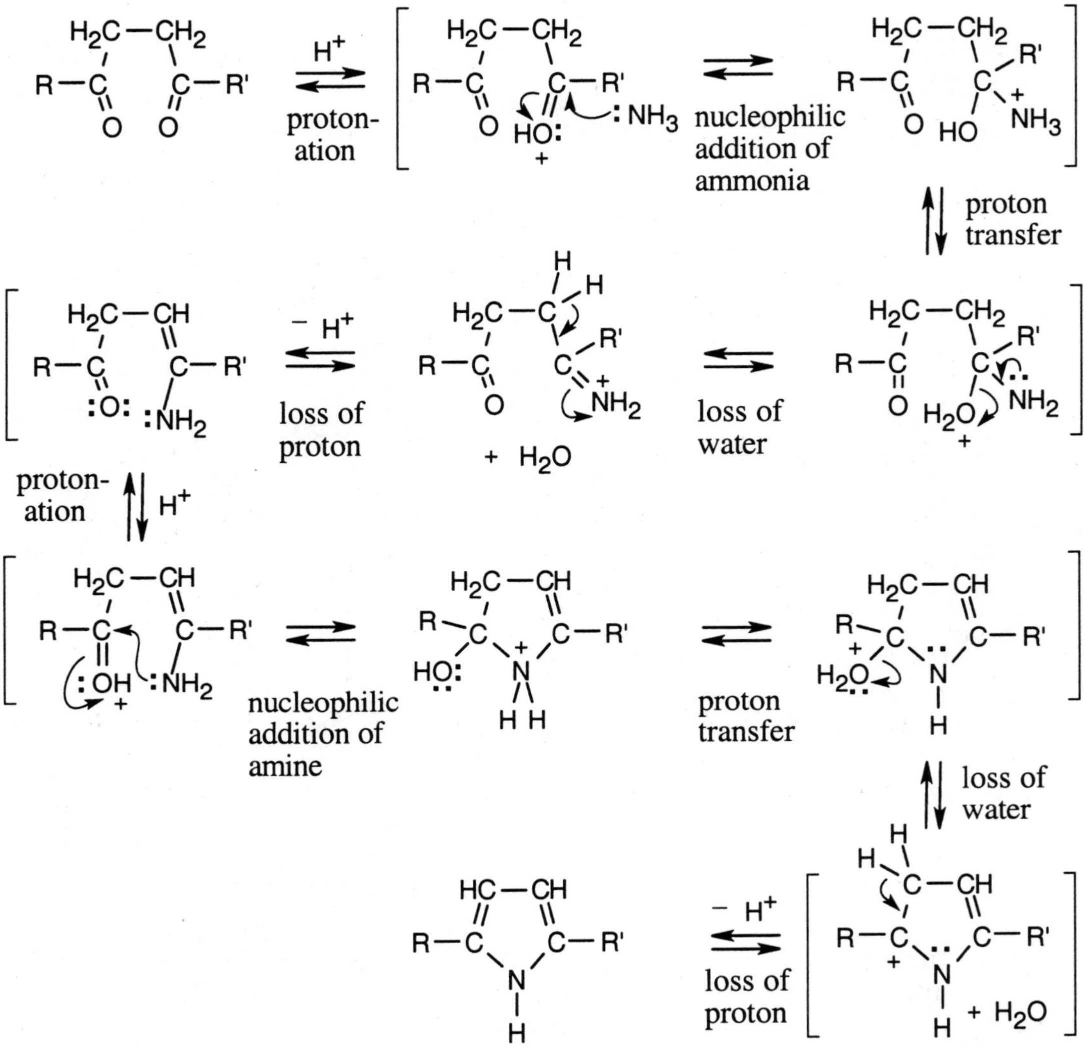

H^+ proton-ation

nucleophilic addition of ammonia

proton transfer

loss of water

$-H^+$ loss of proton

$+ H_2O$

proton-ation H^+

nucleophilic addition of amine

proton transfer

loss of water

$-H^+$ loss of proton

$+ H_2O$

The mechanism consists of the nucleophilic addition of ammonia, first to one of the ketones, and then to the other, with loss of two equivalents of water.

28.32

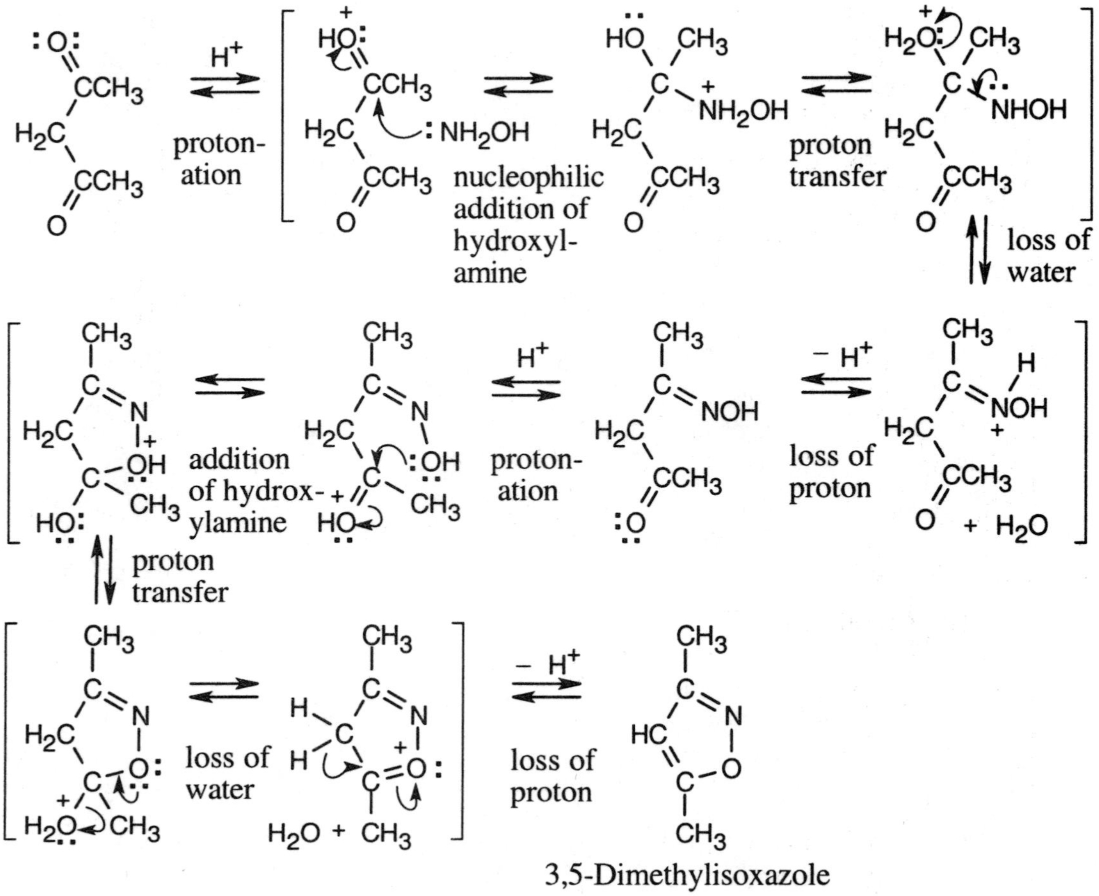

:O: ‖ CCH3 / H2C \ CCH3 // O ⇌ (H+, protonation) [HO:+ ... CCH3 H2C :NH2OH CCH3 O — nucleophilic addition of hydroxylamine ⇌ HO CH3 C +NH2OH H2C CCH3 O ⇌ (proton transfer) H2O+ CH3 C NHOH H2C CCH3 O]

⇅ loss of water

[CH3 C=N H2C +OH C CH3 HO: ⇌ (addition of hydroxylamine) CH3 C=N H2C :OH +C CH3 HO ⇌ (H+, protonation) CH3 C=NOH H2C CCH3 :O ⇌ (− H+, loss of proton) CH3 C=NOH (H, +) H2C CCH3 O + H2O]

⇅ proton transfer

[CH3 C=N H2C O: C + H2O CH3 ⇌ (loss of water) CH3 C=N H C H +C=O: H2O + CH3] ⇌ (− H+, loss of proton) CH3 C=N HC O C CH3

3,5-Dimethylisoxazole

This mechanism is virtually identical to the mechanism illustrated in the previous problem, and involves two nucleophilic additions to carbonyl groups.

28.33

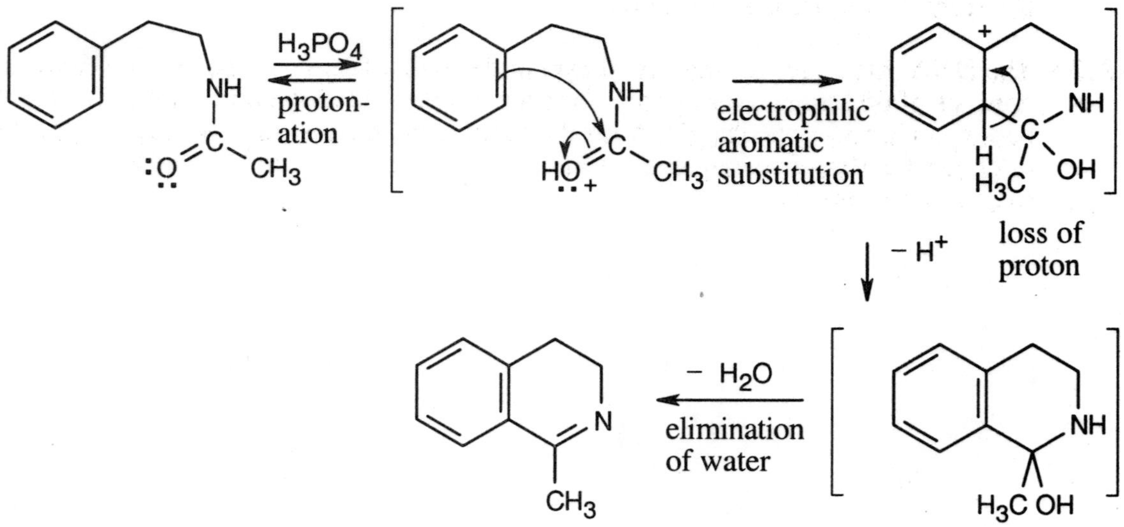

(Ph–CH2CH2–NH–C(=O)–CH3) ⇌ (H_3PO_4, protonation) [Ph–CH2CH2–NH–C(=OH+)–CH3] → (electrophilic aromatic substitution) [cyclized cation, + , H, NH, C, H3C, OH]

↓ − H+ loss of proton

[NH, H3C OH] → (− H_2O, elimination of water) (dihydroisoquinoline, N, CH3)

The cyclization is an electrophilic aromatic substitution.

28.34

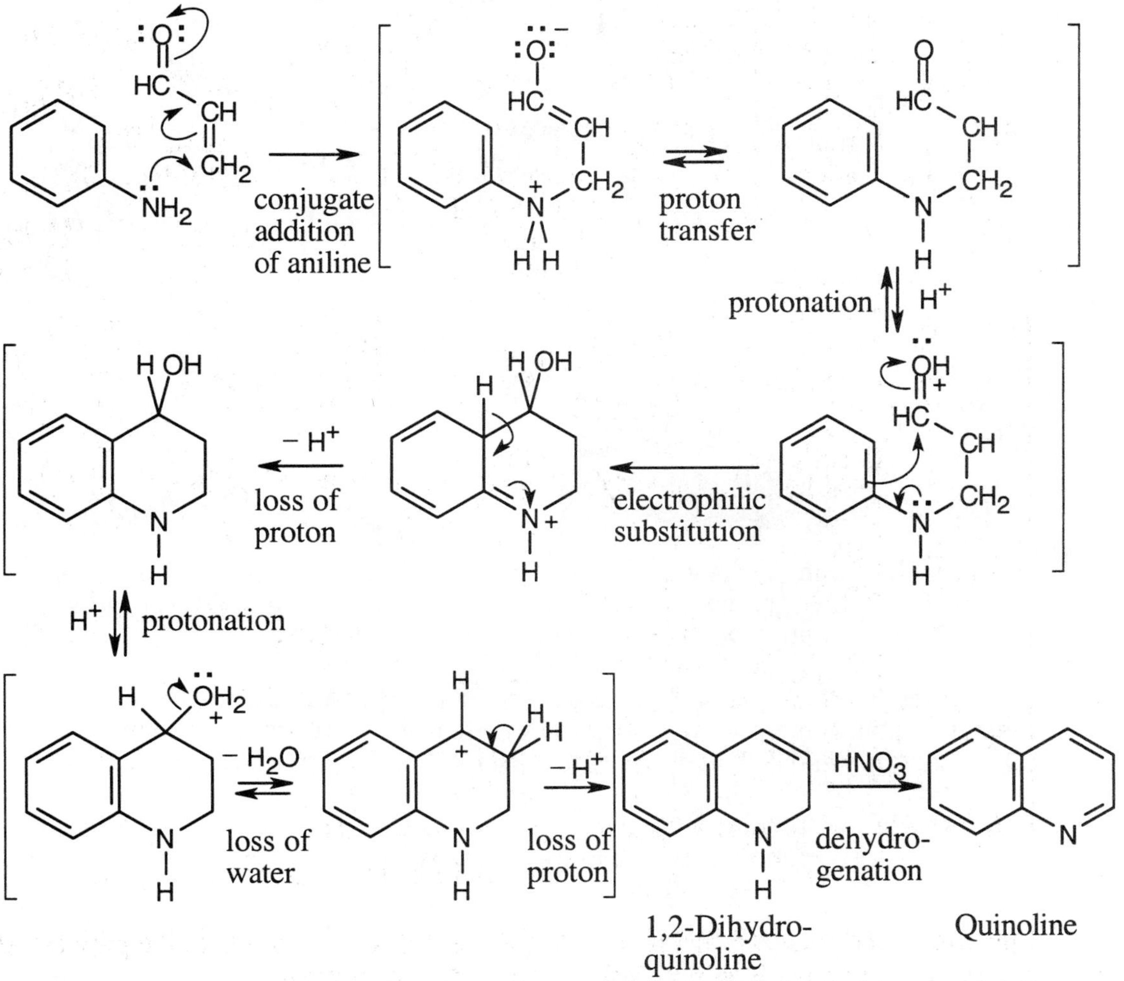

Conjugate addition of aniline to the α,β-unsaturated aldehyde is followed by an internal electrophilic aromatic substitution.

28.35 The DNA that codes for the first polypeptide chain of human insulin (21 amino acids) consists of 66 bases; 3 bases code for each amino acid in the chain, and a 3-base "stop" codon is also needed. The DNA for the second chain consists of 93 bases – the 90 that code for the 30 amino acids and the three-base "stop" codon. Thus, 159 bases are needed to code for human insulin.

28.36 *Position 9:*

Horse amino acid = Gly — Human amino acid = Ser

mRNA codons (5' —> 3'):

Horse				Human					
GGU	GGC	GGA	GGG	UCU	UCC	UCA	UCG	AGU	AGC

DNA bases (template strand 3' —> 5'):

Horse				Human					
CCA	CCG	CCT	CCC	AGA	AGG	AGT	AGC	TCA	TCG

The underlined horse DNA base triplets differ from their human counterparts (also underlined) by only one base.

Position 30:

Horse amino acid = Ala — Human amino acid = Thr

mRNA codons (5' —> 3'):

Horse				Human			
GCU	GCC	GCA	GCG	ACU	ACC	ACA	ACG

DNA bases (template strand 3' —> 5'):

Horse				Human			
CGA	CGG	CGT	CGC	TGA	TGG	TGT	TGC

Each group of three DNA bases from horse insulin has a counterpart in human insulin that differs from it by only one base. It is possible that horse insulin DNA differs from human insulin DNA by only two bases out of 159!

28.37 The percent of A always equals the percent of T, since A and T are complementary. The percent G equals the percent C for the same reason. Thus, sea urchin DNA contains about 32% each of A and T, and about 18% each of G and C.

28.38 Even though the stretch of DNA shown contains UAA in sequence, protein synthesis doesn't stop. The codons are read as 3-base individual units from start to end, and the unit UAA is read as part of two codons, not as a single codon.

28.39 The restriction endonuclease *Alu*I cleaves between the bases G and C in the (5' - 3') sequence AG-CT. It also cleaves between C and G in the (3' - 5') sequence TC-GA. Thus, the sequence in (c), CTCGAG is recognized and cleaved by *Alu*I. The sequence in (a), CAATTC, is also a palindrome and is recognized by a restriction endonuclease.

28.40–28.42

mRNA codon (5'–3'):	(a) AAU	(b) GAG	(c) UCC	(d) CAU
Amino acid:	Asn	Glu	Ser	His
DNA sequence(3'–5'):	TTA	CTC	AGG	GTA
tRNA anticodon(3'–5'):	UUA	CUC	AGG	GUA

The DNA sequence of the template strand is shown.

28.43–28.44 UAC is a codon for tyrosine. It was transcribed from ATG of the template strand of a DNA chain.

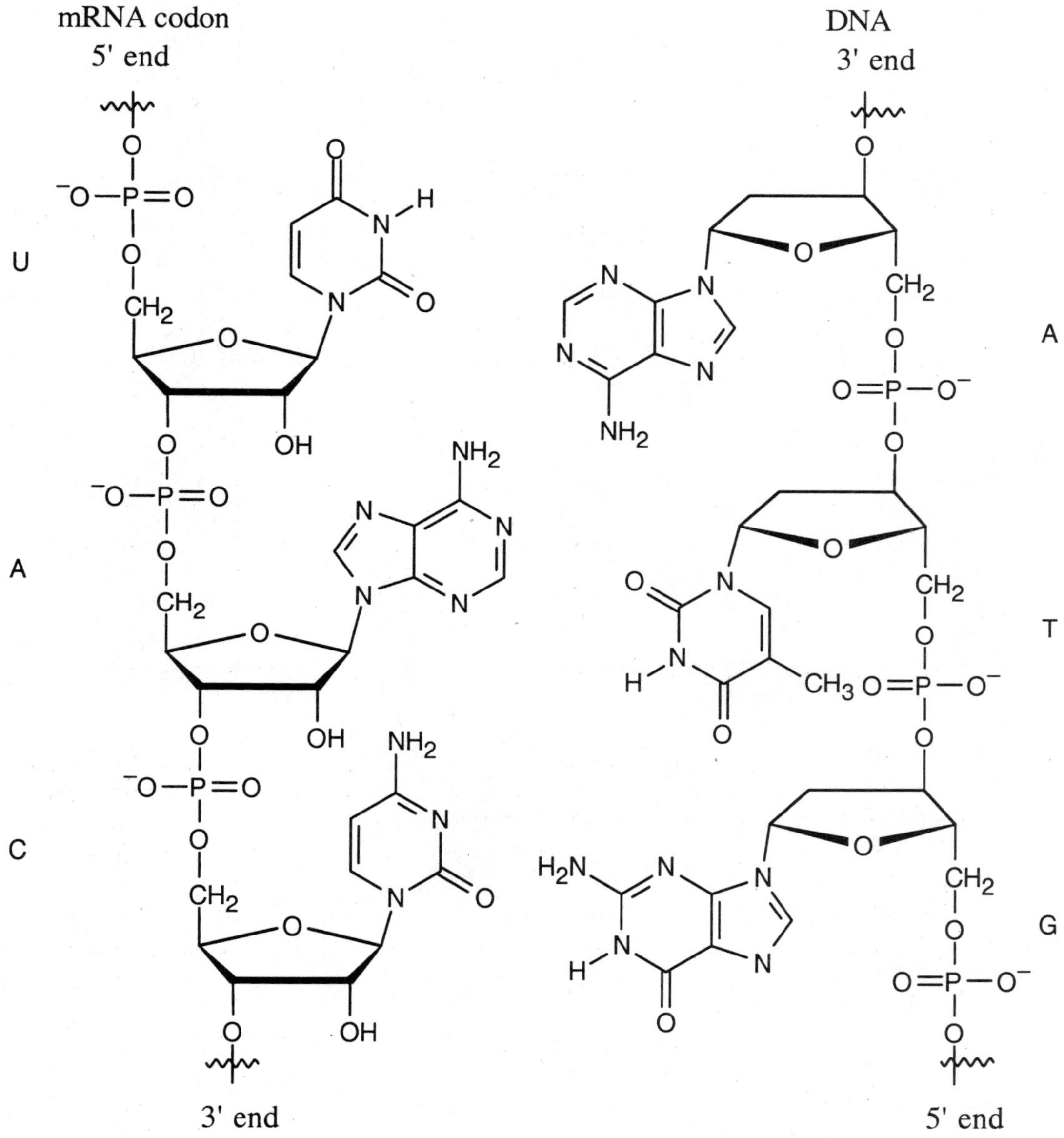

mRNA codon
5' end

U
A
C

3' end

DNA
3' end

A
T
G

5' end

28.45

Tyr	Gly	Gly	Phe	Met	(stop)	is coded by
(UAC)	(GGU)	(GGU)	(UUU)	(AUG)	(UAA)	
(UAU)	(GGC)	(GGC)	(UUC)		(UAG)	
	(GGA)	(GGA)			(UGA)	
	(GGG)	(GGG)				

A total of 2 x 4 x 4 x 2 x 1 x 3 = 194 different mRNA sequences can code for metenkephalin!

28.46 Angiotensin II: Asp—Arg—Val—Tyr—Ile—His—Pro—Phe (stop)

mRNA sequence: (5'–3')	GAU	CGU	GUU	UAU	AUU	CAU	CCU	UUU	UAA
	GAC	CGC	GUC	UAC	AUC	CAC	CCC	UUC	UAG
		CGA	GUA		AUA		CCA		UGA
		CGG	GUG				CCG		
		AGA							
		AGG							

As in the previous problem, many mRNA sequences (13,824) can code for angiotensin II.

28.47 DNA coding strand (5'–3'): CTT— CGA—CCA— GAC—AGC—TTT
mRNA (5'–3'): CUU—CGA—CCA—GAC—AGC—UUU
Amino acid sequence: Leu—Arg—Pro—Asp—Ser—Phe
The mRNA sequence is the complement of the DNA template strand, which is the complement of the DNA coding strand. Thus, the mRNA sequence is a copy of the DNA coding strand, with T replaced by U.

28.48 mRNA sequence (5'–3'): CUA—GAC—CGU—UCC—AAG—UGA
Amino Acid: Leu—Asp—Arg—Ser—Lys (stop)

28.49

	Original Sequence	*Miscopied Sequence*
DNA template strand (3'–5'):	-TAA-CCG-GAT-	-TGA-CCG-GAT-
mRNA sequence (5'–3'):	-AUU-GGC-CUA-	-ACU-GGC-CUA-
Amino acid sequence:	-Ile—Gly—Leu-	-Thr—Gly—Leu-

If this gene sequence were miscopied in the indicated way, an isoleucine in the original protein would be replaced by a threonine in the mutated protein.

28.50 1. First, protect the nucleotides.

(a) Bases are protected by amide formation.

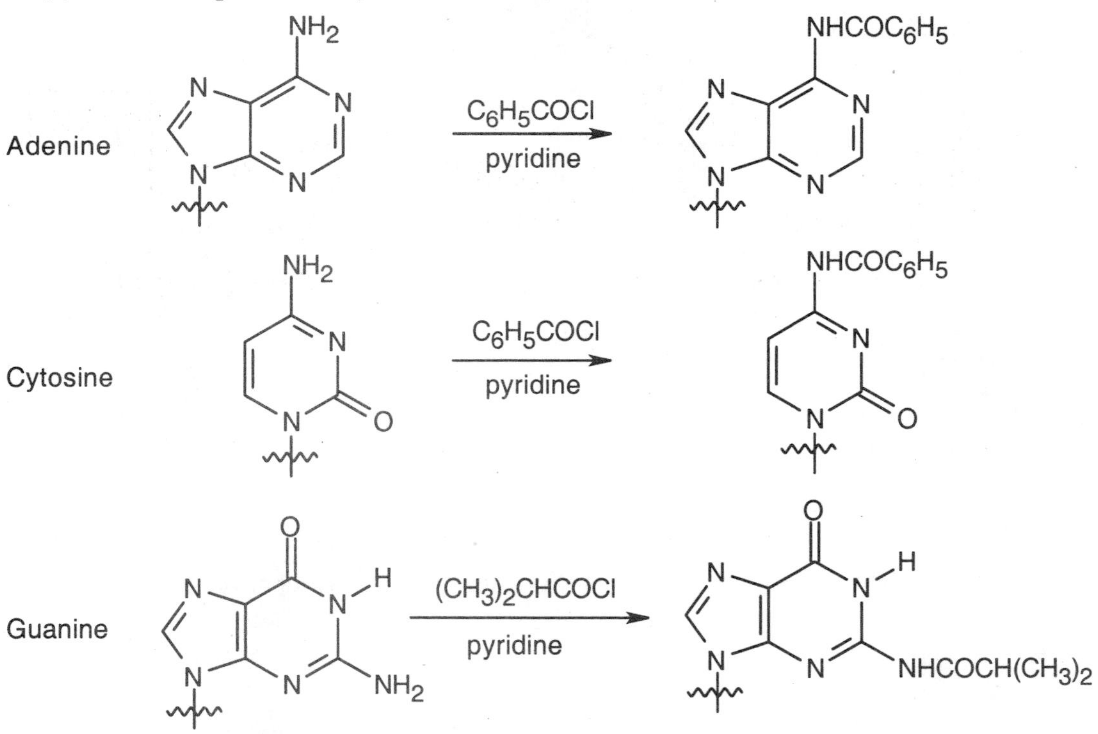

Thymine does not need to be protected.

(b) The 5' hydroxyl group is protected as its *p*–dimethoxytrityl (DMT) ether.

$$\text{HOCH}_2\text{-(Base nucleoside)-OCCH}_2\text{CH}_2\text{CO-C}_6\text{H}_4\text{-NO}_2 \xrightarrow[\text{2. DMTBr}]{\text{1. Base}} \text{DMTOCH}_2\text{-(Base nucleoside)-OCCH}_2\text{CH}_2\text{CO-C}_6\text{H}_4\text{-NO}_2$$

2. Attach a protected 2–deoxycytidine nucleoside to the polymer support.

$$\text{DMTOCH}_2\text{-(Cytosine nucleoside)-OCCH}_2\text{CH}_2\text{CO-C}_6\text{H}_4\text{-NO}_2 + \text{Silica-Si(CH}_2)_3\text{NH}_2 \longrightarrow \text{DMTOCH}_2\text{-(Cytosine nucleoside)-OCCH}_2\text{CH}_2\text{CNH(CH}_2)_3\text{Si-Silica}$$

Let $-\overset{O}{\overset{\|}{C}}CH_2CH_2\overset{O}{\overset{\|}{C}}NH(CH_2)_3Si-\text{Silica} = -\text{Support}$

3. Cleave the DMT ether.

$$\text{DMTOCH}_2\text{-(Cytosine nucleoside)-O-Support} \xrightarrow[\text{CH}_2\text{Cl}_2]{\text{CHCl}_2\text{COOH}} \text{HOCH}_2\text{-(Cytosine nucleoside)-O-Support}$$

4. Couple protected 2'–deoxythymidine to the polymer–2'–deoxycytidine. (The nucleosides have a phosphoramidite group at the 3' position.)

DMTOCH$_2$ Thymine; O; P—N(*i*-Pro)$_2$; NCCH$_2$CH$_2$O + HOCH$_2$ Cytosine; O—Support $\xrightarrow{\text{Tetrazole}}$ DMTOCH$_2$ Thymine; O; P—OCH$_2$ Cytosine; NCCH$_2$CH$_2$O; O—Support

5. Oxidize the phosphite product to a phosphate triester, using iodine.

DMTOCH$_2$ Thymine; O; P—OCH$_2$ Cytosine; NCCH$_2$CH$_2$O; O—Support $\xrightarrow[\text{THF}]{I_2}$ DMTOCH$_2$ Thymine; O; O=P—OCH$_2$ Cytosine; NCCH$_2$CH$_2$O; O—Support

6. Repeat steps 3–5 with protected 2'–deoxyadenosine and protected 2'–deoxyguanosine.

7. Cleave all protecting groups with aqueous ammonia to yield the desired sequence.

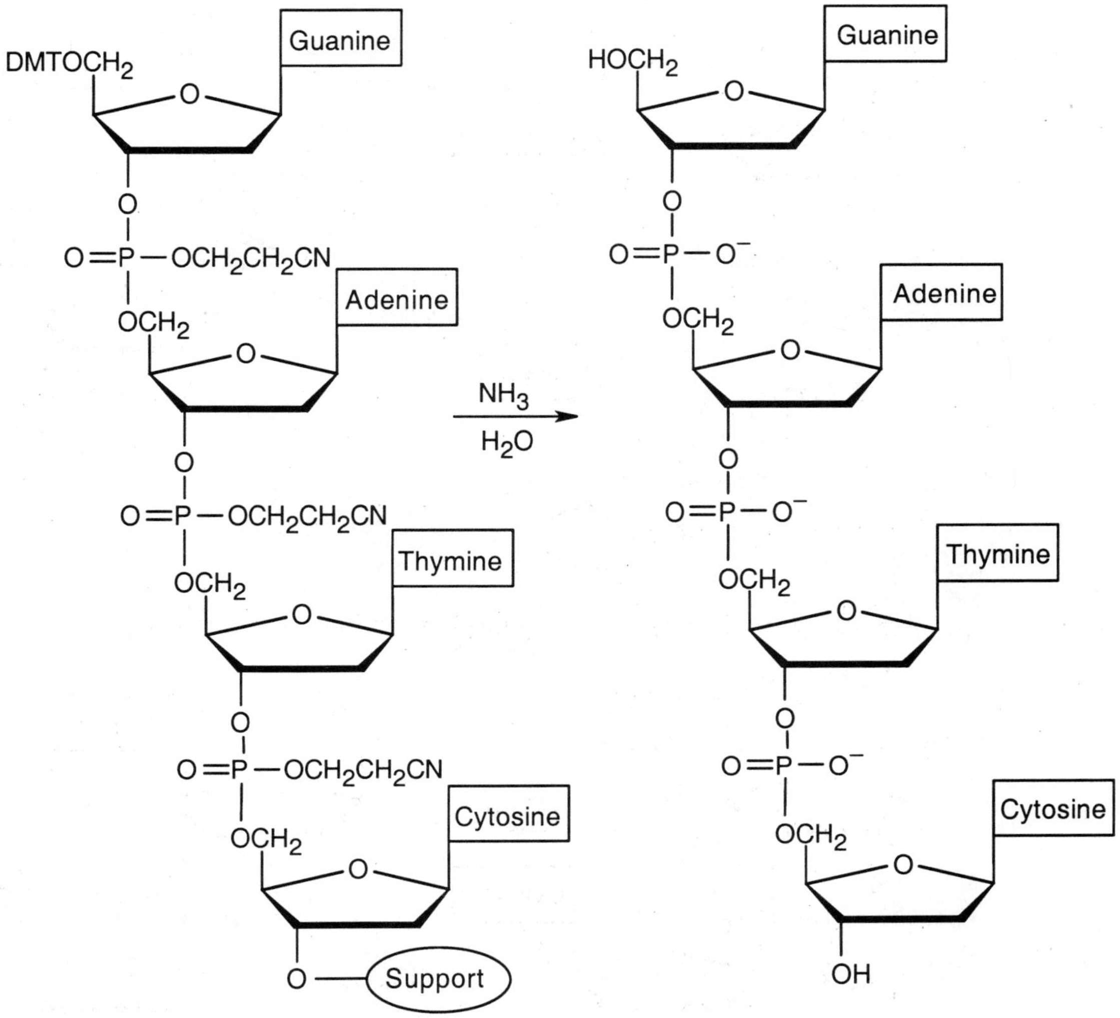

$\xrightarrow[H_2O]{NH_3}$

28.51

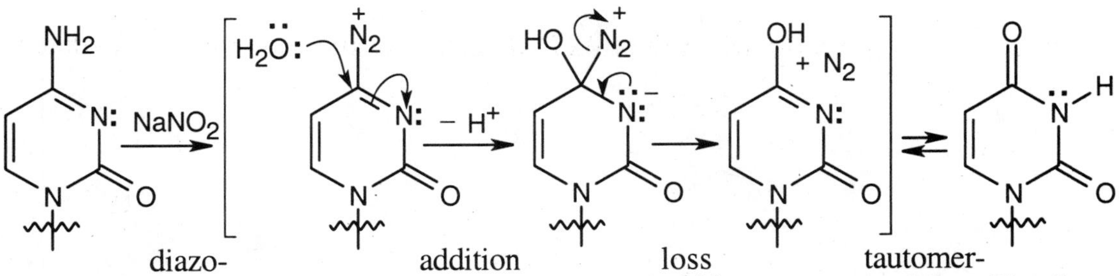

Reaction with nitrous acid diazotizes cytosine, and subsequent reaction with water, followed by tautomerization, yields uracil.

28.52 Both of these cleavages occur by the now-familiar nucleophilic acyl substitution route. A nucleophile adds to the carbonyl group, a proton shifts location, and a second group is eliminated.

Deprotection at 1:

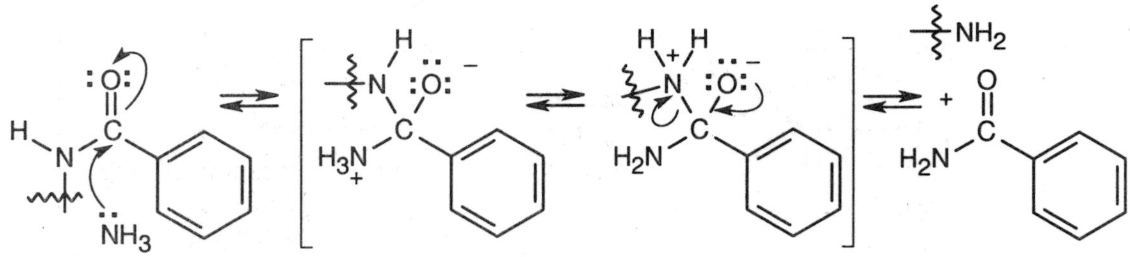

:O:
H
N
C
$\ddot{N}H_3$
⇌ [H N :O: − C $H_3\overset{+}{N}$ ⇌ H H + N :O: − C H_2N] ⇌ NH_2 + O C H_2N

Deprotection at 2:

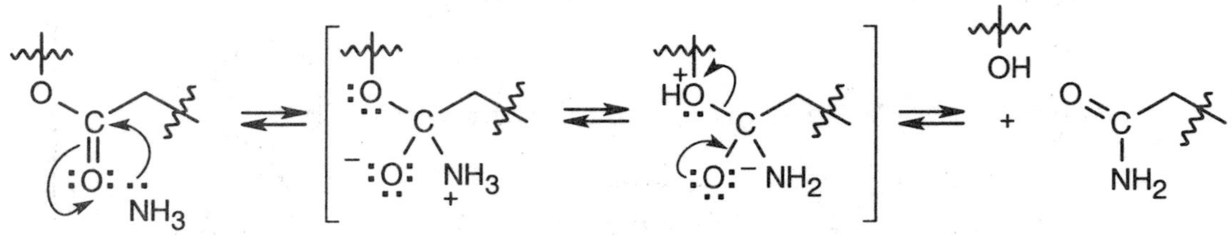

O C :O: NH_3
⇌ [:O C $^-$:O: $\overset{+}{N}H_3$ ⇌ $H\overset{+}{O}$ C :O:$^-$ NH_2] ⇌ OH + O C NH_2

28.53

(DNA)
O
$O=P-O^-$
O
CH_2
O
N N N N
NH_2
H
O
(DNA)

$CH_3OSO_2OCH_3$
methylation
(S_N2 reaction)

[(DNA) O $O=P-O^-$ O CH_2 O N N N N^+ NH_2 CH_3 H O (DNA)]

↕

[(DNA) O $O=P-O^-$ O CH_2 O N N N N $\overset{+}{N}H_2$ CH_3 H O (DNA)]

elimination
of base

[(DNA) O $O=P-O^-$ O CH_2 $\overset{+}{O}$ H $:\ddot{O}H_2$ O (DNA) + N N N N NH_2 CH_3]

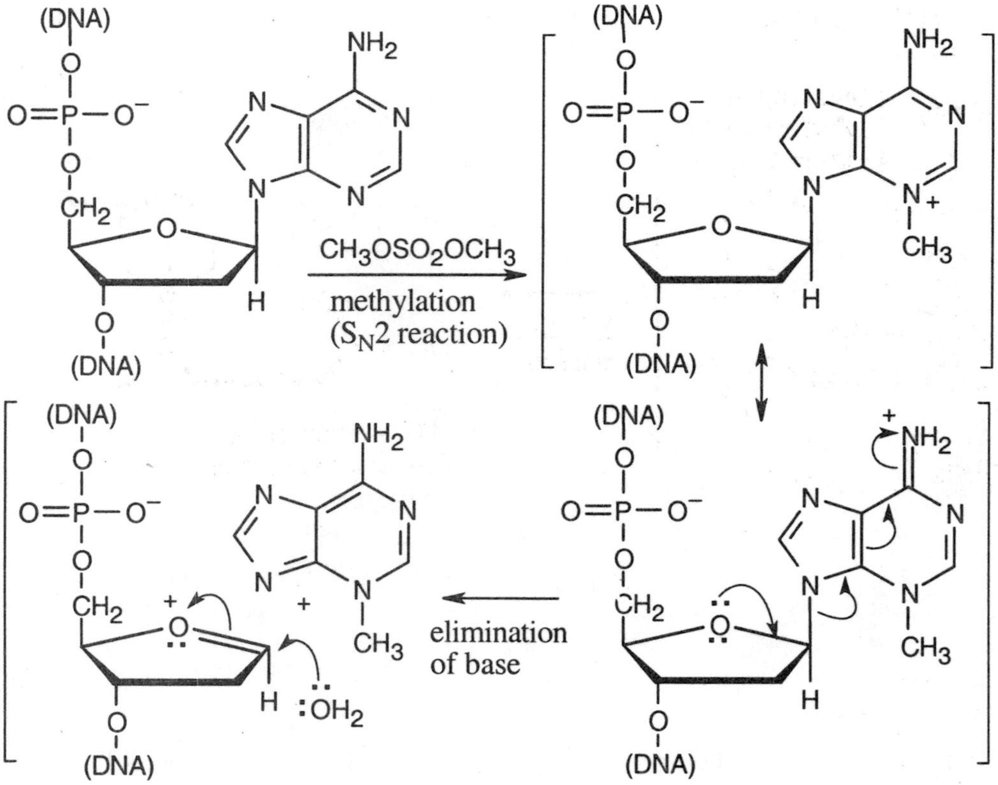

addition of water

proton transfer

opening of ring to form aldehyde

formation of enamine

$+ H_2O$

breaking of DNA chain by elimination

E2 elimination

This intermediate undergoes further reaction.

A Look Ahead

28.54

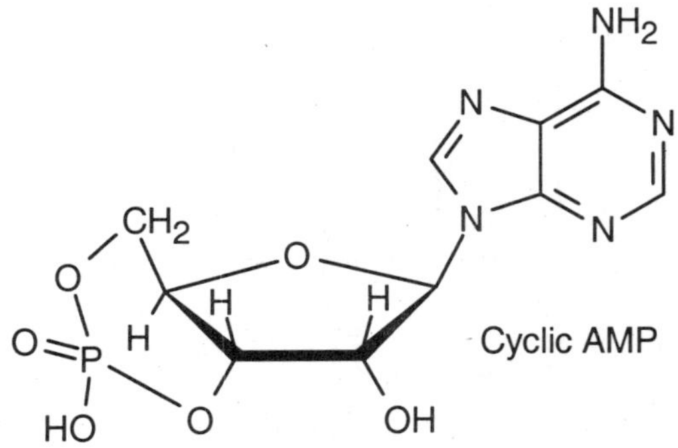

Cyclic AMP

Molecular Modeling

28.55 *N*-Formylpyrrolidine contains a shorter C–N bond (134 pm) than *N*-formylpyrrole. Electrostatic potential maps indicate greater transfer of electron density from N to C–O in *N*-formylpyrrolidine. These results can be described as follows:

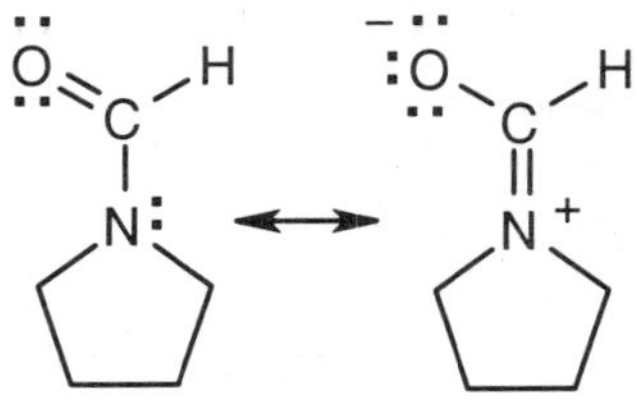

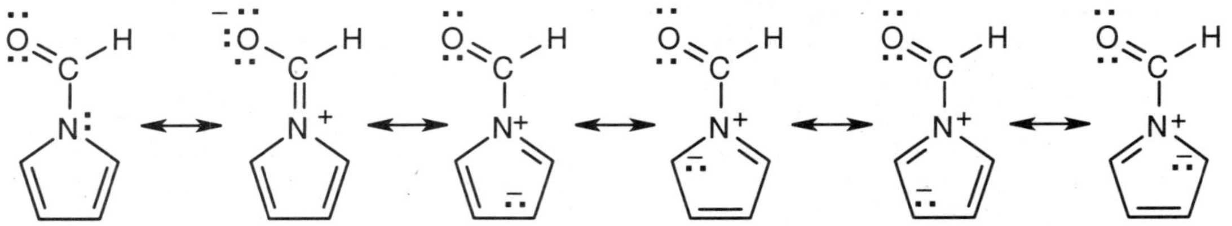

28.56 N7 in *N*9-methylguanine is considerably more negative than N3 in *N*9-methyladenine. Guanine is the stronger nucleophile.

28.57 The binding energies are –92 kJ/mol for the AT base pair, –167 kJ/mol for the GC base pair, –93 kJ/mol for the AG base pair, and –78 kJ/mol for the TC base pair. Only one natural base pair, GC, has a substantially larger binding energy than the unnatural base pairs. This is consistent with the fact that the GC base pair contains three hydrogen bonds, whereas the other pairs, natural and unnatural, contain only two hydrogen bonds. The natural base pairs are flat, and the unnatural base pairs are twisted to minimize steric repulsion (between hydrogens in the AG pair) and unfavorable electrostatic interactions (between oxygens in the TC pair).

28.58 There are 15 base pairs in each strand. The two strands are complementary.
G–C–G–C–G–C–G–C–G–C–G–C–G–C–G in one strand, and
C–G–C–G–C–G–C–G–C–G–C–G–C–G–C in the other strand.

Chapter 29 – The Organic Chemistry of Metabolic Pathways

Chapter Outline

I. Overview of metabolism and biochemical energy (Section 29.1).
  A. Metabolism.
    1. The reactions that take place in the cells of organisms are collectively called metabolism.
      a. The reactions that produce smaller molecules from larger molecules are called catabolism.
      b. The reactions that build larger molecules from smaller molecules are called anabolism.
    2. Catabolism can be divided into four stages:
      a. In digestion, bonds in food are hydrolyzed to yield sugars, fats, and amino acids.
      b. These small molecules are degraded to acetyl CoA.
      c. In the citric acid cycle, acetyl CoA is catabolized to CO_2, and energy is produced.
      d. Energy from the citric acid cycle enters the electron transport chain, where ATP is synthesized.
  B. Biochemical energy.
    1. ATP, a phosphoric acid anhydride, is the storehouse for biochemical energy.
    2. The breaking of a P–O bond of ATP can be coupled with an energetically unfavorable reaction, so that the overall energy change is favorable.
    3. The resulting phosphates are much more reactive than the original compounds.

II. Catabolism (Sections 29.2 – 29.6).
  A. Catabolism of fats (Section 29.2).
    1. Triacylglycerols are first hydrolyzed in the stomach and small intestine to yield glycerol plus fatty acids.
      a. Glycerol is phosphorylated and enters glycolysis.
      b. Fatty acids are degraded by β oxidation, a 4-step spiral that results in the cleavage of an *n*-carbon fatty acid into *n*/2 molecules of acetyl CoA.
      c. Before entering β oxidation, a fatty acid is first converted to its fatty-acyl CoA.
    2. Steps of β oxidation.
      a. Introduction of a double bond conjugated with the carbonyl group.
        i. The reaction is catalyzed by acyl CoA dehydrogenase.
        ii. The enzyme cofactor FAD is also involved.
      b. Conjugate addition of water to form an alcohol.
        The reaction is catalyzed by enoyl CoA hydratase.
      c. Alcohol oxidation.
        i. The reaction is catalyzed by L-3-hydroxyacyl CoA dehydrogenase.
        ii. The cofactor NAD^+ is reduced to $NADH/H^+$ at the same time.
        iii. The mechanism of the reaction resembles a Cannizzaro reaction, followed by conjugate addition of hydride to NAD^+.
      d. Cleavage of acetyl CoA from the chain.
        i. The reaction, which is catalyzed by β-keto thiolase, is a retro-Claisen reaction.
        ii. Nucleophilic addition of coenzyme A to the keto group is followed by loss of acetyl CoA enolate, leaving behind a chain-shortened fatty-acyl CoA.

3. An *n*-carbon fatty acid yields $n/2$ molecules of acetyl CoA after $(n/2 - 1)$ passages of β oxidation.
   a. Since most fatty acids have an even number of carbons, no carbons are left over after β oxidation.
   b. Those with an odd number of carbons require further steps for degradation.

B. Catabolism of carbohydrates (Sections 29.3 – 29.5).
1. Glycolysis (Section 29.3).
   a. Glycolysis is a 10-step series of reactions that converts glucose to pyruvate.
   b. Steps 1–3: Phosphorylation and isomerization.
      i. Glucose is phosphorylated at the 6-position by reaction with ATP.
         The enzyme hexokinase is involved.
      ii. Glucose 6-P is isomerized to fructose 6-P by glucose-6-P isomerase.
      iii. Fructose 6-P is phosphorylated to yield fructose 1,6-bisphosphate.
         ATP and phosphofructokinase are involved.
   c. Steps 4–5: Cleavage and isomerization.
      i. Fructose 1,6 bisphosphate is cleaved to glyceraldehyde 3-phosphate and dihydroxyacetone phosphate.
         The reaction is a reverse aldol reaction catalyzed by aldolase.
      ii. Dihydroxyacetone phosphate is isomerized to glyceraldehyde 3-phosphate.
      iii. The net result is production of two glyceraldehyde 3-phosphates, both of which pass through the rest of the pathway.
   d. Steps 6–8: Oxidation and phosphorylation.
      i. Glyceraldehyde 3-phosphate is both oxidized and phosphorylated to give 3-phosphoglyceroyl phosphate.
         Oxidation occurs via a hemithioacetal to yield a product that forms the mixed anhydride.
      ii. The mixed anhydride reacts with ADP to form ATP and 3-phosphoglycerate
         The enzyme phosphoglycerate kinase is involved.
      iii. 3-Phosphoglycerate is isomerized to 2-phosphoglycerate by phosphoglycerate mutase.
   e. Steps 9–10: Dehydration and dephosphorylation.
      i. 2-Phosphoglycerate is dehydrated by enolase to give PEP.
      ii. Pyruvate kinase catalyzes the transfer of a phosphate group to ADP, with formation of pyruvate.
2. The conversion of pyruvate to acetyl CoA (Section 29.4).
   a. The conversion pyruvate → acetyl CoA is catalyzed by an enzyme complex called pyruvate dehydrogenase complex.
   b. Step 1: Addition of thiamine.
      A nucleophilic ylide group on thiamine pyrophosphate adds to the carbonyl group of pyruvate to yield a tetrahedral intermediate.
   c. Step 2: Decarboxylation.
   d. Step 3: Reaction with lipoamide.
      The enamine product of decarboxylation reacts with lipoamide, displacing sulfur and opening the lipoamide ring.
   e. Step 4: Elimination of thiamine.
   f. Step 5: Acyl transfer.
      i. Acetyl dihydrolipoamide reacts with coenzyme A to give acetyl CoA.
      ii. The resulting dihydrolipoamide is reoxidized to lipoamide by FAD.
      iii. $FADH_2$ is reoxidized to FAD by NAD^+.
   g. Other fates of pyruvate.
      i. In the absence of oxygen, pyruvate is reduced to lactate.
      ii. In bacteria, pyruvate is fermented to ethanol.
3. The citric acid cycle (Section 29.5).
   a. Characteristics of the citric acid cycle.

i. The citric acid cycle is a closed loop.
ii. The intermediates are constantly regenerated.
iii. The cycle operates as long as NAD^+ and $FADH_2$ are available, which means that oxygen must also be available.

b. Steps 1–2: Addition to oxaloacetate.
i. Acetyl CoA adds to oxaloacetate to form citryl CoA, which is hydrolyzed to citrate.
The reaction is catalyzed by citrate synthase.
ii. Citrate is isomerized to isocitrate by aconitase.
The reaction is an E2 dehydration, followed by conjugate addition of water.

c. Steps 3–4: Oxidative decarboxylations.
i. Isocitrate is oxidized by isocitrate dehydrogenase to give a ketone that loses CO_2 to give α-ketoglutarate.
ii. α-Ketoglutarate is transformed to succinyl CoA in a reaction catalyzed by an enzyme complex.

d. Steps 5–6: Hydrolysis and dehydrogenation of succinyl CoA.
i. Succinyl CoA is converted to an acyl phosphate, which transfers a phosphate group to GDP in a reaction catalyzed by succinyl CoA synthase.
ii. Succinate is dehydrogenated by FAD and succinate dehydrogenase to give fumarate.

e. Steps 7–8: Regeneration of oxaloacetate.
i. Fumarase catalyzes the addition of water to fumarate to produce L-malate.
ii. L-malate is oxidized by NAD^+ and malate dehydrogenase to complete the cycle.

B. Catabolism of proteins: Transamination (Section 24.6).
1. The pathway to amino acid catabolism:
a. The amino group is removed by transamination.
b. What remains is converted to a compound that enters the citric acid cycle.
2. Transamination.
a. The $-NH_2$ group of an amino acid adds to the aldehyde group of pyridoxal phosphate to form an imine.
b. The imine rearranges to a different imine.
c. The second imine is hydrolyzed to give an α-keto acid and an amino derivative of pyridoxal phosphate.
d. The pyridoxal derivative transfers its amino group to α-ketoglutarate, to regenerate pyridoxal phosphate and form glutamate.
3. Deamination.
The glutamate from transamination undergoes oxidative deamination to yield ammonium ion and α-ketoglutarate.

III. Anabolism (Sections 29.7 –29.8).
A. Anabolism of fatty acids (Section 29.7).
1. All common fatty acids have an even number of carbons because they are synthesized from acetyl CoA.
2. Steps 1–2: Acyl transfers convert acetyl CoA to more reactive species.
a. Acetyl CoA is converted to acetyl ACP.
b. The acetyl group of acetyl ACP is transferred to a synthase enzyme.
3. Steps 3–4: Carboxylation and acyl transfer.
a. Acetyl CoA reacts with bicarbonate to yield malonyl CoA and ADP.
The coenzyme biotin, a CO_2 carrier, transfers CO_2 in a nucleophilic acyl substitution reaction.
b. Malonyl CoA is converted to malonyl ACP.

4. Step 5: Condensation.
    a. A Claisen condensation forms acetoacetyl CoA from acetyl synthase and malonyl ACP.
    b. The reaction proceeds through an intermediate β-keto acid that loses CO_2 to give acetoacetyl CoA.
5. Steps 6–8: Reduction and dehydrogenation.
    a. The ketone group of acetoacetyl CoA is reduced by NADPH.
    b. The β-hydroxy thiol ester is dehydrated.
    c. The resulting double bond is hydrogenated by NADPH to yield butyryl ACP.
6. The steps are repeated with butyryl synthase and malonyl ACP to give a six-carbon unit.
7. Fatty acids up to palmitic acid are synthesized by this route.
    Elongation of palmitic acid and larger acids occurs with acetyl CoA units as the two-carbon donor.

B. Anabolism of carbohydrates: Gluconeogenesis (Section 29.8).
1. Metabolic pathways and their reverse.
    a. In most cases, the pathway of synthesis isn't the exact reverse of degradation.
    b. If $\Delta G°$ is negative for one route, it must be positive for the exact reverse, which is thus energetically unfavorable.
    c. The metabolic strategy is for one pathway to be related to its reverse but not to be identical.
2. Gluconeogenesis.
    a. Step 1: Carboxylation.
        Pyruvate is carboxylated in a reaction that uses biotin and ATP.
    b. Step 2: Decarboxylation and phosphorylation.
        Concurrent decarboxylation and phosphorylation produce phosphoenolpyruvate.
    c. Steps 3–4: Hydration and isomerization.
        i. Conjugate addition of water gives 2-phosphoglycerate.
        ii. Isomerization produces 3-phosphoglycerate.
    d. Steps 5–7: Phosphorylation, reduction and tautomerization.
        i. Reaction of 3-phosphoglycerate with ATP yields an acyl phosphate.
        ii. The acyl phosphate is reduced by $NADPH/H^+$ to an aldehyde.
        iii. The aldehyde tautomerizes to dihydroxyacetone phosphate.
    e. Step 8: Aldol condensation.
        i. Dihydroxyacetone phosphate and glyceraldehyde 3-phosphate condense to form fructose 1,6-bisphosphate.
        ii. This condensation involves the imine of dihydroxyacetone phosphate, which forms an enamine that takes part in the condensation.
    f. Steps 9-10: Hydrolysis and isomerization.
        i. Fructose 1,6-bisphosphate is hydrolyzed to fructose 6-phosphate.
        ii. Fructose 6-phosphate isomerizes to glucose 6-phosphate.

IV. Some conclusions about biological chemistry (Section 29.9).
1. The mechanisms of biochemical reactions are almost identical to the mechanisms of laboratory reactions.
2. Most metabolic pathways are linear.
    a. Linear pathways make sense when a multifunctional molecule is undergoing transformation.
    b. Cyclic pathways may be more energetically feasible when a molecule is small.
        A cyclic pathway with multifunctional intermediates allows a greater range of reaction choices.

Solutions to Problems

29.1 This reaction is an S_N2 substitution, with ADP as the leaving group.

$$\underset{\text{Glycerol}}{CH_2OH-CHOH-CH_2OH} + {}^{-}O-\overset{O}{\overset{\|}{P}}(O^-)-O-\overset{O}{\overset{\|}{P}}(O^-)-O-\overset{O}{\overset{\|}{P}}(O^-)-OCH_2\text{-ribose(OH, OH)-adenine}(NH_2)\ \text{(ATP)}$$

$$\longrightarrow \underset{\text{Glycerol 1-phosphate}}{CH_2O-\overset{O}{\overset{\|}{P}}(O^-)-O^- \;|\; CHOH \;|\; CH_2OH} + {}^{-}O-\overset{O}{\overset{\|}{P}}(O^-)-O-\overset{O}{\overset{\|}{P}}(O^-)-OCH_2\text{-ribose(OH, OH)-adenine}(NH_2)\ \text{(ADP)}$$

Structure labels: CH_2OH, CHOH, CH_2OH, Glycerol; O, P, O^-, OCH_2, O, N, N, N, N, NH_2, OH, OH, ATP; CH_2O, P, O^-, CHOH, CH_2OH, Glycerol 1-phosphate; O, P, O^-, OCH_2, N, NH_2, OH, OH, ADP

29.2

$$CH_3CH_2-CH_2CH_2-CH_2CH_2-CH_2\overset{O}{\overset{\|}{C}}SCoA$$

Caprylyl CoA

↓ (passage 4)

$$CH_3CH_2-CH_2CH_2-CH_2\overset{O}{\overset{\|}{C}}SCoA + CH_3\overset{O}{\overset{\|}{C}}SCoA$$

Hexanoyl Coa

↓ (passage 5)

$$CH_3CH_2-CH_2\overset{O}{\overset{\|}{C}}SCoA + CH_3\overset{O}{\overset{\|}{C}}SCoA$$

Butanoyl CoA

↓ (passage 6)

$$CH_3\overset{O}{\overset{\|}{C}}SCoA + CH_3\overset{O}{\overset{\|}{C}}SCoA$$

29.3 A fatty acid with *n* carbons yields *n*/2 acetyl CoA molecules after (*n*/2 – 1) passages of the β-oxidation pathway.

(a)

$$CH_3CH_2\overset{\downarrow}{-}CH_2CH_2\overset{\downarrow}{-}CH_2CH_2\overset{\downarrow}{-}CH_2CH_2\overset{\downarrow}{-}CH_2CH_2\overset{\downarrow}{-}CH_2CH_2\overset{\downarrow}{-}CH_2CH_2\overset{\downarrow}{-}CH_2COOH$$

↓ β oxidation

$$8\ CH_3\overset{O}{\overset{\|}{C}}SCoA$$

Seven passages of the β-oxidation pathway are needed.

(b)

$$CH_3CH_2—(CH_2CH_2)_8—CH_2COOH \xrightarrow{\beta \text{ oxidation}} 10\ CH_3\overset{O}{\overset{||}{C}}SCoA$$

Nine passages of the β-oxidation pathway are needed.

29.4 ATP is produced in step 7 (3-phosphoglyceroyl phosphate —> 3-phosphoglycerate) and in step 10 (phosphoenolpyruvate —> pyruvate).

29.5 *Step 1* is a nucleophilic acyl substitution at phosphorus by the –OH group at C6 of glucose, with ADP as the leaving group.
Step 2 is an isomerization, in which the pyranose ring of glucose 6-phosphate opens, keto-enol tautomerism causes isomerization to fructose 6-phosphate, and a furanose ring is formed.
Step 3 is a substitution, similar to the one in step 1, involving the –OH group at C1 of fructose 6-phosphate.
Step 4 is a retro-aldol condensation that cleaves fructose 1,6-bisphosphate to glyceraldehyde 3-phosphate and dihydroxyacetone phosphate.
Step 5 is a isomerization of dihydroxyacetone phosphate to glyceraldehyde 3-phosphate that occurs by keto-enol tautomerization.
In *Step 6*, the aldehyde group of glyceraldehyde 3-phosphate reacts with a thiol group of an enzyme, in a nucleophilic addition reaction, to form a hemithioacetal, which is oxidized by NAD^+ to an acyl thioester. Nucleophilic acyl substitution by phosphate yields the product 3-phosphoglyceroyl phosphate.
Step 7 is a nucleophilic acyl substitution reaction at phosphorus, in which ADP displaces a phosphate group of 3-phosphoglyceroyl phosphate, yielding ATP and 3-phosphoglycerate.
Step 8 is an isomerization of 3-phosphoglycerate to 2-phosphoglycerate.
Step 9 is an E2 elimination of H_2O to form phosphoenolpyruvate.
Step 10 is a displacement reaction at phosphorus that forms ATP and displaces enol pyruvate, which tautomerizes to pyruvate.

29.6

29.7

1 CHO, 2 H–C–OH, 3 HO–C–H, 4 H–C–OH, 5 H–C–OH, 6 CH_2OH (Glucose) → → 1 $CH_2OPO_3^{2-}$, 2 C=O, 3 HO–C–H, 4 H–C–OH, 5 H–C–OH, 6 $CH_2OPO_3^{2-}$ (Fructose 1,6-bisphosphate) —Aldolase→ 1 $CH_2OPO_3^{2-}$, 2 C=O, 3 CH_2OH (Dihydroxyacetone phosphate) + 4 CHO, 5 H–C–OH, 6 $CH_2OPO_3^{2-}$ (Glyceraldehyde 3-phosphate)

Dihydroxyacetone phosphate —Triose phosphate isomerase→ 1 $CH_2OPO_3^{2-}$, 2 CHOH, 3 CHO

3,4 CHO, 2,5 CHOH, 1,6 $CH_2OPO_3^{2-}$ (Glyceraldehyde 3-phosphate) → → 3,4 COO^-, 2,5 C=O, 1,6 CH_3 (Pyruvate) —Pyruvate dehydrogenase complex→ SCoA, 2,5 C=O, 1,6 CH_3 (Acetyl CoA) + 2 CO_2

Carbons 1 and 6 of glucose end up as $-CH_3$ groups of acetyl CoA. and carbons 3 and 4 of glucose end up as CO_2.

29.8 Citrate and isocitrate are tricarboxylic acids.

29.9

Citrate (COO^-, CH_2, HO–C–COO^-, H–C–H, COO^-; B—enzyme) —1.→ [Aconitate (COO^-, CH_2, C=C(COO^-), CH, COO^-) + H_2O: —2.→ COO^-, CH_2, C=C(O^-)(O^-), $H_2\overset{+}{O}$–CH, COO^-] → Isocitrate (COO^-, CH_2, HC–COO^-, HO–CH, COO^-)

Enzyme-catalyzed E2 elimination of H_2O (1) is followed by nucleophilic conjugate addition of water (2) to produce an adduct that isomerizes to isocitrate.

29.10 The first step in the conversion of succinyl CoA to succinate is a nucleophilic acyl substitution by phosphate at the carbonyl carbon of succinyl CoA to form a mixed anhydride.

O=C(SCoA)–CH_2–CH_2–COO^- + $^-$:O–P(=O)(–O^-)–O^- → [$^-$:O–C(SCoA)(–O–P(=O)(–O^-)–O^-)–CH_2–CH_2–COO^-] → O=C(–O–P(=O)(–O^-)–O^-)–CH_2–CH_2–COO^- + $^-$SCoA

Nucleophilic attack by GDP at the phosphate group of the mixed anhydride results in formation of GTP and succinate. To simplify the mechanism, GMP is represented by R.

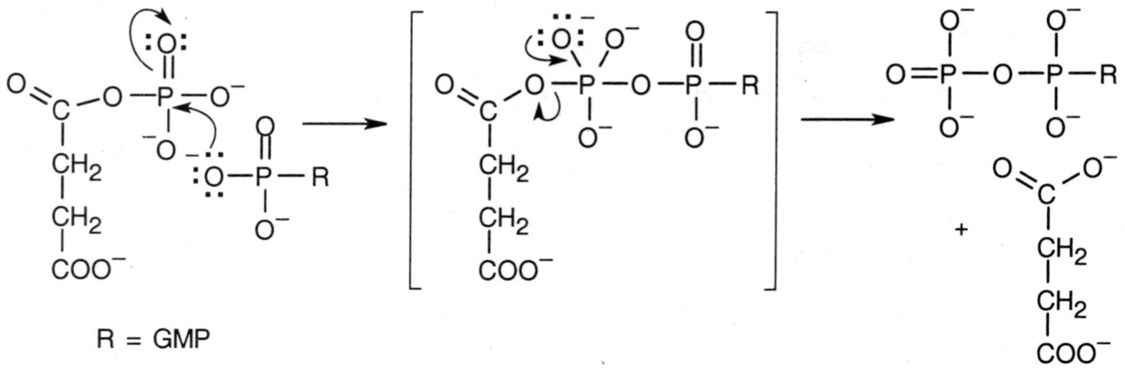

R = GMP

29.11 Position leucine and α-ketoglutarate so that the groups to be exchanged are aligned. This arrangement makes it easy to predict the products of transamination reactions.

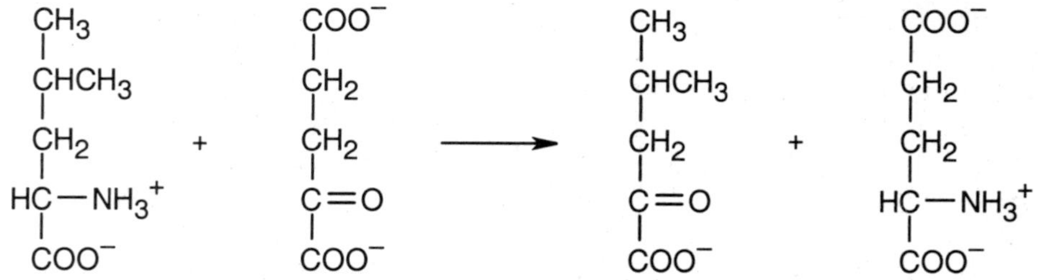

Leucine + α-Ketoglutarate → keto-acid + Glutamate

29.12 You might recognize that β-hydroxybutyryl ACP resembles the β-hydroxy ketones that were described in Chapter 23. These compounds dehydrate readily under both acidic and basic conditions. In this problem, the mechanism will be worked out using base catalysis.

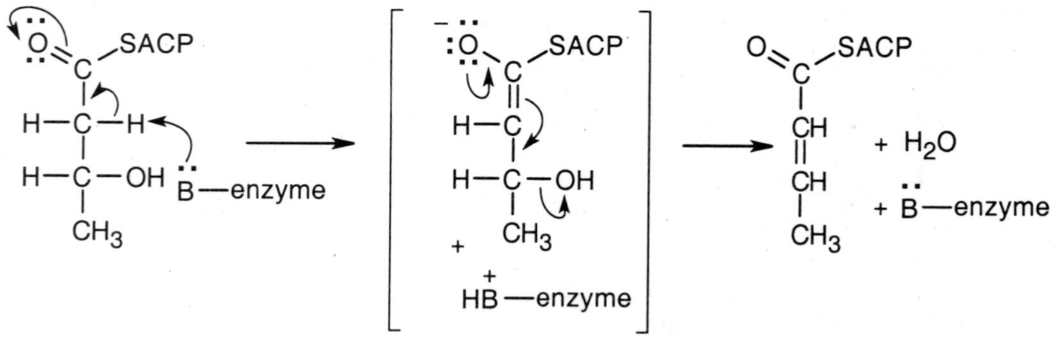

29.13 A fatty acid synthesized from $^{13}CH_3COOH$ has an alternating labeled and unlabeled carbon chain. The carboxylic acid carbon is unlabeled.

$$\overset{*}{C}H_3CH_2\overset{*}{C}H_2CH_2\overset{*}{C}H_2CH_2\overset{*}{C}H_2CH_2\overset{*}{C}H_2CH_2\overset{*}{C}H_2CH_2\overset{*}{C}H_2CH_2\overset{*}{C}H_2COOH$$

29.14 The mechanism consists of addition of hydride to the carbonyl group of 3-phosphoglyceroyl phosphate, followed by loss of phosphate.

$$\text{R–N (NADH/H}^+\text{)} \;+\; {}^{2-}O_3POCH_2CH(OH)C(=O)OPO_3^{2-} \longrightarrow \left[\text{R–N}^+ \text{ (NAD}^+\text{)} \;+\; {}^{2-}O_3POCH_2CH(OH)CH(O^-)OPO_3^{2-} \right] \longrightarrow HOPO_3^{2-} + {}^{2-}O_3POCH_2CH(OH)CHO$$

Glyceraldehyde 3-phosphate

29.15 Note the methyl carbon in the structure of pyruvate shown as the first structure in Figure 29.9. Follow the methyl group through to Steps 6–7 and notice that it has become carbon 3 of glyceraldehyde 3-phosphate, as well as the phosphate-bearing carbon of dihydroxyacetone phosphate. After the aldol condensation shown in Step 8, these two carbons become carbons 1 and 6 of fructose 1,6-bisphosphate and, ultimately, of glucose.

Look back to Problem 29.7, which shows the fate of the glucose carbons in glycolysis. In that problem, we showed that carbons 1 and 6 of glucose became the methyl carbons of pyruvate. Although gluconeogenesis isn't the exact reverse of glycolysis, the fate of the carbons is identical.

Visualizing Chemistry

29.16 The amino acid precursors are valine (a) and methionine (b).

(a)

$$CH_3CH(CH_3)CH(NH_3^+)COO^- + {}^-OOCCH_2CH_2C(=O)COO^- \longrightarrow CH_3CH(CH_3)C(=O)COO^- + {}^-OOCCH_2CH_2CH(NH_3^+)COO^-$$

Valine, α-Ketoglutarate, Glutamate

(b)

$$CH_3SCH_2CH_2CH(NH_3^+)COO^- + {}^-OOCCH_2CH_2C(=O)COO^- \longrightarrow CH_3SCH_2CH_2C(=O)COO^- + {}^-OOCCH_2CH_2CH(NH_3^+)COO^-$$

Methionine, α-Ketoglutarate, Glutamate

29.17 The intermediate is *S*-malate.

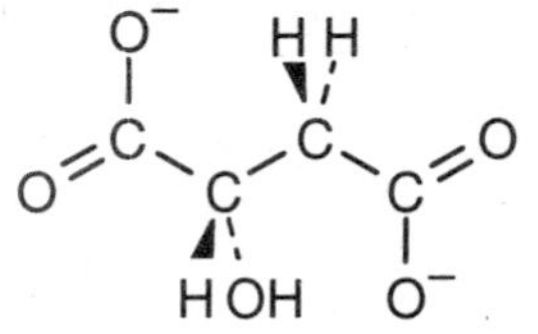

S-Malate

Additional Problems

29.18 Digestion is the breakdown of bulk food in the stomach and small intestine. Hydrolysis of amide, ester and acetal bonds yields amino acids, fatty acids, and simple sugars.

29.19 Metabolism refers to all reactions that take place inside cells. Digestion is a part of metabolism in which food is broken down into small organic molecules.

29.20 Metabolic processes that break down large molecules are known as catabolism. Metabolic processes that assemble larger biomolecules from smaller ones are known as anabolism.

29.21

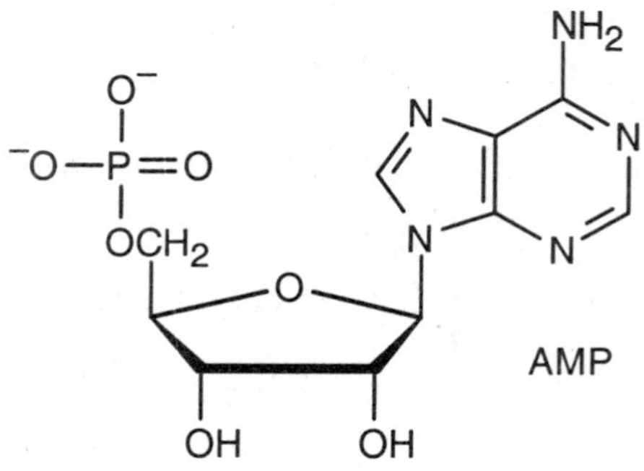

AMP

29.22

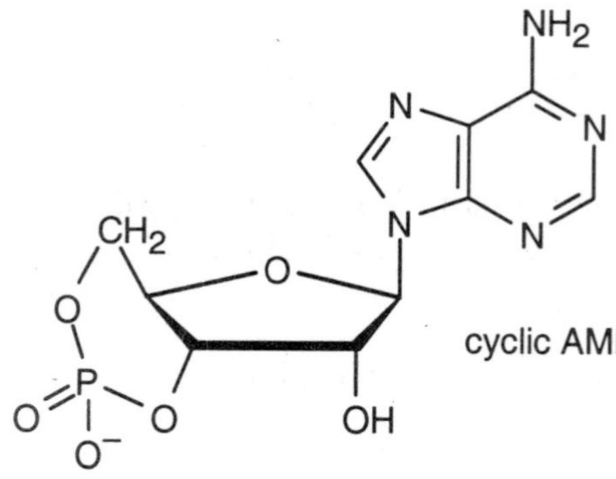

cyclic AMP

29.23 ATP transfers a phosphate group to another molecule in anabolic reactions.

29.24 NAD^+ is a biochemical oxidizing agent that converts alcohols to aldehydes or ketones, yielding NADH and H^+ as byproducts.

29.25 FAD is an oxidizing agent that introduces a conjugated double bond into a biomolecule, yielding $FADH_2$ as a byproduct.

29.26 The exact reverse of an energetically favorable reaction is energetically unfavorable. Since glycolysis is energetically favorable (negative $\Delta G°'$), its exact reverse has a positive $\Delta G°$; and is energetically unfavorable. Instead, glucose is synthesized by gluconeogenesis, an alternate pathway that also has a negative $\Delta G°'$.

29.27

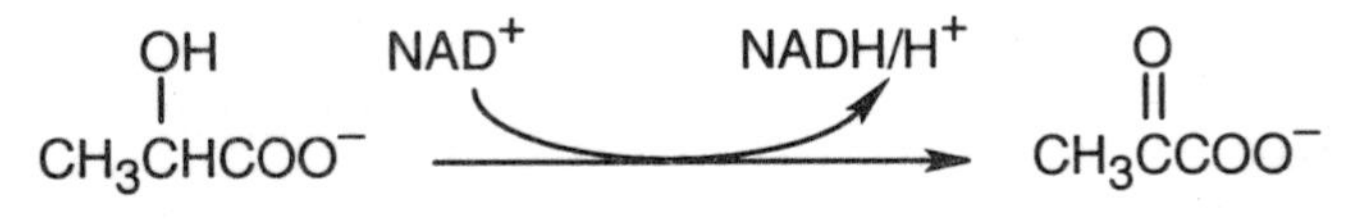

Lactate Pyruvate

NAD^+ is needed to convert lactate to pyruvate because the reaction involves the oxidation of an alcohol.

29.28 (a) One mole of glucose is catabolized to two moles of pyruvate, each of which yields one mole of acetyl CoA. Thus,

1.0 mol glucose —> 2.0 mol acetyl CoA

(b) A fatty acid with n carbons yields $n/2$ moles of acetyl CoA per mole of fatty acid. For palmitic acid ($C_{15}H_{31}COOH$),

$$1.0 \text{ mol palmitic acid} \times \frac{8 \text{ mol acetyl CoA}}{1 \text{ mol palmitic acid}} \longrightarrow 8.0 \text{ mol acetyl CoA}$$

(c) Maltose is a disaccharide that yields two moles of glucose on hydrolysis. Since each mole of glucose yields two moles of acetyl CoA,

1.0 mol maltose —> 2.0 mol glucose —> 4.0 mol acetyl CoA

29.29

	(a) Glucose	(b) Palmitic acid	(c) Maltose
Molecular weight	180.2 amu	256.4 amu	342.3 amu
Moles in 100.0 g	0.5549 mol	0.3900 mol	0.2921 mol
Moles of acetyl CoA produced	2 x 0.5549 mol =1.110 mol	8 x 0.3900 mol = 3.120 mol	4 x 0.2921 mol = 1.168 mol
Grams acetyl CoA produced	898.6 g	2526 g	945.6 g

29.30 Palmitic acid is the most efficient precursor of acetyl CoA on a weight basis.

29.31

$$\text{Glycerol} \xrightarrow{\text{ATP} \curvearrowright \text{ADP}} \text{Glycerol monophosphate} \xrightarrow{\text{NAD}^+ \curvearrowright \text{NADH/H}^+} \text{Glyceraldehyde 3-phosphate}$$

$$\xrightarrow{\text{NAD}^+/\text{P}_i \curvearrowright \text{NADH/H}^+} \text{3-Phosphoglyceroyl phosphate} \xrightarrow{\text{ADP} \curvearrowright \text{ATP}} \text{3-Phosphoglycerate}$$

$$\longrightarrow \text{2-Phosphoglycerate} \xrightarrow{\nearrow H_2O} \text{Phosphoenolpyruvate}$$

$$\xrightarrow{\text{ADP} \curvearrowright \text{ATP}} \text{Pyruvate} \xrightarrow[\text{HSCoA}]{\text{NAD}^+ \curvearrowright \text{NADH/H}^+} \text{Acetyl CoA} + CO_2$$

29.32

$$\underset{\text{Acetoacetyl CoA}}{CH_3\overset{O}{\overset{\|}{C}}CH_2\overset{O}{\overset{\|}{C}}SCoA} + HSCoA \longrightarrow 2\ \underset{\text{Acetyl CoA}}{CH_3\overset{O}{\overset{\|}{C}}SCoA}$$

29.33

(a)

$$CH_3CH_2CH_2CH_2CH_2\overset{O}{\overset{\|}{C}}SCoA \xrightarrow[\text{Acyl CoA dehydrogenase}]{\text{FAD} \curvearrowright \text{FADH}_2} CH_3CH_2CH_2CH{=}CH\overset{O}{\overset{\|}{C}}SCoA$$

(b)

$$CH_3CH_2CH_2CH{=}CH\overset{O}{\overset{\|}{C}}SCoA + H_2O \xrightarrow[\text{hydratase}]{\text{Enoyl CoA}} CH_3CH_2CH_2\overset{OH}{\overset{|}{C}}HCH_2\overset{O}{\overset{\|}{C}}SCoA$$

(c)

$$CH_3CH_2CH_2\overset{OH}{\overset{|}{C}}HCH_2\overset{O}{\overset{\|}{C}}SCoA \xrightarrow[\text{L-3-Hydroxyacyl CoA dehydrogenase}]{\text{NAD}^+ \curvearrowright \text{NADH/H}^+} CH_3CH_2CH_2\overset{O}{\overset{\|}{C}}CH_2\overset{O}{\overset{\|}{C}}SCoA$$

29.34

	Amino acid	*α-Keto acid*
(a)	$CH_3CH(OH)CH(NH_3^+)COO^-$	$CH_3CH(OH)C(=O)COO^-$
(b)	$C_6H_5-CH_2CH(NH_3^+)COO^-$	$C_6H_5-CH_2C(=O)COO^-$
(c)	$H_2NC(=O)CH_2CH(NH_3^+)COO^-$	$H_2NC(=O)CH_2C(=O)COO^-$

29.35 (a) Pyridoxal phosphate is the cofactor associated with transamination.
(b) Biotin is the cofactor associated with carboxylation of a ketone.
(c) Thiamine pyrophosphate is the cofactor associated with decarboxylation of an α-keto acid.

29.36 As we saw in Section 29.1, formation of glucose 6-phosphate from glucose and ATP is energetically favorable (negative $\Delta G^{\circ\prime}$) The reverse reaction, transfer of a phosphate group to ADP from glucose 6-phosphate, is energetically unfavorable and doesn't occur spontaneously. Phosphate transfers to ADP from either 3-phosphoglyceroyl phosphate or phosphoenolpyruvate have negative $\Delta G^{\circ\prime}$ values and are energetically favorable reactions.

In chemical terms, the leaving groups in the reactions of 3-phosphoglyceroyl phosphate (carboxylate) and phosphoenolpyruvate (enolate) are more stable anions than the leaving group in the reaction of glucose (alkoxide), so the reactions are more favorable.

$$RO-P(=O)(O^-)-O^- + :Nu \longrightarrow RO^- + {}^-O-P(=O)(O^-)-Nu$$

	Glucose	3-Phosphoglyceroyl phosphate	Phosphoenolpyruvate
RO^- =	CHO / $H-C-OH$ / $HO-C-H$ / $H-C-OH$ / $H-C-OH$ / CH_2O^-	COO^- / $H-C-OH$ / $CH_2OPO_3^{2-}$	COO^- / $C-O^-$ / $\Vert$ / CH_2

29.37

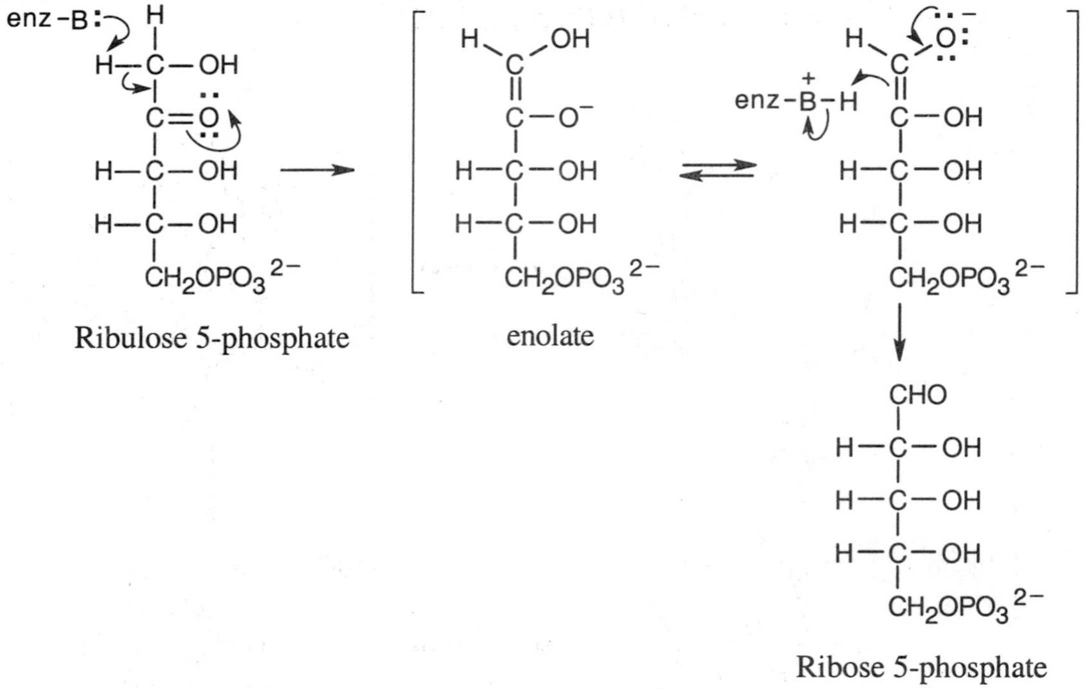

Ribulose 5-phosphate

enolate

Ribose 5-phosphate

The isomerization of ribulose 5-phosphate to ribose 5-phosphate occurs by way of an intermediate enolate.

29.38 This is a reverse aldol reaction, similar to step 4 of glycolysis.

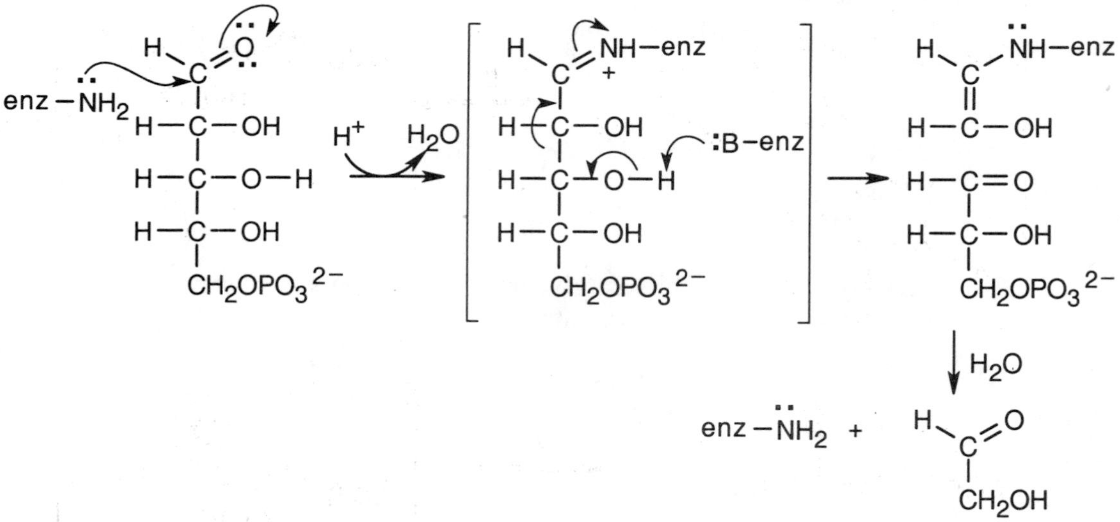

29.39 The steps in the conversion α-ketoglutarate —> succinyl CoA are similar to steps in the conversion pyruvate —> acetyl CoA (shown in Figure 29.5), and the same coenzymes are involved: lipoamide, thiamine, HSCoA and NAD^+.

1 2 3 4 5 6 7 8

NAD+ NADH/H+

An outline of the mechanism: (1) nucleophilic addition of thiamine (2) decarboxylation (3) addition of double bond to lipoamide, with ring opening (4) elimination of thiamine pyrophosphate (5) nucleophilic addition of acetyl CoA to succinyl lipoamide (6) proton transfer (7) elimination of dihydrolipoamide to give succinyl CoA (8) reoxidation of dihydrolipoamide to lipoamide

29.40

2 $CH_3COSCoA$ (Acetyl CoA) —HSCoA→ $CH_3COCH_2COSCoA$ (Acetoacetyl CoA) —H_2O, −HSCoA→ $CH_3COCH_2CO_2^-$ (Acetoacetate)

Acetoacetate —−CO_2→ CH_3COCH_3 (Acetone)

Acetoacetate —NADH/H^+ → NAD^+→ $CH_3CH(OH)CH_2CO_2^-$ (3-Hydroxybutyrate)

Acetone and 3-Hydroxybutyrate: ketone bodies

29.41

$H_3C\text{–}C(=O)\text{SCoA}$ + $^-{:}CH_2\text{–}C(=O)\text{SCoA}$ → $[H_3C\text{–}C(O^-)(\text{SCoA})\text{–}CH_2CSCoA]$ → $CH_3COCH_2COSCoA$ + ^-SCoA

29.42 The first sequence of steps in this mechanism involves formation of the imine (Schiff base) of sedoheptulose 7-phosphate, followed by retro-aldol cleavage to form erythrose 4-phosphate and the Schiff base of dihydroxyacetone.

enz–NH_2 + CH_2OH–C(=O)–CH(OH)–CH(OH)–CH(OH)–CH(OH)–$CH_2OPO_3^{2-}$ → [CH_2OH–C(=$\overset{+}{N}$H–enz)–HO–C–H–H–C–O–H (B–enz)–CH(OH)–CH(OH)–$CH_2OPO_3^{2-}$] → CH_2OH–C(–NH–enz)=C(HO)H + CHO–CH(OH)–CH(OH)–$CH_2OPO_3^{2-}$

Erythrose 4-phosphate

The Schiff base of dihydroxyacetone undergoes an aldol-like condensation with glyceraldehyde 3-phosphate to yield fructose 6-phosphate. This reaction is almost identical to the reaction pictured for Step 8 of gluconeogenesis in Section 29.8.

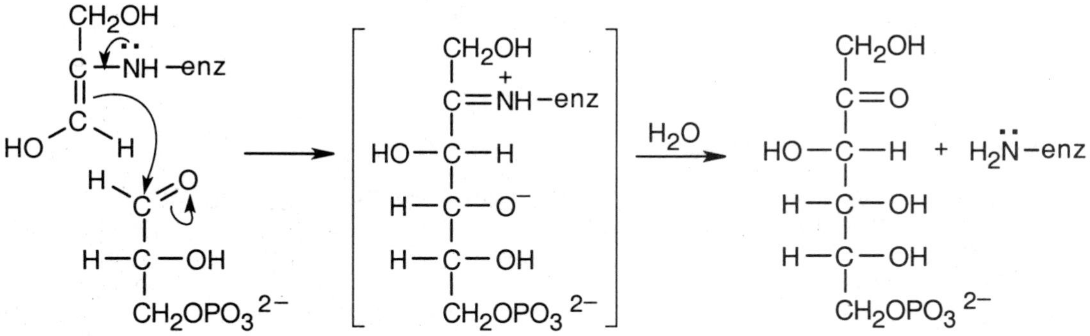

CH_2OH NH–enz HO H H O H–C–OH $CH_2OPO_3^{2-}$ → [CH_2OH C=NH–enz HO–C–H H–C–O⁻ H–C–OH $CH_2OPO_3^{2-}$] $\xrightarrow{H_2O}$ CH_2OH C=O HO–C–H H–C–OH H–C–OH $CH_2OPO_3^{2-}$ + H_2N–enz

Fructose 6-phosphate

29.43

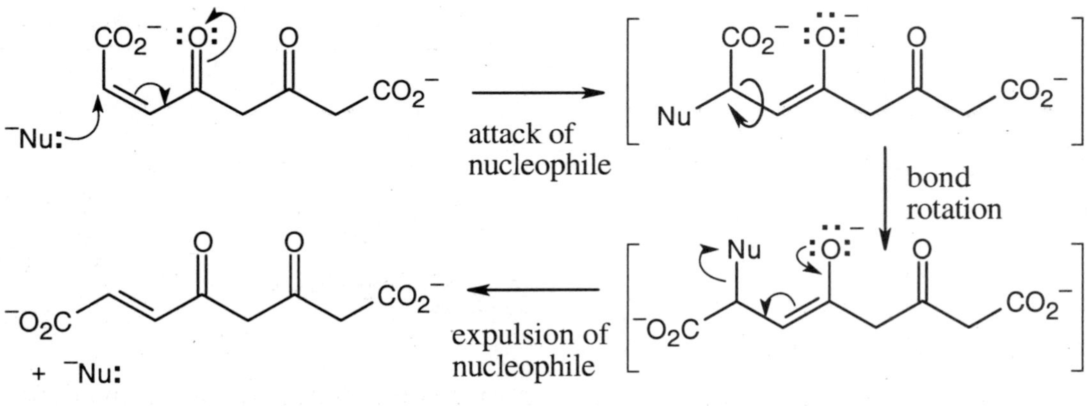

CO_2^- :O: O CO_2^- ⁻Nu: → attack of nucleophile [CO_2^- :O: O Nu CO_2^-] ↓ bond rotation [Nu :O: O ⁻O_2C CO_2^-] ← expulsion of nucleophile ⁻O_2C O O CO_2^- + ⁻Nu:

29.44

:O: O ⁻O_2C CO_2^- ⁻:B–enz → nucleophilic addition of enzyme [:O: :O: ⁻O_2C CO_2^- B-enz] ↓ *retro*-Claisen condensation [O ⁻O_2C B-enz + :O: CO_2^-] $\xleftarrow{H_2O}$ hydrolysis of enzyme-substrate complex ⁻O_2C CO_2^- + ⁻:B–enz + O CO_2^-

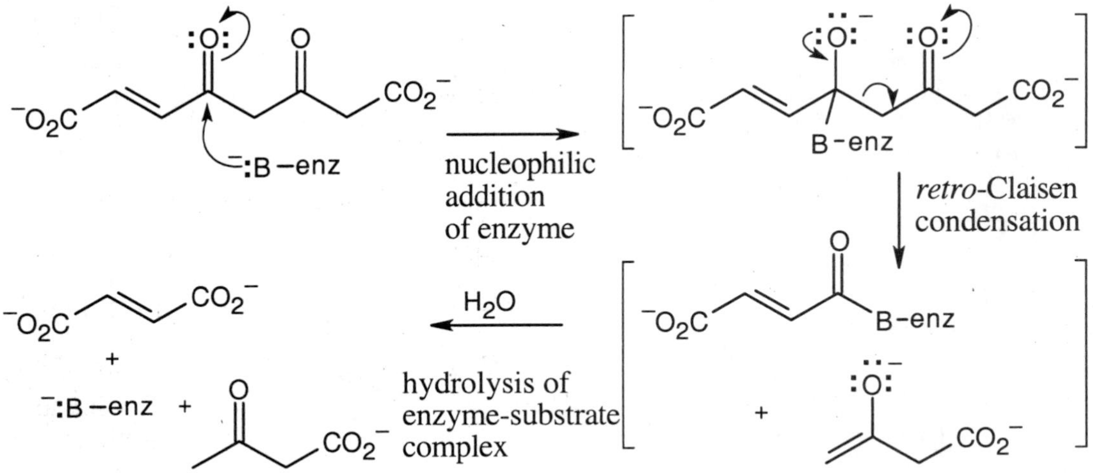

29.45 The first step in the conversion acetoacetate ——> acetyl CoA is the formation of acetoacetyl CoA. This reaction also occurs as the first step in fatty acid catabolism. Although we haven't studied the mechanism, it involves formation of a mixed anhydride.

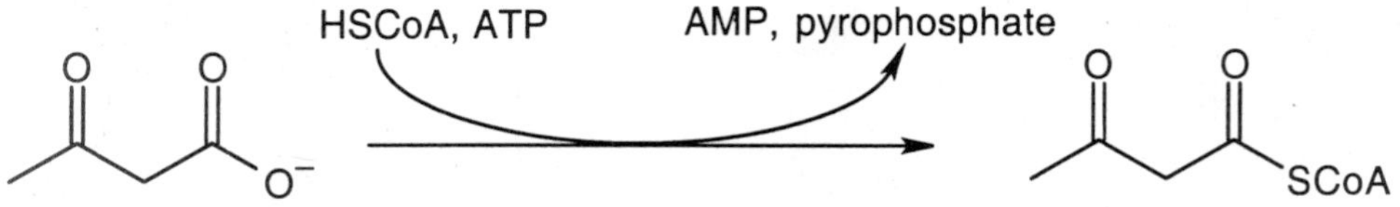

HSCoA, ATP   AMP, pyrophosphate

O O ... O^- ——> O O ... SCoA

The final step is a retro-Claisen reaction, whose mechanism is pictured in Section 29.2 as Step 4 of β-oxidation of fatty acids.

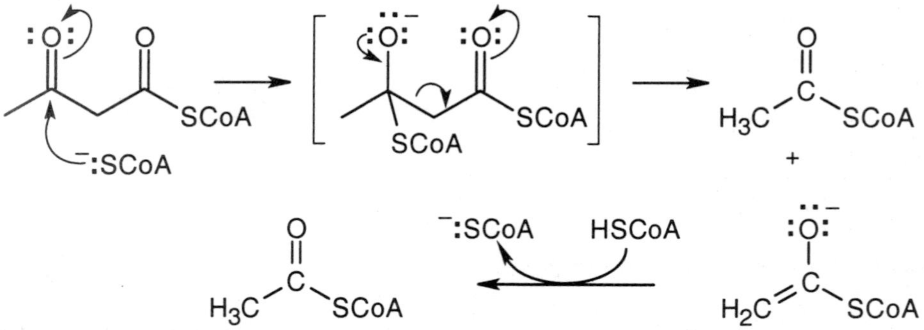

:O: O ... SCoA ——> [:O:⁻ :O: ... SCoA, SCoA] ——> H_3C–C(=O)–SCoA + H_2C=C(–:O:⁻)–SCoA

⁻:SCoA

⁻:SCoA   HSCoA

H_3C–C(=O)–SCoA ←— H_2C=C(–:O:⁻)–SCoA

29.46

(a)

$^{2-}O_3PO$ ... NH_2 + CO_2^- ... O=C ... CO_2^- ——> $^{2-}O_3PO$ ... N=C ... CO_2^- ... CO_2^- + H_2O

H–N+ ... OH ... CH_3

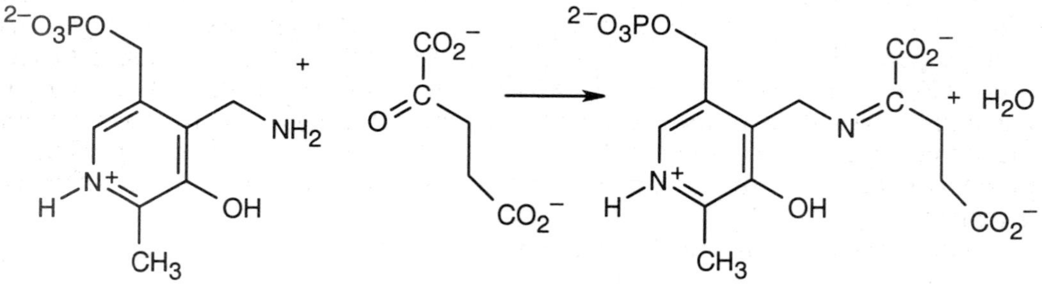

The imine is formed by nucleophilic addition of the amine to the keto group of α-ketoglutarate, followed by loss of water.

(b)

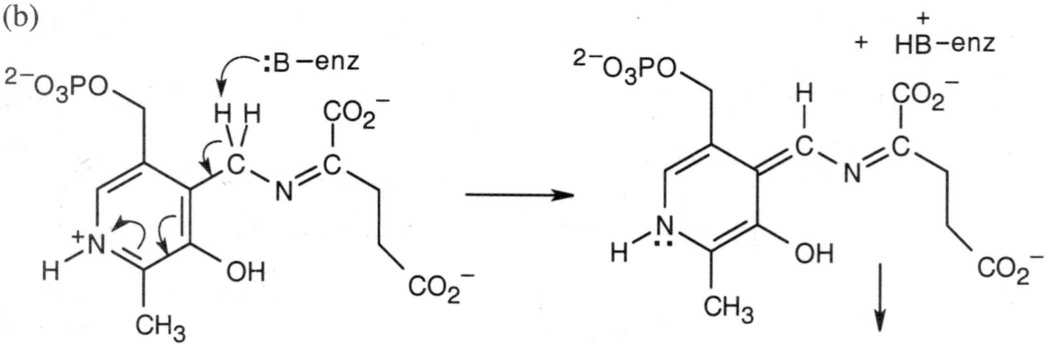

:B–enz

$^{2-}O_3PO$ ... H H ... C ... N=C ... CO_2^- ... CO_2^- ——> $^{2-}O_3PO$ ... H–C ... N=C ... CO_2^- ... CO_2^- + $\overset{+}{H}B$–enz

H–+N ... OH ... CH_3 H–N: ... OH ... CH_3

↓

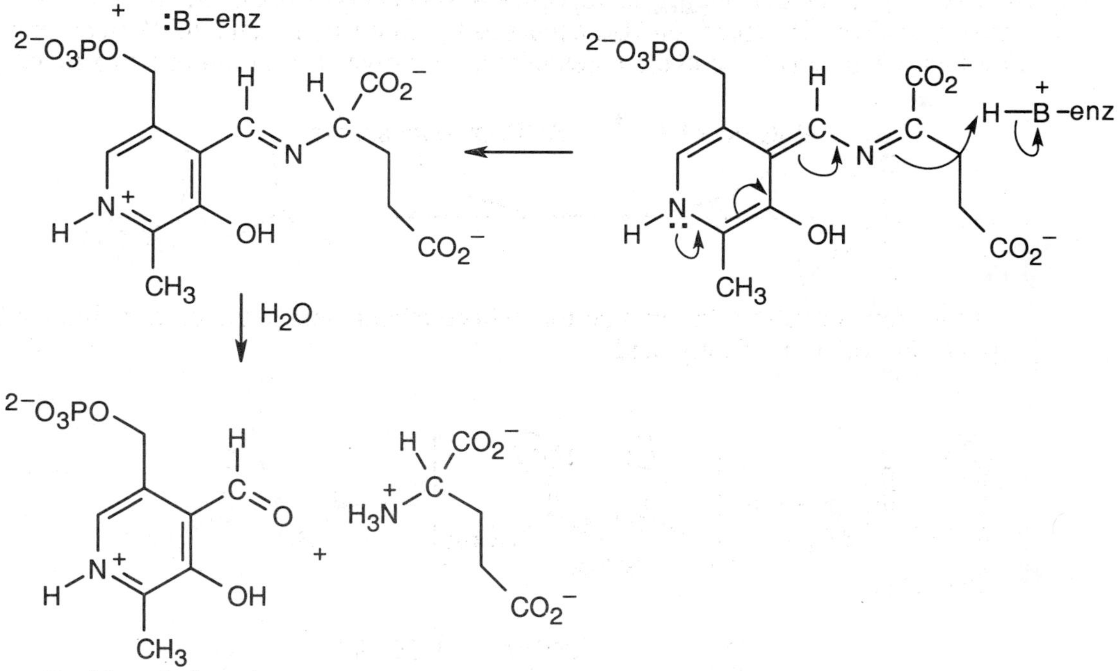

Pyridoxal phosphate Glutamate

Enzyme-catalyzed deprotonation, followed by π bond rearrangement, gives a second imine. This imine undergoes protonation and bond rearrangement to give a third imine, which, when hydrolyzed, yields pyridoxal phosphate and glutamate.

29.47 Now is a good time to use retrosynthetic analysis, which you first encountered in Chapter 8. In this degradative pathway, what might be the precursor to acetyl CoA (the final product)? Pyruvate is a good guess, because we learned how to convert pyruvate to acetyl CoA in Section 29.4. How do we get from serine to pyruvate? A transamination reaction is a possibility. However, the immediate transamination precursor to pyruvate is the amino acid alanine, which differs from serine by one hydroxyl group. Thus, we probably have to design a pathway from serine to pyruvate that takes this difference into account.

Many routes are possible, but here's the simplest:

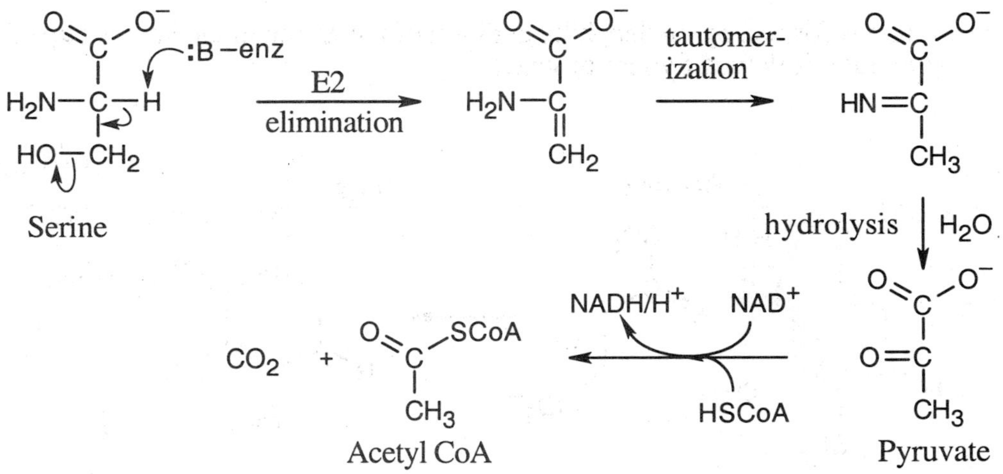

Serine — E2 elimination → tautomerization → hydrolysis (H_2O) → Pyruvate → Acetyl CoA + CO_2

The coenzymes thiamine pyrophosphate and lipoamide are involved in the last step.

29.48

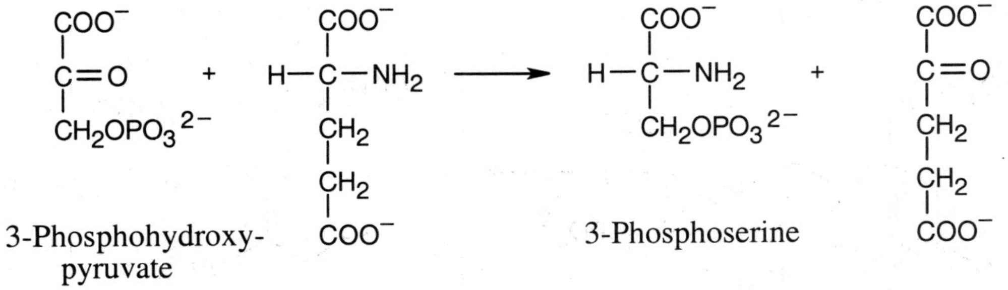

$$\underset{\text{3-Phosphohydroxypyruvate}}{\begin{matrix}COO^-\\ |\\ C{=}O\\ |\\ CH_2OPO_3{}^{2-}\end{matrix}} + \begin{matrix}COO^-\\ |\\ H{-}C{-}NH_2\\ |\\ CH_2\\ |\\ CH_2\\ |\\ COO^-\end{matrix} \longrightarrow \underset{\text{3-Phosphoserine}}{\begin{matrix}COO^-\\ |\\ H{-}C{-}NH_2\\ |\\ CH_2OPO_3{}^{2-}\end{matrix}} + \begin{matrix}COO^-\\ |\\ C{=}O\\ |\\ CH_2\\ |\\ CH_2\\ |\\ COO^-\end{matrix}$$

This reaction is a transamination that requires the coenzyme pyridoxal phosphate as a cofactor. The mechanism, which is described in Figure 29.7 and Problem 29.46, involves two steps. The first step is the nucleophilic addition of glutamate nitrogen to the aldehyde group of pyridoxal phosphate to yield an imine intermediate, which is hydrolyzed to give α-ketoglutarate plus a nitrogen-containing pyridoxal phosphate byproduct.

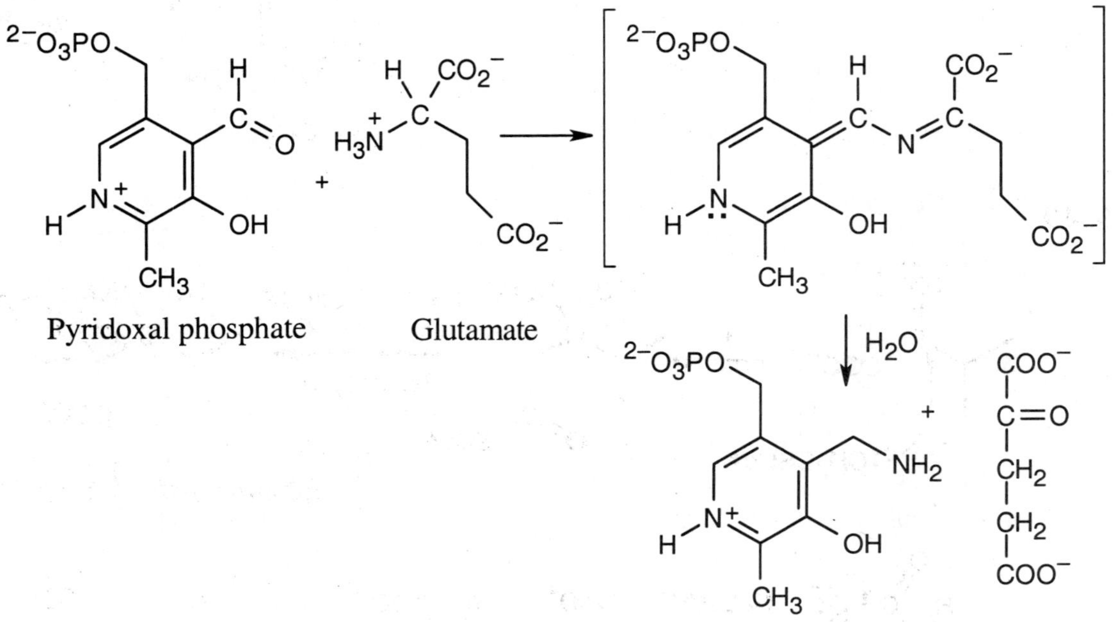

Pyridoxal phosphate + Glutamate → [imine intermediate] $\xrightarrow{H_2O}$ pyridoxamine phosphate (NH_2) + α-ketoglutarate

Pyridoxal phosphate

Glutamate

This byproduct reacts with 3-phosphohydroxypyruvate to give 3-phosphoserine plus regenerated pyridoxal phosphate.

$^{2-}O_3PO$ · NH_2 · H · N^+ · OH · CH_3 · + · COO^- · $C{=}O$ · $CH_2OPO_3{}^{2-}$ → [$^{2-}O_3PO$ · $N{=}C$ · COO^- · $CH_2OPO_3{}^{2-}$ · H · N^+ · OH · CH_3]

↓ H_2O

$^{2-}O_3PO$ · H · $C{=}O$ · H · N^+ · OH · CH_3 + COO^- · $H{-}C{-}NH_2$ · $CH_2OPO_3{}^{2-}$

29.49

$:\ddot{O}:$ · C · COO^- · $^-:CH_2CSCoA$ (with O above C) → 1 → ^-O · COO^- · C · $O{=}C$ · $SCoA$ · H_2O · $HSCoA$ · 2 · hydrolysis → HO · COO^- · C · H · H · COO^- · $:B{-}enz$

E2 elimination ↓ 3

COO^- · $C{=}CHCOO^-$ · $H_2\ddot{O}:$ → 4 → conjugate addition → H · COO^- · C · $CCOO^-$ · H · OH · $NADH/H^+$ · NAD^+ · 5 · oxidation → O · H · $C{-}\ddot{O}:^-$ · C · $CCOO^-$ · O

6 ↓ decarboxylation of a β-keto acid

COO^- · C · O + CO_2

29.50

(a)

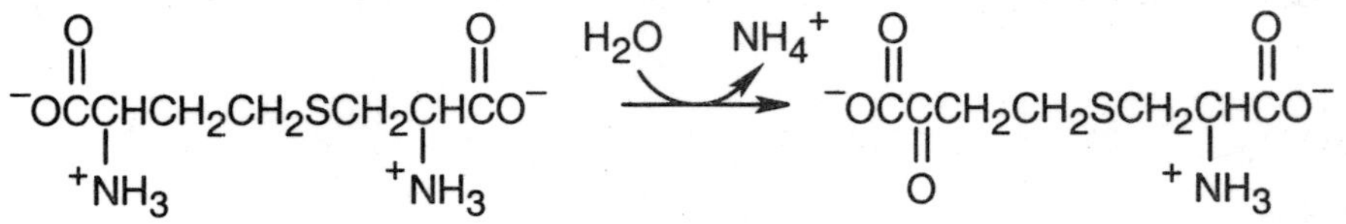

$$^{-}OC(=O)CH(NH_3^{+})CH_2CH_2SCH_2CH(NH_3^{+})CO_2^{-} \xrightarrow{H_2O \;\; NH_4^{+}} {}^{-}OC(=O)C(=O)CH_2CH_2SCH_2CH(NH_3^{+})CO_2^{-}$$

(b)

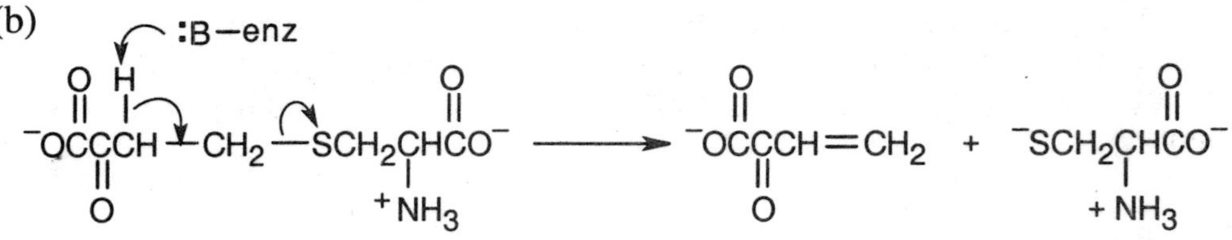

:B–enz

$$^{-}OC(=O)C(=O)CH(H)-CH_2-SCH_2CH(NH_3^{+})CO_2^{-} \longrightarrow {}^{-}OC(=O)C(=O)CH{=}CH_2 + {}^{-}SCH_2CH(NH_3^{+})CO_2^{-}$$

Cysteine

(c)

$$^{-}OC(=O)C(=O)CH{=}CH_2 \xrightarrow{FADH_2 \;\; FAD} {}^{-}OC(=O)C(=O)CH_2CH_3$$

α-Ketobutyrate

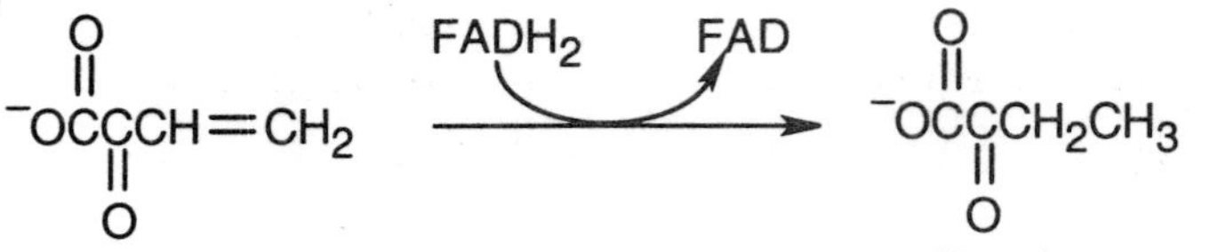

The product of double-bond reduction is α-ketobutyrate. $FADH_2$ is the necessary enzyme cofactor.

Molecular Modeling

29.51 Both reactions shown below are highly exothermic. For oxidation of propane to propanal, $\Delta H° = -214$ kJ/mol, and for oxidation of propanal to glyceraldehyde, $\Delta H° = -220$ kJ/mol.

29.52 Hydroxypropanone enolizes more readily because the enolization energy of hydroxypropanone (41 kJ/mol) is less than the enolization energy of acetone (48 kJ/mol).

29.53 The negative charge in the thiamine model is localized on carbon and is in the σ system. The ketone carbon of pyruvate is most positive. These data suggest the following addition product:

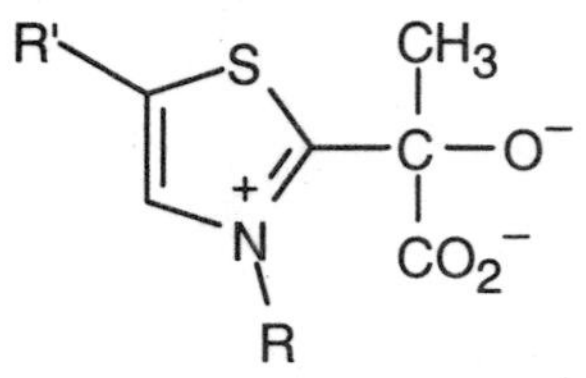

R', S, CH_3, C–O^-, N^+, CO_2^-, R

Review Unit 11: Biomolecules II – Lipids, Nucleic Acids, Metabolic Pathways

Major Topics Covered (with vocabulary):

Lipids:
wax fat oil triacylglycerol fatty acid polyunsaturated fatty acid soap saponification micelle phosphoglyceride sphingolipid lipid bilayer sphingosine sphingomyelin prostaglandin terpene essential oil isoprene rule monoterpene sesquiterpene steroid hormone sex hormone adrenocortical hormone androgen estrogen mineralocorticoid glucocorticoid squalene lanosterol

Heterocycles;
pyrrole thiophene furan pyridine fused-ring heterocycle pyrimidine purine

Nucleic acids and nucleotides:
nucleoside nucleotide deoxyribonucleic acid (DNA) ribonucleic acid (RNA) adenine guanine thymine cytosine 3' end 5' end base pairing double helix complementary pairing major groove minor groove intercalation

Nucleic acids and heredity:
replication semiconservative replication fork DNA ligase transcription mRNA rRNA tRNA coding strand template strand promoter sites exon intron translation codon anticodon

DNA technology:
DNA sequencing Maxam-Gilbert method restriction endonuclease restriction fragment autoradiography dideoxy method DNA synthesis DMT ether phosphoramidite phosphite polymerase chain reaction (PCR)

Metabolic pathways:
metabolism anabolism catabolism digestion phosphoric acid anhydride ATP NAD^+ $NADH/H^+$ β-oxidation pathway glycolysis Schiff base pyruvate acetyl CoA pyruvate dehydrogenase complex thiamine lipoamide citric acid cycle transamination oxidative deamination gluconeogenesis biotin

Types of Problems:

After studying these chapters you should be able to:

– Draw the structures of fats, oils, steroids and other lipids.
– Determine the structure of a fat.
– Predict the products of reactions of fats and steroids.
– Locate the isoprene units in terpenes.
– Understand the mechanism of terpene and steroid biosynthesis.
– Draw the structures and conformations of steroids and other fused-ring systems.

– Draw orbital pictures of heterocycles and explain their acid-base properties.
– Explain orientation and reactivity in heterocyclic reactions, and predict the products of reactions involving heterocycles.

- Formulate mechanisms of reactions involving heterocycles.
- Draw purines, pyrimidines, nucleosides, nucleotides, and representative segments of DNA and their complements.
- List the base sequence that codes for a given amino acid or peptide.
- Deduce an amino acid sequence from a given mRNA sequence (and *vice versa*).
- Draw the anticodon sequence of tRNA, given the mRNA sequence.
- Outline the process of DNA sequencing, and deduce a DNA sequence from an electrophoresis pattern.
- Outline the method of DNA synthesis, and formulate the mechanisms of synthetic steps.

- Explain the basic concepts of metabolism, and understand the energy relationships of biochemical reactions.
- Answer questions relating to the metabolic pathways of carbohydrates, fatty acids and amino acids.
- Formulate mechanisms for metabolic pathways similar to those in the text.

Points to Remember:

* When trying to locate the isoprene units in a terpene, look for an isopropyl group first; at least one should be apparent. After finding it, count 5 carbons, and locate the second isoprene unit. If there are two possibilities for the second unit, choose the one that has the double bond in the correct location.

* In general, the reactions of steroids that are presented in this book are familiar and uncomplicated. Keeping track of the stereochemistry of the tetracyclic ring system is somewhat more complicated.

* In situations where base-pairing occurs, such as replication, transcription or translation, a polynucleotide chain (written with the 5' end on the left and the 3' end on the right) pairs with a second chain (written with the 3' end on the left and the 5' end on the right). Base pairing is complementary, and the two chains are always read in opposite directions.

* When ^{32}P-labeled DNA restriction fragments are treated with reagents that cause splitting, only a few of the many possible splittings occur. If all possible splittings occurred, many very small, unlabeled fragments would be produced, and little useful information would be found.

* Note the difference between transamination and oxidative deamination. Transamination is a reaction in which an amino group of an α-amino acid is transferred to α-ketoglutarate, yielding an α-keto acid and glutamate. In oxidative deamination, glutamate loses its amino group in an NAD^+-dependent reaction that regenerates α-ketoglutarate and produces NH_4^+.

* Look at the steps of glycolysis, and then look at the steps of gluconeogenesis. Several steps in one pathway are the exact reverse of steps in the other pathway because the energy required for these steps is small. Other, high-energy transformations must occur by steps that are not the exact reverse and that require different enzymes. Gluconeogenesis is a metabolic pathway that takes place mainly during fasting and strenuous exercise because dietary sources of carbohydrates are usually available.

* The conversion pyruvate $\rightarrow$ acetyl CoA is catalyzed by pyruvate dehydrogenase complex. The conversion acetyl CoA $\rightarrow$ carbohydrates doesn't occur in animals because they can obtain carbohydrates from food and don't usually need to synthesize carbohydrates. Only plants can, at times, use acetyl CoA to synthesize carbohydrates.

Self-test:

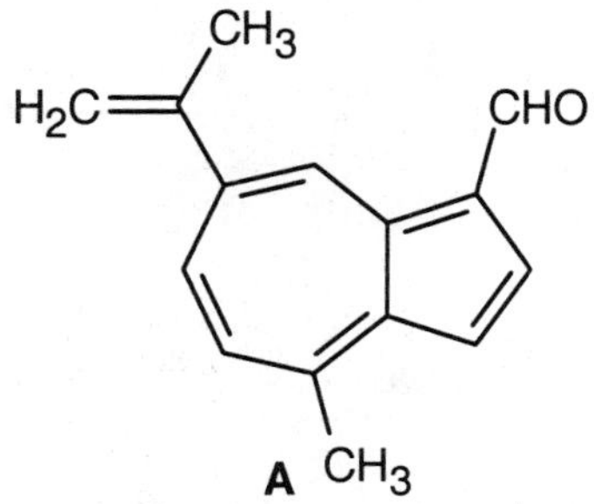

A — Lactaroviolin (an antibiotic)

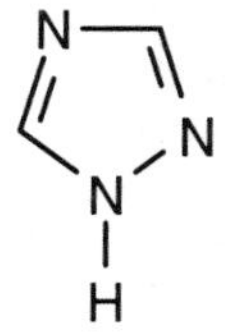

B — 1,2,4-Triazole

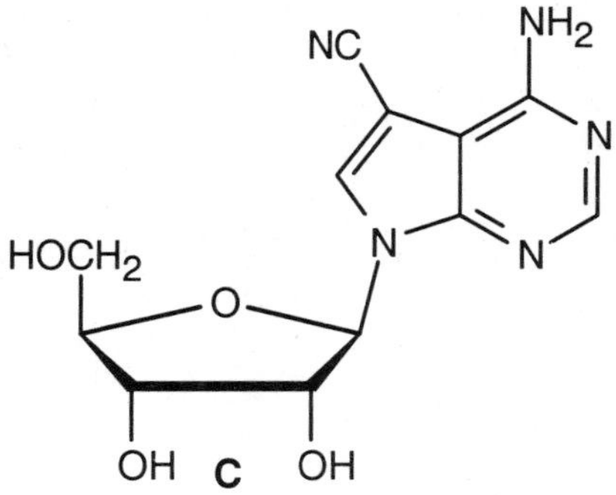

C — Toyocamycin (an antibiotic)

What type of terpene is **A**? Show the location of the isoprene units.

One of the nitrogens of 1,2,4-triazole (**B**) is less basic than the others. Which one is it? What product do you expect when **B** is treated with Br_2, $FeBr_3$?

Toyocamycin (**C**) is related to which nucleoside? What are the differences between **C** and the nucleoside?

3'   5'
–ACG–CCT–TAG–GGC–TTA–GGA–

D

D represents a segment of the template strand of a molecule of DNA. Draw: (a) the coding strand; (b) the mRNA that is synthesized from **D** during transcription; (c) the tRNA anticodons that are complementary to the mRNA codons; (d) the amino acids that form the peptide that **D** codes for.

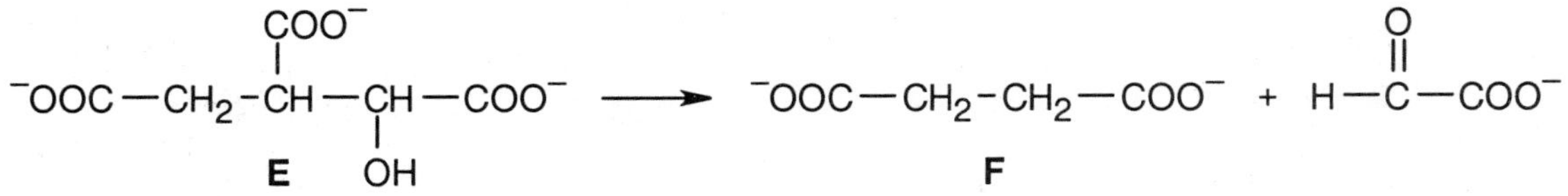

$$^{-}OOC-CH_2-CH(COO^{-})-CH(OH)-COO^{-} \longrightarrow {}^{-}OOC-CH_2-CH_2-COO^{-} + H-C(=O)-COO^{-}$$

E (reactant), **F** (product)

The above reaction is part of a metabolic pathway that occurs in plants. Identify **E** and **F**. What type of reaction is taking place? Do think that NAD^+, FAD, or ATP are needed for this reaction to occur?

Multiple choice:

1. Which type of molecule is most likely to be found in a lipid bilayer?
(a) triacylglycerol (b) prostaglandin (c) sphingomyelin (d) triterpene

2. Which of the following terpenes might have been formed by a tail-to-tail coupling?
(a) monoterpene (b) sesquiterpene (c) diterpene (d) triterpene

3. Prostaglandins and related compounds have all of the following structural features in common except:
(a) cis double bonds (b) a carboxylic acid group (c) a C_{20} chain (d) hydroxyl groups

4. Which type of reaction is least likely to occur?
(a) electrophilic aromatic substitution of pyrrole (b) nucleophilic aromatic substitution of pyrrole (c) electrophilic aromatic substitution of pyridine (d) nucleophilic aromatic substitution of pyridine

5. Which nucleic acid has nonstandard bases, in addition to the usual bases?
(a) DNA (b) mRNA (c) rRNA (d) tRNA

6. Which DNA base doesn't need a protecting group in sequencing or in synthesis?
(a) Thymine (b) Cytosine (c) Adenine (d) Guanine

7. Which amino acid has only one codon?
(a) Tyrosine (b) Arginine (c) Lysine (d) Tryptophan

8. Which of the following enzyme cofactors is not involved in the conversion of pyruvate to acetyl CoA?
(a) Thiamine pyrophosphate (b) Pyridoxal phosphate (c) Lipoamide (d) NAD^+

9. Which of the following steps of the citric acid cycle doesn't produce reduced coenzymes?
(a) Isocitrate $\rightarrow$ α-Ketoglutarate (b) α-Ketoglutarate $\rightarrow$ Succinyl CoA
(c) Fumarate $\rightarrow$ Malate (d) Succinate $\rightarrow$ Fumarate

10. The amino acid aspartate can be metabolized as what citric acid cycle intermediate after transamination?
(a) Oxaloacetate (b) Malate (c) α-Ketoglutarate (d) Succinate

Chapter 30 – Orbitals and Organic Chemistry

Chapter Outline

I. Molecular orbitals (Sections 30.1 – 30.2).
- A. Molecular orbitals of conjugated π systems (Section 30.1).
  1. The *p* orbitals of the sp^2-hybridized carbons of a polyene interact to form a set of π molecular orbitals.
  2. The energies of these orbitals depend on the number of nodes they have.
     - a. The molecular orbitals with fewer nodes are bonding MOs.
     - b. The molecular orbitals with more nodes are antibonding MOs.
  3. A molecular orbital description can be used for any conjugated π system.
     - a. In the ground state, only the bonding orbitals are used.
     - b. On irradiation with UV light, an electron is promoted to an antibonding orbital. This is known as an excited state.
- B. Molecular orbitals and pericyclic reactions (Section 30.2).
  1. The mechanisms of pericyclic reactions can be explained by molecular orbital theory.
     - a. A pericyclic reaction can take place only if the lobes of the reactant MOs have the correct algebraic sign in the transition state.
     - b. If the symmetries of both reactant and product orbitals correlate, the reaction is symmetry-allowed.
     - c. If the symmetries don't correlate, the reaction is symmetry-disallowed. The reaction may still take place, but only by a nonconcerted, high-energy pathway.
  2. A modification of MO theory states that only two MOs need be considered (frontier orbitals):
     - a. The highest occupied molecular orbital (HOMO).
     - b. The lowest occupied molecular orbital (LUMO).

II. Electrocyclic reactions (Sections 30.3 – 30.5).
- A. General description of electrocyclic reactions Section 30.3).
  1. Nature of electrocyclic reactions.
     - a. An electrocyclic reaction involves the cyclization of a conjugated polyene. One π bond is broken, a new σ bond is formed and a cyclic compound results.
     - b. Electrocyclic reactions are reversible.
       - i. The triene-cyclohexadiene equilibrium favors the ring-closed product.
       - ii. The diene-cyclobutene equilibrium favors the ring-opened product.
  2. Stereochemistry of electrocyclic reactions.
     - a. A specific *E,Z* bond isomer yields a specific cyclic stereoisomer under thermal conditions.
     - b. The stereochemical results are opposite when the reactions are carried out under photochemical conditions.
  3. Orbital explanation for outcomes of electrocyclic reactions.
     - a. The signs of the outermost lobes of the interacting orbitals explain these results. For a bond to form, the lobes must be of the same sign.
     - b. The outermost π lobes of the polyene must rotate so that the lobes that form the bonds are of the same sign.
       - i. If the lobes are on the same side of the molecule, the lobes must rotate in opposite directions – disrotatory motion.

ii. If the lobes of the same sign are on opposite sides of the polyene, both lobes must rotate in the same direction – conrotatory motion.

B. Stereochemistry of thermal electrocyclic reactions (Section 30.4).
  1. The stereochemistry of an electrocyclic reaction is determined by the symmetry of the polyene HOMO.
  2. The ground-state electronic configuration is used to identify the HOMO for thermal reactions.
     a. For trienes, the HOMO has lobes of like sign on the same side of the molecule, and ring-closure is disrotatory.
     b. For dienes, ring closing is conrotatory.
  3. In general, polyenes with odd numbers of double bonds undergo disrotatory thermal electrocyclic reactions, and polyenes with even numbers of double bonds undergo conrotatory thermal electrocyclic reactions.

C Stereochemistry of photochemical electrocyclic reactions (Section 30.5).
  1. UV irradiation of a polyene causes excitation of one electron from the ground-state HOMO to the ground-state LUMO.
  2. UV irradiation changes the symmetry of HOMO and LUMO and also changes the reaction stereochemistry.
     a. Photochemical electrocyclic reactions of trienes occur with conrotatory motion.
     b. Photochemical electrocyclic reactions of dienes occur with disrotatory motion.
  3. Thermal and photochemical electrocyclic reactions always take place with opposite stereochemistry.

III. Cycloaddition reactions (Sections 30.6 – 30.7).

A. General description of cycloaddition reactions (Section 30.6).
  1. A cycloaddition reaction is a reaction in which two unsaturated molecules add to give a cyclic product.
  2. Cycloadditions are controlled by the orbital symmetry of the reactants.
     Reactions that are symmetry-disallowed either don't take place or occur by a higher-energy nonconcerted pathway.
  3. The Diels-Alder cycloaddition is an example.
     a. Reaction occurs between a diene and a dienophile to yield a cyclic product.
     b. The products have a specific stereochemistry.
     c. The reaction is known as a [4 + 2] cycloaddition.
  4. Cycloadditions can only occur if the terminal π lobes have the correct stereochemistry.
     a. In suprafacial cycloadditions, a bonding interaction takes place between lobes on the same face of one reactant and lobes on the same face of the other reactant.
     b. Antarafacial cycloadditions occur between lobes on the same face of one reactant and lobes on opposite faces of the other reactant.
     c. Often, antarafacial cycloadditions are symmetry-allowed but geometrically constrained.

B. Stereochemistry of cycloadditions (Section 30.7).
  1. A cycloaddition reaction takes place when a bonding interaction occurs between the HOMO of one reactant and the LUMO of the other reactant.
  2. The symmetries of the terminal lobes of the HOMO and LUMO of the reactants in a [4 + 2] thermal cycloaddition allow the reaction to proceed with suprafacial geometry.
  3. For [2 + 2] cycloadditions:
     a. Orbital symmetry shows that thermal cyclization must occur by an antarafacial pathway.
     b. Because of geometrical constraints, thermal [2 + 2] cycloadditions aren't seen.
     c. Photochemical [2 + 2] cycloadditions take place because the addition can occur by a suprafacial pathway.

4. Thermal and photochemical cycloadditions always take place by opposite stereochemical pathways.

IV. Sigmatropic rearrangements (Sections 30.8 – 30.9).
   A. General description of sigmatropic rearrangements (Section 30.8).
      1. In a sigmatropic rearrangement, a σ-bonded atom or group migrates across a π electron system.
         a. A σ bond is broken, the π bonds move, and a new σ bond is formed in the product.
         b. The σ bonded group can be either at the end or in the middle of the π system.
         c. The notation [3,3] indicates the positions in the groups to which migration occurs.
      2. Sigmatropic rearrangements are controlled by orbital symmetry.
         a. Migration of a group across the same face of the π system is suprafacial rearrangement.
         b. Migration from one face to the other face is antarafacial rearrangement.
         c. Both types of rearrangements are symmetry-allowed, but suprafacial rearrangements are geometrically easier.
   B. Examples of sigmatropic rearrangements (Section 30.8).
      1. The [1,5] migration of a hydrogen atom across two double bonds of a π system is very common.
         Thermal [1,3] hydrogen shifts are unknown.
      2. The Cope rearrangement and the Claisen rearrangement involve reorganization of an odd number of electron pairs and proceed by suprafacial geometry.

V. A summary of rules for pericyclic reactions (Section 30.10).
   A. Thermal reactions with an even number of electron pairs are either conrotatory or antarafacial.
   B. A change from thermal to photochemical, or from even to odd, changes the outcome to disrotatory/suprafacial.
   C. A change of both thermal and even causes no change.

Solutions to Problems

30.1 For ethylene:

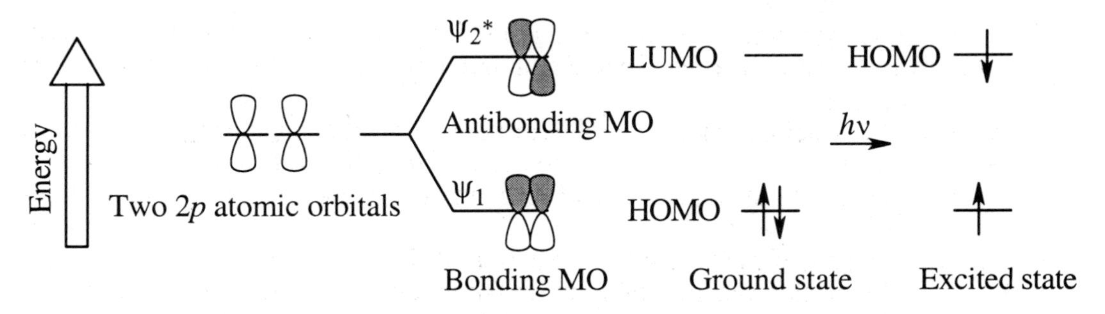

The two π electrons of ethylene occupy ψ_1 in the ground state, making ψ_1 the HOMO and ψ_2^* the LUMO. In the excited state, one electron occupies ψ_1 and the other occupies ψ_2^*, making ψ_2^* the HOMO. Since all orbitals are occupied in the excited state, there is no LUMO.

For 1,3-butadiene:

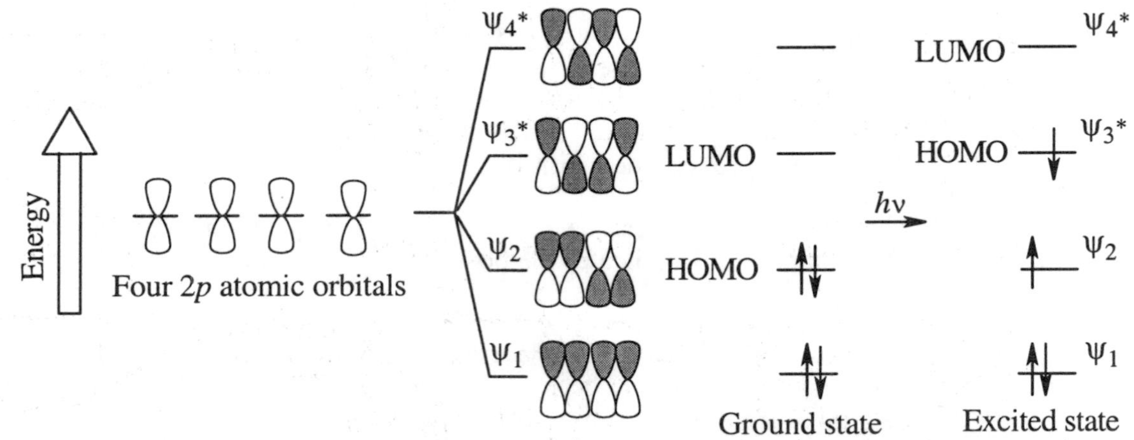

In the ground state, ψ_2 is the HOMO, and ψ_3^* is the LUMO. in the excited state, ψ_3^* is the HOMO, and ψ_4^* is the LUMO.

30.2

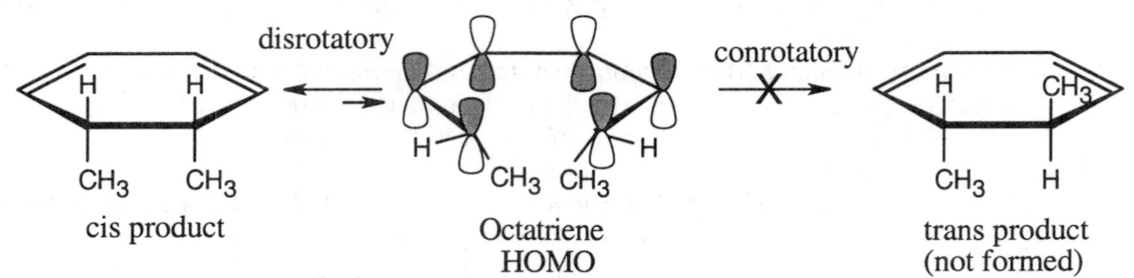

The symmetry of the octatriene HOMO predicts that ring closure will occur by a disrotatory path and that only *cis* product will be formed.

30.3 Note: *Trans*-3,4-dimethylcyclobutene is chiral; the *S,S* enantiomer will be used for this argument.

Path **A**:

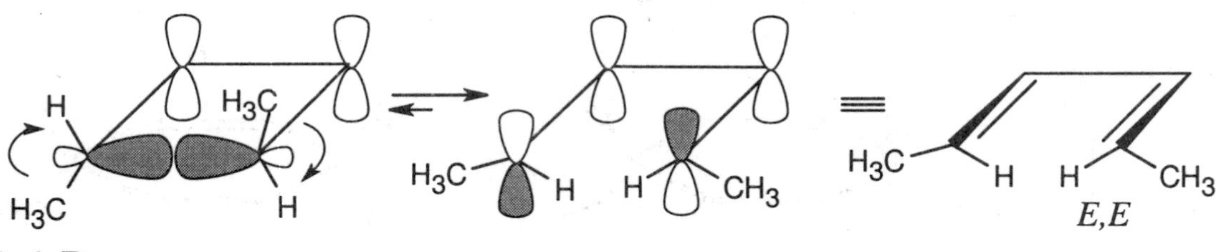

Path **B**:

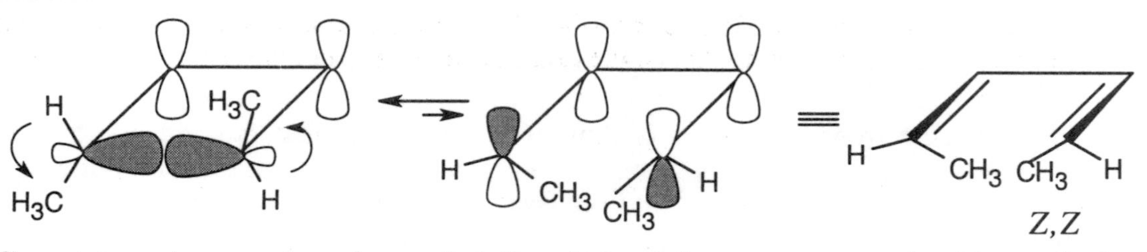

Conrotatory ring opening of *trans*-3,4-dimethylcyclobutene can occur in either a clockwise or a counterclockwise manner. Clockwise opening (path **A**) yields the *E,E* isomer; counterclockwise opening (path **B**) yields the *Z,Z* isomer. Production of (2Z,4Z)-hexadiene is disfavored because of unfavorable steric interactions between the methyl groups in the transition state leading to ring-opened product.

30.4

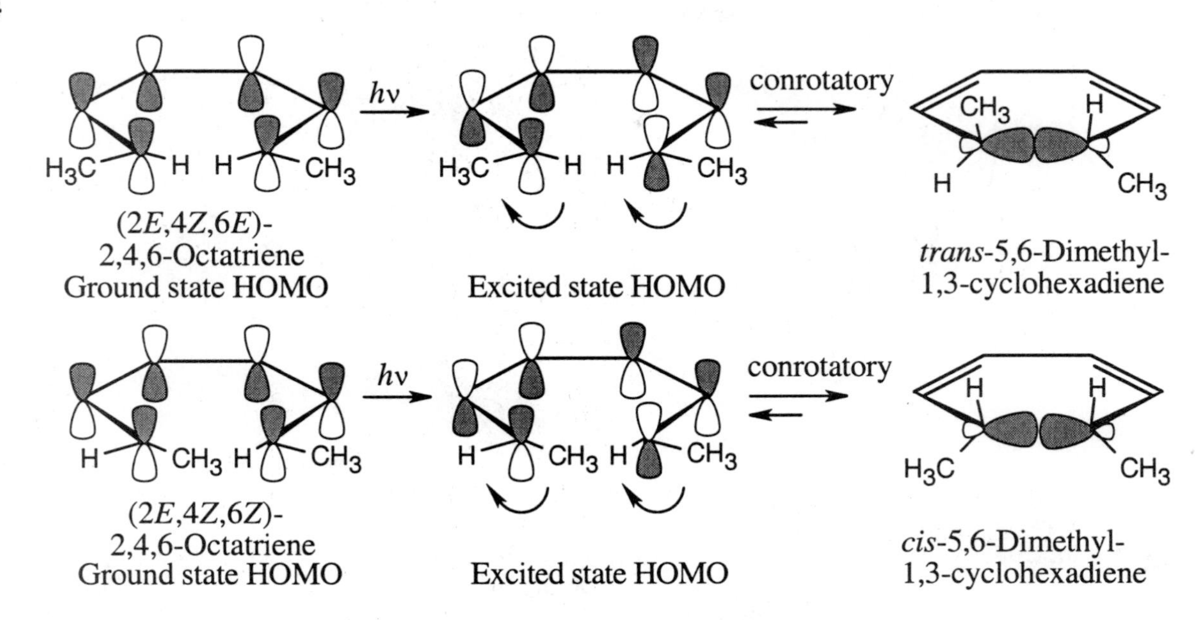

Photochemical electrocyclic reactions of 6 π electron systems always occur in a conrotatory manner.

30.5

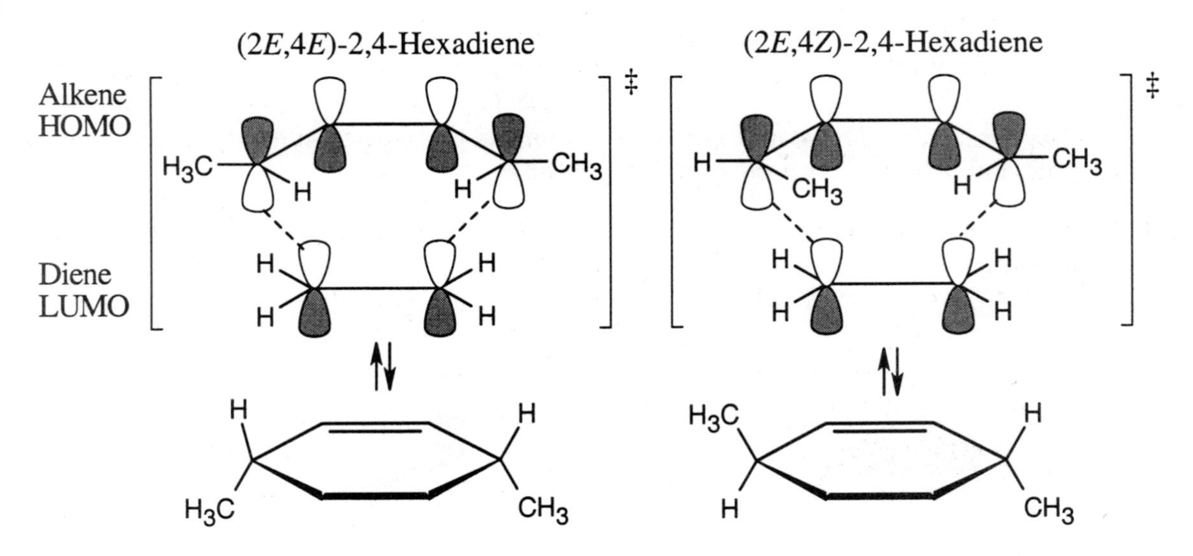

The Diels-Alder reaction is a thermal [4 + 2] cycloaddition, which occurs with suprafacial geometry. The stereochemistry of the diene is maintained in the product.

30.6

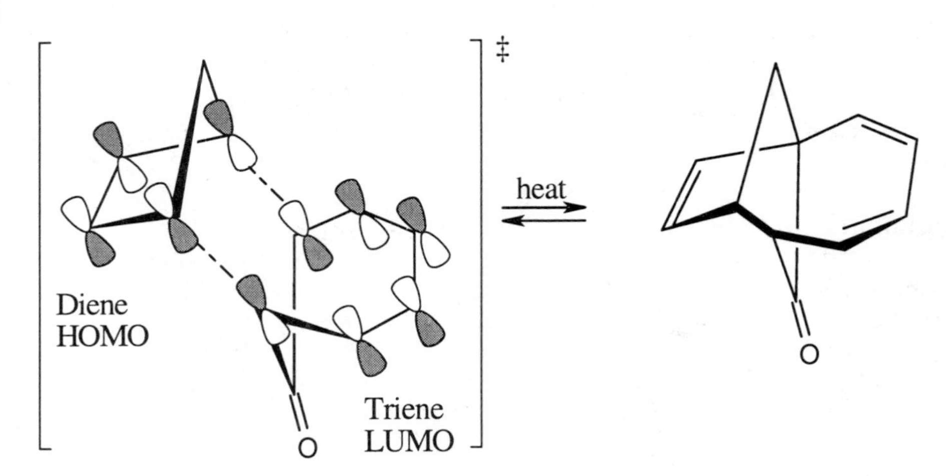

The reaction of cyclopentadiene and cycloheptatrienone is a [6 + 4] cycloaddition. This thermal cycloaddition proceeds with suprafacial geometry since five electron pairs are involved in the concerted process. The π electrons of the carbonyl group do not take part in the reaction.

30.7 This [1,7] sigmatropic reaction proceeds with antarafacial geometry because four electron pairs are involved in the rearrangement.

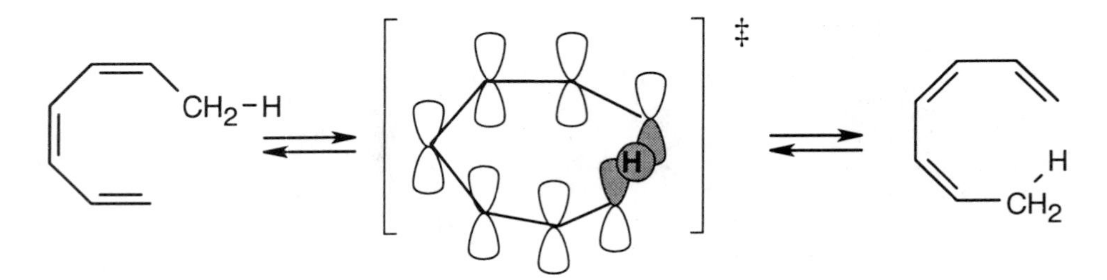

30.8 Scrambling of the deuterium label of 1-deuterioindene occurs by a series of [1,5] sigmatropic rearrangements. This thermal reaction involves three electron pairs—one pair of π electrons from the six-membered ring, the π electrons from the five-membered ring, and two electrons from a carbon-deuterium (or hydrogen) single bond—and proceeds with suprafacial geometry.

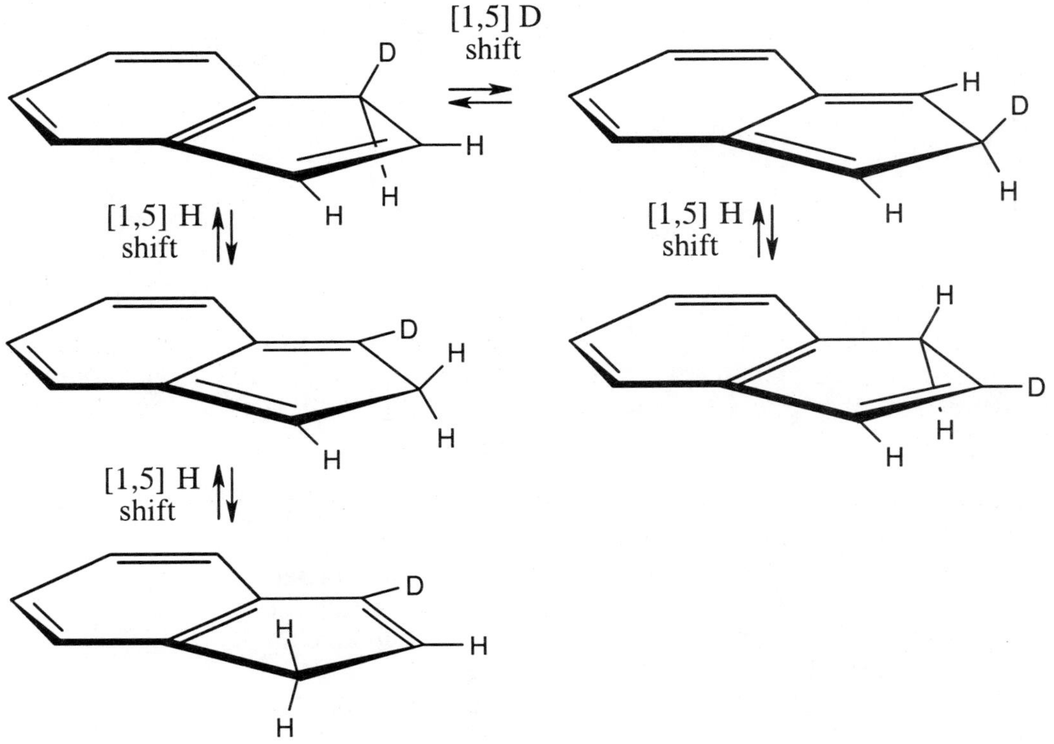

[1,5] D shift

[1,5] H shift

[1,5] H shift

[1,5] H shift

30.9

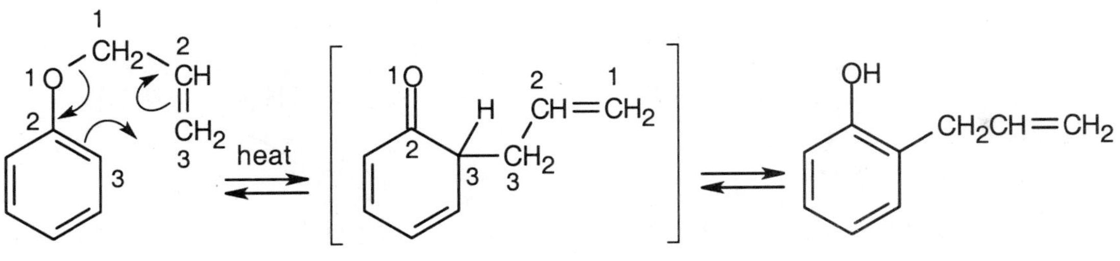

heat

The Claisen arrangement of an unsubstituted allyl phenyl ether is a [3,3] sigmatropic rearrangement in which the allyl group usually ends up in the position *ortho* to oxygen. In this problem both *ortho* positions are occupied by methyl groups. The Claisen intermediate undergoes a second [3,3] rearrangement, and the final product is *p*-allyl phenol.

O
H_3C CH_3
[3,3]
shift
O
CH_3 CH_3
H_3C
[3,3]
shift
O
H_3C CH_3
H
OH
H_3C CH_3

30.10

Type of reaction	*Number of electron pairs*	*Stereochemistry*
(a) Thermal electrocyclic	four	conrotatory
(b) Photochemical electrocyclic	four	disrotatory
(c) Photochemical cycloaddition	four	suprafacial
(d) Thermal cycloaddition	four	antarafacial
(e) Photochemical sigmatropic rearrangement	four	suprafacial

Visualizing Chemistry

30.11

H
CH
CH_2
H
$CH=CH_2$
[3,3]
shift
H
CH
CH_2
H
$CH-CH_2$
≡

This reaction is a [3,3] sigmatropic rearrangement that yields 1,5-cyclodecadiene as a product.

30.12

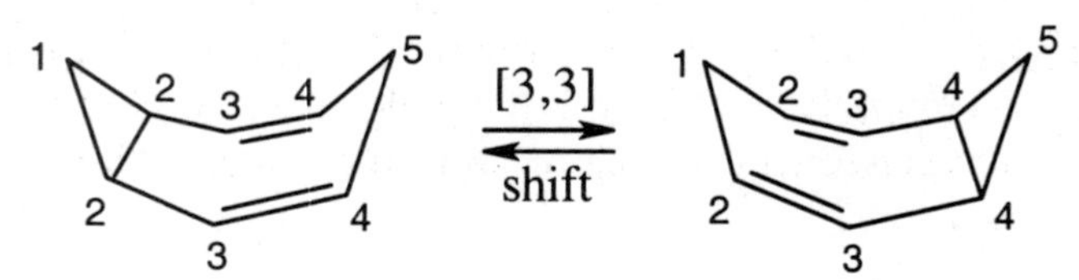

The ^{13}C NMR spectrum of homotropilidene would show five peaks if rearrangement were slow. In fact, rearrangement occurs at a rate that is too fast for NMR to detect. The ^{13}C NMR spectrum taken at room temperature is an average of the two equilibrating forms, in which positions 1 and 5 are equivalent, as are positions 2 and 4. Thus, only three distinct types of carbons are visible in the ^{13}C NMR spectrum of homotropilidene.

Additional Problems

30.13

(a)

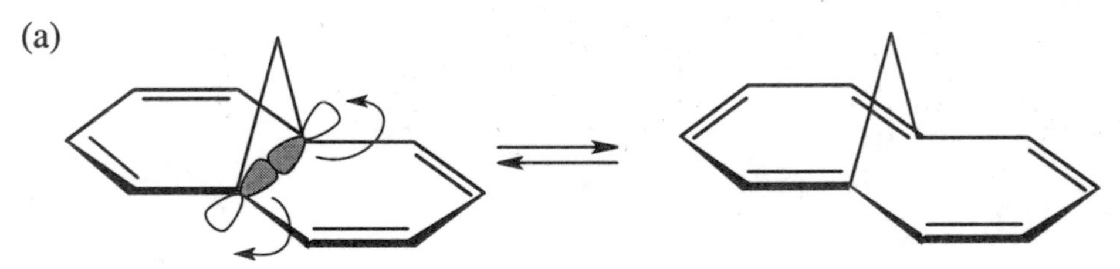

Rotation of the orbitals in the 6 π electron system occurs in a disrotatory fashion. According to the rules in Table 30.1, the reaction should be carried out under thermal conditions.

(b)

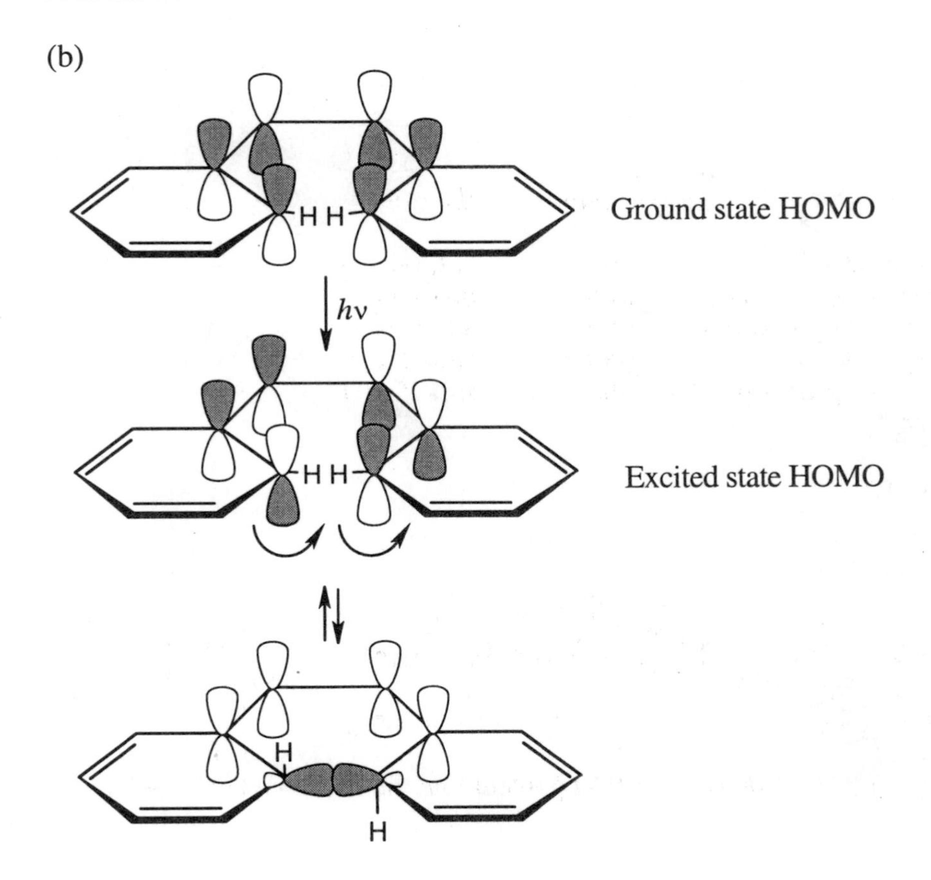

For the hydrogens to be *trans* in the product, rotation must occur in a conrotatory manner. This can happen only if the HOMO has the symmetry pictured. For a 6 π electron system, this HOMO must arise from photochemical excitation of a π electron. To obtain a product having the correct stereochemistry, the reaction must be carried out under photochemical conditions.

30.14 Tables 30.1 – 30.3 may be helpful. The first step is always to find the number of electron pairs involved in the reaction.

Type of reaction	*Number of electron pairs*	*Stereochemistry*
(a) Photochemical [1,5] sigmatropic rearrangement	3	antarafacial
(b) Thermal [4 + 6] cycloaddition	5	suprafacial
(c) Thermal [1,7] sigmatropic rearrangement	4	antarafacial
(d) Photochemical [2 + 6] cycloaddition	4	suprafacial

30.15

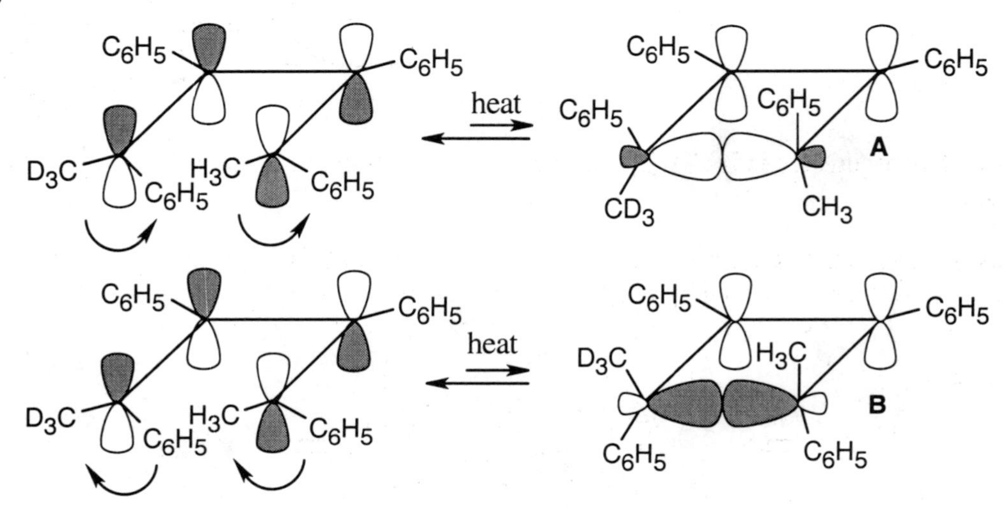

The diene can cyclize by either of two conrotatory paths to form cyclobutenes **A** and **B**. Using **B** as an example:

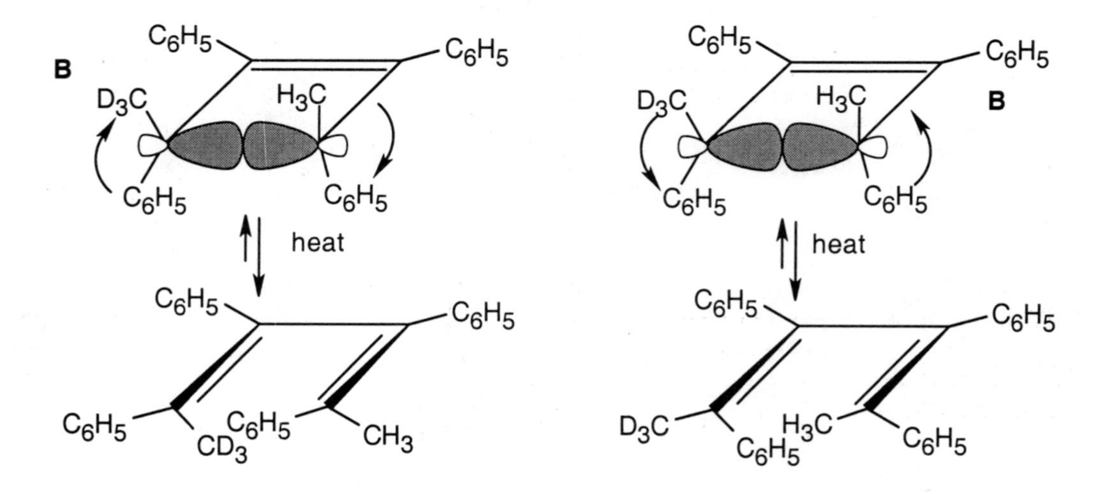

Opening of each cyclobutene ring can occur by either of two conrotatory routes to yield the isomeric dienes.

30.16 A photochemical electrocyclic reaction involving two electron pairs proceeds in a *disrotatory* manner (Table 30.1).

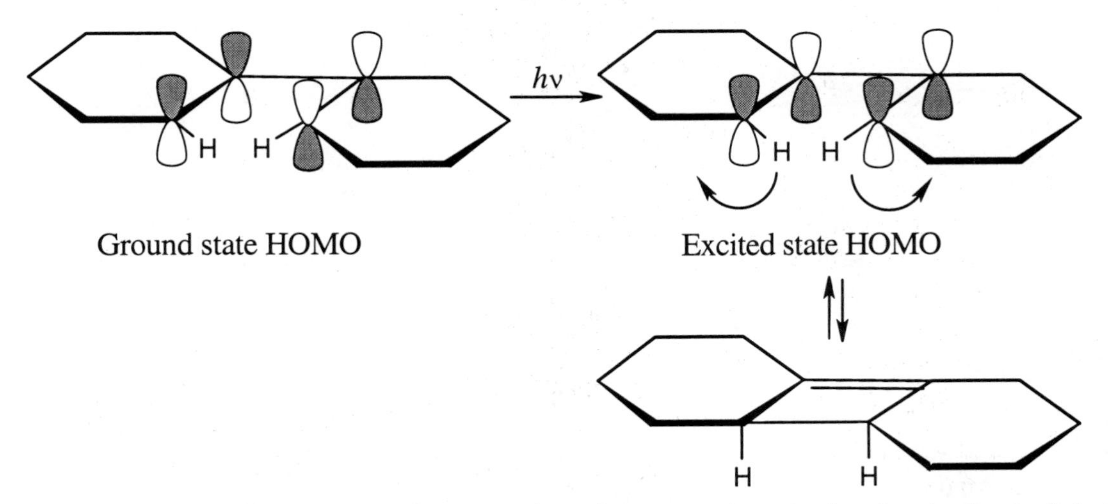

The two hydrogen atoms in the four-membered ring are cis to each other in the cyclobutene product.

30.17 The cyclononatriene is a 6 π electron system that cyclizes by a disrotatory route under thermal conditions. The two hydrogens at the ring junction have a cis relationship.

heat

H

H

H

H

30.18

heat

conrotatory

+

H_3C CH_3 H_3C CH_3

H_3C CH_3

hν

disrotatory

(2*E*,4*Z*,6*Z*,8*E*)-
Decatetraene

H_3C CH_3

Four electron pairs undergo reorganization in this electrocyclic reaction. The thermal reaction occurs with conrotatory motion to yield a pair of enantiomeric *trans*-7,8-dimethyl-1,3,5-cyclooctatrienes. The photochemical cyclization occurs with disrotatory motion to yield the *cis*-7,8-dimethyl isomer.

30.19

heat
conrotatory

H_3C CH_3

CH_3

CH_3

$h\nu$
disrotatory

(2*E*,4*Z*,6*Z*,8*Z*)-
Decatetraene

H_3C CH_3 + H_3C CH_3

30.20

Thermal reaction:

H
H

HOMO

disrotatory

H
H

Photochemical reaction:

H
H

HOMO

conrotatory

H
H

Two electrocyclic reactions, involving three electron pairs each, occur in this isomerization. The thermal reaction is a disrotatory process that yields two *cis*-fused six-membered rings. The photochemical reaction yields the *trans*-fused isomer. The two pairs of π electrons in the eight-membered ring do not take part in the electrocyclic reaction.

30.21

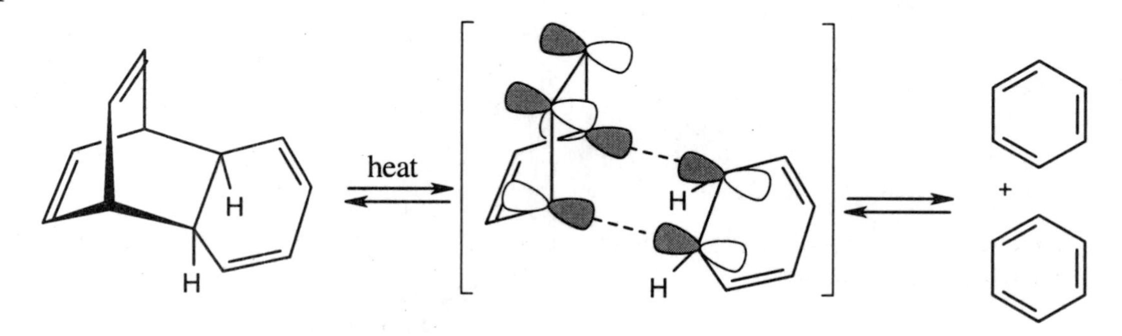

This reaction is a reverse [4 + 2] cycloaddition. The reacting orbitals have the correct symmetry for the reaction to take place by a favorable suprafacial process.

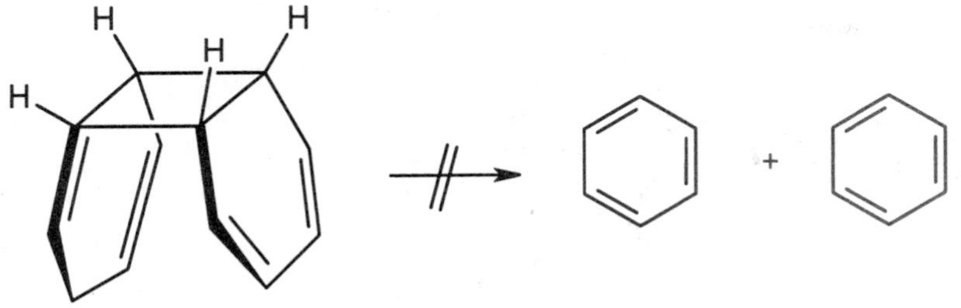

This [2 + 2] reverse cycloaddition is not likely to occur as a concerted process because the antarafacial geometry required for the thermal reaction is not possible for a four π-electron system.

30.22 Ring opening of Dewar benzene is a process involving two electron pairs and, according to Table 30.1, should occur by a conrotatory pathway. However, if you look back to other ring openings of cis-fused cyclobutenes, you will see that conrotatory ring opening produces a diene in which one of the double bonds is trans. Since a trans double bond in a six-membered ring is not likely to be formed, ring opening occurs by a different, higher energy, nonconcerted pathway.

30.23

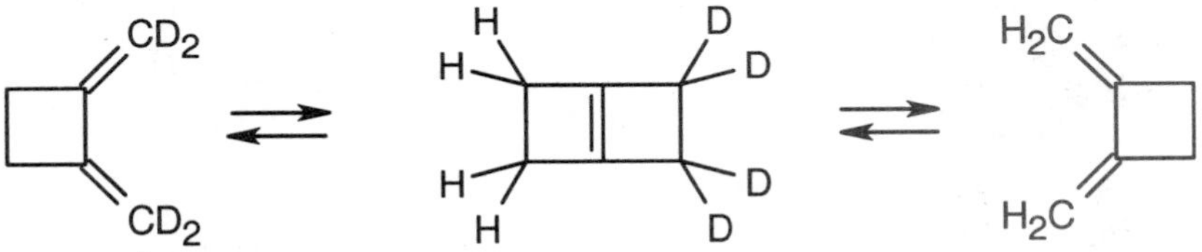

Each electrocyclic reaction involves two pairs of electrons and proceeds in a conrotatory manner.

30.24

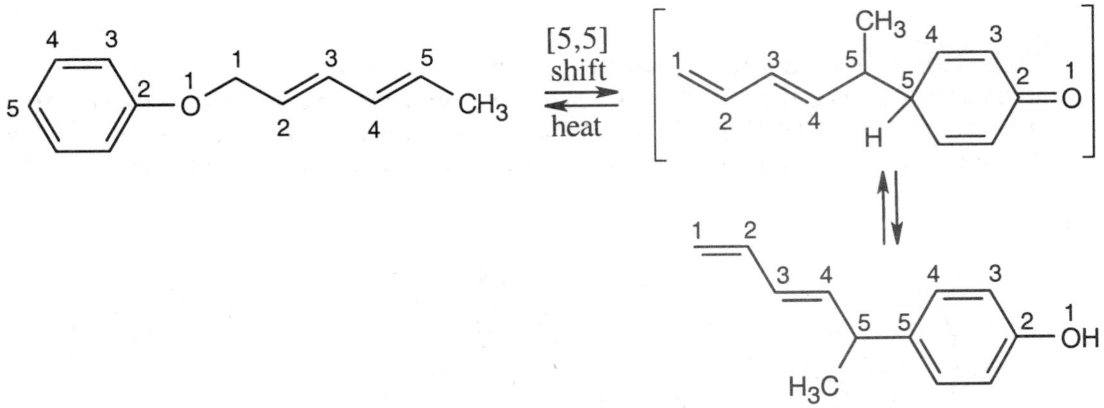

This thermal sigmatropic rearrangement is a suprafacial process since five electron pairs are involved in the reaction.

30.25

conrotatory

H H

conrotatory

H H

The observed product can be formed by a four-electron *pericyclic* process only if the four-membered ring geometry is trans. Ring-opening of the cis isomer by a concerted process would form a severely strained six-membered ring containing a trans double bond. Reaction of the cis isomer to yield the observed product occurs instead by a higher energy, nonconcerted path.

30.26

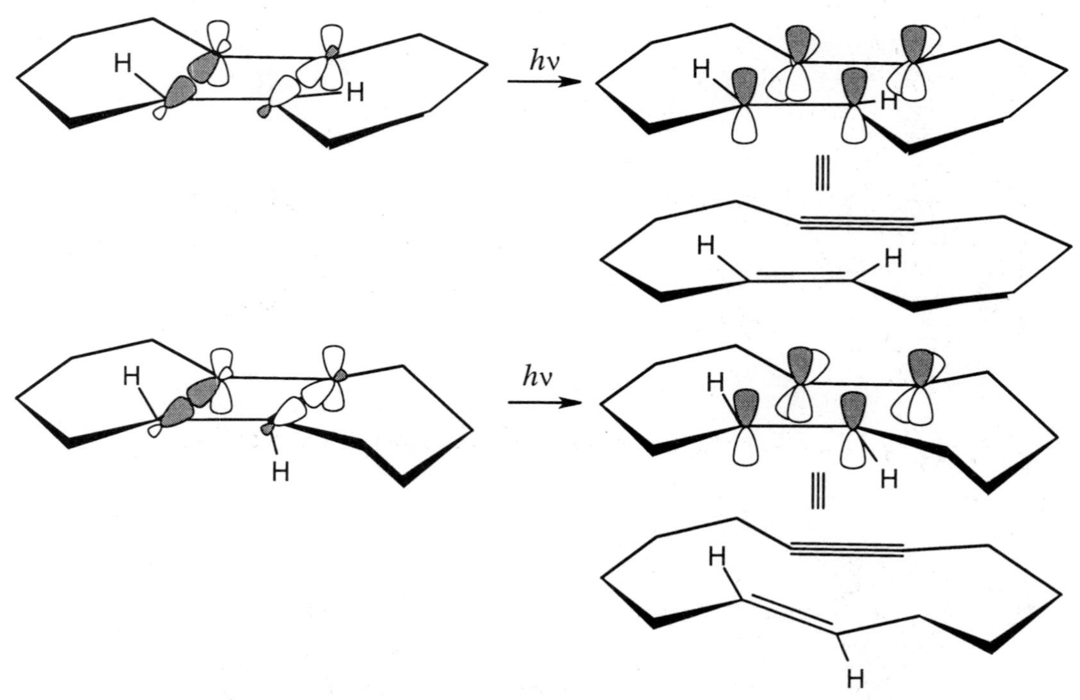

Both reactions are [2 + 2] photochemical electrocyclic reactions, which occur with disrotatory motion.

30.27

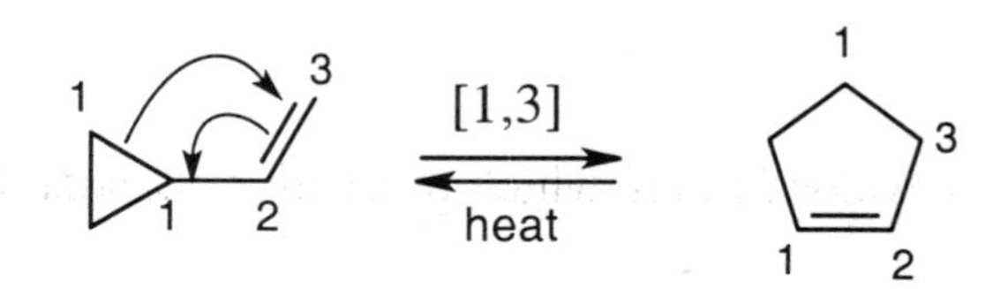

This reaction is a [1,3] sigmatropic rearrangement.

30.28

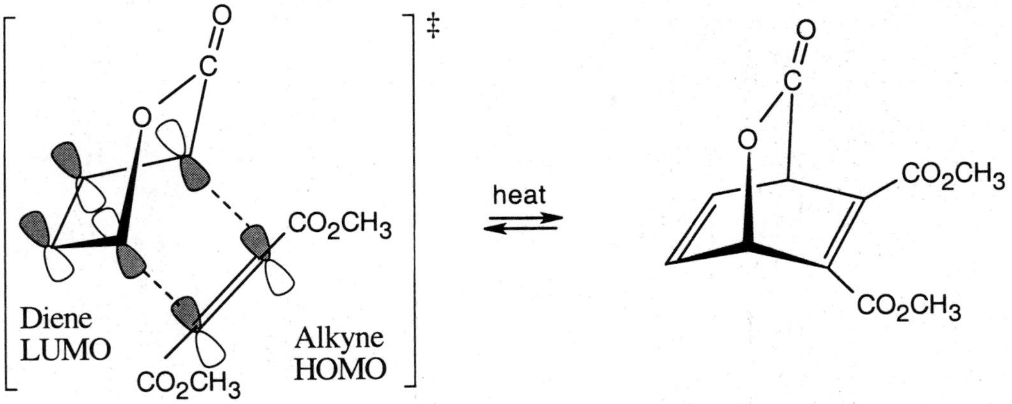

Formation of the bicyclic ring system occurs by a suprafacial [4 + 2] Diels-Alder cycloaddition process. Only one pair of π electrons from the alkyne is involved in the reaction; the carbonyl π electrons are not involved.

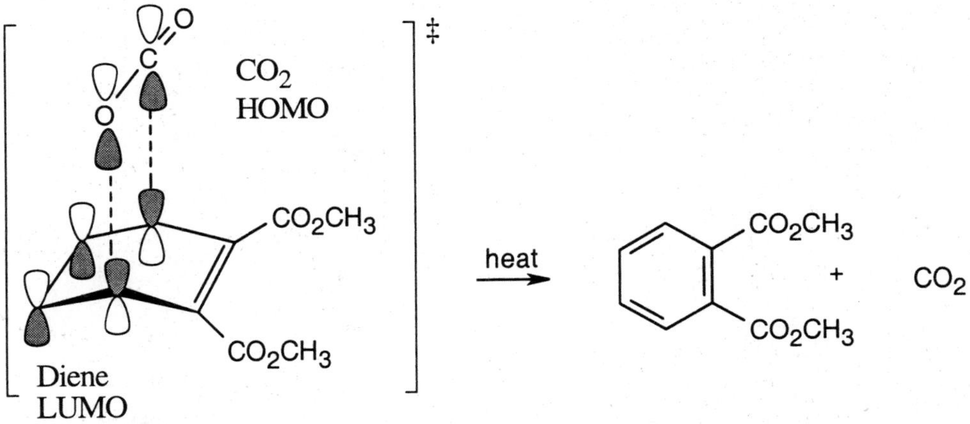

Loss of CO_2 is a reverse Diels-Alder [4 + 2] cycloaddition reaction.

30.29

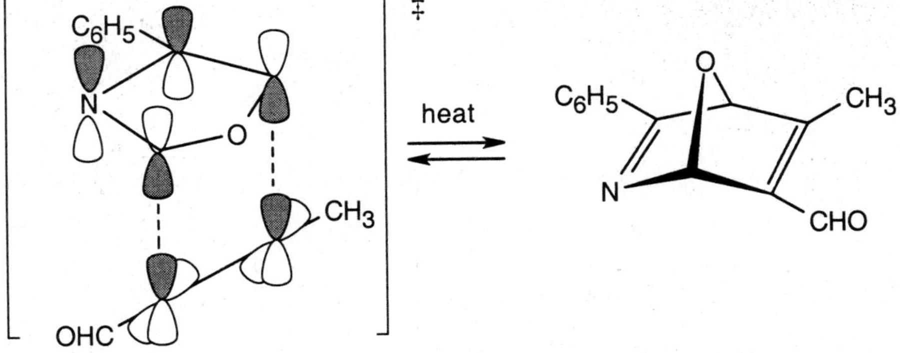

The first reaction is a Diels-Alder [4 + 2] cycloaddition, which proceeds with suprafacial geometry.

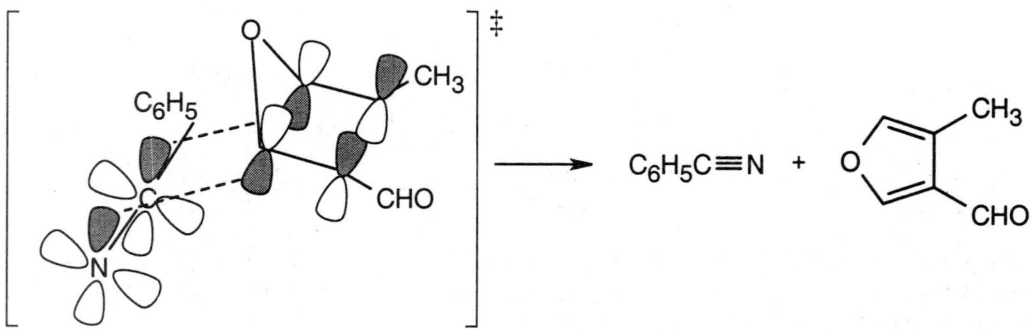

$$\left[\ \text{transition state}\ \right]^{\ddagger} \longrightarrow C_6H_5C{\equiv}N + \text{3-methylfuran-4-carbaldehyde}$$

The second reaction is a reverse Diels-Alder [4 + 2] cycloaddition.

30.30

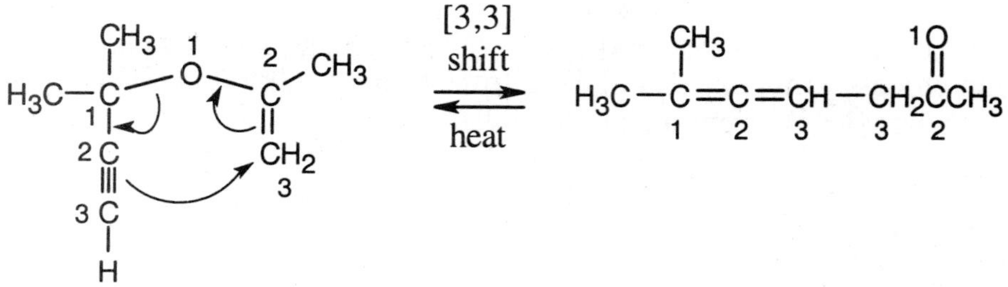

$$H_3C{-}C(CH_3){-}O{-}C(CH_3){=}CH_2,\ C{\equiv}CH \ \underset{\text{heat}}{\overset{[3,3]\ \text{shift}}{\rightleftharpoons}}\ H_3C{-}C(CH_3){=}C{=}CH{-}CH_2CCH_3(=O)$$

An allene is formed by a [3,3] sigmatropic rearrangement.

$$H_3C{-}C(CH_3){=}C{=}CH{-}CH_2\overset{O}{\overset{\|}{C}}CH_3 \ \overset{H^+}{\rightleftharpoons}\ \left[H_3C{-}C(CH_3){=}C{=}CH{-}CH{=}C(OH)CH_3\right] \ \underset{-H^+}{\rightleftharpoons}\ H_3C{-}C(CH_3){=}CH{-}CH{=}CH\overset{O}{\overset{\|}{C}}CH_3$$

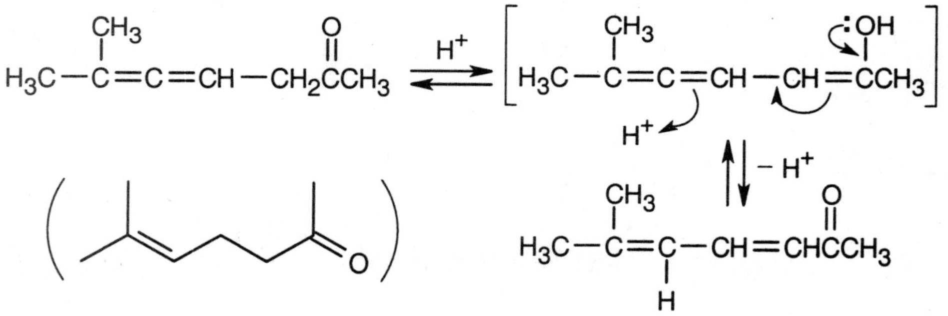

Acid catalyzes isomerization of the allene to a conjugated dienone *via* an intermediate enol.

30.31

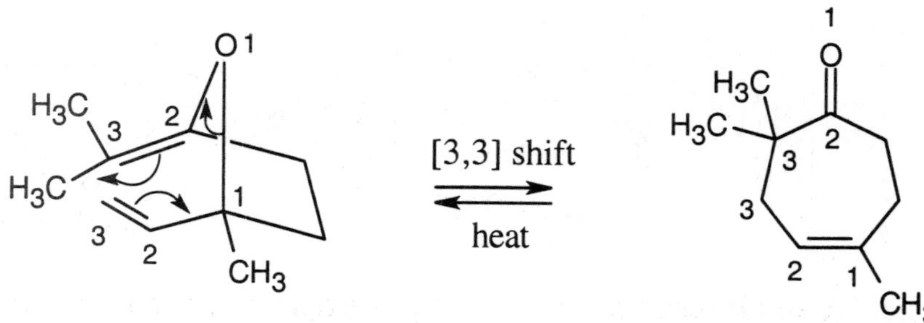

$$\text{bicyclic ether} \ \underset{\text{heat}}{\overset{[3,3]\ \text{shift}}{\rightleftharpoons}}\ \text{karahanaenone}$$

Karahanaenone is formed by a [3,3] sigmatropic rearrangement (Claisen rearrangement).

30.32

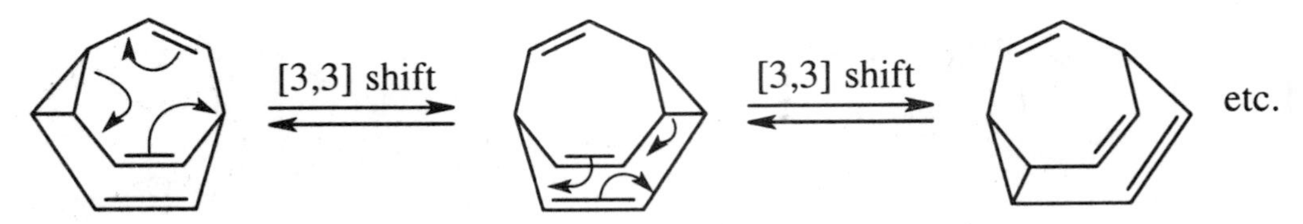

Bullvalene can undergo [3,3] sigmatropic rearrangements in all directions. At 100°, the rate of rearrangement is fast enough to make all hydrogen atoms equivalent, and only one signal is seen in the ^{1}H NMR spectrum.

30.33

Suprafacial shift:

Et
H3C
H
D
CH3

‡
Et
H
CH3
H3C
D
Et
E
CH3
H
S
D
CH3
A

‡
Et
D
H3C
H
CH3
H3C
Z
Et
D
R
H
CH3
B

Antarafacial shift:

Et
H3C
H
D
CH3

‡
Et
D
H3C
H
CH3
Et
E
CH3
D
R
H
CH3
C
not formed

‡
Et
CH3
H3C
H
D
H3C
Z
Et
H
S
D
CH3
D
not formed

The observed products **A** and **B** result from a [1,5] sigmatropic hydrogen shift with suprafacial geometry, and they confirm the predictions of orbital symmetry. **C** and **D** are not formed.

30.34

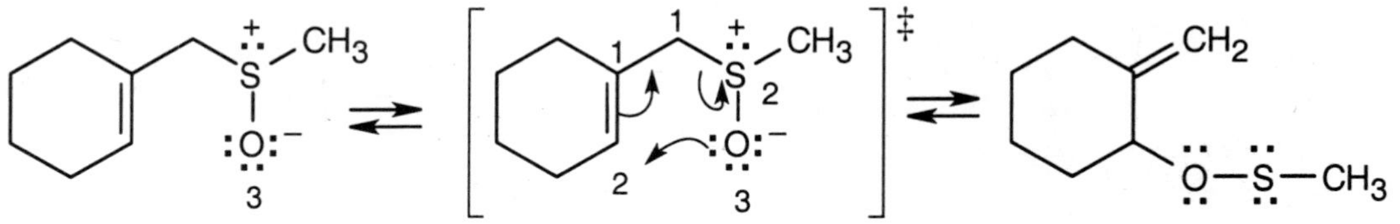

This [2,3] sigmatropic rearrangement involves three electron pairs and should occur with suprafacial geometry.

30.35

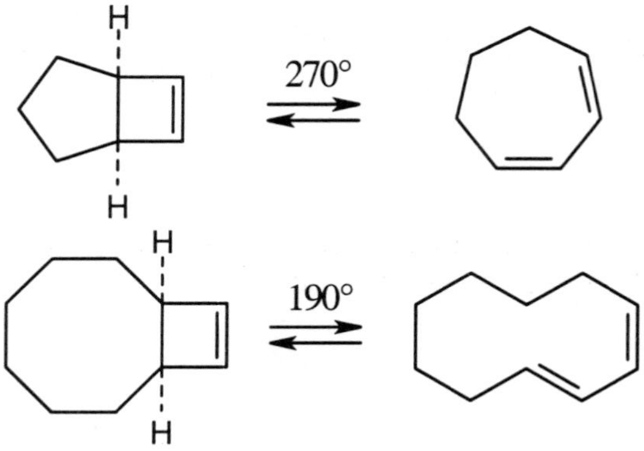

270°

190°

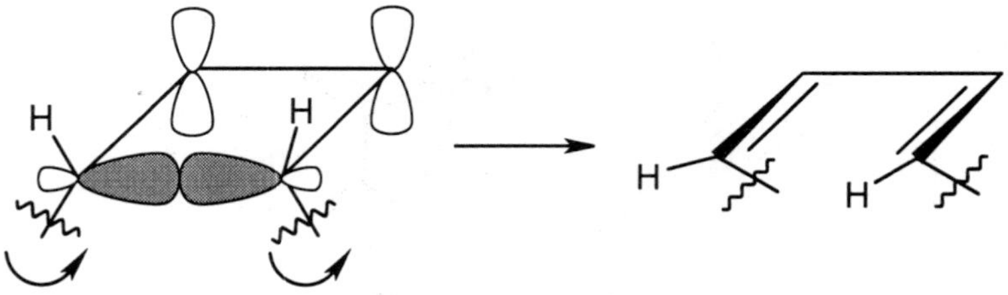

Concerted thermal ring opening of a cis-fused cyclobutene ring yields a product having one cis and one trans double bond. The ten-membered ring product of reaction 2 is large enough to accommodate a trans double bond, but a seven-membered ring containing a trans double bond is highly strained. Opening of the cyclobutene ring in reaction 1 occurs by a higher energy nonconcerted process to yield a seven-membered ring having two cis double bonds.

30.36

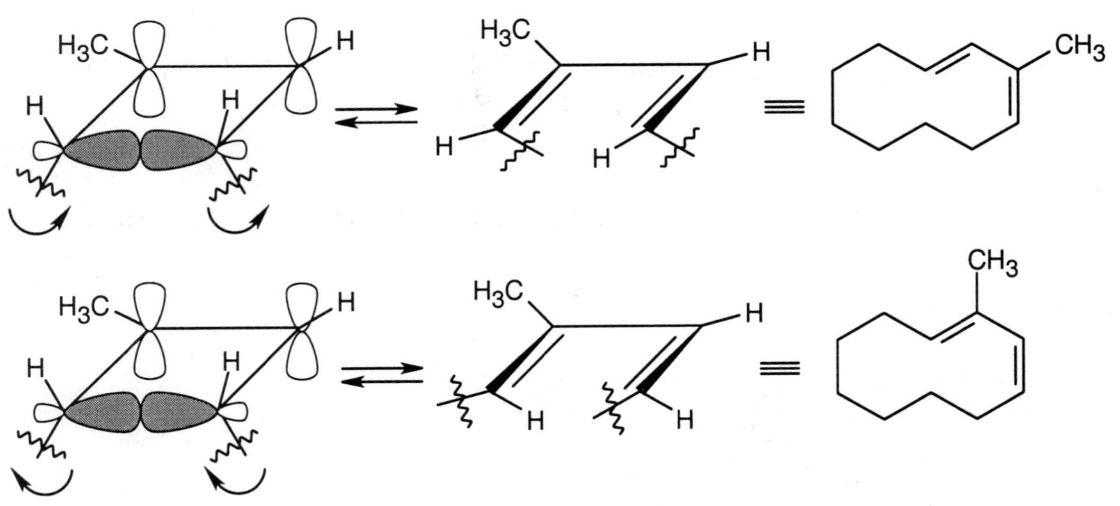

Thermal ring opening of the methylcyclobutene ring can occur by either of two symmetry-allowed conrotatory paths to yield the observed product mixture.

30.37

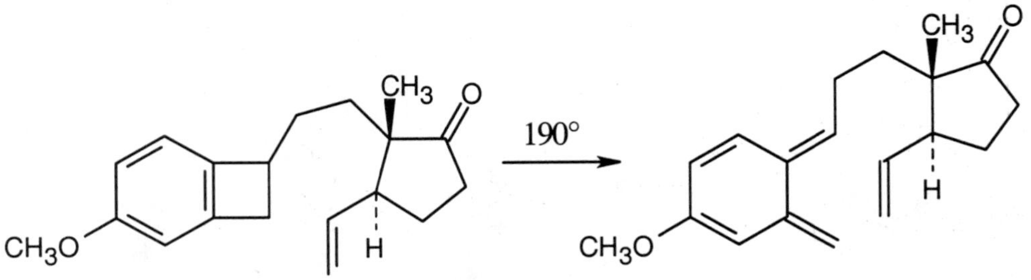

190°

The first reaction is an electrocyclic opening of a cyclobutene ring.

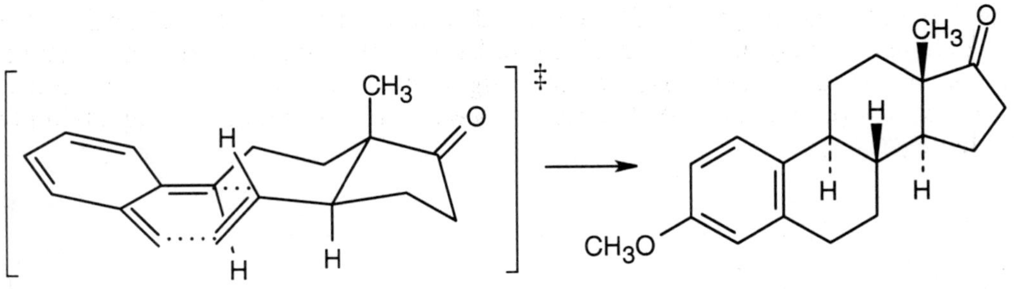

Formation of estrone methyl ether occurs by a Diels-Alder [4 + 2] cycloaddition.

30.38

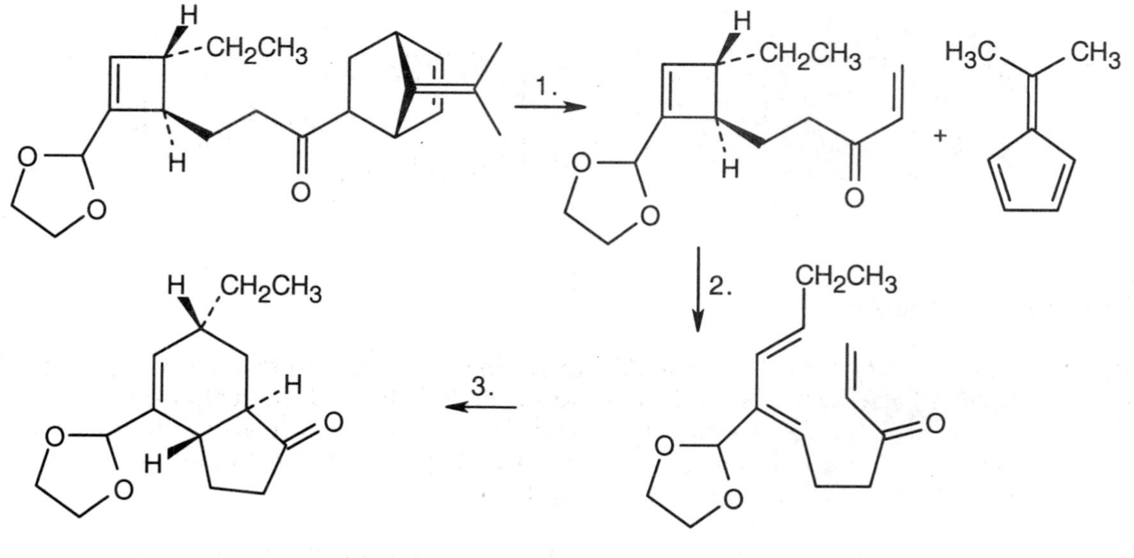

Reaction 1:	Reverse Diels-Alder [4 + 2] cycloaddition;
Reaction 2:	Conrotatory electrocyclic opening of a cyclobutene ring;
Reaction 3:	Diels-Alder [4 + 2] cycloaddition.

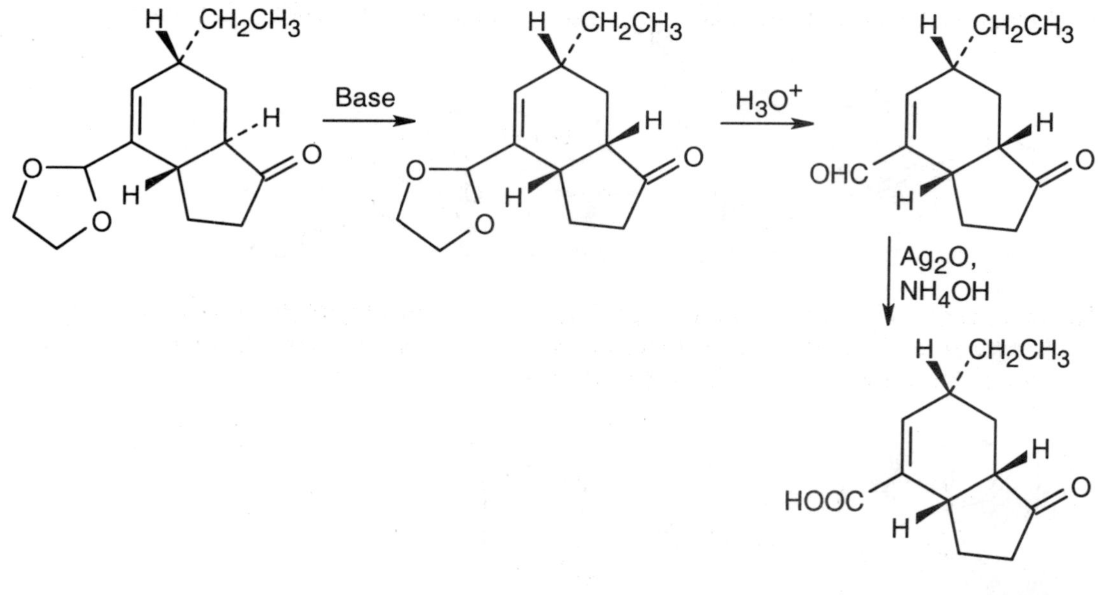

Coronafacic acid

Treatment with base enolizes the ketone and changes the ring junction from *trans* to *cis*. A *cis* ring fusion is less strained when a six-membered ring is fused to a five-membered ring.

30.39

H_3C CH_3 N^+ CH_2 CH H_2C — [3,3] shift / heat → [$\overset{+}{N}(CH_3)_2$, H, $CH_2CH{=}CH_2$] ⇌ $\overset{+}{N}H(CH_3)_2$, $CH_2CH{=}CH_3$

Molecular Modeling

30.40 When R = CH_3, transition state A is lower in energy by 27 kJ/mol, and the product is (*E*)-1,3-pentadiene. When R = CHO, transition state B is lower in energy by 21 kJ/mol, and the product is (*Z*)-2,4-pentadienal. The difference in groups leads to products with different double bond stereochemistry. Since these groups are similar in size, the selectivity must not be due to steric effects.

30.41 In the transition state, cyclopentadiene becomes more electron-poor, and tetracyanoethylene becomes more electron-rich, indicating a transfer of electrons from cyclopentadiene to tetracyanoethylene.

30.42 Arrows identify the diene carbon that is the better electron donor and the dienophile carbon that is the better electron acceptor. The 1,4-disubstituted product is formed more rapidly.

CH_3O + CN → CH_3O, CN

30.43 The barrier for rearrangement is 56 kJ/mol, and rearrangement should be rapid at room temperature. This rearrangement causes equivalence between certain carbons, and 3 signals should be visible in the ^{13}C NMR spectrum.

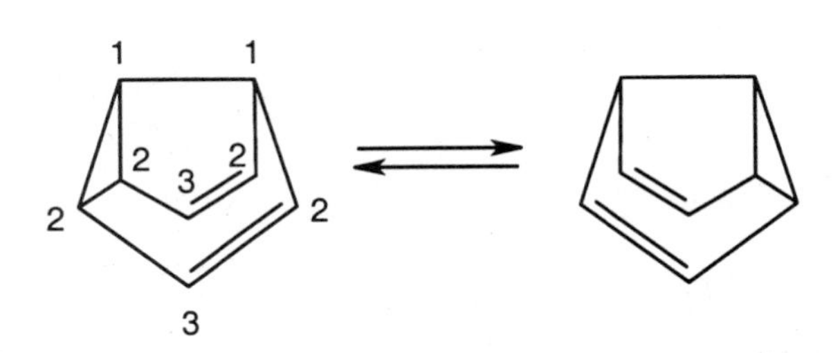

Chapter 31 – Synthetic Polymers

Chapter Outline

I. Chain-growth polymers (Sections 31.1 – 31.3).
   A. General features of chain-growth polymerization reactions (Section 31.1).
      1. How polymerization occurs.
         a. An initiator adds to a carbon-carbon double bond of a vinyl monomer.
         b. The reactive intermediate adds to a second molecule of monomer.
         c. The process is repeated.
      2. Types of polymerization.
         a. A radical initiator leads to radical polymerization.
         b. An acid causes cationic polymerization.
            Acid-catalyzed polymerization is effective only if the vinyl monomers contain electron-donating groups.
         c. Anionic polymerization can be brought about by anionic catalysts.
            i. Vinyl monomers in anionic catalysis must have electron-withdrawing groups.
            ii. Polymerization occurs by Michael addition to the monomer.
            iii. Acrylonitrile, styrene and methyl methacrylate can be polymerized anionically.
            iv. "Super glue" is an example of an anionic polymer.
   B. Stereochemistry of polymerization (Section 31.2).
      1. There are three possible stereochemical outcomes of polymerization of a substituted vinyl monomer.
         a. If the substituents all lie on the same side of the polymer backbone, the polymer is isotactic.
         b. If the substituents alternate along the backbone, the polymer is syndiotactic.
         c. If the substituents are randomly oriented, the polymer is atactic.
      2. The three types of polymers have different properties.
      3. Although radical polymerization can't control stereochemistry, Ziegler-Natta catalysts can yield polymers of desired stereochemical orientation.
         a. Ziegler-Natta catalysts are organometallic-transition metal complexes.
         b. Ziegler-Natta polymers have very little chain-branching.
         c. Ziegler-Natta catalysts are stereochemically controllable.
         d. Polymerization occurs by coordination of the alkene monomer to the complex, followed by insertion into the polymer chain.
      4. Common Ziegler-Natta polymers.
         a. Polyethylene produced by the Ziegler-Natta process (high-density polyethylene) is linear, dense, strong, and heat-resistant.
         b. Other high-molecular-weight polyethylenes have specialty uses.
   C Copolymers (Section 31.3).
      1. Copolymers are formed when two different monomers polymerize together.
      2. The properties of copolymers are different from those of the corresponding monomers.
      3. Types of copolymers.
         a. Random copolymers.
         b. Alternating copolymers.
         c. Block copolymers.
            Block copolymers are formed when an excess of a second monomer is added to a still-active mix.

d. Graft copolymers.
Graft copolymers are made by gamma irradiation of a completed homopolymer to generate a new radical initiation site for further growth of a chain.

II. Step-growth polymers (Section 31.4).
A. Step-growth polymer are formed by reactions in which each bond is formed independently of the others.
B. Most step-growth polymers result from reaction of two difunctional compounds.
Step-growth polymers can also result from polymerization of a single difunctional compound.
C. Types of step-growth polymers.
1. Polyamides.
2. Polycarbonates are formed from carbonates and alcohols or phenols.
3. Polyurethanes.
a. A urethane has a carbonyl group bonded to both an $-NR_2$ group and an –OR group.
b. Most polyurethanes are formed from the reaction of a diisocyanate and a diol.
c. Polyurethanes are used as spandex fibers and insulating foam.
i. Foaming occurs when a small amount of water is added during polymerization, producing bubbles of CO_2.
ii. Polyurethane foams often use a polyol, to increase the amount of cross-linking.

III. Polymer structure and physical properties (Section 31.5).
A. Physical properties of polymers.
1. Because of their large size, polymers experience large van der Waals forces.
These forces are strongest in linear polymers.
2. Many polymers have regions held together by van der Waals forces; these regions are known as crystallites.
a. Polymer crystallinity is affected by the substituents on the chains.
b. T_m is the temperature at which the crystalline regions of a polymer melt.
3. Some polymers have little ordering but are hard at room temperature.
These polymers become soft at a temperature T_g .
B. Polymers can be classified by physical behavior.
1. Thermoplastics.
a. Thermoplastics have a high T_g and are hard at room temperature.
b. Because they become soft at higher temperatures, they can be molded.
c. Plasticizers such as dialkyl phthalates are often added to thermoplastics to keep them from becoming brittle at room temperature.
2. Fibers.
a. Fibers are produced by extrusion of a molten polymer.
b. On cooling, the crystallite regions orient along the axis of the fiber to add tensile strength.
3. Elastomers.
a. Elastomers are amorphous polymers that can stretch and return to their original shape.
b. These polymers have a low T_g and a small amount of cross-linking.
c. The randomly coiled chains straighten out in the direction of the pull, but they return to their random orientation when stretching is done.
d. Natural rubber is an elastomer, but gutta percha is highly crystalline.
4. Thermosetting resins.
a. Thermosetting resins become highly cross-linked and solidify when heated.
b. Bakelite, a phenolic resin formed from phenol and formaldehyde, is the most familiar example.

Solutions to Problems

31.1

Most reactive ⟶ *Least reactive*

$H_2C{=}CHC_6H_5 > H_2C{=}CHCH_3 > H_2C{=}CHCl > H_2C{=}CHCO_2CH_3$

The alkenes most reactive to cationic polymerization contain electron-donating functional groups that can stabilize the carbocation intermediate. The reactivity order of substituents in cationic polymerization is similar to the reactivity order of substituted benzenes in electrophilic aromatic substitution reactions.

31.2

Most reactive ⟶ *Least reactive*

$H_2C{=}CHC{\equiv}N > H_2C{=}CHC_6H_5 > H_2C{=}CHCH_3$

Anionic polymerization occurs most readily with alkenes having electron-withdrawing substituents.

31.3

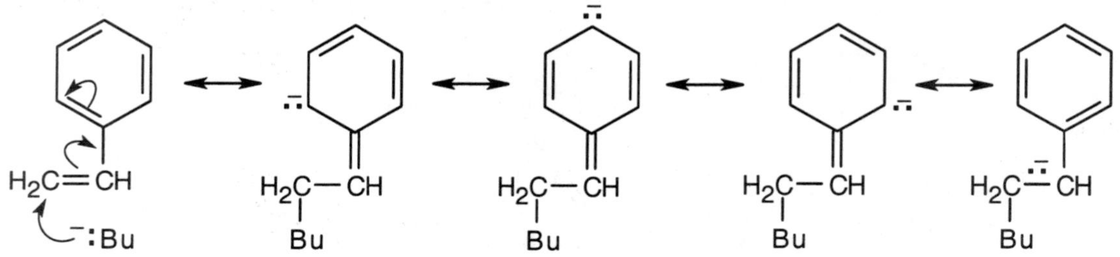

H2C=CH, :Bu ⟷ H2C—CH, Bu ⟷ H2C—CH, Bu ⟷ H2C—CH, Bu ⟷ H2C—CH, Bu

The intermediate anion can be stabilized by resonance involving the phenyl ring.

31.4

$$n\ H_2C{=}CCl_2 \longrightarrow \left(CH_2{-}CCl_2 \right)_n$$

Vinylidene chloride doesn't polymerize in isotactic, syndiotactic or atactic forms because no asymmetric centers are formed during polymerization.

31.5 None of the polypropylenes rotate plane-polarized light. If an optically inactive reagent and an achiral compound react, the product must be optically inactive. For every chirality center generated, an enantiomeric chirality center is also generated, and the resulting polymer mixture is inactive.

31.6

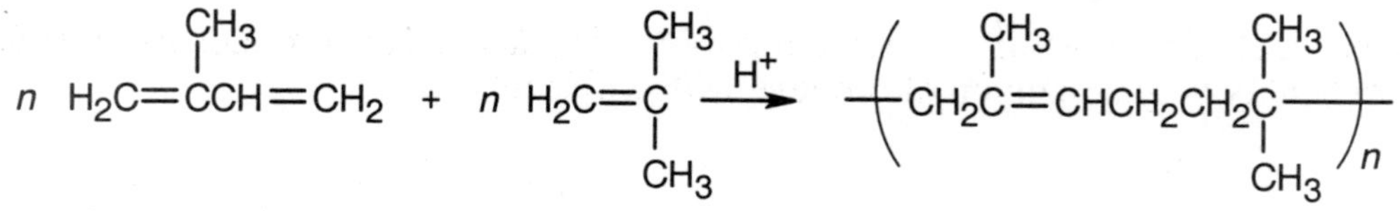

$$n\ H_2C{=}C(CH_3)CH{=}CH_2 + n\ H_2C{=}C(CH_3)_2 \xrightarrow{H^+} \left(CH_2C(CH_3){=}CHCH_2CH_2C(CH_3)_2 \right)_n$$

2-Methyl-1,3-butadiene   2-Methylpropene

31.7

$$-\!\!\left(\mathrm{CH_2CH{=}CHCH_2}\right)\!\!_n \xrightarrow{h\nu} -\!\!\left(\mathrm{CH_2CH{=}CH\dot{C}H}\right)\!\!_n \xrightarrow{m\ \mathrm{H_2C{=}CHC_6H_5}} -\!\!\left(\mathrm{CH_2CH{=}CHCH(CH_2CH(C_6H_5))_m}\right)\!\!_n$$

Irradiation homolytically cleaves an allylic C–H bond because it has the lowest bond energy. The resulting radical adds to styrene to produce a polystyrene graft.

31.8

$$n\ \mathrm{HOCH_2CH_2OH} + n\ \mathrm{HOOC{-}C_6H_4{-}COOH} \longrightarrow -\!\!\left(\mathrm{OCH_2CH_2O{-}OC{-}C_6H_4{-}CO}\right)\!\!_n$$

PET

31.9

$$\mathrm{R{-}N{=}C{=}O} + \mathrm{H{-}OR'} \longrightarrow \left[\mathrm{R{-}\overset{-}{N}{-}C(=O){-}\overset{+}{O}(H)R'}\right] \longrightarrow \mathrm{R{-}NH{-}C(=O){-}OR'}$$

a urethane

31.10

Natural rubber

$\xrightarrow[\text{Pd/C}]{\mathrm{H_2}}$

$\mathrm{CH_3{-}HC{=}CH_2} + \mathrm{H_2C{=}CH_2} \xrightarrow{\text{In}\cdot}$

The product of hydrogenation of natural rubber is atactic. This product also results from the radical copolymerization of propene with ethylene.

31.11

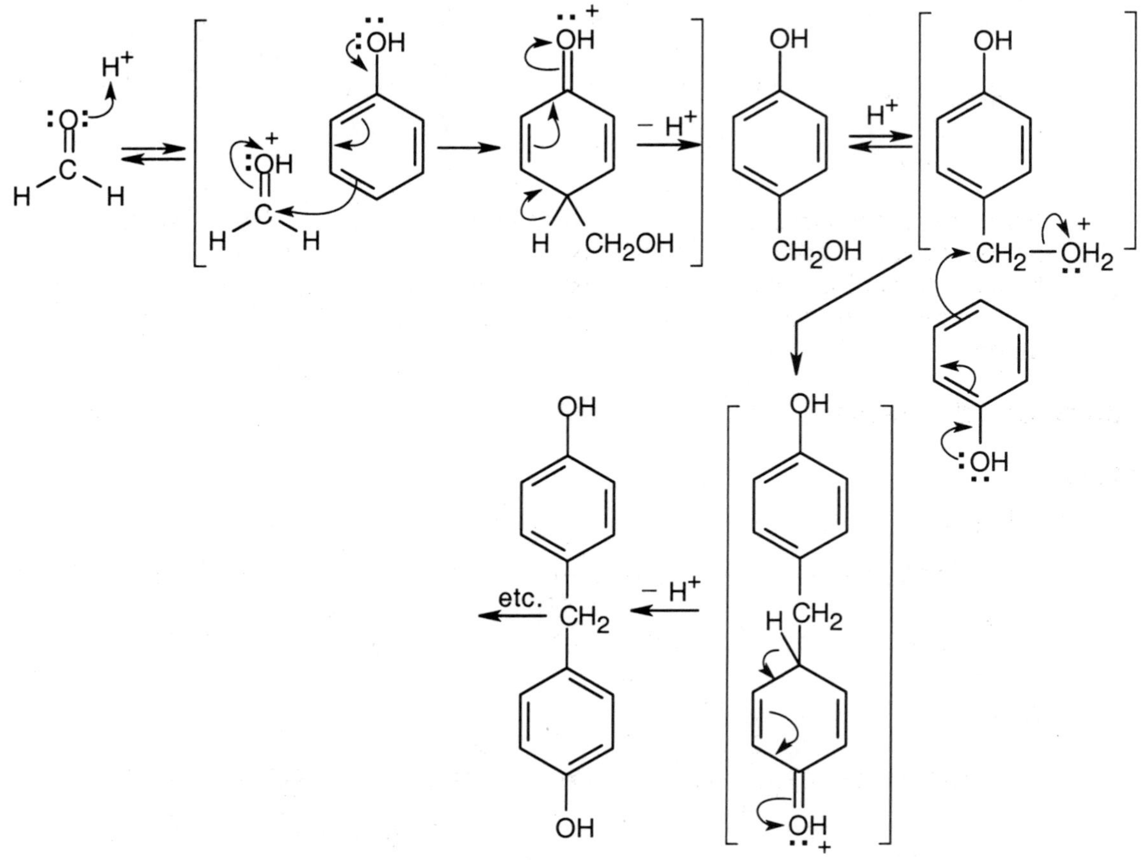

This product can react many times with additional formaldehyde and phenol to yield Bakelite. Reaction occurs at both *ortho* and *para* positions of phenol.

Visualizing Chemistry

31.12

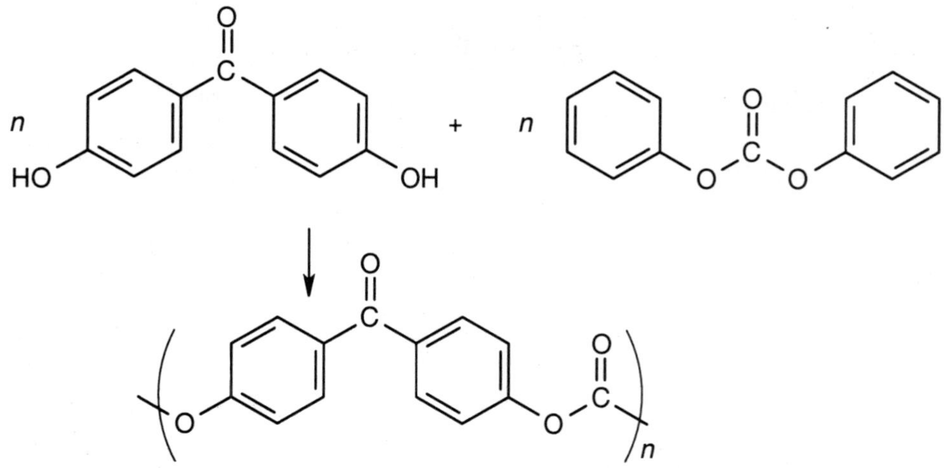

The polymer is a polycarbonate synthesized from the above monomer units.

31.13

(a) n N-vinylpyrrolidone $\longrightarrow$ poly(N-vinylpyrrolidone) (n)

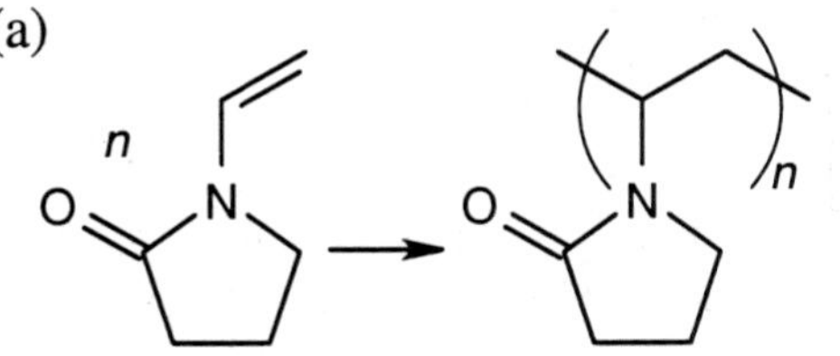

(b) n chloroprene (Cl) $\longrightarrow$ polychloroprene (Cl) (n)

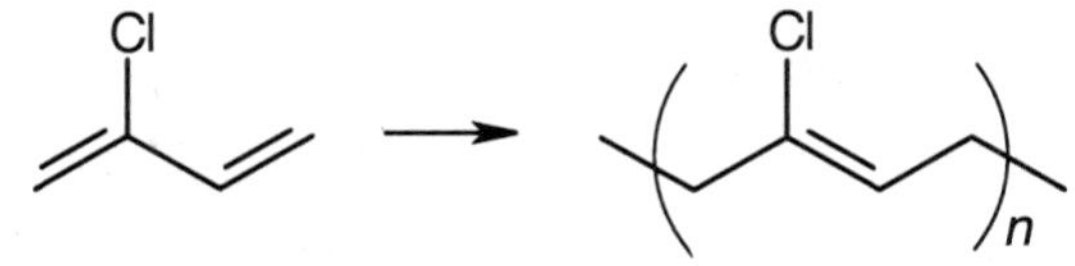

Both of these polymers are chain-growth polymers. To draw the polymer in (a), break the double bond, and draw its extensions, one on each side of the former double bond. In (b), break both double bonds and draw extensions at both ends of the former diene. The remaining double bond migrates to a position between the double bonds of the former diene.

Additional Problems

31.14

(a) $n\ H_2C{=}O \longrightarrow -\!\!(CH_2-O)\!\!-_n$

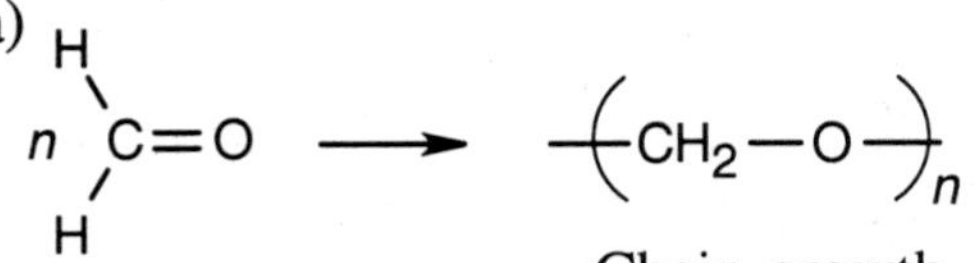

Chain-growth polymer

(b) $n\ F_2C{=}CFCl \longrightarrow -\!\!(CF_2-CFCl)\!\!-_n$

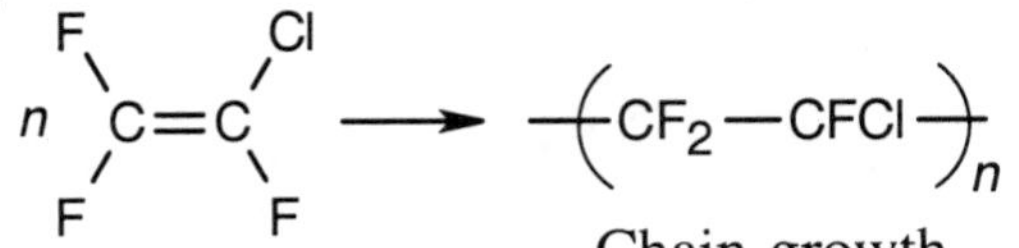

Chain-growth polymer

(c) $n\ H_2NCH_2CH_2CH_2\overset{O}{\overset{\|}{C}}-OH \longrightarrow -\!\!(NHCH_2CH_2CH_2\overset{O}{\overset{\|}{C}})\!\!-_n + n\ H_2O$

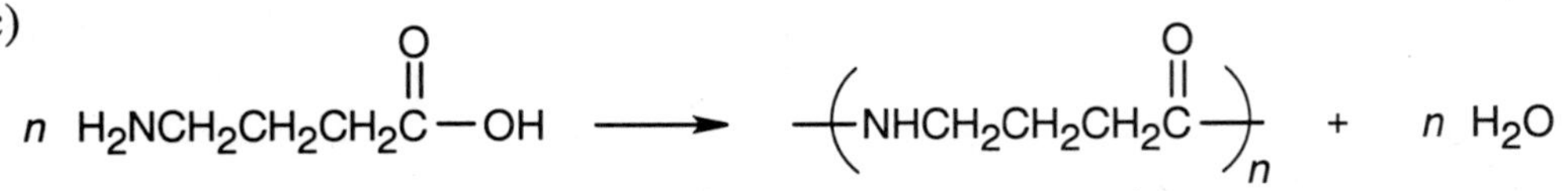

Step-growth polymer

(d) $n\ HO-C_6H_4-\overset{O}{\overset{\|}{C}}-OH \longrightarrow -\!\!(O-C_6H_4-\overset{O}{\overset{\|}{C}})\!\!-_n + n\ H_2O$

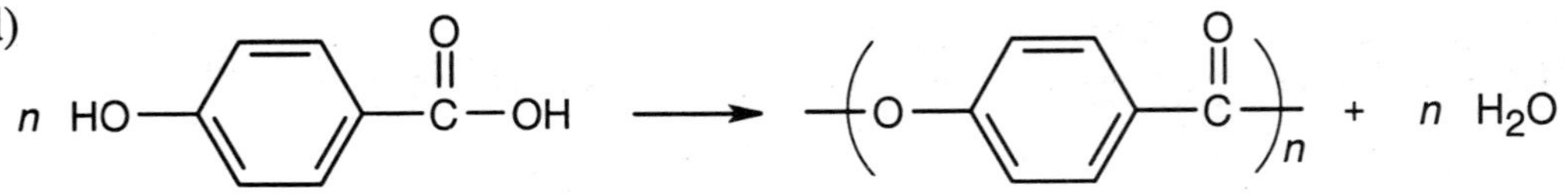

Step-growth polymer

(e) $n\ HO-C_6H_{10}-OH + n\ RO\overset{O}{\overset{\|}{C}}OR \longrightarrow -\!\!(O-C_6H_{10}-O-\overset{O}{\overset{\|}{C}})\!\!-_n + 2n\ ROH$

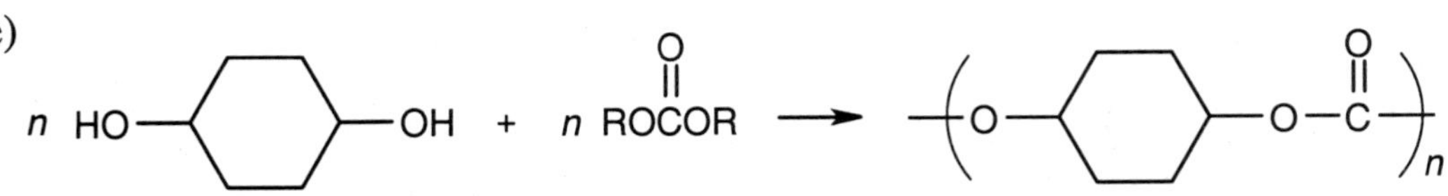

Step-growth polymer

31.15 Remember that isotactic polymers have identical groups on the same side of the polymer backbone. Syndiotactic polymers have alternating identical groups along the polymer backbone. Atactic polymers have a random orientation of groups.

(a)

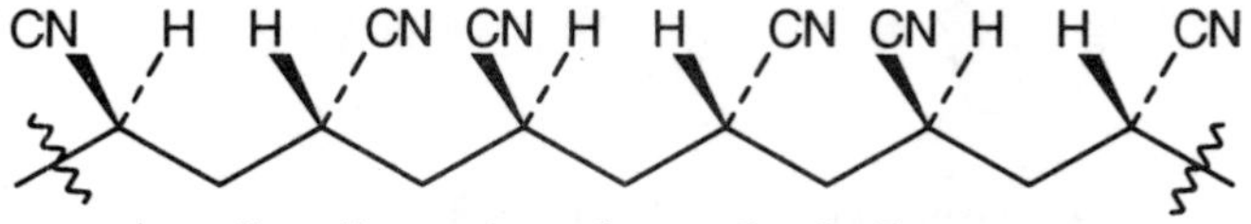

CN H H CN CN H H CN CN H H CN

Syndiotactic polyacrylonitrile

(b)

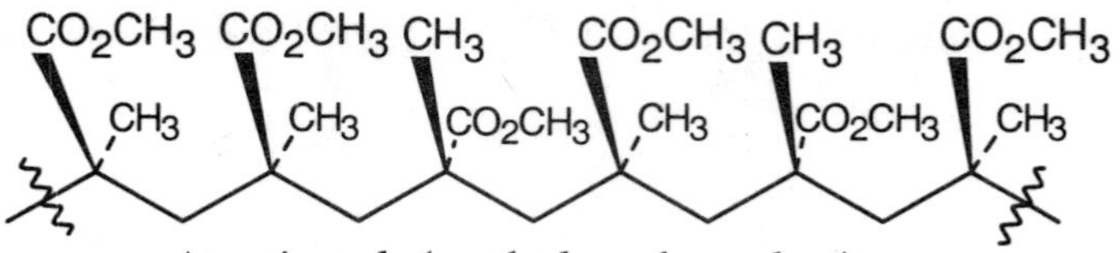

CO_2CH_3 CO_2CH_3 CH_3 CO_2CH_3 CH_3 CO_2CH_3

CH_3 CH_3 CO_2CH_3 CH_3 CO_2CH_3 CH_3

Atactic poly(methyl methacrylate)

(c)

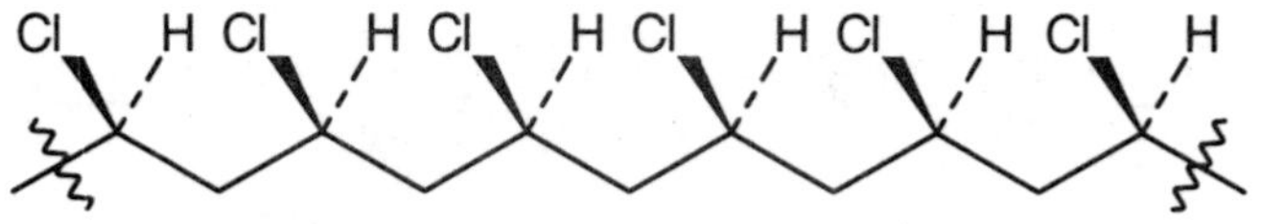

Cl H Cl H Cl H Cl H Cl H Cl H

Isotactic poly(vinyl chloride)

31.16

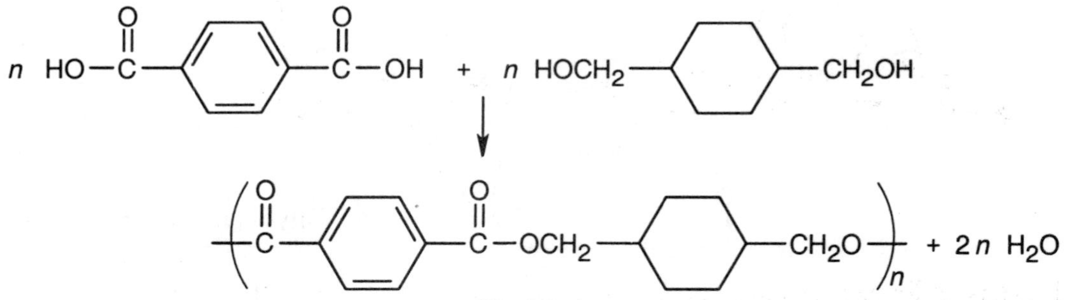

n HO—C(=O)—C_6H_4—C(=O)—OH + n $HOCH_2$—C_6H_{10}—CH_2OH

↓

$\left(C(=O)—C_6H_4—C(=O)—OCH_2—C_6H_{10}—CH_2O \right)_n$ + $2n$ H_2O

Kodel

31.17

n (bis-epoxide diether of resorcinol) + n H_2N—C_6H_4—NH_2

↓ heat

(polymer with OH, OH, N–H, N–H groups)$_n$

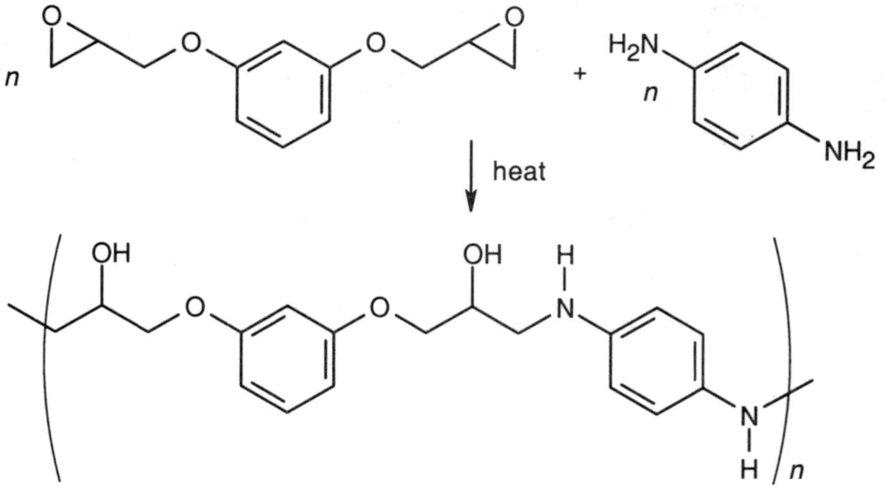

Ring-opening of the epoxide occurs by an S_N2 pathway at the less substituted epoxide carbon.

31.18

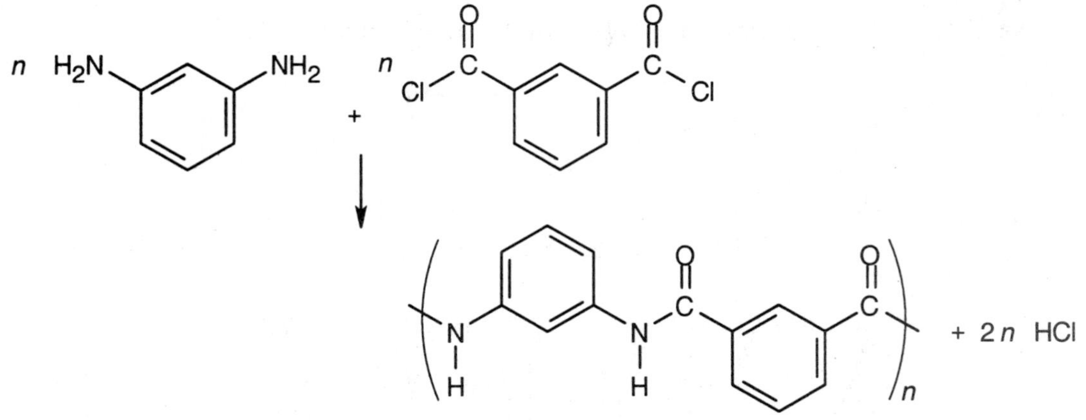

$$n\ H_2N-C_6H_4-NH_2\ (meta) + n\ ClC(=O)-C_6H_4-C(=O)Cl\ (meta) \longrightarrow \left(-NH-C_6H_4-NH-C(=O)-C_6H_4-C(=O)-\right)_n + 2n\ HCl$$

Nomex

31.19

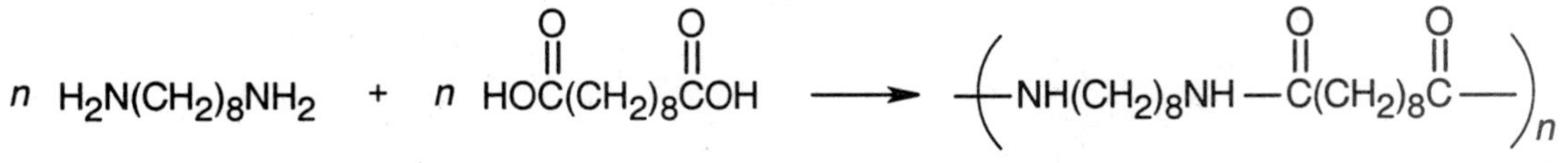

$$n\ H_2N(CH_2)_8NH_2 + n\ HO\overset{O}{\overset{\|}{C}}(CH_2)_8\overset{O}{\overset{\|}{C}}OH \longrightarrow \left(-NH(CH_2)_8NH-\overset{O}{\overset{\|}{C}}(CH_2)_8\overset{O}{\overset{\|}{C}}-\right)_n$$

Nylon 10,10 $+ 2n\ H_2O$

31.20

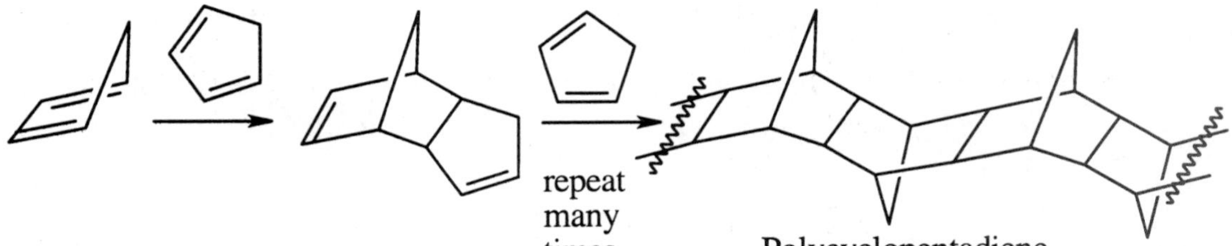

repeat many times

Polycyclopentadiene

31.21 *p*-Divinylbenzene is incorporated into the growing polystyrene chain.

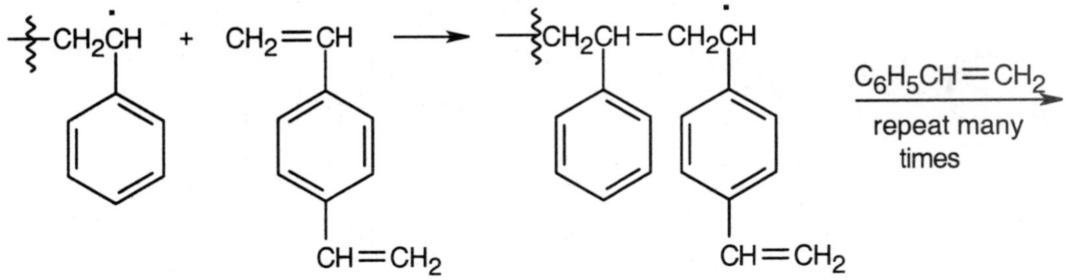

$$\sim CH_2\dot{C}H(C_6H_5) + CH_2=CH(C_6H_4CH=CH_2) \longrightarrow \sim CH_2CH(C_6H_5)-CH_2\dot{C}H(C_6H_4CH=CH_2) \xrightarrow[\text{repeat many times}]{C_6H_5CH=CH_2}$$

Another growing polymer chain reacts with the second double bond of *p*-divinylbenzene.

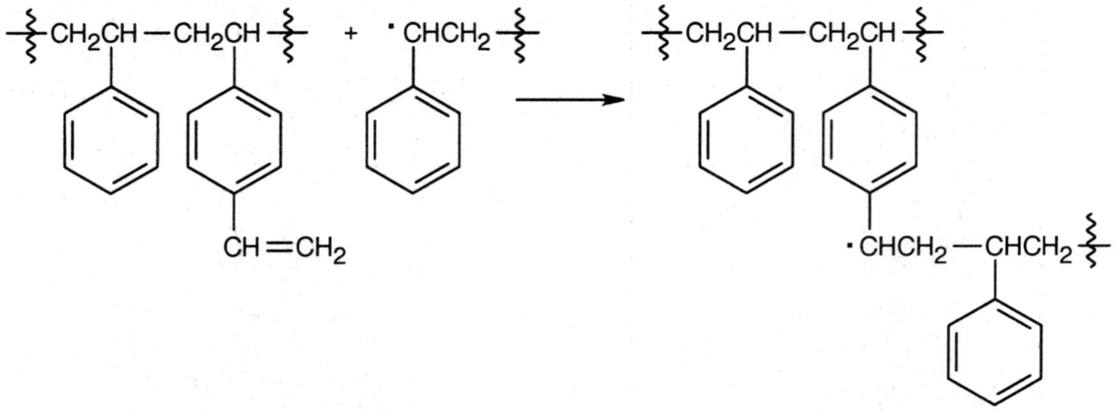

The final product contains polystyrene chains cross-linked by *p*-divinylbenzene units.

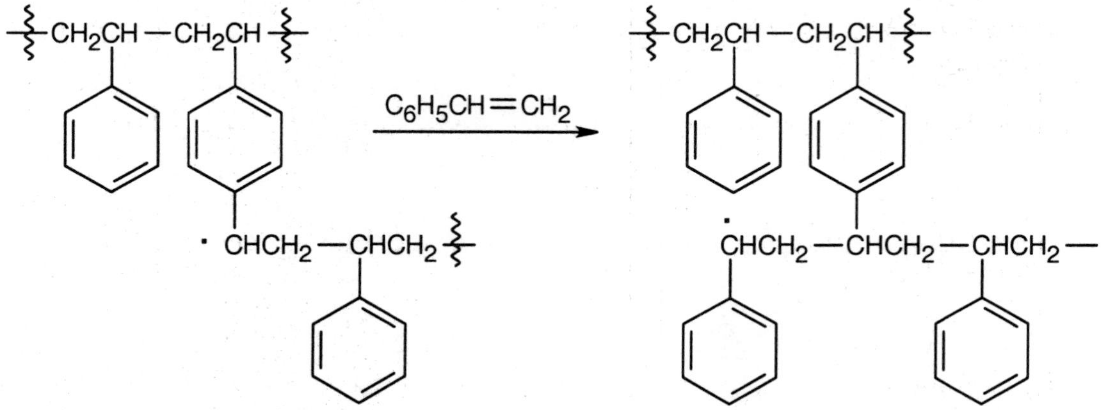

$C_6H_5CH{=}CH_2$

31.22

repeat many times

$HOCH_2CH_2O{+}CH_2CH_2O{+}_n$

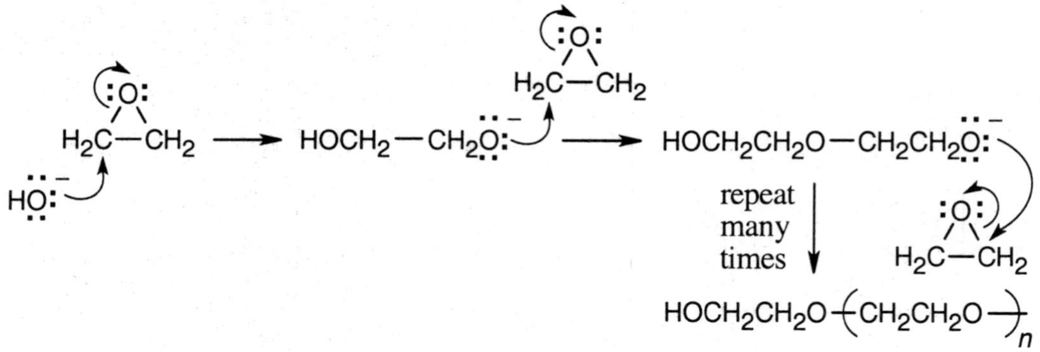

31.23 The white coating on the distillation flask is due to the thermal polymerization of nitroethylene.

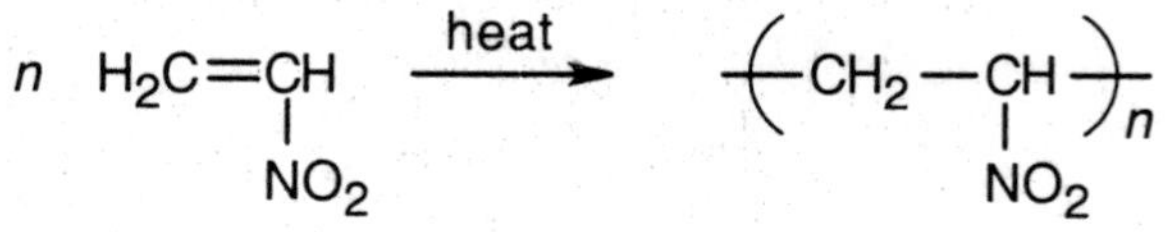

$$n\ H_2C{=}CH(NO_2) \xrightarrow{\text{heat}} {+}CH_2{-}CH(NO_2){+}_n$$

31.24 Poly(vinyl alcohol) is formed by chain-growth polymerization of vinyl acetate, followed by hydrolysis of the acetate groups.

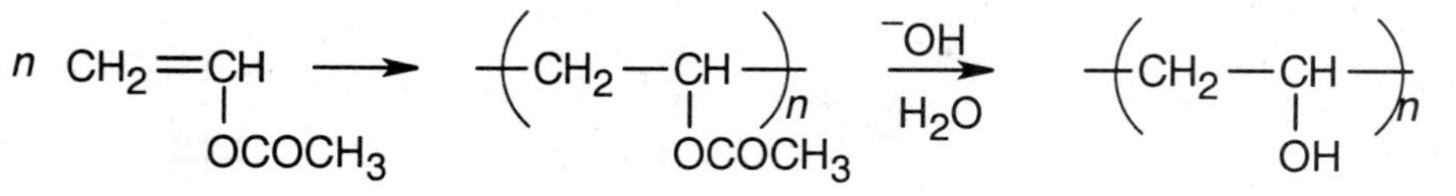

$$n\ CH_2{=}CH(OCOCH_3) \longrightarrow \left(CH_2-CH(OCOCH_3)\right)_n \xrightarrow[H_2O]{^-OH} \left(CH_2-CH(OH)\right)_n$$

Reaction of poly(vinyl alcohol) with butanal produces poly(vinyl butyral).

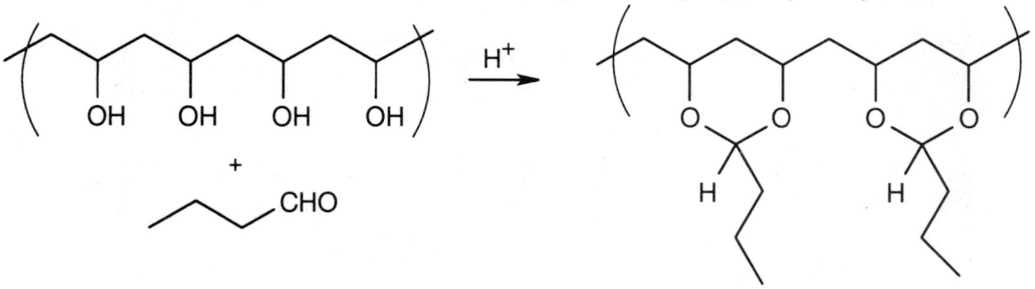

OH OH OH OH + CHO $\xrightarrow{H^+}$ O O O O H H

Poly(vinyl butyral)

31.25

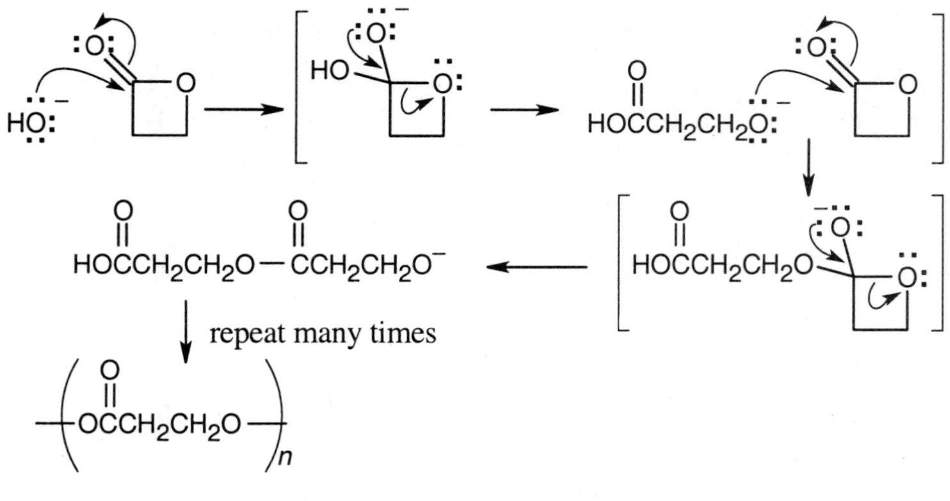

HO^- + :O: O → [HO :O: :O:] → $HOCCH_2CH_2O^-$ + :O: O → [$HOCCH_2CH_2O$:O: :O:] → $HOCCH_2CH_2O-CCH_2CH_2O^-$

repeat many times

$$\left(OCCH_2CH_2O\right)_n$$

The polymer is a polyester.

31.26

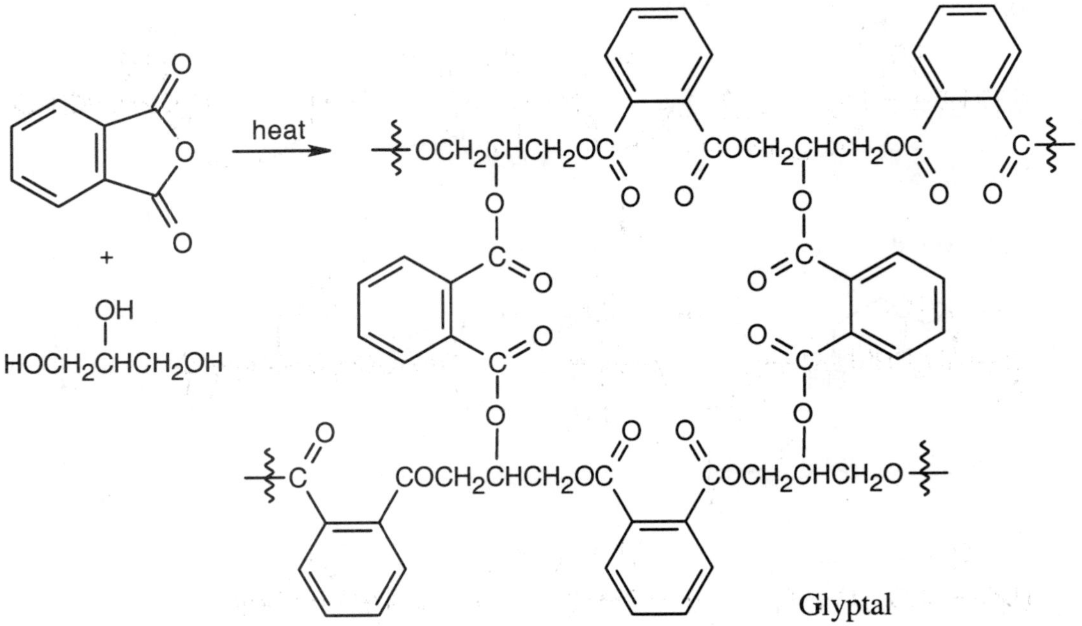

Phthalic anhydride + $HOCH_2CH(OH)CH_2OH$ $\xrightarrow{\text{heat}}$ Glyptal

Use of glycerol as a monomer causes the cross-linking that gives Glyptal its strength.

31.27 Repeated nucleophilic acyl substitution reactions result in the formation of Melmac.

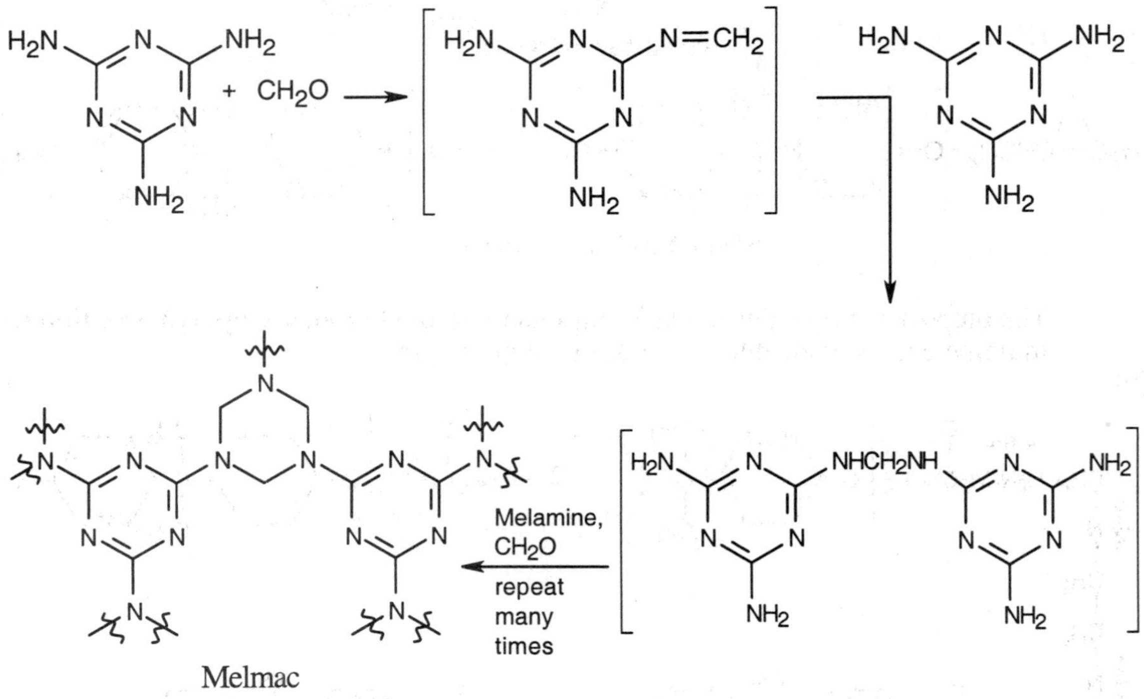

Melamine + CH_2O → [melamine–$N=CH_2$] + melamine → [H_2N-triazine-$NHCH_2NH$-triazine-NH_2] $\xrightarrow{\text{Melamine, } CH_2O \text{, repeat many times}}$ Melmac

31.28

(a)

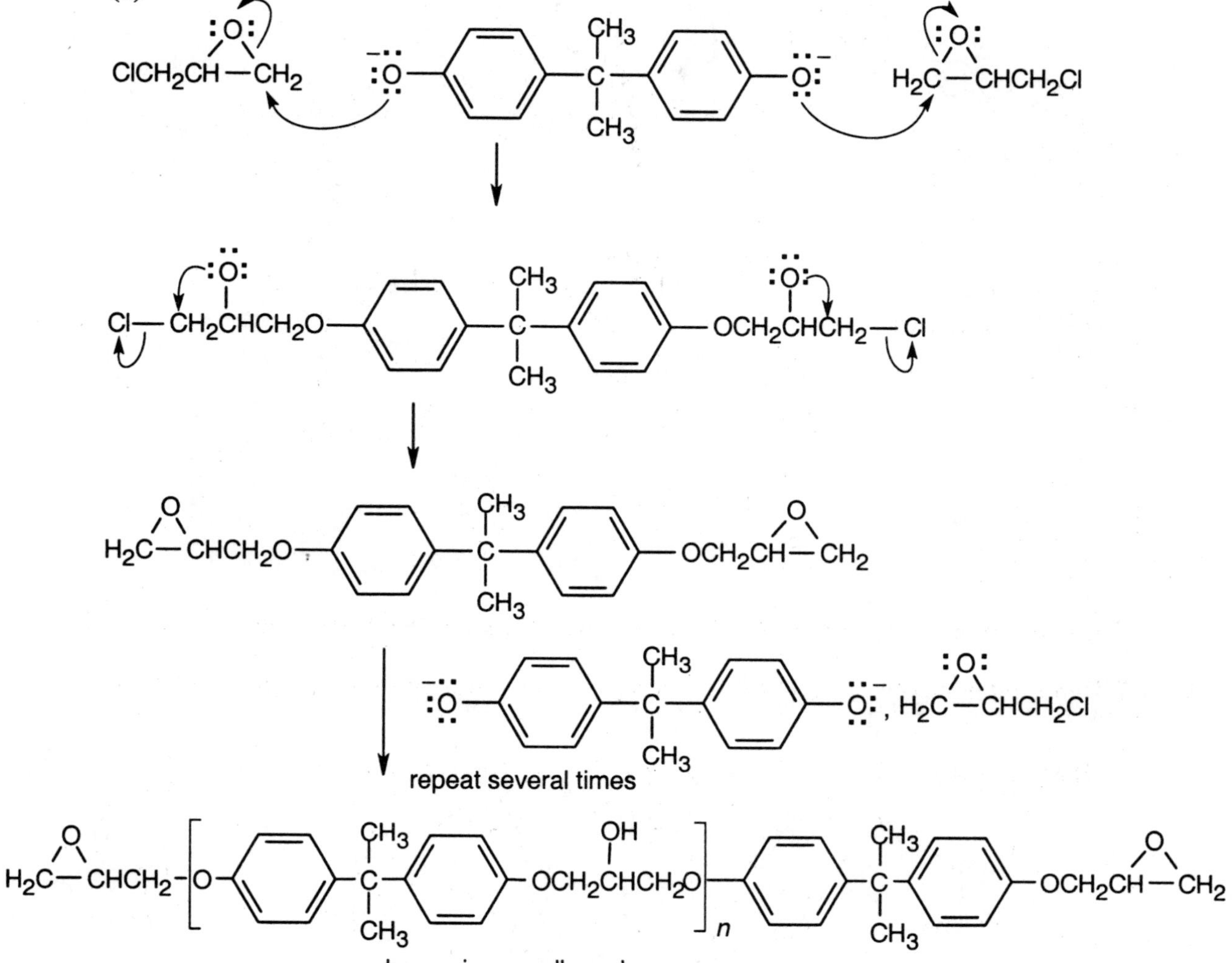

:O:
ClCH$_2$CH—CH$_2$:O$^-$—C$_6$H$_4$—C(CH$_3$)$_2$—C$_6$H$_4$—O$^-$: H$_2$C—CHCH$_2$Cl

Cl—CH$_2$CHCH$_2$O—C$_6$H$_4$—C(CH$_3$)$_2$—C$_6$H$_4$—OCH$_2$CHCH$_2$—Cl

H$_2$C—CHCH$_2$O—C$_6$H$_4$—C(CH$_3$)$_2$—C$_6$H$_4$—OCH$_2$CH—CH$_2$ (epoxide rings)

$^-$:O—C$_6$H$_4$—C(CH$_3$)$_2$—C$_6$H$_4$—O:$^-$, H$_2$C—CHCH$_2$Cl

repeat several times

H$_2$C—CHCH$_2$-[O—C$_6$H$_4$—C(CH$_3$)$_2$—C$_6$H$_4$—OCH$_2$CH(OH)CH$_2$O]$_n$—C$_6$H$_4$—C(CH$_3$)$_2$—C$_6$H$_4$—OCH$_2$CH—CH$_2$

where *n* is a small number

The prepolymer contains epoxide rings and hydroxyl groups. Copolymerization with a triamine occurs at the epoxide ends of the prepolymer.

(b)

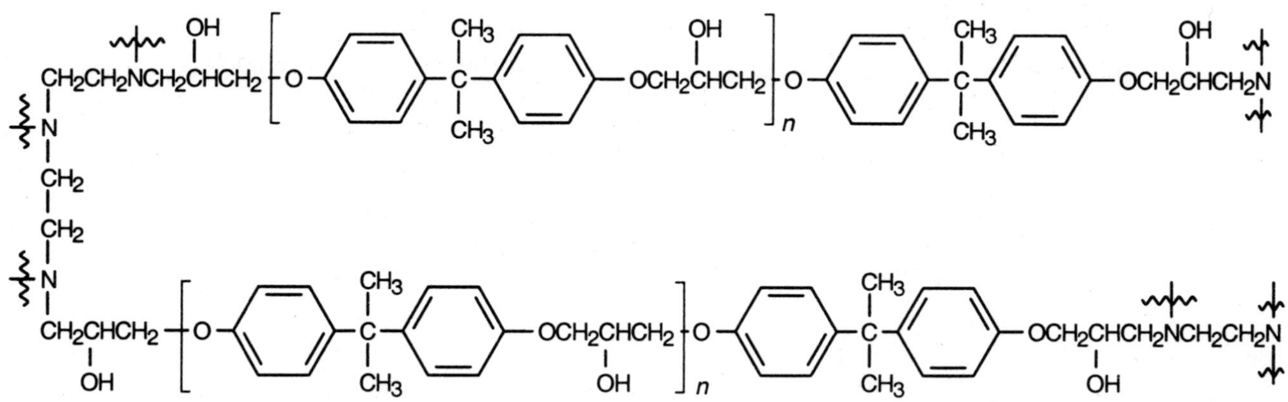

CH$_2$CH$_2$NCH$_2$CH(OH)CH$_2$-[O—C$_6$H$_4$—C(CH$_3$)$_2$—C$_6$H$_4$—OCH$_2$CH(OH)CH$_2$]$_n$-O—C$_6$H$_4$—C(CH$_3$)$_2$—C$_6$H$_4$—OCH$_2$CH(OH)CH$_2$N

N—CH$_2$—CH$_2$—N

CH$_2$CH(OH)CH$_2$-[O—C$_6$H$_4$—C(CH$_3$)$_2$—C$_6$H$_4$—OCH$_2$CH(OH)CH$_2$]$_n$-O—C$_6$H$_4$—C(CH$_3$)$_2$—C$_6$H$_4$—OCH$_2$CH(OH)CH$_2$NCH$_2$CH$_2$N

Cross-linking occurs when the triamine opens epoxide rings on two different chains of the prepolymer.

31.29 (a) The diamine is formed by an electrophilic aromatic substitution reaction of formaldehyde with two equivalents of aniline.

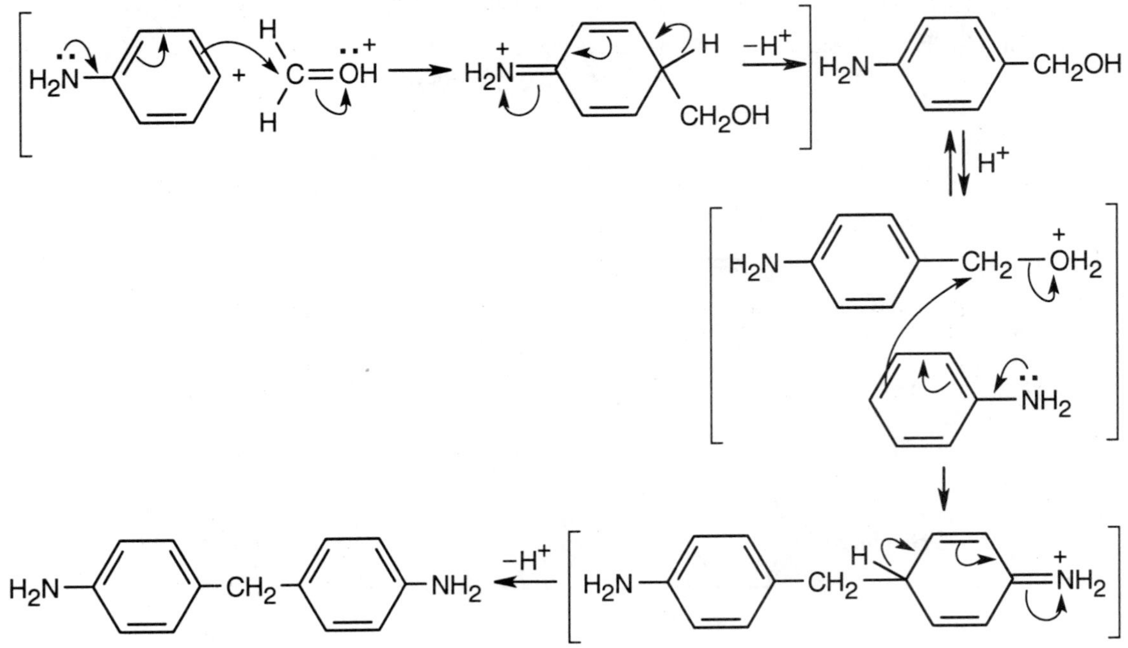

H_2N + H, H, C=$\ddot{O}H^+$ → H_2N^+= H, CH_2OH $-H^+$ → H_2N — CH_2OH

H^+

H_2N — CH_2 — OH_2^+

NH_2

H_2N — CH_2 — NH_2 ← $-H^+$ — H_2N — CH_2 — H — NH_2^+

(b) The diamine reacts with two equivalents of phosgene.

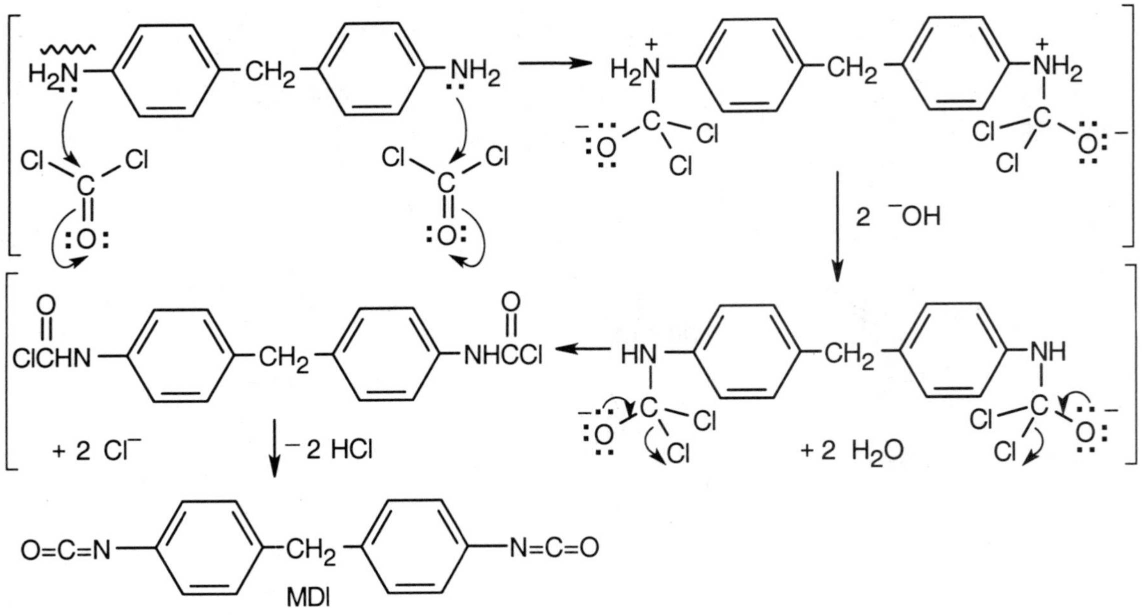

H_2N — CH_2 — NH_2; Cl, Cl, C=O; Cl, Cl, C=O → H_2N^+ — CH_2 — NH_2^+; ^-O—C—Cl, Cl; Cl—C—O^-, Cl

2 ^-OH

ClCHN(C=O) — CH_2 — NHCCl(C=O) ← HN — CH_2 — NH; ^-O—C—Cl, Cl; Cl—C—O^-, Cl

+ 2 Cl^- + 2 H_2O

− 2 HCl

O=C=N — CH_2 — N=C=O

MDI

31.30

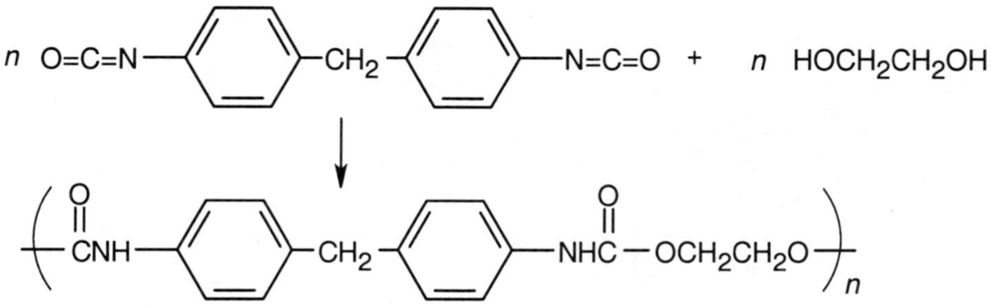

31.31

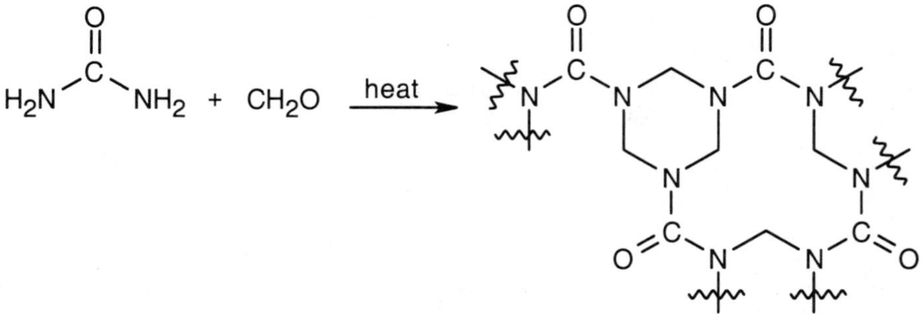

31.32 Step 1: Polystyrene and the phthalimide combine in an electrophilic aromatic substitution reaction

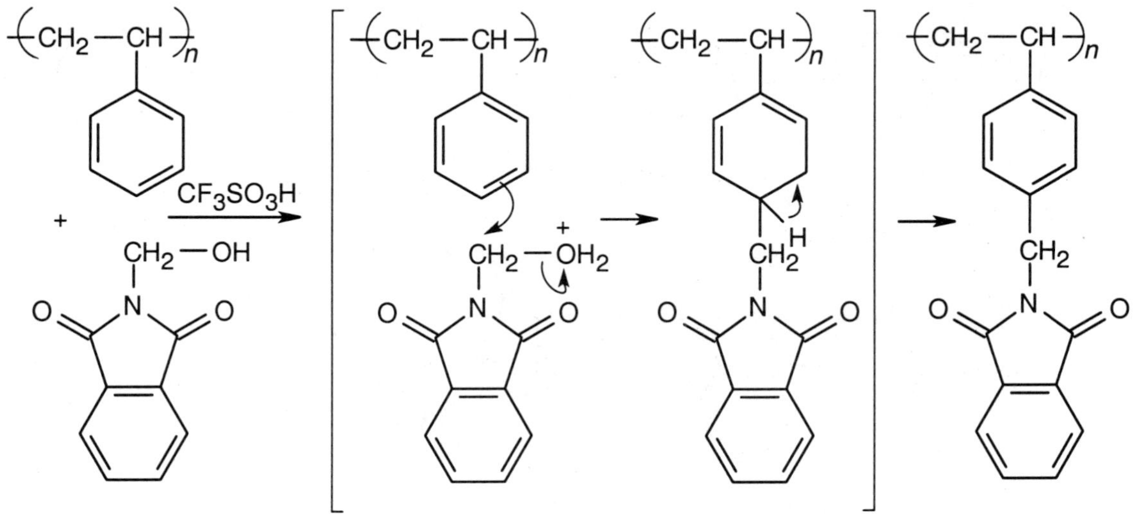

Step 2: The phthalimide is cleaved in a series of steps that involve nucleophilic acyl substitution reactions.

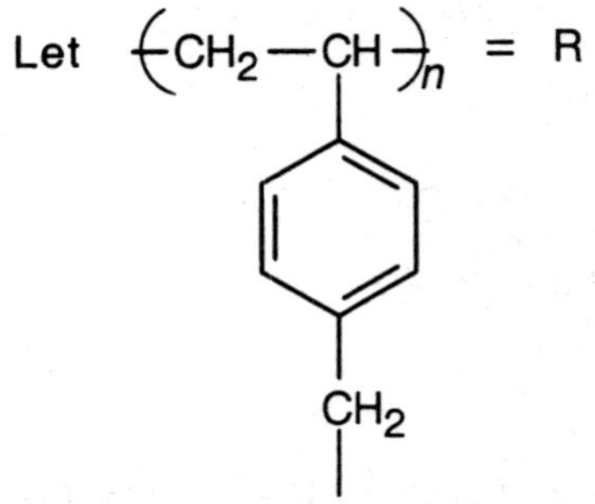

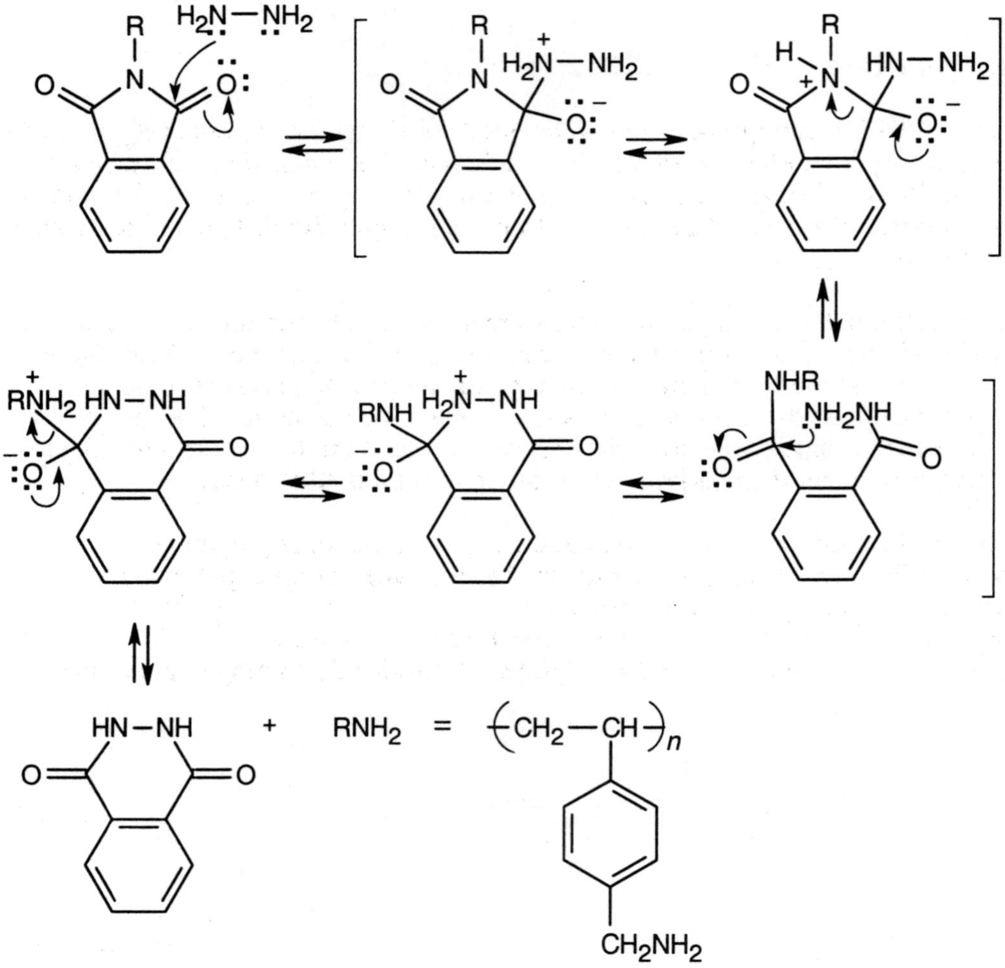

Let $\left(CH_2-CH\right)_n$ = R (with a para-phenylene CH_2 substituent)

H_2N-NH_2; RNH_2; $HN-NH$; $RNH_2 = \left(CH_2-CH\right)_n$ with CH_2NH_2

31.33

$$2\ CH_3CH_2CH_2CHO \xrightarrow[2.\ heat]{1.\ NaOH,\ EtOH} CH_3CH_2CH_2CH{=}C(CH_2CH_3)CHO$$

$$\xrightarrow[2.\ H_3O^+]{1.\ NaBH_4} CH_3CH_2CH_2CH{=}C(CH_2CH_3)CH_2OH \xrightarrow[Pd/C]{H_2} CH_3CH_2CH_2CH_2CH(CH_2CH_3)CH_2OH$$

2-Ethyl-1-hexanol

Aldol self-condensation of butanal, followed by reduction, gives 2-ethyl-1-hexanol.

Molecular Modeling

31.34 Styrene and 2-methylpropene would be good substrates for cationic polymerization because they have electron-rich carbon-carbon double bonds that would be attacked by electrophiles. 2-Propenal and nitroethylene would be good substrates for anionic polymerization because they have electron-poor double bonds that would be attacked by nucleophiles.

31.35 Electrostatic potential maps show that styrene and 2-vinylpyridine are much more effective at delocalizing the developing negative charge than 3-vinylfuran. 3-Vinylfuran would thus be a poor substrate for anionic polymerization. Electrostatic potential maps also show that the heteroatom is the most negative site in the neutral substrates. These substrates are not suitable for cationic polymerization because the electrophilic initiators attack the heteroatom, rather than adding to the carbon-carbon double bond.

31.36 Polymer A: vinyl chloride + vinylidene chloride (random copolymer).
Polymer B: vinyl chloride + vinylidene chloride (alternating copolymer).
Polymer C: acrylonitrile (homopolymer).
Polymer D: acrylonitrile + 1,3-butadiene (block copolymer).
Polymer E: hexafluoropropene + vinylidene fluoride (alternating copolymer).

Review Unit 12: Pericyclic Reactions, Synthetic Polymers

Major Topics Covered (with vocabulary):

Pericyclic reactions:
pericyclic reaction concerted reaction symmetry-allowed symmetry-disallowed frontier orbitals HOMO LUMO electrocyclic reaction disrotatory motion conrotatory motion cycloaddition reaction suprafacial cycloaddition antarafacial cycloaddition sigmatropic rearrangement suprafacial rearrangement antarafacial rearrangement Cope rearrangement Claisen rearrangement

Synthetic polymers:
chain-growth polymer Ziegler-Natta catalyst isotactic syndiotactic atactic homopolymer copolymer block copolymer graft copolymer step-growth polymer polycarbonate polyurethane crystallite melt transition temperature glass transition thermoplastic fiber elastomer thermosetting resin plasticizer

Types of Problems:

After studying these chapters, you should be able to:

- Understand the principles of molecular orbitals, and locate the HOMO and LUMO of conjugated π systems.
- Predict the stereochemistry of thermal and photochemical electrocyclic reactions.
- Know the stereochemical requirements for cycloaddition reactions, and predict the products of cycloadditions.
- Classify sigmatropic reactions by order and predict their products.
- Know the selection rules for pericyclic reactions.

- Locate the monomer units of a polymer; predict the structure of a polymer, given its monomer units.
- Formulate the mechanisms of radical, cationic, anionic, and step-growth polymerizations.
- Understand the stereochemistry of polymerization, and draw structures of atactic, isotactic, and syndiotactic polymers.
- Understand copolymerization, graft polymerization and block polymerization.

Points to Remember:

* Just because a reaction is symmetry-disallowed doesn't mean that it can't occur. Reactions that are symmetry-allowed occur by relatively low-energy, concerted pathways. Reactions that are symmetry-disallowed must take place by higher energy, nonconcerted routes.

* To predict if a reaction is symmetry-allowed, it is only necessary to be concerned with the signs of the outermost lobes.

* The notations in brackets in a sigmatropic rearrangement refer to the positions in the migrating groups to which migration occurs.

* The stereochemical outcome of a concerted reaction run under thermal conditions is always opposite to the stereochemical outcome of the same reaction run under photochemical conditions.

* To show the monomer unit of a chain-growth polymer, find the smallest repeating unit, break the polymer bonds, and draw the monomer with its original double bond in place. To show the monomer unit of a step-growth polymer, find the smallest repeating unit, break the polymer bonds, and draw the monomer unit or units with the small molecules that were displaced by polymerization added to the monomer units.

* Fishhook arrows are used to show movement of single electrons.

Self-test:

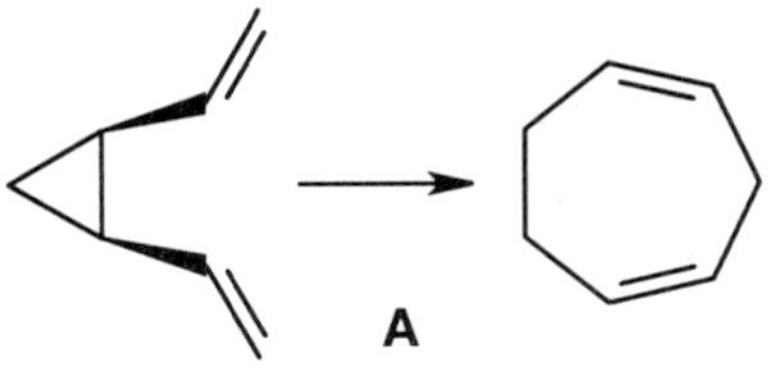

A

$H_2C{=}C(CN)COOCH_2CH(CH_3)CH_3$

B

$H_2C-CH-CH-CH_2$ (two epoxide rings, each O bridging H_2C–CH and CH–CH_2)

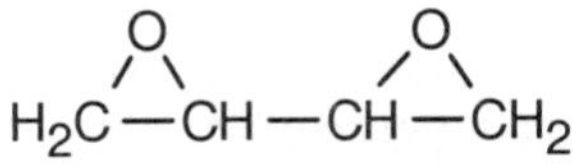

C

What type of reaction is occurring in **A**? Describe it by order and type. If the reaction of the stereoisomer shown proceeds readily, is the reaction being carried out under thermal or photochemical conditions?

Under what type of conditions would you expect monomer **B** to polymerize? (Actually it polymerizes well under all conditions). Is the polymer a chain-growth or a step-growth polymer? Draw a representative segment of the polymer.

Suggest a use for **C** in polymerizations.

Multiple choice:

1. In which orbitals do the outermost lobes have opposite signs on the same side of the π system?
(a) HOMO in the ground state of a 2 π electron system (b) HOMO in the excited state of a 4 π electron system (c) LUMO in the excited state of a 6 π electron system (d) HOMO in the ground state of a 4 π electron system

2. Which reaction is symmetry-disallowed?
(a) disrotatory photochemical ring-opening of a 3 π electron system (b) suprafacial thermal cycloaddition of a 3 π electron system (c) antarafacial thermal sigmatropic rearrangement of a 4 π electron system (d) antarafacial photochemical cycloaddition of a 3 π electron system

3. Which of the following reactions is symmetry-allowed but geometrically constrained?
(a) thermal electrocyclic reaction of a 2 π electron system (b) photochemical cycloaddition of a 2 π electron system (c) thermal sigmatropic rearrangement of a 2 π electron system (d) thermal cycloaddition of a 4 π electron system

4. All of the following sigmatropic rearrangements involve 6 π electrons except:
(a) rearrangement of allyl phenyl ether to *o*-allyl phenol (b) rearrangement of 1,5-heptadiene to 3-methyl-1,5-hexadiene (c) rearrangement of 1,3,5-heptatriene in which a hydrogen atom migrates across the π system (d) rearrangement of homotropilidene

5. Consider the 2 π electron thermal electrocyclic reactions of two double-bond stereoisomers. All of the following are true except:
(a) one reaction is concerted and one isn't (b) the equilibrium lies on the side of the ring-opened product (c) the reaction proceeds with conrotatory motion (d) The ring-closed products are stereoisomers

6. Which of the following monomers is most likely to undergo cationic polymerization?
(a) $H_2C=CF_2$ (b) $H_2C=CH_2$ (c) formaldehyde (d) $H_2C=C(CH_3)_2$

7. Which of the following is not a copolymer?
(a) Saran (b) Nylon 6 (c) Dacron (d) Lexan

8. In which step-growth polymer is an alcohol the byproduct?
(a) polyester (b) polyamide (c) polyurethane (d) polycarbonate

9. Which type of polymer has large regions of oriented crystallites and little or no cross-linking?
(a) a thermoplastic (b) a fiber (c) an elastomer (d) a thermosetting resin

10. A copolymer formed by irradiating a homopolymer in the presence of a second monomer is called a:
(a) random copolymer (b) alternating copolymer (c) graft copolymer (d) block copolymer

Functional-Group Synthesis

The following table summarizes the synthetic methods by which important functional groups can be prepared. The functional groups are listed alphabetically, followed by reference to the appropriate text section and a brief description of each synthetic method.

Acetals, $R_2C(OR')_2$

(Sec. 19.11)	from ketones and aldehydes by acid-catalyzed reaction with alcohols

Acid anhydrides, RCOOCOR'

(Sec. 21.3)	from dicarboxylic acids by heating
(Sec. 21.5)	from acid chlorides by reaction with carboxylate salts

Acid bromides, RCOBr

(Sec. 21.4)	from carboxylic acids by reaction with PBr_3

Acid chlorides, RCOCl

(Sec. 21.3)	from carboxylic acids by reaction with $SOCl_2$

Alcohols, ROH

(Sec. 7.4)	from alkenes by oxymercuration/demercuration
(Sec. 7.5)	from alkenes by hydroboration/oxidation
(Sec. 7.8)	from alkenes by hydroxylation with OsO_4
(Sec. 11.4, 11.5)	from alkyl halides and tosylates by S_N2 reaction with hydroxide ion
(Sec. 18.5)	from ethers by acid-induced cleavage
(Sec. 18.8)	from epoxides by acid-catalyzed ring opening with either H_2O or HX
(Sec. 18.8)	from epoxides by base-induced ring opening
(Sec. 17.5, 19.8)	from ketones and aldehydes by reduction with $NaBH_4$ or $LiAlH_4$
(Sec. 17.6, 19.8)	from ketones and aldehydes by addition of Grignard reagents
(Sec. 20.8)	from carboxylic acids by reduction with either $LiAlH_4$ or BH_3
(Sec. 21.4)	from acid chlorides by reduction with $LiAlH_4$
(Sec. 21.4)	from acid chlorides by reaction with Grignard reagents
(Sec. 21.5)	from acid anhydrides by reduction with $LiAlH_4$
(Sec. 17.5, 21.6)	from esters by reduction with $LiAlH_4$
(Sec. 17.6, 21.6)	from esters by reaction with Grignard reagents

Aldehydes, RCHO

(Sec. 7.8)	from disubstituted alkenes by ozonolysis
(Sec. 7.8)	from 1,2-diols by cleavage with sodium periodate
(Sec. 8.5)	from terminal alkynes by hydroboration followed by oxidation
(Sec. 17.8, 19.2)	from primary alcohols by oxidation
(Sec. 21.4)	from acid chlorides by partial reduction with $LiAl(O\text{-}t\text{-}Bu)_3H$
(Sec. 19.2, 21.6)	from esters by reduction with DIBAH [$HAl(i\text{-}Bu)_2$]
(Sec. 21.8)	from nitriles by partial reduction with DIBAH

Alkanes, RH

(Sec. 7.7)	from alkenes by catalytic hydrogenation
(Sec. 10.8)	from alkyl halides by protonolysis of Grignard reagents
(Sec. 10.9)	from alkyl halides by coupling with Gilman reagents
(Sec. 19.10)	from ketones and aldehydes by Wolff–Kishner reaction

Alkenes, $R_2C=CR_2$

(Sec. 7.1, 11.11)	from alkyl halides by treatment with strong base (E2 reaction)
(Sec. 7.1, 17.7)	from alcohols by dehydration
(Sec. 8.6)	from alkynes by catalytic hydrogenation using the Lindlar catalyst
(Sec. 8.6)	from alkynes by reduction with lithium in liquid ammonia
(Sec. 19.12)	from ketones and aldehydes by treatment with alkylidenetriphenylphosphoranes (Wittig reaction)
(Sec. 22.3)	from α-bromo ketones by heating with pyridine
(Sec. 24.7)	from amines by methylation and Hofmann elimination

Alkynes, RC≡CR

(Sec. 8.3)	from dihalides by base-induced double dehydrohalogenation
(Sec. 8.9)	from terminal alkynes by alkylation of acetylide anions

Amides, $RCONH_2$

(Sec. 21.3)	from carboxylic acids by heating with ammonia
(Sec. 21.4)	from acid chlorides by treatment with an amine or ammonia
(Sec. 21.5)	from acid anhydrides by treatment with an amine or ammonia
(Sec. 21.6)	from esters by treatment with an amine or ammonia
(Sec. 21.8)	from nitriles by partial hydrolysis with either acid or base
(Sec. 26.10)	from a carboxylic acid and an amine by treatment with dicyclohexylcarbodiimide (DCC)

Amines, RNH_2

(Sec. 19.14)	from conjugated enones by addition of primary or secondary amines
(Sec. 21.7, 24.6)	from amides by reduction with $LiAlH_4$
(Sec. 21.8, 24.6)	from nitriles by reduction with $LiAlH_4$
(Sec. 24.6)	from primary alkyl halides by treatment with ammonia
(Sec. 24.6)	from primary alkyl halides by Gabriel synthesis
(Sec. 24.6)	from primary alkyl azides by reduction with $LiAlH_4$
(Sec. 24.6)	from acid chlorides by Curtius rearrangement of acyl azides
(Sec. 24.6)	from primary amides by Hofmann rearrangement
(Sec. 24.6)	from ketones and aldehydes by reductive amination with an amine and $NaBH_3CN$

Amino Acids, $RCH(NH_2)COOH$

(Sec. 26.3)	from α-bromo acids by S_N2 reaction with ammonia
(Sec. 26.3)	from aldehydes by reaction with KCN and ammonia (Strecker synthesis)
(Sec. 26.3)	from α-keto acids by reductive amination
(Sec. 26.3)	from primary alkyl halides by alkylation with diethyl acetamidomalonate

Arenes, Ar–R

(Sec. 16.3)	from arenes by Friedel–Crafts alkylation with an alkyl halide
(Sec. 16.11)	from aryl alkyl ketones by catalytic reduction of the keto group
(Sec. 24.8)	from arenediazonium salts by treatment with hypophosphorous acid

Arylamines, $Ar–NH_2$

(Sec. 16.2, 24.6)	from nitroarenes by reduction with either Fe, Sn, or H_2/Pd.

Arenediazonium salts, $Ar–N_2^+ X^-$

(Sec. 24.8)	from arylamines by reaction with nitrous acid

Arenesulfonic acids $Ar–SO_3H$

(Sec. 16.2)	from arenes by electrophilic aromatic substitution with SO_3/H_2SO_4

Azides, $R–N_3$

(Sec. 11.4, 24.6)	from primary alkyl halides by S_N2 reaction with azide ion

Carboxylic acids, RCOOH

(Sec. 7.8)	from mono- and 1,2-disubstituted alkenes by ozonolysis
(Sec. 16.10)	from arenes by side-chain oxidation with $Na_2Cr_2O_7$ or $KMnO_4$
(Sec. 19.3)	from aldehydes by oxidation
(Sec. 20.6)	from alkyl halides by conversion into Grignard reagents followed by reaction with CO_2
(Sec. 20.6, 21.8)	from nitriles by acid or base hydrolysis
(Sec. 21.4)	from acid chlorides by reaction with aqueous base
(Sec. 21.5)	from acid anhydrides by reaction with aqueous base
(Sec. 21.6)	from esters by hydrolysis with aqueous base
(Sec. 21.7)	from amides by hydrolysis with aqueous base
(Sec. 22.7)	from methyl ketones by reaction with halogen and base (haloform reaction)

Cyanohydrins, RCH(OH)CN

(Sec. 19.7)	from aldehydes and ketones by reaction with HCN

Cycloalkanes

(Sec. 7.6)	from alkenes by addition of dichlorocarbene
(Sec. 7.6)	from alkenes by reaction with CH_2I_2 and Zn/Cu (Simmons–Smith reaction)
(Sec. 16.11)	from arenes by rhodium-catalyzed hydrogenation

Disulfides, RS–SR'

(Sec. 18.11)	from thiols by oxidation with bromine

Enamines, $RCH{=}CRNR_2$

(Sec. 19.9)	from ketones or aldehydes by reaction with secondary amines

Epoxides, $R_2C\overset{O}{—}CR_2$

(Sec. 18.7)	from alkenes by treatment with a peroxyacid
(Sec. 18.7)	from halohydrins by treatment with base

Esters, RCOOR'

(Sec. 21.3)	from carboxylic acid salts by S_N2 reaction with primary alkyl halides
(Sec. 21.3)	from carboxylic acids by acid-catalyzed reaction with an alcohol (Fischer esterification)
(Sec. 21.4)	from acid chlorides by base-induced reaction with an alcohol
(Sec. 21.5)	from acid anhydrides by base-induced reaction with an alcohol
(Sec. 22.8)	from alkyl halides by alkylation with diethyl malonate
(Sec. 22.8)	from esters by treatment of their enolate ions with alkyl halides

Ethers, R–O–R'

(Sec. 16.8)	from activated haloarenes by reaction with alkoxide ions
(Sec. 16.9)	from unactivated haloarenes by reaction with alkoxide ions via benzyne intermediates
(Sec. 18.3)	from primary alkyl halides by S_N2 reaction with alkoxide ions (Williamson ether synthesis)
(Sec. 18.4)	from alkenes by alkoxymercuration/demercuration
(Sec. 18.7)	from alkenes by epoxidation with peroxyacids

Halides, alkyl, R_3C-X

(Sec. 6.8)	from alkenes by electrophilic addition of HX
(Sec. 7.2)	from alkenes by addition of halogen
(Sec. 7.3)	from alkenes by electrophilic addition of hypohalous acid (HOX) to yield halohydrins
(Sec. 8.4)	from alkynes by addition of halogen
(Sec. 8.4)	from alkynes by addition of HX
(Sec. 10.5)	from alkenes by allylic bromination with *N*-bromosuccinimide (NBS)
(Sec. 10.7)	from alcohols by reaction with HX
(Sec. 10.7)	from alcohols by reaction with $SOCl_2$
(Sec. 10.7)	from alcohols by reaction with PBr_3
(Sec. 11.4, 11.5)	from alkyl tosylates by S_N2 reaction with halide ions
(Sec. 16.10)	from arenes by benzylic bromination with *N*-bromosuccinimide (NBS)
(Sec. 18.5)	from ethers by cleavage with either HX
(Sec. 22.3)	from ketones by α-halogenation with bromine
(Sec. 22.4)	from carboxylic acids by α-halogenation with phosphorus and PBr_3 (Hell–Volhard–Zelinskii reaction)

Halides, aryl, Ar–X

(Sec. 16.1, 16.2)	from arenes by electrophilic aromatic substitution with halogen
(Sec. 24.8)	from arenediazonium salts by reaction with cuprous halides (Sandmeyer reaction)

Halohydrins, $R_2CXC(OH)R_2$

(Sec. 7.3)	from alkenes by electrophilic addition of hypohalous acid (HOX)
(Sec. 18.8)	from epoxides by acid-induced ring opening with HX

Imines. $R_2C=NR'$

(Sec. 19.9)	from ketones or aldehydes by reaction with primary amines

Ketones, $R_2C=O$

(Sec. 7.8)	from alkenes by ozonolysis
(Sec. 7.8)	from 1,2-diols by cleavage reaction with sodium periodate
(Sec. 8.5)	from alkynes by mercuric-ion-catalyzed hydration
(Sec. 8.5)	from alkynes by hydroboration/oxidation
(Sec. 16.4)	from arenes by Friedel–Crafts acylation reaction with an acid chloride
(Sec. 17.8, 19.2)	from secondary alcohols by oxidation
(Sec. 19.2, 21.4)	from acid chlorides by reaction with lithium diorganocopper (Gilman) reagents
(Sec. 19.14)	from conjugated enones by addition of lithium diorganocopper reagents
(Sec. 21.8)	from nitriles by reaction with Grignard reagents
(Sec. 22.8)	from primary alkyl halides by alkylation with ethyl acetoacetate
(Sec. 22.8)	from ketones by alkylation of their enolate ions with primary alkyl halides

Nitriles, R–C≡N

(Sec. 11.5, 21.8)	from primary alkyl halides by S_N2 reaction with cyanide ion
(Sec. 21.8)	from primary amides by dehydration with $SOCl_2$
(Sec. 22.8)	from nitriles by alkylation of their α-anions with primary alkyl halides
(Sec. 24.8)	from arenediazonium ions by treatment with CuCN

Nitroarenes, $Ar–NO_2$

(Sec. 16.2)	from arenes by electrophilic aromatic substitution with nitric/sulfuric acids

Organometallics, R–M

(Sec. 10.8)	formation of Grignard reagents from organohalides by treatment with magnesium
(Sec. 10.9)	formation of organolithium reagents from organohalides by treatment with lithium
(Sec. 10.9)	formation of lithium diorganocopper reagents (Gilman reagents) from organolithium reagents by treatment with cuprous halides

Phenols, Ar–OH

(Sec. 16.2, 17.10)	from arenesulfonic acids by fusion with KOH
(Sec. 24.8)	from arenediazonium salts by reaction with Cu_2O and $Cu(NO_3)_2$
(Sec. 16.9)	from aryl halides by nucleophilic aromatic substitution with hydroxide ion

Quinones,

(Sec. 17.11)	from phenols by oxidation with Fremy's salt $[(KSO_3)_2NO]$

Sulfides, R–S–R'

(Sec. 18.11)	from thiols by S_N2 reaction of thiolate ions with primary alkyl halides

Sulfones, $R–SO_2–R'$

(Sec. 18.11)	from sulfides or sulfoxides by oxidation with peroxyacids

Sulfoxides, R–SO–R'

(Sec. 18.11)	from sulfides by oxidation with H_2O_2

Thiols, R–SH

(Sec. 11.5)	from primary alkyl halides by S_N2 reaction with hydrosulfide anion
(Sec. 18.11)	from primary alkyl halides by S_N2 reaction with thiourea, followed by hydrolysis

Functional-Group Reactions

The following table summarizes the reactions of important functional groups. The functional groups are listed alphabetically, followed by a reference to the appropriate text section.

Acetal

1. Hydrolysis to yield a ketone or aldehyde plus alcohol (Sec. 19.11)

Acid anhydride

1. Hydrolysis to yield a carboxylic acid (Sec. 21.5)
2. Alcoholysis to yield an ester (Sec. 21.5)
3. Aminolysis to yield an amide (Sec. 21.5)
4. Reduction to yield a primary alcohol (Sec. 21.5)

Acid chloride

1. Hydrolysis to yield a carboxylic acid (Sec. 21.4)
2. Alcoholysis to yield an ester (Sec. 21.4)
3. Aminolysis to yield an amide (Sec. 21.4)
4. Reduction to yield a primary alcohol (Sec. 21.4)
5. Partial reduction to yield an aldehyde (Sec. 21.4)
6. Grignard reaction to yield a tertiary alcohol (Sec. 21.4)
7. Reaction with a lithium diorganocopper reagent to yield a ketone (Sec. 21.4)

Alcohol

1. Acidity (Sec. 17.3)
2. Oxidation (Sec. 17.8)
   a. Reaction of a primary alcohol to yield an aldehyde or acid
   b. Reaction of a secondary alcohol to yield a ketone
3. Reaction with a carboxylic acid to yield an ester (Sec. 21.3)
4. Reaction with an acid chloride to yield an ester (Sec. 21.4)
5. Dehydration to yield an alkene (Sec. 17.7)
6. Reaction with a primary alkyl halide to yield an ether (Sec. 18.3)
7. Conversion into an alkyl halide (Sec. 17.7)
   a. Reaction of a tertiary alcohol with HX
   b. Reaction of a primary or secondary alcohol with $SOCl_2$
   c. Reaction of a primary or secondary alcohol with PBr_3

Aldehyde

1. Oxidation to yield a carboxylic acid (Sec. 19.3)
2. Nucleophilic addition reactions
   a. Reduction to yield a primary alcohol (Secs. 17.5, 19.8)
   b. Reaction with a Grignard reagent to yield a secondary alcohol (Secs. 17.6, 19.8)
   c. Grignard reaction of formaldehyde to yield a primary alcohol (Sec. 17.6)
   d. Reaction with HCN to yield a cyanohydrin (Sec. 19.7)
   e. Wolff–Kishner reaction with hydrazine to yield an alkane (Sec. 19.10)
   f. Reaction with an alcohol to yield an acetal (Sec. 19.11)
   g. Wittig reaction to yield an alkene (Sec. 19.12)
   h. Reaction with an amine to yield an imine or enamine (Sec. 19.9)
3. Aldol reaction to yield a β-hydroxy aldehyde (Sec. 23.2)
4. Alpha bromination of an aldehyde (Sec. 22.3)

Alkane

1. Radical halogenation to yield an alkyl halide (Secs. 5.3, 10.4)

Alkene

1. Electrophilic addition of HX to yield an alkyl halide (Secs. 6.8–6.12)
   Markovnikov regiochemistry is observed. H adds to the less highly substituted carbon, and X adds to the more highly substituted one.
2. Electrophilic addition of halogen to yield a 1,2-dihalide (Sec. 7.2)
   Anti stereochemistry is observed
3. Oxymercuration/demercuration to yield an alcohol (Sec. 7.4)
   Markovnikov regiochemistry is observed, yielding the more highly substituted alcohol.
4. Hydroboration/oxidation to yield an alcohol (Section 7.5)
5. Hydrogenation to yield an alkane (Sec. 7.7)
6. Hydroxylation to yield a 1,2-diol (Sec. 7.8)
7. Oxidative cleavage to yield carbonyl compounds (Sec. 7.8)
8. Reaction with a peroxyacid to yield an epoxide (Sec. 18.7)
9. Simmons–Smith reaction with CH_2I_2 to yield a cyclopropane (Sec. 7.6)

Alkyne

1. Electrophilic addition of HX to yield a vinylic halide (Sec. 8.4)
2. Electrophilic addition of halogen to yield a dihalide (Sec. 8.4)
3. Mercuric-sulfate-catalyzed hydration to yield a methyl ketone (Sec. 8.5)
4. Hydroboration/oxidation to yield an aldehyde (Sec. 8.5)
5. Alkylation of an alkyne anion (Sec. 8.9)
6. Reduction (Sec. 8.6)
   a. Hydrogenation over Lindlar catalyst to yield a cis alkene
   b. Reduction with Li/NH_3 to yield a trans alkene

Amide

1. Hydrolysis to yield a carboxylic acid (Sec. 21.7)
2. Reduction with $LiAlH_4$ to yield an amine (Sec. 21.7)
3. Dehydration to yield a nitrile (Section 21.8)

Amine

1. S_N2 alkylation of an alkyl halide to yield an amine (Sec. 24.6)
2. Nucleophilic acyl substitution reactions
   a. Reaction with an acid chloride to yield an amide (Sec. 21.4)
   b. Reaction with an acid anhydride to yield an amide (Sec. 21.5)
3. Hofmann elimination to yield an alkene (Sec. 24.7)
4. Formation of an arenediazonium salt (Sec. 24.8)

Arene

1. Oxidation of an alkylbenzene side chain to yield a benzoic acid (Sec. 16.10)
2. Catalytic reduction to yield a cyclohexane (Sec. 16.11)
3. Reduction of an aryl alkyl ketone to yield an arene (Sec. 16.11)
4. Electrophilic aromatic substitution (Secs. 16.1–16.4)
   a. Halogenation (Secs. 16.1–16.2)
   b. Nitration (Sec. 16.2)
   c. Sulfonation (Sec. 16.2)
   d. Friedel–Crafts alkylation (Sec. 16.3)
      Aromatic ring must be at least as reactive as a halobenzene
   e. Friedel–Crafts acylation (Sec. 16.4)

Arenediazonium salt

1. Conversion into an aryl chloride (Sec. 24.8)
2. Conversion into an aryl bromide (Sec. 24.8)
3. Conversion into an aryl iodide (Sec. 24.8)
4. Conversion into an aryl cyanide (Sec. 24.8)
5. Conversion into a phenol (Sec. 24.8)
6. Conversion into an arene (Sec. 24.8)

Arenesulfonic acid

1. Conversion into a phenol (Secs. 16.2, 17.10)

Carboxylic acid

1. Acidity (Secs. 20.3–20.5)
2. Reduction to yield a primary alcohol (Secs. 17.5, 20.8)
   a. Reduction with $LiAlH_4$
   b. Reduction with BH_3
3. Nucleophilic acyl substitution reactions (Sec. 21.3)
   a. Conversion into an acid chloride
   b. Conversion into an acid anhydride
   c. Conversion into an ester
      (1) Fischer esterification
      (2) S_N2 reaction with an alkyl halide

Epoxide

1. Acid-catalyzed ring opening with HX to yield a halohydrin (Sec. 18.8)
2. Ring opening with aqueous acid to yield a 1,2-diol (Sec. 18.8)

Ester

1. Hydrolysis to yield a carboxylic acid (Sec. 21.6)
2. Aminolysis to yield an amide (Sec. 21.6)
3. Reduction to yield a primary alcohol (Secs. 17.6, 21.6)
4. Partial reduction with DIBAH to yield an aldehyde (Sec. 21.6)
5. Grignard reaction to yield a tertiary alcohol (Secs. 17.6, 21.6)
6. Claisen condensation to yield a β-keto ester (Sec. 23.8)

Ether

1. Acid-induced cleavage to yield an alcohol and an alkyl halide (Sec. 18.5)
2. Claisen rearrangement of an allyl aryl ether to yield an *o*-allyl phenol (Secs. 18.6, 30.9)

Halide, alkyl

1. Reaction with magnesium to form a Grignard reagent (Sec. 10.8)
2. Reduction to yield an alkane (Sec. 10.8)
3. Coupling with a diorganocopper reagent to yield an alkane (Sec. 10.9)
4. Nucleophilic substitution (S_N1 or S_N2) (Secs. 11.1–11.9)
5. Dehydrohalogenation to yield an alkene (E1 or E2) (Secs. 11.10–11.14)

Halohydrin

1. Conversion into an epoxide (Sec. 18.7)

Ketone

1. Nucleophilic addition reactions
   a. Reduction to yield a secondary alcohol (Secs. 17.5, 19.8)
   b. Reaction with a Grignard reagent to yield a tertiary alcohol (Secs. 17.6, 19.8)
   c. Wolff–Kishner reaction with hydrazine to yield an alkane (Sec. 19.10)
   d. Reaction with HCN to yield a cyanohydrin (Sec. 19.7)
   e. Reaction with an alcohol to yield an acetal (Sec. 19.11)
   f. Wittig reaction to yield an alkene (Sec. 19.12)
   g. Reaction with an amine to yield an imine or enamine (Sec. 19.9)
2. Aldol reaction to yield a β-hydroxy ketone (Sec. 23.2)
3. Alpha bromination of a ketone (Sec. 22.3)

Nitrile

1. Hydrolysis to yield a carboxylic acid (Sec. 21.8)
2. Reduction to yield a primary amine (Sec. 21.8)
3. Partial reduction with DIBAH to yield an aldehyde (Sec. 21.8)
4. Reaction with a Grignard reagent to yield a ketone (Sec. 21.8)

Nitroarene

1. Reduction to yield an arylamine (Secs. 16.2, 24.6)

Organometallic reagent

1. Reduction by treatment with acid to yield an alcohol (Sec. 10.8)
2. Nucleophilic addition to a carbonyl compound to yield an alcohol (Secs. 17.6, 19.8)
3. Conjugate addition of a lithium diorganocopper to an α,β-unsaturated ketone (Sec. 19.14)
4. Coupling reaction of a lithium diorganocopper reagent with an alkyl halide to yield an alkane (Sec. 10.9)
5. Coupling reaction of a lithium diorganocopper with an acid chloride to yield a ketone (Sec. 21.4)
6. Reaction with carbon dioxide to yield a carboxylic acid (Sec. 20.6)

Phenol

1. Acidity (Sec. 17.4)
2. Reaction with an acid chloride to yield an ester (Sec. 21.4)
3. Reaction with an alkyl halide to yield an ether (Sec. 18.3)
4. Oxidation to yield a quinone (Sec. 17.11)

Quinone

1. Reduction to yield a hydroquinone (Sec. 17.11)

Sulfide

1 Reaction with an alkyl halide to yield a sulfonium salt (Sec. 18.11)
2. Oxidation to yield a sulfoxide (Sec. 18.11)
3. Oxidation to yield a sulfone (Sec. 18.11)

Thiol

1. Reaction with an alkyl halide to yield a sulfide (Sec. 18.11)
2. Oxidation to yield a disulfide (Sec. 18.11)

Reagents in Organic Chemistry

The following table summarizes the uses of some important reagents in organic chemistry. The reagents are listed alphabetically, followed by a brief description of the uses of each and references to the appropriate text sections.

Acetic acid, CH_3COOH: Used as a solvent for the reduction of ozonides with zinc (Section 7.8) and the α-bromination of ketones and aldehydes with Br_2 (Section 22.3).

Acetic anhydride, $(CH_3CO)_2O$: Reacts with alcohols to yield acetate esters (Sections 21.5 and 25.7) and with amines to yield acetamides (Section 21.5).

Aluminum chloride, $AlCl_3$: Acts as a Lewis acid catalyst in Friedel–Crafts alkylation and acylation reactions of aromatic compounds (Sections 16.3 and 16.4).

Ammonia, NH_3: Used as a solvent for the reduction of alkynes by lithium metal to yield trans alkenes (Section 8.6).
- Reacts with acid chlorides and acid anhydrides to yield amides (Sections 21.4 and 21.5).

Borane, BH_3: Adds to alkenes, giving alkylboranes that can be oxidized with alkaline H_2O_2 to yield alcohols (Section 7.5).
- Adds to alkynes, giving vinylic organoboranes that can be oxidized with H_2O_2 to yield aldehydes (Section 8.5).
- Reduces carboxylic acids to yield primary alcohols (Section 20.8).

Bromine, Br_2: Adds to alkenes, yielding 1,2-dibromides (Sections 7.2, 14.5).
- Adds to alkynes yielding either 1,2-dibromoalkenes or 1,1,2,2-tetrabromoalkanes (Section 8.4).
- Reacts with arenes in the presence of $FeBr_3$ catalyst to yield bromoarenes (Section 16.1).
- Reacts with ketones in acetic acid solvent to yield α-bromo ketones (Section 22.3).
- Reacts with carboxylic acids in the presence of PBr_3 to yield α-bromo carboxylic acids (Hell–Volhard–Zelinskii reaction; Section 22.4).
- Reacts with methyl ketones in the presence of NaOH to yield carboxylic acids and bromoform (Haloform reaction; Section 22.7).
- Oxidizes aldoses to yield aldonic acids (Section 25.7).

***N*-Bromosuccinimide (NBS), $(CH_2CO)_2NBr$:** Reacts with alkenes in the presence of aqueous dimethylsulfoxide to yield bromohydrins (Section 7.3).
- Reacts with alkenes in the presence of light to yield allylic bromides (Section 10.5).
- Reacts with alkylbenzenes in the presence of light to yield benzylic bromides; (Section 16.10).

Di-*tert*-butoxy dicarbonate, ($\underline{t}$-BuOCO)$_2$O: Reacts with amino acids to give *t*-BOC protected amino acids suitable for use in peptide synthesis (Section 26.10).

Butyllithium, $CH_3CH_2CH_2CH_2Li$: A strong base; reacts with alkynes to yield acetylide anions, which can be alkylated (Section 8.9).
- Reacts with dialkylamines to yield lithium dialkylamide bases such as LDA [lithium diisopropyl-amide] (Section 22.5).
- Reacts with alkyltriphenylphosphonium salts to yield alkylidenephosphoranes (Wittig reagents (Section 19.12).

Carbon dioxide, CO_2: Reacts with Grignard reagents to yield carboxylic acids (Section 20.6).

Chlorine, Cl_2: Adds to alkenes to yield 1,2-dichlorides (Sections 7.2 and 14.5).
– Reacts with alkanes in the presence of light to yield chloroalkanes by a radical chain reaction pathway (Section 10.4).
– Reacts with arenes in the presence of $FeCl_3$ catalyst to yield chloroarenes (Section 16.2).

***m*-Chloroperoxybenzoic acid, *m*-$ClC_6H_4CO_3H$:** Reacts with alkenes to yield epoxides (Section 18.7).

Chlorotrimethylsilane, $(CH_3)_3SiCl$: Reacts with alcohols to add the trimethylsilyl protecting group (Section 17.9).

Chromium trioxide, CrO_3: Oxidizes alcohols in aqueous acid to yield carbonyl-containing products. Primary alcohols yield carboxylic acids, and secondary alcohols yield ketones (Sections 17.8 and 19.3).

Cuprous bromide, CuBr: Reacts with arenediazonium salts to yield bromoarenes (Sandmeyer reaction; Section 24.8).

Cuprous chloride, CuCl: Reacts with arenediazonium salts to yield chloroarenes (Sandmeyer reaction; Section 24.8).

Cuprous cyanide, CuCN: Reacts with arenediazonium salts to yield substituted benzonitriles (Sandmeyer reaction; Section 24.8).

Cuprous iodide, CuI: Reacts with organolithiums to yield lithium diorganocopper reagents (Gilman reagents; Section 10.9).

Cuprous oxide, Cu_2O: Reacts with arenediazonium salts to yield arenes (Section 24.8).

Dichloroacetic acid, $Cl_2CHCOOH$: Cleaves DMT protecting groups in DNA synthesis (Section 28.16).

Dicyclohexylcarbodiimide (DCC), $C_6H_{11}–N=C=N–C_6H_{11}$: Couples an amine with a carboxylic acid to yield an amide. DCC is often used in peptide synthesis (Section 26.10).

Diethyl acetamidomalonate, $CH_3CONHCH(CO_2Et)_2$: Reacts with alkyl halides in a common method of α-amino acid synthesis (Section 26.3).

Diiodomethane, CH_2I_2: Reacts with alkenes in the presence of zinc–copper couple to yield cyclopropanes (Simmons–Smith reaction; Section 7.6).

Diisobutylaluminum hydride (DIBAH), (*i*-Bu)$_2$AlH: Reduces esters to yield aldehydes (Sections 19.2 and 21.6).
– Reduces nitriles to yield aldehydes (Section 21.8).

2,4-Dinitrophenylhydrazine, 2,4-$(NO_2)_2C_6H_3NHNH_2$: Reacts with ketones and aldehydes to yield 2,4-DNPs that serve as useful crystalline derivatives (Section 19.9).

Ethylene glycol, $HOCH_2CH_2OH$: Reacts with ketones or aldehydes in the presence of an acid catalyst to yield acetals that serve as useful carbonyl protecting groups (Section 19.11).

Ferric bromide, $FeBr_3$: Acts as a catalyst for the reaction of arenes with Br_2 to yield bromoarenes (Section 16.1).

Ferric chloride, $FeCl_3$: Acts as a catalyst for the reaction of arenes with Cl_2 to yield chloroarenes (Section 16.2).

Grignard reagent, RMgX: Reacts with acids to yield alkanes (Section 10.8).
- Adds to carbonyl-containing compounds (ketones, aldehydes, esters) to yield alcohols (Sections 17.6 and 19.8).
- Adds to nitriles to yield ketones (Section 21.8).

Hydrazine, H_2NNH_2: Reacts with ketones or aldehydes in the presence of KOH to yield the corresponding alkanes (Wolff–Kishner reaction; Section 19.10).

Hydrogen bromide, HBr: Adds to alkenes with Markovnikov regiochemistry to yield alkyl bromides (Sections 6.8 and 14.5).
- Adds to alkynes to yield either bromoalkenes or 1,1-dibromoalkanes (Section 8.4).
- Reacts with alcohols to yield alkyl bromides (Sections 10.7 and 17.7).
- Cleaves ethers to yield alcohols and alkyl bromides (Section 18.5).

Hydrogen chloride, HCl: Adds to alkenes with Markovnikov regiochemistry to yield alkyl chlorides (Sections 6.8 and 14.5).
- Adds to alkynes to yield either chloroalkenes or 1,1-dichloroalkanes (Section 8.4).
- Reacts with alcohols to yield alkyl chlorides (Sections 10.7 and 17.7).

Hydrogen cyanide, HCN: Adds to ketones and aldehydes to yield cyanohydrins (Section 19.7).

Hydrogen iodide, HI: Reacts with alcohols to yield alkyl iodides (Section 17.7).
- Cleaves ethers to yield alcohols and alkyl iodides (Section 18.5).

Hydrogen peroxide, H_2O_2: Oxidizes organoboranes to yield alcohols. Used in conjunction with addition of borane to alkenes, the overall transformation effects syn Markovnikov addition of water to an alkene (Section 7.5).
- Oxidizes vinylic boranes to yield aldehydes.(Section 8.5).
- Oxidizes sulfides to yield sulfoxides (Section 18.11).

Hydroxylamine, NH_2OH: Reacts with ketones and aldehydes to yield oximes (Section 19.9).
- Reacts with aldoses to yield oximes as the first step in the Wohl degradation of aldoses (Section 25.7).

Hypophosphorous acid, H_3PO_2: Reacts with arenediazonium salts to yield arenes (Section 24.8).

Iodine, I_2: Reacts with arenes in the presence of CuCl or H_2O_2 to yield iodoarenes (Section 16.2).
- Reacts with methyl ketones in the presence of aqueous NaOH to yield carboxylic acids and iodoform, CHI_3 (Section 22.7).

Iodomethane, CH_3I: Reacts with alkoxide anions to yield methyl ethers (Section 18.3).
- Reacts with carboxylate anions to yield methyl esters (Section 21.6).
- Reacts with enolate ions to yield α-methylated carbonyl compounds (Section 22.8).
- Reacts with amines to yield methylated amines (Section 24.6).

Iron, Fe: Reacts with nitroarenes in the presence of aqueous acid to yield anilines (Section 24.6).

Lindlar catalyst: Acts as a catalyst for the partial hydrogenation of alkynes to yield cis alkenes (Section 8.6).

Lithium, Li: Reduces alkynes in liquid ammonia solvent to yield trans alkenes (Section 8.6).
- Reacts with organohalides to yield organolithium compounds (Section 10.9).

Lithium aluminum hydride, $LiAlH_4$: Reduces ketones, aldehydes, esters, and carboxylic acids to yield alcohols (Sections 17.5, 19.8, and 20.8).
- Reduces amides to yield amines (Section 21.7).
- Reduces alkyl azides to yield amines (Section 24.6).
- Reduces nitriles to yield amines (Sections 21.8 and 24.6).

Lithium diisopropylamide (LDA), $LiN(i\text{-}Pr)_2$: Reacts with carbonyl compounds (aldehydes, ketones, esters) to yield enolate ions (Sections 22.5 and 22.8).

Lithium diorganocopper reagent (Gilman reagent), LiR_2Cu: Couples with alkyl halides to yield alkanes (Section 10.9).
- Adds to α,β-unsaturated ketones to give 1,4-addition products (Section 19.14).
- Reacts with acid chlorides to give ketones (Section 21.4).

Lithium tri-*tert*-butoxyaluminum hydride, $LiAl(O\text{-}t\text{-}Bu)_3H$: Reduces acid chlorides to yield aldehydes (Section 21.4).

Magnesium, Mg: Reacts with organohalides to yield Grignard reagents (Section 10.8).

Mercuric acetate, $Hg(OCOCH_3)_2$: Adds to alkenes in the presence of water, giving α-hydroxy organomercury compounds that can be reduced with $NaBH_4$ to yield alcohols. The overall effect is the Markovnikov hydration of an alkene (Section 7.4).

Mercuric sulfate, $HgSO_4$: Acts as a catalyst for the addition of water to alkynes in the presence of aqueous sulfuric acid, yielding ketones (Section 8.5).

Mercuric trifluoroacetate, $Hg(OCOCF_3)_2$: Adds to alkenes in the presence of alcohol, giving α-alkoxy organomercury compounds that can be reduced with $NaBH_4$ to yield ethers. The overall reaction effects a net addition of an alcohol to an alkene (Section 18.4).

Methyl sulfate, $(CH_3O)_2SO_2$: A reagent used to methylate heterocyclic amine bases during Maxam–Gilbert DNA sequencing (Section 28.15).

Nitric acid, HNO_3: Reacts with arenes in the presence of sulfuric acid to yield nitroarenes (Section 16.2).
- Oxidizes aldoses to yield aldaric acids (Section 25.7).

Nitrous acid, HNO_2: Reacts with amines to yield diazonium salts (Section 24.8).

Osmium tetraoxide, OsO_4: Adds to alkenes to yield 1,2-diols (Section 7.8).
- Reacts with alkenes in the presence of periodic acid to cleave the carbon–carbon double bond, yielding ketone or aldehyde fragments (Section 7.8).

Ozone, O_3: Adds to alkenes to cleave the carbon–carbon double bond and give ozonides, which can be reduced with zinc in acetic acid to yield carbonyl compounds (Section 7.8).

Palladium on barium sulfate, $Pd/BaSO_4$: Acts as a hydrogenation catalyst for nitriles in the Kiliani–Fischer chain-lengthening reaction of carbohydrates (Section 25.7).

Palladium on carbon, Pd/C: Acts as a hydrogenation catalyst for reducing carbon–carbon multiple bonds. Alkenes and alkynes are reduced to yield alkanes (Sections 7.7 and 8.6).
- Acts as a hydrogenation catalyst for reducing aryl ketones to yield alkylbenzenes (Section 16.11).
- Acts as a hydrogenation catalyst for reducing nitroarenes to yield anilines (Section 24.6).

Periodic acid, HIO_4: Reacts with 1,2-diols to yield carbonyl-containing cleavage products (Section 7.8).

Peroxyacetic acid, CH_3CO_3H: Oxidizes sulfoxides to yield sulfones (Section 18.11)

Phenylisothiocyanate, $C_6H_5{-}N{=}C{=}S$: Used in the Edman degradation of peptides to identify N-terminal amino acids (Section 26.8).

Phosphorus oxychloride, $POCl_3$: Reacts with secondary and tertiary and alcohols to yield alkene dehydration products (Section 17.7).

Phosphorus tribromide, PBr_3: Reacts with alcohols to yield alkyl bromides (Section 10.7).
- Reacts with carboxylic acids to yield acid bromides (Section 21.4).
- Reacts with carboxylic acids in the presence of bromine to yield α-bromo carboxylic acids (Hell–Volhard–Zelinskii reaction; Section 22.4).

Platinum oxide (Adam's catalyst), PtO_2: Acts as a hydrogenation catalyst in the reduction of alkenes and alkynes to yield alkanes (Sections 7.7 and 8.6).

Potassium hydroxide, KOH: Reacts with alkyl halides to yield alkenes by an elimination reaction (Sections 7.1 and 11.11).
- Reacts with 1,1- or 1,2-dihaloalkanes to yield alkynes by a twofold elimination reaction (Section 8.3).

Potassium nitrosodisulfonate (Fremy's salt), $K(SO_3)_2NO$: Oxidizes phenols to yield quinones (Section 17.11).

Potassium permanganate, $KMnO_4$: oxidizes alkenes under neutral or acidic conditions to give carboxylic acid double-bond cleavage products (Sections 7.8).
- Oxidizes alkynes to give carboxylic acid triple-bond cleavage products (Section 8.7).
- Oxidizes aromatic side chains to yield benzoic acids (Section 16.10).

Potassium phthalimide, $C_6H_4(CO)_2NK$: Reacts with alkyl halides to yield *N*-alkylphthalimides, which are hydrolyzed by aqueous sodium hydroxide to yield amines (Gabriel amine synthesis; Section 24.6).

Potassium *tert*-butoxide, KO-*t*-Bu: Reacts with alkyl halides to yield alkenes (Sections 11.10 and 11.11).
- Reacts with allylic halides to yield conjugated dienes (Section 14.1).
- Reacts with chloroform in the presence of an alkene to yield a dichlorocyclopropane (Section 7.6).

Pyridine, C_5H_5N: Acts as a basic catalyst for the reaction of alcohols with acid chlorides to yield esters (Section 21.4).
- Acts as a basic catalyst for the reaction of alcohols with acetic anhydride to yield acetate esters (Section 21.5).
- Reacts with α-bromo ketones to yield α,β-unsaturated ketones (Section 22.3).

Pyridinium chlorochromate (PCC), $C_5H_6NCrO_3Cl$: Oxidizes primary alcohols to yield aldehydes and secondary alcohols to yield ketones (Section 17.8).

Pyrrolidine, C_4H_8N: Reacts with ketones to yield enamines for use in the Stork enamine reaction (Sections 19.9 and 23.12).

Rhodium on carbon, Rh/C: Acts as a hydrogenation catalyst in the reduction of benzene rings to yield cyclohexanes (Section 16.11).

Silver oxide, Ag_2O: Oxidizes primary alcohols in aqueous ammonia solution to yield aldehydes (Tollens oxidation; Section 19.3).
- Catalyzes the reaction of monosaccharides with alkyl halides to yield ethers (Section 25.7).
- Reacts with tetraalkylammonium salts to yield alkenes (Hofmann elimination; Section 24.7).

Sodium amide, $NaNH_2$: Reacts with terminal alkynes to yield acetylide anions (Section 8.8).
- Reacts with 1,1- or 1,2-dihalides to yield alkynes by a twofold elimination reaction (Section 8.3).
- Reacts with aryl halides to yield anilines by a benzyne aromatic substitution mechanism (Section 16.9).

Sodium azide, NaN_3: Reacts with alkyl halides to yield alkyl azides (Section 24.6).
- Reacts with acid chlorides to yield acyl azides. On heating in the presence of water, acyl azides yield amines and carbon dioxide (Section 24.6).

Sodium bisulfite, $NaHSO_3$: Reduces osmate esters, prepared by treatment of an alkene with osmium tetraoxide, to yield 1,2-diols (Section 7.8).

Sodium borohydride, $NaBH_4$: Reduces organomercury compounds, prepared by oxymercuration of alkenes, to convert the C–Hg bond to C–H (Section 7.4).
- Reduces ketones and aldehydes to yield alcohols (Sections 17.5 and 19.8).
- Reduces quinones to yield hydroquinones (Section 17.11).

Sodium cyanide, NaCN: Reacts with alkyl halides to yield alkanenitriles (Sections 20.6 and 21.8).

Sodium cyanoborohydride, $NaBH_3CN$: Reacts with ketones and aldehydes in the presence of ammonia to yield an amine by a reductive amination process (Section 24.6).

Sodium dichromate, $Na_2Cr_2O_7$: Oxidizes primary alcohols to yield carboxylic acids and secondary alcohols to yield ketones (Sections 17.8 and 19.2).
- Oxidizes alkylbenzenes to yield benzoic acids (Section 16.10).

Sodium hydride, NaH: Reacts with alcohols to yield alkoxide anions (Section 17.3).

Sodium hydroxide, NaOH: Reacts with arenesulfonic acids at high temperature to yield phenols (Sections 16.2 and 17.10).
- Reacts with aryl halides to yield phenols by a benzyne aromatic substitution mechanism (Section 16.9).
- Reacts with methyl ketones in the presence of iodine to yield carboxylic acids and iodoform (Section 22.7).

Sodium iodide, NaI: Reacts with arenediazonium salts to yield aryl iodides (Section 24.8).

Stannous chloride, $SnCl_2$: Reduces nitroarenes to yield anilines (Sections 16.2 and 24.6).
- Reduces quinones to yield hydroquinones (Section 17.11).

Sulfur trioxide, SO_3: Reacts with arenes in sulfuric acid solution to yield arenesulfonic acids (Section 16.2).

Sulfuric acid, H_2SO_4: Reacts with alcohols and water to yield alkenes (Section 7.4).
- Reacts with alkynes in the presence of water and mercuric sulfate to yield ketones (Section 8.5).
- Catalyzes the reaction of nitric acid with aromatic rings to yield nitroarenes (Section 16.2).
- Catalyzes the reaction of SO_3 with aromatic rings to yield arenesulfonic acids (Section 16.2).

Tetrazole: Acts as a coupling reagent for use in DNA synthesis (Section 28.16).

Thionyl chloride, $SOCl_2$: Reacts with primary and secondary alcohols to yield alkyl chlorides (Section 10.7).
- Reacts with carboxylic acids to yield acid chlorides (Section 21.4).

Thiourea, H_2NCSNH_2: Reacts with primary alkyl halides to yield thiols (Section 18.11).

***p*-Toluenesulfonyl chloride, *p*-$CH_3C_6H_4SO_2Cl$:** Reacts with alcohols to yield tosylates (Sections 11.2 and 17.7).

Trifluoroacetic acid, CF_3COOH: Acts as a catalyst for cleaving *tert*-butyl ethers, yielding alcohols and 2-methylpropene (Section 18.5).
- Acts as a catalyst for cleaving the *t*-BOC protecting group from amino acids in peptide synthesis (Section 26.10).

Triphenylphosphine, $(C_6H_5)_3P$: Reacts with primary alkyl halides to yield the alkyltriphenylphosphonium salts used in Wittig reactions (Section 19.12).

Zinc, Zn: Reduces ozonides, produced by addition of ozone to alkenes, to yield ketones and aldehydes (Section 7.8).
- Reduces disulfides to yield thiols (Section 18.11).

Zinc–copper couple, Zn–Cu: Reacts with diiodomethane in the presence of alkenes to yield cyclopropanes (Simmons–Smith reaction; Section 7.6).

Name Reactions

Acetoacetic ester synthesis (Section 22.8): a multistep reaction sequence for converting a primary alkyl halide into a methyl ketone having three more carbon atoms in the chain.

$$RCH_2X + CH_3\text{–}\overset{O}{\overset{\|}{C}}\text{–}\overset{..-}{C}H\text{–}\overset{O}{\overset{\|}{C}}\text{–}OCH_3 \xrightarrow[\text{2. } H_3O^+\text{, heat}]{\text{1. Heat}} RCH_2\text{–}CH_2\overset{O}{\overset{\|}{C}}CH_3 + CO_2 + CH_3OH$$

Adams' catalyst (Section 7.7): PtO_2, a catalyst used for the hydrogenation of carbon–carbon double bonds.

Aldol condensation reaction (Section 23.2): the nucleophilic addition of an enol or enolate ion to a ketone or aldehyde, yielding a β-hydroxy ketone.

$$2\ R\text{–}\overset{O}{\overset{\|}{C}}\text{–}\overset{|}{\underset{|}{C}}\text{–}H \xrightarrow{\text{NaOH}} R\text{–}\overset{O}{\overset{\|}{C}}\text{–}\overset{|}{\underset{|}{C}}\text{–}\overset{OH}{\underset{R}{C}}\text{–}\overset{|}{\underset{|}{C}}\text{–}H$$

Amidomalonate amino acid synthesis (Section 26.3): a multistep reaction sequence, similar to the malonic ester synthesis, for converting a primary alkyl halide into an amino acid.

$$RCH_2X + {}^-\!:\underset{NHAc}{C}(CO_2Et)_2 \xrightarrow[\text{2. } H_3O^+\text{, heat}]{\text{1. mix}} RCH_2\text{–}\underset{NH_2}{C}H\overset{O}{\overset{\|}{C}}OH + CO_2 + 2\ EtOH$$

Benedict's test (Section 25.7): a chemical test for aldehydes, involving treatment with cupric ion in aqueous sodium citrate.

Cannizzaro reaction (Section 19.13): the disproportionation reaction that occurs when a nonenolizable aldehyde is treated with base.

$$2\ R_3\overset{O}{\overset{\|}{C}}CH \xrightarrow[\text{2. } H_3O^+]{\text{1. } HO^-} R_3\overset{O}{\overset{\|}{C}}COH + R_3CCH_2OH$$

Claisen condensation reaction (Section 23.8): a nucleophilic acyl substitution reaction that occurs when an ester enolate ion attacks the carbonyl group of a second ester molecule. The product is a β-keto ester.

$$2\ R\text{–}CH_2\text{–}\overset{O}{\overset{\|}{C}}\text{–}OCH_3 \xrightarrow[\text{2. } H_3O^+]{\text{1. } HO^-} R\text{–}CH_2\text{–}\overset{O}{\overset{\|}{C}}\text{–}\underset{R}{C}H\text{–}\overset{O}{\overset{\|}{C}}\text{–}OCH_3 + CH_3OH$$

Claisen rearrangement (Sections 18.6 and 30.9): the thermal [3.3] sigmatropic rearrangement of an allyl vinyl ether or an allyl phenyl ether.

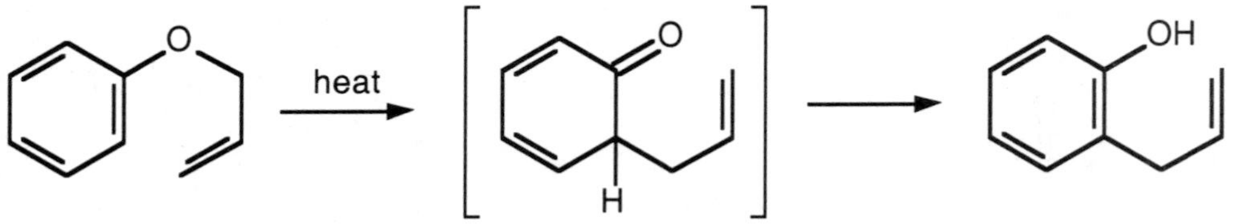

heat

Cope rearrangement (Section 30.9): the thermal [3.3] sigmatropic rearrangement of a 1,5-diene to a new 1,5-diene.

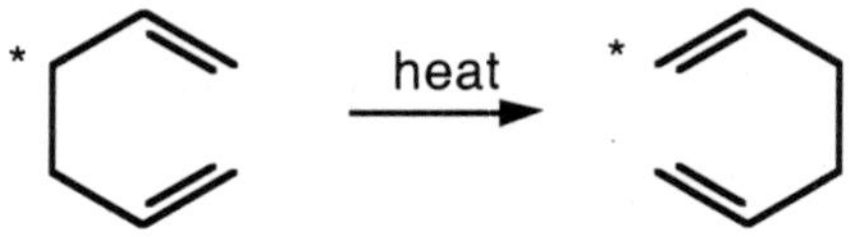

heat

Curtius rearrangement (Section 24.6): the thermal rearrangement of an acyl azide to an isocyanate, followed by hydrolysis to yield an amine.

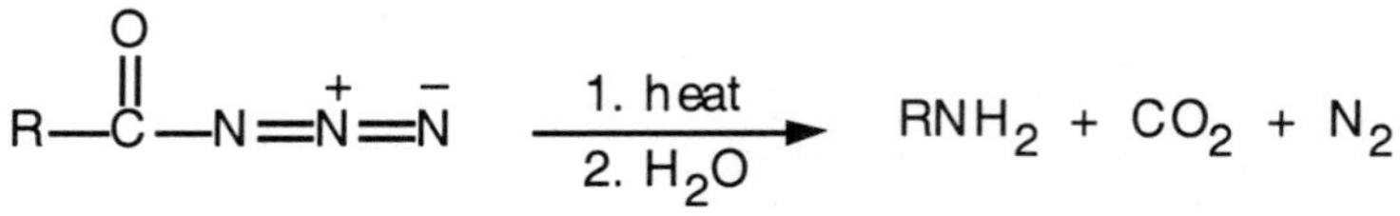

$$R-\overset{O}{\overset{\|}{C}}-N=\overset{+}{N}=\overset{-}{N} \xrightarrow[2.\ H_2O]{1.\ heat} RNH_2 + CO_2 + N_2$$

Diazonium coupling reaction (Section 24.8): the coupling reaction between an aromatic diazonium salt and a phenol or aniline.

$$C_6H_5-\overset{+}{N}\equiv N + C_6H_5-OH \longrightarrow C_6H_5-N=N-C_6H_4-OH$$

Dieckmann reaction (Section 23.10): the intramolecular Claisen condensation reaction of a 1,6- or 1,7-diester, yielding a cyclic β-keto ester.

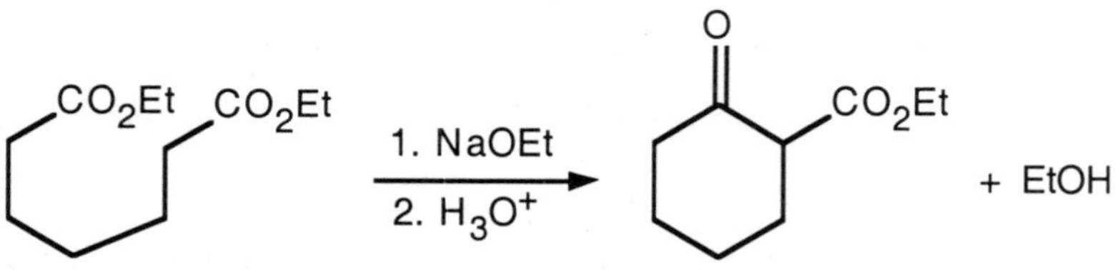

CO_2Et CO_2Et $\xrightarrow[2.\ H_3O^+]{1.\ NaOEt}$ CO_2Et + EtOH

Diels–Alder cycloaddition reaction (Sections 14.8–14.9 and 30.6): the reaction between a diene and a dienophile to yield a cyclohexene ring.

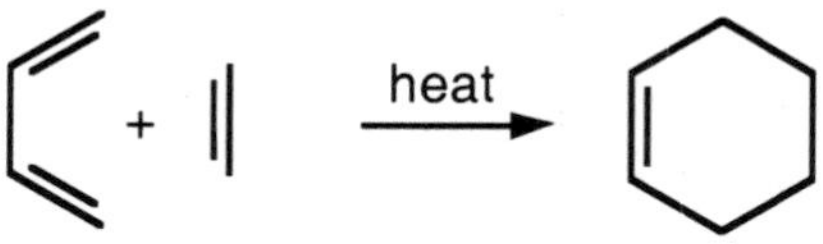

heat

Edman degradation (Section 26.8): a method for cleaving the N-terminal amino acid from a peptide by treatment of the peptide with *N*-phenylisothiocyanate.

$$\text{Ph–N=C=S} + \text{H}_2\text{N–CH(R)–C(=O)–NH–} \longrightarrow \text{phenylthiohydantoin (Ph–N, C=S, N–H, C=O, CHR)} + \text{H}_2\text{N–}$$

Fehling's test (Section 25.7): a chemical test for aldehydes, involving treatment with cupric ion in aqueous sodium tartrate.

Fischer esterification reaction (Section 21.3): the acid-catalyzed reaction between a carboxylic acid and an alcohol, yielding the ester.

$$\text{R–C(=O)–OH} + \text{R'–OH} \xrightarrow{\text{H}^+,\ \text{heat}} \text{R–C(=O)–OR'} + \text{H}_2\text{O}$$

Friedel–Crafts reaction (Section 16.3–16.4): the alkylation or acylation of an aromatic ring by treatment with an alkyl- or acyl chloride in the presence of a Lewis-acid catalyst.

$$\text{C}_6\text{H}_6 \xrightarrow{\text{R–Cl, AlCl}_3} \text{C}_6\text{H}_5\text{–R} \quad \text{Alkylation}$$

$$\text{C}_6\text{H}_6 \xrightarrow{\text{R–C(=O)–Cl, AlCl}_3} \text{C}_6\text{H}_5\text{–C(=O)–R} \quad \text{Acylation}$$

Gabriel amine synthesis (Section 24.6): a multistep sequence for converting a primary alkyl halide into a primary amine by alkylation with potassium phthalimide, followed by hydrolysis.

$$\text{potassium phthalimide (N}^-\text{K}^+) + \text{RCH}_2\text{X} \xrightarrow[\text{2. NaOH, H}_2\text{O}]{\text{1. mix}} \text{C}_6\text{H}_4(\text{CO}_2^-)_2 + \text{RCH}_2\text{NH}_2$$

Gilman reagent (Section 10.9): a lithium dialkylcopper reagent, R_2CuLi, prepared by treatment of a cuprous salt with an alkyllithium. Gilman reagents undergo a coupling reaction with alkyl halides, a 1,4-addition reaction with α,β-unsaturated ketones, and a coupling reaction with acid chlorides to yield ketones.

Glycal assembly method (Section 25.10): a method of polysaccharide synthesis in which a glycal is converted into its epoxide, which is then opened by reaction with an alcohol.

$$\xrightarrow[\text{THF}]{ZnCl_2}$$

Grignard reaction (Section 19.10): the nucleophilic addition reaction of an alkylmagnesium halide to a ketone, aldehyde, or ester carbonyl group.

$$\text{R–Mg–X} + \text{C=O} \xrightarrow[\text{2. } H_3O^+]{\text{1. mix}} \text{C(OH)R}$$

Grignard reagent (Section 10.8): an organomagnesium halide, RMgX, prepared by reaction between an organohalide and magnesium metal. Grignard reagents add to carbonyl compounds to yield alcohols.

Haloform reaction (Section 22.7): the conversion of a methyl ketone to a carboxylic acid and haloform by treatment with halogen and base.

$$R\text{—}\overset{O}{\overset{\|}{C}}\text{—}CH_3 \xrightarrow[\text{2. } H_3O^+]{\text{1. } X_2\text{, NaOH}} R\text{—}\overset{O}{\overset{\|}{C}}\text{—}OH + CHX_3$$

Hell–Volhard–Zelinskii reaction (Section 22.4): the α-bromination of carboxylic acids by treatment with bromine and phosphorus tribromide.

$$\text{—}\overset{H}{\overset{|}{C}}\text{—}\overset{O}{\overset{\|}{C}}\text{—OH} \xrightarrow[\text{2. } H_2O]{\text{1. } Br_2\text{, } PBr_3} \text{—}\overset{Br}{\overset{|}{C}}\text{—}\overset{O}{\overset{\|}{C}}\text{—OH}$$

Hofmann elimination (Section 24.7): a method for effecting the elimination reaction of an amine to yield an alkene. The amine is first treated with excess iodomethane, and the resultant quaternary ammonium salt is heated with silver oxide.

$$R_2N\text{—C—C—H} \xrightarrow[\text{2. } Ag_2O\text{, } H_2O]{\text{1. } CH_3I} \text{C=C} + R_2NCH_3$$

Hofmann rearrangement (Section 24.6): the rearrangement of an *N*-bromoamide to a primary amine by treatment with aqueous base.

$$R{-}\overset{O}{\overset{\|}{C}}{-}NH_2 \xrightarrow[NaOH]{Br_2} \left[R{-}\overset{O}{\overset{\|}{C}}{-}NHBr \right] \longrightarrow RNH_2 + CO_2$$

Kiliani–Fischer synthesis (Section 25.7): a multistep sequence for chain-lengthening an aldose into the next higher homolog.

$$R{-}CHO \xrightarrow[\text{2. } H_2,\ Pd,\ BaSO_4 \quad \text{3. } H_3O^+]{\text{1. HCN}} R{-}CH(OH){-}CHO$$

Koenigs–Knorr reaction (Section 25.7): a method for synthesizing glycosides by reaction of a pyranosyl bromide with an alcohol and Ag_2O.

$$\text{tetra-}O\text{-acetyl pyranosyl bromide } (CH_2OAc,\ AcO,\ AcO,\ OAc,\ Br) \xrightarrow[\text{2. } ^{-}OH,\ H_2O]{\text{1. } ROH,\ Ag_2O} \text{glycoside } (CH_2OH,\ HO,\ HO,\ OH,\ O{-}R)$$

Malonic ester synthesis (Section 22.8): a multistep sequence for converting an alkyl halide into a carboxylic acid with the addition of two carbon atoms to the chain.

$$R{-}CH_2{-}X + {}^{-}{:}\underset{CO_2CH_3}{CH}{-}\overset{O}{\overset{\|}{C}}{-}OCH_3 \xrightarrow[\text{2. } H_3O^+,\ \text{heat}]{\text{1. heat}} RCH_2{-}CH_2\overset{O}{\overset{\|}{C}}OCH_3 + CO_2 + CH_3OH$$

Maxam–Gilbert DNA sequencing (Section 28.15): a chemical method for sequencing long chains of DNA by employing selective cleavage reactions.

McLafferty rearrangement (Section 19.16): a mass spectral fragmentation pathway for carbonyl compounds having a hydrogen three carbon atoms away from the carbonyl carbon.

$$\text{ketone with } \gamma\text{-H} \xrightarrow{e^-} \text{alkene} + H{-}O^{+\cdot}{=}C(CH_3)_2$$

Meisenheimer complex (Section 16.8): an intermediate formed in the nucleophilic aryl substitution reaction of a base with a nitro-substituted aromatic ring.

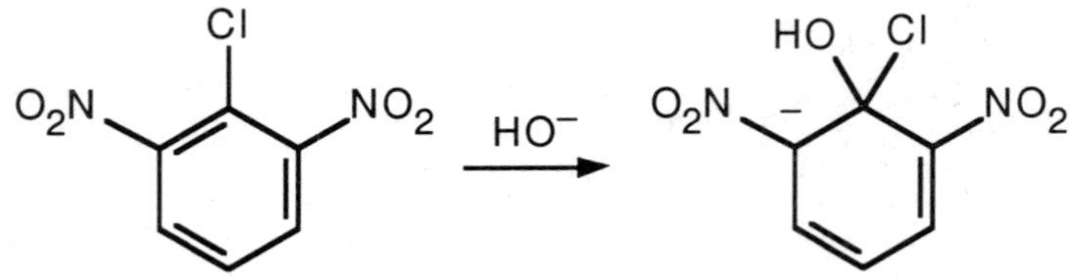

Merrifield solid-phase peptide synthesis (Section 26.11): a rapid and efficient means of peptide synthesis in which the growing peptide chain is attached to an insoluble polymer support.

Michael reaction (Section 23.11): the 1,4-addition reaction of a stabilized enolate anion such as that from a 1,3-diketone to an α,β-unsaturated carbonyl compound.

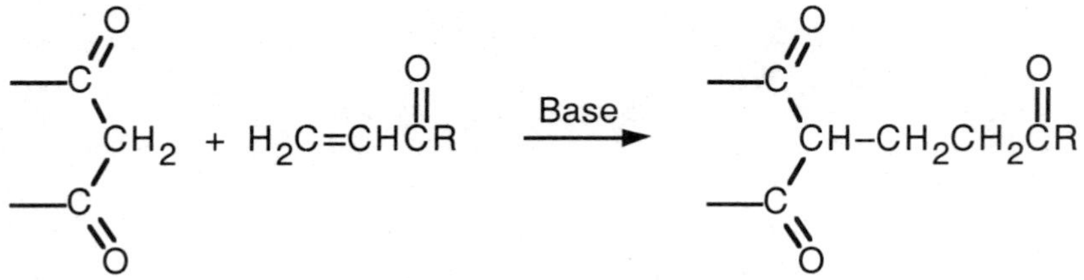

Robinson annulation reaction (Section 23.13): a multistep sequence for building a new cyclohexenone ring onto a ketone. The sequence involves an initial Michael reaction of the ketone followed by an internal aldol cyclization.

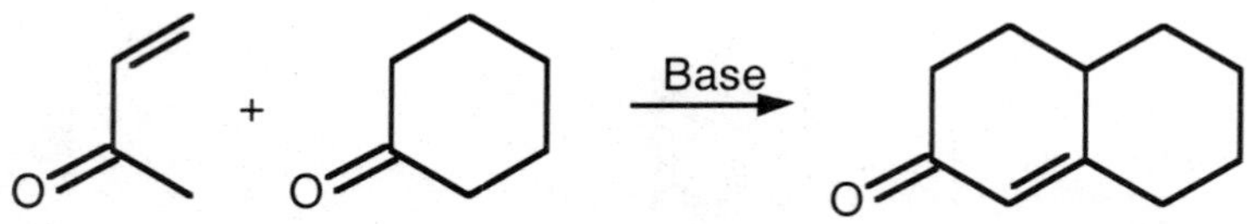

Sandmeyer reaction (Section 24.8): a method for converting aryldiazonium salts into aryl halides by treatment with a cuprous halide.

Sanger dideoxy method (Section 28.15): an enzymatic for DNA sequencing.

Simmons–Smith reaction (Section 7.6): a method for preparing cyclopropanes by treating an alkene with diiodomethane and zinc–copper.

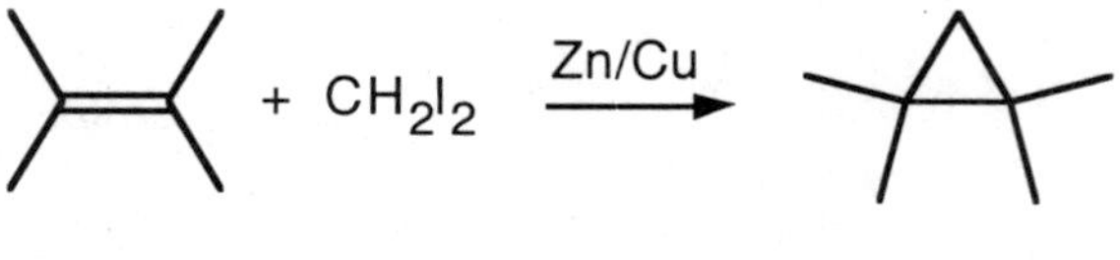

Stork enamine reaction (Section 23.12): a multistep sequence whereby ketones are converted into enamines by treatment with a secondary amine, and the enamines are then used in Michael reactions.

$$\text{(CH}_3)_2\text{C=O} + HNR_2 \longrightarrow CH_2\text{=C(NR}_2)CH_3 \xrightarrow[\text{2. } H_3O^+]{\text{1. } H_2C\text{=CHCOR}'} CH_3COCH_2CH_2CH_2COR'$$

Strecker amino acid synthesis (Section 26.3): a multistep sequence for converting an aldehyde into an amino acid by treatment with ammonium cyanide, followed by hydrolysis.

$$R\text{—}\overset{O}{\overset{\|}{C}}\text{—}H \xrightarrow{NH_3,\ KCN} R\text{—}\overset{NH_2}{\overset{|}{C}}H\text{-}CN \xrightarrow{H_3O^+} R\text{—}\overset{NH_2}{\overset{|}{C}}H\text{-}COOH$$

Tollen's test (Section 19.3): a chemical test for detecting aldehydes by treatment with ammoniacal silver nitrate. A positive test is signaled by formation of a silver mirror on the walls of the reaction vessel.

Walden inversion (Section 11.1): the inversion of stereochemistry at a chirality center during an S_N2 reaction.

$$Nu{:}^- + \text{C—X} \longrightarrow \text{Nu—C} + {:}X^-$$

Williamson ether synthesis (Section 18.3): a method for preparing ethers by treatment of a primary alkyl halide with an alkoxide ion.

$$R\text{–}O^-\ Na^+ + R'CH_2Br \longrightarrow R\text{–}O\text{–}CH_2R' + NaBr$$

Wittig reaction (Section 19.12): a general method of alkene synthesis by treatment of a ketone or aldehyde with an alkylidenetriphenylphosphorane.

$$R\text{—}\overset{O}{\overset{\|}{C}}\text{—}R' + \text{C=PPh}_3 \longrightarrow \text{(R)(R')C=C} + Ph_3P\text{=O}$$

Wohl degradation (Section 25.7): a multistep reaction sequence for degrading an aldose into the next lower homolog.

$$\text{CHO—CH(OH)—R} \xrightarrow[\text{2. } Ac_2O\text{; 3. NaOEt}]{\text{1. } NH_2OH} \text{CHO—R}$$

Wolff–Kishner reaction (Section 19.10): a method for converting a ketone or aldehyde into the corresponding hydrocarbon by treatment with hydrazine and strong base.

$$R-C(=O)-R' \xrightarrow{N_2H_4,\ KOH} R-CH_2-R'$$

Woodward–Hoffmann orbital symmetry rules (Section 30.10): a series of rules for predicting the stereochemistry of pericyclic reactions. Even-electron species react thermally through either antarafacial or conrotatory pathways, whereas odd-electron species react thermally through either suprafacial or disrotatory pathways.

Abbreviations

Å	symbol for Angstrom unit (10^{-8} cm = 10^{-10} m)
Ac–	Acetyl group, $CH_3\overset{O}{\overset{\parallel}{C}}-$
Ar–	aryl group
at. no.	atomic number
at. wt.	atomic weight
$[\alpha]_D$	specific rotation
BOC	*tert*-butoxycarbonyl group, $(CH_3)_3CO\overset{O}{\overset{\parallel}{C}}-$
bp	boiling point
n-Bu	*n*-butyl group, $CH_3CH_2CH_2CH_2-$
sec-Bu	*sec*-butyl group, $CH_3CH_2CH(CH_3)-$
t-Bu	*tert*-butyl group, $(CH_3)_3C-$
cm	centimeter
cm^{-1}	wavenumber, or reciprocal centimeter
D	stereochemical designation of carbohydrates and amino acids
DCC	dicyclohexylcarbodiimide, $C_6H_{11}-N=C=N-C_6H_{11}$
δ	chemical shift in ppm downfield from TMS
Δ	symbol for heat; also symbol for change
ΔH	heat of reaction
dm	decimeter (0.1 m)
DMF	dimethylformamide, $(CH_3)_2NCHO$
DMSO	dimethyl sulfoxide, $(CH_3)_2SO$
DNA	deoxyribonucleic acid
DNP	dinitrophenyl group, as in 2,4-DNP (2,4-dinitrophenylhydrazone)
(*E*)	entgegen, stereochemical designation of double bond geometry
E_{act}	activation energy
E1	unimolecular elimination reaction
E2	bimolecular elimination reaction
Et	ethyl group, CH_3CH_2-
g	gram

hν	symbol for light
Hz	Hertz, or cycles per second (s^{-1})
i-	iso
IR	infrared
J	Joule
J	symbol for coupling constant
K	Kelvin temperature
K_a	acid dissociation constant
kJ	kilojoule
L	stereochemical designation of carbohydrates and amino acids
LAH	lithium aluminum hydride, $LiAlH_4$
Me	methyl group, CH_3–
mg	milligram (0.001 g)
MHz	megahertz (10^6 s^{-1})
mL	milliliter (0.001 L)
mm	millimeter (0.001 m)
mp	melting point
μg	microgram (10^{-6} g)
mμ	millimicron (nanometer, 10^{-9} m)
MW	molecular weight
n-	normal, straight-chain alkane or alkyl group
ng	nanogram (10^{-9} gram)
nm	nanometer (10^{-9} meter)
NMR	nuclear magnetic resonance
–OAc	acetate group, $-OC(=O)CH_3$
PCC	pyridinium chlorochromate
Ph	phenyl group, $-C_6H_5$
pH	measure of acidity of aqueous solution
pK_a	measure of acid strength (= –log K_a)
pm	picometer (10^{-12} m)
ppm	parts per million
n-Pr	*n*-propyl group, $CH_3CH_2CH_2$–
i-Pr	isopropyl group, $(CH_3)_2CH$–

R–	symbol for a generalized alkyl group
(*R*)	*rectus*, designation of chirality center
RNA	ribonucleic acid
(*S*)	*sinister*, designation of chirality center
sec-	secondary
S_N1	unimolecular substitution reaction
S_N2	bimolecular substitution reaction
tert-	tertiary
THF	tetrahydrofuran
TMS	tetramethylsilane nmr standard, $(CH_3)_4Si$
Tos	tosylate group, $-S(=O)_2-C_6H_4-CH_3$
UV	ultraviolet
X–	halogen group (–F, –Cl, –Br, –I)
(*Z*)	zusammen, stereochemical designation of double bond geometry
$\longrightarrow$	chemical reaction in direction indicated
$\rightleftharpoons$	reversible chemical reaction
$\longleftrightarrow$	resonance symbol
$\curvearrowright$	curved arrow indicating direction of electron flow
$\equiv$	is equivalent to
$>$	greater than
$<$	less than
$\approx$	approximately equal to
R-$\wr$	indicates that the organic fragment shown is a part of a larger molecule
▬ (wedge)	single bond coming out of the plane of the paper
-----	single bond receding into the plane of the paper
......	partial bond
δ+, δ–	partial charge
‡	denoting the transition state

Infrared Absorption Frequencies

Functional Group		Frequency (cm^{-1})	Text Section
Alcohol	–O–H	3300–3600 (s)	17.12
	$\rangle$C–O–	1050 (s)	
Aldehyde	–CO–H	2720, 2820 (m)	19.16
aliphatic	$\rangle$C=O	1725 (s)	
aromatic		1705 (s)	
Alkane			12.8
	$\rangle$C–H	2850–2960 (s)	
	$\rangle$C–C$\langle$	800–1300 (m)	
Alkene			12.8
	=C(H)–	3020–3100 (s)	
	$\rangle$C=C$\langle$	1650–1670 (m)	
	$RCH=CH_2$	910, 990 (m)	
	$R_2C=CH_2$	890 (m)	
Alkyne	≡C–H	3300 (s)	12.8
	–C≡C–	2100–2260 (m)	
Alkyl bromide			12.8
	$\rangle$C–Br	500–600 (s)	
Alkyl chloride			12.8
	$\rangle$C–Cl	600–800 (s)	

Amine, *primary*	$-NH_2$	3400, 3500 (s)	24.10
secondary	$>N-H$	3350 (s)	
Ammonium salt	$-N^{+}-H$	2200–3000 (broad)	24.10
Aromatic ring	Ar–H	3030 (m)	15.10
monosubstituted	Ar–R	690–710 (s)	
		730–770 (s)	
o-disubstituted		735–770 (s)	
m-disubstituted		690–710 (s)	
		810–850 (s)	
p-disubstituted		810–840 (s)	
Carboxylic acid	–O–H	2500–3300 (broad)	20.9
associated	$>C=O$	1710 (s)	
free		1760 (s)	
Acid anhydride	$>C=O$	1760, 1820 (s)	21.11
Acid chloride			21.11
aliphatic	$>C=O$	1810 (s)	
aromatic		1770 (s)	
Amide			21.11
aliphatic	$>C=O$	1810 (s)	
aromatic		1770 (s)	
N-substituted		1680 (s)	
N,N-disubstituted		1650 (s)	

Ester			21.11
aliphatic	$\rangle C{=}O$	1735 (s)	
aromatic		1720 (s)	
Ether			18.10
	$-O-C\leftarrow$	1050–1150 (s)	
Ketone			19.16
aliphatic	$\rangle C{=}O$	1715 (s)	
aromatic		1690 (s)	
6-memb. ring		1715 (s)	
5-memb. ring		1750 (s)	
Nitrile			21.11
aliphatic	–C≡N	2250 (m)	
aromatic		2230 (m)	
Phenol	–O–H	3500 (s)	17.12

(s) = strong; (m) = medium intensity

Proton NMR Chemical Shifts

Type of Proton		Chemical Shift (δ)	Text Section
Alkyl, primary	$R{-}CH_3$	0.7–1.3	13.9
Alkyl, secondary	$R{-}CH_2{-}R$	1.2–1.4	13.9
Alkyl tertiary	$R_3C{-}\mathbf{H}$	1.4–1.7	13.9
Allylic	$-\overset{\mid}{C}{=}\overset{\mid}{C}{-}\overset{\mid}{\underset{\mid}{C}}{-}\mathbf{H}$	1.6–1.9	13.9
α to carbonyl	$-\overset{O}{\overset{\parallel}{C}}{-}\overset{\mid}{\underset{\mid}{C}}{-}\mathbf{H}$	2.0–2.3	19.16
Benzylic	$Ar{-}\overset{\mid}{\underset{\mid}{C}}{-}\mathbf{H}$	2.3–3.0	15.10
Acetylenic	$R{-}C{\equiv}C{-}\mathbf{H}$	2.5–2.7	13.9
Alkyl chloride	$Cl{-}\overset{\mid}{\underset{\mid}{C}}{-}\mathbf{H}$	3.0–4.0	13.9
Alkyl bromide	$Br{-}\overset{\mid}{\underset{\mid}{C}}{-}\mathbf{H}$	2.5–4.0	13.9
Alkyl iodide	$I{-}\overset{\mid}{\underset{\mid}{C}}{-}\mathbf{H}$	2.0–4.0	13.9
Amine	$>N{-}\overset{\mid}{\underset{\mid}{C}}{-}\mathbf{H}$	2.2–2.6	24.10
Epoxide	$>\overset{O}{C{-}C}<\mathbf{H}$	2.5–3.5	18.10
Alcohol	$HO{-}\overset{\mid}{\underset{\mid}{C}}{-}\mathbf{H}$	3.5–4.5	17.12
Ether	$RO{-}\overset{\mid}{\underset{\mid}{C}}{-}\mathbf{H}$	3.5–4.5	18.10
Vinylic	$-\overset{\mid}{C}{=}\overset{\mid}{C}{-}\mathbf{H}$	5.0–6.0	13.9
Aromatic	$Ar{-}\mathbf{H}$	6.5–8.0	15.10
Aldehyde	$R{-}\overset{O}{\overset{\parallel}{C}}{-}\mathbf{H}$	9.7–10.0	19.16
Carboxylic acid	$R{-}\overset{O}{\overset{\parallel}{C}}{-}O{-}\mathbf{H}$	11.0–12.0	20.9
Alcohol	$R{-}O{-}\mathbf{H}$	3.5–4.5	17.12
Phenol	$Ar{-}O{-}\mathbf{H}$	2.5–6.0	17.12

Organic Chemical Production
U.S. Chemical Industry – 1998

Ethylene (26,031,000 tons/yr):
prepared by thermal cracking of ethane and propane during petroleum refining; used as starting material for manufacture of polyethylene, ethylene oxide, ethylene glycol, ethylbenzene, 1,2-dichloroethane, and other bulk chemicals.

Propylene (14,307,000 tons/yr):
prepared by steam cracking of light hydrocarbon fractions during petroleum refining; used as starting material for the manufacture of polypropylene, acrylonitrile, propylene oxide, and isopropyl alcohol.

1,2-Dichloroethane (Ethylene dichloride; 12,280,000 tons/yr):
prepared by addition of chlorine to ethylene in the presence of $FeCl_3$ catalyst at 50°C; used as a chlorinated solvent and as starting material for the manufacture of vinyl chloride.

Benzene (7,874,000 tons/yr):
obtained from petroleum by catalytic reforming of hexane and cyclohexane over a platinum catalyst; used as starting material for the synthesis of ethylbenzene, cumene, cyclohexane, and aniline.

Ethylbenzene (6,331,000 tons/yr):
prepared during catalytic reforming in petroleum refining and by an acid-catalyzed Friedel–Crafts alkylation of benzene with ethylene; used almost exclusively for production of styrene.

Styrene (5,695,000 tons/yr):
prepared by high-temperature catalytic dehydrogenation of ethylbenzene; used in the manufacture of polystyrene polymers (thermoplastics, packaging materials).

Ethylene oxide (4,070,000 tons/yr):
prepared by high-temperature air oxidation of ethylene over a silver catalyst; used as starting material for the preparation of ethylene glycol and poly(ethylene glycol).

***p*-Xylene** (3,848,000 tons/yr):
prepared by separation from the mixed xylenes that result during catalytic reforming in gasoline refining; used as starting material for manufacture of the dimethyl terephthalate needed for polyester synthesis.

Cumene (3,357,000 tons/yr):
prepared by a phosphoric-acid-catalyzed Friedel–Crafts reaction between benzene and propylene; used primarily for conversion into phenol and acetone.

1,3-Butadiene (2,033,000 tons/yr):
prepared by steam cracking of gas oil during petroleum refining and by dehydrogenation of butane and butene; used primarily as a monomer component in the manufacture of styrene–butadiene rubber (SBR), polybutadiene rubber, and acrylonitrile–butadiene–styrene (ABS) copolymers.

Acrylonitrile (1,560,000 tons,yr):
prepared by the Sohio ammoxidation process in which propylene, ammonia, and air are passed over a catalyst at 500°C; used in the preparation of acrylic fibers, nitrile rubber, and acrylonitrile–butadiene–styrene (ABS) copolymer.

Aniline (773,000 tons/yr):
prepared by catalytic reduction of nitrobenzene with hydrogen at 350°C; used as starting material for preparing toluene diisocyanate and for the synthesis of dyes and pharmaceuticals.

Isopropyl alcohol (725,000 tons/yr):
prepared by direct high-temperature addition of water to propylene; used in cosmetics formulations, as a solvent and deicer, and as starting material for manufacture of acetone.

***o*-Xylene** (507,000 tons/yr):
obtained by separation from the mixed xylenes that result during catalytic reforming in petroleum refining; used as starting material for preparation of phthalic acid and phthalic anhydride.

2-Ethylhexanol (406,000 tons/yr):
prepared from butanal by aldol condensation and catalytic hydrogenation (the Oxo Process); used in the manufacture of plasticizers, lubricating-oil additives, and detergents.

Nobel Prizes in Chemistry

1901 **Jacobus H. van't Hoff** (The Netherlands):
"for the discovery of laws of chemical dynamics and of osmotic pressure"

1902 **Emil Fischer** (Germany):
"for syntheses in the groups of sugars and purines"

1903 **Svante A. Arrhenius** (Sweden):
"for his theory of electrolytic dissociation"

1904 **Sir William Ramsey** (Britain):
"for the discovery of gases in different elements in the air and for the determination of their place in the periodic system"

1905 **Adolf von Baeyer** (Germany):
"for his researches on organic dyestuffs and hydroaromatic compounds"

1906 **Henri Moissan** (France):
"for his research on the isolation of the element fluorine and for placing at the service of science the electric furnace that bears his name"

1907 **Eduard Buchner** (Germany):
"for his biochemical researches and his discovery of cell-less formation"

1908 **Ernest Rutherford** (Britain):
"for his investigation into the disintegration of the elements and the chemistry of radioactive substances"

1909 **Wilhelm Ostwald** (Germany):
"for his work on catalysis and on the conditions of chemical equilibrium and velocities of chemical reactions"

1910 **Otto Wallach** (Germany):
"for his services to organic chemistry and the chemical industry by his pioneer work in the field of alicyclic substances"

1911 **Marie Curie** (France):
"for her services to the advancement of chemistry by the discovery of the elements radium and polonium"

1912 **Victor Grignard** (France):
"for the discovery of the so-called Grignard reagent, which has greatly helped in the development of organic chemistry"

Paul Sabatier (France):
"for his method of hydrogenating organic compounds in the presence of finely divided metals"

1913 **Alfred Werner** (Switzerland):
"for his work on the linkage of atoms in molecules by which he has thrown new light on earlier investigations and opened up new fields of research especially in inorganic chemistry"

1914 **Theodore W. Richards** (U.S.):
"for his accurate determinations of the atomic weights of a great number of chemical elements"

1915 **Richard M. Willstätter** (Germany):
"for his research on plant pigments, principally on chlorophyll"

1916 No award

1917 No award

1918 **Fritz Haber** (Germany):
"for the synthesis of ammonia from its elements, nitrogen and hydrogen"

1919 No award

1920 **Walther H. Nernst** (Germany):
"for his thermochemical work"

1921 **Frederick Soddy** (Britain):
"for his contributions to the chemistry of radioactive substances and his investigations into the origin and nature of isotopes"

1922 **Francis W. Aston** (Britain):
"for his discovery, by means of his mass spectrograph, of the isotopes of a large number of nonradioactive elements, as well as for his discovery of the whole-number rule"

1923 **Fritz Pregl** (Austria):
"for his invention of the method of microanalysis of organic substances"

1924 No award

1925 **Richard A. Zsigmondy** (Germany):
for his demonstration of the heterogeneous nature of colloid solutions, and for the methods he used, which have since become fundamental in modern colloid chemistry"

1926 **Theodor Svedberg** (Sweden):
"for his work on disperse systems"

1927 **Heinrich O. Wieland** (Germany):
"for his research on bile acids and related substances"

1928 **Adolf O. R. Windaus** (Germany):
"for his studies on the constitution of the sterols and their connection with the vitamins"

1929 **Arthur Harden** (Britain):
Hans von Euler-Chelpin (Sweden):
"for their investigation on the fermentation of sugar and of fermentative enzymes"

1930 **Hans Fischer** (Germany):
"for his researches into the constitution of hemin and chlorophyll, and especially for his synthesis of hemin"

1931 **Frederich Bergius** (Germany):
Carl Bosch (Germany):
"for their contributions to the invention and development of chemical high-pressure methods"

1932 **Irving Langmuir** (U.S.):
"for his discoveries and investigations in surface chemistry"

1933 No award

1934 **Harold C. Urey** (U.S.):
"for his discovery of heavy hydrogen"

1935 **Frederic Joliot** (France):
Irene Joliot-Curie (France):
"for their synthesis of new radioactive elements"

1936 **Peter J. W. Debye** (Netherlands/U.S.):
"for his contributions our knowledge of molecular structure through his investigations on dipole moments and on the diffraction of X rays and electrons in gases"

1937 **Walter N. Haworth** (Britain):
"for his researches into the constitution of carbohydrates and vitamin C"

Paul Karrer (Switzerland):
"for his researches into the constitution of carotenoids, flavins, and vitamins A and B"

1938 **Richard Kuhn** (Germany):
"for his work on carotenoids and vitamins"

1939 **Adolf F. J. Butenandt** (Germany):
"for his work on sex hormones"

Leopold Ruzicka (Switzerland):
"for his work on polymethylenes and higher terpenes"

1940 No award

1941 No award

1942 No award

1943 **Georg de Hevesy** (Hungary):
"for his work on the use of isotopes as tracer elements in researches on chemical processes"

1944 **Otto Hahn** (Germany):
"for his discovery of the fission of heavy nuclei"

1945 **Artturi I. Virtanen** (Finland):
"for his researches and inventions in agricultural and nutritive chemistry, especially for his fodder preservation method"

1946 **James B. Sumner** (U.S.):
"for his discovery that enzymes can be crystallized"

John H. Northrop (U.S.):
Wendell M. Stanley (U.S.):
for their preparation of enzymes and virus proteins in a pure form"

1947 **Sir Robert Robinson** (Britain):
"for his investigations on plant products of biological importance, particularly the alkaloids"

1948 **Arne W. K. Tiselius** (Sweden):
"for his researches on electrophoresis and adsorption analysis, especially for his discoveries concerning the complex nature of the serum proteins"

1949 **William F. Giauque** (U.S.):
"for his contributions in the field of chemical thermodynamics, particularly concerning the behavior of substances at extremely low temperatures"

1950 **Kurt Alder** (Germany):
Otto P. H. Diels (Germany):
"for their discovery and development of the diene synthesis"

1951 **Edwin M. McMillan** (U.S.):
Glenn T. Seaborg (U.S.):
"for their discoveries in the chemistry of the transuranium elements"

1952 **Archer J. P. Martin** (Britain):
Richard L. M. Synge (Britain):
"for their development of partition chromatography"

1953 **Hermann Staudinger** (Germany):
"for his discoveries in the field of macromolecular chemistry"

1954 **Linus C. Pauling** (U.S.):
"for his research into the nature of the chemical bond and its application to the elucidation of the structure of complex substances"

1955 **Vincent du Vigneaud** (U.S.):
"for his work on biochemically important sulfur compounds, especially for the first synthesis of a polypeptide hormone"

1956 **Sir Cyril N. Hinshelwood** (Britain):
Nikolai N. Semenov (U.S.S.R.):
"for their research in clarifying the mechanisms of chemical reactions in gases"

1957 **Sir Alexander R. Todd** (Britain):
"for his work on nucleotides and nucleotide coenzymes"

1958 **Frederick Sanger** (Britain):
"for his work on the structure of proteins, particularly insulin"

1959 **Jaroslav Heyrovsky** (Czechoslovakia):
"for his discovery and development of the polarographic method of analysis"

1960 **Willard F. Libby** (U.S.):
"for his method to use carbon-14 for age determination in archaeology, geology, geophysics, and other branches of science"

1961 **Melvin Calvin** (U.S.):
"for his research on the carbon dioxide assimilation in plants"

1962 **John C. Kendrew** (Britain):
Max F. Perutz (Britain):
"for their studies of the structures of globular proteins"

1963 **Giulio Natta** (Italy):
Karl Ziegler (Germany):
"for their work in the controlled polymerization of hydrocarbons through the use of organometallic catalysts"

1964 **Dorothy C. Hodgkin** (Britain):
"for her determinations by X-ray techniques of the structures of important biochemical substances, particularly vitamin B_{12} and penicillin"

1965 **Robert B. Woodward** (U.S.):
"for his outstanding achievements in the 'art' of organic synthesis"

1966 **Robert S. Mulliken** (U.S.):
"for his fundamental work concerning chemical bonds and the electronic structure of molecules by the molecular orbital method"

1967 **Manfred Eigen** (Germany):
Ronald G. W. Norrish (Britain):
George Porter (Britain):
"for their studies of extremely fast chemical reactions, effected by disturbing the equilibrium with very short pulses of energy"

1968 **Lars Onsager** (U.S.):
"for his discovery of the reciprocal relations bearing his name, which are fundamental for the thermodynamics of irreversible processes"

1969 **Sir Derek H. R. Barton** (Britain):
Odd Hassel (Norway):
"for their contributions to the development of the concept of conformation and its application in chemistry"

1970 **Luis F. Leloir** (Argentina):
"for his discovery of sugar nucleotides and their role in the biosynthesis of carbohydrates"

1971 **Gerhard Herzberg** (Canada):
"for his contributions to the knowledge of electronic structure and geometry of molecules, particularly free radicals"

1972 **Christian B. Anfinsen** (U.S.):
"for his work on ribonuclease, especially concerning the connection between the amino acid sequence and the biologically active conformation"

Stanford Moore (U.S.):
William H. Stein (U.S.):
"for their contribution to the understanding of the connection between chemical structure and catalytic activity of the active center of the ribonuclease molecule"

1973 **Ernst Otto Fischer** (Germany):
Geoffrey Wilkinson (Britain):
"for their pioneering work, performed independently, on the chemistry of the organo-metallic sandwich compounds"

1974 **Paul J. Flory** (U.S.):
"for his fundamental achievements, both theoretical and experimental, in the physical chemistry of macromolecules"

1975 **John Cornforth** (Australia/Britain):
"for his work on the stereochemistry of enzyme-catalyzed reactions"

Vladimir Prelog (Yugoslavia/Switzerland):
"for his work on the stereochemistry of organic molecules and reactions"

1976 **William N. Lipscomb** (U.S.):
"for his studies on the structures of boranes illuminating problems of chemical bonding"

1977 **Ilya Pregogine** (Belgium):
"for his contributions to nonequilibrium thermodynamics, particularly the theory of dissipative structures"

1978 **Peter Mitchell** (Britain):
"for his contribution to the understanding of biological energy transfer through the formulation of the chemiosmotic theory"

1979 **Herbert C. Brown** (U.S.):
"for his application of boron compounds to synthetic organic chemistry"

Georg Wittig (Germany):
"for developing phosphorus reagents, presently bearing his name"

1980 **Paul Berg** (U.S.):
"for his fundamental studies of the biochemistry of nucleic acids, with particular regard to recombinant DNA"

Walter Gilbert (U.S.)
Frederick Sanger (Britain):
"for their contributions concerning the determination of base sequences in nucleic acids"

1981 **Kenichi Fukui** (Japan)
Roald Hoffmann (U.S.):
for their theories, developed independently, concerning the course of chemical reactions"

1982 **Aaron Klug** (Britain):
"for his development of crystallographic electron microscopy and his structural elucidation of biologically important nucleic acid – protein complexes"

1983 **Henry Taube** (U.S.):
"for his work on the mechanisms of electron transfer reactions, especially in metal complexes"

1984 **R. Bruce Merrifield** (U.S.):
"for his development of methodology for chemical synthesis on a solid matrix"

1985 **Herbert A. Hauptman** (U.S.):
Jerome Karle (U.S.):
"for their outstanding achievements in the development of direct methods for the determination of crystal structures"

1986 **John C. Polanyi** (Canada):
"for his pioneering work in the use of infrared chemiluminescence in studying the dynamics of chemical reactions"

Dudley R. Herschbach (U.S.):
Yuan T. Lee (U.S.):
"for their contributions concerning the dynamics of chemical elementary processes"

1987 **Donald J. Cram** (U.S.):
Jean-Marie Lehn (France):
Charles J. Pedersen (U.S.):
"for their development and use of molecules with structure-specific interactions of high selectivity"

1988 **Johann Deisenhofer** (Germany):
Robert Huber (Germany):
Hartmut Michel (Germany):
"for their determination of the structure of the photosynthetic reaction center of bacteria"

1989 **Sidney Altman** (U.S.):
Thomas R. Cech (U.S.):
"for their discovery of catalytic properties of RNA"

1990 **Elias J. Corey** (U.S.):
"for his development of the theory and methodology of organic synthesis"

1991 **Richard R. Ernst** (Switzerland):
"for his contributions to the development of the methodology of high resolution NMR spectroscopy"

1992 **Rudolph A. Marcus** (U.S.):
"for his contributions to the theory of electron-transfer reactions in chemical systems"

1993 **Kary B. Mullis** (U.S.):
"for his development of the polymerase chain reaction"

Michael Smith (Canada):
"for his fundamental contributions to the establishment of oligonucleotide-based site-directed mutagenesis and its development for protein studies"

1994 **George A Olah** (U.S.):
"for pioneering research on carbocations and their role in the chemical reactions of hydrocarbons"

1995 **F. Sherwood Rowland** (U.S.)
Mario Molina (U.S.)
Paul Crutzen (Germany)
"for their work in atmospheric chemistry, particularly concerning the formation and decomposition of ozone"

1996 **Robert F. Curl, Jr.** (U.S.)
Harold W. Kroto (U.K.)
Richard E. Smalley (U.S.)
"for their discovery of carbon atoms bound in the form of a ball (fullerenes)."

1997 **Paul D. Boyer** (U.S.)
John E. Walker (U.K.)
"for having elucidated the mechanism by which ATP synthase catalyzes the synthesis of adenosine triphosphate, the energy currency of living cells"

Jens C. Skou (Denmark)
"for his discovery of the ion-transporting enzyme Na^+–K^+ ATPase, the first molecular pump"

1998 **Walter Kohn** (U.S.)
John A. Pople (U.S.)
"to Walter Kohn for his development of the density-functional theory and to John Pople for his development of computational methods in quantum chemistry"

Answers to Multiple-Choice Questions in Review Units 1–12

Review Unit 1:	1. d	2. b	3. a	4. c	5. d	6. b	7. a	8. d	9. a	10. c
Review Unit 2:	1. c	2. a	3. d	4. c	5. d	6. b	7. a	8. b	9. c	10. b
Review Unit 3:	1. c	2. b	3. a	4. b	5. a	6. d	7. d	8. c	9. d	10. b
Review Unit 4:	1. d	2. b	3. c	4. c	5. b	6. a	7. d	8. c	9. a	10. d
Review Unit 5:	1. b	2. a	3. c	4. a	5. d	6. d	7. c	8. a	9. b	10. c
Review Unit 6:	1. c	2. a	3. b	4. d	5. c	6. c	7. a	8. b	9. d	10. a
Review Unit 7:	1. d	2. c	3. b	4. b	5. a	6. d	7. b	8. a	9. c	10. c
Review Unit 8:	1. a	2. b	3. d	4. b	5. c	6. a	7. c	8. d	9. c	10. b
Review Unit 9:	1. b	2. d	3. a	4. a	5. c	6. d	7. b	8. c	9. b	10. d
Review Unit 10:	1. b	2. a	3. d	4. b	5. c	6. a	7. d	8. b	9. a	10. d
Review Unit 11:	1. c	2. d	3. a	4. b	5. d	6. a	7. d	8. b	9. c	10. a
Review Unit 12:	1. d	2. a	3. c	4. c	5. a	6. d	7. b	8. a	9. b	10. c